Climate Change
Reconsidered II

Fossil Fuels

NIPCC
NONGOVERNMENTAL INTERNATIONAL PANEL
ON CLIMATE CHANGE

About NIPCC and Its Previous Reports

The Nongovernmental International Panel on Climate Change (NIPCC) is an international panel of scientists and scholars who came together to understand the causes and consequences of climate change. NIPCC has no formal attachment to or sponsorship from any government or government agency.

NIPCC seeks to objectively analyze and interpret data and facts without conforming to any specific agenda. This organizational structure and purpose stand in contrast to those of the United Nations' Intergovernmental Panel on Climate Change (IPCC), which *is* government-sponsored, politically motivated, and predisposed to believing that climate change is a problem in need of a U.N. solution.

NIPCC traces its beginnings to an informal meeting held in Milan, Italy in 2003 organized by Dr. S. Fred Singer and the Science & Environmental Policy Project (SEPP). The purpose was to produce an independent evaluation of the available scientific evidence on the subject of carbon dioxide-induced global warming in anticipation of the release of IPCC's *Fourth Assessment Report* (AR4). NIPCC scientists concluded IPCC was biased with respect to making future projections of climate change, discerning a significant human-induced influence on current and past climatic trends, and evaluating the impacts of potential carbon dioxide-induced environmental changes on Earth's biosphere.

To highlight such deficiencies in IPCC's AR4, in 2008 SEPP partnered with The Heartland Institute to produce *Nature, Not Human Activity, Rules the Climate*. In 2009, the Center for the Study of Carbon Dioxide and Global Change joined the original two sponsors to help produce *Climate Change Reconsidered: The 2009 Report of the Nongovernmental International Panel on Climate Change (NIPCC)*, the first comprehensive alternative to the reports of IPCC.

In 2010, a website (www.nipccreport.org) was created to highlight scientific studies NIPCC scientists believed likely would be downplayed or ignored by IPCC during preparation of its next assessment report. In 2011, the three sponsoring organizations produced *Climate Change Reconsidered: The 2011 Interim Report of the Nongovernmental International Panel on Climate Change (NIPCC)*.

In 2013, a division of the Chinese Academy of Sciences translated and published an abridged edition of the 2009 and 2011 NIPCC reports in a single volume. Also in 2013, NIPCC released *Climate Change Reconsidered II: Physical Science,* the first of three volumes bringing the original 2009 report up-to-date with research from the 2011 *Interim Report* plus research as current as the third quarter of 2013. A new website was created (www.ClimateChangeReconsidered.org) to feature the new report and news about its release.

In 2014, the second volume in the *Climate Change Reconsidered II* series, subtitled *Biological Impacts*, was released. The third and final volume, subtitled *Fossil Fuels,* is being released in 2019, and this is its *Summary for Policymakers*.

Climate Change Reconsidered II

Fossil Fuels

Lead Authors

Roger Bezdek, Craig D. Idso, David Legates, S. Fred Singer

Chapter Lead Authors

Dennis Avery, Roger Bezdek, John D. Dunn, Craig D. Idso, David Legates, Christopher Monckton, Patrick Moore, S. Fred Singer, Charles N. Steele, Aaron Stover, Richard L. Stroup

Chapter Contributing Authors

Jerome Arnett, Jr., John Baden, Timothy Ball, Joseph L. Bast, Charles Battig, Edward Briggs, Barry Brill, Kevin Dayaratna, John D. Dunn, James E. Enstrom, Donald K. Forbes, Patrick Frank, Kenneth Haapala, Howard Hayden, Thomas B. Hayward, Jay Lehr, Bryan Leyland, Steve Milloy, Patrick Moore, Willie Soon, Richard Trzupek, Steve Welcenbach, S. Stanley Young

Chapter Reviewers

D. Weston Allen, Mark Alliegro, Charles Anderson, David Archibald, Dennis T. Avery, Timothy Ball, David Bowen, Barry Brill, H. Sterling Burnett, David Burton, William N. Butos, Mark Campbell, Jorge David Chapas, Ian D. Clark, Donald R. Crowe, Weihong Cui, Donn Dears, David Deming, Terry W. Donze, Paul Driessen, John Droz, Jr., James E. Enstrom, Rex J. Fleming, Vivian Richard Forbes, Patrick Frank, Lee C. Gerhard, François Gervais, Albrecht Glatzle, Steve Goreham, Pierre Gosselin, Laurence Gould, Kesten Green, Kenneth Haapala, Hermann Harde, Tom Harris, Howard Hayden, Tom Hennigan, Donald Hertzmark, Ole Humlum, Mary Hutzler, Hans Konrad Johnsen, Brian Joondeph, Richard A. Keen, William Kininmonth, Joseph Leimkuhler, Marlo Lewis, Jr., Bryan Leyland, Anthony R. Lupo, Paul McFadyen, John Merrifield, Alan Moran, Robert Murphy, Daniel W. Nebert, Norman J. Page, Fred Palmer, Garth Paltridge, Jim Petch, Charles T. Rombough, Ronald Rychlak, Tom V. Segalstad, Gary D. Sharp, Jan-Erik Solheim, Willie Soon, Charles N. Steele, David Stevenson, Peter Stilbs, Daniel Sutter, Roger Tattersol, Frank Tipler, Richard Trzupek, Fritz Vahrenholt, Art Viterito, Gösta Walin, Lance Wallace, Thomas F. Walton, James Wanliss, Bernard L. Weinstein. Several additional reviewers and contributors wish to remain anonymous.

Editors

Joseph L. Bast, Diane Carol Bast

NIPCC

NONGOVERNMENTAL INTERNATIONAL PANEL
ON CLIMATE CHANGE

Reviews of *Climate Change Reconsidered II: Physical Science*

"I fully support the efforts of the Nongovernmental International Panel on Climate Change (NIPCC) and publication of its latest report, *Climate Change Reconsidered II: Physical Science,* to help the general public to understand the reality of global climate change."

Kumar Raina, Former Deputy Director General
Geological Survey of India

"*Climate Change Reconsidered II* fulfills an important role in countering the IPCC part by part, highlighting crucial things they ignore such as the Little Ice Age and the recovery (warming) which began in 1800–1850. In contrast to the IPCC, which often ignores evidence of past changes, the authors of the NIPCC report recognize that climatology requires studying past changes to infer future changes."

Syun-Ichi Akasofu, Founding Director & Professor of Physics Emeritus
International Arctic Research Center, University of Alaska Fairbanks

"The work of the NIPCC to present the evidence for natural climate warming and climate change is an essential counter-balance to the biased reporting of the IPCC. They have brought to focus a range of peer-reviewed publications showing that natural forces have in the past and continue today to dominate the climate signal."

Ian Clark, Department of Earth Sciences
University of Ottawa, Canada

"The CCR-II report correctly explains that most of the reports on global warming and its impacts on sea-level rise, ice melts, glacial retreats, impact on crop production, extreme weather events, rainfall changes, etc. have not properly considered factors such as physical impacts of human activities, natural variability in climate, lopsided models used in the prediction of production estimates, etc. There is a need to look into these phenomena at local and regional scales before sensationalization of global warming-related studies."

S. Jeevananda Reddy, Former Chief Technical Advisor
United Nations World Meteorological Organization

"Library shelves are cluttered with books on global warming. The problem is identifying which ones are worth reading. The NIPCC's CCR-II report is one of these. Its coverage of the topic is comprehensive without being superficial. It sorts through conflicting claims made by scientists and highlights mounting evidence that climate sensitivity to carbon dioxide increase is lower than climate models have until now assumed."

Chris de Freitas, School of Environment
The University of Auckland, New Zealand

"Rather than coming from a pre-determined politicized position that is typical of the IPCC, the NIPCC constrains itself to the scientific process so as to provide objective information. If we (scientists) are honest, we understand that the study of atmospheric processes/dynamics is in its infancy. Consequently, the work of the NIPCC and its most recent report is very important."

Bruce Borders, Professor of Forest Biometrics
Warnell School of Forestry and Natural Resources, University of Georgia

"I support [the work of the NIPCC] because I am convinced that the whole field of climate and climate change urgently needs an open debate between several 'schools of thought,' in science as well as other disciplines, many of which jumped on the IPCC bandwagon far too readily. Climate, and even more so impacts and responses, are far too complex and important to be left to an official body like the IPCC."

Sonja A. Boehmer-Christiansen
Reader Emeritus, Department of Geography, Hull University
Editor, *Energy & Environment*

Climate Change Reconsidered II

Fossil Fuels

Published by THE HEARTLAND INSTITUTE
3939 North Wilke Road
Arlington Heights, Illinois 60004 U.S.A.
phone +1 (312) 377-4000
www.heartland.org

1-10 copies	$154 per copy
11-50 copies	$123 per copy
51-100 copies	$98 per copy
101 or more	$79 per copy

Please use the following citation for this report:

Bezdek, R., Idso, C.D, Legates, D., and Singer, S.F. (Eds.) 2019. *Climate Change Reconsidered II: Fossil Fuels.* Nongovernmental International Panel on Climate Change (NIPCC). Arlington Heights, IL: The Heartland Institute.

This print version is black and white. A color version is available for free online at www.climatechangereconsidered.org.

ISBN-13 – 978-1-934791-45-5
ISBN-10 – 1-934791-45-8

2019

1 2 3 4 5 6

Foreword

The release of this volume in the Climate Change Reconsidered (CCR) series ends a five-year pause following the release of *Climate Change Reconsidered II: Biological Impacts.* Several shorter reports were released in the interim, most notably *Why Scientists Disagree about Global Warming* (first edition in 2015, second edition in 2016). While it was never our intention to issue a CCR volume every year, hope did exist that this volume would emerge sooner than it has.

This volume is the product of the Nongovernmental International Panel on Climate Change (NIPCC), a joint project of three nonprofit organizations, The Heartland Institute, the Center for the Study of Carbon Dioxide and Global Change, and the Science and Environmental Policy Project (SEPP). Two of NIPCC's cosponsoring organizations experienced leadership changes since 2014 that affected the release of this new volume.

Joseph Bast and Diane Bast, editors of this volume and previous volumes in the series (as well as the authors of this foreword) retired in early 2018. Joseph Bast was president and CEO of The Heartland Institute and Diane Bast was senior editor, which meant time that might otherwise have been dedicated to this book was devoted to managing a successful corporate succession instead.

Dr. S. Fred Singer, the founder and previously president and chairman of the Science and Environmental Policy Project (SEPP), also handed over the reins to a younger generation of leaders, in his case to Kenneth Haappala the new president and Thomas Sheahan the new chairman. At 93 years young, Dr. Singer insists he is not retiring. He continues to be published in popular and scientific journals and contributed substantially to this volume.

One of the lead authors of the three most recent volumes in the CCR series passed away in January 2016. Dr. Robert Carter's unexpected departure was a heavy blow to everyone – his family, of course, as well as colleagues and friends in Australia, America, and around the world. Just weeks before he passed, Dr. Carter agreed to take this final volume "across the finish line" and was starting to reach out to his extensive global networks of climate scientists for help. So devastated were we that fully a year passed after his death before work resumed on the book. This book is dedicated to his memory.

Amidst all these changes, we also found new partners who made this volume possible. Dr. Roger Bezdek, a distinguished economist specializing in energy and climate issues, stepped in to provide insights and skills that our usual stable of physicists, biologists, and climatologists lacked. With him came nearly two dozen economists and a similar number of engineers with deep expertise in the economic and environmental impacts of fossil fuels and their alternatives, the focus of this book.

Also new for this volume is Dr. David Legates, professor of climatology in the Department of Geography at the University of Delaware and an adjunct professor at the university's Physical Ocean Science and Engineering Program and in the Department of Applied Economics. Dr. Legates is a "scientist's scientist," a recognized authority in his field, and, like Dr. Carter and Dr. Singer, unafraid to speak the truth on the controversial subject of climate change, even at the cost of damaging a promising academic career.

As we said of previous volumes in the CCR series, the sheer size of this volume – 700 pages containing references to thousands of articles and books – suggests what an extraordinary research, writing, and editing endeavor this turned out to be. The topic of this volume – broadly, the benefits and costs of fossil fuels – required reviewing scientific and economic literature on organic chemistry, climate science, public health, economic history, human security, and theoretical studies based on integrated assessment models (IAMs) and cost-benefit analysis. Much of this literature resides outside peer-reviewed academic journals. Consequently readers will see a heavier reliance than in previous volumes on books, government and think tank reports, and sometimes newspaper and magazine reports of news events.

We extend our sincere thanks and appreciation to the 117 scientists, engineers, economists, and other experts who helped write and review this report, as well as the thousands more who conducted the original research that is summarized and cited. Funding for this effort once again came from three family foundations, none of them having any commercial interest in the topic. We thank them for their generosity as well as their patience. No government or corporate funds were solicited or received to support this project.

Diane Carol Bast
Executive Editor
The Heartland Institute

Joseph L. Bast
Director and Senior Fellow
The Heartland Institute

Preface

This new volume, *Climate Change Reconsidered II: Fossil Fuels,* assesses the costs and benefits of the use of fossil fuels with a special focus on concerns related to anthropogenic climate change. It is the fifth volume in the Climate Change Reconsidered (CCR) series produced by the Nongovernmental International Panel on Climate Change (NIPCC).

NIPCC was created by Dr. S. Fred Singer in 2003 to provide an independent review of the reports of the United Nations' Intergovernmental Panel on Climate Change (IPCC). Unlike the IPCC and as its name suggests, NIPCC is a private association of scientists and other experts and nonprofit organizations. It is not a government entity and is not beholden to any political benefactors. This and previous volumes in the CCR series, along with other publications and information about NIPCC, are available for free on NIPCC's website at www.climatechangereconsidered.org.

Summary of Findings

The NIPCC authors, building on previous reports in the CCR series as well as new literature reviews, find that while climate change is occurring and a human impact on climate is likely, there is no consensus on the size of that impact relative to natural variability, the *net* benefits or costs of the impacts of climate change, or whether future climate trends can be predicted with sufficient confidence to guide public policies today. Consequently, concern over climate change is not a sufficient scientific or economic basis for restricting the use of fossil fuels.

The NIPCC authors do something their IPCC counterparts never did: conduct an even-handed cost-benefit analysis of the use of fossil fuels. Despite calling for the end of reliance on fossil fuels by 2100, the IPCC never produced an accounting of the

opportunity cost of restricting or banning their use. That cost, a literature review shows, would be enormous. Estimates of the cost of reducing anthropogenic greenhouse gas (GHG) emissions by the amounts said by the IPCC to be necessary to avoid causing ~2°C warming in the year 2050 range from the IPCC's own estimate of 3.4% to as high as 81% of projected global gross domestic product (GDP) in 2050, the latter estimate nullifying all the gains in human well-being made in the past century. Cost-benefit ratios range from the IPCC's own estimate of 6.8:1 to an alarming 162:1. The costs of specific emission mitigation programs range from 7.4 times to 7,000 times more than the benefits, even assuming the IPCC's faulty science and tenuous associations are correct.

The NIPCC authors conclude, "The global war on energy freedom, which commenced in earnest in the 1980s and reached a fever pitch in the second decade of the twenty-first century, was never founded on sound science or economics. The world's policymakers ought to acknowledge this truth and end that war."

Organization of Inquiry

Since economics can provide insights into the alleged impacts of climate change, this volume begins with a chapter describing how economic principles can be applied to environmental issues. The authors explain how economists use observational data (prices, profits and losses, investment and consumption decisions, etc.) to measure and monetize costs and benefits, to understand how people respond to and solve challenges, and to understand why private as well as government efforts to protect the environment sometimes fail to achieve their objectives.

Chapter 2 updates the literature review of climate science in previous CCR volumes. Most notably, the authors say the IPCC has exaggerated the amount of warming likely to occur if the concentration of atmospheric carbon dioxide (CO_2) were to double and such warming as occurs is likely to be modest and cause no net harm to the global environment or to human well-being. Chapters 3, 4, and 5 catalogue the beneficial impacts of fossil fuels on human prosperity, human health, and the environment. These benefits are enormous – imagine, for a moment, life without electricity, modern medicine, or cars, trucks, and airplanes – yet these benefits are missing from the IPCC's massive tomes. A true accounting of the costs and benefits of ending humanity's reliance on fossil fuels must include the opportunity cost of forgoing these benefits.

Chapters 6 and 7 address two types of costs or harms said to be created by the use of fossil fuels: air pollution and what the IPCC calls threats to "human security." The NIPCC authors show the alleged effects of air pollution have been grossly exaggerated by the U.S. Environmental Protection Agency and the World Health Organization. Similarly, the IPCC's own literature review shows how weak is the case for claiming climate change intensifies "risk factors" such as loss of property and livelihoods, forced migration, and violent conflict. Chapter 8, the final chapter of the book, critiques the IPCC's claim that the cost of reducing the use of fossil fuels is justified by the benefits of a slightly cooler world a century hence. New cost-benefit analyses of climate change, fossil fuels, and regulations demonstrate how adaptation to climate change is invariably the better path than attempting to mitigate it by reducing greenhouse gas emissions.

Acknowledgements

We thank the more-than-100 scientists, scholars, and experts who participated over the course of four years in writing, reviewing, editing, and proofreading this volume. This was a huge undertaking that involved thousands of hours of effort, the vast majority of it unpaid. The result exceeded our hopes, and we trust it meets your expectations.

The NIPCC authors cite thousands of books, scholarly articles, and reports that contradict the IPCC's alarmist narrative. We once again tried to remain true to the facts when representing the findings of others, often by quoting directly and at some length from original sources and describing the methodology used and qualifications that accompanied the stated conclusions. The result may seem tedious at times, but we believe this was necessary and appropriate for a reference work challenging many popular beliefs.

We acknowledge that not every scientist, economist, or historian whose work we cite disagrees with IPCC positions or supports ours, even though their research points in that direction. We recognize there may be some experts we quote who are dismayed to see their work cited in a book written by "skeptics." We ask them to read this book with an open mind and ask themselves how much of what they think they know to be true is based on trust, perhaps misplaced, in claims propagated by the IPCC. Even scientists need to be reminded sometimes that skepticism, not conformity, is the higher value in the pursuit of knowledge.

Craig D. Idso, Ph.D.
Chairman
Center for the Study of Carbon Dioxide and Global Change

David Legates, Ph.D.
Professor, Department of Geography
University of Delaware

Roger Bezdek, Ph.D.
President
Management Information Services, Inc.

S. Fred Singer, Ph.D.
President Emeritus
Science and Environmental Policy Project

Dedication

Photo credit: Hans H.J. Labohm

Robert M. Carter (left) with S. Fred Singer (right), photo
taken in October 2013 in The Hague, Netherlands.

We dedicate this report to the memory of Robert M. Carter, who helped write and edit
previous volumes in the Climate Change Reconsidered series but passed away in 2016 as
the current volume was only beginning to come together. It would have been a far better
work had he lived to help direct our efforts. Bob was a palaeontologist, stratigrapher,
marine geologist, and environmental scientist with a long and distinguished career in the
academy. He was a mentor and friend to hundreds of young scientists and many nonscientists. He proved by personal example that science in the end does not tolerate
corruption, and that what matters most of all is personal integrity.

Abbreviated Table of Contents

Table of Contents

Summary for Policymakers

Introduction

Climate Change Reconsidered II: Fossil Fuels, produced by the Nongovernmental International Panel on Climate Change (NIPCC), assesses the costs and benefits of the use of fossil fuels[1] by reviewing scientific and economic literature on organic chemistry, climate science, public health, economic history, human security, and theoretical studies based on integrated assessment models (IAMs) and cost-benefit analysis (CBA). It is the fifth volume in the Climate Change Reconsidered series (NIPCC 2009, 2011, 2013, 2014) and, like the preceding volumes, it focuses on research overlooked or ignored by the United Nations' Intergovernmental Panel on Climate Change (IPCC).

In its 2013 volume titled *Climate Change Reconsidered II: Physical Science,* NIPCC refuted the scientific basis of the IPCC's claim that dangerous human interference with the climate system is occurring. In its 2014 volume titled *Climate Change Reconsidered II: Biological Impacts,* NIPCC addressed and refuted the IPCC's claim that climate change negatively affects plants, wildlife, and human health.

In this new volume, 117 scientists, economists, and other experts address and refute the IPCC's claim that the impacts of climate change on human well-being and the natural environment justify dramatic reductions in the use of fossil fuels. Specifically, the NIPCC authors critique two recent IPCC reports: *Climate Change 2014: Impacts, Adaptation, and Vulnerability,* the Working Group II contribution to

the IPCC's Fifth Assessment Report (AR5), and *Climate Change 2014: Mitigation of Climate Change,* the Working Group III contribution to AR5 (IPCC, 2014a, 2014b).

The organization of this Summary for Policymakers tracks the organization of the full report. Citations to supporting research and documentation are scant for want of space but can be found at the end of the document. More than 2,000 references appear in the full report.

Part I. Foundations

The most consequential issues in the climate change debate are "whether the warming since 1950 has been dominated by human causes, how much the planet will warm in the 21st century, whether warming is

---

[1] This report follows conventional usage by using "fossil fuels" to refer to hydrocarbons, principally coal, oil, and natural gas, used by humanity to generate power. We recognize that not all hydrocarbons may be derived from animal or plant sources.

'dangerous,' whether we can afford to radically reduce CO_2 emissions, and whether reduction will improve the climate" (Curry, 2015). Addressing these issues requires foundations in environmental economics and climate science. Part I of *Climate Change Reconsidered II: Fossil Fuels* provides those foundations.

1. Environmental Economics

Many environmentalists and climate scientists are not familiar with economic research on environmental issues and have only vague ideas about what economics can bring to the climate change debate. Many economists make a different mistake, accepting unsubstantiated claims that the "science is settled" regarding the causes and consequences of climate change and then limiting their role in the debate to finding the most efficient way to reduce "carbon pollution." Both audiences need to be aware of what economists can bring to the climate change debate.

The most valuable concept economists bring is *opportunity cost,* the value of something that must be given up to acquire or achieve something else. Every choice has a corresponding opportunity cost. By revealing those costs, economics can help policymakers discover cost-effective responses to environmental problems, including climate change (Block, 1990; Markandya and Richardson, 1992; Libecap and Steckel, 2011).

A second key concept is the Environmental Kuznets Curve (EKC), pictured in Figure SPM.1. Fossil fuels and the technologies they power make it possible to use fewer resources and less surface space to meet human needs while also allowing environmental protection to become a positive and widely shared social value and objective. EKCs have been documented for a wide range of countries and air quality, water quality, and other measures of environmental protection (Yandle *et al.*, 2004; Goklany, 2012; Bertinelli *et al.*, 2012).

Economists can help compassionate people reconcile the real-world trade-offs of protecting the environment while using natural resources to produce the goods and services needed by humankind (McKitrick, 2010; Morris and Butler, 2013; Anderson and Leal, 2015). They have demonstrated how committed environmentalists can better achieve their goals by recognizing fundamental economic principles such as discount rates and marginal costs (Anderson and Huggins, 2008). They have shown how entrepreneurs can use private property, price

signals, and markets to discover new ways to protect the environment (Anderson and Leal, 1997; Huggins, 2013).

Figure SPM.1
A typical Environmental Kuznets Curve

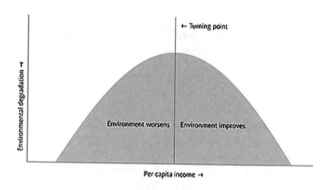

Source: Ho and Wang, 2015, p. 42.

Economists have pointed out the economic and political pitfalls facing renewable and carbon-neutral energies (Morriss *et al.*, 2011; Yonk *et al.*, 2012). Economists have explained how proposals to force a transition away from fossil fuels advanced without an understanding of the true costs and implications of alternative fuels can lead to unnecessary expenses and minimal or even no net reduction in greenhouse gas emissions (Lomborg, 2010; van Kooten, 2013; Heal, 2017; Lemoine and Rudik, 2017).

Economists describe how common resources can be degraded by overuse by "free riders," but also how they can be effectively managed by individuals and nongovernment organizations using their knowledge of local opportunities and costs, the kind of knowledge national and international organizations typically lack (Coase, 1994). These market-based solutions exhibit the sort of spontaneous order that Hayek (1988) often wrote about, a coordination that is not dictated or controlled by a central planner. Ostrom (2010) identified eight design principles shared by entities most successful at managing common-pool resources.

The prosperity made possible by the use of fossil fuels has made environmental protection a social value in countries around the world (Hartwell and Coursey, 2015). The value-creating power of private property rights, prices, profits and losses, and voluntary trade can turn climate change from a possible *tragedy* of the commons into an *opportunity* of the commons (Boettke, 2009). Energy freedom, not government intervention, can balance the interests and needs of today with those of tomorrow. It alone can access the local knowledge needed to find efficient win-win responses to climate change.

2. Climate Science

Chapter 2 provides an overview of the current state of climate science beginning with an explanation of the Scientific Method, which imposes restrictions and duties on scientists intended to ensure the quality, objectivity, utility, and integrity of their work. Key elements of the Scientific Method include experimentation, the testing of competing hypotheses, objective and careful peer review, discerning correlation from causation, and controlling for natural variability. In each of these areas, the IPCC and many scientists whose work is prominent in climate science have been shown to fall short (Essex and McKitrick, 2007; Darwall, 2013; Lewin, 2017; Armstrong and Green, 2018).

Two other topics concerning methodology are the role of consensus in science and ways to manage and communicate uncertainty. Consensus may have a place in science when it is achieved over an extended period of time by independent scientists following the conventions of the Scientific Method. This is not the context in which it is invoked in climate science, and consequently it has been the cause of controversy and polarization of views (Curry, 2012; Lindzen, 2017). Uncertainty is unavoidable in science, but it can be reduced using techniques such as Bayesian inference and honestly communicated to other researchers and the public. Instead of following best practices, the IPCC and its followers make many unmerited declarative statements and issue seemingly confident predictions without error bars (Essex and McKitrick, 2007; Frank, 2015).

The unique chemistry of carbon explains why fossil fuels, composed mainly of carbon and hydrogen, are so widely used as fuel. Kiefer (2013) writes, "Carbon transforms hydrogen from a diffuse and explosive gas that will only become liquid at -423° F [-253° C] into an easily handled, room-temperature liquid with 63% more hydrogen atoms per gallon than pure liquid hydrogen, 3.5 times the volumetric energy density (joules per gallon), and the ideal characteristics of a combustion fuel. ... A perfect combustion fuel possesses the desirable characteristics of easy storage and transport, inertness and low toxicity for safe handling, measured and

adjustable volatility for easy mixing with air, stability across a broad range of environmental temperatures and pressures, and high energy density. Because of sweeping advantages across all these parameters, liquid hydrocarbons have risen to dominate the global economy" (p. 117).

Climate models are a subject of controversy in climate science. General circulation models (GCMs) "run hot," meaning they predict more warming than actually occurred or is likely to occur in the future (Monckton *et al.*, 2015). They hindcast twice as much warming from 1979 to 2016 as actually occurred (Christy, 2018). See Figure SPM.2. Climate models are unable to reproduce many important climate phenomena (Legates, 2014) and are "tuned" to produce results that fall into an "acceptable range" of outputs (Hourdin *et al.*, 2017).

Figure SPM.2
Failure of climate models to hindcast global temperatures, 1979–2015

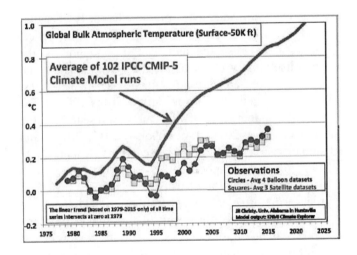

Source: Christy, 2016.

The accuracy of temperature records since pre-industrial times is a second area of controversy. Records from surface stations are known to contain systematic errors due to instrument and recording errors, physical changes in the instrumentation, and database mismanagement, making them too unreliable to form the basis of scientific research, yet they are seldom questioned (Frank, 2015; McLean, 2018). More accurate satellite-based temperature records, which reach back only to 1979, reveal a range of near-global warming of approximately 0.07°C to 0.13°C per decade from 1979 to 2016

(Christy *et al.*, 2018).

Equilibrium climate sensitivity (ECS), a measure of expected warming when CO_2 concentrations in the atmosphere double, is yet another source of controversy in climate science. The IPCC's estimate of ECS is one-third higher than most recent estimates in the scientific literature (Michaels, 2017). There is so much uncertainty in climate models and so many new discoveries being made that a single "true" estimate of ECS is probably impossible to calculate.

Scientists also disagree about whether climate change is negatively affecting human well-being or the natural world. Despite headlines and documentary films claiming the opposite, there is little or no evidence of trends that lie outside natural variability in severe weather events, droughts, forest fires, melting ice, sea-level rise, and adverse effects on plant life. In some cases, the historical record reveals just the opposite: more mild weather and fewer droughts, for example, than in the pre-industrial past. Most plants are known to flourish in a warmer environment with higher levels of CO_2 (Idso and Idso, 2015).

Why do scientists disagree? Partly because *skepticism,* not consensus, is the heart of science. Sources of disagreement can be found in the interdisciplinary character of the issue, fundamental uncertainties concerning climate science (Curry, 2015; Lindzen, 2017), the failure of the IPCC to be an independent and reliable source of research on the subject (IAC, 2010; Laframboise, 2011, 2013), and tunnel vision (bias) among researchers (Kabat, 2008; Berezow and Campbell, 2012).

The final section of Chapter 2 critiques the claim that "97% of scientists agree" that climate change is mostly or entirely the result of the human presence and is dangerous. Surveys, literature reviews, and petitions demonstrate a lively debate is occurring in the scientific community over the basic science and economics of climate change (Solomon, 2010; Curry, 2012; Friends of Science, 2014; Tol, 2014a; Legates *et al.*, 2015; Global Warming Petition Project, n.d.).

In conclusion, fundamental uncertainties arising from insufficient observational evidence and disagreements over how to interpret data and set the parameters of models prevent science from determining whether human greenhouse gas emissions are having effects on Earth's atmosphere that could endanger life on the planet. There is no compelling scientific evidence of long-term trends in global mean temperatures or climate impacts that exceed the bounds of natural variability.

Part II. The Benefits of Fossil Fuels

Part II presents an accounting of the benefits created by the use of fossil fuels. Chapters 3, 4, and 5 address human prosperity, human health benefits, and environmental benefits, respectively.

3. Human Prosperity

The primary reason humans burn fossil fuels is to produce the goods and services that make human prosperity possible. Put another way, humans burn fossil fuels to live more comfortable, safer, and higher-quality lives. Chapter 3 documents the many ways in which fossil fuels contribute to human prosperity.

The role played by fossil fuels in the dramatic rise in human prosperity is revealed by the close correlation between carbon dioxide (CO_2) emissions and world gross domestic product (GDP) shown in Figure SPM.3. Fossil fuels were responsible for such revolutionary technologies as the steam engine and cotton gin, early railroads and steamships, electrification and the electric grid, the internal combustion engine, and the computer and Internet revolution. In particular, the spread of electrification made possible by fossil fuels has transformed the modern world, making possible many of the devices, services, comforts, and freedoms we take for granted (Smil, 2005, 2010; Goklany, 2012; Gordon, 2016).

Figure SPM.3
Relationship between world GDP and CO_2 emissions

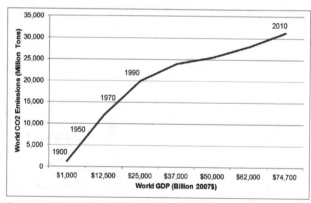

Source: Bezdek, 2014, p. 127.

Today, fossil fuels supply 81% of global primary energy and 78% of U.S. primary energy. They are required to power the revolving turbine electric generators that supply dispatchable energy to electric grids, making electricity available on demand in the quantities needed, not only when the sun shines and the wind blows. Fossil fuels are also essential for fertilizer production and the manufacture of concrete and steel. Access to affordable, plentiful, and reliable energy is closely associated with key measures of global human development, including per-capita GDP, consumption expenditure, urbanization rate, life expectancy at birth, and the adult literacy rate (United Nations Development Program, 2010; Šlaus and Jacobs, 2011). Research reveals a positive relationship between low energy prices and human prosperity (Clemente, 2010; Bezdek, 2014; 2015).

A similar level of human prosperity is not possible by relying on alternative fuels such as solar and wind power. Wind and solar power are intermittent and unreliable, much more expensive than fossil fuels, cannot be deployed without the use of fossil fuels to build them and to provide back-up power, cannot power most modes of transportation, and cannot increase dispatchable capacity sufficiently to meet more than a small part of the rising demand for electricity (Rasmussen, 2010; Bryce, 2010; Smil, 2010, 2016; Stacy and Taylor, 2016).

The contribution of fossil fuels to human prosperity can be estimated in numerous ways, making agreement on a single cost estimate difficult.

However, estimates converge on very high amounts: Coal delivered economic benefits in the United States alone worth between $1.275 trillion and $1.76 trillion in 2015 and supported approximately 6.8 million jobs (Rose and Wei, 2006). Reducing reliance on fossil fuels in the United States by 40% from 2012 to 2030 would cost $478 billion and an average of 224,000 jobs each year (U.S. Chamber of Commerce, 2014).

4. Human Health Benefits

Chapter 4 presents the human health benefits of fossil fuels. Historically, humankind was besieged by epidemics and other disasters that caused frequent widespread deaths and kept the average lifespan to less than 35 years (Omran, 1971). The average lifespan among the ancient Greeks was apparently just 18 years, and among the Romans, 22 years (Bryce, 2014, p. 59, citing Steckel and Rose, 2002). Today, according to the U.S. Census Bureau (2016), "The world average age of death has increased by 35 years since 1970, with declines in death rates in all age groups, including those aged 60 and older. From 1970 to 2010, the average age of death increased by 30 years in East Asia and 32 years in tropical Latin America, and in contrast, by less than 10 years in western, southern, and central Sub-Saharan Africa. … [A]ll regions have had increases in mean age at death, particularly East Asia and tropical Latin America" (pp. 31–3).

Figure SPM.4
Deaths caused by cold vs. heat

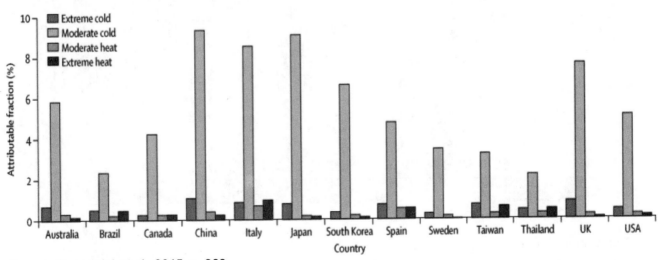

Source: Gasparrini *et al.*, 2015, p. 369.

Fossil fuels have lifted billions of people out of poverty, reducing the negative effects of poverty on human health (Moore and Simon, 2000). They improve human well-being and safety by powering labor-saving and life-protecting technologies such as air conditioning, modern medicine, cars, trucks, and airplanes (Goklany, 2007). Fossil fuels made possible electrification of heating, lighting, manufacturing, and other processes, resulting in protection of human health and extended lives (Bryce, 2014). Fossil fuels also increased the quantity and improved the reliability and safety of the food supply (Moore and White, 2016).

Fossil fuels may also affect human health by contributing to some part of the global warming experienced during the twentieth century or forecast by GCMs for the twenty-first century and beyond. Medical science and observational research in Asia, Australia, Europe, and North America confirm that warming is associated with lower, not higher, temperature-related mortality rates (Keatinge and Donaldson, 2004; Gasparrini et al., 2015; White, 2017). See Figure SPM.4. Research shows warmer temperatures lead to decreases in premature deaths due to cardiovascular and respiratory disease and stroke occurrences (Nafstad et al., 2001; Gill et al., 2012; Song et al., 2018), while warmer temperatures have little if any influence on mosquito- or tick-borne diseases (Murdock et al., 2016).

5. Environmental Benefits

Chapter 5 reviews evidence showing how human use of fossil fuels benefits the environment. The scientific literature on the impacts of warmer temperatures and rising atmospheric CO_2 concentrations on plants finds them to be overwhelmingly positive. This extends to rates of photosynthesis and biomass production and the efficiency with which plants and trees utilize water (Ainsworth and Long, 2005; Bourgault et al., 2017). The result is a remarkable and beneficial Greening of the Earth shown in Figure SPM.5 (Zhu et al. 2016; Campbell et al., 2017; Cheng et al., 2017).

Similarly, the impacts of global warming on terrestrial animals is likely to be net positive. Wildlife benefit from expanding habitats, and real-world data

Figure SPM.5
Greening of the Earth, 1982 to 2009, trend in average observed leaf area index (LAI)

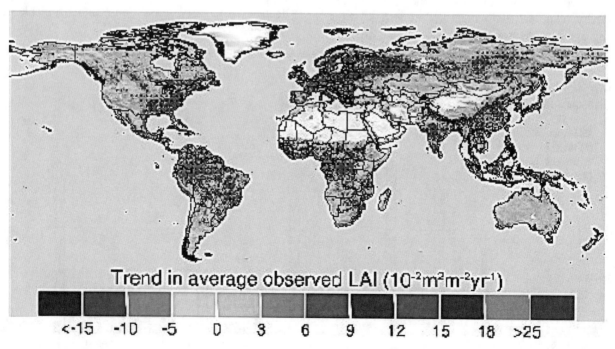

Source: Zhu et al., 2016.

indicate warmer temperatures have not been harmful to wildlife (Willis *et al.*, 2010). Laboratory and field studies of the impact of warmer temperatures and reduced water pH levels (so-called "acidification") on aquatic life find tolerance and adaptation and even examples of benefits (Pandolfi *et al.*, 2011; Baker, 2014).

The fact that carbon and hydrogen are ubiquitous in the natural world helps to explain why the rest of the physical world is compatible with them and even depends on them for life itself (Smil, 2016). The *carbon cycle* minimizes the environmental impact of human emissions of CO_2 by reforming it into other compounds and sequestering it in the oceans, plants, and rocks. According to the IPCC, the residual of the human contribution of CO_2 that remains in the atmosphere after natural processes move the rest to other reservoirs is as little as 0.53% of the carbon entering the air each year and 0.195% of the total amount of carbon thought to be in the atmosphere (IPCC, 2013, p. 471).

The high power density of fossil fuels enable humanity to meet its ever-rising need for food and natural resources while using less surface space, thereby rescuing precious wildlife habitat from development. In 2010, fossil fuels, thermal, and hydropower required less than 0.2% of the Earth's ice-free land, and nearly half that amount was surface covered by water for reservoirs (Smil, 2016, pp. 211–212). Fossil fuels required roughly the same surface area as devoted to renewable energy sources (solar photovoltaic, wind, and liquid biofuels), yet delivered *110 times as much power* (*Ibid.*).

Acid rain, once thought to be a serious environmental threat, is no longer considered one (NAPAP, 1998). Human contributions of oil to the oceans via leakage and spills are trivial in relation to natural sources and quickly disperse and biodegrade (NRC, 2003). The damage caused by oil spills is a net cost of using oil, but not a major environmental problem.

In conclusion, fossil fuels directly benefit the environment by making possible huge (orders of magnitude) advances in efficiency, making it possible to meet human needs while using fewer natural resources. Fossil fuels make it possible for humanity to flourish while still preserving much of the land needed by wildlife to survive. And the prosperity made possible by fossil fuels has made environmental protection both highly valued and financially

possible, producing a world that is cleaner and safer than it would have been in their absence.

Part III. Costs of Fossil Fuels

Part III presents an accounting of the costs of using fossil fuels. Chapters 6 and 7 address impacts on air quality and human security. Chapter 8 reviews the literature on cost-benefit analysis (CBA), integrated assessment models (IAMs), and the "social cost of carbon" (SCC), providing new CBAs for global warming, fossil fuels, and emission mitigation programs.

6. Air Quality

The U.S. Environmental Protection Agency (EPA) claims public health is endangered by exposure to particulate matter (PM), ozone, nitrogen dioxide (NO_2), sulfur dioxide (SO_2), methylmercury, and hydrogen chloride attributed to the combustion of fossil fuels. Other harms include visibility impairment (haze), corrosion of building materials, negative effects on vegetation due to ozone, acid rain, and nitrogen deposition, and negative effects on ecosystems from methylmercury (EPA, 2013).

A review of the evidence shows the EPA and other government agencies exaggerate the public health threat posed by fossil fuels. While the combustion of fossil fuels without pollution abatement technology does release chemicals that could be harmful to humans, other animal life, and plants, the most important issue is not the quantity of emissions but *levels of exposure* (Calabrese and Baldwin, 2003; Calabrese, 2005, 2015). By all accounts, air quality improved in the United States and other developed countries throughout the twentieth century and the trend continues in the twenty-first century (Goklany 2012; EPA, 2018a).

By the EPA's own measures, only 3% of children in the United States live in counties where they might be exposed to what the agency deems "unhealthy air" (EPA, 2018b). Also according to the EPA, 0% of children live in counties in which they might be exposed to harmful levels of carbon monoxide in outdoor air, only 0.1% live in counties where lead exposure might be a threat, 2% live where nitrogen dioxide is a problem, and 3% live where sulfur dioxide is a problem (*Ibid.*). (See Figure SPM.6.)

7

Figure SPM.6
Percentage of children ages 0 to 17 years living in counties with pollutant concentrations above the levels of the current air quality standards, 1999–2016

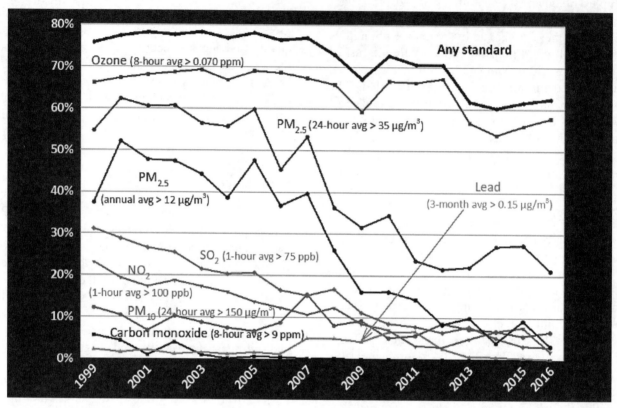

Source: EPA, 2018b, p. 11.

Even these estimates inflate the real public health risk by assuming all children are continuously exposed to the worst air quality measured in the county in which they reside, and by relying on air quality standards that are orders of magnitude lower than medically needed to be protective of human health (Arnett, 2006; Schwartz and Hayward, 2007; Avery, 2010; Belzer, 2017).

The EPA claims PM and ozone remain public health problems in the United States, saying 7% (for PM₁₀) to 21% (for PM₂.₅) of children live in counties where they might be exposed to unhealthy levels of PM and 58% are threatened by ozone. But it is precisely with respect to these two alleged health threats that the EPA's misconduct and violation of sound methodology are most apparent. The agency violated the Bradford Hill Criteria, resisted transparency and accountability for its actions, and even violated the law as it set National Ambient Air Quality Standards (NAAQS) for PM and ozone (Schwartz, 2003; U.S. Senate Committee on Environment and Public Works, 2014; Milloy, 2016).

The EPA's claim that PM kills hundreds of thousands of Americans annually (EPA, 2010, p. G7) is classic scaremongering based on unreliable research (Enstrom, 2005; Milloy and Dunn, 2012; Wolff and Heuss, 2012). The EPA's own measurements show average exposure in the United States to both PM₁₀ and PM₂.₅ has fallen steeply since the 1990s and is now below its NAAQS (EPA, 2018a).

The authors of Chapter 6 conclude that air pollution caused by fossil fuels is unlikely to kill *anyone* in the United States in the twenty-first century, though it may be a legitimate health concern in rapidly growing developing countries that rely on biofuels and burning coal without modern emission control technologies.

7. Human Security

Similar to how the EPA exaggerates the harmful effects of air pollution, the IPCC exaggerates the harmful effects of climate change on "human security," which it defines as "a condition that exists when the vital core of human lives is protected, and when people have the freedom and the capacity to live with dignity" (IPCC, 2014a, p. 759). It collects circumstantial evidence to build a case linking climate change to an almost endless list of maladies, but it never actually tests the null hypothesis that these maladies are due to natural causes. The result is a long and superficially impressive report relying on assumptions and tenuous associations that fall far short of science (Lindzen, 2013; Gleditsch and Nordås, 2014; Tol, 2014b).

Fossil fuels make human prosperity possible (see Chapter 3 and Goklany, 2012). Prosperity in turn, as Benjamin Friedman writes, "more often than not fosters greater opportunity, tolerance of diversity, social mobility, commitment to fairness, and dedication to democracy" (Friedman, 2006, p. 15). All of this serves to protect, not threaten, human security. Prosperity also promotes democracy, and democracies have lower rates of violence and go to war less frequently than any other form of government (Halperin *et al.,* 2004, p. 12).

The cost of wars fought in the Middle East is sometimes attributed to the industrial nations' "addiction to oil." But many of those conflicts have origins and justifications unrelated to oil (Bacevich, 2017; Glaser and Kelanic, 2016; Glaser, 2017). On the verge of becoming a net energy exporter, the United States could withdraw from the region, but it is likely to remain for other geopolitical reasons. If global consumption of oil were to fall as a result of concerns over climate change, the Middle East could become more, not less, violent (Pipes, 2018, p. 21).

Empirical research shows no direct association between climate change and violent conflicts (Salehyan, 2014; Gleditsch and Nordås, 2014). The warming of the second half of the twentieth and early twenty-first centuries coincided with a dramatic decline in the number of fatalities due to warfare. (See Figure SPM.7.) In fact, extensive historical research in China and elsewhere reveals close and positive relationships between a warmer climate and peace and prosperity, and between a cooler climate

Figure SPM.7
Battle-related deaths in state-based conflicts since 1946, by world region

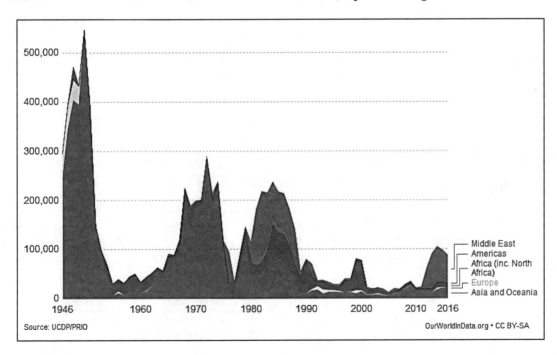

Source: Our World in Data, n.d.

and war and poverty (Yin *et al.*, 2016; Lee *et al.*, 2017). A warmer world is likely to be more prosperous and peaceful than is the world today. Climate change does not pose a military threat to the United States (Kueter, 2012; Hayward *et al.*, 2014). Forcing America's military leaders to utilize costly biofuels, prepare for climate-related humanitarian disasters, and harden military bases for possible changes in weather or sea level attributed to climate change wastes scarce resources and reduces military preparedness (Kiefer, 2013; Smith, 2015).

The authors of Chapter 7 conclude it is probably impossible to attribute to the human impact on climate *any* negative impacts on human security. Deaths and loss of income due to storms, flooding, and other weather-related phenomena are and always have been part of the human condition. Real-world evidence demonstrates warmer weather is closely associated with peace and prosperity, and cooler weather with war and poverty. A warmer world, should it occur, is therefore more likely to bring about peace and prosperity than war and poverty.

8. Cost-Benefit Analysis

Cost-benefit analysis (CBA), sometimes and more accurately called benefit-cost ratio analysis, is an economic tool that can help determine if the financial benefits over the lifetime of a project exceed its costs.

Its use is mandated by executive order for regulations in the United States. In the climate change debate, cost-benefit analysis is used to estimate the net benefits or costs that could result from unabated global warming, from replacing fossil fuels with alternative energy sources, and of particular programs aimed at reducing greenhouse gas emissions or sequestering CO_2. CBA is also employed to estimate the "social cost of carbon."

Chapter 8 starts with a brief tutorial on cost-benefit analysis including its history and use in public policy and the order of "blocks" or "modules" in integrated assessment models (IAMs) (shown in Figure SPM.8). The biggest problem facing the use of IAMs in the climate change debate is the problem of propagation of error, the mounting uncertainty with each step in a complex formula where variables and processes are not known with certainty (Curry, 2011; Frank, 2015, 2016; Heal, 2017). This "cascading uncertainty" makes IAMs "close to useless" for policymakers (Pindyck, 2013). In such cases, the most reliable method of forecasting is not to rely on expert opinion, but to project a simple linear continuation of past trends (Armstrong, 2001).

Two prominent efforts to conduct CBAs of climate change, the U.S. Interagency Working Group on the Social Cost of Carbon (IWG, 2015; since disbanded) and the British Stern Review (Stern, 2007), were severely handicapped by un-

Figure SPM.8
Simplified linear causal chain of an IAM illustrating the basic steps required to obtain SCC estimates

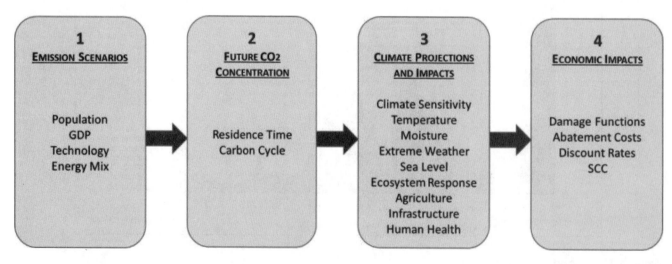

Source: Modified from Parson *et al.*, 2007, Figure ES-1, p. 1.

acknowledged uncertainties, low discount rates, and reliance on the IPCC's flawed climate science (IER, 2014; Byatt, 2006; Mendelsohn, 2006; Tapia Granados and Carpintero, 2013). The complexity of climate science and economics makes conducting any of these CBAs a difficult and perhaps even impossible challenge (Ceronsky *et al.,* 2011). Harvard University Professor of Economics Martin Weitzman remarked, "the economics of climate change is a problem from hell," adding that "trying to do a benefit-cost analysis (BCA) of climate change policies bends and stretches the capability of our standard economist's toolkit up to, and perhaps beyond, the breaking point" (Weitzman, 2015).

Research presented in previous chapters shows how errors or uncertainties in choosing emission scenarios, estimating the amount of carbon dioxide that stays in the atmosphere, the likelihood of increases in flooding and extreme weather, and other

inputs render IAMs unreliable guides for policymakers. Correcting the shortcomings of two of the leading IAMs – the DICE and FUND models – results in a superior analysis that, unsurprisingly, arrives at a very different conclusion, a "social cost of carbon" that is either zero or negative, meaning the social benefits of each additional unit of CO_2 emitted exceed its social costs (Dayaratna *et al.,* 2017).

Figure SPM.9 summarizes evidence presented in previous chapters for all the costs and benefits of fossil fuels. While not exhaustive, the list of impacts in Figure SPM.9 includes most of the topics addressed by the IPCC's Working Group II and can be compared to Assessment Box SPM.2 Table 1 in its Summary for Policymakers (IPCC, 2014a, pp. 21–5). The new review finds 16 of 25 impacts are net benefits, only one is a net cost, and the rest are either unknown or likely to have no net impact.

Figure SPM.9
Impact of fossil fuels on human well-being

Impact	Benefit or Cost	Observations	Chapter References
Acid rain	No net impact	Once feared to be a major environmental threat, the deposition of sulfuric and nitric acid due to smokestack emissions, so-called "acid rain," was later found not to be a threat to forest health and to affect only a few bodies of water, where remediation with lime is an inexpensive solution. The fertilizing effect of nitrogen deposition more than offsets its harms to vegetation. Dramatic reductions in SO_2 and NO_2 emissions since the 1980s mean "acid rain" has no net impact on human well-being today.	5.1, 6.1
Agriculture	Benefit	Fossil fuels have contributed to the enormous improvement in crop yields by making artificial fertilizers, mechanization, and modern food processing techniques possible. Higher atmospheric CO_2 levels are causing plants to grow better and require less water. Numerous studies show the aerial fertilization effect of CO_2 is improving global agricultural productivity, on average by 15%.	3.3, 4.1, 5.2, 5.3, 7.2, 8.2
Air quality	Benefit	Exposure to potentially harmful chemicals in the air has fallen dramatically during the modern era thanks to the prosperity, technologies, and values made possible by fossil fuels. Safe and clean fossil fuels made it possible to rapidly increase energy consumption while improving air quality.	5.2, Chapter 6
Catastrophes	Unknown	No scientific forecasts of possible catastrophes triggered by global warming have been made. CO_2 is not a "trigger" for abrupt climate change. Inexpensive fossil fuel energy greatly facilitates recovery.	7.2, 8.2
Conflict	Benefit	The occurrence of violent conflicts around the world has fallen dramatically thanks to prosperity and the spread of democracy made possible by affordable and reliable energy and a secure food supply.	7.1, 7.3, 8.2
Democracy	Benefit	Prosperity is closely correlated with the values and institutions that sustain democratic governments. Tyranny promoted by zero-sum	7.1

Impact	Benefit or Cost	Observations	Chapter References
		wealth is eliminated. Without fossil fuels, there would be fewer democracies in the world.	
Drought	No net impact	There has been no increase in the frequency or intensity of drought in the modern era. Rising CO_2 lets plants use water more efficiently, helping them overcome stressful conditions imposed by drought.	2.3, 5.3
Economic growth (consumption)	Benefit	Affordable and reliable energy is positively correlated with economic growth rates everywhere in the world. Fossil fuels were indispensable to the three Industrial Revolutions that produced the unprecedented global rise in human prosperity.	Chapter 3, 4.1, 5.2, 7.1, 7.2
Electrification	Benefit	Transmitted electricity, one of the greatest inventions in human history, protects human health in many ways. Fossil fuels directly produce some 80% of electric power in the world. Without fossil fuels, alternative energies could not be built or relied on for continuous power.	Chapter 3, 4.1
Environmental protection	Benefit	Fossil fuels power the technologies that make it possible to meet human needs while using fewer natural resources and less surface space. The aerial CO_2 fertilization effect has produced a substantial net greening of the planet, especially in arid areas, that has been measured using satellites.	1.3, Chapter 5
Extreme weather	No net impact	There has been no increase in the frequency or intensity of extreme weather in the modern era, and therefore no reason to expect any economic damages to result from CO_2 emissions.	2.3, 8.2
Forestry	Benefit	Fossil fuels made it possible to replace horses as the primary means of transportation, saving millions of acres of land for forests. Elevated CO_2 concentrations have positive effects on forest growth and health, including efficiency of water use. Rising CO_2 has reduced and overridden the negative effects of ozone pollution on the photosynthesis, growth, and yield of nearly all the trees that have been evaluated experimentally.	5.3
Human development	Benefit	Affordable energy and electrification, better derived from fossil fuels than from renewable energies, are closely correlated with the United Nations' Human Development Index and advance what the IPCC labels "human capital."	3.2, 4.1, 7.2
Human health	Benefit	Fossil fuels contribute strongly to the dramatic lengthening of average lifespans in all parts of the world by improving nutrition, health care, and human safety and welfare. (See also "Air quality.")	3.2, Chapter 4, 5.2
Human settlements/ migration	Unknown	Forced migrations due to sea-level rise or hydrological changes attributable to man-made climate change have yet to be documented and are unlikely since the global average rate of sea-level rise has not accelerated. Climate change is as likely to decrease as increase the number of people forced to migrate.	7.3
Ocean acidification	Unknown	Many laboratory and field studies demonstrate growth and developmental improvements in aquatic life in response to higher temperatures and reduced water pH levels. Other research illustrates the capability of both marine and freshwater species to tolerate and adapt to the rising temperature and pH decline of the planet's water bodies.	5.5
Oil spills	Cost	Oil spills can harm fish and other aquatic life and contaminate drinking water. The harm is minimized because petroleum is typically reformed by dispersion, evaporation, sinking, dissolution, emulsification, photo-oxidation, resurfacing, tar-ball formation, and biodegradation.	5.1
Other market sectors	No net impact	The losses incurred by some businesses due to climate change, whether man-made or natural, will be offset by profits made by other	1.2, 7.2

Impact	Benefit or Cost	Observations	Chapter References
		businesses taking advantage of new opportunities to meet consumer wants. Institutional adaptation, including of markets, to a small and slow warming is likely.	
Polar ice melting	Unknown	What melting is occurring in mountain glaciers, Arctic sea ice, and polar icecaps is not occurring at "unnatural" rates and does not constitute evidence of a human impact on the climate. Global sea-ice cover remains similar in area to that at the start of satellite observations in 1979, with ice shrinkage in the Arctic Ocean offset by growth around Antarctica.	2.3
Sea-level rise	No net impact	There has been no increase in the rate of increase in global average sea level in the modern era, and therefore no reason to expect any economic damages to result from it. Local sea levels change in response to factors other than climate.	2.3, 8.2
Sustainability	Benefit	Fossil fuels are a sustainable source of energy today and for the foreseeable future. Their impacts do not endanger human health or the environment. A market-based transition to alternative fuels will occur when supply and demand require it.	1.5, 5.2
Temperature-related mortality	Benefit	Cold weather kills more people than warm weather, and fossil fuels enable people to protect themselves from temperature extremes. A world made warmer and more prosperous by fossil fuels would see a net decrease in temperature-related mortality.	4.2
Transportation	Benefit	Fossil fuels revolutionized society by making transportation faster, less expensive, and safer for everyone. The increase in human, raw material, and product mobility was a huge boon for humanity, with implications for agriculture, education, health care, and economic development.	4.1
Vector-borne diseases	No net impact	Warming will have no impact on insect-borne diseases because temperature plays only a small role in the spread of these diseases. The technologies and prosperity made possible by fossil fuels eliminated the threat of malaria in developed countries and could do the same in developing countries regardless of climate change.	4.6
Water resources	Benefit	While access to water is limited by climate and other factors in many locations around the world, there is little evidence warming would have a net negative effect on the situation. Fossil fuels made it possible for water quality in the United States and other industrial countries to improve substantially while improving water use efficiency by about 30% over the past 35 years. Aerial CO_2 fertilization improves plant water use efficiency, reducing the demand for irrigation.	5.2, 5.3

The IPCC's Working Group II says CO_2 emissions must be cut by between 40% and 70% from 2010 levels by 2050 in order to prevent the ~2°C of warming (since pre-industrial times) that would otherwise occur by that year (IPCC, 2014b, pp. 10, 12). Since economic growth is closely related to CO_2 emissions (a proxy for the use of fossil fuels to generate primary energy), the opportunity cost of reducing greenhouse gas (GHG) emissions includes the lost economic prosperity that otherwise would have occurred. Original analysis for this book shows that when this factor is accounted for, reducing GHGs to 70% below 2010 levels by 2050 would

lower world GDP in 2050 by 21% from baseline forecasts. World GDP would be about $231 trillion instead of the $292 trillion now forecast by the World Bank, a loss of $61 trillion.

The IPCC also overlooked the physical limits wind and solar energy face preventing them from generating enough dispatchable energy (available on demand 24/7) to entirely replace fossil fuels, so energy consumption must fall in order for emissions to fall. If global population continues to grow, then per-capita energy consumption must decline even faster. One estimate that takes this factor into account finds reducing GHG emissions by 80% by 2050

Figure SPM.10
Impact of fossil fuels on human health

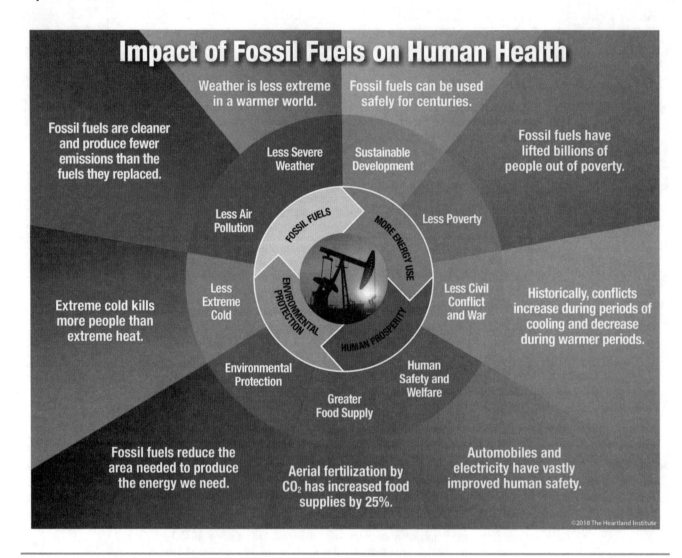

would reduce GDP by 81%, plunging the world into permanent economic recession and undoing all the progress made since 1905 (Tverberg, 2012).

The IPCC estimates the cost of unabated climate change to be between 0.2% and 2% of GDP in 2050 (IPCC, 2014a, p. 663) while the models it relies on produce an average estimate of 0.5%. That is the expected *benefit* of avoiding ~ 2°C of warming by 2050. Since the cost of reducing CO_2 emissions by 70% is approximately 21% of projected GDP that year, the cost-benefit ratio is 42:1 (21 / 0.5). In other words, reducing anthropogenic GHG emissions enough to avoid a 2°C warming by 2050 would cost 42 times as much as the benefits. The estimate by

Tverberg (2012) taking into account the physical limits that prevent alternative energy sources from completely replacing fossil fuels produces an alarming cost-benefit ratio of 162:1 (81 / 0.5).

Cost-benefit analysis can also be applied to greenhouse gas mitigation programs to produce like-to-like comparisons of their cost-effectiveness. The cap-and-trade bill considered by the U.S. Congress in 2009, for example, would have cost 7.4 times more than its benefits, even assuming all of the IPCC's assumptions and claims about climate science were correct. Other bills and programs already in effect have costs exceeding benefits by factors up to 7,000 (Monckton, 2016). In short, even accepting the

IPCC's flawed science and scenarios, there is no justification for adopting GHG emission mitigation programs.

Conclusion

Fossil fuels have benefited humanity by making possible the prosperity that occurred since the first Industrial Revolution, which made possible investments in goods and services that are essential to protecting human health and prolonging human life. Fossil fuels also power the technologies that reduce the environmental impact of a growing human population, saving space for wildlife.

The IPCC and national governments around the world claim the negative impacts of global warming on human health and security, occurring now or likely to occur in the future, more than offset the benefits that come from the use of fossil fuels. This claim lacks any scientific or economic basis. The benefits of fossil fuels are nowhere reported in the IPCC's assessment reports. The analysis conducted here for the first time finds nearly all the impacts of fossil fuel use on human well-being are net positive (benefits minus costs), near zero (no net benefit or cost), or are simply unknown.

The alleged negative human health impacts due to air pollution are exaggerated by researchers who violate the Bradford Hill Criteria and rely too heavily on epidemiological studies finding weak relative risks. The alleged negative impacts on human security due to climate change depend on tenuous chains of causality that find little support in the peer-reviewed literature.

In conclusion, the IPCC and its national counterparts have not conducted proper cost-benefit analyses of fossil fuels, global warming, or regulations designed to force a transition away from fossil fuels. The global war on fossil fuels, which commenced in earnest in the 1980s and reached a fever pitch in the second decade of the twenty-first century, was never founded on sound science or economics. The authors of and contributors to *Climate Change Reconsidered II: Fossil Fuels* urge the world's policymakers to acknowledge this truth and end that war.

References

Ainsworth, E.A. and Long, S.P. 2005. What have we learned from 15 years of free-air CO$_2$ enrichment (FACE)? A meta-analytic review of the responses of photosynthesis, canopy properties and plant production to rising CO$_2$. *New Phytologist* **165**: 351–72.

Anderson, T.L. and Huggins, L.E. 2008. *Greener Than Thou: Are You Really an Environmentalist?* Stanford, CA: Hoover Institution.

Anderson, T.L. and Leal, D.R. 1997. *Enviro-Capitalists: Doing Good While Doing Well.* Lanham, MD: Rowman & Littlefield Publishers, Inc.

Anderson, T.L. and Leal, D.R. 2015. *Free Market Environmentalism for the Next Generation.* New York, NY: Palgrave Macmillan.

Armstrong, J.S. 2001. *Principles of Forecasting – A Handbook for Researchers and Practitioners*. Norwell, MA: Kluwer Academic Publishers.

Armstrong, J.S. and Green, K.C. 2018. Do forecasters of dangerous manmade global warming follow the science? Presented at the International Symposium on Forecasting, Boulder, Colorado, June 18.

Arnett Jr., J.C. 2006. The EPA's fine particulate matter (PM 2.5) standards, lung disease, and mortality: a failure in epidemiology. *Issue Analysis* #4. Washington, DC: Competitive Enterprise Institute.

Avery, G. 2010. Scientific misconduct: The manipulation of evidence for political advocacy in health care and climate policy. *Cato Briefing Papers* No. 117. Washington, DC: Cato Institute. February 8.

Bacevich, A. 2017. *America's War for the Greater Middle East: A Military History*. New York, NY: Random House.

Baker, A.C. 2014. Climate change: many ways to beat the heat for reef corals. *Current Biology* **24**: 10.1016/j.cub.2014.11.014.

Belzer, R. 2017. Testimony before the U.S. House of Representatives Committee on Science, Space, and Technology. February 7.

Berezow, A.B. and Campbell, H. 2012. *Science Left Behind: Feel-Good Fallacies and the Rise of the Anti-Scientific Left*. Philadelphia, PA: PublicAffairs.

Bertinelli, L., Strobl, E., and Zou, B. 2012. Sustainable economic development and the environment: theory and evidence. *Energy Economics* **34** (4): 1105–14.

Bezdek, R.H. 2014. *The Social Costs of Carbon? No, the Social Benefits of Carbon*. Oakton, VA: Management Information Services, Inc.

Bezdek, R.H. 2015. Testimony before the office of administrative hearings for the Minnesota public utilities commission state of Minnesota in the matter of the further investigation into environmental and socioeconomic costs under Minnesota statute 216B.2422, subdivision 3. OAH Docket No. 80-2500-31888, MPUC Docket No. E-999-CI-14-643, June 1.

Block, W.E. (Ed.) 1990. *Economics and the Environment: A Reconciliation*. Toronto, ON: The Fraser Institute.

Boettke, P. 2009. Liberty should rejoice: Elinor Ostrom's Nobel Prize. *The Freeman*. November 18.

Bourgault, M., *et al.* 2017. Yield, growth and grain nitrogen response to elevated CO_2 in six lentil (Lens culinaris) cultivars grown under Free Air CO_2 Enrichment (FACE) in a semi-arid environment. *European Journal of Agronomy* **87:** 50–8.

Bryce, R. 2010. *Power Hungry: The Myths of 'Green' Energy and the Real Fuels of the Future*. New York, NY: PublicAffairs.

Bryce, R. 2014. *Smaller Faster Lighter Denser Cheaper: How Innovation Keeps Proving the Catastrophists Wrong*. New York, NY: PublicAffairs.

Byatt, I., *et al.* 2006. The Stern Review: a dual critique. Part II: economic aspects. *World Economics* **7**: 199–229.

Calabrese, E.J. 2005. Paradigm lost, paradigm found: the re-emergence of hormesis as a fundamental dose response model in the toxicological sciences. *Environmental Pollution* **138** (3): 378–411.

Calabrese, E.J. 2015. On the origins of the linear no-threshold (LNT) dogma by means of untruths, artful dodges and blind faith. *Environmental Research* **142** (October): 432–42.

Calabrese, E.J. and Baldwin, L.A. 2003. Toxicology rethinks its central belief. *Nature* **421** (February): 691–2.

Campbell, J.E., *et al.* 2017. Large historical growth in global terrestrial gross primary production. *Nature* **544**: 84–7.

Ceronsky, M., Anthoff, D., Hepburn, C., and Tol, R.S.J. 2011. Checking the price tag on catastrophe: the social cost of carbon under non-linear climate response. *ESRI Working Paper* No. 392. Dublin, Ireland: Economic and Social Research Institute.

Cheng, L., *et al.* 2017. Recent increases in terrestrial carbon uptake at little cost to the water cycle. *Nature Communications* **8**: 110.

Christy, J.R. 2016. Testimony to the U.S. House Committee on Science, Space & Technology. February 2.

Christy, J. 2017. Testimony before the U.S. House Committee on Science, Space & Technology. March 29.

Christy, J., *et al.* 2018. Examination of space-based bulk atmospheric temperatures used in climate research. *International Journal of Remote Sensing* **39** (11): 3580–607.

Clemente, J. 2010. The statistical connection between electricity and human development. *Power Magazine*. September 1.

Coase, R.H. 1994. *Essays on Economics and Economists*. Chicago, IL: University of Chicago Press.

Curry, J.A. 2011. Reasoning about climate uncertainty. *Climatic Change* **108**: 723.

Curry, J.A. 2012. Climate change: no consensus on consensus. Climate Etc. (website). October 28.

Curry, J. 2015. State of the climate debate in the U.S. Remarks to the U.K. House of Lords, June 15. Climate Etc. (website).

Darwall, R. 2013. *The Age of Global Warming: A History*. London, UK: Quartet Books Ltd.

Dayaratna, K., McKitrick, R., and Kreutzer, D. 2017. Empirically-constrained climate sensitivity and the social cost of carbon. *Climate Change Economics* **8**: 2.

Enstrom, J.E. 2005. Fine particulate air pollution and total mortality among elderly Californians, 1973–2002. *Inhalation Toxicology* **17**: 803–16.

EPA. 2010. U.S. Environmental Protection Agency. *Quantitative Health Risk Assessment for Particulate Matter*. EPA-452/R-10-005. June.

EPA. 2013. U.S. Environmental Protection Agency. *Technical Support Document: Technical Update of the Social Cost of Carbon for Regulatory Impact Analysis under Executive Order 12866*.

EPA. 2018a. U.S. Environmental Protection Agency. Particulate matter ($PM_{2.5}$) trends and particulate matter (PM_{10}) trends (website).

EPA. 2018b. U.S. Environmental Protection Agency. Environments and Contaminants | Criteria Air Pollutants. In: *America's Children and the Environment*. Third edition. Updated January 2018.

Essex, C. and McKitrick, R. 2007. *Taken by Storm: The Troubled Science, Policy, and Politics of Global Warming*. Revised edition. Toronto, ON: Key Porter Books Limited.

Frank, P. 2015. Negligence, non-science, and consensus climatology. *Energy & Environment* **26** (3): 391–415.

Frank, P. 2016. Systematic error in climate measurements: the surface air temperature record. *International Seminars on Nuclear War and Planetary Emergencies 48th Session.* Presentation in Erice, Italy on August 19–25, pp. 337–51.

Friedman, B. 2006. The moral consequences of economic growth. *Society* **43** (January/February): 15–22.

Friends of Science. 2014. *97 Percent Consensus? No! Global Warming Math Myths & Social Proofs.* Calgary, AB: Friends of Science Society.

Gasparrini, A., *et al.* 2015. Mortality risk attributable to high and low ambient temperature: a multicountry observational study. *The Lancet* **386**: 369.

Gill, R.S., Hambridge, H.L., Schneider, E.B., Hanff, T., Tamargo, R.J., and Nyquist, P. 2012. Falling temperature and colder weather are associated with an increased risk of Aneurysmal Subarachnoid Hemorrhage. *World Neurosurgery* **79**: 136–42.

Glaser, C.L. and Kelanic, R.A. (Eds.) 2016. *Crude Strategy: Rethinking the U.S. Military Commitment to Defend Persian Gulf Oil.* Washington, DC: Georgetown University Press.

Glaser, J. 2017. Does the U.S. military actually protect Middle East oil? The National Interest (website). January 9.

Gleditsch, N.P. and Nordås, R. 2014. Conflicting messages? The IPCC on conflict and human security. *Political Geography* **43**: 82–90.

Global Warming Petition Project. n.d. Global warming petition project (website). Accessed July 6, 2018.

Goklany, I.M. 2007. *The Improving State of the World: Why We're Living Longer, Healthier, More Comfortable Lives on a Cleaner Planet.* Washington, DC: Cato Institute.

Goklany, I. 2012. Humanity unbound: how fossil fuels saved humanity from nature and nature from humanity. *Cato Policy Analysis* No. 715. Washington, DC: Cato Institute. December 20.

Gordon, R.J. 2016. *The Rise and Fall of American Growth.* Princeton, NJ: Princeton University Press.

Halperin, M.H., Siegle, J.T., and Weinstein, M.M. 2004. *The Democracy Advantage: How Democracies Promote Prosperity and Peace.* New York, NY: Routledge.

Hartwell, C.A. and Coursey, D.L. 2015. Revisiting the environmental rewards of economic freedom. *Economics and Business Letters* **4** (1): 36–50.

Hayek, F. A. 1988. *The Fatal Conceit: The errors of socialism.* Chicago, IL: University of Chicago Press.

Hayward, T.B, Briggs, E.S., and Forbes, D.K. 2014. *Climate Change, Energy Policy, and National Power.* Chicago, IL: The Heartland Institute.

Heal, G. 2017. The economics of the climate. *Journal of Economic Literature* **55** (3).

Ho, M. and Wang, Z. 2015. Green growth for China? *Resources Magazine.* Resources for the Future. March 3.

Hourdin, F. *et al.* 2017. The art and science of climate model tuning. *Bulletin of the American Meteorological Society* **98** (3): 589–602.

Huggins, L. 2013. *Environmental Entrepreneurship: Markets Meet the Environment in Unexpected Places.* Cheltenham, UK: Edward Elgar.

IAC. 2010. InterAcademy Council. Draft: *Climate Change Assessments: Review of the Processes & Procedures of IPCC.* The Hague, Netherlands: Committee to Review the Intergovernmental Panel on Climate Change. October.

Idso, S.B. and Idso, C.D. 2015. *Mathematical Models vs. Real-World Data: Which Best Predicts Earth's Climatic Future?* Tempe, AZ: Center for the Study of Carbon Dioxide and Global Change.

IER. 2014. Institute for Energy Research. Comment on technical support document: technical update of the social cost of carbon for regulatory impact analysis under executive order no. 12866. February 24.

IPCC. 2013. Intergovernmental Panel on Climate Change. *Climate Change 2013: The Physical Science Basis.* Contribution of Working Group I to the Fifth Assessment Report of the Intergovernmental Panel on Climate Change. New York, NY: Cambridge University Press.

IPCC. 2014a. Intergovernmental Panel on Climate Change. *Climate Change 2014: Impacts, Adaptation, and Vulnerability.* Contribution of Working Group II to the Fifth Assessment Report of the Intergovernmental Panel on Climate Change. New York, NY: Cambridge University Press.

IPCC. 2014b. Intergovernmental Panel on Climate Change. *Climate Change 2014: Mitigation of Climate Change.* Contribution of Working Group III to the Fifth Assessment Report of the Intergovernmental Panel on Climate Change. New York, NY: Cambridge University Press.

IWG. 2015. Interagency Working Group on the Social Cost of Carbon. *Technical Support Document: Technical Update of the Social Cost of Carbon for Regulatory Impact*

Analysis Under Executive Order 12866. Washington, DC. May.

Kabat, G.C. 2008. *Hyping Health Risks: Environmental Hazards in Daily Life and the Science of Epidemiology*. New York, NY: Columbia University Press.

Keatinge, W.R. and Donaldson, G.C. 2004. The impact of global warming on health and mortality. *Southern Medical Journal* **97**: 1093–9.

Kiefer, T.A. 2013. Energy insecurity: the false promise of liquid biofuels. *Strategic Studies Quarterly* (Spring): 114–51.

Kueter, J. 2012. *Climate and National Security: Exploring the Connection*. Washington, DC: George C. Marshall Institute.

Laframboise, D. 2011. *The Delinquent Teenager Who was Mistaken for the World's Top Climate Expert*. Toronto, ON: Ivy Avenue Press.

Laframboise, D. 2013. *Into the Dustbin: Rachendra Pachauri, the Climate Report & the Nobel Peace Prize*. CreateSpace Independent Publishing Platform.

Lee, H.F., Fei, J., Chan, C.Y.S., Pei, Q., Jia, X., and Yue, R.P.H. 2017. Climate change and epidemics in Chinese history: a multi-scalar analysis. *Social Science & Medicine* **174**: 53–63.

Legates, D. 2014. Climate models and their simulation of precipitation. *Energy & Environment* **25** (6–7): 1163–75.

Legates, D.R., Soon, W., Briggs, W.M., and Monckton, C. 2015. Climate consensus and 'misinformation': a rejoinder to agnotology, scientific consensus, and the teaching and learning of climate change. *Science & Education* **24** (3): 299–318.

Lemoine, D. and Rudik, I. 2017. Steering the climate system: using inertia to lower the cost of policy. *American Economic Review* **107** (20).

Lewin, B. 2017. *Searching for the Catastrophe Signal: The Origins of the Intergovernmental Panel on Climate Change*. London, UK: Global Warming Policy Foundation.

Libecap, G.D. and Steckel, R.H. (Eds.) 2011. *The Economics of Climate Change: Adaptations Past and Present*. Chicago, IL: University of Chicago Press.

Lindzen, R.S. 2013. MIT scientist ridicules IPCC climate change report, calls findings 'hilarious incoherence.' *Daily Mail*. September 30.

Lindzen, R.S. 2017. Straight talk about climate change. *Academic Questions* **30**: 419–32.

Lomborg, B. (Ed.) 2010. *Smart Solutions to Climate Change: Comparing Costs and Benefits*. New York, NY: Cambridge University Press.

Markandya, A. and Richardson, J. (Eds.) 1992. *Environmental Economics: A Reader*. New York, NY: St. Martin's Press.

McKitrick, R.R. 2010. *Economic Analysis of Environmental Policy*. Toronto, ON: University of Toronto Press.

McLean, J. 2018. *An Audit of the Creation and Content of the HadCRUT4 Temperature Dataset*. Robert Boyle Publishing.

Mendelsohn, R.O. 2006. A critique of the Stern Report. *Regulation* **29**: 42–6.

Michaels, P. 2017. Testimony. Hearing on At What Cost? Examining the Social Cost of Carbon, before the U.S. House of Representatives Committee on Science, Space, and Technology, Subcommittee on Environment, Subcommittee on Oversight. February 28.

Milloy, S. 2016. *Scare Pollution: Why and How to Fix the EPA*. Lexington, KY: Bench Press Inc.

Milloy, S. and Dunn, J. 2012. Environmental Protection Agency's air pollution research: unethical and illegal? *Journal of American Physicians and Surgeons* **17** (4): 109–10.

Monckton of Brenchley, C. 2016. Is CO_2 mitigation cost effective? In: Easterbrook, D. (Ed.) *Evidence-Based Climate Science*. Second edition. Amsterdam, Netherlands: Elsevier, pp. 175–87.

Monckton, C. *et al.* 2015. Keeping it simple: the value of an irreducibly simple climate model. *Science Bulletin* **60** (15).

Moore, S. and Hartnett White, K. 2016. *Fueling Freedom: Exposing the Mad War on Energy*. Washington, DC: Regnery Publishing.

Moore, S. and Simon, J. 2000. *It's Getting Better All the Time: 100 Greatest Trends of the Last 100 Years*. Washington, DC: Cato Institute.

Morriss, A.P., Bogart, W.T., Meiners, R.E., and Dorchak, A. 2011. *The False Promise of Green Energy*. Washington, DC: Cato Institute.

Morriss, A. and Butler, M. 2013. *Creation and the Heart of Man: An Orthodox Christian Perspective on Environmentalism*. Grand Rapids, MI: Acton Institute.

Murdock, C.C., Sternberg, E.D., and Thomas, M.B. 2016. Malaria transmission potential could be reduced with current and future climate change. *Scientific Reports* **6**: 10.1038/srep27771.

Nafstad, P., Skrondal, A., and Bjertness, E. 2001. Mortality and temperature in Oslo, Norway, 1990–1995. *European Journal of Epidemiology* **17**: 621–7.

NAPAP. 1998. National Acid Precipitation Assessment Program. *Biennial Report to Congress: An Integrated Assessment.* Silver Spring, MD.

NIPCC. 2009. Nongovernmental International Panel on Climate Change. Idso, C.D. and Singer, S.F. (Eds.) *Climate Change Reconsidered: The 2009 Report of the Nongovernmental International Panel on Climate Change (NIPCC).* Chicago, IL: The Heartland Institute.

NIPCC. 2011. Nongovernmental International Panel on Climate Change. Idso, C.D., Carter, R.M., and Singer, S.F. (Eds.) *Climate Change Reconsidered: 2011 Interim Report.* Chicago, IL: The Heartland Institute.

NIPCC. 2013. Nongovernmental International Panel on Climate Change. Idso, C.D., Carter, R.M., and Singer, S.F. (Eds.) *Climate Change Reconsidered: Physical Science.* Chicago, IL: The Heartland Institute.

NIPCC. 2014. Nongovernmental International Panel on Climate Change. Idso, C.D, Idso, S.B., Carter, R.M., and Singer, S.F. (Eds.) *Climate Change Reconsidered II: Biological Impacts.* Chicago, IL: The Heartland Institute.

NRC. 2003. U.S. National Research Council Committee on Oil in the Sea. *Oil in the Sea III: Inputs, Fates, and Effects.* Washington, DC: National Academies Press.

Omran, A.R. 1971. The epidemiologic transition: a theory of the epidemiology of population change. *Milbank Memorial Fund Quarterly* **49** (4 part 1): 509–38.

Ostrom, E. 2010. Polycentric systems for coping with collective action and global environmental change. *Global Environmental Change* **20**: 550–7.

Our World in Data. n.d. War and peace after 1945. Accessed July 6, 2018.

Pandolfi, J.M., Connolly, S.R., Marshall, D.J., and Cohen, A.L. 2011. Projecting coral reef futures under global warming and ocean acidification *Science* **333**: 418–22.

Parson, E., *et al.* 2007. Global-change scenarios: their development and use. *U.S. Department of Energy Publications* 7. Washington, DC.

Pindyck, R.S. 2013. Pricing carbon when we don't know the right price. *Regulation* **36** (2): 43–6.

Pipes, D. 2018. The end of carbon fuels? A symposium of views. *The International Economy* (Spring): 10–21.

Rasmussen, K. 2010. *A Rational Look at Renewable Energy and the Implications of Intermittent Power.* Edition 2.0 (November). South Jordan, UT: Deseret Power.

Rose, A. and Wei, D. 2006. *The Economic Impacts of Coal Utilization and Displacement in the Continental United States, 2015.* State College, PA: Pennsylvania State University. Report prepared for the Center for Energy and Economic Development, Inc. July.

Salehyan, I. 2014. Climate change and conflict: making sense of disparate findings. *Political Geography* **43**: 1–5.

Schwartz, J. 2003. *No Way Back: Why Air Pollution Will Continue to Decline.* Washington, DC: American Enterprise Institute.

Schwartz, J. and Hayward, S. 2007. *Air Quality in America: A Dose of Reality on Air Pollution Levels, Trends, and Health Risks.* Washington, DC: AEI Press.

Šlaus, I. and Jacobs, G. 2011. Human capital and sustainability. *Sustainability* **3** (1): 97–154.

Smil, V. 2005. *Energy at the Crossroads: Global Perspectives and Uncertainties.* Cambridge, MA: The MIT Press.

Smil, V. 2010. *Energy Transitions: History, Requirements and Prospects.* New York, NY: Praeger.

Smil, V. 2016. *Power Density: A Key to Understanding Energy Sources and Uses.* Cambridge, MA: The MIT Press.

Smith, T. 2015. Critique of "climate change adaptation: DOD can improve infrastructure planning and processes to better account for potential impacts." *Policy Brief.* Chicago IL: The Heartland Institute.

Solomon, L. 2010. 75 climate scientists think humans contribute to global warming. *National Post.* December 30.

Song, X., *et al.* 2018. The impact of heat waves and cold spells on respiratory emergency department visits in Beijing, China. *Science of the Total Environment* **615**: 1499–1505.

Stacy, T.F. and Taylor, G.S. 2016. *The Levelized Cost of Electricity from Existing Generation Resources.* Washington, DC: Institute for Energy Research.

Steckel, R.H. and Rose, J.C. (Eds.) 2002. *Backbone of History: Health and Nutrition in the Western Hemisphere.* Cambridge, UK: Cambridge University Press.

Stern, N., *et al.* 2007. *The Economics of Climate Change: The Stern Review.* Cambridge, UK: Cambridge University Press.

Tapia Granados, J.A. and Carpintero, O. 2013. Dynamics and economic aspects of climate change. In: Kang, M.S. and Banga, S.S. (Eds.) *Combating Climate Change: An Agricultural Perspective*. Boca Raton, FL: CRC Press.

Tol, R.J.J. 2014a. Quantifying the consensus on anthropogenic global warming in the literature: a re-analysis. *Energy Policy* **73**: 701.

Tol, R.S.J. 2014b. Bogus prophecies of doom will not fix the climate. *Financial Times*. March 31.

Tverberg, G. 2012. An energy/GDP forecast to 2050. Our Finite World (website). July 26.

United Nations Development Program. 2010. *Human Development Report, 2010*. New York, NY: United Nations Development Program.

U.S. Census Bureau. 2016. *An Aging World: 2015*. International Population Reports P95/16-1. Washington, DC: U.S. Department of Commerce.

U.S. Chamber of Commerce. 2014. *Assessing the Impact of Potential New Carbon Regulations in the United States*. Institute for 21st Century Energy. Washington, DC: U.S. Chamber of Commerce.

U.S. Senate Committee on Environment and Public Works. 2014. *EPA's Playbook Unveiled: A Story of Fraud, Deceit, and Secret Science*. March 19.

van Kooten, G.C. 2013. *Climate Change, Climate Science and Economics*. New York, NY: Springer.

Weitzman, M.L. 2015. A review of William Nordhaus' "The climate casino: risk, uncertainty, and economics for a warming world." *Review of Environmental Economics and Policy* **9**: 145–56.

White, C. 2017. The dynamic relationship between temperature and morbidity. *Journal of the Association of Environmental and Resource Economists* **4**: 1155–98.

Willis, K.J., Bennett, K.D., Bhagwat, S.A., and Birks, H.J.B. 2010. 4°C and beyond: what did this mean for biodiversity in the past? *Systematics and Biodiversity* **8**: 3–9.

Wolff, G.T. and Heuss, J.M. 2012. *Review and Critique of U.S. EPA's Assessment of the Health Effects of Particulate Matter (PM)*. Air Improvement Resource, Inc. Prepared for the American Coalition for Clean Coal Electricity. August 28.

Yandle, B., Vijayaraghavan, M., and Bhattarai, M. 2004. The environmental Kuznets curve: a review of findings, methods, and policy implications. *PERC Research Study* 02-1. Bozeman, MT: PERC.

Yin, J., Fang, X., and Su, Y. 2016. Correlation between climate and grain harvest fluctuations and the dynastic transitions and prosperity in China over the past two millennia. *The Holocene* **26**: 1914–23.

Yonk, R.M., Simmons, R.T, and Steed, B.C. 2012. *Green v. Green: The Political, Legal, and Administrative Pitfalls Facing Green Energy Production*. Oxford, UK: Routledge.

Zhu, Z., *et al.* 2016. Greening of the Earth and its drivers. *Nature Climate Change* **6**: 791–5.

Key Findings

Key findings appear at the top of each chapter and many chapter sections. The following tables collect all the key findings for easier reviewing.

1. Environmental Economics	
Introduction	No one should assume the "science is settled" regarding anthropogenic climate change or that the only role for economists is to recommend the most efficient way to reduce "carbon pollution."
1.1 History	
	Economists have been addressing environmental issues since the discipline was founded in the eighteenth century.
	Economies and ecological systems have many commonalities, with the result that economics and ecology share many key concepts.
	Economists have shown markets can manage access to common-pool resources better than government agencies.
1.2 Key Concepts	
1.2.1 Opportunity Cost	The cost of any choice is the value of forgone uses of the funds or time spent. Economists call this "opportunity cost."
1.2.2 Competing Values	Climate change is not a conflict between people who are selfish and those who are altruistic. People who oppose immediate action to reduce greenhouse gas emissions are just as ethical or moral as those who support such action.
1.2.3 Prices	Market prices capture and make public local knowledge that is complex, dispersed, and constantly changing.
1.2.4 Incentives	Most human action can be understood by understanding the incentives people face. "Moral hazard" occurs when people are able to escape full responsibility for their actions.
1.2.5 Trade	Trade creates value by making both parties better off.
1.2.6 Profits and Losses	Profits and losses direct investments to their highest and best uses.
1.2.7 Unintended Consequences	The art of economics consists in looking not merely at the immediate but at the longer-term effects of any act or policy.
1.2.8 Discount Rates	Discount rates, sometimes referred to as the "social rate of time preference," are used to determine the current value of future costs and benefits.
1.2.9 Cost-benefit Analysis	Cost-benefit analysis, when performed correctly, can lead to better public policy decisions.

1.3 Private Environmental Protection	
1.3.1 Common-pool Resources	Common-pool resources have been successfully protected by tort and nuisance laws and managed by nongovernmental organizations.
1.3.2 Cooperation	Voluntary cooperation can generate efficient solutions to conflicts involving negative externalities.
1.3.3 Prosperity	Prosperity leads to environmental protection becoming a higher social value and provides the resources needed to make it possible.
1.3.4 Local Knowledge	The information needed to anticipate changes and decide how best to respond is local knowledge and the most efficient responses will be local solutions.
1.3.5 Ecological Economics	"Ecological economics" is not a reliable substitute for rigorous mainstream environmental economics.
1.4 Government Environmental Protection	
1.4.1 Property Rights	Governments can protect the environment by helping to define and enforce property rights.
1.4.2 Regulation	Regulations often fail to achieve their objectives due to the conflicting incentives of individuals in governments and the absence of reliable and local knowledge.
	Evidence of "market failure" does not mean government intervention can improve market outcomes.
1.4.3 Bureaucracy	Government bureaucracies predictably fall victim to regulatory capture, tunnel vision, moral hazard, and corruption.
1.4.4 Rational Ignorance	Voters have little incentive to become knowledgeable about many public policy issues. Economists call this "rational ignorance."
1.4.5 Rent-seeking Behavior	Government's ability to promote the goals of some citizens at the expense of others leads to resources being diverted from production into political action. Economists call this "rent-seeking behavior."
1.4.6 Displacement	Government policies that erode the protection of property rights reduce the incentive and ability of owners to protect and conserve their resources. Those policies displace, rather than improve or add to, private environmental protection.
1.4.7 Leakage	"Leakage" occurs when the emissions reduced by a regulation are partially or entirely offset by changes in behavior.
1.5 Future Generations	
1.5.1 Conservation and Protection	Capital markets create information, signals, and incentives to manage assets for long-term value.
1.5.2 Innovation	Markets reward innovations that protect the environment by using less energy and fewer raw materials per unit of output.
1.5.3 Small versus Big Mistakes	Mistakes made in markets tend to be small and self-correcting. Mistakes made by governments tend to be big and more likely to have catastrophic effects.
1.6 Conclusion	
	Climate change is not a problem to be solved by markets or government intervention. It is a complex phenomenon involving choices made by millions or even billions of people producing countless externalities both positive and negative.
	The best responses to climate change are likely to arise from voluntary cooperation mediated by nongovernmental entities using knowledge of local costs and opportunities.
	Energy freedom – allowing markets rather than governments to make important choices about which fuels to use – can turn climate change from a possible *tragedy* of the commons into an *opportunity* of the commons.

2. Climate Science

2.1 A Science Tutorial

2.1.1 Methodology	The Scientific Method is a series of requirements imposed on scientists to ensure the integrity of their work. The IPCC has not followed established rules that guide scientific research.
	Appealing to consensus may have a place in science, but should never be used as a means of shutting down debate.
	Uncertainty in science is unavoidable but must be acknowledged. Many declaratory and predictive statements about the global climate that appear in the IPCC's reports are not warranted by science.
2.1.2 Observations	Surface air temperature is governed by energy flow from the Sun to Earth and from Earth back into space. Whatever diminishes or intensifies this energy flow can change air temperature.
	Levels of carbon dioxide (CO_2) and methane (CH_4) in the atmosphere are governed by processes of the carbon cycle. Exchange rates and other climatological processes are poorly understood.
	The geological record shows temperatures and CO_2 levels in the atmosphere have not been stable, making untenable the IPCC's assumption that they would be stable in the future in the absence of human emissions.
	Water vapor is the dominant greenhouse gas owing to its abundance in the atmosphere and the wide range of spectra in which it absorbs radiation. Carbon dioxide absorbs energy only in a very narrow range of the longwave infrared spectrum.

2.2 Controversies

2.2.1 Temperature Records	Reconstructions of average global surface temperature differ depending on the methodology used. The warming of the twentieth and early twenty-first centuries has not been shown to be beyond the bounds of natural variability.
2.2.2 Climate Models	General circulation models (GCMs) are unable to accurately depict complex climate processes. They do not accurately hindcast or forecast the climate effects of anthropogenic greenhouse gas emissions.
2.2.3 Climate Sensitivity	Estimates of equilibrium climate sensitivity (the amount of warming that would occur following a doubling of atmospheric CO_2 level) range widely. The IPCC's estimate is higher than many recent estimates.
2.2.4 Solar Influence	Solar irradiance, magnetic fields, UV fluxes, cosmic rays, and other solar activity may have greater influence on climate than climate models and the IPCC currently assume.

2.3 Climate Impacts

2.3.1 Severe Weather Events	There is little evidence that the warming of the twentieth and early twenty-first centuries has caused a general increase in severe weather events. Meteorological science suggests a warmer world would see milder weather patterns.
	The link between warming and drought is weak, and by some measures drought decreased over the twentieth century. Changes in the hydrosphere of this type are regionally highly variable and show a closer correlation with multidecadal climate rhythmicity than they do with global temperature.
2.3.2 Melting Ice	The Antarctic ice sheet is likely to be unchanged or is gaining ice mass. Antarctic sea ice is gaining in extent, not retreating. Recent trends in the Greenland ice sheet mass and Arctic sea ice are not outside natural variability.
2.3.3 Sea-level Rise	Long-running coastal tide gauges show the rate of sea-level rise is not accelerating. Local and regional sea levels exhibit typical natural variability.
2.3.4 Harm to Plant Life	The effects of elevated CO_2 on plant characteristics are net positive, including increasing rates of photosynthesis and biomass production.

2.4 Why Scientists Disagree	
2.4.1 Scientific Uncertainties	Fundamental uncertainties and disagreements prevent science from determining whether human greenhouse gas emissions are having effects on Earth's atmosphere that could endanger life on the planet.
2.4.2 An Interdisciplinary Subject	Climate is an interdisciplinary subject requiring insights from many fields of study. Very few scholars have mastery of more than one or two of these disciplines.
2.4.3 Failure of the IPCC	Many scientists trust the Intergovernmental Panel on Climate Change (IPCC) to objectively report the latest scientific findings on climate change, but it has failed to produce balanced reports and has allowed its findings to be misrepresented to the public.
2.4.4 Tunnel Vision	Climate scientists, like all humans, can have tunnel vision. Bias, even or especially if subconscious, can be especially pernicious when data are equivocal and allow multiple interpretations, as in climatology.
2.5 Appeals to Consensus	
2.5.1 Flawed Surveys	Surveys and abstract-counting exercises that are said to show a "scientific consensus" on the causes and consequences of climate change invariably ask the wrong questions or the wrong people. No survey data exist supporting claims of consensus on important scientific questions.
2.5.2 Evidence of Lack of Consensus	Some survey data, petitions, and peer-reviewed research show deep disagreement among scientists on issues that must be resolved before the anthropogenic global warming hypothesis can be accepted.
2.5.3 Petition Project	Some 31,000 scientists have signed a petition saying "there is no convincing scientific evidence that human release of carbon dioxide, methane, or other greenhouse gases is causing or will, in the foreseeable future, cause catastrophic heating of the Earth's atmosphere and disruption of the Earth's climate."
2.5.4 Conclusion	Because scientists disagree, policymakers must exercise special care in choosing where they turn for advice.
2.6 Conclusion	
	Fundamental uncertainties arising from insufficient observational evidence and disagreements over how to interpret data and set the parameters of models prevent science from determining whether human greenhouse gas emissions are having effects on Earth's atmosphere that could endanger life on the planet.
	There is no compelling scientific evidence of long-term trends in global mean temperatures or climate impacts that exceed natural variability.

3. Human Prosperity

3.1 An Energy Tutorial	
3.1.1 Definitions	Some key concepts include energy, power, watts, joules, and power density.
3.1.2 Efficiency	Advances in efficiency mean we live lives surrounded by the latest conveniences, yet we use only about 3.5 times as much energy per capita as did our ancestors in George Washington's time.
3.1.3 Energy Uses	Increased use of energy and greater energy efficiency have enabled great advances in artificial light, heat generation, and transportation.
3.1.4 Energy Sources	Fossil fuels supply 81% of the primary energy consumed globally and 78% of energy consumed in the United States.
3.1.5 Intermittency	Due to the nature of wind and sunlight, wind turbines and solar photovoltaic (PV) cells can produce power only intermittently.
3.2 Three Industrial Revolutions	
3.2.1 Creating Modernity	Fossil fuels make possible such transformative technologies as nitrogen fertilizer, concrete, the steam engine and cotton gin, electrification, the internal combustion engine, and the computer and Internet revolution.

3.2.2 Electrification	Electricity powered by fossil fuels has made the world a healthier, safer, and more productive place.
3.2.3 Human Well-being	Access to energy is closely associated with key measures of global human development including per-capita GDP, consumption expenditure, urbanization rate, life expectancy at birth, and the adult literacy rate.
3.3 Food Production	
3.3.1 Fertilizer and Mechanization	Fossil fuels have greatly increased farm worker productivity thanks to nitrogen fertilizer created by the Haber-Bosch process and farm machinery built with and fueled by fossil fuels.
3.3.2 Aerial Fertilization	Higher levels of carbon dioxide (CO_2) in the atmosphere act as fertilizer for the world's plants.
3.3.3 Economic Value of Aerial Fertilization	The aerial fertilization effect of rising levels of atmospheric CO_2 produced global economic benefits of $3.2 trillion from 1961 to 2011 and currently amount to approximately $170 billion annually.
3.3.4 Future Value of Aerial Fertilization	Over the period 2012 through 2050, the cumulative global economic benefit of aerial fertilization will be approximately $9.8 trillion.
3.3.5 Proposals to Reduce CO_2	Reducing global CO_2 emissions by 28% from 2005 levels, the reduction President Barack Obama proposed in 2015 for the United States, would reduce aerial fertilization benefits by $78 billion annually.
3.4 Why Fossil Fuels?	
3.4.1 Power Density	Fossil fuels have higher power density than all alternative energy sources except nuclear power.
3.4.2 Sufficient Supply	Fossil fuels are the only sources of fuel available in sufficient quantities to meet the needs of modern civilization.
3.4.3 Flexibility	Fossil fuels provide energy in the forms needed to make electricity dispatchable (available on demand 24/7), and they can be economically transported to or stored near the places where energy is needed.
3.4.4 Inexpensive	Fossil fuels in the United States are so inexpensive that they make home heating, electricity, and transportation affordable for even low-income households.
3.5 Alternatives to Fossil Fuels	
3.5.1 Lower Power Density	The low power density of alternatives to fossil fuels is a crippling deficiency that prevents them from ever replacing fossil fuels in most applications.
3.5.2 Limited Supply	Wind, solar, and biofuels cannot be produced and delivered where needed in sufficient quantities to meet current and projected energy needs.
3.5.3 Intermittency	Due to their intermittency, solar and wind cannot power the revolving turbine generators needed to create dispatchable energy.
3.5.4 High Cost	Electricity from new wind capacity costs approximately 2.7 times as much as electricity from existing coal, 3 times more than natural gas, and 3.7 times more than nuclear power.
3.5.5 Future Cost	The cost of alternative energies will fall too slowly to close the gap with fossil fuels before hitting physical limits on their capacity.
3.6 Economic Value of Fossil Fuels	
3.6.1 Energy and GDP	Abundant and affordable energy supplies play a key role in enabling economic growth.
3.6.2 Estimates of Economic Value	Estimates of the value of fossil fuels vary but converge on very high numbers. Coal alone delivered economic benefits worth between $1.3 trillion and $1.8 trillion of U.S. GDP in 2015.
	Reducing global reliance on fossil fuels by 80% by 2050 would probably reduce global GDP by $137.5 trillion from baselines projections.

4. Human Health Benefits	
4.1 Modernity and Public Health	
4.1.1 Technology and Health	Fossil fuels improved human well-being and safety by powering labor-saving and life-protecting technologies such as cars and trucks, plastics, and modern medicine.
4.1.2 Public Health Trends	Fossil fuels play a key and indispensable role in the global increase in life expectancy.
4.2 Morality Rates	
	Cold weather kills more people than warm weather. A warmer world would see a net decrease in temperature-related mortality in virtually all parts of the world, even those with tropical climates.
	Weather is less extreme in a warmer world, resulting in fewer injuries and deaths due to storms, hurricanes, flooding, etc.
4.3 Cardiovascular Disease	
	Higher surface temperatures would reduce the incidence of fatal coronary events related to low temperatures and wintry weather by a greater degree than they would increase the incidence associated with high temperatures and summer heat waves.
	Non-fatal myocardial infarction is also less frequent during unseasonably warm periods than during unseasonably cold periods.
4.4 Respiratory Disease	
	Climate change is not increasing the incidence of death, hospital visits, or loss of work or school time due to respiratory disease.
	Low minimum temperatures are a greater risk factor than high temperatures for outpatient visits for respiratory diseases.
4.5 Stroke	
	Higher surface temperatures would reduce the incidence of death due to stroke in many parts of the world, including Africa, Asia, Australia, the Caribbean, Europe, Japan, Korea, Latin America, and Russia.
	Low minimum temperatures are a greater risk factor than high temperatures for stroke incidence and hospitalization.
4.6 Insect-borne Diseases	
	Higher surface temperatures are not leading to increases in mosquito-transmitted and tick-borne diseases such as malaria, yellow fever, viral encephalitis, and dengue fever.
4.6.1 Malaria	Extensive scientific information and experimental research contradict the claim that malaria will expand across the globe and intensify as a result of CO_2-induced warming.
4.6.2 Dengue Fever	Concerns over large increases in dengue fever as a result of rising temperatures are unfounded and unsupported by the scientific literature, as climatic indices are poor predictors for dengue fever.
4.6.3 Tick-borne Diseases	Climate change has not been the most significant factor driving recent changes in the distribution or incidence of tick-borne diseases.
4.7 Conclusion	
	Fossil fuels directly benefit human health and longevity by powering labor-saving and life-protecting technologies and perhaps indirectly by contributing to a warmer world.

5. Environmental Benefits	
5.1 Fossil Fuels in the Environment	
	Fossil fuels are composed mainly of carbon and hydrogen atoms (and oxygen, in the case of low-grade coal). Carbon and hydrogen appear abundantly throughout the universe and on Earth.

	In addition to mining and drilling, hydrocarbons also enter the environment through natural seepage, industrial and municipal effluent and run-off, leakage from underground storage or wells, and spills and other accidental releases.
	The chemical characteristics of fossil fuels make them uniquely potent sources of fuel. They are more abundant, compact, and reliable, and cheaper and safer to use, than other energy sources.
5.2 Direct Benefits	
5.2.1 Efficiency	The greater efficiency made possible by technologies powered by fossil fuels makes it possible to meet human needs while using fewer natural resources, thereby benefiting the environment.
5.2.2 Saving Land for Wildlife	Fossil fuels make it possible for humanity to flourish while still preserving much of the land needed by wildlife to survive.
5.2.3 Prosperity	The prosperity made possible by fossil fuels has made environmental protection both highly valued and financially possible, producing a world that is cleaner and safer than it would have been in their absence.
5.3. Impact on Plants	
5.3.1 Introduction	
5.3.2 Ecosystem Effects	Elevated CO_2 improves the productivity of ecosystems both in plant tissues aboveground and in the soils beneath them.
5.3.3 Plants under Stress	Atmospheric CO_2 enrichment ameliorates the negative effects of a number of environmental plant stresses including high temperatures, air and soil pollutants, herbivory, nitrogen deprivation, and high levels of soil salinity.
5.3.4 Water Use Efficiency	Exposure to elevated levels of atmospheric CO_2 prompts plants to increase the efficiency of their use of water, enabling them to grow and reproduce where it previously has been too dry for them to exist.
5.3.5 Future Impacts on Plants	The productivity of the biosphere is increasing in large measure due to the aerial fertilization effect of rising atmospheric CO_2.
	The benefits of CO_2 enrichment will continue even if atmospheric CO_2 rises to levels far beyond those forecast by the IPCC.
5.4 Impact on Terrestrial Animals	
	The IPCC's forecasts of possible extinctions of terrestrial animals are based on computer models that have been falsified by data on temperature changes, other climatic conditions, and real-world changes in wildlife populations.
5.4.1 Evidence of Ability to Adapt	Animal species are capable of migrating, evolving, and otherwise adapting to changes in climate that are greater and more sudden than what is likely to result from the human impact on the global climate.
5.4.2 Future Impacts on Terrestrial Animals	Although there likely will be some changes in terrestrial animal population dynamics, few if any will be driven even close to extinction.
5.5 Impact on Aquatic Life	
	The IPCC's forecasts of dire consequences for life in the world's oceans rely on falsified computer models and are contradicted by real-world observations.
5.5.1 Evidence of Ability to Adapt	Aquatic life demonstrates tolerance, adaptation, and even growth and developmental improvements in response to higher temperatures and reduced water pH levels ("acidification").
5.5.2 Future Impacts on Aquatic Life	The pessimistic projections of the IPCC give way to considerable optimism with respect to the future of the planet's marine life.
5.6 Conclusion	
	Combustion of fossil fuels has helped and will continue to help plants and animals thrive leading to shrinking deserts, expanded habitat for wildlife, and greater biodiversity.

6. Air Quality

6.1 An Air Quality Tutorial

6.1.1 Chemistry	The combustion of fossil fuels without air pollution abatement technology releases chemicals known to be harmful to humans, other animal life, and plants.
6.1.2 Exposure	At low levels of exposure, the chemical compounds produced by burning fossil fuels are not known to be toxic.
6.1.3 Trends	Exposure to potentially harmful emissions from the burning of fossil fuels in the United States declined rapidly in recent decades and is now at nearly undetectable levels.
6.1.4 Interpreting Exposure Data	Exposure to chemical compounds produced during the combustion of fossil fuels is unlikely to cause any fatalities in the United States.

6.2 Failure of the EPA

6.2.1 A Faulty Mission	Due to its faulty mission, flawed paradigm, and political pressures on it to chase the impossible goal of zero risk, the U.S. Environmental Protection Agency (EPA) is an unreliable source of research on air quality and its impact on human health.
6.2.2 Violating the Scientific Method	The EPA makes many assumptions about relationships between air quality and human health, often in violation of the Bradford Hill Criteria and other basic requirements of the Scientific Method.
6.2.3 Lack of Integrity and Transparency	The EPA has relied on research that cannot be replicated and violates basic protocols for conflict of interest, peer review, and transparency.
	By conducting human experiments involving exposure to levels of particulate matter and other pollutants it claims to be deadly, the EPA reveals it doesn't believe its own epidemiology-based claims of a deadly threat to public health.
	While the new administration has pledged to improve matters, some current regulations and ambient air quality standards are based on flawed data.

6.3 Observational Studies

6.3.1 Reliance on Observational Studies	Observational studies are easily manipulated, cannot prove causation, and often do not support a hypothesis of toxicity with the small associations found in uncontrolled observational studies.
	Observational studies cited by the EPA fail to show relative risks (RR) that would suggest a causal relationship between chemical compounds released during the combustion of fossil fuels and adverse human health effects.
6.3.2 The Particulate Matter Scare	Real-world data and common sense contradict claims that ambient levels of particulate matter kill hundreds of thousands of Americans and millions of people around the world annually.

6.4 Circumstantial Evidence

	Circumstantial evidence cited by the EPA and other air quality regulators is easily refuted by pointing to contradictory evidence.
	EPA cannot point to any cases of death due to inhaling particulate matter, even in environments where its National Ambient Air Quality Standard (NAAQS) is exceeded by orders of magnitude.
	Life expectancy continues to rise in the United States and globally despite what should be a huge death toll, said to be equal to the entire death toll caused by cancer, attributed by the EPA and WHO to just a single pollutant, particulate matter.

6.5 Conclusion

	It is unlikely that the chemical compounds created during the combustion of fossil fuels kill or harm anyone in the United States, though it may be a legitimate health concern in third-world countries that rely on burning biofuels and fossil fuels without modern emission control technologies.

7. Human Security

7.1 Fossil Fuels

7.1.1 Prosperity	As the world has grown more prosperous, threats to human security have become less common. The prosperity that fossil fuels make possible, including helping produce sufficient food for a growing global population, is a major reason the world is safer than ever before.
7.1.2 Democracy	Prosperity is closely correlated with democracy, and democracies have lower rates of violence and go to war less frequently than any other form of government. Because fossil fuels make the spread of democracy possible, they contribute to human security.
7.1.3 Wars for Oil	The cost of wars fought in the Middle East is not properly counted as one of the "social costs of carbon" as those conflicts have origins and justifications unrelated to oil.
	Limiting access to affordable energy threatens to prolong and exacerbate poverty in developing countries, increasing the likelihood of domestic violence, state failure, and regional conflict.

7.2 Climate Change

7.2.1 The IPCC's Perspective	The IPCC claims global warming threatens "the vital core of human lives" in multiple ways, many of them unquantifiable, unproven, and uncertain. The narrative in Chapter 12 of the Fifth Assessment Report illustrates the IPCC's misuse of language to hide uncertainty and exaggerate risks.
7.2.2 Extreme Weather	Real-world data offer little support for predictions that CO_2-induced global warming will increase either the frequency or intensity of extreme weather events.
7.2.3 Sea-level Rise	Little real-world evidence supports the claim that global sea level is currently affected by atmospheric CO_2 concentrations, and there is little reason to believe future impacts would be distinguishable from local changes in sea level due to non-climate related factors.
7.2.4 Agriculture	Alleged threats to agriculture and food security are contradicted by biological science and empirical data regarding crop yields and human hunger.
7.2.5 Human Capital	Alleged threats to human capital – human health, education, and longevity – are almost entirely speculative and undocumented. There is no evidence climate change has eroded or will erode livelihoods or human progress.

7.3 Violent Conflict

7.3.1 Empirical Research	Empirical research shows no direct association between climate change and violent conflicts.
7.3.2 Methodological Problems	The climate-conflict hypothesis is a series of arguments linked together in a chain, so if any one of the links is disproven, the hypothesis is invalidated. The academic literature on the relationship between climate and social conflict reveals at least six methodological problems that affect efforts to connect the two.
7.3.3 Alleged Sources of Conflict	The scholarly literature does not support the IPCC's claim that climate change intensifies alleged sources of violent conflict including abrupt climate changes, access to water, famine, resource scarcity, and refugee flows.
7.3.4 U.S. Military Policy	Climate change does not pose a military threat to the United States. President Donald Trump was right to remove it from the Pentagon's list of threats to national security.
7.3.5 Conclusion	Predictions that climate change will lead directly or indirectly to violent conflict are not testable. They presume mediating institutions and human capital will not resolve conflicts before they escalate to violence.

7.4 Human History

7.4.1 China	Extensive historical research in China reveals a close and positive relationship between a warmer climate and peace and prosperity, and between a cooler climate and war and poverty.
7.4.2 Rest of the World	The IPCC relies on second- or third-hand information with little empirical backing when commenting on the implications of climate change for conflict.

7.5 Conclusion

	It is probably impossible to attribute to the human impact on climate any negative impacts on

| | human security. Deaths and loss of income due to storms, flooding, and other weather-related phenomena are and always have been part of the human condition. |
| | Real-world evidence demonstrates warmer weather is closely associated with peace and prosperity, and cooler weather with war and poverty. A warmer world, should it occur, is therefore more likely to bring about peace and prosperity than war and poverty. |

8. Cost-Benefit Analysis

8.1 CBA Basics

	Cost-benefit analysis (CBA) is an economic tool that can help determine if the financial benefits over the lifetime of a project exceed its costs.
8.1.1 Use in the Climate Change Debate	In the climate change debate, CBA is used to answer four distinct questions about the costs and benefits of fossil fuels and the costs of measures to mitigate, rather than adapt to, climate change.
8.1.2 Integrated Assessment Models	Integrated assessment models (IAMs) are a key element of cost-benefit analysis in the climate change debate. They are enormously complex and can be programmed to arrive at widely varying conclusions.
8.1.2.1 Background and Structure	A typical IAM has four steps: emission scenarios, future CO_2 concentrations, climate projections and impacts, and economic impacts.
8.1.2.2 Propagation of Error	IAMs suffer from propagation of error, sometimes called cascading uncertainties, whereby uncertainty in each stage of the analysis compounds, resulting in wide uncertainty bars surrounding any eventual results.
8.1.3 IWG Reports	The widely cited "social cost of carbon" calculations produced during the Obama administration by the Interagency Working Group on the Social Cost of Carbon have been withdrawn and are not reliable guides for policymakers.
8.1.4 Stern Review	The widely cited "Stern Review" was an important early attempt to apply cost-benefit analysis to climate change. Its authors focused on worst-case scenarios and failed to report profound uncertainties.

8.2 Assumptions and Controversies

8.2.1 Emission Scenarios	Most IAMs rely on emission scenarios that are little more than guesses and speculative "storylines." Even current greenhouse gas emissions cannot be measured accurately, and technology is likely to change future emissions in ways that cannot be predicted.
8.2.2 Carbon Cycle	IAMs falsely assume the carbon cycle is sufficiently understood and measured with sufficient accuracy as to make possible precise predictions of future levels of carbon dioxide (CO_2) in the atmosphere.
8.2.3 Climate Sensitivity	Many IAMs rely on estimates of climate sensitivity – the amount of warming likely to occur from a doubling of the concentration of atmospheric carbon dioxide – that are too high, resulting in inflated estimates of future temperature change.
8.2.4 Climate Impacts	Many IAMs ignore the extensive scholarly research showing climate change will not lead to more extreme weather, flooding, droughts, or heat waves.
8.2.5 Economic Impacts	The "social cost of carbon" (SCC) derived from IAMs is an accounting fiction created to justify regulation of fossil fuels. It should not be used in serious conversations about how to address the possible threat of man-made climate change.
8.2.5.1 The IPCC's Findings	The IPCC acknowledges great uncertainty over estimates of the "social cost of carbon" and estimates the impact of climate change on human welfare is small relative to many other factors and will barely affect global economic growth rates.
8.2.5.2 Discount Rates	Many IAMs apply discount rates to future costs and benefits that are much lower than the rates conventionally used in cost-benefit analysis and which are mandated by the U.S. Office of Management and Budget (OMB) for use by federal agencies.

8.3 Climate Change

| **8.3.1 The IPCC's Findings** | By the IPCC's own estimates, the cost of reducing emissions in 2050 by enough to avoid a warming of ~2° C would be 6.8 times as much as the benefits would be worth. |

8.3.2 DICE and FUND Models	Changing only three assumptions in two leading IAMs – the DICE and FUND models – reduces the SCC by an order of magnitude for the first and changes the sign from positive to negative for the second.
8.3.3 A Negative SCC	Under very reasonable assumptions, IAMs can suggest the SCC is more likely than not to be negative, even though they have many assumptions and biases that tend to exaggerate the negative effects of GHG emissions.
8.4 Fossil Fuels	
8.4.1 Impacts of Fossil Fuels	Sixteen of 25 possible impacts of fossil fuels on human well-being are net benefits, only one is a net cost, and the rest are either unknown or likely to have no net impact.
8.4.2 Cost of Mitigation	Wind and solar cannot generate enough dispatchable energy (available on demand 24/7) to replace fossil fuels, so energy consumption must fall in order for emissions to fall.
8.4.2.1 High Cost of Reducing Emissions	Transitioning from a world energy system dependent on fossil fuels to one relying on alternative energies would cost trillions of dollars and take decades to implement.
8.4.2.2 High Cost of Reducing Energy Consumption	Reducing greenhouse gas emissions to levels suggested by the IPCC or the goal set by the European Union would be prohibitively expensive.
8.4.3 New Cost-benefit Ratios	The evidence seems compelling that the costs of restricting use of fossil fuels greatly exceed the benefits, even accepting many of the IPCC's very questionable assumptions.
8.5 Regulations	
	Cost-benefit analysis applied to greenhouse gas mitigation programs can produce like-to-like comparisons of their cost-effectiveness.
	The cap-and-trade bill considered by the U.S. Congress in 2009 would have cost 7.4 times more than its benefits, even assuming all of the IPCC's assumptions and claims about climate science were correct.
	Other bills and programs already in effect have costs exceeding benefits by factors up to 7,000. In short, even accepting the IPCC's flawed science and scenarios, there is no justification for adopting expensive emission mitigation programs.
8.6 Conclusion	
	The benefits of fossil fuels far outweigh their costs. Various scenarios of reducing greenhouse gas emissions have costs that exceed benefits by ratios ranging from 6.8:1 to 162:1.

PART I

FOUNDATIONS

1

Environmental Economics

Chapter Lead Authors: Roger Bezdek, Ph.D., Charles N. Steele, Ph.D., Richard L. Stroup, Ph.D.

Contributors: John Baden, Ph.D., Joseph L. Bast

Reviewers: Dennis T. Avery, William N. Butos, Ph.D., Jorge David Chapas, Donald R. Crowe, Donn Dears, Albrecht Glatzle, Dr.Sc.Agr., Kesten Green, Ph.D., Kenneth Haapala, Donald Hertzmark, Ph.D., Brian Joondeph, M.D., M.P.S., Paul McFadyen, Ph.D., John Merrifield, Ph.D., Alan Moran, Ph.D., Robert Murphy, Ph.D., Charles T. Rombough, Ph.D., Ronald Rychlak, J.D., David Stevenson, Daniel Sutter, Ph.D., Richard J. Trzupek, Gösta Walin, Ph.D., Thomas F. Walton, Ph.D., Bernard L. Weinstein, Ph.D.

Citation: Idso, C.D., Legates, D. and Singer, S.F. 2019. Environmental Economics. In: *Climate Change Reconsidered II: Fossil Fuels.* Nongovernmental International Panel on Climate Change. Arlington Heights, IL: The Heartland Institute.

Key Findings

Key findings of this chapter include the following:

Introduction

- No one should assume the "science is settled" regarding anthropogenic climate change or that the only role for economists is to recommend the most efficient way to reduce "carbon pollution."

History

- Economists have been addressing environmental issues since the discipline was founded in the eighteenth century.

- Economies and ecological systems have many commonalities, with the result that economics and ecology share many key concepts.

- Economists have shown markets can manage access to common-pool resources better than government agencies.

Key Concepts

- The cost of any choice is the value of forgone uses of the funds or time spent. Economists call this "opportunity cost."

- Climate change is not a conflict between people who are selfish and those who are altruistic. People who oppose immediate action to reduce greenhouse gas emissions are just as ethical or moral as those who support such action.

- Market prices capture and make public local knowledge that is complex, dispersed, and constantly changing.

- Most human action can be understood by understanding the incentives people face. "Moral hazard" occurs when people are able to escape full responsibility for their actions.

- Trade creates value by making both parties better off.

- Profits and losses direct investments to their highest and best uses.

- The art of economics consists in looking not merely at the immediate but at the longer-term effects of any act or policy.

- Discount rates, sometimes referred to as the "social rate of time preference," are used to determine the current value of future costs and benefits.

- Cost-benefit analysis, when performed correctly, can lead to better public policy decisions.

Private Environmental Protection

- Common-pool resources have been successfully protected by tort and nuisance laws and managed by nongovernmental organizations.

- Voluntary cooperation can generate efficient solutions to conflicts involving negative externalities.

- Prosperity leads to environmental protection becoming a higher social value and provides the resources needed to make it possible.

- The information needed to anticipate changes and decide how best to respond is local knowledge and the most efficient responses will be local solutions.

- "Ecological economics" is not a reliable substitute for rigorous mainstream environmental economics.

Government Environmental Protection

- Governments can protect the environment by helping to define and enforce property rights.

- Regulations often fail to achieve their objectives due to the conflicting incentives of individuals in governments and the absence of reliable and local knowledge.

- Evidence of "market failure" does not mean government intervention can improve market outcomes.

- Government bureaucracies predictably fall victim to regulatory capture, tunnel vision, moral hazard, and corruption.

- Voters have little incentive to become knowledgeable about many public policy issues. Economists call this "rational ignorance."

- Government's ability to promote the goals of some citizens at the expense of others leads to resources being diverted from production to political action. Economists call this "rent-seeking behavior."

- Government policies that erode the protection of property rights reduce the incentive and ability of owners to protect and conserve their resources. Those policies displace, rather than improve or add to, private environmental protection.

- "Leakage" occurs when the emissions reduced by a regulation are partially or entirely offset by changes in behavior.

Future Generations

- Capital markets create information, signals, and incentives to manage assets for long-term value.

- Markets reward innovations that protect the environment by using less energy and fewer raw materials per unit of output.

- Mistakes made in markets tend to be small and self-correcting. Mistakes made by governments tend to be big and are more likely to have catastrophic effects.

Conclusion

- Climate change is not a problem to be solved by markets or government intervention. It is a complex phenomenon involving choices made by millions or even billions of people producing countless externalities both positive and negative.

- The best responses to climate change are likely to arise from voluntary cooperation mediated by nongovernmental entities using knowledge of local costs and opportunities.

- Energy freedom – allowing markets rather than governments to make important choices about which fuels to use – can turn climate change from a possible *tragedy* of the commons into an *opportunity* of the commons.

Introduction

No one should assume the "science is settled" regarding anthropogenic climate change or that the only role for economists is to recommend the most efficient way to reduce "carbon pollution."

Many environmentalists and climate scientists are not familiar with the latest economic research on how common-pool resources, of which the global atmosphere is one, can be managed efficiently. They therefore believe the only thing economists can contribute to the debate over climate change is expertise in finding the most efficient way to reduce "carbon pollution." Many economists allow themselves to be relegated to this role by accepting unsubstantiated claims that the "science is settled" regarding the causes and consequences of climate change. Both audiences need to be aware of basic economic concepts that apply to climate change.

The general acceptance by economists of the findings of the United Nations' Intergovernmental Panel on Climate Change (IPCC) creates the appearance that most economists endorse the theory that man-made emissions of greenhouse gases (GHGs), and carbon dioxide (CO_2) in particular, are causing harm today and possibly a catastrophe in the future. For example, 26 prominent economists signed "The Schelling consensus on climate change policy," which leads with this statement: "Global climate change is one of the greatest problems facing mankind that requires collective action in order to be solved" (Anthoff *et al.,* 2011). Why would economists, who generally do not have backgrounds in physical science and who pride themselves on not presuming to aggregate or order the preferences of others, pledge allegiance to such a dogmatic claim?

Jean Tirole, winner of the 2014 Nobel Prize in economics, wrote in 2017, "Rising sea levels affecting islands and coastal cities, climatic disturbances, heavy rains and extreme droughts, uncertain harvests: we are all aware of the consequences of climate change. … [U]nless the international community acts vigorously, climate change may well compromise, in a dramatic and lasting way, the well-being of future generations" (Tirole, 2017, p. 195). He cites the IPCC and makes reference to the need "to contain the temperature increase to a virtuous 1.5 to 2.0 degrees Celsius" (p. 196). He attributes lack of effort to reduce GHG emissions to "*selfishness with regard to future generations and the free rider problem*" (p. 199). These statements suggest Tirole doesn't know the difference between weather and climate, or between a political organization and a scientific body, and that he thinks one hypothetical construction of global temperature is somehow more "virtuous" than another. Reading such conjecture and moralizing by a Nobel Laureate is disappointing.

Even economists who specialize in climate change fail to take the scientific debate seriously. In a recent book, William Nordhaus, the Sterling Professor of Economics at Yale University, cited the Summary for Policymakers (SPM) of the IPCC's Fourth Assessment Report and two National Academies reports (2001, 2011) and writes, "I could continue with further examples, but the basic findings of expert panels around the world are the same: The processes underlying projections of climate change are established science; the climate is changing unusually rapidly and the earth is warming" (Nordhaus, 2013, p. 296). But climate change in the twentieth century and so far in the twenty-first century was *not* "unusual" and at issue is not whether the planet is warming but how much of that warming is due to anthropogenic causes. His choice of panels rather than peer-reviewed literature is an appeal to authority instead of observational data or the scientific method. He also seems unaware of who writes the summaries for policymakers of the IPCC reports; most are not scientists (Goldenberg, 2014).

In the same book and as part of the same discussion of why he believes the science is settled, Nordhaus accepts the Hadley/NCDC/GISS global average surface temperature record without question or doubt, even though its accuracy has been challenged and since 1979 it has been superseded by superior satellite-based temperature data showing less warming. He cites anecdotes of "melting of glaciers and ice sheets" seemingly unaware that glaciers and ice sheets have waxed and waned for eons and in recent centuries regardless of the amount of CO_2 in the atmosphere. He repeats the IPCC's claim that its computer models cannot account for rising temperatures without a major role for CO_2, so CO_2 must account for rising temperatures … circular reasoning based on unproven presumptions. See also Heal (2017) as an example of an economist who concedes "massive uncertainty" involving climate science and economics, yet considers general circulation models to be a reliable basis for making predictions about future temperatures and climate impacts (pp. 1047, 1052).

Climate scientists have tried to school economists on the actual findings of the climate science community, instead of the distorted portrait created by the IPCC and other government panels, with only limited success. See, for example, Nordhaus (2012) and a reply by three distinguished climate scientists, Cohen, Happer, and Lindzen (2012).

It seems economists have broken what has been called Ray Hyman's Categorical Directive: "Before we try to explain something, we should be sure it actually happened" (Sheaffer, 2009, p. 84). The best available climate science shows the human effect on the global climate is likely too small to be measured against a background of natural variation (NIPCC, 2009, 2013). Most forecasts of future global warming due to human activities are implausible and violate most of the accepted principles of scientific forecasting (Green *et al.*, 2009; Green and Armstrong, 2007). The environmental benefits of a modest global warming are likely to exceed the environmental costs (NIPCC, 2014). Many scientists do not endorse the IPCC's claims of high confidence in predictions of more frequent or severe floods, droughts, hurricanes, and other calamities (Essex and McKitrick, 2007).

The failure of many economists to address the climate issue truthfully and forcefully is surprising. An extensive literature exists describing how interest groups have repeatedly exaggerated environmental threats in order to advance their financial interests or ideological agendas. Green and Armstrong (2011) studied 26 past forecasts of serious environmental harms from human activity and found none of the forecasts was the product of scientific forecasting methods and none proved to be accurate. In 20 of the situations, costly government regulations were imposed with the effect of reducing the welfare of the many while benefiting the few. See also the list of titles in Lehr (2014), the Iron Law of Regulation website, and the references in Section 1.4.5 below. Public choice theory predicts this sort of behavior and documents it across many fields. Public choice theorists are not on the fringe of mainstream economics; at least six won Nobel Prizes for their work: Gary Becker, James M. Buchanan, Ronald Coase, Elinor Ostrom, Vernon Smith, and George Stigler.

Economists have more to offer on climate change than simply advice for designing tax and cap-and-trade schemes. By revealing the costs and benefits of various policy options and market-based alternatives to government regulation, economics can help policymakers discover cost-effective responses to a wide range of environmental problems (Block, 1990; Markandya and Richardson, 1992; Libecap and Steckel, 2011). Environmental economics has become more important as "the quick environmental fixes from command-and-control regulation mainly have been achieved and ... the balance of pollution sources is shifting from large 'point sources' to more diffuse sources that are more difficult and expensive to regulate" (Dietz and Stern, 2002, p. vii). This description certainly applies to global warming, as CO_2 and other greenhouse gases are emitted from billions of sources both anthropogenic and natural.

Economists can help reconcile the real-world tradeoffs of protecting the environment while producing the goods and services needed by humanity by tapping the internal motivation of property owners, the value-creating power of trade, and local knowledge of costs and opportunities (Anderson and Leal, 2015; Morriss and Butler, 2013). They have shown how entrepreneurs can use private property, price signals, and capital markets to protect the environment without relying on government force (Anderson and Leal, 1997; Anderson and Huggins, 2008; Huggins, 2013).

Economists have pointed out the economic, political, legal, and administrative pitfalls facing renewable and carbon-neutral energies (McKitrick, 2010; Morriss et al., 2011; Yonk et al., 2012). Proposals to cap greenhouse gas emissions, "put a price on carbon," and other policies intended to force a transition away from fossil fuels often are advanced without an understanding of the true costs and physical limitations on the supply of alternative fuels. One consequence is their advocates support poorly designed programs that lead to unnecessary expenses, minimal or even no net reductions in emissions, and the unintentional emergence of regulatory hurdles to innovation and future discovery of alternative fuels (McKitrick, 2009; Lomborg, 2010; van Kooten, 2013; Lemoine and Rudik, 2017).

Economists also bring value to the climate change discussion thanks to their expertise in statistical analysis. Darwall (2013) remarks, "economists should be in a better position than others to make their own assessment of the science because much of it is about statistics and modeling" (p. 239). He quotes Ross McKitrick, a Canadian economist, saying, "the typical economist has way more training in data analysis than a typical climatologist," and "once they start reading climate papers they start spotting errors all over the place" (*Ibid.*).

Economists have examined the reasons why poor people and minorities often live in neighborhoods exposed to the highest levels of pollution (Banzhaf, 2012). Understanding how this situation can be the unintended consequence of policies intended to reduce emissions can lead to ideas and proposals that better protect everyone's health and rights.

Economists also can measure and help predict the distributional effects of public policies; e.g., whether the poor are hurt more than the wealthy by policies that seek to reduce greenhouse gas emissions by raising the price of energy (Büchs et al., 2011; Kotkin, 2018). Similarly, economists can determine if poor countries are more vulnerable to climate change than wealthy countries (Mendelsohn et al., 2006).

While some economists are occasionally guilty of the "tunnel vision" described later in this chapter, most are well-schooled in the limits of markets. Fullerton and Stavins (1998) wrote, "many economists – ourselves included – make a living out of analyzing 'market failures' such as environmental pollution. These are situations in which laissez faire policy leads not to social efficiency, but to inefficiency" (p. 5). Market-based approaches to environmental protection, they wrote, "are no panacea" and "the scope of economic analysis is much broader than financial flows" (*Ibid.*, pp. 5–6). On the other hand, economists are more keenly aware than others of the failure of regulation to improve on market results even in cases of "market failure" (Winston, 1993, 2006).

Section 1.1 summarizes the history of environmental economics and introduces free-market environmentalism (FME). Sections 1.2 through 1.4 describe the basic principles and tools of environmental economics based on an earlier book by Richard L. Stroup (Stroup, 2003), one of the coauthors of the present chapter. Stroup's work appears here with the publisher's permission and has been substantially revised and updated with the author's assistance and approval. Section 1.5 describes how markets take into account the interests of future generations. Section 1.6 presents a brief summary and conclusion.

References

Anderson, T.L. and Huggins, L.E. 2008. *Greener Than Thou: Are You Really an Environmentalist?* Stanford, CA: Hoover Institution.

Anderson, T.L. and Leal, D.R. 1997. *Enviro-Capitalists: Doing Good While Doing Well.* Lanham, MD: Rowman & Littlefield Publishers, Inc.

Anderson, T.L. and Leal, D.R. 2015. *Free Market Environmentalism for the Next Generation.* New York, NY: Palgrave Macmillan.

Anthoff, D., *et al.* 2011. Thinking through the climate change challenge. Centre for Economic Policy Research (website). Accessed July 11, 2018.

Banzhaf, S.H. 2012. *The Political Economy of Environmental Justice.* Redwood City, CA: Stanford Economics and Finance, Stanford University Press.

Büchs, M., Bardsley, N., and Duwe, S. 2011. Who bears the brunt? Distributional effects of climate change mitigation policies. *Critical Social Policy* **31** (2): 285–307.

Block, W.E. (Ed.) 1990. *Economics and the Environment: A Reconciliation.* Toronto, ON: The Fraser Institute.

Cohen, R.W., Happer, W., and Lindzen, R.M. 2012. In the climate casino: an exchange. *The New York Times Review of Books.* April 26.

Darwall, R. 2013. *The Age of Global Warming: A History.* London, UK: Quartet Books Ltd.

Dietz, T. and Stern, P.C. (Eds.) 2002. *New Tools for Environmental Protection: Education, Information, and Voluntary Measures.* Washington, DC: National Academies Press.

Essex, C. and McKitrick, R. 2007. *Taken by Storm: The Troubled Science, Policy and Politics of Global Warming.* Second edition. Toronto, ON: Key Porter Books.

Fullerton, D. and Stavins, R. 1998. How do economists really think about the environment? *Discussion Paper* 98-04. Cambridge, MA: Belfer Center for Science and International Affairs, Harvard University. April.

Goldenberg, S. 2014. IPCC report: climate change felt 'on all continents and across the oceans.' *The Guardian.* March 28.

Green, K.C. and Armstrong, J.S. 2007. Global warming: forecasts by scientists versus scientific forecasts. *Energy and Environment* **18**: 997–1021.

Green, K.C. and Armstrong, J.S. 2011. The global warming alarm: forecasts from the structured analogies method. *SSRN Working Paper* 1656056.

Green, K.C. and Armstrong, J.S. 2014. Forecasting global climate change. In: Moran, A. (Ed.) *Climate Change: The Facts 2014.* Melbourne, Australia: IPA, pp. 170–186.

Green, K.C., Armstrong, J.S., and Soon, W. 2009. Validity of climate change forecasting for public policy decision making. *International Journal of Forecasting* **25**: 826–32.

Heal, G. 2017. The economics of the climate. *Journal of Economic Literature* **55** (3).

Huggins, L. 2013. *Environmental Entrepreneurship: Markets Meet the Environment in Unexpected Places.* Cheltenham, UK: Edward Elgar.

Kotkin, J. 2018. Progressive California's growing race challenge. *Orange County Register.* June 30.

Lehr, J. 2014. Replacing the Environmental Protection Agency. *Policy Brief.* Chicago, IL: The Heartland Institute.

Lemoine, D. and Rudik, I. 2017. Steering the climate system: using inertia to lower the cost of policy. *American Economic Review* **107** (20).

Libecap, G.D. and Steckel, R.H. (Eds.) 2011. *The Economics of Climate Change: Adaptations Past and Present.* Chicago, IL: University of Chicago Press.

Lomborg, B. (Ed.) 2010. *Smart Solutions to Climate Change: Comparing Costs and Benefits.* New York, NY: Cambridge University Press.

Markandya, A. and Richardson, J. (Eds.) 1992. *Environmental Economics: A Reader.* New York, NY: St. Martin's Press.

McKitrick, R. 2010. *Economic Analysis of Environmental Policy.* Toronto, ON: University of Toronto Press.

McKitrick, R. (Ed.) 2009. *Critical Topics in Global Warming.* Vancouver, BC: The Fraser Institute.

Mendelsohn, R., Dinar, A., and Williams, L. 2006. The distributional impact of climate change on rich and poor countries. *Environment and Development Economics* **11**: 159–78.

Morriss, A.P., Bogart, W.T., Meiners, R.E., and Dorchak, A. 2011. *The False Promise of Green Energy.* Washington, DC: Cato Institute.

Morriss, A. and Butler, M. 2013. *Creation and the Heart of Man: An Orthodox Christian Perspective on Environmentalism.* Grand Rapids, MI: Acton Institute.

NIPCC. 2009. Nongovernmental International Panel on Climate Change. Idso, C.D. and Singer, S.F. *Climate Change Reconsidered: The 2009 Report of the Nongovernmental International Panel on Climate Change (NIPCC). Chicago,* IL: The Heartland Institute.

NIPCC. 2013. Nongovernmental International Panel on Climate Change. Idso, C.D., Carter, R.M., and Singer, S.F. *Climate Change Reconsidered II: Physical Science.* Nongovernmental International Panel on Climate Change. Chicago, IL: The Heartland Institute.

NIPCC. 2014. Nongovernmental International Panel on Climate Change. Idso, C.D., Idso, S.B., Carter, R.M., and Singer, S.F. *Climate Change Reconsidered II: Biological Impacts.* Nongovernmental International Panel on Climate Change. Chicago, IL: The Heartland Institute.

Nordhaus, W. 2012. Why the global warming skeptics are wrong. *The New York Times Review of Books.* March 22.

Nordhaus, W. 2013. *The Climate Casino: Risk, Uncertainty, and Economics for a Warming World.* New Haven, CT: Yale University Press.

Sheaffer, R. 2009. The fallacy of misplaced rationalism. In: Frazier, K. (Ed.) *Science Under Siege: Defending Science, Exposing Pseudoscience.* New York, NY: Prometheus Books.

Stroup, R.L. 2003. *Eco-nomics: What Everyone Should Know about Economics and the Environment.* Washington, DC: Cato Institute.

Tirole, J. 2017. *Economics for the Common Good.* Princeton, NJ: Princeton University Press, 2017.

van Kooten, G.C. 2013. *Climate Change, Climate Science and Economics: Prospects for an Alternative Energy Future.* New York, NY: Springer.

Winston, C. 1993. Economic deregulation: days of reckoning for microeconomists. *Journal of Economic Literature* **31** (September): 1263–89.

Winston, C. 2006. *Government Failure versus Market Failure: Microeconomics Policy Research and Government Performance.* Washington, DC: AEI-Brookings Joint Center for Regulatory Studies.

Yonk, R.M., Simmons, R.T, and Steed, B.C. 2012. *Green v. Green: The Political, Legal, and Administrative Pitfalls Facing Green Energy Production.* Oxford, UK: Routledge.

1.1 History

Economists have been addressing environmental issues since the discipline was founded in the eighteenth century.

Economists at least since Adam Smith (1776 [1976]) and even before him (see Cantillon, 1755 and the discussion in Rothbard, 1995) have used the tools of economics to address environmental issues. Economics and scholarly interest in the relationship between humans and the natural environment emerged simultaneously due to the same historical events. Writing nearly a century ago, Thomas (1925 [1965]) observed "the first great impulse to a thorough-going development of environmental theories came as a result of the discoveries and colonizing enterprises of the sixteenth and seventeenth centuries known as the Commercial Revolution" (p. 22). Smith was fascinated by what was happening in the American colonies (see "Of Colonies," Book IV, Chapter vii, of *The Wealth of Nations*) and corresponded with Benjamin Franklin while writing his great book.

Economics and ecology emerged as disciplines with more in common than differences. Smith influenced Thomas Malthus (1766–1834), who in turn influenced Charles Darwin, born after Smith's death (Smith lived from 1723 to 1790, Darwin from 1809 to 1882). Darwin attended the University of Edinburgh, where Smith once lectured and was well known. Darwin referred to Smith and cited him in *The Descent of Man*, published in 1871 (Darwin 1871 [1981], p. 81). Smith's insight that markets lead the self-interested individual "by an invisible hand to promote an end which was no part of his intention" (Smith, 1776 [1976], Book 4, Chapter 4) is echoed in Darwin's description of evolution in *The Origin of Species*, a process in which "all organic beings are striving to seize on each place in the economy of nature" (Darwin 1859 [2003]).

Why economists would be interested in the environment was obvious to Thomas: "As economics is almost invariably considered by the economists to include a study of man's exploitation of his physical environment for his own needs, it is not necessary to dwell upon the fact that the study of the physical environment is of the utmost significance for that subject" (Thomas 1925 [1965], p. 9). Malthus, David Ricardo, and John Stuart Mill each addressed the limits on human prosperity posed by the scarcity of land suited to agriculture, coal, and other natural resources.

In 1920, A.C. Pigou recognized the special problem posed by resources owned in common rather than by individuals, observing, "No 'invisible hand' can be relied on to produce a good arrangement of the whole from a combination of separate treatments of the parts. It is therefore necessary that an authority of wider reach should intervene to tackle the collective problems of beauty, of air and light, as those other collective problems of gas and water have been tackled" (Pigou, 1920, p. 195).

By 1931, economists were laying the foundations of what would become natural resource economics (Hotelling, 1931). With some notable exceptions (Mises, [1966] 1998; Knight, 1924), a generation of economists generally accepted Pigou's argument that only governments could solve "collective problems" involving air and water. That changed in 1960 with publication of an essay titled "The Problem of Social Cost" by future Nobel Laureate Ronald Coase (Coase, 1960).

Coase observed that high transaction costs may cause markets to fail to ensure that all of the costs of a person's actions are fully borne by the actor ("internalized"), but transaction costs are ubiquitous (there is no such thing as "zero transaction costs") and positive and negative externalities are resolved everywhere, usually without government intervention. All that is necessary to achieve the most efficient outcomes is for governments to help recognize and enforce the property rights of the parties involved and allow them to negotiate toward a settlement. As Terry Anderson explains it, "Certainly, transaction costs can prevent all costs from being fully accounted for, but unaccounted for costs constitute uncaptured benefits. If water is not owned, and therefore polluted, the entrepreneur who can establish ownership captures the benefits if water quality is improved" (Anderson, 2011). Coase's contribution to the environmental debate is described in more detail in Section 1.3.2.

Environmental economics was strongly influenced by the rise of the modern environmental movement in the 1970s. Publication in 1972 of *The Limits to Growth* by the Club of Rome steered the profession in the direction of forecasting trends in population, technology, and the use of finite natural resources (Meadows *et al.*, 1972). In 1989, *Blueprint for a Green Economy* by David Pearce and coauthors spelled out the public policy implications of environmental values and concerns, calling for recognition of the economic benefits of natural resources, taxes on polluters, and measuring additions and losses to a country's stock of natural resources (Pearce *et al.*, 1989). Many modern-day economists, including Helm (2015), continue to work in this tradition. Baden and Stroup (1981) observed,

> The dawn of the environmental movement coincided with an increased skepticism of private property rights and the market. Many citizen activists blamed self-interest and the institutions that permit its expression for our environmental and natural resource crises. From there it was a short step to the conclusion that management by professional public 'servants,' or bureaucrats, would significantly ameliorate the problems identified in the celebrations accompanying Earth Day 1970 (p. x).

Also during the 1970s, an alternative school of thought called "the new resource economics," or free-market environmentalism (FME), began to emerge (Hardin and Baden, 1977; *Harvard Journal of Law & Public Policy*, 1992; Anderson and Leal, 2015). It advanced critiques of Pigou's dismissal of private solutions to the management of "collective problems" such as pollution by documenting cases where recognizing and enforcing private property rights solved environmental problems without relying on politics and governments. As Anderson commented in 2007, "Secure private property rights that hold people accountable and markets that communicate human values and opportunity costs are the core of FME, and they are as applicable to global warming as they are to land and water conservation" (Anderson, 2007).

According to FME, the market approach to protecting commonly owned resources is to find win-win solutions even when conditions might otherwise cause over-use of a resource by some agents and harm to others. People value and are willing to pay for environmental amenities, meaning there are markets for achieving environmental protection. Since the future value of assets affects their current prices, private ownership of assets creates incentives for conservation and protection that benefit future generations.

FME recognizes that markets are powerful engines of value creation thanks to the incentives created by private property rights, the knowledge generated and communicated by prices and the profit and loss system, and the value created by exchanges in which both parties benefit. Governments routinely fail to manage resources as efficiently as markets due to their isolation from these market forces resulting in moral hazard, careerism, tunnel vision, and other maladies known to afflict government bureaucracies. The best solutions to problems involving common-pool resources, then, are often discovered and implemented by markets rather than by politics and bureaucracies.

FME scholars generally hold that human welfare is the best measure of success in managing natural resources. They stress that protecting the natural environment is a value that grows mainly or even only in the presence of the prosperity that markets make possible. Environmentalists argue to the contrary, that animals and even inanimate objects have "innate" value or make contributions to "sustainability" that should be weighed against any human benefits (Chase, 1995; Davidson, 2000). In contrast to free markets, where values emerge from voluntary transactions, it is not clear who should determine the values that environmentalists argue for or who should bear the cost of attaining them.

Some environmentalists also dispute FME's exposé of repeated government failure, instead attributing favorable trends in air and water quality, for example, entirely to government interventions and not to any market processes, even though many of those trends started before government intervention could have played a role (Simon, 1995; Goklany, 1999; Hayward, 2011, pp. 7ff). These environmentalists hold out hope that concentrating power in the hands of government officials can do more to protect the air, water, and endangered species than giving property owners and others secure property rights and incentives to do the right thing. Finally, some environmentalists blame the free enterprise system for the unequal distribution of wealth among individuals, leading them to subordinate individual liberty to their own goals (Coffman, 1994; Easterbrook, 2003; Buchanan, 2005; Klein, 2014).

Some environmentalists who reject mainstream environmental economics have attempted to create their own school called "ecological economics." The contributions and limits of that effort are described in Section 1.3.5. Not all environmentalists, however, assume a fundamental conflict between free enterprise and environmental protection. Their efforts merit some attention here.

Rothschild (1990) described "the profound similarity" of economies and ecosystems in a book titled *Bionomics: The Inevitability of Capitalism* (p. 213). The titles of sections in his book give an idea of the parallels and their application in both fields: evolution and innovation, organism and organization, energy and values, learning and progress, struggle and competition, feedback loops and free markets, parasitism and exploitation, and mutualism and cooperation. He wrote, "Bionomics is the branch of ecology that examines the economic relations between organisms and their environment. As such, bionomics provides the best starting point for a new way of thinking about the human economy. Cutting through the mind-boggling complexity of the ecosystem, the bionomic perspective illuminates the interplay of forces that maintain stability while spawning change. Problems beyond the reach of orthodox economics are readily understood from the bionomic perspective" (p. 335).

Hawken, Lovins, and Hunter Lovins (2000), in a book titled *Natural Capitalism: Creating the Next Industrial Revolution*, presented what they call "the next industrial revolution," predicting (as well as advocating) a "new type of industrialism, one that differs in its philosophy, goals, and fundamental process from the industrial system that is the standard today" (p. 2). "[N]atural capitalism does not aim to discard market economics," they wrote, "nor reject its valid and important principles or its powerful mechanisms. It does suggest that we should vigorously employ markets for their proper purpose as a tool for solving the problems we face, while better understanding markets' boundaries and limitations" (p. 260).

Hawken *et al.* (2000) were confident markets can address a wide range of environmental challenges, including climate change. They wrote, "The menu of climate-protecting opportunities is so large that over time, they can overtake and even surpass the pace of economic growth. Over the next half-century, even if the global economy expanded by 6- to 8-fold, the rate of releasing carbon by burning fossil fuel could simultaneously decrease by anywhere from one-third to nine-tenths below the current rate" (p. 244).

Nordhaus and Shellenberger (2007) are also environmentalists who endorse market-based approaches to environmental protection. Saying "the two of us had spent all of our professional careers, about thirty years between us, working for the country's largest environmental organizations and foundations, as well as many smaller grassroots ones" (p. 8), they wrote, "As Americans became increasingly wealthy, secure, and optimistic, they started to care more about problems such as air and water pollution and the protection of the wilderness and open space. This powerful correlation between increasing affluence and the emergence of quality-of-life and fulfillment values has been documented in developed and undeveloped countries around the world" (p. 6). "Environmentalists," they observed, "have long misunderstood, downplayed, or ignored the conditions for their own existence. They have tended to view economic growth as the cause [of] but not the solution to ecological crisis" (*Ibid.*).

* * *

This brief overview of the history of environmental economics should lay to rest concerns that economists don't understand ecology or lack the tools to study the best solutions to environmental problems. The conjoined histories of economics and ecology and the extensive commonalities of the subjects mean economists can make valuable contributions to the climate change discussion by identifying market-based solutions to problems arising from pollution and by warning of the shortcomings of relying on government interventions.

References

Anderson, T.L. 2007. Free market environmentalism. *PERC Report* **25** (1).

Anderson, T.L. 2011. Reconciling economics and ecology. *PERC Report* **29** (2).

Anderson, T.L. and Leal, D.R. 2015. *Free Market Environmentalism for the Next Generation*. New York, NY: Palgrave Macmillan.

Baden, J. and Stroup, R.L. (Eds.) 1981. *Bureaucracy vs. Environment: The Environmental Costs of Bureaucratic Governance*. Ann Arbor, MI: The University of Michigan Press.

Buchanan, J.M. 2005. Afraid to be free: dependency as desideratum. *Public Choice* **124**: 19–31.

Cantillon, R. 1755. *Essai sur la nature du commerce en général*.

Chase, A. 1995. *In a Dark Wood: The Fight Over Forests and the Rising Tyranny of Ecology*. Boston, MA: Houghton Mifflin Company.

Coase, R.H. 1960. The problem of social cost. *Journal of Law and Economics* 3: 1–44.

Coffman, M.S. 1994. *Saviors of the Earth? The Politics and Religion of the Environmental Movement*. Chicago, IL: Northfield Publishing.

Darwin, C. 1859 [2003]. *The Origin of Species, 150th Anniversary Edition*. Reprint, New York, NY: Signet.

Darwin, C. 1871 [1981]. *The Descent of Man, and Selection in Relation to Sex*. Reprint, Princeton, NJ: Princeton University Press.

Davidson, E.A. 2000. *You Can't Eat GDP: Economics as if Ecology Mattered*. Cambridge, MA: Perseus Publishing.

Easterbrook, G. 2003. *The Progress Paradox: How Life Gets Better While People Feel Worse*. New York, NY: Random House.

Goklany, I.M. 1999. *Clearing the Air: The Real Story of the War on Air Pollution*. Washington, DC: Cato Institute.

Hardin, G. and Baden, J. 1977. *Managing the Commons*. San Francisco, CA: W.H. Freeman and Company.

Harvard Journal of Law & Public Policy. 1992. Symposium - Free market environmentalism. **15** (2).

Hawken, P., Lovins, A., and Hunter Lovins, L. 2000. *Natural Capitalism: Creating the Next Industrial Revolution*. Washington, DC: U.S. Green Building Council.

Hayward, S. 2011. *Almanac of Environmental Trends*. San Francisco, CA: Pacific Institute for Public Policy Research.

Helm, D. 2015. *Natural Capital: Valuing the Planet*. New Haven, CT: Yale University Press.

Hotelling, H. 1931. The economics of exhaustible resources. *Journal of Political Economy* **39** (2): 137–75.

Klein, N. 2014. *This Changes Everything: Capitalism vs. The Climate*. New York, NY: Simon & Schuster.

Knight, F.H. 1924. Fallacies in the interpretation of social cost. *Quarterly Journal of Economics* **38**: (4).

Meadows, D.H., Meadows, D.L., Randers, J., and Behrens III, W.W. 1972. *The Limits to Growth: A Report for the Club of Rome's Project on the Predicament of Mankind*. New York, NY: Universe Books.

Mises, L. 1966 [1998]. *Human Action: A Treatise on Economics, the Scholar's Edition*. Auburn, AL: Ludwig von Mises Institute.

Nordhaus, T. and Shellenberger, M. 2007. *Break Through: From the Death of Environmentalism to the Politics of Possibility*. Boston, MA: Houghton Mifflin Company.

Pearce, D., Markandya, A., and Barbier, E. 1989. *Blueprint for a Green Economy*. London, UK: Routledge.

Pigou, A.C. 1920. *The Economics of Welfare*. London, UK: Macmillan.

Rothbard, M. 1995. *An Austrian Perspective on the History of Economic Thought*, vol. 1, *Economic Thought before Adam Smith*. Auburn, AL: Ludwig von Mises Institute.

Rothschild, M. 1990. *Bionomics: The Inevitability of Capitalism*. New York, NY: Henry Holt and Company.

Simon, J.L. (Ed.) 1995. *The State of Humanity*. Oxford, UK: Blackwell Publishers, Inc.

Smith, A. 1776 [1976]. *The Wealth of Nations*. Reprint, Indianapolis, IN: Liberty Press.

Thomas, F. 1925 [1965]. *The Environmental Basis of Society: A Study in the History of Sociological Theory*. Reprint, New York, NY: The Century Co.

1.2 Key Concepts

This section introduces nine key economic concepts that shed light on environmental protection. Some important concepts are missing from this section, including population, technology, elasticity of demand, and probably many others. Those concepts can be found in standard textbooks and reference works (e.g., Ward, 2006; Henderson, 2008). While some examples and case studies presented in this section involve climate change, some do not.

References

Henderson, D.R. (Ed.) 2008. *Concise Encyclopedia of Economics*. Second edition. Indianapolis, IN: Library of Economics and Liberty.

Ward, F.A. 2006. *Environmental and Natural Resource Economics*. Upper Saddle River, NJ: Pearson Prentice Hall.

1.2.1 Opportunity Cost

The cost of any choice is the value of forgone uses of the funds or time spent. Economists call this "opportunity cost."

Scarcity is a fundamental fact of life, not just of economics (Becker, 1976; Glaeser and Shleifer, 2014). It is always present in nature, even when human beings are not (Rothschild, 1990). Each population of a species can flourish and expand only until it reaches the limit of available habitat, sunlight, water, and nutrients. Trees grow taller as they compete for sunlight. Some plants spread their leaves horizontally, capturing sunlight while blocking access for other species that might sprout up to compete for water and nutrients. Each successful strategy captures resources, taking them from competing species or populations.

According to Sowell (2007), "the available resources are always inadequate to fulfill all the desires of all the people. Thus there are no 'solutions' … but only trade-offs that still leave many unfulfilled and much unhappiness in the world" (p. 113). Scarcity persists even when supplies increase because people's goals and wants change as they gain control over more resources, giving them the ability to climb what Abraham Maslow famously described as a "hierarchy of needs," rising from physiological needs such as food, clothing, and safety to self-actualization (Maslow, 1943). Maslow's view that basic human needs must be met before higher-level wants and desires become valued has been widely validated in psychology, history, and economics (Abulof, 2017).

The cost of any choice is the value of foregone uses of the funds or time we spend. Economists call this "opportunity cost." Some of these costs are obvious, like the price we pay for a product or service, but others are more subtle and easy to overlook, such as the time we spend learning about which product we want to buy, time spent waiting in line, and the long-term consequences of choices. When governments regulate activities, an estimate of the opportunity cost must include the consequences, many of them unintended, of the new rules.

Advocates of immediate action to reduce greenhouse gas emissions often assume the people of the world can afford to spend more on "low-carbon" fuels or that people can use less energy by being "less wasteful" or making small changes in lifestyle such as riding bicycles to work or replacing incandescent lightbulbs with LED fixtures. This is plainly not the case in developing countries, where limited access to electricity already causes hardships including disease and premature deaths.

The lifestyle change necessary to reduce greenhouse gas emissions as much as called for by the IPCC and various environmental groups would be dramatic. Calculations presented in Chapter 8 show per-capita gross domestic product (GDP) would have to fall by as much as 81% from baseline forecasts, a loss of $238 trillion. To put that figure in perspective,

U.S. GDP in 2017 was only $19.4 trillion and China's GDP was $12.2 trillion (Tverberg, 2012). Most of the economic gains of the past century would be nullified, and with them all the gains in health care, education, transportation, and practically every other part of modern civilization.

Most of the cost of a forced transition away from fossil fuels would be the opportunity cost of using less energy. Even with optimistic assumptions about the rate of innovation and investment, renewables will come up far short of producing the energy needed by a growing global population. Because access to inexpensive and reliable energy is closely correlated with economic growth and human development, a significant reduction in energy supply would cause catastrophic losses in human wellbeing. Imagine a world without cars, trucks, and airplanes, or without aluminum, fertilizer, or the Internet. These are only a few of the things that would have to be surrendered to achieve the emission reduction targets set by the IPCC and the United Nations.

No matter how wealthy the society in which we live, reducing greenhouse gas emissions or investing in adaptation strategies would mean spending less than we otherwise could on schools, public safety, or protecting the environment from threats other than climate change. Ignoring the opportunity costs of climate change actions doesn't make those costs go away. Ignoring them means we cannot prioritize our spending, which leads to wasting scarce resources on activities that produce only small or only hypothetical benefits while passing up opportunities to achieve much greater real benefits.

References

Abulof, U. (Ed.) 2017. Symposium: Rethinking Maslow: human needs in the 21st century. *Society* **54**: 508.

Becker, G.S. 1976. *The Economic Approach to Human Behavior.* Chicago, IL: University of Chicago Press.

Glaeser, E.L. and Shleifer, A. 2014. Obituary. Gary Becker (1930–2014). *Science* **344** (6189): 1233.

Maslow, A.H. 1943. A theory of human motivation. *Psychological Review* **50**: 370–96.

Rothschild, M. 1990. *Bionomics: Economy as Ecosystem.* New York, NY: Henry Holt and Company.

Sowell, T. 2007. *A Conflict of Visions: Ideological Origins of Political Struggles.* Revised Edition. New York, NY: Basic Books.

Tverberg, G. 2012. An energy/GDP forecast to 2050. Our Finite World (website). July 26.

1.2.2 Competing Values

Climate change is not a conflict between people who are selfish and those who are altruistic. People who oppose immediate action to reduce greenhouse gas emissions are just as ethical or moral as those who support such action.

People have differing goals and disagree about which choice is best. Often this disagreement doesn't matter because decisions largely or entirely affect only the person making the choice and those who willingly cooperate with that person. But some decisions affect other people who have not agreed to be affected. In such instances, pursuit of differing goals can lead to conflict. Nowhere is this more evident than in recent environmental matters.

The United States has vast forests but not enough to provide all of the wood, all of the wilderness, and all of the accessible recreation we want. As soon as we log trees, build roads, or improve trails and campsites, we lose some wilderness. Similarly, we have large amounts of fresh water, but if we use water to grow rice in California, the water consumed cannot be used for drinking water in California cities. If we use fire to help a forest renew itself, we will have air pollution downwind while the fire burns. We must make choices about how to allocate our limited resources.

This can be seen in events surrounding the decision of California's San Bernardino County to build a new hospital facility. The county began planning the state-of-the-art complex in 1982. Eleven years later, on the day before groundbreaking in 1993, the U.S. Fish and Wildlife Service determined the Delhi Sands fly, which had been found on the site, was an endangered species. The county was required to spend $4.5 million to move the hospital 250 feet to give the flies a few acres on which to live and a corridor to the nearby sand dunes. This required diverting funds from the county's medical budget to pay for biological studies on accommodating the fly (National Association of Homebuilders *et al.*, 1996; Booth, 1997; Nagle, 1998). Environmentalists who wanted biological diversity were relieved, but county officials were upset at the delay and unexpected costs that taxpayers ultimately would have to bear. To use resources one way sacrifices the use of those resources for other things. There is no escaping this fact. San Bernardino County faced a choice between timely provision of a health care facility and protection of a unique species.

Even environmental goals often conflict. A policy of strict forest preservation (e.g., a wilderness designation) in an old-growth forest does not allow trees to be thinned, although such thinning could minimize forest dieback from insect infestations, disease, or fire. In this case, the goal of preserving the old-growth forest in the short term contradicts the goal of preserving the forest's long-term survival.

Discussions about climate change often frame it as a conflict between people who are selfish and those who care about others, including and perhaps especially future generations. This framing is incorrect. The goals of some individuals are selfish, intended to further only their own welfare, and the goals of others are altruistic, intended to help their fellow man, but in both cases, each person's concern and vision are focused mainly on a narrow set of ends (Sowell, 1980, 2011).

Even the most noble and altruistic goals are typically narrow. Consider two famous examples. The concern felt by Saint Teresa of Calcutta for the indigent and the sick was legendary. So, too, was Sierra Club founder John Muir's love of wilderness and his focus on protecting wilderness for all time. In both cases their goals were widely regarded as noble and altruistic, not narrowly selfish.

Yet one might be tempted to conclude that Saint Teresa would have been willing to sacrifice some of the remaining wilderness in India in order to provide another hospital for the people of Calcutta she cared so much about, and John Muir would have been willing to see fewer hospitals constructed if that helped preserve wilderness. Individuals with unselfish goals, like all other individuals, are narrowly focused. Each individual is willing to see sacrifices made in goals less important to him or her in order to further his or her own narrow purposes.

We know and care most about things that directly affect us, our immediate family, and others close to us. We know much less about things that mostly affect people we never see. When a person acts to achieve his or her narrow set of goals, it doesn't mean the individual cares nothing about others. It just means that for each of us, our strongest interests are narrowly focused. These narrow sets of goals, whatever the mix of selfishness and altruism, correspond to what economists call the "self-interest" of that individual.

People who oppose immediate action to reduce greenhouse gas emissions say they place a higher priority on respecting the rights of others to find their own way in the world, or providing good schools or hospitals or making sure poor people are well provided for, than delaying the uncertain arrival of a small amount of global warming a century hence (Epstein, 2014; Legates and van Kooten, 2014; Carlin, 2015; Moore and Hartnett White, 2016). This is a defendable moral choice, one that doesn't mean they are selfish. People who call themselves environmentalists may care less about the welfare of other people than they do about their own ability to enjoy wilderness or imagine playing a role in bringing about a romantic vision of unspoiled nature (Hulme, 2009). This hardly makes them altruistic.

There are thousands of environmental goals, each competing with others for limited land, water, and other resources. Even without selfishness, the narrow focus of individuals is enough to ensure there will be strong disagreements and competition for scarce natural resources. This narrowness of focus is important for understanding the economics of environmental issues. Depending on the circumstance, narrow goals can lead to tunnel vision, with destructive results, or to satisfying exchanges that make all participants better off.

References

Booth, W. 1997. Flower loving insect becomes symbol for opponents of Endangered Species Act. *Washington Post.* April 4, p. A-1.

Carlin, A. 2015. *Environmentalism Gone Mad: How a Sierra Club Activist and Senior EPA Analyst Discovered a Radical Green Energy Fantasy.* Mount Vernon, WA: Stairway Press.

Epstein, A. 2014. *The Moral Case for Fossil Fuels.* New York, NY: Portfolio/Penguin.

Hulme, M. 2009. *Why We Disagree about Climate Change.* New York, NY: Cambridge University Press.

Legates, D.R. and van Kooten, G.C. 2014. *A Call to Truth, Prudence, and Protection of the Poor 2014: The Case against Harmful Climate Policies Gets Stronger.* Burke, VA: The Cornwall Alliance for the Stewardship of Creation. September.

Moore, S. and Hartnett White, K. 2016. *Fueling Freedom: Exposing the Mad War on Energy.* Washington, DC: Regnery Publishing.

Nagle, J.C. 1998. The Commerce Clause meets the Delhi Sands Flower-Loving Fly. *Michigan Law Review* **97** (174).

National Association of Home Builders. 1996. *National Association of Home Builders, et al., Appellants,* v. *Bruce Babbitt, Secretary, United States Department Of Interior and Mollie Beattie, Director, United States Fish and Wildlife Service, Appellees.* 130 F.3d 1041 (D.C. Cir. 1996).

Sowell, T. 1980. *Knowledge and Decisions.* New York, NY: Basic Books.

Sowell, T. 2011. *Intellectuals and Society.* New York, NY: Basic Books.

1.2.3 Prices

Market prices capture and make public local knowledge that is complex, dispersed, and constantly changing.

The ability of markets to harmonize personal interests and the general welfare is one of the best documented and most firmly established findings of modern economics (Watson, 1969; Becker, 1976; Eatwell *et al.*, 1989; Gwartney *et al.*, 2012). Prices make this harmony possible by capturing and making public local knowledge that is otherwise hidden, dispersed, and constantly changing.

Prices are an expression of agreement between a producer (supply) and a buyer (demand) on the value of a good or service. The price of a good or service is commonly depicted in economics as the intersection of supply and demand curves as shown in Figure 1.2.3.1.

Changes in price reflect a product's increasing or decreasing scarcity compared with other goods and services and present a powerful incentive for buyers and sellers to act on that information (Hayek, 1945; Friedman, 1976; Sowell, 1980; Steil and Hinds, 2009). Without market signals, it would be nearly impossible to evaluate the effect of (or even keep track of) all the factors influencing scarcity in many product uses and locations. Yet each one is relevant to the cost and value of what is preserved, produced, and offered in the marketplace.

Each buyer and each seller may act with little knowledge of what any other person wants or needs. But so long as people are free to choose, market prices direct each person to satisfy the needs of others. Prices encourage producers to provide what consumers want the most, relative to their cost, and to satisfy any particular want in the least costly way.

Figure 1.2.3.1
Prices are determined by supply (S) and demand (D)

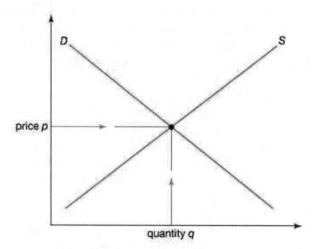

Source: Encyclopedia Britannica, 2013.

Consumers, too, are strongly influenced by prices and they, too, act as if they care about their fellow consumers. When prices increase, they consume less; when prices decrease, they consume more. By economizing when goods are scarce, they allow more for other consumers. They purchase more when the goods are plentiful and there is a lot to go around. Actual and expected offers in the marketplace are guidance from the so-called invisible hand.

Consider energy markets. Each consumer of electricity chooses whether to use electric heat, and how high or low to set the thermostat. In addition to preferences about temperature, these decisions reflect the price of electricity. When prices are high, people will economize, making more electricity available for others. When prices are low, they will consume more. These decisions, in turn, influence the decisions of others, even those made by other industries. For instance, individual consumer choices about electricity consumption affect how much aluminum will be produced and which producers will supply more than others.

Primary aluminum production requires large quantities of electricity. Higher electricity prices raise the price of aluminum compared with substitute metals and especially raise costs for producers that use a lot of electricity per ton of aluminum. Producers who conserve on the use of electricity enjoy a competitive advantage and are likely to produce a larger share of aluminum sold in the market. Thus, even with little or no knowledge of why electricity prices are rising throughout the economy, each consumer makes choices that move sales away from the expensive energy sources and toward conservation or substitute energy sources, and away from inefficient electricity producers and toward more efficient ones.

When resources are not privately owned or are not traded in open markets, the vital flow of information created by prices is missing. That is the case, for example, with national parks in the United States (Leal and Fretwell, 1997). Most of the funds for national parks come from tax dollars appropriated by Congress. Park visitors pay only a small fraction of the cost of the services they receive. Proceeds from national park recreation fees cover only about 10% of the cost of park operations (Regan, 2013). (States do somewhat better; an average of 39% of state park operating costs were recovered by user fees in 2011 (Walls, 2013, p. 5)).

With such a small portion of their budgets coming from user fees, park managers have little information about how much the various services they provide are worth to visitors. To learn what people want, they have to rely on surveys and polls, which can reach only a small number of people and can be misleading. In contrast, for owners of private campgrounds, amusement parks, museums, and other attractions that also draw visitors, information is always flowing and managers always have an incentive to respond to that information. The price they can charge for admission is determined by the value consumers place on their services. Their net profit or loss (as well as news of competitors' profits or losses) directs them to continuously change their budgets to better meet their customers' demonstrated wants.

Turning to the climate change issue, there is no marketplace in which access to the atmosphere is bought and sold; consequently, there is no price system revealing agreement on the value of competing uses. Complicating matters is the fact that human activities contribute only a tiny part to the natural exchange of carbon dioxide between the atmosphere and other reservoirs – the subject of Chapter 5, Section 5.1. Like the managers of public parks, a government agency placed in charge of managing the atmosphere would operate blindly, not knowing how much to charge for use of the atmosphere or how to invest the revenues it might raise by rationing use of the atmosphere.

Some environmentalists and economists see this as a "problem" that could be solved by "putting a price on carbon" (they mean carbon dioxide). But prices efficiently allocate resources only if they are real, that is to say, if they arise from voluntary exchanges among people with defined and enforced private property rights. Assigning a price to a ton of carbon dioxide emissions does not solve the market coordination problem, and since that arbitrary price is likely to be wrong it makes the problem worse. Advocates of a "carbon tax" also assume that an objectively correct or efficient level of taxation can be found, that political leaders would agree to it, and that such a tax could be collected and enforced without creating expenses greater than the forecast benefits of reduced climate change. In reality, there is no such thing as a government agency able to act on a calculation of the "social cost of carbon," even if such a cost were established. A carbon tax is discussed in more detail in Section 1.4.2.

Later chapters explain that the human impact on climate is likely to be too small to be seen against a background of natural variability and the "social cost of carbon" is probably near zero or even negative, if it is knowable at all. These findings suggest that the optimal "carbon tax" is likely to be zero or even negative. In short, it is fruitless to view climate change as a problem in need of a government solution. It is, instead, properly viewed as a natural process, one of many, routinely accommodated by markets without need for government intervention.

The discussion in this section suggests some preliminary implications for the climate change issue:

- Real prices allocate resources to their best and highest uses, even though that is not the intended outcome of individual buyers and sellers.

- Without prices, assets cannot be efficiently managed.

- There is currently no price system that assigns values to competing uses of the atmosphere.

- Real prices reflect agreements by buyers and sellers who are free to choose and cannot be randomly assigned to goods or services by economists or government agencies.

References

Becker, G.S. 1976. *The Economic Approach to Human Behavior*. Chicago, IL: University of Chicago Press.

Eatwell, J., Milgate, M., and Newman, P. 1989. *The New Palgrave: The Invisible Hand*. New York, NY: W.W. Norton & Company, Inc.

Encyclopedia Britannica. 2013. Price (website). Accessed July 31, 2018.

Friedman, M. 1976. *Price Theory*. Chicago, IL: Aldine Publishing Company.

Gwartney, J.D., Stroup, R.L., Sobel, R.S., and Macpherson, D.A. 2012. *Economics: Private and Public Choice*. Fourteenth edition. Boston, MA: Cengage Learning.

Hayek, F. 1945. The use of knowledge in society. *American Economic Review* **35** (4): 519–30.

Leal, D. and Fretwell, H. 1997. Back to the future to save our parks. *PERC Policy Series* PS-10. Bozeman, MT: PERC.

Regan, S. 2013. Funding the National Park System for the next century. Testimony to the U.S. Senate Committee on Energy and Natural Resources. July 25.

Sowell, T. 1980. *Knowledge and Decisions*. New York, NY: Basic Books.

Steil, B. and Hinds, M. 2009. *Money, Markets, and Sovereignty*. New Haven, CT: Yale University Press.

Walls, M. 2013. *Paying for State Parks: Evaluating Alternative Approaches for the 21st Century*. Washington, DC: Resources for the Future.

Watson, D.S. 1969. *Price Theory in Action*. Boston, MA: Houghton Mifflin Company.

1.2.4 Incentives

Most human action can be understood by understanding the incentives people face. "Moral hazard" occurs when people are able to escape full responsibility for their actions.

Nearly everyone would want to save a person who is drowning. But each of us is more likely to try to rescue a person who falls into two feet of water at the edge of a small pond than to try to rescue someone who is struggling in a river near the top edge of Niagara Falls. Economists express this simple reality as a principle: People are more likely to pursue their goals (benefits) when the cost to them is minimal, and they will seek low-cost ways to attain them. These costs and benefits – or penalties and rewards – are called incentives (Becker, 1976; Lazear, 2000).

Knowing the incentives people face makes it possible to understand and sometimes even predict behavior. If a person's goal is to increase his or her income, that person has an incentive to devote long hours to a grueling job or seek to obtain a better-paying one. If the person's goal is to make friends or achieve inner peace, earning a high income is not as important and behavior will be different. The difference between these people is not that some are "greedy" and others not. Their behavior can be explained by understanding the costs and benefits they face in the pursuit of their goals.

As individuals we usually are able to recognize and evaluate the costs of our choices. We are attuned to the relative costs of alternatives available to us, but recognizing and taking into account the costs facing others is more difficult. The costs borne by others generally have less effect on our decisions than the costs we incur directly. When individuals realize they can use resources that properly belong to others for their own benefit, they are tempted to act irresponsibly. Economists call this "moral hazard" (Kotowitz, 1987).

Moral hazard exists in the private marketplace in cases where information asymmetries combine with separation of ownership and control, enabling people to escape full responsibility for their actions. Examples include excessive utilization of health services due to reliance on third-party insurers (Goodman, 2012) and reckless behavior by persons with access to trust funds (Carnegie, 1891; Feldman, 2014). However, as Hülsmann (2006) writes, "there are strong forces at work to eliminate expropriation" when it occurs in free markets. In the examples given, insurers try to limit their financial exposure by not covering treatment for preexisting conditions and by rejecting some claims, and wealthy individuals write trust fund agreements carefully to limit or end access to the funds in the event of misbehavior by the beneficiary. As a result, says Hülsmann, "moral hazard induced expropriation is therefore not only accidental, but also ephemeral in the free market."

A risk of moral hazard arises when one person can over-use a public good (or good held in common) for personal gain even though others may suffer as a result. Writers about climate change often claim manufacturers, energy companies, and people who drive cars and trucks are "dumping carbon pollution" into the atmosphere, exploiting and degrading a common resource without paying a fair price to other stakeholders. If true, this would be an example of moral hazard. Why this is *not* the case is addressed in some detail in Section 1.3 and elsewhere in this book.

In contrast to free markets, political institutions and regulated markets are rife with moral hazard. "Strong forces to eliminate expropriation" are seldom seen. People who work in government or who qualify for government entitlement programs are usually spending someone else's money, not their own, and so have a weaker incentive to spend it wisely. The price of inefficiency, which would be borne by an individual or a business if incurred in the private market, is instead borne by taxpayers or businesses. Consequently, there is little incentive for government agencies to become more efficient. According to Baden and Stroup (1981), "Bureaucrats, like most other people, are predominantly self-interested. Given that an administrator's welfare tends to increase with increments in his budget, many of our resource administrators act as bureaucratic entrepreneurs. Unfortunately, unlike [entrepreneurs] in the private sector, these administrators are not accountable to a bottom line demarcating benefits and costs. Thus, the net 'benefits' of many of their activities are strongly negative" (p. x).

Government agencies have incentives to conceal the basis of their decisions, so they cannot be challenged by the individuals being regulated or taxed. Elected officials have incentives to make campaign promises they have no intention of keeping. Individuals have incentives to fake illnesses or hide assets to qualify for entitlement programs. All these perverse incentives compromise government's ability to deliver services efficiently.

The Endangered Species Act illustrates the harm that can occur when one party determines how another must use an asset. The law gives government officials great latitude in telling landowners what to do if they find an endangered animal such as a red-cockaded woodpecker on their properties. Government officials choose how the animal must be protected, but the landowner must pay the costs. For example, the owner may not be allowed to log land within a certain distance of the bird's colony. In some cases, government officials have prevented plowing land for farming or to create a firebreak. With such power, government officials are likely to be wasteful of some resources (such as land) while ignoring other ways of protecting the species (such as building nest boxes). To the government official, the land is almost

a free good. (See Section 1.4.6 for a more detailed discussion of this example.)

A program the EPA devised for reducing nitrogen based-nutrients that build up in waters such as the Chesapeake Bay offers another example. The agency developed a list of acceptable options among which states could choose, along with an estimated cost per pound of nitrogen removed (Jones *et al.*, 2010, Figure 2). States were required to submit implementation plans describing which solutions they would apply to reduce nitrogen.

One solution was to require storm water retention ponds for new land development projects, with an estimated cost of $92.40 per pound of nitrogen removed. Another option was to plant over-winter cover crops on farm fields, with an estimated cost of $4.70 per pound of nitrogen removed.

Retention ponds reduce immediate runoff but add nutrients to groundwater, the primary source of water pollution. They also impose long-term maintenance issues and attract geese, which add to the nitrogen pollution. Cover crops retain nitrogen, which is then available for the next season's crops, thus removing nitrogen from the water system and making them the superior solution.

Delaware chose to implement the far more expensive, less effective retention pond option because builders need permits and could be forced by government officials to comply, while farmers don't need permits to plant crops. The permitting agency even rejected a builders association's offer to pay into a cover crop fund instead of installing retention ponds (Jones *et al.*, 2010). It was a vivid example of moral hazard at work.

How are incentives relevant to the climate change issue? Without a marketplace in which access to the atmosphere is bought and sold, there are no prices that might make possible the efficient management of the atmosphere for the public good. Individuals have incentives to use the atmosphere to dispose of waste without regard to its possible negative effects on others, unless faced with regulations that prevent such behavior. This system might be judged wrong if actual damage to the public interest were demonstrated *and* if a superior method of rationing use of the atmosphere were available. Both assumptions are severely tested in later chapters.

Some conclusions from this section include the following:

- Human action is determined by incentives people face in the pursuit of their goals.

- In the private sector, incentives align people's actions with the public interest because people generally are held accountable for the consequences of their actions.

- In the public sector, action is often separated from accountability for results, resulting in conduct that may not advance the public interest.

- Incentives concerning the use of the atmosphere are currently distorted by the absence of a price system, but whether this causes social harms or can be corrected is unclear.

References

Baden, J. and Stroup, R.L. (Eds.) 1981. *Bureaucracy vs. Environment: The Environmental Costs of Bureaucratic Governance*. Ann Arbor, MI: The University of Michigan Press.

Becker, G.S. 1976. *The Economic Approach to Human Behavior*. Chicago, IL: University of Chicago Press.

Carnegie, A. 1891. Advantages of poverty. *Nineteenth Century Magazine*. London.

Feldman, A. 2014. Incentive trusts can keep your heirs motivated. *Barron's*. May 17.

Goodman, J. 2012. *Priceless: Curing our Healthcare Crisis*. Oakland, CA: Independent Institute.

Hülsmann, J.G. 2006. The political economy of moral hazard. *Politická ekonomie* **54** (1): 35–47.

Jones, C., Branosky, E., Selman, M., and Perez, M. 2010. How nutrient trading could help restore the Chesapeake Bay. *Working Paper*. Washington, DC: World Resources Institute. February.

Kotowitz, Y. 1987. Moral hazard. In: Eatwell, J., *et al.* (Eds.) *The New Palgrave*. London: Macmillan, pp. 549–51.

Lazear, E.P. 2000. Economic imperialism. *Quarterly Journal of Economics* **115** (1): 99–146.

1.2.5 Trade

Trade creates value by making both parties better off.

The First Theorem of Welfare Economics, also known as Adam Smith's Invisible Hand Theorem, reads "if everyone trades in the competitive marketplace, all mutually beneficial trades will be completed and the resulting equilibrium allocation of resources will be economically efficient." (Pindyck and Rubinfeld, 2000, p. 574). Both sides can gain when goods are exchanged; there does not need to be a winner and a loser. This means voluntary exchanges create value even though no additional goods or services are created. Trade allows people to act on the signals created by prices and the incentives created by the costs and benefits of choices freely made. Trade is the real-world manifestation of markets, the spontaneous order that is created when property rights are protected and people are free to choose how to use what belongs to them (Hayek, 1983). Trade can create value in three ways:

1. Trade channels resources, products, and services from those who value them less to those who value them more. One way to understand this principle is to think about something people really disagree about – say, music. John likes opera. Jane likes rock music. If John has a rock concert ticket and Jane an opera ticket, exchanging the tickets will make both of them better off. Without any change in production, the trade of the opera ticket for the rock concert ticket produces value.

2. Trade enables individuals to direct their resources to activities where they produce the greatest value so they can then trade the fruits of those activities for the items they want for themselves. A farmer in central Montana who grows wheat produces far more than he wants to consume. He trades the wheat for income to buy coffee from Guatemala, shoes from Thailand, and oranges from Florida. The Montana farmer might have been able to grow oranges, but given the cold Montana climate, doing so would have squandered resources. Trade enables people to obtain many things they would not have the proper talent or resources to produce efficiently themselves.

3. Trade enables everyone to gain from the division of labor and economies of scale. Only with trade can individuals specialize narrowly in computer programming, writing books, or playing professional golf, developing highly productive skills that would be impossible to obtain if each

family had to produce everything for itself. Similarly, large automobile factories lower the cost of manufacturing cars so they can be sold at prices within reach of the average worker.

Resource owners gain by trading in three ways: across *uses* (for example, trading out of low-valued crops into ones that earn more money), across *space* (marketing products across geographic distance to different states or nations), and across *time* (using resources now or gaining from conservation or speculation by saving resources until they become more valuable).

Even trade in garbage can create wealth. Consider a city that disposes of garbage in a landfill. If the city is located in an area where underground water lies near the surface, disposing of garbage is dangerous, and very costly measures would have to be taken to protect the water from landfill leakage. Such a city may gain by finding a trading partner with more suitable land where a properly constructed landfill does not threaten to pollute water. The landowner may be willing to accept garbage in return for pay. If so, both parties will be better off.

In some parts of the western United States, rights to divert and use water from rivers and groundwater are bought and sold. Anderson and Libecap (2011) documented 1,766 transactions in 11 states between 1987 and 2008. These transactions allow even water, a resource that meets the definition of a public or common-pool resource, to be traded like a private good, allowing access rights to move to those who value them most highly at a price acceptable to current holders of those rights.

In recent years, more people have been seeking high-quality streams for fly-fishing. They recognize many streams dry up in hot summer months when farmers divert large amounts of water for their fields. To keep more water in streams to keep fish thriving, some fly-fishers are willing to trade cash for the farmers' water rights. And some farmers are happy to part with a portion of the water they have been using in exchange for cash. The Oregon Water Trust (recently renamed The Freshwater Trust) works out trades between individuals committed to protecting salmon and farmers who are willing to give up some of their water. Purkey (2007) wrote,

Consider the story of ranchers Pat and Hedy Voigt. Last year, they reached a permanent, voluntary agreement with one of the [Columbia Basin Water Transactions Program] CBWTP's partners, the Oregon

Water Trust. Between July 21 and September 30, up to 6.5 million gallons of water that they would normally divert each day from the Middle Fork of the John Day River and two of its tributaries will stay in the river, enhancing flows for a distance of 70 miles. In exchange, the Voigts now have the resources to improve irrigation efficiencies on their ranch, even as they benefit one of the largest and best remaining populations of wild spring Chinook and summer steelhead in the lower 48 states.

According to Purkey, similar deals have been struck elsewhere in Oregon and in Idaho, Montana, and Washington. "Across the Columbia Basin, forward-looking landowners are creating innovative strategies that improve their bottom lines and build flexibility into ecosystems facing chronic water shortages. The results of this new model are not only benefiting communities right now but also are helping to prepare the Pacific Northwest for the future" (*Ibid.*).

Trade is important in the climate change discussion for a number of reasons. Access to the atmosphere does not need to be a zero-sum transaction whereby people who produce emissions benefit at the expense of others. Nor do the governments of the world have to agree on the terms and conditions of access for the result to be efficient. Virtually everyone benefits from the energy produced when anthropogenic greenhouse gases are produced as well as from increased agricultural production due to aerial fertilization by carbon dioxide. Positive and negative externalities are exchanged spontaneously in the absence of government policies (or taxes) or even sufficient information to place prices on either one. The result is huge net social benefits documented in Chapters 3, 4, and 5. Interfering with this trade by limiting or even banning the use of fossil fuels would jeopardize these benefits.

It may be objected that future generations do not have a place at the table in the spontaneous marketplace for access to the atmosphere, and since today's emissions may have a negative impact on them this constitutes an inefficiency or injustice. Section 1.5 of this chapter explains how capital markets create incentives for today's investors to protect the interests of future generations, so this concern can be addressed. But consider too that virtually *everything* we do today affects future generations, either for good or for ill, so this can hardly be a justification for government intervention. The Intergovernmental Panel on Climate Change (IPCC) itself admits that the impact of climate change on future generations will be "small relative to the impacts of other drivers (*medium evidence, high agreement*). Changes in population, age, income, technology, relative prices, lifestyle, regulation, governance, and many other aspects of socioeconomic development will have an impact on the supply and demand of economic goods and services that is large relative to the impact of climate change" (IPCC, 2014, p. 662. This suggests climate change does not pose a unique danger that would justify it being treated differently than other challenges.

Common ownership of resources is not a barrier to the use of trade as a way to achieve win-win solutions to conflicts over access. Some of the biggest successes in managing other common-pool resources, such as water described in the examples given above and public lands (grazing rights) in cases described later in this chapter, rely on spontaneous or informal processes with only limited involvement by governments (Ostrom, 2005). The case against attempting to create an artificial marketplace for trading rights to the atmosphere is set forth in Section 1.3 and other parts of this book.

References

Anderson, T. and Libecap, G. 2011. A market solution for our water wars. *Defining Ideas*. Stanford, CA: Hoover Institution.

Hayek, F. 1983. *The Fatal Conceit*. Chicago, IL: University of Chicago Press.

IPCC. 2014. Intergovernmental Panel on Climate Change. *Climate Change 2014: Impacts, Adaptation, and Vulnerability. Part A: Global and Sectoral Aspects.* Contribution of Working Group II to the Fifth Assessment Report of the Intergovernmental Panel on Climate Change. Cambridge, UK and New York, NY: Cambridge University Press.

Ostrom, E. 2005. *Understanding Institutional Diversity*. Princeton, NJ: Princeton University Press.

Pindyck, R.S. and Rubinfeld, D.L. 2000. *Microeconomics*. Fifth Edition. New York, NY: Prentice-Hall.

Purkey, A. 2007. Blue ribbon management for blue ribbon streams. *PERC Reports* **25** (2).

1.2.6 Profits and Losses

Profits and losses direct investments to their highest and best uses.

The profit and loss system is a key element of markets. By allowing investors and producers to keep the profits they earn and suffer any losses they incur, markets ensure that resources are used as efficiently as possible to meet consumer wants and needs (Mises, 1966 [1998], pp. 241–4; Gilder, 1984; Novak, 1991, pp. 104–12).

Profit is a measure of how much value was added to a good or service relative to the cost of resources used. Profits provide a clear index of performance, with high profits indicating resources were purchased at a price much lower than the resulting product was worth to buyers. A large loss indicates the product was worth much less than the resources taken from the rest of the economy to produce it. In this way, profits and losses direct businesses toward activities that most efficiently meet consumer wants and needs

High profits act as a signal to producers to make more products and to potential producers to start making new products. Consumers benefit from the increased supply and competition among producers, which drive down prices. Awareness of profit margins leads to more careful use of natural resources as producers seek to minimize their costs. This can lead to the discovery of new ways to use resources more efficiently.

Hope for profits and fear of losses cause producers to spend untold hours figuring out how to use resources more efficiently. That is why airplanes, batteries, bicycles, bottles, cans, cars, computers and computer chips, printing devices, solar panels, telephones, televisions, and hundreds of other products we use every day are "smaller, faster, lighter, denser, and cheaper" than ever before (Bryce, 2014). The profit and loss system is driving a widespread "dematerialization" process whereby fewer resources and less energy are needed to meet human needs, a trend described and documented in detail in Chapter 5.

Profits reward those who succeed in producing goods and services people are willing to buy at a price higher than the cost of supplying them. Losses have their place, too. They penalize those who have not been able to discover how to create more value than the cost to produce. In effect, people are telling a money-losing firm they want to see that firm's resources go to other products or services more valuable to them.

Large profits are ephemeral. The competition of new entrants, drawn by profits, gradually lowers the sales of existing firms and often their prices as well, reducing profits. Entry continues until profits fall to what economists call normal rates of return. Entry then stops. The first firm to innovate successfully may make above-normal profits (an appropriate reward and critical incentive), but the profits fall as competition heats up.

An entrepreneur seeking to exploit a new profit opportunity usually must (a) discover the new opportunity and (b) find investors willing to take the risk that profits will be made. It may also be necessary to sell potential buyers on the new product or service. All of these activities are costly. But expected profit provides an incentive to persevere for entrepreneurs, investors, and those who must sell the idea to investors and the product to buyers. Expected profit rewards them for making the necessary investments of time, effort, and money to accomplish their tasks. New ideas may need years of effort before they reach fruition. Expected profit is the carrot to attract the needed efforts.

Thanks to the profit and loss system, the most efficient producers win the competition for the use of scarce resources. But this system does not exist in government agencies; unlike private entrepreneurs and investors, government officials typically cannot retain any profits their agencies might earn by being more efficient than competitors, and they do not personally suffer a loss if they are inefficient. Consequently, governments can and do systematically waste resources, taking losses over the long term because they make up the difference by taking money from taxpayers.

Government officials are typically deprived of the signals created by a profit and loss system that might direct them to the most efficient ways to produce a product or deliver a service. When they make poor decisions, they are insulated from the negative consequences because taxpayers, consumers, or regulated businesses must incur the loss. Even if regulators are smart and well-informed, they are unlikely to be smarter or better informed than private investors since they are spending other people's money and not their own.

Framing climate change as a problem requiring a government-led solution necessarily means losing the powerful efficiency-creating power of the profit and loss system. Without profits and losses directing investments in energy sources and technologies, governments must pick winners and losers based on the input of lobbyists, the judgment of bureaucrats influenced by careerism and tunnel vision, and other maladies affecting bureaucracies described in some detail in Section 1.4.3.

A key part of the climate change issue, perhaps more important than any scientific variable or theory, is who should decide what energy sources and technologies ought to be used in light of what we know about climate change. Should those choices be made by individuals and private entities that reap profits or bear losses from their choices, or by government agencies that are immune to such consequences? Vaclav Smil (2010) ended his book *Energy Myths and Realities* with this warning to those who think they can do better than markets at picking an energy source that could replace fossil fuels:

> Do not uncritically embrace unproven new energies and processes just because they fit some preconceived ideological or society-shaping models. Wind turbines or thin-film solar cells may seem to be near-miraculous forms of green salvation, ready to repower America within a decade. But ours is a civilization that was created by fossil fuels, and its social contours and technological foundations cannot be reshaped in a decade or two (p. 163).

References

Bryce, R. 2014. *Smaller Faster Lighter Denser Cheaper: How Innovation Keeps Proving the Catastrophists Wrong.* New York, NY: PublicAffairs.

Gilder, G. 1984. *The Spirit of Enterprise.* New York, NY: Simon and Schuster.

Mises, L. 1996 [1998]. *Human Action: A Treatise on Economics, the Scholar's Edition.* Auburn, AL: Ludwig von Mises Institute.

Novak, M. 1991. *The Spirit of Democratic Capitalism.* Lanham, MD: Madison Books.

Smil, V. 2010. *Energy Myths and Realities: Bringing Science to the Energy Policy Debate.* Washington, DC: American Enterprise Institute.

1.2.7 Unintended Consequences

The art of economics consists in looking not merely at the immediate but at the longer effects of any act or policy.

Hazlitt (1979) wrote, "the whole of economics can be reduced to a single lesson, and that lesson can be reduced to a single sentence. The art of economics consists in looking not merely at the immediate but at the longer effects of any act or policy; it consists in tracing the consequences of that policy not merely for one group but for all groups" (p. 17). Economists are trained to ask, "and then what?"

Unintended consequences are sometimes referred to as "the seen and the unseen." Claude Frédéric Bastiat wrote in 1850, "a law gives birth not only to an effect, but to a series of effects. Of these effects, the first only is immediate; it manifests itself simultaneously with its cause – it is seen. The others unfold in succession – they are not seen: it is well for us if they are foreseen." Bastiat also observed that the difference between a bad and a good legislator is "the one takes account of the visible effect; the other takes account both of the effects which are seen and also of those which it is necessary to foresee."

Overlooking the secondary effects (side effects) of an action is easy, especially if those effects are on other people or will not be experienced soon. When those unintended consequences are negative, they can offset some or all of the benefits of an action. Advocates of a particular goal or state of affairs often are impatient with the sometimes slow pace of markets and voluntary agreements. Passing a law or funding a government program *seems* to be a faster and more direct route to their goal, and this path is often sold to activists by elected officials seeking their campaign support and lobbyists seeking clients. But most government programs fail to achieve their goals precisely because of the unintended consequences economists are trained to look for.

Turning to the issue of climate change, advocates of immediate action to reduce greenhouse gas emissions often overlook the unintended consequences of their recommendations. Reducing emissions by amounts large enough to potentially affect the planet's climate would require large reductions in energy consumption, which would reduce human well-being by diverting resources away from more urgent needs. Because wind and solar power costs two to three times as much as energy derived from the use of fossil fuels, using those alternative energy sources would reduce human

well-being, especially for low-income families that cannot afford to pay more for electricity and home heating (Bezdek, 2010). "Energy poverty" is a critical issue facing developing countries today because access to electricity is crucial to the three dimensions of human development: health, knowledge, and standard of living (Kanagawa and Nakata, 2008).

The money spent today and in the near future on expensive solar and wind power would not be available for other things that contribute to human well-being, such as public health actions to protect people from malaria and other diseases, wells and dams to provide water for agriculture and use in homes, and infrastructure such as electric power plants and power lines (Yadama, 2013; Lomborg, 2006). The full consequences of those missed opportunities would only emerge over time and are largely invisible to today's environmental activists.

Another unintended consequence of reducing greenhouse gas emissions is the negative effect on food production and the natural environment. Carbon dioxide is essential to plant growth, and anthropogenic emissions are thought to be responsible for 70% of the "greening of the Earth" observed from satellites and benefiting more than 25% to 50% of the global vegetated area (Zhu, *et al.*, 2016). (Less than 4% of the globe shows browning (*Ibid.*).) Less greening also means less habitat for wildlife, so efforts to stop or slow global warming could unintentionally lead to the extinction of more species (Goklany, 2015; Hughes *et al.*, 2014). Replacing fossil fuels with wind, solar, and biofuels also would require millions of square miles of wilderness and farmland to be covered with industrial wind turbines, mirrors or photovoltaic panels, or corn planted and harvested to make ethanol. The environmental consequences would be devastating (Kiefer, 2013; Bryce, 2014, p. 212; Smil, 2015, pp. 211–2).

Over the period 2012 through 2050, the cumulative global economic benefit of aerial CO_2 fertilization will be approximately $9.8 trillion (Idso, 2013). Reducing greenhouse gas emissions would mean forfeiting some or all of this benefit. Mariani (2017), in a study described in greater detail in Chapter 5, Section 5.3.6.1, estimates a return of global temperatures and CO_2 levels to pre-industrial conditions would reduce by 18% global production of the four crops (wheat, maize, rice, and soybean) accounting for two-thirds of total global human caloric consumption. Mariani estimates that increases in atmospheric CO_2 to 560 ppm and temperature to +2°C relative to today would improve crop production by 15% above today's values.

Frank *et al.* (2017) note actions to mitigate greenhouse gas emissions could negatively impact food supply in several ways: by diverting agricultural land into land used for energy (e.g. corn from feed to ethanol); by halting or slowing needed land conversion from high-carbon landscapes (forests) into agricultural production; by shifting from more to less greenhouse gas-intensive agricultural commodities (e.g., away from ruminant production); and by adopting greenhouse gas-reducing management practices (e.g., reduced fertilizer application). Figure 1.2.7.1 shows the relative product price change of nine commodities driven by a $150 per ton of CO_2-equivalent (CO_2e) tax (left panel), as well as the overall percent increase in the food price index (right panel, relative to the base year of 2000) for the world and various regions of the world.

In all instances, the CO_2 tax raises the cost of food in all regions. The largest increases (60% to 100%) in the food price index are seen in those regions with less efficient agricultural production systems, such as Oceania, South East Asia, Sub-Saharan Africa, South Asia, and Latin America, while for the world as a whole the price increase is around 38%.

Figure 1.2.7.2 depicts the relationship between greenhouse gas mitigation targets and global average calorie consumption projected for the year 2050. As the figure shows, increasingly ambitious efforts to reduce CO_2 emissions result in greater reductions in daily dietary energy. Using the IPCC's representative concentration pathway (RCP) scenario that limits warming to 1.5°C, for example, a $190 tCO_2e^{-1} carbon tax would reduce daily caloric intake by 285 kcal per capita per day, a 9% decrease. At first glance, such a decline may not appear significant, but as Frank *et al.* note, "this would translate into a rise of 300 million people in the global number of chronically undernourished [individuals]," a 150% increase over the current chronically undernourished population.

Frank *et al.* conclude "a uniform carbon price across sectors does lead to trade-offs with food security at increasingly ambitious stabilization targets. This results from rising food prices driven by the adoption of greenhouse gas abatement strategies [that] limit agricultural land expansion and increase production costs for farmers targeted by the implementation of a carbon price."

Figure 1.2.7.1

Relative price impact of a $150 per tCO₂e carbon tax on emissions from agriculture on global commodity prices and regional food price index

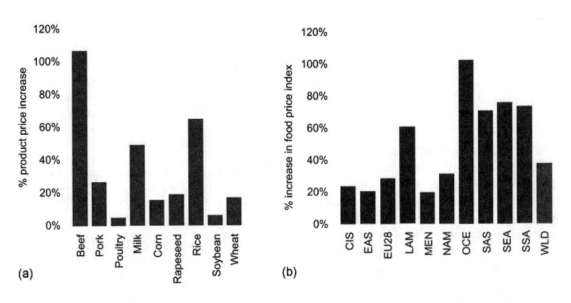

CIS is Commonwealth of Independent States, EAS is East Asia, EU28 is European Union, LAM is Latin America, MEN is Middle East and North Africa, NAM is North America, OCE is Oceania, SAS is South Asia, SEA is South East Asia, SSA is Sub-Saharan Africa, and WLD is World. *Source*: Adapted from Frank *et al.*, 2017.

Figure 1.2.7.2
Per-capita caloric loss caused by proposed carbon taxes

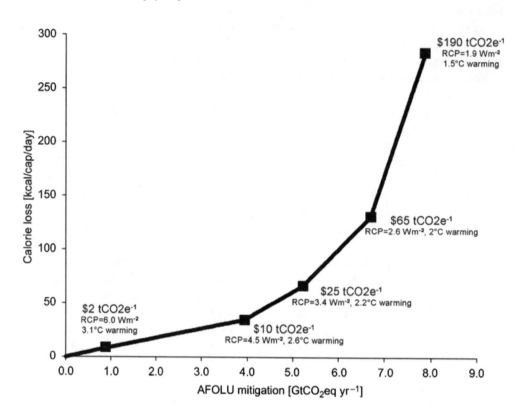

The blue line represents policies where all countries participate to achieve increasingly ambitious climate stabilization targets. Text adjacent to the blue squares indicates the carbon price (tax) associated in achieving climate stabilization for a given representative concentration pathway (RCP) and its associated global temperature reduction in 2050. *Source*: Adapted from Frank *et al.*, 2017.

Advocates of immediate action to reduce the use of fossil fuels probably do not want to increase energy poverty, destroy wildlife habitat, or increase world-wide hunger. These are all unintended consequences of the IPCC's clearly stated goal of reducing and eventually banning fossil fuels (IPCC, 2014, pp. 10, 12). Economists are trained to look for such unintended consequences, to anticipate how changes in incentives lead to changes in behavior which then affect the ability to reach goals. Environmental activists ignore or downplay these consequences due to their tunnel vision. Their vision of a world where energy freedom is replaced with a government-imposed ban on fossil fuels is so compelling they simply refuse to believe it could have a dark side.

References

Bastiat, F. 1850 [1995]. What is seen and what is not seen. In: Cain, S. (Ed.) *Selected Essays on Political Economy*. Reprint, Irvington-on-Hudson, NY: Foundation for Economic Education, pp. 1–50.

Bezdek, R. 2010. Potential economic impact of the EPA endangerment finding on low income groups and minorities. *The Business Review, Cambridge* **16** (1): 127–33.

Bryce, R. 2014. *Smaller Faster Lighter Denser Cheaper: How Innovation Keeps Proving the Catastrophists Wrong*. New York, NY: PublicAffairs.

Frank, S., *et al.* 2017. Reducing greenhouse gas emissions in agriculture without compromising food security? *Environmental Research Letters* **12**: 105004.

Goklany, I.M. 2015. *Carbon Dioxide: The Good News.* London, UK: Global Warming Policy Foundation.

Hazlitt, H. 1979. *Economics in One Lesson.* Norwalk, CT: Arlington House, Inc.

Hughes, A.M., Khandekar, M., and Ollier, C. 2014. *About Face! Why the World Needs More Carbon Dioxide.* Minneapolis, MN: Two Harbors Press.

Idso, C.D. 2013. *The Positive Externalities of Carbon Dioxide.* Tempe, AZ: Center for the Study of Carbon Dioxide and Global Change. October 21.

IPCC. 2014. Intergovernmental Panel on Climate Change. *Climate Change 2014: Impacts, Adaptation, and Vulnerability. Part A: Global and Sectoral Aspects.* Contribution of Working Group II to the Fifth Assessment Report of the Intergovernmental Panel on Climate Change. Cambridge, UK and New York, NY: Cambridge University Press.

Kanagawa, M. and Nakata, T. 2008. Assessment of access to electricity and the socioeconomic impacts in rural areas of developing countries. *Energy Policy* **36**: 2,016–29.

Kiefer, T.A. 2013. Energy insecurity: the false promise of liquid biofuels. *Strategic Studies Quarterly* (Spring): 114–51.

Lomborg, B. 2006. *How to Spend $50 Billion to Make the World a Better Place.* New York, NY: Cambridge University Press.

Mariani, L. 2017. Carbon plants nutrition and global food security. *The European Physical Journal Plus* **132**: 69.

Smil, V. 2016. *Power Density: A Key to Understanding Energy Sources and Uses.* Cambridge, MA: The MIT Press.

Yadama, G.N. 2013. *Fires, Fuel and the Fate of 3 Billion: The State of the Energy Impoverished.* New York, NY: Oxford University Press.

Zhu, Z., *et al.* 2016. Greening of the Earth and its drivers. *Nature Climate Change* **6**: 791–5.

1.2.8 Discount Rates

Discount rates, sometimes referred to as the "social rate of time preference," are used to determine the current value of future costs and benefits.

How do you place a value on a cost or benefit that occurs far in the future? How much should be spent today to avoid a possible but uncertain harm 50 years or 100 years hence?

Discount rates, sometimes referred to as the "social rate of time preference," are used to determine the current value of future benefits or harms. As Kreutzer (2016) wrote, "Discounting is an opportunity cost exercise. The rate should reflect the best alternative return that an investment of the same size could reasonably be expected to generate." Discounting has a long history of use in public policy.

Of the many controversies involved in deciding whether and how to respond to the threat of climate change, none attracts as much attention and condemnation as the choice of the discount rate used to estimate the present value of future impacts. Weitzman (2015) has described this debate over discounting damages as "vigorous," noting "the choice of a discount rate is itself one of the most significant (and controversial) uncertainties in the economics of climate change." And, as Heal and Millner (2014) conclude, there is "no convergence to a single unanimously agreed upon [discount] value in sight."

The rate at which one discounts the value of benefits expected to appear in the future is expressed annually, similar to interest paid on a savings account. "The lower the rate of discount employed, the higher the present value of the estimated future benefits of a public project. Hence, the rate of discount used in evaluating public projects has an important influence on the allocation of resources within the public sector, and may also influence the relative rates of growth of the public and private sectors" (Mikesell, 1977, p. 3).

One method of estimating the current value of future costs and benefits is exponential discounting, which is typically used in finance. It assumes preferences between consuming now or in the future do not change over time, so only the value of time needs to be taken into account. That value, in turn, can be revealed by looking at the interest paid on very safe investments, such as government bonds, for a similar period of time. This method is used by the U.S. Office of Management and Budget (OMB), which defines the "social rate of time preference" as the real rate of return on long-term government debt. It requires cost-benefit analyses be calculated using that published rate and two additional constant rates, a low rate of 3% and a high rate of 7%, to establish a band or range of outputs for decision-making (OMB, 2003). The U.K.'s Treasury uses a standard 3.5% rate (but see below for a recent modification), below the

5% rate typical in the literature (e.g., Nordhaus, 1998; Murphy, 2008; Tol, 2010).

Discount rates are important in understanding the climate change issue because the costs of reducing greenhouse gas emissions mostly occur up-front, in the form of major capital investments in new sources of energy and the infrastructure needed to support them. The benefits of reducing emissions, to the extent they exist, occur far in the future. According to Working Group I's report for the IPCC's Fifth Assessment Report, "cumulative emissions of CO_2 largely determine global mean surface warming by the late twenty-first century and beyond (see Figure SPM.10). Most aspects of climate change will persist for many centuries even if emissions of CO_2 are stopped" (IPCC, 2013, p. 27).

Money spent now to secure benefits far in the future could be used to buy other things that would produce benefits sooner. Some of those benefits, such as food to help feed the world's hungry or clean water in developing countries, are important and may be more important than battling one or two degrees of warming centuries from now (Mendelsohn, 2004; Lomborg, 2006; Lemoine and Rudik, 2017).

Consider the following example: Exponential discounting at the rate of 5% means if we choose to spend $100 today to reduce greenhouse gas emissions by one ton, we lose the opportunity to spend $1,147 50 years from now to reduce emissions then. With advancing technology, that $1,147 spent 50 years from now would likely enable us to reduce emissions by much more than we could with current technology. Since, as the IPCC says, climate is affected by cumulative emissions (ambient CO_2 concentrations) and not annual emissions, early action is difficult to justify.

This example also illustrates that avoiding $1,147 in damages 50 years from now is worth an investment today of about $100, about 9% of the future value. Expressed differently, a dollar of benefit 50 years from now is worth only about 9 cents today. Thus, benefits reaped 50 years in the future need to be worth about 11 times as much as alternative benefits that could be achieved today in order to justify their expense.

Critics of exponential discounting worry that the current generation of investors and emitters won't actually set aside the $100 needed today that would become $1,147 to be used 50 years from now to reduce emissions. What if this modest sum were spent on something else? There is also concern that discount rates of around 5% over-estimate the likely long-term rate of return on investments over so many decades. Do low-probability, high-damage events call for using a lower discount rate? (Ceronsky et al., 2011; Heal, 2017).

Advocates of immediate action to reduce emissions sometimes blame the "greed" or "selfishness" of others for opposition to their plans (Bartholomew and Francis, 2017; Tirole, 2017, p. 196). This could be true, since people sacrificing today are unlikely to live long enough to be among the beneficiaries of a cooler climate 100 years from now. But more likely, people are expressing a reasonable social rate of time preference. Uncertainty grows with time over whether any sacrifice made today will actually benefit future generations. Surveys show the public in the United States cite this uncertainty as the main reason they oppose paying higher taxes on energy to help fight global warming (Ansolabehere and Konisky, 2015). This is not based on ignorance of the issue, but just the opposite. Newspapers and other popular sources of information report regularly on how China, India, and other major emitters are increasing their emissions while the United States is reducing its own, the "leakage" discussed in Section 1.2.10 below. Perhaps physical scientists are less aware of this phenomenon than the less-educated but more-attentive general public.

An alternative to exponential discounting is hyperbolic discounting. Surveys and small-scale experiments show people tend to give more weight to benefits that are very immediate or very distant in the future, and less weight to benefits that might appear at intermediate time scales. This attitude toward time is incorporated into discounting by changing the discount rate chosen for different periods of future time. Its adherents claim it leads to the choice of lower discount rates for events occurring in the far future and therefore makes a stronger case for action today to avoid far-future risks (Farmer and Geanakoplos, 2009; Arrow et al., 2013; Garnaut, 2008).

The U.K.'s Treasury has moved toward hyperbolic discounting by adopting not one but 12 different discount rates taking into account the number of years over which a program operates and whether there is "risk to health and life values" (H.M. Treasury, 2018, Table 8, p. 104).

Choosing the "right" discount rate to use when addressing climate change is addressed again and in greater detail in Chapter 8.

References

Ansolabehere, S. and Konisky, D.M. 2015. *Cheap and Clean: How Americans Think about Energy in the Age of Global Warming*. Cambridge, MA: MIT Press.

Arrow, K.J., *et al.* 2013. How should benefits and costs be discounted in an intergenerational context? *Economics Department Working Paper Series* No. 56-2013. Colchester, UK: University of Essex, Business, Management, and Economics.

Bartholomew and Francis. 2017. Joint message on the world day of prayer for creation by Patriarch Bartholomew and Pope Francis. Greek Orthodox Archdiocese of America. September 1.

Ceronsky, M., Anthoff, D., Hepburn, C., and Tol, R.S.J. 2011. Checking the price tag on catastrophe: the social cost of carbon under non-linear climate response. *ESRI Working Paper* 392. Dublin, Ireland: Economic and Social Research Institute.

Farmer, J.D. and Geanakoplos, J. 2009. Hyperbolic discounting is rational: valuing the far future with uncertain discount rates. *Cowles Foundation Discussion Paper* No. 1719.

Garnaut, R. 2008. *The Garnaut Climate Change Review: Final Report*. Port Melbourne, Australia: Cambridge University Press.

Heal, G. 2017. The economics of the climate. *Journal of Economic Literature* **55** (3).

Heal, G.M. and Millner, A. 2014. Agreeing to disagree on climate policy. *Proceedings of the National Academy of Sciences* **111**: 3695–8.

HM Treasury. 2018. *The Green Book: Central Government Guidance on Appraisal and Evaluation*.

IPCC. 2013. Intergovernmental Panel on Climate Change. *Climate Change 2013: The Physical Science Basis*. Contribution of Working Group I to the Fifth Assessment Report of the Intergovernmental Panel on Climate Change. Cambridge, UK: Cambridge University Press.

Kreutzer, D. 2016. Discounting climate costs. The Heritage Foundation (website). Accessed July 19, 2018.

Lemoine, D. and Rudik, I. 2017. Steering the climate system: using inertia to lower the cost of policy. *American Economic Review* **107** (20).

Lomborg, B. 2006. *How to Spend $50 Billion to Make the World a Better Place*. New York, NY: Cambridge University Press.

Mendelsohn, R. 2004. The challenge of global warming. Opponent paper on climate change. In: Lomborg, B. (Ed.) *Global Crises, Global Solutions*. New York, NY: Cambridge University Press.

Mikesell, R.E. 1977. *The Rate of Discount for Evaluating Public Projects*. Washington, DC: American Enterprise Institute for Public Policy Research.

Murphy, K.M. 2008. *Some Simple Economics of Climate Changes*. Paper presented to the Mont Pelerin Society General Meeting, Tokyo. September 8.

Nordhaus, W.D. 1998. *Economics and Policy Issues in Climate Change*. Washington, DC: Resources for the Future.

OMB. 2003. *Circular A-4: Regulatory Analysis*. Washington, DC: Office of Management and Budget.

Tirole, J. 2017. *Economics for the Common Good*. Princeton, NJ: Princeton University Press.

Tol, R.S.J. 2010. Carbon dioxide mitigation. In: Lomborg, B. (Ed.) *Smart Solutions to Climate Change: Comparing Costs and Benefits*. New York, NY: Cambridge University Press, pp. 74–105.

Weitzman, M.L. 2015. A review of William Nordhaus' The Climate Casino: Risk, Uncertainty, and Economics for a Warming World. *Review of Environmental Economics and Policy* **9**: 145–56.

1.2.9 Cost-benefit Analysis

Cost-benefit analysis, when performed correctly, can lead to better public policy decisions.

Because all people, including those living in wealthy countries, must cope with scarcity, they must choose how much money to spend on environmental improvements and consequently how much less to spend on other goods and services (the opportunity cost of their choices) and in which projects or programs to invest (if any). Cost-benefit analysis (CBA) can help make such choices.

Private cost-benefit analysis is used to determine if the financial benefits to an agent over the lifetime of a project exceed the agent's costs. *Social* cost-benefit analysis attempts to include environmental impacts and other costs and benefits, including unintended consequences, which are not traded in markets and so would not necessarily be taken into account by private economic agents.

CBA is an economic tool that can help determine if the social benefits over the lifetime of a government project exceed its social costs. In the current context, CBA is used to determine in monetary terms the present worth of the social benefits and social costs of using fossil fuels, of mitigation versus unabated global warming, and of environmental regulations. A cost-benefit ratio can be obtained by dividing the projected costs by the projected benefits, or net benefits can be derived by subtracting costs from benefits. Projects earning a cost-benefit ratio less than 1 are possibly worth pursuing. Competing projects can be ranked according to their cost-benefit ratios, net benefits, or cost-effectiveness (Singer, 1979; Dorfman, 1993; Wolka, 2000, p. 8.130; Pearce *et al.*, 2006; van Kooten, 2013).

Economists and other social scientists can identify and attempt to quantify elements on both sides of the cost-benefit equation using observational data regarding supply, demand, prices, and profit generated by millions or billions of voluntary choices taking place in markets around the world and across time. Benefits can include protection of human health from hazards such as air pollution, measured in days or years of life extended, while costs can include slower economic growth (measured in per-capita income or GDP) due to higher taxes or the cost of complying with new regulations. A graph showing a hypothetical cost-benefit analysis for a proposal that would reduce emissions appears as Figure 1.2.9.1.

A variation on CBA is called benefit-cost analysis (BCA), though the two terms are sometimes used interchangeably. Zerbe (2018) writes, "CBA is the traditional approach of valuation, built on the potential compensation test ('PCT') and the avoidance of distributional and other equity considerations. CBA is limited to analyzing only the fair market value of property. BCA recognizes rights and moral sentiments as values insofar as they reflect the willingness to pay ('WTP') to obtain them or the willingness to accept ('WTA') payment for surrendering them." Chapter 8 makes a case for relying on CBA rather than BCA, so this short introduction to the topic focuses on CBA.

In Britain, the use of CBA by governments for all projects (not only environmental projects) is guided by *The Green Book: Central Government Guidance on Appraisal and Evaluation*, originally published in the 1970s by the Treasury and most recently updated in 2018, and a series of supplementary guidance documents listed on page 107 of that book (HM

Figure 1.2.9.1
Cost-benefit analysis of a proposal to reduce emissions

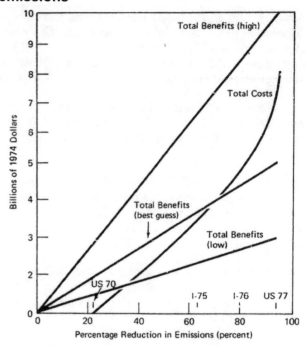

US 70, I-75, I-76, and US 77 are emission reduction scenarios outlined in NAS, 1974. *Source:* Singer, 1979, Figure 2, p. 29.

Treasury, 2018). The entire set of documents constitutes a very fine guide to the issue and is highly recommended, but with apologies to our British colleagues and friends around the world, the rest of this chapter focuses on the application of CBA to environmental issues only in the United States.

The application of cost-benefit analysis to environmental decision-making in the United States dates back to its use by the Army Corps of Engineers in the 1950s, but was developed and applied in earnest starting in the 1970s when new federal air and water protection laws were being implemented (Mishan, 1971; NAS, 1974; Layard, 1974; Maler and Wyzga, 1976; Singer, 1979). The first systematic application of CBA to national regulations in the United States began in 1981 as a result of Executive Order 12291 by President Ronald Reagan (Reagan, 1981).

Under Reagan's executive order, cost-benefit analysis was part of a Regulatory Impact Analysis (RIA), itself part of a broader effort aimed at making regulations more cost-effective and transparent. The effort was controversial, due partly to missteps by the Reagan administration, which "came under harsh criticism from numerous quarters for permitting the Office of Management and Budget (OMB) to delay and block new regulatory initiatives. Critics pointed out the OMB's regulatory review staff was comprised primarily of economists. There were no toxicologists, epidemiologists, or health scientists at OMB to overview EPA proposals" (Graham, 1991, p. 6).

A subsequent executive order issued by President Bill Clinton in 1993 made the Office of Information and Regulatory Affairs (OIRA) within OMB "the repository of expertise concerning regulatory issues, including methodologies and procedures that affect more than one agency, this Executive order, and the President's regulatory policies" (Clinton, 1993). President George W. Bush substantially increased OIRA's authority and staffing and appointed an activist director, the previously quoted John D. Graham.

President Barack Obama, while reaffirming the principles and review process in an executive order issued in 2011 (Obama, 2011), reduced the agency's staff, and it played a smaller role in regulatory policy. Before a congressional committee in 2013, a former deputy director of OIRA testified, "At one time, OIRA had a specific branch of a dozen or so economists who specialized in benefit-cost analysis, and OIRA hired scientific experts in risk analysis. Today it has a few experts scattered among five branches [which] are, for the most part, staffed with overworked although highly competent desk officers" (Morrall, 2013).

Beginning in 2017, it appears President Donald Trump is revitalizing OIRA. Like Bush, he appointed an activist director, Neomi Rao, and has made cutting regulations one of the major themes of his administration. According to Rao (in comments at a Brookings Institution event in early 2018), the federal government issued only three new "significant" regulations in FY 2017 and withdrew more than 15,000 planned rules, reducing regulatory costs by more than $570 million per year and $8 billion in total (Heckman, 2018). In FY 2018, OIRA expects deregulatory actions from federal agencies to outnumber new regulatory actions by a nearly four-to-one ratio, projected to save another $10 billion in compliance costs (*Ibid.*). But just how big a role OIRA plays in this regulation-cutting effort is uncertain.

OIRA has authority to review agency regulations and the analyses used to justify them, as well as to return the regulations to the agencies for reconsideration if it finds the analyses were insufficient. Since 1997, OIRA has issued annual reports to Congress on the benefits and costs of federal regulations. Its 2013 report found "the estimated annual benefits of major Federal regulations reviewed by OMB from October 1, 2002, to September 30, 2012, for which agencies estimated and monetized both benefits and costs, are in the aggregate between $193 billion and $800 billion, while the estimated annual costs are in the aggregate between $57 billion and $84 billion" (OMB, 2013, p. 3). This sounds rigorous and like evidence of a significantly positive overall benefit-cost ratio, but it is not. Williams and Broughel (2013) write,

> Of 37,786 rules finalized in FY2003–FY2012, only 115 rules had estimates of monetized benefits and costs in OIRA's draft report. This is less than one-third of 1% of all final regulations, an abysmal record. Even worse, there are no rules in the report from independent regulatory agencies that have dollar estimates for both benefits and costs.

If OIRA is getting estimates of monetized benefits and costs for less than one-third of 1% of all final regulations, then CBA clearly is not being used aggressively or successfully at the national level. Further evidence of this failure is that there appears to be no correlation between the amount of information provided in a regulatory impact analysis and the net benefits of a regulation (Shapiro and Morrall, 2012). Hahn and Tetlock (2008) also find little evidence that CBA has improved regulatory outcomes.

Requiring the use of CBA apparently doesn't prevent politics – whether ideology or pandering to special interests – from influencing regulatory choices. A study by the Mercatus Center at George Mason University found "the more liberal agencies (Labor, Health and Human Services) got through OIRA with lower-quality analyses in the Obama administration, while the more conservative agencies (Defense, Homeland Security) got through OIRA with lower-quality analyses in the Bush administration" (Morrall, 2013, p. 5, citing Ellig *et al.*, 2013).

Conducting a cost-benefit analysis of climate change is difficult and perhaps impossible due to the enormity of both costs and benefits, their wide dispersal (virtually every person on Earth benefits from the use of fossil fuels and many would benefit from a modest warming), and the long time frame (most of the benefits and costs of climate change might emerge one or even two centuries in the future, if they emerge at all). Economists use integrated assessment models (IAMs) to attempt to monetize the net cost or benefit of climate change, called the "social cost of carbon." Such models are enormously complex and can be programmed to arrive at widely varying conclusions. They are described and critiqued in detail in Chapter 8.

References

Clinton, W.J. 1993. Executive order 12,866. Regulatory planning and review. September 30.

Dorfman, R. 1993. Chapter 18. An introduction to benefit-cost analysis. In: Dorfman, R. and Dorfman, N.S. (Eds.) *Economics of the Environment: Selected Readings.* Third edition. New York, NY: W.W. Norton & Company.

Ellig, J., McLaughlin, P.A., and Morrall, J. 2013. Continuity, change, and priorities: the quality and use of regulatory analysis across US administrations. *Regulation and Governance* **7** (2): 153–73.

Graham, J.D. 1991. *Harnessing Science for Environmental Regulation.* New York, NY: Praeger Publishers.

Hahn, R.W. and Tetlock, P.C. 2008. Has economic analysis improved regulatory decisions? *Journal of Economic Perspectives* **22** (1): 67–84.

Heckman, J. 2018. Trump's 'regulation czar' touts success of 'two-for-one' executive order. Federal News Radio. January 26.

HM Treasury. 2018. *The Green Book: Central Government Guidance on Appraisal and Evaluation.*

Layard, R. (Ed.) 1974. *Cost-Benefit Analysis.* Baltimore, MD: Penguin Books.

Maler, K.G. and Wyzga, R.E. 1976. *Economic Measurement of Environmental Damage.* Paris, France: Organization for Economic Cooperation and Development.

Mishan, E.J. 1971. *Cost-Benefit Analysis.* London, UK: Unwin University Books.

Morrall, J.F. 2013. Reinvigorating, strengthening, and extending OIRA's powers. Testimony to House Committee on the Judiciary, Subcommittee on Regulatory Reform, Commercial, and Antitrust Law. September 30.

NAS. 1974. National Academy of Sciences. Air Quality and Automobile Emission Control. Coordinating Committee on Air Quality Studies. Washington, DC: U.S. Government Printing Office.

Obama, B. 2011. Executive order 13,563. Improving regulation and regulatory review. January 18.

OMB. 2013. Office of Management and Budget. *Draft Report to Congress on the Benefits and Costs of Federal Regulations and Agency Compliance with the Unfunded Mandates Reform Act.* Washington, DC: Office of Management and Budget.

Pearce, D., Atkinson, G., and Mourato, S. 2006. *Cost Benefit Analysis and the Environment: Recent Developments.* Paris, France: Organization for Economic Cooperation and Development.

Reagan, R. 1981. Executive order 12,291. Federal regulation. February 17. Online by Peters, G. and Woolley, J.T. The American Presidency Project.

Shapiro, S. and Morrall, J. 2012. The triumph of regulatory politics: benefit cost analysis and political salience. *Regulation and Governance* **6** (2): 189–206.

Singer, S.F. 1979. *Cost Benefit Analysis as an Aid to Environmental Decision Making.* McLean, VA: The MITRE Corporation.

van Kooten, G.C. 2013. *Climate Change, Climate Science and Economics: Prospects for an Alternative Energy Future.* New York, NY: Springer.

Williams, R. and Broughel, J. 2013. Government report on benefits and costs of federal regulations fails to capture full impact of rules. Mercatus Center (website). December 2.

Wolka, K. 2000. Chapter 8 Analysis and Modeling, Chapter 9 Economics. In: Lehr, J.H. (Ed.) *Standard Handbook of Environmental, Science, Health, and Technology.* New York, NY: McGraw-Hill, pp. 8.122–33.

Zerbe, R.O. 2018. A Distinction between Benefit-Cost Analysis and Cost Benefit Analysis: Moral Reasoning and a Justification for Benefit Cost Analysis.

1.3 Private Environmental Protection

The belief that government action is needed to protect the global atmosphere from "carbon pollution" is based on the flawed assumption that private agents – the people and organizations that use fossil fuels to generate power – are acting without regard to the damages they create that are borne by others. According to this framing, the atmosphere is a common-pool resource in need of effective management and only a government can end this "tragedy of the commons." But environmental economics makes clear that private environmental protection is more common, and more effective, than relying on government intervention.

Section 1.3.1 defines common-pool resources and explains how they have been successfully protected by tort and nuisance laws and managed by nongovernmental organizations that transform a "tragedy of the commons" into "an opportunity of the commons."

Section 1.3.2 explains why positive and negative externalities are ubiquitous and not justifications for government intervention. It describes Coase's theorem, which says socially efficient solutions to conflicts involving externalities can be found so long as both parties are able to negotiate, in effect trading their externalities.

Section 1.3.3 documents how prosperity makes environmental protection a higher public goal and provides the resources needed to achieve it. Economists call this the Environmental Kuznets Curve. Section 1.3.4 describes a huge advantage private environmental protection efforts have over government efforts: their ability to tap local knowledge of values and opportunities. Section 1.3.5 briefly critiques a school of economics called "ecological economics" seeking to justify government intervention rather than working with markets to protect the environment.

1.3.1 Common-pool Resources

Common-pool resources have been successfully protected by tort and nuisance laws and managed by nongovernmental organizations.

Some natural resources such as air, flowing water, and wildlife are held "in common" by the people of a community, nation, or (in the case of the atmosphere) the whole world. They constitute a type of good or service called common-pool resources that is *non-excludable,* meaning non-payers cannot be readily prevented from using or consuming it, and *rivalrous,* meaning consumption or use by one person comes at the expense of others. Common-pool resources can be viewed as one of four types of goods and services that differ according to these two characteristics, as shown in Figure 1.3.1.1.

Common-pool resources are often difficult to protect because "someone has to cover the costs for everybody else. There are too many free riders. Too often, the common resource doesn't get saved" (Avery, 1995, p. 314). Free use of common resources often leads to more demand than can be met by the supply. The classic case is over-grazing on a commons, a pasture open to all herdsmen for cattle grazing (Hardin, 1968; Hardin and Baden, 1977). Each herdsman captures the immediate benefits of grazing another cow even though over-grazing may cause a reduction in next year's grass. The individual herdsman bears only a fraction of the costs – the reduced grazing available next year due to excessive grazing now – because all users share the future costs. If the herdsman removed his cow, he would bear fully the burden of reducing his use and, if someone else adds a cow, still bear some of the cost of over-grazing next year. Thus, each herdsman has an incentive to add cows, even though the pasture may be gradually deteriorating as a result. This situation is known as the "tragedy of the commons."

A similar situation can occur when a fishing territory is open to all fishermen (Anderson and Snyder, 1997; Adler and Stewart, 2013). Each fisherman captures all the benefits of harvesting more fish now, while paying only a small part of the future costs – the reduction of the fish population for future harvests. Ignoring the indirect costs that will occur in the future is easy if the fisherman will not ultimately pay the full, true cost of his or her actions.

In the United States, Canada, and other nations having legal roots in Great Britain, the courts have for centuries provided a way to stop individuals from injuring others by degrading commonly owned resources (Epstein, 1985; Abraham, 2002; Latham *et al.,* 2011; Cushing, 2017). When a victim demonstrates harm has been done or serious harm is threatened, courts can force compensation or issue an injunction to stop the harmful activity. Such harms are called torts or nuisances. Meiners and Yandle (1998) wrote:

> Legal actions can lead to recovery for damages to land as well as to recovery for damages to health or any other benefit attached to our interests in property. A public nuisance is an act that causes inconvenience or damage to public health or order or that obstructs public rights. If a business creates noxious emissions that affect many citizens a public attorney may bring an action on behalf of all affected citizens to have the activity terminated.

Figure 1.3.1.1
Types of goods and services

	Excludable	Non-excludable
Rivalry	**Pure Private** Food, clothing, furniture. Most items of private possession	**Common-pool Resources** Public domain ponds, rivers, wilderness areas, grazing areas, the atmosphere
Non-rivalry	**Club** Swimming pools, toll roads, country clubs, membership organizations, gated communities, private schools	**Pure Public** National defense, public sanitation, crime control, flood defense, contagious disease control

Source: Adapted from Hakim, 2017.

Trespass created rights similar to those against nuisance. If a harmful substance is allowed, intentionally or carelessly, to invade the property of another, whether by land, air, or water, there may be a trespass. If so, the defendant is held responsible for damages.

Since water is often not owned by property owners whose land abuts a lake or a stream, the common law extends protection to water quality through riparian rights. Riparian rights to water are user rights that allow water users to sue those who damage water quality to the point where its use and enjoyment are reduced.

Tort law can even be used to protect the environment from government actions. In the late nineteenth century, the Carmichael family owned a 45-acre farm in Texas, with a river running through it that bordered on the state of Arkansas. The city of Texarkana, Arkansas built a sewage system that deposited sewage in the river in front of the Carmichaels' home. They sued the city in federal court on the grounds that their family and livestock no longer were able to use the river and possibly were exposed to disease.

The court awarded damages to the Carmichaels and granted an injunction against the city, forcing it to stop the sewage dumping. Even though the city of Texarkana was operating properly under state law in building a sewer system, it could not foul the water used by the Carmichaels. Indeed, the judge noted, "I have failed to find a single well-considered case where the American courts have not granted relief under circumstances such as are alleged in this bill against the city" (*Carmichael* v. *City of Texarkana*, 1899).

Reliance on tort and nuisance law to protect the environment declined in the United States beginning in the 1960s and 1970s with passage of regulations that preempted private remedies. While the purpose of these laws was to more effectively achieve the objectives of protecting public health or the natural environment than could be obtained through private legal action, whether they actually had this effect is questionable. Most trends in air and water quality in the United States showed significant improvement *before* the enactment of such laws and little or no change in trends after their adoption (Brubaker, 1995; Simon, 1995; Goklany, 1999; Hayward, 2011, pp. 7ff). For example, Figure 1.3.1.2 shows trends for particulate matter emissions in the United States from 1940 to 1997 and Goklany documents similar trends for carbon monoxide and lead emissions. McKitrick (2015) points out that federal regulations played a complicated and not always positive role in the decline of sulfur dioxide emissions associated with acid deposition. While the intentions may have been good, changes in technology played a bigger role than regulations in reducing emissions.

Climate change activists are attempting to use tort and nuisance laws to protect the atmosphere from "carbon pollution," so far without success. In a recent case where municipalities in California attempted to sue oil companies for their alleged role in causing global warming, federal district Judge William Alsup found for the defendants and dismissed the case. Relevant parts of his opinion read as follows:

Figure 1.3.1.2
U.S. particulate matter emissions, 1940–1997

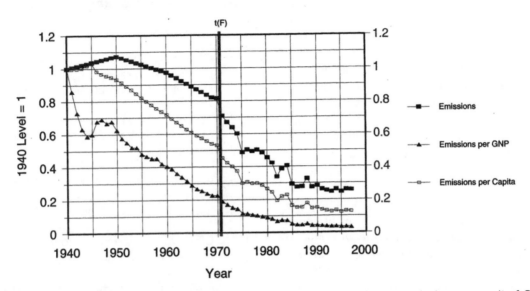

U.S. particulate matter (PM_{10}) emissions relative to 1940 reported as emissions, emissions per unit of GDP, and emissions per capita. The vertical line at 1970 (tF) is time of federalization of environmental regulation. *Source:* Goklany, 1999, Figure 4-6, p. 82.

With respect to balancing the social utility against the gravity of the anticipated harm, it is true that carbon dioxide released from fossil fuels has caused (and will continue to cause) global warming. But against that negative, we must weigh this positive: our industrial revolution and the development of our modern world has literally been fueled by oil and coal. Without those fuels, virtually all of our monumental progress would have been impossible. All of us have benefitted. Having reaped the benefit of that historic progress, would it really be fair to now ignore our own responsibility in the use of fossil fuels and place the blame for global warming on those who supplied what we demanded? Is it really fair, in light of those benefits, to say that the sale of fossil fuels was unreasonable? This order recognizes but does not resolve these questions, for there is a more direct resolution from the Supreme Court and our court of appeals, next considered. ...

In our industrialized and modern society, we needed (and still need) oil and gas to fuel power plants, vehicles, planes, trains, ships, equipment, homes and factories. Our industrial revolution and our modern nation, to repeat, have been fueled by fossil fuels.

This order accordingly disagrees that it could ignore the public benefits derived from defendants' conduct in adjudicating plaintiffs' claims. In the aggregate, the adjustment of conflicting pros and cons ought to be left to Congress or diplomacy (*California* v. *BP et al.,* 2018).

Judge Alsup's opinion reveals the weaknesses in the activists' case and not a shortcoming in the tort law approach to environmental protection. If individuals cannot persuasively demonstrate to the court they are being harmed by pollution, the court will make no attempt to stop that pollution or make those causing it pay damages. Anthropogenic climate change involves billions of people burning fossil fuels and engaging in other activities that may gradually be increasing the concentration of carbon dioxide (CO_2) and other greenhouse gases in the atmosphere. The "polluters" include virtually every person in the world (a human exhales about

2.3 pounds of CO_2 every day), while bona fide "victims" probably have yet to be born. Most alleged victims benefit or will benefit from the prosperity, improved public health, and environmental benefits made possible by fossil fuels. For these reasons (and others having to do with jurisdiction), efforts to compel governments or oil companies to reduce greenhouse gas emissions have failed.

Even in cases where tort law may not prevent over-use of a common resource, solutions are possible without government intervention. Cowen (1988) presented case studies of lighthouses, bees for crop pollination, fire protection, leisure and recreational services, conservation, and public education in which common-pool resources were protected or provided through voluntary agreements and transactions. Often this involves adopting the techniques used by clubs – one of the types of goods and services shown in Figure 1.3.1.1 – to add value to a certain kind of access to a common-pool resource, giving people an incentive to pay for access.

Similarly, Ostrom (1990, 2005, 2010) and her network of researchers documented hundreds of cases where groups avoided the tragedy of the commons without resorting to top-down regulation. Decentralized ensembles of small public and private organizations work together to manage common-pool resources in ways that reflect their knowledge of local opportunities and costs, the knowledge national and international organizations typically lack. They exhibit the sort of spontaneous order that Hayek (1973, 1976, 1979) often wrote about, a coordination that is not dictated or controlled by a central planner. Ostrom identified eight design principles, summarized in Figure 1.3.1.3, shared by entities most successful at managing common-pool resources.

Ostrom's work earned her the Nobel Prize in economics in 2009. Boettke (2009) writes, "Traditional economic theory argues that public goods cannot be provided through the market. Traditional Public Choice theory argues that government often fails to provide solutions. Ostrom shows that decentralized groups can develop various rule systems that enable social cooperation to emerge through voluntary association." According to Boettke, Ostrom showed how nongovernmental organizations can transform disagreements over access to common-pool resources from a "tragedy of the commons" to "an opportunity of the commons."

In these cases and many others, the key concepts of trade, profit-and-loss, and prices allow entrepreneurs to discover what consumers value and then find ways to deliver it despite the hurdles that

Figure 1.3.1.3
Ostrom's eight design principles for effective management of common-pool resources

1. Clear definition of the contents of the common pool resource and effective exclusion of external un-entitled parties;

2. Appropriation and provision of common resources that are adapted to local conditions;

3. Collective-choice arrangements that allow most resource appropriators to participate in the decision-making process;

4. Effective monitoring by monitors who are part of or accountable to the appropriators;

5. A scale of graduated sanctions for resource appropriators who violate community rules;

6. Mechanisms of conflict resolution that are cheap and of easy access;

7. Self-determination of the community recognized by higher-level authorities; and

8. In the case of larger common-pool resources, organization in the form of multiple layers of nested enterprises, with small local common-pool resources at the base level.

Source: Ostrom, 1990.

common ownership places in their paths. This process is described in more detail in the next section.

References

Abraham, K.S. 2002. The relation between civil liability and environmental regulation: an analytical overview. *Washburn Law Journal* **41**: 379, 383–4.

Adler, J.H. and Stewart, N. 2013. Learning how to fish: catch shares and the future of fishery conservation. *UCLA Environmental Law and Policy Review* **31** (1): 150–97.

Anderson, T.L. and Snyder, P.S. 1997. *Water Markets: Priming the Invisible Pump*. Washington, DC: Cato Institute.

Avery, D.T. 1995. *Saving the Planet with Pesticides and Plastic*. Indianapolis, IN: Hudson Institute.

Boettke, P. 2009. Liberty should rejoice: Elinor Ostrom's Nobel Prize. *The Freeman*. November 18.

Brubaker, E. 1995. *Property Rights in the Defense of Nature*. London, UK: Earthscan Publications Ltd.; Toronto, ON: Earthscan Canada.

California v. *BP et al.* 2018. The People of the State of California v. BP P.L.C. et al., No. 3:2017cv06012 - Document 229 (N.D. Cal. 2018). Justia (website). Accessed July 27, 2018.

Carmichael v. *City of Texarkana*. 1899. 94 F. 561. W.D. Ark.

Cowen, T. (Ed.) 1988. *The Theory of Market Failure: A Critical Examination*. Washington, DC: Cato Institute.

Cushing, B. 2017. The role of environmental torts in the future. *Georgetown Environmental Law Review* (website). March 5. Accessed July 29, 2018.

Goklany, I.M. 1999. *Clearing the Air: The Real Story of the War on Air Pollution*. Washington, DC: Cato Institute.

Hakim, S. 2017. Economics and management of privatization. PowerPoint presentation.

Hardin, G. 1968. The tragedy of the commons. *Science* **162**: 1243–8.

Hardin, G. and Baden, J. 1977. *Managing the Commons*. San Francisco, CA: W.H. Freeman and Company.

Hayek, F. 1973, 1976, 1979. *Law, Legislation, and Liberty, Vol. 1: Rules and Order* (1973), *Vol. 2: The Mirage of Social Justice* (1976), *Vol. 3: The Political Order of a Free People* (1979). Chicago, IL: University of Chicago Press.

Hayward, S. 2011. *Almanac of Environmental Trends*. San Francisco, CA: Pacific Institute for Public Policy Research.

Latham, M., Schwartz, V.E., and Appel, C.E. 2011. The intersection of tort and environmental law: where the twains should meet and depart. *Fordham Law Review* **80**: 737.

McKitrick, R. 2015. Clean Power Plan: Acid Rain Part 2? Real Clear Policy (website). August 25. Accessed July 29, 2018.

Meiners, R. and Yandle, B. 1998. The common law: how it protects the environment. *PERC Policy Series* PS-13. Bozeman, MT: PERC.

Ostrom, E. 1990. *Governing the Commons: The Evolution of Institutions for Collective Action*. Cambridge, UK: Cambridge University Press.

Ostrom, E. 2005. *Understanding Institutional Diversity*. Princeton, NJ: Princeton University Press.

Ostrom, E. 2010. Polycentric systems for coping with collective action and global environmental change. *Global Environmental Change* **20**: 550–7.

Simon, J.L. (Ed.) 1995. *The State of Humanity*. Oxford, UK: Blackwell Publishers, Inc.

1.3.2 Cooperation

Voluntary cooperation can generate efficient solutions to conflicts involving negative externalities.

Conflicts regarding the use of scarce resources are often rooted in opposing political viewpoints. Government decisions and regulations tend to favor the side with the most political power or the greatest ability to influence elected officials and regulators. Losers must abide by such outcomes and either pay additional taxes to fund a result they do not support or not receive a service or benefit they are willing to pay for. When politics drive decision-making the process is often a zero-sum game: What one person or interest group gains as a result of the decision, another person or interest group must give up.

Market exchanges, in contrast, produce outcomes that benefit all parties involved (Anderson and McChesney, 2002, and see Section 1.2.5). Even though there is plenty of negotiation and disagreement in the marketplace, the solutions people agree on are ones both parties want – at least compared with available alternatives. A would-be buyer whose offer is rejected does not have to pay. The parties involved are spending their own money so the risk of moral hazard is low.

Cooperation also works when conflicts arise over access to common-pool resources. In the paradigm case, one person's access to a common-pool resource imposes on others a cost, or *negative externality,* that escapes the price mechanism, so the actor is not held accountable for the entire consequences of his action. This can result in a product being overproduced and sold at a price that is less than it should be. But externalities can be traded too, and provided transaction costs are sufficiently low, bargaining can lead to an efficient outcome without government imposing taxes or forcing a reallocation of property. This was an important lesson derived from the work of Ronald Coase, who like Elinor Ostrom mentioned in the previous section, earned a Nobel Prize for his work (Coase, 1960, 1994).

Coase's theorem, as it came to be called, was summarized in different ways by its author and by the many researchers who elaborated on it. The fundamental insight is that every externality necessarily involves two parties, the actor and the person affected by the action, and a solution necessarily involves both parties. In cases involving common-pool resources, the victim likely has several options to avoid damages, such as moving away from the nuisance, choosing to use a different product or service, interfering with the actor's enterprise, suing under tort law, or even threatening to go to elected officials and ask for legislation correcting the injustice. However, all of these options cost time or money, so the victim is probably willing to pay some amount to the actor in return for his using less of the common-pool resource or using it in such a way as to do less damage to others–to reduce the externality. The actor is probably willing to change his behavior if the cost of doing so is less than the amount offered by the victim. The resulting allocation or distribution of the good, according to Coase, will be socially efficient and in terms of resource allocation will be the same regardless of initial assignments of rights/liabilities.

Coase's writing stressed that an efficient outcome of trading in externalities is most likely to occur when "transaction costs" are low, being the costs involved in bringing the parties together and reaching an agreement. Coase knew that in reality those costs are never zero, so the success of negotiating as a solution to negative externalities depends on the design of institutions that can bring the parties together, provide them with the information they need, and make such transactions possible. In cases involving common-pool resources, these are essentially the eight design principles later discovered by Ostrom.

Coase's theorem and Ostrom's design principles tell us cooperation can lead to environmental protection without government intervention. High transaction costs may cause markets to fail to ensure that all of the costs of a person's actions are fully borne by the actor ("internalized"), but the superior solution often is to recognize the property rights of those affected by pollution or other undesirable effects and allow the two parties to negotiate toward a settlement.

Coase was careful to avoid the assumption, common in welfare economics circles, that externalities could be objectively defined or measured outside the very specific circumstances in which they occurred. Since they depend on the actions and judgements of both the actor and the victim, externalities do not exist as objective value-free data. Indeed, who is the "victim" is a matter of perspective, it is not objective. Both parties are exercising rights and deserve compensation if their rights are infringed. As Medema (2011) wrote, "the point to be taken is that there is no such thing as a determinate optimal solution to social cost problems. One can only come to grips with these things on a case-by-case basis, weighing the benefits and costs associated with alternative courses of action and recognizing that both markets/exchange and government activity have associated with them certain costs – often substantial."

Not everyone will get all he or she wants in a negotiation. Those who are not willing to provide any resources will probably be forced outside of negotiations, leaving involved only those who have something to offer. Political decisions do not please everyone either, and people who do not contribute to political candidates or mobilize constituencies to vote are unlikely to have much influence in state and national capitols. The key difference is that in a private setting, those who do not engage in the negotiations or whose offers are rejected do not have to pay for the outcome. In contrast, when a decision is made by a government, taxpayers usually bear the costs, even those who had no say in the decision and who may not benefit from the decision.

Coase's theorem and more generally the success of cooperation in managing common-pool resources around the world has important implications for the climate change issue that will become more apparent in later chapters of this book. For now, consider just these implications:

- Managing Earth's atmosphere like a common-pool resource will require tradeoffs and negotiation among those who benefit from the production and use of fossil fuels (the primary source of anthropogenic greenhouse gases) and those who may suffer from the negative consequences of a warmer world.

- Distinguishing "actors" and "victims" in the climate change phenomenon is difficult or even impossible since everyone emits some greenhouse gases and everyone benefits from the prosperity and technologies made possible by the use of fossil fuels.

- "Social cost" in cases involving common-pool resources is not objectively quantifiable but involves case-specific tradeoffs of rights, costs, and benefits by both actors and victims.

- The efficient solution to climate change is likely to be decentralized, emerging from nongovernment organizations with local knowledge of opportunities and designed to effectively manage common-pool resources, rather than imposed from the top down by national or global government agencies.

References

Anderson, T. and McChesney, F.S. (Eds.) 2002. *Property Rights: Cooperation, Conflict, and Law.* Princeton, NJ: Princeton University Press.

Coase, R.H. 1960. The problem of social cost. *Journal of Law and Economics* **3**: 1–44.

Coase, R.H. 1994. *Essays on Economics and Economists.* Chicago, IL: University of Chicago Press.

Medema, S. 2011. Q&A with Steven Medema on the Coase theorem and environmental economics. Bozeman, MT: PERC. December 1.

1.3.3. Prosperity

Prosperity leads to environmental protection becoming a higher social value and provides the resources needed to make it possible.

Even poor communities are willing to make sacrifices for some basic components of environmental protection, such as access to safe and clean drinking water and sanitary handling of human and animal wastes. As income rises, citizens raise their goals from mere survival to self-realization and spiritual goals (Maslow, 1943; Abulof, 2017). Once basic demands for food, clothing, and shelter are met, people demand cleaner air, cleaner streams, more outdoor recreation, and the protection of wild lands. With higher incomes, citizens place higher priorities on environmental objectives (Ausubel, 1996; Goklany, 2007).

Nordhaus and Shellenberger (2007), quoted earlier in Section 1.1, acknowledged, "As Americans became increasingly wealthy, secure, and optimistic, they started to care more about problems such as air and water pollution and the protection of the wilderness and open space. This powerful correlation between increasing affluence and the emergence of quality-of-life and fulfillment values has been documented in developed and undeveloped countries around the world" (p. 6). They continued, "Environmentalists have long misunderstood, downplayed, or ignored the conditions for their own existence. They have tended to view economic growth as the cause but not the solution to ecological crisis" (*Ibid.*).

Coursey (1992) found the willingness of citizens to spend and sacrifice for a better environment rises more than twice as fast as per-capita income. Conversely, willingness and ability to pay for a better environment falls with falling income. Economists have documented what are called Environmental Kuznets Curves (EKCs) showing how various measures of environmental degradation rise with national per-capita income until a certain tipping point and then begin to fall, often pictured as an inverted U shape (Panayotou, 1993). Figure 1.3.3.1 shows a stylized rendition of the curve.

Figure 1.3.3.1
A typical Environmental Kuznets Curve

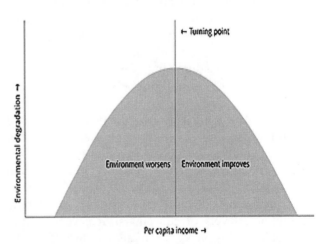

Source: Ho and Wang, 2015, p. 42.

Grossman and Krueger (1995) conducted an extensive literature review of air quality over time and around the world and found ambient air quality tended to deteriorate until average per-capita income reached about $6,000 to $8,000 per year (in 1985 dollars) and then began to sharply improve. Later research confirmed similar relationships for a wide range of countries and air quality, water quality, and other measures of environmental protection (Goklany, 2007, 2012; Criado, *et al.*, 2011; Bertinelli *et al.*, 2012). Yandle *et al.* (2002) surveyed more recent research on EKCs and reported, "Prior to the advent of EKCs, many well-informed people believed that richer economies damaged and even destroyed their natural resource endowments at a faster pace than poorer ones. They thought that environmental quality could only be achieved by escaping the clutches of industrialization and the desire for higher incomes. The EKC's paradoxical relationship cast doubt on this assumption." They found while "there is no single EKC relationship that fits all pollutants for all places and times," the typical inverted U shape is the best way to approximate the link between income and local air pollutants such as oxides of nitrogen, sulfur dioxide, and particulate matter. "The EKC evidence for water pollution is mixed, but there may be an inverted U-shaped curve for biological oxygen demand (BOD), chemical oxygen demand (COD), nitrates, and some heavy metals (arsenic and cadmium). In most cases, the income threshold for improving water quality is much lower than the air pollution improvement threshold."

More recently, Koirala *et al.* (2011) conducted a meta-analysis of 103 empirical EKC studies published between 1992 and 2009 and found EKC-type relationships "for landscape degradation, water pollution, agricultural wastes, municipal-related wastes and several air pollution measures," though not for carbon dioxide emissions. Even sulfur emissions in fast-growing China appear to be declining as the nation prospers (Zilmont *et al.*, 2013).

The relationship between economic growth and environmental impact is more complicated than what is presented here (for more academic background see Grimaud and Rougé, 2005; Grimaud and Tournemaine, 2007; and Schou, 2000, 2002). Factors other than wealth, such as the strength of democratic institutions, levels of educational achievement, and income equality have been shown to affect the environmental impact of prosperity. But these variables are themselves affected by prosperity.

Friedman (2005) documented periods of higher economic growth have led to more tolerance, optimism, and egalitarian perspectives.

The productivity and wealth of nations depend more on their institutions – the law, incentives, and rules in place – than on their natural resources. Countries where private property rights are defined, protected, and can be traded experience significantly greater per-capita wealth, economic growth rates, and rising standards of public health (Gwartney *et al.*, 2014; Miller and Kim, 2015). As might be expected, those countries also experience higher levels of environmental quality. As Hartwell and Coursey wrote, "we find that the correlation between economic freedom and better environmental and public health outcomes remains strong. We conclude that the way forward for environmental policymaking should concentrate on improving property rights and limiting the power of the state, rather than expanding it" (Hartwell and Coursey, 2015, p. 37).

The prosperity made possible by markets creates the resources and change in public values needed to protect the environment. Without markets, a poorer and hungrier world would have little regard for the environment or the interests of future generations, being too busy meeting the more immediate needs of finding food and shelter.

References

Abulof, U. (Ed.) 2017. Symposium: Rethinking Maslow: human needs in the 21st century. *Society* **54**: 508.

Ausubel, J.H. 1996. Liberation of the environment. *Daedalus* **125** (3): 1–17.

Bertinelli, L., Strobl, E., and Zou, B. 2012. Sustainable economic development and the environment: theory and evidence. *Energy Economics* **34** (4): 1105–14.

Coursey, D. 1992. *The Demand for Environmental Quality*. St. Louis, MO: John M. Olin School of Business, Washington University.

Criado, C.O., Valente, S., and Stengos, T. 2011. Growth and pollution convergence: theory and evidence. *Journal of Environmental Economics and Management* **62** (2): 199–214.

Friedman, B. 2005. *The Moral Consequences of Economic Growth*. New York, NY: Vintage Books.

Goklany, I.M. 2007. *The Improving State of the World: Why We're Living Longer, Healthier, More Comfortable*

Lives on a Cleaner Planet. Washington, DC: Cato Institute.

Goklany, I.M. 2012. Humanity unbound: How fossil fuels saved humanity from nature and nature from humanity. *Cato Policy Analysis* #715. Washington, DC: Cato Institute.

Grimaud, A. and Rougé, L. 2005. Polluting non-renewable resources, innovation and growth: welfare and environmental policy. *Resource and Energy Economics* **27**: 109–29.

Grimaud, A. and Tournemaine, F. 2007. Why can an environmental policy tax promote growth through the channel of education? *Ecological Economics* **62**: 27–36.

Grossman, G. and Krueger, A. 1995. Economic growth and the environment. *Quarterly Journal of Economics* **110** (2): 353–77.

Gwartney, J., Lawson, R., and Samida, D. 2014. *Economic Freedom of the World: 2014 Annual Report*. Vancouver, BC: Fraser Institute.

Hartwell, C.A. and Coursey, D.L. 2015. Revisiting the environmental rewards of economic freedom. *Economics and Business Letters* **4** (1): 36–50.

Ho, M. and Wang, Z. 2015. Green growth for China? *Resources*. Washington, DC: Resources for the Future.

Koirala, B.S., Hui, L., and Berrens, R.P. 2011. Further investigation of environmental Kuznets Curve studies using meta-analysis. *International Journal of Ecological Economics and Statistics* **22** (S11).

Maslow, A.H. 1943. A theory of human motivation. *Psychological Review* **50**: 370–96.

Miller, T. and Kim, A.B. 2015. *2015 Index of Economic Freedom*. Washington, DC: The Heritage Foundation.

Nordhaus, T. and Shellenberger, M. 2007. *Break Through: From the Death of Environmentalism to the Politics of Possibility*. New York, NY: Houghton Mifflin Company.

Panayotou, T. 1993. Empirical tests and policy analysis of environmental degradation at different stages of economic development. *Working Paper* WP238. Geneva, Switzerland: International Labour Office, Technology and Employment Programme.

Schou, P. 2000. Polluting non-renewable resources and growth. *Environmental and Resource Economics* **16**: 211–27.

Schou, P. 2002. When environmental policy is super-fluous: growth and polluting resources. *Scandinavian Journal of Economics* **104** (4): 605–20.

Yandle, B., Vijayaraghavan, M., and Bhattarai, M. 2002. The environmental Kuznets curve: A primer. *PERC Research Study* 02-1. Bozeman, MT: PERC.

Zilmont, K., Smith, S.J., and Cofala, J. 2013. The last decade of global anthropogenic sulfur dioxide 2000–2011. *Environmental Research Letters* **8**:1.

1.3.4 Local Knowledge

The information needed to anticipate changes and decide how best to respond is local knowledge and the most efficient responses will be local solutions.

The fact that climate change is a global phenomenon often leads to the assumptions that it is best studied by a global entity, perhaps an Intergovernmental Panel on Climate Change (IPCC), and that it requires a "solution" chosen and implemented by a global government, perhaps the United Nations. Both assumptions are wrong. Institutions may collect massive amounts of data, but this *information* must be processed, interpreted, and added to *knowledge* of local circumstances of time and place in order to lead to the discovery of efficient responses to changes in the world (Hayek, 1945; Kirzner, 1978, 2005; Sowell, 1980; Boettke, 2002; Hess and Ostrom, 2007). Actors operating in the private sector have incentives to gather just enough information – not too much and not too little – because both the costs and the benefits of seeking more information fall upon the actor. Weighing the costs and benefits of more information, the actor won't end up with perfect or complete information but will make a reasonable decision based on the costs and benefits of seeking more knowledge.

Government regulators have very different incentives regarding information and learning. They typically do not bear the cost of information collection and learning, and so will be inclined to demand or require more than is necessary before allowing regulated individuals to act. If damage occurs the regulator could be blamed, so his or her incentive will be to require as much information as possible before allowing a project to go forward. The regulator may ask for study after study to make sure the proposed plan of action will really be safe. The act of accumulating more information becomes an excuse or justification to compel others *not* to act.

Sometimes, government agencies succumb to pressure to find what they believe their superiors want them to find. A good example is the U.S. program that measures surface ambient air temperatures – critical data for understanding climate change. Anthony Watts, a meteorologist, recruited a team of "citizen scientists" to photograph some of the climate-monitoring stations in the U.S. Historical Climatology Network (USHCN) overseen by the National Weather Service, a department of the National Oceanic and Atmospheric Administration (NOAA), to see if those stations complied with NOAA's own quality standards (Watts, 2009). The team eventually surveyed 82.5% of the stations. "We were shocked by what we found," Watts wrote, continuing:

> We found stations located next to the exhaust fans of air conditioning units, surrounded by asphalt parking lots and roads, on blistering-hot rooftops, and near sidewalks and buildings that absorb and radiate heat. We found 68 stations located at wastewater treatment plants, where the process of waste digestion causes temperatures to be higher than in surrounding areas. In fact, we found that 89 percent of the stations – nearly 9 of every 10 – fail to meet the National Weather Service's own siting requirements that stations must be 30 meters (about 100 feet) or more away from an artificial heating or radiating/reflecting heat source (p. 1).

Watts goes on, "In other words, 9 of every 10 stations are likely reporting higher or rising temperatures because they are badly sited. It gets worse. We observed that changes in the technology of temperature stations over time also has caused them to report a false warming trend. We found major gaps in the data record that were filled in with data from nearby sites, a practice that propagates and compounds errors. We found that adjustments to the data by both NOAA and another government agency, NASA, cause recent temperatures to look even higher" (*Ibid.*). The U.S. surface temperature record has long been viewed as the most accurate and complete of the national records relied on by scientists to estimate global temperature trends, so its shortcomings are likely to be shared and even greater in other countries.

A report by the U.S. Government Accountability Office (GAO, 2011) subsequently confirmed Watts' findings and urged NOAA to improve the quality of its surface station network. NOAA agreed with GAO's findings and identified a subset of the USHCN consisting only of supposedly high-quality climate-monitoring stations complying with its siting standards. In 2011, Watts and several colleagues examined "the differences between USHCN temperatures and North American Regional Reanalysis (NARR) temperatures" – that is, the temperature record produced by the subset of higher-quality stations – and found "the most poorly sited stations are warmer compared to NARR than are other stations, and a major portion of this bias is associated with the siting classification rather than the geographical distribution of stations. According to the best-sited stations, the diurnal temperature range in the lower 48 states has no century-scale trend" (Fall *et al.*, 2011).

Similarly and more recently, a doctorate degree was awarded in December 2017 by James Cook University, in Townsville, Australia, to a student whose thesis found scores of flaws in the HadCRUT4 dataset, widely used as the authoritative reconstruction of global temperatures dating back to 1850. That student, now Dr. John McLean, published an updated version of his research in 2018 in which he reported "considerable uncertainty exists about the accuracy of the HadCRUT4" (McLean, 2018, p. 2). "It seems very strange," he wrote, "that man-made warming has been a major international issue for more than 30 years and yet the fundamental data has never been closely examined" (*Ibid*, p. 1).

Whether Watts and McLean are correct or not is obviously important, but not germane to the current point. How could such important data be so unreliable? Why did it take a team of "citizen scientists" in the United States and a graduate student in Australia to expose major flaws in data collection programs created by governments in the United States and United Kingdom and relied on by researchers around the world?

Government officials who oversee government agencies have uses in mind for the data they collect, and those plans affect how data are collected and interpreted. Scott (1998) discovered this while studying failed efforts by governments around the world to force nomadic tribes to settle down, including The Great Leap Forward in China, collectivization in Russia, and compulsory villagization in Tanzania, Mozambique, and Ethiopia. "The more I examined these efforts at sedentarization, the more I came to see them as a state's attempt to make a society legible, to arrange the population in ways that simplified the classic state functions of taxation, conscription, and prevention of rebellion," Scott wrote (p. 2). "In each case, officials took exceptionally complex, illegible, and local social practices, such as land tenure customs or naming customs, and created a standard grid whereby it could be centrally recorded and monitored" (*Ibid.*). Scott refers to this as "seeing like a state."

Climate is certainly "exceptionally complex, illegible." Early on, the IPCC admitted it is "a coupled non-linear chaotic system, and therefore ... long-term prediction of future climate states is not possible" (IPCC, 2001, p. 774). The human impact on climate has been called "one of the most challenging open problems in modern science. Some knowledgeable scientists believe that the climate problem can never be solved" (Essex and McKitrick, 2007). Yet the IPCC now claims future climate conditions *centuries from now* can be predicted with sufficient certainty to make claims about how much greenhouse gases must be reduced and how soon. A mandate to "create a standard grid" out of chaotic observational data could explain the IPCC's lack of interest in natural causes of climate change and reliance on unvalidated computer models. Regarding the latter, Wernick (2014) observed,

> Climate offers a clear case of modeling exercises used to advance political agendas by choosing which data to focus on and how to tweak the (literally) hundreds of parameters in any given model. Whether by design or default, the model tends to vindicate the modeler; for instance, the modeler that selects which natural mechanisms to include and which to neglect when modeling the annual global flux of carbon. Models, and policies to be based on them, ignore the consequences of climate change mitigation strategies, such as costly regressive electricity rates that force even middle-class people to scavenge the forest for fuel, or the benefits of global carbon fertilization. What becomes obscured is the fact that a self-consistent description useful for numerical modeling may not faithfully represent reality, whether physical or social.

The world's political leaders may be motivated by a sincere belief in predictions of catastrophic climate change in centuries to come, but it could also be that collecting extensive data about global energy production and consumption makes the world's energy system "legible" and therefore easier to regulate and tax. The founder of the IPCC and leaders of the UN have not been shy about saying this is their long-term objective (Strong, 1992; UN, 2015; Figueres, 2017).

The IPCC's massive assessment reports and the seemingly endless summits of the parties to the UN's Framework Convention on Climate Change (UNFCCC) reveal the tendency of government bureaucracies to amass information without acquiring the knowledge needed to support action. Proposals from the United Nations feature transfers of billions and even trillions of dollars among international agencies and national governments with seemingly little regard to the scant benefits of such investments. They imply that the goal all along may have been redistribution of income and not preventing or slowing climate change.

The global nature of climate change and the fact that the planet's atmosphere is a global commons obscure the reality that the consequences of climate change are always experienced locally. Consequently, the information needed to anticipate changes and decide how best to respond is *local knowledge* and the most efficient responses will be *local solutions*. It often is forgotten that global estimates of temperature, sea-level rise, and other measures of consequences are model-derived abstractions largely irrelevant to what occurs at specific locations around the world (Essex *et al.,* 2007). For example, changes in sea level at any given site around the world are determined by local and regional changes in shorelines unrelated to estimates of global sea-level rise (Parker and Ollier, 2017). As de Lange and Carter (2014) observe,

> Most coastal hazard is intrinsically local in nature. Other than periodic tsunami and exceptional storms, it is the regular and repetitive local processes of wind, waves, tides and sediment supply that fashion the location and shape of the shorelines of the world. Local relative sea-level is an important determinant too, but in some localities that is rising and in others falling. Accordingly, there is no "one size fits all" sea-level curve or policy that can be applied (p. 33).

What is true of sea-level rise is true of climate impacts more generally. Climate science does not allow us to determine what the local effects of anthropogenic climate change will be. McKitrick (2001) notes,

Anthropogenic additions to the atmosphere will (if they do anything) produce changes in the weather. But weather is a chaotic process, and we have limited expectation of being able to distinguish natural and anthropogenic changes at the local level, even *ex post*. Any damage function we define for the purposes of determining optimal mitigation policy must take for granted a future ability to accurately identify location-specific climate changes and attribute them to anthropogenic causes. If we do not have this ability, climate policy cannot be based on cost-benefit analysis (p. 1).

Since the effect of reducing greenhouse gas emissions, according to the IPCC, will be only to delay the onset of global warming by a few months or years at best, global emission reduction programs are not an effective response to the real on-the-ground consequences of climate change even if one accepts the IPCC's scientific findings. The fact that the impacts of climate change are local explains why even managing the global commons that is the planet's atmosphere is best done by individuals and organizations throughout the world who are experiencing those impacts and not by international organizations based in New York, Paris, or The Hague.

Efforts by the UN and IPCC may actually be preventing other more promising initiatives from advancing, a problem called "displacement" discussed in Section 1.4.6. The record of managing other common-pool resources compiled by Ostrom (1990, 2010) shows top-down and government-led approaches frequently fail while decentralized and often market-based approaches succeed. This insight – that a single top-down solution may be inferior to multiple bottom-up solutions discovered by people with local knowledge and incentives to find the most efficient solutions – is shared by some environmentalists. A group of mostly progressive scholars from Asia, Europe, and North America wrote in 2009:

It is a characteristic of open systems of high complexity and with many ill-understood feed-back effects, such as the global climate classically is, that there are no self-declaring indicators which tell the policy maker when enough knowledge has been accumulated to make it sensible to move into action. Nor, it might be argued, can a policy-maker ever

possess the type of knowledge – distributed, fragmented, private; and certainly not in sufficient coherence or quantity – to make accurate 'top down' directions. Hence, the frequency of failure and unintended consequences (Prins *et al.*, 2009).

The economics of information and knowledge predict neither the IPCC nor the UN will discover the truth about the causes and consequences of climate change or endorse the most efficient response to the phenomenon. Real knowledge and socially optimal responses are most likely to come from the "bottom up," from smaller units of government and private-sector initiatives modeled after those that are successfully managing other common-pool resources.

References

Boettke, P. 2002. Information and knowledge: Austrian economics in search of its uniqueness. *The Review of Austrian Economics* **15** (4): 263–74.

de Lange, W.P. and Carter, R.M. *Sea-level Change: Living with Uncertainty.* London, UK: The Global Warming Policy Foundation.

Essex, C. and McKitrick, R. 2007. *Taken by Storm: The Troubled Science, Policy and Politics of Global Warming.* Second edition. Toronto, ON: Key Porter Books.

Essex, C., McKitrick, R., and Andresen, B. 2007. Does a global temperature exist? *Journal of Non-Equilibrium Thermodynamics* **32** (1): 1–27.

Fall, S., Watts, A., Nielsen-Gammon, J., Jones, E., Niyogi, D., Christy, J., and Pielke Sr., R.A. 2011. Analysis of the impacts of station exposure on the U.S. Historical Climatology Network temperatures and temperature trends. *Journal of Geophysical Research* **116**: D14120. doi:10.1029/2010JD015146.

Figueres, C. 2017. U.N. official admits global warming agenda is really about destroying capitalism. ZeroHedge (website). February 3. Accessed August 2, 2018.

GAO. 2011. Government Accountability Office. *Climate Monitoring: NOAA can improve management of the U.S. Historical Climatology Network.* Washington, DC: Government Accountability Office. August.

Hayek, F. 1945. The use of knowledge in society. *American Economic Review* **35** (4): 519–30.

Hess, C. and Ostrom, E. (Eds.) 2007. *Understanding Knowledge as a Commons: From Theory to Practice.* Cambridge, MA: The MIT Press.

IPCC. 2001. Intergovernmental Panel on Climate Change. *Climate Change 2001: The Scientific Basis.* Contribution of Working Group I to the Third Assessment Report of the Intergovernmental Panel on Climate Change. New York, NY: Cambridge University Press.

Kirzner, I.M. 1978. *Competition and Entrepreneurship.* Chicago, IL: University of Chicago Press.

Kirzner, I.M. 2005. Information-knowledge and action-knowledge. *Economic Journal Watch* **2** (1): 75–81.

McKitrick, R. 2001. Mitigation versus compensation in global warming policy. *Economics Bulletin* **17** (2): 1–6.

McLean, J. 2018. *An Audit of the Creation and Content of the HadCRUT4 Temperature Dataset.* Robert Boyle Publishing.

Ostrom, E. 1990. *Governing the Commons: The Evolution of Institutions for Collective Action.* Cambridge, UK: Cambridge University Press.

Ostrom, E. 2010. Polycentric systems for coping with collective action and global environmental change. *Global Environmental Change* **20**: 550–7.

Parker, A. and Ollier, C.D. 2017. Short-term tide gauge records from one location are inadequate to infer global sea-level acceleration. *Earth Systems and Environment* (December): 1–17.

Prins, G., *et al.* 2009. *How to Get Climate Policy Back on Track.* Oxford, UK: Oxford University Institute for Science Innovation & Society, LSE Mackinder Programme. July 6.

Scott, J.C. 1998. *Seeing Like a State: How Certain Schemes to Improve the Human Condition Have Failed.* New Haven, CT: Yale University Press.

Sowell, T. 1980. *Knowledge and Decisions.* New York, NY: Basic Books.

Strong, M. 1992. Introduction. In: MacNeill, J., Winsemius, P., and Yakushiji, T. (Eds.) *Beyond Interdependence: The Meshing of the World's Economy and the Earth's Ecology.* New York, NY: Oxford University Press, pp. ix–xi.

UN. 2015. United Nations. *Transforming Our World: The 2030 Agenda for Sustainable Development.* New York, NY: United Nations Department of Economic and Social Affairs.

Watts, A. 2009. *Is the U.S. Surface Temperature Record Reliable?* Chicago, IL: The Heartland Institute.

Wernick, I.K. 2014. Living in a material world. *Issues in Science and Technology* **30** (2): 29–31.

1.3.5 Ecological Economics

"Ecological economics" is not a reliable substitute for rigorous mainstream environmental economics.

Environmentalists have attempted to counter the attention given to and impact of the free-market environmentalism movement by creating their own competing school of economics, which they call "ecological economics." Notable authors contributing to this effort include Robert Costanza (1996, 1998, 2004), Herman E. Daly (2000, 2003), Juan Martinez-Alier (1994, 2002), D.J. McCauley (2006), and E.F. Schumacher (1973). This effort should not be confused with efforts by other scholars such as Rothschild (1990) and Hawken, Lovins, and Hunter Lovins (2000) who are critical of how mainstream economists treat environmental topics but not dismissive of the ability of markets and private actors to protect the environment.

While ecological economics has some merits, its origin as an attempt to defend an ideology, rather than to genuinely understand human social action, leads its practitioners to make fundamental errors. One error is to attempt to replace market prices with other means of measuring costs and benefits. The result is reliance on subjective estimates of values often based on survey results, unscientific predictions by experts, or simply popular beliefs. Prices are the essential data of economics precisely because they are an objective account of what people are willing to pay for a good or service.

A second error is uncritically accepting without question the pseudo-science of the environmental movement. For example, a textbook on ecological economics (Common and Stagl, 2005) makes some factually correct statements about climate change, starting with "as a result of the increasing use of fossil fuels in the last two hundred years, the amount of carbon dioxide in the atmosphere has increased. The expert consensus is that this has warmed the planet, and will warm it further. The amount of warming to be expected, by say 2100, is not known with any precision" (pp. 2–3). This is accurate, but the authors go on to write: "But, the expert consensus is that it will be enough to have serious impact on human economic activity and the satisfaction of needs and desires. Beyond 2100, the impacts may be catastrophic" (p. 3). There is no such consensus among either scientists or economists on this matter, as later chapters in this volume will attest. This is simply a statement of environmentalist dogma that prejudices any effort to study the issue objectively. The textbook makes the same mistake on other issues including resource depletion, loss of species, and air and water pollution.

A third error of ecological economics is its slavish devotion to the doctrine of "sustainability." According to Common and Stagl, "The scholars who set up the International Society for Ecological Economics (ISEE) in 1989 were largely motivated by the judgement that the way the world economy was operating was unsustainable" (p. 8). "Sustainability and sustainable development are central concerns of ecological economics," they write, "which has been defined as the science of sustainability, but not of neoclassical economics" (p. 11). And indeed, ISEE's website states as its goal the facilitation of "understanding between economists and ecologists and the integration of their thinking into a trans-discipline aimed at developing a sustainable world" (ISEE, 2015).

Making "a sustainable world" the goal of something purporting to be an academic discipline is problematic at best (Goklany, 2001; Morris, 2002). First, there is no objective definition of sustainability; in particular there is no agreement on what must be sustained and what should be allowed to change and for how long. For example, a recent editorial in *Nature* opined, "'Sustainable development' is a catchphrase that neatly defines what the world must ultimately achieve, but nobody knows precisely what it looks like at full scale" (*Nature*, 2015, p. 407).

Second, sustainability is a political movement generally traced to a political document, *Our Common Future*, produced by an agency of the United Nations in 1987 and often referred to as the Brundtland Report (WCED, 1987). That document and many others in the sustainability literature simply assume that only governments are capable of protecting the environment, rather than treat that postulate as a contestable hypothesis. As the brief history of environmental economics shows, there is extensive research and commentary on how markets often do a better job than governments at protecting natural resources.

Third, sustainability literature relies heavily on forecasts of future population, consumption patterns,

resource availability, emissions, the effects of those emissions, human adaptation to those effects, and more. Those forecasts are apparently made in ignorance of the scientific forecasting literature (Armstrong, 2001; Armstrong and Green, 2018) and of the evidence provided by Simon (1996) showing human ingenuity and free markets solve shortages and deliver more and cheaper resources over the long term. As Thomas Babington Macaulay wrote nearly two centuries ago upon reading similar alarmist prophesies of his day, "By what principle is it that, when we see nothing but improvement behind us, we are to expect nothing but deterioration before us?" (Macaulay, 1830).

These errors make it clear that "ecological economics" cannot be counted on to provide reliable insights into the climate change issue. Mainstream environmental economics has a longer history and superior methodology, is independent of the environmental movement's spin on matters of science and public health, and is not subordinated to a political agenda. As the rest of this chapter demonstrates, environmental economics is a very useful tool in understanding how best to address climate change.

References

Armstrong, J.S. (Ed.) 2001. *Principles of Forecasting: A Handbook for Researchers and Practitioners*. New York, NY: Springer Science+Business Media, LLC.

Armstrong, J.S. and Green, K.C. 2018. Forecasting methods and principles: evidence-based checklists. *Journal of Global Scholars of Marketing Science* **28** (2): 103–59.

Common, M. and Stagl, S. 2005. *Ecological Economics: An Introduction*. Cambridge, UK: Cambridge University Press.

Costanza, R., Segura, O., and Martinez-Alier, J. 1996. *Getting Down to Earth: Practical Applications of Ecological Economics*. Washington, DC: Island Press.

Costanza, R., *et al.* 1998. The value of the world's ecosystem services and natural capital. *Ecological Economics* **25** (1): 3–15.

Costanza, R., Stern, D.I., He, L., and Ma, C. 2004. Influential publications in ecological economics: a citation analysis. *Ecological Economics* **50** (3–4): 261–92.

Daly, H. 2000. *Ecological Economics and the Ecology of Economics: Essays in Criticism.* Cheltenham, UK: Edward Elgar Publishing.

Daly, H. and Farley, J. 2003. *Ecological Economics, Second Edition: Principles and Applications.* Washington, DC: Island Press.

Goklany, I.M. 2001. *The Precautionary Principle: A Critical Appraisal of Environmental Risk Assessment.* Washington, DC: Cato Institute.

Hawken, P., Lovins, A., and Hunter Lovins, L. 2000. *Natural Capitalism: Creating the Next Industrial Revolution.* Washington, DC: US Green Building Council.

ISEE 2015. Cross-discipline approach. International Society for Ecological Economics (website). Accessed July 30, 2018.

Macaulay, T.B. 1830. Southey's Colloquies. In: *Critical and Historical Essays, Volume 1.* Indianapolis, IN: Liberty Fund.

Martinez-Alier, J. 1994. Ecological economics and ecosocialism. In: O'Connor, M. (Ed.) *Is Capitalism Sustainable?* New York, NY: Guilford Press, pp. 23–36.

Martinez-Alier, J. 2002. *The Environmentalism of the Poor: A Study of Ecological Conflicts and Valuation.* Cheltenham, UK: Edward Elgar Publishing.

McCauley, D.J. 2006. Selling out on nature. *Nature* **443** (7): 27–8.

Morris, J. (Ed.) 2002. *Sustainable Development: Promoting Progress or Perpetuating Poverty?* London, UK: Profile Books Ltd.

Nature. 2015. Decoupled ideals. Editorial. **520** (April 23): 407–8.

Rothschild, M. 1990. *Bionomics: The Inevitability of Capitalism.* New York, NY: Henry Holt and Company.

Schumacher, E.F. 1973. *Small Is Beautiful: A Study of Economics as if People Mattered.* London, UK: Blond and Briggs.

Simon, J.L. 1996. *The Ultimate Resource 2.* Princeton, NJ: Princeton University Press.

WCED. 1987. *Our Common Future.* New York, NY: United Nations, World Commission on Environment and Development.

1.4 Government Environmental Protection

Governments may try to protect the environment by helping private parties define and enforce property rights, regulating environmental risks, and owning and managing resources. The first avenue involves the protection of rights and prevention of injury or harm by one person or group against another. In addition to police powers, governments protect rights by recording and maintaining claims, such as records of ownership and sales of land and water rights. These activities help markets function better by upholding the integrity of property rights. This is the subject of Section 1.4.1.

Government efforts to protect the environment by regulation or owning and managing resources have at best a mixed record. As reported previously in Section 1.3, voluntary cooperation, technological changes, and prosperity explain more of the improvement in air and water quality during the twentieth century than regulation. Sometimes government actions are more harmful than beneficial, as demonstrated by the environmental records of the U.S. Forest Service and the former Soviet Union. Why this is the case is the subject of Sections 1.4.2 through 1.4.7.

Understanding why government environmental protection efforts tend to fail is important because most climate change action agendas place nearly full responsibility and discretionary authority in the hands of government officials, as if implementation of mitigation strategies or adaptation programs were less important or perhaps easier than determining "equilibrium climate sensitivity" or the "residency time of carbon dioxide in the atmosphere." As incredible as it might sound, climate science may be the easy part of the climate change issue. What to do about it and how to go about doing it are the more difficult parts.

1.4.1 Property Rights

Governments can protect the environment by helping to define and enforce property rights.

The ability to own and divest property has enormous, but often unrecognizable, effects. A recent cartoon in *The Wall Street Journal* illustrates: A husband and wife are walking out of a home. The man says to the woman, "Their house looks so nice. They must be getting ready to sell it." Pride of ownership and hope that others will place a high value on the things we own are tremendous motivators of conduct that, by an invisible hand, turns out to benefit those around us. There are plenty of other examples: People take better care of their own cars and homes than cars and apartments they rent. People will remove litter from their front yards and carefully mow, weed, and fertilize it, yet will walk past litter and weeds in a nearby park. Property rights explain why they do that.

Property rights hold people accountable for the long-term value of assets they own. Aristotle recognized this point more than 2,000 years ago when he wrote, "What is common to many is taken least care of, for all men have greater regard for what is their own than for what they possess in common with others" (Aristotle 1939, p. 536).

Property rights are traded with mutual consent in markets. Markets are everywhere, from stock exchanges where billions of dollars' worth of ownership interest in capital is traded daily, to farmers' markets that appear along country roads and in urban plazas. Today, more than half the world's population has Internet access, allowing more than 4 billion people to buy and sell goods and services from the comfort of their home with little more than the click of a button on a keyboard (Internet World Stats, 2018). Trade has never been easier, more frequent, or more valuable than it is today.

Governments play a critical role in making these trades possible by protecting individuals' rights to hold and use their properties. A defining characteristic of government is its claim to a "monopoly of the legitimate use of physical force" (Weber, 1918). It can use force (and more often the threat of force) to protect the owners of property rights from trespass, theft, and fraud. Mises (1966 [1998]) colorfully described this realm:

> Beyond the sphere of private property and the market lies the sphere of compulsion and coercion; here are the dams which organized society has built for the protection of private property and the market against violence, malice, and fraud. This is the realm of constraint as distinguished from the realm of freedom. Here are rules discriminating between what is legal and what is illegal, what is permitted and what is prohibited. And here is a grim machine of arms, prisons, and gallows and the men operating it, ready to crush those who dare to disobey (p. 720).

When people are confident in their ownership and the protection of that ownership by their governments, they are more willing to enter a market to produce, sell, or buy goods (Blumenfeld, 1974; Baumol, 2002). But governments often are the biggest violators of enforceable property rights, since such rights restrict their sovereign authority to tax and regulate without limit (Bethell, 1998; Panné *et al.*, 1999). Historian Richard Pipes (1999) found governments rarely create property rights. Although the history of property rights varies from place to place, property rights are usually established informally when land or other natural resources become valuable enough for individuals to utilize them. Later, these informal rights are confirmed or codified as laws by a government entity.

The discovery of gold in California in 1848 illustrates this process. The sudden increase in the value of land briefly led to conflicts among California miners. But soon the miners began to make agreements about how the land and the veins of gold would be divided. Claimants worked mines together, having made contracts spelling out how finds would be allocated. They did this even though there was no effective government in those areas at the time. Later, when the national government came West, it formalized these mining rights and provided legal protection (Anderson and Hill, 2004).

Through most of the history of the United States, the role of governments with respect to land and water was primarily to recognize, record, and protect individual property rights. While the U.S. government claimed ownership to large amounts of land, most of it was gradually settled and became privately owned through various laws such as the Homestead Act of 1862. This policy of divestiture or privatization ended late in the nineteenth century, when the national government decided to keep many western lands.

Once land was privately owned, state governments provided civil courts through which disputes over ownership and incompatible land use could be resolved. As described in Section 1.3.1, among those disputes were disagreements over damage caused by pollution. By enforcing property rights, government courts protected people from excessive pollution, just as they protected individuals from theft and from personal assault (Brubaker, 1995).

Hernando de Soto (2000) discovered the critical role of protecting property rights while studying the informal economy of Peru. He found that through neglect, bureaucratic inertia, and protection of privilege, the Peruvian government had made it impossible for many of its citizens to open businesses. Entrepreneurs had to navigate a labyrinth of onerous requirements in a costly approval process that was nearly impossible to complete. As a result, many people in the poorer sectors operated their enterprises illegally, if they had any business at all. Operating illegally, such persons did not have the basic protection of property rights that governments are generally expected to provide. De Soto concluded that if society is to be cooperative and productive, property rights must be formally recognized so people can plan for the future, knowing they can keep what they earn and any investment they make will not be taken away from them.

Understanding why private property rights are so important and the history of governments both threatening and protecting them is valuable, even critical, for those who would implement a climate change action plan. Defining, trading, and protecting property rights are activities fundamental to a free and prosperous society. While governments are relied upon to use force if necessary to protect these activities, historically governments have been unfriendly and even hostile to private property rights. That hostility has caused some of the greatest human tragedies in history. An action plan to address climate change that dismisses private property rights in favor of giving governments broad and discretionary power is unlikely to succeed.

References

Anderson, T.L. and Hill, P.J. 2004. *The Not So Wild, Wild West*. Stanford, CA: Stanford University Press.

Aristotle. 1939. In: Durant, W. *The Life of Greece*. New York, NY: Simon and Schuster.

Baumol, W.J. 2002. *The Free-Market Innovation Machine: Analyzing the Growth Miracle of Capitalism*. Princeton, NJ: Princeton University Press.

Bethell, T. 1998. *The Noblest Triumph: Property and Prosperity Through the Ages*. New York, NY: St. Martin's Press.

Blumenfeld, S.L. (Ed.) 1974. *Property in a Human Economy: A Selection of Essays Compiled by the Institute for Humane Studies*. LaSalle, IL: Open Court.

Brubaker, E. 1995. *Property Rights in the Defense of Nature*. London, UK: Earthscan Publications Ltd.; Toronto, ON: Earthscan Canada.

De Soto, H. 2000. *The Mystery of Capital*. New York, NY: Basic Books.

Internet World Stats. 2018. Internet usage statistics (website). Accessed July 30, 2018.

Mises, L. 1966 [1998]. *Human Action: A treatise on economics, the scholar's edition*. Auburn, AL: Ludwig von Mises Institute.

Panné, J.L., Paczkowski, A., Bartosek, K., Margolin, J.-L., Werth, N., Courtois, S., Kramer, M. (trans.), and Murphy, J. (trans.). 1999. *The Black Book of Communism: Crimes, Terror, Repression*. Cambridge, MA: Harvard University Press.

Pipes, R. 1999. *Property and Freedom*. New York, NY: Vintage Books.

Weber, M. 1918. Politics as a vocation. Lecture in Munich to the Free Students Union of Bavaria on 28 January.

1.4.2 Regulation

Regulations often fail to achieve their objectives due to the conflicting incentives of individuals in governments and the absence of reliable and local knowledge.

Beginning in the 1970s, several environmental laws were enacted in the United States giving federal agencies sweeping powers to directly control activities that might have negative environmental consequences. The decade saw the passage of the Clean Air Act, Clean Water Act, Endangered Species Act, National Environmental Policy Act, Resource Conservation and Recovery Act, and Toxic Substances Control Act. In 1980, the Comprehensive Environmental Response Compensation and Liability Act, known as Superfund, was enacted to clean up hazardous waste dumps (Easterbrook, 1995; Carlin, 2015).

Standards were set, but were they too tight or too lax? Would the best standards be different in different areas? Technologies are often specified in the regulations formed under such laws. Were they the right technologies? Would they continue to be the right ones? A government agency may have little interest in gathering data to objectively answer these questions because its interests may support more restrictive regulations, regardless of whether they are needed. The information it collects probably will not include valuable knowledge of local circumstances affecting costs and opportunities.

As a result of these federal laws and the mushrooming power of federal regulatory agencies, especially the Environmental Protection Agency (EPA), some potentially dangerous emissions into the air and water were reduced. Those positive outcomes led most citizens to support the environmental laws of the late twentieth century, giving power and influence to environmental activists who moved on to fight other alleged but poorly documented dangers such as acid deposition, particulate matter, and global warming. Fundraising letters produced by environmental groups were especially vivid with alarm and not always accurate regarding the science. The EPA along with state and local agencies adopted regulations that became more and more stringent. The costs imposed on those forced to comply with the new regulations, including taxpayers, grew (Djankov *et al.*, 2002; Trzupek, 2011).

EPA programs such as Superfund and the Clean Air Act allowed government officials to pursue narrow goals without taking into account competing goals or having to provide the kind of cause-and-effect information required in civil litigation. The programs were popular with the public, but when better information was produced, the realized benefits were often much smaller than initially expected because the dangers had been exaggerated and/or the intended solutions did not work in the manner expected (Gots, 1993; Tengs *et al.*, 1995; Graham, 1995; Crews, 2013). Some citizens directly affected by these environmental concerns strongly opposed specific regulations when costs were high or when the effectiveness of regulations could not be demonstrated. Public attention to the high cost of regulation led to a deregulation movement in some areas, with some success (Litan, 2014).

Dawson and Seater (2013) conducted an analysis of the effect of regulation on economic growth in the United States using the number of pages in the *Code of Federal Regulations* as a measure of regulatory burden. They found, "In 2011, nominal GDP was $15.1 trillion. Had regulation remained at its 1949 level, current GDP would have been about $53.9 trillion, an increase of $38.8 trillion. With about 140 million households and 300 million people, an annual loss of $38.8 trillion converts to about $277,100 per household and $129,300 per person. Furthermore, our estimates indicate that the opportunity cost will grow at a rate of about 2% a year (the average reduction in trend over the sample period) if regulation is merely kept at its 2005 level and not increased further" (p. 22). Per-capita GDP in 2011 was approximately $50,000, so but for the presence of federal regulations, average per-capita GDP would have been more than three times as high ($179,300 versus $50,000). The authors note, "our figures are net costs. They are based on the change in total product caused by regulation and so include positive as well as negative effects. Our results thus indicate that whatever positive effects regulation may have on measured output are outweighed by the negative effects" (*Ibid.*). In other words, the lost income is *pure waste* in the sense that it bought nothing of value.

Regulation is expensive because regulators don't have knowledge of local conditions and opportunities and cannot control all the decisions of individuals affected by the rules, leading to inefficiency, circumvention, unintended consequences, and waste (Winston, 2006; Dudley and Brito, 2012). This is particularly problematic in the case of managing greenhouse gases since "carbon emissions are a pervasive result of the use of fossil fuels, and *every decision bearing on the use of fossil fuels will affect these emissions.* Regulatory programs can only be brought to bear on a finite subset of these decisions, where specification of requirement, monitoring, and enforcement are possible" (Montgomery, 1995, p. 37, italics added).

Emission reduction programs relying on command-and-control regulations often are expensive and ineffective because they fail to allow businesses and entrepreneurs to seek out the lowest-cost opportunities to reduce emissions. Tietenberg (1985) surveyed 11 empirical studies comparing the cost of complying with command-and-control regulations to the least-costly methods of achieving the same level of pollution reduction. In all 11 cases, complying with regulations cost more than the least-cost methods, with a mean average ratio of six and a median ratio of four. In other words, command-and-control regulations typically cost between four and six times as much as the least-costly means of reducing emissions by the same amount.

More recently, Nobel Laureate economist Jean Tirole wrote, "It has been empirically verified, however, that top-down policies increase the cost of environmental policies considerably. To judge from experience with other pollutants, introducing a single carbon price might reduce the cost of cutting pollution by at least half in comparison with top-down approaches discriminating between sectors or agents" citing Ellerman *et al.*, 2003; Tietenberg, 2006; and Stavins, 2002 (Tirole, 2017, p. 215).

Taking a regulatory approach to emission reductions has a third deficiency called "new source bias." Costly technology mandates imposed on new plants, machines, buildings, etc. raise the cost of new investments and consequently discourage replacement of existing capital. This slows down the natural turnover of capital, which is responsible for significant advances in energy efficiency. Money that would have gone to new, cleaner goods and services is diverted instead to keeping older, dirtier machines and facilities in use, offsetting some or all of the intended gains. In the case of fuel economy standards for cars and trucks, this "clunker effect" has been estimated to offset 13-16% of the expected fuel savings (Jacobsen and van Benthem, 2015).

Because of the well-known limitations of regulations, some climate change activists promote a "carbon tax" instead. Such a tax would be imposed on the greenhouse gases released during the combustion of coal, oil, and natural gas, raising their prices and giving businesses and consumers incentives to use low-carbon alternatives. A "carbon tax" would be more efficient than regulations only if it replaces existing regulations rather than being in addition to them, yet this is rarely proposed and is likely to be politically impossible. McKitrick (2016a) writes,

[C]arbon pricing only works in the absence of any other emission regulations. If pricing is layered on top of an emission-regulating regime already in place (such as emission caps or feed-in-tariff programs), it will not only fail to produce the desired effects in terms of emission rationing, it will have distortionary effects that cause disproportionate damage in the economy. Carbon taxes are meant to replace all other climate-related regulation, while the revenue from the taxes should not be funnelled into substitute goods, like renewable power (pricing lets the market decide which of those substitutes are worth funding) but returned directly to taxpayers.

Virtually all carbon tax proposals include provisions for giving some of the revenues to the various rent-seekers who make up the global warming movement, a topic addressed in Section 1.4.5 below. Those groups have even opposed "carbon tax" initiatives that don't earmark some of the financial windfall to them (Burnett, 2016). In other words, taxpayers are rightly skeptical that all the revenues raised by a "carbon tax" would be "returned directly to taxpayers."

McKitrick (2016b) also observed, "The economic efficiency of a carbon tax comes not from setting a floor price, but a ceiling price. Policies like the federal biofuels mandate, energy conservation programs, renewables subsidies and coal phaseout rules might reduce carbon dioxide emissions, but they do so at marginal costs of hundreds of dollars per tonne. Adding a carbon tax on top of that does nothing to make the overall policy mix more efficient. But replacing those policies with a carbon tax would. In the process, *it might also lead to higher carbon dioxide emissions,* something that promoters of carbon pricing need to be upfront about" (italics added).

The incidence of a carbon tax would fall on the consumers of virtually all products (since they all require energy to be produced and transported to consumers), so it would act more as a general consumption or sales tax than an environmental tax. As Fullerton (1996) wrote, "Congress can decide who is legally liable to pay a tax, but it cannot legislate the ultimate distribution of burden. A tax on one good may reverberate through the economy in such a way that other prices are affected. An untaxed good may end up with a higher price, and anyone who buys it bears a burden." Nor are taxes free of administrative and compliance costs, or of the opportunity costs incurred when less energy is consumed.

Much of the period after 1970 has been characterized by hostile confrontations between bureaucrats and environmental activists pressing for tougher regulations, and companies and individuals who bear the largest burdens of those regulations resisting (Trzupek, 2011). The discussion of how best to address the threat of climate change takes place in this context, which explains the readiness of many economists, industry groups, and pro-consumer and pro-free enterprise groups to react quickly and negatively to plans that involve new regulations and taxes. This reaction is not knee-jerk or selfish, but based on what has been learned over the past 40 years about the costs and consequences of environmental regulations.

References

Burnett, H.S. 2016. Most environmental groups oppose Washington state's carbon tax initiative. *Environment & Climate News.* August 19.

Carlin, A. 2015. *Environmentalism Gone Mad: How a Sierra Club Activist and Senior EPA Analyst Discovered a Radical Green Energy Fantasy*. Mount Vernon, WA: Stairway Press.

Crews, W. 2013. *Ten Thousand Commandments: A Policymaker's Snapshot of the Federal Regulatory State*. Washington, DC: Competitive Enterprise Institute.

Dawson, J.W. and Seater, J.J. 2013. Federal regulation and aggregate economic growth. *Journal of Economic Growth* **18** (2): 137–77.

Djankov, S., LaPorta, R., Lopez-De-Silanes, F., and Shleifer, A. 2002. The regulation of entry. *Quarterly Journal of Economics* **117**: 1–37.

Dudley, S.E. and Brito, J. 2012. *Regulation: a primer*. Mercatus Center and George Washington University Regulatory Studies.

Easterbrook, G. 1995. *A Moment on the Earth: The Coming Age of Environmental Optimism*. New York, NY: Viking.

Ellerman, D., Harrison, D., and Joskow, P. 2003. *Emissions Trading in the U.S. Experience, Lessons and Considerations for Greenhouse Gases*. Washington, DC: Pew Center on Global Climate Change.

Fullerton, D. 1996. *Why Have Separate Environmental Taxes?* Tax Policy and the Economy **10**. Cambridge, MA: National Bureau of Economic Research.

Gots, R.E. 1993. *Toxic Risks: Science, Regulation, and Perception*. Boca Raton, FL: Lewis Publishers.

Graham, J. 1995. Reform of risk regulation: Achieving more protection at less cost. *Human and Ecological Risk Assessment: An International Journal* **1** (3): 183–206.

Jacobsen, M.R. and van Benthem, A.A. 2015. Vehicle scrappage and gasoline policy. *American Economic Review* **105** (3): 1312–38.

Litan, R. 2014. *Trillion Dollar Economists: How Economists and Their Ideas have Transformed Business*. New York, NY: Wiley.

McKitrick, R. 2016a. A Practical Guide to the Economics of Carbon Pricing. *SPP Research Papers* **9** (28). September.

McKitrick, R. 2016b. Drop all this carbon-tax boosterism; they could easily do more harm than good. *Financial Post*. September 15.

Montgomery, W.D. 1995. *Towards an Economically Rational Response to the Berlin Mandate*. Washington, DC: Charles River Associates. July.

Stavins, R. 2002. Lessons from the American experiment with market-based environmental policies. In: Donahue, J. and Nye, J. (Eds.) *Market-based Governance: Supply Side, Demand Side, Upside, and Downside*. Washington, DC: The Brookings Institution, pp. 173–200.

Tengs, T., Adams, M., Pliskin, J., Fafran, D., Siegel, J., Weinstein, M., and Graham, J. 1995. Five hundred life-saving interventions and their cost-effectiveness. *Risk Analysis* **15** (3): 369–90.

Tietenberg, T.H. 1985. *Emissions Trading: An Exercise in Reforming Pollution Policy*. Washington, DC: Resources for the Future.

Tietenberg, T.H. 2006. *Emissions Trading: Principles and Practice*. Second edition. London, UK: Routledge.

Tirole, J. 2017. *Economics for the Common Good*. Princeton, NJ: Princeton University Press.

Trzupek, R. 2011. *Regulators Gone Wild: How the EPA Is Ruining American Industry*. New York, NY: Encounter Books.

Winston, C. 2006. *Government Failure versus Market Failure: Microeconomics Policy Research and Government Performance*. Washington, DC: AEI-Brookings Joint Center for Regulatory Studies.

1.4.3 Bureaucracy

Government bureaucracies predictably fall victim to regulatory capture, tunnel vision, moral hazard, and corruption.

Government programs often are represented as solutions to social problems without any cost of implementation. But every program requires a bureaucracy to oversee the translation of legislation into regulations, public promotion of the new rules and requirements, enforcement, and regular monitoring of success or failure to achieve goals. Government bureaucracies have been closely studied by economists and found to be rarely efficient (Mises, 1944; Wilson, 1991; Breyer, 1993; Niskanen, 1996; Tullock, 2005). There are several reasons for this.

Many government agencies are given not one but three mandates: to identify, evaluate, and solve a social problem. But combining all three responsibilities in the same entity means the agency has no incentive to decide the social problem does not merit a significant investment of public monies to solve, or that the problem, should it exist, even could be solved. The agency is also charged with measuring its own success and then reporting it to those who control its funding and future existence. The heads of such agencies, no matter how honest or well-intended, cannot objectively evaluate their own performances (Savas, 2000, 2005).

Lobbying by special interests leads to "regulatory capture," the phenomenon of government officials – bureaucrats – reflecting the interests and views of the industries they are supposed to regulate, rather than the consumers they are supposed to protect (Stigler, 1971). The bureaucrats who staff an agency often see future careers on the staffs of the corporations and trade associations that frequent their offices, and they may have been recruited from industry in the first place. Politicians are lobbied by their campaign supporters to place industry insiders on the staff of regulatory agencies, expecting them to be more sympathetic to their concerns.

Even bureaucrats who break away from this pattern are motivated by idealism or careerism to ask for larger budgets and staffs each year (Wildavsky, 1964; Blais and Dion, 1991). Bureaucrats and their staff, therefore, are usually happy to work to expand their programs to deliver benefits to special-interest groups who, in turn, work with politicians to expand their bureau budgets and programs. Hayek (1944) observed that in government "the worst get on top" since their values and skills suit them to winning internecine struggles and persuading others to follow their lead.

Another reason bureaucracies dysfunction can be summarized as tunnel vision. This is the term Supreme Court Justice Stephen Breyer applied to federal regulators, including the EPA (Breyer, 1993). For Breyer, tunnel vision is the tendency of government employees to focus exclusively on the objectives of their agencies, or even the specific programs within their agencies, at the expense of all other concerns. As noted in Section 1.2.2, all people have narrow goals. In the private sector, the rights of other people and competition from other producers bring individual goals into harmony with others. In government, no such invisible hand operates.

Tunnel vision can lead to excessive regulation causing more harm than good (Baden and Stroup, 1981; Greve and Smith, 1992; Nelson, 1995; DeLong, 2002). A notable example in the United States is the 1980 Superfund law intended to clean up abandoned waste sites. Funding came initially from a tax on chemical-producing industries, but the EPA was authorized to obtain compensation from any individual or company it could show had deposited any hazardous waste at the site, no matter how small or innocuous the contribution. Known as "joint and several liability" it enabled the EPA to target companies with the deepest pockets when assessing penalties, even if they were not significant polluters at the site. To obtain this compensation, EPA officials had no responsibility to show wrongdoing, any real damage to others, or even any real and present risk emanating from the site.

Superfund was supposed to cost at most a few billion dollars and be paid for mainly by those whose pollution had caused serious harms or risks. But that was not the result (Wildavsky, 1995, pp. 153–84). In the first 12 years after Superfund was established, the program spent $20 billion, and its costs grew along with delays in its cleanups of hazardous waste sites. Despite the expenditures, the program showed little gain in the way of human health benefits. Hamilton and Viscusi (1996) reported a number of discouraging findings. Among them:

- Most assessed Superfund risks do not pose a threat to human health now; they might do so in the future, but only if people violate common-sense precautions and actually inhabit contaminated sites while disregarding known risks there.

- Even if exposure did occur, there is less than a 1% chance that the risks are as great as the EPA estimates, because of the compounding of extreme assumptions made by the agency.

- Cancer risk is the main concern at Superfund sites, because it has a long latency period and some contaminants at the sites can cause cancer in high-dose exposures. Yet at most of the sites, each cleanup is expected to avert only one-tenth of one case of cancer. Without any cleanup, only 10 of the 150 sites studied were estimated to have one or more expected cases.

- The average cleanup cost per site in the study was $26 million (in 1993 dollars).

- Replacing extreme EPA assumptions with more reasonable ones brought the estimated median cost per cancer case averted to more than $7 billion. At 87 of the 96 sites having the necessary data available, the cost per cancer case averted (only some of which would mean a life saved) was more than $100 million.

- Other national programs in 1996 commonly considered the value of a statistical life to be about $5 million. (Today, the EPA places that value at $7.4 million. See EPA, 2018). Diverting expenditures from most Superfund sites to other sites or other risk-reduction efforts could prolong many more lives or the same number of lives at far less cost.

Hamilton and Viscusi estimate 95% of Superfund expenditures are directed at the last 0.5% of the risk. Many people touched by the program are harmed rather than helped. A designated Superfund site causes property values to fall, residents may be forced to move away, at least temporarily, and people may be badly frightened for no good reason. Consumers and taxpayers foot the enormous bill even though they may never come near a Superfund site.

A third reason bureaucracies turn away from the public interest is moral hazard, explained in Section 1.2.4. Their administrators are tempted to use their expertise, control of information about programs, and monopoly position to push for more authority or a bigger budget. The objective may be to better achieve their agency's objectives, but just as likely it will be to advance personal career objectives or to gain popular recognition for the agency's good works, neither of which advance the public good.

For example, the National Park Service often has used what observers call the Washington Monument strategy: When told to expect budget increases smaller than it would like, the Park Service announces it may have to economize by shortening the hours it can operate the Washington Monument or other popular attractions. In essence, Park Service leaders are presenting a veiled threat, "Give us what we asked for or we will cut back on our most popular services." The tactic was seen in 2013 when the Obama administration ordered hundreds of parks to close, even those not dependent on government funding, during a budget stand-off with Congress (Preston, 2013).

The strategy tends to increase the Park Service budget. The threat of long lines of disgruntled citizens (voters) waiting to get in or expressing outrage at not being able to enter popular attractions is all that is needed to persuade political appointees or congressional committees to increase funding. Private firms rarely if ever use this or similar strategies. Can you imagine Wal-Mart threatening to not sell its most popular product lines unless more customers chose to shop at its stores?

The problem of dysfunctional bureaucracies is not a small one in the climate change discussion. For many years the head of the IPCC – the bureaucracy put in charge of finding a 'scientific consensus" on what should be done about anthropogenic climate change – also worked for the renewable energy industry, a flagrant conflict of interest (Laframboise, 2013). The IPCC's procedures were harshly criticized by an audit conducted by the InterAcademy Council, a respected organization composed of the heads of the world's leading science academies (IAC, 2010). The results of that audit are reported in detail in Chapter 2.

Worse than the IPCC is the United Nations, the IPCC's parent organization and host of the Framework Convention on Climate Change (FCCC), which is tasked with negotiating and then implementing a binding global treaty on climate change. A 2013 report by the Foundation for Defense of Democracies said "The United Nations is a hotbed for corruption and abuse. It is opaque, diplomatically immune, [and] largely unaccountable..." (Dershowitz, 2013). After recounting "the Oil-for-Food scandal, in which the U.N. profited from and covered up for billions in Baghdad kickbacks and corruption" and broken promises of "greater transparency, accountability, an end to Peacekeeper rape, the elimination of redundant mandates, and a more ethical culture," the foundation says "the U.N.'s internal audit division, the Office of Internal Oversight Services, has been roiled with scandals and frictions, including a former chief of the unit accusing the UN Secretary-General of 'deplorable' actions to impede her hiring of investigators, and charging that 'the secretariat is now in a process of decay'" (*Ibid.*).

The UN's problems appear to be structural and not the fault of whoever happens to be the Secretary-General. Allen (2013) wrote, "The United Nations [is] a famously corrupt body in which most votes are controlled by kleptocracies and outright dictatorships. Most of the member-states, as they're called, are rated as either 'not free' or 'partly free' by Freedom House, and both Communist China and Putinist Russia have veto power. And any settlement of the Global Warming issue by the UN would entail massive transfers of wealth from the citizens of wealthy countries to the politicians and bureaucrats of the poorer countries. Other than that, one supposes, the IPCC is entirely trustworthy on the issue. (Well, aside from the fact that the IPCC's climate models predicting Global Warming have already failed.)"

Economists who look at efforts by the IPCC and the UN to address climate change immediately see regulatory capture, tunnel vision, moral hazard, and corruption, all the predictable characteristics of bureaucracies. Environmental activists and many scientists seem unaware of these flaws or willing to excuse them given the presumed gravity of the climate change issue. But the IPCC was *entrusted* to find the truth about climate change science, and the UN was *entrusted* to implement a treaty to manage the global atmosphere. Their obvious shortcomings cannot be irrelevant to the climate change discussion.

After each of its scandals, the UN promised to reform itself. After the scathing audit by the IAC, the IPCC promised to reform itself. Neither has done so because neither *can* do so. Both lack the design principles recommended by Ostrom. The IPCC was never likely to objectively study the climate change issue given its mandate to find a human impact on the global climate. The UN was never likely to negotiate and implement a global program aimed at addressing the challenge of climate change, given the equal voting rights of dictatorships and failed regimes.

References

Allen, S.J. 2013. Climate change violence study exposes ethical problems in science and the media. Capital Research Center (website). August 20.

Baden, J. and Stroup, R. (Eds.). 1981. *Bureaucracy vs. Environment: The Environmental Cost of Bureaucratic Governance*. Ann Arbor, MI: University of Michigan Press.

Blais, A. and Dion, S. (Eds.) 1991. *The Budget-Maximizing Bureaucrat: Appraisals and Evidence*. Pittsburgh, PA: University of Pittsburgh Press.

Breyer, S. 1993. *Breaking the Vicious Circle: Toward Effective Risk Regulation*. Cambridge, MA: Harvard University Press.

DeLong, J.V. 2002. *Out of Bounds, Out of Control: Regulatory Enforcement at the EPA*. Washington, DC: Cato Institute.

Dershowitz, T. 2013. United Nations corruption and the need for reform. Foundation for Defense of Democracy (website). Accessed July 28, 2018.

EPA. 2018. Environmental Protection Agency. What value of statistical life does EPA use? (website). Accessed October 10, 2018.

Greve, M.S. and Smith, F.L. 1992. *Environmental Politics: Public Costs, Private Rewards*. New York, NY: Praeger Publishers.

Hamilton, J. and Viscusi, W.K. 1996. *Calculating Risks*. Cambridge, MA: MIT Press.

Hayek, F. 1944. *The Road to Serfdom*. Chicago, IL: University of Chicago Press.

IAC. 2010. InterAcademy Council. Draft: *Climate Change Assessments: Review of the Processes & Procedures of IPCC*. The Hague, Netherlands: Committee to Review the Intergovernmental Panel on Climate Change. October.

Laframboise, D. 2013. *Into the Dustbin: Rachendra Pachauri, the Climate Report & the Nobel Peace Prize*. CreateSpace Independent Publishing Platform.

Mises, L. 1944. *Bureaucracy*. New Haven, CT: Yale University Press.

Nelson, R.H. 1995. *Public Lands and Private Rights: The Failure of Scientific Management*. Lanham, MD: Rowman & Littlefield Publishers, Inc.

Niskanen Jr., W. 1996. *Bureaucracy and Public Economics*. Cheltenham, UK: Edward Elgar Publishing.

Preston, B. 2013. BREAKING: White House ordering hundreds of privately run, privately funded parks to close. *PJ Media* (website). Accessed July 30, 2018.

Savas, E.S. 2000. *Privatization and Public Private Partnerships*. New York, NY: Chatham House Publishers.

Savas, E.S. 2005. *Privatization in the City: Successes, Failures, Lessons*. Washington, DC: CQ Press.

Stigler, G. 1971. The theory of economic regulation. *Bell Journal of Economics and Management Science* 3: 3–18.

Tullock, G. 2005. *Bureaucracy: The Selected Works of Gordon Tullock, Volume 6*. Indianapolis, IN: Liberty Fund.

Wildavsky, A. 1964. *The Politics of the Budgetary Process*. Boston, MA: Little, Brown.

Wildavsky, A. 1995. *But Is It True? A Citizen's Guide to Environmental Health and Safety Issues*. Cambridge, MA: Harvard University Press.

Wilson, J.Q. 1991. *Bureaucracy: What Government Agencies Do And Why They Do It*. New York, NY: Basic Books.

1.4.4 Rational Ignorance

Voters have little incentive to become knowledgeable about many public policy issues. Economists call this "rational ignorance."

Key to the case for having governments involved in the management of a common-pool resource such as Earth's atmosphere is the belief that doing so gives the general public, or at least the voting public, a say in an important matter. But how much influence do voters have on government policies? And how valuable is their input on an issue as complex and poorly understood as climate change?

Number four of Ostrom's eight design principles for effective management of common-pool resources (listed in Figure 1.3.1.3) is effective monitoring by monitors who are accountable to the people who pay for a program. Voters, by and large, do not and cannot monitor government programs. They are "rationally ignorant" about public policy issues and their elected officials (Downs, 1957; Buchanan and Tullock, 1962; Crain *et al.*, 1988; Olson, 2000). Rational ignorance is making a reasoned choice not to study or master a complex subject because the expected benefits of doing so are not worth the cost (time and effort).

The ballot choice made by a single voter is seldom decisive. Recognizing the outcome almost never hinges on one vote, the individual voter has little incentive to spend time and effort studying issues and candidates in order to cast a more informed vote. This helps to explain why most Americans of voting age cannot name their elected congressional representatives (Haven Insights, 2017), much less identify, understand, or compare the positions of multiple candidates on multiple issues – including the environment.

The likelihood of voters punishing politicians for supporting costly special-interest legislation is low because elected officials cast many votes on many issues, some of them likely to meet with the voter's approval. Since no political candidate is likely to represent the exact interests of a voter, the voter is willing to overlook disagreements. Politicians understand that it can take many disappointments before a voter will choose to vote against an incumbent officeholder.

Those voters who do take the time to learn about issues and vote carefully may nevertheless not represent the broader public interest. A number of factors combine to make special-interest groups more powerful in a representative democracy than their numbers would otherwise suggest (Downs, 1957; Buchanan and Tullock, 1962; Mueller, 1979; Gwartney *et al.*, 2012). Members of an interest group – such as the owners of specific tracts of farmland irrigated with low-cost water – have a strong stake in the outcome of certain political decisions. Thus they have an incentive to hire lobbyists to represent them before Congress and regulatory agencies. They also have an incentive to inform themselves and their allies in local communities and to let legislators know how strongly they feel about an issue of special importance. Many of them encourage voting for political candidates strictly on the basis of whether those candidates support (or promise to support) their specific interests. Such interest groups often are also in a position to provide large financial campaign contributions to candidates who support their positions, and they are sure to remind them of those contributions when votes are on the line.

In contrast, most persons eligible to vote are not attached to any particular special-interest group. For them, examining the issues takes more time and energy than it is worth because they have only a relatively small amount to gain personally from the elimination of special-interest programs or subsidies. For a political candidate, supporting the position of a well-organized, narrowly specialized interest group can generate vocal supporters, campaign workers, and campaign contributions. Supporting the opposition, which is often uninformed, unorganized, and unmotivated, offers politicians little benefit or reward.

Voters who try to become informed and take the time to vote even though their vote won't matter may nevertheless be misinformed. The legacy media (newspapers and broadcast television stations) devote little time and space to the detailed and complicated information necessary for making informed decisions. What sells today are the soft human-interest stories about villains and heroes and dramatic images of shocking, high-risk situations (Sandman *et al.,* 1987; Cohen, 2000; Milloy, 2001). Hard news has largely been reduced to headlines or brief sound-bite-length articles often reporting on the opinions of celebrities, not experts (Ciandella, 2015). The public's loss of respect for legacy media outlets has increased its reliance on new media sources such as cable news (e.g., CNN, Fox, MSNBC), websites (e.g., Drudge Report, Huffington Post), and social media platforms (e.g., Twitter, Facebook), all of which have credibility and bias problems of their own and most of which devote even less space and time to issue analysis.

The media, whether legacy or new, mostly cover the climate change issue by reporting as news the latest claims made by environmental groups, and no wonder: Environmental groups spend billions of dollars a year hyping the possibility of catastrophic climate change. In the United States alone, some 13,716 environmental groups reported combined revenue of $7.4 billion and total assets of $20.6 billion in 2012 (Nichols, 2013). Some of the larger groups include the Environmental Defense Fund (EDF), with $112 million in revenues and $173 million in assets; Natural Resources Defense Council (NRDC), with $97 million in revenue and $248.9 million in assets; and three tax-exempt Greenpeace organizations in the United States with $39.2 million in revenue and $20.6 million in assets.

Voter apathy and rational ignorance, along with the failure of media to report the truth about climate change and many other issues, allow for the creation and maintenance of laws and regulations that do not advance the public interest. Once a law is implemented, a voter will often turn to other matters, especially since the details are complex and unrelated to the voter's everyday activities. Voter apathy and ignorance explain why Superfund was popular when it was passed, why most voters know little or care little about the program's problems, and why voters have not demanded their elected politicians fix the program's flaws or repeal the law. Most voters are likewise ignorant of many other environmental issues.

The same people who fail to act as informed voters often are highly motivated *shoppers* for goods and services. They spend considerable amounts of time and effort evaluating the pros and cons of the choices they make. Imagine you are planning to buy a car next week and also to vote for one of two candidates for the U.S. Senate. In purchasing a car, you have a nearly unlimited number of choices. Do you buy new or used? Sedan or minivan? Honda or Ford? Your options are many and varied. In the voting booth, however, your choice is probably limited to just two candidates. The winning Senate candidate will represent you on hundreds of issues, and it is inconceivable that he or she will agree with you on all of them. Both the car purchase and the Senate vote involve complex tradeoffs. Which of these decisions will command more of your scarce time and energy to research and ultimately choose?

If you spend more time choosing a car than the next Senator, you are not acting irrationally. When it comes to the car, your choice is entirely up to you (or your spouse), and only you are responsible for the costs and reap the full benefits. An uninformed car purchase could be very costly. With respect to the election, if by chance you mistakenly vote for the wrong candidate out of ignorance, the probability is nearly zero that your vote will decide the election. Cumulatively, your vote and those of all others in your state will decide who wins, but your individual choice will not. Thus, a mistake or a poorly informed selection at the ballot box will have little consequence on the actual outcome of an election or on your life.

In light of all this, what is the best way to involve the public in deciding the best ways to respond to climate change? Asking the public to vote on candidates who promise to take one position or another on the issue is unlikely to work for the reasons explained above. Even asking the public to vote on a referendum, say for a "carbon tax" or some emission mitigation plan, won't work because most people won't vote and those who do vote will likely be ill-informed or misinformed.

The more promising route, as our example of the car purchase illustrates, is to engage the public as *shoppers,* individuals seeking to achieve their own goals as efficiently as possible. Respect the value of their time by not insisting they become experts on climate science or vote for candidates who may or may not win and if they do, may or may not deliver on their promises. Instead, give the public opportunities to buy products that are "low-carbon" or otherwise promise to reduce greenhouse gas emissions. Offer investment opportunities in promising new technologies or projects that help people in developing countries adapt to climate change, whether due to natural or anthropogenic causes. Stop subsidizing behavior such as building in floodplains that might lead to higher social costs if sea levels rise or if severe weather events become more frequent.

Treating the public as shoppers rather than voters avoids the problem of rational ignorance, making it the only real market-based response to climate change. Of course it is not without its own difficulties: The global atmosphere remains a common-pool resource, so every individual's private cost-benefit analysis will not include the possible effects of his or her actions on other people. But the "social cost of carbon," if it can be calculated at all, is likely to be zero or even negative (in other words, climate change produces net benefits rather than costs), so its absence from cost-benefit analyses won't be missed. (This is the subject of Chapter 8.)

The tremendous prosperity created by the use of fossil fuels has elevated the value people place on environmental protection as well as their willingness and ability to pay for it. Media exaggeration of the "threat" of climate change probably increases this willingness to pay, helping to transform, as Boettke wrote in 2009 of disagreements over access to common-pool resources generally, from a "tragedy of the commons" to "an opportunity of the commons" (Boettke, 2009).

This option, of treating the public as shoppers rather than voters in the climate change discussion, seldom appears in the academic literature, since writers there simply assume a "social cost of carbon" can be accurately set, is positive, and can be efficiently implemented as a public policy. It seldom appears in the fundraising letters or advocacy reports of environmental groups or in speeches delivered by politicians, either, because it a solution that requires little or no action by governments.

References

Boettke, P. 2009. Liberty should rejoice: Elinor Ostrom's Nobel Prize. *The Freeman.* November 18.

Buchanan, J. and Tullock, G. 1962. *The Calculus of Consent.* Ann Arbor, MI: University of Michigan Press.

Ciandella, M. 2015. *Climate Hypocrites and the Media that Love Them.* Washington, DC: Media Research Center.

Cohen, D. 2000. *Yellow Journalism: Scandal, Sensationalism, and Gossip in the Media.* Brookfield, CT: Twenty-First Century Books.

Crain, W.M., Shughart II, W.F., and Tollison, R.D. 1988. Voters as investors: a rent-seeking resolution of the paradox of voting. In: Rowley, C.K., Tollison, R.D., and Tullock, G. (Eds.) *The Political Economy of Rent-Seeking.* Norwell, MA: Kluwer Academic Publishers, pp. 241–9.

Downs, A. 1957. *An Economic Theory of Democracy.* New York, NY: Harper & Brothers.

Gwartney, J.D., Stroup, R.L., Sobel, R.S., and Macpherson, D.A. 2012. *Economics: Private and Public Choice.* Fourteenth edition. Boston, MA: Cengage Learning.

Haven Insights. 2017. Just 37% of Americans can name their representative (website). May 31. Accessed July 28, 2018.

Milloy, S. 2001. *Junk Science Judo: Self-Defense against Health Scares and Scams.* Washington, DC: Cato Institute.

Mueller, D.C. 1979. *Public Choice.* New York, NY: Cambridge University Press.

Nichols, N. 2013. The tax-exemption rip-off. Townhall (website). October 1.

Olson, M. 2000. *Power and Prosperity: Outgrowing Communist and Capitalist Dictatorships.* New York, NY: Oxford University Press.

Sandman, P.M., Sachsman, D.B., Greenberg, M.R., Gochfeld, M. 1987. *Environmental Risk and the Press: An Exploratory Assessment.* New Brunswick, NJ: Transaction Books.

1.4.5 Rent-seeking Behavior

Government's ability to promote the goals of some citizens at the expense of others leads to resources being diverted from production into political action. Economists call this "rent-seeking behavior."

Government's monopoly on the legitimate use of force gives it the ability to take resources from some people and give them to others. That ability can be used promote transfers that are widely supported and beneficial to most people – what is referred to in the U.S. Constitution as the "general welfare" – but also to advance the specific welfare of a small number of constituents. The beneficiaries of the second type of activity, who economists call rent-seekers, invest resources to convince government officials to take actions that benefit them at the expense of the general public (Olson, 1965, 1984; Rowley *et al.*, 1988; Laband and Sophocleus, 2018). Politicians in turn extract rent from individuals and businesses by threatening to withhold privileges or bestow them on competitors (McChesney, 1987).

The U.S. federal government's program to supply below-cost water to farmers in the arid West illustrates rent-seeking behavior. Using the Central Utah Project's dams and canals, the U.S. Bureau of Reclamation delivered irrigation water from a tributary of the Colorado River to Utah farmers. This transfer of water was highly subsidized by the national treasury. The price to the farmers was only $8 per acre-foot (enough water to cover an acre 1-foot deep) even though the cost of the delivered water was about $400 per acre-foot. Estimates put the value of the water to farmers at about $30 per acre-foot (Anderson and Snyder, 1997). The below-cost water delivery served the landowners and farmers and the small communities where they lived. The high costs (above the amount the farmers paid) were passed on to taxpayers across the nation. Because each individual taxpayer paid only a fraction of the total cost, most taxpayers have never heard of the project and have no idea of the costs they paid.

A more recent example of rent-seeking was documented in a 2017 lawsuit heard by the U.S. Court of Appeals for the District of Columbia Circuit (U.S. Court of Appeals, 2017). The EPA had used an ozone depletion provision of the Clean Air Act to regulate refrigerants for their global warming potential. Use of a $4/pound refrigerant produced by Mexichem Fluor and other manufacturers was banned in favor of a $65/pound option produced by only two companies under patent protection, Honeywell and Chemours. Mexichem Fluor sued the EPA, and the court overturned the regulation. Honeywell and Chemours asked the court to reconsider, which it refused to do. In petitioning the court to reject the request for review, attorneys for Mexichem Fluor noted, "Industry intervenors are rent-seekers trying to use the government to foreclose their competitor's products, not to foster development of new ones" (*Ibid.*, p. 2).

The output-expanding, positive-sum activities of market discovery, innovation, and production are increasingly being replaced by rent-seeking behavior (Tullock, 1987, 2005; Del Rosal, 2011). As transfers dependent on political clout increase, people increasingly redirect their energies to gaining political influence, taking more time, energy, and other resources away from productive activities. Many businesses invest in lobbying because they view it as a cost-effective way to protect their rights and slow the rising costs of complying with environmental regulations. Federal regulation in the United States cost $1.9 trillion in 2017 (Crews, 2018).

Examples of rent-seeking in the heavily subsidized wind and solar power industry are easy to find. NextEra Energy, Inc. is a Florida-based utility that "has grown into a green Goliath, almost entirely under the radar, not through taking on heavy debt to expand or by touting its greenness, but by relentlessly capitalizing on government support for renewable energy, in particular the tax subsidies that help finance wind and solar projects around the country. It then sells the output to utilities, many of which must procure power from green sources to meet state mandates" (Gold, 2018).

Industries producing alternative energies – wind, solar, biofuels, and even nuclear and hydropower – invest hundreds of millions of dollars a year in campaigns to require utilities to purchase their products and taxpayers to pay for their subsidies and tax breaks. Utilities themselves lobby for such policies since rate-of-return regulation gives them a strong incentive to overinvest in capital (Averch and Johnson, 1962). Exelon Nuclear, a division of Exelon Generation that operates the largest fleet of nuclear power plants in the United States, has been vocal in its support of the man-made global warming hypothesis, running full-page ads hyping its "emission free" energy and lobbying for a carbon tax that its fossil-fuel-reliant competitors would have to pay (Snyder and Johnsson, 2013).

Insurance and reinsurance businesses seek to profit from higher insurance rates justified by fears of floods and severe weather (Lloyd's, 2009), even though historically climate extremes are associated with increased profitability for insurance firms as more severe weather creates more interest in their products (Hu and McKitrick, 2015). Banks expect to make billions and even trillions of dollars financing the premature destruction and rebuilding of the world's fossil-fuel-dependent energy system (HSBC, 2016).

Environmental advocacy groups invest in lobbying to advance their own narrow agendas, and their resources rival or exceed those of the business community (Arnold and Gottlieb, 1993; Arnold, 2007; Isaac, 2012). Organizations such as the Environmental Defense Fund (EDF) use fear of catastrophic climate change to raise money using slick direct mail campaigns (Taylor, 2015).

Rent-seeking sometimes makes odd bedfellows, a phenomenon Yandle (1983) labeled "bootleggers and Baptists," a reference to how the two interest groups worked together in the United States to outlaw alcohol sales in some counties or on Sundays. Chesapeake Energy, a company that mostly sells natural gas, gave $26 million to the Sierra Club to attack its rivals in the coal industry (Barringer, 2012). In 2018, ExxonMobil pledged to donate $1 million to a group called Americans for Carbon Dividends which advocates for a "carbon tax" (Pearce, 2018).

Rent-seeking behavior negatively affects more than just the efficient use of resources. The legitimacy of government suffers when the public realizes interest groups and elites use it to their benefit (Codevilla, 2010). Those who lose income without compensation are often upset. For them and for others who do not benefit from "crony capitalism," the public-interest rhetoric of even sincere environmentalists seems hollow (Gilder, 2009, pp. 10–1).

References

Ackerman, B. and Hassler, W. 1981. *Clean Coal/Dirty Air Or How the Clean Air Act Became a Multibillion-Dollar Bail-Out for High-Sulfur Coal Producers and What Should Be Done About It*. New Haven, CT: Yale University Press.

Anderson, T.L. and Snyder, P.S. 1997. *Priming the Invisible Pump*. Bozeman, MT: PERC.

Arnold, R. 2007. *Freezing in the Dark: Money, Power, and Politics and the Vast Left Wing Conspiracy*. Bellevue, WA: Merril Press.

Arnold, R. and Gottlieb, A. 1993. *Trashing the Economy: How Runaway Environmentalism is Wrecking America*. Bellevue, WA: Free Enterprise Press.

Averch, H. and Johnson, L. 1962. Behavior of the firm under regulatory constraint. *American Economic Review* **52**: 1052–69.

Barringer, F. 2012. Answering for taking a driller's cash. *The New York Times*. February 13.

Codevilla, A. 2010. *The Ruling Class: How They Corrupted America and What We Can Do About It*. New York, NY: Beaufort Books.

Crandall, R. 1986. Economic rents as a barrier to deregulation. *Cato Journal* **6** (1): 186–9.

Crews, W. 2018. *Ten Thousand Commandments: A Policymaker's Snapshot of the Federal Regulatory State*. Washington, DC: Competitive Enterprise Institute.

Del Rosal, I. 2011. The empirical measurement of rent seeking costs. *Journal of Economic Surveys* **25** (2): 298–325.

Gilder, G. 2009. *The Israel Test*. Minneapolis, MN: Richard Vigilante Books.

Gold, R. 2018. How a Florida utility became the global king of solar power. *The Wall Street Journal*. June 18.

HSBC. 2016. Statement on Climate Change. October.

Hu, B. and McKitrick, R. 2015. Climatic variations and the market value of insurance firms. *Journal of Insurance Issues* **39** (1).

Isaac, R.J. 2012. *Roosters of the Apocalypse: How the Junk Science of Global Warming Nearly Bankrupted the Western World*. Chicago, IL: The Heartland Institute.

Laband, D. and Sophocleus, J. 2018. Measuring rent-seeking. *Public Choice*: 1–21.

Lloyd's. 2009. *Climate Change and Security: Risks and Opportunities for Business*. Lloyd's 360° Risk Insight and International Institute for Strategic Studies (IISS).

McChesney, F.S. 1987. Rent extraction and rent creation in the economic theory of regulation. *Journal of Legal Studies* **16** (1): 101–18.

Olson, M. 1965. *The Logic of Collective Action*. Cambridge, MA: Harvard Economic Studies.

Olson, M. 1984. *The Rise and Fall of Great Nations*. New Haven, CT: Yale University Press.

Pearce, T. 2018. Oil giant throws $1 million into carbon tax campaign. *The Daily Caller*. October 9.

Rowley, C.K., Tollison, R.D., and Tullock, G. (Eds.) 1988. *The Political Economy of Rent-Seeking*. Norwell, MA: Kluwer Academic Publishers.

Snyder, J. and Johnsson, J. 2013. Exelon falls from green favor as chief fights wind aid. *Bloomberg News*. April 1.

Taylor, J. 2015. Top 10 global warming lies. Chicago, IL: The Heartland Institute. August 1.

Tullock, G. 1987. Rent seeking. In: *The New Palgrave: A Dictionary of Economics, Volume 4*. New York, NY: Palgrave Macmillan, pp. 147–9.

Tullock, G. 2005. *The Rent-Seeking Society: The Selected Works of Gordon Tullock, Volume 5*. Indianapolis, IN: Liberty Fund.

U.S. Court of Appeals. 2017. Joint Response of Petitioners Mexichem Fluor, Inc. and Arkema Inc. to Intervenors' Petitions for Panel Rehearing and Rehearing En Banc. *Mexichem Fluor, Inc., Petitioner*, v. *Environmental Protection Agency, Respondent, The Chemours Company FC, LLC, et al., Intervenors*. Case No. 15-1328. October 18.

Yandle, B. 1983. Bootleggers and Baptists: the education of a regulatory economist. *Regulation* **7** (3): 12–16.

1.4.6 Displacement

Government policies that erode the protection of property rights reduce the incentive and ability of owners to protect and conserve their resources. Those policies displace, rather than improve or add to, private environmental protection.

One of the events that launched the modern environmental movement was a 1969 report that the Cuyahoga River, which flows through the city of Cleveland and empties into Lake Erie, was so polluted that it burned. Of course, the water didn't literally burn, but there was oil and debris on the water; a spark, probably from a train, ignited it. Public outrage that a river could go up in flames galvanized action and helped bring about tougher laws (Meiners *et al.*, 2000).

It turns out the Cuyahoga River fire occurred because efforts to obtain relief from river pollution through the courts had been replaced by command-and-control regulation. A state pollution control board was responsible for issuing permits allowing pollutants to be emitted into the water. According to Meiners *et al.*, the board classified a key stretch of the Cuyahoga as an industrial river, so the companies along its banks did not have to clean up their effluent to any significant degree. In 1965, Bar Realty Corporation, a real estate company, had tried to clean up a Cuyahoga tributary, but the Ohio Supreme Court concluded the state pollution control board, not the courts deciding common-law claims, had the authority – and that board did not require cleanup.

The use of property rights, common law, and market relationships has real advantages. Judges and juries listen to experts on both sides, each bound by rules of evidence and cross-examined by the other, before rendering a decision. Yes, the decisions may not be perfect and the judges and juries may not be experts, but they will likely be much better informed than when they enter an election booth to vote, or when politicians vote as elected representatives.

Evidence from Canada – where, as in the United States, statutory law and government control have been replacing decisions by private owners – suggests common-law protections are stronger than regulatory efforts. Brubaker (1995) reviewed dozens of legal decisions and statutes and found as political control supplanted common law as the favored approach to avoiding pollution, the protection of pollution victims weakened. She wrote:

> Governments have shown that they are not up to the task of preventing resource degradation or pollution; indeed they have often actively encouraged it. ... It is long past time for resources to be shifted away from governments and back to the individuals and communities that have strong interests in their preservation. Such a shift can best be accomplished by strengthening property

rights and by assigning property rights to resources now being squandered by governments (p. 161).

Brubaker's book offers a compelling case that defending property rights historically has been the best way to protect environmental values, more so than relying on governments to do the same thing. Individuals with strong property rights to protect them against those who might cause them harm – governments included – will benefit by finding ways to use those rights effectively to protect themselves and their resources.

Many environmental policies erode property rights. When they do, they often work against the very environmental protection they are intended to provide. The unintended consequences can be dramatic, as illustrated in the case of the Endangered Species Act (Stroup, 1995; Chase, 1995). The intent of this law is to save species presumed to be in danger of extinction, yet only 31 of the approximately 1,800 species it monitors have recovered since the act was passed in 1973 (U.S. FWS, 2014). The law gives federal agents far-reaching powers to control landowners' use of their properties. Those powers have sometimes worked to protect endangered species, but often they have had the opposite effect.

A landowner who provides good habitat for an endangered species, even if by accident, is likely to face restrictions on his or her property rights. Michael Bean, an environmental defense attorney who is sometimes informally credited with authorship of the Endangered Species Act, explained this to a group that included Fish and Wildlife Service (FWS) officials. He said there is "increasing evidence that at least some private landowners are actively managing their lands so as to avoid potential endangered species problems." He emphasized these actions are "not the result of malice toward the environment" but "fairly rational decisions, motivated by a desire to avoid potentially significant economic restraints." He called them a "predictable response to the familiar perverse incentives that sometimes accompany regulatory programs, not just the endangered species program but others" (Bean, 1994).

The case of Benjamin Cone Jr. is a cautionary tale (Welch, 1994). Cone inherited 7,200 acres of land in Pender County, North Carolina. He managed the land primarily for wildlife. He planted chuffa and rye for wild turkey, for example, and the wild turkey made a comeback in Pender County partly due to his efforts. He frequently conducted controlled burns of the property to improve the habitat for quail and deer.

Red-cockaded woodpeckers are listed as an endangered species. They nest in the cavities of old trees and are attracted to places that have both old trees and a clear understory. By clearing the understory to protect quail and deer and by selectively cutting small amounts of timber, Cone may have helped attract the woodpecker to his property.

When Cone intended to sell some timber from his land, the presence of the birds was formally recorded by the U.S. Fish and Wildlife Service. The agency warned Cone not to cut trees or take any other actions that might disturb the birds. FWS did not, however, tell Cone where the nests were. Cone hired a wildlife biologist, who estimated there were 29 birds in 12 colonies. According to the FWS guidelines then in effect for the red-cockaded woodpecker, a circle with a half-mile radius had to be drawn around each colony, within which no timber could be harvested. If Cone harvested the timber, he would be subject to a severe fine, possible imprisonment, or both under ESA. Biologists estimated the presence of the birds put 1,560 acres of Cone's land under the restrictions of the Fish and Wildlife Service.

In response, Cone changed his management techniques. He began to clear-cut 300 to 500 acres every year on the rest of his land. He told an investigator, "I cannot afford to let those woodpeckers take over the rest of the property. I'm going to start massive clear-cutting. I'm going to a 40-year rotation instead of a 75- to 80-year rotation" (Sugg, 1993). By harvesting younger trees, Cone could keep the woodpecker from making new nests in old tree cavities. He also took steps to challenge FWS in court, asking to be compensated for his losses. The agency ultimately avoided that court challenge by negotiating a settlement giving Cone more freedom to use his land.

Cone's experience provides a warning to all landowners under similar circumstances. After Cone informed the owner of neighboring land about possible liabilities in connection with the red-cockaded woodpecker, he noticed the owner clear-cut the property (Welch, 1994). Overall, what has been the result of ESA for the red-cockaded woodpecker? As Bean (1994) has said, "The red-cockaded woodpecker is closer to extinction today than it was a quarter century ago when protection began." Bean recommends the rules be changed to help landowners avoid large reductions in the value of their land from

application of ESA, but no change in the law is currently in sight.

More recently Brian Seasholes, director of Reason Foundation's endangered species project, described how plans by the U.S. Department of the Interior (DOI) to use ESA to protect the greater sage grouse would have had just the opposite effect (Seasholes, 2015a). Although DOI eventually decided against designating the bird as an endangered species, it pursued instead a plan critics (including Seasholes) say would have essentially the same negative effects (Seasholes, 2015b). By using ESA to justify land-use controls that seriously erode the property rights of land owners, the Fish and Wildlife Service has ignored the important positive role that private landowners and institutions historically have played in protecting rare plants and animals.

Displacement is an issue in the climate change discussion for multiple reasons. First, investing in greenhouse gas mitigation efforts displaces investments in other efforts to meet better documented and more urgent needs. Much of the international aid being directed to climate change projects is simply aid that otherwise would have gone to other, presumably valuable, projects (Levi, 2015). Lomborg (2007) wrote of this problem, "This is the real moral problem of the global-warming argument – it means well, but by almost expropriating the public agenda, trying to address the hardest problem, with the highest price tag and the least chance of success, it leaves little space, attention, and money for smarter and more realistic solutions" (p. 123).

Second, seeking a top-down global response displaces more promising national, state, and local responses. Writing about air and water pollution controls before passage of national legislation in the United States, Yandle (1989) observed, "the absence of federal jurisdiction and federal money forced people in states and cities to deal directly with the problem of environmental scarcity. They had no other choice. *As a result, those closest to the problem and most sensitive to the costs resulting from their actions found innovative ways to deal with the problem.* Instead of uniform rules, which are clearly simpler to enforce, local bodies could tailor controls to meet local conditions" (p. 57, italics added). See also Anderson and Hill (1997) and Higgs and Close (2005) for the value of having multiple decentralized solutions to environmental problems.

Third, efforts already underway to mitigate emissions or encourage adaptation are displaced by more expensive and less effective national or international programs. A study of the Joint

Implementation (JI) program, part of the United Nations' Kyoto Protocol, by the Stockholm Environment Institute found "the use of JI offsets may have enabled global GHG emissions to be about 600 million tonnes of carbon dioxide equivalent higher than they would have been if countries had met their emissions domestically" (Kollmuss *et al.*, 2015). The authors analyzed 60 projects and found 73% of the carbon credits did not meet the requirement of "additionality," meaning the projects would have occurred without the added incentive of carbon credits.

More generally, paying to build higher dikes or "harden" infrastructure, or promising to compensate people for flooding or storm damage, discourages potential victims of climate change from taking actions on their own to avoid damages by voluntarily relocating, not investing in improvements in homes or businesses in vulnerable locations, or making their own plans to minimize damages. Because climate change is a slow-moving phenomenon, occurring over decades and centuries, gradual adaptation could be virtually costless as lifestyles and investment patterns change gradually.

Displacement is due to several economic phenomena described earlier in this chapter: moral hazard, unintended consequences, bureaucracy, and rent-seeking. *It is unavoidable* given the incentives faced by and information available to the entities involved. Yet calls for immediate action in response to climate change rarely if ever acknowledge the existence of this problem.

References

Anderson, T.L. and Hill, P.J. (Eds.) 1997. *Environmental Federalism*. Lanham, MD: Rowman & Littlefield Publishers, Inc.

Bean, M. 1994. Transcript of a talk by Michael Bean at a U.S. Fish and Wildlife Service seminar. Arlington, VA: Marymount University. November 3.

Brubaker, E. 1995. *Property Rights in the Defense of Nature*. London, UK: Earthscan Publications Ltd.; Toronto, ON: Earthscan Canada.

Chase, A. 1995. *In a Dark Wood: The Fight Over Forests and the Rising Tyranny of Ecology*. Boston, MA: Houghton Mifflin Company.

Higgs, R. and Close, C.P. (Eds.) 2005. *Re-thinking Green: Alternatives to Environmental Bureaucracy*. Oakland, CA: Independent Institute.

Kollmuss, A., Schneider, L., and Zhezherin, V. 2015. Has Joint Implementation reduced GHG emissions? Lessons learned for the design of carbon market mechanisms. *SEI Working Paper* No. 2015-07. Stockholm, Sweden: Stockholm Environment Institute.

Levi, M. 2015. What matters (and what doesn't) in the G7 climate declaration. Council on Foreign Relations (website). June 10.

Lomborg, B. 2007. *Cool It: The Skeptical Environmentalist's Guide to Global Warming*. New York, NY: Alfred A. Knopf.

Meiners, R., Thomas, S., and Yandle, B. 2000. Burning rivers, common law, and institutional choice for water quality. In: Meiners, R. and Morriss, A. (Eds.) *The Common Law and the Environment*. Lanham, MD: Rowman & Littlefield, pp. 54–85.

Seasholes, B. 2015a. Sage grouse conservation: The proven successful approach. *Policy Brief* No. 131. Los Angeles, CA: Reason Foundation. September.

Seasholes, B. 2015b. The best and worst parts of the decision not to list the sage grouse as endangered. Reason Foundation (website). September 23.

Stroup, R. 1995. The Endangered Species Act: making innocent species the enemy. *PERC Policy Series* PS-3. Bozeman, MT: PERC.

Sugg, I. 1993. Ecosystem Babbitt-babble. *The Wall Street Journal*. April 2, p. A12.

U.S. Fish & Wildlife Service. 2014. *Delisted Species*.

Welch, L. 1994. Property rights conflicts under the Endangered Species Act: protection of the red-cockaded woodpecker. *PERC Working Paper* No. 94-12. Bozeman, MT: PERC.

Yandle, B. 1989. The Political Limits of Environmental Regulation: Tracking the Unicorn. New York, NY: Quorum Books.

1.4.7 Leakage

"Leakage" occurs when the emissions reduced by a regulation are partially or entirely offset by changes in behavior.

In the academic literature on climate change, leakage refers to increases in carbon dioxide or other greenhouse gas emissions occurring outside a state or nation in response to that state or nation's adoption of emission caps or carbon taxes. Leakage can offset much or even all of a state or nation's emission reductions. Leakage can occur for at least four reasons (Niles, 2002):

- Programs that reduce emissions in some countries or industries reduce demand for fossil fuels, allowing businesses in other countries or industries to purchase those fuels at lower prices. This is called the "rebound effect."

- Businesses located in countries or states with lower energy prices and fewer regulations have cost advantages over those in countries and states with high energy prices and burdensome regulations. Consequently, capital migrates from countries and states that impose emission controls to those that do not (Becker and Henderson, 2000; Brunnermeier and Levinson, 2004; Levinson and Taylor, 2008; Hanna, 2010; Stevenson, 2018).

- Changes to behavior occur in response to changes in prices, offsetting some or all of the anticipated emission reduction. For example, higher Corporate Average Fuel Economy (CAFE) standards for new cars and trucks sold in the United States result in people driving more miles and holding onto their older cars longer.

- "Ecological leakage" occurs when secondary and tertiary effects of an effort to reduce emissions produce new emissions that reduce or even entirely cancel out the first round of reductions. For example, production of ethanol from corn resulted in a net *increase* in greenhouse gas emissions due to the energy used to grow the corn and changes in land use prompted by subsidies to producers (Searchinger *et al.*, 2008).

Every study of the Kyoto Protocol, which exempted developing countries from obligations to reduce greenhouse gas emissions, forecast significant leakage. "The imposition of increased energy costs will devastate the U.S. steel industry without a significant decrease in worldwide energy-related emissions from steel making," concluded a study by the Argonne National Laboratory (U.S. Department of Energy, 1997). "Production will simply be shifted to developing countries and may lead to higher levels of overall pollution due to lower standards in those countries" (*Ibid.*). According to a study by WEFA, a consulting firm, 41% of the loss in U.S. GDP due to

the Kyoto Protocol would have come from lost exports and increased imports from developing countries (Novak *et al.*, 1998, p. 30). IPCC (2007) estimated "carbon leakage rates for action under Kyoto range from 5 to 20% as a result of a loss of price competitiveness, but they remain very uncertain." The shift of manufacturing from developed countries such as the United States to developing countries such as China is increasing leakage.

Driven by this global economic transformation, developing countries are dramatically increasing their use of fossil fuels and consequently their share of global greenhouse gas emissions. According to the Netherlands Environmental Assessment Agency, "since 1990, in the EU27, CO_2 emissions decreased from 9.1 to 7.4 tonnes per capita, and in the United States from 19.6 to 16.4 tonnes per capita, they increased in China from 2.1 to 7.1. As such, Chinese citizens, together representing 20% of the world population, on average emitted about the same amount of CO_2 per capita in 2012 as the average European citizen" (Oliver *et al.*, 2013, p. 15).

Not all of the growth in greenhouse gas emissions in developing countries is due to leakage from developed countries. Energy consumption is closely linked to economic growth in both developed and developing countries, and fossil fuels are the least expensive and most reliable source of power for home and commercial applications (Bradley and Fulmer, 2004; BP, 2014; EIA, 2014; Bezdek, 2015). Greenhouse gas emissions therefore will rise if developing countries are successful in raising their populations out of poverty. Global warming policies are among many tax and regulatory policies encouraging investment and manufacturing in developing countries.

Different types of climate programs experience different amounts of leakage. Policies raising costs for energy-intensive industries are likely to have high levels of leakage by driving customers and investors to other countries or industries with lower energy costs. Policies focusing on utilities and with long time frames – giving utilities time to replace older generating capacity with newer, lower-emitting capacity – would have less leakage, since utility customers are relatively immobile.

Imposing higher fuel economy standards on cars and trucks is a good example of leakage (National Research Council, 2001; Lutter and Kravitz, 2003). Mandating higher fuel economy for cars may not reduce the total amount of carbon dioxide emitted if consumers use the fuel savings to drive more miles or drive alone more often. Both changes in behavior occur and historically have cancelled out 20% or more of the fuel savings that might have arisen from U.S. CAFE standards. Summarizing their empirical analysis of gasoline prices and vehicle miles traveled, economists John W. Mayo and John E. Mathis (1988) wrote, "CAFE standards had no independent, statistically significant impact on ... the demand for gasoline."

Federal CAFE standards require car and truck manufacturers to meet national fleet-wide standards for cars and light trucks or pay fines. If one state insists on better fuel economy for the fleet of cars and trucks sold within its borders, manufacturers will oblige by selling only smaller, lighter, and less powerful vehicles in that state, and then sell larger, heavier, and more powerful vehicles in other states, bringing their national corporate average fuel economy back to where it was before the state adopted its standards. Consequently, leakage will simply cancel out whatever emission reductions the stricter state is seeking to make.

Estimates of leakage for national greenhouse gas reduction programs range from 12% (Brown, 1999) to 130% (Babiker, 2005). In other words, reducing carbon emissions by 10 metric tonnes would cause emissions by other countries or states to *increase* between 1.2 and 13 tons. *A net* reduction of 10 tons assuming the lower of the two estimates would require a reduction by the first country or state of 11.4 tons. The second estimate means no reductions by the first country, no matter how high, will lead to a *net reduction* in global emissions since emissions in other countries rise faster than reductions in the first country. In the decades since greenhouse gas reduction programs have been implemented, a large body of research has been created estimating leakage rates by industry, by type of program, and by country (Fischer *et al.,* 2010). While many efforts have been made to discourage leakage, some of them partially successful, it appears to remain an unavoidable part of emission control programs.

Leakage is a classic example of an unintended consequence of government actions, something economists know to look for. Estimates of the effectiveness of greenhouse gas reduction initiatives that don't take leakage into account will over-estimate the benefits of the programs, leading to inaccurate cost-benefit analysis results.

References

Babiker, M.H. 2005. Climate change policy, market structure, and carbon leakage. *Journal of International Economics* **65**: 421.

Becker, R. and Henderson, V. 2000. Effects of air quality regulation on polluting industries. *Journal of Political Economy* **108** (2): 379–421.

Bezdek, R.H. 2015. Economic and social implications of potential UN Paris 2015 global GHG reduction mandates. Oakton, VA: Management Information Services, Inc.

BP. 2014. *BP Energy Outlook 2035.* London, UK: BP p.l.c. January.

Bradley Jr., R.L. and Fulmer, R.W. 2004. *Energy: The Master Resource.* Dubuque, IA: Kendall/Hunt Publishing Company.

Brown, S.P.A. 1999. Global warming policy: Some economic implications. *Policy Report #224.* Dallas, TX: National Center for Policy Analysis.

Brunnermeier, S.B. and Levinson, A. 2004. Examining the evidence on environmental regulations and industry location. *The Journal of Environment and Development* **13** (1): 6–41.

EIA. 2014. U.S. Energy Information Administration. *International Energy Outlook 2014, With Projections to 2040.* September.

Fischer, C., Moore, E., Morgenstern, R., and Arimura, T. 2010. *Carbon policies, competitiveness, and emissions leakage: An international perspective.* Washington, DC: Resources for the Future.

Hanna, R. 2010. U.S. environmental regulation and FDI: evidence from a panel of U.S.-based multinational firms. *American Economic Journal: Applied Economics* **2** (3): 158–89.

IPCC. 2007. Intergovernmental Panel on Climate Change. *Climate Change 2007: Working Group III: Mitigation of Climate Change,* Executive Summary. Intergovernmental Panel on Climate Change. Cambridge, UK: Cambridge University Press.

Levinson, A. and Taylor, M.S. 2008. Unmasking the pollution haven effect. *International Economic Review* **49** (1): 223–54.

Lutter, R. and Kravitz, T. 2003. Do regulations requiring light trucks to be more fuel efficient make economic sense? An evaluation of NHTSA's proposed standards. *Regulatory Analysis* 03-2. Washington, DC: AEI-Brookings Joint Center for Regulatory Studies.

Mayo, J.W. and Mathis, J.E. 1988. The effectiveness of mandatory fuel efficiency standards in reducing the demand for gasoline. *Applied Economics* **20**: 211–9.

National Research Council. 2001. *Effectiveness and Impact of Corporate Average Fuel Economy (CAFE) Standards*. Transportation Research Board. Washington, DC: National Academies Press.

Niles, J.O. 2002. Chapter 13. Tropical forests and climate change. In: Schneider, S.H., Rosencranz, A., and Niles, J.O. (Eds.) *Climate Change Policy: A Survey*. Washington, DC: Island Press, pp. 337–71.

Novak, M., *et al.* 1998. *Global Warming: The High Costs of the Kyoto Protocol*. Philadelphia, PA: WEFA Inc.

Oliver, J.G.J., Janssens-Maenhout, G., Muntean, M., and Peters, J.A.H.W. 2013. *Trends in Global CO2 Emissions; 2013 Report*. The Hague, Netherlands: PBL Netherlands Environmental Assessment Agency; European Union Joint Research Centre.

Searchinger, T., *et al.* 2008. Use of U.S. croplands for biofuels increases greenhouse gases through emissions from land-use change. *Science* **319** (5867): 1238–40.

Stevenson, D.T. 2018. A review of the Regional Greenhouse Gas Initiative. *Cato Journal* **38** (1): 203–23.

U.S. Department of Energy. 1997. *The Impact of Potential Climate Change Commitments on Energy Intensive Industries: A Delphi Analysis*. Argonne, IL: Argonne National Laboratory.

1.5 Future Generations

Previous sections of this chapter showed how markets turn self-interested behavior into behavior that benefits others. In this section, we show how markets create incentives to conserve and protect natural resources for future generations and help ensure the best solutions are the ones adopted.

1.5.1 Conservation and Protection

Capital markets create information, signals, and incentives to manage assets for long-term value.

People sometimes assume private owners have little incentive to protect resources for the future, that they are quite willing to destroy long-term value to realize short-term gain. This line of reasoning suggests only governments can truly preserve a natural resource because the government, unlike the private sector, plans for the long run. This common assumption, however, is largely false (Goklany and Sprague, 1992; Meiners, 1995; Rosegrant *et al.*, 1995; Taylor, 1997; Norton, 1998; Smith, 1999, 2000; Anderson and Leal, 2001; Logomasini and Smith, 2011).

The prices of land and other assets today reflect the future benefits owners expect to receive. In economists' language, today's price is the capitalized value of the future stream of benefits, net of the costs required to protect or produce those benefits.

Just as prices convey information about changing demand and supply all over the globe, a capital market – the buying and selling of capital assets such as land, buildings, bonds, or corporate stock certificates – conveys information about the expected demands, desires, and preferences of people in the future. People who believe a resource will increase in value – that people in the future will value it more highly than people today – can hope to profit by buying it, preserving it, and selling it at a higher price later. Even a shortsighted owner who is personally concerned only with the present will respond to these signals because they change the current value of his or her assets. Of course, the owner can ignore the price signals, but then he or she must deal with the resulting reduction in wealth.

The future value of a resource influences the behavior of its owner. A land owner, for example, will do what the market demands in order to maintain the land's productivity and, where possible, to make investments that improve it. If the land is damaged, its value declines whether the damage occurs through misuse, negligence, trespass, or pollution. If necessary, an owner will go to court against trespassers or polluters to protect the value of property.

Millions of private investors are highly motivated to monitor the performance of private asset managers. When investors in a company's stock view a management decision as a good one, they keep their stock or buy more anticipating the value of the firm will rise. If many investors begin to think this way, their decisions lead the stock's price to rise, increasing shareholder wealth. Similarly, poor decisions lead shareholders to sell the stock and the price tends to fall. Management responds to these capital market signals since they are typically compensated partly with stock options. Managers who fail to keep stock prices stable or rising are likely to be replaced by disappointed stockholders.

The incentive to look to the future is clear for conventional sources of income such as agricultural crops or housing developments, but it also holds true for assets of an environmental nature. Wilderness areas, open spaces, scenic locations, shorelines, and other areas are economic assets that can be and are managed for profit and higher future resale value. In Section 1.2.5 we described a case where rights to water are treated as assets and bought and sold. That market system rewards good stewardship and efficiency.

Consider another example. After a successful and innovative career, television magnate Ted Turner began buying ranches in the West and Southwest (Anderson and Leal, 1997, pp. 4–8; Gunther, 2006). On the Flying D Ranch, south of Bozeman, Montana, he decided not to raise traditional livestock but instead to manage the ranch largely for bison and elk. He decided to increase the number of trophy animals over time. In 2006 he was charging a small number of elk hunters an annual fee of about $12,000 each to spend a week on the ranch trying to shoot a trophy elk. At that time, the ranch was earning roughly $300,000 in additional revenue per year. This added revenue stream raised the resale value of the ranch. It also drove Turner to manage the ranch in a way that is desirable for hunters and encourages the proliferation of diverse wildlife, not just elk, deer, and bison. Admittedly, Turner's motivation was not financial profit; he could afford to lose money and probably could earn more by subdividing and selling the ranch. But the Flying D Ranch and other examples of environmental assets privately managed and earning revenue show how markets allow individuals to achieve their own goals – to "maximize their utility," as economists say – in ways that benefit future generations.

In contrast to private landowners and asset managers, government asset managers receive few if any signals from capital markets. Their property is not for sale and they will not reap the benefits of investments that might improve its long-term value. Government managers are motivated to produce glossy brochures and annual reports highlighting the natural beauty of their latest acquisitions but not to report shortcomings in services and maintenance of parks, wildlands, and other assets already in their possession. Environmental groups realize lack of maintenance of existing parks is a strong argument against acquiring more parkland, so they too are silent on the issue.

The elected officials who oversee the bureaucracies created to manage public assets have strong incentives to promise short-term benefits, such as more recreational opportunities, fewer forest fires,

or more logging on public lands to satisfy well-organized interest groups that make campaign contributions and turn out the vote. But future generations don't vote in the next election, so politicians are free to disregard their interests.

Elected officials often say they care deeply about future generations, since this presumably is what voters want to hear. Perhaps some do. But no one is able to hold them accountable for actually fulfilling their promises. Unlike investors in the private sector, few voters have a financial incentive to monitor the performance of government agencies. Because the assets can't be sold, no one benefits directly from knowing about management changes, so voters choose to remain rationally ignorant. If they vote, they may vote for a candidate on the basis of positions he or she takes on many other issues, or on the basis of misinformation circulated by interest groups.

When an owner or manager in the private sector improves or damages the future value of a natural resource, capital markets change the resource's current price and communicate that change to investors and entrepreneurs in the form of profits and losses, which then motivates decision-makers to take actions that encourage good long-term asset management. A rancher in Montana, for example, can recognize the higher value people are placing on hunting and recreation and profit from it by dedicating some or all of his land to wildlife. Government agencies operate without such information and without the system of rewards and penalties, and so are unlikely to make wise or efficient decisions about managing assets for future generations.

The preceding analysis is relevant to the climate change issue because political entities such as the United Nations or U.S. government should not be assumed to be better stewards of the environment than private parties motivated by profit or by charitable goals. The concerns of future generations are no better protected by politicians and voters today than they are by private asset managers and investors, and probably less so. The best responses to climate change are probably found in the private sector and not in the public sector.

References

Anderson, T.L. and Leal, D.R. 1997. *Enviro-Capitalists: Doing Good While Doing Well*. Lanham, MD: Rowman & Littlefield.

Anderson, T.L. and Leal, D.R. 2001. *Free Market Environmentalism. Revised edition.* Lanham, MD: Rowman & Littlefield.

Goklany, I.M. and Sprague, M.W. 1992. *An Alternative Approach to Sustainable Development.* Washington, DC: Office of Program Development, U.S. Department of Interior.

Gunther, M. 2006. Ted Turner's Montana adventure. *Forbes.* October 4.

Logomasini, A. and Smith, R.J. 2011. Protect endangered species. In: *Liberate to Stimulate: A Bipartisan Agenda to Restore Limited Government and Revive America's Economy.* Washington, DC: Competitive Enterprise Institute.

Meiners, R.E. 1995. Elements of property rights: the common law alternative. In: Yandle, B. (Ed.) *Land Rights: The 1990s Property Rights Rebellion.* Lanham, MD: Rowman & Littlefield, pp. 269–94.

Norton, S.W. 1998. Poverty, property rights, and human well-being: a cross-national study. *Cato Journal* **18** (2): 223–45.

Rosegrant, M.W., Schleyer, M.G., and Yaday, S.N. 1995. Water policy for efficient agricultural diversification: market-based approaches. *Food Policy* **20**: 203–23.

Smith, R.J. 1999. Hawk Mountain Sanctuary Association. *Private Conservation Case Study.* Washington, DC: Competitive Enterprise Institute.

Smith, R.J. 2000. Cypress Bay Plantation, Cummings, South Carolina. *Private Conservation Case Study.* Washington, DC: Competitive Enterprise Institute.

Taylor, C. 1997. The challenge of African elephant conservation. *Conservation Issues* **4** (2): 1, 3–11.

1.5.2 Innovation

Markets reward innovations that protect the environment by using less energy and fewer raw materials per unit of output.

Markets reward innovation, and innovation in turn benefits the environment. The best protection of the atmosphere rests in ensuring that technological innovations continue to increase humanity's ability to meet its material needs without further reducing the land available to wildlife or contaminating the planet's air and water. This is the message of Chapters 3, 4, and 5, but it is introduced briefly here to help explain why economic growth today helps future generations.

Over the past century, new technologies have led to less pollution and to the use of fewer raw materials per unit of output (Simon, 1995; Huber, 1999; Goklany, 2007, 2009; Bryce, 2014; Smil, 2016). This has been true for everything from steel mills (once fiery behemoths belching smoke but now relatively clean, with many using scrap steel as their raw materials) to aluminum cans (which over time have been engineered to become ever thinner and lighter). New technologies have reduced the amount of energy required to produce a dollar of real gross domestic product (GDP) in the United States by two-thirds since 1949. (See Figure 1.5.2.1.)

Innovation is essential to progress, but it means change, and change is always difficult. Choosing to continue doing something the way it has always been done is usually easier than change. Markets reward with profits the creators of innovations that help people meet their goals at lower costs, and penalizes with losses innovations that people don't want or that waste resources (Baumol, 2002).

To have an incentive to innovate, an inventor or entrepreneur must be able to benefit personally from his or her achievement. This incentive comes through private ownership. The owner of a new product or the investors who help him bring it to market can earn large returns in a short period of time by licensing others to use the new product. License-holders in turn can earn larger-than-before returns by using the new product to lower their cost of production or better meet their customers' needs.

The pace of innovation in countries without private property rights is slow, as could be seen by the socialist economies of Eastern Europe before the fall of communism in 1989. The Trabant automobile, produced in East Germany between 1959 and 1989, is a good example. An American auto magazine, *Car and Driver*, brought the Trabant over for a look in 1990 (Ceppos, 1990). On the positive side, the editors reported the car provided basic transportation and was easy to fix (similar things were said about the Model T Fords in the early twentieth century). But the Trabant's top speed was 66 miles per hour, it was noisy, and, the editors said, it had "no discernible handling." It spewed "a plume of oil and gray exhaust smoke" and didn't have a gas gauge. In fact, the Trabant's exhaust was so noxious the Environmental Protection Agency refused to let *Car and Driver* staff drive it on public streets.

Figure 1.5.2.1
U.S. primary energy consumption per real dollar of gross domestic product, 1949–2017

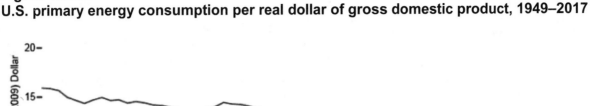

Source: EIA, 2018, Figure 1.7, p. 16.

The Trabant was backward, dirty, and inefficient because its design was the same as when it was first manufactured in 1959. The last model had been introduced in 1964 and since market pressures were absent, the automobile experienced no technological change since then. Cars are much cleaner and still improving today because of market innovation. Today's cars emit a tiny fraction of the pollution emitted by the cars of the early 1970s (Schwartz, 2006; O'Toole, 2012). And while even electric cars require energy from burning fuel in power plants, the emissions from such plants have fallen dramatically, too, as owners have searched out low-sulfur coal and technical devices to reduce pollution. Advances in technology continue to make cars cleaner and safer, just as diesel train engines replaced dirty steam locomotives, and gas and electricity replaced coal for home heating. This story is told in greater detail in Chapters 3, 5, and 6.

Technological change is expected to continue to reduce the energy intensity of the global economy in coming decades, partially offsetting the dramatic rise in demand for energy due to global population growth and rising prosperity (BP, 2014; EIA, 2014; Bezdek, 2015). The environmental consequences of a growing global population would be far worse without innovation, as forests would need to be converted to cropland and emissions of all kinds, not just greenhouse gases, would grow in pace.

The institutions that encourage innovation – property rights and markets – and the freedom and prosperity they make possible must remain in place for future generations to enjoy the safe and clean environment we enjoy today. This is not always made clear in the plans put forward by environmental activists who seem to believe capitalism and protecting the planet's atmosphere are incompatible (Gore, 2006; Klein, 2014).

References

Baumol, W.J. 2002. *The Free-Market Innovation Machine: Analyzing the Growth Miracle of Capitalism.* Princeton, NJ: Princeton University Press.

Bezdek, R.H. 2015. Economic and social implications of potential UN Paris 2015 global GHG reduction mandates. Oakton, VA: Management Information Services, Inc.

BP. 2014. *BP Energy Outlook 2035.* London, UK: BP p.l.c. January.

Bryce, R. 2014. *Smaller Faster Lighter Denser Cheaper: How Innovation Keeps Proving the Catastrophists Wrong.* New York, NY: PublicAffairs.

Ceppos, R. 1990. The car who came in from the cold. *Car and Driver.* December, pp. 89–97.

EIA. 2014. U.S. Energy Information Administration. *International Energy Outlook 2014, With Projections to 2040.* September.

EIA. 2018. U.S. Energy Information Administration. *Monthly Energy Review*. July.

Goklany, I.M. 2007. *The Improving State of the World: Why We're Living Longer, Healthier, More Comfortable Lives on a Cleaner Planet*. Washington, DC: Cato Institute.

Goklany, I.M. 2009. Have increases in population, affluence and technology worsened human and environmental well-being? *The Electronic Journal of Sustainable Development* **1** (3): 3–28.

Gore, A. 2006. *An Inconvenient Truth: The Planetary Emergency of Global Warming and What We Can Do About It*. Emmaus, PA: Rodale Inc.

Huber, P. 1999. *Hard Green: Saving the Environment from the Environmentalists, a Conservative Manifesto*. New York, NY: Basic Books.

Klein, N. 2014. *This Changes Everything: Capitalism vs. The Climate*. New York, NY: Simon & Schuster.

O'Toole, R. 2012. Which is better for the environment: transit or roads? *Issue Brief #111*. Dallas, TX: National Center for Policy Analysis. July 13.

Schwartz, J. 2006. Assessing California vs. federal automobile emissions standards for Pennsylvania. Testimony to the Pennsylvania House of Representatives Committee on Environmental Resources and Energy. February 8.

Simon, J.L. (Ed.) 1995. *The State of Humanity*. Oxford, UK: Blackwell Publishers Inc.

Smil, V. 2016. *Power Density: A Key to Understanding Energy Sources and Uses*. Cambridge, MA: The MIT Press.

1.5.3 Small versus Big Mistakes

Mistakes made in markets tend to be small and self-correcting. Mistakes made by governments tend to be big and more likely to have catastrophic effects.

Free markets are spontaneous orders, self-correcting systems in which many small mistakes are made and quickly corrected. Governments are deliberately created planned institutions whose monopoly on the legitimate use of force allows them to impose the costs and consequences of mistakes on others, sometimes with catastrophic effects (Hayek, 1973, 1976, 1979; Butos and McQuaid, 2001; Hasnas, 2005).

In a market system, inventors and entrepreneurs continuously come up with new products and introduce them to customers, who reject many of them. Most of the businesses launched to sell new goods and services quickly go out of business. In the United States, only about half of newly incorporated businesses survive for five years and only a third survive 10 years or longer (Shane, 2012). But the inventions that do work, the products that do sell, and the businesses that do survive provide the change that transforms the economy and increases wealth over time. Schumpeter (1942) called this the "gale of creative destruction," which he described as the "process of industrial mutation that incessantly revolutionizes the economic structure from within, incessantly destroying the old one, incessantly creating a new one" (pp. 82–3).

Change occurs rapidly in a market system because individuals don't need consensus or majority approval to pursue their ideas as they would if operating under a democratic political process. They are free to invest their own money in new ideas and test them in the marketplace. Successful innovators earn temporary profits, while others must adopt the innovations that work in order to survive in business. The system tolerates many small mistakes in exchange for tapping the wisdom, energy, and aspirations of anyone with an original idea and willingness to work hard.

History is replete with examples of people who have challenged conventional wisdom and produced enormous social benefits. In the 1970s, it looked as though computers would be ever increasing in size and complexity, but a few hobbyists had a different idea. Some innovators put together a crude computer and began selling it as an assemble-it-yourself kit through *Popular Science* magazine. They created the first personal computer, revolutionizing the future of computers and to a large extent changing the way people conduct business and leisure activities.

Such innovations occur in the environmental realm as well, often long before politicians embrace the need for change. The Hawk Mountain Sanctuary in eastern Pennsylvania is a good example of such an innovation (Smith, 1984, 1990, 1999; Anderson and Leal, 1997, pp. 44–6; Furmansky, 2009). Hawk Mountain is a mountain ridge in Pennsylvania that lies along a natural migration route for hawks. In the early 1930s, hunters came to Hawk Mountain from miles around to shoot hawks. At the time, not only was hunting hawks popular, but the biological experts thought hawks and other predatory birds were undesirable and not worth preserving. In fact, the state paid a bounty to those who killed a certain kind of hawk.

Rosalie Edge, a conservationist and activist, opposed the wholesale slaughter of hawks. She tried to convince biologists, state officials, and leaders of the National Audubon Society that hawks have a rightful place in nature and should not be eradicated. Her efforts at persuasion failed, so she took another tack. In 1934, she and some friends came up with enough money to buy an option on Hawk Mountain, and later they bought the mountain. She created a sanctuary for the hawks, forbidding hunts there. Today the 2,000-acre reserve is a prime bird-watching location.

Edge's view that hawks have an important place in nature is now conventional wisdom, but it was radical 84 years ago. Actions taken by far-sighted individuals like Edge can be crucial for environmental protection, since by the time a political majority might be ready to act to save a species, it may be too late. Only because Hawk Mountain was privately owned could Edge exercise her vision of wildlife protection.

The private nonprofit sector historically has been a key component of conservation efforts. Starting late in the nineteenth century, for example, the National Audubon Society was formed to save birds like the snowy egret, which was endangered because women's hats were decorated with egret plumes. In addition to campaigning against wearing such feathers and trying to change some laws, the Audubon Society began to purchase or accept by donation natural areas that would become wildlife preserves. In 2013, Audubon had 44 nature centers, 23 sanctuaries, and 118 million acres of land under conservation (Audubon, 2013).

Mistakes made by private investors and philanthropists mainly affect the actors themselves and only a few others. The mistakes are self-correcting as failed innovations end when their private funding runs out and customers fail to appear, and their failures generate the information needed for later successes. Mistakes made by governments are different in each of these ways. The people affected are often orders of magnitude more than the investors and consumers affected by even a big business's failure. Governments can hide mistakes from public view for many years, and the regulations and subsidies keeping them afloat can send distorted signals to investors and consumers preventing better products or services from being discovered and commercialized. Contemporary examples in the United States of big mistakes by governments include the U.S. Forest Service's policy for many years of suppressing all forest fires (leading to increasingly dangerous wildfires (DeVore, 2018)), fuel economy standards for cars and trucks that result in thousands of highway fatalities every year, and subsidies for ethanol that cost drivers and taxpayers billions of dollars each year but do nothing to benefit the environment.

An infamous example of a government mistake is Lysenkoism, a theory of genetics named after Trofim Lysenko, director of the Soviet Union's Lenin All-Union Academy of Agricultural Sciences (Zubrin, 2013; Ferrara, 2013). Lysenko's theory that plants and animals can pass on to offspring characteristics acquired during their lifetimes was prominent in the Soviet Union of the 1930s. Lysenko rose to power by creating the appearance of being a problem solver, not because he was a highly regarded scientist. Joseph Stalin hailed his pseudoscientific theory because it seemed consistent with the Communist dogma that human nature could be changed by experience.

Once in power, Lysenko used his position to systematically remove from government anyone who challenged his preferred theory, even to the point of ordering the exile and execution of scientists who disagreed with him. Lysenko didn't tolerate disagreement because he didn't need to. He was narrowly focused on what he thought was right, and often that was consistent with advancing his own career. He was given power to suppress dissent and forbid experiments that would have revealed the flaws in his theory.

Lysenko's flawed beliefs contributed to crop failures in the Soviet Union and may have caused millions of deaths. But because it was endorsed by the Communist Party and backed by government force, Lysenkoism remained the official theory of crop genetics in the Soviet Union until the 1960s.

While Lysenkoism may be an extreme example, the growing influence of governments over science is a widely recognized danger (Lindzen, 2012; Curry, 2017). U.S. President Dwight Eisenhower in his Farewell Address of January 17, 1961, warned "against the acquisition of unwarranted influence, whether sought or unsought, by the military-industrial complex. The potential for the disastrous rise of misplaced power exists and will persist." Importantly, Eisenhower went on to say:

The free university, historically the fountainhead of free ideas and scientific discovery, has experienced a revolution in the conduct of research. Partly because of the huge costs involved, a government contract becomes virtually a substitute for intellectual curiosity …

Yet, in holding scientific research and discovery in respect, as we should, we must also be alert to the equal and opposite danger that public policy could itself become the captive of a scientific-technological elite. The prospect of domination of the nation's scholars by Federal employment, project allocations, and the power of money is ever present – and is gravely to be regarded.

Eisenhower's warning seems especially germane to climate science today. The scientific debate about climate change is dominated by government institutions, most notably the Intergovernmental Panel on Climate Change (IPCC), and in the United States by government agencies such as the Environmental Protection Agency (EPA), National Oceanic and Atmospheric Administration (NOAA), and National Aeronautics and Space Administration (NASA). During the Obama administration, all three U.S. agencies took positions on the climate change issue that supported the president's calls for immediate action. Scott Pruitt, the first EPA administrator following the Obama administration, said publicly that Obama had "weaponized" the agency to advance his climate change agenda (Bluey, 2018).

Nearly all of the scientists calling for immediate action to slow or prevent catastrophic climate change debate are government employees or depend on government grants to support their academic careers (Nova, 2009). Many of the "skeptics" are in the private sector, emeritus professors no longer needing grant dollars, or independent scientists with no financial motive to take one side or the other. The possibility that public financing influences the views and public statements of climate change advocates is readily apparent to economists and others who have studied the close-knit climate science community (e.g., Wegman *et al.,* 2006). But efforts to discuss this possible conflict of interest are called "assaults on climate science" by spokespersons for the government science establishment (Gleick *et al.,* 2010, pp. 689–90).

Is concern over anthropogenic climate change the latest Big Mistake by governments around the world? The rest of this book presents extensive evidence that it is. One scientist or a small group of scientists speaking their mind on a controversial issue is unlikely to cause much harm and is to be welcomed. But when the rules of political competition spill over into a scientific controversy and science becomes politicized, the damage to both science and public policy can be huge. Policymakers are well advised to look outside government agencies and beyond government-funded academics to get an accurate presentation of the state of climate science.

References

Anderson, T.L. and Leal, D.R. 1997. *Enviro-Capitalists: Doing Good While Doing Well.* Lanham, MD: Rowman & Littlefield Publishers.

Audubon. 2013. Audubon by the numbers: December 2013 update.

Bluey, R. 2018. The weaponization of the EPA is over: an exclusive interview with Scott Pruitt. The Daily Signal (website). February 25. Accessed July 31, 2018.

Butos, W. and McQuaid, T. 2001. Mind, market, and institutions: the knowledge problem in Hayek's thought. In: Birner, J. and Aimar, T. (Eds.) *The Economics and Social Thought of F.A. Hayek.* London, UK: Routledge, pp. 113–33.

Curry, J.A. 2017. Statement to the Committee on Science, Space, and Technology of the United States House of Representatives. Hearing on climate science: assumptions, policy implications and the scientific method. March 29.

DeVore, C. 2018. California's devastating fires are man-caused – but not in the way they tell us. Forbes (website). July 30.

Eisenhower, D. 1961. Military-industrial complex speech. The Avalon Project (website). New Haven, CT: Yale Law School.

Ferrara, P. 2013. The disgraceful episode of Lysenkoism brings us global warming theory. Forbes (website). April 28.

Furmansky, D.Z. 2009. *Rosalie Edge, Hawk of Mercy: The Activist Who Saved Nature from the Conservationists.* Athens, GA: University of Georgia Press.

Gleick, P.H. *et al.* 2010. Climate change and the integrity of science. Letter. *Science* **328** (May 7).

Hasnas, J. 2005. Hayek, the common law, and fluid drive. *New York University Journal of Law and Liberty* **1** (0): 79–110.

Hayek, F. 1973, 1976, 1979. *Law, Legislation, and Liberty, Vol. 1: Rules and Order* (1973), *Vol. 2: The Mirage of Social Justice* (1976), *Vol. 3: The Political Order of a Free People* (1979). Chicago, IL: University of Chicago Press.

Lindzen, R.S. 2012. Climate science: is it currently designed to answer questions? *Euresis Journal* **2** (Winter): 161–92.

Nova, J. 2009. *Climate Money: The Climate Industry, $79 billion so far – trillions to come*. Haymarket, VA: Science and Public Policy Institute.

Schumpeter, J.A. 1942. *Capitalism, Socialism and Democracy*. London, UK: Routledge.

Shane, S. 2012. Start up failure rates: the definitive numbers. *Small Business Trends* (website). December 12. Citing Census Bureau and Bureau of Labor Statistics (BLS) data.

Smith, R.J. 1984. Special report: The public benefits of private conservation. In: *Environmental Quality, 15th Annual Report of the Council on Environmental Quality*. Washington, DC: Council on Environmental Quality, pp. 363–429.

Smith, R.J. 1990. Private solutions to conservation problems. In: Cowen, T. (Ed.) *The Theory of Market Failure*. Fairfax, VA: George Mason University Press, pp. 341–60.

Smith, R.J. 1999. Hawk Mountain Sanctuary Association. *Private Conservation Case Study*. Washington, DC: Competitive Enterprise Institute.

Wegman, E., Said, Y.H., and Scott, D.W. 2006. *Ad Hoc Committee Report On The 'Hockey Stick' Global Climate Reconstruction*. Washington, DC: U.S. House Committee on Energy and Commerce.

Zubrin, R. 2013. *Merchants of Despair: Radical Environmentalists, Criminal Pseudo-Scientists, and the Fatal Cult of Antihumanism*. Washington, DC: Encounter Books.

1.6 Conclusion

The best responses to climate change are likely to arise from voluntary cooperation mediated by nongovernmental entities using knowledge of local costs and opportunities.

Economics explains how property rights, prices, profits and losses, and exchange lead to the efficient allocation of scarce resources. It reveals how access to common-pool resources – such as wilderness areas, grazing areas, and the atmosphere – can be efficiently managed so long as private property rights are defined and enforced and people are free to negotiate terms. Many institutions have evolved to facilitate such negotiation. Markets produce the prosperity needed to make voluntary environmental protection a social value today, ensuring the necessary resources will be made available to preserve and protect Earth's atmosphere now and for future generations.

Climate change is not a problem to be solved by markets or by government intervention. It is a complex phenomenon involving choices made by millions or even billions of people producing countless externalities both positive and negative. The benefits created by the use of fossil fuels, alleged to be the cause of climate change, have been huge and are well documented; the costs attributed to climate change are less certain but, as will be documented in Chapter 8, are known to be orders of magnitude smaller than the benefits from using fossil fuels.

Climate change presents an opportunity to use the wealth created by fossil fuels today to support an environmental movement based on sound science to study the causes and consequences of climate change and find the responses (plural, because there are likely to be more than one) that maximize private as well as social benefits. The best responses cannot be found in a laboratory by physicists, biologists, or geologists, no matter how brilliant they might be. They must be found in the real world of human action: either in the private sector where decisions are made based on prices and incentives and value is created by trading goods and services; or in the public sector where governments may force compliance with laws shaped by politics and implemented by bureaucracies.

The market approach to climate change involves treating people as shoppers rather than voters. This means allowing them to conduct their own private cost-benefit analyses and then use their local knowledge to discover and craft the best local responses to a global phenomenon. This approach can be called simply "energy freedom."

The freedom-based approach to protecting commonly owned resources is to find win-win solutions even when conditions might otherwise allow some people to over-use the resource and harm others. People value and are willing to pay for environmental amenities, meaning there are markets for finding such solutions. Institutions already exist that can lower the transaction costs that might otherwise stand in the way of such solutions. Since the future value of assets impacts their current prices, private ownership of assets creates incentives for conservation and protection that benefit future generations.

Economics suggests that governments have an important but very limited role to play in environmental protection. They help mainly by recognizing, defining, and enforcing property rights and prosecuting fraud and other criminal acts. Governments historically have done a poor job regulating environmental risks and owning and managing resources such as wilderness areas. Government regulation and ownership often fail to achieve their objectives due to conflicting incentives of individuals in governments (moral hazard), capture of regulatory agencies by special interests, and their inability to collect reliable information or achieve local knowledge.

Efforts to protect Earth's atmosphere by limiting energy freedom and instead empowering governments to restrict the use of fossil fuels or ban them outright fail to work in practice. They erode the protection of private property rights, reducing the incentive and ability of owners to protect and conserve their resources. Taxes, subsidies, and regulations distort the signals sent by prices and profits and losses, resulting in inefficient use of resources. Even when government programs seem to succeed, they often displace rather than improve or add to private environmental protection. Government's ability to promote the goals of some citizens at the expense of others also leads to resources being diverted from production of valuable goods and services into political action (rent seeking) and often outright corruption.

Asking the general public to vote on what to do about climate change is not likely to lead to the most efficient responses. Even voters who are intelligent and well-intentioned often choose to remain ignorant about the issues being voted on by their elected representatives. They realize their individual votes for a candidate are unlikely to affect policies and they often are misinformed by interest groups.

The prosperity made possible by the use of fossil fuels has made environmental protection a social value in countries around the world. The value-creating power of private property rights, prices, profits and losses, and voluntary trade can turn climate change from a possible *tragedy* of the commons into an *opportunity* of the commons. Energy freedom, not government intervention, can balance the interests and needs of today with those of tomorrow. It alone can access the local knowledge needed to find efficient win-win responses to climate change.

2

Climate Science

Chapter Lead Authors: Craig D. Idso, Ph.D., David Legates, Ph.D., S. Fred Singer, Ph.D.

Contributors: Joseph L. Bast, Patrick Frank, Ph.D., Kenneth Haapala, Jay Lehr, Ph.D., Patrick Moore, Ph.D., Willie Soon, Ph.D.

Reviewers: Charles Anderson, Ph.D., Dennis T. Avery, Timothy Ball, Ph.D., David Bowen, Ph.D., David Burton, Mark Campbell, Ph.D., David Deming, Ph.D., Rex J. Fleming, Ph.D., Lee C. Gerhard, Ph.D., François Gervais, Ph.D., Laurence Gould, Ph.D., Kesten Green, Ph.D., Hermann Harde, Ph.D., Howard Hayden, Ph.D., Ole Humlum, Ph.D., Richard A. Keen, Ph.D., William Kininmonth, M.Sc., Anthony Lupo, Ph.D., Robert Murphy, Ph.D., Daniel W. Nebert, M.D., Norman J. Page, Ph.D., Fred Palmer, J.D., Garth Paltridge, Ph.D., D.Sc., FAA, Jim Petch, Ph.D., Jan-Erik Solheim, Peter Stilbs, Ph.D., Roger Tattersol, Frank Tipler, Ph.D., Fritz Vahrenholt, Ph.D., Art Viterito, Ph.D., Lance Wallace, Ph.D.

Citation: Idso, C.D., Legates, D. and Singer, S.F. 2019. Climate Science. In: *Climate Change Reconsidered II: Fossil Fuels.* Nongovernmental International Panel on Climate Change. Arlington Heights, IL: The Heartland Institute.

Key Findings

Key findings of this chapter include the following:

Methodology

- The Scientific Method is a series of requirements imposed on scientists to ensure the integrity of their work. The IPCC has not followed established rules that guide scientific research.

- Appealing to consensus may have a place in science, but should never be used as a means of shutting down debate.

- Uncertainty in science is unavoidable but must be acknowledged. Many declaratory and predictive statements about the global climate are not warranted by science.

Observations

- Surface air temperature is governed by energy flow from the Sun to Earth and from Earth back into space. Whatever diminishes or intensifies this energy flow can change air temperature.

- Levels of carbon dioxide (CO_2) and methane (CH_4) in the atmosphere are governed by processes of the carbon cycle. Exchange rates and other climatological processes are poorly understood.

- The geological record shows temperatures and CO_2 levels in the atmosphere have not been stable, making untenable the IPCC's assumption that they would be stable in the absence of human emissions.

- Water vapor is the dominant greenhouse gas owing to its abundance in the atmosphere and the wide range of spectra in which it absorbs radiation. CO_2 absorbs energy only in a very narrow range of the longwave infrared spectrum.

Controversies

- Reconstructions of average global surface temperature differ depending on the methodology

used. The warming of the twentieth and early twenty-first centuries has not been shown to be beyond the bounds of natural variability.

- General circulation models (GCMs) are unable to accurately depict complex climate processes. They do not accurately hindcast or forecast the climate effects of anthropogenic greenhouse gas emissions.

- Estimates of equilibrium climate sensitivity (the amount of warming that would occur following a doubling of atmospheric CO_2 level) range widely. The IPCC's estimate is higher than many recent estimates.

- Solar irradiance, magnetic fields, UV fluxes, cosmic rays, and other solar activity may have greater influence on climate than climate models and the IPCC currently assume.

Climate Impacts

- There is little evidence that the warming of the twentieth and early twenty-first centuries has caused a general increase in severe weather events. Meteorological science suggests a warmer world would see milder weather patterns.

- The link between warming and drought is weak, and by some measures drought decreased over the twentieth century. Changes in the hydrosphere of this type are regionally highly variable and show a closer correlation with multidecadal climate rhythmicity than they do with global temperature.

- The Antarctic ice sheet is likely to be unchanged or is gaining ice mass. Antarctic sea ice is gaining in extent, not retreating. Recent trends in the Greenland ice sheet mass and Artic sea ice are not outside natural variability.

- Long-running coastal tide gauges show the rate of sea-level rise is not accelerating. Local and regional sea levels exhibit typical natural variability.

- The effects of elevated CO_2 on plant characteristics are net positive, including increasing rates of photosynthesis and biomass production.

Why Scientists Disagree

- Fundamental uncertainties and disagreements prevent science from determining whether human greenhouse gas emissions are having effects on Earth's atmosphere that could endanger life on the planet.

- Climate is an interdisciplinary subject requiring insights from many fields of study. Very few scholars have mastery of more than one or two of these disciplines.

- Many scientists trust the Intergovernmental Panel on Climate Change (IPCC) to objectively report the latest scientific findings on climate change, but it has failed to produce balanced reports and has allowed its findings to be misrepresented to the public.

- Climate scientists, like all humans, can have tunnel vision. Bias, even or especially if subconscious, can be especially pernicious when data are equivocal and allow multiple interpretations, as in climatology.

Appeals to Consensus

- Surveys and abstract-counting exercises that are said to show a "scientific consensus" on the causes and consequences of climate change invariably ask the wrong questions or the wrong people. No survey data exist that support claims of consensus on important scientific questions.

- Some survey data, petitions, and peer-reviewed research show deep disagreement among scientists on issues that must be resolved before the man-made global warming hypothesis can be accepted.

- Some 31,000 scientists have signed a petition saying "there is no convincing scientific evidence that human release of carbon dioxide, methane, or other greenhouse gases is causing or will, in the foreseeable future, cause catastrophic heating of the Earth's atmosphere and disruption of the Earth's climate."

- Because scientists disagree, policymakers must exercise special care in choosing where they turn for advice.

Introduction

A central issue in climate science today is whether human emissions of carbon dioxide, methane, and other "greenhouse gases" are having effects on Earth's atmosphere that could endanger life on the planet. As the size of recent reports by the Intergovernmental Panel on Climate Change (IPCC, 2013, 2014a, 2014b) and the Nongovernmental International Panel on Climate Change (NIPCC, 2009, 2011, 2013, 2014) suggest, climate science is a complex and highly technical subject. Simplistic claims about the relationship between human activity and climate change are misleading.

This chapter focuses on physical and biological sciences. It does not address the impacts of climate change (or fossil fuels) on human prosperity, health, or security or conduct a cost-benefit analysis of climate change or fossil fuels. Those topics are addressed in subsequent chapters. Sometimes science presentations also appear in other chapters, including "tutorials" on air quality (Chapter 6), energy matters (Chapter 7), and integrated assessment models (Chapter 8), but most of the pure science in this book appears in this chapter.

Section 2.1 offers a tutorial describing some of the methodological issues and observational data involved in efforts to understand the causes and consequences of climate change. Section 2.2 describes controversies over four important topics in climate science: temperature records, general circulation models (GCMs), climate sensitivity, and solar influences on climate. Each of these topics is important for discerning and measuring the human impact on the climate.

Section 2.3 examines observational evidence concerning four climate impacts: severe weather events, melting ice, sea-level rise, and effects on plants. Section 2.4 reviews four reasons why scientists disagree: basic scientific uncertainties, the subject's interdisciplinary nature, the failure of the IPCC to win the confidence of many scientists, and tunnel vision (or bias). Section 2.5 looks at claims that a scientific consensus exists on some or all of these issues. A brief summary and conclusion appear in Section 2.6.

Two previous volumes in the *Climate Change Reconsidered* series produced by NIPCC subtitled *Physical Science* (2013) and *Biological Impacts* (2014) contain exhaustive reviews of the scientific literature conducted by lead authors Craig D. Idso, Sherwood Idso, Robert M. Carter, and S. Fred Singer and an international team of some 100 scientists. Combined, they offer more than 2,000 pages of summaries and abstracts of scientific research, nearly all of it appearing in peer-reviewed science journals. Readers seeking a more in-depth treatment of the topics addressed in this chapter are encouraged to read those volumes.

Two reports written in 2018 by teams of scientists led by Jay Lehr were valuable in providing updated references to the scientific literature (Lehr *et al.,* 2018a, 2018b). Sections 2.4 and 2.5 rely on parts of a book titled *Why Scientists Disagree about Global Warming* published by NIPCC in 2015 and revised in 2016 (NIPCC, 2016). Section 2.3.3, on sea-level rise, draws in part from a previous NIPCC special report titled *Data versus Hype: How Ten Cities Show Sea-level Rise Is a False Crisis* (Hedke, 2017).

This chapter provides a comprehensive and balanced account of the latest science on climate change. While acknowledging the extraordinary scientific accomplishment represented by the IPCC's assessment reports, the authors do not hesitate to identify possible errors and omissions. To the extent that this chapter critiques the IPCC's reports, it does so in the spirit of healthy scientific debate and respect.

References

Hedke, D. 2017. Data versus hype: How ten cities show sea-level rise is a false crisis. *Policy Brief.* Arlington Heights, IL: The Heartland Institute. September.

IPCC. 2013. Intergovernmental Panel on Climate Change. *Climate Change 2013: The Physical Science Basis.* Contribution of Working Group I to the Fifth Assessment Report of the Intergovernmental Panel on Climate Change. Cambridge, UK and New York, NY: Cambridge University Press.

IPCC. 2014a. Intergovernmental Panel on Climate Change. *Climate Change 2014: Impacts, Adaptation, and Vulnerability. Part A: Global and Sectoral Aspects.* Contribution of Working Group II to the Fifth Assessment Report of the Intergovernmental Panel on Climate Change.

Cambridge, UK and New York, NY: Cambridge University Press.

IPCC. 2014b. Intergovernmental Panel on Climate Change. *Climate Change 2014: Mitigation of Climate Change.* Contribution of Working Group III to the Fifth Assessment Report of the Intergovernmental Panel on Climate Change. Cambridge, UK and New York, NY: Cambridge University Press.

Lehr, J., *et al.* 2018a. A Critique of the U.S. Global Change Research Program's 2017 Climate Science Special Report. *Policy Study.* Arlington Heights, IL: The Heartland Institute. March.

Lehr, J., *et al.* 2018b. A climate science tutorial prepared for Hon. William Alsup. *Policy Brief.* Arlington Heights, IL: The Heartland Institute. April 13.

NIPCC. 2009. Nongovernmental International Panel on Climate Change. *Climate Change Reconsidered: The 2009 Report of the Nongovernmental International Panel on Climate Change.* Chicago, IL: The Heartland Institute.

NIPCC. 2011. Nongovernmental International Panel on Climate Change. *Climate Change Reconsidered: 2011 Interim Report of the Nongovernmental International Panel on Climate Change.* Chicago, IL: The Heartland Institute.

NIPCC. 2013. Nongovernmental International Panel on Climate Change. *Climate Change Reconsidered II: Physical Science.* Chicago, IL: The Heartland Institute.

NIPCC. 2014. Nongovernmental International Panel on Climate Change. *Climate Change Reconsidered II: Biological Impacts.* Chicago, IL: The Heartland Institute.

NIPCC. 2016. Nongovernmental International Panel on Climate Change. *Why Scientists Disagree about Global Warming.* Arlington Heights, IL: The Heartland Institute.

2.1 A ScienceTutorial

Climate science is confusing and often confused because people claiming to be "climate scientists" are usually specialists in one or a few areas -- including physics, mathematics, computer modeling, and oceanography – who study just one part of the complex climate puzzle. Researchers define concepts and measure values differently and interpret the results through different prisms based on their academic discipline and training. The discipline of climatology is quite new and consequently disagreements start early and new discoveries continuously challenge prevailing wisdom.

For example, recent and current global temperatures are disputed (Christy and McNider, 2017) and some physicists doubt whether the concept of a single global temperature should be used in climate research (Essex *et al.,* 2007). The processes by which carbon dioxide (CO_2) enters the atmosphere and is exchanged with other reservoirs are poorly understood (Falkowski *et al.,* 2000). The role of water vapor and clouds (Chou and Lindzen, 2004; Spencer *et al.,* 2007), ocean currents (D'Aleo and Easterbrook, 2016), solar influences (Ziskin and Shaviv, 2012; Harde, 2017), and CO_2 (Lewis and Curry, 2014; Bates, 2016) in regulating global temperature are all areas of controversy and uncertainty. These are hardly peripheral or unimportant issues.

All this disagreement and uncertainty makes explaining even the basic principles of climate science difficult. Declarative statements usually need to be followed by exceptions, cautions, or alternative interpretations. With these caveats in mind, this "tutorial" presents seven key topics in climate science as shown in Figure 2.1. Later sections of this chapter and later chapters in this book revisit the topics addressed only briefly in this section.

References

Bates, J.R. 2016. Estimating climate sensitivity using two-zone energy balance models. *Earth and Space Science* **3** (5): 207–25.

Chou, M. and Lindzen, R. 2004. Comments on "Examination of the Decadal Tropical Mean ERBS Nonscanner Radiation Data for the Iris Hypothesis." *Journal of Climate* **18** (12): 2123–7.

Christy, J.R. and McNider, R.T. 2017. Satellite bulk tropospheric temperatures as a metric for climate sensitivity. *Asia-Pacific Journal of Atmospheric Sciences* **53** (4): 511–8.

D'Aleo, J.S. and Easterbrook, D.J. 2016. Relationship of multidecadal global temperatures to multidecadal oceanic oscillations. In: Easterbrook, D. (Ed.) *Evidence-Based Climate Science.* Cambridge, MA: Elsevier.

Essex, C., McKitrick, R., and Andresen, B. 2007. Does a global temperature exist? *Journal of Non-Equilibrium Thermodynamics* **32** (1).

Falkowski, P., *et al.* 2000. The global carbon cycle: a test of our knowledge of Earth as a system. *Science* **290** (5490): 291–6.

Figure 2.1
Topics in the tutorial

2.1.1 Methodology

    2.1.1.1 Scientific Method
    2.1.1.2 Consensus
    2.1.1.3 Uncertainty

2.1.2 Observations

    2.1.2.1 Energy Budget
    2.1.2.2 Carbon Cycle
    2.1.2.3 Geological Record
    2.1.2.4 Greenhouse Gases

Harde, H. 2017. Radiation transfer calculations and assessment of global warming by CO_2. *International Journal of Atmospheric Sciences.* Article ID 9251034.

Lewis, N. and Curry, J.A. 2014. The implications for climate sensitivity of AR5 forcing and heat uptake estimates. *Climate Dynamics* **45** (3/4): 1009–23.

Spencer, R., Braswell, W., Christy, J., and Hnilo, J. 2007. Cloud and radiation budget changes associated with tropical intraseasonal oscillations. *Geophysical Research Letters* **34** (15).

Ziskin, S. and Shaviv, N.J. 2012. Quantifying the role of solar radiative forcing over the 20th century. *Advances in Space Research* **50** (6): 762–76.

2.1.1 Methodology

Science is a search for causal explanations of natural events. It is a search for *why* things are the way they are and act the way they do. In Chapter 1, the economic way of thinking was characterized as asking "and then what?" The question leads to careful study of incentives and unintended consequences. In the chapter following this one, engineers are characterized as always asking "how much?" They are keen on measuring energy inputs, power, and efficiency. Scientists, as this section will demonstrate, ask "how do we know?" Skepticism is at the heart of science.

The growth of scientific knowledge proceeds through a process called the *Scientific Method.* It is different from the process called *consensus,* which may have a role in science but is mainly used in politics to determine public policies. The two methods often come into conflict in discussions of climate change. *Uncertainty,* a third concept involving methodology, is unavoidable in science. How to reduce uncertainty and communicate it to the public and policymakers are major sources of disagreement in the climate science community. All three concepts are addressed in this section.

2.1.1.1 Scientific Method

The Scientific Method is a series of requirements imposed on scientists to ensure the integrity of their work. The IPCC has not followed established rules that guide scientific research.

To find reliable answers, scientists use the Scientific Method. Armstrong and Green (2018a) surveyed the literature, citing Hubbard (2016), Munafo *et al.* (2017), and other previously published reviews, to identify practices that have been consistently endorsed by scientists and learned societies. They found general acceptance that the Scientific Method requires scientists to …

1. study important problems,
2. build on prior scientific knowledge,
3. use objective methods,
4. use valid and reliable data,
5. use valid, reliable, and simple methods,
6. use experiments,
7. deduce conclusions logically from prior knowledge and new findings, and
8. disclose all information needed to evaluate the research and to conduct replications.

Armstrong and Green then developed 24 "guidelines for scientists," which appear in Figure 2.1.1.1.1, to ensure compliance with the eight criteria of the Scientific Method. Failure to comply with the Scientific Method often results in what Armstrong and Green call "advocacy research," which they say is characterized by 10 instruments:

1. ignore cumulative scientific knowledge,
2. test a preferred hypothesis against an implausible null hypothesis,
3. show only evidence favoring the preferred hypothesis,
4. do not specify the conditions associated with the hypothesis,
5. ignore important causal variables,
6. use non-experimental data,
7. use data models,
8. use faulty logic,
9. avoid tests of *ex ante* predictive validity, and
10. use *ad hominem* arguments (attack authors and not their reasoning).

Figure 2.1.1.1.1
Armstrong and Green's guidelines for scientists

Selecting a Problem
1. Seek an important problem.
2. Be skeptical about findings, theories, policies, methods, and data, especially absent experimental evidence.
3. Consider replications and extensions of papers with useful scientific findings.
4. Ensure that you can address the problem impartially.
5. If you need funding, ensure that you will nevertheless have control over all aspects of your study.

Designing a Study
6. Acquire existing knowledge about the problem.
7. Develop multiple reasonable hypotheses with specified conditions.
8. Design experiments with specified conditions that can test the predictive validity of hypotheses.

Collecting Data

9. Obtain all valid data.
10. Ensure that the data are reliable.

Analyzing Data

11. Use methods that incorporate cumulative knowledge.
12. Use multiple validated methods.
13. Use simple methods.
14. Estimate effect sizes and prediction intervals.
15. Draw logical conclusions on the practical implications of findings from tests of multiple reasonable hypotheses.

Writing a Scientific Paper

16. Disclose research hypotheses, methods, and data.
17. Cite all relevant scientific papers when presenting evidence.
18. Ensure that summaries of cited prior findings are necessary, explained, and correct.
19. Explain why your findings are useful.
20. Write clearly and succinctly for the widest audience for whom the findings might be useful.
21. Obtain extensive peer review and editing *before* submitting a paper for publication.

Disseminating Findings

22. Provide responses to journal reviewers, including reasons for ignoring their suggestions, and if rejected, appeal to editors if you have useful scientific findings.
23. Consider alternative ways to publish your findings.
24. Inform those who can use your findings.

Source: Armstrong and Green, 2018a, Exhibit 2, pp. 18–19.

Role of Experimentation

Scientific theories differ from observations by being suppositions about what is not observable directly. Only some of their consequences – logical or causal – can be observed. According to Popper (1965), a theory is scientific only if it can be falsified by observational data or experimentation, if not currently then in principle at some future date when the data or tools for further investigation become available. A famous example is how one of the predictions of Albert Einstein's general theory of relativity, first proposed in 1915, was tested and shown to be correct in 1919 by observations gathered during a total solar eclipse (O'Neill, 2017). Popper justified his stance by arguing that humans are fallible – we lack omniscience and so cannot comprehend a theory that might explain everything – and because future observations or experiments could disprove any current theory. Therefore, the best we can do is try to falsify the hypothesis, and by surviving such tests a hypothesis demonstrates it may be close to the truth. Einstein agreed, writing in 1919,

> A theory can thus be recognized as erroneous if there is a logical error in its deductions, or as inadequate if a fact is not in agreement with its consequences. But the truth of a theory can never be proven. For one never knows that even in the future no experience will be encountered which contradicts its consequences; and still other systems of thought are always conceivable which are capable of joining together the same given facts.

This suggests observations in science are useful primarily to falsify hypotheses and cannot prove one is correct. Objecting to Popper's and Einstein's critique of inductive reasoning, Jaynes (2003) writes, "It is not the absolute status of an hypothesis embedded in the universe of all conceivable theories, but the plausibility of an hypothesis relative to a definite set of specified alternatives, that Bayesian inference determines" (p. 310). In other words, the test of a hypothesis is not the impossible standard of omniscience, but rather how well it performs relative to other hypotheses. Bayesian inference is a way of improving the probability that a theory is correct by using Bayes' theorem. Bayes' theorem reads:

$$P(A\backslash B) = \frac{P(B\backslash A)\ P(A)}{P(B)}$$

where A and B are events and $P(B) \neq 0$.

$P(A\backslash B)$ is the likelihood of event A occurring given that B is true.

$P(B\backslash A)$ is the likelihood of event B occurring given that A is true.

$P(A)$ and $P(B)$ are the probabilities of observing A and B independently of each other; this is known as the marginal probability.

Null Hypothesis

Bayes' theorem demonstrates the importance of alternative or "null" hypotheses. A null hypothesis is negative only in the sense that unless it is rebutted, the original hypothesis remains unproven. Failing to disprove the null hypothesis does not mean it is true, only that it survives as a possible alternative explanation. Null hypotheses also need to make specific predictions and be falsifiable. As Jaynes (2003) writes, "we have not asked any definite, well-posed question until we specify the possible alternatives to H0 [null hypothesis]. Then … probability theory can tell us how our hypothesis fares relative to the alternatives that we have specified" (p. 136). Jaynes goes on to write, "This means that if any significance test is to be acceptable to a scientist, we shall need to examine its rationale to see whether it has … some implied if unstated alternative hypotheses. Only when such hypotheses are identified are we in a position to say what the test accomplishes; i.e. what it is testing" (p. 137).

It is relatively easy to assemble reams of "evidence" in favor of a point of view or opinion while ignoring inconvenient facts that would contradict it, a phenomenon called "confirmation bias" and a practice sometimes called "data dredging." The best way to avoid confirmation bias is to entertain alternative hypotheses. Armstrong and Green (2018a) write, "We use the term advocacy to refer to studies that are designed to 'prove' a given hypothesis, as distinct from arguing in favor of an idea. Advocacy studies can be identified operationally by the absence of fair tests of multiple reasonable hypotheses" (p. 7).

The hypothesis implicit in the IPCC's writings, though rarely explicitly stated, is that dangerous global warming is resulting, or will result, from anthropogenic greenhouse gas emissions. As stated, that hypothesis is falsifiable. The null hypothesis is that the warming found in temperature records and changes in polar ice, sea levels, and various weather indices are instances of natural variability or causes unrelated to anthropogenic (human) greenhouse gas emissions. As long as recent average global temperatures, sea-level rise, polar ice melting, etc. are not much different than earlier times when human greenhouse gas emissions were low, the null hypothesis is very reasonable.

Invalidating this null hypothesis requires, at a minimum, direct evidence of human causation of changes in global mean average surface temperature and that recent trends are unprecedented. ("Direct evidence" is knowledge based on observations which, if true, directly prove or disprove a theory without resorting to any assumption or inference. It is distinguished from "circumstantial evidence," which is knowledge that relies on an inference to connect it to a conclusion of fact.) But the IPCC and many other research and advocacy groups make no effort to falsify the null hypothesis. For example, virtually no research dollars are available to study the causes and consequences of *natural* (or what the IPCC calls "internal") climate variability. Rather than investigate the role of ocean currents, solar influences, cosmic rays, and clouds in a fair and balanced way, IPCC researchers dismiss them out of hand as "poorly understood" or "unlikely to have a major effect." Even modest attention to research in these areas would likely force the IPCC to reconsider some of its postulates. The IPCC has used all 10 of Armstrong and Green's instruments of "advocacy research" to defend, rather than test, its hypothesis.

Why doesn't the IPCC study natural causes of climate change? Article 1.2 of the United Nations' Framework Convention on Climate Change, which gave the IPCC its mandate, defines climate change as "a change of climate which is attributed directly or indirectly to human activity that alters the composition of the global atmosphere and which is in addition to natural climate variability observed over comparable time periods" (UNFCCC, 1994). Working Group I of the IPCC has interpreted this as a mandate not to study climate change "in the round" but only a possible *human* impact on climate. As Curry (2013) writes,

> The UNFCCC Treaty provides the rationale for framing the IPCC assessment of climate change and its uncertainties, in terms of identifying dangerous climate change and providing input for decision making regarding CO_2 stabilization targets. In the context of this framing, key scientific questions receive little attention. In detecting and attributing 20th century climate change, the IPCC AR4 all but dismisses natural internal multidecadal variability in the attribution argument. The IPCC AR4 conducted no systematic assessment of the impact of uncertainty in 20th century solar variability on attribution, and indirect solar impacts on climate are little known and remain unexplored in any meaningful way.

Interestingly, the IPCC's Working Group II does not also limit its definition of climate change this way, allowing it to include any impact of climate change "regardless of its cause" in its lengthy catalogue of alleged damages (IPCC, 2014, p. 4, fn. 5). Judging the impacts of man-made climate change to be harmful to human well-being or the environment requires, at a minimum, distinguishing those impacts from impacts that would have occurred in the absence of the human presence. Such a finding also requires balancing all costs and benefits, including the known benefits of higher levels of CO_2 in the atmosphere to ecosystems. Since a steep reduction in the use of fossil fuels is the policy recommendation that arises from a finding that anthropogenic climate change is harmful to humanity, then the costs of living without those energy sources must be weighed as well.

Peer Review

Part of the Scientific Method is independent review of a scientist's work by other scientists who do not have a professional, reputational, or financial stake in whether the hypothesis is confirmed or disproven. Peer review distinguishes academic literature from more popular writing and journalism. Tragically, peer review is in a state of crisis in a wide range of disciplines, affecting even or especially some of the most respected academic journals.

In a series of articles published in leading academic journals, Ioannidis (2005a, 2005b, 2012, 2018; Ioannidis and Trikalinos, 2005) revealed most published research in the health care field cannot be replicated or is likely to be contradicted by later publications His most frequently cited work is titled "Why most published research findings are false." Although the problem is not new (see Mayes *et al.,* 1988), Ioannidis's work generated widespread awareness that peer review is no guarantee of the accuracy or value of a research paper. In fact, he found that the likelihood of research being contradicted is highest when it is published in the most prestigious journals, including *JAMA, Nature,* and *Science.*

Springer, a major publisher of science journals, retracted 16 papers it had published that were simply gibberish generated by a computer program called SCIgen (*Nature,* 2014). In 2016, more than 70% of 1,576 researchers who replied to a survey conducted by *Nature* reported having tried and failed to reproduce another scientist's experiments, and more than half failed to reproduce their own experiments (Baker, 2016). Fifty-two percent agree there is a significant "crisis" of reproducibility. Camerer *et al.* (2018) attempted to replicate 21 experimental social science studies published in *Nature* and *Science* between 2010 and 2015 and found "a significant effect in the same direction as the original study for 13 (62%) studies," while the rest found no effect or an opposite effect. Random chance would have led to a 50% replication rate, so this is a dismal finding. While Camerer *et al.* looked at studies in the social sciences, similar results have been reported in the physical sciences. See Sánchez and Parott (2017) for a review of studies alleging negative health effects of genetically modified foods.

Some journals and academic institutions claim to be engaged in considerable soul-searching and efforts to reform a peer-review process that is plainly broken. However, journals such as *Nature* seem to take the scandal over peer-review corruption seriously only when it concerns issues other than climate science (e.g., Ferguson *et al.,* 2014; Sarewitz, 2016).

This controversy has particular relevance to the climate change debate due to "Climategate," the release by a whistleblower in 2009 and again in 2011 of thousands of emails exchanged among prominent climate scientists discussing their use of the peer-review process to exclude global warming skeptics from journals, punish editors who allowed skeptics' articles to appear, and rush into publication articles refuting or attempting to discredit scientists who disagree with the IPCC's findings (Montford, 2010; Sussman, 2010; Michaels, 2011, Chapter 2). No scientists were punished for their misbehavior and the practice continues today.

The lessons of the peer-review crisis are several. Just because something appears in a peer-reviewed journal does not mean it is credible or reliable. Research that "fails" peer-review or appears in the so-called secondary literature may in fact be credible and reliable. Review by a small cadre of experts behind closed doors is more likely to lead to publication of research that reinforces a prevailing paradigm and overlooks errors, while transparency and open debate lead to generally higher quality research (Raymond, 1999; Luke *et al.*, 2018). This is relevant to Armstrong and Green's Rule #8: Disclose all information needed to evaluate the research and to conduct replications.

Correlation versus Causation

The correlation of two variables does not establish causation, for it is not at all unusual for two trends to co-vary by accident, or in parallel when both are driven by the same outside force. To infer causation one needs a reasoned argument based on some causal theory that has stood up to tests and sits within a framework of theories and "basic statements."

VanCauwenberge (2016) writes, "Sometimes a correlation means absolutely nothing, and is purely accidental (especially when you compute millions of correlations among thousands of variables) or it can be explained by confounding factors. For instance, the fact that the cost of electricity is correlated to how much people spend on education, is explained by a confounding factor: inflation, which makes both electricity and education costs grow over time. This confounding factor has a bigger influence than true causal factors, such as more administrators/ government-funded student loans boosting college tuition."

In the climate change debate, data showing a correlation between observed warming in the Southern Hemisphere between 1963 and 1987 and what was projected to occur by models led some scientists to claim, just days before the Second Conference of the Parties to the United Nations Framework Convention on Climate Change, that this was proof that rising atmospheric CO_2 levels *caused* the temperature to rise (Santer *et al.*, 1996). Michaels

and Knappenberger (1996) quickly pointed out that the observational record actually begins in 1957 and extended to 1995, and when all of the data are used, the warming trend completely disappears. (See Figure 2.1.1.1.2.) Despite this, the theory gained momentum as millions and then billions of dollars were spent searching for a human influence on the climate (Essex and McKitrick, 2007; Darwall, 2013; Lewin, 2017).

Related to the need to distinguish between correlation and causation is the phenomenon of data dredging, also called "data mining" or "p-hacking," whereby large databases are analyzed repeatedly in hopes of finding a calculated probability or p value $\leq$ 0.05, and then selectively reporting the positive results as circumstantial evidence in support of a hypothesis (Goldacre, 2016; Gorman *et al.*, 2017). Reporting positive results greatly increases the odds of being published in academic journals, where articles reporting positive findings outnumber those reporting negative findings by 9:1 or greater (Fanelli, 2012). Advances in data collection and computer processing speeds enable researchers to test thousands and even millions of possible relationships in search of the elusive $p \leq 0.05$ and then to seek a publication willing to accept their findings. This practice is especially apparent in the public health arena where exposure to small doses of chemicals is alleged to be "associated" with negative health effects (see Chapter 6), and also in climatology where small changes in temperatures are alleged to be "associated" with almost countless health and environmental impacts. Data dredging violates the Scientific Method by putting the collection and analysis of data ahead of formulating a reasonable hypothesis and one or more alternative hypotheses.

Control for Natural Variability

To discern the impact of a particular variable or process, scientific experiments attempt to control for natural variability in populations or physical phenomena. Sometimes the "background noise" of natural variability is too great to discern an impact or pattern. When the subject of inquiry is Earth's atmosphere, the largest and most complex phenomenon ever studied by man, it is very difficult to meet the requirements of control for designing, conducting, and interpreting experiments or observational programs.

Figure 2.1.1.1.2
Santer *et al.* (1996) claim to have found a "human influence" on climate and Michaels and Knappenberger (1996) demonstrate no trend

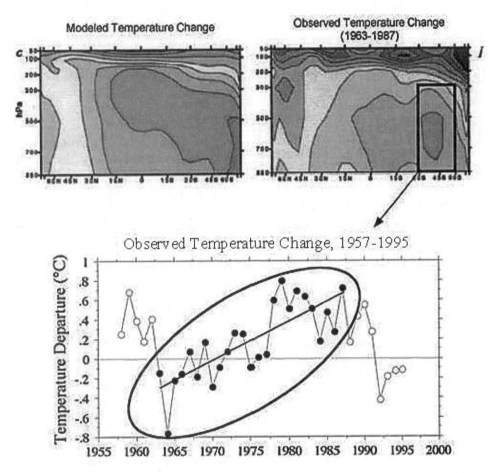

Modeled (upper left) and observed (upper right) temperature changes throughout the atmosphere. Lower image shows time series of temperatures in the region of the highlighted box in the upper right panel for years 1957-1995. Filled circles are years reported by Santer *et al.* (1996). Use of all the available data (open circles) reveals no trend. *Source:* Michaels, 2010, Figure 3, citing Santer *et al.,* 1996 and Michaels and Knappenberger, 1996.

The geological record reveals that we live on a dynamic planet. All aspects of the physical and biological environment are in a constant state of flux (including, of course, temperature). It is wrong to assume no changes would occur in the absence of the human presence. Climate, for example, will be different in 100 years regardless of what humans do. This is a point of contention in the climate change debate because the IPCC seems to assume that global temperatures, solar influences, and exchanges among global carbon reservoirs (to name just three) would remain unchanged, decade after decade and century after century, *but for* the human presence.

Related to this matter, many studies of the impact of climate change on wildlife simply assume temperatures in the area under investigation have risen in pace with estimates of *global* surface temperatures, or that severe weather events have become more frequent, etc., without establishing that the relevant *local* temperature and weather records conform to the postulate. Assertions about specific phenomena should not be made based on global averages. An example of research conducted correctly in this regard is a study of infrastructure needs for a city in California conducted by Pontius (2017). Rather than rely on the mean global average surface temperature, the researcher studied temperature data from the City of Riverside, California from 1901 to 2017. "No evidence of significant climate change beyond natural variability was observed in this temperature record," he reports. "Using a Climate Sensitivity best estimate of 2°C, the increase in temperature resulting from a doubling of atmospheric CO_2 is estimated at approximately 0.009°C/yr which is insignificant compared to natural variability."

Postulates Are Not Science

Postulates, commonly defined as "something suggested or assumed as true as the basis for reasoning, discussion, or belief," can stimulate the search for relevant observations or experiments but more often are merely assertions that are difficult or impossible to test (Kahneman, 2011). For example, most parts of the IPCC's very large assessment reports accept without qualification or acknowledgement of uncertainty the following five postulates:

- The warming of the twentieth century cannot be explained by natural variability.

- The late twentieth century warm peak was of greater magnitude than previous natural peaks.

- Increases in atmospheric CO_2 precede, and then force, parallel increases in temperature.

- Solar influences are too small to explain more than a trivial part of twentieth-century warming.

- A future warming of 2°C or more would be net harmful to the biosphere and human well-being.

All five statements may be true. There is evidence that seems to support all of them, but there is also evidence that contradicts them. In their declarative form, all of these statements are misleading at best and probably untrue. The IPCC expresses "great confidence" and even "extreme confidence" in these postulates, but it did not consider alternative hypotheses or evidence pointing to different conclusions. A true high confidence interval, defined in statistics as the probability that a finding falls within the range of values of the entire population being sampled, cannot be given because

these are statements of opinion and not of fact. Once again, this is a failure to conform to the requirements of the Scientific Method.

* * *

Armstrong and Green (2018b) write, "logical policy requires scientific forecasts of substantive long-term trend in global mean temperatures, major net harmful effects from changing temperatures, and net benefit from proposed policies relative to no action. Failure of any of the three requirements means policy action is unsupported" (p. 29). When they applied a checklist of 20 operational guidelines to the IPCC "business as usual" forecast and to a default no-change model forecast, they found the IPCC scenarios followed none of the guidelines while the no-change model followed 95%. Results like these suggest the IPCC has not been careful to follow the rules of the Scientific Method.

References

Armstrong, J.S. and Green, K.C. 2018a. Guidelines for science: evidence-based checklists. Working Paper Version 502-RG. Philadelphia, PA: The Wharton School, University of Pennsylvania.

Armstrong, J.S. and Green, K.C. 2018b. Do forecasters of dangerous manmade global warming follow the science? Presented at the International Symposium on Forecasting, Boulder, Colorado. June 18.

Baker, M. 2016. 1,500 scientists lift the lid on reproducibility. *Nature* **533**: 452–4.

Camerer, C.F., *et al.* 2018. Evaluating the replicability of social science experiments in *Nature* and *Science* between 2010 and 2015. *Nature Human Behaviour* **2**: 637–44.

Curry, J. 2013. Negotiating the IPCC SPM. Climate Etc. (website). October 1. Accessed November 25, 2018.

Darwall, R. 2013. *The Age of Global Warming: A History.* London, UK: Quartet Books Limited.

Einstein, A. 1919. *Induction and Deduction in Physics.* Berliner Tageblatt, 25 December 1919. Collected Papers of Albert Einstein, Princeton University Press (website). Accessed November 23, 2018.

Essex, C. and McKitrick, R. 2007. *Taken By Storm: The Troubled Science, Policy and Politics of Global Warming.* Revised edition. Toronto, ON: Key Porter Books.

Fanelli, D. 2012. "Positive" results increase down the hierarchy of science. *PLoS ONE* **4** (5): e10068.

Ferguson, C., Marcus, A., and Oransky, I. 2014. Publishing: the peer-review scam. *Nature* **515**: 480–2.

Goldacre, B. 2016. Make journals report clinical trials properly. *Nature* **530** (7588): 7.

Gorman, D., Elkins, A., and Lawley, M. 2017. A systems approach to understanding and improving research integrity. *Science and Engineering Ethics*: 1–19.

Hubbard, R. 2016. *Corrupt Research: The Case for Reconceptualizing Empirical Management and Social Science*. New York, NY: Sage.

Ioannidis, J.P.A. 2005a. Contradicted and initially stronger effects in highly cited clinical research. *Journal of the American Medical Association* **294**: 218–28.

Ioannidis, J.P.A. 2005b. Why most published research findings are false. *PLOS Medicine* **2**: e124.

Ioannidis, J.P.A. 2012. Scientific inbreeding and same-team replication: Type D personality as an example. *Journal of Psychosomatic Research* **73**: 408–10.

Ioannidis, J.P.A. 2018. The proposal to lower P value thresholds to .005. *JAMA* **319**: 1429–30.

Ioannidis, J.P.A. and Trikalinos, T.A. 2005. Early extreme contradictory estimates may appear in published research: the Proteus phenomenon in molecular genetics research and randomized trials. *Journal of Clinical Epidemiology* **58**(6): 543–9.

IPCC. 2014. Intergovernmental Panel on Climate Change. *Climate Change 2014: Impacts, Adaptation, and Vulnerability. Part A: Global and Sectoral Aspects.* Contribution of Working Group II to the Fifth Assessment Report of the Intergovernmental Panel on Climate Change. Cambridge, UK and New York, NY: Cambridge University Press.

Jaynes, E.T. 2003. *Probability Theory: The Logic of Science*. Cambridge, MA: Cambridge University Press.

Kahneman, D. 2011. *Thinking, Fast and Slow*. New York, NY: Macmillan.

Lewin, B. 2017. *Searching for the Catastrophe Signal: The Origins of the Intergovernmental Panel on Climate Change*. London, UK: Global Warming Policy Foundation.

Luke, O., *et al.* 2018. Medical journals should embrace preprints to address the reproducibility crisis. *International Journal of Epidemiology* **47** (5).

Mayes, L.C., Horwitz, R.I., and Feinstein, A.R. 1988. A collection of 56 topics with contradictory results in case-control research. *International Journal of Epidemiology* **17**: 680–5.

Michaels, P.J. 2010. Testimony of Patrick J. Michaels on climate change. Subcommittee on Energy and Environment Committee on Science and Technology, United States House of Representatives. November 17.

Michaels, P. (Ed.) 2011. *Climate Coup: Global Warming's Invasion of Our Government and Our Lives*. Washington, DC: Cato Institute.

Michaels, P.J., and Knappenberger, P.C. 1996. Human effect on global climate? *Nature* **384**: 522-23.

Montford, A.W. 2010. *The Hockey Stick Illusion: Climategate and the Corruption of Science*. London, UK: Stacey International.

Munafo, M.R., *et al.* 2017. A manifesto for reproducible science. *Nature Human Behavior* **1**: 0021.

Nature. 2014. Gibberish papers [news item]. *Nature* **507** (6): 13.

O'Neill, I. 2017. How a Total Solar Eclipse Helped Prove Einstein Right About Relativity. Space.com (website). May 29. Accessed November 23, 2018.

Pontius, F.W. 2017. Sustainable infrastructure: climate changes and carbon dioxide. *American Journal of Civil Engineering* **5** (5): 254–67.

Popper, K. 1965. *Conjectures and Refutations: The Growth of Scientific Knowledge*. Second edition. New York, NY: Harper and Row Publishers.

Raymond, E. 1999. The cathedral and the bazaar. *Knowledge, Technology & Policy* **12**: 23–49.

Sánchez, M.A. and Parott, W.A. 2017. Characterization of scientific studies usually cited as evidence of adverse effects of GM food/feed. *Plant Biotechnology Journal* **15** (10): 1227–34.

Santer, B.D. *et al.* 1996. A search for human influences on the thermal structure of the atmosphere. *Nature* **382**: 39–46.

Sarewitz, D. 2016. Saving science. *The New Atlantis* **49**: 5–40.

Sussman, B. 2010. *Climategate: A Veteran Meteorologist Exposes the Global Warming Scam*. Washington, DC: Worldnet Daily.

UNFCCC. 1994. United Nations Framework Convention on Climate Change (website). Accessed November 23, 2018.

VanCauwenberge, L. 2016. Spurious correlations: 15 examples. Data Science Central (website). January 26.

2.1.1.2 Consensus

Appealing to consensus may have a place in science, but should never be used as a means of shutting down debate.

The meme of "an overwhelming consensus" of scientists favoring one particular view on climate change is very popular, despite its implausibility. For example, Lloyd and Winsberg (2018) write, "While a fair bit of controversy concerning the cause of these phenomena [recent severe weather events] remains in the body politic (especially in the United States), nothing could be further from the truth when it comes to the scientific community. Multiple studies, appearing in peer-reviewed publications, all show similar findings: that roughly 97–98% of actively publishing climate scientists agree with the claim that it is extremely likely that the past century's warming trend is due to human activities" (p. 2). This claim quickly falls apart upon inspection.

That climate change is real or happening is a truism: Climate is always changing. To say it is "due to human activities" begs the questions "how much?" and "how do we know?" Nearly all scientists understand that some of the past century's warming trend was due to natural causes. The issue is how much is natural variability and how much is due to anthropogenic greenhouse gases and changes in land use (mostly agriculture and forestry). Some scientists believe it is difficult or meaningless to ascribe a single temperature to the globe and then to attribute changes to that statistical abstraction to human causes. And whether climate change is dangerous or not is a subjective and political decision, not a scientific concept. Who is at risk? When? And how do the risks created by climate change compare to other risks we face every day?

If the scientific debate were truly over, the range of uncertainty over the impact of carbon dioxide on climate would be smaller than it was in 1979, instead of being virtually the same. Many admissions of uncertainty appear in the IPCC's hefty assessment reports (a topic addressed in the next section and frequently in other chapters), but those reservations and doubts are scrubbed from the often-cited summaries for policymakers (SPMs), an example of scientific malpractice that has been protested by many distinguished scientists (Seitz, 1996; Landsea, 2005; Lindzen, 2012; Tol, 2014; Stavins, 2014). Many scientists look no further than the SPMs and trust them to accurately depict the current state of climate science. They do not. Surveys and abstract-counting exercises purporting to show consensus are critiqued at some length in Section 2.5, so that will not be done here.

Consensus may have a place in science, but only in contexts different than what occur in climate science. It is typically achieved over an extended period of time by independent scientists following the conventions of the Scientific Method, in particular not neglecting the need to entertain competing hypotheses. Consensus emerges from open debate and tolerance of new theories and discoveries; it is not handed down by an international political organization tasked with defending one paradigm. Consensus on basic theories can open up other areas for new research and exploration while leaving the door open to reconsider first principles. Unfortunately, this is not the context in which consensus is invoked in climate science. Curry (2012) wrote,

The manufactured consensus of the IPCC has had the unintended consequences of distorting the science, elevating the voices of scientists that dispute the consensus, and motivating actions by the consensus scientists and their supporters that have diminished the public's trust in the IPCC. Research from the field of science and technology studies are finding that manufacturing a consensus in the context of the IPCC has acted to hyper-politicize the scientific and policy debates, to the detriment of both. Arguments are increasingly being made to abandon the scientific consensus-seeking approach in favor of open debate of the arguments themselves and discussion of a broad range of policy options that stimulate local and regional solutions to the multifaceted and interrelated issues of climate change, land use, resource management, cost effective clean energy solutions, and developing technologies to expand energy access efficiently.

More recently, Curry (2018) writes, "The IPCC and other assessment reports are framed around providing support for the hypothesis of human-caused climate change. As a result, natural processes of climate variability have been relatively neglected in these assessments."

In his book titled *Why We Disagree about Climate Change*, climate scientist Mike Hulme (2009) defends appealing to consensus in climate science by calling climate change "a classic example of ... 'post-normal science,'" which he defines as "the application of science to public issues where 'facts are uncertain, values in dispute, stakes high and decisions urgent'" (quoting Silvio Funtowicz and Jerry Ravetz). Issues that fall into this category, he says, are no longer subject to the cardinal requirements of true science: skepticism, universalism, communalism, and disinterestedness. Instead of experimentation and open debate, post-normal science says "consensus" brought about by deliberation among experts determines what is true, or at least temporarily true enough to direct public policy decisions.

The merits and demerits of post-normalism have been debated (see Carter, 2010; Lloyd, 2018), but it is plainly a major deviation from the rules of the Scientific Method and should be met with skepticism by the scientific community. Claiming a scientific consensus exists tempts scientists to simply sign on to the IPCC's latest reports and pursue the research topics and employ the methodologies approved by the IPCC and its member governments. The result is too little hypothesis-testing in climate science and too much amassing of data that can be "dredged" to support the ruling paradigm. U.S. President Dwight Eisenhower (1961) famously warned of such an outcome of government funding of scientific research in his farewell address:

> The free university, historically the fountainhead of free ideas and scientific discovery, has experienced a revolution in the conduct of research. Partly because of the huge costs involved, a government contract becomes virtually a substitute for intellectual curiosity ...

> Yet, in holding scientific research and discovery in respect, as we should, we must also be alert to the equal and opposite danger that public policy could itself become the captive of a scientific-technological elite. The prospect of domination of the nation's scholars by Federal employment, project allocations, and the power of money is ever present – and is gravely to be regarded.

Some scientists see in this arrangement a license to indulge their political biases. Hulme, for example, writes, "The idea of climate change should be seen as an intellectual resource around which our collective and personal identities and projects can form and take shape. We need to ask not what we can do for climate change, but to ask what climate change can do for us" (*Ibid.,* p. 326). This seems quite distant from the rules of the Scientific Method.

The discipline of climate science is young, leaving many basic questions to be answered. Instead of conforming to a manufactured consensus, climate change science outside the reach of the IPCC is full of new discoveries, new theories, and lively debate. New technologies are bringing new discoveries and surprising evidence. Just a few examples:

- Ilyinskaya *et al.* (2018) estimated CO_2 emissions from Katla, a major subglacial volcanic caldera in Iceland, are "up to an order of magnitude greater than previous estimates of total CO_2 release from Iceland's natural sources" and "further measurements on subglacial volcanoes worldwide are urgently required to establish if Katla is exceptional, or if there is a significant previously unrecognized contribution to global CO_2 emissions from natural sources."

- Martinez (2018) compared changes in electric lights seen from satellites in space coming from free and authoritarian countries to the economic growth rates reported by their governments during the same period. He found "yearly GDP growth rates are inflated by a factor of between 1.15 and 1.3 in the most authoritarian regimes" (see also Ingraham, 2018). The integrated assessment models (IAMs) relied on by the IPCC to estimate the "social cost of carbon" rely on this sort of data being accurate.

- Marbà *et al.* (2018) found seagrass meadows in Greenland could be emerging as a major carbon sink. The meadows "appear to be expanding and increasing their productivity. This is supported by the rapid growth in the contribution of seagrass-derived carbon to the sediment C_{org} [organic carbon] pool, from less than 7.5% at the beginning of 1900 to 53% at present, observed in the studied meadows. Expansion and enhanced productivity of eelgrass meadows in the subarctic Greenland fjords examined here is also consistent with the on average 6.4-fold acceleration of C_{org} burial in sediments between 1940 and present."

■ McLean (2018) conducted what is apparently the first ever audit of the temperature dataset (HadCRUT4) maintained by the Hadley Centre of the UK Met Office and the Climatic Research Unit of the University of East Anglia. He identified some 70 "issues" with the database, including "simple issues of obviously erroneous data, glossed-over sparsity of data, [and] significant but questionable assumptions and temperature data that has been incorrectly adjusted in a way that exaggerates warming" (p. i). Simply cleaning up this database, which is relied on by the IPCC for all its analysis, should be a high priority.

■ Kirkby *et al.* (2016) reported the CLOUD research conducted by CERN (the European Institute for Nuclear Research) provided experimental results supporting the theory that variations in the number of cosmic rays hitting Earth's atmosphere create more or fewer (depending on the strength of the solar magnetic wind) of the low, wet clouds that deflect solar heat back into space. Subsequent to the CLOUD experiment, four European research institutes collaborated on a new climate model giving cosmic rays a bigger role than the models used by the IPCC (Swiss National Science Foundation, 2017).

Kreutzer *et al.* (2016) write, "The idea that the science of climate change is 'settled' is an absurdity, contrary to the very spirit of scientific enquiry. Climate science is in its infancy, and if its development follows anything resembling the normal path of scientific advancement, we will see in the years ahead significant increases in our knowledge, data availability, and our theoretical understanding of the causes of various climate phenomena."

Disagreements among scientists about methodology and the verity of claimed facts make it difficult for unprejudiced lay persons to judge for themselves where the truth lies regarding complex scientific questions. For this reason, politicians and even many scientists look for and eagerly embrace claims of a "scientific consensus" that would free them of the obligation to look at the science and reach an informed opinion of their own. Regarding climate change, that is a poor decision. As Essex and McKitrick (2007) write, "non-scientists [should] stop looking for shortcuts around the hard work of learning the science, and high-ranking scientists [should] stop resorting to authoritarian grandstanding as an easy substitute for the slow work of research, debate, and persuasion" p. 15).

It is too early, the issues are too complex, and new discoveries are too many to declare the debate over. In many ways, the debate has just begun.

References

Carter, R.M. 2010. *Climate: The Counter Consensus.* Stacey International.

Curry, J. 2012. Climate change: no consensus on consensus. Climate Etc. (website). October 28. Accessed November 24, 2018.

Curry, J. 2018. Special report on sea level rise. Climate Etc. (website). Accessed November 27, 2018.

Eisenhower, D. 1961. Transcript of President Dwight D. Eisenhower's Farewell Address (1961). OurDocuments (website). Accessed November 25, 2018.

Essex, C. and McKitrick, R. 2007. *Taken By Storm: The Troubled Science, Policy and Politics of Global Warming.* Revised edition. Toronto, ON: Key Porter Books.

Hulme, M. 2009. *Why We Disagree About Climate Change: Understanding Controversy, Inaction and Opportunity.* New York, NY: Cambridge University Press.

Ilyinskaya, E., *et al.* 2018. Globally significant CO_2 emissions from Katla, a subglacial volcano in Iceland. *Geophysical Research Letters* **45** (9).

Ingraham, C. 2018. Satellite data strongly suggests that China, Russia and other authoritarian countries are fudging their GDP reports. *Washington Post.* May 15.

Kirkby, J., *et al.* 2016. Ion-induced nucleation of pure biogenic particles. *Nature* **533**: 521–6.

Kreutzer, D., *et al.* 2016. The state of climate science: no justification for extreme policies. Washington, DC: The Heritage Foundation. April 22.

Landsea, C. 2005. Resignation Letter of Chris Landsea from IPCC. Climatechangefacts.info (website). Accessed November 20, 2018.

Lindzen, R.S. 2012. Climate science: is it currently designed to answer questions? *Euresis Journal* **2** (Winter): 161–92.

Lloyd, E.A. 2018. The role of "complex" empiricism in the debates about satellite data and climate models. In: Lloyd, E.A. and Winsberg, E. (Eds.) *Climate Modelling: Philosophical and Conceptual Issues.* Cham, Switzerland: Springer Nature.

Lloyd, E.A. and Winsberg, E. 2018. *Climate Modelling: Philosophical and Conceptual Issues.* Cham, Switzerland: Springer Nature.

Marbà, N., Krause-Jensen, D., Masqué, P., and Duarte, C.M. 2018. Expanding Greenland seagrass meadows contribute new sediment carbon sinks. *Scientific Reports* **8**: 14024.

Martinez, L.R. 2018. How much should we trust the dictator's GDP estimates? Chicago, IL: University of Chicago Irving B. Harris Graduate School of Public Policy Studies. May 1.

McLean, J. 2018. *An Audit of the Creation and Content of the HadCRUT4 Temperature Dataset.* Robert Boyle Publishing. October.

Seitz, F. 1996. A major deception on global warming. *The Wall Street Journal.* June 12.

Stavins, R. 2014. Is the IPCC government approval process broken? An Economic View of the Environment (blog). April 25.

Swiss National Science Foundation. 2017. Sun's impact on climate change quantified for first time. Phys.org (website). March 27.

Tol, R.S.J. 2014. IPCC again. Richard Tol Occasional thoughts on all sorts (blog). Accessed June 26, 2018.

2.1.1.3 Uncertainty

Uncertainty in science is unavoidable but must be acknowledged. Many declaratory and predictive statements about the global climate are not warranted by science.

Uncertainty in science is unavoidable. Jaynes (2003, p. 54, fn) writes, "incomplete knowledge is the only working material a scientist has!" But uncertainty can be minimized through experimentation and statistical methods such as Bayesian inference. Jaynes continues, "In scientific inference our job is always to do the best we can with whatever information we have; there is no advance guarantee that our information will be sufficient to lead us to the truth. But many of the supposed difficulties arise from an inexperienced user's failure to recognize and use the safety devices that probability theory as logic always provides" (p. 106).

Kelly and Kolstad (1998) report there are two kinds of uncertainty, *stochastic* and *parametric*. The latter can be expected to decline over time as more is learned about the global climate system and the variables and values used for parameters in general circulation models (GCMs) and integrated assessment models (IAMs) are better constrained. Stochastic uncertainty, on the other hand, can increase, decrease, or remain about the same over time. It is a function of various phenomena that impact economic or geophysical processes but are either not included in the models or included incorrectly. Stochastic uncertainties include the effects of earthquakes, volcanic eruptions, or abrupt economic downturns, such as the global financial crisis of 2007–08. A major element of stochastic uncertainty is the fact that we cannot know the future trend of technology or the economy and are, therefore, always susceptible to surprises.

A third source of uncertainty is *epistemic.* Roy and Oberkampf (2011) define this as "[predictive] uncertainty due to lack of knowledge by the modelers, analysts conducting the analysis, or experimentalists involved in validation. The lack of knowledge can pertain to, for example, modeling of the system of interest or its surroundings, simulation aspects such as numerical solution error and computer round-off error, and lack of experimental data." In describing how to treat such error, Helton *et al.* (2010) note, "the mathematical structures used to represent [stochastic] and epistemic uncertainty must be propagated through the analysis in a manner that maintains an appropriate separation of these uncertainties in the final results of interest."

Uncertainties abound in the climate change debate. For example, there is uncertainty regarding pre-modern-era surface temperatures due to reliance on temperature proxies, such as sediment deposition patterns and oxygen isotopes found in ice cores, and in the modern temperature record due to the placement of temperature stations and changes in technology over time. According to Frank (2016),

Field-calibrations reveal that the traditional Cotton Regional Shelter (Stevenson screen) and the modern Maximum-Minimum Temperature Sensor (MMTS) shield suffer daily average 1σ systematic measurement errors of $\pm0.44°C$ or $\pm0.32°C$, respectively, stemming chiefly from solar and albedo irradiance and insufficient windspeed.

Marine field calibrations of bucket or engine cooling-water intake thermometers revealed typical SST [sea surface temperature] measurement errors of $1\sigma = \pm 0.6°C$, with some data sets exhibiting $\pm 1°C$ errors. These systematic measurement errors are not normally distributed, are not known to be reduced by averaging, and must thus enter into the global average of surface air temperatures. Modern floating buoys exhibit proximate SST error differences of $\pm 0.16°C$.

These known systematic errors combine to produce an estimated lower limit uncertainty of $1\sigma = \pm 0.5°C$ in the global average of surface air temperatures prior to 1980, descending to about $\pm 0.36°C$ by 2010 with the gradual introduction of modern instrumentation (abstract).

Frank (2016) observes that when known uncertainties in the temperature record are more properly accounted, reconstruction of the global temperature record reveals so much uncertainty that "at the 95% confidence interval, the rate or magnitude of the global rise in surface air temperature since 1850 is unknowable." He illustrates the point by calculating error bars due to systematic measurement error and adding them to the widely reproduced graph of global temperatures since 1850 created by the Climatic Research Unit at the University of East Anglia. His graphs are reproduced in Figure 2.1.1.3.1. With the measurement uncertainty so great, it is impossible to know whether human emissions of greenhouse gases have had any impact at all on global air temperature.

The human impact on global average temperature is also uncertain due to our incomplete understanding of the carbon cycle (e.g., exchange rates between CO_2 reservoirs) and the atmosphere (e.g., the behavior of clouds), both described in Section 2.1.2 below. Falkowski et al. (2000) admitted, "Our knowledge is insufficient to describe the interactions

Figure 2.1.1.3.1
2010 HadCRUT temperature record with error bars due to systematic measurement errors

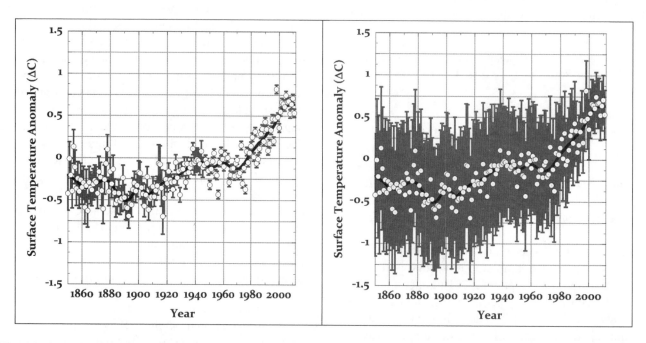

The 2010 global average surface air temperature record obtained from the website of the Climate Research Unit (CRU), University of East Anglia, UK. Left graph shows error bars following the description provided at the CRU website. Right graph shows error bars reflecting uncertainty width due to estimated systematic sensor measurement errors within the land and sea surface records. Source: Frank, 2016, Figure 11, p. 347.

between the components of the Earth system and the relationship between the carbon cycle and other biogeochemical and climatological processes."

Ahlström *et al.* (2017) report "global vegetation models and terrestrial carbon cycle models are widely used for projecting the carbon balance of terrestrial ecosystems. Ensembles of such models show a large spread in carbon balance predictions, ranging from a large uptake to a release of carbon by the terrestrial biosphere, constituting a large uncertainty in the associated feedback to atmospheric CO_2 concentrations under global climate change. ... We conclude that climate bias-induced uncertainties must be decreased to make accurate coupled atmosphere-carbon cycle projections."

Skeie *et al.* (2011) describe another source of uncertainty: measuring human emissions of greenhouse gases. They write, "The uncertainties in present day inventories for [fossil fuel and biofuels black carbon and organic carbon] are about a factor of 2 and there are uncertainties in the rate of change of the emissions and the uncertainties differ in different regions, but are not quantified by Bond *et al.* (2007). Smith *et al.* (2011) found that uncertainties in the regional emissions of SO_2 are far higher than the uncertainties in the global emissions. The 5–95% confidence interval for the global emissions is 9% of the best estimate in 2000 and range between 16% and 7% between 1850 and 2005. The regional uncertainties in the emissions in Former Soviet Union were 20% in 1990 and 30% in China for the year 2000. A formal error propagation for the RF time series of short lived components including the uncertainties in the rate of change and spatial distribution of the emissions is not performed in this study or other published studies. The error estimates for all mechanisms (Fig. 1d) are therefore based on spread found in previous studies."

GCMs, described in Section 2.2.2, and IAMs, described at length in Chapter 8 grapple with the problem of "propagation of error," a term used in statistics referring to how errors or uncertainty in one variable, due perhaps to measurement limitations or confounding factors, are compounded (propagated) when that variable becomes part of a function involving other variables that are also uncertain. Error propagation through sequential calculations is widely used in the physical sciences to reveal the reliability of an experimental result or a calculation from theory. As the number of variables or steps in a function increases, uncertainties multiply until there can be no confidence in the outcomes. In academic literature this is sometimes referred to as "cascading

uncertainties" or "uncertainty explosions." (See Curry and Webster, 2011; and Curry, 2011, 2018.)The IPCC itself illustrated the phenomenon in Working Group II's contribution to the Third Assessment Report (TAR) in the figure reproduced as Figure 2.1.1.3.2. The caption of the image reads, "Range of major uncertainties that are typical in impact assessments, showing the 'uncertainty explosion' as these ranges are multiplied to encompass a comprehensive range of future consequences, including physical, economic, social, and political impacts and policy responses (modified after Jones, 2000, and 'cascading pyramid of uncertainties' in Schneider, 1983" (IPCC, 2001, p. 130).

Figure 2.1.1.3.2
Illustration of cascading uncertainties in climate science

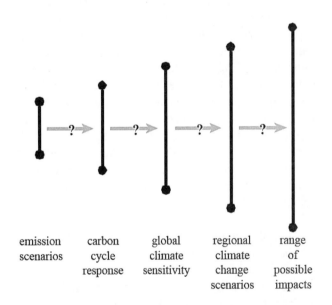

emission scenarios   carbon cycle response   global climate sensitivity   regional climate change scenarios   range of possible impacts

Source: IPCC, 2001, p. 130.

Frank (2015) writes, "It is very well known that climate models only poorly simulate global cloud fraction, among other observables. This simulation error is due to incorrect physical theory. ... [E]ach calculational step delivers incorrectly calculated climate magnitudes to the subsequent step. ... In a sequential calculation, calculational error builds upon initial error in every step, and the uncertainty accumulates with each step" (p. 393). When systematic error is propagated through a model, uncertainty increases with the projection time. When the uncertainty bars become large, no information remains in the projection.

Recognizing this problem, modelers may make their model relatively unresponsive to changes in data or parameter values, creating the appearance of stability until those restraints are questioned and lifted. Hourdin *et al.* (2017), in an article titled "The Art and Science of Climate Model Tuning," write, "Either reducing the number of models or over-tuning, especially if an explicit or implicit consensus emerges in the community on a particular combination of metrics, would artificially reduce the dispersion of climate simulations. *It would not reduce the uncertainty, but only hide it*" (italics added).

Ranges of uncertainty also apply to how to measure alleged climate effects (e.g., loss of livelihood, loss of personal property, forced migration) and how much of the effect to attribute to a specific weather-related event (e.g., flood, drought, hurricane) or to some other non-climate variable (e.g., poverty, civil war, mismanagement of infrastructure). Although considerable progress has been made in climate science and in the understanding of how human activity interacts with and affects the biosphere and economy, significant uncertainties persist in each step of an IAM. As the model progresses through each of these phases, uncertainties surrounding each variable in the chain of computations are compounded one upon another, creating a cascade of uncertainties that peaks upon completion of the final calculation. Tol (2010, p. 79) writes,

> A fifth common conclusion from studies of the economic effects of climate change is that the uncertainty is vast and right- skewed. For example, consider only the studies that are based on a bench-mark warming of 2.5°C. These studies have an average estimated effect of climate change on average output of -0.7% of GDP, and a standard deviation of 1.2% of GDP. Moreover, this standard deviation is only about best estimate of the economic impacts, given the climate change estimates. It does not include uncertainty about future levels of GHG emissions, or uncertainty about how these emissions will affect temperature levels, or uncertainty about the physical consequences of these temperature changes. Moreover, it is quite possible that the estimates are not independent, as there are only a relatively small number of studies, based on similar data, by authors who know each other well.

References

Ahlström, A., Schurgers, G., and Smith, B. 2017. The large influence of climate model bias on terrestrial carbon cycle simulations. *Environmental Research Letters* **12** (1).

Bond, T. C., *et al.* 2007. Historical emissions of black and organic carbon aerosol from energy-related combustion, 1850-2000. *Global Biogeochemical Cycles* **21**, GB2018.

Curry, J.A. 2011. Reasoning about climate uncertainty. *Climatic Change* **108**: 723.

Curry, J.A. 2018. Climate uncertainty & risk. Climate Etc. (blog). July 8. Accessed November 25, 2018.

Curry, J.A. and Webster, P.J. 2011. Climate science and the uncertainty monster. *Bulletin of the American Meteorological Society:* 1667–82.

Falkowski, P., *et al.* 2000. The global carbon cycle: a test of our knowledge of Earth as a system. *Science* **290** (5490): 291–6.

Frank, P. 2015. Negligence, non-science, and consensus climatology. *Energy & Environment* **26** (3): 391-415.

Frank, P. 2016. Systematic error in climate measurements: the global air temperature record. In: Raigaini, R. (Ed.) *The Role of Science in the Third Millennium. International Seminars on Nuclear War and Planetary Emergencies* **48**. Singapore: World Scientific, pp. 337–51.

Helton, J.C., *et al.* 2010. Representation of analysis results involving aleatory and epistemic uncertainty. *International Journal of General Systems* **39** (6): 605–46.

Hourdin, F., *et al.* 2017. The art and science of climate model tuning. *Bulletin of the American Meteorological Society* **98** (3): 589–602.

IPCC. 2001. Intergovernmental Panel on Climate Change. *Climate Change 2001: Impacts, Adaptation and Vulnerability.* Contribution of Working Group II to the Third Assessment Report of the Intergovernmental Panel on Climate Change. New York, NY: Cambridge University Press.

Jaynes, E.T. 2003. *Probability Theory: The Logic of Science.* Cambridge, MA: Cambridge University Press.

Jones, R.N. 2000. Analysing the risk of climate change using an irrigation demand model. *Climate Research* **14**: 89–100.

Kelly, D.L. and Kolstad, C.D. 1998. *Integrated Assessment Models for Climate Change Control*. U.S. Department of Energy grant number DE-FG03-96ER62277. November 1998.

Roy, C.J. and Oberkampf, W.L. 2011. A comprehensive framework for verification, validation, and uncertainty quantification in scientific computing. *Computer Methods in Applied Mechanics and Engineering* **200** (25–28): 2131–44.

Schneider, S.H. 1983. CO₂, climate and society: a brief overview. In: R.S. Chen, E. Boulding, and S.H. Schneider (Eds.), *Social Science Research and Climate Change: An Interdisciplinary Appraisal*. Boston, MA: D. Reidel, pp. 9–15.

Skeie, R.B., *et al.* 2011. Anthropogenic radiative forcing time series from pre-industrial times until 2010. *Atmospheric Chemistry and Physics* **11**: 11827–57.

Smith, S.J., *et al.* 2011. Anthropogenic sulfur dioxide emissions: 1850–2005. *Atmospheric Chemistry and Physics* **11**: 1101–16.

Tol, R.S.J. 2010. Carbon dioxide mitigation. In: B. Lomborg (Ed.) *Smart Solutions to Climate Change: Comparing Costs and Benefits*. Cambridge, MA: Cambridge University Press.

2.1.2 Observations

Science depends on observational data to form and test hypotheses. In climate science, key observational data relate to energy flows in the atmosphere characterized as the energy budget; the movement of carbon among reservoirs, called the carbon cycle; warming and cooling periods seen in the geological and historical records; and sources and behavior of greenhouse gases (GHGs) such as carbon dioxide (CO_2) and methane (CH_4).

2.1.2.1 Energy Budget

Surface air temperature is governed by energy flow from the Sun to Earth and from Earth back into space. Whatever diminishes or intensifies this energy flow can change air temperature.

Figure 2.1.2.1.1 presents a recent effort by the Intergovernmental Panel on Climate Change (IPCC) to characterize and quantify heat flows in Earth's atmosphere. Incoming energy must equal outgoing energy for Earth's temperature to be stable over long periods of time, a balance called radiative equilibrium. Hence, the label "energy budget" is applied to the phenomenon.

The principal source of energy entering Earth's atmosphere is the Sun, providing irradiance, solar wind (plasma), and solar magnetism. The amount of energy reaching any particular point at the top of the atmosphere varies dramatically depending on latitude, season, and diurnal phase (daytime or nighttime). A global average of approximately 340 watts of solar power per square meter (Wm^2, one watt is one joule of energy every second) hits the top of the atmosphere, 100 Wm^2 is reflected back into space by clouds, and approximately 240 Wm^2 enters the atmosphere. Approximately 239 Wm^2 leaves Earth's atmosphere as thermal energy, creating a net "imbalance" of 0.6° [0.2°, 1.0°] Wm^2. That imbalance is the first order cause of rising surface temperatures.

Whatever diminishes or intensifies the energy flow from the Sun to Earth and from Earth back into space can change air temperature. However, the dynamics are not straightforward. Natural variation in the planet's energy budget occurs without human influence, some of the mechanisms are non-linear, and the physics is poorly understood. For example, the changing intensity of the Sun, the planet's changing magnetic field, and galactic cosmic rays all affect incoming solar at the top of the atmosphere. Changes to Earth's albedo (reflectivity) due to changes in snow and ice cover and land use can affect the amount of energy leaving the planet.

The amount of energy reflected back into space by clouds is assumed to be (on average) a constant, but even small changes in cloud cover, cloud brightness, and cloud height – all of which are known to vary spatially and over time and none of which is well modeled – could alter this key variable in the energy budget enough to explain the slight warming of the twentieth century (Lindzen, 2015, p. 55; Hedemann *et al.*, 2017). Surface air temperature at any one place on Earth's terrestrial surface is determined by many factors, only one of which is the small change in the global temperature that presumably emerges from the stylized energy flows shown in Figure 2.1.2.1.1. One such factor is turbulence, which Essex and McKitrick call "one of the most basic and intractable research problems facing humanity. You can't compute it. You can't measure it. But rain falls because of it" (Essex and McKitrick, 2007, p. 20).

Figure 2.1.2.1.1
Global mean energy budget of Earth under present-day climate conditions

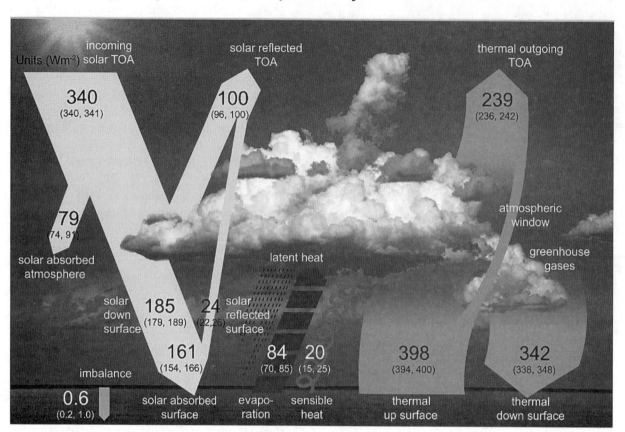

TOA = top of the atmosphere. Imbalance (0.6° [0.2°, 1.0°] Wm² shown at bottom left corner of the image) is thought to be the imbalance causing a net warming of the atmosphere. *Source:* IPCC, 2013, Table 2.11, p. 181.

Earth's rotation produces gyres and flows in two dynamic fluids – the atmosphere and the oceans. Oceans cover more than 70% of Earth's surface and hold approximately 1,000 times as much heat as the atmosphere. This means Earth's surface temperature does not adjust quickly to changes in the atmosphere "due to the ocean's thermal inertia, which is substantial because the ocean is mixed to considerable depths by winds and convection. Thus it requires centuries for Earth's surface temperature to respond fully to a climate forcing" (Hansen *et al.*, 2012). The fluid dynamics of these systems are not well understood. Coupled with rotation, the flows in these two fluids create internal variability in the climate system. The exchange of energy within or between the oceans and the atmosphere can cause one or the other to warm or cool even without any change in the heat provided by the Sun.

El Niño and La Niña cycles dominate the flux of water and energy in the tropical Pacific over periods of two to seven years. These cyclical episodes normally last between nine and 12 months and are part of a complex cycle referred to as the El Niño-Southern Oscillation (ENSO). El Niño, a winter phenomenon (Northern Hemisphere), refers to a period of anomalously warm water in the Central and Eastern Pacific. La Niña is the cold counterpart to the El Niño phenomenon. Both events can have large impacts on global temperatures, rainfall, and storm patterns.

Long-term changes in solar energy entering the top of the atmosphere are caused by changes in the Sun itself as well as Milankovitch cycles – variations in the Earth's orbit due to eccentricity (the changing shape of the Earth's orbit around the Sun), axial tilt (oscillations in the inclination of the Earth's axis in relation to its plane of orbit around the Sun), and precession (the planet's slow wobble as it spins on its axis) on cycles of approximately 100,000, 41,000, and 23,000 years, respectively. Changes in these orbital characteristics affect the seasonal contrasts experienced on Earth. Minimal seasonal contrasts (i.e., cooler summers, warmer winters) are conducive to cooler periods while greater seasonal contrasts promote warmer climate episodes. The prevailing thinking is that warmer winters result in higher snowfall amounts and cooler summers lead to reduced melting of the winter snowpack. The net effect is to raise the planet's reflectivity (albedo), driving temperatures lower over time. The opposite occurs during periods of high seasonal contrast.

The heat energy of fossil fuel combustion is very small compared to the natural heat flux from the Sun and other processes. It is estimated that the total man-made combustion energy amounts to about $0.031\ Wm^2$, averaged over the surface of Earth. The Sun provides $340\ Wm^2$, nearly 11,000 times more. The Sun is responsible for nearly 100% of the heat coming to Earth. A very small fraction is contributed by heat rising through the crust from the molten core.

References

Essex, C. and McKitrick, R. 2007. *Taken by Storm: The Troubled Science, Policy, and Politics of Global Warming.* Toronto, ON: Key Porter Books Limited.

Hansen, J., *et al.* 2012. *Earth's energy imbalance.* Science Briefs (website). U.S. National Aeronautics and Space Administration (NASA). Accessed November 23, 2018.

Hedemann, C., *et al.* 2017. The subtle origins of surface-warming hiatuses. *Nature Climate Change* **7**: 336–9.

IPCC. 2013. Intergovernmental Panel on Climate Change. *Climate Change 2013: The Physical Science Basis.* Contribution of Working Group I to the Fifth Assessment Report of the Intergovernmental Panel on Climate Change. New York, NY: Cambridge University Press.

Lindzen, R.S. 2015. Global warming, models and language. In: Moran, A. (Ed.) *Climate Change: The Facts.* Institute of Public Affairs. Woodville, NH: Stockade Books.

2.1.2.2 Carbon Cycle

Levels of carbon dioxide (CO₂) and methane (CH₄) in the atmosphere are governed by processes of the carbon cycle. Exchange rates and other climatological processes are poorly understood.

Earth's energy budget only partly explains the natural processes that determine surface temperatures. Carbon dioxide (CO_2) and methane (CH_4) are two gases whose rising presence in the atmosphere contributes to rising temperatures. Their concentration in the atmosphere is a function of complex processes characterized in biology as the "carbon cycle." A typical simplified rendering of the cycle appears in Figure 2.1.2.2.1. The Intergovernmental Panel on Climate Change (IPCC) presents a more detailed but no more accurate rendering of the cycle in Figure 6.2 in the Working Group I contribution to its Fifth Assessment Report (IPCC, 2013, p. 474).

Carbon and hydrogen appear abundantly throughout the universe and on Earth. Carbon's unique function as the base element for Earth's biosphere derives from it being the lightest element capable of forming four covalent bonds with atoms of most elements in many variations. ("Covalent bonds" involve the sharing of electron pairs and are stronger than bonds involving single electrons.) The resulting molecules can contain from one to millions of carbon atoms. Carbon is so abundant and apt to bond with other atoms that the discipline of chemistry is divided into *organic chemistry,* which studies only carbon-based compounds, and *inorganic chemistry,* which studies all other compounds. Carbon-based compounds comprise the overwhelming majority of the tens of millions of compounds identified by scientists.

Carbon Reservoirs

Carbon on Earth is stored in four reservoirs: rocks and sediments (lithosphere), oceans and lakes (hydrosphere), vegetation and soil (biosphere), and the air (atmosphere). Contrary to casual assertions and sometimes feigned certainty, the amount of carbon stored in each reservoir and the exchanges among reservoirs are not known with certainty. There is no way to actually measure things as large as the lithosphere or atmosphere, so all estimates of their sizes depend on measurements performed on small parts and fed into models to generate global estimates. These estimates vary widely depending on assumptions made by models. Predictably, estimates of reservoir sizes vary in the literature. The numbers in Figure 2.1.2.2.1 and below from Ruddiman (2008)

Figure 2.1.2.2.1
Schematic of the global carbon cycle

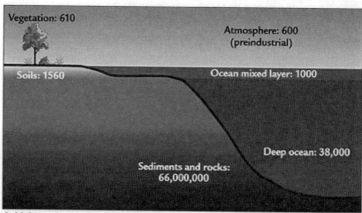

A Major carbon reservoirs (gigatons)

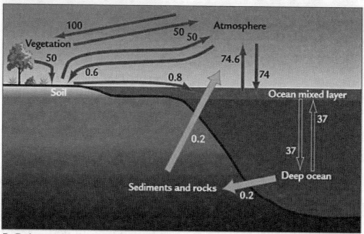

B Carbon exchange rates (gigatons/year)

Source: Modified from Ruddiman, 2008, p. 46. See original for sources.

differ, for example, from those appearing in Falkowski *et al.* (2000), which at the time was thought to be authoritative, and the IPCC's Fifth Assessment Report (IPCC, 2013, pp. 471–4), also thought to be authoritative. According to Ruddiman, carbon is distributed among reservoirs as follows:

Gigatons of Carbon (GtC)	Reservoirs
66,000,000	Rocks and sediment (lithosphere)
39,000	Oceans (hydrosphere)
2,170	Vegetation and soils (biosphere)
600	Air (atmosphere) (preindustrial)

The carbon stored in rocks and sediment dominates the distribution, some 110,000 times as much as the amount in the air. That carbon is mostly in the form of carbonate rocks such as limestone (which is mainly calcium carbonate), marble, chalk, and dolomite. Most of the carbon stored in Earth's mantle was there when Earth formed. Some of the carbon in Earth's crust was deposited there in the form of undecayed biomass produced by the biosphere, buried by the mechanics of plate tectonics, and turned by metamorphism into fossil fuels.

Carbon moves from rocks and sediment into the atmosphere via outgassing from midocean ridges and hotspot volcanoes, leakage of crude oil on the ocean floor, weathering of rocks, upward percolation and migration from deeper in the lithosphere, and human drilling, transporting, and burning of fossil fuels. Just how much carbon is released from the lithosphere in any given year is uncertain since empirical measurements are available for only ~20% of major volcanic gas emission sources. Many researchers suspect computer models of the carbon cycle underestimate the impact of volcanic activity on ocean currents, sea surface temperature, and ice formation and melting (e.g., Viterito, 2017; Smirnov *et al.*, 2017; Kobashi *et al.*, 2017; Slawinska and Robock, 2018; and Ilyinskaya *et al.*, 2018). Wylie (2013) reported,

> In 1992, it was thought that volcanic degassing released something like 100 million tons of CO_2 each year. Around the turn of the millennium, this figure was getting closer to 200 [million]. The most recent estimate, released this February, comes from a team led by Mike Burton, of the Italian National Institute of Geophysics and Volcanology [Burton *et al.*, 2013] – and it's just shy of 600 million tons. It caps a staggering trend: A six-fold increase in just two decades.

Oceans are the second largest reservoir of carbon, containing about 65 times as much as the air. The IPCC and other political and scientific bodies assume roughly 40% of the CO_2 produced by human combustion of fossil fuels is absorbed and sequestered by the oceans, another 15% by plants and animals (terrestrial as well as aquatic), and what is left remains in the air, contributing to the slow increase in atmospheric concentrations of CO_2 during the modern era.

When CO_2 dissolves into water, it reacts with water molecules to form carbonic acid, which increases concentrations of the hydrogen carbonate (HCO_3-) and carbonate (CO_3-_2) ions, which in turn form carbonate rocks, in the process removing CO_2 from the air, a process called weathering. Deep-sea calcium carbonate sediments also neutralize large amounts of CO_2 by reacting with and dissolving it. A hypothetical carbon equilibrium between the atmosphere and the world's oceans would probably show the oceans have assimilated between 80% and 95% of the anthropogenic CO_2 from the atmosphere (World Ocean Review 1, 2010).

Vegetation and soils are the third largest reservoir of carbon, containing approximately 3.6 times as much as the air. The carbon in living matter is derived directly or indirectly from carbon dioxide in the air or dissolved in water. Algae and terrestrial green plants use photosynthesis to convert CO_2 and water into carbohydrates, which are then used to fuel plant metabolism and are stored as fats and polysaccharides. The stored products are then eaten by other organisms, from protozoans to plants to man, which convert them into other forms. CO_2 is added to the atmosphere by animals and some other organisms via respiration. The carbon present in animal wastes and in the bodies of all organisms is also released into the air as CO_2 by decay (DiVenere, 2012).

Uncertainty pervades estimates of the size of the biospheric carbon reservoir. Bastin *et al.* (2017), using high-resolution satellite images covering more than 200,000 plots, found tree-cover and forests in drylands "is 40 to 47% higher than previous estimates, corresponding to 467 million hectares of forest that have never been reported before. This increases current estimates of global forest cover by at least 9%." Their finding, they write, "will be important in estimating the terrestrial carbon sink."

Earth's atmosphere, the fourth carbon reservoir, holds the least carbon – about 600 GtC before human combustion of fossil fuels began to make a measurable contribution. The current best estimate is approximately 870 GtC. Carbon in the atmosphere exists mainly as CO_2 and methane. Pathways for carbon to enter the atmosphere from other reservoirs have already been described. Carbon dioxide leaves the atmosphere by dissolving into bodies of water (oceans, rivers, and lakes), being taken up by plant leaves, branches, and roots during the process of photosynthesis, and being absorbed by the soil.

Carbon dioxide composes approximately 400 parts per million (ppm) of the atmosphere by volume and methane approximately 1,800 parts per billion (ppb). Atmospheric concentrations of both substances have increased since the start of the Industrial Era and both increases are thought to be largely due to human activities, with CO_2 coming from the burning of fossil fuels and methane from agricultural practices and the loss of carbon sinks due to changes in land use. However, uncertainties in measurement and new discoveries cast doubt on this assumption. In any case, compared to natural sources of carbon in the environment, the human contribution is very small.

Exchange Rates

The carbon cycle acts as a buffer to minimize the hypothesized impact of atmospheric CO_2, whether from natural or man-made sources, on surface temperatures. It provides an illustration of Le Chatelier's principle, which states, "when a stress is applied to a chemical system at equilibrium, the equilibrium concentrations will shift in a direction that reduces the effect of the stress" (ChemPRIME, 2011). In the case of CO_2 and the carbon cycle, atmospheric CO_2 is in equilibrium with dissolved CO_2 in the oceans and the biosphere. An increased level of CO_2 in the atmosphere will result in more CO_2 dissolving into water, partly offsetting the rise in atmospheric concentrations as well as forming more carbonic acid, which causes more weathering of rocks, forming calcium ions and bicarbonate ions, which remove more CO_2 from the air. The biosphere will also increase its uptake of CO_2 from the air (the aerial fertilization effect) and sequester some of it in woody plants, roots, peat, and other sediments (Idso, 2018). Atmospheric CO_2 concentrations will then fall, restoring the system to equilibrium.

In contrast to this first order process, where the rate of exchange among reservoirs is directly proportional to the amount of carbon present, the IPCC's carbon cycle model assumes an uptake that scales with the emission rate and not the actual concentration. Such models can never come to a new equilibrium for a slightly increased but constant emission rate due to natural or human influences. The IPCC also contends the time frames on which some parts of the carbon cycle operate, such as the weathering cycle and dissolving into deep oceans, are

too long for them to help offset during the twenty-first century the sudden pulse of CO_2 from the combustion of fossil fuels in the twentieth century. But the fact that some exchange processes operate slowly and others rapidly does not mean, *prima facie*, that an entire pulse of CO_2 cannot be absorbed by the faster-acting parts of the carbon cycle. The different sinks for CO_2 act in parallel and add up to a total uptake as a collective effect, determined by the fastest, not the slowest, sinks (Harde, 2017). The rapid response of the biosphere is seen in the widely documented "greening of the Earth" discussed in Chapter 5, Section 5.3.

The IPCC estimates that between 200 and 220 GtC enters Earth's atmosphere each year. Of that total, 8.9 GtC is anthropogenic (7.8 GtC from fossil fuels and 1.1 GtC from net land use change (agriculture)). The total human contribution, then, is only about 4.3% of total annual releases of carbon into the atmosphere (IPCC, 2013, p. 471, Figure 6.1). The IPCC "assessed that about 15 to 40% of CO_2 emitted until 2100 will remain in the atmosphere longer than 1,000 years" and "the removal of all the human-emitted CO_2 from the atmosphere by natural processes will take a few hundred thousand years (high confidence)," citing Archer and Brovkin (2008) and reproducing the table shown in Figure 2.1.2.2.2 (p. 472).

Human use of fossil fuels contributes only about 3.5% (7.8 Gt divided by 220 Gt) of the carbon entering the atmosphere each year and so, with about 0.5% (1.1 Gt divided by 220 Gt) from net land use change, natural sources account for the remaining 96.0%. The residual of the human contribution the IPCC

Figure 2.1.2.2.2
Time required for natural processes to remove CO_2 from atmosphere (IPCC AR5)

Processes	Time scale (years)	Reactions
Land uptake: Photosynthesis–respiration	$1-10^2$	$6CO_2 + 6H_2O + photons \rightarrow C_6H_{12}O_6 + 6O_2$ $C_6H_{12}O_6 + 6O_2 \rightarrow 6CO_2 + 6H_2O + heat$
Ocean invasion: Seawater buffer	$10-10^3$	$CO_2 + CO_3^{2-} + H_2O \rightleftharpoons 2HCO_3^-$
Reaction with calcium carbonate	10^3-10^4	$CO_2 + CaCO_3 + H_2O \rightarrow Ca^{2+} + 2HCO_3^-$
Silicate weathering	10^4-10^6	$CO_2 + CaSiO_3 \rightarrow CaCO_3 + SiO_2$

Source: IPCC, 2013, p. 472, Box 6.1, Table 1, citing Archer and Brovkin, 2008.

believes remains in the atmosphere after natural processes move the rest to other reservoirs is as little as 1.17 Gt per year (15% of 7.8 Gt), just 0.53% of the carbon entering the atmosphere each year. This is less than two-tenths of 1% (0.195%) of the total amount of carbon thought to be in the atmosphere, per Ruddiman (2008).

The lasting human contribution of carbon emitted to the atmosphere by the use of fossil fuels, according to the IPCC's own estimates, is minuscule, less than 1% of the natural annual flux among reservoirs. As stated earlier, all estimates of the amount of carbon in the four reservoirs and the exchange rates among them are uncertain and constantly being revised in light of new findings. Yet the IPCC assumes exchange rates are estimated with sufficient accuracy to say "It is *extremely likely* that more than half of the observed increase in global average surface temperature from 1951 to 2010 was caused by the anthropogenic increase in greenhouse gas concentrations and other anthropogenic forcings together," while "the contribution of natural forcings is likely to be in the range of −0.1°C to 0.1°C and from natural internal variability is likely to be in the range of −0.1°C to 0.1°C" (IPCC, 2013, p. 17). This seems improbable.

The atmospheric CO_2 trend is a minute residual between titanic sources and sinks that mostly cancel out each other. While measurable in ambient air, the residual is likely to be less than the margin of error in measurements of the reservoirs or natural variability in their exchange rates. The IPCC came close to acknowledging this in the Working Group I contribution to its Third Assessment Report (IPCC, 2001, p. 191), writing "Note that the gross amounts of carbon annually exchanged between the ocean and atmosphere and between the land and atmosphere, represent a sizeable fraction of the atmospheric CO_2 content – and are many times larger than the total anthropogenic CO_2 input. In consequence, an imbalance in these exchanges could easily lead to an anomaly of comparable magnitude to the direct anthropogenic perturbation."

Why does the IPCC assume exchange rates will not continue changing to keep pace with human contributions of CO_2 to the atmosphere, as they have accommodated both natural and anthropogenic changes in the past? A case can be made based on chemistry and biological science that keeping pace should be the null hypothesis, instead of the IPCC's apparent assumptions that some reservoirs already are or soon will be saturated, or that some fraction of anthropogenic CO_2 will remain in the atmosphere

until only very slow natural processes such as weathering and deep ocean sequestration can remove it.

Residence Time

Regarding residence time (the average time carbon spends in a given reservoir), according to Harde (2017, p. 20), "Previous critical analyses facing the IPCC's favored interpretation of the carbon cycle and residence time have been published," citing Jaworowski *et al.* (1992), Segalstad (1998), Dietze (2001), Rörsch *et al.* (2005), Essenhigh (2009), Salby (2012, 2016), and Humlum *et al.* (2013). "Although most of these analyses are based on different observations and methods, they all derive residence times (in some cases also differentiated between turnover and adjustment times) in part *several orders of magnitude shorter than specified in [the Fifth Assessment Report]*. As a consequence of these analyses also a much smaller anthropogenic influence on the climate than propagated by the IPCC can be expected" (italics added).

Harde (2017) derives a residence time of his own, writing, "for the preindustrial period, for which the system is assumed to be in quasi equilibrium, a quite reliable estimate of the average residence time or lifetime can be derived from the simple relation, that under steady state the emission or absorption rate times the average residence time gives the total CO_2 amount in the atmosphere" (p. 21). He calculates a residence time of just three years. Over the industrial era, using the IPCC's own exchange rate estimates, he finds a residence time of 4.1 years. He notes, "a residence time of 4 years is in close agreement with different other independent approaches for this quantity," identifying tests on the fall-out from nuclear bomb testing and solubility data while referencing Sundquist (1985), Segalstad (1998), and Essenhigh (2009).

While the IPCC says it would take longer than one thousand years for oceans and the biosphere to absorb whatever residue of human-produced CO_2 remains after all use of fossil fuels is somehow halted, Harde finds the IPCC's own accounting scheme shows it would take no more than 47.8 years, this derived from the IPCC's own accounting scheme, which considers a slightly increased absorption rate of 2.4%, forced by the instantaneous anthropogenic emission rate of 4.3%. An even more coherent approach presupposing a first order uptake process and no longer distinguishing between a natural and anthropogenic cycle, this results in a unique time scale, the residence time of only four years.

References

Archer, D. and Brovkin, V. 2008. The millennial atmospheric lifetime of anthropogenic CO_2. *Climatic Change* **90** (3): 283–97.

Bastin, J.-F., *et al.* 2017. The extent of forest in dryland biomes. *Science* **356** (6338): 635–8.

Burton, M.R., Sawyer, G.M., and Granieri, D. 2013. Deep carbon emissions from volcanoes. *Reviews in Mineralogy and Geochemistry* **75** (1): 323–54.

ChemPRIME. 2011. Le Chatelier's principle: The carbon cycle and the climate (website). Accessed May 24, 2018.

Dietze, P. 2001. IPCC's most essential model errors (website); Carbon Model Calculations (website). Accessed August 13, 2018.

DiVenere, V. 2012. *The carbon cycle and Earth's climate.* Information sheet for Columbia University Summer Session 2012 Earth and Environmental Sciences Introduction to Earth Sciences.

Essenhigh, R.E. 2009. Potential dependence of global warming on the residence time (RT) in the atmosphere of anthropogenically sourced carbon dioxide. *Energy Fuel* **23**: 2773–84.

Falkowski, P., *et al.* 2000. The global carbon cycle: a test of our knowledge of Earth as a system. *Science* **290** (5490): 291–6.

Harde, H. 2017. Scrutinizing the carbon cycle and CO_2 residence time in the atmosphere. *Global and Planetary Change* **152**: 19–26.

Humlum, O., Stordahl, K., and Solheim, J.E. 2013. The phase relation between atmospheric carbon dioxide and global temperature. *Global and Planetary Change* **100**: 51–69.

Idso, C.D. 2018. CO_2 Science Plant Growth Study Database (website).

Ilyinskaya, E., *et al.* 2018. Globally significant CO_2 emissions from Katla, a subglacial volcano in Iceland. *Geophysical Research Letters* **45** (9).

IPCC. 2001. Intergovernmental Panel on Climate Change. *Climate Change 2001: The Scientific Basis.* Contribution of Working Group I to the Third Assessment Report of the Intergovernmental Panel on Climate Change. New York, NY: Cambridge University Press.

IPCC. 2013. Intergovernmental Panel on Climate Change. *Climate Change 2013: The Physical Science Basis.* Contribution of Working Group I to the Fifth Assessment Report of the Intergovernmental Panel on Climate Change. New York, NY: Cambridge University Press.

Jaworowski, Z., Segalstad, T.V., and Ono, N. 1992. Do glaciers tell a true atmospheric CO_2 story? *Science of the Total Environment* **114**: 227–84.

Kobashi, T., *et al.* 2017. Volcanic influence on centennial to millennial Holocene Greenland temperature change. *Scientific Reports* **7**: Article 1441.

Rörsch, A., Courtney, R.S., and Thoenes, D. 2005. Global warming and the accumulation of carbon dioxide in the atmosphere. *Energy & Environment* **16**: 101–25.

Ruddiman, W.F. 2008. *Earth's Climate: Past and Future.* Second edition. New York, NY: W.H. Freeman and Company.

Salby, M. 2012. *Physics of the Atmosphere and Climate.* New York, NY: Cambridge University Press.

Salby, M. 2016. Atmospheric carbon. Video presentation. University College London. July 18.

Segalstad, T.V. 1998. Carbon cycle modelling and the residence time of natural and anthropogenic atmospheric CO_2: on the construction of the "Greenhouse Effect Global Warming" dogma. In: Bate, R. (Ed.) *Global Warming: The Continuing Debate.* Cambridge, UK: ESEF, pp. 184–219.

Slawinska, J. and Robock, A. 2018. Impact of volcanic eruptions on decadal to centennial fluctuations of Arctic Sea ice extent during the last millennium and on initiation of the Little Ice Age. *Journal of Climate* **31**: 2145–67.

Smirnov, D.A., Breitenbach, S.F.M., Feulner, G., Lechleitner, F.A., Prufer, K.M., Baldini, U.L., Marwan, N., and Kurths, J. 2017. A regime shift in the Sun-Climate connection with the end of the Medieval Climate Anomaly. *Scientific Reports* **7**: 11131.

Sundquist, E.T. 1985. Geological perspectives on carbon dioxide and the carbon cycle. In: Sundquist, E.T. and Broecker, W.S. (Eds.) *The Carbon Cycle and Atmospheric CO_2: Natural Variations Archean to Present.* Washington, DC: American Geophysical Union, pp. 5–59.

Viterito, A. 2017. The correlation of seismic activity and recent global warming: 2016 update. *Environment Pollution and Climate Change* **1** (2): 1000103.

World Ocean Review 1. 2010. The Oceans, the Largest CO_2 Reservoir. Hamburg, Germany: maribus gGmbH.

Wylie, R. 2013. Long invisible, research shows volcanic CO_2 levels are staggering. Live Science (website). October 15. Accessed August 13, 2018.

2.1.2.3 Geological Record

The geological record shows temperatures and CO_2 levels in the atmosphere have not been stable, making untenable the IPCC's assumption that they would be stable in the absence of human emissions.

Estimates of CO_2 concentrations in the atmosphere and local surface temperatures in the distant past can be made by extrapolation from proxy data, which the IPCC defines as "a record that is interpreted, using physical and biophysical principles, to represent some combination of climate-related variations back in time. ... Examples of proxies include pollen analysis, tree ring records, speleothems, characteristics of corals and various data derived from marine sediments and ice cores" (IPCC, 2013,

p. 1460). The most valuable proxy data come from oxygen and hydrogen isotopes in ice and CO_2 in air bubbles preserved in ice cores obtained by drilling in Antarctica and Greenland. Temperature is inferred from the isotopic composition of the water molecules released by melting the ice cores. During colder periods, there will be a higher ratio of ^{16}O to ^{18}O and ^{2}H (also known as deuterium) to ^{1}H in the ice formed than would be found during warm periods. Once again it is important to note that reconstructions of past climatic conditions are not actually data. Like "carbon reservoirs" and "exchange rates," such reconstructions rely on very limited data fed into models and subject to interpretation by scientists.

Proxy data reveal temperatures have varied considerably over the past 600 million years (Lamb, 2011, 2012). Earth's orbital changes, known as Milankovitch cycles and described previously, are the generally accepted explanation for these broad changes in temperatures. Figure 2.1.2.3.1 shows one reconstruction of changes in temperature (blue) and CO_2 levels (purple) for the past 570 million years or

Figure 2.1.2.3.1
Global temperature and atmospheric CO_2 concentration over the past 600 million years

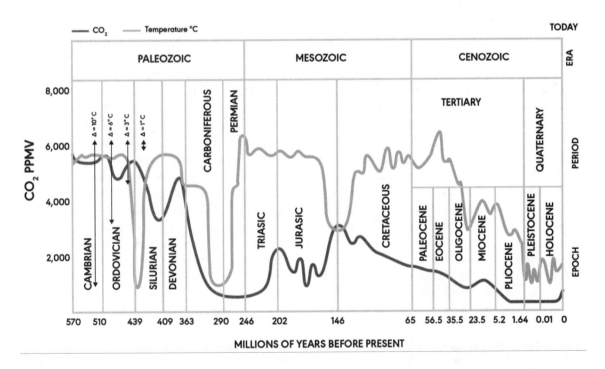

Purple line is CO_2 concentration (ppm); blue line is change in temperature ($\Delta°C$). Horizontal scale is not in constant units. CO_2 scale derived from ratios to levels at around 1911 (300 ppm) calculated by Berner and Kothavala, 2001. *Source*: Adapted from Nahle, 2009 referencing Ruddiman, 2001; Scotese, 2002; and Pagani *et al.*, 2005.

so. Both temperature and and CO_2 are lower today than they have been during most of the era of modern life on Earth since the Cambrian Period. For more than 2.5 million years (the Pleistocene Epoch) the world was in a cold period with long glaciations (ice ages) interrupted by relatively brief warm periods of typically 10,000 to 15,000 years. On this time scale, temperature and CO_2 are completely without correlation. Note "before present" means before 1950, so warming and CO_2 levels since then are not shown.

We have been in the current Holocene Epoch warm period for about 11,500 years. Within the Holocene, there is strong physical evidence for periods of both global warming and cooling, although those periods are less extreme than the Milankovitch-forced glaciation cycles. The current warm period was preceded by the Little Ice Age (1300–1850 AD), which was preceded by the Medieval Warm Period or Medieval Climate Optimum (800–1300 AD), which was preceded by the Dark Ages Cold Period (400–800 AD), which was preceded by the Roman Warm Period or Roman Climate Optimum (250 BC–400 AD). Before that there is evidence of a Minoan Warm Period (~2500 BC) and a thousand-year Holocene Climate Optimum about 6,500 years ago.

Most of the "warm periods" or "climate optimums" are thought to have been at least as warm as Earth's current climate. The Greenland Ice Sheet Project Two (GISP2) used ice cores to estimate temperatures between 1,500 to 10,000 years ago, shown in Figure 2.1.2.3.2 (Alley, 2000). These findings are validated by global glacial advances and retreats, oxygen isotope data from cave deposits, tree ring data, and historic records (Singer and Avery, 2007). Within the past 5,000 years, the Roman Warm Period appears prominently in the GISP2 ice core, about 1,500–1,800 years ago. During that period, ancient Romans wrote of grapes and olives growing farther north in Italy than had been previously thought possible, as well as of there being little or no snow or ice.

Oxygen isotope data from the GISP2 Greenland ice cores show the prominent Medieval Warm Period (MWP) occurring around 900–1300 AD. The MWP was followed by a period of global cooling and the beginning of the Little Ice Age, which spanned the sixteenth to the eighteenth centuries, though some scientists date its start much earlier. The effects of the MWP are also seen in the reconstructions of sea surface temperature near Iceland by Sicre *et al.* (2008), reproduced as Figure 2.1.2.3.3 below. During the MWP in Europe, grain crops flourished, alpine tree lines rose, and the population more than doubled. The Vikings took advantage of the warmer climate to colonize Greenland. The MWP was a global event with proxy data confirming the warm period found in Africa (Lüning *et al.*, 2017), South America (Lüning *et al.,* 2018), North America (McGann, 2008), China (Hong *et al.,* 2009), and many other areas (NIPCC, 2011, Chapter 3). More recent temperature records are discussed in Section 2.2.1. Four observations from the geological record of the carbon cycle, as shown by ice core records and other proxy data, should guide any discussion of the human impact on Earth's climate. First, the concentration of CO_2 in the atmosphere today is below levels that existed during most of the geologic record. Figure 2.1.2.3.1 graphs CO_2 and temperature over geological time with temperature in blue, atmospheric CO_2 concentration in purple, and the trend in CO_2 concentration represented by the purple arrow. Moore (2016, p. 8) writes, "Note the uptick [in CO_2 concentrations] at the far right of the graph representing the reversal of the 600 million-year downward trend due primarily to emissions of CO_2 from the use of fossil fuels for energy. Note that even today, at 400 ppm, CO_2 is still far lower than it has been during most of this 600 million-year history." Figure 2.1.2.3.1 also shows the average level of CO_2 in the atmosphere over the geological span encompassing the evolution and spread of plants, about 300 million years before present, was probably approximately 1,000 ppm, more than twice today's level.

The dramatic fall in atmospheric CO_2 concentrations is significant in the climate change discussion because virtually all species of plant and animal life in the world today arose, evolved, and flourished during periods when atmospheric CO_2 levels were much higher than they are today. Moreover, during the Paleocene-Eocene Thermal Maximum, the average global temperature was 16°C (28.8°F) higher than temperature today. This suggests today's species will survive or even thrive if CO_2 levels rise to several times their current levels and if temperatures increase more than 2°C or 3°C (3.6° or 5.4° F), the increase the IPCC claims would result in unacceptable and irreversible ecological harm.

The second observation from the geological record is that CO_2 levels in the atmosphere are not stable, making untenable the IPCC's assumption that they would be stable in the future in the absence of human emissions. The increase in atmospheric CO_2 concentrations in the modern era, while dramatic when viewed as a trend over thousands or even hundreds of thousands of years, is a brief reversal of

Figure 2.1.2.3.2
Temperatures from Greenland ice cores

A. Greenland GISP2 oxygen isotope curve for the past 10,000 years

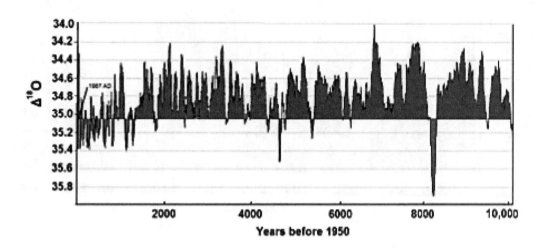

B. Greenland GISP2 oxygen isotope curve for the past 5,000 years.

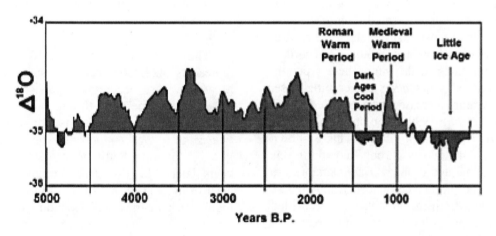

The vertical axis is $\delta^{18}O$, which is a temperature proxy. Horizontal scale for (a) is 10,000 years before 1950 and for (b) is past 5,000 years. The red areas represent temperatures warmer than present (1950). Blue areas are cooler times. Note the abrupt, short-term cooling 8,200 years ago and cooling from about 1500 A.D. to 1950. *Source:* Alley, 2000, plotted from data by Grootes and Stuiver, 1997.

Figure 2.1.2.3.3
Summer sea surface temperature near Iceland

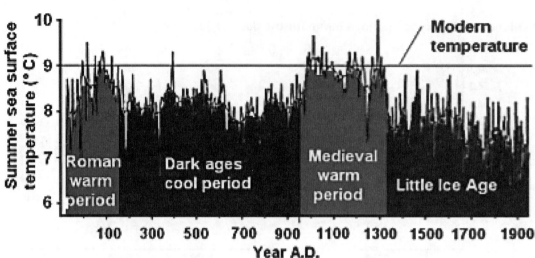

Source: Sicre *et al.*, 2008.

a multi-million-year trend. Today's atmospheric CO_2 concentrations are "unprecedented" only because they are *lower* than at most other points in the record. The third observation from the geological record is that CO_2 concentrations in the atmosphere typically rose several hundred years *after* temperatures rose, indicating temperature increase was not caused by the CO_2 rise (Petit *et al.*, 1999; Monnin *et al.*, 2001; Mudelsee, 2001; Caillon *et al.*, 2003). This is shown in Figure 2.1.2.3.4, where carbon dioxide levels appear as a blue line and changes in temperature are plotted in red. The graph, reproduced from Mearns (2014), shows data from the Vostok ice core drilled in 1995. Mearns explains how the record is created:

> [T]he temperature signal is carried by hydrogen: deuterium isotope abundance in the water that makes the ice whilst the CO_2 and CH_4 signals are carried by air bubbles trapped in the ice. The air bubbles trapped by ice are always deemed to be younger than the ice owing to the time lag between snow falling and it being compacted to form ice. In Vostok, the time lag between snow falling and ice trapping air varies between 2000 and 6500 years. There is therefore a substantial correction applied to bring the gas ages in alignment with the ice ages and the accuracy

of this needs to be born in mind in making interpretations.

The Vostok ice core is 3,310 meters long and represents 422,766 years of snow accumulation. Mearns writes, "There is a persistent tendency for CO_2 to lag temperature throughout and this time lag is most pronounced at the onset of each glacial cycle *'where CO_2 lags temperature by several thousand years'*" (quoting Pettit *et al.*, 1999). Writing three years later, Mearns (2017) comments, "It is quite clear from the data that CO_2 follows temperature with highly variable time lags depending upon whether the climate is warming or cooling. ... The general picture is one of quite strong-co-variance, but in detail there are some highly significant departures where temperature and CO_2 are clearly de-coupled" and "CO_2 in the past played a negligible role [in determining temperature]. It simply responded to bio-geochemical processes caused by changing temperature and ice cover."

For other recent temperature record reconstructions showing the "CO_2 lag" see Soon *et al.* (2015), Davis (2017), and Lüning and Vahrenholt (2017). During periods of glaciation, cooling oceans absorb more CO_2 due to the "solubility pump," which Moore defines as "the high solubility of CO_2 in cold

Figure 2.1.2.3.4
Temperature and CO₂ co-variance in the Vostok ice core

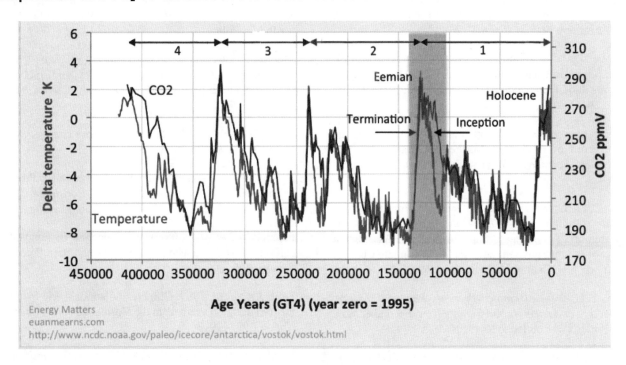

CO₂ and temperature appear well-correlated in a gross sense but there are some significant deviations. At the terminations, the alignment is good but upon descent into the following glaciation there is a time lag between CO₂ and temperature of several thousand years. *Source:* Mearns, 2014.

ocean water at higher latitudes where sinking cold sea-water carries it into the depths of the ocean" (Moore, 2016, p. 10). During warmer inter-glacial periods, oceans absorb less CO₂ or outgas more of it into the air. Plant life absorbs more CO₂ from the air during warm periods than during cold periods, having a countercyclical effect but one that is much smaller given that the ocean reservoir is approximately 65 times as large as the biosphere.

The fourth observation from the geological record is that the rise in CO₂ levels since the beginning of the Industrial Age, whether due to human emissions from the use of fossil fuels and changes to land use or the result of ocean outgassing caused by cyclical warming, could be averting an ecological disaster. As Moore observes, "on a number of occasions during the present Pleistocene Ice Age, CO₂ has dropped during major glaciations to dangerously low levels relative to the requirements of plants for their growth and survival. At 180 ppm, there is no doubt that the growth of many plant

species was substantially curtailed" (Moore, 2016, p. 10, citing Ward, 2005).

"If humans had not begun to use fossil fuels for energy," Moore continues, "it is reasonable to assume that atmospheric CO₂ concentration would have continued to drop as it has for the past 140 million years," perhaps to levels so low during the next glaciation period as to cause "widespread famine and likely the eventual collapse of human civilization. This scenario would not require two million years but possibly only a few thousand" (Moore, 2016, pp. 16–17).

References

Alley, R.B. 2000. The Younger Dryas cold interval as viewed from central Greenland. *Quaternary Science Reviews* **19** (1–5): 213–26.

Berner, R.A. and Kothavala, Z. 2001. GEOCARB III: A revised model of atmospheric CO₂ over Phanerozoic time. *American Journal of Science* **301**: 182–204.

Caillon, N., *et al.* 2003. Timing of atmospheric CO_2 and Antarctic temperature changes across termination III. *Science* **299** (5613): 1728–31.

Davis, W.J. 2017. The relationship between atmospheric carbon dioxide concentration and global temperature for the last 425 million years. *Climate* **5** (76).

Grootes, P.M. and Stuiver, M. 1997. Oxygen 18/6 variability in Greenland snow and ice with 103- to 105-year time resolution. *Journal of Geophysical Research* **102** (C12): 26455–70.

Hong, Y.t., *et al.* 2009. Temperature evolution from the ∫18O record of Hani peat, Northeast China, in the last 14000 years. *Science in China Series D: Earth Sciences* **52**: 952–64.

IPCC. 2013. Intergovernmental Panel on Climate Change. *Climate Change 2013: The Physical Science Basis*. Contribution of Working Group I to the Fifth Assessment Report of the Intergovernmental Panel on Climate Change. New York, NY: Cambridge University Press.

Lamb, H.H. 2011. *Climate: Present, Past and Future, Volume 1: Fundamentals and Climate Now*. Second edition. New York, NY: Routledge Revivals.

Lamb, H.H. 2012. *Climate: Present, Past and Future: Volume 2: Climatic History and the Future*. New York, NY: Routledge.

Lüning, S. and Vahrenholt, F. 2017. Paleoclimatological context and reference level of the 2°C and 1.5°C Paris agreement long-term temperature limits. *Frontiers in Earth Science* **5**: 104.

Lüning, S., Galka, M., and Vahrenholt, F. 2017. Warming and cooling. The medieval climate anomaly in Africa and Arabia. *Paleoceanography and Paleoclimatology* **32** (11): 1219-35.

Lüning, S., *et. al.* 2018. The medieval climate anomaly in South America. *Quaternary International*. In press. https://doi.org/10.1016/j.quaint.2018.10.041.

McGann, M. 2008. High-resolution foraminiferal, isotopic, and trace element records from Holocene estuarine deposits of San Francisco Bay, California. *Journal of Coastal Research* **24**: 1092–109.

Mearns, E. 2014. The Vostok ice core: temperature, CO_2 and CH_4. Energy Matters (website). December 12. Accessed August 13, 2018.

Mearns, E. 2017. The Vostok ice core and the 14,000 year CO_2 time lag. Energy Matters (website). June 14. Accessed August 13, 2018.

Monnin, E., *et al.* 2001. Atmospheric CO_2 concentrations over the last glacial termination. *Science* **291** (5501): 112–14.

Moore, P. 2016. *The Positive Impact of Human CO_2 on the Survival of Life on Earth*. Winnipeg, MB: Frontier Centre for Public Policy.

Mudelsee, M. 2001. The phase relations among atmospheric CO_2 content, temperature and global ice volume over the past 420 ka. *Quaternary Science Reviews* **20**: 583–589.

Nahle, N. 2009. Cycles of global climate change. *Biology Cabinet Journal Online* (website). July.

NIPCC. 2011. Nongovernmental International Panel on Climate Change. *Climate Change Reconsidered: 2011 Interim Report*. Chicago, IL: The Heartland Institute.

Pagani, M., *et al.* 2005. Marked decline in atmospheric carbon dioxide concentrations during the Paleocene. *Science* **309**: 600–3.

Petit, J.R., *et al.* 1999. Climate and atmospheric history of the past 420,000 years from the Vostok ice core, Antarctica. *Nature* **399**: 429–36.

Ruddiman, W.F. 2001. *Earth's Climate: Past and Future*. New York, NY: W.H. Freeman and Co.

Scotese, C.R. 2002. *Analysis of the Temperature Oscillations in Geological Eras*. Paleomar Project (website).

Sicre, M-A., *et al.* 2008. Decadal variability of sea surface temperatures off North Iceland over the last 200 yrs. *Earth Planet Science Letters* **268** (3–4): 137–42.

Singer, S.F. and Avery, D. 2007. *Unstoppable Global Warming—Every 1500 Years*. Lanham, MD: Rowman and Littlefield.

Soon, W., Connolly, R., and Connolly, M. 2015. Re-evaluating the role of solar variability on Northern Hemisphere temperature trends since the 19th century. *Earth-Science Reviews* **150**: 409–52.

Ward, J.K. 2005. Evolution and growth of plants in a low CO_2 world. In: Ehleringer, J., Cerling, T., and Dearning, D. (Eds.) *A History of Atmospheric CO_2 and Its Effects on Plants, Animals, and Ecosystems*. New York, NY: Springer-Verlag.

2.1.2.4 Greenhouse Gases

Water vapor is the dominant greenhouse gas owing to its abundance in the atmosphere and the wide range of spectra in which it absorbs radiation. Carbon dioxide absorbs energy only in a very narrow range of the longwave infrared spectrum.

A central issue in climate science is how much of the energy flowing through the atmosphere is obstructed by atmospheric gases called, erroneously, "greenhouse gases." (The label is erroneous because greenhouses warm the air inside by preventing convection, a process different than how these gases behave in the atmosphere. But the label was coined in 1963 and is the preferred term today.) Many laboratories have conducted repeated tests on the radiative properties of gases for more than a century, with handbooks reporting the results since the 1920s. All gases absorb energy at various wavelengths.

Carbon dioxide (CO_2), the greenhouse gas at the center of the climate change debate, is an invisible, odorless, tasteless, non-toxic gas that is naturally present in the air and essential for the existence of all plants, animals, and humans on Earth. In the photosynthesis process, plants remove CO_2 from the atmosphere and release oxygen, which humans and animals breathe in. CO_2 in the atmosphere does not harm humans directly. In confined spaces, such as in submarines or spacecraft, CO_2 concentrations can build up and threaten human health and safety – but only at concentrations more than 20 times the current trace levels in our atmosphere. Nuclear submarines commonly contain 5,000 parts per million (ppm) of CO_2 after more than a month below the surface (Persson and Wadsö, 2002). The current level of CO_2 in the atmosphere is approximately 405 ppm.

Radiative Properties

A two-atom molecule can spin and oscillate, while a three-atom molecule can also bend, which adds to the possibilities for interactions with radiation. Oxygen (O_2) and nitrogen (N_2) are symmetrical molecules, meaning they are linear and also include only a single element. The molecular stretches of two identical elements do not involve moving charges, so the molecule cannot bend. The chemical elements in CO_2 are different. The C-O stretches of CO_2 include moving charges because the molecular electrons are not symmetrically distributed. These moving atomic charges induce an oscillating electromagnetic (EM) field around the CO_2 molecule. That field can now couple with the EM field of infrared (IR) radiation. The energy quanta associated with the CO_2 bending

mode transition corresponds to a photon of 15 μm longwave infrared radiation (Burch and Williams, 1956; Wilson and Gea-Banacloche, 2012). Similar properties of water vapor explain why it too can absorb radiation in the EM field, but across a much wider range of wavelengths.

When CO_2 absorbs IR radiation, it becomes vibrationally excited. This means the C-O atoms oscillate back-and-forth more quickly and with greater amplitude than they did before the IR was absorbed. The vibrationally excited CO_2 molecule strikes an oxygen (O_2) or nitrogen (N_2) molecule in the air and transfers that vibrational energy to the O_2 or N_2. That energy transfer causes the N_2 or O_2 average velocity – a measure of "translational kinetic energy" – to increase. It is like (to dramatize) slamming a car with a backhoe, causing the car to speed up. That greater translational kinetic energy is also a measure of the "thermal energy" of the gas as measured by its temperature. Water vapor behaves similarly in its respective wavelength spectrum. Briefly put, greenhouse gases transform IR radiation into vibrational energy, and then offload that vibrational energy into air molecules as thermal energy, which is injected into the atmosphere.

Absorption properties vary from gas to gas as shown in Figure 2.1.2.4.1. Some gases, such as nitrogen and oxygen, absorb energy in the ultraviolet spectrum, where wavelengths are shorter than visible light. The greenhouse gases in Earth's atmosphere are mostly transparent to the ultraviolet and visible wavelengths, meaning almost no energy is absorbed by these gases in that part of the spectrum. However, greenhouse gases do absorb energy in the far infrared spectrum (a/k/a longwave infrared or LWIR), at wavelengths much longer than visible light that are invisible to the human eye. If no further radiation of a particular wavelength is absorbed, the wavelength is said to be "saturated."

Figure 2.1.2.4.1 shows how water vapor is the dominant greenhouse gas owing to its abundance in the atmosphere and the wide range of spectra in which it absorbs LWIR radiation. CO_2 absorbs energy only in a very narrow range of the infrared spectrum, a wavelength of 15 μm (micrometers), and is overlapped by the water vapor range. Other greenhouse gases together absorb less than 1% of upgoing LWIR. With increasing altitude water vapor "condenses out" and falls as rain or snow and the concentration of water vapor falls to a few parts per million. When water vapor condenses to liquid water at high altitude, it radiates its excess thermal energy into space. This dynamic process cools and stabilizes

Figure 2.1.2.4.1
Radiation transmitted by the atmosphere and wavelengths absorbed by the five most important greenhouse gases

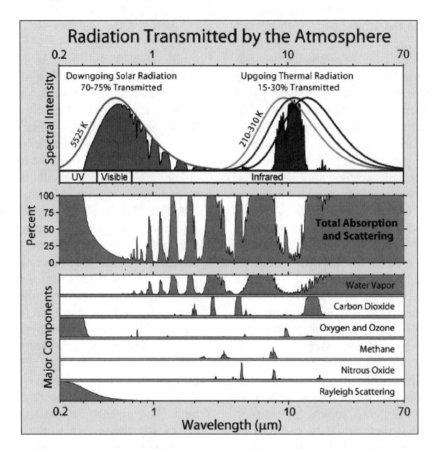

The data used for these figures are based primarily on Spectral Calculator of GATS, Inc. which implements the LINEPAK system of calculating (Gordley *et al.*, 1994) from the HITRAN2004 (Rothman *et al.*, 2004) spectroscopic database. To aid presentation, horizontal scale is not proportional and the absorption spectra were smoothed. Features with a bandwidth narrower than 0.5% of their wavelength may be obscured. *Source:* Wikipedia Commons, n.d.

the climate. At about 10 km (33,000 feet), CO_2, which does not "condense out," becomes the most abundant greenhouse gas.

The ability of greenhouse gases to absorb energy in the longwave infrared spectrum is important because Earth gives off much more LWIR than it receives, so gases that absorb LWIR have a warming effect by preventing the escape of some of the LWIR radiation to space. As shown in Figure 2.1.2.4.1, water vapor emits and absorbs infrared radiation at many more wavelengths than any of the other greenhouse gases, and there is substantially more water vapor in the atmosphere than any of the other greenhouse gases. While this means water vapor can

generate thermal energy, water vapor also has a net *cooling* effect when it forms clouds. Clouds can reflect sunshine back into space during the day, but they can also reflect LWIR downward at night, keeping the surface warmer. Increasing levels of water vapor in the atmosphere, then, by increasing cloud formation, could result in nights getting warmer without increases in day-time high temperatures. Observations confirm this occurred during the twentieth century (Alexander *et al.*, 2006).

The effect of water vapor on surface temperature is especially important because CO_2 is not capable of causing significant warming by itself, due to the narrow range of the spectra it occupies. Scientists aligned with the IPCC claim CO_2 raises global temperature slightly and that rise, in turn, produces an increase in water vapor, which is capable of increasing atmospheric temperature (Chung *et al.*, 2014). But the hydrology of the atmosphere and dynamics of the ocean-atmosphere interface are poorly understood and modeled (Legates, 2014; Christy and McNider, 2017). Whether that effect is large enough to account for the warming of the twentieth century and early twenty-first centuries is a topic of debate and research.

Sources

The greenhouse gases that occur naturally and are also produced by human activities include water vapor, carbon dioxide (CO_2), methane (CH_4), nitrous oxide (N_2O), and ozone (O_3). Other greenhouse gases produced only by human activities include the fluorinated gases such as chlorofluorocarbons (CFCs), hydrochlorofluorocarbons (HCFCs), and hydrofluorocarbons (HFCs). Water vapor is by far the most prevalent greenhouse gas, with an average global ground-level concentration of approximately 14,615 ppm, about 1.5% of the atmosphere near the surface (less in the deserts, more in the tropics) (Harde, 2017). Carbon dioxide is present in the atmosphere at about 405 ppm, methane at 1.8 ppm, and nitrous oxide at 324 ppb. Fluorinated gases are measured in parts per trillion (IPCC, 2013, pp. 167-168).

As reported earlier, the IPCC estimates that between 200 and 220 GtC enters Earth's atmosphere each year. Of that total, 7.8 GtC comes from the combustion of fossil fuels and 1.1 GtC from net land use change (agriculture). The total human contribution, then, is only about 4.3% of total annual releases of carbon into the atmosphere (IPCC, 2013, p. 471, Figure 6.1). Human use of fossil fuels contributes only about 3.5% (7.8 Gt divided by 220 Gt) of the carbon entering the atmosphere each year and so, with about 0.5% (1.1 Gt divided by 220 Gt) from net land use change, natural sources account for the remaining 96.0%.

Carbon dioxide is readily absorbed by water (i.e., the oceans and rain). All else being equal, cold water will absorb more CO_2 than warm water. When the oceans cool they absorb more atmospheric CO_2; when they warm they release CO_2, thereby increasing the CO_2 concentration of the atmosphere. Consequently, over the past million years atmospheric CO_2 levels have varied, with warm "interglacials" causing higher atmospheric CO_2 levels due to oceanic outgassing, and colder glacial periods causing lower atmospheric CO_2 levels due to oceanic absorption. This variation in the CO_2 concentration of the atmosphere over the past 800,000 years is evident in the ice core data presented in Section 2.1.2.3.

Over the past million years, changes in atmospheric CO_2 level are believed to have been primarily due to release or absorption by oceans. However, human emissions related to the use of fossil fuels, manufacture of cement, and changes in land use (agriculture and forestry) are likely to be the main cause of the increase from approximately 280 ppm in 1800 to 405 ppm in 2018. Deforestation is often cited as a contributor of CO_2 into the atmosphere, but the well-documented "greening of the Earth" caused by the increase in atmospheric CO_2 appears to have offset any reduction in global biomass due to agriculture and forestry (De Jong *et al.*, 2012; NIPCC, 2014, pp. 493–508). Figure 2.1.2.4.2 shows the rising concentrations of CO_2 and CH_4 since 1800.

Approximately 40% of the CO_2 released into the atmosphere by human activities is believed to be absorbed by the oceans. Figure 2.1.2.4.3 shows a recreation by Tans (2009) of "cumulative emissions consistent with observed CO_2 increases in the atmosphere and ocean," with uncertainties identified by horizontal lines through diamonds for the years to which they apply. The net terrestrial emissions (from plants, including aquatic plants) are derived "as a residual from better determined terms in the budget" (p. 29). Tans writes, "the atmospheric increase [in CO_2 concentrations] was primarily caused by land use until the early part of the twentieth century, but the net cumulative emissions from the terrestrial biosphere peaked in the late 1930s at ~45 GtC, dwindling to ~20 GtC in the first decade of the twenty-first century" (p. 30).

Tans (2009) also calculated the role fossil fuel emissions may play in determining future CO_2 levels in the atmosphere. He writes, "Instead of adopting the common economic point of view, which, through its emphasis on perpetual growth, implicitly assumes infinite Earth resources, or at least infinite substitutability of resources, let us start with an estimate of global fossil fuel reserves" (p. 32). Using data from the World Energy Council (WEC, 2007), Tans estimates there are 640 GtC of proved reserves and 967 GtC of resources (potential reserves when extraction technology improves). Consumption of fossil fuels rises exponentially at first as the easiest to exploit reserves are used but then declines as more

Figure 2.1.2.4.2
Atmospheric carbon dioxide (CO₂) and methane (CH₄) levels, 1800–present

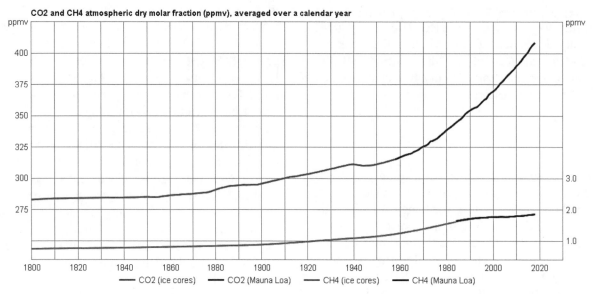

Source: Burton, 2018. See original for data sources.

Figure 2.1.2.4.3
Cumulative emissions and reservoir change

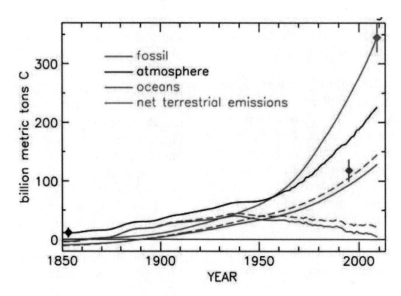

History of cumulative emissions consistent with observed CO_2 increases in the atmosphere and ocean. Uncertainties in the fossil fuel emissions and the accumulations in the atmosphere and ocean are plotted with vertical lines for the years in which they apply. The Hamburg Ocean Carbon Cycle (HAMOCC3) model does not fit the observed cumulative ocean uptake. Therefore, there are two versions of cumulative ocean uptake and net terrestrial emissions: solid lines indicate HAMOCC3 and dashed lines indicate empirical pulse response function. *Source:* Tans, 2009, Figure 2, p. 29.

expensive and energy-intensive methods are needed to extract remaining reserves. Tans applies a logistic model where the rate of extraction is

$$E = dQ/dt = kQ(1 - Q/N)$$

where E is the rate of extraction, Q is cumulative extraction, k is the initial exponential rate of growth, and $(1 - Q/N)$ expresses the increasing difficulty of extraction, which results in slowing of growth. The peak rate of extraction occurs when Q equals half of the total resource. Tans runs the model for two emission scenarios, one assuming 1,000 GtC of fossil fuels will eventually be used and the second, 1,500 GtC. The results are shown in the figure reproduced as Figure 2.1.2.4.4 below.

The Right Climate Stuff

The only scenario presented by the IPCC that does not assume some implementation of worldwide greenhouse gas (GHG) emission controls is RCP8.5, indicating the Representative Concentration Pathway (RCP) scenario would create 8.5 Wm^2 GHG forcing

of global temperature in 2100. RCP8.5 is an extreme outlier in the climate science literature, assuming abnormally high estimates of world population growth and energy use and the absence of any technological improvements in energy efficiency that would lower per-capita growth in CO_2 equivalent (CO_{2eq}) emissions. RCP8.5 would result in more greenhouse gas emissions than 90% of any emissions scenarios published in the technical literature (Riahi *et al.*, 2011). RCP8.5 also does not take into account the fossil-fuel supply and demand changes modeled by Tans (2009) as remaining reserves become more scarce or expensive to develop or the effects of conservation and fuel substitution as prices rise.

Doiron (2016) observes RCP8.5 assumes that by 2100 there would be 930 ppm of CO_2 in the atmosphere. This is 55% more than the 600 ppm of CO_2 that could be generated by burning all the currently known worldwide reserves of coal, oil, and natural gas, according to the U.S. Energy Information Administration's estimates. Working with The Right Climate Stuff (TRCS), a research team composed of retired NASA scientists and engineers, Doiron developed and validated a simple algebraic model for forecasting global mean surface temperature (GMST)

Figure 2.1.2.4.4
Fossil fuel emissions and atmospheric carbon dioxide concentrations, actual and forecast

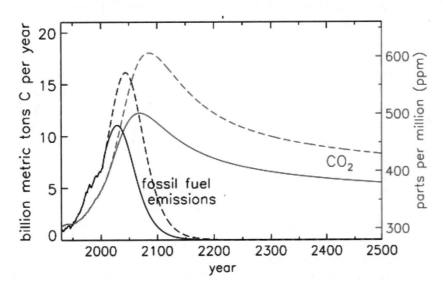

Potential emissions in billion metric tons per year (black lines, left axis) and resulting atmospheric concentration in parts per million (red lines, right axis) for two emission scenarios. Solid lines represent Scenario A (emissions totaling 1,000 GtC), dashed lines represent Scenario B (emissions totaling 1,500 GtC. *Source*: Tans, 2009, Figure 4, p. 32.

using a standard transient climate sensitivity (TCS) variable but basing its forecast of atmospheric CO_2 concentration on historical data and known reserves of fossil fuels. Additionally, the TRCS metric includes a constant, β, based on historical data to account for the warming effects of greenhouse gases other than CO_2 and aerosols. In the TRCS model, TCS with the beta variable is TCS x $(1 + \beta) = 1.8°C$.

The TRCS research team developed two scenarios, RCP6.0 and RCP6.2, projecting CO_2 concentrations of 585 ppm and 600 ppm, respectively, by the year 2100. These scenarios determined a market-driven transition to alternative energy sources would need to *begin* by 2060 to meet the worldwide demand for energy, which will be growing even as reserves of coal, oil, and natural gas are declining and their prices rising. The RCP6.0 and RCP6.2 scenarios project the transition to alternative fuels would be complete by 2130 or 2100, respectively.

The IPCC's Fifth Assessment Report also presented an RCP6.0 scenario, which assumed implementation of modest CO_2 emission controls before 2100. The IPCC and TRCS RCP6.0 scenarios project similar trajectories of CO_2 concentration in the atmosphere by 2100, but the IPCC assumes worldwide controls on the use of fossil fuels would be required to achieve this RCP while the TRCS attributes falling emissions to market forces and a depleting supply of worldwide reserves of coal, oil, and natural gas. The 25-year forecasts for coal, oil, and natural gas consumption published by ExxonMobil (ExxonMobil, 2016) and British Petroleum (BP, 2016) align closely with the fossil-fuel consumption estimates included in the TRCS RCP6.0 and RCP6.2 scenarios.

Figure 2.1.2.4.5 shows the results of the TRCS model using the RCP6.2 emission scenario. The HadCRUT4 temperature anomaly database, which reaches back to 1850, appears on the left side on the vertical axis. The atmospheric CO_2 concentration in ppm is displayed on the right side on the vertical axis. The CO_2 concentration since 1850 and the RCP6.2 projection for the remainder of this century are represented by the green curve, with the sensitivity metric in the model represented by the blue curve. It was found to provide the best fit of the model to the data's long-term temperature increase trends.

The blue curve, the sensitivity metric, threads the narrow path between upper ranges of the temperature data and anomalous data points known to be associated with Super El Niño weather events. Model results with higher values of the sensitivity metric,

represented by the red curve and red dashed curve in Figure 2.1.2.4.5, are clearly too sensitive based on historical CO_2 and temperature measurements. The GMST increase above current conditions for the TRCS model forecasting the RCP6.2 scenario is less than 2°C from 1850 to 2100 and less than 1°C from 2015 to 2100.

* * *

A basic understanding of the Earth's energy budget, carbon cycle, and geological record, and the chemical and radiative properties of greenhouse gases, helps clarify some of the key issues in climate science. Earth's "energy budget" explains how even a small change in the composition of the planet's atmosphere could lead to a net warming or cooling trend, but it also highlights the enormity of the natural processes and their variability compared to the impacts of the human presence, and therefore the difficulty of discerning a human "signal" or influence. The carbon cycle minimizes the impact of human emissions of CO_2 by reforming it into other compounds and sequestering it in the oceans, plants, and rocks. The exact size of any of these reservoirs is unknown, but they necessarily stay in balance with one another – Le Chatelier's principle – by exchanging huge amounts of carbon. According to the IPCC, the residual of the human contribution of CO_2 that remains in the atmosphere after natural processes move the rest to other reservoirs is as little as 0.53% of the carbon entering the air each year and 0.195% of the total amount of carbon thought to be in the atmosphere.

The geological record shows (a) the concentration of CO_2 in the atmosphere today is below levels that existed during most of the record, (b) CO_2 levels in the atmosphere are not stable in the absence of human emissions, (c) CO_2 concentrations in the atmosphere typically rise several hundred years *after* temperatures rise, and (d) the rise in CO_2 levels since the beginning of the Industrial Age may have averted an ecological disaster.

Understanding the atmospheric concentrations and radiative properties of greenhouse gases reveals the important role water vapor plays in Earth's temperature. Water vapor near the surface is present at concentrations approximately 36 times that of CO_2 (14,615 ppm versus 405 ppm) and it absorbs upgoing thermal radiation on a much wider range of wavelengths. However, water vapor's concentration decreases with altitude, making CO_2 the more powerful greenhouse gas at higher levels of the atmos-

Figure 2.1.2.4.5
TRCS validated model with RCP6.2 greenhouse gas and aerosol projections

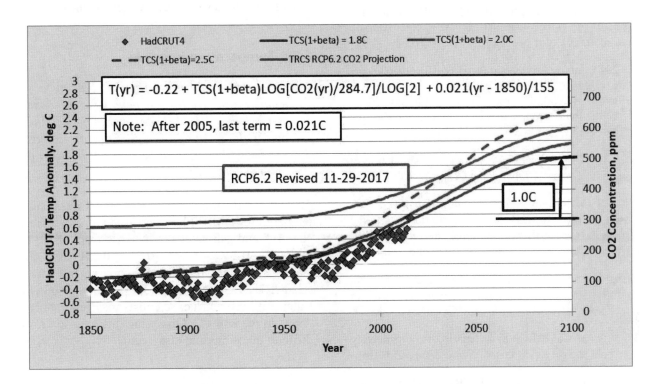

See text for notes. *Source:* Doiron, 2016.

phere. Is a small increase in CO_2 enough to trigger an increase in water vapor sufficient to explain the warming of the twentieth and early twenty-first centuries? This is one important question climate scientists are trying to answer.

References

Alexander, L.V., *et al.* 2006. Global observed changes in daily climate extremes of temperature and precipitation. *Journal of Geophysical Research* **111** (D5).

BP. 2016. BP Energy Outlook to 2035.

Burch, D.E. and Williams, D. 1956. Infrared transmission of synthetic atmospheres II: absorption by carbon dioxide. *Journal of the Optical Society of America* **46**: 237–41.

Burton, D. 2018. Atmospheric carbon dioxide (CO_2) and Methane (CH_4) levels, 1800-present. Sealevel.info (website). Accessed November 23, 2018.

Christy, J.R. and McNider, R.T. 2017. Satellite bulk tropospheric temperatures as a metric for climate sensitivity. *Asia-Pacific Journal of Atmospheric Sciences* **53** (4): 511–18.

Chung, E-S., *et al.* 2014. Upper-tropospheric moistening in response to anthropogenic warming. *Proceedings of the National Academy of Sciences USA* **111** (32): 11636–41.

De Jong, R., Verbesselt, J., Chaepman, M.E., and De Bruin, S. 2012. Trend changes in global greening and browning: contribution of short-term trends to longer-term change. *Global Change Biology* **18**: 642–55.

Doiron, H.H. 2016. *Recommendations to the Trump Transition Team Investigating Actions to Take at the Environmental Protection Agency (EPA): A Report of The Right Climate Stuff Research Team*. November 30, p. 20.

ExxonMobil. 2016. The Outlook for Energy: A View to 2040. ExxonMobil (website). Accessed December 22, 2018.

Gordley, L.L., *et al.* 1994. LINEPAK: Algorithms for modeling spectral transmittance and radiance. *Journal of Quantitative Spectroscopy & Radiative Transfer* **52** (5): 563–80.

Harde, H. 2017. Radiation transfer calculations and assessment of global warming by CO_2. *International Journal of Atmospheric Sciences*: Article 9251034.

IPCC. 2013. Intergovernmental Panel on Climate Change. *Climate Change 2013: The Physical Science Basis.* Contribution of Working Group I to the Fifth Assessment Report of the Intergovernmental Panel on Climate Change. New York, NY: Cambridge University Press.

Legates, D. 2014. Climate models and their simulation of precipitation. *Energy & Environment* **25** (6-7): 1163–75.

NIPCC. 2014. Nongovernmental International Panel on Climate Change. Idso, C.D, Idso, S.B., Carter, R.M., and Singer, S.F. (Eds.) *Climate Change Reconsidered II: Biological Impacts.* Chicago, IL: The Heartland Institute.

Persson, O. and Wadsö, L. 2002. Indoor air quality in submarines. In: Levin, H. (Ed.) *Ninth International Conference on Indoor Air Quality and Climate,* pp. 806–11.

Riahi, K., *et al.* 2011. RCP 8: a scenario of comparatively high greenhouse gas emissions. *Climatic Change* **33** (109).

Rothman, L.S., *et al.* 2004. The HITRAN 2004 molecular spectroscopic database. *Journal of Quantitative Spectroscopy & Radiative Transfer* **96**: 139–204.

Tans, P. 2009. An accounting of the observed increase in oceanic and atmospheric CO_2 and an outlook for the future. *Oceanography* **22**: 26–35.

WEC. 2007. World Energy Council. *Survey of Energy Resources 2007.* London, UK.

Wikipedia Commons. n.d. File: Atmospheric Transmission.png. Wikipedia (website). Accessed November 20, 2018.

Wilson, D.J. and Gea-Banacloche, J. 2012. Simple model to estimate the contribution of atmospheric CO_2 to the earth's greenhouse effect. *American Journal of Physics* **80** (2012): 306–15.

2.2 Controversies

Climate science has made great strides in recent decades thanks especially to satellite data and research laboratories such as CERN, the European Organization for Nuclear Research located on the border between France and Switzerland. However, this progress has not ended controversies that prevent general agreement on some key topics. This section looks at controversies in four areas: temperature records, general circulation models, climate sensitivity, and solar influences on the climate.

2.2.1 Temperature Records

Reconstructions of average global surface temperature differ depending on the methodology used. The warming of the twentieth and early twenty-first centuries has not been shown to be beyond the bounds of natural variability.

The IPCC says it is "certain that global mean surface temperature (GMST) has increased since the late 19th century" and estimates the increase from 1850–1900 to 2003–2012 was 0.78°C [0.72 to 0.85] based on the Hadley Center/Climatic Research Unit dataset (HadCRUT4) (IPCC, 2013, p. 37). While this statement may be true, scientists may reasonably ask "how do we know?" Answering this question reveals difficult questions and disagreements.

How Do We Know?

Recall from Section 2.1.1.3 that Frank (2016) added error bars around the HadCRUT4 temperature record, producing the figure reproduced as Figure 2.1.1.3.1. According to Frank, "these known systematic errors combine to produce an estimated lower limit uncertainty of $1\sigma = \pm 0.5$°C in the global average of surface air temperatures prior to 1980, descending to about ± 0.36°C by 2010 with the gradual introduction of modern instrumentation." The IPCC itself admits its temperature reconstructions are highly uncertain:

> The uncertainty in observational records encompasses instrumental/recording errors, effects of representation (e.g., exposure, observing frequency or timing), as well as effects due to physical changes in the instrumentation (such as station relocations or new satellites). All further processing steps (transmission, storage, gridding, interpolating, averaging) also have their own particular uncertainties. Because there is no unique, unambiguous, way to identify and account for non-climatic artefacts in the vast majority of records, there must be a degree of

uncertainty as to how the climate system has changed (IPCC, 2013, p. 165).

In other words, the IPCC's estimate of 0.78°C [0.72 to 0.85] from 1850–1900 to 2003–2012 is not direct evidence but an estimate based on a long chain of judgements about how to handle data, what data to include and what to leave out, and how to summarize it all into a single "global temperature." The IPCC's temperature estimate cannot be tested because it has no empirical existence. It is a "stylized fact," one of many in climate science.

Serious flaws of the HadCRUT dataset are apparent from emails leaked from the Climatic Research Unit in 2009 (Goldstein, 2009). A file titled "Harry Read Me" contained some 247 pages of email exchanges with a programmer responsible for maintaining and correcting errors in the HadCRUT climate data between 2006 and 2009. Reading only a few of the programmer's comments reveals the inaccuracies, data manipulation, and incompetence that render the database unreliable:

- "Wherever I look, there are data files, no info about what they are other than their names. And that's useless ..." (p. 17).

- "It's botch after botch after botch" (p. 18).

- "Am I the first person to attempt to get the CRU databases in working order?!!" (p. 47).

- "As far as I can see, this renders the [weather] station counts totally meaningless" (p. 57).

- "COBAR AIRPORT AWS [data from an Australian weather station] cannot start in 1962, it didn't open until 1993!" (p. 71).

- "What the hell is supposed to happen here? Oh yeah – there is no 'supposed,' I can make it up. So I have : -)" (p. 98).

- "I'm hitting yet another problem that's based on the hopeless state of our databases. There is no uniform data integrity, it's just a catalogue of issues that continues to grow as they're found" (p. 241).

Also in 2009, after years of denying Freedom of Information Act (FOIA) requests for the HadCRUT dataset on frivolous and misleading grounds (Montford, 2010), Phil Jones, director of the Climatic Research Unit, admitted *the data do not exist.* "We, therefore, do not hold the original raw data but only the value-added (i.e., quality controlled and homogenized) data" (quoted in Michaels, 2009). Recall that "Harry Read Me" was in charge of "quality control" at the time. This means the HadCRUT dataset it is not direct evidence ... evidence that does not rely on inference ... of the surface temperature record prior to the arrival of satellite data in 1979. Relying on such circumstantial evidence to test hypotheses violates the Scientific Method. As Michaels remarked at the time, "If there are no data, there's no science."

Many researchers have identified serious quality control problems with the surface-based temperature record (Balling and Idso, 2002; Pielke Sr., 2007a, 2007b; Watts, 2009; Fall *et al.,* 2011; Frank, 2015, 2016; Parker and Ollier, 2017). Very recently, Hunziker *et al.* (2018) set out to determine the reliability of data from manned weather stations from the Central Andean area of South America by comparing results using an "enhanced" approach to the standard approach for a sample of stations. They found "about 40% of the observations [using the standard approach] are inappropriate for the calculation of monthly temperature means and precipitation sums due to data quality issues. These quality problems, undetected with the standard quality control approach, strongly affect climatological analyses, since they reduce the correlation coefficients of station pairs, deteriorate the performance of data homogenization methods, increase the spread of individual station trends, and significantly bias regional temperature trends." They conclude, "Our findings indicate that undetected data quality issues are included in important and frequently used observational datasets and hence may affect a high number of climatological studies. It is of utmost importance to apply comprehensive and adequate data quality control approaches on manned weather station records in order to avoid biased results and large uncertainties."

McLean (2018) conducted an audit of the HadCRUT4 dataset and found "more than 70 issues of concern" including failure to check source data for errors, resulting in "obvious errors in observation station metadata and temperature data" (p. 88). According to McLean, grid cell values, hemispheric averages, and global averages were derived from too little data to be considered reliable. "So-called 'global' average temperature anomalies have at times been heavily biased toward certain areas of the world and at other times there is a lack of coverage in specific regions." McLean found adjustments to data that "involve heroic assumptions because the necessary information about the conditions was not recorded. They are likely to be flawed." McLean concludes:

> In the opinion of this author, the data before 1950 has negligible real value and cannot be relied upon to be accurate. The data from individual stations might be satisfactory but only if local environments are unchanged and with no manual adjustments to the temperature data. The many issues with the 1850–1949 data make it meaningless to attempt any comparison between it and later data especially in derived values such as averages and the trends in those averages (p. 90).

McLean (2018) is no more satisfied with the record since 1950:

> It is also the opinion of this author that the HadCRUT4 data since 1950 is likewise not fit for purpose. It might be suitable for single-station studies or even small regional studies but only after being deemed satisfactory regards [sic] the issues raised in this report. It is not suitable for the derivation of global or hemispheric averages, not even with wide error margins that can only be guessed at because there are too many points in the data collection and processing that are uncertain and inconsistent (*Ibid.*).

This is the temperature record relied on by the IPCC and climate scientists who claim to know how much the mean global surface temperature has changed since 1850, and hence to know as well what the human impact on climate has been. How much confidence can be placed in their analysis when it is based on such a flawed premise? Sir Fred Hoyle saw this problem more than two decades ago when he wrote: "To raise a delicate point, it really is not very sensible to make approximations … and then to perform a highly complicated computer calculation, while claiming the arithmetical accuracy of the computer as the standard for the whole investigation" (Hoyle, 1996).

In the introduction to his audit, McLean (2018) writes, "it seems very strange that man-made

warming has been a major international issue for more than 30 years and yet the fundamental data has never been closely examined." Indeed.

Compared to What?

The choice of 1850 as the starting point for the timeline IPCC features seems designed to exaggerate the alleged uniqueness of the warming of the twentieth century. The world was just leaving the Little Ice Age, which is generally dated as spanning the sixteenth to the eighteenth centuries. Had the IPCC chosen to start the series before the Little Ice Age, during the Medieval Warm Period (MWP), when temperatures were as warm as they are today (Sicre *et al.*, 2008), there would have been no warming trend to report. The IPCC acknowledges "continental-scale surface temperature reconstructions show with *high confidence*, multidecadal periods during the Medieval Climate Anomaly (950–1250) that were in some regions as warm as the mid-20th century and in others as warm as in the late 20th century" (IPCC, 2013, p. 37). In fact, hundreds of researchers have found the MWP was a global phenomenon (NIPCC, 2011, Chapter 3).

The warming since 1850, if it occurred at all, is meaningful information only if it exceeds natural variability. Proxy temperature records from the Greenland ice core for the past 10,000 years demonstrate a natural range of warming and cooling rates between +2.5°C and -2.5°C/century, significantly greater than rates measured for Greenland or the globe during the twentieth century (Alley, 2000; Carter, 2010, p. 46, Figure 7). The ice cores also show repeated "Dansgaard–Oeschger" events when air temperatures rose at rates of about 10 degrees per century. There have been about 20 such warming events in the past 80,000 years.

Glaciological and recent geological records contain numerous examples of ancient temperatures up to 3°C or more warmer than temperatures as recently as 2015 (Molnár and Végvári, 2017; Ge *et al.*, 2017; Lasher *et al.*, 2017; Simon *et al.*, 2017; Köse *et al.*, 2017; Kawahata *et al.*, 2017; Polovodova Asteman *et al.*, 2018; Badino *et al.*, 2018). During the Holocene, such warmer peaks included the Egyptian, Minoan, Roman, and Medieval warm periods. During the Pleistocene, warmer peaks were associated with interglacial oxygen isotope stages 5, 9, 11, and 31 (Lisiecki and Raymo, 2005). During the Late Miocene and Early Pliocene (6–3 million years ago) temperature consistently attained values 2–3°C above twentieth century values (Zachos *et al.*, 2001).

Phil Jones, director of the Climatic Research Unit, when asked in 2010 (in writing) if the rates of global warming from 1860–1880, 1910–1940, and 1975–1998 were identical, wrote in reply "the warming rates for all 4 periods [he added 1975 – 2009] are similar and not statistically significantly different from each other" (BBC News, 2010). When asked "Do you agree that from 1995 to the present there has been no statistically significant global warming?" Jones answered "yes." His replies contradicted claims made by the IPCC at the time as well as the claim, central to the IPCC's hypothesis, that the warming of the late twentieth and early twenty-first centuries was beyond natural variability.

In its Fifth Assessment Report, the IPCC admits the global mean surface temperature stopped rising from 1997 to 2010, reporting the temperature increase for that period was 0.07°C [-0.02 to 0.18] (IPCC, 2013, p. 37). This "pause" extended 18 years before being interrupted by the major El Niño events of 2010–2012 and 2015–2016. (See Figure 2.2.1 below.) During "the pause" humans released approximately one-third of all the greenhouse gases emitted since the beginning of the Industrial Revolution. If CO_2 concentrations drive global temperatures, their impact surely would have been visible during this period. Either CO_2 is a weaker driver of climate than the IPCC assumes, or natural drivers and variation play a bigger role than it realizes.

Temperatures quickly fell after each El Niño event, though not to previous levels. This step-function increase in temperature (also seen in 1977 in what is known as the Great Pacific Climate Shift) is not the linear increase in global temperatures predicted by general circulation models tuned to assign a leading role to greenhouse gases (Belohpetsky *et al.*, 2017; Jones and Ricketts, 2017). El Niño events are produced by cyclical changes in ocean currents (the El Niño-Southern Oscillation (ENSO) and the Atlantic Multidecadal Oscillation (AMO)) and their known periodicity of 60 to 70 years explains much of the warming and cooling cycles observed in the twentieth century (Douglass and Christy, 2008). The ENSO and AMO are thought to be the result of solar influences (Easterbrook, 2008) dismissed by the IPCC as being too small to have a significant influence on climate. The impact of El Niño events on global temperatures is evidence of natural variability and the strength of natural forcings relative to human greenhouse emissions. (See also D'Aleo and Easterbrook, 2016.)

Satellite Data

NASA research scientists Roy Spencer and John Christy (1990) published a method to use data collected by satellites since December 1978 to calculate global atmospheric temperatures. They now maintain a public database at the University of Alabama – Huntsville (UAH) of nearly 40 years of comprehensive satellite temperature measurements of the atmosphere that has been intensively quality controlled and repeatedly peer-reviewed (Christy and Spencer, 2003a, 2003b; Spencer *et al.,* 2006; Christy *et al.*, 2018). Satellites retrieve data from the entire surface of the planet, something surface-based temperature stations are unable to do, and are accurate to 0.01°C. They are also transparent (available for free on the internet); free from human influences other than greenhouse gas emissions such as change in land use, urbanization, farming, and land clearing, and changing instrumentation and instrument location; and at least somewhat immune to human manipulation. As Santer *et al.* (2014, pp. 185–9) reported, "Satellite TLT [temperature lower troposphere] data have near-global, time-invariant spatial coverage; in contrast, global-mean trends estimated from surface thermometer records can be biased by spatially and temporally non-random coverage changes." Figure 2.2.1.1 presents the latest data from the UAH satellite dataset.

Christy *et al.* (2018) analyzed eight datasets generated by satellites and weather balloons producing bulk tropospheric temperatures beginning in 1979 and ending in 2016, finding the trend is +0.103°C decade^{-1} with a standard deviation among the trends of 0.0109°C. More specifically, they report "a range of near global (+0.07 to +0.13°C decade^{-1}) and tropical (+0.08 to 0.17°C decade^{-1}) trends (1979–2016)." Looking specifically at the tropical (20°S – 20°N) region, where CO_2-forced warming is thought to be most detectable, they find the trend is +0.10 ± 0.03°C decade^{-1}. "This tropical result," they write, "is over a factor of two less than the trend projected from the average of the IPCC climate model simulations for this same period (+0.27°C decade^{-1})." Phrased differently, the only direct evidence available regarding global temperatures since 1979 shows a warming trend of only 0.1°C per decade, or 1°C per century. This is approximately one-third as much as the IPCC's forecast for the twenty-first century, and

Figure 2.2.1.1
Global temperature departure from 1981–2010 average (°C) from 1979 to 2018, UAH satellite-based readings

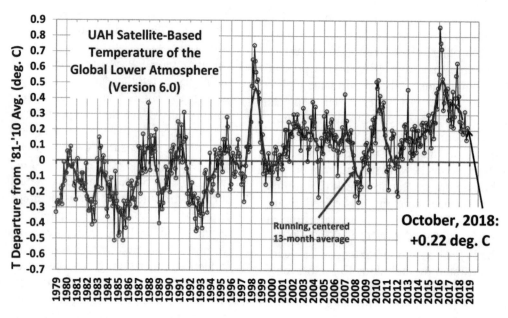

Source: Spencer, 2018.

within the range of natural variability. It is not evidence of a human impact on the global climate.

In 2017, Christy and McNider (2017) used satellite data to "identify and remove the main natural perturbations (e.g. volcanic activity, ENSOs) from the global mean lower tropospheric temperatures (T_{LT}) over January 1979 – June 2017 to estimate the underlying, potentially human-forced trend. The unaltered value is +0.155 K dec^{-1} while the adjusted trend is +0.096 K dec^{-1}, related primarily to the removal of volcanic cooling in the early part of the record." While that trend is "potentially human-forced," it is just as likely to be the result of other poorly understood processes such as solar effects.

Despite the known deficiencies in the HadCRUT surface temperature record and the superior accuracy and global reach of satellite temperature records, the IPCC and many government agencies and environmental advocacy groups continue to rely on the surface station temperature record. One reason for this preference is because some credible surface-based records date back to the 1850s and even earlier. (But recall the conclusion of McLean (2018) that "the data before 1950 has negligible real

value and cannot be relied upon to be accurate.") Another reason becomes clear if one compares the satellite data in Figure 2.2.1.2 to five surface station records shown in Figure 2.2.1.2. Whereas the satellite record shows only 0.1°C per decade warming since 1979, various surface station records (and surface station + satellite in one case) for approximately the same period shows an average warming of about 0.17°C, about 70% higher (Simmons *et al.*, 2016).

Satellite data are valuable for providing a test of the water vapor amplification theory (reported in the next section), demonstrating the inaccuracy of surface station data (particularly their failure to control for urban heat island effects), and exposing the exaggerated claims of those who say global warming (or "climate change") is "already happening." What warming has occurred in the past four decades was too small or slow to have been responsible for the litany of supposed harms attributed to anthropogenic forcing. However, geologists point to how short a period 40 years or even a century are when studying climate, which is known to respond to internal forcings (e.g., ocean currents and solar influences) with multidecadal, centennial, and longer periodicities. Even a century's

Figure 2.2.1.2
Global temperature departure from 1961–1990 average (°C) from 1979 to 2016

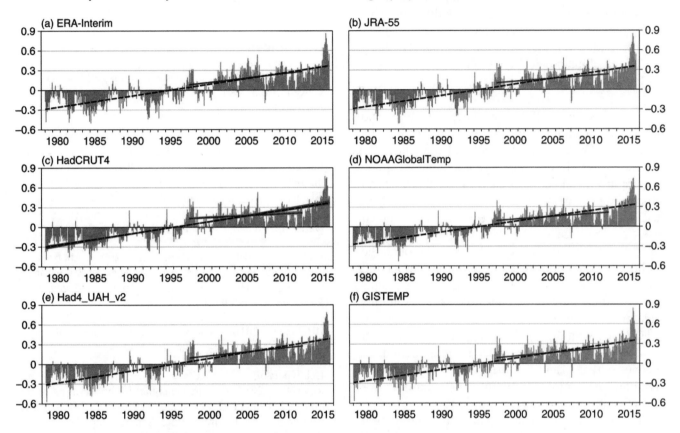

Monthly anomalies in globally averaged surface temperature (°C) relative to 1981–2010, from (a) ERA-Interim, (b) JRA-55, (c) the HadCRUT4 median, (d) NOAAGlobalTemp, (e) Had4_UAH_v2 and (f) GISTEMP, for January 1979 to July 2016. Also shown are least-squares linear fits to the monthly values computed for the full period (black, dashed lines) and for 1998 – 2012 (dark green, solid lines). In the case of HadCRUT4, the corresponding linear fits for each ensemble member are plotted as sets of (overlapping) grey and lighter green lines. *Source:* Simmons *et al.,* 2016.

worth of data would be a mere 1% of the 10,000 years of the Holocene Epoch. How do we know how much of the warming of the twentieth and early twenty-first centuries is due to the human influence?

A Single Global Temperature?

Some scientists challenge the notion that a single temperature can be attributed to the planet's atmosphere. Does it really make sense to create and report the mean average of the temperatures of deserts, corn fields, oceans, and big cities? In a dynamic system where cold weather in one part of the world can mean warmer weather in another, and

where poorly understood weather processes have impacts that dwarf decades and even centuries of long-term climate change, what is the point? As Essex *et al.* (2007) explain, the "statistic called 'global temperature' ... arises from projecting a sampling of the fluctuating temperature field of the Earth onto a single number at discrete monthly or annual intervals." The authors continue,

> While that statistic is nothing more than an average over temperatures, it is regarded as the temperature, as if an average over temperatures is actually a temperature itself, and as if the out-of-equilibrium climate system has only one temperature. But an

average of temperature data sampled from a non-equilibrium field is not a temperature. Moreover, it hardly needs stating that the Earth does not have just one temperature. It is not in global thermodynamic equilibrium – neither within itself nor with its surroundings. It is not even approximately so for the climatological questions asked of the temperature field.

Essex *et al.* (2007) conclude, "there is no physically meaningful global temperature for the Earth in the context of the issue of global warming."

The method of deriving a global average temperature must be arbitrary since there is an infinite number of ways it could be calculated from observational data. Temperatures could be weighed based on the size of the area sampled, but since weather stations are not evenly distributed around the world, the result could be a few dozen stations having more influence on the global or regional "average" than hundreds or thousands of other stations, something McLean (2018) in fact observed in the HadCRUT dataset. Efforts to manipulate and "homogenize" divergent datasets, fill in missing data, remove outliers, and compensate for changes in sampling technology are all opportunities for subjective or just poor decision-making.

The resulting number is an accounting fiction with no real-world counterpart, and therefore is not subject to experimentation or falsification. Essex *et al.* (2007) write, "The resolution of this paradox is not through adoption of a convention. It is resolved by recognizing that it is an abuse of terminology to use the terms 'warming' and 'cooling' to denote upward or downward trends in averages of temperature data in such circumstances. Statistics might go up or down, but the system itself cannot be said to be warming or cooling based on what they do, outside of special circumstances."

Why was the attempt made to infer a single temperature to the planet's atmosphere in the first place? The answer can be found in Chapter 1, where the concept of "seeing like a state" was discussed. Governments need to assign numbers to the things they seek to regulate or tax. Complex realities must be simplified at the cost of misrepresentation and outright falsification to produce the stylized facts that can be used in legislation and then in regulations. Saying the world has a single temperature – 14.9°C (58.82°F) in 2017, according to NASA's Goddard Institute for Space Studies (Hansen *et al.*, 2018) –

violates many of the principles of the Scientific Method, but it fills a need expressed by government officials at the United Nations and in many of the world's capitols.

References

Alley, R.B. 2000. The Younger Dryas cold interval as viewed from central Greenland. *Quaternary Science Reviews* **19**: 213–26.

Badino, F., Ravazzi, C., Vallè, F., Pini, R., Aceti, A., Brunetti, M., Champvillair, E., Maggi, V., Maspero, F., Perego, R., and Orombelli, G. 2018. 8800 years of high-altitude vegetation and climate history at the Rutor Glacier forefield, Italian Alps. Evidence of middle Holocene timberline rise and glacier contraction. *Quaternary Science Reviews* **185**: 41–68.

Balling Jr., R.C. and Idso, C.D. 2002. Analysis of adjustments to the United States Historical Climatology Network (USHCN) temperature database. *Geophysical Research Letters* **29** (10).

BBC News. 2010. Q&A: Professor Phil Jones. BBC News (website). Accessed February 13, 2018.

Belolýpetsky, P.V., Bartsev, S.I., Saltykov, M.Y., and Reýd, P.C. 2017. A staircase signal in the warming of the mid-20th century. Proceedings of the 15th International Conference on Environmental Science and Technology, Rhodes, Greece. August 31–September 2.

Carter, R.M. 2010. *Climate: The Counter Consensus*. London, UK: Stacey International.

Christy, J.R. and Spencer, R.W. 2003a. *Global Temperature Report: 1978–2003*. Huntsville, AL: Earth System Science Center, University of Alabama in Huntsville.

Christy, J.R. and Spencer, R.W. 2003b. Reliability of satellite data sets. *Science* **301** (5636): 1046–9.

Christy, J.R. and McNider, R.T. 2017. Satellite bulk tropospheric temperatures as a metric for climate sensitivity. *Asia-Pacific Journal of Atmospheric Sciences* **53** (4): 511–8.

Christy, J., *et al.* 2018. Examination of space-based bulk atmospheric temperatures used in climate research. *International Journal of Remote Sensing* **39** (11): 3580–607.

D'Aleo, J.S. and Easterbrook, D.J. 2016. Relationship of multidecadal global temperatures to mulidecadal oceanic oscillations. In: Easterbrook, D. (Ed.) *Evidence-based Climate Science*. Cambridge, MA: Elsevier Inc., 191–214.

Douglass, D.H. and Christy, J.R. 2008. Limits on CO_2 climate forcing from recent temperature data of Earth. *Energy and Environment* **20** (1).

Easterbrook, D.J. 2008. Correlation of climatic and solar variations over the past 500 years and predicting global climate changes from recurring climate cycles. In: *Abstracts of 33rd International Geological Congress, Oslo, Norway.*

Essex, C., McKitrick, R., and Andresen, B. 2007. Does a global temperature exist? *Journal of Non-Equilibrium Thermodynamics* **32** (1).

Fall, S., *et al.* 2011. Analysis of the impacts of station exposure on the U.S. Historical Climatology Network temperatures and temperature trends. *Journal of Geophysical Research* **116**.

Frank, P. 2015. Negligence, non-science, and consensus climatology. *Energy & Environment* **26** (3): 391–415.

Frank, P. 2016. Systematic error in climate measurements: the global air temperature record. In: Raigaini, R. (Ed.) *The Role of Science in the Third Millennium. International Seminars on Nuclear War and Planetary Emergencies* 48. Singapore: World Scientific, pp. 337–51.

Ge, Q., Liu, H., Ma, X., Zheng, J., and Hao, Z. 2017. Characteristics of temperature change in China over the last 2000 years and spatial patterns of dryness/wetness during cold and warm periods. *Advances in Atmospheric Sciences* **34** (8): 941–51.

Goldstein, L. 2009. Botch after botch after botch. *Toronto Sun.* November 29.

Hansen, J., Sato, M., Ruedy, R., Schmidt, G.A., Lo, K., and Persin, A. 2018. *Global Temperature in 2017.* New York, NY: Columbia University. January 18.

Hoyle, F. 1996. The great greenhouse controversy. In: Emsley, J. (Ed.) *The Global Warming Debate.* London, UK: The European Science and Environment Forum, 179–89.

Hunziker, S., Brönnimann, S., Calle, J., Moreno, I., Andrade, M., Ticona, L., Huerta, A., and Lavado-Casimiro, W. 2018. Effects of undetected data quality issues on climatological analyses. *Climate of the Past* **14**: 1–20.

IPCC. 2013. Intergovernmental Panel on Climate Change. *Climate Change 2013: The Physical Science Basis.* Contribution of Working Group I to the Fifth Assessment Report of the Intergovernmental Panel on Climate Change. New York, NY: Cambridge University Press.

Jones, R.N. and Ricketts, J.H. 2017. Reconciling the signal and noise of atmospheric warming on decadal timescales. *Earth System Dynamics* **8**: 177–210.

Kawahata, H., Ishizaki, Y., Kuroyanagi, A., Suzuki, A., and Ohkushid, K. 2017. Quantitative reconstruction of temperature at a Jomon site in the Incipient Jomon Period in northern Japan and its implications for the production of early pottery and stone arrowheads. *Quaternary Science Reviews* **157**: 66–79.

Köse, N., Tuncay Güner, H., Harley, G.L., and Guiot, J. 2017. Spring temperature variability over Turkey since 1800CE reconstructed from a broad network of tree-ring data. *Climate of the Past* **13**: 1–15.

Lasher, G.E., Axford, Y., McFarlin, J.M., Kelly, M.A., Osterberg, E.C., and Berkelhammer, M.B. 2017. Holocene temperatures and isotopes of precipitation in Northwest Greenland recorded in lacustrine organic materials. *Quaternary Science Reviews* **170**: 45–55.

Lisiecki, L.E. and Raymo, M.E. 2005. A Pliocene-Pleistocene stack of 57 globally distributed benthic d18O records. *Paleoceanography* **20**: PA1003.

McLean, J. 2018. *An Audit of the Creation and Content of the HadCRUT4 Temperature Dataset.* Robert Boyle Publishing. October.

Michaels, P. 2009. The dog ate global warming. National Review Online (website). September 23. Accessed November 27, 2018.

Molnár, A. and Végvári, Z. 2017. Reconstruction of early Holocene Thermal Maximum temperatures using present vertical distribution of conifers in the Pannon region (SE Central Europe). *The Holocene* **27** (2): 236–45.

Montford, A.W. 2010. *The Hockey Stick Illusion: Climategate and the Corruption of Science.* London, England: Stacey International.

NIPCC. 2011. Nongovernmental International Panel on Climate Change. *Climate Change Reconsidered: 2011 Interim Report of the Nongovernmental International Panel on Climate Change.* Chicago, IL: The Heartland Institute.

Parker, A. and Ollier, C.D. 2017. Discussion of the "hottest year on record" in Australia. *Quaestiones Geographicae* **36** (1).

Pielke Sr., R.A., *et al.* 2007a. Documentation of uncertainties and biases associated with surface temperature measurement sites for climate change assessment. *Bulletin of the American Meteorological Society* **88**: 913–28.

Pielke Sr., R.A., *et al.* 2007b. Unresolved issues with the assessment of multidecadal global land surface temperature trends. *Journal of Geophysical Research* **112** (D24).

Polovodova Asteman, I., Filipsson, H.L., and Nordberg, K. 2018. Tracing winter temperatures over the last two millennia using a NE Atlantic coastal record. *Climate of the Past: Discussions*: https://doi.org/10.5194/cp-2017-160.

Santer, B.D., *et al.* 2014. Volcanic contribution to decadal changes in tropospheric temperature. *Nature Geoscience* **7**: 185–9.

Sicre, M-A., *et al.* 2008. Decadal variability of sea surface temperatures off North Iceland over the last 200 yrs. *Earth Planet Science Letters* **268** (3–4): 137–42.

Simmons, A.J., *et al.* 2016. A reassessment of temperature variations and trends from global reanalyses and monthly surface climatological datasets. *Quarterly Journal of the Royal Meteorological Society* **143**: 742.

Simon, B., Poska, A., Hossann, C., and Tõnno, I. 2017. 14,000 years of climate-induced changes in carbon resources sustaining benthic consumers in a small boreal lake (Lake Tollari, Estonia). *Climatic Change* **145** (1–2): 205–19.

Spencer, R.W., *et al.* 2006. Estimation of tropospheric temperature trends from MSU channels 2 and 4. *Journal of Atmospheric and Oceanic Technology* **23** (3): 417–23.

Spencer, R. 2018. UAH Global Temperature Update for October (blog). November 2. Accessed November 11.

Spencer, R. and Christy, J. 1990. Precise monitoring of global temperature trends from satellites. *Science* **247** (4950): 1558–62.

Watts, A. 2009. *Is the U.S. Surface Temperature Record Reliable?* Chicago, IL: The Heartland Institute.

Zachos, J., Pagani, M., Sloan, L., Thomas, E., and Billups, K. 2001. Trends, rhythms, and aberrations in global climate 65 Ma to present. *Science* **292**: 686–93.

2.2.2 General Circulation Models

General circulation models (GCMs) are unable to accurately depict complex climate processes. They do not accurately hindcast or forecast the climate effects of anthropogenic greenhouse gas emissions.

Working Group III of the IPCC says without mitigation, "global mean surface temperature increases in 2100 from 3.7°C to 4.8°C compared to pre-industrial levels" (IPCC, 2014., p. 8). To arrive at this forecast, the IPCC relies on computer models called general circulation models (GCMs). Relatively few people in the climate science community are experts in building, "tuning," and operating such models, and so the rest of the community accepts their outputs on faith. This is a mistake. Like the history of the flawed HadCRUT temperature record told in the previous section, an examination of how GCMs are created and operate reveals why they are unreliable.

The Map Is Not the Territory

Specialized models, which try to model reasonably well-understood processes like post-glacial rebound (PGR) and radiation transport, are useful because the processes they model are manageably simple and well-understood. Weather forecasting models are also useful, even though the processes they model are very complex and poorly understood, because the models' short-term predictions can be repeatedly tested, allowing the models to be validated and refined. But more ambitious models like GCMs, which attempt to simulate the combined effects of many poorly understood processes, over time periods much too long to allow repeated testing and refinement, are of dubious utility.

Lupo *et al.* (2013) present a comprehensive critique of GCMs in Chapter 1 of *Climate Change Reconsidered II: Physical Science*. The authors write, "scientists working in fields characterized by complexity and uncertainty are apt to confuse the output of models – which are nothing more than a statement of how the modeler believes a part of the world works – with real-world trends and forecasts (Bryson, 1993). Computer climate modelers frequently fall into this trap and have been severely criticized for failing to notice their models fail to replicate real-world phenomena by many scientists, including Balling (2005), Christy (2005), Essex and McKitrick (2007), Frauenfeld (2005), Michaels (2000, 2005, 2009), Pilkey and Pilkey-Jarvis (2007), Posmentier and Soon (2005), and Spencer (2008)."

Confusing models for the real world they are meant to represent is part of the fallacy of a single world temperature discussed by Essex *et al.* (2007) earlier in this section. Similarly, Jaynes (2003) writes,

Common language, or at least, the English language, has an almost universal tendency to disguise epistemological statements by putting them into a grammatical form which suggests to the unwary an ontological statement. A major source of error in current probability theory arises from an unthinking failure to perceive this. To interpret the first kind of statement in the ontological senses is to assert that one's own private thoughts and sensations are realities existing externally in Nature. We call this the "Mind Projection Fallacy," and note the trouble it causes many times in what follows (p. 116).

Mind Projection Fallacy leads to circular arguments. Modelers assume a CO_2 increase causes a temperature increase and so they program that into their models. When asked how they know this happens they say the model shows it, or other models (similarly programmed) show it, or their model doesn't work *unless* CO_2 is assumed to increase temperatures. Modelers may get the benefit of doubt from their colleagues and policymakers owing to the complexity of models and the expense of the supercomputers needed to run them. This does not make them accurate maps of a highly complex territory.

Citing a book by Solomon (2008), Lupo *et al.* (2013) provided the following sample of informed opinion regarding the utility of GCMs:

- Dr. Freeman Dyson, professor of physics at the Institute for Advanced Study at Princeton University and one of the world's most eminent physicists, said the models used to justify global warming alarmism are "full of fudge factors" and "do not begin to describe the real world."

- Dr. Zbigniew Jaworowski, chairman of the Scientific Council of the Central Laboratory for Radiological Protection in Warsaw and a world-renowned expert on the use of ancient ice cores for climate research, said the United Nations "based its global-warming hypothesis on arbitrary assumptions and these assumptions, it is now clear, are false."

- Dr. Hendrik Tennekes, director of research at the Royal Netherlands Meteorological Institute, said "there exists no sound theoretical framework for climate predictability studies" used for global warming forecasts.

- Dr. Antonino Zichichi, emeritus professor of physics at the University of Bologna, and former president of the European Physical Society, said global warming models are "incoherent and invalid."

Dyson (2007) writes, "I have studied the climate models and I know what they can do. The models solve the equations of fluid dynamics, and they do a very good job of describing the fluid motions of the atmosphere and the oceans. They do a very poor job of describing the clouds, the dust, the chemistry, and the biology of fields and farms and forests. They do not begin to describe the real world that we live in."

More recently, Hedemann *et al.* (2017) examined why climate models failed to predict the "surface-warming hiatus" of the early twenty-first century, reporting that other researchers identify model errors in external forcing and heat rearrangements in the ocean. The authors write, "we show that the hiatus could also have been caused by internal variability in the top-of-atmosphere energy imbalance. Energy budgeting for the ocean surface layer over a 100-member historical ensemble reveals that hiatuses are caused by energy-flux deviations as small as 0.08 W m^{-2}, which can originate at the top of the atmosphere, in the ocean, or both. Budgeting with existing observations cannot constrain the origin of the recent hiatus, because the uncertainty in observations dwarfs the small flux deviations that could cause a hiatus."

Coats and Karnauskas (2018) studied whether climate models could hindcast historical trends in the tropical Pacific zonal sea surface temperature gradient (SST gradient) using 41 climate models (83 simulations) and five observational datasets. They found "None of the 83 simulations have a positive trend in the SST gradient, a strengthening of the climatological SST gradient with more warming in the western than eastern tropical Pacific, as large as the mean trend across the five observational data sets. If the observed trends are anthropogenically forced, this discrepancy suggests that state-of-the-art climate models are not capturing the observed response of the tropical Pacific to anthropogenic forcing, with serious implications for confidence in future climate projections." They conclude, "the differences in SST gradient trends between climate models and observational data sets are concerning and motivate the need for process-level validation of the atmosphere-ocean dynamics relevant to climate change in the tropical Pacific."

Dommenget and Rezny (2017) observe that "state-of-the-art coupled general circulation models (CGCMs) have substantial errors in their simulations of climate. In particular, these errors can lead to large uncertainties in the simulated climate response (both globally and regionally) to a doubling of CO_2." The authors use the Globally Resolved Energy Balance (GREB) model to test the impact of "tuning" and find "While tuning may improve model performance (such as reproducing observed past climate), it will not get closer to the 'true' physics nor will it significantly improve future climate change projections. Tuning will introduce artificial compensating error interactions between submodels that will hamper further model development."

Stouffer *et al.* (2017) present a progress report on the Coupled Model Intercomparison Project (CMIP5), the international partnership created to achieve consensus on GCMs for the IPCC. They offer a frank assessment of the shortcomings of the models: "The quantification of radiative forcings and responses was poor, and thus it requires new methods and experiments to address this gap. There are a number of systematic model biases that appear in all phases of CMIP that remain a major climate modeling challenge. These biases need increased attention to better understand their origins and consequences through targeted experiments. Improving understanding of the mechanisms' underlying internal climate variability for more skillful decadal climate predictions and long-term projections remains another challenge for CMIP6."

Hindcasting Failure

There is a relatively simple test of the accuracy of GCMs: run them over a historical interval to see if they can "hindcast" temperature anomalies (departures from an average) recorded in real-world observations. This should be an easy test to pass since the models are "parameterized" to approximate the historical temperature record. (Parameterization is the topic of the next section.) But when this is done, a large and growing divide between the climate model simulations and real-world observations is observed (Douglass, Pearson, and Singer, 2004; Singer, 2011; Singer, 2013).

Christy (2017) performed a test of 102 model runs conducted by GCMs in the CMIP5 against actual temperature observations for the period 1979–2016. Figure 2.2.2.1 shows his results.

Christy's graph shows the growing distance between computer models and observational data from three independent data sources. Almost without exception, the models "run hot," predicting more warming than satellite and weather balloons observe. Christy (2017) summarizes the findings:

The scientific conclusion here, if one follows the Scientific Method, is that the average model trend fails to represent the actual trend of the past 38 years by a highly significant amount. As a result, applying the traditional Scientific Method, one would accept this failure and not promote the model trends as something truthful about the recent past or the future.

More recently, McKitrick and Christy (2018) tested the ability of GCMs to predict temperature change in the tropical 200- to 300-hPa layer (hectopascals, or hPa, is a measure of atmospheric pressure) over the past 60 years, saying this constitutes a strong test of the global warming hypothesis because it meets the four criteria of a valid test: measurability, specificity, independence, and uniqueness. The researchers used model runs using the IPCC's Representative Concentration Pathway 4.5, which employs the best estimate of historical forcings through 2006 and anticipated forcings through 2100. (See Section 8.2.1 of Chapter 8 for a discussion of the IPCC's emission scenarios.)

According to McKitrick and Christy (2018), "the models project on average that the total amount of warming in the target zone since 1958 should have been about 2°C by now, a magnitude well within observational capability, and that the trends should be well established, thus specifying both the magnitude and a timescale" (p. 531). They continue, "simulations in the IPCC AR4 Chapter 9 indicate that, within the framework of mainstream GCMs, greenhouse forcing provides the only explanation for a strong warming trend in the target region" (p. 532). They use a 60-year temperature record composed from radiosondes (instruments carried up into the atmosphere by weather balloons that measure pressure, temperature, and relative humidity) for the period 1958–2017. Figure 2.2.2.2 shows their results.

McKitrick and Christy (2018) also conduct a statistical test (the Vogelsang-Frances F-Test) to determine whether the average temperature prediction of the 102 model runs for the 200- to 300 hPa layer of the atmosphere is statistically different from the average observed temperature. The results are reproduced below as Figure 2.2.2.3. The authors summarize their findings: "The mean restricted trend (without a break term) is 0.325 ± 0.132°C per decade in the models and 0.173 ± 0.056°C per decade in the observations. With a break term included they are 0.389 ± 0.173°C per decade (models) and 0.142 ± 0.115°C per decade (observed)." In other

Figure 2.2.2.1
Climate model forecasts versus observations, 1979–2016

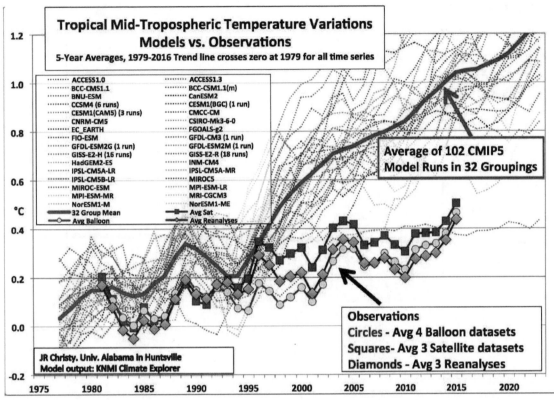

Five-year averaged values of annual mean (1979–2016) tropical bulk T_{MT} as depicted by the average of 102 IPCC CMIP5 climate models (red) in 32 institutional groups (dotted lines). The 1979–2016 linear trend of all time series intersects at zero in 1979. Observations are displayed with symbols: green circles – average of four balloon datasets, blue squares – 3 satellite datasets and purple diamonds – 3 reanalyses. The last observational point at 2015 is the average of 2013–2016 only, while all other points are centered, 5-year averages. *Source*: Christy, 2017.

Figure 2.2.2.2

A test of the tropical 200- to 300-hPa layer of atmosphere warming rate in climate models versus observations

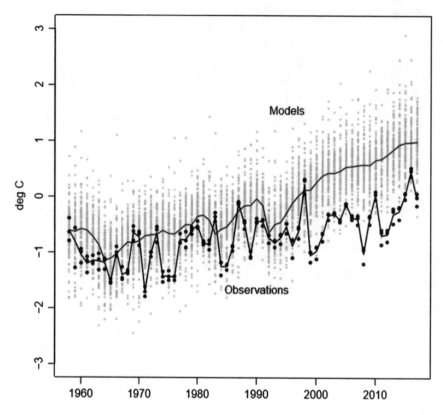

Light red dots show the complete year-by-year array of individual anomaly values from CMIP5. Red line is the annual mean of CMIP5 anomalies. Blue line is the mean of the three observational series, which are shown individually as blue dots. These are positioned so that the year-by-year observational mean starts at the same value as the corresponding model mean. *Source*: McKitrick and Christy, 2018, Figure 3, p. 531.

Figure 2.2.2.3
Test of statistical significance of difference between climate models and observations in tropical 200- to 300-hPa layer of the atmosphere

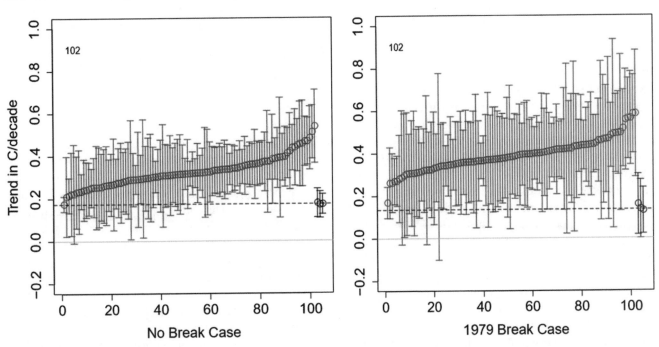

Trend magnitudes (red circles) and 95% confidence intervals (vertical red lines) for 102 values from CMIP5. Observed average trend (blue dashed line) and confidence intervals (vertical blue lines) for average of three observational series (RAOBCORE, RICH, and RATPAC). Numbers in upper left corner indicate number of model trends (out of 102) that exceed observed average trend. *Source*: McKitrick and Christy, 2018, Figure 4, p. 531.

words, the models run hot by about 0.15°C per decade (0.325 – 0.173) and predict nearly twice as much warming in this area of the atmosphere as actually occurred (0.325 / 0.173) during the past 60 years.

Other scientists have confirmed the failure of GCMs to make accurate hindcasts. Green *et al.* (2009) found that when applied to the period of industrialization from 1850 to 1974, the IPCC projection of 3°C per century of warming from human carbon dioxide emissions resulted in errors that were nearly 13 times larger than those from forecasting no change in global mean temperatures for horizons of 91 to 100 years ahead. Monckton *et al.* (2015) found almost without exception, the models "run hot," predicting more warming than satellite and weather balloons observe. Idso and Idso (2015) write, "we find (and document) a total of 2,418 failures of today's top-tier climate models to accurately hindcast a whole host of climatological phenomena. And with this poor record of success,

one must greatly wonder how it is that anyone would believe what the climate models of today project about earth's climate of tomorrow, i.e., a few decades to a century or more from now."

Kravtsov (2017) notes, "identification and dynamical attribution of multidecadal climate undulations to either variations in external forcings or to internal sources is one of the most important topics of modern climate science, especially in conjunction with the issue of human-induced global warming." Using ensembles of twentieth century climate simulations in an attempt to isolate the forced signal and residual internal variability in observed and modeled climate indices, they found "the observed internal variability … exhibits a pronounced multidecadal mode with a distinctive spatiotemporal signature, *which is altogether absent* in model simulations" (italics added).

Roach *et al.* (2018) compared satellite observations with model simulations of the compactness of Antarctic sea ice and the regional distribution of sea ice concentration and found "the simulation of Antarctic sea ice in global climate models often does not agree with observations. ... As a fraction of total sea ice extent, models simulate too much loose, low-concentration sea ice cover throughout the year, and too little compact, high-concentration cover in the summer." Scanlon *et al.* (2018) tested the ability of seven GCMs to accurately hindcast land water storage using data from three Gravity Recovery and Climate Experiment (GRACE) satellite solutions in 186 river basins (~60% of global land area). They found "medians of modeled basin water storage trends greatly underestimate GRACE-derived large decreasing ($\leq$ −0.5 km^3/y) and increasing ($\geq$ 0.5 km^3/y) trends. Decreasing trends from GRACE are mostly related to human use (irrigation) and climate variations, whereas increasing trends reflect climate variations." Specifically, the GRACE satellite detected a large increasing trend in the Amazon while most models estimate decreasing trends, and global land water storage trends are positive for GRACE but negative for models. They conclude, "The inability of models to capture large decadal water storage trends based on GRACE indicates that model projections of climate and human-induced water storage changes may be underestimated."

Of the 102 model runs considered by Christy and McKitrick, only one comes close to accurately hindcasting temperatures since 1979: the INM-CM4 model produced by the Institute for Numerical Mathematics of the Russian Academy of Sciences (Volodin and Gritsun, 2018). That model projects only 1.4°C warming by the end of the century, similar to the forecast made by the Nongovernmental International Panel on Climate Change (NIPCC, 2013) and many scientists, a warming only one-third as much as the IPCC forecasts. Commenting on the success of the INM-CM model compared to the others (as shown in an earlier version of the Christy graphic), Clutz (2015) writes,

(1) INM-CM4 has the lowest CO$_2$ forcing response at 4.1K for 4xCO$_2$. That is 37% lower than multi-model mean.

(2) INM-CM4 has by far the highest climate system inertia: Deep ocean heat capacity in INM-CM4 is 317 W yr m^{-2} K^{-1}, 200% of the mean (which excluded INM-CM4 because it was such an outlier).

(3) INM-CM4 exactly matches observed atmospheric H$_2$O content in lower troposphere (215 hPa), and is biased low above that. Most others are biased high.

So the model that most closely reproduces the temperature history has high inertia from ocean heat capacities, low forcing from CO$_2$ and less water for feedback. Why aren't the other models built like this one?

Parameterization

Weather is defined as the instantaneous state and/or conditions of the atmosphere. Climate is the long-term mean state of the atmospheric conditions, including the variability, extremes, and recurrence intervals, for at least a 30-year period. While these definitions suggest weather and climate are different, each is governed by the same underlying physical causal factors. These factors are represented by seven mathematical equations referred to as the "primitive equations," which form the dynamic core of computer models that attempt to make both weather forecasts and climate projections. Primitive equations represent physical processes for which there are no precise formulations that allow for accurate predictions – processes such as cloud formation, heat exchange between Earth's surface and the atmosphere, precipitation generation, and solar radiation. Therefore, the variables in the equations must be represented by "parameterizations" or ranges.

Computer modelers "tune" these parameters until they get an answer that matches observations or the expectations of the modelers or their peers or funders. An international team of GCM modelers led by Frédéric Hourdin, senior research scientist at the French National Center for Scientific Research (CNRS), revealed how model tuning works in an extraordinary article published in 2017 (Hourdin *et al.* 2017). The authors are high-ranking modelers and yet are frank about the shortcomings of models. They write,

Climate model tuning is a complex process that presents analogy with reaching harmony in music. Producing a good symphony or rock concert requires first a good composition and good musicians who work individually on their score. Then, when playing together, instruments must be tuned, which is a well-defined adjustment of wave frequencies that can be done with the help of electronic devices. But the orchestra harmony is reached also by adjusting to a common tempo as well as by subjective combinations of instruments, volume levels, or musicians' interpretations, which will depend on the intention of the conductor or musicians.

When gathering the various pieces of a model to simulate the global climate, there are also many scientific and technical issues, and tuning itself can be defined as an objective process of parameter estimation to fit a predefined set of observations, accounting for their uncertainty, and a process that can be engineered. However, because of the complexity of the climate system and of the choices and approximations made in each submodel, and because of priorities defined in each climate center, there is also subjectivity in climate model tuning (Tebaldi and Knutti 2007) as well as substantial know how from a limited number of people with vast experience with a particular model (p. 590).

How does "harmony … reached also by adjusting to a common tempo" comply with Armstrong and Green's (2018) list of requirements imposed on researchers by the Scientific Method? Instead of seeking an important problem and being "skeptical about findings, theories, policies, methods, and data, especially absent experimental evidence," modelers adjust, calibrate, and change their methods and findings to be in "harmony" with others. Engineering a research project to fit a predefined set of observations does not sound like addressing the problem impartially. And rather than "exercising control over all aspects of your study," as Armstrong and Green counsel, modelers play a score written by others, surrendering control over their studies to a leader who has "vast experience with a particular model."

"Once a model configuration is fixed," Hourdin *et al.* (2017) continue, "tuning consists of choosing parameter values in such a way that a certain measure of the deviation of the model output from selected observations or theory is minimized or reduced to an acceptable range" (p. 591). One must ask, who determines the acceptable range? Is that range determined by the IPCC or other political and

scientific organizations? What are their conflicts of interest?

"Energy balance tuning," Hourdin *et al.* (2017) write, is "crucial since a change by 1 Wm^2 of the global energy balance typically produces a change of about $0.5 - 1.5$ K in the global-mean surface temperature in coupled simulations depending on the sensitivity of the given model" (pp. 592–3). One can only take from this that variables involving the atmosphere's energy balance are tweaked to make sure the resulting surface temperature forecast is in "an acceptable range." Is that range the IPCC's latest estimate of climate sensitivity or future warming?

Information used to support parameterization, Hourdin *et al.* (2017) write, "can come from theory, from a back-of-the-envelope estimate, from numerical experiments … or from observations" (p. 593). Promising theories and hypotheses can be derived from such myriad and sundry sources, but setting the parameters of computer models running on supercomputers and used to set national and international policies is a different matter. Is there really no quality control on such ad hoc justifications for tuning models? The authors admit that "although tuning is an efficient way to reduce the distance between model and selected observations, it can also risk masking fundamental problems and the need for model improvements" (p. 595). Indeed.

Voosen (2016) also reports on the use of tuning in climate models but blames climate "skeptics" for the secrecy surrounding it. He writes, "Climate models render as much as they can by applying the laws of physics to imaginary boxes tens of kilometers a side. But some processes, like cloud formation, are too fine-grained for that, and so modelers use 'parameterizations': equations meant to approximate their effects. For years, climate scientists have tuned their parameterizations so that the model overall matches climate records. But fearing criticism by climate skeptics, they have largely kept quiet about how they tune their models, and by how much." Given the sausage factory-like environment of climate modeling presented by Hourdin *et al.* (2017), the "skeptics" are right to criticize.

If natural climate forcings and feedbacks are not well understood, then GCMs become little more than an exercise in curve-fitting, changing parameters until the outcomes match the modeler's expectations. As John von Neumann is reported to have once said, "with four parameters I can fit an elephant, and with five I can make him wiggle his trunk" (Dyson, 2004). Of course, conscientious modelers try to avoid abusing their control over parameters, but GCMs provide so much room for subjective judgements that even subconscious bias produces big differences in model outputs. The Scientific Method is meant to protect scientists from such temptation.

Other Specific Shortcomings

The key failure of the climate models is the inability to reproduce warm phases that have repeatedly occurred in a natural way over the past 10,000 years. The models do not contain a meaningful natural climatic driver mechanism, as all natural climate drivers (e.g. sun, volcanoes) are deliberately set to nearly zero. Models without significant natural climate drivers are clearly unable to reproduce the natural temperature perturbations of the pre-industrial past 10,000 years. Climate models need to first pass their hindcast calibration test before they can be said to accurately model present and future climate.

Many important elements of the climate system, including atmospheric pressure, wind, clouds, the distribution of water vapor and carbon dioxide, the condensation of water vapor at ground level, the solar wind, aerosol concentration and distribution, dust concentration and distribution, and the reflectivity of snow and ice are highly uncertain. Lupo *et al.* (2013) cite extensive scholarly research on the following specific problems with climate models:

- Climate models underestimate surface evaporation caused by increased temperature by a factor of 3, resulting in a consequential under-estimation of global precipitation.

- Climate models inadequately represent aerosol-induced changes in infrared (IR) radiation, despite studies showing different mineral aerosols (for equal loadings) can cause differences in surface IR flux between 7 and 25 Wm^{-2}.

- Limitations in computing power restrict climate models from resolving important climate processes; low-resolution models fail to capture many important regional and lesser-scale phenomena such as clouds.

- Model calibration is faulty, as it assumes all temperature rise since the start of the Industrial Revolution has resulted from human activities; in reality, major anthropogenic emissions commenced only in the mid-twentieth century and there is no reason to assume the temperature increase since then is entirely due to human activity.

- Internal climate oscillations (AMO, PDO, etc.) are major features of the historic temperature record; climate models simulate them very poorly.

- Climate models fail to incorporate the effects of variations in solar magnetic field or in the flux of cosmic rays, both of which are known to significantly affect climate.

Christy and McNider (2017) observe, "the mismatch since 1979 between observations and CMIP5 model values suggests that excessive sensitivity to enhanced radiative forcing in the models can be appreciable. The tropical region is mainly responsible for this discrepancy suggesting processes that are the likely sources of the extra sensitivity are (a) the parameterized hydrology of the deep atmosphere, (b) the parameterized heat-partitioning at the ocean atmosphere interface and/or (c) unknown natural variations."

According to Legates (2014, p. 1,165),

GCMs simply cannot reproduce some very important phenomena. For example, hurricanes and most other forms of severe weather (e.g., nor'easters, tornadoes, and thunderstorms) cannot be represented in a GCM owing to the coarse spatial resolution. Other more complex phenomena resulting from interactions among the elements that drive the climate system may be limited or even not simulated at all. Phenomena such as the Pacific Decadal Oscillation, the Atlantic Multidecadal Oscillation, and other complex interrelationships between the ocean and the atmosphere, for example, are inadequately reproduced or often completely absent in climate model simulations. Their absence indicates a fundamental flaw exists in either our understanding of the climate system, the mathematical parameterization of the process, the spatial and temporal limitations imposed by finite computational power, or a combination of all three.

* * *

The outputs of GCMs are only as reliable as the data and theories "fed" into them, which scientists widely recognize as being seriously deficient (Bray and von Storch, 2016; Strengers, *et al.*, 2015). The utility and skillfulness of computer models are dependent on how well the processes they model are understood, how faithfully those processes are simulated in the computer code, and whether the results can be repeatedly tested so the models can be refined (Loehle, 2018). To date, GCMs have failed to deliver on each of these counts.

Clutz (2015), observing how the Russian INM-CM4 climate model most closely reproduces the temperature history since 1850 by incorporating "high inertia from ocean heat capacities, low forcing from CO_2 and less water for feedback," then asked, "Why aren't the other models built like this one?" Why indeed? Unless, as was the case with HadCRUT's flawed temperature record, the purpose of most GCMs is not to accurately model the real climate, but rather to present an image of the climate that meets the needs of the world's political leaders. In this case, the map is definitely not the territory.

References

Armstrong, J.S. and Green, K.C. 2018. Guidelines for science: evidence-based checklists. *Working Paper* version 502-RG. Philadelphia, PA: The Wharton School, University of Pennsylvania.

Balling, R.C. 2005. Observational surface temperature records versus model predictions. In Michaels, P.J. (Ed.) *Shattered Consensus: The True State of Global Warming*. Lanham, MD: Rowman & Littlefield, pp. 50–71.

Bray, D. and von Storch, H. 2016. The Bray and von Storch 5th International Survey of Climate Scientists 2015/2016. *HZG Report* 2016-2. Geesthacht, Germany: GKSS Institute of Coastal Research.

Bryson, R.A. 1993. Environment, environmentalists, and global change: a skeptic's evaluation. *New Literary History* **24**: 783–95.

Christy, J. 2005. Temperature changes in the bulk atmosphere: beyond the IPCC. In Michaels, P.J. (Ed.) *Shattered Consensus: The True State of Global Warming*. Lanham, MD: Rowman & Littlefield, pp. 72–105.

Christy, J. 2017. Testimony before the U.S. House Committee on Science, Space and Technology. March 29.

Christy, J.R. and McNider, R.T. 2017. Satellite bulk tropospheric temperatures as a metric for climate sensitivity. *Asia-Pacific Journal of Atmospheric Sciences* **53** (4): 511–8.

Clutz, R. 2015. Temperatures according to climate models. Science Matters (website). March 24. Accessed November 27, 2018.

Coats, S. and Karnauskas, K.B. 2017. Are simulated and observed twentieth century tropical Pacific sea surface temperature trends significant relative to internal variability? *Geophysical Research Letters* **44**: 9928–37.

Dommenget, D. and Rezny, M. 2018. A caveat note on tuning in the development of coupled climate models. *Journal of Advances in Modeling Earth Systems* **10**: 78–97.

Douglass, D.H., Pearson, B.D., and Singer, S.F. 2004. Altitude dependence of atmospheric temperature trends: climate models versus observation. *Geophysical Research Letters* **31**: L13208.

Dyson, F. 2004. A meeting with Enrico Fermi. *Nature* **427**: 297.

Dyson, F. 2007. Heretical thoughts about science and society. *Edge: The Third Culture*. August.

Essex, C. and McKitrick, R. 2007. *Taken by Storm. The Troubled Science, Policy and Politics of Global Warming*. Toronto, ON: Key Porter Books.

Essex, C., McKitrick, R., and Andresen, B. 2007. Does a global temperature exist? *Journal of Non-Equilibrium Thermodynamics* **32** (1).

Frauenfeld, O.W. 2005. Predictive skill of the El Niño-Southern Oscillation and related atmospheric teleconnections. In Michaels, P.J. (Ed.) *Shattered Consensus: The True State of Global Warming*. Lanham, MD: Rowman & Littlefield, pp. 149–182.

Green, K.C., Armstrong, J.S., and Soon, W. 2009. Validity of climate change forecasting for public policy decision making. *International Journal of Forecasting* **25** (4): 826–32.

Hedemann, C., Mauritsen, T., Jungclaus, J., and Marotzke, J. 2017. The subtle origins of surface-warming hiatuses. *Nature Climate Change* **7**: 336–9.

Hourdin, F., *et al.* 2017. The art and science of climate model tuning. *Bulletin of the American Meteorological Society* **98** (3): 589–602.

Idso, S.B. and Idso, C.D. 2015. *Mathematical Models vs. Real-World Data: Which Best Predicts Earth's Climatic Future?* Tempe, AZ: Center for the Study of Carbon Dioxide and Global Change.

IPCC. 2014. Intergovernmental Panel on Climate Change. *Climate Change 2014: Mitigation of Climate Change.* Contribution of Working Group III to the Fifth Assessment Report of the Intergovernmental Panel on Climate Change. Cambridge, UK and New York, NY: Cambridge University Press.

Jaynes, E.T. 2003. *Probability Theory: The Logic of Science.* Cambridge, MA: Cambridge University Press.

Kravtsov, S. 2017. Pronounced differences between observed and CMIP5-simulated multidecadal climate variability in the twentieth century. *Geophysical Research Letters* **44** (11): 5749–57.

Legates, D. 2014. Climate models and their simulation of precipitation. *Energy & Environment* **25** (6–7): 1163–75.

Loehle, C. 2018. The epistemological status of general circulation models. *Climate Dynamics* **50**: 1719–31.

Lupo, A., *et al.* 2013. Global climate models and their limitations. In: *Climate Change Reconsidered II: Physical Science.* Nongovernmental International Panel on Climate Change. Chicago, IL: The Heartland Institute.

McKitrick, R. and Christy, J. 2018. A test of the tropical 200- to 300-hPa warming rate in climate models. *Earth and Space Science* **5**: 529–36.

Michaels, P.J. 2000. *Satanic Gases: Clearing the Air About Global Warming.* Washington, DC: Cato Institute.

Michaels, P.J. 2005. *Meltdown: The Predictable Distortion of Global Warming by Scientists, Politicians and the Media.* Washington, DC: Cato Institute.

Michaels, P.J. 2009. *Climate of Extremes: Global Warming Science They Don't Want You to Know.* Washington, DC: Cato Institute.

Monckton, C., *et al.* 2015. Why models run hot: results from an irreducibly simple climate model. *Science Bulletin* **60** (1).

NIPCC. 2013. Nongovernmental International Panel on Climate Change. *Climate Change Reconsidered II: Physical Science.* Chicago, IL: The Heartland Institute.

Pilkey, O.H. and Pilkey-Jarvis, L. 2007. *Useless Arithmetic.* New York, NY: Columbia University Press.

Posmentier, E.S. and Soon, W. 2005. Limitations of computer predictions of the effects of carbon dioxide on global temperature. In Michaels, P.J. (Ed.) *Shattered Consensus: The True State of Global Warming.* Lanham, MD: Rowman & Littlefield, pp. 241–81.

Roach, L.A., Dean, S.M., and Renwick, J.A. 2018. Consistent biases in Antarctic sea ice concentration simulated by climate models. *The Cryosphere* **12**: 365–83.

Scanlon, B.R., *et al.* 2018. Global models underestimate large decadal declining and rising water storage trends relative to GRACE satellite data. *Proceedings of the National Academy of Sciences USA* **115**: (6) E1080–E1089.

Singer, S.F. 2011. Lack of consistency between modelled and observed temperature trends. *Energy & Environment* **22**: 375–406.

Singer, S.F. 2013. Inconsistency of modelled and observed tropical temperature trends. *Energy & Environment* **24**: 405–13.

Solomon, L. 2008. *The Deniers: The World Renowned Scientists Who Stood Up Against Global Warming Hysteria, Political Persecution, and Fraud—And those who are too fearful to do so.* Minneapolis, MN: Richard Vigilante Books.

Spencer, R. 2008. *Climate Confusion: How Global Warming Hysteria Leads to Bad Science, Pandering Politicians and Misguided Policies that Hurt the Poor.* New York, NY: Encounter Books.

Stouffer, R.J., *et al.* 2017. CMIP5 scientific gaps and recommendations for CMIP6. *Bulletin of the American Meteorological Society* **98** (1).

Strengers, B., Verheggen, B., and Vringer, K. 2015. Climate science survey questions and responses. PBL Netherlands Environmental Assessment Agency. April 10.

Tebaldi, C. and R. Knutti, 2007. The use of the multi-model ensemble in probabilistic climate projections. *Philosophical Transactions of the Royal Society of London A* **365**: 2053–75.

Volodin, E. and Gritsun, A. 2018. Simulation of observed climate changes in 1850–2014 with climate model INM-CM5. *Earth System Dynamics* **9**: 1235–42.

Voosen, P., 2016. Climate scientists open up their black boxes to scrutiny. *Science* **354**: 401–2.

2.2.3 Climate Sensitivity

Estimates of equilibrium climate sensitivity (the amount of warming that would occur following a doubling of atmospheric CO_2 level) range widely. The IPCC's estimate is higher than many recent estimates.

Climate sensitivity is a metric used to characterize the response of the global climate system to a given forcing. A forcing, in turn, is a chemical or physical process that alters Earth's radiative equilibrium, causing temperatures to rise or fall. Equilibrium climate sensitivity (ECS) is broadly defined as the global mean surface temperature change following a doubling of atmospheric CO_2 concentration and following the passage of time (possibly centuries) required for the atmosphere and oceans to return to equilibrium. Transient climate sensitivity (TCS) is surface temperature change occurring at the time of CO_2 doubling over a period of 70 years (IPCC, 2013, p. 82). Both metrics typically assume a pre-industrial level of CO_2 as the basis of the calculation (e.g., 280 ppm x 2 = 560 ppm).

The current controversy over man-made global warming originated in a 1979 report published by the U.S. National Academy of Sciences called the Charney Report (NRC, 1979). The authors conceded the increase in temperatures from a doubling of atmospheric CO_2 would be modest, probably not measurable at that time. However, they speculated that with water vapor feedback, a doubling of CO_2 would increase atmospheric temperatures sufficiently to result in an increase of surface temperatures by 3 ± 1.5°C. The Charney Report overruled the simple physics calculations of Rasool and Schneider (1971) that had estimated an ECS of 0.6°C (upped to 0.8°C with a simple H_2O feedback).

In the Working Group I contribution to the IPCC's Fifth Assessment Report (AR5), the IPCC claims an ECS in "a range of 2°C to 4.5°C, with the CMIP5 model mean at 3.2°C" (IPCC, 2013, p. 83). This is at odds with the Summary for Policymakers of the same volume, which gives an ECS "in the range 1.5°C to 4.5°C" and says in a footnote on the same page, "No best estimate for equilibrium climate sensitivity can now be given because of a lack of agreement on values across assessed lines of evidence and studies" (p. 16). Either estimate is indistinguishable from the one in the Charney report issued three decades earlier, illustrating the lack of progress made on this key issue of climate science.

The IPCC's ECS Is Too High

The IPCC's estimate is higher than many estimates appearing in the scientific literature, especially those appearing most recently. Christy and McNider (2017), relying on the latest satellite temperature data, write, "If the warming rate of +0.096 K dec^{-1} represents the net T_{LT} response to increasing greenhouse radiative forcings, this implies that the T_{LT} tropospheric transient climate response (ΔT_{LT} at the time CO_2 doubles) is +1.10 ± 0.26 K which is about half of the average of the IPCC AR5 climate models of 2.31 ± 0.20 K. Assuming that the net remaining unknown internal and external natural forcing over this period is near zero, the mismatch since 1979 between observations and CMIP-5 model values suggests that excessive sensitivity to enhanced radiative forcing in the models can be appreciable."

Figure 2.2.3.1 presents a visual representation of estimates of climate sensitivity appearing in scientific research papers published between 2011 and 2016. According to Michaels (2017), the climate sensitivities reported in the literature average ~2.0°C (median) with a range of ~1.1°C (5th percentile) and ~3.5°C (95th percentile). The median is more than one-third lower than the estimate used by the IPCC.

The IPCC ignores mounting evidence that climate sensitivity to CO_2 is much lower than its models assume. Monckton *et al.* (2015) cited 27 peer-reviewed articles "that report climate sensitivity to be below current central estimates." Their list of sources appears in Figure 2.2.3.2.

Figure 2.2.3.1
Equilibrium climate sensitivity estimates from scientific research since 2011 (colored), compared to Roe and Baker (2007) (black)

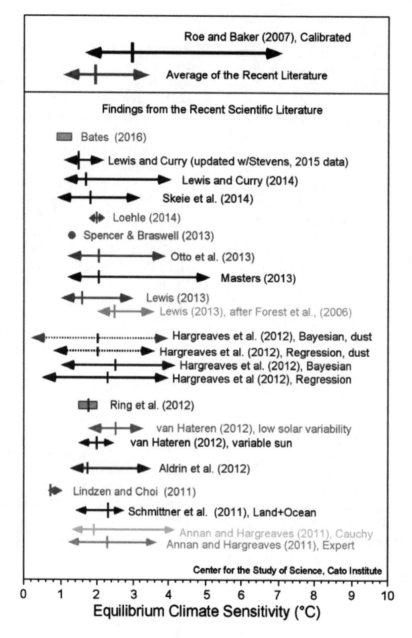

Arrows represent the 5% to 95% confidence bounds for estimates of climate sensitivity released since 2011. Colored vertical lines show the best estimate of climate sensitivity (median of each probability density function or the mean of multiple estimates). Ring *et al.* (2012) present four estimates of the climate sensitivity, and the red box encompasses those estimates. Spencer and Braswell (2013) produce a single ECS value best-matched to ocean heat content observations and internal radiative forcing. Keep in mind that all of these confidence bounds represent spreads about a model mean, and are therefore statements of precision rather than of accuracy. Citations to each of these studies appear in the references section below. *Source:* Michaels, 2017, p. 6.

Figure 2.2.3.2
Research finding climate sensitivity is less than assumed by the IPCC

Michaels, P.J., Knappenberger, P.C., Frauenfeld, O.W., *et al.* 2002. Revised 21st century temperature projections. *Climate Research* **23**: 1–9.

Douglass, D.H., Pearson, B.D., and Singer, S.F. 2004. Altitude dependence of atmospheric temperature trends: climate models versus observation. *Geophysical Research Letters* **31**: L13208. doi: 10.1029/2004GL020103.

Landscheidt, T. 2003. New Little Ice Age instead of global warming? *Energy & Environment* **14** (2): 327–350.

Chylek, P. and Lohmann, U. 2008. Aerosol radiative forcing and climate sensitivity deduced from the Last Glacial Maximum to Holocene transition. *Geophysical Research Letters* **35**: L04804. doi: 10.1029/2007GL032759.

Monckton of Brenchley, C. 2008. Climate sensitivity reconsidered. *Physics & Society* **37**: 6–19.

Douglass, D.H. and Christy, J.R. 2009. Limits on CO_2 climate forcing from recent temperature data of earth. *Energy & Environment* **20**: 1–2.

Lindzen, R.S. and Choi, Y-S. 2009. On the determination of climate feedbacks from ERBE data. *Geophysical Research Letters* **36**: L16705. doi: 10.1029/2009GL039628.

Spencer, R.W. and Braswell, W.D. 2010. On the diagnosis of radiative feedback in the presence of unknown radiative forcing. *Journal of Geophysical Research* **115**: D16109. doi: 10.1029/2009JD013371.

Annan, J.D. and Hargreaves, J.C. 2011. On the generation and interpretation of probabilistic estimates of climate sensitivity. *Climate Change* **104**: 324–436.

Lindzen, R.S. and Choi, Y-S. 2011 On the observational determination of climate sensitivity and its implications. *Asia-Pacific Journal of Atmospheric Sciences* **47**: 377–390.

Monckton of Brenchley, C. 2011. Global brightening and climate sensitivity. In: Zichichi, A. and Ragaini, R. (Eds.) *Proceedings of the 45th Annual International Seminar on Nuclear War and Planetary Emergencies, World Federation of Scientists*. London, UK: World Scientific.

Schmittner, A., Urban, N.M., Shakun, J.D., *et al.* 2011. Climate sensitivity estimated from temperature reconstructions of the last glacial maximum. *Science* **334**: 1385–1388. doi: 10.1126/science.1203513.

Spencer, R.W. and Braswell, W.D. 2011. On the misdiagnosis of surface temperature feedbacks from variations in Earth's radiant-energy balance. *Remote Sensing* **3**: 1603–1613. doi: 10.3390/rs3081603.

Aldrin, M., Holden, M., Guttorp, P., *et al.* 2012. Bayesian estimation of climate sensitivity based on a simple climate model fitted to observations of hemispheric temperature and global ocean heat content. *Environmetrics* **23**: 253–271. doi: 10.1002/env.2140.

Hargreaves, J.C., Annan, J.D., Yoshimori, M., *et al.* 2012. Can the last glacial maximum constrain climate sensitivity? *Geophysical Research Letters* **39**: L24702. doi: 10.1029/2012GL053872.

Ring, M.J., Lindner, D., Cross, E.F., *et al.* 2012. Causes of the global warming observed since the 19th century. *Atmospheric and Climate Sciences* **2**: 401–415. doi: 10.4236/acs.2012.24035.

van Hateren, J.H. 2012. A fractal climate response function can simulate global average temperature trends of the modern era and the past millennium. *Climate Dynamics* **40**: 2651–2670, doi: 10.1007/s00382-012-1375-3.

Lewis, N. 2013. An objective Bayesian improved approach for applying optimal fingerprint techniques to estimate climate sensitivity. *Journal of Climate* **26**: 7414–7429. doi: 10.1175/JCLI-D-12-00473.1.

Masters, T. 2013. Observational estimates of climate sensitivity from changes in the rate of ocean heat uptake and comparison to CMIP5 models. *Climate Dynamics* **42**: 2173–2181. doi: 101007/s00382-013-1770-4.

Otto, A., Otto, F.E.L., Boucher, O., *et al.* 2013. Energy budget constraints on climate response. *Nature Geoscience* **6**: 415–416. diuL19,1938/ngeo1836.

Spencer, R.W. and Braswell, W.D. 2013. The role of ENSO in global ocean temperature changes during 1955–2011 simulated with a 1D climate model. *Asia-Pacific Journal of Atmospheric Sciences* **50**: 229-237. doi: 10.1007/s13143-014-0011-z.

Lewis, N. and Curry, J.A. 2014. The implications for climate sensitivity of AR5 forcing and heat uptake estimates. *Climate Dynamics* **10**. doi: 1007/s00382-014-2342-y.

Loehle, C. 2014. A minimal model for estimating climate sensitivity. *Ecological Modelling* **276**: 80–84. doi: 10.1016/j.ecolmodel.2014.01.006.

McKitrick, R. 2014. HAC-robust measurement of the duration of a trendless subsample in a global climate time series. *Open Journal of Statistics* **4**: 527–535. doi: 10.4236/ojs.2014.47050.

Monckton of Brenchley, C. 2014. Political science: drawbacks of apriorism in intergovernmental climatology. *Energy & Environment* **25**: 1177–1204.

Skeie, R.B., Berntsen, T., Aldrin, M., *et al.* 2015. A lower and more constrained estimate of climate sensitivity using updated observations and detailed radiative forcing time series. *Earth System Dynamics* **5**: 139–175. doi: 10.5194/esd-5-139-2014.

Lewis, N. 2015. Implications of recent multimodel attribution studies for climate sensitivity. *Climate Dynamics* doi: 10.1007/s00382-015-2653-7RSS

Source: Monckton *et al.,* 2015, pp. 1378–1390, footnotes 7 to 33. Some corrections to citations have been made.

Uncertainty

No one actually knows the "true" climate sensitivity value because it is, like so many numbers in the climate change debate, a stylized fact: a single number chosen for the sake of convenience by those who make their living modeling climate change or advocating for government action to slow or stop it. The number is inherently uncertain for much the same reason it is impossible to know how much CO_2 is emitted into the atmosphere or how much of it stays there, which is the enormous size of natural processes relative to the "human signal" caused by our CO_2 emissions. Pindyck (2013) offers some insight with respect to the problems posed by climate feedbacks:

> Here is the problem: the physical mechanisms that determine climate sensitivity involve crucial feedback loops, and the parameter values that determine the strength (and even the sign) of those feedback loops are largely unknown, and for the foreseeable future may even be unknowable. This is not a shortcoming of climate science; on the contrary, climate scientists have made enormous progress in understanding the physical mechanisms involved in climate change. But part of that progress is a clearer realization that there are limits (at least currently) to our ability to pin down the strength of the key feedback loops.

> … We don't know whether the feedback factor f is in fact normally distributed (nor do we know its mean and standard deviation). Roe and Baker [2007] simply assumed a normal distribution. In fact, in an accompanying article in the journal *Science*, Allen and Frame (2007) argued climate sensitivity is in the realm of the "unknowable" (pp. 865, 867).

The IPCC acknowledges there may be natural variability in radiative forcing due to "solar variability and aerosol emissions via volcanic activity" and contends they "are also specified elements in the CMIP5 experimental protocol, but their future time evolutions are not prescribed very precisely" (IPCC, 2013, pp. 1047, 1051). Deferring to the GCMs on which CMIP bases its carbon cycle, the IPCC blandly reports that "some models include the effect" of solar cycles and orbital variations "but most do not." None tries to model volcanic eruptions. "For the other natural aerosols (dust, sea-salt, etc.), no emission or concentration data are recommended. The emissions are potentially computed interactively by the models themselves and many change with climate, or prescribed from separate model simulations carried out in the implementation of CMIP5 experiments, *or simply held constant*" (IPCC, 2013, p. 1051, italics added).

From this description it is likely that most of the GCMs simply assume radiative forcing in the atmosphere would be unchanging for decades or even centuries if not for anthropogenic greenhouse gas emissions. In light of evidence of significant natural variability in global average temperatures on geologic time scales as well as in the modern era, as documented in Section 2.1.2.3, that assumption is unrealistic, and consequently the IPCC's estimate of ECS is also unrealistic.

Earlier in the current chapter, in Section 2.1.1.3, profound uncertainty was documented about re-creations of the global average temperature record

since 1850, including a graph by Frank (2015), who added the missing error bars to a graph produced by the Climatic Research Unit at the University of East Anglia. Frank also applied known sources of uncertainty to projections of *future* temperatures, writing, "CMIP5 climate model simulations of global cloud fraction reveal theory-bias error. Propagation of this cloud forcing error uncovers a [root-sum-squared-error] uncertainty $1\sigma \approx \pm15°C$ in centennially projected air temperature." This single error is so consequential it means "causal attribution of warming is therefore impossible." Frank then applies error bars to the projection of future global temperatures from NASA's Goddard Institute for Space Studies (GISS). His graphic appears here as Figure 2.2.3.3.

After cataloguing a series of errors and uncertainties in the data leading up to a finding of causation (or "attribution"), Frank (2015, p. 406) summarized his findings as follows:

Figure 2.2.3.3
GISS Model II projections of future global averaged surface air temperature anomalies as presented in 1988, (a) without and (b) with uncertainty bars

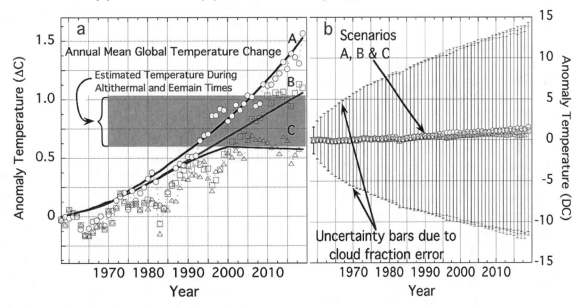

Figure (a) is modified from Hansen, 1988 and Hansen *et al.,*2006; figure (b) has uncertain bars added to account for cloud-forcing error. Note the change in vertical scale to accommodate the uncertainty range of figure (b). *Source:* Frank, 2015, Figure 1, p. 392.

1. The poor resolution of present state-of-the-art CMIP5 GCMs means the response of the terrestrial climate to increased GHGs is far below any level of detection.

2. The poor resolution of CMIP5 GCMs means all past and present projections of terrestrial air temperature can have revealed nothing of future terrestrial air temperature.

3. The lack of any scientific content in consensus proxy paleo-temperature reconstructions means nothing has been revealed of terrestrial paleo-temperatures.

4. The neglected systematic sensor measurement error in the global air temperature record means that neither the rate nor the magnitude of the change in surface air temperatures is knowable. Therefore,

5. Detection and attribution of an anthropogenic cause to climate change cannot have been nor presently can be evidenced in climate observables.

Researchers are continually trying to reduce the uncertainty described by Frank and as they do, estimates of ECS continue to fall. In 2016, the CLOUD research project being conducted by CERN (the European Institute for Nuclear Research) reported the discovery that aerosol particles can form in the atmosphere purely from organic vapors produced naturally by the biosphere (Dunne *et al.*, 2016). In a separate paper (Gordon *et al.*, 2016) CLOUD researchers demonstrated that pure biogenic nucleation was the dominant source of particles in the pristine pre-industrial atmosphere. By raising the baseline aerosol state, this process significantly reduces the estimated aerosol radiative forcing from anthropogenic activities and, in turn, reduces modeled climate sensitivities. The researchers also found evidence of a much bigger role played by cosmic rays in cloud formation than was previously known, a topic covered in the next section.

"This is a huge step for atmospheric science," CLOUD lead-author Ken Carslaw told a writer for the research center's newsletter, the *CERN Courier* (CERN, 2016). "It's vital that we build climate models on experimental measurements and sound understanding, otherwise we cannot rely on them to predict the future. Eventually, when these processes get implemented in climate models, we will have much more confidence in aerosol effects on climate. Already, results from CLOUD suggest that estimates of high climate sensitivity may have to be revised downwards."

Abbot and Marohasy (2017) conducted signal analysis on six temperature proxy datasets and used the resulting component sine waves as input to an artificial neural network (ANN), a form of machine learning. The ANN model was used to simulate the late Holocene period to 1830 CE and then through the twentieth century. The authors report, "the largest deviation between the ANN projections and measured temperatures for six geographically distinct regions was approximately 0.2 °C, and from this an Equilibrium Climate Sensitivity (ECS) of approximately 0.6 °C was estimated. This is considerably less than estimates from the General Circulation Models (GCMs) used by the Intergovernmental Panel on Climate Change (IPCC), and similar to estimates from spectroscopic methods."

If Frank and these other researchers are correct, modelers should not enter into their GCMs or IAMs the IPCC's TCS estimate, since it appears to be too high. But caution is nowhere to be found in the IPCC's discussion of climate sensitivity or in the way it is treated in the popular press or even scholarly research. The IPCC's estimate of equilibrium climate sensitivity of ~3.2°C for a doubling of CO_2 in the atmosphere is accepted as if it were direct evidence or a finding with a high degree of certainty. It is incorporated into IAMs with little debate and no admission of its uncertainty.

* * *

Equilibrium climate sensitivity (ECS) is one of the most important variables in climate science, but it is also the most uncertain and possibly unknowable value. IPCC's estimate of ~3.2°C is not based on direct evidence but on assumptions about Earth's temperature record, the changing composition of its atmosphere, and complex interactions between the atmosphere and oceans that mean discerning the impact of a few trace gases – CO_2 at 405 ppm, CH_4 at only 1.8 ppm, and nitrous oxide at 324 ppb – is likely to be impossible. The current generation of GCMs cannot find a reliable estimate of ECS.

References

Abbot, J. and Marohasy, J. 2017. The application of machine learning for evaluating anthropogenic versus natural climate change. *Journal of Geophysical Research* **14**: 36–46.

Aldrin, M., Holden, M., Guttorp, P., Skeie, R.B., Myhre, G., and Berntsen, T.K. 2012. Bayesian estimation of climate sensitivity based on a simple climate model fitted to observations of hemispheric temperature and global ocean heat content. *Environmetrics* **23** (3): 253–71.

Allen, M.R., and Frame, D.J. 2007. Call off the quest. *Science* **318** (5850): 582-583.

Annan, J.D. and Hargreaves, J.C. 2011. On the generation and interpretation of probabilistic estimates of climate sensitivity. *Climatic Change* **104:** 324–436.

Bates, J.R. 2016. Estimating climate sensitivity using two-zone energy balance models. *Earth and Space Science* **3** (5): 207–25.

CERN. 2016. CLOUD experiment sharpens climate predictions. *CERN Courier.* November 11.

Christy, J.R. and McNider, R.T. 2017. Satellite bulk tropospheric temperatures as a metric for climate sensitivity. *Asia-Pacific Journal of Atmospheric Sciences* **53** (4): 511–518

Dunne, E.M., *et al.* 2016. Global atmospheric particle formation from CERN CLOUD measurements. *Science* **354** (6316).

Frank, P. 2015. Negligence, non-science, and consensus climatology. *Energy & Environment* **26** (3): 391–416.

Gordon, H., *et al.* 2016. Reduced anthropogenic aerosol radiative forcing caused by biogenic new particle formation. *Proceedings of the National Academy of Sciences* **113** (43) 12053-12058.

Hansen, J. 1988. Greenhouse effect and global climate change. Testimony before the US Senate Commission on Energy and Natural Resources.

Hansen, J. *et al.* 2006. Global temperature change. *Proceedings of the National Academy of Sciences USA* **103** (39): 14288–93.

Hargreaves, J.C., Annan, J.D., Yoshimori, M., and Abe-Ouchi, A. 2012. Can the last glacial maximum constrain climate sensitivity? *Geophysical Research Letters* **39**: L24702.

IPCC. 2013. Intergovernmental Panel on Climate Change. *Climate Change 2013: The Physical Science Basis.* Contribution of Working Group I to the Fifth Assessment Report of the Intergovernmental Panel on Climate Change. Cambridge, UK and New York, NY: Cambridge University Press.

Lewis, N. 2013. An objective Bayesian, improved approach for applying optimal fingerprint techniques to estimate climate sensitivity. *Journal of Climate* **26**: 7414–29.

Lewis, N. and Curry, J.A. 2014. The implications for climate sensitivity of AR5 forcing and heat uptake estimates. *Climate Dynamics* **45** (3/4): 1009–23.

Lindzen, R.S. and Choi, Y-S. 2011. On the observational determination of climate sensitivity and its implications. *Asia-Pacific Journal of Atmospheric Science* **47**: 377–90.

Loehle, C. 2014. A minimal model for estimating climate sensitivity. *Ecological Modelling* **276**: 80–4.

Masters, T. 2013. Observational estimates of climate sensitivity from changes in the rate of ocean heat uptake and comparison to CMIP5 models. *Climate Dynamics* **42** (7–8): 2173–218.

Michaels, P. 2017. Testimony. Hearing on At What Cost? Examining the Social Cost of Carbon, before the U.S. House of Representatives Committee on Science, Space, and Technology, Subcommittee on Environment, Subcommittee on Oversight. February 28.

Monckton, C., *et al.* 2015. Keeping it simple: the value of an irreducibly simple climate model. *Science Bulletin* **60** (15).

NRC (National Research Council). 1979. *Carbon Dioxide and Climate: A Scientific Assessment.* Washington, DC: The National Academies Press.

Otto, A., *et al.* 2013. Energy budget constraints on climate response. *Nature Geoscience* **6**: 415–6.

Pindyck, R.S. 2013. Climate change policy: what do the models tell us? *Journal of Economic Literature* **51**: 860–72.

Rasool, S.I. and Schneider, S.H. 1971. Atmospheric carbon dioxide and aerosols: effects of large increases on global climate. *Science* **173**: 138–41.

Ring, M.J., Lindner, D., Cross, E.F., and Schlesinger, M.E. 2012. Causes of the global warming observed since the 19th century. *Atmospheric and Climate Sciences* **2**: 401–15.

Roe, G.H. and Baker, M.B. 2007. Why is climate sensitivity so unpredictable? *Science* **318** (5850): 629–32.

Schmittner, A., *et al.* 2011. Climate sensitivity estimated from temperature reconstructions of the Last Glacial Maximum. *Science* **334**: 1385–8.

Skeie, R.B., Berntsen, T., Aldrin, M., Holden, M., and Myhre, G. 2014. A lower and more constrained estimate of climate sensitivity using updated observations and detailed radiative forcing time series. *Earth System Dynamics* **5**: 139–75.

Spencer, R.W. and Braswell, W.D. 2013. The role of ENSO in global ocean temperature changes during 1955–2011 simulated with a 1D climate model. *Asia-Pacific Journal of Atmospheric Science* **50** (2): 229–37.

van Hateren, J.H. 2012. A fractal climate response function can simulate global average temperature trends of the modern era and the past millennium. *Climate Dynamics* **40** (11–12): 2651–70.

2.2.4 Solar Activity

Solar irradiance, magnetic fields, UV fluxes, cosmic rays, and other solar activity may have greater influence on climate than climate models and the IPCC currently assume.

Solar influences on Earth's climate are a fourth area of controversy in climate science. According to the IPCC (2013), "changes in solar irradiance are an important driver of climate variability, along with volcanic emissions and anthropogenic factors," but they are considered by the IPCC to be too small to explain global temperature fluctuations of more than approximately 0.1°C between minima and maxima (p. 392). Solar influences, according to the IPCC, "cannot explain the observed increases since the late 1970s" (*Ibid.*).

Usoskin (2017) offers an excellent survey of the literature on solar activity. He explains,

> Although scientists knew about the existence of "imperfect" spots on the sun since the early seventeenth century, it was only in the nineteenth century that the scientific community recognized that solar activity varies in the course of an 11-year solar cycle. Solar variability was later found to have many different manifestations, including the fact that the "solar constant," or the total solar irradiance, TSI, (the amount of total incoming solar electromagnetic radiation in all wavelengths per unit area at the top of the atmosphere) is not a constant. The sun appears much more complicated and active than a static hot plasma ball, with a great variety of nonstationary active processes going beyond the adiabatic equilibrium foreseen in the basic theory of sun-as-star. Such transient nonstationary (often eruptive) processes can be broadly regarded as solar activity, in contrast to the so-called "quiet"

sun. Solar activity includes active transient and long-lived phenomena on the solar surface, such as spectacular solar flares, sunspots, prominences, coronal mass ejections (CMEs), etc.

Usoskin (2017) summarizes what he calls the "main features … observed in the long-term evolution of solar magnetic activity":

- Solar activity is dominated by the 11-year Schwabe cycle on an interannual timescale. Some additional longer characteristic times can be found, including the Gleissberg secular cycle, de Vries/Suess cycle, and a quasi-cycle of 2000–2400 years (Hallstatt cycle). However, all these longer cycles are intermittent and cannot be regarded as strict phase-locked periodicities.

- One of the main features of long-term solar activity is that it contains an essential chaotic/stochastic component, which leads to irregular variations and makes solar-activity predictions impossible for a scale exceeding one solar cycle.

- The sun spends about 70% of its time at moderate magnetic activity levels, about 15–20% of its time in a grand minimum, and about 10–15% in a grand maximum.

- Grand minima are a typical but rare phenomena in solar behavior. They form a distinct mode of solar dynamo. Their occurrence appears not periodically, but rather as the result of a chaotic process within clusters separated by 2000–2500 years (around the lows of the Hallstatt cycle). Grand minima tend to be of two distinct types: short (Maunder-like) and longer (Spörer-like).

- The recent level of solar activity (after the 1940s) was very high, corresponding to a prolonged grand maximum, but it has ceased to the normal moderate level. Grand maxima are also rare and irregularly occurring events, though the exact rate of their occurrence is still a subject of debates.

With this background, we can address whether solar activity explains some of the climate changes attributed by the IPCC to anthropogenic forcing.

2.2.4.1 Total Solar Irradiance

Incoming solar radiation is most often expressed as total solar irradiance (TSI), a measure derived from multi-proxy measures of solar activity. Measurements produced from 1979 to 2014 by the now defunct ACRIM satellite, one of 21 observational components of NASA's Earth Observing System, are shown in Figure 2.2.4.1.1. They show TSI rose and fell with solar cycles and ranged between 1,360 and 1,363 Wm^{-2} during the period 1979–2014, a variability of ~3 Wm^{-2} (ACRIM, n.d.). We are currently in the 24th solar cycle since 1755, which began in December 2008 and is expected to end in 2019.

The ACRIM TSI composite shows a small upward pattern from around 1980 to 2000, an increase not acknowledged by the IPCC or incorporated into the models on which it relies (Scafetta and Willson, 2014, Appendix). According to the ACRIM website, "gradual variations in solar luminosity of as little as 0.1% was the likely forcing for the 'Little Ice Age' that persisted in varying degree from the late 14th to the mid-19th centuries." Shapiro *et al.* (2011) estimated the TSI change between the Maunder Minimum and current conditions may have been as large as 6 Wm^{-2}.

Egorova *et al.* (2018, Figure 8b) provide more recent and reliable estimates of about 3.7 to 4.5 Wm^{-2} for the TSI change between the Maunder Minimum and recent activity minima. Those values are still significantly higher than those provided by the CMIP6 estimates (Matthes *et al.*, 2017) expected in

Figure 2.2.4.1.1
Satellite measurement of total solar irradiance (TSI), 1978–2013

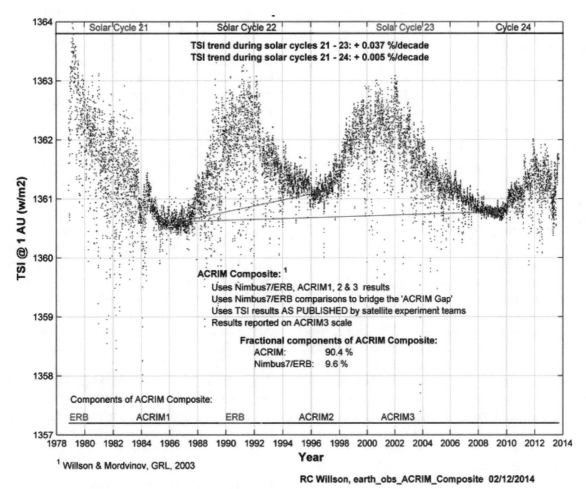

Source: ACRIM. n.d. Total solar irradiance (TSI) monitoring.

the IPCC Sixth Assessment reports forthcoming in 2021–2022.

According to the IPCC, trends in TSI accounted for only +0.05 [-0.01 to +0.10] Wm^{-2} of radiative forcing in 2011 relative to 1750, compared to 1.68 [1.33 to 2.03] Wm^{-2} for CO_2 and 2.29 [1.13 to 3.33] Wm^{-2} for total anthropogenic radiative forcing (IPCC, 2013, p. 14, Figure SPM.5). But changes in solar irradiance and ultraviolet radiation could easily yield a larger influence on global climate than that of CO_2. The absolute forcing of incoming solar radiation is approximately 340 Wm^{-2} at the top of the atmosphere, more than 10 times the forcing of atmospheric CO_2, so even small changes in the absolute forcing of the Sun could result in values larger than the predicted changes in radiative forcing caused by increasing CO_2.

Bond *et al.* (2001) used an accelerator mass spectrometer to study ice-rafted debris found in three North Atlantic deep-sea sediment cores, documenting a characteristic temperature cyclicity with a period of 1,000 to 2,000 years. The climatic oscillations coincided well with solar activity changes as reconstructed based on cosmogenic radionuclides in the Greenland ice cores (beryllium-10, ^{10}Be) and Northern Hemispheric tree rings (carbon-14, ^{14}C). The natural climate cycle occurred throughout the 12,000 years of the Holocene. The last two cold and warm nodes of this oscillation, in the words of Bond *et al.* (2001), were "broadly correlative with the so called 'Little Ice Age' and 'Medieval Warm Period.'" Bond *et al.* concluded, "a solar influence on climate of the magnitude and consistency implied by our evidence could not have been confined to the North Atlantic," suggesting the cyclical climatic effects of the Sun are experienced throughout the world. Soon *et al.* (2014) provide a detailed analysis of the nature of the millennial and bimillennial scales of solar and climatic variability throughout the Holocene.

How do the small changes in solar radiation bring about such significant and pervasive shifts in Earth's global climate? Bond *et al.* (2001) describe a scenario whereby solar-induced changes high in the stratosphere are propagated downward through the atmosphere to Earth's surface, provoking changes in North Atlantic deep water formation that alter the thermohaline circulation of the global ocean. They speculate "the solar signals thus may have been transmitted through the deep ocean as well as through the atmosphere, further contributing to their amplification and global imprint." Concluding their landmark paper, the researchers write the results of their study "demonstrate that the Earth's climate

system is highly sensitive to extremely weak perturbations in the Sun's energy output," noting their work "supports the presumption that solar variability will continue to influence climate in the future."

Research linking changes in solar influences to temperature, sea-level change, precipitation patterns, and other climate impacts is extensive; a summary in Chapter 3 of *Climate Change Reconsidered II: Physical Science* (NIPCC, 2013) is more than 100 pages long. We mention only a few recent studies here. One is Beer *et al.* (2006), who presented the various short- and longer-term scales of solar variability over the past 9,000 years. Beer *et al.* conclude that "comparison [of the solar development] with paleoclimatic data provides strong evidence for a causal relationship between solar variability and climate change."

Scafetta and West (2006a) developed two TSI reconstructions for the period 1900–2000, and their results suggest the Sun contributed 46% to 49% of the 1900–2000 warming of Earth, but with uncertainties of 20% to 30% in their sensitivity parameters. They say the role of the Sun in twentieth-century global warming has been significantly underestimated "because of the difficulty of modeling climate in general and a lack of knowledge of climate sensitivity to solar variations in particular." They also note "theoretical models usually acknowledge as solar forcing only the direct TSI forcing," thereby ignoring "possible additional climate effects linked to solar magnetic field, UV radiation, solar flares and cosmic ray intensity modulations." In a second study published that year, Scafetta and West (2006b) found a "good correspondence between global temperature and solar induced temperature curves during the pre-industrial period, such as the cooling periods occurring during the Maunder Minimum (1645–1715) and the Dalton Minimum (1795–1825)." Scafetta has written and coauthored several papers finding additional evidence of a solar effect on climate (Scafetta, 2008, 2010, 2012a, 2012b, 2013a, 2013b, 2016; Scafetta and West, 2003, 2005, 2007, 2008; Scafetta and Willson, 2009, 2014; Scafetta *et al.* 2017a, 2017b).

Shaviv (2008) found a "very clear correlation between solar activity and sea level" including the 11-year solar periodicity and phase, with a correlation coefficient of r = 0.55. He also found "the total radiative forcing associated with solar cycles variations is about 5 to 7 times larger than those associated with the TSI variations, thus implying the necessary existence of an amplification mechanism, though without pointing to which one." Shaviv argues "the sheer size of the heat flux, and the lack of any phase lag between the flux and the driving force further implies that it cannot be part of an atmospheric feedback and very unlikely to be part of a coupled atmosphere-ocean oscillation mode. It must therefore be the manifestation of real variations in the global radiative forcing." This provides "very strong support for the notion that an amplification mechanism exists. Given that the CRF [cosmic ray flux]/climate links predicts the correct radiation imbalance observed in the cloud cover variations, it is a favorable candidate." Additional work by Shaviv and coauthors with similar results includes Shaviv (2005), Shaviv *et al.* (2014), Howard *et al.* (2015), and Benyamin *et al.* (2017).

Raspopov *et al.* (2008), a team of eight researchers from China, Finland, Russia, and Switzerland, found an approximate 200-year cycle in paleoclimate reconstructions in the Central Asian Mountains that matches well with the solar Suess-de Vries cycle, suggesting the existence of a solar-climate connection. After reviewing additional sets of published palaeoclimatic data from various parts of the world, the researchers concluded the same periodicity is evident in Europe, North and South America, Asia, Tasmania, Antarctica, and the Arctic, as well as "sediments in the seas and oceans," citing 20 independent research papers in support of this statement. They conclude there is "a pronounced influence of solar activity on global climatic processes" related to "temperature, precipitation and atmospheric and oceanic circulation."

de Jager and Duhau (2009) used "direct observations of proxy data for the two main solar magnetic field components since 1844" to derive "an empirical relation between tropospheric temperature variation and those of the solar equatorial and polar activities." When the two researchers applied this relationship to the period 1610–1995, they found a rising linear association for temperature vs. time, upon which were superimposed "some quasi-regular episodes of residual temperature increases and decreases, with semi-amplitudes up to ~0.3°C," and they note "the present period of global warming is one of them." de Jager and Duhau conclude, "the amplitude of the present period of global warming does not significantly differ from the other episodes of relative warming that occurred in earlier

centuries." The late twentieth-century episode of relative warming is merely "superimposed on a relatively higher level of solar activity than the others," giving it the appearance of being unique when it is not.

Qian and Lu (2010) used data from a 400-year solar radiation series based on [10]Be data "to analyze their causality relationship" with the periodic oscillations they had detected in the north Pacific sea surface temperature reconstruction. They determined "the ~21-year, ~115-year and ~200-year periodic oscillations in global-mean temperature are forced by and lag behind solar radiation variability," and the "relative warm spells in the 1940s and the beginning of the 21st century resulted from overlapping of warm phases in the ~21-year and other oscillations." They note "between 1994 and 2002 all four periodic oscillations reached their peaks and resulted in a uniquely warm decadal period during the last 1000 years," representing the approximate temporal differential between the current global warming and the prior Medieval Warm Period.

Soon (2005, 2009) and with coauthors (Soon and Baliunas, 2003; Soon and Legates, 2013; Soon *et al.,* 2000, 2011) has shown close correlations between TSI proxy models and many twentieth-century climate records including temperature records of the Arctic and of China, the sunshine duration record of Japan, and the Equator-to-Pole (Arctic) temperature gradient record. Soon *et al.* (2015) show that the solar models used by the IPCC's climate models were only a small and unrepresentative sample of the models published in the scientific literature. Although several plausible models of solar output have been proposed, the climate models considered only those that showed almost no solar variability since the nineteenth century (see Figure 2.2.4.1.2). The authors then show how solar variability reported in one ignored model (by Hoyt and Schatten (1993), updated by Scafetta and Willson (2014)) closely tracks temperatures in the Northern Hemisphere using a newly reconstructed temperature record from 1881 to 2014 based on primarily rural temperature stations. (See Figure 2.2.4.1.3.) This result is especially significant in that Soon *et al.* (2015) were the first to attempt to avoid the known contamination of non-climatic factors in the surface station records around the world.

Figure 2.2.4.1.2
Solar models considered by IPCC AR5 versus other models in the literature

A. Solar models considered by the IPCC

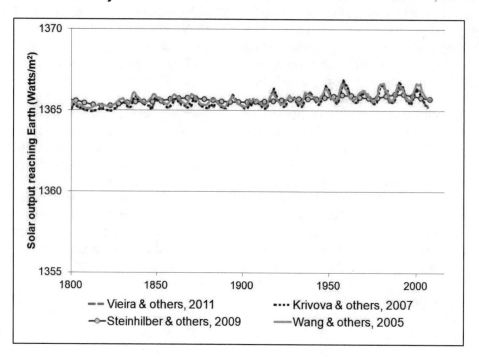

B. Solar models not considered by the IPCC

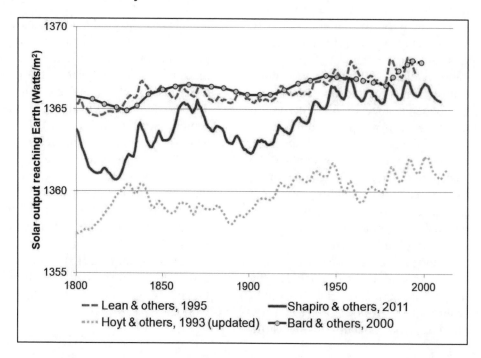

Source: Adapted from Soon *et al.,* 2015, Figure 8, p. 422. See original source for models cited in the figures.

Figure 2.2.4.1.3
Northern Hemisphere rural temperature trends vs. solar output

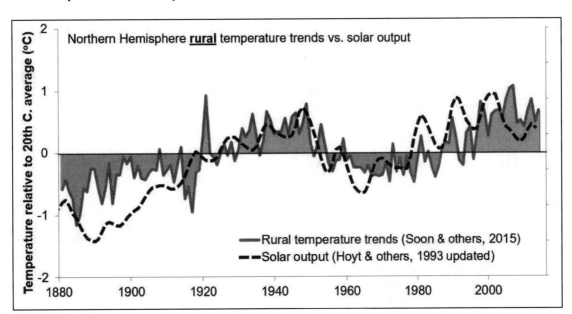

Red and blue represent positive and negative temperature anomalies from twentieth-century average for a Northern Hemisphere temperature reconstruction using primarily rural surface stations (to control for urban heat island effect). Dashed line is solar output according to Hoyt and Schatten (1993) as updated by Scafetta and Willson (2014). *Source:* Adapted from Soon *et al.,* 2015, Figure 27, p. 442.

Several researchers using different methodologies have estimated the percentage of global warming in the modern era that could be attributed to solar activity. Ollila (2017) puts the solar contribution to warming in 2015 at 46% and greenhouse gases at 37%. Harde (2017) estimates that CO_2 contributed 40% and the Sun 60% to global warming over the last century. Booth (2018) estimates that 37% of the warming from 1980 to 2001 was due to solar effects.

Some solar scientists are investigating the possibility of the Sun entering a grand solar minimum (GSM), which could manifest itself within two decades (Lockwood *et al.*, 2011). How an approaching GSM might affect Earth's climate is being studied extensively. Papers by Shindell *et al.* (2001) and others discussed the impact of past low solar activity on regional and global climate. During the last GSM, known popularly as the Maunder Minimum, Earth's climate underwent what has come to be known in climatic terms as the Little Ice Age (LIA), a period that brought the coldest temperatures of the entire Holocene, or current interglacial, in

which we live. Lasting about 200 years (approximately 1650 to 1850), the brunt of the LIA was felt in Europe, which experienced long and extreme winters and cooler summers. Soon and Yaskell (2003) provide a comprehensive discussion of the climatic impact of the Maunder Minimum. Whether the Sun is indeed approaching a new grand solar minimum, however, remains to be seen.

* * *

It is now fairly certain the Sun was responsible for creating multi-centennial global cold and warm periods in the past, and it is quite plausible that modern fluctuations in solar output are responsible for some part of the warming the planet experienced during the past century or so. Besides solar activity changes, other natural climate drivers such as volcanic eruptions and ocean cycles controlled pre-industrial climate change. It is likely that these natural drivers continue to influence modern climate, in addition to yet unquantifiable anthropogenic contributions. A detailed quantitative understanding of the various climatic processes will be possible only once the natural background climate variability is fully understood and successfully calibrated with the known pre-industrial climate change as part of model hindcasts.

References

ACRIM. n.d. Active Cavity Radiometer Irradiance Monitor. Total solar irradiance (TSI) monitoring (website). Accessed December 4, 2018.

ACRIM. n.d. Active Cavity Radiometer Irradiance Monitor. TSI composite time series. (website). Accessed December 4, 2018.

Beer, J., Vonmoos, M., and Muscheler, R. 2006. Solar variability over the past several millennia. *Space Science Reviews* **125**: 67–79.

Benyamin, D., Shaviv, N.J., and Piran, T. 2017. Electron-capture isotopes could constrain cosmic-ray propagation models. *The Astrophysical Journal* **851** (2).

Bond, G., Kromer, B., Beer, J., Muscheler, R., Evans, M.N., Showers, W., Hoffmann, S., Lotti-Bond, R., Hajdas, I., and Bonani, G. 2001. Persistent solar influence on North Atlantic climate during the Holocene. *Science* **294**: 2130–6.

Booth, R.J. 2018. On the influence of solar cycle lengths and carbon dioxide on global temperatures. *Journal of Atmospheric and Solar-Terrestrial Physics* **173**: 96–108.

de Jager, C. and Duhau, S. 2009. Episodes of relative global warming. *Journal of Atmospheric and Solar-Terrestrial Physics* **71**: 194–8.

Egorova, T., Schmulz, W., Rozanov, E., Shapiro, A.I., Usoskin, I., Beer, J., Tagirov, R.V., and Peter, T. 2018. Revised historical solar irradiance forcing. *Astronomy and Astrophysics* **615**: A85.

Harde, H. 2017. Radiation transfer calculations and assessment of global warming by CO_2. *International Journal of Atmospheric Sciences* **2017**: Article 9251034, https://doi.org/10.1155/2017/9251034.

Howard, D., Shaviv, N.J., and Svensmark, H. 2015. The solar and Southern Oscillation components in the satellite altimetry data. *JGR Space Physics* **120** (5).

Hoyt, D.V. and Schatten, K.H. 1993. A discussion of plausible solar irradiance variations, 1700–1992. *Journal of Geophysical Research* **98** (A11): 18,895–906.

IPCC. 2013. Intergovernmental Panel on Climate Change. *Climate Change 2013: The Physical Science Basis*. Contribution of Working Group I to the Fifth Assessment Report of the Intergovernmental Panel on Climate Change. New York, NY: Cambridge University Press.

Lockwood, M., Owens, M.J., Barnard, L., Davis, C.J., and Steinhilber, F. 2011. The persistence of solar activity indicators and the descent of the Sun into Maunder Minimum conditions. *Geophysical Research Letters* **38**.

Matthes, K., Funke, B., Andersson, M.E., and 25 other co-authors. 2017. Solar forcing for CMIP6 (v3.2). *Geoscientific Model Development* **98** (A11): 2247–302.

NIPCC. 2013. Nongovernmental International Panel on Climate Change. Idso, C.D., Carter, R.M., and Singer, S.F. (Eds.) *Climate Change Reconsidered: Physical Science*. Chicago, IL: The Heartland Institute.

Ohmura, A. 2009. Observed decadal variations in surface solar radiation and their causes. *Journal of Geophysical Research* **114**: D00D05, doi:10.1029/2008JD011290.

Ollila, A. 2017. Semi empirical model of global warming including cosmic forces, greenhouse gases, and volcanic eruptions. *Physical Science International Journal* **15** (2): 1–14.

Qian, W.-H. and Lu, B. 2010. Periodic oscillations in millennial global-mean temperature and their causes. *Chinese Science Bulletin* **55**: 4052–57.

Raspopov, O.M., Dergachev, V.A., Esper, J., Kozyreva, O.V., Frank, D., Ogurtsov, M., Kolstrom, T., and Shao, X. 2008. The influence of the de Vries (~200-year) solar cycle on climate variations: results from the Central Asian Mountains and their global link. *Palaeogeography, Palaeoclimatology, Palaeoecology* **259**: 6–16.

Scafetta, N. 2008. Comment on "Heat capacity, time constant, and sensitivity of Earth's climate system" by Schwartz. *Journal of Geophysical Research* **113**: D15104, doi:10.1029/2007JD009586.

Scafetta, N. 2010. Empirical evidence for a celestial origin of the climate oscillations and its implications. *Journal of Atmospheric and Solar-Terrestrial Physics* **72**: 951–70.

Scafetta, N. 2012a. A shared frequency set between the historical mid-latitude aurora records and the global surface temperature. *Journal of Atmospheric and Solar-Terrestrial Physics* **74**: 145–63.

Scafetta, N. 2012b. Testing an astronomically based decadal-scale empirical harmonic climate model versus the IPCC (2007) general circulation climate models. *Journal of Atmospheric and Solar-Terrestrial Physics* **80**: 124–37.

Scafetta, N. 2013a. Solar and planetary oscillation control on climate change: hind-cast, forecast and a comparison with the CMIP5 GCMs. *Energy & Environment* **24** (3-4): 455–96.

Scafetta, N. 2013b. Discussion on climate oscillations: CMIP5 general circulation models versus a semi-empirical harmonic model based on astronomical cycles. *Earth-Science Reviews* **126**: 321–57.

Scafetta, N. 2016. Problems in modeling and forecasting climate change: CMIP5 General Circulation Models versus a semi-empirical model based on natural oscillations. *International Journal of Heat and Technology* **34**, Special Issue 2.

Scafetta, N. and West, B.J. 2003. Solar flare intermittency and the Earth's temperature anomalies. *Physical Review Letters* **90**: 248701.

Scafetta, N. and West, B.J. 2005. Estimated solar contribution to the global surface warming using the ACRIM TSI satellite composite. *Geophysical Research Letters* **32**: 10.1029/2005GL023849.

Scafetta, N. and West, B.J. 2006a. Phenomenological solar contribution to the 1900–2000 global surface warming. *Geophysical Research Letters* **33**: L05708, doi: 10.1029/2005GL025539.

Scafetta, N. and West, B.J. 2006b. Phenomenological solar signature in 400 years of reconstructed Northern Hemisphere temperature record. *Geophysical Research Letters* **33**: L17718, doi: 10.1029/2006GL027142.

Scafetta, N. and West, B.J. 2007. Phenomenological reconstructions of the solar signature in the Northern Hemisphere surface temperature records since 1600. *Journal of Geophysical Research* **112**: D24S03.

Scafetta, N. and West, B.J. 2008. Is climate sensitive to solar variability? *Physics Today* **3**: 50–1.

Scafetta, N. and Willson, R.C. 2009. ACRIM-gap and TSI trend issue resolved using a surface magnetic flux TSI proxy model. *Geophysical Research Letters* **36**: L05701.

Scafetta, N. and Willson, R.C. 2014. ACRIM total solar irradiance satellite composite validation versus TSI proxy models. *Astrophysical Space Science* **350**: 421–42.

Scafetta, N., Mirandola, A., and Bianchini, A. 2017a. Natural climate variability, part 1: observations versus the modeled predictions. *International Journal of Heat and Technology* **35**, Special Issue 1.

Scafetta, N., Mirandola, A., and Bianchini, A. 2017b. Natural climate variability, part 2: interpretation of the post 2000 temperature standstill. *International Journal of Heat and Technology* **35**, Special Issue 1.

Shapiro, A.I., Schmutz, W., Rozanov, E., Schoell, M., Haberreiter, M., Shapiro, A.V., and Nyeki, S. 2011. A new approach to the long-term reconstruction of the solar irradiance leads to a large historical solar forcing. *Astronomy and Astrophysics* **529**: A67.

Shaviv, N.J. 2005. On climate response to changes in the cosmic ray flux and radiative budget. *Journal of Geophysical Research* **110**: 10.1029/2004JA010866.

Shaviv, N.J. 2008. Using the oceans as a calorimeter to quantify the solar radiative forcing. *Journal of Geophysical Research* **113**: A11101, doi:10.1029/2007JA012989.

Shaviv, N.J., *et al.* 2014. Is the solar system's galactic motion imprinted in the Phanerozoic climate? *Science Reports* **4**: 6150.

Shindell, D.T., Schmidt, G.A., Mann, M.E., Rind, D., and Waple, A. 2001. Solar forcing of regional climate change during the Maunder Minimum. *Science* **294**: 2149–52.

Soon, W. W.-H. 2005. Variable solar irradiance as a plausible agent for multidecadal variations in the Arctic-wide surface air temperature record of the past 130 years. *Geophysical Research Letters* **32**: 10.1029/2005GL023429.

Soon, W.W.-H. 2009. Solar Arctic-mediated climate variation on multidecadal to centennial timescales: empirical evidence, mechanistic explanation and testable consequences. *Physical Geography* **30**: 144–84.

Soon, W.W.-H. and Baliunas, S. 2003. Proxy climatic and environmental changes of the past 1000 years. *Climate Research* **23**: 89–110.

Soon, W.W.-H., Connolly, R., and Connolly, M. 2015. Re-evaluating the role of solar variability on Northern Hemisphere temperature trends since the 19th century. *Earth-Science Reviews* **150**: 409–52.

Soon, W.W.-H. and Legates, D.R. 2013. Solar irradiance modulation of Equator-to-Pole (Arctic) temperature gradients: empirical evidence for climate variation on multi-decadal timescales. *Journal of Atmospheric and Solar-Terrestrial Physics* **93**: 45–56.

Soon, W. and Yaskell, S.H. 2003. *The Maunder Minimum and the Variable Sun-Earth Connection*. Singapore: World Scientific Publishing.

Soon, W.W.-H, *et al.* 2000. Climate hypersensitivity to solar forcing? *Annales Geophysicae* **18**: 583–8.

Soon, W.W.-H., *et al.* 2011. Variation in surface air temperature of China during the 20th Century. *Journal of Atmospheric and Solar-Terrestrial Physics* **73**: 2331–44.

Soon, W.W.-H. *et al.,* 2014. A review of Holocene solar-linked climatic variation on centennial to millennial timescales: physical processes, interpretative frameworks and a new multiple cross-wavelet transform algorithm. *Earth-Science Reviews* **134**: 1–15.

Usoskin, Y.G. 2017. A history of solar activity over millennia. *Living Reviews in Solar Physics* **14**: 3.

2.2.4.2 Cosmic Rays

According to the IPCC, "cosmic rays enhance new particle formation in the free troposphere, but the effect on the concentration of cloud condensation nuclei is too weak to have any detectable climate influence during a solar cycle or over the last century (*medium evidence, high agreement*). No robust association between changes in cosmic rays and cloudiness has been identified. In the event that such an association existed, a mechanism other than cosmic ray-induced nucleation of new aerosol particles would be needed to explain it" (IPCC, 2013, p. 573). On this matter, new research has proven the IPCC to be wrong.

The field of galactic cosmic ray (GCR) research begins with the original publication of Svensmark and Friis-Christensen (1997) and later developments are summarized well by Svensmark (2007). Svensmark and his colleagues at the Center for Sun-Climate Research of the Danish National Space Center experimentally determined ions released to the atmosphere by GCRs act as catalysts that significantly accelerate the formation of ultra-small clusters of sulfuric acid and water molecules that constitute the building blocks of cloud condensation nuclei. Svensmark also explains the complex chain of expected atmospheric interactions, in particular how, during periods of greater solar activity, greater shielding of Earth occurs associated with a strong solar magnetic field. That shielding results in fewer cosmic rays penetrating to the lower atmosphere of the Earth, resulting in fewer cloud condensation nuclei being produced and thus fewer and less reflective low-level clouds occurring. More solar radiation is thus absorbed at the surface of Earth, resulting in increasing near-surface air temperatures.

Svensmark provides support for key elements of this scenario with graphs illustrating the close correspondence between global low-cloud amount and cosmic-ray counts over the period 1984–2004. He also notes the history of changes in the flux of galactic cosmic rays estimated since 1700, which correlates well with Earth's temperature history over

the same time period, starting from the latter portion of the Maunder Minimum (1645–1715), when Svensmark says "sunspots were extremely scarce and the solar magnetic field was exceptionally weak," and continuing on through the twentieth century, over which last hundred-year interval, as noted by Svensmark, "the Sun's coronal magnetic field doubled in strength."

Over the past two decades, several studies have uncovered evidence supporting several of the linkages described by Svensmark (e.g., Lockwood *et al.*, 1999; Parker, 1999; Kniveton and Todd, 2001; Carslaw *et al.*, 2002; Shaviv and Veizer, 2003; Veretenenko *et al.*, 2005; Usoskin *et al.*, 2006; Lockwood, 2011; Shapiro *et al.*, 2011; Veretenenko and Ogurtsov, 2012; Georgieva *et al.,* 2012). In 2016, CERN (the European Institute for Nuclear Research) confirmed one of Svensmark's postulates when its large particle beam accelerator, acting on a cloud chamber, revealed that ions from cosmic rays increase the number of cloud condensation nuclei of sizes of at least 50 to 100 nanometers. Kirkby *et al.* (2016) write,

> We find that ions from galactic cosmic rays increase the nucleation rate by one to two orders of magnitude compared with neutral nucleation. Our experimental findings are supported by quantum chemical calculations of the cluster binding energies of representative HOMs [highly oxygenated molecules]. Ion-induced nucleation of pure organic particles constitutes a potentially widespread source of aerosol particles in terrestrial environments with low sulfuric acid pollution.

The CERN experiment documents an important mechanism whereby cosmic rays turn small changes in TSI into larger effects on temperatures. The pre-industrial atmosphere is thought to have been "pristine," without sulfuric acid caused by the combustion of fossil fuels. We now know cosmic rays were seeding clouds then, so their presence or absence due to variation in solar wind explains more of the variability in temperature observed in geological and historical reconstructions than previously thought. Commenting on the finding, Svensmark *et al.* (2017) write, "The mechanism could therefore be a natural explanation for the observed correlations between past climate variations and cosmic rays, modulated by either solar activity or caused by supernova activity in the solar neighborhood on very long time scales where the mechanism will be of profound importance."

An active debate is taking place over the empirical basis for the cosmic ray-low cloud relationship, with scientists raising alternative mechanisms involving the solar wind, aurora, and atmospheric gravity waves (see Soon *et al.,* 2000, 2015; Prikryl *et al.,* 2009a; 2009b; Scafetta 2012a, 2012b; Scafetta and Willson, 2013a, 2013b). The flux of galactic cosmic rays clearly wields an important influence on Earth's climate, likely much more so than that exhibited by the modern increase in atmospheric CO_2. At the very least, these research findings invalidate the IPCC's claim that the GCR-ionization mechanism is too weak to influence global temperatures.

References

Carslaw, K.S., Harrizon, R.G., and Kirkby, J. 2002. Cosmic rays, clouds, and climate. *Science* **298**: 1732–7.

Georgieva, K., Kirov, B., Koucka-Knizova, P., Mosna, Z., Kouba, D., and Asenovska, Y. 2012. Solar influences on atmospheric circulation. *Journal of Atmospheric Solar-Terrestrial Physics* **90–91**: 15–25.

IPCC. 2013. Intergovernmental Panel on Climate Change. *Climate Change 2013: The Physical Science Basis.* Contribution of Working Group I to the Fifth Assessment Report of the Intergovernmental Panel on Climate Change. New York, NY: Cambridge University Press.

Kirkby, J., *et al.* 2016. Ion-induced nucleation of pure biogenic particles. *Nature* **533**: 521–6.

Kniveton, D.R. and Todd, M.C. 2001. On the relationship of cosmic ray flux and precipitation. *Geophysical Research Letters* **28**: 1527–30.

Lockwood, M. 2011. Shining a light on solar impacts. *Nature Climate Change* **1**: 98–9.

Lockwood, M., Stamper, R., and Wild, M.N. 1999. A doubling of the Sun's coronal magnetic field during the past 100 years. *Nature* **399**: 437–9.

Parker, E.N. 1999. Sunny side of global warming. *Nature* **399**: 416–7.

Prikryl, P., Rusin, V., and Rybansky, M. 2009a. The influence of solar wind on extratropical cyclones—Part 1: Wilcox effect revisited. *Annales Geophysicae* **27**: 1–30.

Prikryl, P., Muldrew, D.B., and Sofko, G.J. 2009b. The influence of solar wind on extratropical cyclones—Part 2: a link mediated by auroral atmospheric gravity waves? *Annales Geophysicae* **27**: 31–57.

Scafetta, N. 2012a. Does the Sun work as a nuclear fusion amplifier of planetary tidal forcing? A proposal for a physical mechanism based on the mass-luminosity relation. *Journal of Atmospheric and Solar-Terrestrial Physics* **81–82**: 27–40.

Scafetta, N. 2012b. Multi-scale harmonic model for solar and climate cyclical variation throughout the Holocene based on Jupiter-Saturn tidal frequencies plus the 11-year solar dynamo cycle. *Journal of Atmospheric and Solar-Terrestrial Physics* **80**: 296–311.

Scafetta, N. and Willson, R.C. 2013a. Planetary harmonics in the historical Hungarian aurora record (1523–1960). *Planetary and Space Science* **78:** 38–44.

Scafetta, N. and Willson, R.C. 2013b. Empirical evidences for a planetary modulation of total solar irradiance and the TSI signature of the 1.09-year Earth-Jupiter conjunction cycle. *Astrophysics and Space Science* **348** (1).

Shapiro, A.I., Schmutz, W., Rozanov, E., Schoell, M., Haberreiter, M., Shapiro, A.V., and Nyeki, S. 2011. A new approach to the long-term reconstruction of the solar irradiance leads to a large historical solar forcing. *Astronomy and Astrophysics* **529**: A67.

Shaviv, N. and Veizer, J. 2003. Celestial driver of Phanerozoic climate? *GSA Today* **13** (7): 4–10.

Soon, W., Baliunas, S., Posmentier, E.S., and Okeke, P. 2000. Variations of solar coronal hole area and terrestrial lower tropospheric air temperature from 1979 to mid-1998: Astronomical forcings of change in Earth's climate? *New Astronomy* **4**: 563–79.

Soon, W., Connolly, R., and Connolly, M. 2015. Re-evaluating the role of solar variability on Northern Hemisphere temperature trends since the 19th century. *Earth-Science Reviews* **150**: 409–52.

Svensmark, H. and Friis-Christensen, E. 1997. Variation of cosmic ray flux and global cloud coverage—A missing link in solar-climate relationships. *Journal of Atmospheric and Solar-Terrestrial Physics* **59**: 1225–32.

Svensmark, H. 2007. Cosmoclimatology: a new theory emerges. *Astronomy and Geophysics* **48**: 1–19.

Svensmark, H., Enghoff, M.B., Shaviv, N.J., and Svensmark, J. 2017. Increased ionization supports growth of aerosols into cloud condensation nuclei. *Nature Communications* **8**: Article 2199.

Usoskin, I.G., Solanki, S.K., Taricco, C., Bhandari, N., and Kovaltsov, G.A. 2006. Long-term solar activity reconstructions: direct test by cosmogenic [44]Ti in meteorites. *Astronomy & Astrophysics* **457**: 10.1051/0004-6361:20065803.

Veretenenko, S.V., Dergachev, V.A., and Dmitriyev, P.B. 2005. Long-term variations of the surface pressure in the North Atlantic and possible association with solar activity and galactic cosmic rays. *Advances in Space Research* **35**: 484–90.

Veretenenko, S.V. and Ogurtsov, M. 2012. Regional and temporal variability of solar activity and galactic cosmic ray effects on the lower atmospheric circulation. *Advances in Space Research* **49**: 770–83.

2.2.4.3 Possible Future Impacts

Researchers study the Sun itself as well as the decadal, multidecadal, centennial, and even millennial periods in available climate and solar records in order to extrapolate solar activity into the future. Solheim *et al.* (2012) found significant linear relationships between the average air temperature in a solar cycle and the length of the previous solar cycle for 12 of 13 weather stations in Norway and the North Atlantic, as well as for 60 European stations and for the HadCRUT3N database. For Norway and the other European stations, they found "the solar contribution to the temperature variations in the period investigated is of the order 40%" while "an even higher contribution (63–72%) is found for stations at the Faroe Islands, Iceland and Svalbard," which they note is considerably "higher than the 7% attributed to the Sun for the global temperature rise in AR4 (IPCC, 2007)."

Ludecke *et al.* (2013) considered six periodic components with timescales greater than 30 years in the composite of a six-station temperature record from Central Europe since about 1757, creating a very good reconstruction of the original instrumental records. They project a substantial cooling of the Central European temperature in the next one to two decades but caution their result "does not rule out a warming by anthropogenic influences such as an increase of atmospheric CO_2." In addition, climate system internal oscillations may play a role.

In analyzing the global temperature data records (HadCRUT3 and HadCRUT4, respectively) directly, Loehle and Scafetta (2011) and Tung and Zhou (2013) conclude a large fraction of recent observed warming (60% over 1970–2000 and 40% over the past 50 years) can be accounted for by the natural

upswing of the 60-year climatic ocean cycle during its warming phase. Loehle and Scafetta (2011) proffer that "a 21st Century forecast suggests that climate may remain approximately steady until 2030–2040, and may at most warm 0.5–1.0°C by 2100 at the estimated 0.66°C/century anthropogenic warming rate, which is about 3.5 times smaller than the average 2.3°C/century anthropogenic warming rate projected by the IPCC up to the first decades of the 21st century. However, additional multi-secular natural cycles may cool the climate further."

Scafetta (2016) says his 2011 temperature forecast "has well agreed with the global surface temperature data up to August 2016." He then proposes "a semi-empirical climate model able to reconstruct the natural climatic variability since Medieval times. I show that this model projects a very moderate warming until 2040 and a warming less than 2°C from 2000 to 2100 using the same anthropogenic emission scenarios used by the CMIP5 models. This result suggests that climatic adaptation policies, which are less expensive than the mitigation ones, could be sufficient to address most of the consequences of a climatic change during the 21st century."

In an independent analysis of global temperature data from the Climatic Research Unit (CRU) at the University of East Anglia and the Berkeley Earth Surface Temperature consortium, Courtillot *et al.* (2013) arrive at a new view of the significance of the ~60 year oscillation. They interpret the 60-year period found in the global surface temperature record as "a series of ~30-yr long linear segments, with slope breaks (singularities) in ~1904, ~1940, and ~1974 (±3 yr), and a possible recent occurrence at the turn of the 21st century." Courtillot and his colleagues suggest "no further temperature increase, a dominantly negative PDO index and a decreasing AMO index might be expected for the next decade or two."

By extrapolating present solar cycle patterns into the future, several scientists have suggested a planetary cooling may be expected over the next few decades. The Gleissberg and Suess/de Vries cycles will reach their low points between 2020 and 2040 at a level comparable to what was experienced during the Dalton Minimum. At that time, around 1790–1820, global temperatures were nearly 1°C lower than they are today; conservatively, at least half of that cooling was due to a weaker Sun. Moreover, as Courtillot *et al.* (2013) noted, the Pacific Decadal Oscillation (PDO) is expected to be in a cool phase by 2035, and the Atlantic Multidecadal Oscillation (AMO) will begin to drop around 2020. Such internal climate cycles are generally responsible for about 0.2°C to 0.3°C of the temperature dynamic.

Soon *et al.* (2015) use the new surface temperature record for North America they created using rural stations and the TSI reconstruction created by Hoyt and Schatten (1993) as updated by Scafetta and Willson (2014) to estimate a solar radiative forcing of 1 Wm^{-2} on average causes a change of 1.18°C in surface temperature. The warming from 1881 to 2014 is almost entirely explained by this forcing, leaving CO_2 forcing responsible for a minute "residual" of approximately 0.12°C. They calculate a doubling of CO_2 from its current level of about 400 ppm would cause "at most" 0.44°C of warming (Soon *et al.*, 2015, p. 444).

Finally, Cionco *et al.* (2018) recently provided the first comprehensive boundary conditions for incoming solar radiation that fully account for both the correct orbital solutions (based on Cionco and Soon, 2017) and the intrinsic solar irradiance changes (Velasco Herrera *et al.*, 2015) for the past 2,000 years as well as a forward projection for about 100 years into the future.

* * *

Effects of solar variations on climate are due to changes in radiation reaching the Earth as well as still poorly understood amplifier effects associated with cosmic rays, cloud cover, and stratospheric temperature changes in combination with the ultraviolet (UV) part of the solar spectrum. The CERN experiment in 2016 provided some proof of the mechanism for which the IPCC asked in its Fifth Assessment Report. Numerous case studies of the last decades to millennia have empirically demonstrated a strong link between solar activity and climate. We now know solar influences play a larger role in average surface temperature and other climate indices than the IPCC assumed in 2013 and most climate modelers still assume today. The cautions about avoiding or eliminating non-climatic factors in the world's surface station records from Soon *et al.* (2015) are also important for IPCC authors to note. Estimates of climate sensitivity to a doubling of CO_2 that take the new research into account are lower, and so too are forecasts of future temperature increases. While forecasts differ and are uncertain, one clear conclusion is that the IPCC's prediction of future warming is too high.

References

Cionco, R.G. and Soon, W. 2017. Short-term orbital forcing: a quasi-review and a reappraisal of realistic boundary conditions for climate modeling. *Earth-Science Reviews* **166**: 206–22.

Cionco, R.G., Valentini, J.E., Quaranta, N.E., and Soon, W. 2018. Lunar fingerprints in the modulated incoming solar radiation: in situ insolation and latitudinal insolation gradients as two important interpretative metrics for paleoclimatic data records and theoretical climate modeling. *New Astronomy* **58**: 96–106.

Courtillot, V., Le Mouel, J.-L., Kossobokov, V., Gibert, D., and Lopes, F. 2013. Multi-decadal trends of global surface temperature: a broken line with alternating ~30 yr linear segments? *Atmospheric and Climate Sciences* **3**: 364–71.

Hoyt, D.V. and Schatten, K.H. 1993. A discussion of plausible solar irradiance variations, 1700–1992. *Journal of Geophysical Research* **98** (A11): 18,895–906.

IPCC. 2007. Intergovernmental Panel on Climate Change. *Climate Change 2007: The Physical Science Basis.* Contribution of Working Group I to the Fourth Assessment Report of the Intergovernmental Panel on Climate Change. Cambridge, UK and New York, NY: Cambridge University Press.

Loehle, C. and Scafetta, N. 2011. Climate change attribution using empirical decomposition of climatic data. *The Open Atmospheric Science Journal* **5**: 74–86.

Ludecke, H.-J., Hempelmann, A., and Weiss, C.O. 2013. Multi-periodic climate dynamics: Spectral analysis of long-term instrumental and proxy temperature records. *Climate of the Past* **9**: 447–52.

Scafetta, N. 2016. Problems in modeling and forecasting climate change: CMIP5 General Circulation Models versus a semi-empirical model based on natural oscillations. *International Journal of Heat and Technology* **34**, Special Issue 2.

Scafetta, N. and Willson, R.C. 2014. ACRIM total solar irradiance satellite composite validation versus TSI proxy models. *Astrophysical Space Science* **350**: 421–42.

Solheim, J.-E., Stordahl, K., and Humlum, O. 2012. The long sunspot cycle 23 predicts a significant temperature decrease in cycle 24. *Journal of Atmospheric and Solar-Terrestrial Physics* **80**: 267–84.

Soon, W., Connolly, R., and Connolly, M. 2015. Re-evaluating the role of solar variability on Northern Hemisphere temperature trends since the 19th century. *Earth-Science Reviews* **150**: 409–52.

Tung, K.-K. and Zhou, J. 2013. Using data to attribute episodes of warming and cooling instrumental records. *Proceedings of the National Academy of Sciences USA* **110**: 2058–63.

Velasco Herrera, V.M., Mendoza, B., and Velasco Herrera, G. 2015. Reconstruction and prediction of the total solar irradiance: from the Medieval Warm Period to the 21st century. *New Astronomy* **34**: 221–33.

2.3 Climate Impacts

The Working Group II contribution to the Fifth Assessment Report (AR5) of the United Nations' Intergovernmental Panel on Climate Change (IPCC, 2014) claims climate change causes a "risk of death, injury, and disrupted livelihoods" due to sea-level rise, coastal flooding, and storm surges; food insecurity, inland flooding, and negative effects on fresh water supplies, fisheries, and livestock; and "risk of mortality, morbidity, and other harms during periods of extreme heat, particularly for vulnerable urban populations" (p. 7).

Given the uncertainty that pervades climate science discussed in Section 2.1, and observations showing less warming of the atmosphere than predicted by climate models discussed in Section 2.2, disagreement and uncertainty over the climate impacts of human activity can be expected. This section documents that uncertainty. Observational data on four climate impacts are surveyed here: what the IPCC calls "extreme weather," melting ice, sea-level rise, and effects on plants.

More than two thousand studies on these subjects were reviewed in the Nongovernmental International Panel on Climate Change's *Climate Change Reconsidered II: Physical Science* (NIPCC, 2013) and *Climate Change Reconsidered II: Biological Impacts* (NIPCC, 2014). This section greatly condenses that literature review by leaving out descriptions of research methods and most studies published before 2010. Research is typically presented in chronological order by publication date. Reviews of new research published after 2013 have been added.

References

NIPCC. 2013. Nongovernmental International Panel on Climate Change. Idso, C.D., Carter, R.M., and Singer, S.F. (Eds.) *Climate Change Reconsidered: Physical Science.* Chicago, IL: The Heartland Institute.

NIPCC. 2014. Nongovernmental International Panel on Climate Chage. Idso, C.D, Idso, S.B., Carter, R.M., and Singer, S.F. (Eds.) *Climate Change Reconsidered II: Biological Impacts.* Chicago, IL: The Heartland Institute.

2.3.1 Extreme Weather Events

There is little evidence that the warming of the twentieth and early twenty-first centuries has caused a general increase in "extreme" weather events. Meteorological science suggests a warmer world would see milder weather patterns.

Sutton *et al.* (2018), five British climate scientists, admonished the editors of Nature for repeating the IPCC's claim that climate change is causing extreme weather events, writing in part,

Attribution depends fundamentally on global climate models that can adequately capture regional weather phenomena – including circulation anomalies such as the weak jet stream and large, persistent planetary-scale atmospheric waves that characterized this summer's weather. Accurate simulation of such extremes remains a challenge for today's models. It is not enough to increase the size of the ensemble of simulations if the models themselves have fundamental limitations. Any statement on attribution should therefore always be accompanied by a scientifically robust demonstration of the model's ability to simulate the global and regional weather patterns and the related weather phenomena that lie at the root of extreme events.

Extensive scientific research supports the view expressed by Sutton *et al.* We address six weather phenomena characterized by the IPCC as "extreme weather events" allegedly caused by human greenhouse emissions: high temperatures and heat waves, wildfires, droughts, floods, storms, and hurricanes. In every case, we find the IPCC exaggerates the possibility that such events have or will become more frequent or more intense due to the human presence.

2.3.1.1 Heat Waves

According to the Summary for Policymakers (SPM) of the Working Group I contribution to the IPCC's Fifth Assessment Report, "It is *virtually certain* that there will be more frequent hot and fewer cold temperature extremes over most land areas on daily and seasonal timescales as global mean temperatures increase. It is *very likely* that heat waves will occur with a higher frequency and duration" (IPCC, 2013, p. 20). Regarding past trends, the IPCC writes "There is only *medium confidence* that the length and frequency of warm spells, including heat waves, has increased since the middle of the 20th century mostly owing to lack of data or of studies in Africa and South America. However, it is *likely* that heatwave frequency has increased during this period in large parts of Europe, Asia and Australia" (p. 162). The prediction in the SPM has fed the almost hysterical claims about recent and future heat waves appearing in the media and in "documentary" films. But the IPCC's position overstates the possibility that rising temperatures and heat waves pose a threat to human health.

As explained in Section 2.2.1, the surface station temperature record is too flawed to be used as the basis of scientific research. One of many problems it faces is contamination by heat-emitting and -absorbing activities and structures associated with urbanization. Buildings, roads, parking lots, and loss of green space all combine to raise temperatures, and their effects need to be removed if the purpose of a temperature reconstruction is to measure the effect of anthropogenic greenhouse gases. The IPCC claims to control for this "heat island effect," but researchers have found its adjustments are too small (e.g. McKitrick and Michaels, 2007; Soon *et al.* 2015; Quereda Sala *et al.*, 2017). For example, Zhou and Ren (2011) studied the impact of urbanization on extreme temperature indices for the period 1961–2008 using daily temperature records from the China Homogenized Historical Temperature Datasets compiled by the National Meteorological Information Center of the China Meteorological Administration. They discovered "the contributions of the urbanization effect to the overall trends ranged from 10% to 100%, with the largest contributions coming from tropical nights, daily temperature range, daily maximum temperature and daily minimum temperature," adding "the decrease in daily temperature range at the national stations in North China was caused entirely by urbanization."

A second problem affecting forecasts of future warming is that they fail to consider the cooling effects of the Greening of the Earth phenomenon reported in Section 2.1.2 and in greater detail in Chapter 5. Jeong *et al.* (2010) investigated "the impact of vegetation-climate feedback on the changes in temperature and the frequency and duration of heat waves in Europe under the condition of doubled atmospheric CO_2 concentration in a series of global climate model experiments." Their calculations revealed "the projected warming of 4°C over most of Europe with static vegetation has been reduced by 1°C as the dynamic vegetation feedback effects are included," and "examination of the simulated surface energy fluxes suggests that additional greening in the presence of vegetation feedback effects enhances evapo-transpiration and precipitation, thereby limiting the warming, particularly in the daily maximum temperature." The scientists found "the greening also tends to reduce the frequency and duration of heat waves."

Extensive investigation of historical records and proxy data has found many examples of absolute temperature or variability of temperature exceeding observational data from the twentieth and early twenty-first centuries, lending support to the null hypothesis that recent temperature changes are due to natural causes. For example, Dole *et al.* (2011) ask whether a 2010 summer heat wave in western Russia exceeded natural variability and thus could be evidence of an anthropogenic effect on climate. They used climate model simulations and observational data "to determine the impact of observed sea surface temperatures, sea ice conditions and greenhouse gas concentrations." They found "analysis of forced model simulations indicates that neither human influences nor other slowly evolving ocean boundary conditions contributed substantially to the magnitude of the heat wave." They observed the model simulations provided "evidence that such an intense event could be produced through natural variability alone." "In summary," Dole *et al.* observe, "the analysis of the observed 1880–2009 time series shows that no statistically significant long-term change is detected in either the mean or variability of western Russia July temperatures, implying that for this region an anthropogenic climate change signal has yet to emerge above the natural background variability."

Hiebl and Hofstatter (2012) studied the extent to which temperature variability may have increased in Austria since the late nineteenth century. Using air temperature based on 140 years of data from

Vienna-Hohe Warte, Kremsmunster, Innsbruck-University, Sonnblick, and Graz-University, they found a slow and steady rise in variability during the twentieth century. They also reported a "period of persistently high variability levels before 1900," which leads them to conclude the "relatively high levels of temperature variability during the most recent warm decades from 1990 to 2010 are put into perspective by similar variability levels during the cold late 19th century." They add, "when compared to its inter-annual fluctuations and the evolution of temperature itself, high-frequency temperature variability in the course of the recent 117–139 years appears to be a stable climate feature." Hiebl and Hofstatter conclude concerns about "an increasing number and strength of temperature extremes in terms of deviations from the mean state in the past decades cannot be maintained" and "exaggerated statements seem irresponsible."

Bohm (2012) studied climate data for South Central Europe from 1771–1800 and 1981–2010 and found "the overwhelming majority of seasonal and annual sub-regional variability trends is not significant." Regarding temperature, he reports "most of the variability trends are insignificantly decreasing." In a special analysis of the recent 1981–2010 period that may be considered the first "normal period" under dominant greenhouse-gas-forcing, he found all extremes "remaining well within the range of the preceding ones under mainly natural forcing," and "in terms of insignificant deviations from the long-term mean, the recent three decades tend to be less rather than more variable." Bohm concludes "the … evidence [is clear] that climate variability did rather decrease than increase over the more than two centuries of the instrumental period in the Greater Alpine Region, and that the recent 30 years of more or less pure greenhouse-gas-forced anthropogenic climate were rather less than more variable than the series of the preceding 30-year normal period."

Rusticucci (2012) examined the claim global warming will increase climatic variability, reviewing many studies that have explored this subject throughout South America, particularly as it applies to daily maximum and minimum air temperatures. The Buenos Aires researcher found the most significant trends exist in the evolution of the daily minimum air temperature, with "positive trends in almost all studies on the occurrence of warm nights (or hot extremes of minimum temperature)," as well as negative trends in the cold extremes of the minimum temperature. She states this was the case "in almost all studies." By contrast, she writes, "on

the maximum temperature behavior there is little agreement, but generally the maximum temperature in South America has decreased." Over most of South America there has been a decrease in the extremeness of both daily maximum and minimum air temperatures, with the maximums declining and the minimums rising. These changes are beneficial, as Rusticucci notes cold waves and frost are especially harmful to agriculture, one of the main economic activities in South America. Cold waves and frost days were on the decline nearly everywhere throughout the continent during the warming of the twentieth century.

Deng et al. (2012) used daily mean, maximum, and minimum temperatures for the period 1958–2007 to examine trends in heat waves in the Three Gorges area of China, which comprises the Chongqing Municipality and the western part of Hubei Province, including the reservoir region of the Three Gorges Dam. They found extreme high temperature events showed a U-shaped temporal variation, decreasing in the 1970s and remaining low in the 1980s, followed by an increase in the 1990s and the turn of the twenty-first century, such that "the frequencies of heat waves and long heat waves in the recent years were no larger than the late 1950s and early 1960s." They observe, "coupled with the extreme low frequency in the 1980s, heat waves and long heat waves showed a slight linear decreasing trend in the past 50 years." They note the most recent frequency of heat waves "does not outnumber 1959 or 1961" and "none of the longest heat waves recorded by the meteorological stations occurs in the period after 2003." Deng et al. conclude, citing Tan et al. (2007), "compared with the 1950s and 1960s, short heat waves instead of long heat waves have taken place more often," which, as they describe it, "is desirable, as longer duration leads to higher mortality."

Sardeshmukh et al. (2015) comment on how "it is tempting to seek an anthropogenic component in any recent change in the statistics of extreme weather," but warn fellow scientists that such attribution is likely to be wrong "if the distinctively skewed and heavy-tailed aspects of the probability distributions of daily weather anomalies are ignored or misrepresented." Departures from mean values in temperature record even by "several standard deviations" are "far more common in such a distinctively non-Gaussian world than they are in a Gaussian world. This further complicates the problem of detecting changes in tail probabilities from historical records of limited length and accuracy." Referring to statements in IPCC's AR5 attributing changing extreme weather risks to global warming, the authors write,

> Such statements downplay the fact that there is more to regional climate change than surface warming, and that assessing the changing risks of extreme storminess, droughts, floods, and heat waves requires accurate model representations of multidecadal and longer-term changes in the large-scale modes of natural atmospheric circulation variability and the complex nonlinear climate-weather interactions associated with them. The detection of changes in such modes from the limited observational record is much less clear cut than for surface temperature.

Christy (2012) observed that most of the record highs for heat waves in the United States happened before atmospheric CO_2 levels rose because of human activities. Thirty-eight states set their record highs before 1960, and 23 states' record highs occurred in the 1930s. Also in the United States, the number of days per year during which the temperature broke 100°F (37.8°C) has *declined* considerably since the 1930s, as shown in Figure 2.3.1.1.1. Commenting on this figure, Christy (2016) writes "It is not only clear that hot days have not increased, but it is interesting that in the most recent years there has been a relative dearth of them" (p. 16). The U.S. Environmental Protection Agency's Heat Wave Index confirms the 1930s was the decade in the twentieth century with the most heat waves (EPA, 2016). (See Figure 2.3.1.1.2.)

It is also worth noting that in Canada, the hottest ever recorded temperature of 45°C (113°F) was reached on July 5, 1937 at Yellow Grass, Saskatchewan, on the Canadian Prairies. Also, the deadliest heat wave in Canada occurred July 5–12, 1936, when more than 1,000 people died of heat exhaustion in Manitoba and southern Ontario (Khandekar, 2010). Recent heat waves supposedly "amplified" by higher levels of CO_2 have not reached so high a level or had such tragic consequences.

More recently, Köse *et al.* (2017) used a dataset of 23 tree-ring chronologies to provide a high-resolution spring (March–April) temperature reconstruction over Turkey during the period 1800–2002. The authors report, "the reconstruction is punctuated by a temperature increase during the 20th century; yet extreme cold and warm events during the 19th century seem to eclipse conditions during the 20th century." Similarly, Polovodova Asteman *et al.* (2018) used a ca. 8-m long sediment core from Gullmar Fjord (Sweden) to create a 2000-year record of winter temperatures. They report "the record demonstrates a warming during the Roman Warm Period (~350 BCE – 450 CE), variable bottom water temperatures during the Dark Ages (~450 – 850 CE), positive bottom water temperature anomalies during the Viking Age/Medieval Climate Anomaly (~850 – 1350 CE) and a long-term cooling with distinct multidecadal variability during the Little Ice Age (~1350 – 1850 CE)." Significantly, the temperature reconstruction "also picks up the contemporary warming of the 20th century, which does not stand out in the 2500-year perspective and is of the same magnitude as the Roman Warm Period and the Medieval Climate Anomaly."

Cold Weather

According to the IPCC (2013), "It is *virtually certain* that there will be more frequent hot and fewer cold temperature extremes over most land areas on daily and seasonal timescales as global mean temperatures increase" (p. 20). Contrary to this forecast, many researchers have documented an increase in *cold* weather extremes in many parts of the world since around the beginning of the twenty-first century. Such events confirm the much lower temperature rise revealed by satellite data and rural temperature station records than the unreliable and frequently adjusted HadCRUT surface station record. Cold weather extremes have been observed throughout the Northern Hemisphere and parts of Asia. According to D'Aleo and Khandekar (2016), "Between 1996 and 2015, winter months (January to March) globally have shown no warming in 20 years. Instead, a cooling of 0.9°C (1.5°F)/decade has been identified in the northeastern United States. Cooling of a lesser magnitude has been shown for the lower 48 U.S. states and for winters in the UK for the last 20 years" (p. 107).

Among the papers published in the scientific literature reporting on the phenomenon are those by Benestad (2010), Cattiaux *et al.* (2010), Haigh (2010), Haigh *et al.* (2010), Wang *et al.* (2010), Woollings *et al.* (2010), Seager *et al.* (2010), Taws *et al.*, (2011), Lockwood *et al.* (2011), Sirocko *et al.* (2012), Deser and Phillips (2015), Li *et al.* (2015), Sun *et al.* (2016), and Xie *et al.* (2016). Many of these papers

Figure 2.3.1.1.1
Average number of daily high temperatures at 982 USHCN stations exceeding 100°F (37.8°C) per year, 1895–2014

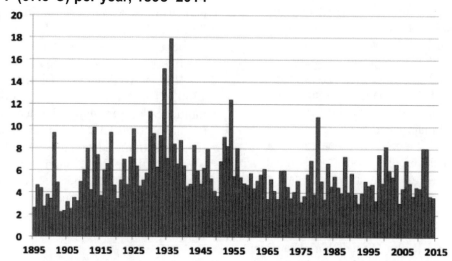

Source: Christy, 2016, citing NOAA data.

Figure 2.3.1.1.2
U.S. heat wave index, 1895–2015

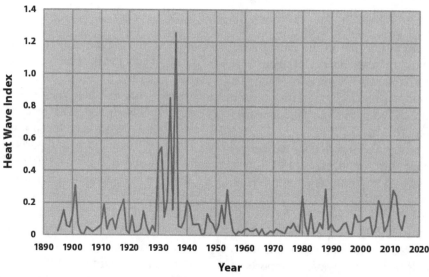

These data cover the contiguous 48 states. An index value of 0.2 could mean that 20% of the country experienced one heat wave, 10% of the country experienced two heat waves, or some other combination of frequency and area resulted in this value. *Source*: EPA, 2016.

suggest reduced solar activity played a prominent role in the observed colder winters.

Brown and Luojus (2018) report, "Early 2018 experienced close to record maximum snow accumulations over Northern Hemisphere and Arctic land areas since satellite passive microwave coverage began in 1979. The Finnish Meteorological Institute confirmed that the 2017/2018 winter has been quite exceptional compared to typical recent winters and is one of the snowier winters in the period since 1979 where passive microwave satellite data have been used to monitor the amount of snow on land. While 2017–2018 is not a record – that title belongs to 1993 with 3649 gigatons of peak snow water storage – close to 3500 gigatons of peak snow water storage were estimated, which ranks as the tenth highest peak snow accumulation since 1979."

Garnett and Khandekar (2018) contend "a colder climate awaits us," noting Canada, China, Europe, Japan, the United States, and other regions of the world have seen at least 25 global cold weather extremes since 2000. They write, "The IPCC-espoused science has highlighted [warm weather extremes] like heat waves, droughts, floods and fires while ignoring the 'cold' reality of the Earth's climate since the new millennium" (p. 435). Some Russian scientists predict cooling in the next few decades, as shown in Figure 2.3.1.1.3. See also Page (2017) and Lüdecke and Weiss (2018) for similar forecasts.

Figure 2.3.1.1.3
Changes in global near-surface air temperature from 1880–2015 and forecast after 2015

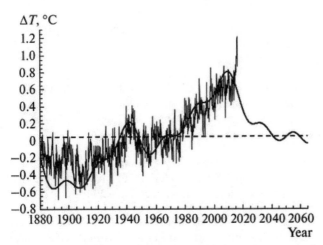

Thin lines represent monthly average ΔT counted from the average value of the global temperature for the period 1901–2000 (dashed line). The solid heavy curve represents calculations performed using spectral data analysis. *Source:* Stozhkov *et al.,* 2017, Figure 1.

As reported in Section 2.2.4, some solar scientists believe the Sun may be entering a grand solar minimum (GSM), which could manifest itself with cooler temperatures (Shindell *et al.,* 2001; Lockwood *et al.,* 2010, 2011; Yndestad and Solheim, 2016). During the last GSM, known as the Maunder Minimum, humanity endured the hardships of the Little Ice Age, a period that brought the coldest temperatures of the Holocene (Soon and Yaskell, 2003). Whether the Sun is indeed approaching a new GSM remains to be seen, but recent temperatures suggest cooling may be as much a concern as warming in coming decades.

References

Benestad, R.E. 2010. Low solar activity is blamed for winter chill over Europe. *Environmental Research Letters* **5**: 021001.

Bohm, R. 2012. Changes of regional climate variability in central Europe during the past 250 years. *The European Physical Journal Plus* **127**: 10.1140/epjp/i2012-12054-6.

Brown, R. and Luojus, K. 2018. 2018 snow assessment. *Global Cryosphere Watch.* World Meteorological Organization (website). Accessed December 5, 2018.

Cattiaux, J., Vautard, R., Cassou, C., Yiou, P., Masson-Delmotte, V., and Codron, F. 2010. Winter 2010 in Europe: a cold extreme in a warming climate. *Geophysical Research Letters* **37**: L20704, doi:10.1029/2010GL044613.

Christy, J. 2012. Testimony to the U.S. Senate Environment and Public Works Committee. August 1.

Christy, J. 2016. Testimony to the U.S. House Committee on Science, Space, and Technology. February 16.

D'Aleo, J.S. and Khandekar, M.L. 2016. Weather extremes. In: Easterbrook, D. (Ed.) *Evidence-based Climate Change.* Cambridge, MA: Elsevier, pp. 103–120.

Deng, H., Zhao, F., and Zhao, X. 2012. Changes of extreme temperature events in Three Gorges area, China. *Environmental and Earth Sciences* **66**: 1783–90.

Deser, C.L. and Phillips, A.S. 2015. Forced and internal components of winter air temperature trends over North America during the past 50 years: mechanisms and implications. *Journal of Climate* **29**: 2237–58, doi:10.1175/JCLI-D-15-0304.1.

Dole, R., Hoerling, M., Perlwitz, J., Eischeid, J., Pegion, P., Zhang, T., Quan, X.-W., Xu, T., and Murray, D. 2011. Was there a basis for anticipating the 2010 Russian heat wave?

Geophysical Research Letters **38**: 10.1029/2010GL046582.

EPA. 2016. U.S. Environmental Protection Agency. Climate change indicators: high and low temperatures (website). Accessed December 6, 2018.

Garnett, E.R. and Khandekar, M.L. 2018. Increasing cold weather extremes since the new millennium: an assessment with a focus on worldwide economic impacts. *Modern Environmental Science and Engineering* **4** (5): 427–38.

Haigh, J.D. 2010. Solar variability and the stratosphere. In: Sobel, A.H., Waugh, D.W., and Polvani, L.M. (Eds.) *The Stratosphere: Dynamics, Transport and Chemistry*. Washington, DC: American Geophysical Union, pp. 173–87.

Haigh, J.D., Winning, A.R., Toumi, R., and Harder, J.W. 2010. An influence of solar spectral variations on radiative forcing of climate. *Nature* **467**: 696–9.

Hiebl, J. and Hofstatter, M. 2012. No increase in multi-day temperature variability in Austria following climate warming. *Climatic Change* **113**: 733–50.

IPCC. 2013. Intergovernmental Panel on Climate Change. *Climate Change 2013: The Physical Science Basis*. Contribution of Working Group I to the Fifth Assessment Report of the Intergovernmental Panel on Climate Change. New York, NY: Cambridge University Press.

IPCC. 2014. Intergovernmental Panel on Climate Change. *Climate Change 2014: Impacts, Adaptation, and Vulnerability. Part A: Global and Sectoral Aspects*. Contribution of Working Group II to the Fifth Assessment Report of the Intergovernmental Panel on Climate Change. New York, NY: Cambridge University Press.

Jeong, S.-J., *et al.* 2010. Potential impact of vegetation feedback on European heat waves in a 2 x CO_2 climate. *Climatic Change* **99**: 625–35.

Khandekar, M.L. 2010. Weather extremes of summer 2010: global warming or natural variability? *Energy & Environment* **21**: 1005–10.

Köse, N., Tuncay Güner, H., Harley, G.L., and Guiot, J. 2017. Spring temperature variability over Turkey since 1800CE reconstructed from a broad network of tree-ring data. *Climate of the Past* **13**: 1–15.

Li, C., *et al.* 2015. Eurasian winter cooling in the warming hiatus of 1998–2012. *Geophysical Research Letters* **42**: 8131–9, doi:10.1002/2015GL065327.

Lockwood, M., Harrison, R.G., Woollings, T., and Solanki, S.K. 2010. Are cold winters in Europe associated with low solar activity? *Environmental Research Letters* **5**: 024001, doi:10.1088/1748-9326/5/2/024001.

Lockwood, M., Owens, M.J., Barnard, L., Davis, C.J., and Steinhilber, F. 2011. The persistence of solar activity indicators and the descent of the Sun into Maunder Minimum conditions. *Geophysical Research Letters* **38**: L22105, doi:10.1029/2011GL049811.

Lüdecke, H.-J. and Weiss, C.-O. 2018. Harmonic analysis of worldwide temperature proxies for 2000 years. *The Open Atmospheric Science Journal* **12**.

McKitrick, R. and Michaels, P.J. 2007. Quantifying the influence of anthropogenic surface processes and inhomogeneities on gridded global climate data. *Journal of Geophysical Research* **112** (24).

Page, N.J. 2017. The coming cooling: usefully accurate climate forecasting for policy makers. *Energy & Environment* **28** (3).

Polovodova Asteman, I., Filipsson, H.L., and Nordberg, K. 2018. Tracing winter temperatures over the last two millennia using a NE Atlantic coastal record. *Climate of the Past: Discussions* **14**: (7).

Quereda Sala, J., Montón Chiva, E., and Quereda Vázquez, V. 2017. Climatic or non-climatic warming in the Spanish mediterranean? *Wulfenia* **24** (10): 195–208.

Rusticucci, M. 2012. Observed and simulated variability of extreme temperature events over South America. *Atmospheric Research* **106**: 1–17.

Sardeshmukh, P.D., Compo, G.P., and Penland, C. 2015. Need for caution in interpreting extreme weather events. *Journal of Climate* **28**: 9166–87.

Seager, R., Kushnir, Y., Nakamura, J., Ting, M., and Naik, N. 2010. Northern Hemisphere winter snow anomalies: ENSO, NAO and the winter of 2009/10. *Geophysical Research Letters* **37**: L14703.

Shindell, D.T., Schmidt, G.A., Mann, M.E., Rind, D., and Waple, A. 2001. Solar forcing of regional climate change during the Maunder Minimum. *Science* **294**: 2149–52.

Sirocko, F., Brunck, H., and Pfahl, S. 2012. Solar influence on winter severity in central Europe. *Geophysical Research Letters* **39**: L16704.

Soon, W., Connolly, R., and Connolly, M. 2015. Re-evaluating the role of solar variability on Northern Hemisphere temperature trends since the 19th century. *Earth-Science Reviews* **150**: 409–52.

Soon, W. and Yaskell, S.H. 2003. *The Maunder Minimum and the Variable Sun-Earth Connection*. Singapore: World Scientific Publishing.

Stozhkov, Y., Bazilevskaya, G., Makhmutov, V., Svirzhevsky, N., Svirzhevskaya, A., Logachev, V., and Okhlopkov, V. 2017. Cosmic rays, solar activity, and changes in the Earth's climate. *Bulletin of the Russian Academy of Sciences: Physics* **81** (2): 252–54.

Sun, L., *et al.* 2016. What caused the recent "Warm Arctic, Cold Continents" trend pattern in winter temperatures? *Geophysical Research Letters* **23** (10).

Sutton, R., Hoskins, B., Palmer, T., Shepherd, T., and Slingo, J. 2018. Attributing extreme weather to climate change is not a done deal. *Nature* **561** (177): 10.

Tan, J., Zheng, Y. Song, G., Kalkstein, L.S., Kalkstein, A.J., and Tang, ZX. 2007. Heat wave impacts on mortality in Shanghai, 1998 and 2003. *International Journal of Biometeorology* **51**: 193–200.

Taws, S.L., Marsh, R., Wells, N.C., and Hirschi, J. 2011. Re-emerging ocean temperature anomalies in late-2010 associated with a repeat negative NAO. *Geophysical Research Letters* **38**: L20601.

Wang, C., Liu, H., and Lee, S.-K. 2010. The record breaking cold temperatures during the winter of 2009/10 in the Northern Hemisphere. *Atmospheric Science Letters* **11**: 161–8.

Woollings, T., Lockwood, M., Masato, G., Bell, C., and Gray, L. 2010. Enhanced signature of solar variability in Eurasian winter climate. *Geophysical Research Letters* **37**: L20805.

Xie, Y., Huang, J., and Liu, Y. 2016. From accelerated warming to warming hiatus in China. *International Journal of Climatology* **37** (4): 1758–73.

Yndestad, H. and Solheim, J.E. 2016. The influence of solar system oscillation on the variability of the total solar irradiance. *Astronomy* **51**: 10.1016/j.newast.2016.08.020.

Zhou, Y.Q. and Ren, G.Y. 2011. Change in extreme temperature event frequency over mainland China, 1961–2008. *Climate Research* **50**: 125–39.

2.3.1.2 Wildfires

According to model-based predictions, larger and more intense wildfires will become more frequent because of CO_2-induced global warming. Many scientists have begun to search for a link between fire and climate, often examining past trends to see if they support the models' projections. While some studies find fires were more common during the Medieval Warm Period, and so might well increase if warming resumes in the twenty-first century, others find little or no impact on fires and some even find a declining trend during the twentieth century.

A warmer world with higher levels of CO_2 in the atmosphere produces more vegetation and consequently more fuel for fires. While this could be interpreted as "climate change will cause more forest fires," this hardly supports the meme that this is a net environmental harm, since rising temperatures and CO_2 levels are responsible for the increased mass of trees and other plants being burned. For example, Turner *et al.* (2008) determined "climatically-induced variation in biomass availability was the main factor controlling the timing of regional fire activity during the Last Glacial-Interglacial climatic transition, and again during Mid-Holocene times, with fire frequency and magnitude increasing during wetter climatic phases." In addition, they report spectral analysis of the Holocene part of the record "indicates significant cyclicity with a periodicity of ~1500 years that may be linked with large-scale climate forcing."

Riano *et al.* (2007) conducted "an analysis of the spatial and temporal patterns of global burned area with the Daily Tile US National Oceanic and Atmospheric Administration-Advanced Very High-Resolution Radiometer Pathfinder 8 km Land dataset between 1981 and 2000." For several areas of the world this investigation revealed there were indeed significant upward trends in land area burned. Some parts of Eurasia and western North America, for example, had annual upward trends as high as 24.2 pixels per year, where a pixel represents an area of 64 km^2. These increases in burned area, however, were offset by equivalent decreases in burned area in tropical Southeast Asia and Central America. Consequently, observe Riano *et al.*, "there was no significant global annual upward or downward trend in burned area." They also note "there was also no significant upward or downward global trend in the burned area for any individual month." In addition, they found "latitude was not determinative, as divergent fire patterns were encountered for various land cover areas at the same latitude."

Marlon *et al.* (2008) observe "large, well-documented wildfires have recently generated worldwide attention, and raised concerns about the impacts of humans and climate change on wildfire regimes." The authors used "sedimentary charcoal records spanning six continents to document trends in both natural and anthropogenic biomass burning [over] the past two millennia." They found "global biomass burning declined from AD 1 to ~1750, before rising sharply between 1750 and 1870," after which it "declined abruptly." In terms of attribution, they note the initial long-term decline in global biomass burning was due to "a long-term global cooling trend," while they suggest the rise in fires that followed was "linked to increasing human influences." With respect to the final decline in fires that took place after 1870, however, they note it occurred "despite increasing air temperatures and population." As for what may have overpowered the tendency for increased global wildfires that would "normally" have been expected to result from the warming of the Little Ice Age-to-Current Warm Period transition, the nine scientists attribute "reduction in the amount of biomass burned over the past 150 years to the global expansion of intensive grazing, agriculture and fire management."

McAneney *et al.* (2009) assembled a database of building losses in Australia since 1900 and found "the annual aggregate numbers of buildings destroyed by bushfire since 1926 ... is 84," but "most historical losses have taken place in a few extreme fires." Nevertheless, they observe "the most salient result is that the annual probability of building destruction has remained almost constant over the last century," even in the face of "large demographic and social changes as well as improvements in fire fighting technique and resources." McAneney *et al.* conclude, "despite predictions of an increasing likelihood of conditions favoring bushfires under global climate change, we suspect that building losses due to bushfires are unlikely to alter materially in the near future."

Girardin *et al.* (2009) investigated "changes in wildfire risk over the 1901–2002 period with an analysis of broad-scale patterns of drought variability on forested eco-regions of the North American and Eurasian continents." The seven scientists report "despite warming since about 1850 and increased incidence of large forest fires in the 1980s, a number of studies indicated a decrease in boreal fire activity in the last 150 years or so." They find "this holds true for boreal southeastern Canada, British Columbia, northwestern Canada and Russia." With respect to this long-term "diminishing fire activity," Girardin *et al.* observe "the spatial extent for these long-term changes is large enough to suggest that climate is likely to have played a key role in their induction." The authors further note, "the fact that diminishing fire activity has also been detected on lake islands on which fire suppression has never been conducted provides another argument in support of climate control."

Brunelle *et al.* (2010) collected sediments during the summers of 2004 and 2005 from a drainage basin located in southeastern Arizona (USA) and northeastern Sonora (Mexico), from which samples were taken "for charcoal analysis to reconstruct fire history." Their results "show an increase in fire activity coincident with the onset of ENSO, and an increase in fire frequency during the Medieval Climate Anomaly [MCA]." During this latter period, from approximately AD 900 to 1260, "background charcoal reaches the highest level of the entire record and fire peaks are frequent," and "the end of the MCA shows a decline in both background charcoal and fire frequency, likely associated with the end of the MCA-related drought in western North America (Cook *et al.*, 2004)." Brunelle *et al.* speculate that if the region of their study warms in the future, "warming and the continuation of ENSO variability will likely increase fire frequency (similar to the MCA) while extreme warming and the shift to a persistent El Niño climate would likely lead to the absence of fires, similar to >5000 cal yr BP."

Wallenius *et al.* (2011) "studied *Larix*-dominated forests of central Siberia by means of high-precision dendro-chronological dating of past fires." They found "in the 18th century, on average, 1.9% of the forests burned annually, but in the 20th century, this figure was only 0.6%," and "the fire cycles for these periods were 52 and 164 years, respectively." In addition, they report "a further analysis of the period before the enhanced fire control program in the 1950s revealed a significant lengthening in the fire cycle between the periods 1650–1799 and 1800–1949, from 61 to 152 years, respectively." They note "a similar phenomenon has been observed in Fennoscandia, southern Canada and the western United States, where the annually burned proportions have decreased since the 19th century (Niklasson and Granstrom, 2000; Weir *et al.*, 2000; Heyerdahl *et al.*, 2001; Bergeron *et al.*, 2004)." They also found "in these regions, the decrease has been mostly much steeper, and the current fire cycles are several hundreds or thousands of years."

Girardin *et al.* (2013) write that many people have supposed that "global wildfire activity resulting from human-caused climatic change is a threat to communities living at wildland-urban interfaces world-wide and to the equilibrium of the global carbon cycle." The eight researchers note "broadleaf deciduous stands are characterized by higher leaf moisture loading and lower flammability and rate of wildfire ignition and initiation than needleleaf evergreen stands," citing the work of Paatalo (1998),

Campbell and Flannigan (2000), and Hely *et al.* (2001). And they therefore speculate that the introduction of broadleaf trees into dense needleleaf evergreen landscapes "could decrease the intensity and rate of spread of wildfires, improving suppression effectiveness, and reducing wildfire impacts," citing Amiro *et al.* (2001) and Hirsch *et al.* (2004).

Girardin *et al.* (2013) integrated into a wildfire modeling scheme information about millennial-scale changes in wildfire activity reconstructed from analyses of charred particles found in the sediments of 11 small lakes located in the transition zone between the boreal mixed-wood forests and the dense needle leaf forests of eastern boreal Canada. They report their assessment of millennial-scale variations of seasonal wildfire danger, vegetation flammability, and fire activity suggests "feedback effects arising from vegetation changes are large enough to offset climate change impacts on fire danger."

Asking whether such vegetation changes occur in the real world, the Canadian and French scientists cite the work of McKenney *et al.* (2011) and Terrier *et al.* (2013) suggesting that "future climate warming will lead to increases in the proportion of hardwood forests in both southern and northern boreal landscapes." They note this change in landscapes likely will have other benefits as well, such as "the higher albedo and summer evapotranspiration from deciduous trees, which would cool and counteract regional warming (Rogers *et al.*, 2013), and the increase in the resilience of forests to climatic changes (Drobyshev *et al.*, 2013)."

Yang *et al.* (2014) note fire is a critical component of the biosphere that "substantially influences land surface, climate change and ecosystem dynamics." To accurately predict fire regimes in the twenty-first century, they write, "it is essential to understand the historical fire patterns and recognize the interactions among fire, human and environmental factors." They "developed a 0.5° x 0.5° data set of global burned area from 1901 to 2007 by coupling the Global Fire Emission Database version 3 with a process-based fire model and conducted factorial simulation experiments to evaluate the impacts of human, climate and atmospheric components."

The seven scientists found "the average global burned area was about 442×10^4 km^2/yr during 1901–2007," with "a notable declining rate of burned area globally (1.28×10^4 km^2/yr)." They also found "burned area in the tropics and extra-tropics exhibited a significant declining trend, with no significant trend

detected at high latitudes." They report "factorial experiments indicated that human activities were the dominant factor in determining the declining trend of burned area in the tropics and extra-tropics" and "climate variation was the primary factor controlling the decadal variation of burned area at high latitudes." They note elevated CO_2 and nitrogen deposition "enhanced burned area in the tropics and southern extra-tropics" but "suppressed fire occurrence at high latitudes."

According to Hanson and Odion (2014), "there is widespread concern about an increase in fire severity in the forests of the western United States," citing Agee and Skinner (2005), Stephens and Ruth (2005), and Littell *et al.* (2009); but they write prior studies of the subject have "provided conflicting results about current trends of high-severity fire," possibly due to the fact that they "have used only a portion of available fire severity data, or considered only a portion of the Sierra Nevada." Using remote sensing data obtained from satellite imagery to assess high-severity fire trends since 1984, Hanson and Odion analyzed the entire region included within the Sierra Nevada Ecosystem Project (SNEP, 1996), which includes all of the Sierra Nevada and the southern Cascade mountains located within California, USA.

The two researchers report they could find "no trend in proportion, area or patch size of high-severity fire," while also noting "the rate of high-severity fire has been lower since 1984 than the estimated historical rate." They conclude that predictions of excessive, high-severity fire throughout the Sierra Nevada in the future "may be incorrect."

Noting "forest fires are a serious environmental hazard in southern Europe," Turco *et al.* (2016) write that "quantitative assessment of recent trends in fire statistics is important for assessing the possible shifts induced by climate and other environmental/ socioeconomic changes in this area." They analyzed "recent fire trends in Portugal, Spain, southern France, Italy and Greece, building on a homogenized fire database integrating official fire statistics provided by several national/EU agencies." The nine researchers from Greece, Italy, and Spain report, "during the period 1985–2011, the total annual burned area (BA) displayed a general decreasing trend" (see Figure 2.3.1.2.1) and "BA decreased by about 3020 km^2 over the 27-year-long study period (i.e. about -66% of the mean historical value)." They note "these results are consistent with those obtained on longer time scales when data were available" and

Figure 2.3.1.2.1
Total annual burned area (BA) and number of fires (NF) in Mediterranean Europe, 1985–2011

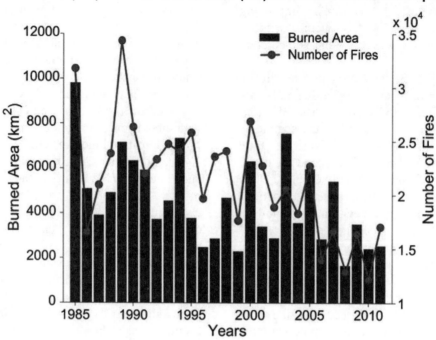

Source: Adapted from Turco *et al.*, 2016.

"similar overall results were found for the annual number of fires (NF), which globally decreased by about 12,600 in the study period (about -59%)."

Calder *et al.* (2015) used 12 lake-sediment charcoal records taken within and surrounding the Mount Zirkel Wilderness, a mountainous area of subalpine forests in northern Colorado, USA, to reconstruct the history of wildfire in the region over the past two millennia. The researchers found "warming of ~0.5 °C ~1,000 years ago increased the percentage of our study sites burned per century by ~260% relative to the past ~400 y," confirming that wildfires during the Medeival Warm Period were much more frequent and intense than they are in the modern era. The authors report, "only 15% of our study area burned in the past 80 y and only 30% of the area in a 129,600-ha study area in Yellowstone National Park burned from 1890 to 1988. Using Yellowstone National Park fire history as a baseline for comparison, our minimum estimate of 50% of sites burned within a century at the beginning of the MCA (Medieval Climate Anomaly) exceeds any century-scale estimate of Yellowstone National Park burning for the past 750 y."

Zhang *et al.* (2016) studied trends in forest fires and carbon emissions in China from 1988 to 2012. They note, "As one of the largest potential mechanisms for release of carbon from forest ecosystems, fires can have substantial impacts on the net carbon balance of ecosystems through emissions of carbon into the atmosphere and changes in net ecosystem productivity in postfire environments (Li *et al.*, 2014; Liu *et al.*, 2014). For 1960 to 2000, global estimates of approximately 2.07–2.46 Pg carbon emission per year due to fire have been reported (Schultz *et al.*, 2008; van der Werf *et al.*, 2010; Randerson *et al.*, 2012). These emissions represent about 26%–31% of current global CO_2 emissions from fossil fuels and industrial processes (Raupach *et al.*, 2007)." Using data reported in the *China Agriculture Yearbooks* from 1989 to 2013, the researchers were able to identify 169,100 forest fires that occurred in China, an average of 6,764 fires per year. "During the entire period, no significant temporal trends of fire numbers and burned area were observed." The frequency of fires declined during the 1990s, rose during the 2000s, and fell precipitously after 2008. They conclude, "The results indicated that no significant increases in fire occurrence and carbon emissions were observed during the study period at the national level."

Introducing their study of a somewhat unusual subject, Meigs *et al.* (2016) write "in western North America, recent widespread insect outbreaks and wildfires have sparked acute concerns about potential insect-fire interactions," noting "although previous research shows that insect activity typically does not increase wildfire likelihood, key uncertainties remain regarding insect effects on wildfire severity." The five U.S. researchers developed "a regional census of large wildfire severity following outbreaks of two prevalent bark beetle and defoliator species – mountain pine beetle (*Dendroctonus ponderosae*) and western spruce budworm (*Choristoneura freemani*) – across the U.S. Pacific Northwest." They report, "in contrast to common assumptions of positive feedbacks, we find that insects generally reduce the severity of subsequent wildfires," noting "specific effects vary with insect type and timing," but "both insects decrease the abundance of live vegetation susceptible to wildfire at multiple time lags," so that "by dampening subsequent burn severity, native insects could buffer rather than exacerbate fire regime changes expected due to land use and climate change." In light of these findings, they recommend "a precautionary approach when designing and implementing forest management policies intended to reduce wildfire hazard and increase resilience to global change."

Noting the economic and ecological costs of wildfires in the United States have risen in recent decades, Balch *et al.* (2017) studied more than 1.5 million government records of wildfires that had to be either extinguished or managed by state or federal agencies from 1992 to 2012. The six scientists report "humans have vastly expanded the spatial and seasonal 'fire niche' in the coterminous United States, accounting for 84% of all wildfires and 44% of total area burned." They note "during the 21-year time period, the human-caused fire season was three times longer than the lightning-caused fire season" and humans "added an average of 20,000 wildfires per year across the United States."

Earl and Simmonds (2017) examined the spatial and temporal patterns of fire activity for Australia over the period 2001–2015 using satellite data from the MODerate resolution Imaging Spectroradiometer (MODIS) sensors on the Terra and Aqua satellites, which they write allows for "a more consistent and comprehensive evaluation" of fire trends. They derived fire count numbers from an active fire algorithm, allowing them to calculate seasonal and annual fire activity on a 0.1° x 0.1° grid box scale (~1000 km^2). Results for Australia as a whole are presented in Figure 2.3.1.2.2. Annual fire numbers for Australia have decreased across the past 15 years of study, although the decrease is not statistically significant. Seasonally, the most abundant fire season was in the spring (48% of all fires), followed by winter (21%), summer (16%), and fall (15%). Summer was the only season found to exhibit a statistically significant trend ($p < 0.05$), showing a decline over the period of record.

With respect to possible climatic drivers of annual fire number statistics, Earl and Simmonds (2017) conducted a series of analyses to explore their relationship with large-scale climate indices, including ENSO, the Indian Ocean Dipole, and a precipitation dataset covering the continent. Their results revealed some significant relationships across both space and time. The authors did *not* conduct an analysis between the fire records and temperature, which omission we have remedied in Figure 2.3.1.2.3. As illustrated there, a statistically significant relationship exists between Australian temperature and annual fire counts, such that a 1°C temperature increase results in a 2.39 x 10^5 decline in annual fire count. Consequently, these data would appear to contradict claims that rising temperatures will increase fire frequency.

Figure 2.3.1.2.2
Annual number of Australian fires over the period 2001–2015

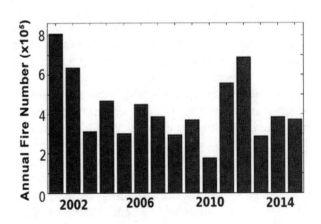

Source: Adapted from Earl and Simmonds, 2017.

Figure 2.3.1.2.3
Relationship between the annual number of Australian fires and Australian temperature anomalies over the period 2001–2015

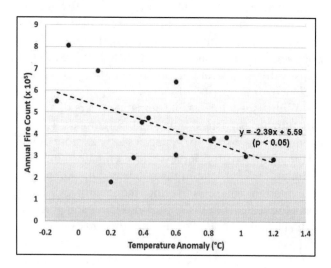

$$y = -2.39x + 5.59$$
$$(p < 0.05)$$

Source: Adapted from Earl and Simmonds, 2017.

Stirling (2017) studied the circumstances around the Fort McMurray, Alberta wildfire of 2016, said at the time to be "the costliest [insured] natural disaster in Canadian history," to see if claims that the fire was the result of global warming had a scientific basis. She begins by noting "Wildfires are an expected and essential occurrence in the vast boreal forests of Canada. Fires are essential for the regrowth of certain coniferous species whose seeds can only be released from the pine cones under the intense heat of a wildfire. But unusually dry periods between snow-melt and spring rain, careless campers or reckless intentional activities, and uncontrollable natural conditions like high winds and aging conifers can turn otherwise manageable wildfires into catastrophic, extreme events within hours or days if appropriate personnel, wildfire fighting equipment, and sufficient budget are not immediately available."

Stirling (2017) identifies the proximate causes of the fire and its destructiveness as a dry spring, sparse snow cover, inadequate resources for appropriate management of the fire hazard being "at the ready," and the high ratio of aging conifers. "Aging trees die from the bottom up," she explains, "with lower branches remaining on the stem. These become 'ladder fuels,' literally offering a small fire a way to race up the tree to the crown." Her review of 96 years of Fort McMurray temperature records of the monthly average daytime highs found "no apparent warming trend." (Recall from Section 2.1.1 that a common mistake made in climate research is to assume that global averages and trends accurately describe local and regional circumstances.) Stirling also observes that some 80% of forest fires are caused by human interaction with wilderness, which is "directly proportional to humans building into and extending activity in forested areas." She also notes fossil fuels allow "humans to escape wildfires in mass evacuations by car, truck and plane, and to fight wildfires with fossil-fueled air craft like water bombers, helicopters and motorized water pumps and vehicles."

Wildfires in the United States have been the focus of public attention in recent years due to drought conditions and forest mismanagement mainly in California. Nationally, the average annual number of forest fires did not increase between 1983, the first year for which comparable data are available, and 2017 (NIFC, n.d.). The number of acres burned did increase, though that figure is highly variable and has not increased since 2007. Both trends appear in Figure 2.3.1.2.4 below.

References

Agee, J.K. and Skinner, C.N. 2005. Basic principles of forest fuel reduction treatments. *Forest Ecology and Management* **211**: 83–96.

Amiro, B.D., Stocks, B.J., Alexander, M.E., Flannigan, M.D., and Wotton, B.M. 2001. Fire, climate change, carbon and fuel management in the Canadian Boreal forest. *International Journal of Wildland Fire* **10**: 405–13.

Balch, J.K., Bradley, B.A., Abatzoglou, J.T., Nagy, R.C., Fusco, E.J., and Mahood, A.L. 2017. Human-started wildfires expand the fire niche across the United States. *Proceedings of the National Academy of Sciences USA* **114**: 2946–51.

Bergeron, Y., Gauthier, S., Flannigan, M., and Kafka, V. 2004. Fire regimes at the transition between mixedwood and coniferous boreal forest in northwestern Quebec. *Ecology* **85**: 1916–32.

Brunelle, A., Minckley, T.A., Blissett, S., Cobabe, S.K., and Guzman, B.L. 2010. A ~8000 year fire history from an Arizona/Sonora borderland cienega. *Journal of Arid Environments* **24**: 475–81.

Figure 2.3.1.2.4
Number of U.S. wildland fires and acres burned per year, 1983–2017

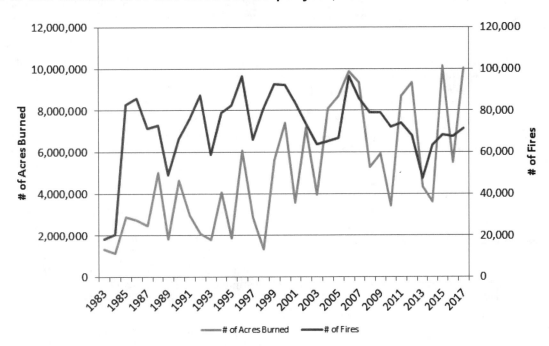

Source: National Interagency Fire Center (NIFC), n.d.

Calder, W.J., *et al.* 2015. Medieval warming initiated exceptionally large wildfire outbreaks in the Rocky Mountains. *Proceedings of the National Academy of Sciences USA* **112** (43): 13,261–6.

Campbell, I.D. and Flannigan, M.D. 2000. Long-term perspectives on fire-climate-vegetation relationships in the North American boreal forest. In: Kasischke, E.A. and Stocks, B.J. (Eds.) *Fire, Climate Change, and Carbon Cycling in the Boreal Forests.* New York, NY: Springer-Verlag, pp. 151–72.

Cook, E.R., Woodhouse, C., Eakin, C.M., Meko, D.M., and Stahle, D.W. 2004. Long-term aridity changes in the western United States. *Science* **306**: 1015–8.

Drobyshev, Y., Gewehr, S., Berninger, F., and Bergeron, Y. 2013. Species specific growth responses of black spruce and trembling aspen may enhance resilience of boreal forest to climate change. *Journal of Ecology* **101**: 231–42.

Earl, N. and Simmonds, I. 2017. Variability, trends, and drivers of regional fluctuations in Australian fire activity. *Journal of Geophysical Research: Atmospheres* **122**: 7445–60.

Girardin, M.P., Ali, A.A., Carcaillet, C., Blarquez, O., Hely, C., Terrier, A., Genries, A., and Bergeron, Y. 2013.

Vegetation limits the impact of a warm climate on boreal wildfires. *New Phytologist* **199**: 1001–11.

Girardin, M.P., Ali, A.A., Carcaillet, C., Mudelsee, M., Drobyshev, I., Hely, C., and Bergeron, Y. 2009. Heterogeneous response of circumboreal wildfire risk to climate change since the early 1900s. *Global Change Biology* **15**: 2751–69.

Hanson, C.T. and Odion, D.C. 2014. Is fire severity increasing in the Sierra Nevada, California, USA? *International Journal of Wildland Fire* **23**: 1–8.

Hely, C., Flannigan, M., Bergeron, Y., and McRae, D. 2001. Role of vegetation and weather on fire behavior in the Canadian mixed-wood boreal forest using two fire behavior prediction systems. *Canadian Journal of Forest Research* **31**: 430–41.

Heyerdahl, E.K., Brubaker, L.B., and Agee, J.K. 2001. Spatial controls of historical fire regimes: a multiscale example from the interior west, USA. *Ecology* **82**: 660–78.

Hirsch, K., Kafka, V., and Todd, B. 2004. Using forest management techniques to alter forest fuels and reduce wildfire size: an exploratory analysis. In: Engstrom, R.T., Galley, K.E.M., and de Groot, W.J. (Eds.) *Fire in*

Temperate, Boreal, and Montane Ecosystems. Tallahassee, FL: Tall Timber Research Station, pp. 175–84.

Li, F., *et al.* 2014. Quantifying the role of fire in the Earth system – Part 2: impact on the net carbon balance of global terrestrial ecosystems for the 20th century. *Biogeosciences* **11**: 1345–60.

Littell, J.S., McKenzie, D., Peterson, D.L., and Westerling, A.L. 2009. Climate and wildfire area burned in western US ecoprovinces, 1916–2003. *Ecological Applications* **19**: 1003–21.

Liu, D., *et al.* 2014. The contribution of China's Grain to Green Program to carbon sequestration. *Landscape Ecology* **29**: 1675–88.

Marlon, J.R., Bartlein, P.J., Carcaillet, C., Gavin, D.G., Harrison, S.P., Higuera, P.E., Joos, F., Power, M.J., and Prentice, I.C. 2008. Climate and human influences on global biomass burning over the past two millennia. *Nature Geoscience* **1**: 697–702.

McAneney, J., Chen, K., and Pitman, A. 2009. 100-years of Australian bushfire property losses: is the risk significant and is it increasing? *Journal of Environmental Management* **90**: 2819–22.

McKenney, D.W., Pedlar, J.H., Rood, R.B., and Price, D. 2011. Revisiting projected shifts in the climate envelopes of North American trees using updated general circulation models. *Global Change Biology* **17**: 2720–30.

Meigs, G.W., Zald, H.S.J., Campbell, J.L., Keeton, W.S., and Kennedy, R.E. 2016. Do insect outbreaks reduce the severity of subsequent forest fires? *Environmental Research Letters* **11**: 10.1088/1748-9326/11/4/045008.

NIFC. n.d. National Interagency Fire Center. Total Wildland Fires and Acres (1926–2017) (website). Accessed December 6, 2018.

Niklasson, M. and Granstrom, A. 2000. Numbers and sizes of fires: long-term spatially explicit fire history in a Swedish boreal landscape. *Ecology* **81**: 1484–99.

Paatalo, M.-L. 1998. Factors influencing occurrence and impacts of fires in northern European forests. *Silva Fennica* **32**: 185–202.

Randerson, J.T., *et al.* 2012. Small fire contributions to global burned area and biomass burning emissions. *Journal of Geophysical Research* **117**.

Raupach, M.R., *et al.* 2007. Global and regional drivers of accelerating CO_2 emissions. *Proceedings of the National Academy of Sciences USA* **104** (24): 10,288–93.

Riano, D., Moreno Ruiz, J.A., Isidoro, D., and Ustin, S.L. 2007. Global spatial patterns and temporal trends of burned

area between 1981 and 2000 using NOAA-NASA Pathfinder. *Global Change Biology* **13**: 40–50.

Rogers, B.M., Randerson, J.T., and Bonan, G.B. 2013. High-latitude cooling associated with landscape changes from North American boreal forest fires. *Biogeosciences* **10**: 699–718.

Schultz, M.G., *et al.* 2008. Global wildland fire emissions from 1960 to 2000. *Global Biogeochemical Cycles* **22**: GB2002.

SNEP. 1996. *Sierra Nevada Ecosystem Project, Final Report to Congress, Vol. I.* Davis, CA: University of California at Davis, Centers for Water and Wildland Resources.

Stephens, S.L. and Ruth, L.W. 2005. Federal forest-fire policy in the United States. *Ecological Applications* **15**: 532–42.

Stirling, M. 2017. Fort McMurray wildfire 2016: conflating human-caused wildfires with human-caused global warming. Available at SSRN: https://ssrn.com/abstract=2929576.

Terrier, A., Girardin, M.P., Perie, C., Legendre, P., and Bergeron, Y. 2013. Potential changes in forest composition could reduce impacts of climate change on boreal wildfires. *Ecological Applications* **23**: 21–35.

Turco, M., Bedia, J., Di Liberto, F., Fiorucci, P., von Hardenberg, J., Koutsias, N., Llasat, M.-C., Xystrakis, F., and Provenzale, A. 2016. Decreasing fires in Mediterranean Europe. *PLoS ONE* **11** (3): e0150663.

Turner, R., Roberts, N., and Jones, M.D. 2008. Climatic pacing of Mediterranean fire histories from lake sedimentary microcharcoal. *Global and Planetary Change* **63**: 317–24.

van der Werf, G.R., *et al.* 2010. Global fire emissions and the contribution of deforestation savanna forest agricultural and peat fires (1997–2009). *Atmospheric Chemistry and Physics* **10**: 11,707–35.

Wallenius, T., Larjavaara, M., Heikkinen, J., and Shibistova, O. 2011. Declining fires in Larix-dominated forests in northern Irkutsk district. *International Journal of Wildland Fire* **20**: 248–54.

Weir, J.M.H., Johnson, E.A., and Miyanishi, K. 2000. Fire frequency and the spatial age mosaic of the mixed-wood boreal forest in western Canada. *Ecological Applications* **10**: 1162–77.

Yang, J., Tian, H., Tao, B., Ren, W., Kush, J., Liu, Y., and Wang, Y. 2014. Spatial and temporal patterns of global burned area in response to anthropogenic and environmental factors: reconstructing global fire history for the 20th and early 21st centuries. *Journal of Geophysical Research: Biogeosciences* **119**: 249–63.

Zhang, Y.D.., *et al.* 2016. Historical trends of forest fires and carbon emissions in China from 1988 to 2012. *Journal of Geophysical Research: Biogeosciences* **121**: 2506–17.

2.3.1.3 Droughts

The link between warming and drought is weak, and by some measures drought decreased over the twentieth century. Changes in the hydrosphere of this type are regionally highly variable and show a closer correlation with multidecadal climate rhythmicity than they do with global temperature.

Higher surface temperatures are said to result in more frequent, severe, and longer-lasting droughts. The IPCC expresses doubt, however, that this is or will become a major problem. In the Working Group I contribution to the Fifth Assessment Report, the authors write "compelling arguments both for and against significant increase in the land area affected by drought and/or dryness since the mid-20th century have resulted in a *low confidence* assessment of observed and attributable large-scale trends" and "*high confidence* that proxy information provides evidence of droughts of greater magnitude and longer duration than observed during the 20th century in many regions" (IPCC, 2013, p. 112).

The historical record is replete with accounts of megadroughts lasting for several decades to centuries that occurred during the Medieval Warm Period (MWP), dwarfing modern-day droughts (e.g., Seager *et al.*, 2007; Cook *et al.*, 2010). Atmospheric CO_2 concentrations were more than 100 ppm lower during the Medieval Warm Period than they are today. The clear implication is that natural processes operating during the MWP were responsible for droughts that were much more frequent and lasted much longer than those observed in the twentieth and twenty-first centuries or even forecast for the rest of the twenty-first century by all but the most unrealistic climate models.

Minetti *et al.* (2010) examined a regional inventory of monthly droughts for the portion of South America located south of approximately 22°S latitude, dividing the area of study into six sections (the central region of Chile plus five sections making

up most of Argentina). They note "the presence of long favorable tendencies [1901–2000] regarding precipitations or the inverse of droughts occurrence are confirmed for the eastern Andes Mountains in Argentina with its five sub-regions (Northwest Argentina, Northeast Argentina, Humid Pampa, West-Centre Provinces and Patagonia) and the inverse over the central region of Chile." From the middle of 2003 to 2009, however, they report "an upward trend in the occurrence of droughts with a slight moderation over the year 2006." They additionally note the driest single-year periods were 1910–1911, 1915–1916, 1916–1917, 1924–1925, and 1933–1934, suggesting twentieth century warming has not promoted an abnormal increase in droughts in the southern third of South America.

Sinha *et al.* (2011) observed "proxy reconstructions of precipitation from central India, north-central China (Zhang *et al.*, 2008), and southern Vietnam (Buckley *et al.*, 2010) reveal a series of monsoon droughts during the mid 14th–15th centuries that each lasted for several years to decades," and "these monsoon megadroughts have no analog during the instrumental period." They note "emerging tree ring-based reconstructions of monsoon variability from SE Asia (Buckley *et al.*, 2007; Sano *et al.*, 2009) and India (Borgaonkar *et al.*, 2010) suggest that the mid 14th–15th century megadroughts were the first in a series of spatially widespread megadroughts that occurred during the Little Ice Age" and "appear to have played a major role in shaping significant regional societal changes at that time."

Wang *et al.* (2011) estimated soil moisture and agricultural drought severities and durations in China for the period 1950–2006, identifying a total of 76 droughts. Wang *et al.* report "climate models project that a warmer and moister atmosphere in the future will actually lead to an enhancement of the circulation strength and precipitation of the summer monsoon over most of China (e.g., Sun and Ding, 2010) that will offset enhanced drying due to increased atmospheric evaporative demand in a warmer world (Sheffield and Wood, 2008)." Tao and Zhang (2011) provide some support for this statement, finding the net effect of physiological and structural vegetation responses to expected increases in the atmosphere's CO_2 content will lead to "a decrease in mean evapotranspiration, as well as an increase in mean soil moisture and runoff across China's terrestrial ecosystem in the 21st century," which should act to lessen, or even offset, the "slightly drier" soil moisture conditions modeled by Wang *et al.*

Buntgen *et al.* (2011) "introduce and analyze 11,873 annually resolved and absolutely dated ring-width measurement series from living and historical fir (*Abies alba* Mill.) trees sampled across France, Switzerland, Germany and the Czech Republic, which continuously span the AD 962–2007 period," and which "allow Central European hydroclimatic springtime extremes of the industrial era to be placed against a 1000 year-long backdrop of natural variations." The nine researchers found "a fairly uniform distribution of hydroclimatic extremes throughout the Medieval Climate Anomaly, Little Ice Age and Recent Global Warming." Such findings, Buntgen *et al.* state, "may question the common belief that frequency and severity of such events closely relates to climate mean states."

Kleppe *et al.* (2011) reconstructed the duration and magnitude of extreme droughts in the northern Sierra Nevada region of the Fallen Leaf Lake (California, USA) watershed to estimate paleo-precipitation near the headwaters of the Truckee River-Pyramid Lake watershed of eastern California and northwestern Nevada. The six scientists found "submerged Medieval trees and geomorphic evidence for lower shoreline corroborate a prolonged Medieval drought near the headwaters of the Truckee River-Pyramid Lake watershed," and water-balance calculations independently indicated precipitation was "less than 60% normal." They note these findings "demonstrate how prolonged changes of Fallen Leaf's shoreline allowed the growth and preservation of Medieval trees far below the modern shoreline." In addition, they note age groupings of such trees suggest similar megadroughts "occurred every 600–1050 years during the late Holocene."

Pederson *et al.* (2012) attempt to put the southeastern United States' recent drought variability in a long-term perspective by reconstructing historic drought trends in the Apalachicola-Chattahoochee-Flint river basin over the period 1665–2010 using a dense and diverse tree-ring network. This network, they write, "accounts for up to 58.1% of the annual variance in warm-season drought during the 20th century and captures wet eras during the middle to late 20th century." The 12 researchers found the Palmer Drought Severity Index reconstruction for their study region revealed "recent droughts are not unprecedented over the last 346 years" and "droughts of extended duration occurred more frequently between 1696 and 1820," when most of the world was in the midst of the Little Ice Age.

They also found their results "confirm the findings of the first reconstruction of drought in the southern Appalachian Mountain region, which indicates that the mid-18th and early 20th centuries were the driest eras since 1700," citing Stahle *et al.* (1988), Cook *et al.* (1988), and Seager *et al.* (2009).

Chen *et al.* (2012) used the standard precipitation index to characterize drought intensity and duration throughout the Southern United States (SUS) over the past century. According to the nine researchers, there were "no obvious increases in drought duration and intensity during 1895–2007." Instead, they found "a slight (not significant) decreasing trend in drought intensity." They note "although reports from IPCC (2007) and the U.S. Climate Report (Karl *et al.*, 2009) indicated that it is likely that drought intensity, frequency, and duration will increase in the future for the SUS, we did not find this trend in the historical data." They also note, although "the IPCC (2007) and U.S. Climate Report predicted a rapid increase in air temperature, which would result in a higher evapotranspiration thereby reducing available water," they "found no obvious increase in air temperature for the entire SUS during 1895–2007."

According to Hao *et al.* (2014), the global areal extent of drought fell from 1982 to 2012 across all five levels used to rank drought conditions. A figure illustrating their findings is reproduced as Figure 2.3.1.3.1.

Khandekar (2014) notes the Canadian prairies "are very drought prone, … but since 2005, droughts have been replaced by floods, with 2010 and 2014 ranking among the wettest summers on record" (p. 10, citing Garnett and Khandekar, 2010). And with respect to India, he notes "most climate models have achieved only limited success in simulating and predicting the features of the monsoon, for example extreme rainfall events and the associated flooding or extended monsoon-season dry spells." Noting there is "an urgent need to develop more skillful algorithms for the short-term prediction of regional and localized floods and droughts" (p. 8), he concludes, "Reducing greenhouse gas emissions (to reduce floods or other extreme weather events in future) is a meaningless exercise and will do nothing to influence future climate extremes" (*Ibid.*)

Delworth *et al.* (2015) observe "portions of western North America have experienced prolonged drought over the last decade" and acknowledge "the underlying causes of the drought are not well established in terms of the role of natural variability versus human-induced radiative forcing changes, such as from increasing greenhouse gases." They

Figure 2.3.1.3.1
Global areal extent of five levels of drought for 1982–2012

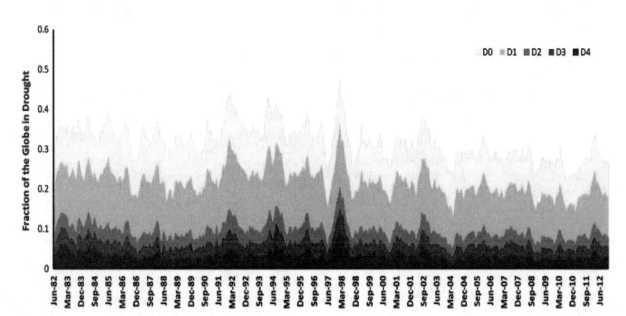

Fraction of the global land in D0 (abnormally dry), D1 (moderate), D2 (severe), D3 (extreme), and D4 (exceptional) drought condition (Data: Standardized Precipitation Index data derived from MERRA-Land). *Source:* Hao *et al.*, 2014.

further note the drought "has occurred at the same time as the so-called global warming hiatus, a decadal period with little increase in global mean surface temperature." Could the two events be related? The authors pose the hypothesis that the hiatus *caused* the drought, proposing as the mechanism enhanced easterly winds in the Pacific (unrelated to anthropogenic forcing) citing Kosaka and Xie (2013) and England *et al.* (2014). The five authors, all of them affiliated with NOAA, conducted an experiment with three climate models using observational data pertaining to the Pacific winds and the drought, and "find a clear link" between the hiatus and the drought. According to the model results, "tropical wind anomalies account for 92% of the simulated North American drought during the recent decade, with 8% from anthropogenic radiative forcing changes." They predict drought conditions will continue so long as the Pacific wind anomaly continues.

McCabe *et al.* (2017) studied monthly runoff for 2,109 hydrologic units (HUs) in the coterminous United States from 1901–2014, recording the frequency of drought as indicated by the HU runoff

percentile dropping to the 20th percentile or lower. A drought was considered to end when the HU runoff percentile exceeded the 20th percentile. Among their findings, and the one most relevant to the current discussion, is "for most of the continental United States, drought frequency appears to have *decreased* during the 1901 through 2014 period" (italic added).

Ault *et al.* (2018) conducted a rare test of a null hypothesis in the global warming debate, that megadroughts in the western United States "are inevitable and occur purely as a consequence of internal climate variability." They test the hypothesis using a linear inverse model (LIM) constructed from global sea surface temperature anomalies and self-calibrated Palmer Drought Severity Index data for North America. They find "Despite being trained only on seasonal data from the late twentieth century, the LIM produces megadroughts that are comparable in their duration, spatial scale, and magnitude to the most severe events of the last 12 centuries. The null hypothesis therefore cannot be rejected with much confidence when considering these features of megadrought, meaning that similar events are possible today, even without any changes to boundary conditions."

Drought conditions in the United States reached their lowest level in 2017 since the United States Drought Monitor (USDM) began keeping records in 2000. (The USDM is a partnership between the University of Nebraska-Lincoln, the U.S. Department of Agriculture, and the U.S. National Oceanic and Atmospheric Administration (NOAA), see USDM, n.d.). The long drought in California ended that year, or at least was interrupted, and more than 100 inches of snow fell in parts of the Sierra Nevada mountain range. However, drought conditions over the entire United States increased their range in 2018, reaching levels last seen in 2014. As shown in Figure 2.3.1.3.2, there has been a generally decreasing trend in drought area in the continental United States for the past 10 years.

The findings of Kleppe *et al.* (2011) and many others whose works they cite suggest the planet during the Medieval Warm Period experienced less precipitation and longer and more severe drought than have been experienced to date in the modern era. In addition, their data suggest such dry conditions have occurred regularly, in cyclical fashion, "every 650–1150 years during the mid- and late-Holocene." These observations suggest there is nothing unusual, unnatural, or unprecedented about the global number or intensity of droughts during the modern era.

References

Ault, T.R., St. George, S., Smerdon, J.E., Coats, S., Mankin, J.S., Carrillo, C.M., Cook, B.I., and Stevenson, S. 2018. A robust null hypothesis for the potential causes of megadrought in Western North America. *Journal of Climate* **31**: 3–24.

Borgaonkar, H.P., Sikdera, A.B., Rama, S., and Panta, G.B. 2010. El Niño and related monsoon drought signals in 523-year-long ring width records of teak (*Tectona grandis* L.F.) trees from south India. *Palaeogeography, Palaeo-climatology, Palaeoecology* **285**: 74–84.

Buckley, B.M., Anchukaitis, K.J., Penny, D., Fletcher, R., Cook, E.R., Sano, M., Nam, L.C., Wichienkeeo, A., Minh, T.T., and Hong, T.M. 2010. Climate as a contributing factor in the demise of Angkor, Cambodia. *Proceedings of the National Academy of Sciences USA* **107**: 6748–52.

Figure 2.3.1.3.2
United States areal extent of five levels of drought for 2000–2018

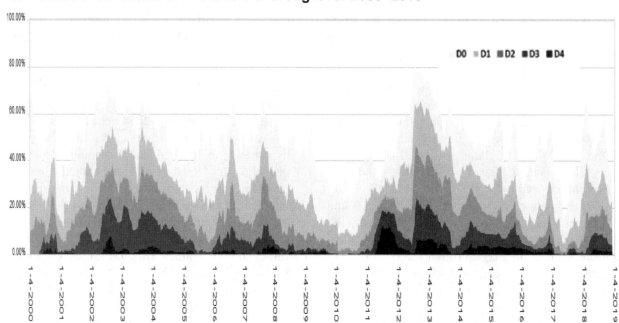

Fraction of the United States land area in D0 (abnormally dry), D1 (moderate), D2 (severe), D3 (extreme), and D4 (exceptional) drought condition. *Source:* USDM, n.d.

Buckley, B.M., Palakit, K., Duangsathaporn, K., Sanguantham, P., and Prasomsin, P. 2007. Decadal scale droughts over northwestern Thailand over the past 448 years: links to the tropical Pacific and Indian Ocean sectors. *Climate Dynamics* **29**: 63–71.

Buntgen, U., Brazdil, R., Heussner, K.-U., Hofmann, J., Kontic, R., Kyncl, T., Pfister, C., Chroma, K., and Tegel, W. 2011. Combined dendro-documentary evidence of Central European hydroclimatic springtime extremes over the last millennium. *Quaternary Science Reviews* **30**: 3947–59.

Chen, G., Tian, H., Zhang, C., Liu, M., Ren, W., Zhu, W., Chappelka, A.H., Prior, S.A., and Lockaby, G.B. 2012. Drought in the Southern United States over the 20th century: variability and its impacts on terrestrial ecosystem productivity and carbon storage. *Climatic Change* **114**: 379–97.

Cook, E.R., Kablack, M.A., and Jacoby, G.C. 1988. The 1986 drought in the southeastern United States—how rare an event was it? *Journal of Geophysical Research* **93**: 14,257–60.

Cook, E.R., Seager, R., Heim Jr., R.R., Vose, R.S., Herweijer, C., and Woodhouse, C. 2010. Mega-droughts in North America: placing IPCC projections of hydroclimatic change in a long-term palaeoclimate context. *Journal of Quaternary Science* **25**: 48–61.

Delworth, T., *et al.* 2015. A link between the hiatus in global warming and North American drought. *Journal of Climate* **28** (9): 3834–45.

England, M.H., *et al.* 2014. Recent intensification of wind-driven circulation in the Pacific and the ongoing warming hiatus. *Nature Climate Change* **4**: 222–7.

Garnett, E.R. and Khandekar, M.L. 2010. Summer 2010: wettest on the Canadian Prairies in 60 years! A preliminary assessment of causes and consequences. *Canadian Meteorological and Oceanographic Society Bulletin* **38**: 204–8.

Hao, Z., *et al.* 2014. Global integrated drought monitoring and prediction system. *Scientific Data* **1**: Article 140001.

IPCC. 2007. Intergovernmental Panel on Climate Change. *Climate Change 2007: The Physical Science Basis.* Contribution of Working Group I to the Fourth Assessment Report of the Intergovernmental Panel on Climate Change. Cambridge, UK: Cambridge University Press.

IPCC. 2013. Intergovernmental Panel on Climate Change. *Climate Change 2013: The Physical Science Basis.* Contribution of Working Group I to the Fifth Assessment Report of the Intergovernmental Panel on Climate Change. New York, NY: Cambridge University Press.

Khandekar, M. 2014. Floods and droughts in the Indian Monsoon: natural variability trumps human impact. *GWPF Briefing* 12. London, UK: The Global Warming Policy Foundation.

Karl, T.R., Melillo, J.M., and Peterson, T.C. 2009. *Global Climate Change Impacts in the United States.* Cambridge, UK: Cambridge University Press.

Kleppe, J.A., Brothers, D.S., Kent, G.M., Biondi, F., Jensen, S., and Driscoll, N.W. 2011. Duration and severity of Medieval drought in the Lake Tahoe Basin. *Quaternary Science Reviews* **30**: 3269–79.

Kosaka, Y. and Xie, S.-P. 2013. Recent global-warming hiatus tied to equatorial Pacific surface cooling. *Nature* **501**: 403–7.

McCabe, G.J., Wolock, D.M., and Austin, S.H. 2017. Variability of runoff-based drought conditions in the conterminous United States. *International Journal of Climatology* **37**: 1014–21, doi:10.1002/joc.4756

Minetti, J.L., Vargas, W.M., Poblete, A.G., de la Zerda, L.R., and Acuña, L.R. 2010. Regional droughts in southern South America. *Theoretical and Applied Climatology* **102**: 403–15.

Pederson, N., Bell, A.R., Knight, T.A., Leland, C., Malcomb, N., Anchukaitis, K.J., Tackett, K., Scheff, J., Brice, A., Catron, B., Blozan, W., and Riddle, J. 2012. A long-term perspective on a modern drought in the American Southeast. *Environmental Research Letters* **7**: 10.1088/1748-9326/7/1/014034.

Sano, M., Buckley, B.M., and Sweda, T. 2009. Tree-ring based hydroclimate reconstruction over northern Vietnam from *Fokienia hodginsii*: eighteenth century mega-drought and tropical Pacific influence. *Climate Dynamics* **33**: 331–40.

Seager, R., Graham, N., Herweijer, C., Gorodn, A.L., Kushnir, Y., and Cook, E. 2007. Blueprints for medieval hydroclimate. *Quaternary Science Reviews* **26**: 2322–36.

Seager, R., Tzanova, A., and Nakamura, J. 2009. Drought in the southeastern United States: causes, variability over the last millennium, and the potential for future hydroclimate change. *Journal of Climate* **22**: 5021–45.

Sheffield, J. and Wood, E.F. 2008. Projected changes in drought occurrence under future global warming from multi-model, multi-scenario, IPCC AR4 simulations. *Climate Dynamics* **31**: 79–105.

Sinha, A., Stott, L., Berkelhammer, M., Cheng, H., Edwards, R.L., Buckley, B., Aldenderfer, M., and Mudelsee, M. 2011. A global context for megadroughts in monsoon Asia during the past millennium. *Quaternary Science Reviews* **30**: 47–62.

Stahle, D.W., Cleaveland, M.K., and Hehr, J.G. 1988. North Carolina climate changes reconstructed from tree rings: AD 372–1985. *Science* **240**: 1517–19.

Sun, Y. and Ding, Y.-H. 2010. A projection of future changes in summer precipitation and monsoon in East Asia. *Science in China Series D: Earth Sciences* **53**: 284–300.

Tao, F. and Zhang, Z. 2011. Dynamic response of terrestrial hydrological cycles and plant water stress to climate change in China. *Journal of Hydrometeorology* **12**: 371–93.

USDM. n.d. United States Drought Monitor (website). Accessed December 5, 2018.

Wang, A., Lettenmaier, D.P., and Sheffield, J. 2011. Soil moisture drought in China, 1950–2006. *Journal of Climate* **24**: 3257–71.

Zhang, P.Z., *et al.* 2008. A test of climate, sun, and culture relationships from an 1810-year Chinese cave record. *Science* **322**: 940–42.

2.3.1.4 Floods

Climate model simulations generally predict a future with more frequent and severe floods in response to CO_2-induced global warming. Confirming such predictions has remained an elusive task, according to the Working Group I contribution to the IPCC's Fifth Assessment Report. The authors write, "there continues to be a lack of evidence and thus *low confidence* regarding the sign of trend in the magnitude and/or frequency of floods on a global scale over the instrumental record" (IPCC, 2013, p. 112). That conclusion is not in fact due to a "lack of evidence," but the presence of real-world data contradicting the narrative the IPCC attempts to present. A large body of scientific research shows CO_2-induced global warming has not increased the frequency or magnitude of floods nor is it likely to do so in the future.

Two problems confront those claiming that climate change has caused more flooding in recent decades. The first is the failure to control for increases in impervious surfaces (roads, parking lots, buildings, etc.) near rivers, which result in more, and more rapid, run-off during heavy rains. Bormann *et al.* (2011) studied data from 78 river gauges in Germany and found a significant impact by changes in nearby surfaces. When impervious surfaces increase in area, they note, "runoff generation can be expected to increase and infiltration and groundwater recharge decrease," which can lead to increases in river flow and a potential for more frequent and extreme floods. They conclude these facts "should be emphasized in the recent discussion on the effect of climate change on flooding."

A second problem is controlling for the increasing value and vulnerability of property located near river and lake shorelines, an issue addressed by many scholars including Pielke and Landsea (1998), Crompton and McAneney (2008), Pielke *et al.* (2008), Barredo (2009, 2010), and Neumayer and Barthel (2011). Barredo *et al.* (2012) write "economic impacts from flood disasters have been increasing over recent decades" more often due to the rising value of properties located near water than to climate change. The authors examined "the time history of insured losses from floods in Spain between 1971 and 2008" to see "whether any discernible residual signal remains after adjusting the data for the increase in the number and value of insured assets over this period of time." They found "the absence of a significant positive trend in the adjusted insured flood losses in Spain," suggesting "the increasing trend in the original losses is explained by socio-economic factors, such as the increases in exposed insured properties, value of exposed assets and insurance penetration." "The analysis rules out a discernible influence of anthropogenic climate change on insured losses," they write, a finding that "is consistent with the lack of a positive trend in hydrologic floods in Spain in the last 40 years."

Many researchers have documented past floods that are larger than any in the industrial era, meaning natural variability cannot be ruled out as the cause of even major and unusual floods in the modern era. Zhang *et al.* (2009) found coolings of 160- to 170-year intervals dominated climatic variability in the Yangtze Delta in China over the past millennium, and these cooling periods promoted locust plagues by enhancing temperature-associated drought/flood events. The six scientists state "global warming might not only imply reduced locust plague[s], but also reduced risk of droughts and floods for entire China," noting these findings "challenge the popular view that global warming necessarily accelerates natural and biological disasters such as drought/flood events and outbreaks of pest insects." They contend their results are an example of "benign effects of global warming on the regional risk of natural disasters."

Benito *et al.* (2010) reconstructed flood frequencies of the Upper Guadalentin River in southeast Spain using "geomorphological evidence, combined with one-dimensional hydraulic modeling and supported by records from documentary sources at Lorca in the lower Guadalentin catchment." The combined palaeoflood and documentary records indicate past floods were clustered during particular time periods: AD 950–1200 (10), AD 1648–1672 (10), AD 1769–1802 (9), AD 1830–1840 (6), and AD 1877–1900 (10), where the first time interval coincides with the Medieval Warm Period and the latter four fall within the Little Ice Age. By calculating mean rates of flood occurrence over each of the five intervals, a value of 0.40 floods per decade during the Medieval Warm Period and an average value of 4.31 floods per decade over the four parts of the Little Ice Age can be determined. The latter value is more than ten times greater than the mean flood frequency experienced during the Medieval Warm Period.

Villarini and Smith (2010) "examined the distribution of flood peaks for the eastern United States using annual maximum flood peak records from 572 U.S. Geological Survey stream gaging stations with at least 75 years of observations." This work revealed "only a small fraction of stations exhibited significant linear trends," and "for those stations with trends, there was a split between increasing and decreasing trends." They also note "no spatial structure was found for stations exhibiting trends." Thus, they conclude, "there is little indication that human-induced climate change has resulted in increasing flood magnitudes for the eastern United States."

Villarini *et al.* (2011) similarly analyzed data from 196 U.S. Geological Survey streamflow stations with a record of at least 75 years over the midwestern United States. Most streamflow changes they observed were "associated with change-points (both in mean and variance) rather than monotonic trends," and they indicate "these non-stationarities are often associated with anthropogenic effects," which they identify as including "changes in land use/land cover, changes in agricultural practice, and construction of dams and reservoirs." "In agreement with previous studies (Olsen *et al.*, 1999; Villarini *et al.*, 2009)," they conclude, "there is little indication that anthropogenic climate change has significantly affected the flood frequency distribution for the Midwest U.S."

Stewart *et al.* (2011) derived "a complete record of paleofloods, regional glacier length changes (and associated climate phases) and regional glacier advances and retreats (and associated climate transitions) … from the varved sediments of Lake Silvaplana (ca. 1450 BC–AD 420; Upper Engadine, Switzerland)," indicating "these records provide insight into the behavior of floods (i.e. frequency) under a wide range of climate conditions." They found "an increase in the frequency of paleofloods during cool and/or wet climates and windows of cooler June–July–August temperatures" and the frequency of flooding "was reduced during warm and/or dry climates." Reiterating that "the findings of this study suggest that the frequency of extreme summer–autumn precipitation events (i.e. flood events) and the associated atmospheric pattern in the Eastern Swiss Alps was not enhanced during warmer (or drier) periods," Stewart *et al.* acknowledge "evidence could not be found that summer–autumn floods would increase in the Eastern Swiss Alps in a warmer climate of the 21st century."

Hirsch and Ryberg (2012) compared global mean carbon dioxide concentration (GMCO$_2$) to a streamflow dataset consisting of long-term (85- to 127-year) annual flood series from 200 stream gauges deployed by the U.S. Geological Survey in basins with little or no reservoir storage or urban development (less than 150 persons per square kilometer in AD 2000) throughout the coterminous United States. The authors determine whether the patterns of the statistical associations between the two parameters were significantly different from what would be expected under the null hypothesis that flood magnitudes are independent of GMCO$_2$. The authors report "in none of the four regions defined in this study is there strong statistical evidence for flood magnitudes increasing with increasing GMCO$_2$." One region, the southwest, showed a statistically significant *negative* relationship between GMCO$_2$ and flood magnitudes. Hirsch and Ryberg conclude "it may be that the greenhouse forcing is not yet sufficiently large to produce changes in flood behavior that rise above the 'noise' in the flood-producing processes." It could also mean the "anticipated hydrological impacts" envisioned by the IPCC and others are simply incorrect.

Zha *et al.* (2012) conducted a paleohydrological field investigation in the central portion of the Jinghe River, the middle and upper reaches of which are located in a semiarid zone with a monsoonal climate, between Binxian county and Chunhua county of Shaanxi Province, China. Their analysis revealed during the mid-Holocene climatic optimum, the climate was warm-humid, the climate system was stable, and "there were no flood records identified in the middle reaches of the Yellow river." Thereafter, however, they report "global climatic cooling events occurred at about 4200 years BP, which was also well recorded by various climatic proxies in China." These observations led them to conclude, "the extraordinary floods recorded in the middle reaches of the Jinghe River were linked to the global climatic events" – all of which were global *cooling* events.

Wilhelm *et al.* (2012) analyzed the sediments of Lake Allos, a 1-km-long by 700-m-wide high-altitude lake in the French Alps (44°14'N, 6°42'35'E), by means of both seismic survey and lake-bed coring, revealing the presence of 160 graded sediment layers over the past 1,400 years. Comparisons of the most recent of these layers with records of historic floods suggest the sediment layers are representative of significant floods that were "the result of intense meso-scale precipitation events." Of special interest is their finding of "a low flood frequency during the Medieval Warm Period and more frequent and more intense events during the Little Ice Age." Wilhelm *et al.* additionally state "the Medieval Warm Period was marked by very low hydrological activity in large rivers such as the Rhone, the Moyenne Durance, and the Tagus, and in mountain streams such as the Taravilla lake inlet." Of the Little Ice Age, they write, "research has shown higher flood activity in large rivers in southern Europe, notably in France, Italy, and in smaller catchments (e.g., in Spain)."

Sagarika *et al.* (2014) examined variability and trends in seasonal and water year (October through September) streamflow for 240 stream gauges considered to be minimally impaired by human influences in the coterminous United States for the years 1951–2010. They report finding positive trends in streamflow for many sites in the eastern United States and negative trends in streamflow for sites in the Pacific Northwest.

McCabe and Wolock (2014) examined "spatial and temporal patterns in annual and seasonal minimum, mean, and maximum daily streamflow values" using a database drawn from 516 reference stream gauges located throughout the coterminous United States for the period 1951–2009. Cluster analysis was used to classify the stream gauges into 14 groups based on similarity in their temporal patterns of streamflow. They found "some small magnitude trends over time" which "are only weakly associated with well-known climate indices. We conclude that most of the temporal variability in flow is unpredictable in terms of relations to climate

indices and infer that, for the most part, future changes in flow characteristics cannot be predicted by these indices."

Hao *et al.* (2016) studied trends in floods in China over a 2,000-year record to see if drought and flood events coincided with known warm and cold periods. They begin by observing "there has been no significant trend in the mean precipitation over the whole country" from 1951 to 2009, despite a measured increase in temperature of 1°C during that period. The authors used a 2,000-year temperature series created by Ge *et al.* (2013) using 28 proxies including historical documents, tree rings, ice cores, lake sediments, and stalagmites, producing a record with a time resolution finer than 10 years. A data set of precipitation anomalies and grading system for the severity of droughts and flood disasters created by Zhang (1996) was then compared to the temperature record. The results showed only weak correlations and a random assortment of positive as well as negative associations depending on region. The results "showed that there has been no fixed spatial pattern of precipitation anomalies during either cold or warm periods in Eastern China over the past 2000 years." Which means neither drought conditions nor flooding in China correlate with changes in mean average global temperature.

Macdonald and Sangster (2017) lament that "one of the greatest challenges presently facing river basin managers is the dearth of reliable long-term data on the frequency and severity of extreme floods," and set out to address that challenge by presenting "the first coherent large-scale national analysis undertaken on historical flood chronologies in Britain, providing an unparalleled network of sites (Fig. 1), permitting analysis of the spatial and temporal distribution of high-magnitude flood patterns and the potential mechanisms driving periods of increased flooding at a national scale (Britain) since AD 1750." The authors report, "The current flood-rich period (2000) is of particular interest with several extreme events documented in recent years, though it should be noted from a historical perspective that these are not unprecedented, with several periods with comparable [Flood Index] scores since ca. 1750, it remains unclear at present whether the current period (2000) represents a short or long flood-rich phase." They conclude, "The apparent increase in flooding witnessed over the last decade appears in consideration to the long-term flood record not to be unprecedented; whilst the period since 2000 has been considered as flood-rich, the period 1970–2000 is 'flood poor', which may partly explain why recent floods are often perceived as extreme events. The much publicised (popular media) apparent change in flood frequency since 2000 may reflect natural variability, as there appears to be no shift in long-term flood frequency."

Hodgkins *et al.* (2017) studied major floods (25–100 year return period) from 1961 to 2010 in North America and from 1931 to 2010 in Europe to see if such events had become more frequent over time. More than 1,200 flood gauges were studied in diverse catchments from North America and Europe; only minimally altered catchments were used, and trends were assessed on a variety of flood characteristics and for a variety of specific regions. "Overall," they write, "the number of significant trends in major-flood occurrence across North America and Europe was approximately the number expected due to chance alone. Changes over time in the occurrence of major floods were dominated by multidecadal variability rather than by long-term trends. There were more than three times as many significant relationships between major-flood occurrence and the Atlantic Multidecadal Oscillation than significant long-term trends."

* * *

Summarizing, the historical record suggests no global trend toward increasing flooding events in the modern era, while proxy data give a contradictory picture of major floods due to natural causes, more flooding during cool periods than during warm periods or *vice versa,* or (as in the case of China) no correlation at all between floods and temperature. This being the case, it is unlikely that human CO_2 emissions are currently causing a global increase in floods or that warmer temperatures forecast for the rest of the twenty-first century would trigger such an increase.

References

Barredo, J.I. 2009. Normalized flood losses in Europe: 1970–2006. *Natural Hazards and Earth System Sciences* **9**: 97–104.

Barredo, J.I. 2010. No upward trend in normalized windstorm losses in Europe: 1970–2008. *Natural Hazards and Earth System Sciences* **10**: 97–104.

Barredo, J.I., Sauri, D., and Llasat, M.C. 2012. Assessing trends in insured losses from floods in Spain 1971–2008. *Natural Hazards and Earth System Sciences* **12**: 1723–9.

Benito, G., Rico, M., Sanchez-Moya, Y., Sopena, A., Thorndycraft, V.R., and Barriendos, M. 2010. The impact of late Holocene climatic variability and land use change on the flood hydrology of the Guadalentin River, southeast Spain. *Global and Planetary Change* **70**: 53–63.

Bormann, H., Pinter, N., and Elfert, S. 2011. Hydrological signatures of flood trends on German rivers: flood frequencies, flood heights and specific stages. *Journal of Hydrology* **404**: 50–66.

Crompton, R.P. and McAneney, K.J. 2008. Normalized Australian insured losses from meteorological hazards: 1967–2006. *Environmental Science and Policy* **11**: 371–8.

Ge, Q.S., *et al.* 2013. Temperature changes over the past 2000 yr in China and comparison with the Northern hemisphere. *Climate of the Past* **9**: 1153–60.

Hao, Z., *et al.* 2016. Spatial patterns of precipitation anomalies in eastern China during centennial cold and warm periods of the past 2000 years. *International Journal of Climatology* **36** (1): 467–75.

Hirsch, R.M. and Ryberg, K.R. 2012. Has the magnitude of floods across the USA changed with global CO_2 levels? *Hydrological Sciences Journal* **57**: 10.1080/02626667.2011.621895.

Hodgkins, G.A., *et al.* 2017. Climate-driven variability in the occurrence of major floods across North America and Europe. *Journal of Hydrology* **552**: 704–17.

IPCC. 2013. Intergovernmental Panel on Climate Change. *Climate Change 2013: The Physical Science Basis.* Contribution of Working Group I to the Fifth Assessment Report of the Intergovernmental Panel on Climate Change. New York, NY: Cambridge University Press.

Macdonald, N. and Sangster, H. 2017. High-magnitude flooding across Britain since AD 1750. *Hydrology and Earth System Sciences* **21**: 1631–50.

McCabe, G.J. and Wolock, D. 2014. Spatial and temporal patterns in conterminous United States streamflow characteristics. *Geophysical Research Letters* **41**: 19.

Neumayer, E. and Barthel, F. 2011. Normalizing economic loss from natural disasters: a global analysis. *Global Environmental Change* **21**: 13–24.

Olsen, J.R., Stedinger, J.R., Matalas, N.C., and Stakhiv, E.Z. 1999. Climate variability and flood frequency estimation for the Upper Mississippi and Lower Missouri

Rivers. *Journal of the American Water Resources Association* **35**: 1509–23.

Pielke Jr., R.A., Gratz, J., Landsea, C.W., Collins, D., Saunders, M.A., and Musulin, R. 2008. Normalized hurricane damage in the United States: 1900–2005. *Natural Hazards Review* **31**: 29–42.

Pielke Jr., R.A. and Landsea, C.W. 1998. Normalized hurricane damage in the United States: 1925–95. *Weather and Forecasting* **13**: 621–31.

Sagarika, S., Kalrab, A., and Ahmada, S. 2014. Evaluating the effect of persistence on long-term trends and analyzing step changes in streamflows of the continental United States. *Journal of Hydrology* **517**: 36–53.

Stewart, M.M., Grosjean, M., Kuglitsch, F.G., Nussbaumer, S.U., and von Gunten, L. 2011. Reconstructions of late Holocene paleofloods and glacier length changes in the Upper Engadine, Switzerland (ca. 1450 BC–AD 420). *Palaeogeography, Palaeoclimatology, Palaeoecology* **311**: 215–23.

Villarini, G., Serinaldi, F., Smith, J.A., and Krajewski, W.F. 2009. On the stationarity of annual flood peaks in the continental United States during the 20th century. *Water Resources Research* **45**: 10.1029/2008WR007645.

Villarini, G. and Smith, J.A. 2010. Flood peak distributions for the eastern United States. *Water Resources Research* **46**: 10.1029/2009WR008395.

Villarini, G., Smith, J.A., Baeck, M.L., and Krajewski, W.F. 2011. Examining flood frequency distributions in the Midwest U.S. *Journal of the American Water Resources Association* **47**: 447–63.

Wilhelm, B., *et al.* 2012. 1400 years of extreme precipitation patterns over the Mediterranean French Alps and possible forcing mechanisms. *Quaternary Research* **78**: 1–12.

Zha, X., Huang, C., Pang, J., and Li, Y. 2012. Sedimentary and hydrological studies of the Holocene palaeofloods in the middle reaches of the Jinghe River. *Journal of Geographical Sciences* **22**: 470–8.

Zhang, P.Y. 1996. *Historical Climate Change of China.* Jinan, China: Shandong Science and Technology Press, pp. 307–38 (in Chinese).

Zhang, Z., *et al.* 2009. Periodic temperature-associated drought/flood drives locust plagues in China. *Proceedings of the Royal Society B* **276**: 823–31.

2.3.1.5 Storms

The IPCC writes "it is *likely* that the number of heavy precipitation events over land has increased in more regions than it has decreased in since the mid-20th century, and there is *medium confidence* that anthropogenic forcing has contributed to this increase. … For the near and long term, CMIP5 projections confirm a clear tendency for increases in heavy precipitation events, in the global mean seen in the AR4 [Fourth Assessment Report], but there are substantial variations across regions. Over most of the mid-latitude land masses and over wet tropical regions, extreme precipitation will *very likely* be more intense and more frequent in a warmer world" (IPCC, 2013, p. 112). Climate models generally predict more storms as the human impact on climate increases during the twenty-first century (e.g., Seeley and Romps, 2014).

Referring to extreme or intense storms, Dezileau *et al.* (2011) note the key question is, "are they linked to global warming or are they part of natural climate variability?" They write "it is essential to place such events in a broader context of time, and trace the history of climate changes over several centuries," because "these extreme events are inherently rare and therefore difficult to observe in the period of a human life." Analyzing regional historical archives and sediment cores extracted from two Gulf of Aigues-Mortes lagoons in the northwestern part of the occidental Mediterranean Sea for bio- and geo-indicators of past storm activities there, they were able to assess "the frequency and intensity of [extreme] events during the last 1500 years" as well as "links between past climatic conditions and storm activities." They found evidence of four "catastrophic storms of category 3 intensity or more," which occurred at approximately AD 455, 1742, 1848, and 1893, all before human greenhouse gases could have been a factor.

Dezileau *et al.* (2011) write, "the apparent increase in intense storms around 250 years ago lasts to about AD 1900," whereupon "intense meteorological activity seems to return to a quiescent interval after (i.e. during the 20th century AD)." They add, "interestingly, the two periods of most frequent superstorm strikes in the Aigues-Mortes Gulf (AD 455 and 1700–1900) coincide with two of the coldest periods in Europe during the late Holocene (Bond cycle 1 and the latter half of the Little Ice Age.)" The authors suggest "extreme storm events are associated with a large cooling of Europe," and they calculate the risk of such storms occurring during that cold period "was higher than today by a factor of 10," noting "if this regime came back today, the implications would be dramatic."

Barredo (2010) examined large historical windstorm event losses in Europe over the period 1970–2008 for 29 European countries. After adjusting the data for "changes in population, wealth, and inflation at the country level and for inter-country price differences using purchasing power parity," the researcher, employed by the Institute for Environment and Sustainability, European Commission-Joint Research Centre in Ispra, Italy, reports "the analyses reveal no trend in the normalized windstorm losses and confirm increasing disaster losses are driven by society factors and increasing exposure," adding "increasing disaster losses are overwhelmingly a consequence of changing societal factors."

Page *et al.* (2010), working with sediment cores extracted from Lake Tutira on the eastern end of New Zealand's North Island, developed a 7,200-year history of the frequency and magnitude of storm activity based on analyses of sediment grain size, diatom, pollen, and spore types and concentrations, carbon and nitrogen concentrations, and tephra and radiocarbon dating. They report millennial-scale cooling periods tend to "coincide with periods of increased storminess in the Tutira record, while warmer events match less stormy periods." Their research shows the sudden occurrence of a string of years, or even decades, of unusually large storms is something that can happen at almost any time without being driven by human activities such as the burning of fossil fuels.

Gascon *et al.* (2010) conducted a study they describe as "the first to document the climatology of major cold-season precipitation events that affect southern Baffin Island [Canada]." They examined the characteristics and climatology of the 1955–2006 major cold-season precipitation events at Iqaluit, the capital of Nunavut, located on the southeastern part of Baffin Island in the northwestern end of Frobisher Bay, basing their work on analyses of hourly surface meteorological data obtained from the public archives of Environment Canada. The three researchers detected a "non-significant decrease" in autumn and winter storm activity over the period of their study. The authors' results are depicted in Figure 2.3.1.5.1.

Figure 2.3.1.5.1
Cold-season occurrences of major precipitation events at Iqaluit, Nunavut, Canada

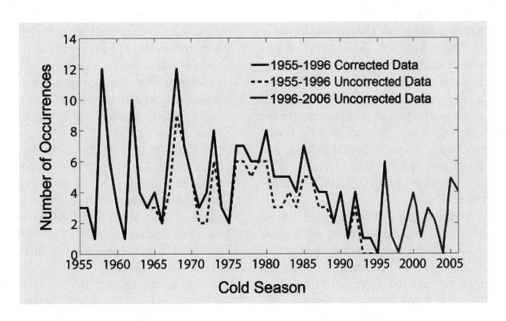

Source: Adapted from Gascon *et al.*, 2010.

Alexander *et al.* (2011) analyzed storminess across southeast Australia using extreme (standardized seasonal 95th and 99th percentiles) geostrophic winds deduced from eight widespread stations possessing sub-daily atmospheric pressure observations dating back to the late nineteenth century, finding "strong evidence for a significant reduction in intense wind events across SE Australia over the past century." They note "in nearly all regions and seasons, linear trends estimated for both storm indices over the period analyzed show a decrease," while "in terms of the regional average series," they write, "all seasons show statistically significant declines in both storm indices, with the largest reductions in storminess in autumn and winter."

Mallinson *et al.* (2011) employed optically stimulated luminescence dating of inlet-fill and flood tide delta deposits from locations in the Outer Banks barrier islands of North Carolina, USA to provide a "basis for understanding the chronology of storm impacts and comparison to other paleoclimate proxy data" in the region over the past 2,200 years. Analyses of the cores revealed "the Medieval Warm Period (MWP) and Little Ice Age (LIA) were both characterized by elevated storm conditions as indicated by much greater inlet activity relative to today." They write, "given present understanding of atmospheric circulation patterns and sea-surface temperatures during the MWP and LIA, we suggest that increased inlet activity during the MWP responded to intensified hurricane impacts, while elevated inlet activity during the LIA was in response to increased nor'easter activity." The group of five researchers state their data indicate, relative to climatic conditions of the Medieval Warm Period and Little Ice Age, there has more recently been "a general decrease in storminess at mid-latitudes in the North Atlantic," reflecting "more stable climate conditions, fewer storm impacts (both hurricane and nor'easter), and a decrease in the average wind intensity and wave energy field in the mid-latitudes of the North Atlantic."

Li *et al.* (2011), citing "unprecedented public concern" with respect to the impacts of climate change, set out to examine the variability and trends of storminess for the Perth, Australia metropolitan coast. They conducted an extensive set of analyses using observations of wave, wind, air pressure, and water level over the period 1994–2008. The results of their analysis, in their view, would serve "to validate or invalidate the climate change hypothesis" that rising CO_2 concentrations are increasing the

frequency and severity of storms. As shown in Figure 2.3.1.5.2, all storm indices showed significant interannual variability over the period of record, and "no evidence of increasing (decreasing) trends in extreme storm power was identified to validate the climate change hypotheses for the Perth region."

Sorrel *et al.* (2012) note the southern coast of the English Channel in northwestern France is "well suited to investigate long-term storminess variability because it is exposed to the rapidly changing North Atlantic climate system, which has a substantial influence on the Northern Hemisphere in general." They present "a reappraisal of high-energy estuarine and coastal sedimentary records," finding "evidence for five distinct periods during the Holocene when storminess was enhanced during the past 6,500 years." The six scientists write, "high storm activity occurred periodically with a frequency of about 1,500 years," with the last extreme stormy period "coinciding with the early to mid-Little Ice Age." They note "in contrast, the warm Medieval Climate Optimum was characterized by low storm activity (Sorrel *et al.*, 2009; Sabatier *et al.*, 2012)."

Khandekar (2013) observed that "many excellent studies on thunderstorm climatology (e.g., Changnon, 2001) have used over 100 years of data to document that 'thunderstorms and related activity in the U.S. peaked during the 1920s and 1930s and since then have declined in the late 1990s.'" See Changnon and Kunkel (2006) and Changnon (2010) for more recent presentations of this research. Khandekar also reports research conducted by Hage (2003), who "extracts data from several thousand Prairie-farm newsletters and reconstructs windstorm activity from 1880 to 1995 for the Canadian Prairies." The study (by Hage) concludes that "severe windstorms and associated thunderstorm activity peaked during the early part of the twentieth century and has since then declined steadily."

Yang *et al.* (2015) report that over the period December 2013–February 2014, "there was a pronounced reduction of extratropical storm (ETS) activity over the North Pacific Ocean and the west coast of the United States of America (USA), and a substantial increase of ETS activity extending from central Canada down to the midwestern USA." In hopes of explaining the "extreme North America winter storm season of 2013/14," they used the Geophysical Fluid Dynamics Laboratory Forecast-Oriented Low Ocean Resolution model to conduct a series of simulations. The authors' modeling exercise found "no statistically significant change" in ETS over mid-America or the Pacific coastal

Figure 2.3.1.5.2
Annual storm trends for Perth, Australia, 1994–2008

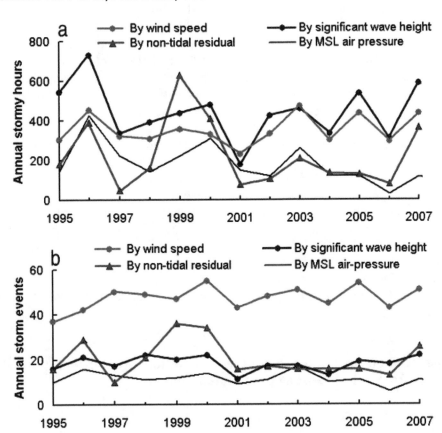

Annual storm trends for Perth, Australia defined by (a) stormy hours and (b) number of storm events, as determined by wind speed, significant wave height, non-tidal residual water level, and mean sea level pressure. Adapted from Li *et al., 2011.*

region from 1940–2040 could be attributed to anthropogenic forcing, while a "significant decrease starting [in] 2000" was found for mid-Canada. They conclude, "thus, the impact of antorpogenic forcing prescribed on this model did not contribute to the 2013-14 extreme ETS events over North America."

Degeai *et al.* (2015) studied sediments in a lagoon in southern France to find the "sedimentological signature" of ancient storms. The study area is located in the Languedoc region along the continental shelf of the Gulf of Lions in the Northwestern Mediterranean. "The many lagoons in this coastal plain give an excellent opportunity to find sedimentary sequences recording the palaeostorm events," they write. They found "phases of high storm activity occurred during cold periods, suggesting a climatically-controlled mechanism for the occurrence of these storm periods." They also

found a "new 270-year solar-driven pattern of storm cyclicity" and 10 major storm periods with a mean duration of 96 ± 54 years. Phases of higher storm activity occurred generally during the cold episodes of the Little Ice Age, the Dark Ages Cold Period, and the Iron Age Cold Period. Extreme storm waves were recorded on the French Mediterranean coast to the East of the Rhone delta during the Little Ice Age. Periods of low storm activity occurred during periods that coincide with the Roman Warm Period and the Medieval Warm Period.

Zhang *et al.* (2017a) note "understanding the trend of localized severe weather under the changing climate is of great significance but remains challenging which is at least partially due to the lack of persistent and homogeneous severe weather observations at climate scales while the detailed physical processes of severe weather cannot be resolved in global climate models." They created a database of "continuous and coherent severe weather reports from over 500 manned stations" across China and discovered "a significant decreasing trend in severe weather occurrence across China during the past five decades. The total number of severe weather days that have either thunderstorm, hail and/or damaging wind decrease about 50% from 1961 to 2010. It is further shown that the reduction in severe weather occurrences correlates strongly with the weakening of East Asian summer monsoon which is the primary source of moisture and dynamic forcing conducive for warm-season severe weather over China."

Turning from China to the United States, the number of extreme rainfall events – defined as a greater-than-normal proportion of one-day precipitation originating from the highest 10th percentile of one-day precipitation – are alleged to be increasing in the United States. A database often cited is Step 4 of the Climate Exchange Index maintained by the National Oceanic and Atmospheric

Administration (NOAA, 2018). But most or even all of the increase is likely due to a change in instrumentation and analysis methodology rather than a change in weather. Figure 2.3.1.5.3 plots data from 1910 through 2014 and finds from 1990 through 1992, a near zero slope exists with a percent of explained variance (R^2) of only 0.007%. Then, from 1995 to 2014, another insignificant slope with an R^2 of 0.24% is observed. Thus, the data do not exhibit a trend, but rather a discontinuity occurring between 1992 and 1995.

Climate forcing caused by CO_2 is predicted to produce a linear increase in surface temperature and other climate indices, not sudden discontinuities such as that shown in NOAA's extreme precipitation record. The more likely cause of the jump is that between 1992 and 1995, the National Weather Service (NWS) changed the way it measures precipitation at its "first-order" weather stations network, replacing manual observation gauges with electronic devices that are equipped with wind shields, are closer to the ground, and are "corrected" by

Figure 2.3.1.5.3
Percentage of the contiguous United States with a much greater-than-normal proportion of precipitation derived from extreme 1-day precipitation events

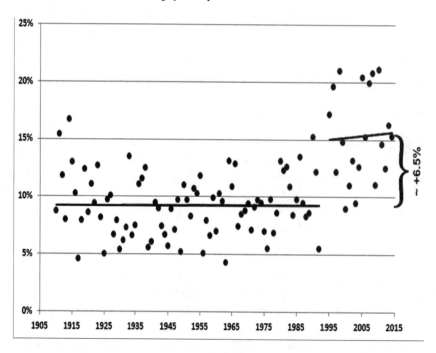

Annual time-series of the proportion of the contiguous United States with a greater-than-normal proportion of its one-day precipitation originating from the highest 10th percentile of one-day precipitation (Climate Extreme Index Step 4). The expected value is 10%. Separate regression lines are plotted for the period from 1910 to 1992 ($R^2 = 7 \times 10^{-5}$) and from 1995 to 2014 ($R^2 = 0.0024$). *Source:* Adapted from Gleason *et al.*, 2008, Figure 2, p. 2129.

NWS staff to account for some known biases. The sudden jump in extreme precipitation reports is probably just an artifact of this change in data-collection methodology. This is independently confirmed by evidence that the record of floods across the United States shows no evidence of increasing extreme rainfall events since 1950 (Hirsch and Ryberg, 2012; McCabe and Wolock, 2014).

As mentioned at the beginning of this section, climate models generally predict more extreme rainfall events. However, the models do a poor job simulating such events and therefore are unreliable guides to the future. Zhang *et al.* (2017b) note, "meeting the demand for robust projections for extreme short-duration rainfall is challenging, however, because of our poor understanding of its past and future behaviour. The characterization of past changes is severely limited by the availability of observational data. Climate models, including typical regional climate models, do not directly simulate all extreme rainfall producing processes, such as convection." The authors report on efforts to improve the models by focusing on precipitation–temperature relationships, but those relationships are tenuous and of limited use for projecting future precipitation extremes.

* * *

In conclusion, storms, like floods, sometimes appear to be more frequent and more intense during periods of global *cooling,* not warming, and are unrelated to anthropogenic forcing. Climate models are not a reliable guide to the frequency or intensity of future extreme rainfall events. If the human presence is causing the climate to warm, in many parts of the world this may produce weather that is calm, not stormier.

References

Alexander, L.V., Wang, X.L., Wan, H., and Trewin, B. 2011. Significant decline in storminess over southeast Australia since the late 19th century. *Australian Meteorological and Oceanographic Journal* **61**: 23–30.

Barredo, J.I. 2010. No upward trend in normalized windstorm losses in Europe: 1970–2008. *Natural Hazards and Earth System Sciences* **10**: 97–104.

Changnon, S.A. 2001. *Thunderstorms across the Nation: An Atlas of Storms, Hail and Their Damages in the 20th Century*. Mahomet, IL: Changnon Climatologist.

Changnon, S.A. 2010. *An Atlas of Windstorms in the United States*. Champaign, IL: University of Illinois.

Changnon, S.A. and Kunkel, K.E. 2006. *Severe Storms in the Midwest*. Champaign, IL: Midwestern Regional Climate Center, Illinois State Water Survey.

Degeai, J., Devillers, B., Dezileau, L., Oueslati, H., and Bony, G. 2015. Major storm periods and climate forcing in the Western Mediterranean during the Late Holocene. *Quaternary Science Reviews* **129**: 37–56.

Dezileau, L., Sabatier, P., Blanchemanche, P., Joly, B., Swingedouw, D., Cassou, C., Castaings, J., Martinez, P., and Von Grafenstein, U. 2011. Intense storm activity during the Little Ice Age on the French Mediterranean coast. *Palaeogeography, Palaeoclimatology, Palaeo-ecology* **299**: 289–97.

Gascon, G., Stewart, R.E., and Henson, W. 2010. Major cold-season precipitation events at Iqaluit, Nunavut. *Arctic* **63**: 327–37.

Gleason, K.L., Lawrimore, J.H., Levinson, D.H., Karl, T.R., and Karoly, D.J. 2008. A revised U.S. climate extremes index. *Journal of Climate* **21**: 2124–37.

Hage, K.D. 2003. On destructive Canadian Prairie windstorms and severe winters. *Natural Hazards* **29**: 207–28.

Hirsch, R.M. and Ryberg, K.R. 2012. Has the magnitude of floods across the USA changed with global CO_2 levels? *Hydrological Sciences Journal* **57**: 10.1080/02626667.2011.621895.

IPCC. 2013. Intergovernmental Panel on Climate Change. *Climate Change 2013: The Physical Science Basis*. Contribution of Working Group I to the Fifth Assessment Report of the Intergovernmental Panel on Climate Change. New York, NY: Cambridge University Press.

Khandekar, M.L. 2013. Are extreme weather events on the rise? *Energy & Environment* **24**: 537–49.

Li, F., Roncevich, L., Bicknell, C., Lowry, R., and Ilich, K. 2011. Interannual variability and trends of storminess, Perth, 1994–2008. *Journal of Coastal Research* **27**: 738–45.

Mallinson, D.J., Smith, C.W., Mahan, S., Culver, S.J., and McDowell, K. 2011. Barrier island response to late Holocene climate events, North Carolina, USA. *Quaternary Research* **76**: 46–57.

McCabe, G.J. and Wolock, D. 2014. Spatial and temporal patterns in conterminous United States streamflow characteristics. *Geophysical Research Letters* **41**: 19.

NOAA. 2018. U.S. National Oceanic and Atmospheric Administration. A single index for measuring extremes (website). August. Accessed December 9, 2018.

Page, M.J., *et al.* 2010. Storm frequency and magnitude in response to Holocene climate variability, Lake Tutira, North-Eastern New Zealand. *Marine Geology* **270**: 30–44.

Sabatier, P., Dezileau, L., Colin, C., Briqueu, L., Bouchette, F., Martinez, P., Siani, G., Raynal, O., and Von Grafenstein, U. 2012. 7000 years of paleostorm activity in the NW Mediterranean Sea in response to Holocene climate events. *Quaternary Research* **77**: 1–11.

Seeley, J. and Romps, D. 2014. The effect of global warming on severe thunderstorms in the United States. *Journal of Climate* **28** (6): 2443–58.

Sorrel, P., Debret, M., Billeaud, I., Jaccard, S.L., McManus, J.F., and Tessier, B. 2012. Persistent non-solar forcing of Holocene storm dynamics in coastal sedimentary archives. *Nature Geoscience* **5**: 892–6.

Sorrel, P., Tessier, B., Demory, F., Delsinne, N., and Mouaze, D. 2009. Evidence for millennial-scale climatic events in the sedimentary infilling of a macrotidal estuarine system, the Seine estuary (NW France). *Quaternary Science Reviews* **28**: 499–516.

Yang, X., *et al.* 2015. Extreme North America winter storm season of 2013/14: roles of radiative forcing and the global warming hiatus. *Bulletin of the American Meteorological Society* **96** (12): S25–S28.

Zhang, Q., Ni, X., and Zhang, F. 2017a. Decreasing trend in severe weather occurrence over China during the past 50 years. *Scientific Reports* **7**: Article 42310.

Zhang, X., Zwiers, F.W., Li, G., Wan, H., and Cannon, A.J. 2017b. Complexity in estimating past and future extreme short-duration rainfall. *Nature Geoscience* **10**: 255–9.

2.3.1.6 Hurricanes

For many years, nearly all climate model output suggested hurricanes (in the Atlantic and Northeastern Pacific Basin, although the terms typhoon, severe cyclonic storm, or tropical cyclone are used elsewhere) should become more frequent and intense as planetary temperatures rise. Scientists worked to improve the temporal histories of these hurricane characteristics for various ocean basins around the world to evaluate the plausibility of such projections. In nearly all instances, the research revealed such trends do not exist.

As a result of these findings, the IPCC revised its conclusion on hurricanes, stating in the Fifth Assessment Report, "there is *low confidence* in long-term (centennial) changes in tropical cyclone activity, after accounting for past changes in observing capabilities ... and there is *low* confidence in attribution of changes in tropical cyclone activity to human influence owing to insufficient observational evidence, lack of physical understanding of the links between anthropogenic drivers of climate and tropical cyclone activity and the low level of agreement between studies as to the relative importance of internal variability, and anthropogenic and natural forcings" (IPCC, 2013, p. 113). Or maybe no relationship exists between the human presence and hurricanes.

The prediction of more frequent or more intense hurricanes is significant due to the destruction they may cause to the environment and homes and businesses in exposed coastal areas. However, much of that damage is likely to be due to continued human migration to coastal areas and not to more hurricanes. According to Pielke *et al.* (2005), by 2050 "for every additional dollar in damage that the Intergovernmental Panel on Climate Change expects to result from the effects of global warming on tropical cyclones, we should expect between $22 and $60 of increase in damage due to population growth and wealth," citing the findings of Pielke *et al.* (2000). They state, "the primary factors that govern the magnitude and patterns of future damages and casualties are how society develops and prepares for storms rather than any presently conceivable future changes in the frequency and intensity of the storms." The authors note many continue to claim a significant hurricane–global warming connection for advocating anthropogenic CO_2 emission reductions that "simply will not be effective with respect to addressing future hurricane impacts," additionally noting "there are much, much better ways to deal with the threat of hurricanes than with energy policies (e.g., Pielke and Pielke, 1997)."

Klotzbach *et al.* (2018) updated earlier research by Pielke *et al.,* this time covering trends in hurricanes making landfall in the continental United States (CONUS) since 1990. They "found no significant trends in landfalling hurricanes, major hurricanes, or normalized damage consistent with what has been found in previous studies." They report hurricane activity is influenced by El Niño–Southern Oscillation on the interannual time scale and by the Atlantic Multidecadal Oscillation on the multidecadal time scale. "Despite a lack of trend in observed CONUS landfalling hurricane activity since 1900," they write, "large increases in inflation-adjusted hurricane-related damage have been observed, especially since the middle part of the twentieth century. We demonstrate that this increase in damage is strongly due to societal factors, namely, increases in population and wealth along the U.S. Gulf and East Coasts."

The Scientific Debate

Many of the first major advances in understanding hurricane genesis, tracks, and cyclical patterns were produced by William W. Gray and his many students and colleagues at Colorado State University, located in Fort Collins, Colorado, USA. Klotzbach *et al.* (2017) described Gray's contributions in a tribute published after Gray's death in 2016:

> Gray pioneered the compositing approach to observational tropical meteorology through assembling of global radiosonde datasets and tropical cyclone research flight data. In the 1970s, he made fundamental contributions to knowledge of convective–larger-scale interactions. Throughout his career, he wrote seminal papers on tropical cyclone structure, cyclogenesis, motion, and seasonal forecasts. His conceptual development of a seasonal genesis parameter also laid an important framework for both seasonal forecasting as well as climate change studies on tropical cyclones. His work was a blend of both observationally based studies and the development of theoretical concepts.

In the 1990s, Gray connected the natural cycles in Atlantic basin hurricane activity with variability in air and sea surface temperatures (SST) (Gray, 1990; Gray *et al.,* 1997), and then to variability in the Atlantic Multidecadal Oscillation (AMO) (Goldenberg *et al.,* 2001). Gray identified the source of AMO variability as natural changes in the thermohaline circulation, the large-scale movement of water through Earth's oceans thought to be propelled by changes in temperatures and density and surface winds (Klotzbach and Gray, 2008). When the AMO is in its warm (positive) phase, conditions are more favorable to the formation of hurricanes as characterized by above-average far north and tropical Atlantic SSTs, below-average tropical Atlantic sea-level pressures, and reduced levels of tropical Atlantic vertical wind shear. When the AMO is in its cool phase, the atmosphere is drier, has more inhibiting vertical wind shear, and cannot sustain deep convection as readily. Figure 2.3.1.6.1 shows this natural variability from 1880 to 2015 (Klotzbach *et al.*, 2015). Recent research by Barcikowska *et al.* (2017) confirms many of Gray's observations, with the authors finding "observational SST and atmospheric circulation records are dominated by an almost 65-yr variability component" and "the recently observed (1970s–2000s) North Atlantic warming and eastern tropical Pacific cooling might presage an ongoing transition to a cold North Atlantic phase with possible implications for near-term global temperature evolution."

Gray did not support the theory that human sulfate aerosols in the atmosphere were masking the effect of CO_2 and warmer temperatures on hurricane genesis. Gray's view on this matter was strenghthened by the CERN cloud particle experiment reported in Section 2.2.4, which found "ion-induced nucleation of pure organic particles constitutes a potentially widespread source of aerosol particles in terrestrial environments with low sulfuric acid pollution" (Kirkby *et al.*, 2016). Instead, Gray argued (consistent with the views of the authors of this section) that climate models are unable to distinguish natural from anthropogenic forcings and misrepresent the role of water vapor feedback. He also also argued that rising temperatures from around 1970 to 2000 were due not to anthropogenic causes but to a long-term weakening in the strength of the Atlantic thermohaline circulation possibly due to solar influences (Gray, 2012).

Sobel *et al.* (2016) summarized the views of climate modelers as follows: "Theory and numerical simulations suggest that human emissions of greenhouse gases, acting on their own, should have already caused a small increase in tropical cyclone (TC) intensities globally. The same theory and simulations indicate that we should not expect to be able to discern this increase in recent historical observations because of the confounding influences of aerosol forcing (which acts to oppose greenhouse gas forcing) and large natural variability (which compromises trend detection). Current expectations for the future are that aerosol forcing will remain level or decrease while greenhouse gas forcing continues to increase, leading to considerable increases in TC intensity as the climate warms further." In light of the CERN experiment and other research on cosmic rays described in Section 2.2.4, this expectation is not justified.

Figure 2.3.1.6.1
Tropical cyclone Atlantic multidecadal variability

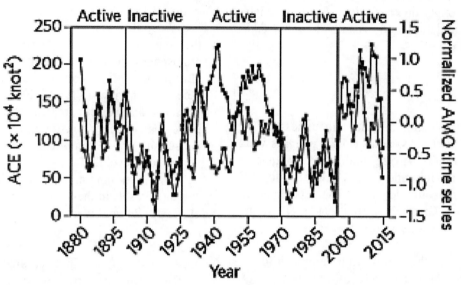

Three-year-averaged accumulated cyclone energy (ACE) in the Atlantic basin (green line) and three-year-averaged standardized normalized Atlantic multidecadal oscillation (AMO) (blue line) from 1880–2014 with predicted value for 2015 (red squares). The 2015 AMO value is the January–June-averaged value. The year listed is the third year being averaged (for example, 1880 is the 1878–1880 average). Correlation between the two time series is 0.61. *Source:* Klotzbach *et al.*, 2015.

Sobel *et al.* (2016) acknowledge "the validity of data from the earlier periods in the longest-term observational data sets has been strongly questioned," "large natural variability, including substantial components with decadal and longer frequencies, further confounds trend detection in records," and "robustly detectable trends in basin-average PI [potential impact] are found only in the North Atlantic, where both surface warming and, to some extent, tropical tropopause cooling have contributed to an increase in PI between 1980 and 2013." Still,

they argue "it would be inappropriate to go on to conclude that there is no human influence on TCs at present. To draw that conclusion would be a type II statistical error, conflating absence of evidence with evidence of absence." But the evidence offered by Gray and others in this section suggests there is indeed "evidence of absence," and hurricane activity is unlikely to be affected by human influences on climate.

Uncertainty in the Hurricane Database

Landsea and Franklin (2013) studied the hurricane database (called HURDAT) used by the National Hurricane Center to report the intensity, central pressure, position, and three measures of radii (size) of Atlantic and eastern North Pacific basin tropical and subtropical cyclones, with the goal of measuring changes in their uncertainty over the past decade. They observe that "given the widespread use of HURDAT for meteorological, engineering, and financial decision making, it is surprising that very little has been published regarding the uncertainties inherent in the database."

Readers may note the similarity between this comment and one made by McLean (2018) reported in Section 2.2.1. McLean examined the HadCRUT4 surface station temperature record and wrote, "It seems very strange that man-made warming has been a major international issue for more than 30 years and yet the fundamental data has never been closely examined." McLean uncovered so many errors and methodological problems with the HadCRUT4 database that it is plainly not suited for scientific research. Landsea and Franklin are only slightly less critical of the HURDAT2 database.

Landsea and Franklin explain "a best track is defined as a subjectively smoothed representation of a tropical cyclone's history over its lifetime, based on a poststorm assessment of all available data." While based on observational data, a best track is a stylized fact, a subjective interpretation of several different and "often contradictory" datasets by a small group of specialists tasked with assigning numbers to an extremely complex and ultimately unknowable set of natural processes. "Because the best tracks are subjectively smoothed," the authors write, "they will not precisely recreate a storm's history, even when that history is known to great accuracy." So how accurate are best tracks?

Landsea and Franklin compared two surveys completed by the specialists employed by the National Hurricane Center, six in 1999 and 10 in 2010, asking them to assign uncertainty values to each of six quantities used to produce a best track. The results of the comparison did show progress in reducing uncertainty in the ten years that passed between surveys, but the amount of uncertainty remaining was surprising. The quantity with the least uncertainty is position, ranging from 7.5% for U.S. landfalling cyclones to 12.5% for satellite-only monitoring. Intensity and central pressure have uncertainties ranging from 17.5% to 20% for satellite-only and 10% to 12.5% for both satellite-aircraft monitoring and at landfall in the United States. For wind radii, the relative uncertainty for cyclones making a U.S. landfall is around 25% to 30%, and for those being observed by satellite only, 35% to 52.5%.

Finally, Landsea and Franklin note the best tracks database "goes back to 1851, but it is far from being complete and accurate for the entire century and a half." As one looks further back in time, "in addition to larger uncertainties, biases become more pronounced as well with tropical cyclone frequencies being underreported and the tropical cyclone intensities being underanalyzed. That is, some storms were missed and many intensities are too low in the preaircraft reconnaissance era (before 1944 for the western half of the basin) and in the presatellite era (before 1972 for the entire basin)."

* * *

Source: Landsea, C.W. and Franklin, J.L. 2013. Atlantic hurricane database uncertainty and presentation of a new database format. *Monthly Weather Review* **141**: 3576–92.

Searching for a Trend

Maue (2011) obtained global TC life cycle data from the IBTrACS database of Knapp *et al.* (2010), which contains six-hourly best-track positions and intensity estimates for the period 1970–2010, from which he calculated the accumulated cyclone energy (ACE) metric (Bell *et al.*, 2000), analogous to the power dissipation index (PDI) used by Emanuel (2005) in his attempt to link hurricanes with global warming. Maue found "in the pentad since 2006, Northern Hemisphere and global tropical cyclone ACE has decreased dramatically to the lowest levels since the late 1970s." He also found "the global frequency of tropical cyclones has reached a historical low." Maue noted "there is no significant linear trend in the frequency of global TCs," in agreement with the analysis of Wang *et al.* (2010). "[T]his current period of record inactivity," as Maue describes it, suggests the long-held contention that global warming increases the frequency and intensity of tropical storms is simply not true. Maue has continuously updated his analysis, with the latest results shown in Figure 2.3.1.6.2.

Villarini *et al.* (2011) used a statistical model developed by Villarini *et al.* (2010), in which "the frequency of North Atlantic tropical storms is modeled by a conditional Poisson distribution with a rate of occurrence parameter that is a function of tropical Atlantic and mean tropical sea surface temperatures (SSTs)," to examine "the impact of different climate models and climate change scenarios on North Atlantic and U.S. landfalling tropical storm activity." The five researchers report their results "do not support the notion of large increases in tropical storm frequency in the North Atlantic basin over the twenty-first century in response to increasing greenhouse gases." They also note "the disagreement among published results concerning increasing or decreasing North Atlantic tropical storm trends in a warmer climate can be largely explained (close to half of the variance) in terms of the different SST projections (Atlantic minus tropical mean) of the different climate model projections."

Figure 2.3.1.6.2

Cyclonic energy, globally and Northern Hemisphere, from 1970 through October 2018

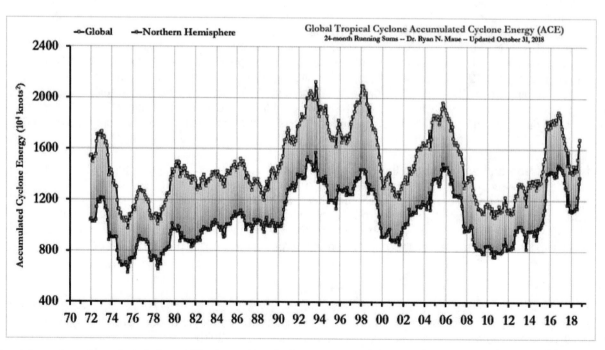

Last four decades of global and Northern Hemisphere Accumulated Cyclone Energy (ACE): 24-month running sums. Note that the year indicated represents the value of ACE through the previous 24 months for the Northern Hemisphere (bottom line/gray boxes) and the entire globe (top line/blue boxes). The area in between represents the Southern Hemisphere total ACE. *Source:* Maue, 2018.

Vecchi and Knutson (2011) conducted an analysis of the characteristics of Atlantic hurricanes whose peak winds exceeded 33 meters/second for the period 1878–2008 based on the HURDAT database, developing a new estimate of the number of hurricanes that occurred in the pre-satellite era (1878–1965) based on analyses of TC storm tracks and the geographical distribution of the tracks of the ships that reported TC encounters. The two researchers report "both the adjusted and unadjusted basin-wide hurricane data indicate the existence of strong interannual and decadal swings." Although "existing records of Atlantic hurricanes show a substantial increase since the late 1800s," their analysis suggests "this increase could have been due to increased observational capability." They write, "after adjusting for an estimated number of 'missed' hurricanes (including hurricanes that likely would have been mis-classified as tropical storms), the secular change since the late-nineteenth century in Atlantic hurricane frequency is nominally negative – though not statistically significant." The two researchers from NOAA's Geophysical Fluid Dynamics Laboratory contend their results "do not support the hypothesis that the warming of the tropical North Atlantic due to anthropogenic greenhouse gas emissions has caused Atlantic hurricane frequency to increase."

Ying *et al.* (2011), working with tropical cyclone best track and related observational severe wind and precipitation datasets created by the Shanghai Typhoon Institute of the China Meteorological Administration, identified trends in observed TC characteristics over the period 1955 to 2006 for the whole of China and four sub-regions. They found over the past half-century there have been changes in the frequency of TC occurrence in only one sub-region, where they determined "years with a high frequency of TC influence have significantly become less common." They also note, "during the past 50 years, there have been no significant trends in the days of TC influence on China" and "the seasonal rhythm of the TC influence on China also has not changed." They found "the maximum sustained winds of TCs affecting the whole of China and all sub-regions have decreasing trends" and "the trends of extreme storm precipitation and 1-hour precipitation were all insignificant." Thus, for the whole of China and essentially all of its component parts, major measures of TC impact have remained constant or slightly decreased.

Sun *et al.* (2011) analyzed data pertaining to TCs for the period 1951–2005 over the northwestern Pacific and South China Sea, obtained from China's Shanghai Typhoon Institute and the National Climate Center of the China Meteorological Administration. They determined the frequency of all TCs affecting China "tended to decrease from 1951 to 2005, with the lowest frequency [occurring] in the past ten years." In addition, the average yearly number of super typhoons was "three in the 1950s and 1960s" but "less than one in the past ten years." They write "the decrease in the frequency of super typhoons, at a rate of 0.4 every ten years, is particularly significant (surpassing the significance test at the 0.01 level)," adding "there is a decreasing trend with the extreme intensity of these TCs during the period of influence in the past 55 years." The authors' findings are shown in Figure 2.3.1.6.3.

Xiao *et al.* (2011) "developed a Tropical Cyclone Potential Impact Index (TCPI) based on the air mass trajectories, disaster information, intensity, duration and frequency of tropical cyclones," using observational data obtained from the China Meteorological Administration's *Yearbook of Tropical (Typhoon) Cyclones in China* for the years 1951–2009 plus the *Annual Climate Impact Assessment* and *Yearbook of Meteorological Disasters* in China, also compiled by the China Meteorological Administration, but for the years 2005–2009. The five researchers report "China's TCPI appears to be a weak decreasing trend over the period [1949–2009], which is not significant overall, but significant in some periods."

Hoarau *et al.* (2012) analyzed intense cyclone activity in the northern Indian Ocean from 1980 to 2009 based on a homogenous reanalysis of satellite imagery. The three French researchers conclude "there has been no trend towards an increase in the number of categories 3–5 cyclones over the last 30 years," noting "the decade from 1990 to 1999 was by far the most active with 11 intense cyclones while 5 intense cyclones formed in each of the other two decades"; i.e., those that preceded and followed the 1990s. They state there has "not been a regular increase in the number of cyclone 'landfalls' over the last three decades (1980–2009)."

Zhao *et al.* (2018) studied TC frequency in the western North Pacific, "the most active basin over the global oceans, experienc[ing] on average about 26 TCs each year, accounting for nearly 1/3 of the global annual total TC counts," during 1979–2014. They note an "abrupt shift" occurred in 1998, after which TC frequency dropped precipitously. The mean annual TC frequency fell from 20 during 1979–1997 to only 15.5 from 1998–2014. "A similar reduc-

Figure 2.3.1.6.3
Declining number of typhoons and frequency of super typhoons affecting China (1951–2005)

A. Annual number of typhoons

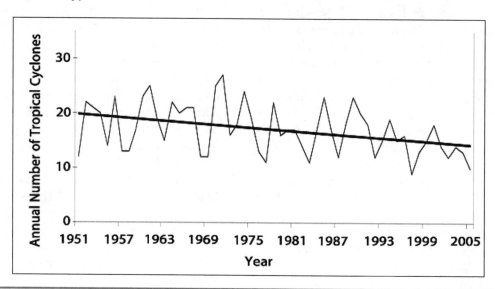

B. Annual number of super typhoons

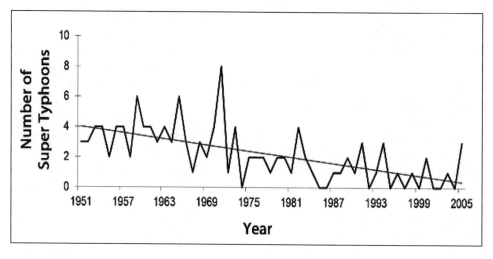

Source: Sun *et al.,* 2011.

tion of TC frequency was observed in the WNP basin and of global TC activity was observed by Liu and Chan (2013) and Maue (2011)."

Before the active 2017 season, the United States had not experienced landfall of a Category 3 or greater hurricane in nearly 12 years, the longest such "hurricane drought" in the United States since the 1860s (Truchelut and Staeling, 2017; Landsea, 2018). Landsea (2015) explained why hurricanes making

landfall on the U.S. coast is representative of global hurricane activity:

Hurricanes striking the continental United States compose a sizable percentage (23%) of all Atlantic basin hurricanes since 1972, the first year for reliable all Atlantic basin hurricane frequency owing to the invention of Dvorak satellite intensity technique

(Dvorak 1975) coupled with available satellite imagery for the basin. The rather lengthy coastline of the United States tends to experience more hurricane strikes in busy seasons, but not every active year causes more U.S. landfalls because of variability in genesis locations and steering flow. The linear correlation coefficient of U.S. hurricanes with all Atlantic basin hurricanes is 0.49 for the years 1972–2014 (statistically significant beyond the 99.8% level after accounting for serial correlation). Thus, while the sample size per season of U.S. hurricanes is substantially smaller than for all Atlantic basin hurricanes, the U.S. hurricane time series reflects some of the same variability as seen in the whole basin.

Commenting on claims by Kunkel *et al.* (2013) that Atlantic basinwide activity had risen in recent years, Landsea (2015) writes, "The long U.S. landfall record is an indication that this recent upward phase of activity in the Atlantic basin was preceded by quiet and active periods of similar magnitude. Furthermore, because of the use of over 100 years of reliable U.S. hurricane records, one can conclude that there has been no long-term century-scale increase in U.S. hurricane frequencies."

References

Barcikowska, M.J., Knutson, T.R., and Zhang, R. 2017. Observed and simulated fingerprints of multidecadal climate variability and their contributions to periods of global SST stagnation. *Journal of Climate* **30**: 721–37.

Bell, G.D., Halpert, M.S., Schnell, R.C., Higgins, R.W., Lawrimore, J., Kousky, V.E., Tinker, R., Thiaw, W., Chelliah, M., and Artusa, A. 2000. Climate assessment for 1999. *Bulletin of the American Meteorology Society* **81**: S1–S50.

Dvorak, V.F. 1975. Tropical cyclone intensity analysis and forecasting from satellite imagery. *Monthly Weather Review* **103**: 420–30.

Emanuel, K. 2005. Increasing destructiveness of tropical cyclones over the past 30 years. *Nature* **436**: 686–8.

Goldenberg, S.B., Landsea, C.W., Mestas-Nuñez, A.M., and Gray, W.M. 2001. The recent increase in Atlantic hurricane activity: causes and implications. *Science* **293**: 474–9.

Gray, W.M. 1990: Strong association between West African rainfall and U.S. landfall of intense hurricanes. *Science* **249**: 1251–6.

Gray, W.M., Sheaffer, J.D., and Landsea, C.W. 1997. Climate trends associated with multidecadal variability of Atlantic hurricane activity. In: Diaz, H.F. and Pulwarty, R.S. (Eds.) *Hurricanes: Climate and Socioeconomic Impacts*. Berlin, Germany: Springer-Verlag, pp. 15–53.

Gray, W.M. 2012. The physical flaws of the global warming theory and deep ocean circulation changes as the primary climate driver. Presentation to the Seventh International Conference on Climate Change. Chicago, IL: The Heartland Institute.

Hoarau, K., Bernard, J., and Chalonge, L. 2012. Intense tropical cyclone activities in the northern Indian Ocean. *International Journal of Climatology* **32**: 1935–45.

IPCC. 2013. Intergovernmental Panel on Climate Change. *Climate Change 2013: The Physical Science Basis*. Contribution of Working Group I to the Fifth Assessment Report of the Intergovernmental Panel on Climate Change. New York, NY: Cambridge University Press.

Kirkby, J., *et al.* 2016. Ion-induced nucleation of pure biogenic particles. *Nature* **533**: 521–6.

Klotzbach, P.J. and Gray, W.M. 2008. Multidecadal variability in North Atlantic tropical cyclone activity. *Journal of Climate* **21**: 3932–5.

Klotzbach, P., Gray, W., and Fogarty, C. 2015. Active Atlantic hurricane era at its end? *Nature Geoscience* **8**: 737–8.

Klotzbach, P.J., *et al.* 2017. The science of William M. Gray: his contributions to the knowledge of tropical meteorology and tropical cyclones. *Bulletin of the American Meteorological Society* **98**.

Klotzbach, P.J., Bowen, S.G., Pielke Jr., R., and Bell, M. 2018. Continental U.S. hurricane landfall frequency and associated damage observations and future risks. *Bulletin of the American Meteorological Society* **99** (7): 1359–77.

Knapp, K.R., Kruk, M.C., Levinson, D.H., Diamond, H.J., and Neumann, C.J. 2010. The International Best Track Archive for Climate Stewardship (IBTrACS): unifying tropical cyclone best track data. *Bulletin of the American Meteorological Society* **91**: 363–76.

Kunkel, K.E., *et al.* 2013. Monitoring and understanding trends in extreme storms: state of knowledge. *Bulletin of the American Meteorological Society* **94**: 499–514.

Landsea, C.W. 2015. Comments on "Monitoring and understanding trends in extreme storms: state of knowledge." *Bulletin of the American Meteorological Society* **96**: 1175–6.

Landsea, C. 2018. What is the complete list of continental U.S. landfalling hurricanes? (website). Hurricane Research Division, Atlantic Oceanographic and Meteorological Laboratory, National Oceanic and Atmospheric Administration. Accessed June 29, 2018.

Landsea, C.W. and Franklin, J.L. 2013. Atlantic hurricane database uncertainty and presentation of a new database format. *Monthly Weather Review* **141**: 3576–92.

Liu, K.S. and Chan, J.C.L. 2013. Inactive period of western North Pacific tropical cyclone activity in 1998–2011. *Journal of Climate* **26**: 2614–30.

Maue, R.N. 2011. Recent historically low global tropical cyclone activity. *Geophysical Research Letters* **38**: 10.1029/2011GL047711.

Maue, R.N. 2018. Global Tropical Cyclone Activity (website). Accessed November 16, 2018.

McLean, J. 2018. *An Audit of the Creation and Content of the HadCRUT4 Temperature Dataset*. Robert Boyle Publishing. October.

Pielke Jr., R.A. and Pielke Sr., R.A. 1997. *Hurricanes: Their Nature and Impacts on Society*. Hoboken, NJ: John Wiley and Sons.

Pielke Jr., R.A., Pielke Sr., R.A., Klein, R., and Sarewitz, D. 2000. Turning the big knob: energy policy as a means to reduce weather impacts. *Energy and Environment* **11**: 255–76.

Pielke Jr., R.A., Landsea, C., Mayfield, M., Laver, J., and Pasch, R. 2005. Hurricanes and global warming. *Bulletin of the American Meteorological Society* **86**: 1571–5.

Sobel, A.H., *et al.* 2016. Human influence on tropical cyclone intensity. *Science* **353** (6296): 242–6.

Sun, L.-h., Ai, W.-x., Song, W.-l., and Wang, Y.-m. 2011. Study on climatic characteristics of China-influencing tropical cyclones. *Journal of Tropical Meteorology* **17**: 181–6.

Truchelut, R.E. and Staehling, E.M. 2017. An energetic perspective on United States tropical cyclone landfall droughts. *Geophysical Research Letters* **44**: 12,013–9, https://doi.org/10.1002/2017GL076071.

Vecchi, G.A. and Knutson, T.R. 2011. Estimating annual numbers of Atlantic hurricanes missing from the HURDAT database (1878–1965) using ship track density. *Journal of Climate* **24**: 1736–46.

Villarini, G., Vecchi, G.A., Knutson, T.R., Zhao, M., and Smith, J.A. 2011. North Atlantic tropical storm frequency response to anthropogenic forcing: projections and sources of uncertainty. *Journal of Climate* **24**: 3224–38.

Villarini, G., Vecchi, G.A., and Smith, J.A. 2010. Modeling of the dependence of tropical storm counts in the North Atlantic basin on climate indices. *Monthly Weather Review* **138**: 2681–705.

Wang, C., Liu, H., and Lee, S.-K. 2010. The record breaking cold temperatures during the winter of 2009/10 in the Northern Hemisphere. *Atmospheric Science Letters* **11**: 161–8.

Xiao, F., Yin, Y., Luo, Y., Song, L., and Ye, D. 2011. Tropical cyclone hazards analysis based on tropical cyclone potential impact index. *Journal of Geographical Sciences* **21**: 791–800.

Ying, M., Yang, Y.-H., Chen, B-D., and Zhang, W. 2011. Climatic variation of tropical cyclones affecting China during the past 50 years. *Science China Earth Sciences* **54**: 10.1007/s11430-011-4213-2.

Zhao, H., Wu, L., and Raga, G.B. 2018. Inter-decadal change of the lagged inter-annual relationship between local sea surface temperature and tropical cyclone activity over the western North Pacific. *Theoretical and Applied Climatology* **134** (1–2): 707–20.

2.3.2 Melting Ice

The Antarctic ice sheet is likely to be unchanged or is gaining ice mass. Antarctic sea ice is gaining in extent, not retreating. Recent trends in the Greenland ice sheet mass and Artic sea ice are not outside natural variability.

According to the Working Group I contribution to the IPCC's Fifth Assessment Report, "over the last two decades, the Greenland and Antarctic ice sheets have been losing mass, glaciers have continued to shrink almost worldwide, and Arctic sea ice and Northern Hemisphere spring snow cover have continued to decrease in extent (*high confidence*)" (IPCC, 2013, p. 9). The IPCC gives estimates of the average rate of net ice loss during the late twentieth and early twenty-first centuries for glaciers around the world, the Greenland and Antarctic ice sheets, and mean Arctic sea ice extent, noting "mean Antarctic sea ice extent *increased* at a rate in the range of 1.2 to 1.8% per decade between 1979 and 2012" (italics added).

A tutorial on the cryosphere – those places on or near Earth's surface so cold that water is present around the year as snow or ice in glaciers, ice sheets, and sea ice – appears in Chapter 5 of *Climate Change Reconsidered II: Physical Science* (NIPCC, 2013) and will not be repeated here. This section focuses narrowly on the IPCC's claim to know with "high confidence" that Antarctica is in fact losing mass and that melting in the Arctic is due to anthropogenic rather than natural forcing. The possible effects of ice melting on sea-level rise are the subject of Section 2.3.3.

Any discussion of the issue must begin by observing that glaciers, ice sheets, and sea ice around the world continuously advance and retreat due to the net difference between accumulation and ablation. Both the acquisition and loss of ice and snow are determined by physical processes that include regional temperature fluctuations, precipitation variability, solar cycles, ocean current cycles, wind, and local geological conditions. Globally, a general pattern exists of ice retreating since the end of the Little Ice Age, a period when many glaciers reached or approached their maximum extents of the Holocene. The melting observed during the modern era precedes any possible anthropogenic forcing and observed melting today is not proof of a human impact on climate.

Complicating this discussion is the fact, well-known among experts in the field but not widely communicated to other researchers or the public, that measurements of mass balance (the net balance between the mass gained by snow deposition – accumulation – and the loss of mass by melting, calving, or other processes – ablation) vary depending on the techniques used. Inference, not direct observation, is required to produce such estimates. Computer models have a particularly poor record of hindcasting ice sheets and sea ice (Rosenblum and Eisenman, 2017). Even if recent trends represent a change from historical patterns, this would not be *prima facie* evidence of an anthropogenic problem. Melting ice produces human and ecological benefits as well as incurring costs, a fact well understood by many millions of people whose supplies of fresh water rely on melting glaciers. No effort has yet been made to weigh those real benefits against the imagined costs.

2.3.2.1 Antarctic Ice Sheet and Sea Ice

We start in Antarctica because it is massively large compared to the Arctic ice sheet, accounting for 90% of the world's ice. Even very small increases in the Antarctic ice sheet mass balance are enough to offset or compensate for melting at the opposite pole.

There are conflicting estimates and no single authoritative measure of Antarctica's mass balance or its changes over geological time or human history. Until recently, most researchers believed Antarctica's ice sheet experienced little net gain or loss since satellite data first became available in 1992 (Tedesco and Monaghan, 2010; Quinn and Ponte, 2010; Zwally and Giovinetto, 2011). More recently, modelers using the IPCC's dubious forecasts of temperature and precipitation and perhaps looking for a human signal find a small net loss by the ice sheet (IMBIE Team, 2018).

Climate models generally predict that a warmer climate would result in more snowfall over Antarctica. This is due to the ability of warmer air (but still below freezing) to transport more moisture across the Antarctic continent. By itself, increased snowfall would increase the Antarctic ice sheet so much it would cause a drop in global sea level of 20 to 43 millimeters (0.8 – 1.7 inches) in 2100 and 73 to 163 mm (2.9 – 6.4 inches) in 2200, compared with today (Ligtenberg *et al.*, 2013). However, models also predict this increase would be more than offset by increases in surface melt, ice discharge, ice-shelf collapses, and ocean-driven melting. One key complication that current climate models failed to account for is the significant geothermal heating beneath the ice sheet (Schroeder *et al.*, 2014) as well as some 138 volcanoes, 91 of which have not been previously identified, recently discovered beneath the West Antarctic (van Wyk de Vries *et al.*, 2018). Due to the complexity of the processes involved, the future and even current ice sheet mass balances of Antarctica are unknown.

While the Antarctic ice sheets were once thought to have been stable over long periods of geological time, this thinking is now known to be incorrect. The editors of *Nature Geoscience* (2018) write, "first came sediment and model evidence that the West Antarctic ice sheet collapsed during previous interglacial periods and under Pliocene warmth. Then came erosional data showing that several regions of the East Antarctic ice sheet also retreated and advanced throughout the Pliocene. An extended record of ice-sheet extent from elsewhere on the East Antarctic coast now paints a more complicated picture of the sensitivity of this ice sheet to warming."

According to the editors of *Nature Geoscience*, Antarctic ice sheet changes occur very slowly. "In terms of immediate sea-level rise, it is reassuring that it seems to require prolonged periods of time lasting hundreds of thousands to millions of years to induce even partial retreat [of the East Antarctic ice sheet]. Nevertheless, we must not take its stability completely for granted – we cannot be sure how the East Antarctic ice sheet will respond to rates of warming that might exceed one to two degrees in a few thousand years."

Antarctic melting is not to be feared even "in a few thousand years" if temperatures in the Antarctic do not rise substantially. Stenni *et al.* (2017) report that their new reconstruction of Antarctic temperature "confirm[s] a significant cooling trend from 0 to 1900 CE (current era) across all Antarctic regions where records extend back into the 1st millennium, with the exception of the Wilkes Land coast and Weddell Sea coast regions. Within this long-term cooling trend from 0–1900 CE we find that the warmest period occurs between 300 and 1000 CE, and the coldest interval from 1200 to 1900 CE. Since 1900 CE, significant warming trends are identified for the West Antarctic Ice Sheet, the Dronning Maud Land coast and the Antarctic Peninsula regions, and these trends are robust across the distribution of records that contribute to the unweighted isotopic composites and also significant in the weighted temperature reconstructions. Only for the Antarctic Peninsula is this most recent century-scale trend unusual in the context of natural variability over the last 2000-years."

Growing evidence suggests temperatures in the region rose in the modern era prior to about 1998 but have since stopped and may now be cooling once more. Carrasco (2013) reported finding a decrease in the warming rate from stations on the western side of the Antarctic Peninsula between 2001 and 2010, as well as a slight cooling trend for King George Island (in the South Shetland Islands just off the peninsula). Similarly, in an analysis of the regional stacked temperature record over the period 1979–2014, Turner *et al.* (2016) reported a switch from warming (1979–1997) to cooling (1999–2014). While warming on the Antarctic Peninsula (typically measured at the Faraday/Vernadsky station) is often cited as proof that Antarctica is warming, temperatures elsewhere on the enormous continent suggest a different story.

More recently, Oliva *et al.* (2017) updated the study by Turner *et al.* (2016) "by presenting an updated assessment of the spatially-distributed temperature trends and interdecadal variability of mean annual air temperature and mean seasonal air temperature from 1950 to 2015, using data from ten stations distributed across the Antarctic Peninsula region." They found the "Faraday/Vernadsky warming trend is an extreme case, circa twice those of the long-term records from other parts of the northern Antarctic Peninsula." They also note the presence of significant decadal-scale variability among the 10 temperature records, which they linked to large-scale atmospheric phenomena such as ENSO, the Pacific Decadal Oscillation, and the Southern Annular Mode. Perhaps most important is their confirmation that "from 1998 onward, a turning point has been observed in the evolution of mean annual air temperatures across the Antarctic Peninsula region, changing from a warming to a cooling trend." This cooling has amounted to a 0.5° to 0.9°C decrease in temperatures in most of the Antarctic Peninsula region, the only exception being three stations located in the southwest sector of the peninsula that experienced a slight delay in their thermal turning point, declining only over the shorter period of the past decade. Oliva *et al.* (2017) cite independent evidence from multiple other sources in support of the recent cooling detected in their analysis, including an "increase in the extent of sea ice, positive mass-balance of peripheral glaciers and thinning of the active layer of permafrost."

Colder temperatures in Antarctica appear to have halted net melting on the continent. Lovell *et al.* (2017) set out to determine the temporal changes in the glacial terminus positions of 135 outlet glaciers (91 marine- and 44 land-terminating) spanning approximately 1,000 kilometers across three major drainage basins along the coastline of East Antarctica (Victoria Land, Oates Land, and George V Land). This was accomplished by comparing terminus position changes in seven satellite images over the period 1972–2013. In describing their findings, Lovell *et al.* write, "between 1972 and 2013, 36% of glacier termini in the entire study area advanced and 25% of glacier termini retreated, with the remainder showing no discernible change outside of the measurement error (± 66 m or ± 1.6 m yr^{-1}) and classified as 'no change.'" Although there were some regional differences in glacier termini changes, the authors found no correlation with those changes and changes in air temperature or sea ice trends. Instead, they write, "sub-decadal glacier terminus variations in these regions over the last four decades were more closely linked to non-climatic drivers, such as terminus type and geometry, than any obvious climatic or oceanic forcing."

Similarly, Fountain *et al.* (2017) analyzed changes in glacier extent along the western Ross Sea in Antarctica over the past 60 years. The authors used digital scans of paper maps based on aerial imagery acquired by the U.S. Geological Survey, along with modern-day satellite imagery from a variety of platforms, to calculate changes in the terminus positions, ice speed, calving rates, and ice front advance and retreat rates from 34 glaciers in this region over the period 1955–2015. The authors report "no significant spatial or temporal patterns of terminus position, flow speed, or calving emerged, implying that the conditions associated with ice tongue stability are unchanged," at least over the past six decades. However, they also report "the net change for all the glaciers, weighted by glacier width at the grounding line, has been [one of] *advance*" (italics added) with an average rate of increase of +12 $\pm$ 88 m yr^{-1}. Over a period during which the bulk of the modern rise in atmospheric CO_2 has occurred, not only have the majority of glaciers from this large region of Antarctica not retreated, they have collectively grown.

Engel *et al.* (2018) analyzed surface mass-balance records from two glaciers on James Ross Island, located off the northeastern edge of the Antarctic Peninsula. The first glacier, Whisky Glacier, is a land-terminating valley glacier, while the second, Davies Dome, is an ice dome. According to the authors, "because of their small volume, these glaciers are expected to have a relatively fast dynamic response to climatic oscillations and their mass balance is also considered to be a sensitive climate indicator," citing the work of Allen *et al.* (2008). The researchers found that over the period of study (2009–2015), Davis Dome and Whisky Glacier experienced cumulative mass *gains* of 0.11 $\pm$ 0.37 and 0.57 $\pm$ 0.67 meters of water equivalent, respectively; their annual surface mass balances were *positive* in every year except 2011/2012.

Engel *et al.* (2018) write their findings "indicate a change from surface mass loss that prevailed in the region during the first decade of the 21st century to predominantly positive surface mass balance after 2009/2010." They also note the positive mass balances observed on Davis Dome and Whisky Glacier "coincide with the surface mass-balance records from Bahía del Diablo Glacier on nearby Vega Island, Bellingshausen Ice Dome on King George Island and Hurd and Johnsons glaciers on Livingston Island," which records reveal "a regional change from a predominantly negative surface mass balance in the first decade of the 21st century to a positive balance over the 2009–2015 period." Their findings appear in Figure 2.3.2.1.1. The authors also noted "a significant decrease in the warming rates reported from the northern Antarctic Peninsula since the end of the 20th century" which "is also consistent with the regional trend of climate cooling on the eastern side of the Antarctic Peninsula."

Moving from the Antarctic ice sheet to the *sea ice* surrounding the continent, Comiso *et al.* (2017) report "the Antarctic sea ice extent has been slowly increasing contrary to expected trends due to global warming and results from coupled climate models." They note record high levels of sea ice extent were reported in 2012 and 2014, and "the positive trend is confirmed with newly reprocessed sea ice data that addressed inconsistency issues in the time series." The authors produce a new sea ice record "to show that the positive trend in sea ice extent is real using an updated and enhanced version of the sea ice data" in response to concerns expressed by Eisenman *et al.* (2014) that the trend might be an artifact of inconsistency in the processing of data before and after January 1992. That problem was fixed when the entire dataset was reprocessed, as reported by Comiso and Nishio (2008). To further improve the dataset, Comiso *et al.* (2017) correct inconsistencies among sensors, "the tie point for open water was made dynamic; and the threshold for the lower limit for ice was relaxed to allow retrieval of ice at 10% ice concentration. Further adjustments in brightness temperature T_B were made to improve consistency in the retrieval of ice concentration, ice extent, and ice area from the different sensors." After making these improvements, the authors report a positive trend of +19.9 $\pm$ 2.0 10^3 km^2/year, or +1.7 $\pm$ 0.2% / decade. The results of their new analysis are shown in Figure 2.3.2.1.2.

Comiso *et al.* (2017) write, "The positive trend, however, should not be regarded as unexpected despite global warming and the strong negative trend in the Arctic ice cover because the distribution of global surface temperature trend is not uniform. In the Antarctic region the trend in surface temperature is about 0.1°C decade^{-1} while the trend is 0.6°C decade^{-1} in the Arctic and 0.2°C decade^{-1} globally since 1981." Actually, the satellite record shows a global temperature trend from 1979 to 2016 of only 0.1°C decade^{-1}, just half of their estimate (Christy *et al.*, 2018), which makes the increasing extent of Southern Hemispheric sea ice even more expected.

Figure 2.3.2.1.1
Surface mass-balance records for glaciers around the northern Antarctic Peninsula

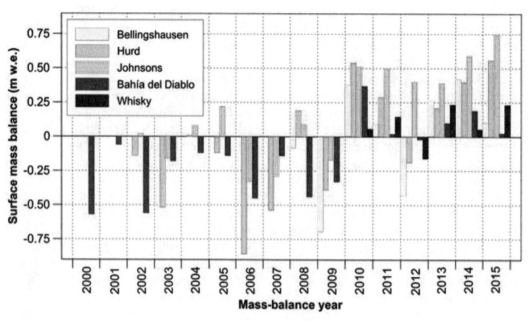

Source: Engel *et al.,* 2018.

Figure 2.3.2.1.2
Monthly averages of the Antarctic sea ice extent, November 1978 to December 2015

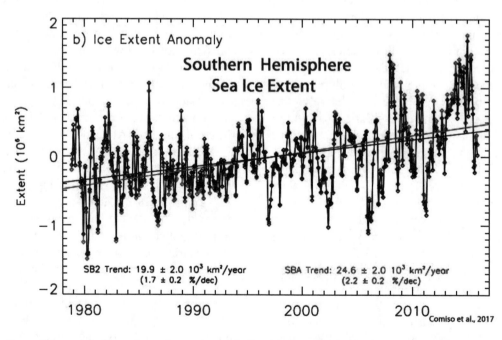

Time series of monthly anomalies of sea ice extents derived using the newly enhanced SB2 data (black) and the older SBA data (red) from November 1978 to December 2015. The trend lines using SB2 and SBA data are also shown and the trend values with statistical errors are provided. Adapted from Comiso *et al.*, 2017, Figure 3b.

Purich *et al.* (2018) believe the expansion of Antarctic sea ice can be explained by the addition of fresh water to the Southern Ocean surface, a process called "surface freshening." "The majority of CMIP5 models underestimate or fail to capture this historical surface freshening," they write, "yet little is known about the impact of this model bias on regional ocean circulation and hydrography." The authors use GCMs to model the addition of freshwater to the Southern Ocean and find it "causes a surface cooling and sea ice increase under preindustrial conditions, because of a reduction in ocean convection and weakened entrainment of warm subsurface waters into the surface ocean."

* * *

Despite climate model predictions, the Antarctic ice sheet is likely to be unchanged or is gaining ice mass while Antarctic sea ice is gaining in extent, not retreating. A long regional cooling trend in surface temperatures appears to have ended around the beginning of the twentieth century, probably due to internal variability unrelated to the human presence. The subsequent warming trend may have ended in the past decade.

References

Allen, R.J., Siegert, M.J., and Payne, T. 2008. Reconstructing glacier-based climates of LGM Europe and Russia - Part 1: numerical modelling and validation methods. *Climate of the Past* **4**: 235–48.

Carrasco, J.F. 2013. Decadal changes in the near-surface air temperature in the western side of the Antarctic Peninsula. *Atmospheric and Climate Sciences* **3**: 275–81.

Christy, J., *et al.* 2018. Examination of space-based bulk atmospheric temperatures used in climate research. *International Journal of Remote Sensing* **39** (11): 3580–607.

Comiso, J.C. and Nishio, F. 2008: Trends in the sea ice cover using enhanced and compatible AMSR-E, SSM/I, and SMMR data. *Journal of Geophysical Research* **113**: C02S07.

Comiso, J., Gerstena, R.A., Stock, L.V., Turner, J., Perez, G.J., and Cho, K. 2017. Positive trend in the antarctic sea ice cover and associated changes in surface temperature. *Journal of Climate* **30** (6): 2251–67.

Eisenman, I., Meier, W.N., and Norris, J.R. 2014. A spurious jump in the satellite record: has Antarctic sea ice expansion been overestimated? *Cryosphere* **8**: 1289–96.

Engel, Z., Láska, K., Nývlt, D., and Stachoň, Z. 2018. Surface mass balance of small glaciers on James Ross Island, north-eastern Antarctic Peninsula, during 2009–2015. *Journal of Glaciology* **64**: 349–61.

Fountain, A.G., Glenn, B., and Scambos, T.A. 2017. The changing extent of the glaciers along the western Ross Sea, Antarctica. *Geology* **45**: 927–30.

IMBIE Team. 2018. Mass balance of the Antarctic Ice Sheet from 1992-2017. *Nature* **558**: 219–22.

IPCC. 2013. Intergovernmental Panel on Climate Change. *Climate Change 2013: The Physical Science Basis.* Contribution of Working Group I to the Fifth Assessment Report of the Intergovernmental Panel on Climate Change. New York, NY: Cambridge University Press.

Ligtenberg, S.R.M., Berg, W.J., Broeke, M.R., Rae, J.G.L., and Meijgaard, E. 2013. Future surface mass balance of the Antarctic ice sheet and its influence on sea level change, simulated by a regional atmospheric climate model. *Climate Dynamics* **41**: 867–84.

Lovell, A.M., Stokes, C.R., and Jamieson, S.S.R. 2017. Sub-decadal variations in outlet glacier terminus positions in Victoria Land, Oates Land and George V Land, East Antarctica (1972–2013). *Antarctic Science* **29**: 468–83.

Nature Geoscience. 2018. A history of instability. Editorial. *Nature Geoscience* **11** (83).

NIPCC. 2013. Nongovernmental International Panel on Climate Change. Idso, C.D., Carter, R.M., and Singer, S.F. (Eds.) *Climate Change Reconsidered: Physical Science.* Chicago, IL: The Heartland Institute.

Oliva, M., *et al.* 2017. Recent regional climate cooling on the Antarctic Peninsula and associated impacts on the cryosphere. *Science of the Total Environment* **580**: 210–23.

Purich, A., England, M.H., Caia, W., Sullivan, A., and Durack, P.J. 2018. Impacts of broad-scale surface freshening of the southern ocean in a coupled climate model. *Journal of Climate* **31**: 2613–32.

Quinn, K.J. and Ponte, R.M. 2010. Uncertainty in ocean mass trends from GRACE. *Geophysical Journal International* **181**: 762–8.

Rosenblum, E. and Eisenman, I. 2017. Sea ice trends in climate models only accurate in runs with biased global warming. *Journal of Climate* **30**: 6265–78.

Stenni, B., *et al.* 2017. Antarctic climate variability on regional and continental scales over the last 2,000 years. *Climate of the Past* **13** (11): 1609–34.

Tedesco, M. and Monaghan, A.J. 2010. Climate and melting variability in Antarctica. *EOS, Transactions, American Geophysical Union* **91**: 1–2.

Turner, J., *et al.* 2016. Absence of 21st century warming on Antarctic Peninsula consistent with natural variability. *Nature* **535**.

Zwally, H.J. and Giovinetto, M.B. 2011. Overview and assessment of Antarctic Ice-Sheet mass balance estimates: 1992–2009. *Surveys in Geophysics* **32**: 351–76.

2.3.2.2 Arctic Ice Sheet and Sea Ice

According to Kryk *et al.* (2017), "Greenland is the world's largest, non-continental island located between latitudes 59 ° and 83 ° N, and longitudes 11 ° and 74 ° W. Greenland borders with Atlantic Ocean to the East, with the Arctic Ocean to the North and Baffin Bay to the West. Three-quarters of Greenland is solely covered by the permanent ice sheet." Models cited by the IPCC find "the average rate of ice loss from the Greenland ice sheet has *very likely* substantially increased from 34 [-6 to 74] Gt yr^{-1} over the period 1992 to 2001 to 215 [157 to 274] Gt yr^{-1} over the period 2002 to 2011" (IPCC, 2013, p. 9). Also according to the IPCC, "the annual mean Arctic sea ice extent decreased over the period 1979 to 2012 with a rate that was *very likely* in the range of 3.5 to 4.1% per decade. ..."

While it is easy to find alarming accounts of rising temperatures and melting ice in the Arctic region even in respected science journals (e.g., Lang *et al.*, 2017), such accounts neglect to report natural variability in the historical and geological record against which recent trends must be compared. MacDonald *et al.* (2000) used radiocarbon-dated macrofossils to document how "Over most of Russia, forest advanced to or near the current arctic coastline between 9000 and 7000 yr B.P. (before present) and retreated to its present position by between 4000 and 3000 yr B.P. ... During the period of maximum forest extension, the mean July temperatures along the northern coastline of Russia may have been 2.5° to 7.0°C warmer than modern. The development of forest and expansion of treeline likely reflects a number of complimentary environmental conditions, including heightened summer insolation, the demise

of Eurasian ice sheets, reduced sea-ice cover, greater continentality with eustatically lower sea level, and extreme Arctic penetration of warm North Atlantic waters."

Miller *et al.* (2005) summarized the main characteristics of the glacial and climatic history of the Canadian Arctic's Baffin Island since the Last Glacial Maximum by presenting biotic and physical proxy climate data derived from six lacustrine sediment cores recovered from four sites on Baffin Island. This work revealed that "glaciers throughout the Canadian Arctic show clear evidence of Little Ice Age expansion, persisting until the late 1800s, followed by variable recession over the past century." They also report that wherever the Little Ice Age advance can be compared to earlier advances, "the Little Ice Age is the most extensive Late Holocene advance," and "some glaciers remain at their Little Ice Age maximum." Since the Little Ice Age in the Canadian Arctic spawned the region's most extensive glacial advances of the entire Holocene, it is only to be *expected* that the region should be experiencing significant melting as the planet recovers from that historic cold era.

Also working with sediment cores recovered from three mid-Arctic lakes on the Cumberland Peninsula of eastern Baffin Island, Frechette *et al.* (2006) employed radiocarbon dating of macrofossils contained in the sediment, together with luminescence dating, to isolate and study the portions of the cores pertaining to the interglacial that preceded the Holocene, which occurred approximately 117,000 to 130,000 years ago, reconstructing the past vegetation and climate of the region during this period based on pollen spectra derived from the cores. This work revealed that "in each core," as they describe it, "last interglacial sediments yielded remarkably high pollen concentrations, and included far greater percentages of shrub (*Betula* and *Alnus*) pollen grains than did overlying Holocene sediments." They then infer "July air temperatures of the last interglacial to have been 4 to 5°C warmer than present on eastern Baffin Island." This clearly reveals that Arctic region temperatures today are not unprecedented. In a companion study, Francis *et al.* (2006) estimated "summer temperatures during the last interglacial were higher than at any time in the Holocene, and 5 to 10°C higher than present."

A major review of the literature conducted in 2006 (CAPE-Last Interglacial Project Members, 2006) reported "quantitative reconstructions of LIG [Last Interglaciation] summer temperatures suggest that much of the Arctic was 5°C warmer during the LIG than at present." With respect to the impacts of this warmth, they note Arctic summers of the LIG "were warm enough to melt all glaciers below 5 km elevation except the Greenland Ice Sheet, which was reduced by ca 20–50% (Cuffey and Marshall, 2000; Otto-Bliesner *et al.*, 2006)." In addition, they write "the margins of permanent Arctic Ocean sea ice retracted well into the Arctic Ocean basin and boreal forests advanced to the Arctic Ocean coast across vast regions of the Arctic currently occupied by tundra."

Frauenfeld *et al.* (2011) used the known close correlations of total annual observed melt extent across the Greenland ice sheet, summer temperature measurements from stations located along Greenland's coast, and variations in atmospheric circulation across the North Atlantic to create a "near-continuous 226-year reconstructed history of annual Greenland melt extent dating from 2009 back into the late eighteenth century." Their graph of the record appears as Figure 2.3.2.2.1. The researchers found "the recent period of high-melt extent is similar in magnitude but, thus far, shorter in duration, than a period of high melt lasting from the early 1920s through the early 1960s. The greatest melt

extent over the last 2-1/4 centuries occurred in 2007; however, this value is not statistically significantly different from the reconstructed melt extent during 20 other melt seasons, primarily during 1923–1961." Similarly, Bjørk *et al.* (2018) found the rates at which Greenland's peripheral glaciers are currently retreating were exceeded during the "early twentieth century post-Little-Ice-Age retreat."

Vasskog *et al.* (2015) provides a summary of what is (and is not) presently known about the history of the Greenland ice sheet (GrIS) over the previous glacial-interglacial cycle. The authors find that during the last interglacial period (130–116 ka BP), global temperatures were 1.5°–2.0°C warmer than the peak warmth of the present interglacial, or Holocene, in which we are now living. They estimate the GrIS was "probably between ~7[%] and 60% smaller than at present," and that melting contributed to a rise in global sea level of "between 0.5 and 4.2 m." Comparing the present interglacial to the past interglacial, atmospheric CO_2 concentrations are currently 30% higher yet global temperatures are 1.5°–2°C cooler, GrIS volume is from 7% to 60% larger, and global sea level is at least 0.5-4.2 m lower, none of which observations signal catastrophe for the present.

Figure 2.3.2.2.1
Reconstructed history of the total ice melt extent index over Greenland, 1784–2009

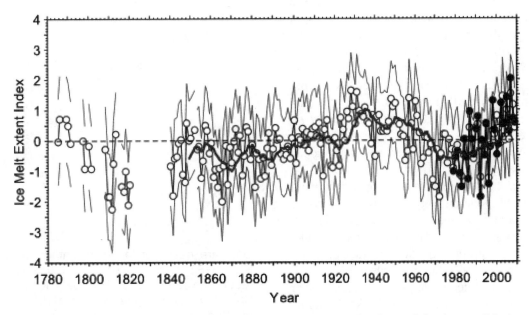

Observed values of the ice melt index (blue solid circles), reconstructed values of the ice melt index (gray open circles), the 10-year trailing moving average through the reconstructed and fitted values (thick red line), and the 95% upper and lower confidence bounds (thin gray lines). Source: Frauenfeld *et al.*, 2011, Figure 2.

Lusas *et al.* (2017) observe, "Prediction of future Arctic climate and environmental changes, as well as associated ice-sheet behavior, requires placing present-day warming and reduced ice extent into a long-term context." Using calibrated radiocarbon dating of organic remains in small lakes near the Istorvet ice cap in East Greenland, they discover "deglaciation of the region before ~10,500 years BP, after which time the ice cap receded rapidly to a position similar to or less extensive than present." The record "suggests that the ice cap was similar to or smaller than present throughout most of the Holocene. This restricted ice extent suggests that climate was similar to or warmer than present, in keeping with other records from Greenland that indicate a warm early and middle Holocene."

Mangerud and Svendsen (2017) observe that remains of shallow marine mollusks that are today extinct close to Svalbard, a Norwegian group of islands located in the Arctic Ocean north of continental Norway, are found in deposits there dating to the early Holocene. The presence of "the most warmth-demanding species found, *Zirfaea crispata,*" indicates August temperatures on Svalbard were 6°C warmer at around 10,200 – 9,200 YBP, when this species lived there. Another mussel, *Mytilus edulis,* returned to Svalbard in 2004 "following recent warming, and after almost 4000 years of absence, excluding a short re-appearance during the Medieval Warm Period 900 years ago." Based on their study of these mollusk remains, the authors conclude "a gradual cooling brought temperatures to the present level at about 4.5 cal. ka BP. The warm early-Holocene climate around Svalbard was driven primarily by higher insolation and greater influx of warm Atlantic water, but feedback processes further influenced the regional climate." These findings, like those of Kryk *et al.* (2017), make it clear that today's temperatures in this part of the Arctic are not unprecedented and indeed are cool compared to past periods.

Hofer *et al.* (2017) observe that the loss of mass by the GrIS has generally been attributed to rising temperatures and a decrease in surface albedo, but "we show, using satellite data and climate model output, that the abrupt reduction in surface mass balance since about 1995 can be attributed largely to a coincident trend of decreasing summer cloud cover enhancing the melt-albedo feedback. Satellite observations show that, from 1995 to 2009, summer cloud cover decreased by $0.9 \pm 0.3\%$ per year. Model output indicates that the GrIS summer melt increases by 27 ± 13 gigatons (Gt) per percent reduction in summer cloud cover, principally because of the impact of increased shortwave radiation over the low albedo ablation zone." The authors attribute the reduction in cloud cover to "a state shift in the North Atlantic Oscillation promoting anticyclonic conditions in summer [which] suggests that the enhanced surface mass loss from the GrIS is driven by synoptic-scale changes in Arctic-wide atmospheric circulation," not anthropogenic forcing.

As with changing views about the Antarctic ice sheet, past estimates of Greenland's mass balance and changes to the same showed little net loss and possibly small gains (Johannessen *et al.,* 2005; Zwally *et al.,* 2011; Jezek, 2012). More recently, computer models as well as satellite data appear to show a steady loss of mass from the Greenland ice sheet between 2002 and 2017, as shown in Figure 2.3.2.2.2. However, "Glacier area measurements from LANDSAT and ASTER, available since 1999 for 45 of the widest and fastest-flowing marine-terminating glaciers, reveal a pattern of continued relative stability since 2012–13" (Box *et al.,* 2018).

Figure 2.3.2.2.2
Change in total mass (Gt) of the Greenland ice sheet from GRACE satellite measurements, 2002–2017

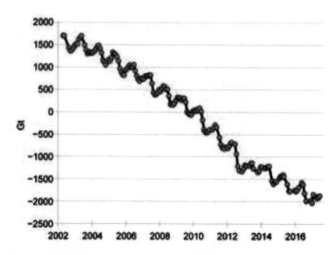

Data are based on an unweighted average of JPL RL05, GFZ RL05, and the CSR RL05 solutions, which reduce noise in the GRACE data for 2017. *Source:* Box *et al.,* 2018, Figure 5.12, p. S154, citing Sasgen *et al.,* 2012.

Turning from the GrIS to Arctic *sea ice*, Darby *et al.* (2001) developed a 10,000-year multi-parameter environmental record from a thick sequence of post-glacial sediments obtained from cores extracted from the upper continental slope off the Chukchi Sea Shelf in the Arctic Ocean. They uncovered "previously unrecognized millennial-scale variability in Arctic Ocean circulation and climate" along with evidence suggesting "in the recent past, the western Arctic Ocean was much warmer than it is today." More specifically, they write, "during the middle Holocene the August sea surface temperature fluctuated by 5°C and was 3–7°C warmer than it is today," and they report their data reveal "rapid and large (1–2°C) shifts in bottom water temperature," concluding that "Holocene variability in the western Arctic is larger than any change observed in this area over the last century."

Van Kooten (2013, pp. 232–3) observes, "Historically, the Arctic is characterized by warm periods when there were open seas and the Arctic sea ice did not extend very far to the south. Ships' logs identify ice-free passages during the warm periods of 1690–1710, 1750–1780 and 1918–1940, although each of these warm periods was generally preceded and followed by colder temperatures, severe ice conditions and maximum southward extent of the ice (e.g., during 1630–1660 and 1790–1830)." Van Kooten continues, "there must have been little ice in the Davis Strait west of Greenland as the Vikings

Kryk *et al.* (2017) note "Arctic temperatures are very variable, making it difficult to identify long-term trends, particularly on a regional scale," and "only until recently, the area of the North Atlantic, including SW Greenland region, was one of the few areas in the world where cooling was observed, however in the period 1979–2005 the trend reversed and strong warming was observed." The authors use changes in diatom species composition in Godthåbsfjord region, SW Greenland, to create a reconstruction of sea surface temperature and sea ice concentration (SIC) from 1600–2010 showing, among other things, that current temperatures and SIC are not at all unprecedented. Their findings are reproduced as Figure 2.3.2.2.3. See also Werner *et al.* (2017) for a review of four Arctic summer temperature reconstructions and the authors' original reconstruction, all showing temperature peaks higher than those observed in recent years.

Figure 2.3.2.2.3
Southwest Greenland sea surface temperatures and sea ice concentration, 1600–2010

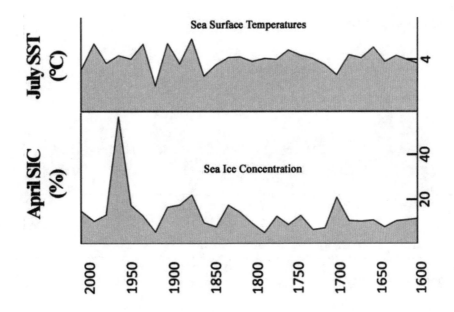

SST = sea surface temperature, SIC = sea ice concentration. *Source:* Adapted from Kryk *et al.*, 2017, Figure 3.

The melting trend recorded by satellites since 1979 was preceded by a period of expanding ice extent from 1943 to about 1970 (e.g., Suo *et al.,* 2013), so looking only at the record since 1979 exaggerates the appearance of unusual melting. Not until ~2005 did the recent melting period reach the low point previously reached in 1943. Given uncertainties in measurement and changes in technology, it is possible today's sea ice extent is still not the lowest since 1900. See, for example, the Arctic sea ice dataset created by Connolly *et al.* (2017) reproduced as Figure 2.3.2.2.4.

Commenting on their findings, Connolly *et al.* (2017) write, "if we also consider the full envelope of the associated confidence intervals, we cannot rule out the possibility that similarly low sea ice extents occurred during the 20th century. That is, the upper bounds of the estimates for all years since 2004 are still greater than the lower bounds for several years in the early 20th century." They also note "this late-1970s reversal in sea ice trends was *not* captured by the hindcasts of the recent CMIP5 climate models used for the latest IPCC reports, which suggests that current climate models are still quite poor at modelling past sea ice trends."

Slawinska and Robock (2018) used the Community Earth System Model to study the impacts of volcanic perturbations and the phase of the Atlantic Multidecadal Oscillation on the extent of sea ice and other climate indices. The authors show "at least in the Last Millennium Ensemble, volcanic eruptions are followed by a decadal-scale positive response of the Atlantic multidecadal overturning circulation, followed by a centennial-scale enhancement of the Northern Hemispheric sea ice extent. It is hypothesized that a few mechanisms, not just one, may have to play a role in consistently explaining such a simulated climate response at both decadal and centennial time scales." Of particular relevance to the topic being addressed here, the authors contend "prolonged fluctuations in solar irradiance associated with solar minima potentially amplify the enhancement of the magnitude of volcanically triggered anomalies of Arctic sea ice extent." As such natural sources of variability are better understood, the possible role of anthropogenic greenhouse gases in sea ice extent necessarily becomes smaller.

* * *

The Greenland ice sheet and neighboring glaciers and sea ice vary in mass and extent over time due to natural forces unrelated to the human presence. Recent temperatures and melting are not outside the range of natural variability found even in the past century. If there is anything strange or unusual about

Figure 2.3.2.2.4
Annual Arctic sea ice extent trends from 1900 to 2017

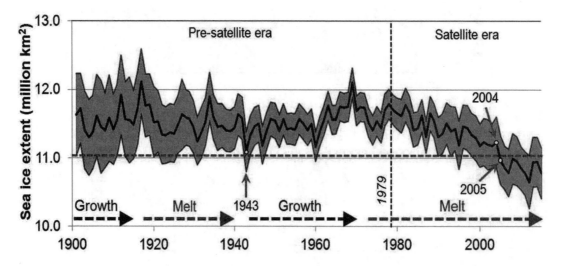

Periods of net sea ice growth and melt are indicated at the bottom of the figure. Satellite data became available in 1979. *Source:* Connolly *et al.,* 2017.

current Arctic temperatures it is that they are *lower* than what they were during the maximum warmth of the current interglacial and, even more so, the prior interglacial. If the Arctic behaves anything like the Antarctic in this regard, one can extend this comparison back in time through *three more* interglacials, all of which were also warmer than the current one (Petit *et al.*, 1999; Augustin *et al.*, 2004). Furthermore, it should be evident that the region's current life forms, including most notably polar bears, fared just fine during these much warmer interglacials, or else they would not be here today.

References

Augustin, L., *et al.* 2004. Eight glacial cycles from an Antarctic ice core. *Nature* **429**: 623–8.

Bjørk, A.A., et al. 2018. Changes in Greenland's peripheral glaciers linked to the North Atlantic Oscillation. *Nature Climate Change* **8**: 48–52.

Box, J.E., *et al.* 2018. Greenland Ice sheet. In: State of the climate in 2017. Special Supplement. *Bulletin of the American Meteorological Society* **99** (8).

CAPE-Last Interglacial Project Members. 2006. Last Interglacial Arctic warmth confirms polar amplification of climate change. *Quaternary Science Reviews* **25**: 1383–1400.

Connolly, R., Connolly, M., and Soon, W. 2017. Re-calibration of Arctic sea ice extent datasets using Arctic surface air temperature records. *Hydrological Sciences Journal* **62** (8).

Cuffey, K.M. and Marshall, S.J. 2000. Substantial contribution to sea-level rise during the last interglacial from the Greenland ice sheet. *Nature* **404**: 591–4.

Darby, D., Bischof, J., Cutter, G., de Vernal, A., Hillaire-Marcel, C., Dwyer, G., McManus, J., Osterman, L., Polyak, L., and Poore, R. 2001. New record shows pronounced changes in Arctic Ocean circulation and climate. *EOS, Transactions, American Geophysical Union* **82**: 601, 607.

Francis, D.R., Wolfe, A.P., Walker, I.R., and Miller, G.H. 2006. Interglacial and Holocene temperature reconstructions based on midge remains in sediments of two lakes from Baffin Island, Nunavut, Arctic Canada. *Palaeogeography, Palaeoclimatology, Palaeoecology* **236**: 107–24.

Frauenfeld, O.W., Knappenberger, P.C., and Michaels, P.J. 2011. A reconstruction of annual Greenland ice melt extent, 1784–2009. *Journal of Geophysical Research* **116**.

Frechette, B., Wolfe, A.P., Miller, G.H., Richard, P.J.H., and de Vernal, A. 2006. Vegetation and climate of the last interglacial on Baffin Island, Arctic Canada. *Palaeogeography, Palaeoclimatology, Palaeoecology* **236**: 91–106.

Hofer, S., Tedstone, A.J., Fettweis, X., and Bamber, J.L. 2017. Decreasing cloud cover drives the recent mass loss on the Greenland Ice Sheet. *Science Advances* **3** (6): e1700584.

IPCC. 2013. Intergovernmental Panel on Climate Change. *Climate Change 2013: The Physical Science Basis.* Contribution of Working Group I to the Fifth Assessment Report of the Intergovernmental Panel on Climate Change. New York, NY: Cambridge University Press.

Jezek, K.C. 2012. Surface elevation and velocity changes on the south-central Greenland ice sheet: 1980–2011. *Journal of Glaciology* **58**: 1201–11.

Johannessen, O.M., Khvorostovsky, K., Miles, M.W., and Bobylev, L.P. 2005. Recent ice-sheet growth in the interior of Greenland. *Science* **310**: 1013–6.

Kryk, A., Krawczyk, D.W., Seidenkrantz, M.-S., and Witkowski, A. 2017. Holocene oceanographic variation in the Godthåbsfjord region, SW Greenland, using diatom proxy. Presented to the 11th International Phycological Congress, University of Szczecin, Szczecin, Poland. August 13–19.

Lang, A., Yang, S., and Kaas, E. 2017. Sea ice thickness and recent Arctic warming. *Geophysical Research Letters* **44** (1): 409–18.

Lusas, A.R., Hall, B.L., Lowell, T.V., Kelly, M.A., Bennike, O., Levy, L.B., and Honsaker, W. 2017. Holocene climate and environmental history of East Greenland inferred from lake sediments. *Journal of Paleolimnology* **57** (4): 321–41.

MacDonald, G.M., *et al.* 2000. Holocene treeline history and climate change across northern Eurasia. *Quaternary Research* **53** (3): 302–11.

Mangerud, J. And Svendsen, J.I. 2017. The Holocene Thermal Maximum around Svalbard, Arctic North Atlantic; molluscs show early and exceptional warmth. *The Holocene* **28** (1).

Miller, G.H., Wolfe, A.P., Briner, J.P., Sauer, P.E., and Nesje, A. 2005. Holocene glaciation and climate evolution of Baffin Island, Arctic Canada. *Quaternary Science Reviews* **24**: 1703–21.

Otto-Bliesner, B.L., Marshall, S.J., Overpeck, J.T., Miller, G.H., Hu, A., and CAPE Last Interglacial Project members. 2006. Simulating Arctic climate warmth and icefield retreat in the last interglaciation. *Science* **311**: 1751–3.

Petit, J.R., *et al.* 1999. Climate and atmospheric history of the past 420,000 years from the Vostok ice core, Antarctica. *Nature* **399**: 429–36.

Sasgen, I., *et al.* 2012. Timing and origin of recent regional ice-mass loss in Greenland. *Earth and Planetary Science Letters* **333–334**: 293–303.

Slawinska, J. and Robock, A. 2018. Impact of volcanic eruptions on decadal to centennial fluctuations of Arctic Sea ice extent during the last millennium and on initiation of the Little Ice Age. *Journal of Climate* **31**: 2145–67.

Suo, L., *et al.* 2013. External forcing of the early 20th century Arctic warming. *Tellus A: Dynamic Meteorology and Oceanography* **65** (1): 20578.

van Kooten, G.C. 2013. *Climate Change, Climate Science and Economics.* New York, NY: Springer.

Vasskog, K., Langebroek, P.M., Andrews, J.T., Nilsen, J.E.Ø., and Nesje, A. 2015. The Greenland Ice Sheet during the last glacial cycle: current ice loss and contribution to sea-level rise from a palaeoclimatic perspective. *Earth-Science Reviews* **150**: 45–67.

Werner, J.P., Divine, D.V., Charpentier Ljungqvist, F., Nilsen, T., and Francus, P. 2017. Spatio-temporal variability of Arctic summer temperatures over the past two millennia: an overview of the last major climate anomalies. *Climate of the Past: Discussions*: doi:10.5194/cp-2017-29.

Zwally, H.J., *et al.* 2011. Greenland ice sheet mass balance: distribution of increased mass loss with climate warming: 2003–07 versus 1992–2002. *Journal of Glaciology* **57**: 88–102.

2.3.2.3 Non-polar Glaciers

During the past 25,000 years (late Pleistocene and Holocene) glaciers around the world have fluctuated broadly in concert with changing climate, at times shrinking to positions and volumes smaller than today. Many non-polar glaciers have been retreating since the Little Ice Age, a natural occurrence unrelated to the human presence. This fact notwithstanding, mountain glaciers around the world show a wide variety of responses to local climate variation and do not respond to global temperature change in a simple, uniform way.

Tropical mountain glaciers in both South America and Africa have retreated in the past 100 years because of reduced precipitation and increased solar radiation; retreat of some glaciers elsewhere began at the end of the Little Ice Age in 1850, when no impact from anthropogenic CO_2 was possible. The

data on global glacial history and ice mass balance do not support the claims made by the IPCC that CO_2 emissions are causing most glaciers today to retreat and melt. Chapter 5 of *Climate Change Reconsidered II: Physical Science* (NIPCC, 2013) reported many examples of glaciers that were either stable or advancing in the modern era. Rather than attempt a similarly comprehensive literature review again, we mention only a few examples reported most recently in the literature.

Yan *et al.* (2017) studied sea ice on the Bohai Sea, a gulf of the Yellow Sea on the northeastern coast of China. Despite being one of the busiest seaways in the world, the sea is becoming ice-covered during winter months with growing frequency. The recent global winter cooling trend reported in Section 2.3.1 has affected the Bohai Sea, producing media reports of record-setting cold and ice (*Beijing Review*, 2010, 2014; *China Daily*, 2012; *China Travel Guide*, 2013; *Daily Mail*, 2016). Yan *et al.* observe, "Despite the backdrop of continuous global warming, sea ice extent has been found not to consistently decrease across the globe, and instead exhibit heterogeneous variability at middle to high latitudes." Using satellite imagery, Yan *et al.* reveal an upward trend of $1.38 \pm 1.00\%$ yr^{-1} (R = 1.38, i.e. at a statistical significance of 80%) in Bohai Sea ice extent over the 28-year period. The researchers also report a decreasing mean ice-period average temperature based on data from 11 meteorological stations around the Bohai Sea. Their results are shown in Figure 2.3.2.3.1.

Sigl *et al.* (2018) note "Starting around AD 1860, many glaciers in the European Alps began to retreat from their maximum mid-19th century terminus positions, thereby visualizing the end of the Little Ice Age in Europe," confirming this retreat began before human greenhouse gas emissions could have been a principal driving factor. The authors note some researchers nevertheless contend radiative forcing by increasing deposition of industrial black carbon to snow might account for "the abrupt glacier retreats in the Alps." To test this hypothesis, they used "sub-annually resolved concentration records of refractory black carbon (rBC; using soot photometry) as well as distinctive tracers for mineral dust, biomass burning and industrial pollution from the Colle Gnifetti ice core in the Alps from AD 1741 to 2015. These records allow precise assessment of a potential relation between the timing of observed acceleration of glacier melt in the mid-19th century with an increase of rBC deposition on the glacier caused by the industrialization of Western Europe."

Figure 2.3.2.3.1
Increasing sea ice extent on Bohai Sea in northeastern China, 1988–2015

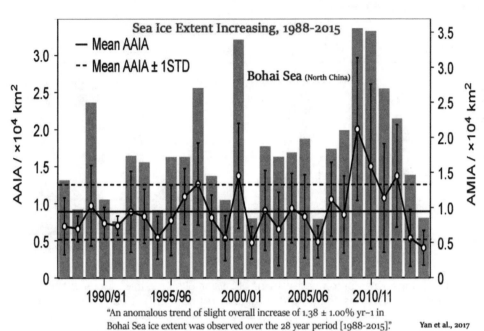

Left axis and solid lines are annual average ice area (AAIA). Right axis and shaded columns are annual maximum ice area (AMIA). Source: Adapted from Yan *et al.*, 2017.

Sigl *et al.* (2018) establish that industrial black carbon (BC) deposition began in 1868–1884 (5%–95% range) with the highest change point probability in 1876. "The median timing of industrial BC deposition at the four Greenland ice-core sites is AD 1872 (ToE analysis) or AD 1891 (Bayesian change-point), respectively, in good agreement with the Alpine ice cores," they report. By that time, they report, "the majority of Alpine glaciers had already experienced more than 80 % of their total 19th century length reduction, casting doubt on a leading role for soot in terminating of the Little Ice Age." Their plot of mean glacier length retreat and advance rates, shown in Figure 2.3.2.3.2, shows significant natural variability correlated with ambient temperature (not shown in the graph) but unrelated to soot emissions or human greenhouse gas emissions.

More evidence that natural variability in glacier advances and retreats exceeds that witnessed in the modern era comes from Oppedal *et al.* (2018). The authors produce a 7,200-year-long reconstruction of advances and retreats of the Diamond glacier, a "cirque glacier" (a glacier formed in a bowl-shaped depression on the side of or near mountains) on north-central South Georgia, an island south of the Antarctic Convergence. The authors infer glacier activity "from various sedimentary properties including magnetic susceptibility (MS), dry bulk density (DBD), loss-on-ignition (LOI) and geochemical elements (XRF), and tallied to a set of terminal moraines." They plot their findings in the figure reproduced here as Figure 2.3.2.3.3.

Oppedal *et al.* (2018) also found the study site was "deglaciated prior to 9900 ± 250 years ago when Neumayer tidewater glacier retreated up-fjord." Significantly, one of the periods when the glacier "was close to its Maximum Holocene extent" was "in the Twentieth century (likely 1930s)." Clearly, the extent of this glacier has varied considerably over the past 7,000 years and the retreat since 1930 is well within the bounds of natural variation. As for a mechanism explaining the waxing and waning of this glacier, Oppedal *et al.* write, "glacier fluctuations are largely in-phase with reconstructed Patagonian glaciers, implying that they respond to centennial climate variability possibly connected to corresponding modulations of the Southern Westerly Winds."

Figure 2.3.2.3.2
Nineteenth century glacier retreat in the Alps preceded the emergence of industrial black carbon deposition on high-alpine glaciers

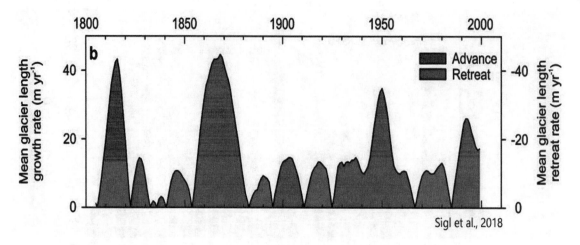

Mean glacier length change rate (smoothed with an 11-year filter) of the glacier stack length record indicating phases of average glacier advances (blue) and of glacier retreat (red). *Source:* Adapted from Sigl *et al.,* 2018, Figure 8b.

Figure 2.3.2.3.3
Glacier reconstruction for Diamond glacier on South Georgia island

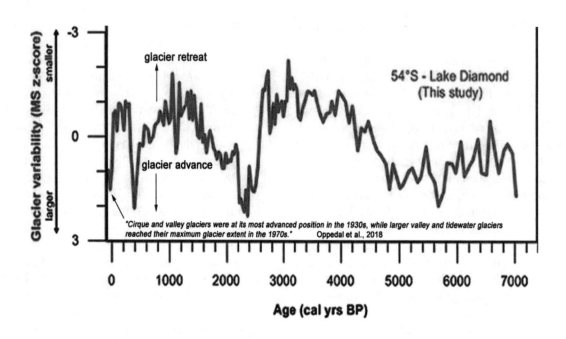

Source: Adapted from Oppedal *et al.*, 2018.

* * *

On the matter of whether trends in melting ice in recent decades can be attributed to the human presence, the IPCC writes, "anthropogenic forcings are *very likely* to have contributed to Arctic sea ice loss since 1979" and "ice sheets and glaciers are melting, and anthropogenic influences are likely to have contributed to the surface melting of Greenland since 1993 and to the retreat of glaciers since the 1960s" (IPCC, 2013, p. 870). But such melting is not occurring at the rates reported by the IPCC, and melting is not unusual by historical or geological time standards. Computer models *assume* attribution to the human presence and then, in a circular fashion, are cited as proof of such attribution.

In conclusion, the Antarctic ice sheet is likely to be unchanged or is gaining ice mass. Antarctic sea ice is gaining in extent, not retreating. The Greenland ice sheet and Artic sea ice are losing mass but historically show variability exceeding the changes seen in the late twentieth and early twenty-first centuries. Any significant warming, whether anthropogenic or natural, will melt ice. To claim anthropogenic global warming is occurring based on such information is to confuse the consequences of warming with its cause.

References

Beijing Review. 2010. Still Life. February 11.

Beijing Review. 2014. Frozen Sea. February 5.

China Daily. 2012. Bohai Sea freezes into massive ice rink. February 2.

China Travel Guide. 2013. "Real" 2013 of Huanghai and Bohai Sea region "frozen" polar spectacle. July 24.

Daily Mail. 2016. Chinese coast frozen as sea turns into an amazing winter wonderland. January 31.

IPCC. 2013. Intergovernmental Panel on Climate Change. *Climate Change 2013: The Physical Science Basis.* Contribution of Working Group I to the Fifth Assessment Report of the Intergovernmental Panel on Climate Change. New York, NY: Cambridge University Press.

NIPCC. 2013. Nongovernmental International Panel on Climate Change. Idso, C.D., Carter, R.M., and Singer, S.F. (Eds.) *Climate Change Reconsidered: Physical Science.* Chicago, IL: The Heartland Institute.

Oppedal, L.T., Bakke, J., Paasche, Ø., Werner, J.P., and van der Bilt, W.G.M. 2018. Cirque Glacier on South Georgia shows centennial variability over the last 7000 years. *Frontiers in Earth Science* 6 (2).

Sigl, M., Abram, N.J., Gabrieli, J., Jenk, T.M., Osmont, D., and Schwikowski, M. 2018. 19th century glacier retreat in the Alps preceded the emergence of industrial black carbon deposition on high-alpine glaciers. *The Cryosphere* 12: 3311–31.

Yan, Y., *et al.* 2017. Multidecadal anomalies of Bohai Sea ice cover and potential climate driving factors during 1988–2015. *Environmental Research Letters* 12.

2.3.3 Sea-level Rise

Long-running coastal tide gauges show the rate of sea-level rise is not accelerating. Local and regional sea levels exhibit typical natural variability.

According to the Working Group I contribution to the IPCC's Fifth Assessment Report (IPCC, 2013), "it is very likely that the global mean rate [of sea level rise] was 1.7 [1.5 to 1.9] mm yr^{-1} between 1901 and 2010 for a total sea level rise of 0.19 [0.17 to 0.21] m" (p. 1139) and "it is very likely that the rate of global mean sea level rise during the 21st century will exceed the rate observed during 1971–2010 for all Representative Concentration Pathway (RCP) scenarios due to increases in ocean warming and loss of mass from glaciers and ice sheets" (p. 1140).

Also according to the IPCC (2013), mass loss from the Greenland and Antarctic ice sheets over the period 1993–2010 expressed as sea-level equivalent was "about 5.9 mm (including 1.7 mm from glaciers around Greenland) and 4.8 mm, respectively," and ice loss from glaciers between 1993 and 2009 (excluding those peripheral to the ice sheets) was 13 mm (p. 368). The total is 23.7 mm (5.9 + 4.8 + 13), which is slightly less than 1 inch.

Like ice melting, sea-level rise is a research area that has recently come to be dominated by computer models. Whereas researchers working with datasets built from long-term coastal tide gauges typically report a slow linear rate of sea-level rise, computer modelers assume a significant anthropogenic forcing and tune their models to find or predict an acceleration of the rate of rise. This section reviews recent research to determine if there is any evidence of such an acceleration and then examines claims that islands and coral atolls are being inundated by rising seas.

2.3.3.1 Recent Sea-level Trends

The recent Pleistocene Ice Age slowly ended 20,000 years ago with an initially slow warming and a concomitant melting of ice sheets. As a result, sea level rose nearly 400 feet to approximately the present level. For the past thousand years it is generally believed that globally averaged sea-level change has been less than seven inches per century, a rate that is functionally negligible because it is frequently exceeded by coastal processes such as erosion and sedimentation. Local and regional sea levels continue to exhibit typical natural variability – in some places rising and in others falling – unrelated to changes in the global average sea level.

Measuring changes in sea level is difficult due to the roles and impacts of gravity variations, density of the water due to salinity differences, temperature of the water, wind, atmospheric pressure differences, changes in land level and land uses, and uncertainty regarding new meltwater from glaciers. The change in technologies used to measure sea level with the arrival of satellite altimetry created discontinuities in datasets resulting in conflicting estimates of sea levels and their rates of change (e.g., Chen *et al.*, 2013; Cazenave *et al.*, 2014). While some researchers infer from satellite data rates of sea-level rise of 3 mm yr^{-1} or even higher (Nerem *et al.*, 2018), the accuracy of those claims have been severely criticized (Church *et al.*, 2010; Zhang and Church, 2012; Parker, 2015; Parker and Ollier, 2016; Mörner, 2017; Roach *et al.*, 2018). Others have spliced together measurements from different locations at different times (Church and White, 2006). In fact, all the (very slight) acceleration reported by Church and White (2006) occurred prior to 1930 – when CO_2 levels were under 310 ppm (Burton, 2012).

Many researchers place the current rate of global sea-level rise at or below the IPCC's historic estimate for 1901–2020 of 1.7 mm/year. Parker and Ollier (2016) averaged all the tide gauges included in the Permanent Service for Mean Sea Level (PSMSL), a repository for tide gauge data used in the measurement of long-term sea-level change based at the National Oceanography Centre in Liverpool, England, and found a trend of about + 1.04 mm/year for 570 tide gauges of any length. When they selected tide gauges with more than 80 years of recording, they found the average trend was only + 0.25 mm/year. They also found no evidence of acceleration in either dataset.

Parker and Ollier (2017) described six datasets they characterized as especially high quality:

- The 301 stations of the PSMSL database having a range of years greater than or equal to 60 years, "PSMSL-301."

- Mitrovica's 23 gold standard tide stations with minimal vertical land motion suggested by Douglas, "Mitrovica-23."

- Holgate's nine excellent tide gauge records of sea-level measurements, "Holgate-9."

- The 199 stations of the NOAA database (global and the USA) having a range of years greater than or equal to 60 years, "NOAA-199."

- The 71 stations of the NOAA database (USA only), having a range of years greater than or equal to 60 years, "US 71."

- The eight tide gauges of California, USA of years range larger than 60 years, "California-8."

According to Parker and Ollier (2017), "all consistently show a small sea-level rate of rise and a negligible acceleration." The average trends and accelerations for these data sets are:

- $+ 0.86 \pm 0.49$ mm/year and $+ 0.0120 \pm 0.0460$ mm/year^2 for the PSMSL-301 dataset.

- $+ 1.61 \pm 0.21$ mm/year and $+ 0.0020 \pm 0.0173$ mm/year^2 for the Mitrovica-23 dataset.

- $+ 1.77 \pm 0.17$ mm/year and $+ 0.0029 \pm 0.0118$ mm/year^2 for the Holgate-9 dataset.

- $+ 1.00 \pm 0.46$ mm/year and $+ 0.0052 \pm 0.0414$ mm/year^2 for the NOAA-199 dataset.

- $+ 2.12 \pm 0.55$ mm/year and $- 0.0077 \pm 0.0488$ mm/year^2 for the US 71 dataset.

- $+ 1.19 \pm 0.29$ mm/year and $+ 0.0014 \pm 0.0266$ mm/year^2 for the California-8 dataset.

Bezdek (2017) notes "one region in the USA identified as being particularly susceptible to sea-level rise is the Chesapeake Bay region, and it has been estimated that by the end of the century Norfolk, Virginia could experience sea-level rise of 0.75 meters to more than 2.1 meters." The author's research revealed that water intrusion was in fact a "serious problem in much of the Chesapeake Bay region" but "due not to 'sea level rise' but primarily to land subsidence due to groundwater depletion and, to a lesser extent, subsidence from glacial isostatic adjustment. We conclude that water intrusion will thus continue even if sea levels decline." The author goes on to recommend water management policies that have been "used successfully elsewhere in the USA and other nations to solve water intrusion problems."

Wang and Zhou (2017) studied two tide gauge stations in the Pearl River Estuary on the coast of China (Macau and Hong Kong), applying a "peaks-over-threshold model of extreme value theory to statistically model and estimate secular parametric trends of extreme sea level records." Tide gauge data for Macau and Hong Kong spanned the period 1925–2010 and 1954–2014, respectively. In describing their findings, the two Chinese researchers note there are "evident decadal variations in the intensity and frequency of extremes in [the] sea level records," but "none of the parameters (intensity and frequency) of daily higher high-water height extremes in either Macau or Hong Kong has a significant increasing or decreasing trend." Similar results were obtained upon examination of trends of extremes in tidal residuals, where Wang and Zhou again report "none of the parameters presents a significant trend in recent decades."

Watson (2017) notes "some 28 of the 30 longest records in the Permanent Service for Mean Sea Level (PSMSL) global data holdings are European, extending as far back as 1807 (Brest, France). Such records provide the world's best time series data with which to examine how kinematic properties of the trend might be changing over time." He chose 83 tide gauge records with a minimum of 80 years reporting, at the locations shown in Figure 2.3.3.1.1, and used "a recently developed analytical package titled 'msltrend' specifically designed to enhance estimates of trend, real-time velocity, and acceleration in the relative mean sea-level signal derived from long annual average ocean water level time series."

Figure 2.3.3.1.1
Location of tide gauge records in Europe with at least 80 years of reporting

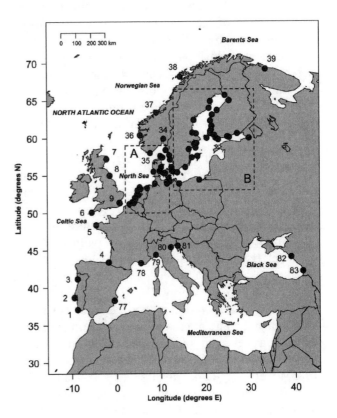

Source: Watson, 2017, Figure 1, p. 24.

Even though "the msltrend package has been specifically designed to enhance substantially estimates of trend, real-time velocity, and acceleration in relative mean sea level derived from contemporary ocean water level data sets," Watson (2017) reports (with apparent surprise), "Key findings are that at the 95% confidence level, no consistent or compelling evidence (yet) exists that recent rates of rise are higher or abnormal in the context of the historical records available across Europe, nor is there any evidence that geocentric rates of rise are above the global average. It is likely a further 20 years of data will distinguish whether recent increases are evidence of the onset of climate change–induced acceleration." Watson (2017), like many other researchers, observed "the quasi 60-year oscillation identified in all oceanic basins of the world (Chambers, Merrifield, and Nerem, 2012)."

Frederikse *et al.* (2018) comment on how "different sea level reconstructions show a spread in sea level rise over the last six decades," citing among the reasons for disagreement "vertical land motion at tide-gauge locations and the sparse sampling of the spatially variable ocean." The authors create a new reconstruction of sea level from 1958 to 2014 using tide-gauge records, observations of vertical land motion, and estimates of ice-mass loss, terrestrial water storage, and barotropic atmospheric forcing and find "a trend of 1.5 ± 0.2 mm yr^{-1} over 1958–2014 (1σ), compared to 1.3 ± 0.1 mm yr^{-1} for the sum of contributors," an acceleration of 0.07 ± 0.02 mm yr^{-2}.

Ahmed *et al.* (2018) used a geographic information system (GIS) and remote sensing techniques to study land erosion (losses) and accretion (gains) for the entire coastal area of Bangladesh for the period 1985–2015. Because it is a low-lying river delta especially vulnerable to sea-level rise, concerns are often expressed that it could be a victim of global warming-induced sea-level rise (e.g., Cornwall, 2018). Ahmed *et al.* find "the rate of accretion in the study area is slightly higher than the rate of erosion. Overall land dynamics indicate a net gain of 237 km^2 (7.9 km^2 annual average) of land in the area for the whole period from 1985 to 2015." Rather than sinking beneath rising seas, Bangladesh is actually *growing into the sea.*

Contrary to the IPCC's statement that it is "very likely" sea-level rise is accelerating, Burton (2018) reports the highest quality coastal tide gauges from around the world show no evidence of acceleration since the 1920s or before, and therefore no evidence of being affected by rising atmospheric CO_2 levels. Figure 2.3.3.1.2 shows three coastal sea-level measurement records (in blue), all more than a century long, in each case juxtaposed with atmospheric CO_2 levels (in green).

The mean sea-level (MSL) trend at Honolulu, Hawaii, USA is +1.48 mm/year; at Wismar, Germany is +1.42 mm/year; and at Stockholm, Sweden is -3.75 mm/year. The first two graphs are typical sea-level trends from especially high-quality measurement records located on opposite sides of the Earth at sites that are little affected by distortions like tectonic instability, vertical land motion, and ENSO. The trends are nearly identical, and perfectly typical: only about 6 inches per century, a rate that has not increased in more than nine decades. At Stockholm, sea-level rise is negative due to regional vertical land motion. To see how typical these trends are, as well as to observe natural variability for reasons already presented, see the entire 375 tide stations for which NOAA did long-term trend analysis at http://sealevel.info/MSL_global_thumbnails5.html.

Local sea-level trends vary considerably because they depend not only on the average global trend, but also on tectonic movements of adjacent land. In many places vertical land motion, either up or down, exceeds the very slow global sea-level trend. Consequently, at some locations sea level is rising much faster than the global rate, and at other locations sea level is falling. Figure 2.3.3.1.3 shows sea level since 1930 at Grand Isle, Louisiana, USA and Skagway, Alaska, USA.

* * *

The best available data show dynamic variations in Pacific sea level in accord with El Niño-La Niña cycles, superimposed on a natural long-term rise in the volume of water in Earth's oceans (called the eustatic rise) (Australian Bureau of Meteorology, 2011; Scafetta, 2013). Though the range of natural variation has yet to be fully described, evidence is lacking for any recent changes in global sea level that lie outside natural variation.

Figure 2.3.3.1.2
Coastal measurement of sea-level rise (blue) in three cities vs. aerial CO₂ concentration (green)

A. Honolulu, Hawaii, USA

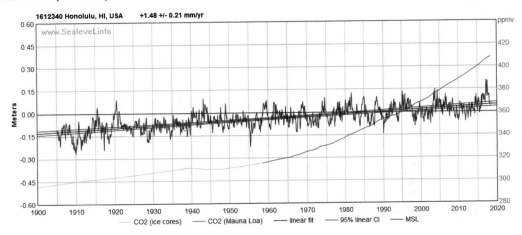

B. Wismar, Germany

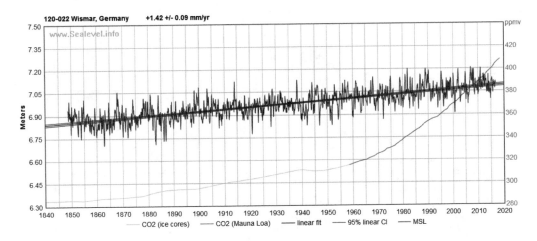

C. Stockholm, Sweden

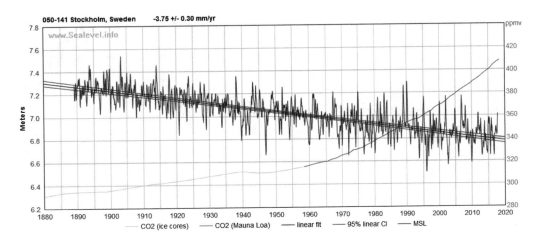

Mean sea level at Honolulu, HI, USA (NOAA 1612340, 760-031, PSMSL 155), Wismar, Germany (NOAA 120-022, PSMSL 8), and Stockholm, Sweden (NOAA 050-141, PSMSL 78). Monthly mean sea level in meters (blue, left axis) without the regular seasonal fluctuations due to coastal ocean temperatures, salinities, winds, atmospheric pressures, and ocean currents. CO_2 concentrations in ppmv (green, right axis). The long-term linear trend (red) and its 95% confidence interval (grey). The plotted values are relative to the most recent mean sea-level data established by NOAA CO-OPS. *Source:* Burton, 2018.

Figure 2.3.3.1.3
Sea level from 1930, for Grand Isle, Louisiana, USA and Skagway, Alaska, USA

A. Grand Isle, Louisiana, USA

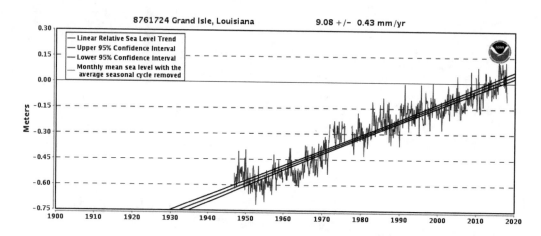

B. Skagway, Alaska, USA

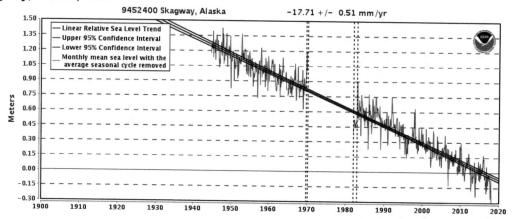

See previous figure for notes. *Source*: Burton, 2018, using NOAA data.

References

Ahmed, A., Drake, F., Nawaz, R., and Woulds, C. 2018. Where is the coast? Monitoring coastal land dynamics in Bangladesh: an integrated management approach using GIS and remote sensing techniques. *Ocean & Coastal Management* **151**: 10–24.

Australian Bureau of Meteorology. 2011. The South Pacific sea-level and climate monitoring program. Sea-level summary data report, July 2010–June 2011.

Bezdek, R.H. 2017. Water intrusion in the Chesapeake Bay region: is it caused by climate-induced sea level rise? *Journal of Geoscience and Environment Protection* **5**: 252–63.

Burton, D. 2012. Comments on 'Assessing future risk: quantifying the effects of sea level rise on storm surge risk for the southern shores of Long Island, New York. *Natural Hazards* **63**: 1219.

Burton, D. 2018. Mean sea level at Honolulu, HI, USA (NOAA 1612340, 760-031, PSMSL 155, mean sea level at Wismar, Germany (NOAA 120-022, PSMSL 8, and mean sea level at Stockholm, Sweden (NOAA 050-141, PSMSL 78. Sea Level Info (website). Accessed December 11, 2018.

Cazenave, A., *et al.* 2014. The rate of sea-level rise. *Nature Climate Change* **4**: 358–61.

Chambers, D.P., Merrifield, M.A., and Nerem, R.S. 2012 Is there a 60-year oscillation in global mean sea level? *Geophysical Research Letters* **39** (18).

Chen, J.L., Wilson, C.R., and Tapley, B.D. 2013. Contribution of ice sheet and mountain glacier melt to recent sea level rise. *Nature Geoscience* **6**: 549–52.

Church, J.A. and White, N.J. 2006. A 20th century acceleration in global sea-level rise. *Geophysical Research Letters* **33**: L01602.

Church, J.A., *et al.* 2010. Sea-level rise and variability: synthesis and outlook for the future. In: Church, J.A., Woodworth, P.L, Aarup, T., and Wilson, W.S. (Eds.) *Understanding Sea-Level Rise and Variability*. Chichester, UK: Wiley-Blackwell, pp. 402–19.

Cornwall, W. 2018. As sea levels rise, Bangladeshi islanders must decide between keeping the water out—or letting it in. *Science Social Sciences*: doi:10.1126/science.aat4495, March 1.

Frederikse, T., Jevrejeva, S., Riva, R.E.M., and Dangendorf, S. 2018. A consistent sea-level reconstruction and its budget on basin and global scales over 1958–2014. *Journal of Climate* **31** (3): 1267–80.

IPCC. 2013. Intergovernmental Panel on Climate Change. *Climate Change 2013: The Physical Science Basis*. Contribution of Working Group I to the Fifth Assessment Report of the Intergovernmental Panel on Climate Change. New York, NY: Cambridge University Press.

Mörner. N.-A. 2017. Sea level manipulation. *International Journal of Engineering Science Invention* **6** (8): 48–51.

Nerem, R.S., *et al.* 2018. Climate-change–driven accelerated sea-level rise detected in the altimeter era. *Proceedings of the National Academy of Sciences USA* **115** (9): 2022–5.

Parker, A. 2015. Accuracy and reliability issues in the use of global positioning system and satellite altimetry to infer the absolute sea level rise. *Journal of Satellite Oceanography and Meteorology* **1** (1): 13–23.

Parker, A. and Ollier, C.D. 2016. Coastal planning should be based on proven sea level data. *Ocean & Coastal Management* **124**: 1–9.

Parker A. and Ollier, C.D. 2017. California sea level rise: evidence based forecasts vs. model predictions. *Ocean & Coastal Management* **149**: 198–209.

Roach, L.A., Dean, S.M., and Renwick, J.A. 2018. Consistent biases in Antarctic sea ice concentration simulated by climate models. *The Cryosphere* **12**: 365–83.

Scafetta, N. 2013. Multi-scale dynamical analysis (MSDA) of sea level records versus PDO, AMO, and NAO indexes. *Climate Dynamics*: 10.1007/s00382-013-1771-3.

Wang, W. and Zhou, W. 2017. Statistical modeling and trend detection of extreme sea level records in the Pear River Estuary. *Advances in Atmospheric Sciences* **34**: 383–96.

Watson, P.J. 2017. Acceleration in European mean sea level? A new insight using improved tools. *Journal of Coastal Research* **33** (1): 23–38.

Zhang, X. and Church, J.A. 2012. Sea level trends, interannual and decadal variability in the Pacific Ocean. *Geophysical Research Letters* **39**: 10.1029/2012GL053240.

2.3.3.2 Islands and Coral Atolls

Small islands and Pacific coral atolls (an island made of coral that encircles a lagoon partially or completely) are thought to be particularly at risk of harm from sea-level rise due to their low elevation and fragile shorelines. Since global sea levels have risen slowly but steadily since 1900, and that rise is alleged to have accelerated since 1990, its negative effects should be visible as a loss of surface area, but repeated studies have found this is not the case.

Island researchers generally have found that atoll shorelines are most affected by direct weather and infrequent high tide events due to El Niño-Southern Oscillation events and the impacts of increasing human populations. Pacific island ecologies are very resilient to hurricanes and floods since they happen so frequently, and plants and animals have learned to adapt and recover (Smithers and Hoeke, 2014; Mann and Westphal, 2016). Most flooding results not from sea-level rise, but from spring tides or storm surges in combination with development pressures such as borrow pit digging (a hole where soil, gravel, or sand has been dug for use at another location) or groundwater withdrawal. Persons emigrating from the islands generally do so for social and economic reasons rather than in response to environmental threats.

Biribo and Woodroffe (2013) write, "low-lying reef islands on atolls appear to be threatened by impacts of observed and anticipated sea-level rise," noting that "widespread flooding in the interior of Fongafale on Funafuti Atoll, in Tuvalu, is often cited as confirmation that 'islands are sinking' (Patel, 2006)." To see if this was true, the two scientists examined changes in shoreline position on most of the reef islands on Tarawa Atoll, the capital of the Republic of Kiribati, by analyzing "reef-island area and shoreline change over 30 years determined by comparing 1968 and 1998 aerial photography using geographical information systems."

Biribo and Woodroffe (2013) determined that the reef islands of Tarawa Atoll "substantially increased in size, gaining about 450 ha, driven largely by reclamations on urban South Tarawa, accounting for 360 ha (~80% of the net change)." Of the 40 islands of North Tarawa, where population is absent or sparse, they report that "25 of the reef islands in this area showed no change at the level of detection, 13 showed net accretion and only two displayed net erosion." In addition, they indicate that "similar reports of reef island area increase have been observed on urban Majuro, in the Marshall Islands, again mainly related to human activity," citing Ford (2012). And they say "a recent analysis of changes in area of 27 reef islands from several Pacific atolls for periods of 35 or 61 years concluded that they were growing (Webb and Kench, 2010)," likely "as a result of more prolific coral growth and enhanced sediment transport on reef flats when the sea is higher," under which conditions they note that "shorelines will actually experience accretion, thus increasing reef island size (Kinsey and Hopley, 1991)."

Introducing their study, Kench et al. (2015) write, "low-lying coral reef islands are coherent accumulations of sand and gravel deposited on coral reef surfaces that provide the only habitable land in atoll nations such as Kiribati, Tuvalu, and the Marshall Islands in the Pacific Ocean, and the Maldives in the Indian Ocean." And they write that in extreme cases, "rising sea level is expected to erode island coastlines," forcing "remobilization of sediment reservoirs and promoting island destabilization," thereby making them "unable to support human habitation and rendering their populations among the first environmental refugees," citing Khan et al. (2002) and Dickinson (2009). But will this ever really happen?

One phenomenon that suggests it could occur is the high rate of sea-level rise ($5.1 \pm 0.7 \text{mm/yr}^{-1}$) and the consequent changes in shoreline position that have occurred over the past 118 years at 29 islands of Funafuti Atoll in the tropical Pacific Ocean. However, Kench et al. (2015) write, "despite the magnitude of this rise, no islands have been lost," noting, in fact, that "the majority have enlarged, and there has been a 7.3% increase in net island area over the past century (AD 1897–2013)." They add "there is no evidence of heightened erosion over the past half-century as sea-level rise accelerated," noting that "reef islands in Funafuti continually adjust their size, shape, and position in response to variations in boundary conditions, including storms, sediment supply, as well as sea level." The scientists conclude that "islands can persist on reefs under rates of sea-level rise on the order of 5 mm/year," which is a far greater rate-of-rise than what has been observed over the past half-century of significant atmospheric CO_2 enrichment.

Ford and Kench (2015) used historic aerial photographs and recent high-resolution satellite imagery to determine "shoreline changes on six atolls and two mid-ocean reef islands in the Republic of the Marshall Islands." This work revealed, "since the middle of the 20th century more shoreline has accreted than eroded, with 17.23% showing erosion, compared to 39.74% accretion and 43.03% showing no change." Consequently, they determine "the net result of these changes was the growth of the islands examined from 9.09 km^2 to 9.46 km^2 between World War Two (WWII) and 2010." In light of these findings, Ford and Kench conclude that "governments of small island nations need to acknowledge that island shorelines are highly dynamic and islands have persisted and in many cases grown in tandem with sea level rise."

Purkis *et al.* (2016) observed that "being low and flat, atoll islands are often used as case studies against which to gauge the likely impacts of future sea-level rise on coastline stability." The authors examined remotely sensed images from Diego Garcia, an atoll island situated in the remote equatorial Indian Ocean, to determine how its shoreline has changed over the past five decades (1963–2013), during which time sea level in the region has been rising more than 5 mm per year, over at least the last 30 years, based on data they obtained from the National Oceanographic Data Center. According to the four scientists, "the amount of erosion on Diego Garcia over the last 50 years is almost exactly balanced by the amount of accretion, suggesting the island to be in a state of equilibrium." Commenting on the significance of this finding, Purkis *et al.* write their study "constitutes one of the few that have documented island shoreline dynamics at timescales relevant to inform projections of future change."

Testut *et al.* (2016) acquired baseline data on both absolute and relative sea-level variations and shoreline changes in the Scattered Islands region of the Indian Ocean, based on aerial image analysis, satellite altimetry, field observations, and *in situ* measurements derived from the 2009 and 2011 Terres Australes et Antarctiques Francaises scientific expeditions. They discovered "Grande Glorieuse Island has increased in area by 7.5 ha between 1989 and 2003, predominantly as a result of shoreline accretion," which "occurred over 47% of shoreline length." They also note "topographic transects and field observations show that the accretion is due to sediment transfer from the reef outer slopes to the reef flat and then to the beach."

Duvat *et al.* (2017) studied shoreline change in atoll reef islands of the Tuamotu Archipelago in French Polynesia by examining aerial photographs and satellite images of 111 atoll reef islands from the area taken over the past 50 years. According to the researchers, their findings bring "new irrefutable evidences on the persistence of reef islands over the last decades." Over the past three to five decades, the total net land area of the studied atolls "was found to be stable, with 77% of the sample islands maintaining their area, while 15% expanded and 8% contracted." Furthermore, they note that seven out of the eight islands that decreased in area were very small in area (less than 3 hectares), whereas "all of the 16 islands larger than 50 hectares were stable in area."

McAneney *et al.* (2017) created a 122-year record of major flooding depths at the Rarawai Sugar Mill on the Ba River in the northwest of the Fijian Island of Viti Levu. "Reconstructed largely from archived correspondence of the Colonial Sugar Refining Company, the time series comprises simple measurements of height above the Mill floor." The authors report their findings as follows: "It exhibits no statistically significant trends in either frequency or flood heights, once the latter have been adjusted for average relative sea-level rise. This is despite persistent warming of air temperatures as characterized in other studies. There is a strong dependence of frequency (but not magnitude) upon El Niño-Southern Oscillation (ENSO) phase, with many more floods in La Niña phases. The analysis of this long-term data series illustrates the difficulty of detecting a global climate change signal from hazard data...."

Summarizing her own review of the scientific literature on sea-level rise, Curry (2018) writes, "Tide gauges show that sea levels began to rise during the 19th century, after several centuries associated with cooling and sea level decline. Tide gauges also show that rates of global mean sea level rise between 1920 and 1950 were comparable to recent rates. Recent research has concluded that there is no consistent or compelling evidence that recent rates of sea level rise are abnormal in the context of the historical records back to the 19th century that are available across Europe."

Kench *et al.* (2018) recount the "dispiriting and forlorn consensus" that rising sea levels will inundate atoll islands and argue there is "a more nuanced set of options to be explored to support adaptation in atoll states. Existing paradigms are based on flawed assumptions that islands are static landforms, which will simply drown as the sea level rises. There is growing evidence that islands are geologically dynamic features that will adjust to changing sea level and climatic conditions," citing Webb and Kench (2010), Ford (2013), McLean and Kench (2015), and Duvat and Pillet (2017). The authors test the theory that rising sea levels were inundating atoll islands by analyzing "shoreline change in all 101 islands in the Pacific atoll nation of Tuvalu." "Surprisingly," they write, "we show that all islands have changed and that the dominant mode of change has been island expansion, which has increased the land area of the nation." The nation saw a net increase in land area of 73.5 ha (2.9%) despite sea-level rise. While 74% of the islands gained land area, 27% decreased in size. "Expansion of islands on reef surfaces indicates a net addition of sediment," the researchers write. "Implications of increased sediment volumes are profound as they suggest positive sediment generation balances for these islands and maintenance of an active linkage between the reef sediment production regime and transfer to islands, which is critical for ongoing physical resilience of islands."

Finally, Kench *et al.* (2018) report "direct anthropogenic transformation of islands through reclamation or associated coastal protection works/development has been shown to be a dominant control on island change in other atoll nations. However, in Tuvalu direct physical interventions that modify coastal processes are small in scale because of much lower population densities. Only 11 of the study islands have permanent habitation and, of these, only two islands sustain populations greater than 600. Notably, there have been no large-scale reclamations on Tuvaluan islands within the analysis window of this study (the past four decades)." These results, Kench *et al.* write, "challenge perceptions of island loss, showing islands are dynamic features that will persist as sites for habitation over the next century, presenting alternate opportunities for adaptation that embrace the heterogeneity of island types and their dynamics."

* * *

Small islands and Pacific coral atolls are not being inundated by rising seas due to anthropogenic climate change. Direct evidence reveals many islands and atolls are increasing, not decreasing, in area as natural process lead to more prolific coral growth and enhanced sediment transport on reef flats. Combined with evidence that sea levels are not rising at unusual or unprecedented rates around the world, this means the IPCC's concern over rising sea levels is without merit.

References

Biribo, N. and Woodroffe, C.D. 2013. Historical area and shoreline change of reef islands around Tarawa Atoll, Kiribati. *Sustainability Science* **8**: 345–62.

Curry, J. 2018. Special report on sea level rise. Climate Etc. (website). Accessed November 27, 2018.

Dickinson, W.R. 2009. Pacific atoll living: how long already and until when? *GSA Today* **19**: 124–32.

Duvat, V.K.E. and Pillet, V. 2017. Shoreline changes in reef islands of the Central Pacific: Takapoto Atoll, Norther Tuamotu, French Polynesia. *Geomorphology* **282**: 96–118.

Ford, M. 2012. Shoreline changes on an urban atoll in the central Pacific Ocean: Majuro Atoll, Marshall Islands. *Journal of Coastal Research* **28**: 11–22.

Ford, M. 2013. Shoreline changes interpreted from multi-temporal aerial photographs and high resolution satellite images: Wotje Atoll, Marshall Islands. *Remote Sensing of the Environment* **135**: 130–40.

Ford, M.R. and Kench, P.S. 2015. Multi-decadal shoreline changes in response to sea level rise in the Marshall Islands. *Anthropocene* **11**: 14–24.

Kench, P.S., Ford, M.R., and Owen, S.D. 2018. Patterns of island change and persistence offer alternate adaptation pathways for atoll nations. *Nature Communications* **9**: Article 605.

Kench, P.S., Thompson, D., Ford, M.R., Ogawa, H., and McLean, R.F. 2015. Coral islands defy sea-level rise over the past century: records from a central Pacific atoll. *Geology* **43**: 515–8.

Khan, T.M.A., Quadir, D.A., Murty, T.S., Kabir, A., Aktar, F., and Sarker, M.A. 2002. Relative sea level changes in Maldives and vulnerability of land due to abnormal coastal inundation. *Marine Geodesy* **25**: 133–43.

Kinsey, D.W. and Hopley, D. 1991. The significance of coral reefs as global carbon sinks–response to Greenhouse. *Palaeogeography, Palaeoclimatology, Palaeoecology* **89**: 363–77.

Mann, T. and Westphal, H. 2016. Multi-decadal shoreline changes on Taku Atoll, Papua New Guinea: observational evidence of early reef island recovery after the impact of storm waves. *Geomorphology* **257**: 75–84.

McAneney, J., van den Honert, R., and Yeo, S. 2017. Stationarity of major flood frequencies and heights on the Ba River, Fiji, over a 122-year record. *International Journal of Climatology* **37** (S1): 171–8.

McLean, R.F. and Kench, P.S. 2015. Destruction or persistence of coral atoll islands in the face of 20th and 21st century sea-level rise? *WIREs Climate Change* **6**: 445–63.

Patel, S.S. 2006. Climate science: a sinking feeling. *Nature* **440**: 734–6.

Purkis, S.J., Gardiner, R., Johnston, M.W., and Sheppard, C.R.C. 2016. A half-century of coastline change in Diego Garcia, the largest atoll island in the Chagos. *Geomorphology* **261**: 282–98.

Smithers, S.G. and Hoeke, R.K. 2014. Geomorphological impacts of high-latitude storm waves on low-latitude reef islands–observations of the December 2008 event on Nukutoa, Takuu, Papua New Guinea. *Geomorphology* **222**: 106–21.

Testut, L., Duvat, V., Ballu, V., Fernandes, R.M.S., Pouget, F., Salmon, C., and Dyment, J. 2016. Shoreline changes in a rising sea level context: the example of Grande Glorieuse, Scattered Islands, Western Indian Ocean. *Acta Oecologica* **72**: 110–9.

Webb, A.P. and Kench, P.S. 2010. The dynamic response of reef islands to sea-level rise: evidence from multi-decadal analysis of island change in the Central Pacific. *Global and Planetary Change* **72**: 234–46.

2.3.4 Harm to Plant Life

The effects of elevated CO_2 on plant characteristics are net positive, including increasing rates of photosynthesis and biomass production.

According to the Working Group II contribution to the IPCC's Fifth Assessment Report, a "large fraction of both terrestrial and freshwater species faces increased extinction risk under projected climate change during and beyond the 21st century, especially as climate change interacts with other stressors, such as habitat modification, over-exploitation, pollution, and invasive species (*high confidence*)" (IPCC 2014, pp. 14–15). Like so many of the IPCC's other predictions, this one ignores natural variability and extensive data on plant and animal life that contradict it.

Chapter 5 addresses the impact of climate change and fossil fuels on the environment in great detail. In this section we focus narrowly on the science concerning the effects of rising levels of carbon dioxide (CO_2) on plant life. This section updates the literature review in Chapter 2 of *Climate Change Reconsidered II: Biological Impacts* (NIPCC, 2014) examining the effect of elevated CO_2 on plant characteristics. The key findings of that report are presented in Figure 2.3.4.1. There is a host of other effects of significance, including the efficiency with which plants and trees utilize water, which are addressed in Chapter 5.

Figure 2.3.4.1
Key Findings: Impacts on plant characteristics

- Atmospheric CO_2 enrichment (henceforth referred to as "rising CO_2") enhances plant growth, development, and ultimate yield (in the case of agricultural crops) by increasing the concentrations of plant hormones that stimulate cell division, cell elongation, and protein synthesis.

- Rising CO_2 enables plants to produce more and larger flowers, as well as other flower-related changes having significant implications for plant productivity and survival, almost all of which are positive.

- Rising CO_2 increases the production of glomalin, a protein created by fungi living in symbiotic association with the roots of 80% of the planet's vascular plants, where it is having a substantial positive impact on the biosphere.

- Rising CO_2 likely will affect many leaf characteristics of agricultural plants, with the majority of the changes leading to higher rates and efficiencies of photosynthesis and growth as well as increased resistance to herbivory and pathogen attack.

- Rising CO_2 stimulates photosynthesis in nearly all plants, enabling them to produce more nonstructural carbohydrates that can be used to create important carbon-based secondary compounds, one of which is lignin.

- Rising CO_2 leads to enhanced plant fitness, flower pollination, and nectar production, leading to increases in fruit, grain, and vegetable yields of agricultural crops as well as productivity increases in natural vegetation.

- As rising CO_2 causes many plants to increase biomass, the larger plants likely will develop more extensive root systems enabling them to extract greater amounts of mineral nutrients from the soil.

- Rising CO_2 causes plants to sequentially reduce the openness of their stomata, thus restricting unnecessary water loss via excessive transpiration, while some plants also reduce the density (number per area) of stomates on their leaves.

- Rising CO_2 significantly enhances the condensed tannin concentrations of most trees and grasses, providing them with stronger defenses against various herbivores both above and below ground. This in turn reduces the amount of methane, a potent greenhouse gas, released to the atmosphere by ruminants browsing on tree leaves and grass.

- As the atmosphere's CO_2 content rises, many plant species may not experience photosynthetic acclimation even under conditions of low soil nitrogen. In the event that a plant cannot balance its carbohydrate sources and sinks, CO_2-induced acclimation provides a way of achieving that balance by shifting resources away from the site of photosynthesis to enhance sink development or other important plant processes.

Source: Chapter 2. "Plant Characteristics," *Climate Change Reconsidered II: Biological Impacts.* Nongovernmental International Panel on Climate Change. Chicago, IL: The Heartland Institute, 2014.

Introduction

"It should be considered good fortune that we are living in a world of gradually increasing levels of atmospheric carbon dioxide" writes Wittwer (1995). He adds, "the rising level of atmospheric CO_2 is a universally free premium, gaining in magnitude with time, on which we can all reckon for the foreseeable future." Similarly, Benyus (2002) writes, "Organisms don't think of CO_2 as a poison. Plants and organisms that make shells, coral, think of it as a building block."

The geological history briefly recounted in Section 2.1.2.3 shows life flourished when atmospheric CO_2 concentrations were at double and triple their current levels. Section 2.3.1.3 showed how the frequency and intensity of drought has not increased in the modern era. Instead, the Earth has experienced a significant "greening" observed by satellites during the past 30 years, with approximately 20% of Earth's surface becoming greener, including 36% of Africa, while only 3% of Earth has browned (Myneni, 2015; see also Bastos *et al.,* 2017; Brandt, 2017; Zeng *et al.,* 2018). As atmospheric CO_2 concentrations increase, so does plant growth.

The positive relationship between atmospheric CO_2 concentrations and plant growth is well known by botanists, biologists, and agronomists. Kimball (1983a, 1983b), for example, conducted two of the earliest analyses of the peer-reviewed scientific literature dealing with plant responses to atmospheric CO_2 enrichment. From 770 individual plant responses, he determined that a 300 parts per million (ppm) rise in the air's CO_2 content boosted the productivity of most herbaceous plants by approximately 33%. Other reviews conducted soon afterwards by Cure and Acock (1986), Mortensen (1987), and Lawlor and Mitchell (1991) produced similar results. On the basis of research such as this, commercial greenhouses use CO_2 generators to elevate the level of CO_2 in their facilities to 800 to 1,200 ppm (e.g., Ontario.ca, 2009). Recall that the ambient concentration of CO_2 in 2018 was about 405 ppm and the pre-industrial level is thought to have been about 280 ppm. An increase of 300 ppm is therefore an approximate doubling of the pre-industrial level.

Perhaps the largest such review was that of Idso (1992), who analyzed papers published over the decade subsequent to the reviews of Kimball. This comprehensive assessment of the pertinent literature incorporated a total of 1,087 observations of plant responses to atmospheric CO_2 enrichment obtained from 342 peer-reviewed scientific journal articles. Idso determined that 93% of the plant responses to atmospheric CO_2 enrichment were positive, 5% were negligible, and only 2% were negative. The mean growth response curve of the plants investigated in these studies is depicted in Figure 2.3.4.2. The database of experiments was published in a peer-reviewed journal in 1994 (Idso and Idso, 1994) and has been maintained and continuously updated by the Center for the Study of Carbon Dioxide and Global Change (Idso, 2018). See Idso (2012, 2013) for more recent surveys of the literature, and appendices to *Climate Change Reconsidered II: Biological Impacts* (NIPCC, 2014, pp. 1,045–62).

Figure 2.3.4.2
Positive impact of CO_2 on plants and trees

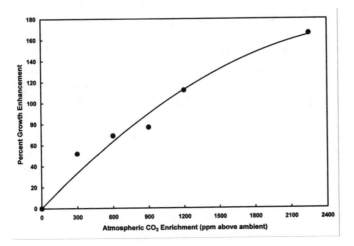

Source: NIPCC, 2014, p. 1, Figure 1, citing Idso, 1992.

Additional research shows the positive effects of this "aerial fertilization" are even greater when combined with rising temperatures, the mechanisms by which are explained later in this section. Figure 2.3.4.3 shows how the net photosynthetic rate for one type of tree (bigtooth aspen) exposed to elevated CO_2 increases as temperature rises, and how higher CO_2 levels enable the tree to produce glucose at temperatures considerably higher than it would otherwise tolerate. Many other plants show a similar response.

Figure 2.3.4.3
Positive impact of CO_2 is enhanced by warmer temperatures

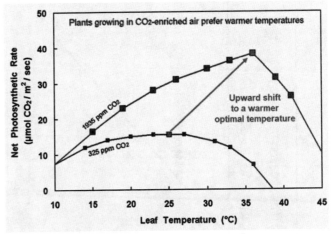

Relationship of leaf temperature and CO_2 exchange rate, a proxy for net photosynthetic rate, for two levels of exposure to CO_2, 1,935 ppm (large boxes) and 325 ppm (small boxes), for bigtooth aspen (*Populus grandidentata*). *Source:* Adapted from Jurik *et al.*, 1984, Figure 1, p. 1023.

Warmer water temperatures similarly benefit some (but not all) kinds of sea life. Figure 2.3.4.4 shows how calcification rates for *Montastraea annularis*, a species of coral, rise with seawater temperatures. This species is the most abundant species of reef-building coral in the Caribbean.

Literature Review

Chapter 5 surveys literature on the impacts of rising temperatures and atmospheric CO_2 levels on ecosystems, plants under stress, water use efficiency, and the future impacts of climate change on plants. Here we focus on the more narrow topic of plant responses to atmospheric CO_2 enrichment.

Weiss *et al.* (2010) grew rooted shoot cuttings of the cacti *Hylocereus undatus* (red pitaya) and *Selenicereus megalanthus* (yellow pitaya) for one full year (August 2006 to August 2007) in vented chambers maintained at either ambient or elevated atmospheric CO_2 concentrations (380 or 1,000 ppm, respectively) in a cooled greenhouse, where the plants were fertilized twice weekly. The researchers

Figure 2.3.4.4
Positive impact of warmer temperatures on reef-building coral in the Caribbean

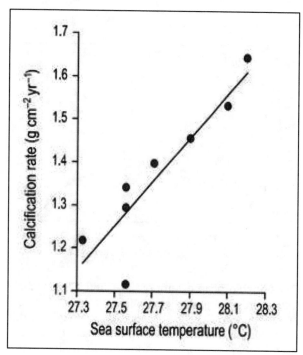

Yearly mean calcification rate of *Montastraea annularis* vs. mean annual sea surface temperature for the several sites studied by Carricart-Ganivet (2004) (blue circles) and two sites studied by Carricart-Ganivet and Gonzalez-Diaz (2009) (red circles). The line that has been fit to the data is described by: Calcification Rate = 0.51 SST - 12.85 (r^2 = 0.82, p < 0.002). Adapted from Carricart-Ganivet and Gonzalez-Diaz, 2009.

measured net photosynthesis on four days in mid-April and made final biomass determinations at the conclusion of the study. In addition, they conducted a second one-year study of eight-year-old plants to investigate the fruit development responses of mature plants to atmospheric CO_2 enrichment; this work was done in open-top chambers maintained in the same greenhouse.

Weiss *et al.* (2010) report, "*H. undatus* plants enriched with CO_2 demonstrated 52%, 22%, 18% and 175% increases, relative to plants measured in ambient CO_2, in total daily net CO_2 uptake, shoot elongation, shoot dry mass, and number of reproductive buds, respectively," while corresponding responses for *S. megalanthus* were 129%, 73%, 68%, and 233%. They also found a slight (7%) increase in the fruit fresh mass of *H. undatus* and a much greater 63% increase in the fruit fresh mass of *S. megalanthus* due to the CO_2 enrichment. They conclude their experiments demonstrate "the vast potential of possible increases in the yields of CAM [crassulacean acid metabolism] crops under CO_2 enrichment."

Jiang *et al.* (2012) studied the relationship between CO_2 enrichment and brassinosteroids (BRs), "naturally occurring plant steroid hormones that are ubiquitously distributed in the plant kingdom." They report "BRs play prominent roles in various physiological processes including the induction of a broad spectrum of cellular responses, such as stem elongation, pollen tube growth, xylem differentiation, leaf epinasty, root inhibition, induction of ethylene biosynthesis, proton pump activation, regulation of gene expression and photosynthesis, and adaptive responses to environmental stress." They also note, "as potent plant growth regulators, BRs are now widely used to enhance plant growth and yield of important agricultural crops."

Working with well-watered and fertilized cucumber plants that had reached the three-leaf growth stage in pots within controlled environment growth chambers maintained at either ambient (380 ppm) or enriched (760 ppm) atmospheric CO_2 concentrations and with or without being sprayed with a solution of brassinosteroids (0.1 μM 24-epibrassinolide), Jiang *et al.* (2012) measured rates of net photosynthesis, leaf area development, and shoot biomass production over a period of one additional week. They determined that their doubling of the air's CO_2 concentration resulted in a 44.1% increase in CO_2 assimilation rate and the BR treatment "also significantly increased CO_2 assimilation under ambient atmospheric CO_2 conditions and the increase was close to that by CO_2 enrichment."

Jiang *et al.* (2012) report the combined treatment of "plants with BR application under CO_2-enriched conditions showed the highest CO_2 assimilation rate, which was increased by 77.2% relative to the control." Likewise, "an elevation in the atmospheric CO_2 level from 380 to 760 ppm resulted in a 20.5% and 16.0% increase in leaf area and shoot biomass accumulation, respectively," while the plants that received the BR application "exhibited 22.6% and 20.6% increases in leaf area and shoot biomass accumulation, respectively." The combined treatment of "CO_2 enrichment and BR application further improved the plant growth, resulting in 49.0% and 40.2% increases in leaf area and shoot biomass, relative to that of the control, respectively."

Goufo *et al.* (2014) note crop plants need *phenolic compounds* "for structural support, constitutive and induced protection and defense against weeds, pathogens and insects." They point out carbon dioxide is one of the four major raw materials plants need to produce phenolic compounds, the other three being water, nutrients, and light. They conducted a two-year field study of a japonica rice variety (*Oryza sativa* L. cv. Ariete), employing open-top chambers maintained at either 375 or 550 ppm CO_2 over two entire life cycles of the crop, during which time numerous plant samples were collected at five growth stages and assessed for many plant-produced substances, including phenolics. They found all plant organs had higher levels of phenolic acids and flavonoids in response to "CO_2 enrichment during the maturity stages."

Goufo *et al.* (2014) explain "phenolic compounds are emerging as important defense compounds in rice," particularly noting the phenolic compound tricin "inhibits the growth of *Echinochloa colonum*, *Echinochloa crusgalli*, *Cyperus iris* and *Cyperus difformis*," which they say "are the most noxious weeds in rice fields." They add that several flavonoids "have also been found to exhibit antibiotic activities against the soil-borne pathogenic fungi *Rhizoctonia solani* and *Fusarium oxysporum*," which they say are "the causal agents of rice seedling rot disease." They suggest the ongoing rise in the atmosphere's CO_2 concentration may "increase plant resistance to specific weeds, pests and pathogens."

Zhang *et al.* (2017) grew cotton in climate-controlled growth chambers under two temperature (27/20°C (80.6/68°F) or 34/27°C (93.2/80.6°F) day/night) and CO_2 (400 or 800 ppm) regimes. The plants were sampled and various parameters measured to observe the impact of elevated CO_2 and elevated temperature. At the end of the experiment (105 days after sowing), they report, elevated temperature enhanced dry matter by 43% under ambient CO_2 conditions and by 60% under elevated CO_2 conditions (Figure 2.3.4.5, left panel). Elevated CO_2 also enhanced dry matter, by 17% under ambient temperature conditions and 31% under elevated temperatures. The highest increase in dry matter content was noted in the elevated temperature and elevated CO_2 treatment, suggesting to the authors that "[CO_2] enrichment could enhance the effect of rising temperature on dry matter content."

Reporting on other parameters, Zhang *et al.* (2017) note that although early measurements made at 45 days after sowing revealed calculable differences, by 75 days after sowing and on through the end of their experiment they found no significant effect of elevated CO_2 on leaf nitrogen or carbon/nitrogen (C:N) ratio at either ambient or elevated temperature. Leaf soluble sugars, total starch, and total foliar nonstructural carbohydrates, by contrast, all experienced marked increases from elevated CO_2 under ambient and elevated temperatures. Elevated CO_2 also tended to enhance total leaf phenolic concentrations, while elevated temperature tended to reduce them (Figure 2.3.4.5, right panel).

Sgherri *et al.* (2017) write, lettuce is "an important source of phytochemicals such as phenolic compounds," which compounds (including antioxidants), "have been recognized as phytonutrients able to lower the incidence of some types of cancer and cardiovascular diseases (Hooper and Cassidy, 2006)." Noting plant phytochemical composition and antioxidant activity can be altered by environmental factors, such as rising atmospheric CO_2 and salinity stress, they investigated the effects of elevated CO_2 and salinity stress on the phytochemical composition of two lettuce cultivars, Blonde of Paris Batavia (a green leaf cultivar) and Oak Leaf (a red leaf cultivar).

Sgherri *et al.* (2017) grew the two lettuce cultivars under both ambient (400 ppm) and elevated (700 ppm) CO_2 for 35 days after sowing. They then subjected a portion of plants in each CO_2 treatment to salt stress by adding Hoagland solution supplemented with 200 millimeters sodium chloride (NaCl) each day until harvest. Upon harvest, they took measurements to ascertain plant growth and phytonutrient differences. Under ambient CO_2 growth conditions, Sgherri *et al.* report salinity stress caused yield reductions of 5% and 10% in the green and red lettuce cultivars, respectively. Under normal salt conditions, elevated CO_2 stimulated yields, inducing gains of 29% and 38% in the green and red cultivars, respectively. In the combined treatment of elevated CO_2 and salinity stress, the positive impacts of elevated CO_2 ameliorated the negative impacts of salt stress. With respect to phytochemicals, as shown in Figure 2.3.4.6, both salt stress and elevated CO_2 increased plant antioxidant capacity, total phenols, and total flavonoids. Sgherri *et al.* conclude, "the application of moderate salinity or elevated CO_2, alone or in combination, can induce the production of some phenolics that increase the health benefits of lettuce."

Figure 2.3.4.5
Impact of elevated CO₂ and temperature on cotton plants

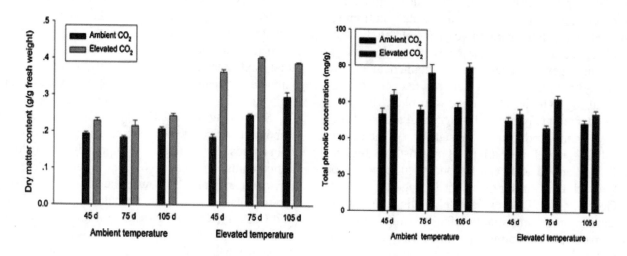

Ambient temperature is 27/20°C (80.6/68°F) day/night and elevated temperature is 34/27°C (93.2/80.6°F) day/night. Ambient CO₂ concentration is 400 ppm and elevated is 800 ppm. *Source*: Zhang *et al.*, 2017.

Li *et al.* (2017) observe that more frequent droughts are among the predicted effects of future climate change. While the beneficial effects of elevated levels of CO₂ on winter wheat (*Triticum aestivum L.*) plants exposed to drought conditions for a single generation are well documented, "the transgenerational effect of e[CO₂] [elevated CO₂] in combination of drought on stomatal behavior, plant water consumption and water use efficiency (WUE) have not been investigated." The researchers harvested seeds from plants after two generations (2014–2015) continuously grown in ambient CO₂ (a[CO₂], 400 µmol l⁻¹) and e[CO₂] (800 µmol l⁻¹) and sowed them in four- liter pots, and the plants were grown separately in greenhouse cells with either a[CO₂] or e[CO₂]. At stem elongation stage, in each of the cells half of the plants were subjected to progressive drought stress until all the plant available soil water was depleted, and the other half were well-watered and served as controls. "The results," the researchers report, "showed that transgenerational exposure of the winter wheat plants to e[CO₂] could attenuate the negative impact of drought stress on dry biomass (DM) and WUE. The modulations of multi-generational e[CO2] on leaf abscisic acid concentration, stomatal conductance, and leaf water status could have contributed to the enhanced DM

and WUE. These findings provide new insights into the response of wheat plants to a future drier and CO₂-enriched environment."

Nakano *et al.* (2017) studied the possible impact of future higher levels of CO₂ in the atmosphere on rice (*Oryza sativa* L.) grain yield. They grew "a chromosome segment substitution line (CSSL) and a near-isogenic line (NIL) producing high spikelet numbers per panicle (CSSL-*GN1* and NIL-*APO1*, respectively) under free-air CO₂ enrichment (FACE) conditions and examined the effects of a large sink capacity on grain yield, its components, and growth-related traits under increased atmospheric CO₂ concentrations." They found that under ambient conditions, CSSL-*GN1* and NIL-*APO1* exhibited a similar grain yield to Koshihikari rice, but under FACE conditions, CSSL-*GN1* and NIL-*APO1* had an equal or higher grain yield than Koshihikari because of the higher number of spikelets and lower reduction in grain filling. "Thus," they conclude, "the improvement of source activity by increased atmospheric CO₂ concentrations can lead to enhanced grain yield in rice lines that have a large sink capacity. Therefore, introducing alleles that increase sink capacity into conventional varieties represents a strategy that can be used to develop high-yielding varieties under increased atmospheric CO₂ concentra-

Figure 2.3.4.6
Two lettuce cultivars subjected to salt treatment under ambient or elevated CO_2

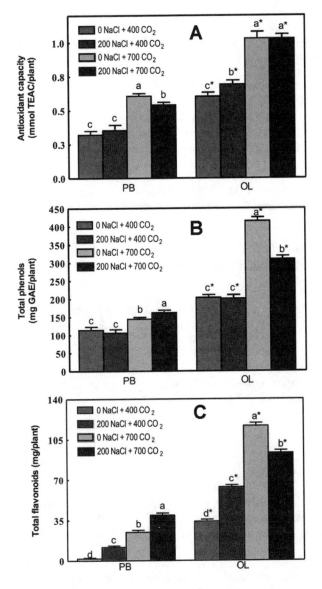

Panel (A) reports change in antioxidant capacity, (B) change in total phenols, and (C) change in total flavonides in Blonde of Paris green-leaf (PB) and Oak Leaf red-leaf (OL). TEAC is trolox equivalent antioxidant capacity; GAE is gallic acid equivalent. Dark green is control plot (no additional sodium chloride (NaCl) and ambient CO_2 (400 ppm)), purple is 200 mm of NaCl added every day at ambient CO_2, bright green is no NaCl and 700 ppm CO_2 and blue is 200 NaCl and 700 CO_2. *Source:* Adapted from Sgherri *et al.*, 2017.

tions, such as those predicted in the near future."

Pau *et al.* (2018) analyzed "a 28-year record of tropical flower phenology in response to anthropogenic climate and atmospheric change." They found "compared to significant climatic factors, CO_2 had on average an approximately three-, four-, or fivefold stronger effect than rainfall, solar radiation, and the Multivariate ENSO Index, respectively." That effect is invariably positive, resulting in increased flowering of mid-story trees and shrub species and a lengthening of flowering duration for canopy and mid-story trees. The authors conclude, "Given that atmospheric CO_2 will likely continue to climb over the next century, a long-term increase in flowering activity may persist in some growth forms until checked by nutrient limitation or by climate change through rising temperatures, increasing drought frequency and/or increasing cloudiness and reduced insolation."

* * *

In conclusion, the effects of elevated CO_2 on plant characteristics are net positive, including increasing rates of photosynthesis and production of biomass and phenolics. Thousands of laboratory and field experiments reveal why plants benefit from higher CO_2 levels and higher temperatures. By focusing narrowly on the effects of elevated CO_2 on plants, this section provides the scientific basis for discussions in Chapters 5, 7, and 8 on the impacts on ecosystems, agriculture, and human well-being.

References

Bastos, A., *et al.* 2017. Was the extreme Northern Hemisphere greening in 2015 predictable? *Environmental Research Letters* **12**: 044016.

Benyus, J.M. 2002. *Biomimicry: Innovation Inspired by Nature.* New York, NY: Harper Perennial.

Brandt, M. 2017. Human population growth offsets climate-driven increase in woody vegetation in sub-Saharan Africa. *Nature Ecology & Evolution* **1**: 0081.

Carricart-Garnivet, J.P. 2004. Sea surface temperature and the growth of the West Atlantic reef-building coral *Montastraea annularis. Journal of Experimental Marine Biology and Ecology* **302**: 249–60.

Carricart-Ganivet, J.P. and Gonzalez-Diaz, P. 2009. Growth characteristics of skeletons of *Montastraea annularis*

(Cnidaria: Scleractinia) from the northwest coast of Cuba. *Ciencias Marinas* **35**: 237–43.

Cure, J.D. and Acock, B. 1986. Crop responses to carbon dioxide doubling: a literature survey. *Agricultural and Forest Meteorology* **38**: 127–145.

Goufo, P., Pereira, J., Moutinho-Pereira, J., Correia, C.M., Figueiredo, N., Carranca, C., Rosa, E.A.S., and Trindade, H. 2014. Rice (*Oryza sativa* L.) phenolic compounds under elevated carbon dioxide (CO_2) concentration. *Environmental and Experimental Botany* **99**: 28–37.

Hooper, L. and Cassidy, A. 2006. A review of the health care potential of bioactive compounds. *Journal of the Science of Food and Agriculture* **86**: 1805–13.

Idso, K.E. 1992. Plant responses to rising levels of atmospheric carbon dioxide. *Climatological Publications Scientific Paper* No. 23. Tempe, AZ: Office of Climatology, Arizona State University.

Idso, C.D. 2012. *The State of Earth's Terrestrial Biosphere: How is It Responding to Rising Atmospheric CO_2 and Warmer Temperatures?* Tempe, AZ: Center for the Study of Carbon Dioxide and Global Change.

Idso, C.D. 2013. *The Positive Externalities of Carbon Dioxide: Estimating the Monetary Benefits of Rising Atmospheric CO_2 Concentrations on Global Food Production.* Tempe, AZ: Center for the Study of Carbon Dioxide and Global Change.

Idso, C.D. 2018. CO2Science Plant Growth Study Database (website).

Idso, K.E. and Idso, S.B. 1994. Plant responses to atmospheric CO_2 enrichment in the face of environmental constraints: A review of the past 10 years' research. *Agricultural and Forest Meteorology* **69**: 153-203.

IPCC. 2014. Intergovernmental Panel on Climate Change. *Climate Change 2014: Impacts, Adaptation, and Vulnerability. Part A: Global and Sectoral Aspects.* Contribution of Working Group II to the Fifth Assessment Report of the Intergovernmental Panel on Climate Change. New York, NY: Cambridge University Press.

Jiang, Y.-P., Cheng, F., Zhou, Y.-H., Xia, X.-J., Shi, K., and Yu, J.-Q. 2012. Interactive effects of CO_2 enrichment and brassinosteroid on CO_2 assimilation and photosynthetic electron transport in *Cucumis sativus*. *Environmental and Experimental Botany* **75**: 98–106.

Jurik, T.W., Webber, J.A., and Gates, D.M. 1984. Short-term effects of CO_2 on gas exchange of leaves of bigtooth aspen (*Populus grandidentata*) in the field. *Plant Physiology* **75**: 1022–6.

Kimball, B.A. 1983a. Carbon dioxide and agricultural yield: an assemblage and analysis of 330 prior observations. *Agronomy Journal* **75**: 779–88.

Kimball, B.A. 1983b. *Carbon Dioxide and Agricultural Yield: An Assemblage and Analysis of 770 Prior Observations.* Phoenix, AZ: U.S. Water Conservation Laboratory.

Lawlor, D.W. and Mitchell, R.A.C. 1991. The effects of increasing CO_2 on crop photosynthesis: a review of field studies. *Plant, Cell and Environment* **14**: 807–18.

Li, Y., Li., X., Yu, J., and Liu, F. 2017. Effect of the transgenerational exposure to elevated CO_2 on the drought response of winter wheat: stomatal control and water use efficiency. *Environmental and Experimental Botany* **136**: 78–84.

Mortensen, L.M. 1987. Review: CO_2 enrichment in greenhouses. Crop responses. *Scientia Horticulturae* **33**: 1–25.

Myneni, R.B. 2015. *The Greening Earth*. Final Workshop for the Arctic Biomass Project. Svalbard, Norway, October 20–23.

Nakano, H., *et al.* 2017. Quantitative trait loci for large sink capacity enhance rice grain yield under free-air CO_2 enrichment conditions. *Scientific Reports* **7**: 1827.

NIPCC. 2014. Nongovernmental International Panel on Climate Change. Idso, C.D, Idso, S.B., Carter, R.M., and Singer, S.F. (Eds.) *Climate Change Reconsidered II: Biological Impacts.* Chicago, IL: The Heartland Institute.

Ontario.ca. 2009. Ontario Ministry of Agriculture, Food, and Rural Affairs. Carbon dioxide in greenhouses (website). Accessed December 7, 2018.

Pau, S., *et al.* 2018. Long-term increases in tropical flowering activity across growth forms in response to rising CO_2 and climate change. *Global Change Biology* **24**: 5.

Sgherri, C., Pérez-López, U., Micaelli, F., Miranda-Apodaca, J., Mena-Petite, A., Muñoz-Rueda, A., and Quartacci, M.F. 2017. Elevated CO_2 and salinity are responsible for phenolics-enrichment in two differently pigmented lettuces. *Plant Physiology and Biochemistry* **115**: 269–78.

Weiss, I., Mizrahi, Y., and Raveh, E. 2010. Effect of elevated CO_2 on vegetative and reproductive growth characteristics of the CAM plants *Hylocereus undatus* and *Selenicereus megalanthus*. *Scientia Horticulturae* **123**: 531–6.

Wittwer, S.H. 1995. *Food, Climate, and Carbon Dioxide: The Global Environment and World Food Production.* Boca Raton, FL: CRC Press.

Zeng, Z., *et al.* 2018. Impact of Earth greening on the terrestrial water cycle. *Journal of Climate* **31**: 2633–50.

Zhang, S., Fu, W., Zhang, Z., Fan, Y., and Liu, T. 2017. Effects of elevated CO$_2$ concentration and temperature on some physiological characteristics of cotton (*Gossypium hirsutum* L.) leaves. *Environmental and Experimental Botany* **133**: 108–7.

2.4 Why Scientists Disagree

Previous sections in this chapter revealed considerable disagreement among scientists on basic matters of climatology, including how to reduce and report uncertainty, natural versus anthropogenic impacts on the planet's energy budget and carbon cycle, and whether temperature reconstructions and computer models are reliable tools for scientific research. *Why do scientists disagree on so many matters?*

It should be said at the outset that disagreement is not uncommon in science. Today's "truth" is not the same as yesterday's "truth" or tomorrow's "truth," because truth, like science, is never settled. The Scientific Method assures that every theory can be challenged by experiment and a better understanding of complex physical processes that are currently poorly understood or unknown. Still, disagreements over matters of science regarding climate seem particularly stubborn and involve matters fundamental to our understanding of climate, not only on the "frontiers" or periphery of scientific research. This section offers four explanations for why this is the case:

- Fundamental uncertainties arise from insufficient observational evidence and disagreements over how to interpret data and set the parameters of models.

- Climate is an interdisciplinary subject requiring insights from many fields. Very few scholars have mastery of more than one or two of these disciplines.

- Many scientists trust the United Nations' Intergovernmental Panel on Climate Change (IPCC) to objectively report the latest scientific findings on climate change, but it has failed to produce balanced reports and has allowed its findings to be misreported to the public.

- Climate scientists, like all humans, can have tunnel vision. Bias, even or especially if subconscious, can be especially pernicious when data are equivocal and allow multiple interpretations, as in climatology.

2.4.1 Scientific Uncertainties

Fundamental uncertainties and disagreements prevent science from determining whether human greenhouse gas emissions are having effects on Earth's atmosphere that could endanger life on the planet.

The first and most obvious reason why scientists disagree is because the human impact on climate remains a puzzle. Essex and McKitrick (2007) write, "Climate is one of the most challenging open problems in modern science. Some knowledgeable scientists believe that the climate problem can never be solved." Bony *et al.* (2015) write, "Fundamental puzzles of climate science remain unsolved because of our limited understanding of how clouds, circulation and climate interact." Reporting in *Nature* on Bony's 2015 study, Schiermeier (2015) wrote, "There is a misconception that the major challenges in physical climate science are settled. 'That's absolutely not true,' says Sandrine Bony, a climate researcher at the Laboratory of Dynamic Meteorology in Paris. 'In fact, essential physical aspects of climate change are poorly understood'" (p. 140). See also Stevens and Bony (2013); Stouffer *et al.* (2017), and Collins *et al.* (2018).

Uncertainty was the topic of an earlier section of this chapter (Section 2.1.1.3) and examples of uncertainty appear in other sections. Here is a brief summary of areas where uncertainty prevents climate scientists from attaining what the IPCC refers to as "high agreement" or even "medium agreement."

Methodological Uncertainty

Efforts to predict future climate conditions rely on complex general circulation models (GCMs) and even more complex integrated assessment models (IAMs). Such models introduce uncertainty into climate science and therefore are a source of disagreement for four reasons:

- *Parametric uncertainty* involves the proper setting of parameters and their variability in a simulation. So little is known about climate processes that parameterization for climate models is a subjective process only weakly constrained by observational data and best practices, causing wide variation in model outputs (Lupo *et al.,* 2013; Hourdin, 2017).

- *Stochastic uncertainties* arise from random events that cannot be predicted (Kelly and Kolstad, 1998). In climatology, they include abrupt economic downturns, changes in the Sun, volcanic eruptions, and wars. Any of these events could have a greater effect on climate than decades of forcing by CO_2.

- *Epistemic uncertainty* is due to what modelers do not know: the behavior of atmospheric and ocean processes; missing, erroneous, or unknowingly adjusted data; unacknowledged variabilities and uncertainties in data; and more (Roy and Oberkampf, 2011; Loehle, 2018).

- *Propagation of error* means errors or uncertainty in one variable, due perhaps to measurement limitations or confounding factors, are compounded (propagated) when that variable becomes part of a function involving other variables that are also uncertain, leading to "cascading uncertainties" or "uncertainty explosions" (Curry and Webster, 2011; Frank, 2015; Curry, 2018).

Temperature Record

Fundamental to the theory of anthropogenic climate change is an accurate reconstruction of a record of Earth's surface temperature. Yet this is probably the source of greatest uncertainty in climate science. The IPCC itself admits its temperature reconstructions are highly uncertain:

> The uncertainty in observational records encompasses instrumental/recording errors, effects of representation (e.g., exposure, observing frequency or timing), as well as effects due to physical changes in the

instrumentation (such as station relocations or new satellites). All further processing steps (transmission, storage, gridding, interpolating, averaging) also have their own particular uncertainties. Because there is no unique, unambiguous, way to identify and account for non-climatic artefacts in the vast majority of records, there must be a degree of uncertainty as to how the climate system has changed (IPCC, 2013, p. 165).

McLean (2018) conducted an audit of the HadCRUT4 dataset and found "more than 70 issues of concern," including failure to check source data for errors, resulting in "obvious errors in observation station metadata and temperature data" (p. 88). He found the dataset "has been incorrectly adjusted in a way that exaggerates warming." Emails from a programmer responsible for maintaining and correcting errors in the HadCRUT climate data between 2006 and 2009 reveal inaccuracies, data manipulation, and incompetence that render the dataset unreliable (Goldstein, 2009; Montford, 2010). In 2009, in response to an academic's request for the HadCRUT dataset, Phil Jones, director of the Climatic Research Unit at the University of East Anglia, admitted, "We, therefore, do not hold the original raw data but only the value-added (i.e., quality controlled and homogenized) data" (Michaels, 2009).

Studies of the positioning of weather stations in the United States – thought to have the best network of such stations in the world – found extensive violations of siting rules leading to contamination by urban heat islands (Pielke, 2007a, 2007b). The IPCC claims to control for heat island effects but researchers have found its adjustments are too small (e.g. McKitrick and Michaels, 2007; Soon *et al.* 2015; Quereda Sala *et al.*, 2017).

The warming since 1850 is meaningful information only if it exceeds natural variability. Proxy temperature records from the Greenland ice cores for the past 10,000 years demonstrate a natural range of warming and cooling rates between +2.5°C and -2.5°C/century, significantly greater than rates measured for Greenland or the globe during the twentieth century (Alley, 2000; Carter, 2010; Lamb, 2011, 2012). The ice cores also show repeated "Dansgaard–Oeschger" events when air temperatures rose at rates of about 10°C per century. There have been about 20 such warming events in the past 80,000 years.

In its Fifth Assessment Report (AR5), the IPCC admits the global mean surface temperature stopped rising from 1997 to 2010, reporting the temperature increase for that period was 0.07°C [-0.02 to 0.18] (IPCC, 2013, p. 37). This "pause" was interrupted by the major El Niño events of 2010–2012 and 2015–2016. During "the pause" humans released approximately one-third of all the greenhouse gases emitted since the beginning of the Industrial Revolution. If atmospheric CO_2 concentrations drive global temperatures, their impact surely would have been visible during this period. Either CO_2 is a weaker driver of climate than the IPCC assumes, or natural drivers and variation play a bigger role than it realizes (Davis *et al.*, 2018).

Energy Budget

Climate models wrongly assume that global temperatures, solar influences, and exchanges among global carbon reservoirs would remain unchanged decade after decade and century after century, *but for* the human presence. But the ACRIM total solar irradiance (TSI) composite shows a small upward pattern from around 1980 to 2000, an increase not acknowledged by the IPCC or incorporated into the models on which it relies (Scafetta and Willson, 2014). The absolute forcing of incoming solar radiation is approximately 340 Wm^{-2} at the top of the atmosphere, more than 10 times the forcing of all atmospheric CO_2, so even small changes in the absolute forcing of the Sun could result in values larger than the much smaller predicted changes in radiative forcing caused by human greenhouse gas emissions.

Two TSI reconstructions for the period 1900–2000 by Scafetta and West (2006) suggest the Sun contributed 46% to 49% of the 1900–2000 warming of Earth, but with uncertainties of 20% to 30% in their sensitivity parameters. Close correlations exist between TSI proxy models and many twentieth-century climate records including temperature records of the Arctic and of China, the sunshine duration record of Japan, and the Equator-to-Pole (Arctic) temperature gradient record (Soon, 2005, 2009; Ziskin and Shaviv, 2012). The solar models used by the IPCC report less variability than other reconstructions published in the scientific literature (Soon *et al.*, 2015), leading the IPCC to understate the importance of solar influences. Some solar scientists are investigating the possibility of the Sun entering a grand solar minimum (GSM), which

could manifest itself within two decades (Lockwood *et al.*, 2011).

Small changes in cloud cover, cloud brightness, and cloud height – all of which are known to vary spatially and over time and none of which is well modeled – could alter the planet's energy budget enough to explain the slight warming of the twentieth century (Lindzen, 2015). The role of water vapor and clouds on reflectivity and the planet's energy budget is not accurately modeled (Chou and Lindzen, 2004; Spencer *et al.*, 2007; Lindzen and Choi, 2011).

Research conducted by CERN (the European Institute for Nuclear Research) in 2016 provided experimental results supporting the theory that variations in the number of cosmic rays hitting Earth's atmosphere create more or fewer (depending on the strength of the solar magnetic wind) of the low, wet clouds that reflect solar heat back into space (Kirkby *et al.*, 2016). This could be the mechanism for converting small changes in TSI into larger changes in surface temperature, a mechanism the IPCC contends is missing (Svensmark, 2007; Svensmark *et al.*, 2017). The CLOUD experiment also found pure biogenic nucleation can produce aerosols in the pristine pre-industrial atmosphere, creating clouds (Gordon *et al.*, 2016). "The results from CLOUD suggest that estimates of high climate sensitivity may have to be revised downwards" (CERN, 2016).

The role of ocean currents in determining temperature and precipitation is probably understated by climate models (D'Aleo and Easterbrook, 2016). El Niño and La Niña cycles dominate the flux of water and energy in the tropical Pacific over periods of two to seven years. These cyclical episodes have large impacts on global temperatures, rainfall, and storm patterns relative to the forcing of CO_2. Partly due to the failure to accurately model these processes, climate sensitivity to a doubling of CO_2 is probably overstated by climate models (Lewis and Curry, 2014; Bates, 2016; Christy and McNider, 2017).

Carbon Cycle

The carbon cycle is not sufficiently understood or measured with sufficient accuracy to make declarative statements about the human contribution of CO_2 to the atmosphere, how long it resides there, and how it affects exchange rates among the planet's four carbon reservoirs (lithosphere, oceans, biosphere, and atmosphere) (Falkowski *et al.* 2000; Harde, 2017a, 2017b). Empirical measurements are available for only ~20% of major volcanic gas emission sources (Burton *et al.*, 2013). CO_2 emissions from volcanoes could be "a significant previously unrecognized contribution to global CO_2 emissions from natural sources" (Ilyinskaya *et al.*, 2018; see also Viterito, 2017 and Smirnov *et al.*, 2017). More than 80% of our ocean is unmapped, unobserved, and unexplored (NOAA, 2018). Seagrass meadows in Greenland could be emerging as a major carbon sink (Marbà *et al.*, 2018). New satellite images have increased estimates of global forest cover by at least 9%, "requiring revision to the biospheric carbon sink" (Bastin *et al.*, 2017).

Anthropogenic greenhouse gas emissions are minuscule compared to the exchanges among these carbon reservoirs, meaning even small mismeasurements or uncertainty regarding natural processes could account for all the forcing attributed to anthropogenic CO_2. Human greenhouse gas emissions thought to remain in the atmosphere each year constitute just two-tenths of 1% (0.195%) of the total amount of carbon thought to be in the atmosphere (IPCC, 2014; Ruddiman, 2008). Even human greenhouse gas emissions are not measured with precision. Nearly half of global economic activity takes place in the informal market and is unlikely to be accurately accounted for (Jutting and de Laiglesia, 2009). Authoritarian regimes inflate yearly GDP growth rates by a factor between 1.15 and 1.3, meaning GCMs are relying on false data (Martinez, 2018).

Computer Model Problems

Climate models fail to accurately hindcast past temperatures and consistently "run hot," predicting warmer temperatures than are likely to occur (Monckton *et al.*, 2015; Idso and Idso, 2015; Hope *et al.*, 2017; McKitrick and Christy, 2018). Computer modelers "tune" their models until they reach a result that matches observations or the expectations of the modelers, their peers, or funders (Hourdin *et al.*, 2017). Modeling turbulence is "one of the most basic and intractable research problems facing humanity. You can't compute it. You can't measure it. But rain falls because of it" (Essex and McKitrick, 2007, p. 20). The hydrology of the atmosphere and dynamics of the ocean-atmosphere interface are poorly understood and modeled, yet even small errors in this area have major effects on models (Legates, 2014; Christy and McNider, 2017).

Lupo *et al.* (2013) cite extensive scholarly research finding climate models underestimate surface evaporation caused by increased temperature by a factor of 3; inadequately represent aerosol-induced changes in infrared radiation; are unable to capture many important regional and lesser-scale phenomena such as clouds; assume all temperature rise since the start of the Industrial Revolution has resulted from human activities when in reality, major anthropogenic emissions commenced only in the mid-twentieth century; poorly simulate internal climate oscillations such as the AMO and PDO; and fail to incorporate the effects of variations in solar magnetic field or in the flux of cosmic rays, both of which are known to significantly affect climate. Forecasts of future warming fail to consider the cooling effects of the Greening of the Earth phenomenon (Jeong *et al.*, 2010).

Modelers assume a CO_2 increase causes a temperature increase and so they program that into their models. When asked how they know this happens they say the model shows it, or other models (similarly programmed) show it, or their model doesn't work *unless* CO_2 is assumed to increase temperatures. Modelers may get the benefit of the doubt from their colleagues and policymakers owing to the complexity of models and the expense of the supercomputers needed to run them. That does not make them accurate maps of a highly complex territory.

Climate Impacts

Measurement of climate impacts is severely handicapped by missing and unreliable data, smoothed and "homogenized" databases, overlooked variability, and the substitution of global and stylized facts for regional and local observational data. Uncertainty leads to widely varying claims about whether "climate change" is already happening and can be attributed to the human presence. Some examples of still unresolved issues in this area include:

- Are the number and intensity of heat waves rising and cold days falling globally as forecast by the IPCC? (Li *et al.*, 2015; Sardeshmukh *et al.*, 2015; EPA, 2016; Sun *et al.*, 2016)

- Have there been increasing trends in storms, floods, droughts, or hurricanes in the modern era? (IPCC, 2013, p. 112; Hao *et al.,* 2014, 2016; Sutton *et al.,* 2018)

- How much of rising property damage caused by extreme weather events is due to population growth and the increasing value and vulnerability of property located near river and lake shorelines, and not to anthropogenic climate change? (Pielke and Landsea, 1998; Crompton and McAneney, 2008; Pielke *et al.,* 2008; Barredo, 2009, 2010; Neumayer and Barthel, 2011).

- Do the well-known benefits of aerial CO_2 fertilization and warmer temperatures more than offset the hypothetical negative effects of climate change on plant life? (Idso, 2012, 2013; Myneni, 2015; Bastos *et al.,* 2017; Brandt, 2017; Zeng *et al.,* 2018)

- What impact will physical limits on the supply of fossil fuels and market forces, such as rising prices, have on future greenhouse gas emissions? (Tans, 2009; Doiron, 2016; Wang *et al.,* 2017)

* * *

In short, scientists disagree because so much about the climate is still unknown. This simple truth is not publicized because uncertainty discourages action (Samieson, 1996; Shackley and Wynne, 1996), and so climate activists coach scientists to conceal it (Moser and Dilling, 2007). Kreutzer *et al.* (2016) write, "The idea that the science of climate change is 'settled' is an absurdity, contrary to the very spirit of scientific enquiry. Climate science is in its infancy, and if its development follows anything resembling the normal path of scientific advancement, we will see in the years ahead significant increases in our knowledge, data availability, and our theoretical understanding of the causes of various climate phenomena."

References

Alley, R.B. 2000. The Younger Dryas cold interval as viewed from central Greenland. *Quaternary Science Reviews* 19: 213–26.

Barredo, J.I. 2009. Normalized flood losses in Europe: 1970–2006. *Natural Hazards and Earth System Sciences* 9: 97–104.

Barredo, J.I. 2010. No upward trend in normalized windstorm losses in Europe: 1970–2008. *Natural Hazards and Earth System Sciences* 10: 97–104.

Bastin, J.-F., *et al.* 2017. The extent of forest in dryland biomes. *Science* 356 (6338): 635–8.

Bastos, A., *et al.* 2017. Was the extreme Northern Hemisphere greening in 2015 predictable? *Environmental Research Letters* 12: 044016.

Bates, J.R. 2016. Estimating climate sensitivity using two-zone energy balance models. *Earth and Space Science* 3 (5): 207–25.

Bony, S., *et al.* 2015. Clouds, circulation and climate sensitivity. *Nature Geoscience* 8: 261–8.

Brandt, M. 2017. Human population growth offsets climate-driven increase in woody vegetation in sub-Saharan Africa. *Nature Ecology & Evolution* 1: 0081.

Burton, M.R., Sawyer, G.M., and Granieri, D. 2013. Deep carbon emissions from volcanoes. *Reviews in Mineralogy and Geochemistry* 75 (1): 323–54.

Carter, R.M. 2010. *Climate: The Counter Consensus.* London, UK: Stacey International.

CERN. 2016. CLOUD experiment sharpens climate predictions. *CERN Courier.* November 11.

Chou, M. and Lindzen, R. 2004. Comments on "Examination of the decadal tropical mean ERBS nonscanner radiation data for the iris hypothesis." *Journal of Climate* 18 (12): 2123–7.

Christy, J.R. and McNider, R.T. 2017. Satellite bulk tropospheric temperatures as a metric for climate sensitivity. *Asia-Pacific Journal of Atmospheric Sciences* 53 (4): 511–8.

Collins, M., *et al.* 2018. Challenges and opportunities for improved understanding of regional climate dynamics. *Nature Climate Change* 8: 101–8.

Crompton, R.P. and McAneney, K.J. 2008. Normalized Australian insured losses from meteorological hazards: 1967–2006. *Environmental Science and Policy* 11: 371–8.

Curry, J.A. 2018. Climate uncertainty & risk. Climate Etc. (blog). July 8. Accessed November 25, 2018.

Curry, J.A. and Webster, P.J. 2011. Climate science and the uncertainty monster. *Bulletin of the American Meteorological Society* 92 (12): 1667–82.

D'Aleo, J.S. and Easterbrook, D.J. 2016. Relationship of multidecadal global temperatures to multidecadal oceanic oscillations. In: Easterbrook, D. (Ed.) *Evidence-Based Climate Science*. Cambridge, MA: Elsevier.

Davis, W.J., Taylor, P.J., and Barton Davis, W. 2018. The Antarctic Centennial Oscillation: a natural paleoclimate cycle in the Southern Hemisphere that influences global temperature. *Climate* **6** (1): 3.

Doiron, H.H. 2016. *Recommendations to the Trump Transition Team Investigating Actions to Take at the Environmental Protection Agency (EPA): A Report of The Right Climate Stuff Research Team*. November 30, p. 20.

EPA. 2016. U.S. Environmental Protection Agency. Climate change indicators: high and low temperatures (website). Accessed December 6, 2018.

Essex, C. and McKitrick, R. 2007. *Taken by Storm. The Troubled Science, Policy and Politics of Global Warming*. Toronto, ON: Key Porter Books.

Falkowski, P., *et al.* 2000. The global carbon cycle: a test of our knowledge of Earth as a system. *Science* **290** (5490): 291–6.

Frank, P. 2015. Negligence, non-science, and consensus climatology. *Energy & Environment* **26** (3): 391–415.

Goldstein, L. 2009. Botch after botch after botch. *Toronto Sun*. November 29.

Gordon, H., *et al.* 2016. Reduced anthropogenic aerosol radiative forcing caused by biogenic new particle formation. *Proceedings of the National Academy of Sciences USA* **113** (43): 12053–8.

Hao, Z., *et al.* 2014. Global integrated drought monitoring and prediction system. *Scientific Data* **1**: Article 140001.

Hao, Z., *et al.* 2016. Spatial patterns of precipitation anomalies in eastern China during centennial cold and warm periods of the past 2000 years. *International Journal of Climatology* **36** (1): 467–75.

Harde, H. 2017a. Radiation transfer calculations and assessment of global warming by CO_2. *International Journal of Atmospheric Sciences*: Article 9251034.

Harde, H. 2017b. Scrutinizing the carbon cycle and CO_2 residence time in the atmosphere. *Global and Planetary Change* **152**: 19–26.

Hope, A.P., Canty, T.P., Salawitch, R.J., Tribett, W.R., and Bennett, B.F. 2017. Forecasting global warming. In: Salawitch, R.J., Canty, T.P., Hope, A.P., Tribett, W.R., and Bennett, B.F. (Eds.) *Paris Climate Agreement: Beacon of Hope*. New York, NY: Springer Climate, pp. 51–113.

Hourdin, F., *et al.* 2017. The art and science of climate model tuning. *Bulletin of the American Meteorological Society* **98** (3): 589–602.

Idso, C.D. 2012. *The State of Earth's Terrestrial Biosphere: How is It Responding to Rising Atmospheric CO_2 and Warmer Temperatures?* Tempe, AZ: Center for the Study of Carbon Dioxide and Global Change.

Idso, C.D. 2013. *The Positive Externalities of Carbon Dioxide: Estimating the Monetary Benefits of Rising Atmospheric CO_2 Concentrations on Global Food Production*. Tempe, AZ: Center for the Study of Carbon Dioxide and Global Change.

Idso, S.B. and Idso, C.D. 2015. *Mathematical Models vs. Real-World Data: Which Best Predicts Earth's Climatic Future?* Tempe, AZ: Center for the Study of Carbon Dioxide and Global Change.

Ilyinskaya, E., *et al.* 2018. Globally significant CO_2 emissions from Katla, a subglacial volcano in Iceland. *Geophysical Research Letters* **45** (9).

IPCC. 2013. Intergovernmental Panel on Climate Change. *Climate Change 2013: The Physical Science Basis.* Contribution of Working Group I to the Fifth Assessment Report of the Intergovernmental Panel on Climate Change. New York, NY: Cambridge University Press.

IPCC. 2014. Intergovernmental Panel on Climate Change. *Climate Change 2014: Impacts, Adaptation, and Vulnerability. Part A: Global and Sectoral Aspects.* Contribution of Working Group II to the Fifth Assessment Report of the Intergovernmental Panel on Climate Change. New York, NY: Cambridge University Press.

Jamieson, D. 1996. Scientific uncertainty and the political process. *Annals of the American Academy of Political and Social Science* **545**: 35–43.

Jeong, S.-J., *et al.* 2010. Potential impact of vegetation feedback on European heat waves in a 2 x CO_2 climate. *Climatic Change* **99**: 625–35.

Jutting, J. and de Laiglesia, J.R. (Eds.) 2009. *Is Informal Normal? Towards More and Better Jobs in Developing Countries*. Paris, France: Organization for Economic Cooperation and Development.

Kelly, D.L. and Kolstad, C.D. 1998. *Integrated Assessment Models for Climate Change Control*. U.S. Department of Energy grant number DE-FG03-96ER62277. November.

Kirkby, J., *et al.* 2016. Ion-induced nucleation of pure biogenic particles. *Nature* **533**: 521–6.

Kreutzer, D., *et al.* 2016. The state of climate science: no justification for extreme policies. Washington, DC: The Heritage Foundation. April 22.

Lamb, H.H. 2011. *Climate: Present, Past and Future, Volume 1: Fundamentals and Climate Now*. Second edition. New York, NY: Routledge Revivals.

Lamb, H.H. 2012. *Climate: Present, Past and Future: Volume 2: Climatic History and the Future*. New York, NY: Routledge.

Legates, D. 2014. Climate models and their simulation of precipitation. *Energy & Environment* **25** (6–7): 1163–75.

Lewis, N. and Curry, J.A. 2014. The implications for climate sensitivity of AR5 forcing and heat uptake estimates. *Climate Dynamics* **45** (3/4): 1009–23.

Li, C., *et al.* 2015. Eurasian winter cooling in the warming hiatus of 1998–2012. *Geophysical Research Letters* **42**: 8131–9, doi:10.1002/2015GL065327.

Lindzen, R.S. 2015. Global warming, models and language. In: Moran, A. (Ed.) *Climate Change: The Facts*. Institute of Public Affairs. Woodville, NH: Stockade Books.

Lindzen, R.S. and Choi, Y-S. 2011. On the observational determination of climate sensitivity and its implications. *Asia-Pacific Journal of Atmospheric Science* **47**: 377–90.

Lockwood, M., Owens, M.J., Barnard, L., Davis, C.J., and Steinhilber, F. 2011. The persistence of solar activity indicators and the descent of the Sun into Maunder Minimum conditions. *Geophysical Research Letters* **38**: L22105, doi:10.1029/2011GL049811.

Loehle, C. 2018. The epistemological status of general circulation models. *Climate Dynamics* **50**: 1719–31.

Lupo, A., *et al.,* 2013. Global climate models and their limitations. In: *Climate Change Reconsidered II: Physical Science*. Nongovernmental International Panel on Climate Change. Chicago, IL: The Heartland Institute.

Marbà, N., Krause-Jensen, D., Masqué, P., and Duarte, C.M. 2018. Expanding Greenland seagrass meadows contribute new sediment carbon sinks. *Scientific Reports* **8**: 14024.

Martinez, L.R. 2018. How much should we trust the dictator's GDP estimates? Chicago, IL: University of Chicago Irving B. Harris Graduate School of Public Policy Studies. May 1.

McKitrick, R. and Christy, J. 2018. A test of the tropical 200- to 300-hPa warming rate in climate models. *Earth and Space Science* **5**: 529–36.

McKitrick, R. and Michaels, P.J. 2007. Quantifying the influence of anthropogenic surface processes and inhomogeneities on gridded global climate data. *Journal of Geophysical Research* **112** (24).

McLean, J. 2018. *An Audit of the Creation and Content of the HadCRUT4 Temperature Dataset*. Robert Boyle Publishing. October.

Michaels, P.J. 2009. *Climate of Extremes: Global Warming Science They Don't Want You to Know*. Washington, DC: Cato Institute.

Monckton, C., *et al.* 2015. Why models run hot: results from an irreducibly simple climate model. *Science Bulletin* **60** (1).

Montford, A.W. 2010. *The Hockey Stick Illusion: Climategate and the Corruption of Science*. London, UK: Stacey International.

Moser, S.C. and Dilling, L. (Eds.) 2007. *Creating a Climate for Change: Communicating Climate Change and Facilitating Social Change*. New York, NY: Cambridge University Press.

Neumayer, E. and Barthel, F. 2011. Normalizing economic loss from natural disasters: a global analysis. *Global Environmental Change* **21**: 13–24.

NOAA. 2018. U.S. National Oceanic and Atmospheric Administration. How much of the ocean have we explored? (website). July 11. Accessed December 9, 2018.

Myneni, R.B. 2015. *The Greening Earth*. Final Workshop for the Arctic Biomass Project. Svalbard, Norway. October 20–23.

Pielke Jr., R.A., Gratz, J., Landsea, C.W., Collins, D., Saunders, M.A., and Musulin, R. 2008. Normalized hurricane damage in the United States: 1900–2005. *Natural Hazards Review* **31**: 29–42.

Pielke Jr., R.A. and Landsea, C.W. 1998. Normalized hurricane damage in the United States: 1925–95. *Weather and Forecasting* **13**: 621–31.

Pielke Sr., R.A., *et al.* 2007a. Documentation of uncertainties and biases associated with surface temperature measurement sites for climate change assessment. *Bulletin of the American Meteorological Society* **88**: 913–28.

Pielke Sr., R.A., *et al.* 2007b. Unresolved issues with the assessment of multidecadal global land surface temperature trends. *Journal of Geophysical Research* **112** (D24).

Quereda Sala, J., Montón Chiva, E., and Quereda Vázquez, V. 2017. Climatic or non-climatic warming in the Spanish mediterranean? *Wulfenia* **24** (10): 195–208.

Roy, C.J. and Oberkampf, W.L. 2011. A comprehensive framework for verification, validation, and uncertainty quantification in scientific computing. *Computer Methods in Applied Mechanics and Engineering* **200** (25–28): 2131–44.

Ruddiman, W.F. 2008. *Earth's Climate: Past and Future.* Second edition. New York, NY: W.H. Freeman and Company.

Sardeshmukh, P.D., Compo, G.P., and Penland, C. 2015. Need for caution in interpreting extreme weather events. *Journal of Climate* **28**: 9166–87.

Scafetta, N. and West, B.J. 2006. Phenomenological solar contribution to the 1900–2000 global surface warming. *Geophysical Research Letters* **33**: L05708, doi: 10.1029/2005GL025539.

Scafetta, N. and Willson, R.C. 2014. ACRIM total solar irradiance satellite composite validation versus TSI proxy models. *Astrophysical Space Science* **350**: 421–2.

Schiermeier, Q. 2015. IPCC report under fire. *Nature* **508** (7496): 298..

Shackley, S. and Wynne, B. 1996. Representing uncertainty in global climate change science and policy: boundary-ordering devices and authority. *Science, Technology and Human Values* **21**: 275–302.

Smirnov, D.A., Breitenbach, S.F.M., Feulner, G., Lechleitner, F.A., Prufer, K.M., Baldini, U.L., Marwan, N., and Kurths, J. 2017. A regime shift in the Sun-climate connection with the end of the Medieval Climate Anomaly. *Scientific Reports* **7**: 11131.

Soon, W. W.-H. 2005. Variable solar irradiance as a plausible agent for multidecadal variations in the Arctic-wide surface air temperature record of the past 130 years. *Geophysical Research Letters* **32**: 10.1029/2005GL023429.

Soon, W. W.-H. 2009. Solar Arctic-mediated climate variation on multidecadal to centennial timescales: empirical evidence, mechanistic explanation and testable consequences. *Physical Geography* **30**: 144–84.

Soon, W., Connolly, R., and Connolly, M. 2015. Re-evaluating the role of solar variability on Northern Hemisphere temperature trends since the 19th century. *Earth-Science Reviews* **150**: 409–52.

Spencer, R., Braswell, W., Christy, J., and Hnilo, J. 2007. Cloud and radiation budget changes associated with tropical intraseasonal oscillations. *Geophysical Research Letters* **34** (15).

Stevens, B. and Bony, S. 2013. What are climate models missing? *Science* **340** (6136): 1053–4.

Stouffer, R.J., *et al.* 2017. CMIP5 Scientific Gaps and Recommendations for CMIP6. *Bulletin of the American Meteorological Society* **98** (1).

Sun, L., *et al.* 2016. What caused the recent "Warm Arctic, Cold Continents" trend pattern in winter temperatures? *Geophysical Research Letters* **23** (10).

Sutton, R., Hoskins, B., Palmer, T., Shepherd, T., and Slingo, J. 2018. Attributing extreme weather to climate change is not a done deal. *Nature* **561** (177): 10.1038/d41586-018-06631-7.

Svensmark, H. 2007. Cosmoclimatology: a new theory emerges. *Astronomy and Geophysics* **48**: 1–19.

Svensmark, H., Enghoff, M.B., Shaviv, N.J., and Svensmark, J. 2017. Increased ionization supports growth of aerosols into cloud condensation nuclei. *Nature Communications* **8**: Article 2199.

Tans, P. 2009. An accounting of the observed increase in oceanic and atmospheric CO_2 and an outlook for the future. *Oceanography* **22**: 26–35.

Viterito, A. 2017. The correlation of seismic activity and recent global warming: 2016 update. *Environment Pollution and Climate Change* **1** (2): 1000103.

Wang, J., Feng, L., Tang, X., Bentley, Y., and Höök, M. 2017. The implications of fossil fuel supply constraints on climate change projections: a supply-side analysis. *Futures* **86** (February): 58–72.

Zeng, Z., *et al.* 2018. Impact of Earth greening on the terrestrial water cycle. *Journal of Climate* **31**: 2633–50.

Ziskin, S. and Shaviv, N.J. 2012. Quantifying the role of solar radiative forcing over the 20th century. *Advances in Space Research* **50** (6): 762–76.

2.4.2 An Interdisciplinary Topic

Climate is an interdisciplinary subject requiring insights from many fields of study. Very few scholars have mastery of more than one or two of these disciplines.

"Global warming is a topic that sprawls in a thousand technical directions," write Essex and McKitrick (2007, p. 17). They continue, "There is no such thing as an 'expert' on global warming, because no one can master all the relevant subjects. On the subject of climate change everyone is an amateur on many if not most topics, including ourselves."

The collaboration of experts from such fields as agronomy, astronomy, astrophysics, biology, botany, cosmology, economics, geochemistry, geology, history, oceanography, paleontology, physics, scientific forecasting, and statistics, among other disciplines, explains at least some of the confusion and disagreement on basic issues of climatology. Geologists, for example, view time in millennia and eons and are aware of past fluctuations in both global temperatures and carbon dioxide concentrations in the atmosphere, with the two often moving in opposite directions. Solar physicists think of time in terms of seconds and even nanoseconds and study action at an atomic level. Few geologists understand solar physics while few physicists understand geology.

In their attempts to summarize and simplify their research findings for researchers in other fields, scientists and others resort to "stylized facts," generalizations that may accurately convey the gist of their scientific research but are necessarily inaccurate, overly simplified, and often unsuited for use as inputs to the work of other researchers. The "mean global surface temperature" is one such stylized fact, the use of which in climate science has caused extensive controversy and disagreement. Historical records of total solar irradiance (TSI), greenhouse gas "inventories" and emission scenarios, global sea-level rise, and hurricane "best tracks" are just a few more of the many examples discussed earlier in this chapter and in other chapters of this book.

Related to this is the fact that scientists are often optimistic about the resilience or safety of the environment in their own area of research and expertise, but are pessimistic about risks with which they are less familiar. Simon (1999) wrote,

This phenomenon is apparent everywhere. Physicians know about the extraordinary progress in medicine that they fully expect to continue, but they can't believe in the same sort of progress in natural resources. Geologists know about the progress in natural resources that pushes down their prices, but they worry about food. Even worse, some of those who are most optimistic about their own areas point with alarm to other issues to promote their own initiatives. The motive is sometimes self-interest (pp. 47–8).

Physical scientists who think they can discern a human impact on climate are apt to call for actions to diminish or end such an impact, the logical solution to the "problem" they have found. But when doing so they step outside their area of expertise and express only informed opinions. First, can they actually predict future human carbon dioxide emissions, and then future CO_2 levels in the atmosphere, and then the impact of that concentration on the global mean surface temperature, and then the impact of that change on weather, sea level, plants, and human well-being? And on each of these impacts, where? when? And how do we know? How many physicists, geologists, or chemists know enough to endorse the science behind each step in this chain argument? The answer is "none."

Second and equally important, climate change is not a "problem" simply because physicists or climatologists say it is. What to do about climate change is a question whose answer lies in the domain of social science, not physical science. Who benefits and who is hurt if climate change is allowed to proceed unabated? Whose rights are violated and whose should be protected? What institutions, laws, and precedents exist that govern situations like this? Who has the right to decide, to intervene, and to enforce action? What are the probabilities of success of such interventions based on past experience? Physicists, climatologists, and other physical scientists may have opinions on all these matters, but they have no expertise.

Economists, of course, do not have all the answers, either. They typically enter the debate late in the chained argument and are asked to "monetize" climate impacts, recommend a discount rate used to value benefits that occur beyond 50 or 60 years in the future, and give advice on the best way to reduce "carbon pollution." These matters were addressed in Chapter 1 and will be again in Chapter 8, but it is worth noting here that economists give conflicting advice on every point. Some call for immediate action (e.g., Stiglitz, 2018), while others say the benefits exeed the costs in the short and medium run, so "climate change would appear to be an important issue primarily for those who are concerned about the distant future, faraway lands, and remote probabilities" (Tol, 2018).

Monetizing climate impacts requires evaluating risks, which in turn requires choosing one of hundreds or thousands of competing climate impact scenarios. Economists must then estimate or assume how quickly and successfully humans can substitute renewable energies for fossil fuels or adapt to climate changes or (more likely) some combination of both. There is no agreement among economists on this matter, or on the right discount rate to use; some recommend 7% while others recommend rates as low as 0%. The choice of discount rates means the difference between action today and no action until 2050 or even beyond. Rather than recommend a carbon tax or some other government "solution" to climate change, some economists recommend ways to tap local knowledge to turn climate change into a win-win opportunity (Goklany, 2007; Lomberg, 2010; van Kooten, 2013; Morris, 2015; Kahn, 2010, 2016–17; Olmstead and Rhode, 2016–17; Zycher, 2017; Anderson *et al.,* 2018; Tol, 2018).

The Rise of Computer Models

"*Climatology*," writes Ball (2015), "is the study of weather patterns of a place or region, or the change of weather patterns over time. *Climate science* is the study of one component piece of climatology." Many of the practitioners of climate science are not climatologists or even very familiar with the subject. They come from other disciplines – physics, geology, biology, or even engineering, economics, and law – and lend their expertise to the effort to solve the puzzle that is Earth's climate. "The analogy I've used for decades," Ball writes,

> is that climatology is a puzzle of thousands of pieces; climate science is one piece of the puzzle. A practical approach to assembling the puzzle is to classify pieces into groups. The most basic sorting identifies the corner pieces, the edge pieces and then color. Climatologists say the four corner pieces, which are oceans, atmosphere, lithosphere, and the cosmos are not even fully identified or understood. Climate scientists tend to hold one piece of the puzzle and claim it is the key to everything.

Because it is a young discipline – many of its pioneers are still writing or only recently deceased – climatology very quickly fell victim to the growth of specialization in the academy. To bridge the distances between specialties, academics have turned to systems analysis, sometimes defined as "the analysis of the requirements of a task and the expression of those requirements in a form that permits the assembly of computer hardware and software to perform the task." In climate science,

general circulation models (GCMs) and integrated assessment models (IAMs) are used to "perform the task" as well as handle the immense amounts of data being generated by new satellites and spectrometer analysis of various temperature and weather proxies.

Computer models are useful, but they cannot solve disagreements when there is no agreement on a general theory of climate (Essex and McKitrick, 2007). When an expert in one field, say physics, presents an estimate of the sensitivity of the climate to rising carbon dioxide levels, an expert in another field, say geology, can quickly challenge her understanding of the carbon cycle, and rightly so. The physicist probably accepts uncritically estimates of the size of carbon reservoirs and exchange rates offered by, say, the IPCC, and is unaware of the significant uncertainty and error bars surrounding those estimates. If she knew, her calculation of climate sensitivity would likely be much different. Many geologists are unfamiliar with the extensive literature on the impact of rising levels of atmospheric CO_2 on photosynthesis and plant growth, so they stand to be corrected by biologists, botanists, and agronomists. And then it might take an economist to estimate how the application of technology will change agriculture and forestry in the future, affecting the rate of exchange between the biosphere and atmosphere and once again affecting the estimate of climate sensitivity. As reported in Section 2.1.1.3, uncertainties propagate and confusion and disagreement rise exponentially.

GCMs and IAMs can be tuned according to the knowledge, opinions, and biases of their modelers, and then run with an infinite combination of databases reflecting underlying uncertainties in observational data (Hourdin *et al.*, 2017). No wonder 102 GCMs produce the wide array of hindcasts and forecasts of temperature, shown in the graph from McKitrick and Christy (2018) reproduced as Figure 2.2.2.1 earlier in this chapter. It brings to mind the famous tale of a group of blind men touching various parts of an elephant, each arriving at a very different idea of what it is like: to one it is like a tree, to another, a snake, and to a third, a wall. A wise man tells the group, "You are all right. An elephant has all the features you mentioned."

The role of "wise man" in climate science falls to computer modelers, but as Ball (2015) remarks, they are seldom climatologists. He writes,

> After a discussion with a computer modeler in 1998, I realized the limitations of his weather and climate knowledge. Despite this,

I watched modelers take over as climate scientists and become keynote speakers at most climate conferences. It became so technologically centered that whoever had the biggest fastest computers were the 'state of the art' climate experts. I recall the impact of the Cray computer on climate science. The idiocy continues today with the belief that the only limitation to the models is computer capacity and speed.

References

Anderson, S.E. *et al.* 2018. The critical role of markets in climate change adaptation. NBER Working Paper No. 24645, May.

Ball, T. 2015. Generalization, specialization and climatology. Watts Up with That (website). June 29. Accessed December 8, 2018.

Essex, C. and McKitrick, R. 2007. *Taken by Storm. The Troubled Science, Policy and Politics of Global Warming.* Toronto, ON: Key Porter Books.

Goklany, I.M. 2007. *The Improving State of the World.* Washington, DC: Cato Institute.

Hourdin, F., *et al.* 2017. The art and science of climate model tuning. *Bulletin of the American Meteorological Society* **98** (3): 589–602.

Kahn, M.E. 2010. *Climatopolis: How Our Cities will Thrive in the Hotter Future.* New York, NY: Basic Books.

Kahn, M.E. 2016-17. Climatopolis revisited. *PERC Reports* **35**: 2.

Lomborg, B. 2010. (Ed.) *Smart Solutions to Climate Change: Comparing Costs and Benefits.* Cambridge, MA: Cambridge University Press.

McKitrick, R. and Christy, J. 2018. A Test of the tropical 200- to 300-hPa warming rate in climate models. *Earth and Space Science* **5**: 529–36.

Morris, J. 2015. Assessing the social costs and benefits of regulating carbon emissions. *Policy Study* No. 445. Los Angeles, CA: Reason Foundation.

Olmstead, A.L. and Rhode, P.W. 2016–17. Amber waves of change. *PERC Reports* **35**: 2.

Simon, J. 1999. *Hoodwinking the Nation.* New Brunswick, NJ: Transaction Publishers.

Stiglitz, J.E. 2018. Expert report prepared for plaintiffs and attorneys for plaintiffs, *Juliana v. U.S.A.,* in the United States District Court, District of Oregon, Case No. 6:15-cv-01517-TC, submitted June 28.

Tol, R.S.J. 2018. The economic impacts of climate change. *Review of Environmental Economics and Policy* **12**: 1.

van Kooten, G.C. 2013. *Climate Change, Climate Science and Economics.* New York, NY: Springer.

Zycher, B. 2017. *The Deeply Flawed Conservative Case for a Carbon Tax.* Washington, DC: American Enterprise Institute.

2.4.3 Failure of the IPCC

Many scientists trust the United Nations' Intergovernmental Panel on Climate Change (IPCC) to objectively report the latest scientific findings on climate change, but it has failed to produce balanced reports and has allowed its findings to be misreported to the public.

The Intergovernmental Panel on Climate Change (IPCC) was created in 1988 by the United Nations as a joint project of two of its agencies, the Environmental Programme (UNEP) and the World Meteorological Organization (WMO). Its initial Statement of Principles reads, "The role of the IPCC is to assess on a comprehensive, objective, open and transparent basis the scientific, technical and socio-economic information relevant to understanding the scientific basis of risk of human-induced climate change, its potential impacts, and options for adaptation and mitigation" (IPCC, 1988).

The IPCC provides research support to the UN's Framework Convention on Climate Change (UNFCCC), an entity that arose from the 1992 United Nations Conference on Environment and Development (UNCED), also known as the Rio Earth Summit. The IPCC supports the Conference of the Parties (COP), which meets annually to oversee implementation of the 1997 Kyoto Protocol and 2015 Paris Accord.

The regular and special reports of the IPCC dominate the climate change debate, and rightly so. The five multi-volume "assessments" produced to date total more than 10,000 pages, much of it consisting of dense literature reviews in which hundreds and possibly thousands of climate scientists and policy advocates take part. The hefty tomes identify hundreds of scientists as lead authors, contributors, or reviewers and are presented by the IPCC as representing the "consensus" of the scientific community on the climate change issue. Many of the world's science academies, membership organizations, and leading journals have endorsed the IPCC's findings and give its spokespersons extensive attention. In 2007, the IPCC was a co-recipient (with former U.S. Vice President Al Gore) of the Nobel Peace Prize.

While there is no disputing the fact that the IPCC has summarized immense amounts of high-quality and consequential research, researchers and policymakers should understand that the organization labors under political and institutional constraints that undermine the credibility of its work, and that an audit of its procedures found disturbing violations of proper scientific procedures.

Political and Institutional Restraints

The IPCC was created by the United Nations to build a scientific case for giving it authority to regulate the planet's atmosphere as a collective good or resource commons, a kind of resource defined and described in Chapter 1, Section 1.3. Much of what the IPCC does and does not do can be explained by the "seeing like a state" phenomenon described by James C. Scott (1998) and explained in Chapter 1, Section 1.3.4.

The story of the creation of the IPCC is told well in such books as *The Age of Global Warming: A History* (Darwall, 2013) and *Searching for the Catastrophe Signal: The Origins of the Intergovernmental Panel on Climate Change* (Lewin, 2017) and will not be repeated here. Article 1.2 of the Framework Convention on Climate Change (UNFCCC, 1992), which provides the IPCC's mandate, defines climate change as "a change of climate which is attributed directly or indirectly to human activity that alters the composition of the global atmosphere and which is in addition to natural climate variability observed over comparable time periods." Working Group I of the IPCC has interpreted this as a mandate to not study climate change "in the round" or to look at natural as well as man-made influences on climate. Instead, it seeks to find and report only a possible human impact on climate, and thereby make a scientific case for adopting national and international policies that could reduce that impact. Missing from the 1,535-page

Working Group I contribution to IPCC's Fifth Assessment Report (IPCC, 2013) is a serious analysis of natural causes of climate variability. The section titled "Natural Radiative Forcing Changes: Solar and Volcanic" runs only six pages, and "The Carbon Dioxide Fertilization Effect" merits only a single page. A balanced report would have devoted hundreds of pages to each topic.

Similarly, Working Group II of the IPCC views its assignment as being to catalogue all possible *harms* of "climate change," including those arising from natural as well as anthropogenic causes, and it hardly mentions the benefits of a warmer climate, increasing atmospheric CO_2, or the fossil fuels it argues should be banned by the end of the twenty-first century (IPCC, 2014). Yet it is apparent, even obvious, that the moderate warming experienced in the late-twentieth and early twenty-first centuries has been beneficial to many parts of the world. Increasing crop yields, the retreat of deserts, and the reduced toll of cold days and nights are well documented in the peer-reviewed literature but almost entirely absent from the IPCC's work.

A second institutional constraint on the IPCC is the inevitable consequence of asking a committee to deal with a complex and controversial issue. The incentive structure of committees leads them to allow the inclusion of declaratory statements and confident predictions so long as they are accompanied by caveats and admissions of uncertainty and dissenting views. The result is a schizophrenic style that veers from declarations of "extreme confidence" to admissions of complete doubt and uncertainty, often before a paragraph ends but sometimes in a different chapter or volume of the assessment report. An example, provided by Gleditsch and Nordås (2014), is in Chapter 22 of the Working Group II contribution to AR5 (on Africa), which "states explicitly that 'causality between climate change and violent conflict is difficult to establish' (p. 5), yet goes on to say on the same page that 'the degradation of natural resources as a result of both overexploitation and climate change will contribute to increased conflicts over the distribution of these resources.'" Once the pattern is noticed, it becomes apparent on many pages of every working group report.

The IPCC is able to present conflicting opinions side-by-side and yet avoid nonsense by making ample use of what Gleditsch and Nordås call "expressions of uncertainty." Such words as may, might, can, and could, and such phrases as "has a potential to," "is a potential cause of," and "is sensitive to" appear on average 1.2 times per page in the Working Group I Summary for Policymakers (SPM) and 1.3 times per page in the Working Group II SPM. They write, "The frequent use of 'may' terms might have been justified as a way of indicating that 'under certain circumstances, a relationship is likely.' But this does not work well if those circumstances are not specified. On the whole, it would probably be best to avoid the use of terms like 'may' in academic writing except to state conjectures. Misrepresentation of the scientific basis is a real hazard when using such terminology" (p. 88).

A third institutional constraint is that the IPCC's "members" are not scientists but national governments that are members of the United Nations. The WMO, for example, has "191 member States and Territories" (WMO, n.d.). The national governments that created the IPCC also fund it, staff it, select the scientists who participate in its work, and importantly, revise and rewrite the reports after the scientists have concluded their work. In 2014, a reporter for *Science* described the political interference on display in events leading up to the release of the Working Group III contribution to AR5: "Although the underlying technical report from WG III was accepted by the IPCC, final, heated negotiations among scientific authors and diplomats led to a substantial deletion of figures and text from the influential 'Summary for Policymakers' (SPM). … [S]ome fear that this redaction of content marks an overstepping of political interests, raising questions about division of labor between scientists and policy-makers and the need for new strategies in assessing complex science. Others argue that SPM should explicitly be coproduced with governments" (Wible, 2014). The subtitle of the article is "Did the 'Summary for Policymakers' become a summary *by* policy-makers?"

Serving the State

The IPCC's first report (IPCC, 1990) found "increases in atmospheric concentrations of greenhouse gases may lead to irreversible change in the climate which could be detectable by the end of this century" (p. 53). Every assessment report since then has claimed with rising certainty that there is a "discernable human impact" on the climate and that steps must be taken to avoid a global climate crisis, even though the IPCC's estimates of climate sensitivity to CO_2 have stayed largely unchanged

since 1990 and declined in the peer-reviewed literature, temperatures have risen only half as much as the IPCC predicted, and few if any of the negative climate impacts predicted by the IPCC have been observed. *Why, as direct evidence increasingly pointed away from a climate change crisis, has the IPCC's rhetoric become more extreme?* The problems of "tunnel vision" and "moral hazard" afflicting government bureaucracies, described in Chapter 1, Section 1.4.3, provide an explanation for this trend. So too does the phenomenon of "seeing like a state," whereby government agencies produce data and stylized facts they believe will satisfy their political overseers.

Many admissions of uncertainty appear in the IPCC's hefty assessment reports, including AR5, a fact established earlier in this chapter and repeated in other chapters, but the IPCC's purpose and agenda work to ensure that uncertainty is not broadly advertised. The opening words of the foreword to the Working Group I contribution to AR5, for example, read: "'Climate Change 2013: the Physical Science Basis' presents clear and robust conclusions in the global assessment of climate change science – not the least of which is that the science now shows with 95 percent certainty that human action is the dominant cause of observed warming since the mid-20th century" (p. v). The authors – the Secretary General of the WMO and Executive Director of the UNEP – continue in this same declarative tone, "warming in the climate system is unequivocal, with many of the observed changes unprecedented over decades to millennia."

Only people deeply familiar with AR5 realize what the Secretary General and Executive Director chose *not* to say. The "95 percent certainty" is not an expression of statistical significance but only a rhetorical expression of the strength of opinions, a number quite literally arrived at by a show of hands around a table. It is not derived from a poll of the scientists who contributed to the volume but only the opinion of a few individuals, including nonscientists, who were involved in the writing and rewriting of the Summary for Policymakers. The only survey done of contributors to AR5 found a majority do not endorse this statement (Fabius Maximus, 2015).

Similarly, the IPCC found the warming is "unequivocal" except when it is not, such as during the "pause" from 1997 to 2010 when the real scientists who helped write AR5 acknowledge there was no warming at all ("0.07°C [-0.02 to 0.18])" (IPCC, 2013, p. 37). Similarly, "observed changes [are] unprecedented" except, as this chapter documents, for temperatures, extreme weather events, polar ice melting, and sea-level rise. What is left?

The doubts and uncertainty that the Scientific Method requires be revealed so other researchers know what is scientific fact and what is conjecture or speculation do appear in AR5, and in other IPCC assessment reports, but they are scrubbed from the Summaries for Policymakers (SPMs). This scientific malpractice has been protested by many distinguished scientists (Seitz, 1996; Landsea, 2005; Lindzen, 2012; Tol, 2014; Stavins, 2014). Unfortunately, many scientists look no further than the SPMs and trust them to accurately depict the current state of climate science.

Tol (2014) commented on how the AR5 Summary for Policymakers, "drafted by the scholars of the IPCC, is rewritten by delegates of the governments of the world," each with political agendas that lead to interference with the report's scientific content. Even the scientists who participate are biased: "The IPCC does not guard itself against selection bias and groupthink. Academics who worry about climate change are more likely to publish about it, and more likely to get into the IPCC. Groups of like-minded people reinforce their beliefs. The environment agencies that comment on the draft IPCC report will not argue that their department is obsolete. The IPCC should therefore be taken out of the hands of the climate bureaucracy and transferred to the academic authorities" (Tol, 2014).

The IAC Audit

It is often remarked that nearly all of the world's national science academies have "endorsed" the IPCC's findings. Such a claim is superficially true but scientifically meaningless: none of these academies surveyed its members to see if they agree with everything contained in the IPCC's massive reports. Even if *most* of the members say they approve of *most* of what the IPCC writes in its latest report, what does this say about the views of the much smaller community of climate scientists, engineers, economists, and others who specialize in climatology and have informed opinions on it? Voting is not an effective way of separating sound science from pseudoscience.

Also very telling is that when the presidents of those same institutions conducted an audit of the IPCC's practices, they found ample grounds to doubt the organization's credibility. The InterAcademy

Council (IAC), made up of the presidents of the world's leading national science academies, audited the IPCC in 2010 (IAC, 2010). Among its findings:

- *Fake confidence intervals*: The IAC was highly critical of the IPCC's method of assigning "confidence" levels to its forecasts, singling out "… the many statements in the Working Group II Summary for Policymakers that are assigned high confidence but are based on little evidence. Moreover, the apparent need to include statements of 'high confidence' (i.e., an 8 out of 10 chance of being correct) in the Summary for Policymakers led authors to make many vaguely defined statements that are difficult to refute, therefore making them of 'high confidence.' Such statements have little value" (p. 61).

- *Use of gray sources*: Too much reliance on unpublished and non-peer-reviewed sources. Three sections of the IPCC's 2001 climate assessment cited peer-reviewed material only 36%, 59%, and 84% of the time (p. 63).

- *Political interference*: Line-by-line editing of the summaries for policymakers during "grueling Plenary session that lasts several days, usually culminating in an all-night meeting. Scientists and government representatives who responded to the Committee's questionnaire suggested changes to reduce opportunities for political interference with the scientific results …" (p. 64).

- *The use of secret data*: "An unwillingness to share data with critics and enquirers and poor procedures to respond to freedom-of-information requests were the main problems uncovered in some of the controversies surrounding the IPCC (Russell *et al.*, 2010; PBL, 2010). Poor access to data inhibits users' ability to check the quality of the data used and to verify the conclusions drawn …" (p. 68).

- *Selection of contributors is politicized*: Politicians decide which scientists are allowed to participate in the writing and review process: "political considerations are given more weight than scientific qualifications" (p. 14).

- *Chapter authors exclude opposing views*: "Equally important is combating confirmation bias – the tendency of authors to place too much weight on their own views relative to other views (Jonas *et al.*, 2001). As pointed out to the Committee by a presenter and some questionnaire respondents, alternative views are not always cited in a chapter if the Lead Authors do not agree with them ..." (p. 18).

- *Need for independent review*: "Although implementing the above recommendations would greatly strengthen the review process, it would not make the review process truly independent because the Working Group Co-chairs, who have overall responsibility for the preparation of the reports, are also responsible for selecting Review Editors. To be independent, the selection of Review Editors would have to be made by an individual or group not engaged in writing the report, and Review Editors would report directly to that individual or group (NRC, 1998, 2002)" (p. 21).

The quotations above are from a publicly circulated draft of the IAC's final report, still available online (see reference). The final report was heavily edited to water down and perhaps hide the extent of problems uncovered by the investigators, itself evidence of misconduct. The IPCC accepted the IAC's findings and promised to make changes … too late to affect AR5.

Some climate scientists spoke out early and forcefully against the problems they saw as compromising the integrity of the IPCC, but their voices were difficult to hear amid a steady drumbeat of alarmism from media outlets. As a result, many scientists, economists, and policymakers have been misled into thinking the IPCC is the final or even only authority on climate change. German meteorologist and physicist Klaus-Ekhart Pul said in an interview in 2012, "Ten years ago I simply parroted what the IPCC told us. One day I started checking the facts and data – first I started with a sense of doubt but then I became outraged when I discovered that much of what the IPCC and the media were telling us was sheer nonsense and was not even supported by any scientific facts and measurements. To this day I still feel shame that as a scientist I made presentations of their science without first checking it" (translated by Gosselin, 2012).

* * *

This is a harsh criticism of an organization that, as noted at the beginning of this section, has summarized large amounts of high-quality and consequential research. It is not our purpose to disparage the individuals involved, and a careful reading of this section will show that we did not do so. The IPCC's failures arise from its mission, its oversight by two government agencies, and the inevitable dynamics of assigning a committee to address a complex problem. Our point is a simple one: The IPCC is not the final word on climate science.

References

Darwall, R. 2013. *The Age of Global Warming: A History.* London, UK: Quartet Books.

Fabius Maximus. 2015. New study undercuts the key IPCC finding (website).

Gleditsch, N.P. and Nordås, R. 2014. Conflicting messages? The IPCC on conflict and human security. *Political Geography* **43**: 82–90.

Gosselin, P. 2012. The belief that CO_2 can regulate climate is "sheer absurdity" says prominent German meteorologist. NoTricksZone (website). Accessed November 20, 2018.

IAC. 2010. InterAcademy Council. Draft: *Climate Change Assessments: Review of the Processes & Procedures of IPCC*. The Hague, Netherlands: Committee to Review the Intergovernmental Panel on Climate Change. October.

IPCC. 1988. Intergovernmental Panel on Climate Change. Principles Governing IPCC Work. Approved October 1988, amended 2003, 2006, 2012, and 2013.

IPCC. 1990. Intergovernmental Panel on Climate Change. *First Assessment Report*. United Nations World Meteorological Association/Environment Programme.

IPCC. 2013. Intergovernmental Panel on Climate Change. *Climate Change 2013: The Physical Science Basis*. Contribution of Working Group I to the Fifth Assessment Report of the Intergovernmental Panel on Climate Change. New York, NY: Cambridge University Press.

IPCC. 2014. Intergovernmental Panel on Climate Change. *Climate Change 2014: Impacts, Adaptation, and Vulnerability. Part A: Global and Sectoral Aspects*. Contribution of Working Group II to the Fifth Assessment Report of the Intergovernmental Panel on Climate Change. Cambridge, UK and New York, NY: Cambridge University Press.

Jonas, E., Schulz-Hardt, S., Frey, D., and Thelen, N. 2001. Confirmation bias in sequential information search after preliminary decisions. *Journal of Personality and Social Psychology* **80** (4): 557.

Landsea, C. 2005. Resignation Letter of Chris Landsea from IPCC. Climatechangefacts.info (website). Accessed November 20, 2018.

Lewin, B. 2017. *Searching for the Catastrophe Signal: The Origins of the Intergovernmental Panel on Climate Change*. London, UK: The Global Warming Policy Foundation.

Lindzen, R.S. 2012. Climate science: is it currently designed to answer questions? *Euresis Journal* **2** (Winter): 161–92.

NRC. 1998. National Research Council. *Peer Review in Environmental Technology Development Programs*. Washington, DC: The National Academies Press.

NRC. 2002. National Research Council. *Knowledge and Diplomacy: Science Advice in the United Nations System*. Washington, DC: The National Academies Press.

PBL. 2010. *Assessing an IPCC Assessment: An Analysis of Statements on Projected Regional Impacts in the 2007 Report*. The Hague, Netherlands: Netherlands Environmental Assessment Agency.

Russell, M., *et al.* 2010. *The Independent Climate Change E-mails Review*. Report to the University of East Anglia.

Scott, J.C. 1998. *Seeing Like a State: How Certain Schemes to Improve the Human Condition Have Failed*. New Haven, CT: Yale University Press.

Seitz, F. 1996. A major deception on global warming. *The Wall Street Journal*. June 12.

Stavins, R. 2014. Is the IPCC government approval process broken? An Economic View of the Environment (blog). April 25.

Tol, R.S.J. 2014. IPCC again. Richard Tol Occasional thoughts on all sorts (blog). Accessed June 26, 2018.

UNFCC. 1992. United Nations Framework Convention on Climate Change. Bonn, Germany.

WMO. n.d. United Nations World Meteorological Organization. Who we are (website). Accessed December 9, 2018.

Wible, B. 2014. IPCC lessons from Berlin. *Science* **345** (6192 July): 34.

2.4.4 Tunnel Vision

Climate scientists, like all humans, can have tunnel vision. Bias, even or especially if subconscious, can be especially pernicious when data are equivocal and allow multiple interpretations, as in climatology.

Bias, or what might better be called "tunnel vision" (Breyer, 1993), is another reason for disagreement among scientists and other writers on climate change. Scientists, no less than other human beings, bring their personal beliefs and interests to their work and sometimes make decisions based on them that direct their attention away from research findings that would contradict their opinions. Bias is often subconscious but, if recognized, can be overcome by careful adherence to procedures or by being guided by professional ethics, but sometimes it leads to outright corruption.

Essex and McKitrick (2007) write, "Journalists have taken to writing 'exposé' articles that never seem to expose the substance of the scientific arguments at issue, but instead grub around for connections, however tenuous, between scientists and the petroleum industry … but always leaving just-barely unstated the libelous premise that such people would falsify their research or misrepresent their real views for some filthy lucre" (p. 52). "Charges of 'sowing doubt' mean nothing," they add. "They can be hurled at any opponent: my idea is the right one and yours is just sowing doubt. But the lessons of half a millennium have taught us over and over that doubt is part of the lifeblood of scientific advance. We need it. It's worth fighting for" (p. 55).

To obtain funding (and more funding), it helps scientists immensely to have the public – and thus Congress and potentially private funders – worried about the critical nature of the problems they study. This incentive makes it less likely researchers will interpret existing knowledge or present their findings in a way that reduces public concern (Lichter and Rothman, 1999; Kellow, 2007; Kabat, 2008). As a result, scientists often gravitate toward emphasizing worst-case scenarios, though there may be ample evidence to the contrary. This bias of alarmism knows no political bounds, affecting scientists of all political stripes (Berezow and Campbell, 2012; Lindzen, 2012).

Freedman (2010) identifies a long list of reasons why experts are often wrong, including pandering to audiences or clients, lack of oversight, reliance on flawed evidence provided by others, and failure to take into account important confounding variables. Scientists, especially those in charge of large research projects and laboratories, may have a financial incentive to seek more funding for their programs. They are not immune to having tunnel vision regarding the importance of their work and employment. Each believes his or her mission is more significant and essential relative to other budget priorities.

Park *et al.* (2014), in a paper published in *Nature*, summarized research on publication bias, careerism, data fabrication, and fraud to explain how scientists converge on false conclusions. They write, "Here we show that even when scientists are motivated to promote the truth, their behaviour may be influenced, and even dominated, by information gleaned from their peers' behaviour, rather than by their personal dispositions. This phenomenon, known as herding, subjects the scientific community to an inherent risk of converging on an incorrect answer and raises the possibility that, under certain conditions, science may not be self-correcting."

Some journalists seem to recognize only one possible source of bias, and that is funding from "the fossil fuel industry." The accusation often permeates conversations of the subject, perhaps second only to the "consensus" claim, and the two are often paired, as in "only scientists paid by the fossil fuel industry dispute the overwhelming scientific consensus." The accusation doesn't carry any weight for many reasons:

■ There has never been any evidence of a climate scientist accepting money from industry to take a position or change his or her position in the climate debate (Singer, 2010; Cook, 2014).

■ Vanishingly few global warming skeptics have ever been paid by the fossil fuel industry, certainly not more than a tiny fraction of the 31,478 American scientists who signed the Global Warming Petition or the hundreds of meteorologists and climate scientists reported in Section 2.1 who tell survey-takers they do not agree with the IPCC.

■ Funding of alarmists by government agencies, liberal foundations, environmental advocacy groups, and the alternative energy industry exceeds funding from the fossil fuel industry by as much as four orders of magnitude (Nova, 2009; Butos and McQuade, 2015). Does government and interest-group funding of alarmists not also have a "corrupting" influence on its recipients?

■ The most prominent organizations supporting global warming skepticism get little if any money from the fossil fuel industry. Their support comes overwhelmingly from individuals (and their foundations) motivated by concern over the corruption of science and the enormous costs it is imposing on the public.

Curry (2015) worries more about the influence of government grants than private funding on scientists: "I am very concerned that climate science is becoming biased owing to biases in federal funding priorities and the institutionalization by professional societies of a particular ideology related to climate change. Many scientists, and institutions that support science, are becoming advocates for UN climate policies, which is leading scientists into overconfidence in their assessments and public statements and into failures to respond to genuine criticisms of the scientific consensus. In short, the climate science establishment has become intolerant to disagreement and debate, and is attempting to marginalize and de-legitimize dissent as corrupt or ignorant."

The extensive funding of global warming alarmism by government agencies, corporations, and liberal foundations and the negative effect it has on the public's understanding of the issue were described in Chapter 1, Sections 1.4.4 and 1.4.5. Darwall (2018) describes how political, business, and advocacy group interests converge to form a "Climate Industrial Complex" with immense resources at its disposal. He writes, "there is another Moore's law. The second one says the richer you are, the more likely you are to support green causes. ... Climate change is ethics for the wealthy: It legitimizes great accumulations of wealth. Pledging to combat it immunizes climate-friendly corporate leaders and billionaires from being targeted as memmbers of the top one-tenth of the top one percent" (pp. 211-12). If supporting renewable energy harmed their interests, Darwall writes, these major donors to the global warming movement would "drop it in a nanosecond."

* * *

Scientists disagree about climate change for many reasons, and often different reasons. Skepticism, the heart of science, means questioning orthodoxy and always asking, *how do we know?* Disagreement over climate science is to be expected and even encouraged for reasons presented in this section. As Essex and McKitrick (2007) write, "We need it. It's worth fighting for."

References

Berezow, A.B. and Campbell, H. 2012. *Science Left Behind: Feel-Good Fallacies and the Rise of the Anti-Scientific Left.* Philadelphia, PA: PublicAffairs.

Breyer, S. 1993. *Breaking the Vicious Circle: Toward Effective Risk Regulation.* Cambridge, MA: Harvard University Press.

Butos, W.N. and McQuade, T.J. 2015. Causes and consequences of the climate science boom. *The Independent Review* **20** (2 Fall): 165–96.

Cook, R. 2014. Merchants of Smear. *Policy Brief.* Arlington Heights, IL: The Heartland Institute.

Curry, J. 2015. Is Federal Funding Biasing Climate Research? Climate Etc. (website). May 6. Accessed May 21, 2018.

Darwall, R. 2018. *Green Tyranny: Exposing the totalitarian roots of the climate industrial complex.* New York, NY: Encounter Books.

Essex, C. and McKitrick, R. 2007. *Taken by Storm: The Troubled Science, Policy, and Politics of Global Warming.* Toronto, ON: Key Porter Books Limited.

Freedman, D.H. 2010. *Wrong: Why Experts Keep Failing Us – And How to Know When Not to Trust Them.* New York, NY: Little, Brown and Company.

Kabat, G.C. 2008. *Hyping Health Risks: Environmental Hazards in Daily Life and the Science of Epidemiology.* New York, NY: Columbia University Press.

Kellow, A. 2007. *Science and Public Policy: The Virtuous Corruption of Virtual Environmental Science.* Northampton, MA: Edward Elgar Publishing.

Lichter, S.R. and Rothman, S. 1999. *Environmental Cancer – A Political Disease?* New Haven, CT: Yale University Press.

Lindzen, R.S. 2012. Climate science: is it currently designed to answer questions? *Euresis Journal* **2** (Winter): 161–92.

Nova, J. 2009. *Climate Money.* Washington, DC: Science and Public Policy Institute.

Park, I-U., Peacey, M.W., and Munafo, M.R. 2014. Modelling the effects of subjective and objective decision making in scientific peer review. *Nature* **506** (6): 93–6.

Singer, S.F. 2010. A response to 'the climate change debates.' *Energy & Environment* **21** (7): 847–51.

2.5 Appeals to Consensus

As explained in Section 2.1.1.2, scientific disagreements are not resolved by a show of hands. While consensus is a method used in politics to achieve change, the Scientific Method is used by scientists to reduce uncertainty and errors with the goal of understanding "why things are the way they are and act the way they do." This chapter makes clear the presence of disagreement on a wide range of important scientific topics in climate science, including most notably reconstructions of temperature trends in the past, the reliability of climate models, the sensitivity of climate to a doubling of CO_2, and the role of solar influences. Along the way we discovered uncertainty and disagreement over scores of other important matters.

Regrettably, claims of a "scientific consensus" on the causes and consequences of climate change have been used to shut down debate and provide cover to those who want political action. This section rebuts such claims by disclosing flaws in surveys allegedly finding a consensus and describing evidence of a distinct *absence* of consensus on many of the most important topics in climate science.

2.5.1 Flawed Surveys

Surveys and abstract-counting exercises that are said to show a "scientific consensus" on the causes and consequences of climate change invariably ask the wrong questions or the wrong people. No survey data exist that support claims of consensus on important scientific questions.

Claims of a "scientific consensus" on the causes and consequences of climate change rely on a handful of essays reporting the results of surveys or efforts to count the number of articles published in peer-reviewed journals that appear to endorse or reject the positions of the IPCC. The U.S. National

Aeronautics and Space Administration (NASA) on its website cites four sources supporting its claim that "Multiple studies published in peer-reviewed scientific journals show that 97% or more of actively publishing climate scientists agree: Climate-warming trends over the past century are extremely likely due to human activities" (NASA, 2018). As this section reveals, those surveys and abstract-counting exercises do not support that claim.

2.5.1.1 Oreskes, 2004

The most frequently cited source for a "consensus of scientists" is a 2004 essay for the journal *Science* written by historian Naomi Oreskes (Oreskes, 2004). Oreskes reported examining abstracts from 928 papers reported by the Institute for Scientific Information database published in scientific journals from 1993 and 2003, using the keywords "global climate change." Although not a scientist, she concluded 75% of the abstracts either implicitly or explicitly supported the IPCC's view that human activities were responsible for most of the observed warming over the previous 50 years while none directly dissented.

Oreskes' essay appeared in a "peer-reviewed scientific journal," as NASA reports on its website, but the essay itself was not peer-reviewed. It was an opinion essay and the editors had not asked to see her database. This opinion essay became the basis of a book, *Merchants of Doubt* (Oreskes and Conway, 2010), and then an academic career built on claiming that global warming "deniers" are a tiny minority within the scientific community, and then a movie based on her book released in 2015. Her 2004 claims were repeated in former Vice President Al Gore's movie, *An Inconvenient Truth*, and in his book with the same title (Gore, 2006).

Oreskes did not distinguish between articles that acknowledged or assumed some human impact on climate, however small, and articles that supported the IPCC's more specific claim that human emissions are responsible for more than 50% of the global warming observed during the past 50 years. The abstracts often are silent on the matter, and Oreskes apparently made no effort to go beyond those abstracts. Her definition of consensus also is silent on whether man-made climate change is dangerous or benign, a rather important question.

Oreskes' literature review overlooked hundreds of articles in peer-reviewed journals written by prominent global warming skeptics including John

Christy, Sherwood Idso, Richard Lindzen, and Patrick Michaels. More than 1,350 such articles (including articles published after Oreskes' study was completed) are identified in an online bibliography (Popular Technology.net, 2014).

Oreskes' methodology was flawed by assuming a nonscientist could decern the findings of scientific research by reading only the abstracts of published papers. Even trained climate scientists are unable to do so because abstracts do not accurately reflect their articles' findings. According to Park *et al.* (2014), abstracts routinely overstate or exaggerate research findings and contain claims that are irrelevant to the underlying research. Park *et al.* find "a mismatch between the claims made in the abstracts, and the strength of evidence for those claims based on a neutral analysis of the data, consistent with the occurrence of herding." They note abstracts often are loaded with "keywords" to ensure they are picked up by search engines and thus cited by other researchers.

Oreskes' methodology is further flawed, as are the other surveys and abstract-counting exercises discussed in this section, by surveying the opinions and writings of scientists and often nonscientists who may write about climate but are by no means experts on or even casually familiar with the science dealing with attribution – that is, attributing a specific climate effect (such as a temperature increase) to a specific cause (such as rising atmospheric CO_2 levels). Most articles simply reference or assume to be true the claims of the IPCC and then go on to address a different topic, such as the effect of ambient temperature on the life-cycle of frogs or correlations between temperature and outbreaks of influenza. Attribution is the issue the surveys ask about, but they ask people who have never studied the issue. The number of scientists actually knowledgeable about this aspect of the debate may be fewer than 100 in the world. Several are prominent skeptics (John Christy, Richard Lindzen, Patrick Michaels, and Roy Spencer, to name only four) and many others may be.

Monckton (2007) finds numerous other errors in Oreskes' essay, including her use of the search term "global climate change" instead of "climate change," which resulted in her finding fewer than one-thirteenth of the estimated corpus of scientific papers on climate change published over the stated period. Monckton also points out Oreskes never stated how many of the 928 abstracts she reviewed actually endorsed her limited definition of "consensus." Medical researcher Klaus-Martin Schulte used the same database and search terms as Oreskes to examine papers published from 2004 to

February 2007 and found fewer than half endorsed the "consensus" and only 7% did so explicitly (Schulte, 2008). His study is described in more detail in Section 2.5.2.1.

References

Gore, A. 2006. *An Inconvenient Truth: The Planetary Emergency of Global Warming and What We Can Do About It*. Emmaus, PA: Rodale Press.

Monckton, C. 2007. *Consensus? What Consensus? Among Climate Scientists, the Debate Is Not Over*. Washington, DC: Science and Public Policy Institute.

NASA. 2018. National Aeronautics and Space Administration. Scientific consensus: Earth's climate is warming (website).

Oreskes, N. 2004. Beyond the ivory tower: the scientific consensus on climate change. *Science* **306** (5702): 1686.

Oreskes, N. and Conway, E.M. 2010. *Merchants of Doubt: How a Handful of Scientists Obscured the Truth on Issues from Tobacco Smoke to Global Warming*. New York: NY: Bloomsbury Press.

Park, I.-U., Peacey, M.W., and Munafo, M.R. 2014. Modelling the effects of subjective and objective decision making in scientific peer review. *Nature* **506** (7486): 93–6.

Popular Technology.net. 2014. 1350+ peer-reviewed papers supporting skeptic arguments against ACC/AGW alarmism (website). February 12.

Schulte, K-M. 2008. Scientific consensus on climate change? *Energy & Environment* **19** (2): 281–6.

2.5.1.2 Doran and Zimmerman, 2009

Doran and Zimmerman (2009) reported conducting a survey that found "97% of climate scientists agree" that mean global temperatures have risen since before the 1800s and that humans are a significant contributing factor. The researchers had sent an online survey to 10,257 Earth scientists working for universities and government research agencies, generating responses from 3,146 people. The survey asked only two questions:

Q1. When compared with pre-1800s levels, do you think that mean global temperatures have generally risen, fallen, or remained relatively constant?

Q2. Do you think human activity is a significant contributing factor in changing mean global temperatures?

Overall, 90% of respondents answered "risen" to question 1 and 82% answered "yes" to question 2. The authors achieved their 97% figure by reporting the "yes" answers from only 79 of their respondents who "listed climate science as their area of expertise and who also have published more than 50% of their recent peer-reviewed papers on the subject of climate change." That is, Doran and Zimmerman applied *ex post facto* criteria to exclude 10,178 of their 10,257 sample population. Commenting on the survey, Solomon (2010) wrote:

> The two researchers started by altogether excluding from their survey the thousands of scientists most likely to think that the Sun, or planetary movements, might have something to do with climate on Earth – out were the solar scientists, space scientists, cosmologists, physicists, meteorologists and astronomers. That left the 10,257 scientists in disciplines like geology, oceanography, paleontology, and geochemistry. … The two researchers also decided that scientific accomplishment should not be a factor in who could answer – those surveyed were determined by their place of employment (an academic or a governmental institution). Neither was academic qualification a factor – about 1,000 of those surveyed did not have a Ph.D., some didn't even have a master's diploma.

Most "skeptics" of man-made global warming would answer those two questions the same way as alarmists would. The controversy in the science community is not over whether the climate warmed since the Little Ice Age or whether there is a human impact on climate, but "whether the warming since 1950 has been dominated by human causes, how much the planet will warm in the 21st century, whether warming is 'dangerous,' whether we can afford to radically reduce CO_2 emissions, and whether reduction will improve the climate" (Curry, 2015). The IPCC has expressed informed opinions on all these subjects, but those opinions often are at odds with extensive scientific research.

The survey by Doran and Zimmerman fails to produce evidence that would back up claims of a "scientific consensus" about the causes or consequences of climate change. They simply asked the wrong people the wrong questions. The "97%" figure so often attributed to their survey refers to the opinions of only 79 scientists, hardly a representative sample of scientific opinion.

References

Curry, J. 2015. State of the climate debate in the U.S. Remarks to the U.K. House of Lords, June 15. Climate Etc. (website).

Doran, P.T. and Zimmerman, M.K. 2009. Examining the scientific consensus on climate change. *EOS* **90** (3): 22–3.

Solomon, L. 2010. 75 climate scientists think humans contribute to global warming. *National Post.* December 30.

2.5.1.3 Anderegg et al., 2010

The third source cited by NASA as proof of a "scientific consensus" is Anderegg *et al.* (2010), who report using Google Scholar to identify the views of the most prolific writers on climate change. The authors found "(i) 97–98% of the climate researchers most actively publishing in the field support the tenets of ACC [anthropogenic climate change] outlined by the Intergovernmental Panel on Climate Change, and (ii) the relative climate expertise and scientific prominence of the researchers unconvinced of ACC are substantially below that of the convinced researchers." Like Oreskes (2014), Anderegg *et al.* was not peer reviewed. It was an "invited paper," which allowed its authors to bypass peer review.

This is not a survey of scientists, whether "all scientists" or specifically climate scientists. Instead, Anderegg *et al.* simply counted the number of articles found on the internet written by 908 scientists. This counting exercise is the same flawed methodology utilized by Oreskes, falsely assuming abstracts of papers accurately reflect their findings. Further, Anderegg *et al.* did not determine how many of the 908 authors believe global warming is harmful or that the science is sufficiently established to be the basis for public policy. Anyone who cites this study in defense of these views is mistaken.

Anderegg *et al.* also didn't count as "skeptics" the scientists whose work exposes gaps in the man-made global warming theory or contradicts claims that climate change will be catastrophic. Avery (2007) identified several hundred scientists who fall into this category, even though some profess to "believe" in global warming.

Looking past the "97–98%" claim, Anderegg *et al.* found the average skeptic has been published about half as frequently as the average alarmist (60 versus 119 articles). Most of this difference was driven by the hyper-productivity of a handful of alarmist climate scientists: The 50 most prolific alarmists were published an average of 408 times, versus only 89 times for the skeptics. The extraordinary publication rate of alarmists should raise a red flag. It is unlikely these scientists actually participated in most of the experiments or research contained in articles bearing their names. The difference in productivity between alarmists and skeptics can be explained by several factors other than merit:

- Publication bias: Articles reporting statistically significant correlations are much more likely to get published than those that do not (Fanelli, 2012);

- Heavy government funding of the search for one result but little or no funding for other results: The U.S. government alone paid $64 billion to climate researchers during the four years from 2010 to 2013, virtually all of it explicitly assuming or intended to find a human impact on climate and virtually nothing on the possibility of natural causes of climate change (Butos and McQuade, 2015, Table 2, p. 178);

- Resumé padding: It is increasingly common for academic articles on climate change to have multiple and even a dozen or more authors, inflating the number of times a researcher can claim to have been published (Hotz, 2015). Adding a previously published researcher's name to the work of more junior researchers helps ensure publication (as was the case with Anderegg *et al.* (2010) and Doran and Zimmerman (2009), in both cases the primary authors were college students);

- Differences in the age and academic status of global warming alarmists versus skeptics: Climate scientists who are skeptics tend to be older and more are emeritus than their counterparts on the alarmist side; skeptics are thus under less pressure and often are simply less eager to publish.

So what, exactly, did Anderegg *et al.* discover? That a small clique of climate alarmists had their names added to hundreds of articles published in academic journals, something that probably would have been impossible or judged unethical just a decade or two ago. Anderegg *et al.* simply assert those "top 50" are more credible than scientists who publish less, but they make no effort to prove this and there is ample evidence they are not (Solomon, 2008). Once again, Anderegg *et al.* did not ask if authors believe global warming is a serious problem or if science is sufficiently established to be the basis for public policy. Anyone who cites this study as evidence of scientific support for such views is misrepresenting the paper.

References

Anderegg, W.R.L., Prall, J.W., Harold, J., and Schneider, S.H. 2010. Expert credibility in climate change. *Proceedings of the National Academy of Sciences USA* **107** (27): 12107–9.

Avery, D.T. 2007. 500 scientists whose research contradicts man-made global warming scares. The Heartland Institute (website). September 14.

Butos, W.N. and McQuade, T.J. 2015. Causes and consequences of the climate science boom. *The Independent Review* **20** (2): 165–96.

Doran, P.T. and Zimmerman, M.K. 2009. Examining the scientific consensus on climate change. *EOS* **90** (3): 22–3.

Fanelli, D. 2012. "Positive" results increase down the hierarchy of science. *PLoS ONE* **4** (5): e10068.

Hotz, R.L. 2015. How many scientists does it take to write a paper? Apparently thousands. *The Wall Street Journal.* August 10.

Oreskes, N. 2004. Beyond the ivory tower: the scientific consensus on climate change. *Science* **306** (5702): 1686.

Solomon, L. 2008. *The Deniers: The World Renowned Scientists Who Stood Up Against Global Warming Hysteria, Political Persecution, and Fraud—And those who are too fearful to do so.* Minneapolis, MN: Richard Vigilante Books.

2.5.1.4 Cook et al., 2013

NASA's fourth source proving a "scientific consensus" is an abstract-counting exercise by Cook *et al.* (2013). The authors reviewed abstracts of peer-reviewed papers from 1991 to 2011 and found 97% of those that stated a position either explicitly or implicitly affirmed that human activity is responsible for some warming. The study was quickly critiqued by Legates *et al.* (2015), who found "just 0.3% endorsement of the standard definition of consensus: that most warming since 1950 is anthropogenic." They note "only 41 papers – 0.3% of all 11,944 abstracts or 1.0% of the 4,014 expressing an opinion, and not 97.1% – had been found to endorse the standard or quantitative hypothesis."

Scientists whose work questions the consensus, including Craig Idso, Nils-Axel Mörner, Nicola Scafetta, and Nir J. Shaviv, protested that Cook misrepresented their work (Popular Technology.net, 2012, 2013). Richard Tol, a lead author of the IPCC reports, said of the Cook report, "the sample of papers does not represent the literature. That is, the main finding of the paper is incorrect, invalid and unrepresentative" (Tol, 2013). On a blog of *The Guardian*, a British newspaper that had reported on the Cook report, Tol (2014) explained:

> Any conclusion they draw is not about 'the literature' but rather about the papers they happened to find. Most of the papers they studied are not about climate change and its causes, but many were taken as evidence nonetheless. Papers on carbon taxes naturally assume that carbon dioxide emissions cause global warming – but assumptions are not conclusions. Cook's claim of an increasing consensus over time is entirely due to an increase of the number of irrelevant papers that Cook and Co. mistook for evidence.

Montford (2013) revealed the authors of Cook *et al.* were marketing the expected results of the paper before the research itself was conducted; changed the definition of an endorsement of the global warming hypothesis mid-stream when it became apparent the abstracts they were reviewing did not support their original (IPCC-based) definition; and gave incorrect guidance to the volunteers recruited to read and score abstracts. Montford concludes "the consensus referred to is trivial" since the paper "said nothing about global warming being dangerous" and "the project was not a scientific investigation to determine the extent of agreement on global warming, but a public relations exercise."

* * *

Friends of Science, a group of Canadian retired Earth and atmospheric scientists, reviewed the four surveys and abstract-counting exercises summarized above (Friends of Science, 2014). They conclude, "these surveys show there is no 97% consensus on human-caused global warming as claimed in these studies. None of these studies indicate any agreement with a catastrophic view of human-caused global warming" (p. 4). We concur.

References

Cook, J., Nuccitelli, D., Green, S.A., Richardson, M., Winkler, B., Painting, R., Way, R., Jacobs, P., and Skuce, A. 2013. Quantifying the consensus on anthropogenic global warming in the scientific literature. *Environmental Research Letters* **8** (2).

Friends of Science. 2014. *97 Percent Consensus? No! Global Warming Math Myths & Social Proofs*. Calgary, AB: Friends of Science Society.

Legates, D.R., Soon, W., Briggs, W.M., and Monckton, C. 2015. Climate consensus and 'misinformation': a rejoinder to agnotology, scientific consensus, and the teaching and learning of climate change. *Science & Education* **24** (3): 299–318.

Montford, A. 2013. Consensus? What consensus? *GWPF Note* No. 5. London, UK: Global Warming Policy Foundation.

Popular Technology.net. 2012. The truth about Skeptical Science (website). March 18.

Popular Technology.net. 2013. 97% Study falsely classifies scientists' papers, according to the scientists that published them (website). May 21.

Tol, R. 2013. Open letter to Professor Peter Høj, president and vice-chancellor, University of Queensland. August 2013.

Tol, R. 2014. The claim of a 97% consensus on global warming does not stand up. The Guardian (blog). June 6.

2.5.2 Evidence of Lack of Consensus

Some survey data, petitions, and the peer-reviewed literature show deep disagreement among scientists on scientific issues that must be resolved before the man-made global warming hypothesis can be accepted.

In contrast to the surveys and abstract-counting exercises described above, which try but fail to find a consensus in support of the claim that global warming is man-made and dangerous, many authors and surveys have found widespread disagreement or even that a majority of scientists oppose the alleged consensus. These surveys and studies generally suffer the same methodological errors as afflict the ones described above, but they suggest that even playing by the alarmists' rules, the results demonstrate disagreement rather than consensus on key issues.

2.5.2.1 Klaus-Martin Schulte, 2008

Schulte (2008), a practicing physician, observed, "Recently, patients alarmed by the tone of media reports and political speeches on climate change have been voicing distress, for fear of the imagined consequences of anthropogenic 'global warming.'" Concern that his patients were experiencing unnecessary stress "prompted me to review the literature available on 'climate change and health' via PubMed" and then to attempt to replicate Oreskes' 2004 report.

"In the present study," Schulte writes, "Oreskes' research was brought up to date by using the same search term on the same database to identify abstracts of 539 scientific papers published between 2004 and mid-February 2007." According to Schulte, "The results show a tripling of the mean annual publication rate for papers using the search term 'global climate change', and, at the same time, a significant movement of scientific opinion away from the apparently unanimous consensus which Oreskes had found in the learned journals from 1993 to 2003. Remarkably, the proportion of papers explicitly or implicitly rejecting the consensus has risen from zero in the period 1993–2003 to almost 6% since 2004. Six papers reject the consensus outright."

Schulte also found "Though Oreskes did not state how many of the papers she reviewed explicitly endorsed the consensus that human greenhouse-gas emissions are responsible for more than half of the past 50 years' warming, only 7% of the more recent papers reviewed here were explicit in endorsing the consensus even in the strictly limited sense she had defined. The proportion of papers that now explicitly or implicitly endorse the consensus has fallen from 75% to 45%."

Schulte's findings demonstrate that if Oreskes' methodology were correct and her findings for the period 1993 to 2003 accurate, then scientific publications in the more recent period 2004–2007 show a strong tendency away from the consensus Oreskes claimed to have found. We doubt the utility of the methodology used by both Oreskes and Schulte. Nevertheless, it is useful to note the same methodology applied during two time periods seems to reveal a significant shift from consensus to open debate on the causes of climate change.

Reference

Schulte, K-M. 2008. Scientific consensus on climate change? *Energy & Environment* **19** (2): 281–6.

2.5.2.2 Bray and von Storch, 2015–2016

Surveys by German scientists Dennis Bray and Hans von Storch conducted in 1996, 2003, 2008, 2010, and 2015-16 have consistently found climate scientists have deep doubts about the reliability of the science underlying claims of man-made climate change. Questions about climate science in the surveys, which have stayed largely the same over the years in order to discern trends, ask respondents to rank their agreement with a statement on a scale from 1 to 7, with 1 = "very inadequate," "poor," or "none" (depending on the wording of the question) and 7 = "very adequate," "very good," or "a very high level." Histograms then show the probability distribution of answers for each question.

Bast and Taylor (2007) analyzed the results of the 2003 survey and found only 55.8% of respondents agreed with the statement that "climate change is mostly the result of anthropogenic (manmade) causes" and more scientists "strongly disagree" than "strongly agree." When climate scientists were asked if "climate models can accurately predict climate conditions in the future," only a third (35.1%) agreed, while 18.3% were uncertain and nearly half (46.6%) disagreed. Most histograms showed bell-shaped distributions suggesting disagreement and uncertainty rather than agreement and confidence. Bast (2010) analyzed the Bray and von Storch 2010 survey and once again found bell-shaped distributions for about a third of the questions addressing scientific issues. Bast writes, "The remaining two-thirds are divided almost equally between distributions that lean toward skepticism and those that lean toward alarmism." He concludes,

"There is certainly no consensus on the science behind the global warming scare."

The latest survey by Bray and von Storch (2016) was conducted in 2015 and released in 2016. The survey was sent by email to 3,879 individuals who were mostly contributors to past IPCC reports and writers appearing in 10 top-ranked peer-reviewed climate journals. Complete and partial responses were received from 651 respondents, a 17% response rate. All but 55 of the respondents (8.5%) reported working for universities or government agencies, suggesting as could be expected by the sampling procedure that proponents of the IPCC's views were over-represented in the sample. Given that the number of scientists active in fields related to climate number in the thousands, this small number of responses cannot be a representative sample of scientific opinion.

Surprisingly, given the skewed sample that was surveyed, only 48% of respondents said they agree "very much" with the statement that "most of recent or near future climate change is, or will be, the result of anthropogenic causes." Recall that the IPCC (2013) wrote in its Fifth Assessment Report that "It is *extremely likely* that more than half of the observed increase in global average surface temperature from 1951 to 2010 was caused by the anthropogenic increase in greenhouse gas concentrations and other anthropogenic forcings together" (p. 17). The IPCC defines "extremely likely" as "95–100% probability" (p. 36). It appears a majority of even the career climate scientists surveyed by Bray and von Storch disagree with the IPCC.

On whether "climate models accurately simulate the climatic conditions for which they are calibrated," a plurality (41.53%) ranked this a 5 and equal numbers (20%) ranked it a 4 or a 6, a bell-shaped distribution that skews toward the alarmist direction but still shows deep disagreement. Only 4% said they agree "very much" with the statement. The IPCC (2013) expressed *very high confidence* that its computer model simulations agree with the observed trend from 1951 to 2012 (p. 15).

On other questions there are once again bell-shaped distributions showing *most* scientists are unsure about basic questions of climate science. For example, 80% of respondents gave a 4 or less when asked how well atmospheric models can deal with the influence of clouds and precipitation and 76% give a 4 or less to the ability of climate models to simulate a global mean value of precipitation values for the next 50 years. Only 9% ranked the ability of climate models to simulate a global mean value for temperature for the next 50 years as "very good." The IPCC (2013) contends it can forecast temperatures to 2100 and beyond with *"high confidence"* (p. 21). Interestingly, 42% of respondents agreed with the statement that "the collective authority of a consensus culture of science paralyzes new thought," with 9% saying they "strongly agree."

Setting aside its small and skewed sample, Bray and von Storch's results should be easy to interpret. On most questions, most scientists are somewhere in the middle, somewhat convinced that man-made climate change is occurring but concerned about the lack of reliable data and other fundamental uncertainties. Very few scientists share the "95–100% probability" claimed by the IPCC.

References

Bast, J. and Taylor, J. 2007. Scientific consensus on global warming. Chicago, IL: The Heartland Institute.

Bast, J. 2010. Analysis: New international survey of climate scientists. *Policy Brief.* Chicago, IL: The Heartland Institute.

Bray, D. and von Storch, H. 2016. The Bray and von Storch 5th International Survey of Climate Scientists 2015/2016. *HZG Report* 2016-2. Geesthacht, Germany: GKSS Institute of Coastal Research.

IPCC. 2013. Intergovernmental Panel on Climate Change. *Climate Change 2013: The Physical Science Basis.* Contribution of Working Group I to the Fifth Assessment Report of the Intergovernmental Panel on Climate Change. New York, NY: Cambridge University Press.

2.5.2.3 Verheggen et al., 2014, 2015

Verheggen *et al.* (2014) and Strengers, Verheggen, and Vringer (2015) reported the results of a survey they conducted in 2012 of contributors to the IPCC reports, authors of articles appearing in scientific literature, and signers of petitions on global warming (but not the Global Warming Petition Project, described below). By the authors' own admission, "signatories of public statements disapproving of mainstream climate science … amounts to less than 5% of the total number of respondents," suggesting the sample is heavily biased toward pro-"consensus" views. Nevertheless, this survey found fewer than half of respondents agreed with the IPCC's most recent claims.

A total of 7,555 people were contacted and 1,868 questionnaires were returned, for a response rate of 29%. Verheggen *et al.* asked specifically about agreement or disagreement with the IPCC's claim in its Fifth Assessment Report (AR5) that it is "virtually certain" or "extremely likely" that net anthropogenic activities are responsible for more than half of the observed increase in global average temperatures in the past 50 years.

When asked "What fraction of global warming since the mid-20th century can be attributed to human induced increases in atmospheric greenhouse gas (GHG) concentrations?," 64% chose fractions of 51% or more, indicating agreement with the IPCC AR5. (Strengers, Verheggen, and Vringer, 2015, Figure 1a.1) When those who chose fractions of 51% or more were asked, "What confidence level would you ascribe to your estimate that the anthropogenic GHG warming is more than 50%?," 65% said it was "virtually certain" or "extremely likely," the language used by the IPCC to characterize its level of confidence (Strengers, Verheggen, and Vringer, 2015, Figure 1b).

The math is pretty simple: Two-thirds of the respondents to this survey – a sample heavily biased toward the IPCC's point of view by including virtually all its editors and contributors – agreed with the IPCC on the impact of human emissions on the climate, and two-thirds of those who agreed were as confident as the IPCC in that finding. Sixty-five percent of 64% is 41.6%, so fewer than half of the survey's respondents support the IPCC. More precisely – since some responses were difficult to interpret – 42.6% (797 of 1,868) of respondents were highly confident that more than 50% of the warming is human-caused.

This survey, like the Bray and von Storch surveys previously described, shows the IPCC's position on global warming is the minority view of the science community. Since the sample was heavily biased toward contributors to the IPCC reports and academics most likely to publish, one can assume a survey of a larger universe of scientists would reveal even less support for the IPCC's position. Verheggen *et al.* reported their findings only in tables in a report issued a year after their original publication rather than explain them in the text of their peer-reviewed article. It took the efforts of a blogger to call attention to the real data (Fabius Maximus, 2015). Once again, the data reveal no scientific consensus.

References

Fabius Maximus. 2015. New study undercuts the key IPCC finding (website).

Strengers, B., Verheggen, B., and Vringer, K. 2015. Climate science survey questions and responses. PBL Netherlands Environmental Assessment Agency. April 10.

Verheggen, B., Strengers, B., Cook, J., van Dorland, R., Vringer, K., Peters, J., Visser, H., and Meyer, L. 2014. Scientists' views about attribution of global warming. *Environmental Science & Technology* **48** (16): 8963–71. doi: 10.1021/es501998e.

2.5.2.4 Survey of Meteorologists

One way to test with a poll or survey whether a purported consensus position on a particular topic of science is correct is to ask experts in neighboring fields for their opinions. For example, if the objective is to determine whether acupuncture is a legitimate health therapy, it would not make sense to poll acupuncturists, who have an obvious emotional as well as financial stake in the matter. A survey that found 97% consensus on this issue among them would be meaningless. On the other hand, a survey of experts in neighboring disciplines such as doctors, physical therapists, and nurses would make sense. Many of them know enough about the subject to make an informed judgement and they are less likely to be biased by professional interests.

In the case of climatology, the alleged consensus position is that computer models can accurately predict a human impact on the climate decades and even centuries in the future. Climate scientists, in particular climate modelers, may be too wed to this postulate to view it objectively, leaving it up to experts of neighboring fields to decide its validity. One such "neighboring discipline" is meteorology.

The leadership of the American Meteorological Society (AMS) has long supported the IPCC and the anthropogenic global warming theory, but its members have been much more skeptical. According to its website, "It is clear from extensive scientific evidence that the dominant cause of the rapid change in climate of the past half century is human-induced increases in the amount of atmospheric greenhouse gases" (AMS, 2012). As part of its campaign to "educate" its members on the issue, the AMS contracts with an advocacy organization called the Center for Climate Change Communication to conduct an annual survey of its members' views.

The latest AMS survey, Maibach *et al.* (2017), generated 465 complete and partial responses from AMS members. The survey asked respondents to complete the following sentence, "Do you think that the climate change that has occurred over the past 50 years has been caused by...," by choosing among seven options ranging from "largely or entirely due to human activity (81–100%)" to "there has been no climate change over the past 50 years." As could be expected, only 1% of respondents chose the "no climate change" option. Much more interesting is that only 15% chose the "largely or entirely by human activity (81–100%)," while 34% said "mostly by human activity (60–80%)" and the rest, 51%, said "more or less equally by human activity and natural events," "mostly by natural events," or "largely or entirely by natural events."

The AMS survey also asked "over the next 50 years, to what extent can additional climate change be averted if mitigation measures are taken worldwide (i.e., substantially reducing emissions of carbon dioxide and other greenhouse gases)?" Only 1% said "almost all additional climate change can be averted," the same percentage as "don't think there will be additional climate change over the next 50 years." Sixty-nine percent said either "a moderate amount" or "a small amount" of climate change could be averted, and 13% said "almost no additional climate change can be averted." This is consistent with the view that most climate change is due to natural causes and not human activity.

The AMS survey also asked, "which of the following best describes the impact(s) of local climate change in your media market over the past 50 years?" If climate change is already happening and causing harms, as the IPCC and its many allies have been saying for three decades, meteorologists would be in a good position to know since many of them are paid to report on it every day. But 49% said "the impacts have been approximately equally mixed between beneficial and harmful," and another 12% said the impacts have been "primarily beneficial" or "exclusively beneficial." Only 3% said "the impacts have been exclusively harmful." Compare this to the long lists of alleged "damages" caused by climate change and near absence of benefits reported in the IPCC's latest reports.

It is disappointing that even 15% of meteorologists apparently believe climate change is mostly caused by human activity and that 36% think the impacts of climate change have been primarily harmful. The evidence reported earlier in this chapter makes it clear that both views are probably wrong.

But the fact that a majority of meteorologists do not subscribe to the IPCC's claims of *very high confidence* is independent confirmation that the alleged consensus is not generally supported by the science community.

References

AMS. 2012. American Meteorological Society. Climate Change. An Information Statement of the American Meteorological Society (website). Adopted by AMS Council 20 August 2012. Accessed December 7, 2018.

Maibach, E., *et al.* 2017. *A 2017 National Survey of Broadcast Meteorologists: Initial Findings.* Fairfax, VA: George Mason University, Center for Climate Change Communication.

2.5.3 Global Warming Petition Project

Some 31,000 scientists have signed a petition saying "there is no convincing scientific evidence that human release of carbon dioxide, methane, or other greenhouse gases is causing or will, in the foreseeable future, cause catastrophic heating of the Earth's atmosphere and disruption of the Earth's climate."

The Global Warming Petition Project (2015) is a statement about the causes and consequences of climate change signed by 31,478 American scientists, including 9,021 with Ph.D.s. The full statement reads:

> We urge the United States government to reject the global warming agreement that was written in Kyoto, Japan in December, 1997, and any other similar proposals. The proposed limits on greenhouse gases would harm the environment, hinder the advance of science and technology, and damage the health and welfare of mankind.
>
> There is no convincing scientific evidence that human release of carbon dioxide, methane, or other greenhouse gases is causing or will, in the foreseeable future, cause catastrophic heating of the Earth's atmosphere and disruption of the Earth's

climate. Moreover, there is substantial scientific evidence that increases in atmospheric carbon dioxide produce many beneficial effects upon the natural plant and animal environments of the Earth.

This is a strong statement of dissent from the perspective advanced by the IPCC. The fact that more than 15 times as many scientists have signed it as are alleged to have "participated" in some way or another in the research, writing, and review of the IPCC's assessments (IPCC, n.d.) is significant. These scientists actually endorse the statement that appears above. By contrast, fewer than 100 of the scientists (and nonscientists) who are listed in the appendices to the IPCC reports actually participate in the writing of the all-important Summary for Policymakers or the editing of the final report to comply with the summary, and therefore could be said to endorse the main findings of that report. The survey by Verheggen *et al.* (2014) reported above shows many or even most of the scientists who participate in the IPCC do not endorse its declarative statements and unqualified predictions.

The Global Warming Petition Project has been criticized for including names of suspected nonscientists, including names submitted by environmental activists for the purpose of discrediting the petition. But the organizers of the project painstakingly reconfirmed the authenticity of the names in 2007, and a complete directory of those names appeared as an appendix to *Climate Change Reconsidered: Report of the Nongovernmental International Panel on Climate Change (NIPCC)*, published in 2009 (NIPCC, 2009). For more information about The Petition Project, including the text of the letter endorsing it written by the late Dr. Frederick Seitz, past president of the National Academy of Sciences and president emeritus of Rockefeller University, visit the project's website at www.petitionproject.org.

References

Global Warming Petition Project. 2015. Global warming petition project (website). Accessed July 12, 2018.

IPCC. n.d. Intergovernmental Panel on Climate Change. IPCC Participants (website). Accessed December 7, 2018.

NIPCC. 2009. Nongovernmental International Panel on Climate Change. *Climate Change Reconsidered: 2009 Report of the Nongovernmental International Panel on*

Climate Change (NIPCC). Chicago, IL: The Heartland Institute.

Verheggen, B., Strengers, B., Cook, J., van Dorland, R., Vringer, K., Peters, J., Visser, H., and Meyer, L. 2014. Scientists' views about attribution of global warming. *Environmental Science & Technology* **48** (16): 8963–71.

2.5.4 Conclusion

The most important fact about climate science, often overlooked, is that scientists disagree about the environmental impacts of the combustion of fossil fuels.

As this section makes apparent, the surveys and abstract-counting exercises that are said to show a "scientific consensus" on the causes and consequences of climate change invariably ask the wrong questions or the wrong people. No survey data exist that support claims of consensus on important scientific questions. At best, there is broad agreement that the planet may have warmed in the late twentieth century and that a human impact could be discernible, but even these statements are no more than expressions of opinion unless the terms are carefully qualified and defined. There is no consensus on the following matters:

- anthropogenic greenhouse gas emissions are responsible for most or all of the warming of the twentieth and early twenty-first centuries;

- climate models can accurately forecast temperatures and precipitation 50 or more years into the future;

- mean surface temperatures or their rate of change in the twentieth and twenty-first centuries exceed those observed in the historical and geological record;

- climate impacts such as storms, floods, the melting of ice, and sea-level rise are unprecedented since before the beginning of the industrial age; and

- rising CO_2 levels and temperatures would have a negative impact on plant life.

In each of these areas, at issue here is not whether some scientists are right and others wrong, but that considerable and valid evidence exists *on all sides*. As befits a young academic discipline, scientists are still learning how much they do not know. Consensus may have a place in science, and it may someday arrive on key issues in climate science, but that day has not yet arrived.

Phil Jones, director of the Climatic Research Unit at the University of East Anglia, when asked if the debate on climate change is over, told the BBC, "I don't believe the vast majority of climate scientists think this. This is not my view" (BBC News, 2010). When asked, "Do you agree that according to the global temperature record used by the IPCC, the rates of global warming from 1860–1880, 1910–1940 and 1975–1998 were identical?" Jones replied,

Temperature data for the period 1860–1880 are more uncertain, because of sparser coverage, than for later periods in the 20th Century. The 1860–1880 period is also only 21 years in length. As for the two periods 1910–40 and 1975–1998 the warming rates are not statistically significantly different … I have also included the trend over the period 1975 to 2009, which has a very similar trend to the period 1975–1998. So, in answer to the question, the warming rates for all four periods are similar and not statistically significantly different from each other.

Finally, when asked "Do you agree that from 1995 to the present there has been no statistically significant global warming" Jones answered "yes." Each of his statements contradicts claims made by the IPCC at the time, claims that are still being repeated today. It was an honest admission by Jones of the lack of scientific consensus on one of the most complex and controversial scientific issues of the day.

Sarewitz (2016) observes that "the vaunted scientific consensus around climate change – which largely rests on fundamental physics that has been well understood for more than a century – applies only to a narrow claim about the discernible human impact on global warming. The minute you get into questions about the rate and severity of future impacts, or the costs of and best pathways for addressing them, no semblance of consensus among experts remains" (p. 30).

References

BBC News. 2010. Q&A: Professor Phil Jones. February 13.

Sarewitz, D. 2016. Saving science. *The New Atlantis* **49**: 5–40.

2.6 Conclusion

Because scientists disagree, policymakers must exercise special care in choosing where they turn for advice.

Climate is an exciting field of study for scientists from many fields and types of training. While much has been learned, there is still more that is unknown than known about climate processes.

The "science tutorial" at the beginning of this chapter attempted to explain the main methodological issues and most important types of observations that together account for much of the controversy within climate science. One recurring theme has been the presence of uncertainty – in temperature records, the carbon cycle, the energy budget, human greenhouse gas emissions, and many more key areas – but also the promise that uncertainty can be reduced through rigorous adherence to the Scientific Method. Regrettably, the disciplines of the Scientific Method have not been consistently applied in all areas of climate science, resulting in polarization, intolerance, and expressions of false certainty.

The good news in this chapter is that the feared negative impacts of climate change – more frequent and intense extreme weather events, more melting ice at the poles, rapidly rising sea levels, and harm to plant life – are unlikely to emerge. There is no compelling scientific evidence of long-term trends in any of these areas that exceed the bounds of natural variability. Climate science suggests a warmer world with higher levels of atmospheric CO_2 is likely to see fewer extreme weather events and a continuation of the Greening of the Earth witnessed in the past four decades. Drawing from an extensive review of the scientific evidence, our conclusion is that the human effect on the global climate is very small and the impacts of that effect are likely to be benign.

We understand why scientists disagree on this matter. Fundamental uncertainties arise from the lack of a comprehensive physical theory of climate, from the large errors and low resolution of climate models, from insufficient or inaccurate observational evidence, and from disagreements over how to interpret data and how to set the parameters of models. Climate is an interdisciplinary subject requiring insights from many fields. Very few scholars have mastery of more than one or two of these disciplines. The Intergovernmental Panel on Climate Change (IPCC) is widely viewed as an independent source of science on all causes of climate change, but it is not. It is agenda-driven, and many people, scientists included, have been misled by its work. Finally, climate scientists, like all humans, can have tunnel vision. Sources of bias include careerism, grant-seeking, political views, and confirmation bias.

Fundamental uncertainties and disagreements prevent science from determining whether human greenhouse gas emissions are having effects on Earth's atmosphere that could endanger life on the planet. Because scientists disagree, policymakers must exercise special care in choosing where they turn for advice. Rather than rely exclusively on the IPCC, policymakers should seek out advice from independent, nongovernment organizations and scientists whose views are less likely to be affected by political and financial conflicts of interest. Policymakers should resist pressure from lobby groups to silence scientists who question the authority of the IPCC to speak for "climate science."

The distinguished British biologist Conrad Waddington (1941) wrote, "It is … important that scientists must be ready for their pet theories to turn out to be wrong. Science as a whole certainly cannot allow its judgment about facts to be distorted by ideas of what ought to be true, or what one may hope to be true." That statement merits reflection by those who continue to assert the fashionable belief, in the face of direct evidence to the contrary, that anthropogenic greenhouse gas emissions are causing or will cause dangerous global warming.

Reference

Waddington, C.H. 1941. *The Scientific Attitude*. London, UK: Penguin Books.

PART II

BENEFITS OF
FOSSIL FUELS

Introduction to Part II

Part I provided the fundamental economics and science needed to understand proposals to severely restrict the use of fossil fuels in order to slow or stop climate change. Chapter 1, addressing environmental economics, found:

> The prosperity made possible by the use of fossil fuels has made environmental protection a social value in countries around the world. The value-creating power of private property rights, prices, profits and losses, and voluntary trade can turn climate change from a possible *tragedy* of the commons into an *opportunity* of the commons. Energy freedom, not government intervention, can balance the interests and needs of today with those of tomorrow. It alone can access the local knowledge needed to find efficient win-win responses to climate change.

Chapter 2, addressing climate science, found:

> Fundamental uncertainties arising from insufficient observational evidence and disagreements over how to interpret data and set the parameters of models prevent science from determining whether human greenhouse gas emissions are having effects on Earth's atmosphere that could endanger life on the planet. There is no compelling scientific evidence of long-term trends in global mean temperatures or climate impacts that exceed the bounds of natural variability.

In the face of such economic and scientific findings, many experts recommend a "no regrets" strategy of relying on policies that generate value even if climate change turns out not to be a major problem (NCPA, 1991; Adler *et al.*, 2000; Goklany, 2001; Lomborg, 2008; Murray and Burnett, 2009; Carter, 2010; The Hartwell Group, 2010, 2011; van Kooten, 2013; Vahrenholt and Lüning, 2015; Bailey, 2015; Moore and Hartnett White, 2016). Such a strategy might include ending subsidies to development in floodplains and improving the design and construction of levees and flood walls (to reduce flood damage), improving forest management (to reduce forest fires), reducing urban traffic congestion (to reduce emissions from cars and trucks), and improving emergency response systems (to minimize the loss of life during natural disasters). A majority of voters may support policies that protect the environment and save human lives while also addressing the possibility of harmful climate change.

"No regrets" is not the strategy advocated by the IPCC and its many allies in the environmental movement. They advocate instead for immediate major reductions in the use of fossil fuels, hoping this would reduce the level of CO_2 in the atmosphere, which in turn they hope would slow or stop future climate changes, seemingly without regard to economic and scientific facts that suggest otherwise. Relying on invalidated climate models, they claim CO_2 emissions must be reduced by 80% by 2050 to avoid a climate catastrophe (Long and Greenblatt, 2012; National Research Council, 2013; World Energy Council, 2013; IPCC, 2014). If we reject the "no regrets" option, either because of genuine disagreement over economics and science or ideological fervor, we are not relieved of the obligation to weigh the cost of our decision. In particular,

- Can wind turbines, solar photovoltaic (PV) panels, and biofuels meet the world's growing

Citation: Idso, C.D., Legates, D. and Singer, S.F. 2019. Human Prosperity. In: *Climate Change Reconsidered II: Fossil Fuels.* Nongovernmental International Panel on Climate Change. Arlington Heights, IL: The Heartland Institute.

need for dispatchable energy to produce electricity, heat homes, and power manufacturing and transportation?

- How much more would energy cost if fossil fuels were banned or phased out? What impact would that have on human prosperity and health?

- What would be the opportunity cost of such a transition? In other words, what other opportunities to advance human well-being would be foregone?

Answering these questions requires an accurate accounting of the benefits of our current reliance on fossil fuels. For that reason, Part II (Chapters 3, 4, and 5) surveys the three largest benefits of fossil fuels: human prosperity, human health benefits, and environmental protection. Part III (Chapters 6, 7, and 8) will survey the costs of fossil fuels and conduct cost-benefit analyses of fossil fuels, climate change, and policies proposed to prevent or delay the onset of anthropogenic climate change.

Chapter 3 reports the contribution fossil fuels make to human prosperity. The contribution is large: One study projected the "existence value" of coal production, transportation, and consumption for electric power generation in the United States at $1.275 trillion (in 2015 dollars) and estimated coal supported 6.8 million U.S. jobs (Rose and Wei, 2006). An additional benefit is the value of increased food production due to rising levels of atmospheric CO_2, a phenomenon called aerial fertilization. Its worth is estimated to have been $3.2 trillion from 1961 to 2011 and is currently approximately $170 billion annually (Idso, 2013). Chapter 3 also explains why alternatives to fossil fuels – wind turbines, solar PV panels, and biofuels –cannot sustain the prosperity made possible by fossil fuels.

Chapter 4 reports the human health benefits of fossil fuels. Fossil fuels are responsible for the prosperity that makes possible better nutrition, housing, and working conditions, and cleaner air and water, contributing to the dramatic improvements in human longevity and decline in the incidence of diseases and premature death. The marginally warmer temperatures observed in some parts of the world at the end of the twentieth and beginning of the twenty-first centuries have further contributed to human health by preventing millions of premature deaths globally from illnesses or health effects related to colder temperatures (Gasparrini *et al.*, 2015).

Chapter 5 describes the environmental protection made possible by fossil fuels. These benefits go beyond meeting human needs and providing the goods and services that contribute to human flourishing and modernity. As Nobel Laureate Amartya Sen wrote in 2015, "We can have many reasons for our conservation efforts, not all of which need to be parasitic on our own living standards (or need-fulfillment), and some of which may turn precisely on our sense of values and on our acknowledgment of our reasons for taking fiduciary responsibility for other creatures on whose lives we can have a powerful influence" (Sen, 2014). Fossil fuels make it possible to feed a growing global population without massive deforestation or air and water pollution. The aerial fertilization effect further benefits forests and terrestrial species and promotes biodiversity.

References

Adler, J., Crews, C.W., Georgia, P., Lieberman, B., Melugin, J., and Seiver, M-L. 2000. *Greenhouse Policy Without Regrets: A Free Market Approach to the Uncertain Risks of Climate Change.* Washington, DC: Competitive Enterprise Institute.

Bailey, R. 2015. *The End of Doom: Environmental Renewal in the Twenty-first Century.* New York, NY: Thomas Dunne Books/St. Martin's Press.

Carter, R.M. 2010. *Climate: The Counter Consensus.* London, UK: Stacey International.

Gasparrini, A., *et al.* 2015. Mortality risk attributable to high and low ambient temperature: a multi-country observational study. *Lancet* **386**: 369–75.

Goklany, I.M. 2001. *The Precautionary Principle: A Critical Appraisal of Environmental Risk Assessment.* Washington, DC: Cato Institute.

Idso, C.D. 2013. *The Positive Externalities of Carbon Dioxide.* Tempe, AZ: Center for the Study of Carbon Dioxide and Global Change. October 21.

IPCC. 2014. *Climate Change 2014: Impacts, Adaptation, and Vulnerability.* Contribution of Working Group II to the Fifth Assessment Report of the Intergovernmental Panel on Climate Change. Geneva, Switzerland: World Meteorological Organization.

Lomborg, B. 2008. *Cool It: The Skeptical Environmentalist's Guide to Global Warming.* New York, NY: Alfred A. Knopf.

Long, J.C.S. and Greenblatt, J. 2012. The 80% solution: radical carbon emission cuts for California. *Issues in Science and Technology.* September.

Moore, S. and Hartnett White, K. 2016. *Fueling Freedom: Exposing the Mad War on Energy.* New York, NY: Regnery.

Murray, I. and Burnett, H.S. 2009. 10 Cool Global Warming Policies. *Policy Report #321.* Dallas, TX: National Center for Policy Analysis.

National Research Council. 2013. *Transitions to Alternative Vehicles and Fuels.* Washington, DC: National Academies Press.

NCPA (National Center for Policy Analysis). 1991. *Progressive Environmentalism: A Pro-human, Pro-science, Pro-free Enterprise Agenda for Change, Task Force Report.* Dallas, TX: National Center for Policy Analysis.

Rose, A. and Wei, D. 2006. *The Economic Impacts of Coal Utilization and Displacement in the Continental United States, 2015.* Report prepared for the Center for Energy and Economic Development, Inc. State College, PA: Pennsylvania State University. July.

Sen, A. 2014. Global warming is just one of many environmental threats that demand our attention. *New Republic.* August 22.

The Hartwell Group. 2010. *The Hartwell Paper: A New Direction for Climate Policy after the Crash of 2009.* Institute for Science, Innovation & Society, University of Oxford; LSE Mackinder Programme. London, UK: London School of Economics and Political Science.

The Hartwell Group. 2011. *Climate Pragmatism: Innovation, Resilience and No Regrets.* Toronto, ON: The Hartwell Group. June.

Vahrenholt, F. and Lüning, S. 2015. *The Neglected Sun: Why the Sun Precludes Climate Catastrophe.* Second English Edition. Arlington Heights, IL: The Heartland Institute.

van Kooten, G.C. 2013. *Climate Change, Climate Science and Economics: Prospects for an Alternative Energy Future.* New York, NY: Springer.

World Energy Council. 2013. *Goal of Fossil Fuel Independence by 2050.*

3

Human Prosperity

Chapter Lead Authors: Roger Bezdek, Ph.D., Craig Idso, Ph.D.

Contributors: Joseph L. Bast, Howard Hayden, Ph.D.

Reviewers: David Archibald, Timothy Ball, Ph.D., Barry Brill, OPE, JP, H. Sterling Burnett, Ph.D., Ian D. Clark, Ph.D., Weihong Cui, Donn Dears, David Deming, Ph.D., Terry W. Donze, Paul Driessen, J.D., John Droz, Jr., Vivian Richard Forbes, Lee C. Gerhard, Ph.D., Steve Goreham, Pierre Gosselin, Kesten Green, Ph.D., Mary Hutzler, Hans Konrad Johnsen, Ph.D., Joseph Leimkuhler, Bryan Leyland, Alan Moran, Ph.D., Robert Murphy, Ph.D., Tom V. Segalstad, Ph.D., Peter Stilbs, Ph.D., Richard J. Trzupek, Fritz Vahrenholt, Ph.D., Gösta Walin, Ph.D.

Citation: Idso, C.D., Legates, D. and Singer, S.F. 2019. Human Prosperity. In: *Climate Change Reconsidered II: Fossil Fuels.* Nongovernmental International Panel on Climate Change. Arlington Heights, IL: The Heartland Institute.

Key Findings

Key findings of this chapter include the following:

Energy Tutorial

- Some key concepts include energy, power, watts, joules, and power density.

- Advances in efficiency mean we live lives surrounded by the latest conveniences, yet we use only about 3.5 times as much energy per capita as did our ancestors in George Washington's time.

- Increased use of energy and greater energy efficiency have enabled great advances in artificial light, heat generation, and transportation.

- Fossil fuels supply 81% of the primary energy consumed globally and 78% of energy consumed in the United States.

- Due to the nature of wind and sunlight, wind turbines and solar photovoltaic (PV) cells can produce power only intermittently.

Three Industrial Revolutions

- Fossil fuels make possible such transformative technologies as nitrogen fertilizer, concrete, the steam engine and cotton gin, electrification, the internal combustion engine, and the computer and Internet revolution.

- Electricity powered by fossil fuels has made the world a healthier, safer, and more productive place.

- Access to energy is closely associated with key measures of global human development including per-capita GDP, consumption expenditure, urbanization rate, life expectancy at birth, and the adult literacy rate.

Food Production

- Fossil fuels have greatly increased farm worker productivity thanks to nitrogen fertilizer created by the Haber-Bosch process and farm machinery built with and fueled by fossil fuels.

- Higher levels of carbon dioxide (CO_2) in the atmosphere act as fertilizer for the world's plants.

- The aerial fertilization effect of rising levels of atmospheric CO_2 produced global economic benefits of $3.2 trillion from 1961 to 2011 and currently amount to approximately $170 billion annually.

- The economic value of CO_2 fertilization of crops over the period 2012–2050 is forecast to be $9.8 trillion.

- Reducing global CO_2 emissions by 28% from 2005 levels, the reduction President Barack Obama proposed in 2015 for the United States, would reduce aerial fertilization benefits by $78 billion annually.

Why Fossil Fuels?

- Fossil fuels have higher power density than all alternative energy sources except nuclear power.

- Fossil fuels are the only sources of fuel available in sufficient quantities to meet the needs of modern civilization.

- Fossil fuels provide energy in the forms needed to make electricity dispatchable (available on demand 24/7) and they can be economically transported to or stored near the places where energy is needed.

- Fossil fuels in the United States are so inexpensive that they make home heating, electricity, and transportation affordable for even low-income households.

Alternatives to Fossil Fuels

- The low power density of alternatives to fossil fuels is a crippling deficiency that prevents them from ever replacing fossil fuels in most applications.

- Wind, solar, and biofuels cannot be produced and delivered where needed in sufficient quantities to meet current and projected energy needs.

- Due to their intermittency, solar and wind power cannot power the revolving turbine generators needed to create dispatchable energy.

- Electricity from new wind capacity costs approximately 2.7 times as much as existing coal, 3 times more than combined cycle gas, and 3.7 times more than nuclear power.

- The cost of alternative energies will fall too slowly to close the gap with fossil fuels before hitting physical limits on their capacity.

Economic Value of Fossil Fuels

- Abundant and affordable energy supplies play a key role in enabling economic growth.

- Estimates of the value of fossil fuels vary but converge on very high numbers. Coal alone delivered economic benefits worth between $1.3 trillion and $1.8 trillion of U.S. GDP in 2015.

- Reducing global reliance on fossil fuels by 80% by 2050 would probably reduce global GDP by $137.5 trillion from baseline projections.

Introduction

This chapter documents the economic benefits of fossil fuels, generally measured as per-capita income or gross domestic product (GDP). Later chapters focus on the human health and environmental benefits. Parts of this chapter originally appeared in reports by Roger H. Bezdek (Bezdek, 2014) and Craig D. Idso (Idso, 2013) which have been substantially updated and revised.

Section 3.1 offers a primer on energy, similar to the tutorials on climate science in Chapter 2, fossil fuels in Chapter 5, and cost-benefit analysis in Chapter 8. Section 3.2 describes the indispensable role played by fossil fuels in creating the modern world. Billions of lives were improved and continue to be improved every day by having access to safe, reliable, and affordable energy. Electricity, overwhelmingly generated with fossil fuels, has improved human well-being in countless ways.

Section 3.3 describes how fossil fuels improved agricultural productivity thanks to nitrogen fertilizer created by the Haber-Bosch process, agricultural machinery built with and fueled by fossil fuels, and the aerial fertilization effect of rising levels of atmospheric CO_2. Section 3.4 explains why fossil fuels are uniquely suited to meeting the energy demands of modern civilization due to their density, sufficient supply, flexibility of use, and low cost.

Section 3.5 explains why alternative fuels – wind turbines, solar photovoltaic (PV) panels, and biofuels (primarily wood and ethanol) – cannot replace fossil fuels as the primary source of energy for human use. Section 3.6 surveys the economic literature estimating the economic value of fossil fuels. It finds coal alone contributed between $1.3 trillion and $1.8 trillion to the U.S. economy in 2015 and reducing global reliance on fossil fuels by 80% by 2050 would probably reduce global GDP by $2.7 trillion a year. Section 3.7 provides a brief summary and conclusion.

References

Bezdek, R.H. 2014. *The Social Costs of Carbon? No, the Social Benefits of Carbon*. Oakton, VA: Management Information Services, Inc.

Idso, C.D. 2013. *The Positive Externalities of Carbon Dioxide: Estimating the Monetary Benefits of Rising Atmospheric CO$_2$ Concentrations on Global Food Production*. Tempe, AZ: Center for the Study of Carbon Dioxide and Global Change. October 21.

3.1 An Energy Tutorial

Other sections of this chapter deal at length with the economics of energy, and especially how abundant, inexpensive energy leads to productivity, higher GDP, and other economic benefits. This first section provides useful background by reviewing the science and technology behind the engines and machines that produce and use energy. It defines key terms, explains how efficiency is measured, presents basic facts about the leading uses and sources of energy, and explains the differences between dispatchable and intermittent power and why it matters.

3.1.1 Definitions

Some key concepts include energy, power, watts, joules, and power density.

Energy is the capacity or power to do work, such as lifting or moving an object by the application of force. Energy comes in many forms, such as kinetic (energy due to motion), potential (energy due to location), electrical, mechanical, chemical, thermal, and nuclear. Energy can be converted from one form to another by such processes as combustion and letting water descend through a water turbine to drive a generator.

Power is the amount of energy converted from one form to another divided by the time interval of the conversion; in other words, the rate of conversion of energy from one form to another. Power is energy divided by time; energy is power multiplied by time. The International System of Units uses the familiar *watt* (abbreviated W) as the unit of power. One watt is one *joule* (J), the unit of energy, per one second (s), the unit of time. One joule is one watt-second. A *gigajoule* (GJ) is one billion (10^9) joules. An *exajoule* (EJ) is 10^{18} joules.

Power density is energy flow per unit of time, which can be measured in joules per second (watts) divided by a unit of space, as in watts per square meter or W/m^2.

A secondary unit of energy is the kilowatt-hour, which is 1,000 watts multiplied by 1 hour (3,600 seconds) = 3,600,000 J. In the United States, one kWh of electricity (sometimes written kWh_e or kWhe) costs about 14 cents. One thermal kWh (kWh_t or kWht) from gasoline costs about 8 cents. It is wise, when discussing energy policy, to stick to watts and joules with occasional use of watt-hours. Very simple ideas become difficult when there is a profusion of units: a ton of coal, gallon of gasoline, British Thermal Unit (BTU), million-ton-of-carbon-equivalent (MTCE), barrel of oil, cord of wood, calories, kilocalories, foot-pounds, and so forth. For example, what is the efficiency of an engine that consumes one gallon of gasoline to produce 13 horsepower-hours? Conversion factors can be found at the online Engineering Toolbox and in Hayden (2015).

3.1.2 Efficiency

Advances in efficiency mean we live lives surrounded by the latest conveniences, yet we use only about 3.5 times as much energy per capita as did our ancestors in George Washington's time.

All conversions of energy from one form to another are characterized by an efficiency factor. When the output shaft of a car's engine is transferred to the wheels, the efficiency is very high, well above 95%; the limitation is simply friction. On the other hand, when heat is used to produce mechanical energy, the efficiency is much lower; the car's engine typically has an efficiency of around 25%.

A schematic design of a heat engine – which could be a steam engine, a gasoline engine, an aircraft's jet engine, etc. – is shown in Figure 3.1.2.1.

**Figure 3.1.2.1
Energy flow in a heat engine**

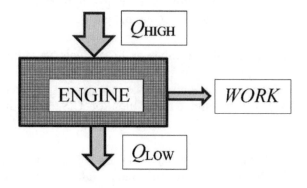

Source: Hayden, 2015, p. 41, Figure 17.

Some heat energy Q_{HIGH} flows from a hot source into an engine, which does some work *WORK*. The Second Law of Thermodynamics demands that some heat energy Q_{LOW} will flow into the lower-temperature surroundings. The efficiency of the heat-to-work engine is *WORK* divided by Q_{HIGH}. Generally speaking, the higher the temperature of the source of heat, and the lower the temperature of the surroundings (usually out of our control), the higher will be the efficiency of the engine.

The first steam engine, designed by Newcomen, was used to pump water out of a coal mine. It was huge, with a 28-inch (71 cm) diameter piston that traveled up and down a distance of 8 feet (2.4 m). It had about as much power as today's garden tractor and an efficiency of 0.05%. By way of comparison, General Electric's 9H, a 50 Hz combined-cycle gas turbine, feeds as much as 530 megawatts (MW) into the UK's electric grid with thermal efficiency of nearly 60% (Langston, 2018).

Today we have cars, planes, trucks, railroad cars, electric lights, electric motors, computers, the Internet, aluminum, refrigeration, furnaces, air conditioning, and all sorts of conveniences, yet we use only about 3.5 times as much energy per capita as did our ancestors in George Washington's time, as shown in Figure 3.1.2.2. Our modern conveniences were not available to our colonial ancestors, yet they used around 3,000 (thermal) watts (100 GJ) per capita. Mostly, the energy was consumed for home heating, cooking with firewood, and lighting with candles. (Energy from horses is not considered in this brief discussion.) The vast improvement in lifestyle occurred with such a small increase in per-capita energy consumption due to vast improvements in energy efficiency.

Figure 3.1.2.2
Annual U.S. per-capita energy consumption in GJ per capita

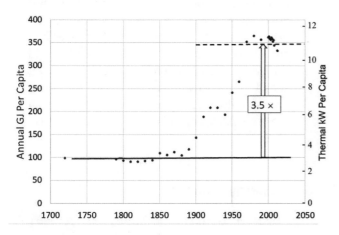

Source: Adapted from Hayden, 2015, Figure 13, p. 21.

The process of improving efficiency will continue, but with less breathtaking results. For example, going from a Rumford fireplace at 9% efficiency to a modern furnace of 90% efficiency is a dramatic change. If it were possible to achieve 100% efficiency, the change would be less dramatic. Turning the shaft of a generator to produce electricity is already well above 98% efficiency. The friction of railroad cars is already so low that to keep the train moving straight on a level track requires a forward force equal to about 50-millionths of the weight of the train, meaning 1/10 pound keeps a ton rolling (Federal Rail Administration, 2009).

For heat engines, it has long been known that one path to higher efficiency lies in developing materials that can withstand high temperatures. Pratt and Whitney developed a method for producing single-crystal superalloys that is being used to create gas turbine blades with no grain boundaries between crystals, where cracks or corrosion develop (Langston, 2018). Steam engines and internal combustion engines will see further improvements in efficiency.

3.1.3 Energy Uses

Increased use of energy and greater energy efficiency have enabled great advances in artificial light, heat generation, and transportation.

The main uses of energy in a modern civilization are for light, heat (including home heating and heat for electricity generation and manufacturing processes), and transportation.

Light

In the novels of Jane Austen, parties were scheduled to coincide with the full moon so guests could travel at night in their horse-drawn carriages. Suffice it to say that our ancestors lived in a dark world. The common source of light early on was the candle, which produces about 0.17 lumens (a unit of the amount of visible light) per watt. By comparison, an incandescent light bulb with a filament produces about 30 lumens per watt, and a light-emitting diode produces about 200.

Candles were overtaken by whale oil in cities located near coasts or with adequate roads and infrastructure to import the oil, which burns with a very clean flame in lamps. The slaughter of whales led to a shortage that made the price soar, and kerosene, a product refined from petroleum, replaced it. According to Beckmann (1977), "sperm oil rose from 43¢ a gallon in 1823 to $2.25 a gallon, and whale oil rose from 23¢ in 1832 to $1.45 a gallon. … By 1861 [the price of kerosene] had fallen to 10¢ a barrel [*sic*, it was 10¢ a gallon], and within two years kerosene had pushed out all other lighting fuels from the market, including tallow candles (as a permanent source of light). Though not everyone had been able to afford whale oil, practically everyone was able to buy kerosene."

Kerosene was then overtaken by electric lighting made possible by fossil fuels and hydroelectric generation dams. Thomas Edison's Pearl Street Station in New York City generated direct-current electricity in 1882 and by 1884 was powering about 10,000 lamps and providing some local buildings with some heat. The rest of that story is told in Section 3.2.2 and so won't be repeated here.

Heat

In the late 1700s, houses were heated by fireplaces that were very smoky and very inefficient. Count Rumford (Benjamin Thompson), who was the first scientist to prove that heat and mechanical energy were related, invented a way to improve fireplaces on both counts (Brown, 1981). The efficiency of a Rumford fireplace was about 9%. At about the same time, Benjamin Franklin invented the cast-iron stove, for which the efficiency was probably 15% to 20%. Modern home furnaces have efficiencies of roughly 90%. Of course, people in those early days were not measuring efficiency. They simply had to cut, split, and stack lots of firewood for cooking and home-heating requirements.

The British denuded their forests, using the best timber for building the ships that made the kingdom powerful at sea, and much of the rest for the manufacture of glass windows for mansions. Part of the reason for the rebellion against King George was that he claimed the best trees on American lands owned by people in the colonies.

Transportation

Until the advent of steam locomotives in the early 1800s, all transportation was by foot, animals, or ships with sails. The vast majority of people never traveled more than a few tens of kilometers (a few dozen miles) from where they were born. Even in the early decades of the 1900s, two major problems in

big cities were the removal of horse manure and disposal of dead horses.

The first primitive locomotive was built in 1812, and the first practical one was Stephenson's 8-ton "LOCOMOTION No. 1" built in 1825 for the Stockton & Darlington Railroad. It was capable of pulling 90 tons of coal at 15 mph. Today's coal trains move 10,000 tons at more than 60 mph. It was not until about 1850 that train travel became common. The trains were powered mostly by coal for the next century. City trains (streetcars and trolleybuses) were powered by overhead power lines, but only after electricity became widespread.

Trains, streetcars, and trolleybuses can take you to a station near where you want to go, but that station may be a long walk from your final destination. There will be stops along the way to allow other passengers to board or leave, and the trains operate on schedules that may not coincide with yours. For reasons of convenience, cars and trucks became the default means of transportation except in a few cities with high population densities. Cars and trucks delivered unprecedented mobility, opening up innumerable opportunities for commerce, recreation, and individual freedom. O'Toole (2009) writes,

> No matter where you are in the United States, you owe almost everything you see around you to mobility. If you live in a major city, your access to food, clothing, and other goods imported from outside the city depends on mobility. If you live in a rural area, your access to the services enjoyed by urban dwellers, such as electrical power and communications lines that are installed and served by trucks, depends on mobility. If you spend your vacations hiking in the most remote wilderness areas, your ability to reach the trailheads depends on your mobility (p. 6).

3.1.4 Energy Sources

Fossil fuels supply 81% of the primary energy consumed globally and 78% of energy consumed in the United States.

Already during the second half of the seventeenth century, a shortage of wood was leading to rising prices and restrictions on the harvesting of forest trees in England and elsewhere in Europe. The abundance of trees in the Americas assured wood's prominence longer, but by around 1900 coal had overtaken wood as the world's primary energy supply, and fossil fuels have dominated ever since, as shown in Figure 3.1.4.1.

Fossil Fuels

The world's energy supply increased dramatically from 1900 to 2009, with nearly all the increase supplied by coal, oil, and natural gas, as shown by Figure 3.1.4.1. According to the International Energy Agency (IEA, n.d.), 81% of total world energy consumption was supplied by fossil fuels in 2016. Biofuels and waste supplied about 9.8%, nuclear provided 4.9%, hydroelectric provides 2.5%, and wind, solar and other renewables combined contributed only 1.6%. See Figure 3.1.4.2. It is important to note these are stylized facts, a simplified presentation summarizing data that are incomplete, derived from models, and known to have inaccuracies. The presentation illustrates the ability of coal, oil, and natural gas to increase rapidly in supply relative to renewables – primarily wind turbines and solar PV cells – which contribute very little to global energy supplies.

The history of U.S. energy consumption from 1635 to the present is shown in Figure 3.1.4.3. Until 1850, virtually all energy came from firewood. Now our energy comes also from petroleum, coal, natural gas, nuclear power stations, hydro, wind, geothermal, solar thermal, and solar photovoltaics. The vast

Figure 3.1.4.1
The world's total primary energy supply for 1900–2009

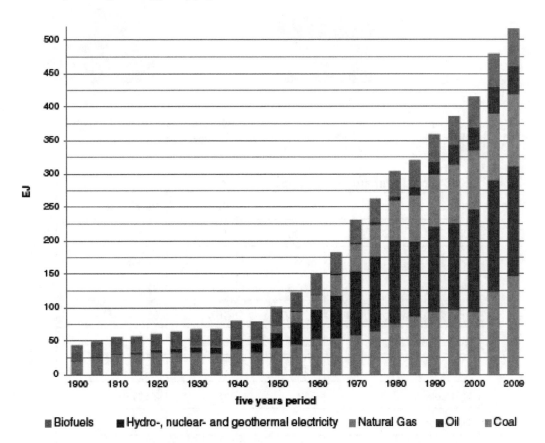

Source: Bithas and Kalimeris, 2016, Figure 2.1, p. 8.

majority of the increase in energy demand is due to the increase in population, which has increased by a factor of 100 since the founding of the nation. In 2017, fossil fuels accounted for 78% of primary energy production in the United States; nuclear and nonhydroelectric renewables each contributed about 9.5%, and hydroelectric produced 3% (EIA, 2018, Table 1.2, p. 5).

Figure 3.1.4.4 shows the complex energy flow in the United States, with sources on the left. The widths of the lines are proportional to the amounts of energy flowing in the directions indicated. The light gray areas represent "rejected energy," which is primarily the Q_{LOW} from heat engines explained in Section 3.1.1. By far the main sources of energy for producing electricity are coal, natural gas, and nuclear. Petroleum is not used much for electricity

Figure 3.1.4.2
Global primary energy supply by fuel, 1971 and 2016

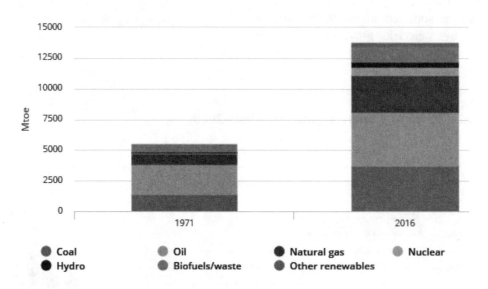

2016 global primary energy supply by fuel			
Source	**Mtoe**	**EJ**	**Percentage**
Oil	4,390	184	31.9%
Coal	3,731	156	27.1%
Natural gas	3,035	127	22.1%
Fossil fuels subtotal	11,156	467	81.1%
Biofuels and waste	1,349	56	9.8%
Nuclear	680	28	4.9%
Hydroelectric	349	15	2.5%
Other renewables	226	9	1.6%
Total	**13,760**	**575**	**100.0%**

Primary energy sources for the world in millions of tons of oil equivalent. Mtoe = megatonnes (million tons) oil equivalent, EJ = exajoules. *Source:* IEA, n.d.

Figure 3.1.4.3
U.S. energy consumption from 1635–2013 by energy source

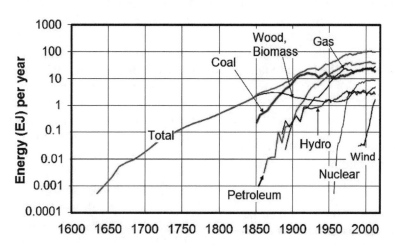

Note the vertical axis is a logarithmic scale, equal differences in order of magnitude are represented by equal distances from the value of 1. *Source:* Hayden, 2015, Figure 9, p. 18.

Figure 3.1.4.4
Sources and uses of energy in the United States in 2016

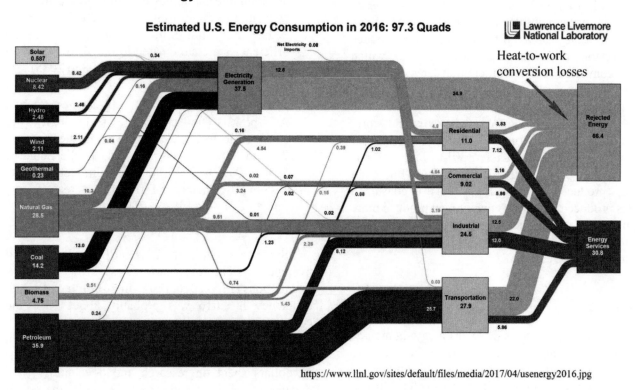

A quad is a quadrillion BTU, or 1.055×10^{18} J = 1.055 EJ. *Source:* Lawrence Livermore National Laboratory (LLNL), 2017.

production, but is the source for almost all of our transportation needs. Natural gas is the most versatile fuel, providing energy for electricity production and heating, cooking, and process heat for homes, commercial establishments, and industries. The world as a whole consumes 576 EJ, about 5.8 times as much energy as the 100 EJ consumed by the United States. Coal supplies 162 EJ, 28% of the world's total energy, versus coal's 37% share in the United States.

Bioenergy

Biofuels (mainly wood) supply about 10% of the global energy supply. Ethanol and biodiesel are often proposed as "climate friendly" alternatives to gasoline and diesel fuel, which are derived from petroleum. Biofuels are severely limited by their lack of power density, a matter discussed at some length in Section 3.4.1 below and again in Chapter 5, Section 5.2.2. Here in a nutshell is the engineering science behind that problem.

Chlorophyll absorbs about 6.6% of the sunlight falling on it. Of that amount, some energy is used to combine carbon (stolen from CO_2), hydrogen (stolen from H_2O), nitrogen and various minerals into green leaves. All in all, 90% of that absorbed solar energy is used up in the plant itself. The best plants that can be grown in large areas of the United States produce about 10 tons of drymatter per acre per year, which when converted into biofuel translates into 1.2 thermal watts per square meter of land (Bomgardner, 2013). This yield diminishes to only 0.069 and 0.315 W/m^2 for biofuels produced from soy and corn, respectively, when energy is deducted to account for farming and processing (Kiefer, 2013). Full accounts of the ethanol production process generally find more energy (produced by using fossil fuels) is consumed than is produced, meaning ethanol may produce more greenhouse gas emissions than the oil it replaces (Searchinger *et al.*, 2008; Melillo *et al.*, 2009; Mosnier *et al.*, 2013).

Solar Energy

There are three main uses of solar energy: home and workplace heating, conversion of solar heat to electricity, and direct conversion of sunlight to electricity by using photovoltaic (PV) cells. Most of the emphasis has been on PV cells, so we focus on them here. Higher PV efficiency means less collection area is needed to produce the same energy output. Less obvious, perhaps, is that higher efficiency sometimes comes at a dramatically higher cost.

Figure 3.1.4.5 shows progress in improving module efficiencies for PV cells from 1975 to 2016. (A module is better known as a panel, a collection of cells pre-wired and packaged for modular installation.) These modules often are on the cutting edge of research and not yet ready for commercial applications. Some, for example, are the size of your fingernail. Efficiencies as high as 46% have been obtained but only for very small, very expensive cells. For large-scale photovoltaics, efficiencies tend to be around 15% and have not been improving much over time.

The electricity generated by PV cells must be converted from direct current (DC) to alternating current (AC) before sale on wholesale electricity markets or for direct use in an AC household system, resulting in a loss of energy reported as "system efficiency." The U.S. Energy Information Administration (2010) made the following forecasts of improvements in module and system efficiencies:

- *Module Efficiency.* Module efficiencies for crystalline technologies operating in the field are estimated to range from 14% in 2008 to 20% in 2035. For thin-film technologies, module efficiencies are anticipated to range from 10% to 14% over this same time span (2008 to 2035).

- *System Efficiency.* System efficiencies (DC to AC power) for crystalline technologies are expected to increase from levels in the range of 78% to 82% in 2008, to levels in the range of 86% to 90% in 2035. For thin-film technologies, system efficiencies are forecast to increase from a range of 77% to 81% in 2008, to a range of 86% to 90% in 2035.

Six years later, in 2016, EIA still assumed average system efficiency was 80% for new installations (EIA, 2016, p. 18, fn 26). In fact, PV cell efficiency in 2016 was probably about 15% and conversion to grid voltage was about 80%, making an overall efficiency of only 12%. There is every reason to expect efficiency to improve, but despite billions of dollars spent on research and development and decades of subsidies and tax breaks, so far the rate of increase has not been rapid.

Figure 3.1.4.5
Progress in improving best research-cell efficiencies from 1975 to 2016

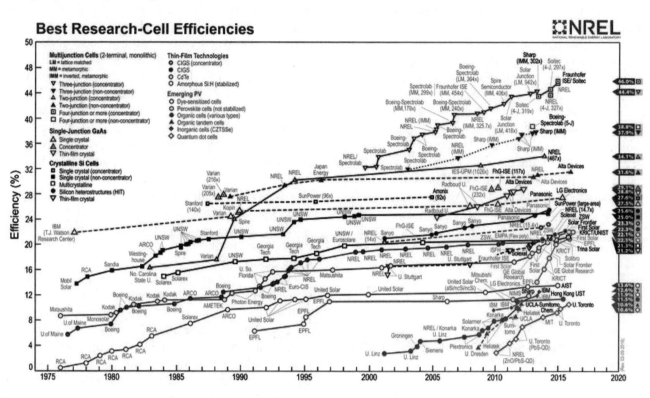

Source: National Renewable Energy Laboratory, 2018.

Wind

Wind turbines extract kinetic energy from the wind. Wind turbines rotate slowly, measured in rotations per minute (RPM), but the tips of the blades move very rapidly. As an example, the Vestas V80-2.0 MW turbine turns at 16.7 revolutions per minute, doing a full rotation in 3.6 seconds. But the tips of the 80-meter-diameter turbine move at about 70 meters per second (about 155 miles per hour) and can reach up to 80 m/s (about 180 mph).

Because wind turbines are designed around the properties of the wind, the tip speed is a multiple of the wind speed, regardless of the rotor diameter. Accordingly, the RPM decreases with rotor diameter. Generators, on the other hand, work best at high rotation rate, typically hundreds to thousands of RPM. Gearboxes are required to convert the ponderous rotation of the wind turbine rotor into the high RPM of the generator. To date, gearboxes have been the main cause of wind turbine failures.

The annual capacity factor (CF) of wind – the average annual power divided by the nameplate power of the generator – is a matter of engineering design. A small generator on a large-diameter turbine will have a high CF. A large generator turned by a small-diameter turbine will have a small CF. The current best engineering compromise is a 35% CF. Wind turbines cause the air to slow down. This coupled with the need to avoid turbulence means wind turbines have to be spaced some distance apart, typically about 10 rotor diameters. The power produced by the wind turbine is proportional to the area swept by the rotors: If you double the diameter, you quadruple the power. But the distance between adjacent turbines must double in both directions, thereby quadrupling the land area. Consequently, the power produced per unit area of land is independent of the size of the wind turbines.

For arrays of industrial wind turbines there are two useful numbers to know: the CF is 35% by design, and the year-round average power per unit area, or power density, is about 1.2 W/m^2 (12 kW/ha, 5 kW/acre).

The wind energy arriving at the wind turbine per unit of time is proportional to the *cube* – the third power – of the wind speed. As a consequence, the power generated by a wind turbine varies dramatically with wind speed. Figure 3.1.4.6 shows the power curve for the Vestas unit discussed above; the curves for any model of industrial wind turbine by any manufacturer is similar in shape. The only significant difference is the peak power. At wind speeds below about 4.5 m/s (10 mph), no power is produced. At 6 m/s (13.4 mph), the power is 200 kW. At 12 m/s (27 mph), the power is 1600 kW. When the wind speed doubles from 6 m/s to 12 m/s, the power output increases by a factor of 8. Above about 14 m/s (31.3 mph), the generated power is 2 MW until the speed reaches 25 m/s (56 mph), after which the system must be shut down to avoid tearing itself apart.

Figure 3.1.4.6
Power produced by the Vestas V80-2.0 MW versus wind speed

Power curve V80-2.0 MW

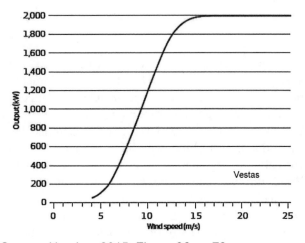

Source: Hayden, 2015, Figure 36, p. 72.

3.1.5 Intermittency

Due to the nature of wind and sunlight, wind turbines and solar photovoltaic (PV) cells can produce power only intermittently.

Wind turbines and solar photovoltaic (PV) cells are unable to produce a steady stream of energy into an electric grid and sometimes produce no power at all for hours, days, or even weeks. Both are unable to produce *dispatchable* energy, defined as energy available to an electric grid on demand. PV cell output drops when clouds, rain, or dust reduce the amount of sunlight reaching the panel, during seasons that tend to be cloudy or rainy, and of course every day from nightfall until sunrise. Figure 3.1.5.1 records actual solar energy production in southeastern Australia on a typical day, June 19, 2018. Evident from the figure are high levels of variability during daylight hours and, as expected, zero power production at night.

Figure 3.1.5.2 illustrates the dramatic volatility of wind power on a typical day (June 19, 2018) in southeastern Australia. Wind turbines were constantly ramping up and down, from 100% to 0% often in just minutes.

Data plotted in Figures 3.1.5.1 (solar) and 3.1.5.2 (wind) are from the energy markets and systems in southeastern Australia operated by the Australian Energy Market Operator (AEMO). AEMO also documents the very small amounts of energy generated by wind and solar relative to total energy consumption in southeastern Australia, shown in Figure 3.1.5.3. Fossil fuels, the only dispatchable energy source available in sufficient amounts to meet demand, makes up for the shortfalls when the sun doesn't shine and the wind doesn't blow. Similar patterns are apparent in all industrialized countries of the world.

In many countries, calm periods sometimes last longer than a week. It is probably reasonable to assume that the installed capacity of wind power needs to be something like twice the system maximum demand and the storage system needs to be able to receive this surplus energy at a rate greater than the system demand. So if the system demand is 1,000 MW, 3,000 – 4,000 MW of generating capacity is needed supported by about 2,500 MW of storage. So a 1,000 MW system needs to have a connected generating capacity of about 6,000 MW. This will be extremely expensive.

Energy can be stored primarily in three ways: as chemical energy in batteries, as gravitational potential energy behind dams, and as heat, typically heated water. While each method is widely used and has valuable applications, all have limitations making them unable to store more than a small fraction of the energy used on an hourly or daily basis. Without back-up power produced by fossil fuels, a renewables-

Figure 3.1.5.1
Solar energy production in southeastern Australia on June 19, 2018

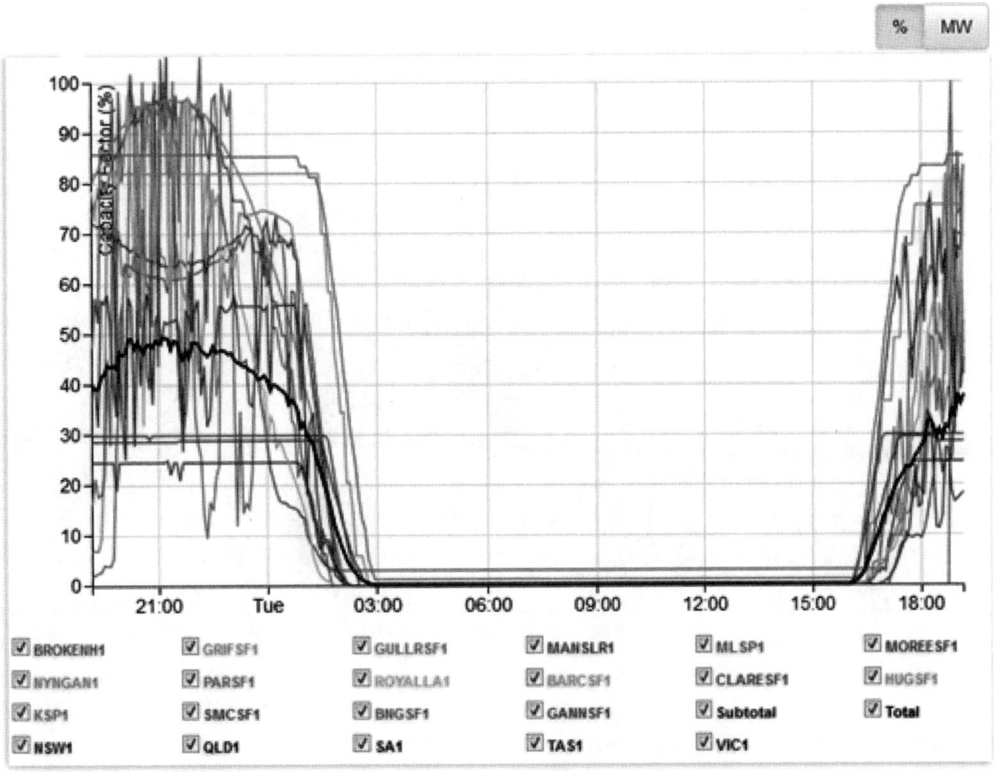

The default, capacity factor graph shows the output as a percentage of registered capacity. Alternatively, you may view the actual output in megawatts.

Source: AEMO, 2018.

only energy system would require a vast amount of storage. The only technology available at the moment that can provide this is pumped storage hydropower. An installation consists of a lower lake and an upper lake 300 to 800 m above the lower lake. The two lakes are connected by a pipeline with a power station at the lower lake that can either generate from water from the upper lake or pump water from the lower lake to the upper lake.

Conventional pumped hydro power stations store water for between 6 and 10 hours and normally generate during the morning and evening peak demand periods and pump the early hours of the morning and the middle of the day. There might be six storage schemes in the world that can provide storage for longer than a few days. In order to support solar and wind generation a pumped storage scheme would need to have far more capacity than any pumped hydro system operating today, enough to

power a grid for days or weeks at a time. It would be extremely difficult to find a site for such a station because it needs huge basins less than about 10 miles apart, one of which is hundreds of metres higher than the other. It is also necessary to find a source of make-up water because the evaporation from two lakes is likely to be quite large.

The economics of such a scheme will be dubious because it likely involves submerging thousands of acres of land plus an investment of thousands or millions of tons of concrete and steel to build the dam, hydroelectric turbines, power lines, etc. A very large quantity of water has to be pumped up to the upper lake using expensive wind and solar power, held there for days, weeks or months, and then used to generate electricity with an overall loss of about 20%. The reality is that there are few suitable sites available and those that exist are likely to be remote from solar and wind power sources, thus incurring

Figure 3.1.5.2
Wind energy production in southeastern Australia on June 19, 2018

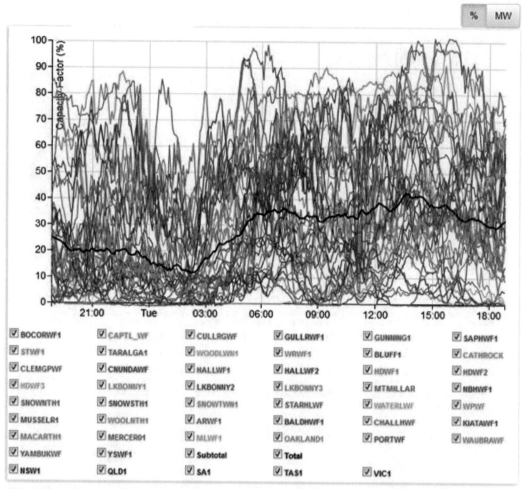

This graph depicts performance of wind farms connected to the electricity grid in south-eastern Australia over a 24-hour period.

The default, capacity factor graph shows the output as a percentage of registered capacity. On average wind farms in south-east Australia operate at a capacity factor of around 30-35%.

Source: AEMO, 2018.

Figure 3.1.5.3
Energy production by source in southeastern Australia on June 19, 2018

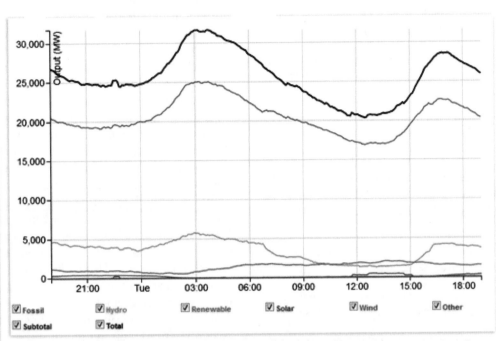

This graph shows the amount of energy being contributed into the grid by each type of generator. It also shows the total which essentially amounts to energy being demanded of the grid at any given time.

Source: AEMO, 2018.

very large transmission losses.

In addition to the storage problem, the variability of wind and solar power creates problems for electrical grid operators that may be unsolvable at high penetration rates. This topic is addressed in Section 3.5.3.

References

AEMO. 2018. Australian Energy Market Operator. Wind Energy (website). Accessed June 19, 2018.

Beckmann, P. 1977. *Pages from US Energy History.* Golem Press.

Bithas, K. and Kalimeris, P. 2016. *Revisiting the Energy-Development Link.* SpringerBriefs in Economics. DOI 10.1007/978-3-319-20732-2_2.

Bomgardner, M.M. 2013. Seeking biomass feedstocks that can compete: biofuels and biobased chemical makers hope to win with cellulosic sugars. *C & E News.* August 12.

Brown, S.C. 1981. *Benjamin Thompson, Count Rumford.* Cambridge, MA: MIT Press.

EIA. 2010. U.S. Energy Information Administration. Photovoltaic (PV) Cost and Performance Characteristics for Residential and Commercial Applications, Final Report.

EIA. 2016. U.S. Energy Information Administration. Wind and Solar Data and Projections from the U.S. Energy Information Administration: Past Performance and Ongoing Enhancements.

EIA. 2018. U.S. Environmental Protection Agency. *Monthly Energy Review November 2018.*

Federal Rail Administration. 2009. *Comparative Evaluation of Rail and Truck Fuel Efficiency on Competitive Corridors, Final Report.* November 19.

Hayden, H.C. 2015. *Energy: A Textbook.* Pueblo West, CO: Vales Lake Publishing.

IEA. n.d. International Energy Agency. World Energy Balances. Accessed November 4, 2018.

Kiefer, T.A. 2013. Energy insecurity: the false promise of liquid biofuels. *Strategic Studies Quarterly* (Spring): 114–51.

Langston, L.S. 2018. Single-crystal turbine blades earn ASME milestone status. *Machine Design.* March 16.

Lawrence Livermore National Laboratory (LLNL). 2017. Energy Flow Charts (website). Accessed November 4, 2018.

Melillo, J., *et al.* 2009. Indirect emissions from biofuels: how important? *Science* **326** (5958): 1397–9.

Mosnier, A.P., *et al.* 2013. The net global effects of alternative U.S. biofuel mandates: fossil fuel displacement, indirect land use change, and the role of agricultural productivity growth. *Energy Policy* **57** (June): 602–14.

National Renewable Energy Laboratory. 2018. U.S. Department of Energy. Research Cell Efficiency Records (website). Accessed November 4, 2018.

O'Toole, R. 2009. *Gridlock: Why We're Stuck in Traffic and What to Do About It.* Washington, DC: Cato Institute.

Searchinger, T., *et al.* 2008. Use of US croplands for biofuels increases greenhouse gases through emissions from land-use change. *Science* **319**: 1238–40.

3.2 Three Industrial Revolutions

The primary reason humans burn fossil fuels is to produce the goods and services that make human prosperity possible. Put another way, we burn fossil fuels to live more comfortable, safer, and higher-quality lives. The close connection between fossil fuels and human prosperity is revealed by the history of the Industrial Revolution and analysis of more recent technological innovations.

3.2.1 Creating Modernity

Fossil fuels make possible such transformative technologies as nitrogen fertilizer, concrete, the steam engine and cotton gin, electrification, the internal combustion engine, and the computer and Internet revolution.

Prior to the widespread use of fossil fuels, humans expended nearly as much energy (calories) producing food and finding fuel (primarily wood and dung) to warm their dwellings as their primitive technologies were able to produce. Back-breaking work to provide bare necessities was required from sun-up to sun-down, by children as well as adults, leaving little time for any other activity. The result was a vicious cycle in which the demands of the immediate present prevented investing the time and capital needed to think about and discover ways to improve productivity and therefore the future (Simon, 1981; Bradley and Fulmer, 2004; Epstein, 2014).

According to Goklany (2012), "For most of its existence, mankind's well-being was dictated by disease, the elements and other natural factors, and the occasional conflict. Virtually everything required – food, fuel, clothing, medicine, transport, mechanical power – was the direct or indirect product of living nature" (p. 2). Generations of farmers and craftsmen used the same tools and worked the same land as their ancestors. Progress, whether measured by lifespan, population, or per-capita income, was almost nonexistent. The main sources of non-labor power in that era were windmills, waterwheels, and grass-fed horses, none of which could be easily scaled up. Prosperity came slowly to humanity. According to Maddison (2006):

- "Over the past millennium, world population rose 22–fold. Per capita income increased 13–fold, world GDP nearly 300–fold. This contrasts sharply with the preceding millennium, when world population grew by only a sixth, and there was no advance in per capita income.

- "From the year 1000 to 1820 the advance in per capita income was a slow crawl – the world average rose about 50 per cent. Most of the growth went to accommodate a fourfold increase in population.

- "Since 1820, world development has been much more dynamic. Per capita income rose more than eightfold, population more than fivefold.

- "Per capita income growth is not the only indicator of welfare. Over the long run, there has been a dramatic increase in life expectation. In the year 1000, the average infant could expect to live about 24 years. A third would die in the first year of life, hunger and epidemic disease would ravage the survivors.

- "There was an almost imperceptible rise [in life expectancy] up to 1820, mainly in Western Europe. Most of the improvement has occurred since then. Now the average infant can expect to survive 66 years" (p. 19).

The increasing use of fossil fuels was responsible for the astonishing change in human well-being starting around 1800. Gordon (2012, 2016) analyzed economic growth in the United States over the past several hundred years and identified fossil fuels as the power source that drove not one but *three* Industrial Revolutions. The first (1750 to 1830) resulted from the invention of the steam engine and cotton gin and proceeded through the development of the early railroads and steamships, although much of the impact of railroads on the American economy came later, between 1850 and 1900. The *Second Industrial Revolution* (1870 to 1900) was the most important, with the invention of electricity generation, lights, motors, and the internal combustion engine, and widespread access to running water with indoor plumbing. Both of the first two revolutions required about 100 years for their full effects to percolate through the economy.

During the two decades 1950–70 the benefits of the Second Industrial Revolution were still transforming the economy, including air conditioning, home appliances, and the interstate highway system. After 1970, productivity growth from this second revolution slowed markedly as the new inventions had reached every corner of the country. The *Third Industrial Revolution* (1970 to present) is marked by the computer and Internet revolution. Its beginnings can be traced back to around 1960, but it really took off and reached a climax during the dot-com era of the late 1990s. It continues to revolutionize science, medicine, manufacturing, and transportation.

As documented in Section 3.1, fossil fuels provided the energy required by nearly all of the revolutionary technologies Gordon identified, from the steam engine and cotton gin of the past to high-tech manufacturing and the mobile computer devices of today (Ayres and Warr, 2009). See Figure 3.1.4.1 in Section 3.1 for estimates of the sources of the world's total primary energy supply for 1900–2009. Figure 3.2.1.1 shows the rapid increase in the use of coal in the United States beginning around 1850, then oil in 1900, followed by natural gas in 1920. Wood remained the main source of fuel in the United States until about 1883, when it was overtaken by coal. Energy consumption followed somewhat similar trajectories in many other countries, with differences determined by natural resource endowments, assignment of private property rights to natural resources, and government policies.

Figure 3.2.1.1 History of energy consumption in the United States, 1775–2009

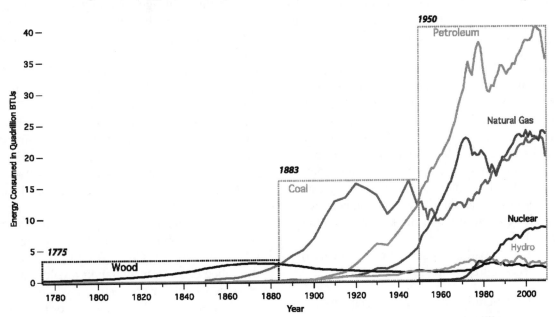

Source: EIA, 2011.

In 2016, 81% of global energy consumption is supplied by fossil fuels (IEA, 2018.). Approximately 63% of electricity worldwide was produced by the combustion of fossil fuels (coal, oil, or natural gas), while nuclear accounted for 20% and all renewable energies (solar, wind, biomass, geothermal, and hydroelectric) combined accounted for the remaining 17% (EIA, n.d.). Wind energy generated 6.3% of electricity and solar generated 1.3%.

Key to the ongoing technological developments of the three Industrial Revolutions was the fact that initial technologies accelerated the generation of ideas that facilitated even better technologies through, among other things, greater accumulation of human capital (via greater populations, time-expanding illumination, and time-saving machinery) and more rapid exchange of ideas and knowledge (via greater and faster trade and communications) (Smil, 1994, 2005; Bradley, 2000). The benefits continue to accumulate today as cleaner-burning fossil fuels bring electricity to developing countries and replace wood and dung as sources of indoor heating (Yadama, 2013).

Without cheap and reliable energy, there would be less food (and what food we had would be less fresh, less nutritious, and less safe), no indoor plumbing, no air conditioning, no labor-saving home appliances such as washing machines and clothes driers, no agricultural machinery, few hospitals, and no speedy ambulances to take us to a hospital when we need urgent medical care. Sterilizing medical devices would be extremely difficult without electricity. Natural gas is also the fuel stock of plastics; without it, the hospital we might succeed in finding would have no syringes, no tubes, and no bags for plasma.

Goklany (2012) summarized the benefits as follows: "Americans currently have more creature comforts, they work fewer hours in their lifetimes, their work is physically less demanding, they devote more time to acquiring a better education, they have more options to select a livelihood and live a more fulfilling life, they have greater economic and social freedom, and they have more leisure time and greater ability to enjoy it." Goklany's research shows these trends are also evident in other industrialized nations (Goklany, 2007).

Fossil fuels made possible the growth of America's largest cities. Platt (1991) observed,

> Although generally ignored by scholars, energy fuels constitute a natural resource that has had a major impact on regional

economies, including the growth of their urban centers. With the shift from wood to coal, the Midwest's virtually unlimited supply of that fuel became crucial to maintaining transportation rates on a par with or lower than those in cities farther east. Vast fields of bituminous (soft) coal throughout Illinois would allow Chicago's commerce and manufacturing to develop in step with those of its counterparts in the East. In contrast, regions of relative coal scarcity such as the Southwest would lag behind in manufacturing while high transportation rates added an extra tax on their commerce (p. 7).

Platt emphasizes the role played by coal in attracting industry to the Chicago area:

> The importance of this natural resource to the growth of the industrial cities of the Midwest cannot be overstressed. A "glut" of cheap coal would act as a magnet attracting a wide array of energy-intensive industries to locate and flourish in Chicago, the transportation hub of the region. … And it was these very energy-intensive industries that represented the vanguard of the industrial revolution. The rise of big business and the jobs it created were in large part responsible for the city's phenomenal growth in the late nineteenth century (*Ibid.*, pp. 7, 9).

Calling Chicago "the city that coal built," Platt wrote: "Whereas the region's grain, hogs, and timber fed the growth of the first city, its abundance of cheap coal fueled the second wave of industrial development." Coal-gas-powered gas lamps, inaugurated in 1850, were 12 to 14 times more powerful than the standard candle or oil lamp. Gas streetlamps made nightlife possible, improved safety for travelers and protection against muggers, and lowered the odds of accidental fires. Safe indoor lighting came within reach of the non-rich for the first time in history. The same story can be told of all the world's cities.

Fossil fuels also made possible a vast expansion of human mobility (Rae, 1965; Lomasky, 1997; O'Toole, 2001; Cox, 2006). O'Toole (2009) described eight "transportation revolutions," only one of which could have occurred without fossil fuels. They were, in chronological order, steamboats, canals, roads across mountains, railroads, horse-powered streetcars, automobiles, superhighways, and jetliners (*Ibid.*, pp. 8–20). Fossil fuels were essential to creating the steel and powering the factories used to create streetcars, so even horse-powered streetcars would have been rare without fossil fuels. Increased mobility produced major economic benefits as employers were able to draw workers from a larger area and workers were able to choose among a larger number of potential employers without having to relocate their families. O'Toole cites research in France that found for every 10% increase in travel speeds, the pool of workers available to employers increased by 15%. He noted, "This gives employers access to more highly skilled workers, which in turn increases worker productivity by 3%." Research in California, he says, found "doubling the distance workers can commute to work increases productivity by 25%," citing Prud'homme and Lee (1999) and Cervero (2001) (*Ibid.*, p. 5).

Collier (2007) explained how global economic development today depends on the continued availability of fossil fuel-powered transportation. Some of the poorest countries in the world are landlocked and face high transportation costs, preventing them from being able to participate in the global economy. He mentions Burkina Faso, Central African Republic, Chad, Malawi, and Uganda. For these countries, "air freight offers a potential lifeline into European markets. The key export products are likely to be high-value horticulture…" (p. 180). He further observes that coastal resource-poor countries are unable to access international markets because they lack ports and airports to compete with China and other first-movers. "Breaking out of limbo," he says, requires "big-push aid" for "raising export infrastructure up to globally competitive levels" (p. 182). More than aid, they need affordable and reliable fossil fuels to build and utilize this infrastructure.

References

Ayres, R.U. and Warr, B. 2009. *The Economic Growth Engine: How Energy and Work Drive Material Prosperity*. Northampton, MA: Edward Elgar.

Bradley Jr., R.L. 2000. *Julian Simon and the Triumph of Energy Sustainability*. Washington, DC: American Legislative Exchange Council.

Bradley Jr., R.L. and Fulmer, R.W. 2004. *Energy: The Master Resource*. Dubuque, IA: Kendall/Hunt Publishing.

Cervero, R. 2001. Efficient urbanization: economic performance and the shape of the metropolis. *Urban Studies* **38** (1): 1651–71.

Collier, P. 2007. *The Bottom Billion: Why the Poorest Countries Are Failing and What Can Be Done About It*. New York, NY: Oxford University Press.

Cox, W. 2006. *War on the Dream: How Anti-Sprawl Policy Threatens the Quality of Life*. Lincoln, NE: iUniverse.

EIA. 2011. U.S. Energy Information Administration. History of energy consumption in the United States, 1775–2009. Today in Energy (website). Accessed May 22, 2018.

EIA. n.d. U.S. Energy Information Administration. Frequently Asked Questions: What is U.S. electricity generation by energy source? (website). Accessed March 7, 2018.

Epstein, A. 2014. *The Moral Case for Fossil Fuels*. New York, NY: Portfolio/Penguin.

Goklany, I.M. 2007. *The Improving State of the World: Why We're Living Longer, Healthier, More Comfortable Lives on a Cleaner Planet*. Washington, DC: Cato Institute.

Goklany, I.M. 2012. Humanity unbound: How fossil fuels saved humanity from nature and nature from humanity. *Cato Policy Analysis* #715. Washington, DC: Cato Institute.

Gordon, R.J. 2012. Is U.S. economic growth over? Faltering innovation confronts the six headwinds. *Working Paper* No. 18315, National Bureau of Economic Research.

Gordon, R.J. 2016. *The Rise and Fall of American Growth: The U.S. Standard of Living Since the Civil War*. Princeton, NJ: Princeton University Press.

IEA. 2018. International Energy Agency. World Energy Balances (website). Accessed November 4, 2018.

Lomasky, L.E. 1997. Autonomy and automobility. *The Independent Review* **2** (1): 5–28.

Maddison, A. 2006. *The World Economy. Volume 1: A Millennial Perspective*. Paris, France: Development Centre of the Organisation for Economic Co-operation and Development.

Maddison, A. 2010. Angus Maddison. Historical Statistics (website). Groningen, The Netherlands: Groningen Growth and Development Center, University of Groningen.

O'Toole, R. 2001. *The Vanishing Automobile and Other Urban Myths: How Smart Growth Will Harm American Cities*. Bandon, OR: The Thoreau Institute.

O'Toole, R. 2009. *Gridlock: Why We're Stuck in Traffic and What to Do About It*. Washington, DC: Cato Institute.

Platt, H.L. 1991. *The Electric City: Energy and the Growth of the Chicago Area, 1880–1930*. Chicago, IL: University of Chicago Press.

Prud'homme, R. and Lee, C-W. 1999. Size, sprawl, speed, and the efficiency of cities. *Urban Studies* **36** (11): 1849–58.

Rae, J.B. 1965. *The American Automobile: A Brief History*. Chicago, IL: University of Chicago Press.

Simon, J. (Ed.) 1981. *The Ultimate Resource*. Princeton, NJ: Princeton University Press.

Smil, V. 1994. *Energy in World History*. Boulder, CO: Westview Press.

Smil, V. 2005. *Energy at the Crossroads: Global Perspectives and Uncertainties*. Cambridge, MA: MIT Press.

Yadama, G.N. 2013. *Fires, Fuels, and the Fate of 3 Billion: The State of the Energy Impoverished*. New York, NY: Oxford University Press.

3.2.2 Electrification

Electricity powered by fossil fuels has made the world a healthier, safer, and more productive place.

Fossil fuels' greatest contribution to human prosperity is making electricity affordable and dispatchable. In 2000, the U.S. National Academy of Engineering (NAE) announced "the 20 engineering achievements that have had the greatest impact on quality of life in the 20th century." The achievements were nominated by 29 professional engineering societies and ranked by "a distinguished panel of the nation's top engineers" chaired by H. Guyford Stever, former director of the National Science Foundation and science advisor to the president. The experts ranked electrification the number one achievement. "[E]lectrification powers almost every pursuit and enterprise in modern society," NAE reported. "It has literally lighted the world and impacted countless areas of daily life, including food production and processing, air conditioning and heating, refrigeration, entertainment, transportation, communication, health care, and computers" (NAE, 2000).

NAE contrasted modern life with life before electricity, saying "One hundred years ago, life was a constant struggle against disease, pollution, deforestation, treacherous working conditions, and enormous cultural divides unbreachable with current communications technologies. By the end of the 20th century, the world had become a healthier, safer, and more productive place, primarily because of engineering achievements" (*Ibid*.). Constable and Somerville, in a book published in 2003 by the National Academies Press, commented on the extraordinary engineering achievements electricity launched:

The greatest engineering achievements of the twentieth century led to innovations that transformed everyday life. *Beginning with electricity*, engineers have brought us a wide range of technologies, from the mundane to the spectacular. Refrigeration opened new markets for food and medicine. Air conditioning enabled population explosions in places like Florida and Arizona. The invention of the transistor, followed by integrated circuits, ushered in the age of ubiquitous computerization, impacting everything from education to entertainment. The control of electromagnetic radiation has given us not only radio and television, but also radar, x-rays, fiber optics, cell phones, and microwave ovens. The airplane and automobile have made the world smaller, and highways have transformed the landscape (Constable and Somerville, 2003, p. 9, Box 1, italics added).

Fossil fuels brought electricity to the homes and workplaces of billions of people around the world. Bryce (2014) wrote,

Edison's breakthrough designs at the Pearl Street plant [the world's first coal-fired electricity generating plant] allowed humans to reproduce the lightning of the sky and use it for melting, heating, lighting, precision machining, and a great many other uses. Electric lights meant workers could see better and therefore make more precise drawings and fittings. Electricity allowed steel producers to operate their furnaces with greater precision, which led to advances in metallurgy. Electric power allowed factories to operate drills and other precision equipment at speeds unimaginable on the old pulley-driven systems, which relied on waterwheels or steam power. As Henry Ford wrote in 1930, without electricity "there could be nothing of what we call modern industry" (pp. 30–31).

Electrification made its first contribution to modernity by bringing light to cities. Government regulators routinely granted monopolies to coal gas companies for street, business, and residential lighting, and those companies formed cartels to keep prices high even as technology improved and supplies increased. The arrival of the electric arc lamp and then Edison's incandescent lamp "set off separate revolutions in the technology of making gaslights and in the business practices of local utilities," triggering "intense competition to sign up customers and extend service territories, including working-class neighborhoods" (Platt, 1991, p. 16).

Manufacturers began to adopt the new technology in 1890. At first they connected electric generators to existing steam engines on their premises to power lights. Then electric motors were mounted on ceilings, upper floors, and in attics to replace the rotating shafts that had been hung from ceilings and connected to machines on shop floors by long belts. Then electric motors were moved to the shop floor, often at individual workstations, making the belts unnecessary. The final step was connecting the factory to a central electric generation station that would replace the on-site steam engines.

The results of the switch to central electric power stations "were revolutionary," Platt wrote (p. 216). "From the eve of the war [WWI] to the onset of the Great Depression, industrial power use [in the Chicago area] increased tenfold, or a spectacular 68% annually over the fifteen-year period. Energy consumption by commercial and residential customers also grew at a vigorous rate of almost 30% a year, while public transportation lagged behind with an anemic annual rate of 3.5%" (p. 217). It was, he wrote, "the birth of the machine age" (p. 226).

Electricians wired homes first for electric lights and then outlets to power everything from stoves and refrigerators to space heaters, radios, clocks, toasters, washing machines and clothes dryers, and vacuum cleaners. Every aspect of daily life was changed. "By the late twenties the use of more and more electricity, gas, and oil in everyday life had become so ubiquitous as to wrap urban America in an 'invisible world' of energy. Even the shock wave of the Great Depression could not halt the steady rise in household consumption of electricity, preserving the new standard of living" (Platt, 1991, pp. 235–6).

Electricity had a powerful effect on culture. Suddenly, millions of people were listening to radios and then watching television, hearing news and music and reading newspapers printed on electric printing presses. "Popular culture" emerged for the first time, knitting together communities once separated by distance and unaware of the music, ideas, and lifestyles of people who lived farther away than a day's journey on horseback or in a horse-drawn carriage. While the greater mobility made possible by cars and trucks caused a radical decentralization of authority and society itself (with the creation of suburbs), the new electric media brought the nation closer together by creating a shared body of knowledge and entertainment. "For the first time, Midwestern farmers, Italian immigrants, the suburban elite, small children, and myriad others were all spending leisure time in the same pursuit" (Platt, 1991, p. 286).

While Chicago and other cities in the Midwest benefited from their ample supplies of coal, cities in the South were benefitting from another invention made possible by electricity: climate control, most importantly air conditioning. Willis Carrier originally developed climate control to facilitate ink drying in the printing industry in New York City in the early 1900s, but his signature technology soon produced nearly incalculable benefits to society. Air conditioning made factory work tolerable in the South, reduced infant mortality, eliminated malaria, and allowed developers to build skyscrapers and apartment buildings. Air conditioning industrialized and urbanized the South, lifting it out of its post-Civil War depression (Arsenault, 1984).

In the United States, many of the central changes in society since World War II would not have been possible without air conditioning in homes and workplaces. Arizona, Florida, Georgia, New Mexico, Southern California, and Texas all experienced above-average growth during the latter half of the twentieth century – which would have been impossible without air conditioning. Air conditioning was crucial for the explosive postwar growth of Sunbelt cities like Houston, Las Vegas, Miami, and Phoenix. Without it people simply could not live and thrive in such hot locations.

Air conditioning launched new forms of architecture and altered the ways Americans live, work, and play. From suburban tract houses to glass skyscrapers, indoor entertainment centers, high-tech manufacturers' clean rooms, and pressurized modules for space exploration, many of today's structures and products would not exist without the invention of climate control. As the technology of climate control developed, so also did the invention of more sophisticated products that required increasingly precise temperature, humidity, and filtration controls – consumer products such as computer chips and CDs must be manufactured in "clean rooms" that provide dust-free environments. The development of the entire information technology (IT) industry could not have occurred without the cooling technologies first pioneered by air conditioning.

Electricity propelled the transportation revolution begun by fossil fuels by making possible headlights for cars and trucks, street lighting, traffic lights, airlines, mass transit, and telecommuting. It revolutionized health care by making possible modern hospitals and clinics, and agriculture by allowing the refrigeration of produce. Electricity created the "global village" via advances in communication including the telephone, radio, television, fax machines, cell phones, computers, the Internet, satellites, email, social media, and more. Electricity powered by fossil fuels, in short, created the modern age.

References

Arsenault, R. 1984. The end of the long hot summer: the air conditioner and Southern culture. *The Journal of Southern History* **50**: 597–628.

Bryce, R. 2014. *Smaller Faster Lighter Denser Cheaper: How Innovation Keeps Proving the Catastrophists Wrong*. New York, NY: PublicAffairs.

Constable, G. and Somerville, B. 2003. *A Century of Innovation: Twenty Engineering Achievements that Transformed our Lives*. Washington, DC: The National Academies Press.

NAE. 2000. National Academy of Engineering reveals top engineering impacts of the 20th century: Electrification cited as most important. *News Release*. Washington, DC: National Academy of Engineering. February 22.

Platt, H.L. 1991. *The Electric City: Energy and the Growth of the Chicago Area, 1880–1930*. Chicago, IL: University of Chicago Press.

3.2.3. Human Well-being

Access to energy is closely associated with key measures of global human development including per-capita GDP, consumption expenditure, urbanization rate, life expectancy at birth, and the adult literacy rate.

The prosperity made possible by the use of fossil fuels enabled societies to invest in education, health care, housing, and other essential goods and services that lead to major improvements in human well-being. According to Moore and Harnett White (2016), "The story of human advancement is the story of the discovery of cheap, plentiful, and versatile energy. Fossil fuels are the ignition switch to modern life" (p. xiii). Alternative sources of energy such as wind turbines, hydroelectric dams, and biofuels were replaced by fossil fuels with superior properties. "It wasn't until man harnessed fossil fuels – predominantly oil, gas, and coal – that industrialization achieved unprecedented productivity. ... Energy, in short, is the wellspring of mankind's greatest advances" (*Ibid.*, p. xiv). Similarly, Epstein (2014) writes,

[T]he benefits of cheap, reliable energy to power the machines that civilization runs on are enormous. They are just as fundamental to life as food, clothing, shelter, and medical care – indeed, all of these require cheap, reliable energy. By failing to consider the benefits of fossil fuel energy, the experts didn't anticipate the spectacular benefits that energy brought about in the last thirty years (p. 16).

Tucker (2008) adds, "Coal is the most important fossil fuel in history. The Industrial Revolution would never have occurred without it. In fact, for all intents and purposes, coal *was* the Industrial Revolution. Only a few nations have ever industrialized without shifting most of their energy dependence to coal, as the experience of China and India proves again today" (p. 61).

That access to affordable and reliable energy is the key to human well-being throughout the world can be demonstrated by the close correlations between energy consumption and GDP. Bezdek (2014) plotted global CO_2 emissions data from the U.S. Energy Information Administration and International Energy Agency and global GDP data from the U.S. Bureau of Economic Analysis to produce the graph shown in Figure 3.2.3.1.

Other scholars compare per-capita energy consumption to rank on the United Nations' Human Development Index (HDI) scorecard, a summary composite index measuring on a scale of 0 to 1 a nation's average achievement in three dimensions of human development: health, knowledge, and standard of living. Health is measured by life expectancy at birth; knowledge is measured by a combination of the adult literacy rate and the combined primary, secondary, and tertiary education gross enrollment ratio; and standard of living is measured by GDP per capita (UNDP, 2015). United Nations member states are listed and ranked each year according to these measures.

Kanagawa and Nakata (2008) examined data from 120 countries and found countries with higher per-capita electricity consumption showed higher scores with respect to the HDI. Similarly, statistical analysis by Clemente (2010) found countries using at least 2,000 kWh of electricity per capita a year have a significantly higher HDI than those that use less. Other researchers using different indices arrive at similar conclusions:

- Niu *et al.* (2013) found electricity consumption is closely correlated with five basic human development indicators: per-capita GDP, consumption expenditure, urbanization rate, life expectancy at birth, and the adult literacy rate.

- Manheimer (2012) plotted yearly per-capita energy use versus yearly per-capita GDP in the year 2000 for a number of countries, producing the graph reproduced as Figure 3.2.3.2 below. He observed "the two are very strongly correlated; there are no rich countries that use little energy per capita. Countries high up on the graph have more educated populations who live more pleasant, longer lives, and who live in cleaner environments than countries lower down on the graph."

- Mazur (2011) found electricity consumption is essential for improvement and well-being in less-developed countries, especially in populous nations such as China and India.

- Ghali and El-Sakka (2004) report per-capita energy and electricity consumption are highly correlated with economic development and other indicators of modern lifestyle.

Epstein (2014) illustrated the close correlation between global CO_2 emissions produced by the combustion of fossil fuels with rising human life expectancy, per-capita GDP, and global population in the four graphs shown in Figure 3.2.3.3. Numerous scholars have documented the close relationship between the cost of energy (typically electricity but sometimes petroleum and natural gas) and GDP growth in the United States and globally. Their work is reported in Section 3.5.1.

The disparity in access to electricity around the world is staggering. Approximately 3.9 billion people – 12 times the population of the United States and almost half the population of the world – have either no electricity or rely on biomass, coal, or kerosene for cooking (IEA, 2017). The average consumer in Germany, for example, uses 15 times as much power each year as the average citizen of India. In Europe, virtually no household lacks access to electricity. By contrast, in India, more than 400 million people have no electricity, 600 million cook with wood or dung, and more than one billion have no refrigeration (*Ibid.*). The consequences of these differences in electricity access are revealed in a comparison of each country's HDI score. In Germany, a newborn can expect to live until age 79, while in India its life expectancy is 64, 15 years less. In Germany, primary education completion and literacy rates are about 100%; in India, they hover around 70%. In Germany, GDP per capita is $34,401; in India it is $2,753. Consequently, Germany's HDI is 0.947, while India's is just 0.612.

Figure 3.2.3.1
Relationship between world GDP and annual CO₂ emissions

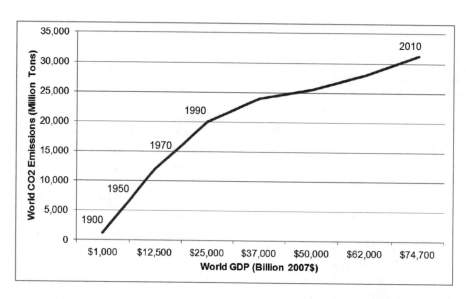

Source: Bezdek, 2014, p. 127, citing data from U.S. Energy Information Administration, International Energy Agency, and U.S. Bureau of Economic Analysis to 2007.

Figure 3.2.3.2
Per-capita GDP and per-capita energy consumption

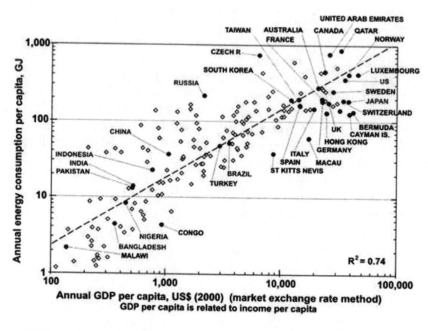

Source: Manheimer, 2012. Author says "Chart compiled by D. Lightfoot from information available from Energy Information Administration, *International Energy Annual* 2003; see also www.mcgill.ca/gec3/gec3members/lightfoot]."

Figure 3.2.3.3
Fossil fuel use and human progress, 0 AD – 2000 AD

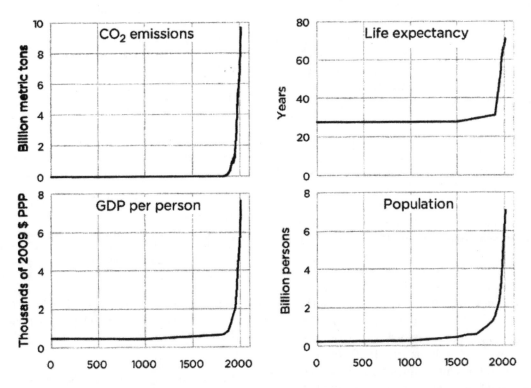

Source: Epstein, 2014, Figure 3.1, p. 77, citing Boden, Marland, and Andres, 2010; Bolt and van Zanden, 2013; and World Bank, World Development Indicators (WDI) Online Data, April 2014.

The connection between prosperity and public health can be illustrated by comparing Ethiopia and the Netherlands. According to ifitweremyhome.com, a website that allows international comparisons on a wide range of characteristics, Ethiopia has about three times as much good farmland per person as the Netherlands. (Good farmland has fertile soils, good weather, and enough rainfall to support substantial crop production.) About the same percentage of each country's land area is covered by forests. But residents of Ethiopia consume, on average, 99.52 percent less electricity and 99.27 percent less oil than those of the Netherlands. The result is dramatic: The average Ethiopian makes 97.7 percent less money than the average Netherlander. If you lived in Ethiopia instead of the Netherlands you would:

- be 10.5 times more likely to have HIV/AIDS;

- have a 17 times higher chance of dying in infancy;

- die 23.75 years sooner; and

- spend 99.25 percent less money on health care.

These numbers reveal an almost unimaginable difference in the quality of life between these two countries. Of course Ethiopia is not the only developing country in the world facing severe challenges. Lomborg (2007) noted, "[I]t is obvious that there are many other and more pressing issues for the third world, such as almost four million people dying [annually] from malnutrition, three million from HIV/AIDS, 2.5 million from indoor and outdoor air pollution, more than two million from lack of micronutrients (iron, zinc, and vitamin A), and almost two million from lack of clean drinking water" (p. 42). The lack of access to affordable energy and the prosperity it makes possible, not climate change, threatens the health of millions of people in developing countries. Lomborg also wrote,

317

... local surveys in that country [Tanzania] show the biggest concerns are the lack of capital to buy seeds, fertilizers, and pesticides; pests and animal diseases; costly education; high HIV-infection rates; malaria; and low-quality health services. I believe we have to dare to ask whether we help Tanzanians best by cutting CO_2, which would make no difference to the glaciers, or through HIV policies that would be cheaper, faster, and have much greater effect (*Ibid.*, p. 57).

These examples make it clear that the prosperity made possible by fossil fuels was not equally shared by all the peoples of the world. Still, a rising tide lifts all boats. Pinkovskiy and Sala-i-Martin (2009) estimated the income distribution for 191 countries between 1970 and 2006 and found,

Using the official $1/day line [the United Nations' definition of poverty], we estimate that world poverty rates have fallen by 80% from 0.268 in 1970 to 0.054 in 2006. The corresponding total number of poor has fallen from 403 million in 1970 to 152 million in 2006. Our estimates of the global poverty count in 2006 are much smaller than found by other researchers. We also find similar reductions in poverty if we use other poverty lines. We find that various measures of global inequality have declined substantially and measures of global welfare increased by somewhere between 128% and 145%.

In conclusion, the close correlation between energy consumption and many measures of quality of life show the value of fossil fuels isn't just something to read about in history books. Billions of lives are improved *every day* by having access to safe and affordable energy produced from fossil fuels.

References

Bezdek, R.H. 2014. *The Social Costs of Carbon? No, the Social Benefits of Carbon.* Oakton, VA: Management Information Services, Inc.

Boden, T.A., Marland, G., and Andres, R.J. 2010. Global, regional, and national fossil-fuel CO_2 emissions. Carbon Dioxide Information Analysis Center (CDIAC). Oak Ridge, TN: Oak Ridge National Laboratory, U.S. Department of Energy.

Bolt, J. and van Zanden, J.L. 2013. The first update of the Maddison Project: re-estimating growth before 1820. *Maddison Project Working Paper* 4.

Clemente, J. 2010. The statistical connection between electricity and human development. *Power Magazine.* September 1.

Epstein, A. 2014. *The Moral Case for Fossil Fuels.* New York, NY: Portfolio/Penguin.

Ghali, K.H. and El-Sakka, M.I.T. 2004. Energy use and output growth in Canada: a multivariate cointegration analysis. *Energy Economics* **26**: 225–38.

IEA. 2017. International Energy Agency. *World Energy Outlook 2017.*

Kanagawa, M. and Nakata, T. 2008. Assessment of access to electricity and the socioeconomic impacts in rural areas of developing countries. *Energy Policy* **36**: 2,016–29.

Lomborg, B. 2007. *Cool It: The Skeptical Environmentalist's Guide to Global Warming.* New York, NY: Alfred A. Knopf.

Manheimer, W.M. 2012. American physics, climate change, and energy. *Physics & Society* **41** (2): 14.

Mazur, A. 2011. Does increasing energy or electricity consumption improve quality of life in industrial nations? *Energy Policy* **39** (2): 568–72.

Moore, S. and Hartnett White, K. 2016. *Fueling Freedom: Exposing the Mad War on Energy.* Washington, DC: Regnery Publishing.

Niu, S., Jia, Y., Wang, W., He, R., Hu, L., and Liu, H. 2013. Electricity consumption and human development level: a comparative analysis based on panel data for 50 countries. *International Journal of Electrical Power & Energy Systems* **53**: 338–47.

Pinkovskiy, M. and Sala-i-Martin, X. 2009. Parametric estimations of the world distribution of income. *Working Paper* No. 15433. Cambridge, MA: National Bureau of Economic Research.

Tucker, W. 2008. *Terrestrial Energy: How Nuclear Power Will Lead the Green Revolution and End America's Energy Odyssey.* Baltimore, MD: Bartleby Press.

UNDP. 2015. United Nations Development Program. International Human Development indicators.

World Bank. 2014. World Development Indicators (website). Accessed June 8, 2014.

3.3 Food Production

The United Nations' Intergovernmental Panel on Climate Change (IPCC) projects the net impact of climate change on global agriculture will be negative, although it seems far from confident in its prediction. According to the Working Group II contribution to its Fifth Assessment Report:

> For the major crops (wheat, rice, and maize) in tropical and temperate regions, climate change without adaptation is projected to negatively impact production for local temperature increases of 2°C or more above late-20th-century levels, although individual locations may benefit (*medium confidence*). Projected impacts vary across crops and regions and adaptation scenarios, with about 10% of projections for the period 2030–2049 showing yield gains of more than 10%, and about 10% of projections showing yield losses of more than 25%, compared to the late 20th century. After 2050 the risk of more severe yield impacts increases and depends on the level of warming (IPCC, 2014, pp. 17–18).

There are numerous problems with the IPCC's forecast that make it unreliable. The prediction is for "local temperature increases of 2°C or more above late-20th-century levels," which the IPCC's models do not predict will occur globally until the *end* of the twenty-first century. This means the IPCC's forecast is irrelevant for eight decades or, as is more likely, even longer if its forecasts are wrong, as the climate science reviewed in Chapter 2 suggests. Extensive biological research suggests plants would *benefit* from a warming of less than 2°C, yet the IPCC is silent about that benefit of climate change.

The IPCC assumes no adaptation by the world's farmers, even though adaptation is already taking place as farmers continuously choose crops and hybrids and change such parameters as when to plant, fertilize, and harvest to maximize their output. *This mistake alone invalidates the IPCC's predictions.* No credible expert on global agriculture believes farmers will fail to adjust their practices to accommodate and benefit from climate changes as they occur. The slow pace of climate change predicted by the IPCC's own models suggests such gradual adaptation could be accomplished easily.

Note as well that the IPCC makes its prediction with only "medium confidence," which presumably means "better than a 50% chance." This is little more than a guess and not a scientific forecast. Finally, the forecast oddly focuses on the tails of the distribution of possible outcomes, where apparently only 10% predict positive effects and 10% predict negative effects. One supposes 80% predict no net impact, but this is not what the IPCC's opening sentence implies or the message the media took from its report.

For all these reasons, the IPCC's forecasts regarding global food production are not credible. *So what is more likely to occur?* We know that fossil fuels revolutionized agriculture, making it possible for an ever-smaller part of the population to raise food sufficient to feed a growing global population without devastating nature or polluting air and water. The aerial fertilization effect of higher levels of atmospheric CO_2 has further increased food production. Contradicting forecasts of global famine and starvation by such popular figures as Paul Ehrlich and John Holdren (Ehrlich, 1971; Ehrlich, Ehrlich and Holdren, 1977), the world's farmers increased their production of food at a faster rate than population growth, as shown in Figure 3.3.1.

Growing global food production is resulting in less hunger and starvation worldwide. In 2015, the Food and Agriculture Organization of the United Nations (FAO) reported "the number of hungry people in the world has dropped to 795 million – 216 million fewer than in 1990–92 – or around one person out of every nine" (FAO, 2015). In developing countries, the share of the population that is undernourished (having insufficient food to live an active and healthy life) fell from 23.3 percent 25 years earlier to 12.9 percent. A majority of the 129 countries monitored by FAO reduced undernourishment by half or more since 1996 (*Ibid.*).

Section 3.3.1 explains how fossil fuels created and today sustain the fertilization and mechanization that made possible the Green Revolution so plainly visible in Figure 3.3.1. Section 3.3.2 explains the phenomenon of aerial fertilization: Rising levels of atmospheric CO_2 promote plant growth, increasing agricultural yields beyond levels farmers would otherwise achieve. Sections 3.3.3 and 3.3.4 calculate the current and future value of aerial fertilization. Section 3.3.5 estimates the value of global food production that would have been lost had the world adopted and actually achieved the goal President Barack Obama set for reducing U.S. greenhouse gas emissions.

Figure 3.3.1
Global population, CO₂ emissions, and food production from 1961 to 2010, normalized to a value of unity at 1961

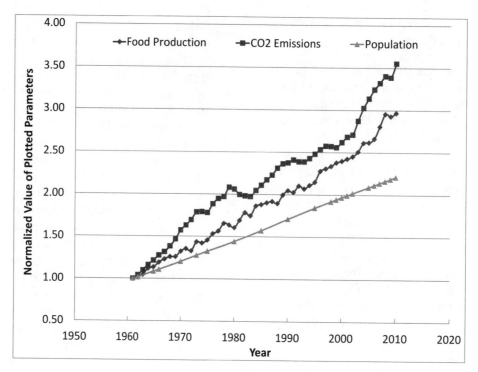

On the x axis, a "normalized value" of 2 represents a value that is twice the amount reported in 1961. Food production data represent the total production values of the 45 crops that supplied 95% of the total world food production over the period 1961–2011, using sources and a methodology described later in this section. *Source*: Idso, 2013, p. 24, Figure 8.

Extensive documentation regarding the positive effects of fossil fuels, CO₂, and higher surface temperatures on plants and animals appears in Chapter 5. To avoid needless repetition, it is referenced but not presented in this chapter.

References

Ehrlich, P. 1971. *The Population Bomb*. Revised and Expanded Edition. New York, NY: Ballantine Books.

Ehrlich, P., Ehrlich, A., and Holdren, J. 1977. *Ecoscience: Population, Resources, Environment*. Third Edition. New York, NY: W.H. Freeman.

FAO. 2015. Food and Agriculture Organization of the United Nations. *The State of Food Insecurity in the World 2015*. Rome, Italy.

Idso, C.D. 2013. *The Positive Externalities of Carbon Dioxide: Estimating the Monetary Benefits of Rising Atmospheric CO₂ Concentrations on Global Food Production*. Tempe, AZ: Center for the Study of Carbon Dioxide and Global Change.

IPCC. 2014. Intergovernmental Panel on Climate Change. *Climate Change 2014: Impacts, Adaptation, and Vulnerability*. Contribution of Working Group II to the Fifth Assessment Report of the Intergovernmental Panel on Climate Change. Geneva, Switzerland: World Meteorological Organization.

3.3.1 Fertilizer and Mechanization

Fossil fuels have greatly increased farm worker productivity thanks to nitrogen fertilizer created by the Haber-Bosch process and farm machinery built with and fueled by fossil fuels.

Cars and trucks dramatically improved the quantity and quality of food while reducing its cost in several ways: by improving productivity in fields with artificial fertilizer and increasingly specialized vehicles for sowing, cultivating, and reaping; by speeding the delivery of food from fields to processing plants and grocery stores; by inducing more competition among grocers and farmers by greatly expanding the range of businesses competing for a consumer's business; and by allowing food crops to be grown on land that would have been devoted to grazing or raising feed for horses. Historian Harold Platt (1991) wrote,

> The application of massive amounts of energy to every step in the commercial food chain was chiefly responsible for the revolution in what Americans ate. The war brought recent innovations to the manufacture of artificial fertilizers to technological maturity, helped ice makers kill off the natural ice business, turned shoppers toward the new cash-and-carry supermarkets, and made processed foods socially acceptable among the middle classes. During the 1920s, the food industry made intensive use of heat and refrigeration to offer a wider variety of better-tasting canned and baked goods as well as fresh fruits, dairy products, vegetables, and meats year round. "Foods formerly limited to the well-to-do," Hoover's economic experts noted in 1929, "have come more and more within the reach of the masses" (p. 221).

Gasoline-powered tractors similarly transformed agriculture with life-saving consequences. Thanks in large part to productivity gains made possible by tractors and increasingly specialized gasoline-powered vehicles, the percentage of the U.S. working population engaged in agriculture fell from about 80 to 90 percent in 1800 to just 1.5 percent in 2011 (Goklany, 2012, p. 19). Other developed countries witnessed the same trend. Agricultural labor has always been more hazardous than occupations in manufacturing and other industries, hence this migration to other occupations has saved countless lives.

The gasoline-powered tractor was invented in 1892, and farmers swiftly began replacing their horses and mules with the new technology. By the start of the twenty-first century, U.S. farmers were using some five million tractors (McKnight and Meyers, 2007, p. 12, citing Dimitri et al., 2005). Tractors brought their own risks –30,000 people in the United States were killed from the early twentieth century to 1971 by farm tractor overturns – but continued technological innovation is addressing that problem, too. "Roll-over protection structures" on new tractors reduced the annual number of deaths from tractor overturns from about 500 in 1966 to 200 deaths per year by 1985 (Ibid.).

One of the greatest achievements in human history was the discovery of a way to make ammonia from natural gas, thereby enabling farmers to add ammonia to their soil and dramatically increase crop yields. Ammonia (NH_3) is a potent organic fuel for most soil bacteria and plants (see Kiefer, 2013, citing Mylona et al., 1995; Matiru and Dakora, 2004; Sanguinetti et al., 2008; and Hayat et al., 2010). Ammonia is added to soil naturally by symbiotic soil and root bacteria, but at a slower rate than plants are able to use.

The discovery was made in 1909 by Fritz Haber and Carl Bosch, and the process is now known as the Haber-Bosch process. Natural gas and atmospheric nitrogen are converted into ammonia using an iron catalyst at high temperature and pressure. In 2014, U.S. farmers applied 19 million tons of man-made ammonia-based fertilizer to their fields (USDA, 2018), helping to make possible the "Green Revolution," an enormous increase in yields around the world beginning in the early 1960s due mainly to the use of cultivars that more responsive to nitrogen fertilizer, chemical pesticides, and irrigation. The Green Revolution brought to an end the conversion of wildlife habitat into cropland (Ausubel et al., 2013). Today, more land is being converted from cropland to forests and prairies than vice versa each year (Ibid.). This point is explained and documented in Chapter 5, Section 5.2.

Following the Green Revolution is what some call the "Gene Revolution" (Davies, 2003), the application of biotechnology to food crops resulting in a second wave of yield improvements. This wave, while initiated by breakthroughs in genetic engineering and related fields of research, will rely on energy-intensive technologies to produce the fertilizers, pesticides, irrigation, and dissemination of information needed for new ideas to be widely implemented in fields throughout the world.

References

Ausubel, J., Wernick, I., and Waggoner, P. 2013. Peak farmland and the prospect for land sparing. *Population and Development Review* **38**: 221–42.

Davies, W. 2003. An historical perspective from the Green Revolution to the Gene Revolution. *Nutrition Reviews* **61** (suppl 6): S124–S134.

Dimitri, C., Effland, A., and Conklin, N. 2005. *The 20th Century Transformation of U.S. Agriculture and Farm Policy*. Washington, DC: U.S. Department of Agriculture, Economic Research Service.

Goklany, I.M. 2012. Humanity unbound: How fossil fuels saved humanity from nature and nature from humanity. *Cato Policy Analysis* #715. Washington, DC: Cato Institute.

Hayat, R., Ali, S., Amara, U., Khalid, R., and Ahmed, I. 2010 Soil beneficial bacteria and their role in plant growth promotion: a review. *Annals of Microbiology* **60** (4): 579–98.

Kiefer, T.A. 2013. Energy insecurity: the false promise of liquid biofuels. *Strategic Studies Quarterly* (Spring): 114–51.

Matiru, V.N. and Dakora, F.D. 2004. Potential use of Rhizobial bacteria as promoters of plant growth for increased yield in landraces of African cereal crops. *African Journal of Biotechnology* **3** (1): 1–7.

McKnight, R.H. and Meyers, M.L. 2007. *The History of Occupational Safety and Health in U.S. Agriculture*. Lexington, KY: University of Kentucky College of Public Health.

Mylona, P., Pawlowski, K., and Bisseling, T. 1995. Symbiotic nitrogen fixation. *Plant Cell* **7** (July): 869–85.

Platt, H.L. 1991. *The Electric City: Energy and the Growth of the Chicago Area, 1880–1930*. Chicago, IL: University of Chicago Press.

Sanguinetti, G., Noirel, J., and Wright, P.C. 2008. MMG: a probabilistic tool to identify submodules of metabolic pathways. *Bioinformatics* **24** (8): 1078–84.

USDA. 2018. U.S. Department of Agriculture. *Fertilizer Use and Price* (website). Table 4. U.S. consumption of selected nitrogen materials. Economic Research Service. Accessed May 23, 2018.

3.3.2 Aerial Fertilization

Higher levels of carbon dioxide (CO_2) in the atmosphere act as fertilizer for the world's plants.

Since CO_2 is the basic "food" of essentially all terrestrial plants, the more of it there is in the atmosphere, the bigger and better they grow. At locations across the planet, the increase in the atmosphere's CO_2 concentration has stimulated vegetative productivity (Zhu *et al.*, 2016; Cheng *et al.*, 2017). Long-term studies confirm the findings of shorter-term experiments, demonstrating numerous growth-enhancing, water-conserving, and stress-alleviating effects of elevated atmospheric CO_2 on plants growing in both terrestrial and aquatic ecosystems (Idso and Idso, 1994; Ainsworth and Long, 2005; Bunce, 2005, 2012, 2013, 2014, 2016; Bourgault *et al.*, 2017; Sanz-Sáez *et al.*, 2017; Sultana *et al.*, 2017). Chapter 5 summarizes extensive research in support of this finding.

Since the start of the Industrial Revolution, it can be calculated on the basis of the work of Mayeux *et al.* (1997) and Idso and Idso (2000) that the 120 ppm increase in atmospheric CO_2 concentration increased agricultural production per unit land area by 70% for C_3 cereals, 28% for C_4 cereals, 33% for fruits and melons, 62% for legumes, 67% for root and tuber crops, and 51% for vegetables. A nominal doubling of the atmosphere's CO_2 concentration will raise the productivity of Earth's herbaceous plants by 30% to 50% (Kimball, 1983; Idso and Idso, 1994), while the productivity of its woody plants will rise by 50% to 80% (Saxe *et al.* 1998; Idso and Kimball, 2001).

Claims that global warming will reduce global food output are frequently made (e.g., Challinor *et al.*, 2014), but these forecasts invariably are based on computer models not validated by real-world data. Crop yields have continued to rise globally despite predictions and claims of higher temperatures, more droughts, etc. As Sylvan Wittwer (1995), the father of agricultural research on this topic, so eloquently put it more than two decades ago:

> The rising level of atmospheric CO_2 could be the one global natural resource that is progressively increasing food production and total biological output, in a world of otherwise diminishing natural resources of land, water, energy, minerals, and fertilizer. It is a means of inadvertently increasing the productivity of farming systems and other

photosynthetically active ecosystems. The effects know no boundaries and both developing and developed countries are, and will be, sharing equally … [for] the rising level of atmospheric CO_2 is a universally free premium, gaining in magnitude with time, on which we all can reckon for the foreseeable future.

The relationship described by Wittwer is illustrated in Figure 3.3.1, showing anthropogenic CO_2 emissions, food production, and human population all experienced rapid and interlinked growth over the past five decades.

References

Ainsworth, E.A. and Long, S.P. 2005. What have we learned from 15 years of free-air CO_2 enrichment (FACE)? A meta-analytic review of the responses of photosynthesis, canopy properties and plant production to rising CO_2. *New Phytologist* **165**: 351–72.

Bourgault, M., Brand, J., Tausz-Posch, S., Armstrong, R.D., O'Leary, G.L., Fitzgerald, G.J., and Tausz, M. 2017. Yield, growth and grain nitrogen response to elevated CO_2 in six lentil (*Lens culinaris*) cultivars grown under Free Air CO_2 Enrichment (FACE) in a semi-arid environment. *European Journal of Agronomy* **87**: 50–8.

Bunce, J.A. 2005. Seed yield of soybeans with daytime or continuous elevation of carbon dioxide under field conditions. *Photosynthetica* **43**: 435–8.

Bunce, J.A. 2012. Responses of cotton and wheat photosynthesis and growth to cyclic variation in carbon dioxide concentration. *Photosynthetica* **50**: 395–400.

Bunce, J.A. 2013. Effects of pulses of elevated carbon dioxide concentration on stomatal conductance and photosynthesis in wheat and rice. *Physiologia Plantarum* **149**: 214–21.

Bunce, J.A. 2014. Limitations to soybean photosynthesis at elevated carbon dioxide in free-air enrichment and open top chamber systems. *Plant Science* **226**: 131–5.

Bunce, J.A. 2016. Responses of soybeans and wheat to elevated CO_2 in free-air and open top chamber systems. *Field Crops Research* **186**: 78–85.

Challinor, A.J., Watson, J., Lobell, D.B., Howden, S.M., Smith, D.R., and Chhetri, N. 2014. A meta-analysis of crop yield under climate change and adaptation. *Nature Climate Change* **4**: 287–91.

Cheng, L., *et al.* 2017. Recent increases in terrestrial carbon uptake at little cost to the water cycle. *Nature Communications* **8**: 110.

Idso, K.E. and Idso, S.B. 1994. Plant responses to atmospheric CO_2 enrichment in the face of environmental constraints: a review of the past 10 years' research. *Agricultural and Forest Meteorology* **69**: 153–203.

Idso, C.D. and Idso, K.E. 2000. Forecasting world food supplies: the impact of the rising atmospheric CO_2 concentration. *Technology* **7S**: 33–55.

Idso, S.B. and Kimball, B.A. 2001. CO_2 enrichment of sour orange trees: 13 years and counting. *Environmental and Experimental Botany* **46**: 147–53.

Kimball, B.A. 1983. Carbon dioxide and agricultural yield: an assemblage and analysis of 430 prior observations. *Agronomy Journal* **75**: 779–88.

Mayeux, H.S., Johnson, H.B., Polley, H.W., and Malone, S.R. 1997. Yield of wheat across a subambient carbon dioxide gradient. *Global Change Biology* **3**: 269–78.

Sanz-Sáez, A., Koester, R.P., Rosenthal, D.M., Montes, C.M., Ort, D.R., and Ainsworth, E.A. 2017. Leaf and canopy scale drivers of genotypic variation in soybean response to elevated carbon dioxide concentration. *Global Change Biology* **23**: 3908–20.

Saxe, H., Ellsworth, D.S., and Heath, J. 1998. Tree and forest functioning in an enriched CO_2 atmosphere. *New Phytologist* **139**: 395–436.

Sultana, H., Armstrong, R., Suter, H., Chen, D., and Nicolas, M.E. 2017. A short-term study of wheat grain protein response to post-anthesis foliar nitrogen application under elevated CO_2 and supplementary irrigation. *Journal of Cereal Science* **75**: 135–7.

Wittwer, S.H. 1995. *Food, Climate, and Carbon Dioxide: The Global Environment and World Food Production.* Boca Raton, FL: Lewis Publishers.

Zhu, Z., *et al.* 2016. Greening of the Earth and its drivers. *Nature Climate Change* **6**: 791–5.

3.3.3 Economic Value of Aerial Fertilization

The aerial fertilization effect of rising levels of atmospheric CO_2 produced global economic benefits of $3.2 trillion from 1961 to 2011 and currently amount to approximately $170 billion annually.

Calculating the economic value of aerial fertilization begins with the United Nations' Food and Agriculture Organization (FAO) database, called FAOSTAT, of historic annual global crop yield and production data and the monetary value associated with that production for more than 160 crops grown and used world-wide since 1961 (FAO, 2013). No data are available prior to that time, so the present analysis is limited to the 50-year time window of 1961–2011.

More than half of the crops in the FAOSTAT database each account for less than 0.1% of the world's total food production. The analysis below focuses only on those crops that account for 95% of global food production. This was accomplished by taking the average 1961–2011 production contribution of the most important crop, adding to that the contribution of the second most important crop, and continuing in like manner until 95% of the world's total food production was reached. The results of this procedure produced the list of 45 crops shown in Figure 3.3.3.1.

Other data needed to estimate the economic value of aerial fertilization are annual global atmospheric CO_2 values since 1961 and plant-specific CO_2 growth response factors. The annual global CO_2 data were obtained from the IPCC report titled *Annex II: Climate System Scenario Tables – Final Draft Underlying Scientific-Technical Assessment* (IPCC, 2013). The plant-specific CO_2 growth response factors – which represent the percent growth enhancement expected for each crop listed in Figure 3.3.3.1 in response to a known rise in atmospheric CO_2 – were acquired from the online Plant Growth Database (PGD) maintained by the Center for the Study of Carbon Dioxide and Global Change at www.co2science.org/ (Idso, 2013b).

Figure 3.3.3.1
The 45 crops that supplied 95% of the global food production from 1961 to 2011

Crop	% of Total Production	Crop	% of Total Production
Sugar cane	20.492	Rye	0.556
Wheat	10.072	Plantains	0.528
Maize	9.971	Yams	0.523
Rice, paddy	9.715	Groundnuts, with shell	0.518
Potatoes	6.154	Rapeseed	0.494
Sugar beet	5.335	Cucumbers and gherkins	0.492
Cassava	3.040	Mangoes, mangosteens, guavas	0.406
Barley	2.989	Sunflower seed	0.398
Vegetables fresh nes	2.901	Eggplants (aubergines)	0.340
Sweet potatoes	2.638	Beans, dry	0.331
Soybeans	2.349	Fruit Fresh Nes	0.321
Tomatoes	1.571	Carrots and turnips	0.320
Grapes	1.260	Other melons (inc.cantaloupes)	0.302
Sorghum	1.255	Chillies and peppers, green	0.274
Bananas	1.052	Tangerines, mandarins, clem.	0.264
Watermelons	0.950	Lettuce and chicory	0.262
Oranges	0.935	Pumpkins, squash and gourds	0.248
Cabbages and other brassicas	0.903	Pears	0.243
Apples	0.886	Olives	0.241
Coconuts	0.843	Pineapples	0.230
Oats	0.810	Fruit, tropical fresh nes	0.230
Onions, dry	0.731	Peas, dry	0.228
Millet	0.593		
Sum of All Crops = 95.2%			

"Nes" is "not elsewhere specified." "Clem." is clementines. *Source*: Idso, 2013a, Table 1, p. 8.

Figure 3.3.3.2
Mean percentage yield increases produced by a 300 ppm increase in atmospheric CO_2 concentration for crops accounting for 95 percent of global food production

Crop	% Biomass Change	Crop	% Biomass Change
Sugar cane	34.0%	Rye	38.0%
Wheat	34.9%	Plantains	44.8%
Maize	24.1%	Yams	47.0%
Rice, paddy	36.1%	Groundnuts, with shell	47.0%
Potatoes	31.3%	Rapeseed	46.9%
Sugar beet	65.7%	Cucumbers and gherkins	44.8%
Cassava	13.8%	Mangoes, mangosteens, guavas	36.0%
Barley	35.4%	Sunflower seed	36.5%
Vegetables fresh nes	41.1%	Eggplants (aubergines)	41.0%
Sweet potatoes	33.7%	Beans, dry	61.7%
Soybeans	45.5%	Fruit Fresh Nes	72.3%
Tomatoes	35.9%	Carrots and turnips	77.8%
Grapes	68.2%	Other melons (inc.cantaloupes)	4.7%
Sorghum	19.9%	Chillies and peppers, green	41.1%
Bananas	44.8%	Tangerines, mandarins, clem.	29.5%
Watermelons	41.5%	Lettuce and chicory	18.5%
Oranges	54.9%	Pumpkins, squash and gourds	41.5%
Cabbages and other brassicas	39.3%	Pears	44.8%
Apples	44.8%	Olives	35.2%
Coconuts	44.8%	Pineapples	5.0%
Oats	34.8%	Fruit, tropical fresh nes	72.3%
Onions, dry	20.0%	Peas, dry	29.2%
Millet	44.3%		

"Nes" is "not elsewhere specified." "Clem." is clementines. *Source*: Idso, 2013a, Table 2, p. 9.

The PGD was used to calculate the mean crop growth response to a 300-ppm increase in atmospheric CO_2 concentration, a simulation often used in experiments, for each crop listed in Figure 3.3.3.1. In cases where no CO_2 enrichment data appeared in the database, the mean responses of similar plants or groups of plants were utilized. Also, there were some instances where the plant category in the FAO database represented more than one plant in the PGD. For example, the designation *Oranges* represents a single crop category in the FAO database, yet there were two different types of oranges listed in PGD (*Citrus aurantium* and *Citrus reticulata x C. paradisi x C. reticulata*). To produce a single number to represent the CO_2-induced growth response for the *Oranges* category, a weighted average from the growth responses of both orange species listed in the PGD was calculated. This procedure was repeated in other such circumstances.

The final results for all crops appear in Figure 3.3.3.2 above.

Figure 3.3.3.2 reveals the significant impact a hypothetical rise of 300 ppm in atmospheric CO_2 concentrations would have on yields of the world's 45 most important crops. The increases range from less than 10% for pineapples and "other melons" to more than 60% for sugar beets, grapes, beans, fruits, and carrots and turnips.

Determining the monetary benefit of atmospheric CO_2 enrichment on historic crop production begins by calculating the *increased annual yield* for each crop due to each year's increase in atmospheric CO_2 concentration above the baseline value of 280 ppm that existed at the beginning of the Industrial Revolution. Illustrating this process for wheat, in 1961 the global yield of wheat from the FAOSTAT database was 10,889 hectograms per hectare (Hg/Ha), the atmospheric CO_2 concentration was

317.4 ppm, representing an increase of 37.4 ppm above the 280 ppm baseline, and the CO_2 growth response factor for wheat as listed in Figure 3.3.3.2 is 34.9% for a 300 ppm increase in CO_2. To determine the impact of the 37.4 ppm rise in atmospheric CO_2 on 1961 wheat yields, the wheat-specific CO_2 growth response factor of 34.9% per 300 ppm CO_2 increase (mathematically written as 34.9%/300 ppm) is multiplied by the 37.4 ppm increase in CO_2 that has occurred since the Industrial Revolution. The resultant value of 4.35% indicates the degree by which the 1961 yield was enhanced above the baseline yield value corresponding to an atmospheric CO_2 concentration of 280 ppm.

The 1961 yield is then divided by this relative increase (1.0435) to determine the baseline yield in Hg/Ha (10,889/1.0435 = 10,435). The resultant baseline yield amount of 10,435 Hg/Ha is subtracted from the 1961 yield total of 10,889 Hg/Ha, revealing that 454 Hg/Ha of the 1961 yield was due to the 37.4 ppm rise in CO_2 since the start of the Industrial Revolution. Similar calculations are then made for each of the remaining years in the 50-year period, as well as for each of the 44 remaining crops accounting for 95% of global food production.

The next step is to determine what percentage of the *total annual yield* of each crop in each year was due to CO_2. This is accomplished by taking the results calculated in the previous step and dividing them by the corresponding total annual yields. For example, using the calculations for wheat from above, the 454 Hg/Ha yield due to CO_2 in 1961 was divided by the total 10,889 Hg/Ha wheat yield for that year, revealing that 4.17% of the total wheat yield in 1961 was due to the historical rise in atmospheric CO_2. Again, such percentage calculations were completed for all crops for each year in the 50-year period 1961–2011.

Knowing the annual percentage influences of CO_2 on all crop yields (production per Ha), the next step is to determine how that influence is manifested in total *crop production value*. This is accomplished by multiplying the CO_2-induced yield percentage increases by the corresponding annual *production* of each crop, and by then multiplying these data by the gross production *value* (in constant 2004–2006 U.S.

dollars) of each crop per metric ton, which data were obtained from the FAOSTAT database. The end result of these calculations becomes an estimate of the *annual monetary benefit* of atmospheric CO_2 enrichment (above the baseline of 280 ppm) on crop production since 1961. These findings appear in Figure 3.3.3.3.

As can be seen from Figure 3.3.3.3, the benefit of Earth's rising atmospheric CO_2 concentration on global food production is enormous. Such benefits over the period 1961–2011 amounted to at least $1.8 billion for each of the 45 crops examined; and for nine of the crops the monetary increase due to CO_2 over this period was well over $100 billion. The largest of these benefits is noted for rice, wheat, and grapes, which saw increases of $579 billion, $274 billion, and $270 billion, respectively.

Figure 3.3.3.4 plots the rise in the annual total monetary value of the CO_2 benefit for all 45 crops over the 50-year period from 1961 to 2011. The curve rises because the CO_2 effect each year must be examined relative to the baseline value of 280 ppm. Thus, the CO_2 benefit is getting larger each year as the atmospheric CO_2 level rises. At 410 ppm presently, the CO_2 effect is 40 percent greater now than it was around the turn of the twentieth century. Whereas the annual value of the CO_2 benefit amounted to approximately $18.5 billion in 1961, by the end of the record it had grown to more than $140 billion annually. Projecting the line forward to 2015 (not shown in the figure) puts the annual benefit at approximately $170 billion. Summing these annual benefits across the 50-year time period of 1961–2011, as is done in Figure 3.3.3.3, shows the cumulative CO_2-induced benefit on global food production since 1961 is $3.2 trillion.

In conclusion, aerial fertilization by higher levels of CO_2 increased the monetary value of crop production by approximately $170 billion in 2015 and the benefit is rising every year. The cumulative economic value of aerial fertilization since 1961 is more than $3.2 trillion. This is a major benefit to human prosperity and well-being due to the use of fossil fuels.

Figure 3.3.3.3
Annual average monetary value of CO$_2$ aerial fertilization on global crop production from 1961–2011 (in constant 2004–2006 U.S. dollars)

Crop	Production Rank	Monetary Benefit of CO$_2$	Crop	Production Rank	Monetary Benefit of CO$_2$
Rice, paddy	4	$579,013,089,273	Carrots and turnips	35	$36,439,812,318
Wheat	2	$274,751,908,146	Cucumbers and gherkins	29	$33,698,222,461
Grapes	13	$270,993,488,618	Watermelons	16	$32,553,055,795
Maize	3	$182,372,524,324	Pears	41	$31,577,067,767
Soybeans	11	$148,757,417,756	Fruit Fresh Nes	34	$29,182,817,600
Potatoes	5	$147,862,516,739	Fruit, tropical fresh nes	44	$28,837,991,342
Vegetables fresh nes	9	$143,295,147,644	Millet	23	$24,748,422,190
Tomatoes	12	$140,893,704,588	Eggplants (aubergines)	32	$22,794,746,004
Sugar cane	1	$107,420,713,630	Cassava	7	$21,850,017,436
Apples	19	$98,329,393,797	Onions, dry	22	$20,793,394,925
Sugar beet	6	$69,247,223,819	Sorghum	14	$20,579,850,257
Barley	8	$63,046,887,462	Tangerines, mandarins, clem.	38	$18,822,174,419
Bananas	15	$58,264,644,460	Coconuts	20	$17,949,253,896
Yams	26	$56,163,446,226	Sunflower seed	31	$17,585,395,685
Groundnuts, with shell	27	$51,076,843,461	Plantains	25	$17,384,141,669
Olives	42	$50,604,186,875	Lettuce and chicory	39	$15,029,691,577
Oranges	17	$50,173,178,154	Pumpkins, squash and gourds	40	$13,140,422,653
Beans, dry	33	$47,240,266,167	Oats	21	$12,615,396,815
Mangoes, mangosteens, guavas	30	$40,731,776,757	Rye	24	$8,981,587,998
Sweet potatoes	10	$39,889,080,598	Peas, dry	45	$5,667,935,087
Chillies and peppers, green	37	$39,813,008,532	Other melons (inc.cantaloupes)	36	$2,477,799,109
Rapeseed	28	$38,121,172,234	Pineapples	43	$1,779,091,848
Cabbages and other brassicas	18	$37,501,047,431	Sum of all crops = $3,170,050,955,544		

"Nes" is "not elsewhere specified." "Clem." is clementines. *Source:* Adapted from Idso, 2013a, Table 3, p. 11.

Figure 3.3.3.4
Annual monetary value of CO$_2$ aerial fertilization on global crop production for 45 crops from 1961 to 2011

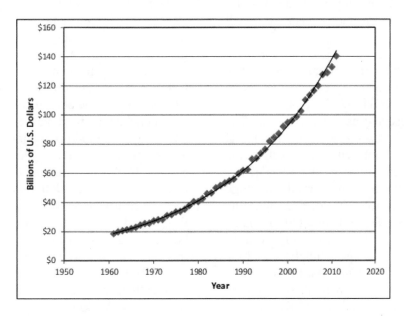

Source: Idso, 2013a, Figure 1, p. 12.

References

Idso, C.D. 2013a. *The Positive Externalities of Carbon Dioxide: Estimating the Monetary Benefits of Rising Atmospheric CO_2 Concentrations on Global Food Production.* Tempe, AZ: Center for the Study of Carbon Dioxide and Global Change. October 21.

Idso, C.D. 2013b. Plant Growth Database. CO_2 Science (website). Tempe AZ: Center for the Study of Carbon Dioxide and Global Change.

FAO. 2013. Food and Agriculture Organization of the United Nations. FAO Statistics Database. Rome, Italy.

IPCC. 2013. Intergovernmental Panel on Climate Change. Annex II: Climate System Scenario Tables - Final Draft Underlying Scientific-Technical Assessment. In: *Climate Change 2013: The Physical Science Basis*, contribution of Working Group I to the IPCC Fifth Assessment Report (AR5). Cambridge, UK and New York, NY: Cambridge University Press.

3.3.4 Future Value of Aerial Fertilization

Over the period 2012 through 2050, the cumulative global economic benefit of aerial fertilization will be approximately $9.8 trillion.

Future monetary benefits of rising atmospheric CO_2 concentrations on crop production also can be estimated. The methodology for doing so is slightly different from that used in calculating the historic values. In explaining these methods, sugar cane will serve as the example.

First, the 1961–2011 historic yield data for sugar cane are plotted as the blue line in Figure 3.3.4.1. The portion of each year's annual yield due to rising atmospheric CO_2, as per calculations described in Section 3.3.3, are presented as the green line. The annual yield due to rising CO_2 is subtracted from total annual yield to generate the red line, which is the contribution of everything else that tended to influence crop yield over that time period. Although many factors play a role in determining the magnitude of this latter effect, it is referred to here as the *techno-intel effect*, as it derives primarily from continuing advancements in agricultural technology and scientific research that expands our knowledge or intelligence base. For the most part, these advances were part of the three Industrial Revolutions discussed in Section 3.2.

As depicted in Figure 3.3.4.1, the relative influence of atmospheric CO_2 on the total yield of sugar cane is increasing with time. This fact is further borne out in Figure 3.3.4.2, where techno-intel yield values are plotted as a percentage of total sugar cane yield. Whereas the influence of technology and intelligence accounted for approximately 96% of the observed yield values in the early 1960s, by the end of the record in 2011 it accounted for only 89%.

The three trends revealed in Figure 3.3.4.1 can be projected forward to the year 2050 using a second-order polynomial fitted to the data. The results are depicted in Figure 3.3.4.3. By knowing the annual total yield, as well as the portion of the annual total yield that is due to the techno-intel effect between 2012 and 2050, the part of the total yield that is due to CO_2 can be calculated by subtracting the difference between them. These values appear in the figure as the dashed green line.

Linear trends for each crop's 1961–2011 production data were next extended forward in time to provide projections of annual production values through 2050. As with the historic calculations discussed in the previous section, these production values were multiplied by the corresponding annual percentage influence of CO_2 on 2012–2050 projected crop yields. The resultant values were then multiplied by an estimated gross production value (in constant 2004–2006 U.S. dollars) for each crop per metric ton. As there are several potential unknowns that may influence the future production value assigned to each crop, a simple 50-year average of the observed gross production values was applied over the period 1961–2011. The ensuing monetary values for each of the 45 crops over the period 2012 through 2050 are listed in Figure 3.3.4.4.

The economic benefit of aerial fertilization by CO_2 can be expressed as an annual benefit per ton of CO_2 emitted by the combustion of fossil fuels. This is accomplished by dividing the annual dollar benefit of CO_2 on global food production by annual global CO_2 emissions. The resultant values are plotted in Figure 3.3.4.5. The social benefit was near $2 per ton of CO_2 emitted during the 1960s and 1970s. Thereafter, it rose in linear fashion to a value of $4.14 at the end of the record. Although comparisons of the social benefits and costs of fossil fuels are not discussed in this chapter (they are taken up in Chapter 8), we note our estimate of the annual benefit of aerial fertilization in 2010, $4.14, is similar to EPA's Interagency Working Group's (IWG) 2010 estimate, $4.70, of the "social cost of carbon" using a 5% discount rate (IWG, 2010). This is remarkable because

Figure 3.3.4.1
Sources of increasing sugar cane yields from 1961 to 2011

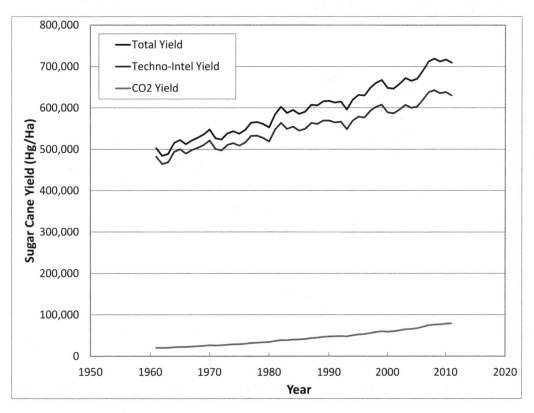

Source: Idso, 2013, Figure 2, p. 13.

Figure 3.3.4.2
Percentage of the total annual yield of sugar cane from 1961 to 2011 attributable to the techno-intel effect

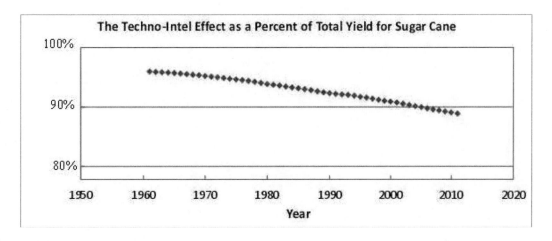

Source: Idso, 2013, Figure 3, p. 14.

Figure 3.3.4.3
Historical and projected increases in total yield and the portion of the total yield due to the techno-intel and CO₂ effects from 2012 to 2050

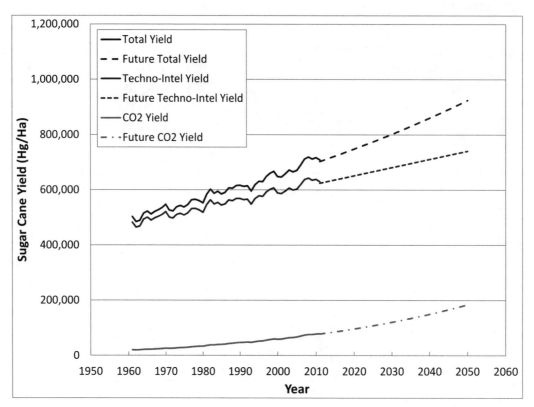

Source: Idso, 2013, Figure 4, p. 13.

Figure 3.3.4.4
Monetary benefit (in 2004–2006 \$) of Earth's rising atmospheric CO_2 concentration on 45 crops for the period 2012–2050

Crop	Production Rank	Monetary Benefit of CO_2	Crop	Production Rank	Monetary Benefit of CO_2
Rice, paddy	4	\$1,847,162,847,355	Beans, dry	33	\$121,672,752,990
Wheat	2	\$731,810,134,138	Eggplants (aubergines)	32	\$121,040,127,404
Soybeans	11	\$622,840,779,401	Sugar beet	6	\$118,016,992,389
Vegetables fresh nes	9	\$603,158,136,300	Pears	41	\$106,648,093,649
Maize	3	\$582,352,695,047	Fruit Fresh Nes	34	\$96,939,989,779
Tomatoes	12	\$538,622,004,026	Tangerines, mandarins, clem.	38	\$94,049,613,976
Grapes	13	\$507,943,670,190	Fruit, tropical fresh nes	44	\$92,676,868,053
Sugar cane	1	\$366,333,858,080	Onions, dry	22	\$83,094,062,469
Apples	19	\$306,866,752,703	Sweet potatoes	10	\$70,623,018,596
Potatoes	5	\$268,944,859,065	Cassava	7	\$66,454,408,155
Yams	26	\$206,504,638,016	Pumpkins, squash and gourds	40	\$65,141,087,416
Bananas	15	\$200,878,216,972	Lettuce and chicory	39	\$54,406,821,316
Rapeseed	28	\$176,560,583,707	Coconuts	20	\$52,278,524,212
Cucumbers and gherkins	29	\$165,126,686,871	Sunflower seed	31	\$50,554,512,301
Oranges	17	\$165,014,960,801	Plantains	25	\$45,996,854,219
Chillies and peppers, green	37	\$162,527,401,900	Millet	23	\$43,337,359,355
Olives	42	\$157,323,187,194	Sorghum	14	\$38,314,226,074
Groundnuts, with shell	27	\$148,440,689,387	Other melons (inc.cantaloupes)	36	\$11,163,081,357
Watermelons	16	\$144,909,503,686	Peas, dry	45	\$10,484,435,272
Barley	8	\$127,842,645,165	Pineapples	43	\$6,926,670,057
Carrots and turnips	35	\$126,282,174,308	Rye	24	\$5,804,121,850
Mangoes, mangosteens, guavas	30	\$124,067,842,115	Oats	21	\$4,904,374,119
Cabbages and other brassicas	18	\$122,664,616,192	Sum of all crops = \$9,764,706,877,630		

Source: Adapted from Idso, 2013, Table 4, p. 17.

Figure 3.3.4.5
Economic benefits of aerial fertilization of CO$_2$ in \$ per ton of CO$_2$ emissions, 1961 to 2010

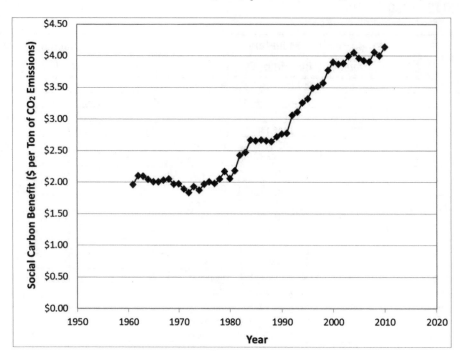

Source: Calculations from data in Idso, 2013.

it means the economic benefits of aerial fertilization *alone* will offset nearly all the projected social costs forecast by IWG

Figure 3.3.4.4 reveals a tremendous future economic benefit of Earth's rising atmospheric CO$_2$ concentration. Over the period 2012 through 2050, the cumulative benefit is \$9.8 trillion, much larger than the \$3.2 trillion that was observed in the longer 50-year historic period of 1961–2011.

By incorporating the additional CO$_2$-induced productivity benefits realized by the timber industry, along with those experienced outside the human timber and agricultural industries – i.e., the rest of the plants existing and sustaining wild nature – it is likely that this CO$_2$-induced productivity benefit is sufficient to completely overpower all the hypothetical human welfare damages forecast by the IPCC.

References

Idso, C.D. 2013. *The Positive Externalities of Carbon Dioxide: Estimating the Monetary Benefits of Rising Atmospheric CO$_2$ Concentrations on Global Food*

Production. Tempe, AZ: Center for the Study of Carbon Dioxide and Global Change. October 21.

IWG. 2010. Interagency Working Group. *Technical support document: Social cost of carbon for regulatory impact analysis under Executive Order 12866*. Washington, DC: U.S. Environmental Protection Agency.

3.3.5 Proposals to Reduce CO$_2$ Emissions

Reducing global CO$_2$ emissions by 28% from 2005 levels, the reduction President Barack Obama proposed in 2015 for the United States, would reduce aerial fertilization benefits by \$78 billion annually.

In 2015, the Obama administration proposed reducing CO$_2$ emissions by 28% below 2005 levels (Showstack, 2015). While that proposal would have applied only to the United States, other countries are contemplating similar or larger emission reductions. *What effect would a global 28% reduction of CO$_2$*

emissions have on the aerial fertilization benefits discussed above?

Globally, 29.7 billion tons of CO_2 were emitted in 2005. A 28% reduction would drop annual emissions to 21.4 billion tons, a value last seen more than 30 years ago, in 1987. As shown in Figure 3.3.4.5, the social benefit of CO_2 from increased agricultural productivity amounted to $2.65 per ton of CO_2 emitted at that time, meaning the world would lose a minimum of $1.49 per ton of CO_2 in benefits ($4.14, the value in 2010, minus $2.65, the value in 1987), or $78 billion annually.

The decline in aerial fertilization by CO_2 caused by mandated emission reductions could cause food shortages in countries that presently have only limited food supplies, causing malnutrition and starvation, and possibly igniting conflict and war. There is no reason to believe advocates of reducing the use of fossil fuels have taken this into consideration.

* * *

The world's rising population and prosperity since the start of widespread use of fossil fuels have led to rising CO_2 emissions and likely contribute to rising CO_2 concentrations in the atmosphere. This development has benefited food production, creating an economic value calculated here of $3.2 trillion from 1961 to 2011, annual benefits as of 2015 of approximately $170 billion, and cumulative anticipated benefits worth $9.8 trillion over the period 2012 through 2050.

The economic benefits of aerial fertilization alone will offset nearly all the social costs forecast by climate change activists, even granting their highly dubious assumptions and methodologies. Reducing global CO_2 emissions by 28% from 2005 levels, the target President Barack Obama proposed for the United States in 2015, would reduce aerial fertilization benefits by $78 billion annually.

References

Idso, C.D. 2013. *The Positive Externalities of Carbon Dioxide: Estimating the Monetary Benefits of Rising Atmospheric CO_2 Concentrations on Global Food Production*. Tempe, AZ: Center for the Study of Carbon Dioxide and Global Change. October 21.

Showstack, R. 2015. White House submits greenhouse gas emission targets. *EOS: Earth & Space Science News* **96**: 6–7.

3.4 Why Fossil Fuels?

Fossil fuels – coal, oil, and natural gas – replaced alternative energy sources because they have a higher power density than any substitute except nuclear power and are in abundant supply, flexible, and inexpensive.

Fossil fuels have four qualities that made them uniquely suited to fuel the three industrial revolutions that created modernity: they are (1) able to deliver more power per unit of space (energy density) than any competing fuel except nuclear power, (2) available in sufficient supply to meet human needs, (3) flexible enough to support dispatchable power generation in a wide range of circumstances, and (4) so inexpensive that they make electricity and transportation affordable for even low-income households. These qualities enabled fossil fuels to displace other resources that were less dense, in shorter supply, less flexible, and more expensive. These qualities also explain why fossil fuels continue to dominate the global energy supply today.

3.4.1 Power Density

Fossil fuels have higher power density than all alternative energy sources except nuclear power.

Power density was defined in Section 3.1 as energy flow per unit of time, which can be measured in joules per second (watts) divided by a unit of space, as in watts per square meter or W/m^2. When energy sources are ranked by their relative power density, as shown in Figure 3.4.1.1, it quickly becomes clear that fossil fuels dominate all fuels except nuclear power. A natural gas well, for example, is nearly 50 times more power-dense than a wind turbine, more than 100 times as dense as a biomass-fueled power plant, and 1,000 times as dense as corn ethanol. Coal (not shown in the figure) has an energy density 50% to 75% that of oil, still far superior to solar, wind, and biofuels (Layton, 2008; Smil, 2010).

According to Smil (2016), "fossil fuels are enormously concentrated transformations of biomass, and hence the power densities associated with their extraction are unrivaled by any other form of terrestrial energy" (p. 97). Smil also notes, "Obviously, the higher the density of an energy resource, the lower are its transportation (as well as

Figure 3.4.1.1
Relative power density

W/m²	Energy Sources
56	nuclear
53	natural gas well
28	gas stripper well
27	oil stripper well (10 barrels/day)
6.7	solar PV
5.5	oil stripper well (2 barrels/day)
1.2	wind turbine
0.4	biomass-fueled power plant
0.05	corn ethanol

Source: Bryce, 2010, p. 93. See sources in original.

storage) costs, and this means that its production can take place farther away from the centers of demand. Crude oil has, at ambient pressure and temperature, the highest energy density of all fossil fuels (42 Gj/t), and hence it is a truly global resource, with production ranging from the Arctic coasts to equatorial forests and hot deserts" (*Ibid.,* p. 12).

High power density explains why a basket of coal light enough for a single person to carry can heat a home for an entire day and night even in the cold of winter, and why the lights did not go out in New England states in the United States during the exceptionally frigid winter of 2013–2014. It explains how a car can travel more than 300 miles on a 13-gallon tank of gasoline, and how a pipe less than one inch in diameter can provide enough natural gas to meet the cooking, heating, and hot water needs of even large homes. High power density explains why jet airplanes powered by kerosene can make non-stop ocean-crossing trips and how ships can make similarly long trips without having to stop at ports. High power density means fossil fuels can be conveniently stockpiled near where they will be used, making them less vulnerable to supply interruptions (National Coal Council, 2014; U.S. Department of Energy, 2017). All of these features produce huge economic benefits.

The uranium used in nuclear reactors has an energy density even higher than fossil fuels (80,620 GJ/kg), but the facilities needed to harness that power reduce its power density to closer to that of fossil fuels, as shown in Figure 3.4.1.1. Unjustified public concern over the safety of nuclear power, fueled by environmental advocacy groups and yellow journalism, has slowed or stopped the expansion of nuclear power in the United States and in most other parts of the world, though not in China (Hibbs, 2018).

References

Bryce, R. 2010. *Power Hungry: The Myths of "Green" Energy and the Real Fuels of the Future.* New York, NY: PublicAffairs.

Hibbs, M. 2018. *The Future of Nuclear Power in China.* Washington, DC: Carnegie Endowment for World Peace.

Layton, B.E. 2008. A comparison of energy densities of prevalent energy sources in units of joules per cubic meter. *International Journal of Green Energy* **5**: 438–55.

National Coal Council. 2014. Reliable and resilient, the value of our existing coal fleet: an assessment of measures to improve reliability and efficiency while reducing emissions. May.

Smil, V. 2010. *Power Density Primer: Understanding the Spatial Dimension of the Unfolding Transition to Renewable Electricity Generation (Part I – Definitions).* May 8.

Smil, V. 2016. *Power Density: A Key to Understanding Energy Sources and Uses.* Cambridge, MA: The MIT Press.

U.S. Department of Energy. 2017. Staff Report to the Secretary on Electricity Markets and Reliability. August.

3.4.2 Sufficient Supply

Fossil fuels are the only sources of fuel available in sufficient quantities to meet the needs of modern civilization.

Bithas and Kalimeris (2016) write,

The milestone that determined the transition from the organic economy to the fossil fuel economy, the invention that characterized the era called "The Industrial Revolution," was the steam engine. The unique process that the steam engine initiated was the conversion of chemical energy (heat) into mechanical energy (motion) (McNeill, 2000). The biomass energy stocks accumulated in the earth's crust for hundreds of millions of years were now available to serve human needs for the first time in mankind's history, to such an extent that the dawn of the fossil fuel era was about to begin (p. 7).

Three figures appearing earlier in this chapter, Figures 3.1.4.1, 3.1.4.3, and 3.2.1.1, illustrated how fossil fuels were able to produce the enormous amounts of energy required globally and in the United States since the beginning of the industrial age. According to the U.S. Energy Information Administration (EIA, 2018), fossil fuels supplied 78% of total U.S. primary energy in 2017 and according to the International Energy Agency (IEA, n.d.) they supplied 81% of global energy use in 2016.

Fossil fuels quickly supplanted wood as the preeminent form of energy, rescuing millions of acres of forests from logging. Fossil fuels supply, as wood never could, the vast amount of energy needed by businesses using new labor-saving technologies and urban centers needing fuels for home and business heating, cooling, and lighting. Without ample supplies of coal, electrification of many processes from manufacturing to home heating, cooking, and laundry would not have taken place. Wood, wind turbines, and biofuels (and more recently solar PV panels) could not and still cannot provide more than a small fraction of total energy needs.

The demand for energy is expected to grow dramatically in the years ahead. According to the International Energy Agency (IEA, 2017), even in its "New Policies Scenario" which assumes subsidies and tax policies that discriminate against fossil fuels and raise the price of energy, global energy needs still expand by 30% between today and 2040. "This is the equivalent of adding another China and India to today's global demand," the authors write. "A global economy growing at an average rate of 3.4% per year, a population that expands from 7.4 billion today to more than 9 billion in 2040, and a process of urbanisation that adds a city the size of Shanghai to the world's urban population every four months are key forces that underpin our projections. The largest contribution to demand growth – almost 30% – comes from India, whose share of global energy use rises to 11% by 2040 (still well below its 18% share in the anticipated global population)." Figure 3.4.2.1 illustrates where the biggest increases in energy demand are expected to occur between 2016 and 2040. Note that according to the IEA, energy demand

in the United States, Europe, and Japan is projected to decline.

The growing population and per-capita incomes of a prosperous world underscore the importance of having an ample supply of high-quality energy. However, since supplies of fossil fuels are thought to be exhaustible (though there are theories to the contrary, see Gold (1992, 1999) and Colman *et al.* (2017)), some fear the possibility of eventual depletion. Similar fears were raised by economist William Stanley Jevons in an 1865 book ominously titled *The Coal Question; An Inquiry Concerning the Progress of the Nation, and the Probable Exhaustion of Our Coal Mines.* During the 1970s, environmental advocacy groups such as the Sierra Club and Club of Rome and even national governments proclaimed fossil fuels would run out or be in short supply by the turn of the century (Holdren, 1971; Meadows *et al.,* 1972; Joint Economic Committee, 1980). Pessimists who have followed Jevons' lead are still prominent (e.g., Gore, 1992, 2007; Klare, 2012), but their predictions have repeatedly been found to be wrong (e.g., Simon, 1999; Bailey, 2015; Pinker, 2018; and many others). Commenting on such predictions, Clayton (2013) wrote,

> The logic appears unimpeachable at first glance. But it's wrong. The prices of raw materials have not traveled the path this story would predict for any traded commodity once inflation is factored in, over long stretches of time. One of the most powerfully counter-intuitive and empirically conclusive findings in economic history is that the real prices of nearly all major resources have actually trended lower over very long periods of time, even if they're produced at higher and higher rates. (Oil, once OPEC got involved, is the glaring exception. But even oil prices since OPEC came about haven't simply climbed higher and higher as global consumption has grown.) Though non-renewable commodity prices can rise steeply over years or even decades when supply and demand conditions warrant, over the centuries they've tended to decline after adjusted for inflation.

According to the U.S. Energy Information Administration, as of December 31, 2014, total world proven recoverable reserves of coal were about 1.2 trillion short tons, enough to last for centuries at projected rates of demand. In the United States alone,

Figure 3.4.2.1
Change in primary energy demand, 2016–2040 (Mtoe)

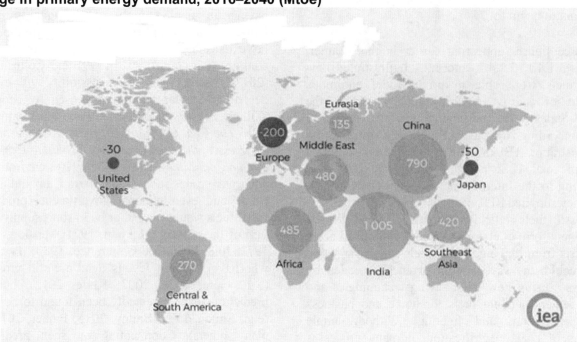

Source: IEA, 2017.

estimated recoverable reserves of coal totaled 254,896 million short tons, enough to last about 348 years. EIA estimates that as of January 1, 2016, there were an estimated 6,879 trillion cubic feet (Tcf) of total world proven reserves of gross natural gas. The United States had 2,462 Tcf of technically recoverable resources of dry (consumer-grade) natural gas, enough to last about 90 years, with advancing technology (such as the combination of horizontal drilling and hydraulic fracturing) and higher prices likely to make that reserve last for decades or even centuries (EIA, n.d.). In short, if humanity ever stops using fossil fuels it will not be because the supply ran out.

One sign of fossil fuels' continued abundance is its relatively stable price. According to the U.S. Energy Information Administration (EIA, 2018), "U.S. energy expenditures declined for the fifth consecutive year [in 2016], reaching $1.0 trillion in 2016, a 9% decrease in real terms from 2015. Adjusted for inflation, total energy expenditures in 2016 were the lowest since 2003. Expressed as a percent of gross domestic product (GDP), total energy expenditures were 5.6% in 2016, the lowest since at least 1970." In the nearly half-century since Holdren, Meadows, and others warned of an

imminent energy crisis, total U.S. energy consumption rose 44% (from 67.8 quadrillion Btu in 1970 to 97.4 in 2016), yet spending on energy as a percentage of GDP did not increase at all. If fossil fuels – responsible for some 78% of U.S. energy supply – were becoming scarce, their prices would be rising relative to other goods and services.

It is not only fossil fuels whose supply is probably inexhaustible. According to Clayton, "Raw materials prices show a secular deterioration relative to manufactured goods over long stretches of time. Since 1871, the *Economist* industrial commodity-price index has sunk to roughly half its value in real terms, seeing average annual compound growth of - 0.5% per year over the ensuing 140 years. Even after the boom years of the 2000s – in 2008, for instance, as commodity indexes soared, the *Economist* index never climbed more than halfway above where it stood 163 years earlier, in real terms" (*Ibid.*). As explained in Chapter 1, the prices of scarce goods do not fall over time. Fossil fuels are becoming more, not less, abundant with time.

"The exhaustion of fossil fuels on the global scale is not imminent," wrote McNeill (2000). "Predictions of dearth have proved false since the 1860s. Indeed, quantities of proven reserves of coal, oil, and natural gas tended to grow faster than production in the twentieth century. Current predictions, which will be revised, imply several decades before oil or gas should run out, and several centuries before coal might. We can continue to live off the accumulated geologic capital of the eons for some time to come – if we can manage or accept the pollution caused by fossil fuels."

References

Bailey, R. 2015. *The End of Doom: Environmental Renewal in the Twenty-first Century*. New York, NY: St. Martin's Press.

Bithas, K. and Kalimeris, P. 2016. *Revisiting the Energy-Development Link*. SpringerBriefs in Economics. DOI 10.1007/978-3-319-20732-2_2.

Clayton, B. 2013. Bad news for pessimists everywhere. Energy, Security, and Climate (blog). Council on Foreign Relations. March 22.

Colman, D.R., Poudel, S., Stamps, B.W., Boyd, E.S., and Spear, J.R. 2017. The deep, hot biosphere: twenty-five years of retrospection. *Proceedings of the National Academy of Sciences USA* **114** (27): 6895–903.

EIA. 2018. U.S. Energy Information Administration. In 2016, U.S. energy expenditures per unit GDP were the lowest since at least 1970. Today in Energy (website). July 30.

EIA. 2018. U.S. Energy Information Administration. *Monthly Energy Review November 2018* (website). Accessed November 28, 2018.

EIA. n.d. U.S. Energy Information Administration. Frequently asked questions (website). Accessed May 22, 2018.

Gold, T. 1992. The deep, hot biosphere. *Proceedings of the National Academy of Sciences USA* **89**: 6045–9.

Gold, T. 1999. *The Deep Hot Biosphere: The Myth of Fossil Fuels*. New York, NY: Springer.

Gore, A. 1992. *Earth in the Balance*. New York, NY: Houghton Mifflin.

Gore, A. 2007. *The Assault on Reason*. New York, NY: Penguin Press.

Holdren, J. 1971. *Energy: A Crisis in Power*. San Francisco, CA: Sierra Club.

IEA. 2017. International Energy Agency. World Energy Outlook 2017.

IEA. 2018. International Energy Agency. World Energy Balances (website). Accessed November 4, 2018.

Jevons, W.S. 1865. *The Coal Question; An Inquiry Concerning the Progress of the Nation, and the Probable Exhaustion of Our Coal Mines*. London, UK: Macmillan & Co.

Joint Economic Committee. 1980. *The Global 2000 Report*. Congress of the United States. Washington, DC: U.S. Government Printing Office.

Klare, M. 2012. *The Race for What's Left: The Global Scramble for the World's Last Resources*. New York, NY: Macmillan.

McNeill, J.R. 2000. *Something New Under the Sun: An Environmental History of the Twentieth-Century World*. New York, NY: W.W. Norton & Company.

Meadows, D.H., *et al.* 1972. *The Limits to Growth*. New York, NY: New American Library.

Pinker, S. 2018. *Enlightenment Now: The Case for Reason, Science, Humanism, and Progress*. New York, NY: Viking.

Simon, J. 1999. *Hoodwinking the Nation*. New Brunswick, NJ: Transaction Publishers.

3.4.3 Flexibility

Fossil fuels provide energy in the forms needed to make electricity dispatchable (available on demand 24/7) and they can be economically transported to or stored near the places where energy is needed.

Following their high power density and sheer abundance, the third reason fossil fuels have been the fuel of choice since the beginning of the Industrial Revolution is their flexibility. Fossil fuels can be economically transported to or stored near the places where energy is needed and they can power technologies able to generate electricity on demand 24 hours a day, seven days a week. This feature is extremely valuable because modern economies require a constant supply of electricity 24/7, not just when the sun shines and the wind blows (Clack *et al.*, 2017). Electric grids need to be continuously balanced – energy fed into the grid must equal energy leaving the grid – which requires dispatchable (on-demand) energy and spinning reserves (Backhaus and

Chertkov, 2013; Dears, 2015). Today, only fossil fuels and nuclear power can provide dispatchable power in sufficient quantities to keep grids balanced.

Coal, the fossil fuel that takes the solid form, can be safely mined, processed, transported in railcars, and stored in outdoor piles until it needs to be used. Its inexpensive storage capacity makes it the fuel of choice for electricity generation (Stacy and Taylor, 2016). Even natural gas is more vulnerable to supply interruptions than is coal, and both are more reliable than alternatives except for nuclear energy (U.S. Department of Energy, 2017; Bezdek, 2017).

Oil, the liquid fossil fuel, is ideal for autonomous transportation vehicles such as cars, trucks, airplanes, and ships. Smil (2016) was quoted earlier in this chapter saying "crude oil has, at ambient pressure and temperature, the highest energy density of all fossil fuels (42 Gj/t), and hence it is a truly global resource, with production ranging from the Arctic coasts to equatorial forests and hot deserts." Oil's superior properties are apparent when modern forms of transportation are compared to those powered by wind (schooners) and biofuels (horses and horse-drawn carriages). It is also superior to hydrogen, which sometimes is proposed as a substitute for gasoline for transportation uses. Hydrogen gas is highly flammable and will explode at concentrations in air ranging from 4% to 75% by volume in the presence of a flame or a spark. Because hydrogen is so light it must be stored under pressure, introducing more cost, weight, and risk, and this is difficult to do because hydrogen embrittles many metals. A typical automobile gas tank holds 15 gallons of gasoline weighing 90 pounds, while the corresponding hydrogen tank would need to hold 60 gallons and would need to be insulated (McCarthy, 2005).

Natural gas, the fossil fuel in a gaseous state, is ideal for home heating and cooking since it burns so cleanly that it causes little indoor air pollution. Natural gas is typically compressed to about 15 times atmospheric pressure for pipeline distribution over many hundreds of miles, making it instantly available when needed to produce electricity or meet other energy needs. Pipeline pressure is reduced to about 30% over atmospheric pressure at a customer's home, making it safe for use by furnaces, water heaters, and stoves. Pipelines allow natural gas to be economically transported to areas that are far removed from well sites and where on-site storage of coal or oil would be uneconomical. The unique features of natural gas make it superior to coal or oil for specific applications, while offering the high energy density, abundant supply, and "always on"

availability that make it superior to other alternatives (Hayden, 2015).

High-pressure natural gas lines, transporting gas over long distances, have much lower loss of energy per unit of energy transported than high-voltage electric lines. A gas line is often buried in the ground, with a narrow safety zone around it, whereas high-voltage power systems require wide clearances in forests and rural areas above ground.

References

Backhaus, S. and Chertkov, M. 2013. Getting a grip on the electrical grid. *Physics Today* **66**: 42–8.

Bezdek, R. 2017. Death of U.S. coal industry greatly exaggerated, part two. Public Utilities Fortnightly (website). October 15.

Clack, C.T.M., *et al.* 2017. Evaluation of a proposal for reliable low-cost grid power with 100% wind, water, and solar. *Proceedings of the National Academy of Sciences USA* **114** (26) 6722–7.

Dears, D. 2015. *Nothing to Fear: A Bright Future for Fossil Fuels.* The Villages, FL: Critical Thinking Press.

Hayden, H.C. 2015. *Energy: A Textbook.* Pueblo West, CO: Vales Lake Publishing.

McCarthy, J. 2005. Progress and Its Sustainability, Hydrogen (website). Accessed May 23, 2018.

Smil, V. 2016. *Power Density: A Key to Understanding Energy Sources and Uses.* Cambridge, MA: The MIT Press.

U.S. Department of Energy. 2017. Staff Report to the Secretary on Electricity Markets and Reliability. August.

3.4.4 Inexpensive

Fossil fuels in the United States are so inexpensive that they make home heating, electricity, and transportation affordable for even low-income households.

The most dense, abundant, and flexible energy source in the world would be little used if it came at a price so high that few people could afford to use it. Fossil fuels do not suffer from this hypothetical problem. Coal, oil, and natural gas are often the least expensive sources of energy for many applications. Despite the enormous contribution of energy to industry and quality of life, total energy expenditures in the United States were only 5.6% of GDP in 2016 (EIA, 2018). The U.S. average energy price was $15.92 per million British thermal units (MBtu) in 2016. Expenditures on electricity accounted for 74% of residential expenditures, 80% of commercial expenditures, and 37% of industrial expenditures. (*Ibid.*).

Electricity for home and industrial uses in the United States, where fossil fuels produce 78% of electricity, is less expensive than in many other parts of the world, where taxes, regulations, and forced reliance on alternative energies have artificially inflated its price. According to the National Coal Council (2014), "in 2013 the average price of residential and industrial electricity in the U.S. was one-half to one-third the price of electricity in Germany, Denmark, Italy, Spain, the UK, and France" (p. 1 referencing Table B-1).

Except in areas near hydroelectric dams and in some cases nuclear power facilities, electricity generated by fossil fuels is almost always less expensive than alternatives. Since this is a contentious issue, it is discussed in detail in Section 3.5.4. Here we can focus on why fossil fuels are able to generate electricity so much less expensively than alternatives (except nuclear power).

The efficiencies of converting natural resources to energy and then using that energy differ dramatically from place to place and depend on many variables. Using hydroelectric power to generate electricity in the Pacific Northwest, for example, is more efficient than using coal or natural gas due to its abundant availability, while coal and natural gas are better choices in the Midwest where hydroelectric power opportunities are more limited. Nevertheless, it is possible to estimate and rank fuels by their average or typical *energy return on investment* (EROI), which is the amount of useful energy a fuel yields divided by how much energy is required to produce it. This calculation reveals the superior efficiency of fossil fuels compared to alternative energies and the reason they are so much more affordable.

Kiefer (2013) conducted a thorough literature review of the EROIs for 12 fuels in the United States. Figure 3.4.4.1 reproduces the graph showing his results, with EROI scores on the vertical axis and the amount of energy each fuel produced in 2010 on the horizontal. According to Kiefer,

[C]urrent petroleum diesel and gasoline production EROIs are variously estimated between 10:1 and 20:1. A conservative approach least favorable to petroleum is to postulate an 8:1 EROI, which represents the lowest value calculated since 1920. An 8:1 EROI means that one barrel of liquid fuel energy input can support the exploration, drilling, extraction, and refining of enough crude oil to make eight new barrels of liquid fuel energy – which for petroleum happens to come with a bonus of one barrel of chemical feedstock for plastics, lubricants, organic compounds, industrial chemicals, and asphalt (p. 124).

Figure 3.4.4.1 illustrates the high efficiency of coal (for electricity production), natural gas, and petroleum relative to that of any other source of energy save hydroelectric, the supply of which is limited by geography and opposition to the construction of new dams, and nuclear, to which opposition is also fierce. Wind and solar are seen as having highly variable EROIs, extending below 1:1 at their low points (meaning they consume more energy during production than they release when used) and reaching the EROIs achieved by fossil fuels only in their best circumstances. Ethanol and biodiesel fuels barely reach a 3:1 EROI and often are below 1:1. The figure also demonstrates, by their position to the right of all other fuels, how fossil fuels dominate the supply of energy in the United States.

More evidence of the affordability of fossil fuels can be seen in Figure 3.4.4.2, which plots electricity prices in the 50 U.S. states against the percentage of electric power produced with coal in each of those states. Except for a few states where hydropower produces inexpensive energy, the price of electricity is lowest in states where coal is the preeminent source of electric power.

* * *

In conclusion, fossil fuels produce 81% of the primary energy consumed globally and 78% in the U.S due to four characteristics: power density, abundant supply, flexibility, and low cost. These are the reasons fossil fuels were indispensable to the creation of Modernity, to the electrification of the world, and to the dramatic improvement in human well-being.

Figure 3.4.4.1
Energy return on investment (EROI) of U.S. energy sources

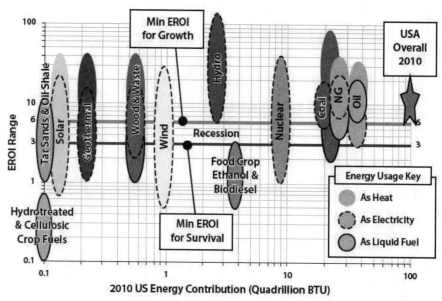

Min EROI for Growth (6:1) is minimum EROI historically required by the U.S. economy to avoid economic recessions. Min EROI for Survival (3:1) is the minimum quality a raw energy feedstock must have to overcome production costs and conversion losses and still deliver positive net energy to modern civilization. Note the vertical axis is a logarithmic scale, equal differences in order of magnitude are represented by equal distances from the value of 1. *Source:* Kiefer, 2013, p. 120. Sources appear in author's footnotes 21–24 on p. 143.

Figure 3.4.4.2
Relationship between coal generation and retail electricity prices by state

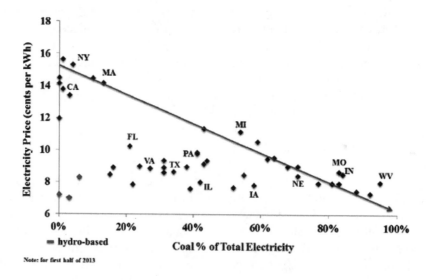

Source: Bezdek, 2014, p. 10, citing U.S. Energy Information Administration, *Electric Power Monthly*, August 2013.

References

Bezdek, R.H. 2014. *The Social Costs of Carbon? No, the Social Benefits of Carbon*. Oakton, VA: Management Information Services, Inc.

EIA. 2018. U.S. Energy Information Administration. In 2016, U.S. energy expenditures per unit GDP were the lowest since at least 1970. Today in Energy (website). July 30.

Kiefer, T.A. 2013. Energy insecurity: the false promise of liquid biofuels. *Strategic Studies Quarterly* (Spring): 114–51.

National Coal Council. 2014. Reliable and resilient, the value of our existing coal fleet: an assessment of measures to improve reliability and efficiency while reducing emissions. May.

Stacy, T.F. and Taylor, G.S. 2016. *The Levelized Cost of Electricity from Existing Generation Resources*. Washington, DC: Institute for Energy Research.

3.5 Alternatives to Fossil Fuels

Could today's level of global prosperity be sustained without fossil fuels? The United Nations' Intergovernmental Panel on Climate Change (IPCC) claims that avoiding a climate catastrophe requires "substantial cuts in anthropogenic GHG emissions by mid-century through large scale changes in energy systems and potentially land use (*high confidence*)" and "emissions levels near zero GtCO$_2$eq or below in 2100" (IPCC, 2014, pp. 10, 12). (Emissions can supposedly fall to below zero through the use of "carbon dioxide removal technologies.") The IPCC estimates the cost of reducing emissions to meet these goals would be 1% to 4% (median: 1.7%) of global GDP in 2030, 2% to 6% (median: 3.4%) in 2050, and 3% to 11% (median: 4.8%) in 2100 relative to consumption in baseline scenarios (*Ibid.,* pp. 15–16, text and Table SPM.2).

The IPCC's belief that human greenhouse gas emissions must be reduced to "near zero or below" to avoid a climate catastrophe is simply wrong, as shown by the science reviewed in Chapter 2 and elsewhere in this volume. There is no impending climate crisis that requires such action. Also wrong is the IPCC's claim that the cost of such a draconian reduction in the use of fossil fuels would be only a few percentage points of baseline global GDP. Modern civilization relies on quantities and qualities of energy that only fossil fuels can deliver.

Alternative energies such as wind turbines, solar PV cells, and biofuels do not have the features that made fossil fuels the fuel of choice for the past two centuries – high density, abundant supply, flexibility, and low cost. The *apparent* cost of a forced transition would be far more than the IPCC's estimates, and the *opportunity* cost would be greater still.

Reference

IPCC. 2014. Intergovernmental Panel on Climate Change. *Climate Change 2014: Mitigation of Climate Change*. Contribution of Working Group III to the Fifth Assessment Report of the Intergovernmental Panel on Climate Change. New York, NY: Cambridge University Press.

3.5.1 Low Power Density

The low power density of alternatives to fossil fuels is a crippling deficiency that prevents them from ever replacing fossil fuels in most applications.

"The fundamental problem with both wind and biofuels," writes Bryce, "is that they are not dense. Producing significant quantities of energy from either wind or biomass simply requires too much land. The problem is not one of religious belief, it's simple math and basic physics" (Bryce, 2014, p. 212). "The punch line," he writes, is this:

> [E]ven if we ignore wind energy's incurable intermittency, its deleterious impact on wildlife, and how 500-foot-high wind turbines blight the landscape and harm the landowners who live next to them, its paltry power density simply makes it unworkable. Wind-energy projects require too much land and too much airspace. In the effort to turn the low power density of the wind into electricity, wind turbines standing about 150 meters high [492 feet] must sweep huge expanses of air. (A 6-megawatt offshore turbine built by Siemens sports turbine blades with a total diameter of 154 meters [505 feet] that sweep an area of 18,600 square meters [200,209 square feet]. That sweep area is nearly three times the area of a regulation soccer pitch.) By sweeping those enormous expanses of air, wind turbines are

killing large numbers of bats and birds (*Ibid.*, p. 212).

Bryce estimates replacing U.S. coal-fired generation capacity in 2011 (300 gigawatts) with wind turbines at 1 watt per square meter would have required 300 billion square meters, roughly 116,000 square miles (Bryce, 2014, pp. 217–218). Driessen, using a number of conservative assumptions, estimated using windmills to produce the same amount of energy as is currently produced globally by fossil fuels would require 14.4 million onshore turbines requiring some 570 million acres (890,625 square miles), an area equal to 25% of the entire land area of the United States (30% of the lower 48 states) (Driessen, 2017).

A study of the use of biofuels to replace fossil fuels conducted by the UK's Energy Research Centre and published in 2011 found that replacing half of current global primary energy supply with biofuels would require an area ranging from twice to ten times the size of France. Replacing the entire current global energy supply with biofuels would require …

an area of high yielding agricultural land the size of China. … In addition these estimates assume that an area of grassland and marginal land larger than India (>0.5Gha) is converted to energy crops. The area of land allocated to energy crops could occupy over 10% of the world's land mass, equivalent to the existing global area used to grow arable crops (Slade *et al.*, 2011, p. vii).

Kiefer (2013) wrote, "Biofuel production is a terribly inefficient use of land, and this can best be illustrated with power density, a key metric for comparing energy sources" (p. 131). Biodiesel and ethanol produced from soy and corn have power densities of only 0.069 and 0.315 W/m^2 respectively, "300 times worse than the 90 W/m^2 delivered by the average US petroleum pumpjack well on a two-acre plot of land" (*Ibid.*). Replacing the energy used by the United States each year just for transportation "would require more than 700 million acres of corn. This is 37% of the total area of the continental United States, more than all 565 million acres of forest, and more than triple the current amount of annually harvested cropland. Soy biodiesel would require 3.2 billion acres – one billion more than all US territory including Alaska" (*Ibid.*).

The power density estimates cited above probably underestimate the advantage fossil fuels have over renewable energies by not taking into account the resources needed to build wind turbines or the difficulty of transporting ethanol, which is corrosive and cannot be transported through pipelines. Concerning the former, Bryce (2010) reports,

[E]ach megawatt of wind power capacity requires about 870 cubic meters of concrete and 460 tons of steel. For comparison, each megawatt of power capacity in a combined-cycle gas turbine power plant … requires about 27 cubic meters of concrete and 3.3 tons of steel. In other words, a typical megawatt of reliable wind power capacity requires about 32 times as much concrete and 139 times as much steel as a typical natural gas-fired power plant (p. 90).

Wind turbines are designed to last approximately 35 years, and there is some evidence they frequently do not last that long (Hughes, 2012). It is unlikely that wind turbines generate enough energy in their lifetimes to recover the enormous amounts of energy used to create their enormous pads and the infrastructure needed to bring their power to businesses and consumers (*Ibid.*). Facts like these prompted even James Hansen, an outspoken global warming activist and former director of the NASA Goddard Institute for Space Studies, to admit in 2011 that "suggesting that renewables will let us phase rapidly off fossil fuels in the U.S., China, India, or the world as a whole is almost the equivalent of believing in the Easter Bunny and Tooth Fairy" (Hansen, 2011, p. 5).

References

Bryce, R. 2010. *Power Hungry: The Myths of "Green" Energy and the Real Fuels of the Future*. New York, NY: PublicAffairs.

Bryce, R. 2014. *Smaller Faster Lighter Denser Cheaper: How Innovation Keeps Proving the Catastrophists Wrong*. New York, NY: PublicAffairs.

Driessen, P. 2017. Revisiting wind turbine numbers. Townhall (website). September 2. Accessed May 18, 2018.

Hansen, J. 2011. Baby Lauren and the Kool-Aid. Climate Science, Awareness, and Solutions (blog). July 29.

Hughes, G. 2012. *Why is wind power so expensive? An economic analysis.* London, UK: Global Warming Policy Foundation.

Kiefer, T.A. 2013. Energy insecurity: the false promise of liquid biofuels. *Strategic Studies Quarterly* (Spring): 114–51.

Slade, R., Saunders, R., Gross, R., and Bauen, A. 2011. *Energy from Biomass: The Size of the Global Resource.* London, UK: Imperial College Centre for Energy Policy and Technology and UK Energy Research Centre.

3.5.2 Limited Supply

Wind, solar, and biofuels cannot be produced and delivered where needed in sufficient quantities to meet current and projected energy needs.

Combined, geothermal, wind, solar, and other non-hydro, non-biofuels contributed only 1.6% of global energy supplies in 2016 (IEA, n.d.). With the best locations for hydroelectric facilities already in use and public opposition to the building of new facilities, proposals to achieve a 100% renewable energy world rely on fantastic increases in energy from wind, solar, and biofuels or equally fantastic reductions in per-capita energy consumption, and more often a combination of the two.

Renewable energy's tiny installed capacity is not due to a lack of public taxpayer support; globally, hundreds of billions of dollars have been spent subsidizing solar and wind power generation. In the United States, many states even have laws mandating the public utilities pay premium prices to purchase power from solar and wind companies. Nor is the limited capacity of renewables a problem that can be solved by pouring more money into research and development, as was demonstrated in Section 3.1 and Figure 3.1.4.5. Rather, it is due to physical limitations inherent to renewable energies:

- *Low density:* renewable energy's low energy density would require unacceptably large areas be covered with wind turbines or solar panels, or planted in soy or corn destined to become ethanol, in order to meet even a fraction of total energy demand in the United States or in most other countries. Low density means limited supply because many areas cannot accommodate the massive industrial wind facilities or arrays of

solar panels envisioned by their advocates (Bryce, 2010).

- *Intermittency:* wind and solar power are intermittent, making their output of low or even no value to electric grid operators seeking dispatchable energy available 24/7. Wind and solar power require redundant coal or natural gas back-up generation capacity for when the wind does not blow and the sun does not shine, effectively doubling its cost (E.ON Netz, 2005). Ethanol must be trucked to refineries and end users because it corrodes pipelines, and its low BTU content makes it an undesirable transportation fuel. Wind and solar cannot supply power for ships or airplanes or in emergency situations during and after floods, winter storms, or natural disasters.

- *Expensive:* Advocates of solar and wind power have claimed for decades that they are closing the price gap with fossil fuels, yet power from new investment in solar and wind (plus back-up power from natural gas) still costs approximately three times as much as power from existing long-lived coal plants (Stacy and Taylor, 2016). Claims of cost parity invariably hide subsidies and tax breaks, ignore intermittency and the cost of integrating solar and wind into electric grids, and attribute fictional "social costs" to fossil fuels while ignoring the very real social costs imposed by wind turbines on people and on wildlife (Bezdek, 2014).

Clack *et al.* (2017), critiquing a report by Mark Jacobson *et al.* (2015) claiming wind, solar, and hydropower could completely replace fossil fuels, illustrated how unrealistic Jacobson *et al.*'s forecast is with the graphic reproduced as Figure 3.5.2.1 below. The graphic shows how achieving "100% decarbonization" in the United States would require a 14-fold increase in wind, solar, and hydroelectric capacity additions (measured as watts per year per capita) versus the U.S. historical average *every year* from 2015 to 2047 and beyond. That expansion of capacity is not just unprecedented in the United States (as well as German) history, it is *six times* as much as has ever been added in any one year in U.S. history.

Figure 3.5.2.1
Wind, solar, and hydroelectric capacity additions required in United States to achieve 100% decarbonization versus historical trends in United States, Germany, and China

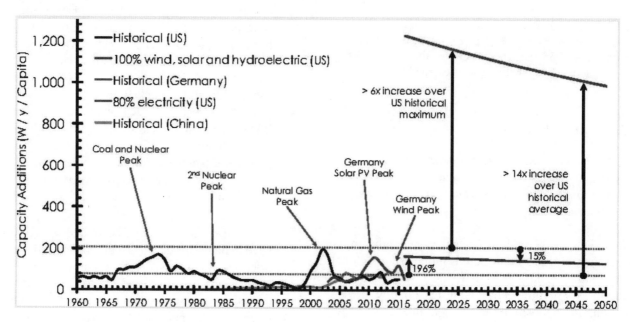

Source: Clack *et al.,* 2017, Figure 4.

References

Bezdek, R.H. 2014. *The Social Costs of Carbon? No, the Social Benefits of Carbon.* Oakton, VA: Management Information Services, Inc.

Bryce, R. 2010. *Power Hungry: The Myths of "Green" Energy and the Real Fuels of the Future.* New York, NY: PublicAffairs.

Clack, C.T.M., *et al.* 2017. Evaluation of a proposal for reliable low-cost grid power with 100% wind, water, and solar. *Proceedings of the National Academy of Sciences USA* **114** (26) 6722–7.

E.ON Netz. 2005. *Wind Power Report 2005.* Bayreuth, Germany.

IEA. n.d. International Energy Agency. World Energy Balances (website). Accessed November 4, 2018.

Jacobson, M.Z., Delucchi, M.A., Cameron, M.A., and Frew, B.A. 2015. Low-cost solution to the grid reliability problem with 100% penetration of intermittent wind, water, and solar for all purposes. *Proceedings of the National Academy of Sciences USA* **112** (49):15060–5.

Stacy, T.F. and Taylor, G.S. 2016. *The Levelized Cost of Electricity from Existing Generation Resources.* Washington, DC: Institute for Energy Research.

3.5.3 Intermittency

Due to their intermittency, solar and wind power cannot power the revolving turbine generators needed to create dispatchable energy.

Modern economies require a constant supply of electricity 24/7, not just when the sun shines and the wind blows. The grid needs to be continuously balanced – energy fed into the grid must equal energy leaving the grid – which requires dispatchable (on-demand) energy and spinning reserves (Backhaus and Chertkov, 2013). Today, only fossil fuels and nuclear power can provide dispatchable power in sufficient quantities to keep grids balanced. Clack *et al.* observed in 2017,

Wind and solar are variable energy sources, and some way must be found to address the issue of how to provide energy if their immediate output cannot continuously meet instantaneous demand. The main options are to (*i*) curtail load (i.e., modify or fail to satisfy demand) at times when energy is not available, (*ii*) deploy very large amounts of energy storage, or (*iii*) provide supplemental energy sources that can be dispatched when needed. It is not yet clear how much it is possible to curtail loads, especially over long durations, without incurring large economic costs. There are no electric storage systems available today that can affordably and dependably store the vast amounts of energy needed over weeks to reliably satisfy demand using expanded wind and solar power generation alone. These facts have led many U.S. and global energy system analyses to recognize the importance of a broad portfolio of electricity generation technologies, including sources that can be dispatched when needed (p. 6722).

Section 3.1 explained why wind and solar power are inevitably intermittent and Figures 3.1.5.1, 3.1.5.2, and 3.1.5.3 showed the impact this variability has on energy supplies in southeastern Australia. South Australia, a state in the southern central part of Australia with a population in 2013 of 1.677 million, relies on renewable energy sources for 53% of its electric power generation. In 2016 it experienced a blackout caused by a series of tornadoes that lasted 12 days, and it has since experienced numerous more, albeit shorter, blackouts due to the closure of coal-fired power plants unable to compete with subsidized wind power and the unreliable nature of its industrial wind turbine installations (Orr and Palmer, 2018). South Australia's reliance on renewable energy has also led to dramatically higher energy prices, the highest in the world and some three times higher than in the United States (Potter and Tillett, 2017).

E-ON Netz, a global company that operates industrial wind facilities in Germany, the UK, and elsewhere, reported in 2005 that "Wind energy is only able to replace traditional power stations to a limited extent. Their dependence on the prevailing wind conditions means that wind power has a limited load factor even when technically available. It is not possible to guarantee its use for the continual cover of electricity consumption. Consequently, traditional power stations with capacities equal to 90% of the installed wind power capacity must be permanently online in order to guarantee power supply at all times" (E-ON Netz, 2005). This is a remarkably candid admission of a flaw in wind power which, in the absence of government subsidies and mandates, would render it useless as a supplier of energy for electricity production.

The intermittency of wind and solar means greater reliance on them requires a correspondingly larger investment in back-up generating capacity powered by fossil fuels or nuclear power, or not-yet-invented energy storage systems. Given the vagaries of wind and solar, such a storage system would have to be large enough to store all the energy that will be demanded for many days, possibly weeks, until the wind and solar systems come back online. The technology to safely and economically store such large amounts of electricity does not exist, at least not outside the few areas where large bodies of water and existing dams make pumped-storage hydroelectricity possible. The frequent announcements of "breakthroughs" in battery technology have not resulted in commercial products capable of even a small fraction of the storage needed to transition away from fossil fuels.

To use Australia once again as an example, in November 2017, the world's largest lithium ion battery was installed in South Australia to help avoid blackouts caused by the variability of wind power (BBC, 2017). The battery cost $50 million and, according to its creator, when fully charged can power up to 30,000 homes for one hour. While hailed by some as a milestone in the effort to accommodate the intermittency of wind and solar power, this battery proves just the opposite. While sufficient to smooth out small interruptions in power supply for short periods of time, it is clearly not scalable. To understand why, consider the following:

- One hour of back-up energy is trivial compared to the amount of time wind and solar power are unavailable and during and following storms and natural disasters, when solar and wind power are most likely to be unavailable.

- 30,000 homes is trivial compared to the number of homes affected by blackouts due to south Australia's reliance on renewable energy. The city of Adelaide has a population of 1.3 million living in 515,000 private homes, 17 times more than could be served (for one hour) by the new battery.

- The high cost of Musk's battery relative to the benefit becomes apparent with a little math. The $50 million battery could power just 3.4 homes for one year at a cost of $14.6 million per home per year or $1.2 million per month per home. A battery large enough to power all of Adelaide for just one hour would cost $2.2 billion.

- For roughly the same investment, South Australia (or Adelaide) could buy a state-of-the-art 650 MW Ultra Supercritical coal-powered plant (EIA, 2017, Table 1), which could provide continuous power for the entire city of Adelaide every year for decades to come (650 MW x 750 homes/MW = 487,500 homes, see California ISO Glossary, n.d).

In addition to the storage problem, the variability of wind and solar power creates problems for electrical grid operators that may be unsolvable at high penetration rates. Clack *et al.* (2017) write: "In a system where variable renewable resources make up over 95% of the U.S. energy supply, renewable energy forecast errors would be a significant source of uncertainty in the daily operation of power systems. The LOADMATCH model does not show the technical ability of the proposed system ... to operate reliably given the magnitude of the architectural changes to the grid and the degree of uncertainty imposed by renewable resources." Hirth (2013) estimated the integration costs of wind energy to be up to 50% of total generation costs at penetration rates of 30% to 40%. Gowrisankaran *et al.* (2016) estimated the social cost for 20% solar generation is $46.00 per MWh due to intermittency. See also IER (2014) for a discussion of the "energy duck curve," whereby increasing penetration by solar power during the day creates a need to rapidly ramp up power from fossil fuel generation in the evening as workers return to their homes and start using lights and electrical appliances.

A power system relying on wind and solar power for more than 20% to 40% of total power needs begins to experience serious problems with frequency stability, voltage stability and clearing of faults in the power system. For frequency stability a power system must have turbine generators with a very large flywheel effect. If there is insufficient flywheel effect a small disturbance will result in the system

frequency increasing or decreasing very rapidly, leading to a system collapse. Wind power has a very small flywheel effect and solar power and batteries have none. Conventional rotating turbine generators make a major contribution to voltage stability because, when needed, they can rapidly import or export what is known as "reactive power" that is essential for system voltage stability. Wind and solar power installations cannot do this to the same extent.

If the conductors in a major transmission line break and fall to the ground it is essential to isolate the faulty section of line rapidly to avoid system collapse. Conventional generators provide the high currents for short periods needed to maintain system voltage and indicate that a fault has occurred and which line needs to be isolated. Wind, solar power and batteries cannot provide the necessary high currents. So if a fault occurs in a system dominated by wind and solar power the chances are that there will be a massive drop of voltage followed by a system collapse.

Restoring the power after a collapse in a system supplied by wind and solar power is almost impossible. Conventional rotating turbine machines are needed to supply the step changes in electricity demand as the system is restored block by block. Wind and solar power cannot do this. For technical reasons turbines at a pumped storage scheme cannot do this either and there is always a risk that the pumped storage lakes will be empty when a system collapse occurs.

Open cycle gas turbines that can respond rapidly to fluctuations caused by changing loads and changing generation from wind and solar are the only practical option. When they are operating in this back-up mode, carbon dioxide emissions are increased compared to operating at a steady high load and, anyway, open cycle gas turbines are substantially less efficient than combined cycle gas turbines. Unfortunately combined cycle gas turbines cannot change output in the time scale needed.

A system that gets more than 50% of its energy from wind and solar and uses gas turbines when wind and solar output is low or absent also incurs huge losses. A system with a demand of 1000 MW would require 4000 – 5000 MW of wind and solar power and at least 800 MW of gas turbines. For quite a large proportion of the time wind and solar would be capable of providing all the energy demanded – but it would be unable to do so because of system stability problems referred to above. When the wind and solar generation was at a maximum, there would be 2000 – 3000 MW of surplus power available that would have to be dumped in one way or another. As a result, the capacity factor of the wind and solar plants would be much reduced, thus further increasing the cost of energy from them.

In short, dispatchable baseload power is physically, conceptually, and economically different from unpredictable bursts of power. The latter has very few practical uses unless accompanied by storage. The low worth of intermittent power has been disguised in NW Europe where shortfalls can be backed up by imports and surplus production is readily exported at good prices. International cost-benefit assessments are heavily influenced by the unique experience of this region, which has the world's greatest penetration of renewables. However, more than 90% of countries do not import or export electricity.

Since renewables cannot replace fossil fuel or nuclear generation stations, *both* renewable generators and traditional generators must be built and maintained. This forces an overcapacity of generation, with approximately half of all capacity being idle much of the time, depressing the wholesale price of electricity. From 1990 to 2014, Europe built 70% more capacity, the majority of it renewables, while demand for electricity increased by only 26%. Wholesale electricity prices paid to generators continue to decline. Today, no new power plant, either renewable or conventional, can be built in Europe without a government subsidy.

References

Backhaus, S. and Chertkov, M. 2013. Getting a grip on the electrical grid. *Physics Today* **66**: 42–8.

BBC. 2017. Tesla mega-battery in Australia activated December 1. Australia BBC News. December 1.

California ISO Glossary. n.d. Factsheet. Accessed November 5, 2018.

Clack, C.T.M., *et al.* 2017. Evaluation of a proposal for reliable low-cost grid power with 100% wind, water, and solar. *Proceedings of the National Academy of Sciences USA* **114** (26) 6722–7.

EIA. 2017. U.S. Energy Information Administration. Updated Capital Cost Estimates for Utility Scale Electricity Generating Plants. Accessed November 5, 2018.

E.ON Netz. 2005. *Wind Power Report 2005*. Bayreuth, Germany.

Gowrisankaran, G., *et al.* 2016. Intermittency and the value of renewable energy. *Journal of Political Economy* **124** (4): 1187–234.

Hirth, L. 2013. The market value of variable renewables: the effect of solar and wind power variability on their relative price. RSCAS 2013/36. Fiesole, Italy: Robert Schuman Centre for Advanced Studies, Loyola de Palacio Programme on Energy Policy.

IER. 2014. Institute for Energy Research. Solar Energy's Duck Curve. October 27. Accessed November 5, 2018.

Orr, I. and Palmer, F. 2018. How the premature retirement of coal-fired power plants affects energy reliability, affordability. *Policy Study* No. 145. Arlington Heights, IL: The Heartland Institute. February.

Potter, B. and Tillett, A. 2017. Australian households pay highest power prices in the world. *Financial Review*. August 5.

3.5.4 High Cost

Electricity from new wind capacity costs approximately 2.7 times as much as electricity from existing coal, 3 times more than natural gas, and 3.7 times more than nuclear power.

Advocates of rapid decarbonization say the cost of wind and solar power is falling relative to fossil fuels and either has already reached parity or soon will. This has been the claim and the promise since the 1970s, yet electricity generated by wind turbines and solar PV cells is still much more expensive than power from coal, natural gas, and nuclear-powered generators. Claims to the contrary invariably feature methodological errors that ignore or under-estimate real costs while heaping imaginary costs onto fossil fuel generation. Of course, energy costs vary among countries, regions, and states in the United States due to many factors – including local climate conditions, existing infrastructure, population density, and regulations and taxes – so no estimate applies to all areas and circumstances and all estimates are to some degree inaccurate.

Levelized cost of electricity (LCOE)

To accurately compare the cost of producing electricity with each type of fuel requires a methodology that takes into account "capital costs, fuel costs, fixed and variable operations and maintenance (O&M) costs, financing costs, and an assumed utilization rate for each plant type" (EIA, 2018a). The results of such comparisons, called the levelized cost of electricity (LCOE), are estimates of "per-megawatthour cost (in discounted real dollars) of building and operating a generating plant over an assumed financial life and duty cycle" (*Ibid.*). The U.S. Energy Information Administration (EIA) calculates the LCOE for 15 energy sources in the United States, divided into "dispatchable technologies" (generally available regardless of time of day or season) and "non-dispatchable technologies" (generally available only during daytime (solar) or when weather allows wind turbines to operate), for *new* generation resources only. The results of EIA's latest calculations, for facilities entering service in 2022, appear in Figure 3.5.4.1.

According to EIA's projections, the LCOE of new coal generation with 30% carbon capture and storage (CCS) is $130.10/MWh, while the LCOE of new coal with 90% CCS is $119.10/MWh. New natural gas generation is substantially less, ranging from $49.00/MWh for advanced combined-cycle to $98.70/MWh for conventional combustion turbine. Wind and solar energy are categorized as non-dispatchable technologies since their intermittent nature makes them unworkable as a source of baseload power. Off-shore wind at $138.00/MWh and solar thermal at $165.10/MWh, without tax credits, are more expensive than even the most expensive uses of fossil fuels, although tax credits bring the LCOE of both below the LCOE of new coal with 30% CCS, which does not benefit from tax credits. Surprisingly, EIA puts new on-shore wind costs at $59.10/MWh and new solar photovoltaic (PV) costs at $63.20/MWh without subsidies, making them competitive with most fossil fuels. If these figures are accurate, it is difficult to understand why government policies subsidize these facilities at all.

EIA's analysis is valuable, but it is frequently misinterpreted in the climate change debate. The analysis looks only at *future* construction, ignoring the enormous current investment in *existing* long-lived fossil-fuel generation capacity. As Stacy and Taylor (2016) write,

> The approach taken by the federal Energy Information Administration (EIA) … ignores the cost of electricity from all of our existing

resources and publishes LCOE calculations for new generation resources only. If no existing generation sources were closed before the end of their economic life, EIA's approach would provide sufficient information to policymakers on the costs of different electricity policies. … However, in the current context of sweeping environmental regulations on conventional generators – coupled with mandates and subsidies for intermittent resources – policies are indeed forcing existing generation sources to close early. Federal policies alone threaten to shutter 110 gigawatts of coal and nuclear generation capacity (p. 1).

The point is an important one. Coal-powered stations with abundant fuel do not simply disappear at the end of a nominal 50-year life. Their infrastructure, technology, and hardware are continuously replaced and upgraded in much the same way as an electrical grid. They will continue producing for as long as their short-run marginal costs remain competitive with the long-run marginal costs of new generators. Even decommissioned plants are likely to be replaced in situ by new power plants, whose costs will be significantly lower than those of a theoretical "greenfield" site. Climate policy is said to be "urgent" and there is much rhetoric to the effect that the next 30-odd years will be crucial (e.g., Lovins *et al.*, 2011). During that period, the generating capacity of most countries (especially developed countries) will be entirely dominated by existing sites, meaning the EIA's LCOEs will have virtually no application.

A second misinterpretation is assuming EIA's LCOE calculations take into account the intermittency of solar and wind power, and consequently the need for those facilities to maintain additional reserve capacity of dispatchable back-up generation units. As explained in Sections 3.1.4 and 3.5.3, every wind turbine and solar panel needs a fossil fuel-powered generator of nearly equal capacity standing behind it ready to generate power when the wind does not blow or the sun does not shine (Rasmussen, 2010; E.ON Netz, 2005). Joskow (2011) noted:

Figure 3.5.4.1
Estimated levelized cost of electricity (unweighted average) for new generation resources entering service in the United States in 2022 (2017 $/MWh)

Plant type	Capacity factor (%)	Levelized capital cost	Levelized fixed O&M	Levelized variable O&M	Levelized transmission cost	Total system LCOE	Levelized tax credit[1]	Total LCOE including tax credit
Dispatchable technologies								
Coal with 30% CCS[2]	85	84.0	9.5	35.6	1.1	130.1	NA	130.1
Coal with 90% CCS[2]	85	68.5	11.0	38.5	1.1	119.1	NA	119.1
Conventional CC	87	12.6	1.5	34.9	1.1	50.1	NA	50.1
Advanced CC	87	14.4	1.3	32.2	1.1	49.0	NA	49.0
Advanced CC with CCS	87	26.9	4.4	42.5	1.1	74.9	NA	74.9
Conventional CT	30	37.2	6.7	51.6	3.2	98.7	NA	98.7
Advanced CT	30	23.6	2.6	55.7	3.2	85.1	NA	85.1
Advanced nuclear	90	69.4	12.9	9.3	1.0	92.6	NA	92.6
Geothermal	90	30.1	13.2	0.0	1.3	44.6	-3.0	41.6
Biomass	83	39.2	15.4	39.6	1.1	95.3	NA	95.3
Non-dispatchable technologies								
Wind, onshore	41	43.1	13.4	0.0	2.5	59.1	-11.1	48.0
Wind, offshore	45	115.8	19.9	0.0	2.3	138.0	-20.8	117.1
Solar PV[3]	29	51.2	8.7	0.0	3.3	63.2	-13.3	49.9
Solar thermal	25	128.4	32.6	0.0	4.1	165.1	-38.5	126.6
Hydroelectric[4]	64	48.2	9.8	1.8	1.9	61.7	NA	61.7

Notes:

1 The tax credit component is based on targeted federal tax credits such as the PTC [Production Tax Credit] or ITC [Investment Tax Credit] available for some technologies. It reflects tax credits available only for plants entering service in 2022 and the substantial phase out of both the PTC and ITC as scheduled under current law. Technologies not eligible for PTC or ITC are indicated as NA or not available. The results are based on a regional model, and state or local incentives are not included in LCOE calculations.

2 Because Section 111(b) of the Clean Air Act requires conventional coal plants to be built with CCS [carbon capture and storage] to meet specific CO_2 emission standards, two levels of CCS removal are modeled: 30%, which meets the NSPS [New Source Performance Standards], and 90%, which exceeds the NSPS but may be seen as a build option in some scenarios. The coal plant with 30% CCS is assumed to incur a 3 percentage-point increase to its cost of capital to represent the risk associated with higher emissions.

3 Costs are expressed in terms of net AC power available to the grid for the installed capacity.

4 As modeled, hydroelectric is assumed to have seasonal storage so that it can be dispatched within a season, but overall operation is limited by resources available by site and season.

CCS=carbon capture and sequestration. CC=combined-cycle (natural gas). CT=combustion turbine. PV=photovoltaic.

Source: EIA, 2018a, Table 1b, p. 5.

Economic evaluations of alternative electric generating technologies typically rely on comparisons between their expected "levelized cost" per MWh supplied. I demonstrate that this metric is inappropriate for comparing intermittent generating technologies like wind and solar with dispatchable generating technologies like nuclear, gas combined cycle, and coal. It overvalues intermittent generating technologies compared to dispatchable base load generating technologies. It also likely overvalues wind generating technologies compared to solar generating technologies.

On the one hand, the LCOEs assume a tabula rasa, such as might occur in remote areas of less developed countries, where there are no sunk costs or stranded investments in fossil fuel production. But on the other hand, the LCOEs also assume fossil fuel back-up or spinning reserve exists to supply back-up power at zero cost when needed. The assumptions of the scenario are internally inconsistent. Instead, the capital cost of coal or natural gas back-up power should be added to the LCOE of wind and solar power along with fuel costs when they are called up to provide power. Crucially, back-up power capacity will not be maintained or built because "imposed costs" described later will render it uneconomic if the renewables have dispatch priority.

Another problem encountered when using EIA's LCOE estimates is the transmission costs for electricity from solar and wind do not rise fast enough to reflect the higher integration costs (as high as 50% of total generation costs) at penetration rates of 30% to 40% (Gowrisankaran et al., 2016; Hirth, 2013). A fourth problem is that EIA's LCOEs account for some but not all of the many subsidies, tax breaks, and regulatory protections renewable energies enjoy over fossil fuels. This is described in more detail below. A fifth problem is EIA already adds 3% to the cost of capital for coal-fired power and coal-to-liquids plants, equivalent to an emissions fee of $15 per metric ton of carbon dioxide emissions (EIA, 2018a, p. 3). Unless there is a national $15 per metric ton tax on carbon emissions, such an "adjustment" prejudges the answer to the very question LCOEs are often invoked to answer.

Sixth and finally, LCOE estimates are sensitive to assumptions about realized capacity factors: the average power a facility delivers divided by its rated peak power (also known as "nameplate capacity").

Realized capacity factors vary for individual power plants, for different times of day and times of year, and for different locations (Boccard, 2009; Pomykacz and Olmsted, 2014). EIA (2018b) reported the following average realized capacity factors for power plants across the United States in June 2018:

CC Natural Gas	54.8
Coal	53.5
Wind	36.7
Solar PV	27.0
Solar Thermal	21.8

On average, coal and combined cycle natural gas have realized capacity factors substantially greater than wind and solar. When LCOE estimates are being made, an error of several percentage points in the capacity factor assumptions for coal and CC natural gas facilities would have a small effect in terms of the percentage change of overall capacity and hence energy costs. A similar error in the case of renewable energy, like wind, could mean a considerable increase or decrease in LCOE. For example, an error by 10 percentage points for wind would translate into 30% lower realized capacity factor and 30% higher LCOE, since the main costs of wind are fixed costs, "fuel" costs are zero.

Stacy and Taylor (2016) produced "much-needed cost comparisons between existing resources that face early closure and the new resources favored by current policy to replace them." They used data from documents (known as Form 1) submitted to the Federal Energy Regulatory Commission (FERC) to estimate the LCOE of existing fossil fuel generation. They also added to the LCOE of alternative energies "the amount intermittent resources increase the LCOE for conventional resources by reducing their utilization rates without reducing their fixed costs. We refer to these as 'imposed costs,' and we estimate them to be as high as $25.9 per megawatt-hour of intermittent generation when we model combined cycle natural gas energy displaced by wind, and as high as $40.6 per megawatt-hour of intermittent generation when we model combined cycle and combustion turbine natural gas energy displaced by PV solar" (Stacy and Taylor, 2016, p. 1). Their findings appear in Figure 3.5.4.2.

According to Stacy and Taylor's analysis, existing conventional coal generation resources in 2015 had an LCOE of only $39.9/MWh and natural gas of $34.4/MWh, far below the cost of electricity from new wind ($107.40/MWh) and new solar PV ($140.30/MWh). The results, the authors write, show the sharp contrast between the high cost of electricity from new generation resources and the average low cost from the existing fleet. Existing coal-fired power plants, for example, generate reliable electricity at an LCOE-E [LCOE-Existing] of $39.9 per megawatt-hour on average. Compare that to the LCOE of a new coal plant, which is $95.1 per megawatt-hour according to EIA estimates. This analysis also shows that, on average, continuing to operate existing natural gas, nuclear, and hydroelectric resources is far less costly than building and operating new plants to replace them. Existing generating facilities produce electricity at a substantially lower levelized cost than new plants of the same type (p. 1).

Despite the appearance of precision, this calculation produces only an approximation of values that are highly uncertain and are not observed in the real world. The methodology and its use here are complicated by the different year-dollars used by the EIA (2017) and Stacy and Taylor (2013), and while Stacy and Taylor used numbers from EIA's *Annual Energy Outlook 2015,* EIA reduced the cost of wind by 20% and the cost of solar PV by 50% in the latest, 2018, *Outlook.* Such changes reflect the rising benefits of taxes, mandates, and subsidies and their growing adverse effects on fossil fuel capacity factors and operating costs.

The difference between the LCOE of existing and new generation is salient to the current public policy debate because existing coal power plants are long-lived facilities, with useful lives estimated at 50 years and likely to be longer, compared to only 25 years or less for wind and solar installations (Pomykacz and Olmsted, 2014; IER, 2018). Moreover, electricity demand in the United States is essentially flat, with net generation falling 2.5% between 2008 and 2017 (from 4,119,388 thousand MWh to 4,014,804 thousand MWh) (EIA, 2018c). According to Stacy and Taylor, "absent mandates for new generation and the onset of new federal environmental regulations forcing some coal fired generating capacity to retire, almost no new generation capacity would have been necessary" between 2004 and 2014 (Stacy and Taylor, 2016, p. 6). The situation is similar in much of Europe, though not in developing countries. Recall from Section 3.4 that global energy needs are expected to rise by 30% between today and 2040 and total electricity generation will rise even more.

Comparing the LCOE for existing fossil-fuel generation to new wind capacity + imposed costs finds wind costs 2.69 times as much as coal (107.4/39.9), 3.12 times as much as CC gas (107.4/34.4), and 3.69 times as much as nuclear

Figure 3.5.4.2
Levelized cost of electricity (LCOE) of existing and new generation resources, in 2013 $ / MWh

Generator Type	LCOE of EXISTING Generation (at actual 2015 Capacity Factors and Fuel Costs)	LCOE of NEW Generation (at actual 2015 Capacity Factors and Fuel Costs)
Dispatchable Full-Time-Capable Resources		
Conventional Coal	39.9	N/A
Conventional Combined Cycle Gas (CC gas)	34.4	55.3
Nuclear	29.1	90.1
Hydro	35.4	122.2
Dispatchable Peaking Resources		
Conventional Combustion Turbine Gas (CT gas)	88.2	263.0
Intermittent Resources – As Used in Practice		
Wind including cost imposed on CC gas	N/A	107.4 + other costs
PV Solar including cost imposed on CC and CT gas	N/A	140.3 + other costs

Source: Stacy and Taylor, 2016.

power (107.4/29.1). These are sizeable differences indeed, and important since a change of even 10% in the cost of electricity in the United States results in a loss of approximately 1.3% of GDP, about $253 billion in 2017, as reported in Section 3.6.1 below.

Stacy and Taylor conclude, "Electricity from the existing generating fleet is less expensive than from its available new replacements, and existing generators whose construction cost repayment and recovery obligations have been substantially or entirely met are often the least-cost producers in their resource fleet. Cost trends extracted from Form 1 indicate the fleet average cost of electricity from existing resources is on track to remain a lower cost option than new generation resources for at least a decade – and possibly far longer" (p. 35).

Stacy and Taylor's analysis concentrated on the EIA's LCOE estimate, but other estimates are similarly flawed. For example, an LCOE calculated by Lazard, a financial services company with offices in New York City and London, is often cited as providing proof that solar and wind power are achieving or have already achieved parity with fossil fuels, but the following disclaimer appears in bold print on the first page of the latest report:

> Other factors would also have a potentially significant effect on the results contained herein, but have not been examined in the scope of this current analysis. These additional factors, among others, could include: capacity value vs. energy value; stranded costs related to distributed generation or otherwise; network upgrade, transmission or congestion costs or other integration-related costs; significant permitting or other development costs, unless otherwise noted; and costs of complying with various environmental regulations (e.g., carbon emissions offsets, emissions control systems) (Lazard, 2017, p. 1).

On the next page of its report, Lazard admits to not taking into account "reliability or intermittency-related considerations (e.g., transmission and back-up generation costs associated with certain Alternative Energy technologies)" (*Ibid.*, p. 2). It is precisely the stranded costs, integration expenses, and "intermittency-related considerations" that cause wind and solar power to incur some of their largest

costs. Any LCOE that fails to take these matters into account is inaccurate and useless for public policy purposes.

Subsidies

Renewable energies sometimes appear to be cost-competitive with fossil fuels due to the extremely wide and complicated web of government policies biased in favor of renewable energy. Consumers may be told they can sign up for "100% renewable energy" without any increase in their monthly utility bill, not realizing that renewable portfolio mandates on utilities force all ratepayers to subsidize their choice by paying higher rates. In the United States, state governments add a layer of subsidies and tax credits to those provided by the national government. Schleede (2010) notes EIA does not account properly for five-year double declining balance accelerated depreciation, state and local tax breaks, state mandates, and more that make wind and solar appear to be less costly than they really are.

The U.S. Energy Information Administration regularly reports on federal subsidies to energy producers and consumers. In its latest report (EIA, 2018d), EIA estimated subsidies to producers totaled $7.5 billion in 2016. Additional subsidies for smart grid and transmission, conservation, and to end users totaled $7.45 billion. Of the subsidies directly to producers, renewables received 89%, coal received 17%, and nuclear received 5%. Natural gas and petroleum liquids producers paid $940 million more than they received via energy-specific tax provisions (expensing of exploration, development, and refining costs), which EIA reports as a *negative net subsidy* to the industry of -$773 million. This huge subsidy imbalance between renewables and other fuels is even more apparent when the subsidies are calculated on a per-unit-of-output basis. According to EIA, hydroelectric power received no federal subsidies in 2016. Remaining renewables generated 7.9 quadrillion Btu of primary energy in 2016 (EIA, 2018e). On a per-Btu basis, the subsidies to renewables were 10 times larger than for coal power and 2 times larger than for coal power. Natural gas and petroleum liquids received no net subsidy, and indeed recorded a *negative* subsidy of -$15 million per quadrillion Btu of energy produced. See Figure 3.5.4.3.

Figure 3.5.4.3
Direct U.S. federal financial interventions and subsidies, 2016

Beneficiary	2016 subsidies and support (millions)	Primary energy production (quadrillion Btu)	% of subsidies	Subsidy per quadrillion Btu (millions)	Subsidy per Btu to non-hydro renewables: other fuels
Non-hydro renewables	$6,682	7.856	88.67%	$850.56	--
Coal	$1,262	14.667	16.75%	$86.04	10:1
Nuclear	$365	8.427	4.84%	$43.31	2:1
Natural gas and petroleum liquids	($773)	50.94	-10.26%	($15.17)	--
Total	$7,536	81.89	100.00%	$92.03	--

Source: EIA, 2018d, Table 2, p. 5 and Table 3, p. 9; EIA, 2018e, Table 1.2, p. 5.

The massive subsidization of renewables relative to fossil fuels in the United States is not new. During the years 2011–2016, renewable energy (solar, wind, biomass, geothermal, and hydro) received $89 billion in federal incentives, nearly four times the federal incentives for oil and natural gas combined (Bezdek, 2017). See Figure 3.5.4.4. Notably, oil and gas supplied more than 61% of U.S. energy needs whereas wind and solar provided less than 3%. Thus, per unit of energy, renewables are massively subsidized compared to oil and gas. In much of the world, renewables are even more heavily favored than fossil fuels than in the United States.

Stiglitz (2018) claims below-cost federal leases are driving prices and giving fossil fuels an advantage. But less than half of U.S. coal production is from federal leases (BLM, 2018a) and less than one-third for both oil and gas production (Humphries, 2016). Oil and gas production from federal lands is declining while U.S. production has increased about 60% since 2008. This would not be the case if federal leases were underpriced. Stiglitz claims leases are sold "at prices far below what the competitive equilibrium price would be," but lease winners bear risks (geological characteristics of reservoir, cost of extraction, and the future retail price for output are all uncertain) so the "competitive equilibrium price" is undefined. Underprices leases would create profitable opportunities for other businesses to bid for them, thus driving up lease prices. Further, winners pay royalties on oil and gas recovered, so lease price is not even the most relevant issue (BLM, 2018b).

Figure 3.5.4.4
U.S. federal tax incentives for oil, natural gas, and renewables, 2011–2016

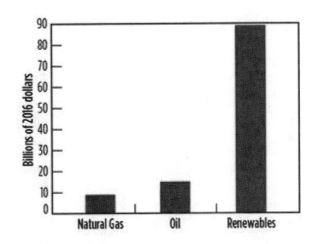

Source: Bezdek, 2017.

Because oil and natural gas are traded internationally, their prices are set by the world market. The U.S. government and its lessees are bargaining over distribution of rents, not over the price of the product. The United States produces about 15% of world petroleum and 20% of natural gas (EIA, 2018f). This haggling over how rents will be split is highly unlikely to be driving down world prices. Finally, proof that solar and wind are not cost-competitive is seen when investment in new installations virtually stops when subsidies are interrupted. New wind and solar power can compete with fossil fuels only if utilities are required to buy it, often at prices that are two and three times as high as the price of coal and natural gas-generated power.

Other Costs

Stacy and Taylor (2016) found electricity produced by solar and wind generators in the United States cost approximately three times as much as electricity produced with fossil fuels. Their calculation of the LCOE for solar and wind is likely still too low because it does not take into account all the subsidies described above or other costs. Some unaccounted costs are deterioration of wind turbine output over time, negative environmental and neighborhood effects, and opportunity costs.

The performance and capacity factors of wind turbines deteriorate over time. A seminal study analyzed the rate of aging of a national fleet of wind turbines using public data for the actual and theoretical ideal load factors from the UK's 282 industrial wind facilities (Staffell and Green, 2014). It found:

- Load factors declined with age, at a rate similar to that of other rotating machinery.

- Onshore wind installations' output declines 16% a decade.

- Performance declines with age occurred in all wind installations and all generations of turbines.

- Decreasing output over a wind installation's life increased the levelized cost of electricity.

The study determined this degradation rate was consistent for different vintages of turbines and for individual wind installations, from those built in the early 1990s to early 2010s.

The Renewable Energy Foundation, an organization that actually advocates in favor of renewable energy facilities, also conducted a comprehensive study of the available capacity factors over time for wind turbines in the UK and came to similar findings. Using monthly observations for 282 onshore installations in the UK with an age range of zero to 19 years, it found "the normalized load factor

for UK onshore wind farms declines from a peak of about 24 percent at age one to 15 percent at age 10 and 11 percent at age 15" (Hughes 2012). In other words, the capacity factors for wind generators decline significantly every year after installation.

Other costs attributable to renewable energy but not counted in the LCOE exercises include *environmental harms* such as killing birds and bats. According to Smallwood (2013), "I estimated 888,000 bat and 573,000 bird fatalities/year (including 83,000 raptor fatalities) at 51,630 megawatt (MW) of installed wind-energy capacity in the United States in 2012." Since wind turbine capacity in the United States has grown since then, it is certain bird and bat kills have increased apace.

According to Hambler (2013), "Because wind farms tend to be built on uplands, where there are good thermals, they kill a disproportionate number of raptors. In Australia, the Tasmanian wedge-tailed eagle is threatened with global extinction by wind farms. In North America, wind farms are killing tens of thousands of raptors including golden eagles and America's national bird, the bald eagle. In Spain, the Egyptian vulture is threatened, as too is the Griffon vulture – 400 of which were killed in one year at Navarra alone. Norwegian wind farms kill over ten white-tailed eagles per year and the population of Smøla has been severely impacted by turbines built against the opposition of ornithologists."

According to Taylor (2015), the Ivanpah solar power plant in the Mojave Desert in California killed 3,500 birds in its first year of operation. According to the Institute for Energy Research (IER, 2015), "The [Ivanpah] facility is estimated to have killed 83 different species of birds. The most commonly killed birds were mourning doves (14 percent of fatalities), followed by yellow-rumped warblers, tree swallows, black-throated sparrows and yellow warblers. Of the birds that died from known causes, about 47 percent died from being toasted by the heat of the solar flux. Just over half of the known deaths were attributed to collisions."

Tang *et al.* (2017) reported that construction of the wind turbines in the area of China they studied elevated both day (by 0.45-0.65°C) and night (by 0.15-0.18°C) temperatures, which increase, they say, "suppressed soil moisture and enhanced water stress in the study area." As a result, they calculated an approximate 14.5%, 14.8%, and 8.9% decrease in leaf area index (LAI), enhanced vegetation index (EVI), and normalized difference vegetation index (NDVI), respectively, over the period of study, as well as "an inhibiting [wind farm] effect of 8.9% on summer gross primary production (GPP) and 4.0% on annual net primary production (NPP)." These several findings led Tang *et al.* to conclude that their research "provides significant observational evidence that wind farms can inhibit the growth and productivity of the underlying vegetation."

Increased use of biofuels (primarily wood and ethanol) also has negative environmental consequences that often go unreported. Di Fulvio *et al.* (2019) studied the ecological impact of land use changes expected to be made in the European Union to meet its goal of reducing greenhouse gas emissions by 80% by 2050. They estimate such changes would result in the extinction of approximately 1% of the total of all global species by 2050. Models used to predict extinctions are notoriously inaccurate, as explained in Chapter 5, but there is little doubt that the massive expansion of acreage devoted to the production of ethanol and other biofuels would have a negative effect on many species.

Another uncounted cost of renewable energy is *negative neighborhood effects,* such as those caused by wind turbines on crop yields and property values. Linowes (2013) reports on soil compacting, destruction of irrigation piping and crops, and the end of aerial spraying of insecticides in fields near wind turbines. A study by a London School of Economics economist of some two million home sales in England and Wales from 2000 to 2011 found "Wind farms reduce house prices in postcodes where the turbines are visible; and they reduce prices relative to postcodes close to wind farms where the turbines are not visible. Averaging over wind farms of all sizes, prices fall by around 5–6% within 2km, by less than 2% in the range 2–4km and by less than 1% at 14km, which is the limit of likely visibility" (Gibbons, 2014).

There are many accounts of possible negative health effects due to the low-frequency sound and vibrations produced by wind turbines. Frequencies below 200Hz can be generated by thunder, volcano eruptions, earthquakes, or storms, all events that can cause anxiety or fear. It is possible humans are "wired" to respond this way, making nearby wind turbines a nuisance or worse. The Sahlgrenska Academy Institute of Medicine at the University of Gothenburg in Sweden has conducted extensive research on the issue (Sahlgrenska Academy, n.d.).

Opportunity cost

Research papers claiming to show the feasibility of a rapid and inexpensive transition from fossil fuels, such as Jacobson *et al.* (2015) and an earlier paper by Jacobson and a coauthor (Jacobson and Delucchi, 2009), fail to take into account the opportunity cost of abandoning the existing energy generation infrastructure. The sheer size of the global energy market makes replacing that infrastructure massively expensive and time consuming. The electric grids in the United States and around the world represent investments of trillions of dollars and require hundreds of billions of dollars a year in new investment simply to maintain, improve, and keep pace with population and consumption growth. They also generate hundreds of billions of dollars in revenue each year. Replacing them with more advanced grids and long-distance high-voltage power lines that could accommodate disbursed solar and wind energy or hydropower located far from urban centers would cost several times total past investments *in addition to* ongoing investments in modernization and expansion of the existing grid until it can be replaced, and would require decades to plan and implement. Given competing interests, decentralized government decision-making, already high levels of government indebtedness, and strong NIMBY (not in my backyard) opposition to new infrastructure projects around the world, such proposals for a 100% renewable future are no more than academic exercises.

Smil (2010) notes the global oil industry "handles about 30 billion barrels annually or 4 billion tons" and operates about 3,000 large tankers and more than 300,000 miles of pipelines. "Even if an immediate alternative were available, writing off this colossal infrastructure that took more than a century to build would amount to discarding an investment worth well over $5 trillion – and it is quite obvious that its energy output could not be replaced by any alternative in a decade or two" (Smil, 2010, p. 140). Later, Smil writes the cost of a transition "would be easily equal to the total value of U.S. gross domestic product (GDP), or close to a quarter of the global economic product" (*Ibid.,* p. 148).

A second opportunity cost is living without affordable and convenient energy. If renewable energies cannot produce the quantity and quality of energy needed to sustain current and future levels of human prosperity, then the quality of life for millions and potentially billions of people will be diminished. The cost of renewables therefore includes not being able to own or use a car or truck, live in a single-family dwelling, or work more than a short distance from home. It could mean reduced access to fresh and affordable food, clean water and sanitation, quality health care, and educational and recreational opportunities. A shortage of energy, even if prices are government-controlled and access to energy is rationed, would be profoundly costly.

* * *

In summary, the cheapest form of energy in most locations in the developed world is continued production from existing facilities with significant remaining lifespans. Those facilities are predominantly powered by fossil fuels. Energy produced by solar PV cells, wind turbines, and ethanol can contribute to the world's energy mix but they lack power density and sufficient supply, are not dispatchable when needed, and are too costly to meet more than a small fraction of the world's energy needs.

Without fossil fuels, most homes and businesses not located near a nuclear power plant or a river able to produce hydropower would be without electricity. While wind turbines and solar PV cells can generate power in some places and under some circumstances, only fossil fuels can produce enough energy to forge steel, make concrete, power locomotives and ocean-crossing ships and airplanes, and many other components of modern industry. Biofuels, such as ethanol, cannot replace more than a small fraction of petroleum used around the world for transportation. Indeed, without fossil fuels it would be impossible to manufacture wind turbines and solar PV cells, or build the massive concrete foundations for wind turbines or modern hydroelectric dams, or plant and irrigate and harvest corn or soybeans in sufficient quantity to power more than a percent or two of a modern civilization's daily energy consumption. There would also be no high-voltage power lines or towers to transport electricity generated by solar panels or wind turbines, and no batteries (or dams, in the case of hydropower) to store power for when it is needed.

References

Bezdek, R.H. 2017. Oil and gas in the capitals. *World Oil* **238** (5).

BLM. 2018a. U.S. Bureau of Land Management. Frequently asked questions about the Federal Coal Leasing Program (website). U.S. Bureau of Land Management, Department of the Interior. Accessed November 6, 2018.

BLM. 2018b. U.S. Bureau of Land Management. National Fluids Lease Sale Program, Frequently Asked Questions (website). U.S. Bureau of Land Management, Department of the Interior. Accessed November 6, 2018.

Di Fulvio, F., *et al.* 2019. Spatially explicit LCA analysis of biodiversity losses due to different bioenergy policies in the European Union. *Science of the Total Environment* **651**: 1505–16.

EIA. 2018a. U.S. Energy Information Administration. *Levelized cost and levelized avoided cost of new generation resources in the Annual Energy Outlook 2018.* March.

EIA. 2018b. U.S. Energy Information Administration. *Electric Power Monthly* (website). Table 6.7.A. Capacity Factors for Utility Scale Generators Primarily Using Fossil Fuels, January 2013-April 2018 and Table 6.7.B. Capacity Factors for Utility Scale Generators Not Primarily Using Fossil Fuels. June 25.

EIA. 2018c. U.S. Energy Information Administration. *Electricity Data Browser* (website). Net general for all sectors, annual.

EIA. 2018d. U.S. Energy Information Administration. Direct Federal Financial Interventions and Subsidies in Energy in Fiscal Year 2016. Washington, DC.

EIA. 2018e. U.S. Energy Information Administration. Monthly Energy Review, November 2018. Washington, DC.

EIA. 2018f. U.S. Energy Information Administration. United States remains the world's top producer of petroleum and natural gas hydrocarbons. Today in Energy (website). May 21.

E.ON Netz. 2005. *Wind Power Report 2005.* Bayreuth, Germany.

Gibbons, S. 2014. *Gone with the Wind: Valuing the Visual Impacts of Wind Turbines through House Prices.* London, UK: London School of Economics & Political Sciences and Spatial Economics Research Centre.

Gowrisankaran, G., *et al.* 2016. Intermittency and the value of renewable energy. *Journal of Political Economy* **124** (4): 1187–1234.

Hambler, C. 2013. Wind farms vs wildlife. The Spectator (website). January 5.

Hirth, L. 2013. The market value of variable renewables: the effect of solar and wind power variability on their relative price. RSCAS 2013/36. Fiesole, Italy: Robert Schuman Centre for Advanced Studies, Loyola de Palacio Programme on Energy Policy.

Hughes, G. 2012. *The Performance of Wind Farms in the United Kingdom*. London, UK: Renewable Energy Foundation.

Humphries, M. 2016. U.S. Crude Oil and Natural Gas Production in Federal and Nonfederal Areas. Congressional Research Service. Washington, DC: U.S. Government Printing Office.

IER. 2015. Institute for Energy Research. License to kill: wind and solar decimate birds and bats. April 29.

IER. 2018. Institute for Energy Research. The mounting solar panel waste problem. September 12.

Jacobson, M.Z. and Delucchi, M.A. 2009. A plan to power 100 percent of the planet with renewables. *Scientific American* **303** (5): 58–65.

Jacobson, M.Z., *et al.* 2015. Low-cost solution to the grid reliability problem with 100% penetration of intermittent wind, water, and solar for all purposes. *Proceedings of the National Academy of Sciences USA* **112** (49):15060–5.

Joskow, P.L. 2011. Comparing the costs of intermittent and dispatchable electricity generating technologies. *American Economic Review* **101** (3): 238–41.

Lazard. 2017. *Lazard's Levelized Cost of Energy Analysis – Version 11.0*. November.

Linowes, L. 2013. The incompatibility of wind and crop 'farming.' MasterResource (blog). July 1.

Lovins, A.B. *et al.* 2011. *Reinventing Fire: Bold Business Solutions for the New Energy Era*. White River Junction, VT: Chelsea Green Publishing.

Pomykacz, M. and Olmsted, C. 2014. The appraisal of power plants. *The Appraisal Journal* (Summer): 216–30.

Rasmussen, K. 2010. *A Rational Look at Renewable Energy and the Implications of Intermittent Power*. Edition 2.0. November. South Jordan, UT: Deseret Power.

Sahlgrenska Academy. n.d. Institute of Medicine at the University of Gothenburg. Low frequency noise and human response (website). Accessed November 6, 2018.

Schleede, G.R. 2010. *The True Cost of Electricity from Wind Is Always Underestimated and Its Value Is Always Overestimated*. Washington, DC: Science & Public Policy Institute. February 10.

Smallwood, K.S. 2013. Comparing bird and bat fatality-rate estimates among North American wind-energy projects. *Wildlife Society Bulletin* **37** (1).

Smil, V. 2010. *Energy Myths and Realities: Bringing Science to the Energy Policy Debate*. Washington, DC: American Enterprise Institute.

Stacy, T.F. and Taylor, G.S. 2016. *The Levelized Cost of Electricity from Existing Generation Resources*. Washington, DC: Institute for Energy Research.

Staffell, I. and Green, R. 2014. How does wind farm performance decline with age? *Renewable* Energy 66 (June): 775–86.

Stiglitz, J.E. 2018. Expert report of Joseph E. Stiglitz. Prepared for plaintiffs and attorneys for plaintiffs in the case *of Juliana et al.* v. *United States of America*. United States District Court, District of Oregon.

Tang, B., *et al.* 2017. The observed impacts of wind farms on local vegetation growth in northern China. *Remote Sensing* **9**: 332, doi:10.3390/rs9040332.

Taylor, P. 2015. 3,500 birds died at Ivanpah 'power towers' in 1st year – report. *Greenwire*. April 24.

3.5.5 Future Cost

The cost of alternative energies will fall too slowly to close the gap with fossil fuels before hitting physical limits on their capacity.

The research summarized above showed why alternative energies such as wind and solar cannot completely supplant fossil fuels and why their true cost is extremely high relative to fossil fuels. *How will this change in the future?* Short of a world government imposing its will by decree, will alternative energies *ever* replace fossil fuels?

Levelized cost of electricity (LCOE) exercises assume the rate of price reduction previously experienced for wind and solar electricity generation will continue indefinitely. This is most unlikely as the reductions of the last decade are due chiefly to falling manufacturing costs of turbines and PV panels – as those items graduated from bespoke (tailor-made) development to mass production. They also benefited from large government subsidies, especially in China and Germany, which have proven to be unsustainable (Reed, 2017; Reuters, 2018). Turbines and panels now comprise relatively minor components of the life-cycle costs of installations. The major remaining components (e.g., labor and raw materials) are likely to be much more resistant to progress.

Cost trends for renewable energy are usually summarized in the form of experience curves, a statistical relation between the installed capacity or total output and the unit costs of production. The curves reflect "learning rates" defined as the percentage decline in unit costs for each doubling of output or capacity. The experience curves approach dates back to the study by Wright (1936) documenting unit costs decreased by 10% to 15% every time production of an airplane doubled. Experience curves have been documented in a variety of other industries by Bruce Henderson of the Boston Consulting Group (Henderson, 1970) and other authors since then. The Stern Report (2006), which greatly affected environmental policy in the United Kingdom, used learning rates estimated by the International Energy Agency (IEA, 2000) in its analysis of options for clean energy production. The United Nations' Intergovernmental Panel on Climate Change (IPCC) also used learning rates in its Fourth Assessment Report (IPCC, 2007). More recently, Upstill and Hall (2018) estimated the learning rate for carbon sequestration and storage to be 6.3%.

Figure 3.5.5.1 provides a summary of the learning rates reported in the literature based on works of McDonald and Schrattenholzer (2001), Neij *et al.* (2003), and Junginger *et al.* (2004). The average of study results for wind energy suggests the per-unit cost of electricity from that source falls by about 16% (with a wide range from 4% to 32%) for each doubling of capacity. Solar exhibits higher learning rates of 15% to 25% for each doubling, with ethanol production estimated to have learning rates around 20% for each doubling of production.

An important limitation on the use of experience curves is the fact that one industry does not learn and become more efficient over time while its competitors are frozen in time and learn nothing at all. Electricity producers using coal, natural gas, and nuclear power are all climbing experience curves of their own and becoming more efficient over time. To use experience curves to predict when new wind energy + natural gas back-up will be cost-competitive with existing coal, for example, would require knowing the shape of the curves for all three energy producers and then estimating the difference, the net progress wind + natural gas would make over coal, over time. It is entirely possible coal will keep pace with the productivity gains of wind energy + natural gas in the coming decades – especially if the

regulatory environment were to change so as not to disfavor coal (Orr and Palmer, 2018) – meaning new wind + natural gas would never become cost-competitive.

The wind energy industry faces physical limits on its ability to improve the efficiency of its turbines, regardless of learning, as explained in Section 3.1.2. Generally the industry has been lowering costs by increasing the height of the towers, but industrial wind facilities are facing increasing opposition from land owners and communities. Taller turbines will mean larger set-backs from houses and communities, constraining their ability to increase capacity. For many nations of Europe, recent building of wind systems has been offshore, despite the higher expense, because opposition to onshore facilities has been too high. The 20,000 land-based turbines in Denmark and Germany may not be possible to replace when they reach their end of life because of lack of subsidies and community opposition. These realities suggest the cost of electricity from wind power probably will fall by less than 16% (the average from studies listed in Figure 3.6.1) for each doubling of capacity.

Recall that Stacy and Taylor found the LCOE of new wind energy in 2015 was 2.69 times as much as coal, 3.12 times as much as CC gas, and 3.69 times as much as nuclear power. Assume that the per-unit price of electricity from wind energy will fall 16% relative to the price of coal, natural gas, and nuclear energy for every doubling in wind's output. How many doublings of wind capacity would have to occur before the per unit cost of wind equals or is less than coal, natural gas, or nuclear energy? The math is easy. Wind producers would need to double their output six times (64 times current output) to be price-competitive with existing coal, seven times (128 times current output) to be competitive with CC gas, and eight times (256 times current output) to be competitive with nuclear power. These are, of course, impossible output numbers. In the United States, sometime around the fourth doubling wind energy would hypothetically produce all of the electricity needed to meet demand without fossil fuels. (Four doublings from a base of 254 billion kWh would be 4,064 billion kWh. Total U.S. electricity production in 2017 was approximately 4,015 billion kWh (EIA, 2018)). Of course, wind by itself cannot power an electric grid. Given its current learning rate, the wind energy industry would have to produce four times the entire energy consumption of the United States before it will have lowered its cost-per-unit to the current cost of coal, eight times to be price-

Figure 3.5.5.1
Learning rates in different renewable energy technologies

Type of Energy	Region	Period	Dependent Variable	Explanatory Variable	Learning Rate	Source
Electricity from biomass	EU	1980–1995	sp. prod. cost ($/kWh)	cum. prod. (TWh)	0.15	IEA (2000)
Ethanol	Brazil	1979–1995	sp. sale price ($/boe)	cum. prod. (cubic meters)	0.2	Goldemberg (1996)
Ethanol	Brazil	1978–1995	sp. sale price ($/boe)	cum. prod. (cubic meters)	0.22	IEA (2000)
Solar PV	EU	1985–1995	sp. prod. cost (ECU/kWh)	cum. prod. (TWh)	0.35	IEA (2000)
Solar PV modules	EU	1976–1996	sale price ($/W peak)	cum. sales (MW)	0.21	IEA (2000)
Solar PV modules	World	1976–1992	sale price ($/W peak)	cum. sales (MW)	0.18	IEA (2000)
Solar PV modules	World	1968–1998	sp. inv. price ($/W peak)	cum. cap. (MW)	0.2	Harmon (2000)
Solar PV panels	US	1959–1974	sp. sale price ($/W peak)	cum. cap. (MW)	0.22	Maycock and Wakefield (1975)
Wind	Germany	1990–1998	specific investment price ($/kW)	cum. cap. (MW)	0.08	Durstewitz (1999)
Wind power	Denmark	1982–1997	sp. inv. price ($/kW)	cum. cap. (MW)	0.04	IEA (2000)
Wind power	EU	1980–1995	sp. prod. cost ($/kWh)	cum. prod. (TWh)	0.18	IEA (2000)
Wind power	Germany	1990–1998	sp. inv. price ($/kW)	cum. cap. (MW)	0.08	IEA (2000)
Wind power	US	1985–1994	sp. prod. cost ($/kWh)	cum. prod. (TWh)	0.32	IEA (2000)
Wind power	World	1992–2001	turnkey investment costs for UK and Spain	global installed cap. (MW)	0.15–0.18	Junginger et al. (2004)
Wind turbines	Denmark	1982–1997	specific investment price ($/kW)	cum. cap. (MW)	0.08	Neij (1999)
Wind turbines	Denmark	1981–2000	levelized production cost ($/kW)	cum. cap. produced (MW)	0.17	Neij et al. (2003)

Sources: Authors' summaries and interpretations as well as those by McDonald and Schrattenholzer (2001), Neij *et al.* (2003), and Junginger *et al.* (2004).

competitive with CC gas, and 16 times to be competitive with nuclear power.

A similar calculation could be done for solar power, but the point has been made. Despite optimistic assumptions about learning rates and economies of scale, wind and solar power will *never* achieve price parity with fossil fuels and nuclear power. They will reach the physical limits of their technologies long before their prices fall to that of competing fuels.

In 2008, the U.S. Department of Energy set a more realistic goal for wind + natural gas back-up to provide 20% of U.S. electricity needs (USDOE, 2008). Wind energy would achieve this with between one and two doublings, leaving its per-unit cost still between two and three times as high as power from coal, natural gas, and nuclear power. This finding contradicts the very optimistic forecasts of the wind industry, which were the basis for the USDOE report and parroted in the popular press.

The situation in countries other than the United States is different. Fossil fuels are not as abundant (often due to government policies, not differences in natural endowments) and international trade in electricity may substitute for the need for energy storage to offset the intermittency of wind and solar power. In many European countries, taxes are the largest part of the cost of energy, not production costs, so tax reform could keep energy affordable even as reliance on more expensive renewable fuels is increased. Issues of energy security also inform the international debate but are not addressed by the LCOE exercise.

References

EIA. 2018. U.S. Energy Information Administration. Frequently Asked Questions: What is U.S. electricity generation by energy source? (website). March 7.

Henderson, B.D. 1970. *The Product Portfolio: Perspectives*. Boston, MA: The Boston Consulting Group.

IEA. 2000. International Energy Agency. *Experience Curves for Energy Technology Policy*. Paris, France.

IPCC. 2007. Intergovernmental Panel on Climate Change. *Climate Change 2007: Synthesis Report*. Contribution of Working Groups I, II and III to the Fourth Assessment Report of the Intergovernmental Panel on Climate Change. Geneva, Switzerland: Intergovernmental Panel on Climate Change.

Junginger, M., Faaij, A., and Turkenburg, W. 2004. Cost reduction prospects for offshore wind farms. *Wind Engineering* **28** (1): 92–118.

McDonald, A. and Schrattenholzer, L. 2001. Learning rates for energy technologies. *Energy Policy* **29** (4): 255–61.

Neij, L., Anderson, P.D., Durstewitz, M., Helby, P., Hoppe-Kilpper, M., and Morthorst, P.E. 2003. *Experience curves: A tool for energy policy assessment*. Lund, Sweden: Environmental and Energy Systems Studies, Lund University.

Orr, I. and Palmer, F. 2018. How to prevent the premature retirement of coal-fired power plants. *Policy Study* No. 148. Arlington Heights, IL: The Heartland Institute. February.

Reed, S. 2017. Germany's shift to green power stalls, despite huge investments. *New York Times* (October 7).

Reuters. 2018. Chinese solar projects facing closure amid subsidy backlog: government report. October 12.

Stern, N. 2006. *Stern Review on the Economics of Climate Change*. Cambridge, UK: Cambridge University Press.

Upstill, G. and Hall, P. 2018. Estimating the learning rate of a technology with multiple variants: the case of carbon storage. *Energy Policy* **121**: 498–505.

USDOE. 2008. U.S. Department of Energy. *20% Wind Energy by 2030: Increasing Wind Energy's Contribution to U.S. Electricity Supply*. DOE/GO-102008-2567. May.

Wright, T.P. 1936. Factors affecting the cost of airplanes. *Journal of Aeronautical Sciences* **3** (4): 122–8.

3.6 Economic Value of Fossil Fuels

The late Julian Simon, perhaps the leading resource economist of his time, wrote in 1981,

> Energy is the master resource, because energy enables us to convert one material into another. As natural scientists continue to learn more about the transformation of materials from one form to another with the aid of energy, energy will be even more important. Therefore, if the cost of usable energy is low enough, all other important resources can be made plentiful, as H.E. Goeller and A.M. Weinberg showed. …
>
> On the other hand, if there were to be an absolute shortage of energy – that is, if there were no oil in the tanks, no natural gas in the pipelines, no coal to load onto the railroad cars – then the entire economy would come to a halt. Or if energy were available but only at a very high price, we would produce much smaller amounts of most consumer goods and services (p. 162).

More recently, Aucott and Hall (2014) write, "it was cheap energy that led to robust growth and made industrial economies rich. In our view it has been a serious failure on the part of traditional economics to consider the importance of energy only as related to its cost share rather than its absolute necessity, growth in use, and power to create the infrastructure and activities that support and drive industrial economies. In our view, the physical importance of energy makes it different from other commodities; the role of energy cannot be adequately equated strictly to traditional financial factors" (p. 6561). They add, "If the price of energy goes up, almost everything costs more, and this ripples through the economy. Fertilizer may be a useful analogy. Adding

50 kg of nitrogen per hectare can change the yield of corn by several tons per ha. This is because nitrogen is typically a 'limiting nutrient.' It may be that energy is the 'limiting nutrient' of the economy" (p. 6568).

Energy alone is not sufficient to create the conditions for prosperity, but it is absolutely necessary. It is impossible to operate a factory, run a store, grow crops, or deliver goods to consumers without using some form of energy. Access to reliable energy is particularly crucial to human development as electricity has, in practice, become indispensable for lighting, clean water and sanitation, refrigeration, and the running of household appliances.

Since fossil fuels provide 81% of the primary energy consumed in the world, its economic value must be considerable. Monetizing that value – expressing it in dollars, pounds, or euros – is not a simple task. There are many efforts reported in the literature, each covering different parts of the world, different time periods, or different fuels, and each with different assumptions leading to different conclusions. This section first documents the close association between the cost of energy and gross domestic product (GDP), then summarizes six studies illustrating six different methodologies, and finally offers a comparison of the estimates.

References

Aucott, M. and Hall, C. 2014. Does a change in price of fuel affect GDP growth? An examination of the U.S. data from 1950–2013. *Energies* 7: 6558–70.

Simon, J. (Ed.) 1981. *The Ultimate Resource*. Princeton, NJ: Princeton University Press.

3.6.1 Energy and GDP

Abundant and affordable energy supplies play a key role in enabling economic growth.

The job losses and price increases resulting from increased energy costs reduce incomes as firms, households, and governments spend more of their budgets on electricity and less on other items, such as home goods and services. Virtually all economists agree there is a negative relationship between energy price increases and economic activity, though there are differences of opinion in regard to the mechanisms through which price impacts are felt.

Following is a sample of informed opinion on the issue:

- "Economic growth in the past has been driven primarily not by 'technological progress' in some general and undefined sense, but specifically by the availability of ever cheaper energy – and useful work – from coal, petroleum, or gas" (Ayres and Warr, 2009).

- "The theoretical and empirical evidence indicates that energy use and output are tightly coupled, with energy availability playing a key role in enabling growth. Energy is important for growth because production is a function of capital, labor, and energy, not just the former two or just the latter as mainstream growth models or some biophysical production models taken literally would indicate" (Stern, 2010).

- "The bottom line is that an enormous increase in energy supply will be required to meet the demands of projected population growth and lift the developing world out of poverty without jeopardizing current standards of living in the most developed countries" (Brown *et al.*, 2011).

Aucott and Hall (2014) found "a threshold exists in the vicinity of 4%; if the percent of GDP spent on fuels is greater than this, poorer economic performance has been observed historically" (p. 6567). Bildirici and Kayikci (2012a, 2012b; 2013) found causal relationships between electricity consumption and economic growth in the Commonwealth of Independent States countries and in-transition countries in Europe. Lee and Lee (2010) analyzed the demand for energy and electricity in OECD countries and found a statistically valid relationship between electricity consumption and economic growth. Baumeister *et al.* (2010) examined the economic consequences of oil shocks across a set of industrialized countries over time and found energy costs and GDP are negatively correlated.

Blumel *et al.* (2009) used Chilean data to estimate the long-run impact of increased electricity and energy prices on the nation's economy. Kerschner and Hubacek (2008) reported significant correlations between energy and GDP in a study of the potential economic effects of peak oil, although they noted sectoral impacts are more significant. Sparrow (2008) analyzed the impacts of coal utilization in Indiana and estimated electricity costs significantly affect economic growth in the state.

Figure 3.6.1.1 presents three decades of rigorous research on the relationship between GDP and energy and electricity prices. Some studies looked only at oil prices, others at all types of energy, and some only at electricity prices. These studies support price-GDP elasticity estimates of about -0.17 for oil, -0.13 for electricity, -0.14 for all sources of energy, and -0.15 for all the studies in the table.

A price-GDP elasticity of -0.1 implies a 10% increase in the price, *ceteris paribus*, will result in a 1% decrease in GDP or, in the case of a state, Gross State Product (GSP). Thus, for example, the elasticity estimate for electricity of -0.13 means a 10% increase in the price of electricity in the United States results in a loss of approximately 1.3% of GDP, about $253 billion in 2017 (BEA, 2018). Estimates of the impacts of oil shocks and other energy price perturbations have produced different results, with smaller time-series econometric models producing energy price change-output elasticities of -2.5% to -11%, while large disaggregated macro models estimate much smaller impacts, in the range of -0.2% to -1.0% (Brown and Hunnington, 2010).

Knowing the energy price-GDP elasticity enables us to determine the impact of higher energy costs on human prosperity and the value of fossil fuels. Those calculations appear in the following sections.

References

Anderson, K.P. 1982. *Industrial Location and Electric Utility Price Competition.* New York, NY: National Economic Research Associates, Inc.

Aucott, M. and Hall, C. 2014. Does a change in price of fuel affect GDP growth? An examination of the U.S. data from 1950–2013. *Energies* 7: 6558–70; doi:10.3390/en7106558.

Ayres, R.U. and Warr, B. 2009. *The Economic Growth Engine: How Energy and Work Drive Material Prosperity.* Northampton, MA: Edward Elgar.

Baumeister, C., Peersman, G., and Van Robays, I. 2010. *The Economic Consequences of Oil Shocks: Differences Across Countries and Time.* Ghent, Belgium: Ghent University.

BEA. U.S. Bureau of Economic Analysis. 2018. Gross Domestic Product, 4th quarter and annual 2017 (advance estimate). *News Release.* Washington, DC: Bureau of Economic Analysis, U.S. Department of Commerce.

Bildirici, M. and Kayikci, F. 2012a. Economic growth and electricity consumption in former Soviet republics. *Energy Economics* 34 (3): 747–53.

Bildirici, M. and Kayikci, F. 2012b. Economic growth and electricity consumption in former Soviet republics. *IDEAS.* St. Louis, MO: Federal Reserve Bank of St. Louis.

Bildirici, M. and Kayikci, F. 2013. Economic growth and electricity consumption in emerging countries of Europa: an ARDL analysis. *Economic Research - Ekonomska Istrazivanja* 25 (3): 538–59.

Blumel, G., Espinoza, R.A., and de la Luz Domper, G.M. 2009. Does energy cost affect long run economic growth? Time series evidence using Chilean data. Las Condes, Chile: Instituto Libertad y Desarrollo Facultad de Ingeniería, Universidad de los Andes. March 22.

Brown, J.H., *et al.* 2011. Energetic limits to economic growth. *BioScience* 61 (1).

Brown, S.P.A. and Huntington, H.G. 2010. *Estimating U.S. Oil Security Premiums.* Washington, DC: Resources for the Future. June.

Brown, S.P.A. and Yucel, M.K. 1999. Oil prices and U.S. aggregate economic activity: a question of neutrality. *Economic and Financial Review.* Second Quarter. Dallas, TX: Federal Reserve Bank of Dallas.

Considine, T. 2006. *Coal: America's Energy Future.* Volume II. Appendix: Economic Benefits of Coal Conversion Investments. Prepared for the National Coal Council. March.

Deloitte Consulting Pty (Ltd). 2017. An overview of electricity consumption and pricing in South Africa. Report prepared for Eskom Holdings SOC Ltd. February 24.

Global Insight. 2006. *The Impact of Energy Price Shocks on the UK Economy: A Report to the Department of Trade and Industry.* May 18.

Hewson, T. and Stamberg, J. 1996. *At What Cost? Manufacturing Employment Impacts from Higher Electricity Prices.* Arlington, VA: Energy Ventures Analysis.

Hooker, M.A. 1996. What happened to the oil price-macroeconomy relationship? *Journal of Monetary Economics* 38: 195–213.

Hu, Y. and Wang, S.Y. 2015. The relationship between energy consumption and economic growth: evidence from China's industrial sectors. *Energies* 8: 9392–406.

Huntington, H.G., Barrios, J.J., and Arora, V. 2017. Review of key international demand elasticities for major

Figure 3.6.1.1
Summary of energy- and electricity price-GDP elasticity estimates

Year Analysis Published	Author	Type of Energy	Elasticity Estimate
2017	Deloitte Consulting Pty (Ltd).	Electricity	~ -0.1
2017	Huntington, Barrios, and Arora	Energy	-0.024 to -0.17
2017	Lu, Wen-Cheng	Electricity	-0.07
2015	Hu & Wang	Energy	-.087 to -0.10
2011	Inglesi-Lotz and Blignaut	Electricity	-1.10
2010	Lee and Lee	Energy and electricity	-0.01 and -0.19
2010	Brown and Huntington	Oil	-0.01 to -0.08
2010	Baumeister, Peersman, and Robays	Oil	-0.35
2009	Blumel, Espinoza, and de la Luz Domper	Energy and electricity	-0.085 to -0.16
2008	Kerschner and Hubacek	Oil	-0.03 to -0.17
2008	Sparrow	Electricity	-0.3
2007	Maeda	Energy	-0.03 to -0.075
2007	Krishna Rao	Energy	-0.3 to -0.37
2007	Lescaroux	Oil	-0.1 to -0.6
2006	Rose and Wei	Electricity	-0.1
2006	Oxford Economic Forecasting	Energy	-0.03 to -0.07
2006	Considine	Electricity	-0.3
2006	Global Insight	Energy	-0.04
2004	IEA	Oil	-0.08 to -0.13
2002	Rose and Yang	Electricity	-0.14
2002	Klein and Kenny	Electricity	-0.06 to -0.13
2001	Rose and Ranjan	Electricity	-0.14
2001	Rose and Ranjan	Energy	-0.05 to -0.25
1999	Brown and Yucel	Oil	-0.05
1996	Hewson and Stamberg	Electricity	-0.14
1996	Rotemberg and Woodford	Energy	-0.25
1996	Joutz and Gardner	Energy	-0.072
1996	Hewson and Stamberg	Electricity	-0.5 and -0.7
1996	Hooker	Energy	-0.07 to -0.29
1995	Lee, Ni, and Ratti	Oil	-0.14
1982	Anderson	Electricity	-0.14
1981	Rasche and Tatom	Energy	-0.05 to -0.11

Sources: See References for citations. Authors' interpretation of study results. This table necessarily over-simplifies some complicated findings and leaves out caveats and findings unrelated to the present purpose.

industrializing economies. U.S. Energy Information Administration. December.

IEA. 2004. International Energy Agency. Analysis of the impact of high oil prices on the global economy. Paris, May.

Inglesi-Lotz, R. and Blignaut, J.N. 2011. South Africa's electricity consumption: a sectoral decomposition analysis. *Applied Energy* **88** (12): 4779–84.

Joutz, F. and Gardner, T. 1996. Economic growth, energy prices, and technological innovation. *Southern Economic Journal* **62** (3): 653–66.

Kerschner, C. and Hubacek, K. 2008. Assessing the suitability of input-output analysis for enhancing our understanding of potential economic effects of peak-oil. Leeds, UK: Sustainability Research Institute, School of Earth and Environment, University of Leeds.

Klein, D. and Kenny, R. 2002. *Mortality Reductions from Use of Low-cost Coal-fueled Power: An Analytical Framework*. Mclean, VA: Twenty-First Century Strategies and Duke University. December.

Krishna Rao, P.V. 2007. *Surviving in a World with High Energy Prices*. New York, NY: Citigroup Energy Inc. September 19.

Lee, C-C. and Lee, J-D. 2010. A panel data analysis of the demand for total energy and electricity in OECD countries. *The Energy Journal* **31** (1): 1–23.

Lee, K., Ni, S., and Ratti, R.A. 1995. Oil shocks and the macroeconomy: the role of price variability. *Energy Journal* **16:** 39–56.

Lescaroux, F. 2007. An interpretative survey of oil price-GDP elasticities. *Oil & Gas Science and Technology* **62** (5): 663–71.

Lu,W-C. 2017. Electricity consumption and economic growth: evidence from 17 Taiwanese industries. *Sustainability* **9:** 50.

Maeda, A. 2007. On the world energy price-GDP relationship. Presented at the 27th USAEE/IAEE North American Conference, Houston, Texas, September 16–19.

Oxford Economic Forecasting. 2006. DTI energy price scenarios in the Oxford models. May.

Rasche, R.H. and Tatom, J.A. 1981. Energy price shocks, aggregate supply, and monetary policy: the theory and international evidence. In: Brunner, K. and Meltzer, A.H. (Eds.) *Supply Shocks, Incentives, and National Wealth*. Carnegie-Rochester Conference Series on Public Policy No. 14. Amsterdam, Netherlands: North-Holland Publishing Co.

Rose, A. and Ranjan, R. 2001. *The Economic Impact of Coal Utilization in Wisconsin*. State College, PA: Pennsylvania State University, Department of Energy, Environmental, and Mineral Economics. August.

Rose, A. and Wei, D. 2006. *The Economic Impacts of Coal Utilization and Displacement in the Continental United States, 2015*. Report prepared for the Center for Energy and Economic Development, Inc. State College, PA: Pennsylvania State University. July.

Rose, A. and Yang, B. 2002. *The Economic Impact of Coal Utilization in the Continental United States*. Alexandria, VA: Center for Energy and Economic Development.

Rotemberg, J.J. and Woodford, M. 1996. Imperfect competition and the effects of energy price increases on the economy. *Journal of Money, Credit, and Banking* **28** (4): 550–77.

Sparrow, F.T. 2008. Measuring the contribution of coal to Indiana's economy. *CCTR Briefing*. West Lafayette, IN: Center for Coal Technology Research, Purdue University. December 12.

Stern, D.I. 2010. *The Role of Energy in Economic Growth*. The United States Association for Energy Economics and the International Association for Energy Economics. USAEE-IAEE WP 10-055. November.

3.6.2 Estimates of Economic Value

Estimates of the value of fossil fuels vary but converge on very high numbers. Coal alone delivered economic benefits worth between $1.3 trillion and $1.8 trillion of U.S. GDP in 2015.

Reducing global reliance on fossil fuels by 80% by 2050 would probably reduce global GDP by $137.5 trillion from baseline projections.

There are at least six ways to calculate the past and present economic value of fossil fuels:

1. *Comparing LCOEs:* Estimates of the levelized cost of electricity (LCOE) can be combined with energy price-GDP elasticity to estimate the cost of replacing fossil fuels with alternative fuels.

2. *Existence value:* The existence value of fossil fuels is the value of economic activity

specifically attributable to their low cost relative to alternative fuels.

3. *Historical relationships:* The historical relationships between electricity costs or per-capita energy consumption, on the one hand, and GDP and other measures of prosperity on the other hand, can be used to forecast the cost of switching to higher-priced and less-abundant alternative energies.

4. *Bottom-up estimates:* Bottom-up calculations use data concerning the cost of existing and new production capacity and transmission infrastructure, premature retirement of existing resources, and economic models to estimate the incremental cost of reducing reliance on fossil fuels.

5. *Macroeconomic models:* Specific policies designed to reduce the use of fossil fuels are entered into macroeconomic models to produce estimates of their impacts on GDP, employment, economic growth, and more.

6. *Modeled as a tax increase:* Since increases in energy costs have effects similar to tax hikes, proposals such as the Obama administration's Clean Power Plan can be treated as though they were taxes on carbon dioxide emissions and entered into macroeconomic models as such.

In most of these methodologies, the economic value of fossil fuels appears as the *cost avoided* by not shifting to higher-priced, less-reliable forms of energy. This section describes a single example of each of these six methodologies and then compares the results of all six studies. There are many more studies than the ones summarized here, but the ones chosen are authoritative or representative of the findings of others.

3.6.2.1 Comparing LCOEs

Combining the levelized cost of electricity (LCOE) for electricity produced by coal and alternative energies reported in Section 3.5.4 with the electricity-GDP elasticity estimate reported in Section 3.6.1 allows us to estimate the present value, measured in GDP, of the current level of reliance on fossil fuels by the United States. The same technique allows us to consider a number of scenarios under which some or all of the electricity currently generated by coal is replaced by wind energy, the most cost-competitive alternative energy other than nuclear power. The results of such an analysis are reported in Figure 3.6.2.1.1.

Figure 3.6.2.1.1 offers stylized facts answering the question, "Given the current cost differences between existing coal resources and new wind energy resources, how would replacing some or all of current coal powered generators with wind turbines affect the price of electricity, and how would that affect GDP?" Another way of framing the question is, "How much do consumers *benefit* from relying on coal rather than wind to produce the electricity they use?" This is not a forecast of the actual cost of converting from coal to wind energy since (a) replacing 100% of coal generation with wind power is not physically possible, (b) such a conversion could not take place in a single year, (c) the length of time allowed to retire existing coal resources and build new wind energy would substantially affect the cost of such a conversion, and (d) the methodology assumes no changes in the output and cost of other energy sources (nuclear, hydro, and others) that would be strongly affected by an overall increase in the cost of electricity. This is also not the only way to calculate an existence value. Another way, illustrated by a study by Rose and Wei (2006), is presented in Section 3.6.2.2 below.

It is also important to note the cost of replacing coal with wind is likely to be more logarithmic than what the table shows. New wind tends to be increasingly expensive as the best sites have been selected already and major expansions of wind capacity would likely require positioning wind turbines offshore, which is considerably more expensive than onshore installations. Costs due to the intermittency of wind power escalate as its market penetration rises and sectors that rely on continuous high-quality energy must somehow accommodate intermittent power instead.

Figure 3.6.2.1.1
Cost of replacing coal power with wind energy in the United States, LCOE method

Generator Type	LCOE/ MWh [1]	% of production (2015) [2]	Current annual Output (TWh) [2]	Annual Cost (millions)	Replace 20% of coal power with wind energy (millions)	Replace 40% of coal power with wind energy (millions)	Replace 100% of coal with wind energy (millions)
Coal (existing)	$39.90	39.0%	1,596	$63,680	($12,738)	($25,476)	($61,297)
Wind (new)*	$107.4	4.4%	180	--	$34,288	$68,576	$171,439
Other	--	56.6%	2,317	--	--	--	--
Totals	--	100.0%	4,093	--	--	--	--
Net Cost	--	--	--	--	$21,550	$43,099	$110,142
% Change in Cost	--	--	--	--	33.83%	67.67%	172.93%
% Change in COE [3]	--	--	--	--	13.20%	26.39%	67.44%
Loss of GDP [4]	--	--	--	--	($307,059)	($614,118)	($1,569,409)
% of GDP lost	--	--	--	--	(1.72%)	(3.43%)	(8.77%)

Notes and sources:

* Including cost imposed on CC gas for back-up power generation.
[1] Stacy and Taylor, 2016.
[2] EIA, 2015.
[3] Coal's % of production x % change in cost = change in average cost of electricity (COE).
[4] Best estimate of electricity price-GDP elasticity in the United States of -0.13, from Figure 3.7.1.1, x 2015 GDP estimate of $17.9 trillion from BEA, 2015.

With these caveats in mind, the numbers in Figure 3.6.2.1.1 allow us to make the following statements:

- Coal in 2015 provided 39% of U.S. electricity (1,596 TWh in 2015) at a levelized cost of approximately $64 billion.

- If 20% of the power generated by coal were generated instead by wind with natural gas back-up generation, the annual net cost would increase by $22 billion and electricity prices would rise 13%, causing a loss of GDP of approximately $307 billion, 1.72% of U.S. GDP in 2015.

- If 40% of the power generated by coal were generated instead by wind + gas back-up, the annual net cost would increase by $43 billion and electricity prices would rise 26%, causing a loss of GDP of approximately $614 billion, 3.43% of U.S. GDP in 2015.

- If it were physically possible for 100% of the power generated by coal to be generated instead by wind + gas back-up, the annual net cost would increase by $110 billion and electricity prices would rise 67%, causing a loss of GDP of approximately $1.6 trillion, 8.77% of GDP in 2015.

These are enormous costs. Coal today, compared to the next best alternative energy (other than nuclear power), provides a direct annual benefit in the United States of about $110 billion, and by lowering electricity rates it increases GDP by approximately $1.6 trillion a year, about 9% of total U.S. GDP. The results would be similar if the comparison used natural gas rather than coal, since their LCOEs are similar ($39.9 for coal and $34.4 for natural gas). The LCOE of solar PV cells is 30% higher than wind power ($140.3 versus $107.4) so substituting solar for fossil fuels would cost even more.

As discussed in Section 3.5, the learning rate and economies of scale for wind energy would reduce this cost if coal were phased out only as wind energy became economically competitive on a non-subsidized basis, but this would cap wind's penetration at about 10% or less of U.S. electricity needs, a level it has already achieved. This is not the proposal made by the United Nations' Intergovernmental Panel on Climate Change (IPCC) or its followers. The analysis in Section 3.5 shows wind energy is so far from being cost-competitive with coal that even optimistic forecasts of costs falling 16% relative to coal with every doubling of output would mean wind would hit the physical limits of its production capacity and the country's need for electricity long before its cost fell to the level of coal or natural gas. This means a forced transition from affordable fossil fuels to alternative energies would impose considerable costs, resulting in lost income and slower economic growth.

Our analysis confirms the concerns expressed by many experts about the impact of anti-coal policies pursued by the Obama administration. For example, Ann Norman, a senior research fellow with the National Center for Policy Analysis, wrote in 2014: "Losing coal would not be as much of a problem if we had a cost-effective, large-scale energy alternative available. But the environmentalist left will not touch nuclear power (an energy source that produces no carbon emissions), and renewables are unreliable and expensive, hardly suited to replace coal" (Norman, 2014).

References

BEA. 2015. U.S. Bureau of Economic Analysis. National income and product accounts, Gross Domestic Product: second quarter 2015 (third estimate) corporate profits: second quarter 2015 (revised estimate). *News Release.* Washington, DC: Bureau of Economic Analysis, U.S. Department of Commerce.

EIA. 2015. U.S. Energy Information Administration. What is U.S. electricity generation by energy source? (website).

Norman, A. 2014. Power grid reliability as coal plants retire. NCPA: Energy and Environment: Clearing the Air (blog). April 23.

Rose, A. and Wei, D. 2006. *The Economic Impacts of Coal Utilization and Displacement in the Continental United States, 2015.* Report prepared for the Center for Energy and Economic Development, Inc. State College, PA: Pennsylvania State University. July.

Stacy, T.F. and Taylor, G.S. 2016. *The Levelized Cost of Electricity from Existing Generation Resources.* Washington, DC: Institute for Energy Research.

3.6.2.2 Existence Values

Rose and Wei (2006) conducted a sophisticated analysis of the existence impacts of coal-fueled electricity generation and the likely impact on GDP, household income, and employment of displacing 33% and 66% of projected coal generation by alternative energy resources over a 10-year period beginning in 2006 and ending in 2015. Their analysis took into account the positive economic effects associated with alternative investments in oil, natural gas, nuclear, and renewable energy supplies.

A method of capturing the locational attractiveness of a good or service is not to claim the entirety of output of its direct and indirect users, but only an amount relating to the price advantage of the input over its competitors. Rose and Wei calculated a price differential between coal and alternative fuels in electricity production, then estimated how much economic activity was attributable to this cost saving. They used an economy-wide elasticity of output with respect to energy prices of -0.10, meaning the availability of coal-fueled electricity at a price 10% lower than that of its nearest competitor is responsible for increasing total state or regional economic activity by 1.0%.

Explaining the implications of this methodology, the authors write, "Essentially, we are measuring the economic activity attributable to relatively cheaper coal in contrast to what would take place if a state were dependent on more expensive alternatives, which we assume would be a combination of oil/gas, renewable, and nuclear electricity" (Rose and Wei, 2006, p. 13).

The authors first estimated the level of coal-based electricity generation in each of the lower-48 states in 2015 based on projections made in 2006 by the EIA and EPA. They used IMPLAN input-output tables to estimate the direct and indirect (multiplier) economic output, household income, and jobs created by coal-fueled electricity generation in each state. Two estimates were produced: (1) upper-range ("high") prices for coal substitutes (nuclear, natural gas, and renewables) and (2) a lower-range ("low") price substitutes scenario. The authors' findings are summarized in Figure 3.6.2.2.1. They summarized their findings (depicted in the gray-shaded three cells) as follows: "Our analysis shows that, in 2015, U.S. coal production, transportation and consumption for electric power generation will contribute more than $1 trillion (2005 $) of gross output directly and indirectly to the economy of the lower-48 United States. Based on an average of two energy price scenarios … we calculate that $362 billion of household income and 6.8 million U.S. jobs will be attributable to the production, transportation and use of domestic coal to meet the nation's electric generation needs" (p. 4).

The authors then evaluated the impacts of two scenarios in which alternative energies replaced 66% and 33% of coal generation over the ten-year period from 2006 to 2015. The found "the average impacts of displacing 66% of coal-fueled generation in 2015 [would be a] $371 billion (2005 $) reduction in gross economic output; $142 billion reduction of annual household incomes; and 2.7 million job losses" (p. 4). The average impacts of displacing 33% of coal-based generation in 2015 were estimated to be "$166 billion (2005 $) reduction in gross economic output; $64 billion reduction of annual household incomes;

Figure 3.6.2.2.1
Regional summary of the "existence" value of U.S. coal utilization in electric generation, 2015 (in billions of 2005 dollars and millions of jobs)

Region	High-Price Alternatives	Low-Price Alternatives	Average
Southeast			
Output	$309	$166	$238
Earnings	$106	$55	$80
Jobs	2.2	1.1	1.6
Northeast			
Output	$145	$65	$105
Earnings	$56	$24	$40
Jobs	0.9	0.4	0.6
Midwest			
Output	$409	$199	$304
Earnings	$137	$65	$101
Jobs	2.4	1.2	1.8
Central			
Output	$305	$149	$227
Earnings	$106	$50	$78
Jobs	2.1	1.0	1.5
West			
Output	$213	$135	$174
Earnings	$78	$48	$63
Jobs	1.5	0.9	1.2
48 States			
Output	$1,381	$714	$1,047
Earnings	$482	$242	$362
Jobs	9.0	4.6	6.8

Source: Rose and Wei (2006), Table S-I, p. 6.

and 1.2 million job losses" (p. 5). Using an inflation calculator we can convert Rose and Wei's 2005 $ to 2015 $, making them comparable to the LCOE-derived estimates reported in Section 3.6.2. Here are the results:

- Coal in 2015 will contribute $1,275 billion of GDP directly and indirectly to the economy of the lower-48 United States and $445 billion of household income.

- If 33% of the power generated by coal were generated instead by gas, nuclear power, and renewables, GDP would decline by $204 billion, annual household incomes would fall by $79 billion, and 1.2 million jobs would be lost.

- If 66% of the power generated by coal were generated instead by alternatives, GDP would fall $456 billion, annual household incomes would drop by $175 billion, and 2.7 million jobs would be lost.

Rose and Wei's estimates are lower than what the LCOE exercise reported in Section 3.6.2.1 would have found for 33% and 66% substitution scenarios. There are many reasons for the difference: use of a lower energy cost-GDP elasticity rate; this study's time frame (10 years) versus the static analysis in Section 3.6.2.1; the exclusion of Alaska and Hawaii; including natural gas and nuclear power as possible substitutes for coal; and possibly not including the added burden of wind and solar power on natural gas power generators producing back-up power.

Reference

Rose, A. and Wei, D. 2006. *The Economic Impacts of Coal Utilization and Displacement in the Continental United States, 2015.* Report prepared for the Center for Energy and Economic Development, Inc. State College, PA: Pennsylvania State University. July.

3.6.2.3 Historical Relationships

Section 3.6.1 reviewed the extensive literature on the close relationships between energy prices and GDP. That literature enables us to forecast the impact on GDP of rising energy prices due to substitution by renewables assuming past correlations continue. An example of this methodology applied to global rather than U.S.-only energy is Tverberg (2012). Using estimates of energy consumption by Vaclav Smil, BP, EIA, and other sources, Tverberg plotted a plausible scenario in which governments force fossil fuel consumption to fall by 80% by 2050, the target endorsed by the European Union (EU). She projects non-fossil fuel power sources to rise more rapidly than their historical rate but not fast enough to offset the loss of fossil fuel power, requiring a decrease in global energy consumption of 50%. She then divides energy consumption by global population estimates and forecasts from the United Nations to estimate actual and projected per-capita energy consumption over the period. The results are shown in Figure 3.6.2.3.1.

Next, Tverberg created a database of the annual rate of change in global energy consumption, population, and GDP for 11 multi-year periods since 1920 relying "on population and GDP estimates of Angus Maddison and energy estimates of Vaclav Smil, supplemented by more recent data (mostly for 2008 to 2010) by BP, the EIA, and USDA Economic Research Service." Tverberg applied regression analysis to the data and found a 10% increase in per-capita energy consumption correlates with an 8.9% increase in per-capita GDP. Applying this finding to her scenario of a 50% reduction in energy consumption by 2050 created what Tverberg called a "a best-case estimate of future GDP if a decrease in energy supply of the magnitude shown were to take place." Her results are sobering:

- World per-capita energy consumption in 2050 would fall to what it was in 1905.

- Global per-capita GDP would decline by 42% from its 2010 level.

- Global GDP would be some $137.5 trillion (in 2015$) less in 2050 than baseline projections.

- The average global economic growth rate from 2012 to 2050 would be -0.59%.

A common baseline forecast for annual global GDP growth is 3% (PricewaterhouseCoopers LLP (2015). Tverberg's forecast could thus be stylized as an annual loss of 3.59% GDP from what it otherwise would have been. Since world GDP was approximately $74.4 trillion in 2015, the loss that year would have been $2.67 trillion.

Figure 3.6.2.3.1

A forecast of global per-capita energy consumption assuming 80% reduction in fossil fuel consumption by 2050

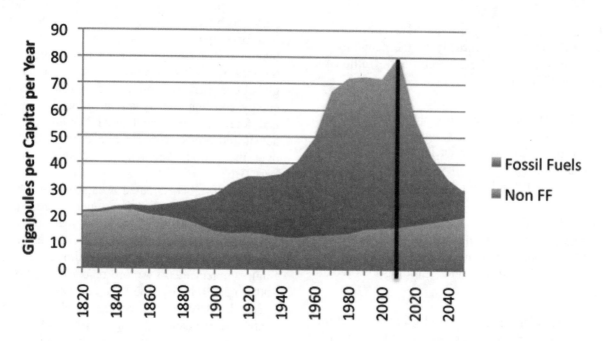

Assuming governments force consumption of fossil fuels to decrease by 80% by 2050, the goal set by the EU, and use of non-fossil fuels increases but not by enough to make up the entire gap, reducing total energy consumption by 50%. Amounts before the black vertical line are actual; after the black line are forecast in this scenario. *Source*: Tverberg, 2012.

Tverberg contends her estimates represent an optimistic "best-case" scenario since "the issue of whether we can really continue transitioning to a service economy when much less fuel in total is available is also debatable. If people are poorer, they will cut back on discretionary items. Many goods are necessities: Food, clothing, basic transportation. Services tend to be more optional – getting one's hair cut more frequently, attending additional years at a university, or sending grandma to an Assisted Living Center. So the direction for the future may be toward a mix that includes fewer, rather than more, services, and so will be more energy intensive" (*Ibid.*).

"If our per capita energy consumption drops to the level it was in 1905," Tverberg wrote, "can we realistically expect to have robust international trade, and will other systems hold together? While it is easy to make estimates that make the transition sound easy, when a person looks at the historical data, making the transition to using less fuel looks quite difficult, even in a best-case scenario." She concludes

that such a worldwide reduction in reliance on fossil fuels is "very unlikely" (*Ibid.*).

References

PricewaterhouseCoopers LLP. 2015. The World in 2050: Will the shift in global economic power continue? London, UK.

Tverberg, G. 2012. An energy/GDP forecast to 2050. Our Finite World (website). July 26.

3.6.2.4 Bottom-up Estimates

In 2014, the U.S. Chamber of Commerce commissioned IHS Energy and IHS Economics, a global consulting firm, to produce a bottom-up estimate of the incremental cost of the Obama administration's proposal "to reduce gross U.S. greenhouse gas emissions by 42% below the 2005 level by 2030 (as stated in the administration's 2010 submission to the UN Framework Convention on Climate Change associating the U.S. with the Copenhagen Accord)" (U.S. Chamber of Commerce, 2014, p. 3). The administration's proposal eventually became the Clean Power Plan, a regulation with somewhat different targets, so the Chamber's analysis is not directly applicable to that regulation.

The authors establish a "Reference Case" describing energy trends expected in the absence of the administration's proposal. In the Reference Case, natural gas expands its market share due to "technological advancements in drilling techniques, a resulting reduction in unit production costs, and an expanded domestic resource base estimated at 3,400 trillion cubic feet (Tcf) – enough to supply demand at current levels for more than 100 years. …" (p. 20).

Also in the Reference Case, "generator retirements from 2011 through 2030 total 154 gigawatts (GW), with 85 GW of coal-fired power plants retiring in this time frame" (p. 17), reducing coal's share of electricity production from about 40% in 2013 to about 29% in 2030. U.S. total energy and electricity demand are forecast to grow an average of 1.4% per year and U.S. GDP is projected to grow 2.5% per year. The authors note, "prior to the mid-1980s, electricity demand grew more quickly than GDP; during the 1960s, electric demand grew twice as fast as GDP. Since 1980, electricity demand has grown more slowly than GDP. During the previous decade, for every 1% increase in GDP, electricity demand grew roughly 0.6%" (p. 19). They attribute the "progressively weaker relationship" to "the countervailing effect of rising retail electricity prices and a continued strong emphasis on energy efficiency policies at both the U.S. federal and state levels" (*Ibid.*).

The authors create a "Policy Case" to forecast the impacts of the proposal to reduce gross U.S. greenhouse gas emissions by 42% below the 2005 level by 2030. They predict "baseload coal plant retirements would jump sharply in the Policy Case, with an additional 114 gigawatts – about 40% of existing capacity – being shut down by 2030 compared with the Reference Case" (p. 3). They then do a bottom-up calculation of the incremental cost of such a forced transition away from coal, including "the costs for new incremental generating capacity, necessary infrastructure (transmission lines and natural gas and CO_2 pipelines), [and] decommissioning stranded asset costs.…" (p. 4). These costs are partially offset by lower fuel use and

operation and maintenance expenses incurred by coal-fired electricity producers. Their findings are summarized in Figure 3.6.2.4.1.

Figure 3.6.2.4.1
Cost to reduce U.S. greenhouse gas emissions by 42% below the 2005 level by 2030

Incremental cost item	Incremental cost (billions)
Power plant construction	$339
Electric transmission	$16
Natural gas infrastructure	$23
CCS pipelines	$25
Coal plant decommissioning	$8
Coal unit efficiency upgrades	$3
Coal unit stranded costs	$30
Demand-side energy efficiency	$106
Operations and maintenance costs	-$5
Fuel costs	-$66
Total incremental costs	$478

Source: U.S. Chamber of Commerce, 2014, Table ES-1, p. 4. Note in the original reads: "Please see Appendix C for power generation addition unit costs and more detail on the calculation of natural gas pipelines, transmission, CCS pipelines, coal plant decommissioning, and coal unit stranded assets."

As shown in Figure 3.6.2.4.1 the incremental cost of reducing existing coal generation output by 40%, net of savings, would be $478 billion (in constant 2012 dollars) by 2030. Most of the costs incurred by the coal industry would be passed on to consumers in the form of higher electricity rates. The authors estimate "the Policy Case will cause U.S. consumers to pay nearly $290 billion more for electricity between 2014 and 2030, or an average of $17 billion more per year" (p. 5). Average annual GDP during the same period is projected to be $51 billion less than in the Reference Case, "with a peak decline of nearly $104 billion in 2025" (p. 7). Average annual employment would be an average of 224,000 fewer per year, "with a peak decline in employment of 442,000 jobs in 2022."

The U.S. Chamber of Commerce's estimate of

the cost of reducing reliance on coal by 42% below the 2005 level by 2030, the pledge made by the Obama administration in 2010, $478 billion, is not far from what the less sophisticated LCOE methodology used in Section 3.6.2.1 would predict ($644 billion, not shown in Figure 3.6.2.1.1). The major difference is likely that the LCOE methodology assumed coal would be replaced by wind power and not a combination of wind power, gas, and other less-expensive alternatives.

Reference

U.S. Chamber of Commerce. 2014. *Assessing the Impact of Potential New Carbon Regulations in the United States.* Institute for 21st Century Energy. Washington, DC: U.S. Chamber of Commerce.

3.6.2.5 Macroeconomic Models

In 2014, the U.S. Energy Information Administration (EIA, 2015a) used a macroeconomic model to estimate the cost of the Clean Power Plan (CPP), a regulation setting targets for reductions in greenhouse gas emissions associated with electrical generation to 25% below 2005 levels by 2020 and 30% below by 2030. The regulation has since been rescinded, but the analysis of the regulation provides an example of the use of macroeconomic models to forecast the cost of reducing reliance on fossil fuels. The report used EIA's National Energy Modeling System (NEMS), "a modular economic modeling system used by EIA to develop long-term projections of the U.S. energy sector, currently through the year 2040." Data on existing energy costs, supply, and demand were taken from the *Annual Energy Outlook 2015,* a database maintained by EIA (EIA, 2015b).

The modeling exercise is complicated because CPP relied on states to implement emission reduction programs and gives them some latitude in the choice of tools to achieve the reductions. States did not need to begin to reduce CO_2 emissions until 2020 and were expected to reach performance goals, measured in pounds of CO_2 emitted per megawatthour of electricity generated from affected electric generating units, by 2030. Emission reduction targets and methods approved by the EPA to attain them are described in an EPA document titled "Best System of Emissions Reduction (BSER)." Those methods, which the EPA calls "building blocks," are:

1. Improving the thermal efficiency of individual affected sources (heat rate improvement);

2. Dispatching the generating fleet to substitute less-carbon-intensive affected sources for more-carbon-intensive affected sources (re-dispatch for reduced emissions);

3. Expanding the use of low- or zero-carbon generation in order to displace affected sources (low- and zero-carbon capacity expansion).

The modeling exercise is further complicated because Congress asked for consideration of nine scenarios (e.g., extension of the Clean Power Plan targets beyond 2030, treatment of future nuclear capacity similar to the treatment of renewable capacity, and sensitivities for expenditures and effectiveness of energy efficiency programs). It is not necessary for our purposes to review all of EIA's findings. Instead, we focus on what EIA calls its "Base Policy case." The agency found:

> Increased investment in new generating capacity as well as increased use of natural gas for generation lead to electricity prices that are 3% to 7% higher on average from 2020–25 in the Clean Power Plan cases, versus the respective baseline cases. … While prices return to near-baseline levels by 2030 in many regions, prices remain at elevated levels in some parts of the country. … Economic activity indicators, including real gross domestic product (GDP), industrial shipments, and consumption, are reduced relative to baseline under the Clean Power Plan. Across cases that start from the AEO2015 Reference case, the reduction in cumulative GDP over 2015–40 ranges from 0.17%–0.25%, with the high end reflecting a tighter policy beyond 2030.

EIA seemed to trivialize the impact of CPP on GDP in its summary by focusing on early years, before the costs are significant, and later years, when it claims (implausibly) that technological advances will lower the cost of alternative energies to below the cost of fossil fuels. Most other researchers use an electricity-GDP elasticity of around -0.13 and would say an increase in electricity prices of 3% to 7% would reduce annual GDP between 0.39% and 0.91% below baseline forecasts, between two and four times EIA's estimate, about $69.8 billion to $162.9 billion a year using an estimated U.S. 2015 GDP of $17.9 trillion.

Kevin D. Dayaratna, senior statistician and research programmer at The Heritage Foundation (Dayaratna, 2015), "unpacked" EIA's findings and produced a series of tables presenting the *annual* impact of CPP on manufacturing employment, overall employment, GDP, annual income for a family of four, and annual household electricity expenditures for four of EIA's nine cases. Two of his tables, for impacts on GDP and overall employment, appear below as Figures 3.6.2.5.1 and 3.6.2.5.2.

Dayaratna's tables show EIA's analysis found significant costs, but still less than previous methodologies would predict: more than $100 billion a year in GDP is lost in each of eight years (2021–2028), cumulative GDP loss amounts to $25 trillion to $30 trillion, and job losses total more than 100,000 in nine years (2021–2029). EIA's forecast of positive impacts beginning in 2031 are counterintuitive, to say the least, given the low learning rates and physical limitations confronting solar and wind power and the almost certain decrease in energy consumption due to the inability of alternatives to meet population-driven rising electricity demand.

One reason EIA's analysis finds relatively low costs is because it accepts the U.S. Environmental Protection Agency's assumption that energy conservation efforts will reduce the rate of increase in electricity consumption below historical levels without imposing costs on consumers. That assumption has been severely criticized by the Electric Reliability Coordinating Council (ERCC), a group of energy companies:

> There is no doubt that our economy is becoming more energy efficient, but EPA's claims about future improvements are simply wishful thinking. We are not aware of any serious analysis showing, as EPA claims, that it will save you money by increasing your electricity rates. The efficiency promises made by environmentalist groups such as the NRDC [who have led the call for this regulatory proposal] are beyond what any state, no matter how green, has achieved and are wholly unrealistic. Further, the economy remains in doldrums, with growth stunted over the last five years. If economic recovery picks up – which the Administration believes is likely – counting on appreciably less energy use will not be an option. What happens if policies rely on energy efficiency beyond what is viable given economic conditions? The result is energy rationing (ERCC, 2014).

Economic growth in the United States did in fact increase after 2014, validating the ERCC's concern.

References

Dayaratna, K.D. 2015. The economic impact of the Clean Power Plan. Testimony before the U.S. House Committee on Science, Space, and Technology. June 24.

EIA. 2015a. U.S. Energy Information Administration. Analysis of the impacts of the Clean Power Plan. May.

EIA. 2015b. U.S. Energy Information Administration. *Annual Energy Outlook 2015.* Washington, DC: U.S. Department of Energy.

ERCC. 2014. Electric Reliability Coordinating Council. ERCC answers seven questions you should have about EPA's proposed rule on carbon emissions from existing power plants (website). June 2.

Figure 3.6.2.5.1
Impact of Clean Power Plan on U.S. GDP, 2015–2040

Year	Clean Power Plan (CPP)	CPP Policy Extension	CPP Policy with New Nuclear	CPP Policy with Biomass CO^2
2015	-$226,562,000	-$3,906,000	-$31,250,000	-$224,609,000
2016	-$3,052,735,000	-$4,857,422,000	-$2,546,875,000	-$3,001,954,000
2017	-$4,068,360,000	-$7,039,063,000	-$3,580,078,000	-$4,035,157,000
2018	-$3,375,000,000	-$5,656,250,000	-$4,134,766,000	-$3,210,937,000
2019	-$2,232,422,000	-$7,986,329,000	-$2,347,657,000	-$3,605,469,000
2020	-$61,351,563,000	-$68,333,985,000	-$57,271,485,000	-$69,414,063,000
2021	-$116,789,062,000	-$104,718,750,000	-$110,716,796,000	-$130,425,781,000
2022	-$106,982,422,000	-$95,548,828,000	-$99,708,985,000	-$122,193,359,000
2023	-$117,937,500,000	-$102,208,984,000	-$113,904,296,000	-$131,744,140,000
2024	-$134,919,922,000	-$129,205,078,000	-$133,191,407,000	-$145,988,281,000
2025	-$147,900,391,000	-$152,810,547,000	-$143,474,610,000	-$151,742,188,000
2026	-$131,402,344,000	-$140,197,266,000	-$124,250,000,000	-$132,585,937,000
2027	-$119,218,750,000	-$126,933,594,000	-$111,230,469,000	-$116,541,015,000
2028	-$101,009,766,000	-$102,050,781,000	-$101,613,281,000	-$91,500,000,000
2029	-$69,599,609,000	-$68,685,547,000	-$72,375,000,000	-$59,783,203,000
2030	-$28,572,266,000	-$38,830,078,000	-$32,189,453,000	-$18,908,203,000
2031	$4,818,359,000	-$24,000,000,000	$5,240,234,000	$12,169,922,000
2032	$25,599,609,000	-$6,478,516,000	$33,091,797,000	$35,785,156,000
2033	$42,017,578,000	$7,806,640,000	$52,261,719,000	$46,839,843,000
2034	$45,316,406,000	$3,943,359,000	$59,785,156,000	$45,500,000,000
2035	$32,140,625,000	-$17,144,531,000	$48,419,922,000	$41,345,703,000
2036	$15,488,281,000	-$36,867,188,000	$32,669,921,000	$20,746,093,000
2037	$6,117,187,000	-$49,640,625,000	$20,839,844,000	$5,699,219,000
2038	$1,623,047,000	-$58,640,625,000	$10,347,656,000	-$12,435,547,000
2039	-$7,621,093,000	-$68,392,578,000	$701,172,000	-$20,111,328,000
2040	-$12,064,453,000	-$66,421,874,000	$1,589,844,000	-$16,769,531,000

Source: Author's calculations based on U.S. Energy Information Administration, "Analysis of the Impacts of the Clean Power Plan: Macroeconomic," http://www.eia.gov/oiaf/aeo/tablebrowser/ (accessed June 22, 2015).

Figures in 2009 chain-weighted U.S. dollars (a method of adjusting real dollar amounts for inflation over time). CPP Policy Extension = a scenario in which the Clean Power Plan is extended after 2030 with additional targets. CPP Policy with New Nuclear = a scenario in which new nuclear power installations are accorded the same treatment as new eligible renewables in the compliance calculation. CPP Policy with Biomass CO^2 = a scenario assuming that the emission rate for biomass fuel is 195 pounds CO_2 per MMBtu in place of EIA's Reference case assumption that biomass is carbon-neutral. *Source*: Dayaratna, 2015, Table 3.

Figure 3.6.2.5.2
Impact of Clean Power Plan on overall employment, 2015–2040 (negative numbers represent lost jobs due to slower economic growth)

Year	Clean Power Plan (CPP)	CPP Policy Extension	CPP Policy with New Nuclear	CPP Policy with Biomass CO^2
2015	-1,206	351	213	-641
2016	-16,663	-26,886	-13,702	-16,388
2017	-25,772	-42,481	-22,476	-26,703
2018	-25,482	-44,403	-27,465	-28,061
2019	4,410	41,489	-3,067	1,725
2020	-57,755	-39,261	-46,097	-84,366
2021	-282,913	-183,823	-264,496	-350,708
2022	-234,329	-139,465	-206,054	-324,371
2023	-189,423	-114,548	-165,253	-258,331
2024	-211,365	-211,166	-189,392	-264,084
2025	-378,602	-452,668	-330,384	-390,458
2026	-453,460	-537,445	-399,994	-423,004
2027	-479,034	-536,194	-427,948	-425,873
2028	-422,989	-426,819	-423,660	-339,371
2029	-277,939	-264,175	-314,606	-192,703
2030	-78,506	-83,221	-121,551	9,964
2031	140,549	64,148	107,666	210,083
2032	293,762	187,958	295,578	345,978
2033	375,579	269,927	408,599	441,223
2034	387,955	281,433	449,509	439,117
2035	329,147	210,174	404,159	408,356
2036	254,502	113,861	313,233	305,390
2037	179,932	33,920	222,305	194,901
2038	147,323	1,251	163,421	111,572
2039	110,611	-15,183	122,604	58,762
2040	90,821	26,032	127,472	91,934

Source: Author's calculations based on U.S. Energy Information Administration, "Analysis of the Impacts of the Clean Power Plan: Macroeconomic," http://www.eia.gov/oiaf/aeo/tablebrowser/ (accessed June 22, 2015).

Figures in 2009 chain-weighted U.S. dollars (a method of adjusting real dollar amounts for inflation over time). CPP Policy Extension = a scenario in which the Clean Power Plan is extended after 2030 with additional targets. CPP Policy with New Nuclear = a scenario in which new nuclear power installations are accorded the same treatment as new eligible renewables in the compliance calculation. CPP Policy with Biomass CO^2 = a scenario assuming that the emission rate for biomass fuel is 195 pounds CO_2 per MMBtu in place of EIA's Reference case assumption that biomass is carbon-neutral. *Source*: Dayaratna, 2015, Table 2.

3.6.2.6 Modeled As a Tax Increase

The U.S. Environmental Protection Agency (EPA) estimated the Clean Power Plan (CPP) would reduce CO_2 emissions at an average cost of $30 a ton, "an average of the estimates for each building block, weighted by the total estimated cumulative CO_2 reductions for each of these building blocks over the 2022–2030 period" (EPA, 2014, p. 446). The EPA estimates the cost of "building block three," "Expanding the use of low- or zero-carbon generation in order to displace affected sources (low- and zero-carbon capacity expansion)," to be $37 per ton on average from 2022 to 2030 (p. 769). This latter figure is most salient to the current analysis.

Dayaratna *et al.* (2014) explained why CPP can be modeled as a tax increase: "Taxing CO_2-emitting energy incentivizes businesses and consumers to change production processes, technologies, and behavior in a manner comparable to the Clean Power Plan regulatory scheme. Modeling comparable tax changes as a substitute for estimating the macroeconomic impact of complex regulatory schemes is a widely accepted practice."

Dayaratna *et al.* go on to treat CPP as a $37/ton carbon dioxide tax. The authors "employed the Heritage Energy Model (HEM), a derivative of the [EIA's] National Energy Model System 2014 Full Release (NEMS). This model includes modules covering a variety of energy markets and integrates with the IHS Global Insight macroeconomic model. … We modeled the impact of a revenue-neutral carbon tax starting at $37 per ton in 2015 through 2030."

"The costs," Dayaratna *et al.* wrote, "turn out to be substantial." If implemented, CPP would reduce cumulative GDP "by more than $2.5 trillion between now and 2030. Employment would track nearly 300,000 jobs below the no-carbon-regulation baseline in an average year, with some years seeing an employment deficit of more than 1 million jobs." The researchers also found CPP, modeled as a tax on carbon dioxide emissions, would cause a peak employment shortfall of more than 1 million jobs and total income loss of more than $7,000 per person (inflation-adjusted) by the year 2030. They point out that EIA "analyzed the economic impact of a carbon tax using essentially the same model and found similarly devastating results. Comparing the EIA's $25-carbon-tax estimate with the baseline shows more than $2 trillion in lost GDP from 2014 to 2030 and a peak employment differential of 1 million lost

jobs," referencing EIA's *Annual Energy Outlook 2014* (EIA, 2014).

The Heritage Foundation analysis is valuable, but like other methodologies described in this section it has drawbacks and limits. A carbon tax may be more efficient than the arbitrary caps, timelines, and technology mandates contained in CPP, so modeling CPP as a tax underestimates the cost of displacing fossil fuels and consequently their current value. The carbon tax in the model was assumed to be "revenue neutral," meaning its revenues would be offset by reductions in other tax collections, and consequently its impact on GDP would be less. Examples of new taxes that were "revenue neutral" are difficult to find in human history (see Chapter 1, Section 1.4 for some reasons why this is the case), so it is fair to guess that a new carbon tax would have a larger negative effect on economic growth than forecast by either EIA or Dayaratna *et al.*

References

Dayaratna, K.D., Loris, N., and Kreutzer, D.W. 2014. The Obama administration's climate agenda: Underestimated costs and exaggerated benefits. *Backgrounder* #2975. Washington, DC: Heritage Foundation. November 13.

EIA. 2014. U.S. Energy Information Administration. *Annual Energy Outlook 2014*. Table labeled Macroeconomic Indicators.

EPA. 2014. U.S. Environmental Protection Agency. Clean Power Plan final rule. Washington, DC.

3.6.3 Comparison of Estimates

The six studies summarized in this section are difficult to compare or reconcile since they vary in what was measured and over what time periods. For example, the first estimate using LCOEs looked only at the case of replacing existing coal resources in the United States with new wind energy *ceteris paribus*, with no time frame and no consideration of the effects on other sources of electricity generation. Rose and Wei looked at only the lower-48 states and envisioned a ten-year transition (from 2006 to 2015) away from coal to natural gas, nuclear, and renewable fuels. Tverberg looked at global costs of reducing fossil fuel consumption by 80% and estimated effects in the year 2050. Figure 3.6.3.1 presents a summary of the findings in a table that makes the results easier to interpret.

Figure 3.6.3.1
Summary of six estimates of the cost of replacing fossil fuels with alternatives, measured as percentage of GDP, U.S.-only unless specified as global

Authors	Methodology	Time period	Current value of fossil fuels	Replace 20% by 2020 and 32% by 2025	Replace 33%	Replace 40%	Replace 66%	Replace 80% (global)
NIPCC (2018)	Comparison of LCOEs in the U.S.	2015	+$1.57 trillion GDP	-$307 billion GDP (20%) -$491 billion GDP (32%) (annual)	-$506 billion GDP (annual)	-$614 billion GDP (annual)	-$1.01 trillion GDP (annual)	--
Rose and Wei (2006)	Existence value of coal in the U.S.	2006–2015	+$1.275 trillion GDP (cumulative) +6.8 million jobs	--	-$166 billion GDP (annual) -1.2 million jobs	--	-$371 billion GDP (annual) -2.7 million jobs	--
Tverberg (2012) (global)	Historical relationship of energy consumption and global GDP	2012–2050	--	--	--	--	--	-$137.5 trillion global GDP (2050)
U.S. Chamber of Commerce (2014)	Cost of Clean Power Plan, bottom-up estimate	2015–2030	--	--	--	-$478 billion GDP (cumulative) -224,000 jobs per year (average)	--	--
EIA (2015)	Cost of Clean Power Plan, macroeconomic model	2015–2030	--	- $1.1 trillion GDP (cumulative) -196,00 jobs per year (average)	--	--	--	--
Davaratna, Loris, and Kreutzer (2014)	Cost of Clean Power Plan, modeled as a tax increase	2015–2030	--	-$2.5 trillion GDP (cumulative) - 300,000 jobs per year (average)	--	--	--	--

This table significantly simplifies the findings of six reports and therefore leaves out many caveats and other findings. In some cases these are static estimates that do not reflect the likely incremental cost of replacing fossil fuels over time. *Sources*: See References. NIPCC (2018) refers to this volume of *Climate Change Reconsidered II*s.

A few generalizations can be offered:

- Use of coal in the United States delivered economic benefits worth between $1.275 trillion and $1.57 trillion in 2015. Fossil fuels support approximately 6.8 million jobs in the United States.

- Replacing 20% of the energy produced with fossil fuels in the United States with wind power would cost approximately $300 billion and replacing 32% would cost approximately $491 billion. Achieving these reductions by 2020 and 2025, the stated goals of the Clean Power Plan (CPP), would cost between $1.1 trillion and $2.5 trillion in cumulative lost GDP and destroy between 196,000 and 300,000 jobs each year between 2015 and 2030.

- Replacing 33% of the energy produced by coal in the United States in 2006 with alternatives (including natural gas and nuclear power) by 2015 would have cost $166 billion a year and 1.2 million jobs. Replacing this same amount of coal generation with wind power would cost $506 billion a year.

- Replacing 40% of the energy produced by coal in the United States in 2012 with alternatives by 2030, the goal proposed by the Obama administration in 2010, would cost $478 billion and an average of 224,000 jobs each year. Replacing it with wind power would cost $614 billion a year.

- Replacing 66% of the energy produced by coal in the United States in 2006 with alternatives by 2015 would have cost $371 billion a year and 2.7 million jobs. Replacing it with wind power would have cost $1.0 trillion a year.

- Reducing global reliance on fossil fuels by 80% by 2050 would cause the loss of $137.5 trillion of global GDP in 2050.

Given the great variation in and independence of the methodologies used to reach these conclusions, as well as known and unknown limitations and flaws in several of the studies, it may be surprising the results are at least somewhat consistent. Fossil fuels deliver economic benefits to the United States of between

$1.275 trillion (for coal alone) and $1.76 trillion (for all fossil fuels) a year in added GDP and some 6.8 million jobs (for coal alone). Continued reliance on fossil fuels in the year 2050 would be worth approximately 42% of global GDP, about $137.5 trillion in today's dollars.

Relying on fossil fuels and using alternative energies only as they become cost-competitive would *save* consumers the enormous expenses documented by these studies. Reducing our dependency on fossil fuels is costly, measured as hundreds of billions of dollars of GDP and hundreds of thousands of jobs annually. As the world's population continues to grow and billions of people rise out of poverty, using abundant and affordable fossil fuels is more important than ever.

References

Dayaratna, K.D., Loris, N., and Kreutzer, D.W. 2014. The Obama administration's climate agenda: Underestimated costs and exaggerated benefits. *Backgrounder #2975*. Washington, DC: Heritage Foundation. November 13.

EIA. 2015. U.S. Energy Information Administration. Analysis of the impacts of the Clean Power Plan. May.

Rose, A. and Wei, D. 2006. *The Economic Impacts of Coal Utilization and Displacement in the Continental United States, 2015*. Report prepared for the Center for Energy and Economic Development, Inc. State College, PA: Pennsylvania State University. July.

Tverberg, G. 2012. An energy/GDP forecast to 2050. Our Finite World (website). July 26.

U.S. Chamber of Commerce. 2014. *Assessing the Impact of Potential New Carbon Regulations in the United States*. Institute for 21st Century Energy. Washington, DC: U.S. Chamber of Commerce.

3.7 Conclusion

Despite much compelling evidence of the progress made in human well-being thanks to the use of fossil fuels, sometimes we wish for the "good old days." But our ancestors didn't think of horse-drawn carriages, open-hearth fires in homes, and stifling heat during warm summer nights that way. To them, safe and affordable transportation, clean and reliable home heating, and air conditioning would have been unmitigated blessings leading to tremendous improvements in their quality of life.

Fossil fuels make a dramatic contribution to public health by reducing poverty by supporting the technologies we rely on to keep us safe and well, making electrification of many processes possible, and helping to create a safe and plentiful food supply. Replacing fossil fuels with alternative energies that are more costly or less reliable would mean losing many of these benefits.

Fossil fuels created the modern era. They raised the standard of living, dramatically improved human health, increased human lifespan, and helped elevate billions of persons out of poverty. What we recognize today as modernity – modern cities, fast and affordable transportation, television and the Internet – are all products of fossil fuels. In the words of energy historian Vaclav Smil (2005), "The most fundamental attribute of modern society is simply this: Ours is a high energy civilization based largely on combustion of fossil fuels."

The research presented in this chapter supports Smil's observation. Renewable fuels such as wind turbines, solar PV cells, and ethanol cannot replace fossil fuels. None is sufficiently energy-dense, available in sufficient quantities, dispatchable (always available on demand), or affordable to play more than a small role in meeting the world's growing energy needs. Multiple methodologies aimed at monetizing the benefits of fossil fuels place their value at trillions of dollars a year.

In Chapter 1, "opportunity cost" was defined as the value of foregone uses of the funds or time spent following a choice. Every choice has an *apparent* cost, say the higher price of electricity produced by choosing to rely on wind or solar power instead of coal or natural gas. Research presented in Sections 3.4.4 and 3.4.5 found electricity generated by new wind capacity in the United States costs approximately 2.7 times as much as coal, 3 times as much as combined cycle gas, and nearly four times as much as nuclear power. This is the apparent cost of the choice, but the *opportunity* cost is far greater. The high prices and intermittency of alternative energies raise the cost of electricity, slowing economic growth, and their limited supply raises the prospect of living with much less energy. *What would that look like?*

Deming (2013) speculated about "what would happen to the U.S. today if the fossil fuel industry went on a strike of indefinite duration?" Some of the consequences he described include:

- "With no diesel fuel, the trucking industry would grind to a halt. Almost all retail goods in the U.S. are delivered by trucks. Grocery shelves would begin to empty. Food production at the most basic levels would also stop. Without gasoline, no farm machinery would function, nor could pesticides or fertilizers be produced on an industrial scale. The U.S. cannot feed 315 million people with an agricultural technology based on manure and horse-drawn plows. After two weeks mass starvation would begin.

- "Locomotives once ran on coal but today are powered by diesel engines. With no trains or trucks running there would be no way to deliver either raw materials or finished products. All industrial production and manufacturing would stop. Mass layoffs would ensue. At this point, it would hardly matter. With virtually all transportation systems out, the only people who could work would be those who owned horses or were capable of walking to their places of employment.

- "42% of electric power in the U.S. is produced by burning coal. With natural gas also out of the picture, we would lose another 25%. … With two-thirds of the electric power gone, the grid would shut down entirely. [Probably not entirely… electricity would still be available in some areas near dams and nuclear power plants.] No electricity also means no running water and no flush toilets. When the bottled water ran out, people would drink from streams and ponds and epidemic cholera would inevitably follow.

- "Hospitals could continue to function for a few days on backup generators. But with no diesel fuel being produced, the backups would also fail. Emergency surgeries would have to be conducted by daylight in rooms with windows. Because kerosene is a petroleum byproduct, lighting by kerosene lamps would not be an option. Even candles today are made of paraffin, another petroleum byproduct. It is doubtful if sufficient beeswax could be found to manufacture enough candles to light the 132 million homes in the U.S.

- "With no electricity, little to no fuel, and no way to transport either people or commodities, the U.S. would revert to the eighteenth century within a matter of days to weeks. The industrial revolution would be reversed. The gross domestic product would shrink by more than 95%. Depending on the season and location, people would begin to either freeze or swelter in their homes."

This dark tale of a future without fossil fuels may be easy to criticize, but it is hardly less scientific or less credible than the even darker predictions of a climate Armageddon coming from the United Nations' Intergovernmental Panel on Climate Change (IPCC) and many advocacy groups that echo its views. Deming's narrative has the virtue of relying on actual data consistent with what is reported in this chapter and the predictable consequences of abruptly

ending the use of fossil fuels, whereas IPCC's forecasts rest on assumptions and computer models. Unlike the IPCC and its allies, Deming did not claim to be making a scientific forecast. If only for that reason, Deming seems to be the more trustworthy of the parties.

References

Deming, D. 2013. What if Atlas Shrugged? LewRockwell.com (website). Accessed October 24, 2018.

Smil, V. 2005. *Energy at the Crossroads: Global Perspectives and Uncertainties*. Cambridge, MA: MIT Press.

4

Human Health Benefits

Chapter Lead Authors: Roger Bezdek, Ph.D., Craig D. Idso, Ph.D.

Contributors: Jerome C. Arnett, Jr., M.D., Charles Battig, M.D., John D. Dunn, M.D., J.D., Steve Milloy, J.D., S. Stanley Young, Ph.D.

Reviewers: D. Weston Allen, M.B.B.S., F.R.A.C.G.P., Dip.Phys.Med., Mark Alliegro, Ph.D., James E. Enstrom, Ph.D., Steve Goreham, Kesten Green, Ph.D., Daniel W. Nebert, M.D., Gary D. Sharp, Ph.D.

Key Findings

Key findings of this chapter include the following:

Modernity and Public Health

- Fossil fuels improved human well-being and safety by powering labor-saving and life-protecting technologies such as cars and trucks, plastics, and modern medicine.

- Fossil fuels play a key and indispensable role in the global increase in life expectancy.

Mortality Rates

- Cold weather kills more people than warm weather. A warmer world would see a net decrease in temperature-related mortality in virtually all parts of the world, even those with tropical climates.

- Weather is less extreme in a warmer world, resulting in fewer injuries and deaths due to storms, hurricanes, flooding, etc.

Citation: Idso, C.D., Legates, D. and Singer, S.F. 2019. Human Health Benefits. In: *Climate Change Reconsidered II: Fossil Fuels.* Nongovernmental International Panel on Climate Change. Arlington Heights, IL: The Heartland Institute.

Cardiovascular Diseases

■ Higher surface temperatures would reduce the incidence of fatal coronary events related to low temperatures and wintry weather by a greater degree than they would increase the incidence associated with high temperatures and summer heat waves.

■ Non-fatal myocardial infarction is also less frequent during unseasonably warm periods than during unseasonably cold periods.

Respiratory Disease

■ Climate change is not increasing the incidence of death, hospital visits, or loss of work or school time due to respiratory disease.

■ Low minimum temperatures are a greater risk factor than high temperatures for outpatient visits for respiratory diseases.

Stroke

■ Higher surface temperatures would reduce the incidence of death due to stroke in many parts of the world, including Africa, Asia, Australia, the Caribbean, Europe, Japan, Korea, Latin America, and Russia.

■ Low minimum temperatures are a greater risk factor than high temperatures for stroke incidence and hospitalization.

Insect-borne Diseases

■ Higher surface temperatures are not leading to increases in mosquito-transmitted and tick-borne diseases such as malaria, yellow fever, viral encephalitis, and dengue fever.

■ Extensive scientific information and experimental research contradict the claim that malaria will expand across the globe and intensify as a result of CO_2-induced warming.

■ Concerns over large increases in dengue fever as a result of rising temperatures are unfounded and

unsupported by the scientific literature, as climatic indices are poor predictors for dengue fever.

■ Climate change has not been the most significant factor driving recent changes in the distribution or indicence of tick-borne diseases.

Conclusion

■ Fossil fuels directly benefit human health and longevity by powering labor-saving and life-protecting technologies and perhaps indirectly by contributing to a warmer world.

Introduction

Fossil fuels directly benefit human health and longevity by powering labor-saving and life-protecting technologies such as cars and trucks, plastics, and modern medicine. They may also indirectly benefit human health by contributing to some part of the increase in surface temperatures that the United Nations' Intergovernmental Panel on Climate Change (IPCC) claims occurred during the twentieth and early twenty-first centuries and may continue for the rest of the twenty-first century and beyond. How much warming has occurred and will occur, and how much can be attributed to the combustion of fossil fuels, are unsolved scientific puzzles, as explained in Chapter 2. Whereas the IPCC predicts a global temperature increase of between 2°C and 4°C by 2100 (compared to the 1850–1900 average) (IPCC, 2013, p. 20), the Nongovernmental International Panel on Climate Change (NIPCC) says its best-guess forecast is of ~0.3 to 1.1°C (NIPCC, 2013).

In the Working Group II contribution to its Fifth Assessment Report, the IPCC admits "at present the world-wide burden of human ill-health from climate change is relatively small compared with effects of other stressors and is not well quantified," but it also claims "impacts from recent climate-related extremes, such as heat waves, droughts, floods, cyclones, and wildfires, reveal significant vulnerability and exposure of some ecosystems and many human systems to current climate variability (*very high confidence*)" (IPCC, 2014, p. 6). It further claims to have "high confidence" that climate change will contribute to eight "risk factors" including "risk of severe ill-health" and "mortality and morbidity during periods of extreme heat" (*Ibid.*, p. 13).

As has been common throughout its history, the IPCC's claims have been repeated by legacy news media, politicians, environmental activists, and subsidy-seekers in the renewable energy industry, while its more cautious findings, qualifications, and admissions of uncertainty are unreported or even hidden. This has led to widespread fear of the health effects of global warming (Schulte, 2008) and even political attack ads claiming people are dying of "carbon pollution" (WMC, 2015).

Independent researchers who should know better have also overlooked the IPCC's errors and admissions of uncertainties. This failure is illustrated by an otherwise commendable effort to quantify the health and other effects of global warming by Richard S.J. Tol, a professor of economics at the University of Sussex and professor of the economics of climate change at the Vrije Universiteit Amsterdam who is otherwise an outspoken critic of the IPCC (Tol, 2013). Tol developed the Climate Framework for Uncertainty, Negotiation, and Distribution (FUND) model, which he said "is a fully integrated model, including scenarios of population, economy, energy use, and emissions; a carbon cycle and simple climate model; and a range of impact models" (Tol, 2011, p. 4).

Tol's model forecasts the decline in the number of deaths due to cold temperatures would exceed the increase in the number of deaths due to warm temperatures in the year 2055 and beyond, a finding supported by extensive research summarized in this chapter. But Tol also contends "Climate change has caused the premature deaths of a substantial number of people over the 20th century – on average 7.5 per million per year. In 2000, according to FUND, 90,000 people died because of climate change" (p. 13). These numbers, drawn from the IPCC and public health advocacy groups, are not empirical data and should not be treated as though they were. They are derived from computer models that assume more local warming than actually occurred, assume causation when medical evidence says otherwise, and contradict actual public health data showing falling numbers of deaths due to respiratory and cardiovascular diseases and insect-borne diseases such as malaria.

Chapter 3 already explained some of the human health benefits produced by the prosperity made possible by fossil fuels. In this chapter, those benefits are explained in greater detail. Section 4.1 documents the direct human health benefits due to the prosperity and technologies made possible by fossil fuels. Section 4.2 documents how medical science and observational research in Asia, Europe, and North America confirm that global warming is associated with lower, not higher, temperature-related mortality rates. Sections 4.3, 4.4, and 4.5 report research showing warmer temperatures lead to decreases in premature deaths due to cardiovascular and respiratory disease and stroke occurrences. Section 4.6 finds global warming has little if any influence on mosquito-borne diseases such as malaria and dengue fever or tick-borne diseases. Section 4.7 is a brief summary and conclusion.

The health benefits of climate change were the subject of Chapter 7 of a previous volume in the *Climate Change Reconsidered II* series, subtitled *Biological Effects* (NIPCC, 2014). The authors of that chapter provided a more comprehensive survey of the literature than is presented here, including older studies excluded from this review and detailed summaries of the methodologies utilized by the authors. A summary of that chapter appears below as Figure 4.1. This new chapter features research published as recently as 2018.

Figure 4.1
Impacts on human health

- Warmer temperatures lead to a decrease in temperature-related mortality, including deaths associated with cardiovascular disease, respiratory disease, and strokes. The evidence of this benefit comes from research conducted in every major country of the world.

- In the United States the average person who died because of cold temperature exposure lost in excess of 10 years of potential life, whereas the average person who died because of hot temperature exposure likely lost no more than a few days or weeks of life.

- In the United States, some 4,600 deaths are delayed each year as people move from cold northeastern states to warm southwestern states. Between 3% and 7% of the gains in longevity experienced over the past three decades was due simply to people moving to warmer states.

- Cold-related deaths are far more numerous than heat-related deaths in the United States, Europe, and almost all countries outside the tropics. Coronary and cerebral thrombosis account for about half of all cold-related mortality.

- Warmer temperatures are reducing the incidence of cardiovascular diseases related to low temperatures and wintry weather by a much greater degree than they increase the incidence of cardiovascular diseases associated with high temperatures and summer heat waves.

- A large body of scientific examination and research contradicts the claim that malaria will expand across the globe and intensify as a result of CO_2-induced warming.

- Concerns over large increases in vector-borne diseases such as dengue as a result of rising temperatures are unfounded and unsupported by the scientific literature, as climatic indices are poor predictors for dengue disease.

- Whereas temperature and climate largely determine the geographical distribution of ticks, they are not among the significant factors determining the incidence of tick-borne diseases.

Source: Chapter 7. "Human Health," *Climate Change Reconsidered II: Biological Impacts* (Chicago, IL: The Heartland Institute, 2014).

References

IPCC. 2013. *Climate Change 2013: The Physical Science.* Contribution of Working Group I to the Fifth Assessment Report of the Intergovernmental Panel on Climate Change. New York, NY: Cambridge University Press.

IPCC. 2014. *Climate Change 2014: Impacts, Adaptation, and Vulnerability. Part A: Global and Sectoral Aspects.* Contribution of Working Group II to the Fifth Assessment Report of the Intergovernmental Panel on Climate Change. New York, NY: Cambridge University Press.

NIPCC. 2013. Nongovernmental International Panel on Climate Change. *Climate Change Reconsidered II: Physical Science.* Chicago, IL: The Heartland Institute.

NIPCC. 2013. Nongovernmental International Panel on Climate Change. *Climate Change Reconsidered II: Biological Impacts.* Chicago, IL: The Heartland Institute.

Schulte, K-M. 2008. Scientific consensus on climate change? *Energy & Environment* **19** (2): 281–6.

Tol, R.S.J. 2011. The economic impact of climate change in the 20th and 21st centuries. *Assessment Paper.* Copenhagen Consensus on Human Challenges.

Tol, R.S.J. 2013. Open letter to Professor Peter Høj, president and vice-chancellor, University of Queensland. August 2013.

WMC. 2015 Wisconsin Manufacturers and Commerce. Wisconsin chamber decries outrageous environmentalist ad attacking Sen. Johnson. *News Release.* September 3.

4.1 Modernity and Public Health

Fossil fuels improved public health in developed countries around the world by making electrification, safe transportation, plastics, and modern medicine possible. None of these technological advances would have occurred without fossil fuels providing the abundant, convenient, and affordable energy needed to power them or the critical feedstock used to create them. Proof of fossil fuels' success in advancing public health can be seen in the rising lifespans of people living all around the world, but especially those living with the most abundant supplies of energy.

4.1.1 Technology and Health

Fossil fuels improved human well-being and safety by powering labor-saving and life-protecting technologies such as cars and trucks, plastics, and modern medicine.

Fossil fuels produce the quantity and quality of energy needed to fuel technologies that produce enormous human health benefits. Electricity, whose widespread use is possible only with fossil fuels, made and continues to make the biggest contributions to public health. Electricity makes food safer, more plentiful, and more affordable by making possible refrigeration from fields to grocery stores and freezers for long-term storage of food, and powering the modern canning industry, dramatically reducing waste. Electricity promotes good health by greatly facilitating the sharing of information, allowing people in need of help to call family members or emergency medical services, patients to communicate with doctors and other medical professionals between visits to hospitals and clinics, and doctors to share insights and discoveries with one another across long distances. Electricity continues to revolutionize health care with electronic medical records, robotic surgery, and remote diagnosis.

Safe and clean transportation is a second example of health benefits made possible by fossil fuels. The development and widespread use of cars and trucks averted a public health crisis. According to Smith (1990), "a horse produces approximately 45 pounds of manure each day. In high-density urban environments, massive tonnages accumulated, requiring constant collection and disposal. Flies, dried dung dust, and the smell of urine filled the air, spreading disease and irritating the lungs. On rainy days, one walked through puddles of liquid wastes. Occupational diseases in horse-related industries were common" (p. 25). According to Smith, New York City was disposing of 15,000 dead horses every year in the 1890s, a task that posed major threats to public health. Tenner (1997) offered a sobering reprise of the bad old days of horses in the streets causing diseases and deaths by accidents, and the public health revolution made possible by automobiles:

Less remembered today than the sanitary problems caused by horses were the safety hazards they posed. Horses and horse-drawn vehicles were dangerous, killing more riders, passengers, and pedestrians than is generally

appreciated. Horses panicked. In frequent urban traffic snarls, they bit and kicked some who crossed their path. Horse-related accidents were an important part of surgical practice in Victorian England and no doubt in North America as well. In the 1890s in New York, per capita deaths from wagons and carriage accidents nearly doubled. By the end of the century they stood at nearly six per hundred thousand of population. Added to the five or so streetcar deaths, the rate of about 110 per million is close to the rates of motor vehicle deaths in many industrial countries in the 1980s. On the eve of motorization, the urban world was not such a gentle place.

The automobile was an answer to disease and danger. In fact, private internal-combustion transportation was almost utopian (pp. 333–4).

A third way fossil fuels-enabled technology has contributed to human health is by allowing the use of plastic products instead of metal, glass, or wooden products (Avery, 2000; North and Halden, 2013). Natural gas is the primary feedstock of plastic products made in the United States while oil is used by most Eurpean and Asian producers. See, for example, the representation of the production process of polyvinyl chloride (PVC) in Figure 4.1.1.1. According to the American Chemistry Council (ACC, 2015), falling prices of natural gas and oil in the United States made possible by the development of shale oil produced "a flood of new investment in U.S. plastics capacity announced since 2010. New factories are being built to produce more plastic resins, about half of that new resin production will be exported. In addition, with newly-available supplies of low-cost resin, producers of plastic products are building new production facilities" (p. 3).

While it is fashionable to minimize the benefits plastics bring to society, they are undeniably immense. Andrady and Neal (2009) write, "Plastics deliver unparalleled design versatility over a wide range of operating temperatures. They have a high strength-to-weight ratio, stiffness and toughness, ductility, corrosion resistance, bio-inertness, high thermal/electrical insulation, non-toxicity and outstanding durability at a relatively low lifetime cost compared with competing materials; hence plastics are very resource efficient."

Figure 4.1.1.1
Production process of polyvinyl chloride (PVC) in the United States

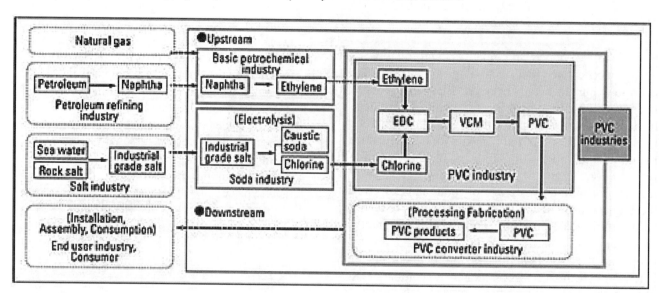

Source: PVC.org, n.d.

Plastic components make up a growing percentage of the total composition of airplanes, cars, ships, buildings, and many other important technologies of the modern age. PVC is a maintenance-free and non-combustible material that is now essential for buildings, furniture, piping, and upholstery. Computers, cellphones, and the Internet itself would be impossible without plastics. A new technology – 3D printing – is using plastics to further revolutionize the manufacturing of consumer products.

Plastics contribute directly to improving public health in many ways. Plastic film helps protect food and other products in inventory and during shipping while dramatically reducing weight and cost relative to other types of containers. Plastic containers, plates, and cups replace glass containers that can cause injuries when broken. Plastic bottles with childproof caps reduce instances of accidental poisoning. Plastic airbags and seatbelts save lives every day by protecting passengers of cars and trucks from impact during accidents. Plastic insulation of electric wiring dramatically reduces the incidence of home and business fires and death and injury by electrocution.

Plastics have extensive applications in medicine, including disposable surgical gloves, masks, gowns, syringes, and petri dishes; flexible tubing and bags for plasma; tamper-resistent packaging for drugs; and parts for innumerable medical devices. Plastic contact lenses restore vision to millions of people without the inconvenience of glasses, and plastic makes glasses lighter, break-resistant, and more affordable. Plastic prostheses are lighter, last longer, cost less than alternatives, and look more life-like. North and Halden (2013) write, "Plastics are cost-effective, require little energy to produce, and are lightweight and biocompatible. This makes them an ideal material for single-use disposable devices, which currently comprise 85% of medical equipment. Plastics can also be soft, transparent, flexible, or biodegradable and many different types of plastics function as innovative materials for use in engineered tissues, absorbable sutures, prosthetics, and other medical applications."

Plastics infused with antibiotics, called antimicrobial plastic, can help stop the spread of diseases in hospitals, a major global public health threat. Surfaces containing antimicrobial plastic repel or kill bacteria on surfaces that doctors and patients touch, such as furniture in emergency and examination rooms. Sterile plastics can replace glass and steel containers used to store medicines and medical waste. Plastics speed the invention and wider use of new medical devices by making prototypes dramatically less expensive to create and modify. Plastic joints are typically longer-lasting than metal joints, thereby reducing pain and the need for repeat surgeries. Because plastic devices are cheaper to produce than the metal or wood products they replace, they reduce the cost of many steps in the patient care cycle, thereby lowering the cost of and increasing public access to health care.

A fourth way fossil fuels contribute to human health is by enabling the world's farmers to increase their output faster than population, resulting in less hunger and starvation around the world. In 2015, the Food and Agriculture Organization of the United Nations (FAO) reported "the number of hungry people in the world has dropped to 795 million – 216 million fewer than in 1990–92 – or around one person out of every nine" (FAO, 2015). In developing countries, the share of the population that is undernourished (having insufficient food to live an active and healthy life) fell from 23.3 percent 25 years earlier to 12.9 percent. A majority of the 129 countries monitored by the FAO reduced undernourishment by half or more since 1996 (*Ibid.*).

Chapter 3, Section 3.3.1, explained how fossil fuels created and sustain the use of fertilizers and machines that make possible the Green Revolution and the more recent "Gene Revolution"; documented how rising levels of atmospheric CO_2 promote plant growth, increasing agricultural yields beyond levels farmers would otherwise achieve; and estimated the current and future value of aerial fertilization. Despite all this good news, concern has been expressed that increases in the *quantity* of food produced has come at a cost in *lower quality* food. There is some evidence that while aerial fertilization promotes crop yields, it may lower the level of key nuitrients relative to total plant mass, making crops less nutritious. What does the latest science say about this?

It is possible to contrive growing conditions in which something other than CO_2 limits plant growth and health, or in which a shortage of some soil nutrient causes better crop yields to be accompanied by reduced levels of some nutrient, but such contrived conditions are easily avoided through normal agricultural fertilization practices. In real-world greenhouses, additional CO_2 is dramatically beneficial for agriculture at levels far beyond what are likely to be reached in the outdoor atmosphere, and the nutrient value of such crops grown with extra CO_2 is not significantly different from other crops.

Dong *et al.* (2018) write, "a comprehensive review of recent studies explaining and targeting the key role of the effect of elevated CO_2 on vegetable quality is lacking." To remedy this knowledge gap,

the team of five researchers performed a meta-analysis of existing studies on the topic. In all, they examined 57 published works, which included CO_2 enrichment studies on root vegetables (carrot, radish, sugar beet, and turnip), stem vegetables (broccoli, celery, Chinese kale, ginger, onion, potato, and scallion), leafy vegetables (cabbage, Chinese cabbage, chives, fenugreek, Hongfengcat, lettuce, oily sowthistle, palak, and spinach) and fruit vegetables (cucumber, hot pepper, strawberry, sweet pepper, and tomato). The specific focus of their analysis was to examine measurements of nutritional quality on the vegetables, including measurements of soluble sugars, organic acids, protein, nitrates, antioxidants, and minerals.

The results of the analysis, shown in Figure 4.1.1.2, reveal elevated CO_2 "increased the concentrations of fructose, glucose, total soluble sugar, total antioxidant capacity, total phenols, total flavonoids, ascorbic acid, and calcium in the edible part of vegetables by 14.2%, 13.2%, 17.5%, 59.0%, 8.9% 45.5%, 9.5%, and 8.2%, respectively, but [that it] decreased the concentrations of protein, nitrate, magnesium, iron, and zinc by 9.5%, 18.0%, 9.2%, 16.0%, and 9.4%. The concentrations of titratable acidity, total chlorophyll, carotenoids, lycopene, anthocyanins, phosphorus, potassium, sulfur, copper, and manganese were not affected."

In commenting on their findings, Dong *et al.* say that "overall, elevated CO_2 promotes the accumulation of antioxidants in vegetables, thus improving vegetable quality," while adding that the CO_2-induced stimulation of total antioxidant capacity, total phenols, total flavonoids, ascorbic acid, and chlorophyll b indicate "an improvement of beneficial compounds in vegetables."

For those concerned about the decreases in protein, nitrate, magnesium, iron, and zinc that were also observed in the meta-analysis, these slight declines can be reduced, if not reversed, through the application of several management approaches that were investigated and discussed by the authors, including "(1) selecting vegetable species or cultivars that possess greater ability in carbon fixation and synthesis of required quality-related compounds; (2) optimizing other environmental factors (e.g., moderate CO_2 concentrations, moderate light intensity, increased N availability, or increased fertilization of Fe or Zn) to promote carbon fixation and nutrient uptake interactively when growing plants under elevated CO_2; (3) harvesting vegetable products earlier in cases of over maturity and reduced benefit of elevated CO_2 to vegetative growth; and (4)

Figure 4.1.1.2

Effects of elevated CO2 on the concentrations of various plant compounds and minerals in vegetables

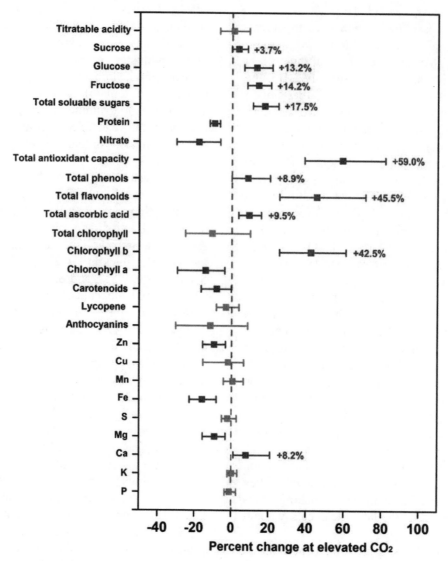

Data are means of percent change (relative to ambient CO_2) with 95% confidence intervals. Green squares and error bars represent positive changes, blue negative and grey indicate no significant change. *Source:* Adapted from Dong *et al.*, 2018.

combining elevated CO_2 with mild environmental stress (e.g., ultraviolet-B radiation or salinity) in instances when this enhances vegetable quality and might counteract the dilution effect or direct metabolic pathways toward the synthesis of health-beneficial compounds."

The findings of Dong *et al.* clearly show that CO_2-induced plant nutritional enhancements outweigh CO_2-induced plant nutritional declines.

Thus, it can reasonably be concluded that rising atmospheric CO_2 concentrations will yield future health benefits to both human and animal plant consumers.

A fifth contribution of fossil fuels to human health, ironic in light of current debates, is to protect mankind from the climate. As Goklany wrote, "these technologies, by lowering humanity's reliance on living nature, inevitably ensured that human well-being is much less subject to whims of nature (as expressed through the weather, climate, disease, and other natural disasters)" (Goklany, 2012, p. 1). Writing in 2011, Goklany noted,

[E]xtreme weather events ... now contribute only 0.07% to global mortality. Mortality from extreme weather events has declined even as all-cause mortality has increased, indicating that humanity is coping better with extreme weather events than it is with far more important health and safety problems. The decreases in the numbers of deaths and death rates reflect a remarkable improvement in society's adaptive capacity, likely due to greater wealth and better technology, enabled in part by use of hydrocarbon fuels. Imposing additional restrictions on the use of hydrocarbon fuels may slow the rate of improvement of this adaptive capacity and thereby worsen any negative impact of climate change. At the very least, the potential for such an adverse outcome should be weighed against any putative benefit arising from such restrictions (p. 4.).

Epstein (2014) pointed out, "Climate is no longer a major cause of death, thanks in large part to fossil fuels" (p. 126). "Historically, drought is the *number-one climate related cause of death*. Worldwide it has gone down by 99.98% in the last eighty years for many energy-related reasons: oil-powered drought-relief convoys, more food in general because of more prolific, fossil-fuel-based agriculture, and irrigation systems" (*Ibid.*). Environmentalists, Epstein wrote, have the issue backward: "[W]e don't take a safe climate and make it dangerous; we take a dangerous climate and make it safe. High-energy civilization, not climate, is the driver of climate livability. No matter what, climate will always be naturally hazardous – and the key question will always be whether we have the adaptability to handle it, or better yet, master it" (*Ibid.*).

References

ACC. 2015. American Chemistry Council. The rising competitive advantage of U.S. plastics. Washington, DC.

Andrady, A.L. and Neal, M.A. 2009. Applications and societal benefits of plastics. *Philosophical Transactions of the Royal Society of London B Biological Science* **364** (1526): 1977–84.

Avery, D. 2000. *Saving the Planet with Plastic and Pesticides.* Second edition. Indianapolis, IN: Hudson Institute.

Dong, J., Gruda, N., Lam, S.K., Li, X. and Duan, Z. 2018. Effects of elevated CO_2 on nutritional quality of vegetables: a review. *Frontiers in Plant Science* **9**: Article 924.

Epstein, A. 2014. *The Moral Case for Fossil Fuels.* New York, NY: Portfolio/Penguin.

FAO. 2015. Food and Agriculture Organization. *The State of Food Insecurity in the World 2015.* Rome, Italy.

Goklany, I.M. 2011. *Wealth and Safety: The Amazing Decline in Deaths from Extreme Weather in an Era of Global Warming, 1900–2010.* Los Angeles, CA: Reason Foundation.

Goklany, I.M. 2012. Humanity unbound: How fossil fuels saved humanity from nature and nature from humanity. *Cato Policy Analysis* #715. Washington, DC: Cato Institute.

North, E.J. and Halden, R.U. 2013. Plastics and environmental health: the road ahead. *Review of Environmental Health* **28** (1): 1–8.

PVC.org. n.d. How is PVC made? (website) Accessed November 8, 2018.

Smith, F.L. 1990. Auto-nomy: The liberating benefits of a safer, cleaner, and more mobile society. *Reason* (August/September).

Tenner, E. 1997. *Why Things Bite Back: Technology and the Revenge of Unintended Consequences.* New York, NY: VintageBooks.

4.1.2 Public Health Trends

Fossil fuels play a key and indispensable role in the global increase in life expectancy.

Historically, humankind was besieged by epidemics and other disasters that caused frequent widespread deaths and kept the average lifespan to less than 35 years (Omran, 1971). The average lifespan among the ancient Greeks was apparently just 18 years, and among the Romans, 22 years (Bryce, 2014 p. 59, citing Steckel and Rose, 2002). The discovery of uses for fossil fuels in the late-eighteenth and early-nineteenth century dramatically changed the world, a story told in Chapter 3.

The evidence of progress in public health in the United States and other developed countries in the twentieth and early-twenty-first centuries is overwhelming (Lehr, 1992; Lomborg, 2001). Economist Julian Simon edited a series of volumes (Simon, 1981, 1995, 1998; Simon and Kahn, 1984) and coauthored a book published posthumously (Moore and Simon, 2000) providing extensive data showing long-term trends for everything from mortality and longevity to food supplies, air and water quality, and the affordability of housing. Nearly every trend showed dramatic improvement over time. As Simon wrote in the introduction to his 1995 book, "Most important, fewer people are dying young. And life expectancy in the rich countries has increased most sharply in the older age cohorts, among which many thought there was no improvement. Perhaps most exciting, the quantities of education that people obtain all over the world are sharply increasing, which means less tragic waste of human talent and ambition" (Simon, 1995, p. 2). The trend Simon observed in 1995 continues today. According to the U.S. Census Bureau:

- "The world average age of death has increased by 35 years since 1970, with declines in death rates in all age groups, including those aged 60 and older (Institute for Health Metrics and Evaluation, 2013; Mathers *et al.,* 2015). From 1970 to 2010, the average age of death increased by 30 years in East Asia and 32 years in tropical Latin America, and in contrast, by less than 10 years in western, southern, and central Sub-Saharan Africa (Institute for Health Metrics and Evaluation, 2013; Figure 4-1) ;

- "In the mean age at death between 1970 and 2010 across different WHO regions, all regions have had increases in mean age at death, particularly East Asia and tropical Latin America;

- "Global life expectancy at birth reached 68.6 years in 2015. A female born today is expected to live 70.7 years on average and a male 66.6 years. The global life expectancy at birth is projected to increase almost 8 years, reaching 76.2 years in 2050" (U.S. Census Bureau, 2015, pp. 31–33).

The U.S. historical record reveals the close correlation between prosperity and public health and longevity. As the nation grew richer thanks to fossil fuels, the incidence of nearly every disease in the United States fell dramatically, as shown in Figure 4.1.2.1.

Figure 4.1.2.1
Incidence of selected diseases in the United States, 1912–1997

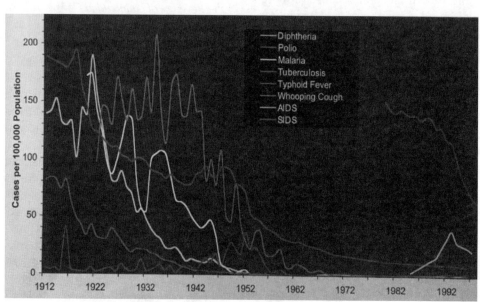

Source: Moore and Simon, 2000, p. 35.

As Moore and Simon write, "Before 1900, major killers included such infectious diseases as tuberculosis, smallpox, diphtheria, polio, influenza, and bronchitis. Just three infectious diseases – tuberculosis, pneumonia, and diarrhea – accounted for almost half of all deaths in 1900. Now few Americans die from these diseases, and many diseases have been completely eradicated due to a medley of modern medicines" (Moore and Simon, 2000, p. 34).

The economic growth created by abundant and affordable energy, documented in Chapter 3, Section 3.6.1, is closely correlated with better health and longevity. Brenner (1984), in a study for the Joint Economic Committee of the U.S. Congress, found a one-percentage-point increase in the unemployment rate (e.g., from 5% to 6%) would lead to a 2% increase in the age-adjusted mortality rate. The growth of real income per capita also showed a significant correlation to decreases in mortality rates (except for suicide and homicide), hospitalization for mental illness, and property crimes. The European Commission has supported similar research showing comparable results throughout the European Union (Brenner, 2000, 2003).

Both in the United States and Europe, Brenner found that changes to the economic status of individuals result in changes to their health and lifespan, with decreased real income per capita and increased unemployment leading to increased mortality. Econometric analyses of time-series data have measured the relationship between changes in the economy and changes in health outcomes, and studies have determined declines in real income per capita and increases in unemployment led to elevated mortality rates over a subsequent period of six years (Brenner, 2005). The loss of disposable income also reduces the amount families can spend on critical health care, especially among the poorest and least healthy (Keeney, 1990; Lutter and Morrall, 1994; Viscusi, 1994; Viscusi and Zeckhauser, 1994; Hjalte *et al.*, 2003).

The U.S. Environmental Protection Agency (EPA) has acknowledged "People's wealth and health status, as measured by mortality, morbidity, and other metrics, are positively correlated. Hence, those who bear a regulation's compliance costs may also suffer a decline in their health status, and if the costs are large enough, these increased risks might be greater than the direct risk-reduction benefits of the regulation" (EPA, 1995). The U.S. Office of Management and Budget, Food and Drug Administration, and Occupational Safety and Health Administration use methodologies similar to the EPA's to assess the degree to which their regulations induce premature death among those who bear the costs of federal mandates (OMB, 1993).

The global correlation between prosperity and life expectancy is represented by the "Preston curve" shown in Figure 4.1.2.2. It has been closely studied since the 1970s (Preston, 1975, 2007; Deaton, 2003, 2004; Bloom and Canning, 2007). The data make very clear that the people who live in impoverished societies live shorter lives, with the relationship strongest at per-capita GDP levels of less than $30,000. At that stage of development investments in nutrition, clean water and sanitation, and public safety have their greatest impact on public health. The flattening of the curve at higher income levels reveals how other factors then affect lifespan, with the leading factors being educational attainment, spending on health care services, and personal savings. All of these factors contributing to longevity – nutrition, clean water and sanitation, public safety, education, health care services, and personal income – are positively impacted by abundant and affordable energy, impacts documented at length in Chapter 3 and earlier in this chapter. Fossil fuels played a key role in increasing global wealth and longevity.

It took eight millennia for the average global life expectancy to rise from 20 years to the high 20s. Since the discovery of fossil fuels, life expectancy soared to 75 years and longer in developed countries. Life expectancy increased for all age groups, from infancy to old age, as the three Industrial Revolutions brought improved nutrition, cleaner air and water, and safer work conditions to virtually every person in developed countries and to many in developing countries.

References

Bloom, D.E. and Canning, D. 2007. Commentary: the Preston Curve 30 years on: still sparking fires. *International Journal of Epidemiology* **36** (3).

Brenner, H. 1984. *Estimating the Effects of Economic Change on National Health and Social Well-Being*. Washington, DC: Joint Economic Committee, U.S. Congress.

Brenner, H. 2000. *Estimating the Social Cost of Unemployment and Employment Policies in the European Union and the United States*. Luxembourg: European Commission Directorate-General for Employment, Industrial Relations, and Social Affairs.

Figure 4.1.2.2
The Millenium Preston Curve
Life expectancy and per-capita GDP in 2000

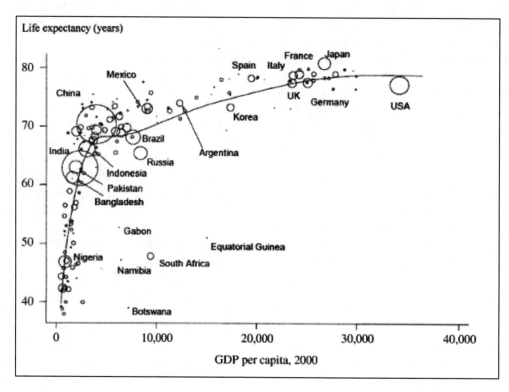

Dots and circles represent countries. Circles have diameter proportional to population size. GDP per capita is in purchasing power parity dollars. *Source:* Deaton, 2004.

Brenner, H. 2003. *Unemployment and Public Health in Countries of the European Union.* Luxembourg: European Commission Directorate-General for Employment, Industrial Relations, and Social Affairs.

Brenner, H. 2005. Health benefits of low-cost energy: an econometric study. *EM* (November): 28–33.

Bryce, R. 2014. *Smaller Faster Lighter Denser Cheaper: How Innovation Keeps Proving the Catastrophists Wrong.* New York, NY: PublicAffairs.

Deaton, A. 2003. Health, inequality, and economic development. *Journal of Economic Literature* **41**: 113–58.

Deaton, A. 2004. Health in an age of globalization. In: *Brookings Trade Forum 2004.* New York, NY: Brookings Institution Press.

EPA. 1995. U.S. Environmental Protection Agency. *On the Relevance of Risk-Risk Analysis to Policy Evaluation.* Washington, DC: Economic Analyis and Innovations Division. August 16.

Hjalte, K., *et al.* 2003. Health analysis – an alternative method for economic appraisal of health policy and safety regulation: Some empirical Swedish estimates. *Accident Analysis & Prevention* **35** (1): 37–46.

Institute for Health Metrics and Evaluation. 2013. *The Global Burden of Disease: Generating Evidence, Guiding Policy.* Seattle, WA: IHME.

Keeney, R.L. 1990. Mortality risks induced by economic expenditures. *Risk Analysis* **10** (1): 147–59.

Lehr, J. (Ed.) 1992. *Rational Readings on Environmental Concerns.* New York, NY: Wiley.

Lomborg, B. 2001. *The Skeptical Environmentalist: Measuring the Real State of the World.* Cambridge, MA: Cambridge University Press.

Lutter, R. and Morrall, J.F. 1994. Health-health analysis: a new way to evaluate health and safety regulation. *Journal of Risk and Uncertainty* **8** (1): 43–66.

Mathers, C.A., Stevens, G.A., Boerma, T., White, R.A., and Tobias, M.I. 2015. Causes of international increases in older age life expectancy. *The Lancet* **385** (9967): 540–8.

Moore, S. and Simon, J. 2000. *It's Getting Better All the Time: 100 Greatest Trends of the Last 100 Years.* Washington, DC: Cato Institute.

OMB. 1993. U.S. Office of Management and Budget. *Regulatory Impact Analysis: A Primer.* Circular A-4.

Omran, A. 1971. The epidemiologic transition: a theory of the epidemiology of population change. *The Milbank Memorial Fund Quarterly* **49** (4): 509–38.

Preston, S.H. 1975. The changing relation between mortality and level of economic development. *Population Studies* **29** (2).

Preston, S.H. 2007. The changing relation between mortality and level of economic development. *International Journal of Epidemiology* **36** (3): 484–90.

Simon, J. (Ed.) 1981. *The Ultimate Resource.* Princeton, NJ: Princeton University Press.

Simon, J. (Ed.) 1995. *The State of Humanity.* Cambridge, MA: Blackwell Publishers Inc.

Simon, J. 1998. *The Ultimate Resource 2.* New York, NY: Princeton University Press.

Simon, J. and Kahn, H. (Eds.) 1984. *The Resourceful Earth: A Response to 'Gobal 2000'.* Oxford, UK: Basil Blackwell Publisher Limited.

Steckel, R.H. and Rose, J.C. (Eds.) 2002. *Backbone of History: Health and Nutrition in the Western Hemisphere.* Cambridge, UK: Cambridge University Press.

U.S. Census Bureau. 2016. *An Aging World: 2015.* International Population Reports P95/16-1. Washington, DC: U.S. Department of Commerce.

Viscusi, W.K. 1994. Risk-risk analysis. *Journal of Risk and Uncertainty* **8** (1): 5–17.

Viscusi, W.K. and Zeckhauser, R.J. 1994. The fatality and injury costs of expenditures. *Journal of Risk and Uncertainty* **8** (1): 19–41.

4.2 Mortality Rates

Cold weather kills more people than warm weather. A warmer world would see a net decrease in temperature-related mortality in virtually all parts of the world, even those with tropical climates.

As stated in the introduction to this chapter, the United Nations' Intergovernmental Panel on Climate Change (IPCC) claims the carbon dioxide (CO_2) produced during the combustion of fossil fuels affects human health *indirectly* by causing an increase in surface temperatures that creates a "risk of severe ill-health" and "mortality and morbidity during periods of extreme heat" (IPCC, 2014, p. 13). Section 4.2.1 provides the basis in medical science for why warmer temperatures can be expected to *reduce* rather than increase mortality rates. Sections 4.2.2 through 4.2.6 summarize observational research conducted globally and specifically in Asia, Australia, Europe, and North America confirming the hypothesis.

Reference

IPCC. 2014. Intergovernmental Panel on Climate Change. *Climate Change 2014: Impacts, Adaptation, and Vulnerability. Part A: Global and Sectoral Aspects.* Contribution of Working Group II to the Fifth Assessment Report of the Intergovernmental Panel on Climate Change. New York, NY: Cambridge University Press.

4.2.1 Medical Science

Carbon dioxide (CO_2) is invisible, odorless, and nontoxic, and it does not seriously affect human health until the CO_2 content of the air reaches approximately 15,000 ppm, more than 37 times greater than the current concentration of atmospheric CO_2 (Luft *et al.*, 1974). A long-term rise in the atmosphere's CO_2 content, in the recent past or in the future, will have no direct adverse human health consequences. Even extreme model projections do not indicate anthropogenic activities will raise the air's CO_2 concentration above 1,000 to 2,000 ppm.

The medical literature shows, overwhelmingly, that warmer temperatures and a smaller difference between daily high and low temperatures, as occurred during the twentieth and early twenty-first centuries, reduce mortality rates (the subject of this section) as well as illness and mortality due to cardiovascular and respiratory disease and stroke occurrence (the subject of later sections). Humans are increasingly adapting to changes in temperature (and weather generally) thanks to the spread of technologies such as air conditioning, more efficient home heating, better insulation, and improvements in clothing and transportation. See, for example, the studies by Matzarakis *et al.* (2011) and Matthies and Menne (2009) and the additional sources they cite. This means over time, even the relatively small number of deaths caused by exposure to heat waves is declining, making the small temperature increase that may occur during the coming century highly unlikely to cause any deaths.

Medical science explains why colder temperatures often cause diseases and sometimes fatalities whereas warmer temperatures are associated with health benefits. Keatinge and Donaldson (2001) explain that "cold causes mortality mainly from arterial thrombosis and respiratory disease, attributable in turn to cold-induced hemo-concentration and hypertension [in the first case] and respiratory infections [in the second case]." McGregor (2005) notes "anomalous cold stress can increase blood viscosity and blood pressure due to the activation of the sympathetic nervous system which accelerates the heart rate and increases vascular resistance (Collins *et al.*, 1985; Jehn *et al.*, 2002; Healy, 2003; Keatinge *et al.*, 1984; Mercer, 2003; Woodhouse *et al.*, 1993)" due to vasoconstriction to reduce blood flow and heat loss at the surface, adding, "anomalously cold winters may also increase other risk factors for heart disease such as blood clotting or fibrinogen concentration, red blood cell count per volume and plasma cholesterol."

Wang *et al.* (2013) write, "a large change in temperature within one day may cause a sudden change in the heart rate and circulation of elderly people, which all may act to increase the risk of cardiopulmonary and other diseases, even leading to fatal consequences." This is significant for the climate change debate because, as Wang *et al.* also observe, "it has been shown that a rise of the minimum temperature has occurred at a rate three times that of the maximum temperature during the twentieth century over most parts of the world, which has led to a decrease of the diurnal temperature range (Karl *et al.*, 1984, 1991)." Robeson (2002) demonstrated, based on a study of 50 years of daily temperatures at more than 1,000 U.S. weather stations, that temperature variability declines with greenhouse warming, and at a very substantial rate, so this aspect of a warmer world would lead to a reduction in temperature-related deaths. Braganza *et al.* (2004) reported, "observed reductions in DTR over the last century are large." Alexander *et al.* (2006) found a global trend toward warmer nights and a much smaller trend toward warmer days for the

period 1951–2003, concluding "these results agree with earlier global studies … which imply that rather than viewing the world as getting hotter it might be more accurate to view it as getting less cold." See also Easterling *et al.* (1997) and Seltenrich (2015).

Keatinge and Donaldson (2004) report coronary and cerebral thrombosis account for about half of all cold-related deaths, and respiratory diseases account for approximately half of the rest. They say cold stress causes an increase in arterial thrombosis "because the blood becomes more concentrated, and so more liable to clot during exposure to cold." As they describe it, "the body's first adjustment to cold stress is to shut down blood flow to the skin to conserve body heat," which "produces an excess of blood in central parts of the body," and to correct for this effect, "salt and water are moved out from the blood into tissue spaces," leaving behind "increased levels of red cells, white cells, platelets and fibrinogen" that lead to increased viscosity of the blood and a greater risk of clotting. The British scientists report the infections that cause respiratory-related deaths spread more readily in cold weather because people "crowd together in poorly ventilated spaces when it is cold." In addition, they say "breathing of cold air stimulates coughing and running of the nose, and this helps to spread respiratory viruses and bacteria." The "train of events leading to respiratory deaths," they continue, "often starts with a cold or some other minor infection of the upper airways," which "spreads to the bronchi and to the lungs," whereupon "secondary infection often follows and can lead to pneumonia." They also note cold stress "tends to suppress immune responses to infections," and respiratory infections typically "increase the plasma level of fibrinogen, and this contributes to the rise in arterial thrombosis in winter."

Keatinge and Donaldson also note "cold spells are closely associated with sharp increases in mortality rates," and "deaths continue for many days after a cold spell ends." On the other hand, they report, "increased deaths during a few days of hot weather are followed by a lower than normal mortality rate," because "many of those dying in the heat are already seriously ill and even without heat stress would have died within the next 2 or 3 weeks." With respect to the implications of global warming for human mortality, Keatinge and Donaldson state, "since heat-related deaths are generally much fewer than cold-related deaths, the overall effect of global warming on health can be expected to be a beneficial one." They report, "the rise in temperature of 3.6°F expected over the next 50 years would increase heat-related deaths in Britain by about 2,000 but reduce cold-related deaths by about 20,000."

Keatinge and Donaldson's (2004) reference to deaths that typically would have occurred shortly even without excess heat is a phenomenon researchers call "displacement" or "harvesting." A study from Germany found "cold spells lead to excess mortality to a relatively small degree, which lasts for weeks," while "the mortality increase during heat waves is more pronounced, but is followed by lower than average values in subsequent weeks" (Laschewski and Jendritzky, 2002). The authors say the latter observation suggests people who died from short-term exposure to heat possibly "would have died in the short term anyway." They found the mean duration of above-normal mortality for the 51 heat episodes that occurred from 1968 to 1997 was 10 days, with a mean increase in mortality of 3.9%, after which there was a mean decrease in mortality of 2.3% for 19 days. The net effect of the two perturbations was an overall decrease in mortality of 0.2% over the full 29-day period.

References

Alexander, L.V., et al. 2006. Global observed changes in daily climate extremes of temperature and precipitation. *Journal of Geophysical Research* **111** (D5).

Braganza, K., Karoly, D.J., and Arblaster, J.M. 2004. Diurnal temperature range as an index of global climate change during the twentieth century. *Geophysical Research Letters* **31** (13).

Collins, K.J., Easton, J.C., Belfield-Smith, H., Exton-Smith, A.N., and Pluck, R.A. 1985. Effects of age on body temperature and blood pressure in cold environments. *Clinical Science* **69**: 465–70.

Easterling, D.R., et al. 1997. Maximum and minimum temperature trends for the globe. *Science* **277**: 364–7.

Healy, J.D. 2003. Excess winter mortality in Europe: a cross country analysis identifying risk factors. *Journal of Epidemiology and Public Health* **57**: 784–9.

Jehn, M., Appel, L.J., Sacks, F.M., and Miller III, E.R. 2002. The effect of ambient temperature and barometric pressure on ambulatory blood pressure variability. *American Journal of Hypertension* **15**: 941–5.

Karl, T.R., et al. 1984. A new perspective on recent global warming: asymmetric trends of daily maximum and minimum temperature. *Bulletin of the American Meteorological Society* **74**: 1007–23.

Karl, T.R., et al. 1991. Global warming: evidence for asymmetric diurnal temperature change. *Geophysical Research Letters* **18**: 2253–6.

Keatinge, W.R. and Donaldson, G.C. 2001. Mortality related to cold and air pollution in London after allowance for effects of associated weather patterns. *Environmental Research* **86**: 209–16.

Keatinge, W.R. and Donaldson, G.C. 2004. The impact of global warming on health and mortality. *Southern Medical Journal* **97**: 1093–9.

Keatinge, W.R., et al. 1984. Increases in platelet and red cell counts, blood viscosity, and arterial pressure during mild surface cooling: factors in mortality from coronary and cerebral thrombosis in winter. *British Medical Journal* **289**: 1404–8.

Laschewski, G. and Jendritzky, G. 2002. Effects of the thermal environment on human health: an investigation of 30 years of daily mortality data from SW Germany. *Climate Research* **21**: 91–103.

Luft, U.C., Finkelstein, S., and Elliot, J.C. 1974. Respiratory gas exchange, acid-base balance, and electrolytes during and after maximal work breathing 15 mm Hg $PICO_2$. In: Nahas, G. and Schaefer, K.E. (Eds.) *Carbon Dioxide and Metabolic Regulations*. New York, NY: Springer-Verlag, pp. 273–81.

Matthies, F. and Menne, B. 2009. Prevention and management of health hazards related to heatwaves. *International Journal of Circumpolar Health* **68**: 8–22.

Matzarakis, A., Muthers, S., and Koch, E. 2011. Human biometeorological evaluation of heat-related mortality in Vienna. *Theoretical and Applied Climatology* **105**: 1–10.

McGregor, G.R. 2005. Winter North Atlantic Oscillation, temperature and ischaemic heart disease mortality in three English counties. *International Journal of Biometeorology* **49**: 197–204.

Mercer, J.B. 2003. Cold—an underrated risk factor for health. *Environmental Research* **92**: 8–13.

Robeson, S.M. 2002. Relationships between mean and standard deviation of air temperature: implications for global warming. *Climate Research* **22**: 205–13.

Seltenrich, N. 2015. Between extremes: health effects of heat and cold. *Environmental Health Perspectives* **123**: A276–A280.

Wang, M-z., Zheng, S., He, S-l., Li, B., Teng, H-j., Wang, S-g., Yin, L., Shang, K-z., and Li, T-s. 2013. The association between diurnal temperature range and emergency room admissions for cardiovascular, respiratory, digestive and genitourinary disease among the elderly: a time series study. *Science of the Total Environment* **456–457**: 370–5.

Woodhouse, P.R., Khaw, K., and Plummer, M. 1993. Seasonal variation of blood pressure and its relationship to ambient temperature in an elderly population. *Journal of Hypertension* **11**: 1267–74.

4.2.2 Global

Gasparrini *et al.* (2015a) analyzed more than 74 million deaths in 384 locations across 13 countries between 1985 and 2012, finding 20 times more people die from cold-related rather than heat-related weather events, and extreme cold weather is much deadlier. They write, "Our findings show that temperature is responsible for advancing a substantial fraction of deaths, corresponding to 7.71% of mortality in the selected countries within the study period. Most of this mortality burden was caused by days colder than the optimum temperature (7.29%), compared with days warmer than the optimum temperature (0.42%). Furthermore, most deaths were caused by exposure to moderately hot and cold temperatures, and the contribution of extreme days was comparatively low, despite increased RRs [relative risks]." They also found "the optimum temperature at which the risk is lowest was well above the median, and seemed to be increased in cold regions." A figure illustrating their findings is reproduced below as Figure 4.2.2.1.

In a second paper, Gasparrini *et al.* (2015b) collected data for more than 20.2 million heat-related deaths that occurred in Australia, Canada, Japan, South Korea, Spain, the United Kingdom, and the United States during the summer months between 1985 and 2012. They report "mortality risk due to heat appeared to decrease over time in several countries, with relative risks associated with high temperatures significantly lower in 2006 compared with 1993 in the United States, Japan and Spain"; there was "a non-significant decrease in Canada"; "temporal changes were difficult to assess in Australia and South Korea due to low statistical power"; and they "found little evidence of variation in the United Kingdom," while "in the United States, the risk seemed to be completely abated in 2006 for summer temperatures below their 99th percentile."

They concluded there was "a statistically significant decrease in the relative risk for heat-related mortality in 2006 compared with 1993 in the majority of countries included in the analysis.

Seltenrich (2015) writes, "while isolated heat waves pose a major health risk and grab headlines when they occur, recent research has uncovered a more complex and perhaps unexpected relationship between temperature and public health," which is, as he continues, that "on the whole, far more deaths occur in cold weather than in hot." Seltenrich reports that "an analysis by the Centers for Disease Control and Prevention of U.S. temperature-related deaths between 2006 and 2010 showed that 63% were attributable to cold exposure, while only 31% were attributable to heat exposure," citing *National Health Statistics Report No. 76* of the National Center for Health Statistics of the U.S. Centers for Disease Control and Prevention. "In Australia and the United Kingdom, cold-related mortality between 1993 and 2006 exceeded heat-related mortality by an even greater margin, and is likely to do so through at least the end of the century," he writes, citing Vardoulakis *et al.* (2014).

Arbuthnott *et al.* (2016) examined "variations in temperature related mortality risks over the 20th and 21st centuries [to] determine whether population adaptation to heat and/or cold has occurred." A search of 9,183 titles and abstracts dealing with the subject returned 11 studies examining the effects of ambient temperature over time and six studies comparing the effect of heatwaves at specific points in time. Of the first 11 studies, with respect to the hot end of the temperature spectrum, Arbuthnott *et al.* report "all except one found some evidence of decreasing susceptibility." At the cold end of the temperature spectrum, they say "there is little consistent evidence for decreasing cold related mortality, especially over the latter part of the last century." With respect to the impacts of specific heatwave events on human health, Arbuthnott *et al.* state that four of the six papers included in this portion of their analysis revealed "a decrease in expected mortality," again signaling there has been a decrease in the vulnerability of the human populations studied over time. As for the cause(s) of the observed temperature-induced mortality declines, the authors acknowledge their methods are incapable of making that determination. However, they opine that it may, in part, be related to physiological acclimatization (human adaptation) to temperature.

Figure 4.2.2.1
Deaths caused by cold vs. heat

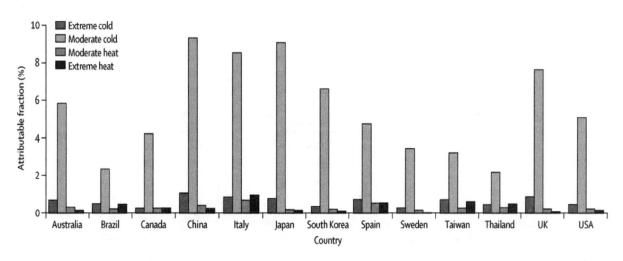

Source: Gasparrini *et al.*, 2015a, p.369.

Son *et al.* (2016) examined how mortality in Sao Paulo, Brazil, was affected by extremes of heat and cold over the 14.5-year period from 1996 to 2010, using "over-dispersed generalized linear modeling and Bayesian hierarchical modeling." They found "cold effects on mortality appeared higher than heat effects in this subtropical city with moderate climatic conditions."

Guo *et al.* (2014) obtained daily temperature and mortality data from 306 communities located in 12 countries (Australia, Brazil, Canada, China, Italy, Japan, Korea, Spain, Taiwan, Thailand, the United Kingdom, and the United States) within the time period 1972–2011. In order to "obtain an easily interpretable estimate of the effects of cold and hot temperatures on mortality," they "calculated the overall cumulative relative risks of death associated with cold temperatures (1st percentile) and with hot temperatures (99th percentile), both relative to the minimum-mortality temperature [75th percentile]" (see Figure 4.2.2.2). Despite the "widely ranging climates" they encountered, they report "the minimum-mortality temperatures were close to the 75th percentile of temperature in all 12 countries, suggesting that people have adapted to some extent to their local climates."

References

Arbuthnott, K., Hajat, S., Heaviside, C., and Vardoulakis, S. 2016. Changes in population susceptibility to heat and cold over time: assessing adaptation to climate change. *Environmental Health* **15**: 33.

Gasparrini, A., *et al.* 2015a. Mortality risk attributable to high and low ambient temperature: a multi-country observational study. *Lancet* **386**: 369–75.

Gasparrini, A., *et al.* 2015b. Temporal variation in heat-mortality associations: a multi-country study. *Environmental Health Perspectives* **123**: 1200–7.

Guo, Y., *et al.* 2014. Global variation in the effects of ambient temperature on mortality. *Epidemiology* **25**: 781–9.

Seltenrich, N. 2015. Between extremes: health effects of heat and cold. *Environmental Health Perspectives* **123**: A276–A280.

Son, J.-Y., Gouveia, N., Bravo, M.A., de Freitas, C.U., and Bell, M.L. 2016. The impact of temperature on mortality in a subtropical city: effects of cold, heat, and heat waves in Sao Paulo, Brazil. *International Journal of Biometeorology* **60**: 113–21.

Vardoulakis, S., Dear, K., Hajat, S., Heaviside, C., Eggen, B., and McMichael, A.J. 2014. Comparative assessment of the effects of climate change on heat- and cold-related

Figure 4.2.2.2
The pooled overall cumulative relation between temperature and deaths over lags of 0–21 days in 12 countries/regions

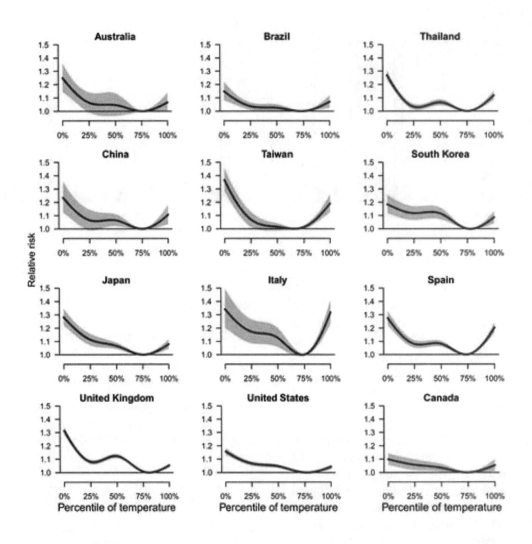

Source: Adapted from Guo *et al.*, 2014.

mortality in the United Kingdom and Australia. *Environmental Health Perspectives* **122**: 1285–92.

4.2.3 Asia

Behar (2000) studied sudden cardiac death (SCD) and acute myocardial infarction (AMI) in Israel, concentrating on the role temperature may play in the incidence of these health problems. Behar notes "most of the recent papers on this topic have concluded that a peak of SCD, AMI and other

cardiovascular conditions is usually observed in low temperature weather during winter." He cites an Israeli study by Green *et al.* (1994), which reported between 1976 and 1985 "mortality from cardio-vascular disease was higher by 50% in mid-winter than in mid-summer, both in men and women and in different age groups," even though summer temperatures in the Negev, where much of the work was conducted, often exceed 30°C (86°F) and winter temperatures typically do not drop below 10°C (50°F). Behar concludes these results "are reassuring for populations living in hot countries."

Kan *et al.* (2003) investigated the association between temperature and mortality in Shanghai, China, finding a V-like relationship between total mortality and temperature that had a minimum mortality risk at 26.7°C. Above this optimum temperature, they observed that total mortality increased by 0.73% for each degree Celsius increase, while for temperatures below the optimum value, total mortality increased by 1.21% for each degree Celsius decrease. The net effect of a warming in Shanghai, China, therefore, would likely be decreased mortality on the order of 0.5% per degree Celsius increase in temperature, or perhaps more.

Kan *et al.* (2007) examined the association between diurnal temperature range (DTR, defined as daily maximum temperature minus daily minimum temperature) and human mortality, using daily weather and mortality data from Shanghai over the period January 1, 2001 to December 31, 2004. They say their data suggest "even a slight increase in DTR is associated with a substantial increase in mortality." Their results suggest that in addition to the reduction in human mortality typically provided by the increase in daily mean temperature, the accompanying decrease in DTR also should have been tending to reduce human mortality.

Ma *et al.* (2011) investigated the impact of heat waves and cold spells on hospital admissions in Shanghai, China. The four researchers report the number of excess (above normal) hospital admissions during an eight-day heat wave was 352 whereas during a 10-day cold spell there were 3,725 excess admissions. Ma *et al.* conclude "the cold spell seemed to have a larger impact on hospital admission than the heat wave in Shanghai."

Cheng and Kan (2012) analyzed mortality, air pollution, temperature, and covariate data over the period January 1, 2001 through December 31, 2004 in Shanghai. They report they "did not find a significant interaction between air pollution and higher temperature [>85th percentile days]," but "the interaction between PM_{10} [particulate matter, 10 micrometers or smaller] and extreme low temperature [<15th percentile days] was statistically significant for both total and cause-specific mortality." Compared to normal temperature days (15th to 85th percentile), they found a 10-μg/m3 increase in PM_{10} on extreme low temperature days led to all-cause mortality rising from 0.17% to 0.40%. They add, "the interaction pattern of O_3 with low temperature was similar," noting their finding of "a stronger association between air pollution and daily mortality on extremely cold days confirms those of three

earlier seasonal analyses in Hong Kong, Shanghai and Athens," citing Touloumi *et al.* (1996), Wong *et al.* (1999, 2001), and Zhang *et al.* (2006).

Guo *et al.* (2012) examine the nonlinear and delayed effects of temperature on cause-specific and age-specific mortality employing data from 1999 to 2008 for Chiang Mai, Thailand, with a population of 1.6 million people. Controlling for season, humidity, ozone, and particulate matter (PM_{10}) pollution, the three researchers found "both hot and cold temperatures resulted in an immediate increase in all mortality types and age groups," but "the hot effects on all mortality types and age groups were short-term, while the cold effects lasted longer." The cold effects were greater, with more people dying from them than from the effects of heat.

Lindeboom *et al.* (2012) used daily mortality and weather data for the period 1983–2009 pertaining to Matlab, Bangladesh to measure lagged effects of weather on mortality, controlling for time trends and seasonal patterns. The four researchers report "mortality in the Matlab surveillance area shows overall weak associations with rainfall, and stronger negative association with temperature." They determined there was "a 1.4% increase in mortality with every 1°C decrease in mean temperature at temperatures below 29.2°C," but only "a 0.2% increase in mortality with every 1°C increase in mean temperature."

Wang *et al.* (2013) evaluated the short-term effect of DTR on emergency room (ER) admissions among elderly adults in Beijing. The nine researchers report "significant associations were found between DTR and four major causes of daily ER admissions among elderly adults in Beijing." They state "a 1°C increase in the 8-day moving average of DTR (lag 07) corresponded to an increase of 2.08% in respiratory ER admissions and 2.14% in digestive ER admissions," and "a 1°C increase in the 3-day and 6-day moving average of DTR (lag 02 and lag 05) corresponded to a 0.76% increase in cardiovascular ER admissions, and a 1.81% increase in genitourinary ER admissions, respectively."

Wu *et al.* (2013) assessed the health effects of temperature on mortality in four subtropical cities of China (Changsha, Guangzhou, Kunming, and Zhuhai). The 11 researchers report a U-shaped relationship between temperature and mortality was found in the four cities, indicating "mortality is usually lowest around a certain temperature and higher at lower or higher temperatures." Although "both low and high temperatures were associated with increased mortality in the four subtropical

Chinese cities," Wu *et al.* state the "cold effect was more durable and pronounced than the hot effect."

Yang *et al.* (2013) examined the effects of DTR on human mortality rates using daily meteorological data for the period January 1, 2003 through December 31, 2010 from a single station located in the heart of the urban area of Guangzhou City (the largest metropolis in Southern China). They found "a linear DTR-mortality relationship, with evidence of increasing mortality with DTR increase," where "the effect of DTR occurred immediately and lasted for four days," such that over that time period, a 1°C increase in DTR was associated with a 0.47% increase in non-accidental mortality. In addition, they report there was a joint adverse effect with temperature "when mean temperature was below 22°C [71.6°F], indicating that high DTR enhanced cold-related mortality." In light of their findings, the eight researchers speculate the expected "decrease in DTR in future climate scenarios might lead to two benefits: one from decreasing the adverse effects of DTR [which is reduced due to greater warming at night than during the day], and the other from decreasing the interaction effect with temperature [which is expected to rise with greenhouse warming]."

Onozuka and Hagihara (2015) acquired data on daily emergency ambulance dispatches in Japan's 47 prefectures from 2007 to 2010, which they used to determine relationships between medical emergency transport and temperature. They found the fraction of ambulance dispatches attributable to low temperatures was 6.94% for all causes, while that attributable to high temperatures was 1.01% for all causes. They report "the majority of temperature-related emergency transport burden was attributable to lower temperature," which burden was almost seven times greater than that attributable to higher temperatures.

Huang *et al.* (2015) analyzed community-specific daily mortality data for the period January 1, 2006 to December 31, 2011, obtained from the Chinese Center for Disease Control and Prevention, together with community-specific daily meteorological data for the same period, obtained from the China National Weather Data Sharing System. They found temperature-mortality relationships were "approximately V-shaped or U-shaped, with a minimum mortality temperature (MMT)," above and below which human mortality increased. For each of the 66 communities they studied, they calculated "the change in mortality risk for a 1°C decrease in temperature below the MMT (cold effect) and for a

1°C temperature increase above the MMT (heat effect)." This work revealed that a 1°C temperature increase above the MMT resulted in a mean increase of 1.04% in human mortality for the 66 communities, while a 1°C temperature decrease below the MMT resulted in a mean increase of 3.44% in human mortality, demonstrating that cooling below the minimum mortality temperature was 3.31 times more deadly than was warming above it.

Chau and Woo (2015) examined summer (June–August) versus winter (December–February) excess mortality trends among the older population (65 years and older) of Hong Kong citizens over the 35-year period 1976–2010. They performed statistical analyses that searched for relationships between various measures of extreme meteorological data and recorded deaths due to cardiovascular and respiratory-related causes. They report there was an average rise in mean temperature of "0.15°C per decade in 1947–2013 and an increase of 0.20°C per decade in 1984–2013." They also note that over the 35-year period of their analysis "winter became less stressful," with fewer extreme cold spells. Summers, on the other hand, became "more stressful as the number of Hot Nights in summer increased by 0.3 days per year and the number of summer days with very high humidity (daily relative humidity over 93%) increased by 0.1 days per year." Given such observations, it would be expected under the global warming hypothesis that cold-related deaths should have declined and heat-related deaths should have increased across the length of the record. As shown in Figure 4.2.3.1, cold-related death rates did indeed decline (by 49.3%), from approximately 21 deaths per 1,000 persons in 1976 to 10.6 deaths in 2010. Heat-related death rates, however, did not increase. Rather, they too declined, from 13.2 per 1,000 persons in 1976 to 8.10 in 2010 (a decrease of 38.8%). Thus, both cold- and heat-related death rates declined over the 35-year period of study. The authors concluded, "Hong Kong has not observed an increase in heat-related deaths as predicted in the Western literature."

Ma *et al.* (2015) used a distributed lag non-linear model to determine the community-specific effects of extreme hot and cold temperatures on non-accidental mortality during 2006–2011 in 66 Chinese cities, after which they conducted a multivariate meta-analysis that enabled them to pool the individual estimates of community-specific effects. They found a U-shaped relationship whereby both daily maximum and minimum temperatures were associated with increased mortality risk compared to that of the overall mean temperature, but the relative risk (RR) at the mean daily minimum temperature was significantly greater than the RR at the mean daily maximum temperature. Typically experienced extreme cold throughout China, they conclude, is much more deadly than is typically experienced extreme heat, which the 14 researchers note "is consistent with previous studies," including Guo *et al.* (2011), Guo *et al.* (2013), Chen *et al.* (2013), Wu *et al.* (2013), and Xie *et al.* (2013).

Ng *et al.* (2016) analyzed "daily total [from natural causes], cardiovascular and respiratory disease mortality and temperature data from 1972 to 2010 for 47 prefectures." They report their data "show a general decrease in excess heat-related mortality over the past 39 years despite increasing temperatures [of approximately 1°C]," demonstrating, in their words, "that some form of adaptation to extreme temperatures has occurred in Japan." More specifically, their data revealed a national reduction of 20, 21, and 46 cases of deaths per 1,000 due to natural, cardiovascular, and respiratory causes, corresponding to respective drops of 69, 66, and 81%. Ng *et al.* write, an "increase of AC [air conditioning] prevalence was not associated

with a reduction of excess mortality over time," yet they note "prefectures and populations with improved economic status documented a larger decline of excess mortality," adding that "healthcare resources were associated with fewer heat-related deaths in the 1970s, but the associations did not persist in the more recent period (i.e., 2006–2010)."

Wang *et al.* (2016) collected daily mortality and meteorological data from 66 communities across China over the period 2006–2011. They analyzed the data to discern relationships between cold spell characteristics and human mortality, finding cold spells significantly increased human mortality risk in China. They found the combined cumulative excess mortality risk (CER) for all of China when defining cold spells with a 5th and 2.5th percentile temperature intensity threshold was 28.5 and 39.7%, respectively. However, there were notable geographic differences: CER was tempered and near zero in the colder/higher latitudes, but increased to 58.7 and 92.9% at the corresponding 5th and 2.5th percentile temperature intensity thresholds for the warmest and most southern latitude. Such geographic differences in mortality risk, according to the authors, are likely the product of better physiological and behavioral

Figure 4.2.3.1
Summer and winter age-standardized mortality rate (per 1,000 population) for adults age 65 and older in Hong Kong over the period 1976–2010

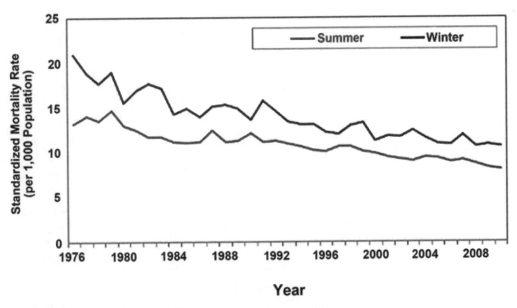

Source: Adapted from Chau and Woo, 2015.

acclimatization of the northerly populations to cold weather.

Wang *et al.* also report that the strength of the temperature/mortality relationship was modified by cold spell characteristics and human-specific factors, such that there was a significant increase in non-accident mortality during cold spells that were longer, stronger, and/or earlier in the season. In addition, mortality rates were found to increase by age and decrease by education level; the older and less educated tended to experience the greatest risk of death. The health status of an individual was also a factor. Those with respiratory illnesses had higher CER rates than those suffering from cardiovascular or cerebrovascular diseases.

Cui *et al.* (2016) set out to examine which end of the temperature spectrum (hot or cold weather) exerts a greater deleterious effect on human health. They focused their analysis on the largest city within the Sichuan Province of China, Chengdu, with a population of 14.65 million. They gathered daily meteorological and death record data for the city for the period January 1, 2011 through December 31, 2014 and estimated the relationship between daily mortality and ambient temperature using a distribution lag model with a quasi-Poisson regression, controlling for long-time trends and day of the week. They also calculated the relative risk of mortality, defined as the risk of death attributable to heat or cold above or below the optimum temperature at which minimum mortality occurred. They found the "total fraction of deaths caused by both heat and cold was 10.93%." However, they note "the effect of cold was significant and was responsible for most of the burden," whereas "the effect of heat was small and non-significant." The effect of cold temperatures was calculated to be ten times larger than that of warm temperatures (9.96% vs. 0.97%). Figure 4.2.3.2 illustrates the far greater impact of cold temperatures.

Chung *et al.* (2017) note "understanding how the temperature-mortality association worldwide changes over time is crucial to addressing questions of human adaptation under climate change." They investigated the temporal change in this relationship for 15 cities in three countries from Northeast Asia (Japan, Korea, and Taiwan) over the past four decades, during which time temperatures increased in all cities. They utilized a generalized linear model with splines, allowing them to investigate a nonlinear association between temperature and mortality, as well as a non-

Figure 4.2.3.2

Exposure-response relationship of temperature and non-accidental deaths in Chengdu, China (2011–2014)

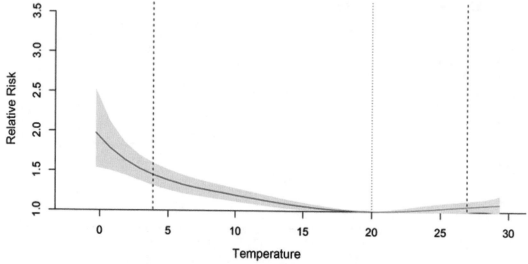

The blue part of the curve is the exposure-response association (with 95% empirical confidence interval, shaded grey) of cold, and the red one presents the heat. The dotted line at 20°C is minimum mortality temperature and the dashed lines are the 2.5th and 97.5th percentile. *Source*: Adapted from Cui *et al.*, 2016.

linear change in this association over time. Additionally, their analysis was stratified by cause-specific mortality (from cardiovascular, respiratory, and non-cardiorespiratory) and age group (under age 65, 65–75 years, and greater than 75 years of age).

Their analysis revealed that cold-related mortality risk remained relatively constant over time, with only one of the 15 locations exhibiting a trend that was statistically significant. In contrast, all of the study locations revealed declines in heat-related mortality (weighted average of approximately -16% change), only three of which declines were not statistically significant. Chung *et al.* also report the temporal pattern of decreasing heat-related mortality differed by age and cause of death, where the oldest segment of the population and respiratory-caused deaths experienced the largest decreases.

Chung *et al.*'s findings dispel two claims: that global warming will enhance heat-related deaths and that the elderly population will suffer the most. In direct contradiction of these assertions, the results of this study clearly demonstrate that populations are adapting to warmer temperatures much better than to colder temperatures (by a factor of about 10), as evidenced by declining trends in mortality risk over time, and the elderly are not suffering a disproportionate number of heat-related deaths. Whereas trends in heat-related deaths were higher in the older populations at the beginning of the records, they have disproportionately declined and, Chung *et al.* report, "converged to become similar among the three age groups in later years." Wang *et al.* (2017) studied 122 communities across mainland China, using daily non-accidental mortality and meteorological data for the period January 1, 2007 through December 31, 2012 and a quasi-Poisson regression with a distributed lag nonlinear model to estimate the relationship between daily mean temperature and mortality in each of the communities they studied. They pooled their data into one of five temperature zones from which they analyzed the temperature-mortality relationship at the regional and national level. They report that both high and low temperatures increase the risk of mortality, but that the risk is higher and lasts longer at the cold edge of the temperature spectrum. Qualitatively, the relative risk of mortality due to cold was 1.63 versus 1.15 for heat.

To further illustrate the greater danger of extreme cold, the average relative risk of mortality due to extreme cold and heat for each of the five temperature regions are presented in Figure 4.2.3.3.

References

Behar, S. 2000. Out-of-hospital death in Israel: should we blame the weather? *Israel Medical Association Journal* **2**: 56–7.

Chau, P.H. and Woo, J. 2015. The trends in excess mortality in winter vs. summer in a sub-tropical city and its association with extreme climate conditions. *PLoS ONE* **10**: e0126774.

Chen, R., Wang, C., Meng, X., Chen, H., Thach, T.Q., Wong, C.-M., and Kan, H. 2013. Both low and high temperature may increase the risk of stroke mortality. *Neurology* **81**: 1064–70.

Cui, Y., *et al.* 2016. Heat or cold: which one exerts greater deleterious effects on health in a basin climate city? Impact of ambient temperature on mortality in Chengdu, China. *International Journal of Environmental Research and Public Health* **13**: 1225.

Green, M.S., Harari, G., and Kristal-Boneh, E. 1994. Excess winter mortality from ischaemic heart disease and stroke during colder and warmer years in Israel. *European Journal of Public Health* **4**: 3–11.

Guo, Y., Barnett, A.G., Pan, X., Yu, W., and Tong, G. 2011. The impact of temperature on mortality in Tianjin, China: a case-crossover design with a distributed lag nonlinear model. *Environmental Health Perspectives* **119**: 1719–25.

Guo, Y., Li, S., Zhang, Y., Armstrong, B., Jaakkola, J.J.K., Tong, S.K., and Pan, X. 2013. Extremely cold and hot temperatures increase the risk of ischaemic heart disease mortality: epidemiological evidence from China. *Heart* **99**: 195–203.

Guo, Y., Punnasiri, K., and Tong, S. 2012. Effects of temperature on mortality in Chiang Mai city, Thailand: a time series study. *Environmental Health* **11**: 36.

Huang, Z., *et al.* 2015. Individual-level and community-level effect modifiers of the temperature-mortality relationship in 66 Chinese communities. *BMJ Open* **5**.

Kan, H., *et al.* 2007. Diurnal temperature range and daily mortality in Shanghai, China. *Environmental Research* **103**: 424–31.

Kan, H-D., Jia, J., and Chen, B-H. 2003. Temperature and daily mortality in Shanghai: a time-series study. *Biomedical and Environmental Sciences* **16**: 133–9.

Lindeboom, W., Alam, N., Begum, D., and Streatfield, P.K. 2012. The association of meteorological factors and mortality in rural Bangladesh, 1983–2009. *Global Health Action* **5**: 61–73.

403

Figure 4.2.3.3

Pooled mortality risks of extreme cold and heat for five temperature zones

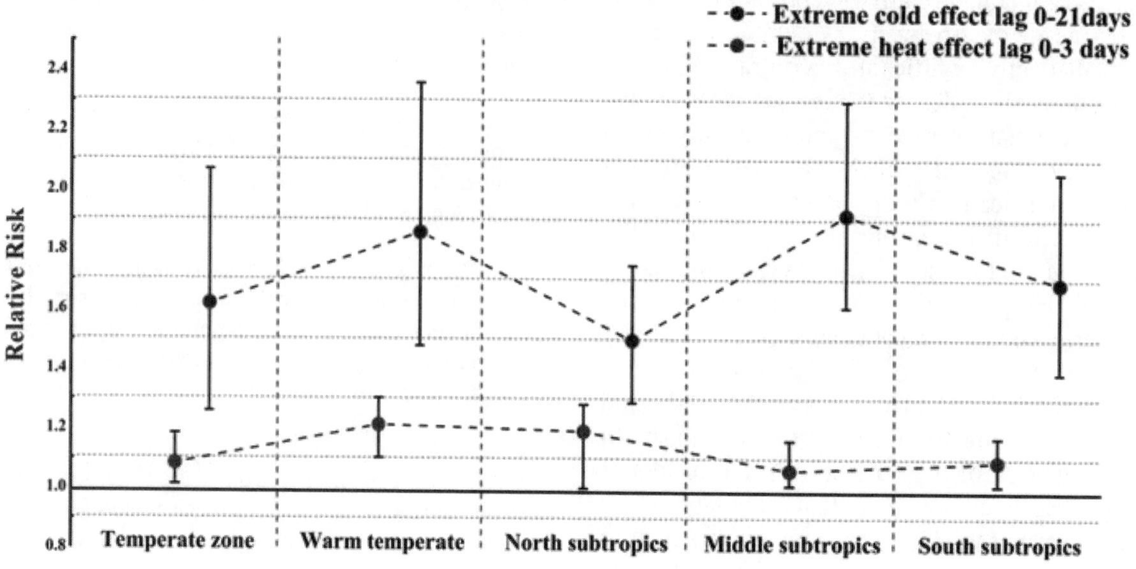

Source: Wang *et al.*, 2017.

Ma, W., *et al.* 2015. The temperature-mortality relationship in China: an analysis from 66 Chinese communities. *Environmental Research* **137**: 72–7.

Ma, W., Xu, X., Peng, L., and Kan, H. 2011. Impact of extreme temperature on hospital admission in Shanghai, China. *Science of the Total Environment* **409**: 3634–7.

Ng, C.F.S., Boeckmann, M., Ueda, K., Zeeb, H., Nitta, H., Watanabe, C., and Honda, Y. 2016. Heat-related mortality: effect of modification and adaptation in Japan from 1972 to 2010. *Global Environmental Change* **39**: 234–43.

Onozuka, D. and Hagihara, A. 2015. All-cause and cause-specific risk of emergency transport attributable to temperature. *Medicine* **94**: e2259.

Touloumi, G., Samoli, E., and Katsouyanni, K. 1996. Daily mortality and "winter type" air pollution in Athens, Greece: a time series analysis within the APHEA project. *Journal of Epidemiology and Community Health* **50** (Supplement 1): 47–51.

Wang, C., Zhang, Z., Zhou, M., Zhang, L., Yin, P., Ye, W., and Chen, Y. 2017. Nonlinear relationship between extreme temperature and mortality in different temperature zones: a systematic study of 122 communities across the mainland of China. *Science of the Total Environment* **586**: 96–106.

Wang, L., *et al.* 2016. The impact of cold spells on mortality and effect modification by cold spell characteristics. *Scientific Reports* **6**: 38380.

Wang, M-z., *et al.* 2013. The association between diurnal temperature range and emergency room admissions for cardiovascular, respiratory, digestive and genitourinary disease among the elderly: a time series study. *Science of the Total Environment* **456–457**: 370–5.

Wong, C.M., Ma, S., Hedley, A.J., and Lam, T.H. 1999. Does ozone have any effect on daily hospital admissions for circulatory diseases? *Journal of Epidemiology and Community Health* **53**: 580–1.

Wong, C.M., Ma, S., Hedley, A.J., and Lam, T.H. 2001. Effect of air pollution on daily mortality in Hong Kong. *Environmental Health Perspectives* **109**: 335–40.

Wu, W., *et al.* 2013. Temperature-mortality relationship in four subtropical Chinese cities: a time-series study using a distributed lag non-linear model. *Science of the Total Environment* **449**: 355–62.

Xie, H., *et al.* 2013. Short-term effects of the 2008 cold spell on mortality in three subtropical cities in Guangdong province, China. *Environmental Health Perspectives* **121**: 210–6.

Yang, J., Liu, H.-Z., Ou, C.-Q., Lin, G.-Z., Zhou, Q., Shen, G.-C., Chen, P.-Y., and Guo, Y. 2013. Global climate change: impact of diurnal temperature range on mortality in Guangzhou, China. *Environmental Pollution* **175**: 131–6.

Zhang, Y., *et al.* 2006. Ozone and daily mortality in Shanghai, China. *Environmental Health Perspectives* **114**: 1227–32.

4.2.4 Australia

Bennett *et al.* (2014) studied the ratio of summer to winter deaths against a background of rising average annual temperatures over a period of four decades in Australia, finding this summer/winter "death ratio" had increased from a value of 0.71 to 0.86 since 1968, due to summer deaths rising faster than winter deaths.

Bennett *et al.* also note "the change [the increase in summer/winter death ratio] has so far been driven more by reduced winter mortality [due to reductions in extreme cold] than by increased summer mortality [due to increases in extreme warmth]," as well as the fact that the greater number of typical winter-season deaths "is largely explained," in their words, "by infectious disease transmission peaks during winter and the exacerbation of chronic diseases, especially cardiovascular and respiratory conditions," citing Cameron *et al.* (1985).

In a study of Adelaide, Brisbane, Melbourne, Perth, and Sydney, Australia, Huang *et al.* (2015) split "seasonal patterns in temperature, humidity and mortality into their stationary (seasonal) and non-stationary (unseasonal) parts," where "a stationary seasonal pattern is consistent from year-to-year, and a non-stationary pattern varies from year-to-year," with the aim to determine "how unseasonal patterns in temperature and humidity in winter and summer were associated with unseasonal patterns in death." Working with mortality data for more than 1.5 million deaths from January 1, 1988 to December 31, 2009, the researchers found there were "far more deaths in winter," such that "death rates were 20–30% higher in a winter than a summer" (see Figure 4.2.4.1). They note "this seasonal pattern is consistent across much of the world, and many countries suffer 10% to 30% excess deaths in winter," citing the work of Healy (2003) and Falagas *et al.* (2009). They also report that winters that were colder or drier than a typical winter had significantly increased death risks, whereas "summers that were warmer or more humid than average showed no increase in death risks."

Utilizing a database of natural hazard event impacts known as PerilAUS, produced by Risk Frontiers, an independent research center sponsored by the insurance industry and located at Australia's Macquarie University, Coates *et al.* (2014) derived "a lower-bound estimate of heat-associated deaths in Australia since European settlement." The estimate for "extreme heat events," also often referred to as "heat waves," from the time of European settlement in 1844 to 2010 was at least 5,332, while from 1900 to 2010 it was 4,555.

The five researchers also determined "both deaths and death rates (per unit of population) fluctuate widely but show an overall decrease with time." In South Australia, for example, where the death rate has been the highest, they report "the decadal death rate has fallen from 1.69 deaths per 100,000 population in the 1910s to 0.26 in the 2000s," a decline of nearly 85%. Although "the elderly are significantly more vulnerable to the risk of heat-associated death than the general population, and this vulnerability increases with age," they find "death rates amongst seniors also show a decrease with time." That finding is in harmony with Bobb *et al.* (2014), who found much the same thing for the elderly in the United States, where between 1987 and 2005, the decline in death rate due to heat "was largest among those ≥ 75 years of age."

References

Bennett, C.M., Dear, K.B.G., and McMichael, A.J. 2014. Shifts in the seasonal distribution of deaths in Australia, 1968–2007. *International Journal of Biometeorology* **58**: 835–42.

Bobb, J.F., Peng, R.D., Bell, M.L., and Dominici, F. 2014. Heat-related mortality and adaptation to heat in the United States. *Environmental Health Perspectives* **122**: 811–6.

Cameron, A.S., Roder, D.M., Esterman, A.J., and Moore, B.W. 1985. Mortality from influenza and allied infections in South Australia during 1968–1981. *Medical Journal of Australia* **142**: 14–7.

Coates, L., Haynes, K., O'Brien, J., McAneney, J., and Dimer de Oliveira, F. 2014. Exploring 167 years of vulnerability: an examination of extreme heat events in Australia 1844–2010. *Environmental Science & Policy* **42**: 33–44.

Falagas, M.E., *et al.* 2009. Seasonality of mortality: the September phenomenon in Mediterranean countries. *Canadian Medical Association Journal* **181**: 484–6.

Figure 4.2.4.1
Stationary seasonal patterns of mortality (standardized to January) in five Australian cities (1988–2009)

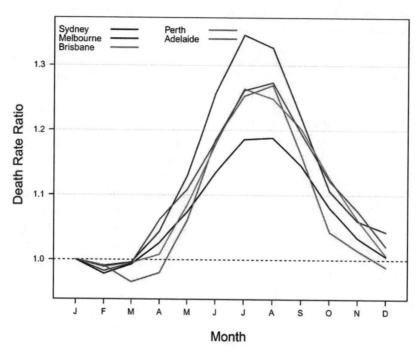

Australian spring/summer is September to February; fall/winter is March to August. *Source*: Adapted from Huang *et al.,* 2015.

Healy, J.D. 2003. Excess winter mortality in Europe: a cross country analysis identifying key risk factors. *Journal of Epidemiology and Community Health* **57**: 784–9.

Huang, C., Chu, C., Wang, X., and Barnett, A.G. 2015. Unusually cold and dry winters increase mortality in Australia. *Environmental Research* **136**: 1–7.

4.2.5 Europe

Keatinge and Donaldson (2001) analyzed the effects on human mortality of temperature, wind, rain, humidity, and sunshine during high pollution days in the greater London area over the period 1976–1995. They observed simple plots of mortality rate versus daily air temperature revealed a linear increase as temperatures fell from 15°C (59°F) to near 0°C (32°F). Mortality rates at temperatures above 15°C, however, were "grossly alinear," as they describe it, showing no trend. Only low temperatures were found to have a significant effect on immediate and long-term mortality. They conclude, "the large, delayed

increase in mortality after low temperature is specifically associated with cold and is not due to associated patterns of wind, rain, humidity, sunshine, SO_2, CO_2 or smoke."

Kysely and Huth (2004) calculated deviations of the observed number of deaths from the expected number of deaths for each day of the year in the Czech Republic for the period 1992–2000. They found "the distribution of days with the highest excess mortality in a year is clearly bimodal, showing a main peak in late winter and a secondary one in summer." Regarding the smaller number of summer heat-wave-induced deaths, they also found "a large portion of the mortality increase is associated with the harvesting effect, which consists in short-term shifts in mortality and leads to a decline in the number of deaths after hot periods (e.g. Rooney *et al.*, 1998; Braga *et al.*, 2002; Laschewski and Jendritzky, 2002)." For the Czech Republic, they report, "the mortality displacement effect in the severe 1994 heat waves can be estimated to account for about 50% of the total number of victims." As they describe it, "people who would have died in the short term even in the absence of oppressive weather conditions made up about half of the total number of deaths."

Diaz *et al.* (2005) examined the effect of extreme winter temperature on mortality in Madrid, Spain for people older than 65, using data from 1,815 winter days over the period 1986–1997, during which time 133,000 deaths occurred. They found that as maximum daily temperature dropped below 6°C (42.8°F), which they describe as an unusually cold day (UCD), "the impact on mortality also increased significantly." They also found the impact of UCDs increased as the winter progressed, with the first UCD of the season producing an average of 102 deaths/day at a lag of eight days and the sixth UCD producing an average of 123 deaths/day at a lag of eight days.

Laaidi *et al.* (2006) conducted an observational population study in six regions of France between 1991 and 1995 to assess the relationship between temperature and mortality in areas of widely varying climatic conditions and lifestyles. In all cases they found "more evidence was collected showing that cold weather was more deadly than hot weather." These findings, the researchers say, are "broadly consistent with those found in earlier studies conducted elsewhere in Europe (Kunst *et al.*, 1993; Ballester *et al.*, 1997; Eurowinter Group, 1997; Keatinge *et al.*, 2000a, 2000b; Beniston, 2002; Muggeo and Vigotti, 2002), the United States (Curriero *et al.*, 2002) and South America (Gouveia *et al.*, 2003)." They also say their findings "give grounds for confidence in the near future," stating even a 2°C warming over the next half century "would not increase annual mortality rates."

Analitis *et al.* (2008) analyzed short-term effects of cold weather on mortality in 15 major European cities using data from 1990–2000 and found "a 1°C decrease in temperature was associated with a 1.35% increase in the daily number of total natural deaths and a 1.72%, 3.30% and 1.25% increase in cardiovascular, respiratory, and cerebro-vascular deaths, respectively." In addition, they report "the increase was greater for the older age groups" and the cold effect "persisted up to 23 days, with no evidence of mortality displacement." They conclude their results "add evidence that cold-related mortality is an important public health problem across Europe and should not be overlooked by public health authorities because of the recent focus on heat-wave episodes."

Christidis *et al.* (2010) compiled the numbers of daily deaths from all causes for men and women 50 years of age or older in England and Wales for the period 1976–2005, and then compared the death results with surface air temperature data. As expected, during the hottest portion of the year, warming led to increases in death rates, whereas during the coldest portion of the year warming led to decreases in death rates. The three scientists report there were only 0.7 death per million people per year due to warming in the hottest part of the year, but a decrease of fully 85 deaths per million people per year due to warming in the coldest part of the year, for a phenomenal lives-saved to lives-lost ratio of 121.4.

Fernandez-Raga *et al.* (2010) obtained data from weather stations situated in eight of the provincial capitals in the Castile-Leon in Spain for the period 1980–1998, and they obtained contemporary mortality data for deaths associated with cardiovascular, respiratory, and digestive system diseases. For all three of the disease types studied, they found "the death rate is about 15% higher on a winter's day than on a summer's day," which they describe as "a result often found in previous studies," citing Fleming *et al.* (2000), Verlato *et al.* (2002), Grech *et al.* (2002), Law *et al.* (2002), and Eccles (2002). Their data, plotted in Figure 4.2.5.1, clearly demonstrate the people of the Castile-Leon region of Spain are much more likely to die from a cardiovascular disease in the extreme cold of winter than in the extreme heat of summer. The same holds true with respect to dying from respiratory and digestive system diseases.

Wichmann *et al.* (2011) investigated the association between the daily three-hour maximum apparent temperature (which reflects the physiological experience of combined exposure to humidity and temperature) and deaths due to cardiovascular disease (CVD), cerebrovascular disease (CBD), and respiratory disease (RD) in Copenhagen over the period 1999–2006. During the warm half of the year (April–September), they found a rise in temperature had an inverse or protective effect with respect to CVD mortality (a 1% decrease in death in response to a 1°C increase in apparent temperature). This finding is unusual but also has been observed in Dublin, Ireland, as reported by Baccini *et al.* (2008, 2011). Wichmann *et al.* found no association with RD and CBD mortality. At the other end of the thermal spectrum, during the cold half of the year, all three associations were inverse or protective. This finding, according to the researchers, is "consistent with other studies (Eurowinter Group, 1997; Nafstad *et al.*, 2001; Braga *et al.*, 2002;

O'Neill *et al.*, 2003; Analitis *et al.*, 2008)."

Matzarakis *et al.* (2011) studied the relationship between heat stress and all-cause mortality in the densely populated city of Vienna, Austria. Based on data from 1970–2007, and after adjusting the long-term mortality rate to account for temporal variations in the size of the population of Vienna, temporal changes in life expectancy, and the changing age structure of Vienna's population, the three researchers found a significant relationship between heat stress and mortality. However, over this 38-year period, "some significant decreases of the sensitivity were found, especially in the medium heat stress levels," they report. These decreases in sensitivity, they write, "could indicate active processes of long-term adaptation to the increasing heat stress." In the discussion section of their paper, they write such sensitivity changes "were also found for other regions," citing Davis *et al.* (2003b), Koppe (2005), Tan *et al.* (2007), and Donaldson and Keatinge (2008). In the conclusion of their paper, they refer to these changes as "positive developments."

Kysely and Plavcova (2012) write, "there is much concern that climate change may be associated with large increases in heat-related mortality," but "growing evidence has been emerging that the relationships between temperature extremes and mortality impacts are nonstationary," and "most of these studies point to declining heat-related mortality in developed countries, including the US, Australia, the UK, the Netherlands and France (Davis *et al.*, 2002, 2003a, 2003b; Bi and Walker, 2001; Donaldson *et al.*, 2001; Garssen *et al.*, 2005; Carson *et al.*, 2006; Fouillet *et al.*, 2008; Sheridan *et al.*, 2009)." This is true, they note, despite "aging populations and prevailing rising trends in temperature extremes."

Kysely and Plavcova then examined "temporal changes in mortality associated with spells of large positive temperature anomalies (hot spells) in extended summer season in the population of the Czech Republic (Central Europe) during 1986–2009." They found declining mortality trends in spite of rising temperature trends, just the opposite of what the IPCC claims will occur in response to global

Figure 4.2.5.1
Monthly deaths in the Castile-Leon Region of Spain attributable to cardiovascular disease

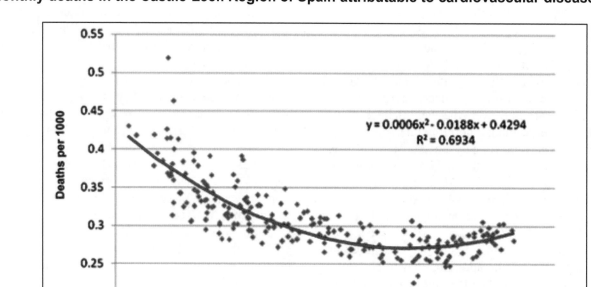

Source: Adapted from Fernandez-Raga *et al.*, 2010.

warming. The Czech scientists add, "the finding on reduced vulnerability of the population remains unchanged if possible confounding effects of within-season acclimatization and mortality displacement are taken into account," and "neither does it depend on the changing age structure of the population, since similar (and slightly more pronounced) declines in the mortality impacts are found in the elderly (age group 70+ years) when examined separately."

Carmona et al. (2016) determined the impact of daily minimum temperatures on mortality in each of Spain's 52 provincial capitals; they report this effort revealed relative cold-induced mortality increases of 1.13 due to natural causes, 1.18 due to circulatory causes, and 1.24 due to respiratory causes, all of which they found to be "slightly greater than those obtained to date for heat." They note, "from a public health standpoint, there is a need for specific cold wave prevention plans at a regional level which would enable mortality attributable to low temperatures to be reduced."

Minimum mortality temperature (MMT) is defined as the temperature at which the lowest mortality rate for a given location occurs over a given time period of examination. Above and below this value mortality rates increase as temperatures rise and fall, respectively. Todd and Valleron (2015) note MMT values in comparative studies from different latitudes are repeatedly shown to be higher in locations where the mean summer temperature is higher. Such observations, according to the two French researchers, have led to the interpretation that humans are capable of adapting to local climatic conditions. However, they additionally note that "drawing conclusions from this geographic observation about the possible adaptability of human populations to future climate change requires observing that, similarly, MMT at a given location changes over time when climate changes." With this caveat in mind, Todd and Valleron investigated whether MMT for a given location does indeed change over time as climate changes.

Todd and Valleron examined the change in MMT in France over the 42-year period from 1968–2009 and over three 14-year subsets: 1968–1981, 1982–1995, and 1996–2009. Their data included 228 0.5 x 0.5 degree latitude/longitude grid squares of daily mean temperature and individual death certificate information for persons > 65 years old who died in France over the 42-year period of examination (approximately 16.5 million persons).

Their results indicate MMT was strongly correlated with, and had a positive linear relationship

with, mean summer temperature over the entire period. They also determined that mean MMT increased from 17.5°C in the first 14-year period (1968–2981) to 17.8°C and 18.2°C in the second (1982–1995) and third (1996–2009) time periods examined. Todd and Valleron conclude their "spatiotemporal analysis indicated some human adaptation to climate change, even in rural areas."

Ballester et al. (2016) analyzed a host of climate variables against daily regional counts of mortality from 16 European countries over the period 1998–2005. They report their analyses "highlight the strong association between year-to-year fluctuations in winter mean temperature and mortality, with higher seasonal cases during harsh winters." Exceptions were noted for Belgium, the Netherlands, and the United Kingdom, which lack of correlation was likely explained by socioeconomic factors (e.g., higher housing efficiency, better health care, reduced economic and fuel poverty, etc.). Upon further analysis, Ballester et al. determined that, despite the lack of mortality association in those three countries, "it can be concluded that warmer winters will contribute to the decrease in winter mortality everywhere in Europe."

Citing the IPCC, Díaz et al. (2018) note climate models predict heat waves will become more frequent and intense in the future. However, they say the impact of such events on human health "is not so clear," as human adaptation, improved health services, and the implementation of advance warning systems can minimize the impacts of heat waves on health. They investigated whether there has been a temporal change in the relative risk of human mortality in response to these and other mitigating factors.

Díaz et al. examined the relationship between temperature and mortality for three time periods (1983–1992, 1993–2003, and 2004–2013) using data from ten Spanish provinces, carried out for the summer period only (June–September) in each year. They found "there has been a sharp decrease in mortality attributable to heat over the past 10 years" in Spain. More specifically, they found an identical relative risk of mortality due to heat of 1.15 across the first (1983–1992) and second (1993–2003) time periods, which thereafter experienced a statistically significant decline to 1.01 during the third period (2004–2013).

Díaz et al. write their work shows "a drastic decrease in the impact of heat, with a decline in attributable risk per degree of $T_{threshold}$ values from 14% to 1%; a decrease of around 93%," which they note "is similar to the decline in the heat-related mortality rate found in Australia (Coates *et al.*, 2014) of 85% and above, though also around 70% in the U.S. (Barreca *et al.*, 2016)."

The temporal variation in relative risk observed by Díaz *et al.* calls into question the results of numerous model-based studies that project the future impact of heat on mortality to be constant over time. As observed by Díaz *et al.* and others in different parts of the world (Coates *et al.*, 2014; Petkova *et al.*, 2014; Gasparrini *et al.*, 2015; Åström *et al.*, 2016, Barreca *et al.*, 2016) such projections are likely to overestimate the true impact of heat on human mortality. If results observed for Spain apply to the rest of the world, humankind has little to fear, for its relative risk of mortality declined to such a point in the most recent period (1.01; 95% confidence interval of 1.00 to 1.01) that there was no significant heat-related mortality risk.

References

Analitis, A., *et al.* 2008. Effects of cold weather on mortality: results from 15 European cities within the PHEWE project. *American Journal of Epidemiology* **168**: 1397–1408.

Åström, D.O., Tornevi, A., Ebi, K.L., Rocklöv, J., and Forsberg, B. 2016. Evolution of minimum mortality temperature in Stockholm, Sweden, 1901–2009. *Environmental Health Perspectives* **124**: 740–4.

Baccini, M., *et al.* 2008. Heat effects on mortality in 15 European cities. *Epidemiology* **19**: 711–9.

Baccini, M., Tom, K., and Biggeri, A. 2011. Impact of heat on mortality in 15 European cities: attributable deaths under different weather scenarios. *Journal of Epidemiology and Community Health* **65**: 64–70.

Ballester, F., Corella, D., Perez-Hoyos, S., and Saez, M. 1997. Mortality as a function of temperature. A study in Valencia, Spain, 1991–1993. *International Journal of Epidemiology* **26**: 551–61.

Ballester, J., Rodó, X., Robine, J.-M., and Herrmann, F.R. 2016. European seasonal mortality and influenza incidence due to winter temperature variability. *Nature Climate Change* **6**: 927–31.

Barreca, A., Clay, K., Deschenes, O., Greenstone, M., and Shapiro, J.S. 2016. Adapting to climate change: the remarkable decline in the US temperature-mortality relationship over the twentieth century. *Journal of Political Economy* **124**: 105–9.

Beniston, M. 2002. Climatic change: possible impact on human health. *Swiss Medical Weekly* **132**: 332–7.

Bi, P. and Walker, S. 2001. Mortality trends for deaths related to excessive heat (E900) and excessive cold (E901), Australia, 1910–1997. *Environmental Health* **1**: 80–6.

Braga, A.L.F., Zanobetti, A., and Schwartz, J. 2002. The effect of weather on respiratory and cardiovascular deaths in 12 US cities. *Environmental Health Perspectives* **110**: 859–63.

Carmona, R., Diaz, J., Miron, I.J., Ortiz, C., Leo, I., and Linares, C. 2016. Geographical variation in relative risks associated with cold waves in Spain: the need for a cold wave prevention plan. *Environment International* **88**: 103–11.

Carson, C., Hajat, S., Armstrong, B., and Wilkinson, P. 2006. Declining vulnerability to temperature-related mortality in London over the 20th century. *American Journal of Epidemiology* **164**: 77–84.

Christidis, N., Donaldson, G.C., and Stott, P.A. 2010. Causes for the recent changes in cold- and heat-related mortality in England and Wales. *Climatic Change* **102**: 539–53.

Coates, L., Haynes, K., O'Brien, J., McAneney, J., and De Oliveira, F.M. 2014. Exploring 167 years of vulnerability: an examination of extreme heat events in Australia 1844–2010. *Environmental Science & Policy* **42**: 33–44.

Curriero, F.C., Heiner, K.S., Samet, J.M., Zeger, S.L., Strug, L., and Patz, J.A. 2002. Temperature and mortality in 11 cities of the Eastern United States. *American Journal of Epidemiology* **155**: 80–7.

Davis, R.E., Knappenberger, P.C., Novicoff, W.M., and Michaels, P.J. 2002. Decadal changes in heat-related human mortality in the Eastern US. *Climate Research* **22**: 175–84.

Davis, R.E., Knappenberger, P.C., Novicoff, W.M., and Michaels, P.J. 2003a. Decadal changes in summer mortality in U.S. cities. *International Journal of Biometeorology* **47**: 166–75.

Davis, R.E., Knappenberger, P.C., Michaels, P.J., and Novicoff, W.M. 2003b. Changing heat-related mortality in the United States. *Environmental Health Perspectives* **111**: 1712–8.

Diaz, J., Garcia, R., Lopez, C., Linares, C., Tobias, A., and Prieto, L. 2005. Mortality impact of extreme winter temperatures. *International Journal of Biometeorology* **49**: 179–83.

Donaldson, G.C. and Keatinge, W.R. 2008. Direct effects of rising temperatures on mortality in the UK. In: Kovats, R.S. (Ed.) *Health Effects of Climate Change in the UK 2008: An Update of the Department of Health Report 2001/2002*. London, UK: Department of Health, pp. 81–90.

Donaldson, G.C., Kovats, R.S., Keatinge, W.R., and McMichael, A.J. 2001. Heat- and cold-related mortality and morbidity and climate change. In: Maynard, R.L. (Ed.) *Health Effects of Climate Change in the UK*. London, UK: Department of Health, pp. 70–80.

Eccles, R. 2002. An explanation for the seasonality of acute upper respiratory tract viral infections. *Acta Oto-Laryngologica* **122**: 183–91.

Eurowinter Group. 1997. Cold exposure and winter mortality from ischaemic heart disease, cerebrovascular disease, respiratory disease, and all causes in warm and cold regions of Europe. *The Lancet* **349**: 1341–6.

Fernandez-Raga, M., Tomas, C., and Fraile, R. 2010. Human mortality seasonality in Castile-Leon, Spain, between 1980 and 1998: the influence of temperature, pressure and humidity. International *Journal of Biometeorology* **54**: 379–92.

Fleming, D.M., Cross, K.W., Sunderland, R., and Ross, A.M. 2000. Comparison of the seasonal patterns of asthma identified in general practitioner episodes, hospital admissions, and deaths. *Thorax* **55**: 662–5.

Fouillet, A., *et al.* 2008. Has the impact of heat waves on mortality changed in France since the European heat wave of summer 2003? A study of the 2006 heat wave. *International Journal of Epidemiology* **37**: 309–17.

Garssen, J., Harmsen, C., and de Beer, J. 2005. The effect of the summer 2003 heat wave on mortality in the Netherlands. *Euro Surveillance* **10**: 165–8.

Gasparrini, A., *et al.* 2015. Temporal variation in heat-mortality associations: a multicountry study. *Environmental Health Perspectives* **123**: 1200–7.

Gouveia, N., Hajat, S., and Armstrong, B. 2003. Socioeconomic differentials in the temperature-mortality relationship in Sao Paulo, Brazil. *International Journal of Epidemiology* **32**: 390–7.

Grech, V., Balzan, M., Asciak, R.P., and Buhagiar, A. 2002. Seasonal variations in hospital admissions for asthma in Malta. *Journal of Asthma* **39**: 263–8.

Keatinge, W.R. and Donaldson, G.C. 2001. Mortality related to cold and air pollution in London after allowance for effects of associated weather patterns. *Environmental Research* **86**: 209–16.

Keatinge, W.R., *et al.* 2000a. Winter mortality in relation to climate. *International Journal of Circumpolar Health* **59**: 154–9.

Keatinge, W.R., *et al.* 2000b. Heat related mortality in warm and cold regions of Europe: observational study. *British Medical Journal* **321**: 670–3.

Koppe, C. 2005. Gesundheitsrelevante Bewertung von thermischer Belastung unter Berucksichtigung der kurzfristigen Anpassung der Bevolkerung an die lokalen Witterungsverhaltnisse. Albert-Ludwigs-University of Freiburg, Germany.

Kunst, A.E., Looman, W.N.C., and Mackenbach, J.P. 1993. Outdoor temperature and mortality in the Netherlands: a time-series analysis. *American Journal of Epidemiology* **137**: 331–41.

Kysely, J. and Huth, R. 2004. Heat-related mortality in the Czech Republic examined through synoptic and 'traditional' approaches. *Climate Research* **25**: 265–74.

Kysely, J. and Plavcova, E. 2012. Declining impacts of hot spells on mortality in the Czech Republic, 1986–2009: adaptation to climate change? *Climatic Change* **113**: 437–53.

Laaidi, M., Laaidi, K., and Besancenot, J.-P. 2006. Temperature-related mortality in France, a comparison between regions with different climates from the perspective of global warming. *International Journal of Biometeorology* **51**: 145–53.

Laschewski, G. and Jendritzky, G. 2002. Effects of the thermal environment on human health: an investigation of 30 years of daily mortality data from SW Germany. *Climate Research* **21**: 91–103.

Law, B.J., Carbonell-Estrany, X., and Simoes, E.A.F. 2002. An update on respiratory syncytial virus epidemiology: a developed country perspective. *Respiratory Medicine Supplement B* **96**: S1–S2.

Matzarakis, A., Muthers, S., and Koch, E. 2011. Human biometeorological evaluation of heat-related mortality in Vienna. *Theoretical and Applied Climatology* **105**: 1–10.

Muggeo, V.M.R. and Vigotti, M.A. 2002. Modelling trend in break-point estimation: an assessment of the heat tolerance and temperature effects in four Italian cities. In: Stasinopoulos, M. and Touloumi, G. (Eds.) *Proceedings of the 17th International Workshop on Statistical Modelling*. University of North London, Chania, Greece, pp. 493–500.

Nafstad, P., Skrondal, A., and Bjertness, E. 2001. Mortality and temperature in Oslo, Norway, 1990–1995. *European Journal of Epidemiology* **17**: 621–7.

O'Neill, M.S., Zanobetti, A., and Schwartz, J. 2003. Modifiers of the temperature and mortality association in seven US cities. *American Journal of Epidemiology* **157**: 1074–82.

Petkova, E.P., Gasparrini, A., and Kinney, P.L. 2014. Heat and mortality in New York City since the beginning of the 20th century. *Epidemiology* **25**: 554–60.

Rooney, C., McMichael, A.J., Kovats, R.S., and Coleman, M.P. 1998. Excess mortality in England and Wales, and in Greater London, during the 1995 heat wave. *Journal of Epidemiology and Community Health* **52**: 482–6.

Sheridan, S.C., Kalkstein, A.J., and Kalkstein, L.S. 2009. Trends in heat-related mortality in the United States, 1975–2004. *Natural Hazards* **50**: 145–60.

Tan, J., Zheng, Y., Tang, X., Guo, C., Li, L., Song, G., Zhen, X., Yuan, D., Kalkstein, A., and Chen, H. 2007. Heat wave impacts on mortality in Shanghai 1998 and 2003. *International Journal of Biometeorology* **51**: 193–200.

Todd, N. and Valleron, A.-J. 2015. Space-time covariation of mortality with temperature: a systematic study of deaths in France, 1968–2009. *Environmental Health Perspectives* **123**: 659–64.

Verlato, G., Calabrese, R., and De Marco, R. 2002. Correlation between asthma and climate in the European Community Respiratory Health Survey. *Archives of Environmental Health* **57**: 48–52.

Wichmann, J., Anderson, Z.J., Ketzel, M., Ellermann, T., and Loft, S. 2011. Apparent temperature and cause-specific mortality in Copenhagen, Denmark: a case-crossover analysis. *International Journal of Environmental Research and Public Health* **8**: 3712–27.

4.2.6 North America

Goklany and Straja (2000) examined trends in United States death rates over the period 1979–1997 due to excessive hot and cold weather. They report there were no trends in deaths due to either extreme heat or extreme cold in the entire population or in the older, more-susceptible age groups, those aged 65 and over, 75 and over, and 85 and over. Deaths due to extreme cold in these older age groups exceeded those due to extreme heat by as much as 80 to 125%. With respect to the absence of trends in death rates attributable to either extreme heat or cold, Goklany and Straja say this "suggests that adaptation and technological change may be just as important determinants of such trends as more obvious meteorological and demographic factors."

Davis *et al.* (2002) studied changes in the impact of high temperatures on daily mortality rates over a period of four decades in six major metropolitan areas (Atlanta, Boston, Charlotte, Miami, New York City, and Philadelphia) along a north-south transect in the eastern United States. They found few significant weather-mortality relationships for any decade or demographic group in the three southernmost cities examined, where warmer weather is commonplace. In the three northernmost cities, however, there were statistically significant decreases in population-adjusted mortality rates during hot and humid weather between 1964 and 1994. The authors write, "these statistically significant reductions in hot-weather mortality rates suggest that the populace in cities that were weather-sensitive in the 1960s and 1970s have become less impacted by extreme conditions over time because of improved medical care, increased access to air conditioning, and biophysical and infrastructural adaptations." They further note, "this analysis counters the paradigm of increased heat-related mortality rates in the eastern US predicted to result from future climate warming."

Davis *et al.* (2003) evaluated "annual excess mortality on days when apparent temperatures – an index that combines air temperature and humidity – exceeded a threshold value for 28 major metropolitan areas in the United States from 1964 through 1998." They found "for the 28-city average, there were 41.0 ± 4.8 excess heat-related deaths per year (per standard million) in the 1960s and 1970s, 17.3 ± 2.7 in the 1980s, and 10.5 ± 2.0 in the 1990s," a remarkable decline. They conclude, "heat-related mortality in the United States seems to be largely preventable at present."

Davis *et al.* (2004) examined the seasonality of mortality due to all causes, using monthly data for 28 major U.S. cities from 1964 to 1998, then calculated the consequences of a future 1°C warming of the conglomerate of those cities. At all locations studied, they report "warmer months have significantly lower mortality rates than colder months." They calculate "a uniform 1°C warming results in a net mortality *decline* of 2.65 deaths (per standard million) per metropolitan statistical area" (italics added). The primary implication of Davis *et al.*'s findings, in their words, "is that the seasonal mortality pattern in US cities is largely independent of the climate and thus insensitive to climate fluctuations, including changes related to increasing greenhouse gases."

O'Neill *et al.* (2005) assessed the influence of air pollution and respiratory epidemics on empirical associations between apparent temperature, which "represents an individual's perceived air temperature," and daily mortality in Mexico's largest and third-largest cities: Mexico City and Monterrey, respectively. They found "in Mexico City, the 7-day temperature mortality association has a hockey stick shape with essentially no effect of higher temperatures," whereas in Monterrey the function they fit to the data "shows a U-shape," with "a higher mortality risk at both ends of the distribution," although the effect is much weaker at the high-temperature end of the plot than at the low-temperature end, and the absolute value of the slope of the mortality vs. temperature relationship is smaller across the high-temperature range of the data.

Interestingly, the researchers also found that "failure to control for respiratory epidemics and air pollution resulted in an overestimate of the impact of hot days by 50%," whereas "control for these factors had little impact on the estimates of effect of cold days." They note "most previous assessments of effects of heat waves on hot days have not controlled for air pollution or epidemics." Stedman (2004) made a similar claim after analyzing the impact of air pollutants present during a 2003 heat wave in the United Kingdom, claiming to have found 21% to 38% of the total excess deaths claimed to be due to high temperatures were actually the result of elevated concentrations of ozone and PM_{10} (particulate matter of diameter less than 10μm). Likewise, Fischer *et al.* (2004) claimed 33% to 50% of the deaths attributed to the same heat wave in the Netherlands were caused by concurrent high ozone and PM_{10} concentrations.

O'Neill *et al.*, Stedman, and Fischer *et al.* are correct in pointing to factors other than temperature contributing to deaths during periods of very warm or very cold temperatures. However, attributing fatalities to air pollution in developed countries is a dubious exercise at best. As explained in detail in Chapter 6, epidemiological studies purporting to show such attribution are easily manipulated, cannot prove causation, and often do not support a hypothesis of toxicity with the small associations in uncontrolled observational studies. Exaggeration of effects and certainty has been the rule in that field, just as has been in the climate debate.

Deschenes and Moretti (2009) analyzed the relationship between weather and mortality, based on "data that include the universe of deaths in the United States over the period 1972–1988," in which they "match each death to weather conditions on the day of death and in the county of occurrence." They discovered "hot temperature shocks are indeed associated with a large and immediate spike in mortality in the days of the heat wave," but "almost all of this excess mortality is explained by near-term displacement." As a result, "in the weeks that follow a heat wave, we find a marked decline in mortality hazard, which completely offsets the increase during the days of the heat wave," so "there is virtually no lasting impact of heat waves on mortality." In the case of cold temperature days, they also found "an immediate spike in mortality" but "there is no offsetting decline in the weeks that follow," so "the cumulative effect of one day of extreme cold temperature during a thirty-day window is an increase in daily mortality by as much as 10%."

Vuteovici *et al.* (2014) studied the impact of variations of diurnal temperature on daily mortality of residents of Montreal aged 65 years and older during the period 1984–2007, finding "a 5.12% increase in the cumulative effects on mortality for an increase of the diurnal temperature range from 6°C to 11°C." When the diurnal temperature range increased from 11°C to 17.5°C, they found an 11.27% increase in mortality.

Petkova *et al.* (2014) "examined adaptation patterns by analyzing daily temperature and mortality data spanning more than a century in New York City," where using a distributed-lag nonlinear model they analyzed the heat-mortality relationship in people 15 years of age or older during two periods – 1900–1948 and 1973–2006 – in order to "quantify population adaptation to high temperatures over time." The three researchers report that "during the first half of the century, the decade-specific relative risk of mortality at 29°C vs. 22°C ranged from 1.30 in the 1910s to 1.43 in the 1900s." Since 1973, however, they found "there was a gradual and substantial decline in the relative risk, from 1.26 in the 1970s to 1.09 in the 2000s." In addition, they say "age-specific analyses indicated a greater risk for people of age 65 years and older in the first part of the century," but "less evidence for enhanced risk among this older age group in more recent decades." Petkova *et al.*'s discovery that the excess mortality originally experienced at high temperatures fell substantially over the course of the century they studied is indicative, in their words, of "population adaptation to heat in recent decades," which they attribute primarily to "the rapid spread and widespread availability of air conditioning."

Bobb *et al.* (2014) note increasing temperatures are anticipated to have health impacts but "little is known about the extent to which the population may be adapting." They examined "the hypothesis that if adaptation is occurring, then heat-related mortality would be deceasing over time," using "a national database of daily weather, air pollution, and age-stratified mortality rates for 105 U.S. cities (covering 106 million people) during the summers of 1987–2005." They found, "on average across cities, the number of deaths (per 1,000 deaths) attributable to each 10°F increase in same-day temperature decreased from 51 in 1987 to 19 in 2005" (see Figure 4.2.6.1). They report "this decline was largest among those ≥ 75 years of age, in northern regions, and in cities with cooler climates." In addition, they write that "although central air conditioning (AC) prevalence has increased, we did not find statistically significant evidence of larger temporal declines among cities with larger increases in AC prevalence." They conclude the U.S. population has "become more resilient to heat over time."

White (2017) used daily hospital visit and meteorological data to examine the dynamic relationship between temperature and morbidity in California over the period 2005–2014. He determined that the 31-day cumulative impact of a cold day with a mean temperature under 40°F (4.5°C) resulted in an 11% net increase in total morbidity (defined by the number of emergency department visits), which value rose to 17% when the cumulative impact was extended another month. (See Figure 4.2.6.2.)

The most influential disease category driving cumulative cold temperature morbidity was respiratory disease (including influenza and pneumonia), which far exceeded any other cause and amounted to approximately 6 of the 8.5 increased hospital visits per 100,000 persons that were due to cold temperatures. Stratifying the effect of cold temperatures by age group, White further reports the greatest risk of morbidity fell within the youngest age group (children under 5 years of age), which group experienced a 27.7% increase in hospital visits above the mean daily visit rate – a value four times as large as that observed for the least affected group (non-elderly adults aged 25–64).

With respect to the warm end of the temperature spectrum, White found the 31-day cumulative impact of a hot day (with mean temperature above 80°F (26.7°C) was to increase human morbidity by 5.1%. Several disease categories contributed to this overall relationship, including injuries, nervous system, and

Figure 4.2.6.1
Excess U.S. deaths (per 1,000) attributable to each 10°F increase in the same day's summer temperature, 1987–2005

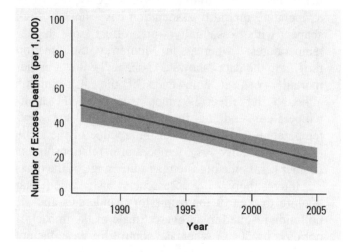

Shaded region represents 95% confidence intervals.
Source: Adapted from Bobb *et al.*, 2014.

Figure 4.2.6.2
Cumulative (one month) effect of temperature (°F) on emergency department visits in California (2005–2014)

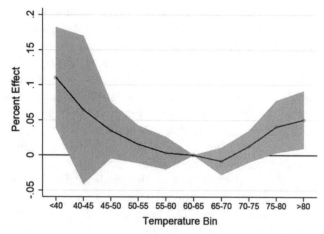

Shaded regions represent 95% confidence intervals.
Source: White, 2017.

genitourinary problems. As with cold weather, hot weather morbidity was also influenced by age, and the effect was more pronounced in the youngest age categories, where morbidity risk was 8.6% and 9.5% for the under 5 and 5–14 age groups, respectively. White also reported finding evidence of an early harvesting effect (an initial morbidity increase that is offset by a later decrease) on the elderly during hot days. The over 64 age group, for example, experienced an initial contemporaneous increase of 3.7% in morbidity on days where the mean temperature was above 80°F. However, the cumulative morbidity effect fell to 2.7% at the end of 31 days.

White also calculated the health care cost impacts associated with both hot and cold morbidity, reporting that a day over 80°F mean temperature is associated with an approximate $8,000 increase in hospital costs per 100,000 persons, whereas cost impacts from a mean daily temperature below 40°F amount to a much larger $12,000 per 100,000 people. Lastly, White also estimated the potential effects of climate change on human morbidity and its associated health care costs using future temperature projections derived from climate model output produced under the IPCC's RCP8.5 greenhouse gas emission scenario. Results of that exercise revealed "the effects of climate change on hospital visits and costs in California to be negligible."

Allen and Sheridan (2018) investigated the relationship between all-cause mortality and extreme temperature events for 50 metropolitan areas in the United States with populations from 1.0 million to 19.5 million for the period 1975–2004. They assessed mortality impacts in response to three-day means of daily apparent temperature for a cumulative 14-day period following hot and cold events. They also subdivided their calculations to discern such effects from both heat and cold events (95th and 5th percentiles) and extreme heat and cold events (97.5th and 2.5th percentiles) that occurred both early and late in the summer and winter seasons, respectively. They also determined the impact on mortality of the length of heat and cold spells (short = events lasting two days or less, long = events lasting for three or more days).

Allen and Sheridan found the highest relative risk of mortality is for extreme cold events that occur early in the winter season and last for two days or less, which risk is double that observed for extreme heat events that occur in the summer. They also found the cumulative relative risk of mortality values from both short and long temperature events decline over the course of the winter and summer seasons, suggesting there is a seasonal human adaptation to extreme weather events occurring at both ends of the temperature spectrum.

References

Allen, M.J. and Sheridan, S.C. 2018. Mortality risks during extreme temperature events (ETEs) using a distributed lag non-linear model. *International Journal of Biometeorology* **62**: 57–67.

Bobb, J.F., Peng, R.D., Bell, M.L., and Dominici, F. 2014. Heat-related mortality and adaptation to heat in the United States. *Environmental Health Perspectives* **122**: 811–6.

Davis, R.E., Knappenberger, P.C., Michaels, P.J., and Novicoff, W.M. 2003. Changing heat-related mortality in the United States. *Environmental Health Perspectives* **111**: 1712–8.

Davis, R.E., Knappenberger, P.C., Michaels, P.J., and Novicoff, W.M. 2004. Seasonality of climate-human mortality relationships in US cities and impacts of climate change. *Climate Research* **26**: 61–76.

Davis, R.E., Knappenberger, P.C., Novicoff, W.M., and Michaels, P.J. 2002. Decadal changes in heat-related human mortality in the eastern United States. *Climate Research* **22**: 175–84.

Deschenes, O. and Moretti, E. 2009. Extreme weather events, mortality, and migration. *The Review of Economics and Statistics* **91**: 659–81.

Fischer, P.H., Brunekreef, B., and Lebret, E. 2004. Air pollution related deaths during the 2003 heat wave in the Netherlands. *Atmospheric Environment* **38**: 1083–5.

Goklany, I.M. and Straja, S.R. 2000. U.S. trends in crude death rates due to extreme heat and cold ascribed to weather, 1979–97. *Technology* **7S**: 165–73.

O'Neill, M.S., Hajat, S., Zanobetti, A., Ramirez-Aguilar, M., and Schwartz, J. 2005. Impact of control for air pollution and respiratory epidemics on the estimated associations of temperature and daily mortality. *International Journal of Biometeorology* **50**: 121–9.

Petkova, E.P., Gasparrini, A., and Kinney, P.L. 2014. Heat and mortality in New York City since the beginning of the 20th century. *Epidemiology* **25**: 554–60.

Stedman, J.R. 2004. The predicted number of air pollution related deaths in the UK during the August 2003 heatwave. *Atmospheric Environment* **38**: 1087–90.

Vuteovici, M., Goldberg, M.S., and Valois, M.-F. 2014. Effects of diurnal variations in temperature on non-accidental mortality among the elderly population of Montreal, Quebec, 1984–2007. *International Journal of Biometeorology* **58**: 843–52.

White, C. 2017. The dynamic relationship between temperature and morbidity. *Journal of the Association of Environmental and Resource Economists* **4**: 1155–98.

4.3 Cardiovascular Disease

Higher surface temperatures would reduce the incidence of fatal coronary events related to low temperatures and wintry weather by a greater degree than they would increase the incidence associated with high temperatures and summer heat waves.

Non-fatal myocardial infarction is also less frequent during unseasonably warm periods than during unseasonably cold periods.

Cardiovascular diseases (CVDs) affect the heart or blood vessels. They include arrhythmia, arteriosclerosis, congenital heart disease, coronary artery disease, diseases of the aorta and its branches, disorders of the peripheral vascular system, endocarditis, heart valve disease, hypertension, orthostatic hypotension, and shock. According to the Working Group II contribution to the Intergovernmental Panel on Climate Change (IPCC) Fifth Assessment Report, "Numerous studies of temperature-related morbidity, based on hospital admissions or emergency presentations, have reported increases in events due to cardiovascular, respiratory, and kidney diseases (Hansen *et al.*, 2008; Knowlton *et al.*, 2009; Lin and Chan, 2009) and the impact has been related to the duration and intensity of heat (Nitschke *et al.*, 2011) (IPCC, 2014, p. 721). The IPCC overlooks the fact that cooler temperatures cause an even larger number of premature deaths, with the result that a warmer world would experience fewer deaths in total.

Nafstad *et al.* (2001) examined the effects of temperature on mortality due to all forms of cardiovascular disease for citizens of New South Wales, Australia, over the period 1990 to 1995. Their analysis showed the average daily number of cardiovascular-related deaths was 15% higher in the winter months (October–March) than in the summer months (April–September), leading them to conclude

"a milder climate would lead to a substantial reduction in average daily number of deaths." This confirmed an earlier finding by Enqueslassie *et al.* (1993) of the Hunter region of New South Wales which covered July 1, 1985 to June 30, 1990 and found "fatal coronary events and non-fatal definite myocardial infarction were 20–40% more common in winter and spring than at other times of year." With respect to daily temperature effects, they found "rate ratios for deaths were significantly higher for low temperatures," noting "on cold days coronary deaths were up to 40% more likely to occur than at moderate temperatures."

Hajat and Haines (2002) analyzed data obtained between January 1992 and September 1995 for registered patients aged 65 and older from several medical practices in London, England. They found the number of general practitioner consultations was higher in the cool-season months (October–March) than in the warm-season months (April–September) for all CVDs.

Braga *et al.* (2002) determined both the acute effects and lagged influence of temperature on cardiovascular-related deaths in a study of both "hot" and "cold" cities in the United States: Atlanta, Georgia; Birmingham, Alabama; and Houston, Texas comprised the "hot" group, and Canton, Ohio; Chicago, Illinois; Colorado Springs, Colorado; Detroit, Michigan; Minneapolis-St. Paul, Minnesota; New Haven, Connecticut; Pittsburgh, Pennsylvania; and Seattle and Spokane, Washington comprised the "cold" group. They found in the hot cities neither hot nor cold temperatures had much impact on mortality related to CVDs. In the cold cities, on the other hand, both high and low temperatures were associated with increased CVD deaths. The effect of cold temperatures persisted for days, whereas the effect of high temperatures was restricted to the day of the death or the day before. For all CVD deaths, the hot-day effect was five times smaller than the cold-day effect. In addition, the hot-day effect included some "harvesting," where Braga *et al.* observed a deficit of deaths a few days later, which they did not observe for the cold-day effect.

Gouveia *et al.* (2003) determined the number of cardiovascular-related deaths in adults aged 15–64 in Sao Paulo, Brazil over the period 1991–1994 increased by 2.6% for each 1°C decrease in temperature below 20°C (68°F), while they found no evidence for heat-induced deaths due to temperatures rising above 20°C. In the elderly (65 years of age and above), a 1°C warming above 20°C led to a 2% increase in deaths, but a 1°C cooling below 20°C led to a 6.3% increase in deaths, more than three times as many cardiovascular-related deaths due to cooling than to warming in the elderly.

McGregor *et al.* (2004) obtained and analyzed data on ischaemic heart disease (IHD) and temperature for five English counties aligned on a north-south transect (Tyne and Wear, West Yorkshire, Greater Manchester, West Midlands, and Hampshire) for the period 1974–1999. They determined "the seasonal cycles of temperature and mortality are inversely related," and "the first harmonic accounts for at least 85% (significant at the 0.01 level) of the variance of temperature and mortality at both the climatological and yearly time scales." They also report "years with an exaggerated mortality peak are associated with years characterized by strong temperature seasonality," and "the timing of the annual mortality peak is positively associated with the timing of the lowest temperatures."

Chang *et al.* (2004) studied the effects of monthly mean temperature on rates of hospitalization for arterial stroke and acute myocardial infarction (AMI) among young women aged 15–49 from 17 countries in Africa, Asia, Europe, Latin America, and the Caribbean. These efforts revealed "among young women from 17 countries, the rate of hospitalized AMI, and to a lesser extent stroke, was higher with lower mean environmental air temperature." They report, "on average, a 5°C reduction in mean air temperature was associated with a 7 and 12% increase in the expected hospitalization rates of stroke and AMI, respectively." Finally, they note, "lagging the effects of temperature suggested that these effects were relatively acute, within a period of a month."

Bartzokas *et al.* (2004) "examined the relationship between hospital admissions for cardiovascular (cardiac in general including heart attacks) and/or respiratory diseases (asthma etc.) in a major hospital in Athens [Greece] and meteorological parameters for an 8-year period." Over the study period, "there was a dependence of admissions on temperature" and low temperatures were "responsible for a higher number of admissions," they found. Specifically, "there was a decrease of cardiovascular or/and respiratory events from low to high values [of temperature], except for the highest temperature class in which a slight increase was recorded."

Nakaji *et al.* (2004) evaluated seasonal trends in deaths in Japan from 1970 to 1999 and recorded mean monthly temperature. The nine researchers note Japan has "bitterly cold winters," and their analysis

indicates the numbers of deaths due to infectious and parasitic diseases – including tuberculosis, respiratory diseases including pneumonia and influenza, diabetes, digestive diseases, and cerebrovascular and heart diseases – rise to a maximum during that cold time of year. Of the latter two categories, they found peak mortality rates due to heart disease and stroke were one-and-a-half to two times greater in winter (January) than in August and September, when mortality rates for those conditions are at their yearly minimums.

Sharovsky *et al.* (2004) investigated "associations between weather (temperature, humidity, and barometric pressure), air pollution (sulfur dioxide, carbon monoxide, and inhalable particulates), and the daily death counts attributed to myocardial infarction" in Sao Paulo, Brazil, where 12,007 deaths were observed from 1996 to 1998. As mean daily temperature dropped below 18°C, death rates rose in essentially linear fashion to attain a value at 12°C (the typical lower limit of observed temperature in Sao Paulo) more than 35% greater than the minimum baseline value registered between 21.6°C and 22.6°C. Sharovsky *et al.* say "myocardial infarction deaths peak in winter not only because of absolute low temperature but possibly secondary to a decrease relative to the average annual temperature."

Kovats *et al.* (2004) analyzed patterns of temperature-related hospital admissions and deaths in Greater London during the mid-1990s. For the three-year period 1994–1996, cardiovascular-related deaths were approximately 50% greater during the coldest part of the winter than during the peak warmth of summer, whereas respiratory-related deaths were nearly 150% greater in the depths of winter cold than at the height of summer warmth. With respect to heat waves, the mortality impact of the notable heat wave of July 29 to August 3, 1995 was so tiny it could not be discerned among the random scatter of plots of three-year-average daily deaths from cardiovascular and respiratory problems versus day of year.

Carder *et al.* (2005) investigated the relationship between outside air temperature and deaths due to all non-accident causes in the three largest cities of Scotland (Glasgow, Edinburgh, and Aberdeen) between January 1981 and December 2001. They observed "an overall increase in mortality as temperature decreases," which "appears to be steeper at lower temperatures than at warmer temperatures," while "there is little evidence of an increase in mortality at the hot end of the temperature range." The seven scientists found, for temperatures below 11°C, a 1°C drop in the daytime mean temperature on any one day was associated with an increase in cardiovascular-caused mortality of 3.4% over the following month. At any season of the year a decline in air temperature in the major cities of Scotland leads to increases in deaths due to cardiovascular causes, whereas there is little or no such increase in mortality associated with heat waves.

Cagle and Hubbard (2005) examined the relationship between temperature and cardiac-related deaths in King County, Washington (USA) over the period 1980–2000. They determined there was an average of 2.86 cardiac-related deaths per day for all days when the maximum temperature fell within the broad range of 5–30°C. For days with maximum temperatures less than 5°C, the death rate rose by 15% to a mean value of 3.30, whereas on days with maximum temperatures greater than 30°C, death rates *fell* by 3% to a mean value of 2.78. In addition, "the observed association between temperature and death rate is not due to confounding by other meteorological variables," and "temperature continues to be statistically significantly associated with death rate even at a 5-day time lag."

Tam *et al.* (2009) employed daily mortality data in Hong Kong for the years 1997 to 2002 to examine the association between diurnal temperature range (DTR) and cardiovascular disease among the elderly (age 65 and older). They report "a 1.7% increase in mortality for an increase of 1°C in DTR at lag days 0–3" and describe these results as being "similar to those reported in Shanghai." The four researchers state "a large fluctuation in the daily temperature – even in a tropical city like Hong Kong – has a significant impact on cardiovascular mortality among the elderly population."

Cao *et al.* (2009) assessed the relationship between DTR and coronary heart disease (CHD) deaths of elderly people (66 years of age or older) occurring in Shanghai between January 1, 2001 and December 31, 2004. They found "a 1°C increase in DTR (lag = 2) corresponded to a 2.46% increase in CHD mortality on time-series analysis, a 3.21% increase on unidirectional case-crossover analysis, and a 2.13% increase on bidirectional case-crossover analysis," and "the estimated effects of DTR on CHD mortality were similar in the warm and cool seasons." The seven scientists conclude their "data suggest that even a small increase in DTR is associated with a substantial increase in deaths due to CHD."

Bayentin *et al.* (2010) analyzed the standardized daily hospitalization rates for ischemic heart disease (IHD) and their relationship with climatic conditions up to two weeks prior to the day of admission to determine the short-term effects of climate conditions on the incidence of IHD over the period 1989–2006 for 18 health regions of Quebec. The authors report "a decline in the effects of meteorological variables on IHD daily admission rates" that "can partly be explained by the changes in surface air temperature," which they describe as warming "over the last few decades."

Toro *et al.* (2010) used data on 7,450 cardiovascular-related deaths in Budapest, Hungary between 1995 and 2004 to find potential relationships between those deaths and daily maximum, minimum, and mean temperature, air humidity, air pressure, wind speed, and global radiation. The six Hungarian scientists report "on the days with four or more death cases, the daily maximum and minimum temperatures tend to be lower than on days without any cardiovascular death events," "the largest frequency of cardiovascular death cases was detected in cold and cooling weather conditions," and "we found a significant negative relationship between temperature and cardiovascular mortality."

Bhaskaran *et al.* (2010) explored the short-term relationship between ambient temperature and risk of heart attacks (myocardial infarction) in England and Wales by analyzing 84,010 hospital admissions from 2003 to 2006. They found a broadly linear relationship between temperature and heart attacks that was well characterized by log-linear models without a temperature threshold, such that each 1°C reduction in daily mean temperature was associated with a 2.0% cumulative increase in risk of myocardial infarction over the current and following 28 days. They also report heat had no detrimental effect, as an increased risk of myocardial infarction at higher temperatures was not detected.

Kysely *et al.* (2011) used a database of daily mortality records in the Czech Republic that cover the 21-year period 1986–2006 – which, in their words, "encompasses seasons with the hottest summers on record (1992, 1994, 2003) as well as several very cold winters (1986/87, 1995/96, 2005/06)" – to compare the effects of hot and cold periods on cardiovascular mortality. The four Czech scientists report "both hot and cold spells are associated with significant excess cardiovascular mortality," but "the effects of hot spells are more direct (unlagged) and typically concentrated in a few days of a hot spell, while cold spells are associated with indirect (lagged) mortality impacts persisting after a cold spell ends." Although they report "the mortality peak is less pronounced for cold spells," they determined "the cumulative magnitude of excess mortality is larger for cold than hot spells." They conclude, "in the context of climate change, substantial reductions in cold-related mortality are very likely in mid-latitudinal regions, particularly if the increasing adaptability of societies to weather is taken into account (cf. Christidis *et al.*, 2010)," and "it is probable that reductions in cold-related mortality will be more important than possible increases in heat-related mortality."

Lim *et al.* (2012) assessed the effects of increasing DTR on hospital admissions for the most common cardiovascular and respiratory diseases in the four largest cities of Korea (Seoul, Incheon, Daegu, and Busan) for the period 2003–2006. According to the three South Korean researchers, the data showed "the area-combined effects of DTR on cardiac failure and asthma were statistically significant," and the DTR effects on asthma admissions were greater for the elderly (75 years or older) than for the non-elderly group. "In particular," they write, "the effects on cardiac failure and asthma were significant with the percentage change of hospital admissions per 1°C increment of DTR at 3.0% and 1.1%, respectively."

Wanitschek *et al.* (2013), noting Austria's 2005/2006 winter was very cold whereas the 2006/2007 winter was extraordinarily warm, studied the cases of patients who were suffering acute myocardial infarctions and had been referred for coronary angiography (CA). They compared the patients' risk factors and in-hospital mortality rates between these two consecutive winters and found nearly identical numbers of CA cases (987 vs. 983), but 12.9% of the CA cases in the colder winter were acute, while 10.4% of the cases in the warmer winter were acute. They conclude, "the average temperature increase of 7.5°C from the cold to the warm winter was associated with a decrease in acute coronary angiographies ..."

Vasconcelos *et al.* (2013) studied the health-related effects of a daily human-biometeorological index known as the Physiologically Equivalent Temperature (PET), which is based on the input parameters of air temperature, humidity, mean radiant temperature, and wind speed, as employed by Burkart *et al.* (2011), Grigorieva and Matzarakis (2011), and Cohen *et al.* (2012), focusing their attention on Lisbon and Oporto Counties in Portugal over the period 2003–2007. The five Portuguese researchers report there was "a linear relationship between daily mean PET, during winter, and the risk of myocardial infarction, after adjustment for confounding factors," thus confirming "the thermal environment, during winter, is inversely associated with acute myocardial infarction morbidity in Portugal." They observed "an increase of 2.2% of daily hospitalizations per degree fall of PET, during winter, for all ages." In Portugal and many other countries where low winter temperatures "are generally under-rated compared to high temperatures during summer periods," Vasconcelos *et al.* conclude cold weather is "an important environmental hazard" that is much more deadly than the heat of summer.

Hart (2015) writes, "warm temperatures are thought to be associated with increased death rates," citing the works of Longstreth (1991) and Zanobettia *et al.* (2012). Rather than accepting the assumption at face value, Hart conducted a statistical analysis to test it. Using linear multiple regression analysis of data from all 67 counties in Alabama (USA), he analyzed the relationship between daily mean air temperature and land elevation (both as predictor variables) and death rates from cancer and heart disease (the two response variables) over the periods 2006–2010 and 2008–2010, respectively. Hart reports there was no "statistically significant adverse health effects for either predictor with these response variables" (see Figure 4.3.1). However, as evident in the figure, his analysis did reveal an inverse relationship between temperature and heart disease death rates, such that a one degree Fahrenheit rise in temperature would have the effect of reducing heart disease death rates by 12 persons per 100,000 over the temperature range analyzed in his study. The findings, Hart writes, "contradict dire predictions of adverse health consequences as a result of global warming," yet are "consistent with a previous report that indicated a beneficial association between warmer temperatures and decreased mortality (Idso *et al.*, 2014)."

Ponjoan *et al.* (2017) analyzed the effects of both heat waves and cold spells on emergency hospitalizations due to cardiovascular diseases in Catalonia (a region of Spain in the Mediterranean basin) over the period 2006–2013. They used the self-controlled case series statistical methodology to assess the relative incidence rate ratios (IRRs) of hospitalizations during the hot and cold waves in comparison to reference time periods with normal temperature exposure. Heat waves were defined as a period of at least three days in July and August in which the daily maximum temperatures were higher than the 95th percentile of daily maximum temperature for those two months. Cold waves were similarly defined as periods of at least three days in January and February when daily minimum temperatures were lower than the 95th percentile of daily minimum temperatures for those two months. IRRs were adjusted for age, time interval, and air pollution. The number of hospitalizations due to cardiovascular diseases during January and February over the period of study was 22,611, whereas there were only 17,017 during July and August.

Ponjoan *et al.* found the incidence of cardiovascular hospitalizations increases during cold spells by 20%, rising to 26 and 29% when a lag of three and seven days, respectively, are added to the cold spell. In contrast, Ponjoan *et al.* report, "the effect of heatwaves on overall cardiovascular hospitalizations was not significantly different from the null," adding "no significant differences were observed when stratifying by sex, age or cardiovascular type categories" for either heat waves or cold spells. This latter finding challenges the oft-repeated concern that the elderly will disproportionately suffer more health maladies than younger people during such temperature departures (either hot or cold).

In a review of the research literature, Claeys *et al.* (2017) note "acute myocardial infarctions (AMIs) are the leading cause of mortality worldwide and are usually precipitated by coronary thrombosis, which is induced by a ruptured or eroded atherosclerotic plaque that leads to a sudden and critical reduction in blood flow," citing Davies and Thomas (1985), Nichols *et al.* (2013), and a 2014 report of the American Heart Association that was produced by a team of 44 researchers. Claeys *et al.* note, "the majority of the temperature-related mortality has been shown to be attributable much more to cold, when compared with extreme hot weather," citing the work of Gasparrini *et al.* (2015) and reporting that "for each 10°C decrease in temperature, there was a 9% increase in the risk of AMI."

Zhang et al. (2017) studied daily meteorological data and records of all registered deaths in Wuhan, central China, between 2009 and 2012, performing a series of statistical procedures to estimate the exposure-response impact of DTR on human mortality (including non-accidental deaths and those due to cardiorespiratory, cardiovascular, respiratory, stroke, and ischemic heart disease causes) and years of life lost (YLL). They found a 1°C increase in DTR at lag 0–1 days significantly increased daily non-accidental mortality by 0.65% and cardiovascular-specific mortality by 1.12%. The relationships between DTR and deaths due to other investigated causes (cardiorespiratory, respiratory, stroke, and ischemic heart disease) were not significant. Nor was there any significant relationship between DTR and YLL for any of the cause-specific mortalities

Figure 4.3.1

Scatter plots of heart disease death rates and mean temperature (left panel) and cancer death rates and mean temperature (right panel) for all 67 counties in Alabama

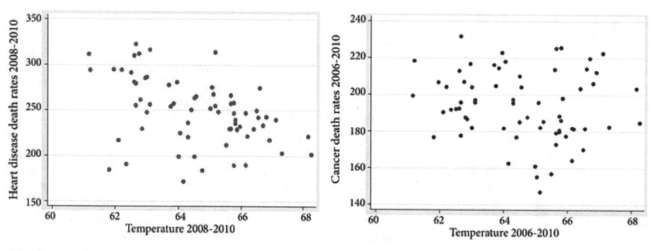

Source: Hart, 2015.

investigated. In stratifying their findings by subgroups for the two mortality categories that did have a significant relationship with DTR (non-accidental and cardiovascular deaths), Zhang *et al.* found that, "compared with males and younger persons, females and the elderly suffered more significantly and substantially from both increased mortality and YLL in relation to a high DTR." They also found that those who had obtained a higher degree of education were more susceptible to increased mortality and YLL from a DTR increase than those who were less educated, an unusual finding for which they suggested further research would be required.

Multiple researchers have reported a decline in DTR in recent decades at locations all across the globe in conjunction with a rise in global temperatures. For China, Shen *et al.* (2014) recently calculated a country-wide DTR decrease at a mean rate of 0.157°C/decade based on an analysis of 479 weather stations over the period 1962–2011; they cited the works of a number of other authors who also determined that the "DTR decreased significantly in China over the past several decades, including Karl *et al.* (1991, 1993), Kukla and Karl (1993), Dai *et al.* (1997, 1999), Liu *et al.* (2004), Ye *et al.* (2010), Zhou and Ren (2011), Wang and Dickinson (2013), Xia (2013), and Wang *et al.* (2014). Consequently, it would appear that if global temperatures continue to rise and the DTR continues to decline, there will likely be fewer cases of non-accidental and cardiovascular deaths in the future.

Daisuke Onozuka and Akihito Hagihara (Onozuka and Hagihara, 2017), two researchers at the Kyushu University Graduate School of Medical Sciences in Fukuoka, Japan, state that out-of-hospital cardiac arrest (OHCA) is "an on-going public health issue with a high case fatality rate and associated with both patient and environmental factors," including temperature. Recognizing the concern that exists over the potential impacts of climate change on human health, the two scientists investigated the population-attributable risk of OHCA in Japan due to temperature, and the relative contributions of low and high temperatures on that risk, for the period 2005–2014.

Onozuka and Hagihara obtained OHCA data on more than 650,000 cases in the 10-year period from all 47 Japanese prefectures. Using climate data acquired from the Japan Meteorological Agency, they conducted a series of statistical analyses to determine the temperature-related health risk of OHCA.

They found "temperature accounted for a substantial fraction of OHCAs, and ... most of [the] morbidity burden was attributable to low temperatures." Nearly 24% of all OHCAs were attributable to non-optimal temperature, and low temperature was responsible for 23.64%. The fraction of OHCAs attributed to high temperature was just 0.29% – a morbidity burden two orders of magnitude smaller than that due to low temperature.

Onozuka and Hagihara also examined the impact of extreme versus moderate temperatures, as well as the effects of gender and age on OHCA risk. With respect to extreme versus moderate temperatures, the two scientists report "the effect of extreme temperatures was substantially less than that of moderate temperatures." For gender, they determined the attributable risk of OHCA was higher for females (26.86%) than males (21.12%). For age, they found that the elderly (75–110 years old) had the highest risk at 28.39%, followed by the middle-aged (65–74 years old, 25.24% attributable risk), and then the youngest section of the population (18–64 years old, 17.93% attributable risk).

Onozuka and Hagihara's analysis reveals moderately cold temperatures carry an inherently far greater risk for OHCA than moderately warm temperatures; extremely cold or extremely warm temperatures are responsible for only a small fraction of attributable risk of OHCA; and female and elderly portions of the population are more prone to the temperature-related effects on OHCA. It would appear modest global warming would yield great benefits for the Japanese by reducing OHCA risk/morbidity, particularly for women and the elderly.

These several studies clearly demonstrate global warming is beneficial to humanity, reducing the incidence of cardiovascular diseases related to low temperatures and wintry weather by a much greater degree than it increases the incidence of cardiovascular diseases associated with high temperatures and summer heat waves.

References

Bartzokas, A., Kassomenos, P., Petrakis, M., and Celessides, C. 2004. The effect of meteorological and pollution parameters on the frequency of hospital admissions for cardiovascular and respiratory problems in Athens. *Indoor and Built Environment* **13**: 271–5.

Bayentin, L., El Adlouni, S., Ouarda, T.B.M.J., Gosselin, P., Doyon, B., and Chebana, F. 2010. Spatial variability of climate effects on ischemic heart disease hospitalization rates for the period 1989–2006 in Quebec, Canada. *International Journal of Health Geographics* **9**: 5.

Bhaskaran, K., Hajat, S., Haines, A., Herrett, E., Wilkinson, P., and Smeeth, L. 2010. Short term effects of temperature on risk of myocardial infarction in England and Wales: time series regression analysis of the Myocardial Ischaemia National Audit Project (MINAP) registry. *British Medical Journal* **341**: c3823.

Braga, A.L.F., Zanobetti, A., and Schwartz, J. 2002. The effect of weather on respiratory and cardiovascular deaths in 12 U.S. cities. *Environmental Health Perspectives* **110**: 859–63.

Burkart, K., Khan, M., Kramer, A., Breitner, S., Schneider, A., and Endlicher, W. 2011. Seasonal variations of all-cause and cause-specific mortality by age, gender, and socioeconomic condition in urban and rural areas of Bangladesh. *International Journal for Equity in Health* **10**: 32.

Cagle, A. and Hubbard, R. 2005. Cold-related cardiac mortality in King County, Washington, USA 1980–2001. *Annals of Human Biology* **32**: 525–37.

Cao, J., Cheng, Y., Zhao, N., Song, W., Jiang, C., Chen, R., and Kan, H. 2009. Diurnal temperature range is a risk factor for coronary heart disease death. *Journal of Epidemiology* **19**: 328–32.

Carder, M., McNamee, R., Beverland, I., Elton, R., Cohen, G.R., Boyd, J., and Agius, R.M. 2005. The lagged effect of cold temperature and wind chill on cardiorespiratory mortality in Scotland. *Occupational and Environmental Medicine* **62**: 702–10.

Chang, C.L., Shipley, M., Marmot, M., and Poulter, N. 2004. Lower ambient temperature was associated with an increased risk of hospitalization for stroke and acute myocardial infarction in young women. *Journal of Clinical Epidemiology* **57**: 749–57.

Christidis, N., Donaldson, G.C., and Stott, P.A. 2010. Causes for the recent changes in cold- and heat-related mortality in England and Wales. *Climatic Change* **102**: 539–53.

Claeys, M.J., Rajagopalan, S., Nawrot, T.S., and Brook, R.D. 2017. Climate and environmental triggers of acute myocardial infarction. *European Heart Journal* **38**: 955–60.

Cohen, P., Potchter, O., and Matzarakis, A. 2012. Daily and seasonal climatic conditions of green urban open spaces in the Mediterranean climate and their impact on human comfort. *Building and Environment* **51**: 285–95.

Dai, A., DelGenio, A.D., and Fung, I.Y. 1997. Clouds, precipitation and diurnal temperature range. *Nature* **386**: 665–6.

Dai, A., Trenberth, K.E., and Karl, T.R. 1999. Effects of clouds, soil moisture, precipitation, and water vapor on diurnal temperature range. *Journal of Climate* **12**: 2451–73.

Davies, M.J. and Thomas, A.C. 1985. Plaque fissuring - the cause of acute myocardial infarction, sudden ischaemic death, and crescendo angina. *British Heart Journal* **53**: 363–73.

Enquselassie, F., Dobson, A.J., Alexander, H.M., and Steele, P.L. 1993. Seasons, temperature and coronary disease. *International Journal of Epidemiology* **22**: 632–6.

Gasparrini, A., *et al.* 2015. Mortality risk attributable to high and low ambient temperature: a multi-country observational study. *Lancet* **386**: 369–75.

Gouveia, N., Hajat, S., and Armstrong, B. 2003. Socioeconomic differentials in the temperature-mortality relationship in Sao Paulo, Brazil. *International Journal of Epidemiology* **32**: 390–7.

Grigorieva, E. and Matzarakis, A. 2011. Physiologically equivalent temperature as a factor for tourism in extreme climate regions in the Russian far east: preliminary results. *European Journal of Tourism, Hospitality and Recreation* **2**: 127–42.

Hajat, S. and Haines, A. 2002. Associations of cold temperatures with GP consultations for respiratory and cardiovascular disease amongst the elderly in London. *International Journal of Epidemiology* **31**: 825–30.

Hansen, A.L., *et al.* 2008. The effect of heat waves on hospital admissions for renal disease in a temperate city of Australia. *International Journal of Epidemiology* **37** (6): 1359–65.

Hart, J. 2015. Association between air temperature and deaths due to cancer and heart disease in Alabama. *Applied Scientific Reports* **2**: 1.

IPCC. 2014. Intergovernmental Panel on Climate Change. *Climate Change 2014: Impacts, Adaptation, and Vulnerability. Part A: Global and Sectoral Aspects.* Contribution of Working Group II to the Fifth Assessment Report of the Intergovernmental Panel on Climate Change. New York, NY: Cambridge University Press.

Karl, T.R., *et al.* 1991. Global warming: evidence for asymmetric diurnal temperature change. *Geophysical Research Letters* **18**: 2253–6.

Karl, T.R., *et al*. 1993. Asymmetric trends of daily maximum and minimum temperature. *Bulletin of the American Meteorological Society* **74**: 1007–23.

Knowlton, K.M., *et al.* 2009. The 2006 California heat wave: impacts on hospitalizations and emergency department visits. *Environmental Health Perspectives* **117** (1): 61–7.

Kovats, R.S., Hajat, S., and Wilkinson, P. 2004. Contrasting patterns of mortality and hospital admissions during hot weather and heat waves in Greater London, UK. *Occupational and Environmental Medicine* **61**: 893–8.

Kukla, G. and Karl, T.R. 1993. Nighttime warming and the greenhouse effect. *Environmental Science and Technology* **27**: 1468–74.

Kysely, J., Plavcova, E., Davidkovova, H., and Kyncl, J. 2011. Comparison of hot and cold spell effects on cardiovascular mortality in individual population groups in the Czech Republic. *Climate Research* **49**: 113–29.

Lim, Y.-H., Hong, Y.-C., and Kim, H. 2012. Effects of diurnal temperature range on cardiovascular and respiratory hospital admissions in Korea. *Science of the Total Environment* **417–418**: 55–60.

Lin, R.T. and Chan, C.C. 2009. Effects of heat on workers' health and productivity in Taiwan. *Global Health Action* **2**.

Liu, B., Xu, M., Henderson, M., Qi, Y., and Li, Y. 2004. Taking China's temperature: daily range, warming trends, and regional variations, 1955–2000. *Journal of Climate* **17**: 4453–62.

Longstreth, J. 1991. Anticipated public health consequences of global climate change. *Environmental Health Perspectives* **96**: 139–44.

McGregor, G.R., Watkin, H.A., and Cox, M. 2004. Relationships between the seasonality of temperature and ischaemic heart disease mortality: implications for climate based health forecasting. *Climate Research* **25**: 253–63.

Nafstad, P., Skrondal, A., and Bjertness, E. 2001. Mortality and temperature in Oslo, Norway, 1990–1995. *European Journal of Epidemiology* **17**: 621–7.

Nakaji, S., *et al.* 2004. Seasonal changes in mortality rates from main causes of death in Japan (1970–1999). *European Journal of Epidemiology* **19**: 905–13.

Nichols, M., Townsend, N., Scarborough, P., and Rayner, M. 2013. Cardiovascular disease in Europe: epidemiological update. *European Heart Journal* **34**: 3028–34.

Nitschke, M., *et al.* 2011. Impact of two recent extreme heat episodes on morbidity and mortality in Adelaide, South Australia: a case-series analysis. *Environmental Health* **10** (1): 42–51.

Onozuka, D. and Hagihara, A. 2017. Out-of-hospital cardiac arrest risk attributable to temperature in Japan. *Scientific Reports* **7**: 39538.

Ponjoan, A., *et al.* 2017. Effects of extreme temperatures on cardiovascular emergency hospitalizations in a Mediterranean region: a self-controlled case series study. *Environmental Health* **16**: 32.

Sharovsky, R., Cesar, L.A.M., and Ramires, J.A.F. 2004. Temperature, air pollution, and mortality from myocardial infarction in Sao Paulo, Brazil. *Brazilian Journal of Medical and Biological Research* **37**: 1651–7.

Shen, X., Liu, B., Li, G., Wu, Z., Jin, Y., Yu, P., and Zhou, D. 2014. Spatiotemporal change of diurnal temperature range and its relationship with sunshine duration and precipitation in China. *Journal of Geophysical Research: Atmospheres* **119**: 13,163–79.

Tam, W.W.S., Wong, T.W., Chair, S.Y., and Wong, A.H.S. 2009. Diurnal temperature range and daily cardiovascular mortalities among the elderly in Hong Kong. *Archives of Environmental and Occupational Health* **64**: 202–6.

Toro, K., Bartholy, J., Pongracz, R., Kis, Z., Keller, E., and Dunay, G. 2010. Evaluation of meteorological factors on sudden cardiovascular death. *Journal of Forensic and Legal Medicine* **17**: 236–42.

Vasconcelos, J., Freire, E., Almendra, R., Silva, G.L., and Santana, P. 2013. The impact of winter cold weather on acute myocardial infarctions in Portugal. *Environmental Pollution* **183**: 14–8.

Wang, F., Zhang, C., Peng, Y., and Zhou, H. 2014. Diurnal temperature range variation and its causes in a semiarid region from 1957 to 2006. *International Journal of Climatology* **34**: 343–54.

Wang, K. and Dickinson, R.E. 2013. Contribution of solar radiation to decadal temperature variability over land. *Proceedings of the National Academy of Sciences USA* **110**: 14,877–82.

Wanitschek, M., Ulmer, H., Sussenbacher, A., Dorler, J., Pachinger, O., and Alber, H.F. 2013. Warm winter is associated with low incidence of ST elevation myocardial infarctions and less frequent acute coronary angiographies in an alpine country. *Herz* **38**: 163–70.

Xia, X. 2013. Variability and trend of diurnal temperature range in China and their relationship to total cloud cover and sunshine duration. *Annals of Geophysics* **31**: 795–804.

Ye, J., Li, F., Sun, G., and Guo, A. 2010. Solar dimming and its impact on estimating solar radiation from diurnal temperature range in China, 1961–2007. *Theoretical and Applied Climatology* **101**: 137–42.

Zanobettia, A., O'Neill, M.S., Gronlund, C.J., and Schwartz, J.D. 2012. Summer temperature variability and long-term survival among elderly people with chronic disease. *Proceedings of the National Academy of Sciences USA* **1097**: 6608–13.

Zhang, Y., Yu, C., Yang, J., Zhang, L., and Cui, F. 2017. Diurnal temperature range in relation to daily mortality and years of life lost in Wuhan, China. *International Journal of Environmental Research and Public Health* **14**: 891.

Zhou, Y.Q. and Ren, G.Y. 2011. Change in extreme temperature events frequency over Mainland China during 1961–2008. *Climatic Research* **50**: 125–39.

4.4 Respiratory Disease

Climate change is not increasing the incidence of death, hospital visits, or loss of work or school time due to respiratory disease.

Respiratory diseases affect the organs and tissues that make gas exchange possible in humans and other higher organisms. They range from the common cold, allergies, and bronchiolitis to life-threatening conditions including asthma, chronic obstructive pulmonary disease (COPD), pneumonia, and lung cancer. Sudden acute respiratory disease (SARS) is a condition in which breathing becomes difficult and oxygen levels in the blood drop lower than normal.

Respiratory diseases are widespread. Non-fatal respiratory diseases impose enormous social costs due to days lost from work and school (Mourtzoukou and Falagas, 2007). Contrary to claims made by the IPCC, real-world data reveal unseasonable cold temperatures cause more deaths and hospital admissions due to respiratory disease than do unseasonable warm temperatures.

Before reviewing the literature on respiratory diseases generally, some background information on asthma may be useful. Childhood asthma affects more than 300 million people worldwide (Baena-Cagnani and Badellino, 2011) and has been increasing by 50% per decade in many countries (Beasley *et al.,* 2000). Like so many government bureaucracies, the U.S. Centers for Disease Control

and Prevention (CDC, 2014) has sought to associate a subject of public concern with climate change, but alternative explanations include increasing hygiene (Liu, 2007), antibiotic use (Kozyraskyj *et al.,* 2007), and the pasteurisation of cow's milk (Ewaschuk *et al.,* 2011; Loss *et al.,* 2017). Rising atmospheric CO_2 concentrations and temperatures may increase ragweed pollen numbers and perhaps other pollens associated with respiratory allergies (Wayne *et al.,* 2002), but ragweed pollen allergenicity can vary four-fold (Lee *et al.,* 1979) and pollen numbers are spatially and temporally highly variable (Weber, 2002). All of this suggests there is no simple relationship between climate or surface temperature and asthma, and therefore no reason to believe somehow slowing or stopping climate change would reduce the incidence of asthma. Asthma is discussed at greater length in Chapter 6.

Some of the studies cited earlier in this chapter on lower death rates due to warmer temperatures and cardiovascular disease also identified specific reductions in fatalities due to respiratory diseases, so their research also appears in this section. Keatinge and Donaldson (2001), for example, studied the effects of temperature on mortality in people over 50 years of age in the greater London area over the period 1976–1995. Simple plots of mortality rate versus daily air temperature revealed a linear increase in mortality as the air temperature fell from 15°C to near 0°C (59°F–32°F). Mortality rates at temperatures above 15°C, on the other hand, showed no trend. The authors say it is because "cold causes mortality mainly from arterial thrombosis and *respiratory disease*, attributable in turn to cold-induced hemo-concentration and hypertension and *respiratory infections*" (italics added).

Nafstad *et al.* (2001) studied the association between temperature and daily mortality in citizens of Oslo, Norway over the period 1990–1995. They found the mean daily number of respiratory-related deaths was considerably higher in winter (October–March) than in summer (April–September). Winter deaths associated with respiratory diseases were 47% more numerous than summer deaths. They conclude, "a milder climate would lead to a substantial reduction in average daily number of deaths."

Hajat and Haines (2002) examined the relationship between cold temperatures and the number of visits by the elderly to general practitioners for asthma, lower respiratory diseases other than asthma, and upper respiratory diseases other than allergic rhinitis as obtained for registered patients aged 65 and older from several London practices between January 1992 and September 1995. They found the mean number of consultations was higher in cool-season months (October–March) than in warm-season months (April–September) for all respiratory diseases. At mean temperatures below 5°C, the relationship between respiratory disease consultations and temperature was linear, and stronger at a time lag of six to 15 days. A 1°C decrease in mean temperature below 5°C was associated with a 10.5% increase in all respiratory disease consultations.

Braga *et al.* (2002) conducted a time-series analysis of both the acute and lagged influence of temperature and humidity on mortality rates in 12 U.S. cities, finding no clear evidence for a link between humidity and respiratory-related deaths. With respect to temperature, they found respiratory-related mortality increased in cities with more variable temperature. This phenomenon, they write, "suggests that increased temperature variability is the most relevant change in climate for the direct effects of weather on respiratory mortality."

Gouveia *et al.* (2003) extracted daily counts of deaths from all causes, except violent deaths and neonatal deaths (up to one month of age), from Sao Paulo, Brazil's mortality information system for the period 1991–1994 and analyzed them for effects of temperature. For respiratory-induced deaths, death rates due to a 1°C cooling were twice as great as death rates due to a 1°C warming in adults and 2.8 times greater in the elderly.

Nakaji *et al.* (2004) evaluated seasonal trends in deaths due to various diseases in Japan, using nationwide vital statistics from 1970 to 1999 and concurrent mean monthly air temperature data. They found the numbers of deaths due to respiratory diseases, including pneumonia and influenza, rise to a maximum during the coldest time of the year. The team of nine scientists concludes, "to reduce the overall mortality rate and to prolong life expectancy in Japan, measures must be taken to reduce those mortality rates associated with seasonal differences."

Bartzokas *et al.* (2004) "examined the relationship between hospital admissions for cardiovascular (cardiac in general including heart attacks) and/or respiratory diseases (asthma etc.) in a major hospital in Athens [Greece] and meteorological parameters for an 8-year period." Over the study period, they found, "there was a dependence of admissions on temperature," and low temperatures were "responsible for a higher number of admissions." Specifically, "there was a decrease of cardiovascular or/and respiratory events from low to high values [of temperature], except for the highest temperature class in which a slight increase was recorded."

Kovats *et al.* (2004) studied patterns of temperature-related hospital admissions and deaths in Greater London during the mid-1990s. For the three-year period 1994–1996, they found respiratory-related deaths were nearly 150% greater in the depth of winter cold than at the height of summer warmth. They also found the mortality impact of the heat wave of July 29 to August 3, 1995 (which boosted daily mortality by just over 10%) was so tiny it could not be discerned among the random scatter of plots of three-year-average daily deaths from cardiovascular and respiratory problems versus day of year. Similarly, in a study of temperature effects on mortality in three English counties (Hampshire, West Midlands, and West Yorkshire), McGregor (2005) found "the occurrence of influenza ... helps elevate winter mortality above that of summer."

Carder *et al.* (2005) investigated the relationship between outside air temperature and deaths due to all non-accident causes in the three largest cities of Scotland (Glasgow, Edinburgh, and Aberdeen) between January 1981 and December 2001. The authors observed "an overall increase in mortality as temperature decreases," which "appears to be steeper at lower temperatures than at warmer temperatures," and "there is little evidence of an increase in mortality at the hot end of the temperature range." Specifically regarding respiratory disease, they found "for temperatures below 11°C, a 1°C drop in the daytime mean temperature on any one day was associated with an increase in respiratory mortality of 4.8% over the following month."

Donaldson (2006) studied the effect of annual mean daily air temperature on the length of the yearly season for respiratory syncytial virus (RSV), which causes bronchiolitis, in England and Wales for 1981–2004. Reporting "climate change may be shortening the RSV season," Donaldson found "the seasons associated with laboratory isolation of respiratory syncytial virus (for 1981–2004) and RSV-related emergency department admissions (for 1990–2004) ended 3.1 and 2.5 weeks earlier, respectively, per 1°C increase in annual central England temperature (P = 0.002 and 0.043, respectively)." Since "no relationship was observed between the start of each season and temperature," he reports, "the RSV season has become shorter." He concludes, "these findings imply a health benefit of global warming in England and Wales associated with a reduction in the duration of the RSV season and its consequent impact on the health service."

Frei and Gassner (2008) studied hay fever prevalence in Switzerland from 1926 to 1991, finding it rose from just under 1% of the country's population to just over 14%, but from 1991 to 2000 it leveled off, fluctuating about a mean value on the order of 15%. The authors write, "several studies show that no further increase in asthma, hay fever and atopic sensitization in adolescents and adults has been observed during the 1990s and the beginning of the new century," citing Braun-Fahrlander et al. (2004) and Grize et al. (2006). They write, "parallel to the increasing hay fever rate, the pollen amounts of birch and grass were increasing from 1969 to 1990," but "subsequently, the pollen of these plant species decreased from 1991 to 2007." They say this finding "is more or less consistent with the changes of the hay fever rate that no longer increased during this period and even showed a tendency to decrease slightly." Nearly identical findings were presented a year later (Frei, 2009). Although some have claimed rising temperatures and atmospheric CO_2 concentrations will lead to more pollen and more hay fever (Wayne et al., 2002), the analyses of Frei (2009) and Frei and Gassner (2008) suggest that is not true of Switzerland.

Jato et al. (2009) collected airborne samples of Poaceae pollen in four cities in Galicia (Northwest Spain) – Lugo, Ourense, Santiago, and Vigo – noting "the global climate change recorded over recent years may prompt changes in the atmospheric pollen season (APS)." The four researchers report "all four cities displayed a trend towards lower annual total Poaceae pollen counts, lower peak values and a smaller number of days on which counts exceeded 30, 50 and 100 pollen grains/m³." The percentage decline in annual pollen grain counts between 1993 and 2007 in Lugo was approximately 75%, and in Santiago the decline was 80%, as best as can be determined from the graphs of the researchers' data. In addition, they write, "the survey noted a trend towards delayed onset and shorter duration of the APS." Thus, even though there was a "significant trend towards increasing temperatures over the months prior to the onset of the pollen season," according to the Spanish scientists, Poaceae pollen became far less of a negative respiratory health factor in the four cities over the decade and a half of their study.

Miller et al. (2012) extracted from the U.S. National Health Interview Survey for 1998 to 2006 annual prevalence data for frequent otitis media (defined as three or more ear infections per year), respiratory allergy, and non-respiratory seizures in children. They also obtained average annual temperatures for the same period from the U.S. Environmental Protection Agency. They found "annual temperature did not influence the prevalence of frequent otitis media," "annual temperature did not influence prevalence of respiratory allergy," and "annual temperature and sex did not influence seizure prevalence." They conclude their findings "may demonstrate that average temperature is not likely to be the dominant cause of the increase in allergy burden or that larger changes in temperatures over a longer period are needed to observe this association." They further conclude, "in the absence of more dramatic annual temperature changes, we do not expect prevalence of otitis media to change significantly as global warming may continue to affect our environment."

Lin et al. (2013) used data on daily area-specific deaths from all causes, circulatory diseases, and respiratory diseases in Taiwan, developing relationships between each of these cause-of-death categories and a number of cold-temperature related parameters for 2000–2008. The five researchers discovered "mortality from [1] all causes and [2] circulatory diseases and [3] outpatient visits of respiratory diseases has a strong association with cold temperatures in the subtropical island, Taiwan." In addition, they found "minimum temperature estimated the strongest risk associated with outpatient visits of respiratory diseases."

Xu et al. (2013a) found that reported cases of childhood asthma increased 0–9 days after a diurnal temperature range (DTR) above 10°C, and also found a 31% increase in emergency department admissions per 5°C increment in DTR.

Xu et al. (2013b) found in Brisbane, Queensland, that "a 1°C increase in diurnal temperature range was associated with a 3% increase of Emergency Department Admissions for childhood diarrhea." Their conclusion that "the incidence of childhood diarrhea may increase if climate variability increases as predicted" reveals gross ignorance of the science on greenhouse warming. This is another example of the statement on page 2 about "independent researchers who should know better."

Liu et al. (2015) examined the association between temperature change and clinical visits for childhood respiratory tract infections (RTIs) in Guangzhou, China based on outpatient records of clinical visits for pediatric RTIs between January 1, 2012 and December 31, 2013, where temperature change was defined as the difference between the mean temperatures of two consecutive days, and where a distributed lag non-linear model was employed to examine the impact of the observed temperature changes. Their analyses of the weather and hospital data revealed "a large temperature decrease was associated with a significant risk for an RTI, with the effect lasting for ~10 days." In addition, they found that "children aged 0–2 years, and especially those aged <1 year, were particularly vulnerable to the effects of temperature drop," noting an extreme temperature decrease "was significantly associated with increased pediatric outpatient visits for RTI's in Guangzhou."

Xie *et al.* (2017) investigated the relationship between acute bronchitis and diurnal temperature range (DTR), which they considered to be "a meteorological indicator closely associated with urbanization, global climate change and [reflective of] the stability of the weather." They examined 14,055 cases of acute bronchitis among children aged 0–14 in Hefei, China over the winter months (December–February) of 2010 through 2013. Their analysis indicated the risk of bronchitis increased as the DTR increased, which relationship was greatest at a 3-day lag, where a 1°C increase in DTR led to a 1.0% increase in the number of daily cases for childhood bronchitis.

Song *et al.* (2018) analyzed daily meteorological data and daily emergency hospital visits in the Haidian district of Beijing, China between 2009 and 2012. They studied the relative risk of respiratory morbidity for both heat waves and cold spells of ≥ 2, ≥ 3, and ≥ 4 days of duration in which the average daily temperature fell within the 95th through 99th percentiles and 1st through 5th percentiles, respectively.

Song *et al.* found the relationship between ambient temperature and respiratory emergency department visits followed a U-shaped curve, where the minimum relative risk value of 1.0 was observed at a mean daily temperature of 21.5°C, a full 6.0°C warmer than the mean average temperature of the entire study period (15.5°C). Given that this minimum-morbidity temperature is much higher than the mean temperature over the period of study, some form of human adaptability or respiratory morbidity acclimation to warmer weather appears to be taking place.

Song *et al.*'s work also revealed that the relative risk (RR) of daily respiratory morbidity due to cold spells is typically much greater (RR of ~1.7 vs RR of ~1.4) than that due to heat waves (see Figure 4.4.1).

The only exception is the RR for heat waves lasting four or more days at the 99th percentile threshold. That data point is represented by only one heat wave in the entire study and may be an aberration skewing the results, given the closer relationship in RR values observed at the 95th through 98th percentiles.

In light of these several findings, it would appear the most effective policies for reducing respiratory emergency department visits would be targeted towards the higher relative risks observed at the cold end of the temperature spectrum.

References

Baena-Cagnani, C. and Badellino, H. 2011. Diagnosis of allergy and asthma in childhood. *Current Allergy and Asthma Reports* **11**: 71–7.

Bartzokas, A., Kassomenos, P., Petrakis, M., and Celessides, C. 2004. The effect of meteorological and pollution parameters on the frequency of hospital admissions for cardiovascular and respiratory problems in Athens. *Indoor and Built Environment* **13**: 271–5.

Beasley, R., *et al.* 2000. Prevalence and etiology of asthma. *Journal of Allergy and Clinical Immunology* **105**: 466–72.

Braga, A.L.F., Zanobetti, A., and Schwartz, J. 2002. The effect of weather on respiratory and cardiovascular deaths in 12 U.S. cities. *Environmental Health Perspectives* **110**: 859–63.

Braun-Fahrlander, C., *et al.* 2004. No further increase in asthma, hay fever and atopic sensitization in adolescents living in Switzerland. *European Respiratory Journal* **23**: 407–13.

Carder, M., McNamee, R., Beverland, I., Elton, R., Cohen, G.R., Boyd, J., and Agius, R.M. 2005. The lagged effect of cold temperature and wind chill on cardiorespiratory mortality in Scotland. *Occupational and Environmental Medicine* **62**: 702–10.

CDC. 2014. Centers for Disease Control and Prevention. Allergens (website). December 11. Accessed June 12, 2018.

Donaldson, G.C. 2006. Climate change and the end of the respiratory syncytial virus season. *Clinical Infectious Diseases* **42**: 677–9.

Ewaschuk, J.B., *et al.* 2011: Effect of pasteurization on immune components of milk: implications for feeding preterm infants. *Applied Physiology, Nutrition and Metabolism* **36** (2): 175–82.

Figure 4.4.1
Relative risk in daily respiratory morbidity due to cold spells and heat waves in Beijing, China

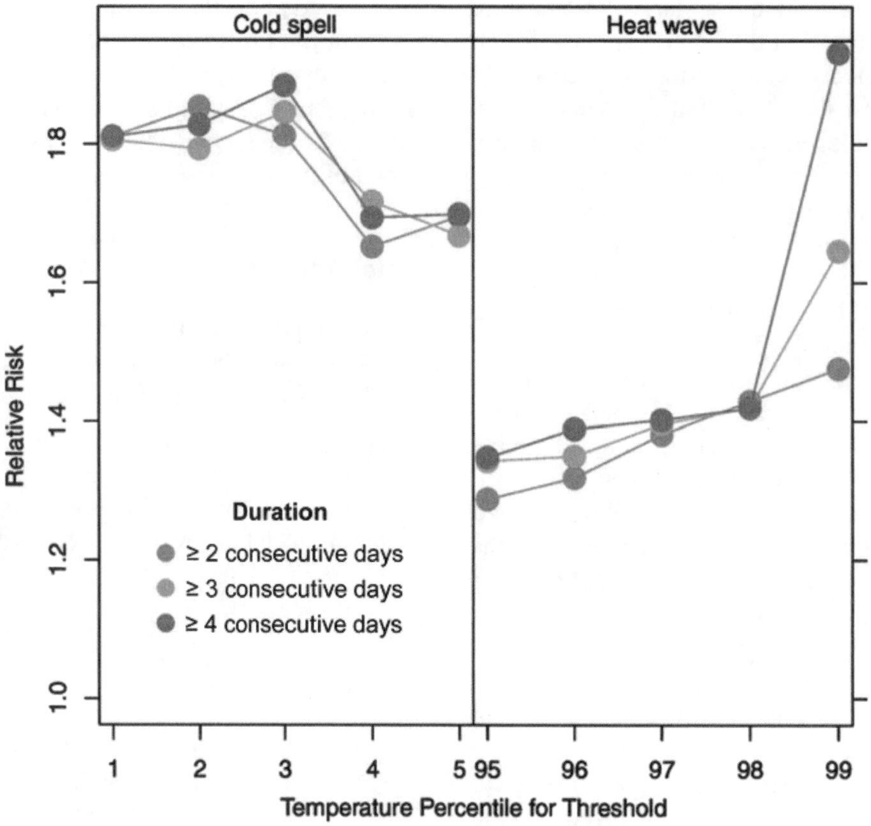

Source: Song *et al.*, 2018.

Frei, T. 2009. Trendwende bei der pollinose und dem pollenflug? *Allergologie* **32**: 123–7.

Frei, T. and Gassner, E. 2008. Trends in prevalence of allergic rhinitis and correlation with pollen counts in Switzerland. *International Journal of Biometeorology* **52**: 841–7.

Gouveia, N., Hajat, S., and Armstrong, B. 2003. Socioeconomic differentials in the temperature-mortality relationship in Sao Paulo, Brazil. *International Journal of Epidemiology* **32**: 390–7.

Grize, L., *et al*. 2006. Trends in prevalence of asthma, allergic rhinitis and atopic dermatitis in 5–7-year old Swiss children from 1992 to 2001. *Allergy* **61**: 556–62.

Hajat, S. and Haines, A. 2002. Associations of cold temperatures with GP consultations for respiratory and cardiovascular disease amongst the elderly in London. *International Journal of Epidemiology* **31**: 825–30.

Jato, V., Rodriguez-Rajo, F.J., Seijo, M.C., and Aira, M.J. 2009. Poaceae pollen in Galicia (N.W. Spain): characterization and recent trends in atmospheric pollen season. *International Journal of Biometeorology* **53**: 333–44.

Keatinge, W.R. and Donaldson, G.C. 2001. Mortality related to cold and air pollution in London after allowance for effects of associated weather patterns. *Environmental Research* **86**: 209–16.

Kovats, R.S., Hajat, S., and Wilkinson, P. 2004. Contrasting patterns of mortality and hospital admissions during hot weather and heat waves in Greater London, UK. *Occupational and Environmental Medicine* **61**: 893–8.

Kozyrskyj, A.L., *et al*. 2007: Increased risk of childhood asthma from antibiotic use in early life. *Chest Journal* **131** (6): 1753–9.

Lee, Y.S., Dickinson, D.B., Schlager, D., and Velu, J.G. 1979: Antigen E content of pollen from individual plants of short ragweed (*Ambrosia artemisiifolia*). *Journal of Allergy and Clinical Immunology* **63**: 336–9.

Lin, Y.-K., Wang, Y.-C., Lin, P.-L., Li, M.-H., and Ho, T.-J. 2013. Relationships between cold-temperature indices and all causes and cardiopulmonary morbidity and mortality in a subtropical island. *Science of the Total Environment* **461–462**: 627–35.

Liu, A.H. 2007: Hygiene theory and allergy and asthma prevention. *Paediatric and Neonatal Epidemiology* **21** (3): 2–7.

Liu, Y., *et al.* 2015. Association between temperature change and outpatient visits for respiratory tract infections among chidlren in Guangzhou, China. *International Journal of Environmental Research and Public Health* **12**: 439–54.

Loss, G., *et al.* 2011: The protective effect of farm milk consumption on childhood asthma and atopy: the GABRIELA study. *The Journal of Allergy and Clinical Immunology* **128** (4): 766–73.

McGregor, G.R. 2005. Winter North Atlantic Oscillation, temperature and ischaemic heart disease mortality in three English counties. *International Journal of Biometeorology* **49**: 197–204.

Miller, M.E., Shapiro, N.L., and Bhattacharyya, N. 2012. Annual temperature and the prevalence of frequent ear infections in childhood. *American Journal of Otolaryngology - Head and Neck Medicine and Surgery* **33**: 51–5.

Mourtzoukou, E.G. and Falagas, M.E. 2007. Exposure to cold and respiratory tract infections. *International Journal of Tuberculosis and Lung Disease* **11**: 938–43.

Nafstad, P., Skrondal, A., and Bjertness, E. 2001. Mortality and temperature in Oslo, Norway. 1990–1995. *European Journal of Epidemiology* **17**: 621–7.

Nakaji, S., *et al.* 2004. Seasonal changes in mortality rates from main causes of death in Japan (1970–1999). *European Journal of Epidemiology* **19**: 905–13.

Song, X., *et al.* 2018. The impact of heat waves and cold spells on respiratory emergency department visits in Beijing, China. *Science of the Total Environment* **615**: 1499–1505.

Wayne, P., *et al.* 2002. Production of allergenic pollen by ragweed (Ambrosia artemisiifolia L.) is increased in CO_2-enriched atmospheres. *Annals of Allergy, Asthma, and Immunology* **88**: 279–82.

Weber, R.W. 2002: Mother Nature strikes back: global warming, homeostasis, and implications for allergy. *Annals of Allergy, Asthma & Immunology* **88**: 251–2.

Xie, M., *et al.* 2017. Effect of diurnal temperature range on the outpatient visits for acute bronchitis in children: a time-series study in Hefei, China. *Public Health* **144**: 103–8.

Xu, Z., Huang, C., Su, H., Turner, L.R., Qiao, Z., and Tong, S. 2013a. Diurnal temperature range and childhood asthma: a time-series study. *Environmental Health* **12**: 10.1186/1476-069X-12-12.

Xu, Z., Huang, C., Turner, L.R., Su, H., Qiao, Z., and Tong, S. 2013b. Is diurnal temperature range a risk factor for childhood diarrhea? *PLoS One* **8**: e64713.

4.5 Stroke

Higher surface temperatures would reduce the incidence of death due to stroke in many parts of the world, including Africa, Asia, Australia, the Caribbean, Europe, Japan, Korea, Latin America, and Russia.

Low minimum temperatures are a greater risk factor than high temperatures for stroke incidence and hospitalization.

Strokes are either ischemic or hemorrhagic. An ischemic stroke occurs when blood flow to part of the brain is cut off, due to a clot forming either on an atherosclerotic plaque (fatty deposit) in a cerebral artery or in the heart or blood vessels leading to the brain, breaking off and travelling to the brain. By contrast, the most common causes of hemorrhagic stroke are high blood pressure and brain aneurysms. An aneurysm is a dilated segment of artery due to a weakness or thinness in the blood vessel wall, which leads to excessive ballooning, leakage of blood, or rupture. The result is blood seeping into or around the brain tissue, causing damage to brain cells.

According to the IPCC, rising atmospheric CO_2 concentrations due to the combustion of fossil fuels cause surface temperatures to rise, which then causes increased deaths due to strokes. However, as was the case with cardiovascular disease and respiratory disease, examination of real-world data reveals unseasonably cold temperatures cause more deaths and hospital admissions due to stroke than do unseasonably warm temperatures.

Feigin *et al.* (2000) examined the relationship between the incidence of stroke and ambient temperatures over the period 1982–1993 in

Novosibirsk, Siberia, which has one of the highest stroke incidence rates in the world. Based on analyses of 2,208 patients with sex and age distributions similar to those of Russia as a whole, the researchers found a statistically significant association between stroke occurrence and low ambient temperature. In the case of ischemic stroke (IS), which accounted for 87% of all strokes, they determined "the risk of IS occurrence on days with low ambient temperature [was] 32% higher than that on days with high ambient temperature." They conclude the "very high stroke incidence in Novosibirsk, Russia may partially be explained by the highly prevalent cold factor there."

Hong *et al.* (2003) investigated the association between the onset of ischemic stroke and prior episodic decreases in temperature in 545 patients who suffered strokes in Incheon, Korea from January 1998 to December 2000. They report "decreased ambient temperature was associated with risk of acute ischemic stroke," with the strongest effect being seen on the day after exposure to cold weather, further noting "even a moderate decrease in temperature can increase the risk of ischemic stroke." They also found "risk estimates associated with decreased temperature were greater in winter than in the summer," which suggests "low temperatures as well as temperature changes are associated with the onset of ischemic stroke." Finally, they explain the reason for the 24- to 48-hour lag between exposure to cold and the onset of stroke "might be that it takes some time for the decreasing temperature to affect blood viscosity or coagulation."

Nakaji *et al.* (2004) evaluated seasonal trends in deaths due to various diseases in Japan using nationwide vital statistics from 1970 to 1999 together with mean monthly temperature data. They found the peak mortality rate due to stroke was two times greater in winter (January) than at the time of its yearly minimum (August and September).

Chang *et al.* (2004) analyzed data from the World Health Organization (WHO) Collaborative Study of Cardiovascular Disease and Steroid Hormone Contraception (WHO, 1995) to determine the effects of monthly mean temperature on rates of hospitalization for arterial ischemic stroke and acute myocardial infarction among women aged 15 to 49 from 17 countries in Africa, Asia, the Caribbean, Europe, and Latin America. They found among these women, a 5°C reduction in mean air temperature was associated with a 7% increase in the expected hospitalization rate due to stroke, and this effect was

relatively acute, within a period of about a month, the scientists write.

Gill *et al.* (2012) write, "in the past two decades, several studies reported that meteorologic changes are associated with monthly and seasonal spikes in the incidence of aneurysmal subarachnoid hemorrhage (aSAH)," and "analysis of data from large regional databases in both hemispheres has revealed increased seasonal risk for aSAH in the fall, winter and spring," citing among other sources Feigin *et al.* (2001), Abe *et al.* (2008), and Beseoglu *et al.* (2008). Gill *et al.* identified the medical records of 1,175 patients at the Johns Hopkins Hospital in Baltimore, Maryland (USA) who were admitted with a radiologically confirmed diagnosis of aSAH between January 1, 1991 and March 1, 2009. The six scientists report both "a one-day decrease in temperature and colder daily temperatures were associated with an increased risk of incident aSAH," and "these variables appeared to act synergistically" and were "particularly predominant in the fall, when the transition from warmer to colder temperatures occurred." Gill *et al.* add their study "is the first to report a direct relationship between a temperature decrease and an increased risk of aSAH," and "it also confirms the observations of several reports of an increased risk of aSAH in cold weather or winter," citing Nyquist *et al.* (2001) and other sources.

References

Abe, T., Ohde, S., Ishimatsu, S., Ogata, H., Hasegawa, T., Nakamura, T., and Tokuda, Y. 2008. Effects of meteorological factors on the onset of subarachnoid hemorrhage: a time-series analysis. *Journal of Clinical Neuroscience* **15**: 1005–10.

Beseoglu, K., Hanggi, D., Stummer, W., and Steiger, H.J. 2008. Dependence of subarachnoid hemorrhage on climate conditions: a systematic meteorological analysis from the Dusseldorf metropolitan area. *Neurosurgery* **62**: 1033–8.

Chang, C.L., Shipley, M., Marmot, M., and Poulter, N. 2004. Lower ambient temperature was associated with an increased risk of hospitalization for stroke and acute myocardial infarction in young women. *Journal of Clinical Epidemiology* **57**: 749–57.

Feigin, V.L., Anderson, C.S., Anderson, N.E., Broad, J.B., Pledger, M.J., and Bonita, R. 2001. Is there a temporal pattern to the occurrence of subarachnoid hemorrhage in the southern hemisphere? Pooled data from 3 large, population-based incidence studies in Australasia, 1981 to 1997. *Stroke* **32**: 613–9.

Feigin, V.L., Nikitin, Yu.P., Bots, M.L., Vinogradova, T.E., and Grobbee, D.E. 2000. A population-based study of the associations of stroke occurrence with weather parameters in Siberia, Russia (1982–92). *European Journal of Neurology* **7**: 171–8.

Gill, R.S., Hambridge, H.L., Schneider, E.B., Hanff, T., Tamargo, R.J., and Nyquist, P. 2012. Falling temperature and colder weather are associated with an increased risk of Aneurysmal Subarachnoid Hemorrhage. *World Neurosurgery* **79**: 136–42.

Hong, Y-C., Rha, J-H., Lee, J-T., Ha, E-H., Kwon, H-J., and Kim, H. 2003. Ischemic stroke associated with decrease in temperature. *Epidemiology* **14**: 473–8.

Nakaji, S., *et al.* 2004. Seasonal changes in mortality rates from main causes of death in Japan (1970–1999). *European Journal of Epidemiology* **19**: 905–13.

Nyquist, P.A., Brown Jr., R.D., Wiebers, D.O., Crowson, C.S., and O'Fallon, W.M. 2001. Circadian and seasonal occurrence of subarachnoid and intracerebral hemorrhage. *Neurology* **56**: 190–3.

WHO. 1995. World Health Organization. WHO collaborative study of cardiovascular disease and steroid hormone contraception. A multi-national case-control study of cardiovascular disease and steroid hormone contraceptives: description and validation of methods. *Journal of Clinical Epidemiology* **48**: 1513–47.

4.6 Insect-borne Diseases

Higher surface temperatures are not leading to increases in mosquito-transmitted and tick-borne diseases such as malaria, yellow fever, viral encephalitis, and dengue fever.

The IPCC's Fifth Assessment Report (AR5) backs down from previous predictions that global warming would facilitate the spread of mosquito-borne diseases including malaria and dengue fever and tick-borne diseases. The full report from Working Group II on the subject (IPCC, 2014a, Chapter 11, pp. 722–6) repeatedly admits there is no evidence that climate change has affected the range of vector-borne diseases. However, the *Summary for Policymakers* inexplicably warns, "Throughout the 21st century, climate change is expected to lead to increases in ill-health in many regions and especially in developing countries with low income, as compared to a baseline without climate change (*high confidence*)." Among the "examples" given is "vector-borne diseases (*medium confidence*)" (IPCC, 2014b, pp. 19–20).

In a research report in *Science*, Rogers and Randolph (2000) note "predictions of global climate change have stimulated forecasts that vector-borne diseases will spread into regions that are at present too cool for their persistence." However, the effect of warmer temperatures on insect-borne diseases is complex, sometimes working in favor of and sometimes against the spread of a disease. For example, ambient temperature has historically not determined the range of insect-borne diseases and human adaptation to climate change overwhelms the role of climate. Even those who support the IPCC admit, "It's a little bit tricky to make a solid prediction" (Irfan, 2011, quoting Marm Kilpatrick).

Gething *et al.* (2010), writing specifically about malaria, may have put it best when they said there has been "a decoupling of the geographical climate-malaria relationship over the twentieth century, indicating that non-climatic factors have profoundly confounded this relationship over time." They note "non-climatic factors, primarily direct disease control and the indirect effects of a century of urbanization and economic development, although spatially and temporally variable, have exerted a substantially greater influence on the geographic extent and intensity of malaria worldwide during the twentieth century than have climatic factors." As for the future, they conclude climate-induced effects "can be offset by moderate increases in coverage levels of currently available interventions."

This section investigates the reliability of the IPCC's claim with respect to the three main kinds of insect-borne diseases: malaria, dengue fever, and tick-borne diseases. According to scientific examination and research on this topic, there is little support for the claims appearing in the latest IPCC *Summary for Policymakers*.

References

Gething, P.W., Smith, D.L., Patil, A.P., Tatem, A.J., Snow, R.W., and Hay, S.I. 2010. Climate change and the global malaria recession. *Nature* **465**: 342–5.

IPCC. 2014a. Intergovernmental Panel on Climate Change. Chapter 11: Human health: Impacts, adaptations, and co-benefits. In: *Climage Change 2014: Impacts, Adaptation, and Vulnerabilities.* Contribution of Working Group II to the Fifth Assessment Report of the Intergovernmental Panel on Climate Change. New York, NY: Cambridge University Press.

IPCC. 2014b. Intergovernmental Panel on Climate Change. *Summary for Policymakers*. In: *Climate Change 2014: Impacts, Adaptation, and Vulnerabilities*. Contribution of Working Group II to the Fifth Assessment Report of the Intergovernmental Panel on Climate Change. New York, NY: Cambridge University Press.

Irfan, U. 2011. Climate change may make insect-borne diseases harder to control. Scientific American (website). November 21. Accessed May 21, 2018.

Rogers, D.J. and Randolph, S.E. 2000. The global spread of malaria in a future, warmer world. *Science* **289**: 1763–6.

4.6.1 Malaria

Extensive scientific information and experimental research contradict the claim that malaria will expand across the globe and intensify as a result of CO_2-induced warming.

Jackson *et al.* (2010) say "malaria is one of the most devastating vector-borne parasitic diseases in the tropical and subtropical regions of the world," noting it affects more than 100 countries. According to the World Health Organization (WHO, 2017), an estimated 216 million cases of malaria occurred worldwide in 2016, down from 237 million cases in 2010. Ninety-one countries reported indigenous malaria cases with Africa accounting for 90% of all cases. Approximately 445,000 people died from malaria that year.

According to Reiter (2000), claims that malaria will become more widespread due to CO_2-induced global warming ignore other important factors and disregard known facts. A historical analysis of malaria trends, for example, reveals this disease was an important cause of illness and death in England during a period of colder-than-present temperatures throughout the Little Ice Age. Its transmission began to decline only in the nineteenth century, during a warming phase when, according to Reiter, "temperatures were already much higher than in the Little Ice Age." In short, malaria was prevalent in Europe during some of the coldest centuries of the past millennium, and it has only recently undergone widespread decline, when temperatures have been warming, Clearly, there are other factors at work that are more important than temperature. Such factors include the quality of public health services, public awareness, irrigation and agricultural activities, land use practices, civil strife, natural disasters, ecological

change, population change, use of insecticides, and the movement of people (Reiter, 2000; Reiter, 2001; Hay *et al.*, 2002a, 2002b).

Nevertheless, concerns have lingered about the possibility of widespread future increases in malaria due to global warming. These concerns are generally rooted in climate models that typically use only one, or at most two, climate variables in making their predictions of the future distribution of the disease over Earth, and they generally do not include any of the non-climatic factors listed in the preceding paragraph. When more variables are included, a less-worrisome future is projected. In one modeling study, for example, Rogers and Randolph (2000) employed five climate variables and obtained very different results. Briefly, they used the present-day distribution of malaria to determine the specific climatic constraints that best define that distribution, after which the multivariate relationship they derived from this exercise was applied to future climate scenarios derived from climate models in order to map potential future geographical distributions of the disease. Their study revealed very little change: a 0.84% increase in potential malaria exposure under the "medium-high" scenario of global warming and a 0.92% *decrease* under the "high" scenario. Rogers and Randolph explicitly state their quantitative model "contradicts prevailing forecasts of global malaria expansion" and "highlights the use of multivariate rather than univariate constraints in such applications."

Hay *et al.* (2002a) investigated long-term trends in meteorological data at four East African highland sites that experienced significant increases in malaria cases over the past couple of decades, reporting "temperature, rainfall, vapour pressure and the number of months suitable for *P. falciparum* transmission have not changed significantly during the past century or during the period of reported malaria resurgence," thus these factors could not be responsible for the observed increases in malaria cases. Likewise, Shanks *et al.* (2000) examined trends in temperature, precipitation, and malaria rates in western Kenya over the period 1965–1997, finding no linkages among the variables.

Small *et al.* (2003) examined trends in a climate-driven model of malaria transmission between 1911 and 1995, using a spatially and temporally extensive gridded climate dataset to identify locations in Africa where the malaria transmission climate suitability index had changed significantly over this time interval. They found "climate warming, expressed as a systematic temperature increase over the 85-year period, does not appear to be responsible for an increase in malaria suitability over any region in Africa." They conclude "research on the links between climate change and the recent resurgence of malaria across Africa would be best served through refinements in maps and models of precipitation patterns and through closer examination of the role of nonclimatic influences."

Kuhn *et al.* (2003) analyzed the determinants of temporal trends in malaria deaths within England and Wales in 1840–1910 and found "a 1°C increase or decrease was responsible for an increase in malaria deaths of 8.3% or a decrease of 6.5%, respectively," which explains "the malaria epidemics in the 'unusually hot summers' of 1848 and 1859." Nevertheless, the long-term near-linear temporal decline in malaria deaths over the period of study, the researchers write, "was probably driven by nonclimatic factors," among which they identify increasing livestock populations (which tend to divert mosquito biting away from humans), decreasing acreages of marsh wetlands (where mosquitoes breed), as well as "improved housing, better access to health care and medication, and improved nutrition, sanitation, and hygiene." Kuhn *et al.* say "the projected increase in proportional risk is clearly insufficient to lead to the reestablishment of endemicity."

Zhou *et al.* (2004) employed a nonlinear mixed-regression model study that focused on the numbers of monthly malaria outpatients of the past 10 to 20 years in seven East African highland sites and their relationships to the numbers of malaria outpatients during the previous time period, seasonality, and climate variability. They state, "for all seven study sites, we found highly significant nonlinear, synergistic effects of the interaction between rainfall and temperature on malaria incidence, indicating that the use of either temperature or rainfall alone is not sensitive enough for the detection of anomalies that are associated with malaria epidemics." Githeko and Ndegwa (2001), Shanks *et al.* (2002), and Hay *et al.* (2002b) reached the same conclusion. In addition, climate variability – not just temperature or not just warming – contributed less than 20% of the temporal variance in the number of malaria outpatients, and at only two of the seven sites studied.

Rogers and Randolph (2006) conducted a major review of the potential impacts of global warming on vector-borne diseases, focusing on recent upsurges of malaria in Africa, asking, "Has climate change already had an impact?" They demonstrate "evidence for increasing malaria in many parts of Africa is

overwhelming, but the more likely causes for most of these changes to date include land-cover and land-use changes and, most importantly, drug resistance rather than any effect of climate," noting "the recrudescence of malaria in the tea estates near Kericho, Kenya, in East Africa, where temperature has not changed significantly, shows all the signs of a disease that has escaped drug control following the evolution of chloroquine resistance by the malarial parasite."

Childs *et al.* (2006) present a detailed analysis of malaria incidence in northern Thailand based on a quarter-century monthly time series (January 1977 through January 2002) of total malaria cases in the country's 13 northern provinces. Over this period, when the IPCC claims the world warmed at a rate and to a level unprecedented over the prior one to two millennia, Childs *et al.* report there was an approximately constant rate of *decline* in total malaria incidence (from a mean monthly incidence in 1977 of 41.5 cases per hundred thousand people to 6.72 cases per hundred thousand people in 2001). Noting "there has been a steady reduction through time of total malaria incidence in northern Thailand, with an average decline of 6.45% per year," they say this result "reflects changing agronomic practices and patterns of immigration, as well as the success of interventions such as vector control programs, improved availability of treatment and changing drug policies."

Zell *et al.* (2008) conducted a similar review of the literature and determined "coupled ocean/atmosphere circulations and continuous anthropogenic disturbances (increased populations of humans and domestic animals, socioeconomic instability, armed conflicts, displaced populations, unbalanced ecosystems, dispersal of resistant pathogens etc.) appear to be the major drivers of disease variability," and "global warming at best contributes."

Reiter (2008) came to similar conclusions, writing, "simplistic reasoning on the future prevalence of malaria is ill-founded; malaria is not limited by climate in most temperate regions, nor in the tropics, and in nearly all cases, 'new' malaria at high altitudes is well below the maximum altitudinal limits for transmission." He further states, "future changes in climate may alter the prevalence and incidence of the disease, but obsessive emphasis on 'global warming' as a dominant parameter is indefensible; the principal determinants are linked to ecological and societal change, politics and economics."

Hulden and Hulden (2009) analyzed malaria statistics collected in Finland from 1750 to 2008 via correlation analyses between malaria frequency per million people and all variables that have been used in similar studies throughout other parts of Europe, including temperature data, animal husbandry, consolidation of land by redistribution, and household size. Over the entire period, "malaria frequency decreased from about 20,000–50,000 per 1,000,000 people to less than 1 per 1,000,000 people," they report. The two Finnish researchers conclude, "indigenous malaria in Finland faded out evenly in the whole country during 200 years with limited or no counter measures or medication," making that situation "one of the very few opportunities where natural malaria dynamics can be studied in detail." Their study indicates "malaria in Finland basically was a sociological disease and that malaria trends were strongly linked to changes in the human household size and housing standard."

Russell (2009) studied the "current or historic situations of the vectors and pathogens, and the complex ecologies that might be involved" regarding malaria, dengue fever, the arboviral arthritides (Ross River and Barmah Forest viruses), and the arboviral encephalitides (Murray Valley encephalitis and Kunjin viruses) in Australia and found "there might be some increases in mosquito-borne disease in Australia with a warming climate, but with which mosquitoes and which pathogens, and where and when, cannot be easily discerned." He concludes, "of itself, climate change as currently projected, is not likely to provide great cause for public health concern with mosquito-borne disease in Australia."

Nabi and Qader (2009) considered the climatic conditions that impact the spread of malaria – temperature, rainfall, and humidity – and the host of pertinent nonclimatic factors that play important roles in its epidemiology: the presence or absence of mosquito control programs, the availability or non-availability of malaria-fighting drugs, changing resistances to drugs, the quality of vector control, changes in land use, the availability of good health services, human population growth, human migrations, international travel, and standard of living. The two researchers report "global warming alone will not be of a great significance in the upsurge of malaria unless it is accompanied by a deterioration in other parameters like public health facilities, resistance to anti-malarial drugs, decreased mosquito control measures," etc. They say "no accurate prediction about malaria can truly be made," because "it is very difficult to estimate what the other factors will be like in the future."

Jackson et al. (2010) linked reported malaria cases and deaths from the years 1996 to 2006 for 10 countries in western Africa (Benin, Burkina Faso, Cote d'Ivoire, Gambia, Ghana, Liberia, Mali, Senegal, Sierra Leone, and Togo) with corresponding climate data from the U.S. National Oceanic and Atmospheric Administration's National Climatic Data Center. They searched for transitive relationships between the weather variables and malaria rates via spatial regression analysis and tests for correlation. Jackson et al. report their analyses showed "very little correlation exists between rates of malaria prevalence and climate indicators in western Africa." This result, as they describe it, "contradicts the prevailing theory that climate and malaria prevalence are closely linked and also negates the idea that climate change will increase malaria transmission in the region."

Haque et al. (2010) analyzed monthly malaria case data for the malaria endemic district of Chittagong Hill Tracts in Bangladesh from January 1989 to December 2008, looking for potential relationships between malaria incidence and various climatic parameters (rainfall, temperature, humidity, sea surface temperature, and the El Niño-Southern Oscillation), as well as the normalized difference vegetation index (NDVI), a satellite-derived measure of surface vegetation greenness. The six scientists report, "after adjusting for potential mutual confounding between climatic factors there was no evidence for any association between the number of malaria cases and temperature, rainfall and humidity," and "there was no evidence of an association between malaria cases and sea surface temperatures in the Bay of Bengal and [the El Niño-Southern Oscillation index for Niño Region 3]."

Gething et al. (2010) compared historical and contemporary maps of the range and incidence of malaria and found endemic/stable malaria is likely to have covered 58% of the world's land surface around 1900 but only 30% by 2007. They report, "even more marked has been the decrease in prevalence within this greatly reduced range, with endemicity falling by one or more classes in over two-thirds of the current range of stable transmission." They write, "widespread claims that rising mean temperatures have already led to increases in worldwide malaria morbidity and mortality are largely at odds with observed decreasing global trends in both its endemicity and geographic extent." Rather, "the combined natural and anthropogenic forces acting on the disease throughout the twentieth century have resulted in the great majority of locations undergoing a net reduction in transmission between one and three orders of magnitude larger than the maximum future increases proposed under temperature-based climate change scenarios."

Stern *et al.* (2011) examined trends in temperature and malaria for the Highlands of East Africa, which span Rwanda, Burundi, and parts of Kenya, Tanzania, and Uganda, to resolve controversies over whether the area has warmed and malaria has become more prevalent. They report temperature has increased significantly in the region, yet "malaria in Kericho and many other areas of East Africa has decreased during periods of unambiguous warming."

Nkurunziza and Pilz (2011) assessed the impact of an increase in temperature on malaria transmission in Burundi, a landlocked country in the African Great Lakes region of East Africa. They found "an increase in the maximum temperature will cause an increase in minimum temperature," and "the increase in the latter will result in a decreasing maximum humidity, leading to a decrease in rainfall." These results, they continue, "suggest that an increased temperature will result in a shortening of the life span of mosquitoes (due to decreasing humidity) and decrease in the capacity of larva production and maturation (due to decreasing rainfall)." Thus, "the increase in temperature will not result in an increased malaria transmission in Burundi," which is "in good agreement with some previous works on the topic," citing as examples WHO, WMO, UNEP (2003), Lieshout *et al.* (2004), and Thomas (2004). In a final statement on the matter, Nkurunziza and Pilz note that in regions with endemic malaria transmission, such as Burundi, "the increase in temperature may lead to unsuitable climate conditions for mosquitoes survival and, hence, probably to a decreasing malaria transmission."

Béguin *et al.* (2011) estimated populations at risk of malaria (PAR) based on climatic variables, population growth, and GDP per capita (GDPpc). GDPpc is an approximation for per-capita income ("income" for short) for 1990, 2010, and 2050, based on sensitivity analyses for the following three scenarios: (1) a worst-case scenario, in which income declines to 50% of its 2010 values by 2050; (2) a "growth reduction" scenario, in which income declines by 25% in 2030 and 50% in 2050, relative to the A1B scenario (socioeconomic change plus climate change); and (3) a scenario in which income stays constant at 2010 values. The results are presented in Figure 4.6.1.1. The authors observe, "under the A1B climate scenario, climate change has much weaker effects than GDPpc increase on the geographic distribution of malaria." This result is consistent with the few studies that have considered the impact of climate change and socioeconomic factors on malaria. (See, e.g., Tol and Dowlatabadi, 2001; Bosello *et al.*, 2006). With respect to malaria, therefore, climate change is a relatively minor factor compared to economic development.

Paaijmans *et al.* (2012) examined the effects of temperature on the rodent malaria Plasmodium yoelii and the Asian malaria vector Anopheles stephensi. The three U.S. researchers found "vector competence (the maximum proportion of infectious mosquitoes, which implicitly includes parasite survival across the incubation period) tails off at higher temperatures, even though parasite development rate increases." Moreover, "the standard measure of the parasite incubation period (i.e., time until the first mosquitoes within a cohort become infectious following an infected blood-meal) is incomplete because parasite development follows a cumulative distribution, which itself varies with temperature. Finally, "including these effects in a simple model dramatically alters estimates of transmission intensity and reduces the optimum temperature for transmission." Therefore, in regard to "the possible effects of climate warming," they conclude "increases in temperature need not simply lead to increases in transmission."

Paaijmans *et al.* conclude their results "challenge current understanding of the effects of temperature on malaria transmission dynamics," and they note their findings imply "control at higher temperatures might be more feasible than currently predicted."

Using a high-resolution computer model that incorporates "climate-driven hydrology as a determinant of mosquito populations and malaria transmission," Yamana et al. (2016) found the impact of future climate change on West African malaria transmission will be "negative at best, and positive but insignificant at worst," confirming that "no major increases in the frequency or the severity of malaria outbreaks in West Africa are expected as a result of climate change."

Murdock *et al.* (2016) investigated "how increases in temperature from optimal conditions (27°C to 30°C and 33°C) interact with realistic diurnal temperature ranges (DTR: ± 0°C, 3°C and 4.5°C) to affect the ability of key vector species from Africa and Asia (*Anopheles gambiae* and *An. Stephensi*) to transmit the human malaria parasite,

Figure 4.6.1.1

Effects of climate change and socioeconomic factors on the projected future global distribution of malaria

Model type	Population at Risk 2030 [billions]	Population at Risk 2050 [billions]
Socioeconomic changes only (no climate change)	3.52	1.74
Socioeconomic and climatic changes (A1B scenario)	3.58 [3.55–3.60]	1.95 [1.93–1.96]
Socioeconomic changes and CC (slower growth scenario)	3.82 [3.39–3.84]	3.42 [3.28–3.45]
No growth scenario, only CC	4.61 [4.54–4.67]	5.20 [5.11–5.25]
Pessimistic growth scenario and CC	5.18 [5.07–5.30]	6.27 [6.19–6.32]

CC refers to climate change scenarios developed by the authors; mean temperature of the coldest month and mean precipitation of the wettest month were modeled. *Source*: Béguin *et al.*, 2011.

Plasmodium falciparum." They report "the effects of increasing temperature and DTR on parasite prevalence, parasite intensity, and mosquito mortality decreased overall vectorial capacity for both mosquito species" (see Figure 4.6.1.2). They also note "increases of 3°C from 27°C reduced vectorial capacity by 51–89% depending on species and DTR, with increases in DTR alone potentially halving transmission," and "at 33°C, transmission potential was further reduced for *An. Stephensi* and blocked completely in *An. Gambiae*." The researchers concluded that rather than *increasing* malaria transmission, any current or future warming should actually diminish malaria transmission potential in what are currently high transmission settings.

Zhao *et al.* (2016) quantified the impact of several factors that led to freeing Europe from endemic malaria transmission during the twentieth century. They analyzed spatial datasets representing climatic, land use, and socioeconomic factors thought to be associated with the decline of malaria in twentieth century Europe and integrated the data with historical malaria distribution maps in order to quantify changes and differences across the continent before, during, and after malaria elimination. Their goal was to understand which factors significantly influence malaria transmission and decline, as well as which factors continue to play a role in limiting the risks of its re-establishment.

Of the nine factors analyzed by Zhao *et al.*, three of them were climate-related (temperature,

precipitation, and frost day frequency), each of which is "often considered to have an effect of malaria transmission," they note. The three European researchers, however, found "indicators relating to socio-economic improvements such as wealth, life expectancy and urbanization were strongly correlated with the decline of malaria in Europe, whereas those describing climatic and land use changes showed weaker relationships." More often than not, they found, changes in climate tended to run counter to observed trends in malaria; i.e., climate changes were thought to lead to an increased number of cases yet the actual numbers declined. It would appear that socioeconomic and land use factors are more than capable of compensating for any unfavorable changes in climate that may lead to malaria transmission and outbreaks. As long as countries continue to focus on improving these more important factors, malaria trends will continue to remain little influenced by future climate change, model projections notwithstanding.

References

Béguin, A., Hales, S., Rocklöv, J., Åström, C., Louis, V.R., and Sauerborn, R. 2011. The opposing effects of climate change and socio-economic development on the global distribution of malaria. *Global Environmental Change* **21**: 1209–14.

Figure 4.6.1.2
Effects of temperature and diurnal temperature range (DTR) on vectorial capacity

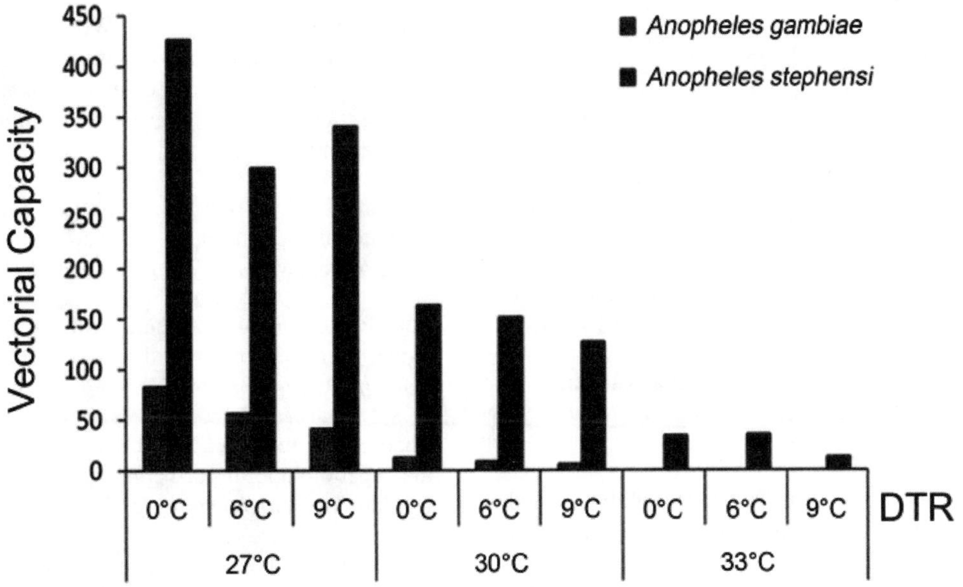

Source: Murdock *et al.*, 2016.

Bosello, F., Roson, R., and Tol, R.S.J. 2006. Economy-wide estimates of the implications of climate change: human health. *Ecological Economics* **58** (3): 579–91.

Childs, D.Z., Cattadori, I.M., Suwonkerd, W., Prajakwong, S., and Boots, M. 2006. Spatiotemporal patterns of malaria incidence in northern Thailand. *Transactions of the Royal Society of Tropical Medicine and Hygiene* **100**: 623–31.

Gething, P.W., Smith, D.L., Patil, A.P., Tatem, A.J., Snow, R.W., and Hay, S.I. 2010. Climate change and the global malaria recession. *Nature* **465**: 342–5.

Githeko, A.K. and Ndegwa, W. 2001. Predicting malaria epidemics in the Kenyan highlands using climate data: a tool for decision makers. *Global Change and Human Health* **2**: 54–63.

Haque, U., Hashizume, M., Glass, G.E., Dewan, A.M., Overgaard, H.J., and Yamamoto, T. 2010. The role of climate variability in the spread of malaria in Bangladeshi highlands. *PLoS One* **16**: 5.

Hay, S.I., Cox, J., Rogers, D.J., Randolph, S.E., Stern, D.I., Shanks, G.D., Myers, M.F., and Snow, R.W. 2002a. Climate change and the resurgence of malaria in the East African highlands. *Nature* **415**: 905–9.

Hay, S.I., Rogers, D.J., Randolph, S.E., Stern, D.I., Cox, J., Shanks, G.D., and Snow, R.W. 2002b. Hot topic or hot air? Climate change and malaria resurgence in East African highlands. *Trends in Parasitology* **18**: 530–4.

Hulden, L. and Hulden, L. 2009. The decline of malaria in Finland—the impact of the vector and social variables. *Malaria Journal* **8**: 94.

Jackson, M.C., Johansen, L., Furlong, C., Colson, A., and Sellers, K.F. 2010. Modelling the effect of climate change on prevalence of malaria in western Africa. *Statistica Neerlandica* **64**: 388–400.

Kuhn, K.G., Campbell-Lendrum, D.H., Armstrong, B., and Davies, C.R. 2003. Malaria in Britain: past, present, and future. *Proceedings of the National Academy of Sciences USA* **100**: 9,997–10,001.

Lieshout, M.V., Kovats, R.S., Livermore, M.T.J., and Martens, P. 2004. Climate change and malaria: analysis of the SRES climate and socio-economic scenarios. *Global Environmental Change* **14**: 87–99.

Murdock, C.C., Sternberg, E.D., and Thomas, M.B. 2016. Malaria transmission potential could be reduced with current and future climate change. *Scientific Reports* **6**: 10.1038/srep27771.

Nabi, S.A. and Qader, S.S. 2009. Is global warming likely to cause an increased incidence of malaria? *Libyan Journal of Medicine* **4**: 18–22.

Nkurunziza, H. and Pilz, J. 2011. Impact of increased temperature on malaria transmission in Burundi. *International Journal of Global Warming* **3**: 77–87.

Paaijmans, K.P., Blanford, S., Chan, B.H.K., and Thomas, M.B. 2012. Warmer temperatures reduce the vectorial capacity of malaria mosquitoes. *Biology Letters* **8**: 465–8.

Reiter, P. 2000. From Shakespeare to Defoe: malaria in England in the Little Ice Age. *Emerging Infectious Diseases* **6**: 1–11.

Reiter, P. 2001. Climate change and mosquito-borne disease. *Environmental Health Perspectives* **109**: 141–61.

Reiter, P. 2008. Global warming and malaria: knowing the horse before hitching the cart. *Malaria Journal* **7** (Supplement 1): S1–S3.

Rogers, D.J. and Randolph, S.E. 2000. The global spread of malaria in a future, warmer world. *Science* **289**: 1763–6.

Rogers, D.J. and Randolph, S.E. 2006. Climate change and vector-borne diseases. *Advances in Parasitology* **62**: 345–81.

Russell, R.C. 2009. Mosquito-borne disease and climate change in Australia: time for a reality check. *Australian Journal of Entomology* **48**: 1–7.

Shanks, G.D., Biomndo, K., Hay, S.I., and Snow, R.W. 2000. Changing patterns of clinical malaria since 1965 among a tea estate population located in the Kenyan highlands. *Transactions of the Royal Society of Tropical Medicine and Hygiene* **94**: 253–5.

Shanks, G.D., Hay, S.I., Stern, D.I., Biomndo, K., and Snow, R.W. 2002. Meteorologic influences on *Plasmodium falciparum* malaria in the highland tea estates of Kericho, Western Kenya. *Emerging Infectious Diseases* **8**: 1404–8.

Small, J., Goetz, S.J., and Hay, S.I. 2003. Climatic suitability for malaria transmission in Africa, 1911–1995. *Proceedings of the National Academy of Sciences USA* **100**: 15,341–5.

Stern, D.I., Gething, P.W., Kabaria, C.W., Temperley, T.H., Noor, A.M., Okiro, E.A., Shanks, G.D., Snow, R.W., and Hay, S.I. 2011. Temperature and malaria trends in Highland East Africa. *PLoS One* **6** (9): 10.1371/journal.pone.0024524.

Thomas, C. 2004. Malaria: a changed climate in Africa? *Nature* **427**: 690–1.

Tol, R.S.J. and Dowlatabadi, H. 2001. Vector-borne diseases, development & climate change. *Integrated Assessment* **2**: 173–81.

WHO.2017. World Health Organization. *World Malaria Report 2017*. Geneva, Switzerland.

WHO, WMO, UNEP. 2003. World Health Organization, World Meteorological Organization, United Nations Environment Program. *Climate Change and Human Health—Risks and Responses: Summary*. Geneva, Switzerland.

Yamana, T.K., Bomblies, A., and Eltahir, E.A.B. 2016. Climate change unlikely to increase malaria burden in West Africa. *Nature Climate Change* **6**: 1009–15.

Zell, R., Krumbholz, A., and Wutzler, P. 2008. Impact of global warming on viral diseases: what is the evidence? *Current Opinion in Biotechnology* **19**: 652–60.

Zhao, X., Smith, D.L., and Tatem, A.J. 2016. Exploring the spatiotemporal drivers of malaria elimination in Europe. *Malaria Journal* **15**: 122.

Zhou, G., Minakawa, N., Githeko, A.K., and Yan, G. 2004. Association between climate variability and malaria epidemics in the East African highlands. *Proceedings of the National Academy of Sciences USA* **101**: 2375–80.

4.6.2 Dengue Fever

Concerns over large increases in dengue fever as a result of rising temperatures are unfounded and unsupported by the scientific literature, as climatic indices are poor predictors for dengue fever.

According to Ooi and Gubler (2009a), "dengue/dengue hemorrhagic fever is the most important vector-borne viral disease globally," with more than half the world's population living in areas deemed to be at risk of infection. Kyle and Harris (2008) note "dengue is a spectrum of disease caused by four serotypes of the most prevalent arthropod-borne virus affecting humans today," and "its incidence has increased dramatically in the past 50 years," to where "tens of millions of cases of dengue fever are estimated to occur annually, including up to 500,000 cases of the life-threatening dengue hemorrhagic fever/dengue shock syndrome."

Some of the research papers summarized in previous sections address dengue fever as well as malaria. With a few worthy exceptions, we do not repeat those summaries in this section. The most important exceptions are papers written or coauthored by Paul Reiter (2001, 2003, 2010a, 2010b), one of the world's premier authorities on the subject. Reiter analyzed the history of malaria and dengue fever in an attempt to determine whether the incidence and range of influence of these diseases would indeed increase in response to higher global surface temperatures. His reviews established what is now widely accepted among experts in the field: that the natural history of these vector-borne diseases is highly complex, and the interplay of climate, ecology, vector biology, and a number of other factors defies definition by the simplistic analyses utilized in the computer models relied on by environmental activists and the IPCC.

That there has in fact been a resurgence of these diseases in parts of the world is true, but as Reiter (2001) notes, it is "facile to attribute this resurgence to climate change." This he shows via a number of independent analyses that clearly demonstrate factors associated with politics, economics, and human activity are the principal determinants of the spread of these diseases. He describes these factors as being "much more significant" than climate in promoting disease expansion. Reiter took up the subject again in 2003 with 19 other scientists as coauthors (Reiter *et al.*, 2003), and again in 2010.

Tuchman *et al.* (2003) conducted an empirical study of the impact of a doubling of atmospheric CO_2 concentrations (from 360 to 720 ppm) on development rates and survivorship of four species of detritivorous mosquito larvae eating leaf litter from *Populus tremuloides* (Michaux) trees. They report larval mortality was 2.2 times higher for *Aedes albopictus* (Skuse) mosquitos that were fed leaf litter that had been produced in the high-CO_2 chambers than it was for those fed litter that had been produced in the ambient-air chambers. In addition, they found larval development rates of *Aedes triseriatus* (Say), *Aedes aegypti* (L.), and *Armigeres subalbatus* (Coquillett) were slowed by 78%, 25%, and 27%, respectively. The researchers suggest "increases in lignin coupled with decreases in leaf nitrogen induced by elevated CO_2 and subsequent lower bacterial productivity [on the leaf litter in the water] were probably responsible for [the] decreases in survivorship and/or development rate of the four species of mosquitoes." Concerning the significance of these findings, Tuchman *et al.* write, "the indirect impacts of an elevated CO_2 atmosphere on mosquito larval survivorship and development time could potentially be great," because longer larval development times could result in fewer cohorts of

mosquitoes surviving to adulthood. With fewer mosquitoes, there should be lower levels of mosquito-borne diseases.

Kyle and Harris (2008) write "there has been a great deal of debate on the implications of global warming for human health," but "at the moment, there is no consensus." However, "in the case of dengue," they report, "it is important to note that even if global warming does not cause the mosquito vectors to expand their geographic range, there could still be a significant impact on transmission in endemic regions," because "a 2°C increase in temperature would simultaneously lengthen the lifespan of the mosquito and shorten the extrinsic incubation period of the dengue virus, resulting in more infected mosquitoes for a longer period of time." Nevertheless, they state there are "infrastructure and socioeconomic differences that exist today and already prevent the transmission of vector-borne diseases, including dengue, even in the continued presence of their vectors," citing Reiter (2001).

Wilder-Smith and Gubler (2008) conducted a review of the scientific literature, noting "the past two decades saw an unprecedented geographic expansion of dengue" and "global climate change is commonly blamed for the resurgence of dengue," but they add, "there are no good scientific data to support this conclusion." The two researchers report, "climate has rarely been the principal determinant of [mosquitoes'] prevalence or range," and "human activities and their impact on local ecology have generally been much more significant." They cite as contributing factors "urbanization, deforestation, new dams and irrigation systems, poor housing, sewage and waste management systems, and lack of reliable water systems that make it necessary to collect and store water," further noting "disruption of vector control programs, be it for reasons of political and social unrest or scientific reservations about the safety of DDT, has contributed to the resurgence of dengue around the world."

In addition, Wilder-Smith and Gubler write "large populations in which viruses circulate may also allow more co-infection of mosquitoes and humans with more than one serotype of virus," which would appear to be borne out by the fact that "the number of dengue lineages has been increasing roughly in parallel with the size of the human population over the last two centuries." Most important, perhaps, is "the impact of international travel," of which they say "humans, whether troops, migrant workers, tourists, business travelers, refugees, or others, carry the virus into new geographic areas," and these movements "can lead to epidemic waves." The two researchers conclude, "population dynamics and viral evolution offer the most parsimonious explanation for the observed epidemic cycles of the disease, far more than climatic factors."

Ooi and Gubler (2009b) examined "the history of dengue emergence" in order to determine "the major drivers for the spread of both the viruses and mosquito vectors to new geographic regions." The two researchers note "frequent and cyclical epidemics are reported throughout the tropical world, with regular importation of the virus via viremic travelers into both endemic and non-endemic countries." They state, "there is no good evidence to suggest that the current geographic expansion of the dengue virus and its vectors has been or will be due to global warming."

Russell *et al.* (2009) showed the dengue vector (the *Aedes aegypti* mosquito) "was previously common in parts of Queensland, the Northern Territory, Western Australia and New South Wales," and it had "in the past, covered most of the climatic range theoretically available to it," adding "the distribution of local dengue transmission has [historically] nearly matched the geographic limits of the vector." This being the case, they conclude the vector's current absence from much of Australia "is not because of a lack of a favorable climate." Thus, they reason "a temperature rise of a few degrees is not alone likely to be responsible for substantial increases in the southern distribution of *A. aegypti* or dengue, as has been recently proposed." Instead of futile attempts to limit dengue transmission by controlling the world's climate, therefore, the medical researchers recommend "well resourced and functioning surveillance programs, and effective public health intervention capabilities, are essential to counter threats from dengue and other mosquito-borne diseases."

Johansson *et al.* (2009) studied the association between the El Niño Southern Oscillation (ENSO) and dengue incidence in Puerto Rico (1986–2006), Mexico (1985–2006), and Thailand (1983–2006) using wavelet analysis as a tool to identify time- and frequency-specific associations. The three researchers report they "did not find evidence of a strong, consistent relationship in any of the study areas," and Rohani (2009), who wrote a Perspective piece on their study, states the three researchers found "no systematic association between multi-annual dengue outbreaks and El Niño Southern Oscillation." Thus,

as stated in the "Editors' Summary" of the Johansson *et al.* paper, their findings "provide little evidence for any relationship between ENSO, climate, and dengue incidence."

Shang *et al.* (2010) analyzed dengue cases in Taiwan at their onset dates of illness from 1998 to 2007, in order to "identify correlations between indigenous dengue and imported dengue cases (in the context of local meteorological factors) across different time lags." The researchers write, "the occurrence of indigenous dengue was significantly correlated with temporally-lagged cases of imported dengue (2–14 weeks), higher temperatures (6–14 weeks), and lower relative humidity (6–20 weeks)," and "imported and indigenous dengue cases had a significant quantitative relationship in the onset of local epidemics." The six Taiwanese researchers conclude, "imported dengue are able to serve as an initial facilitator, or spark, for domestic epidemics" while "meteorology alone does not initiate an epidemic." Rather than point to global warming, they state unequivocally that "an increase in viremic international travelers has caused global dengue hemorrhagic fever case numbers to surge in the past several decades."

Reiter (2010a) observed "the introduction and rapidly expanding range of *Aedes albopictus* in Europe is an iconic example of the growing risk of the globalization of vectors and vector-borne diseases," and "the history of yellow fever and dengue in temperate regions confirms that transmission of both diseases could recur, particularly if *Aedes aegypti*, a more effective vector, were to be re-introduced." He states "conditions are already suitable for transmission." Much more important than a rise or fall of a couple degrees of temperature, Reiter says, is "the quantum leap in the mobility of vectors and pathogens that has taken place in the past four decades, a direct result of the revolution of transport technologies and global travel."

Carbajo *et al.* (2012) evaluated the relative contributions of geographic, demographic, and climatic variables to the recent spread of dengue in Argentina. They found dengue spatial occurrence "was positively associated with days of possible transmission, human population number, population fall and distance to water bodies." When considered separately, the researchers write, "the classification performance of demographic variables was higher than that of climatic and geographic variables." Thus, although useful in estimating annual transmission risk, Carbajo *et al.* conclude temperature "does not fully describe the distribution of dengue occurrence at the country scale," and "when taken separately, climatic variables performed worse than geographic or demographic variables."

Williams *et al.* (2014) used a dynamic life table simulation model and statistically downscaled daily values for future climate to assess "climate change induced changes to mosquito bionomics," focusing on "female mosquito abundance, wet weight, and the extrinsic incubation period for dengue virus in these mosquitoes." They based their work on simulations of *Ae. aegypti* populations for current (1991–2011) and future (2046–2065) climate conditions for the city of Cairns, Queensland (which has historically experienced the most dengue virus transmission in all of Australia), as derived from the MPI ECHAM 5 climate model for the IPCC-proposed B1 and A2 emission scenarios.

Their work revealed "*Aedes aegypti* abundance is predicted to increase under the B1, but decrease under the A2, scenario," and "mosquitoes are predicted to have a smaller body mass in a future climate." Williams *et al.* say "it is therefore unclear whether dengue risk would increase or decrease in tropical Australia with climate change." They conclude their findings "challenge the prevailing view that a future, warmer climate will lead to larger mosquito populations and a definite increase in dengue transmission."

These several observations indicate concerns over large increases in dengue fever as a result of rising temperatures are unfounded and unsupported by the scientific literature, as climatic indices are poor predictors for dengue fever.

References

Carbajo, A.E., Cardo, M.V., and Vezzani, D. 2012. Is temperature the main cause of dengue rise in non-endemic countries? The case of Argentina. *International Journal of Health Geographics* **11**: 26.

Johansson, M.A., Cummings, D.A.T., and Glass, G.E. 2009. Multiyear climate variability and dengue-El Niño Southern Oscillation, weather and dengue incidence in Puerto Rico, Mexico, and Thailand: a longitudinal data analysis. *PLoS Medicine* **6**: e1000168.

Kyle, J.L. and Harris, E. 2008. Global spread and persistence of dengue. *Annual Review of Microbiology* **62**: 71–92.

Ooi, E.-E. and Gubler, D.J. 2009a. Dengue virus-mosquito interactions. In: Hanley, K.A. and Weaver, S.C. (Eds.) *Frontiers in Dengue Virus Research*. London, UK: Caister Academic Press, pp. 143–55.

Ooi, E.-E. and Gubler, D.J. 2009b. Global spread of epidemic dengue: the influence of environmental change. *Future Virology* **4**: 571–80.

Reiter, P. 2001. Climate change and mosquito-borne disease. *Environmental Health Perspectives* **109**: 141–61.

Reiter, P. 2010a. Yellow fever and dengue: A threat to Europe? *Eurosurveillance* **15** (10).

Reiter, P. 2010b. A mollusc on the leg of a beetle: human activities and the global dispersal of vectors and vector-borne pathogens. In: Relman, D.A., Choffnes, E.R., and Mack, A. (Rapporteurs) *Infectious Disease Movement in a Borderless World*. Washington, DC: The National Academies Press, pp. 150–65.

Reiter, P., *et al.* 2003. Texas lifestyle limits transmission of Dengue virus. *Emerging Infectious Diseases* **9**: 86–9.

Rohani, P. 2009. The link between dengue incidence and El Niño Southern Oscillation. *PLoS Medicine* **6**: e1000185.

Russell, R.C., Currie, B.J., Lindsay, M.D., Mackenzie, J.S., Ritchie, S.A., and Whelan, P.I. 2009. Dengue and climate change in Australia: predictions for the future should incorporate knowledge from the past. *Medical Journal of Australia* **190**: 265–8.

Shang, C.-S., Fang, C.-T., Liu, C.-M., Wen, T.-H., Tsai, K.-H., and King, C.-C. 2010. The role of imported cases and favorable meteorological conditions in the onset of dengue epidemics. *PLoS* **4**: e775.

Tuchman, N.C., *et al.* 2003. Nutritional quality of leaf detritus altered by elevated atmospheric CO_2: effects on development of mosquito larvae. *Freshwater Biology* **48**: 1432–9.

Wilder-Smith, A. and Gubler, D.J. 2008. Geographic expansion of Dengue: the impact of international travel. *Medical Clinics of North America* **92**: 1377–90.

Williams, C.R., Mincham, G., Ritchie, S.A., Viennet, E., and Harley, D. 2014. Bionomic response of Aedes aegypti to two future climate change scenarios in far north Queensland, Australia: implications for dengue outbreaks. *Parasites & Vectors* **7**: 447.

4.6.3 Tick-borne Diseases

Climate change has not been the most significant factor driving the recent temporal patterns in the epidemiology of tick-borne diseases.

Sarah Randolph of the University of Oxford's Department of Zoology is a leading scholar on tick-borne diseases. She and fellow Oxford faculty member David Rogers observed in 2000 that tick-borne encephalitis (TBE) "is the most significant vector-borne disease in Europe and Eurasia," having "a case morbidity rate of 10–30% and a case mortality rate of typically 1–2% but as high as 24% in the Far East" (Randolph and Rogers, 2000). The disease is caused by a flavivirus (TBEV) maintained in natural rodent-tick cycles; humans may be infected if bitten by an infected tick or by drinking untreated milk from infected sheep or goats.

Early discussions on the relationship of TBE to global warming predicted the disease would expand its range and become more of a threat to humans in a warmer world. However, Randolph and Rogers (2000) note, "like many vector-borne pathogen cycles that depend on the interaction of so many biotic agents with each other and with their abiotic environment, enzootic cycles of TBEV have an inherent fragility," so "their continuing survival or expansion cannot be predicted from simple univariate correlations." Confining their analysis to Europe, Randolph and Rogers first matched the present-day distribution of TBEV to the present-day distributions of five climatic variables: monthly mean, maximum, and minimum temperatures, plus rainfall and saturation vapor pressure, "to provide a multivariate description of present-day areas of disease risk." They applied this understanding to outputs of a general circulation model of the atmosphere that predicted how these five climatic variables may change in the future.

The results indicate the distribution of TBEV might expand both north and west of Stockholm, Sweden in a warming world. For most other parts of Europe, however, the two researchers say "fears for increased extent of risk from TBEV caused by global climate change appear to be unfounded." They report, "the precise conditions required for enzootic cycles of TBEV are predicted to be disrupted" in response to global warming, and the new climatic state "appears to be lethal for TBEV." This finding, they write, "gives the lie to the common perception that a warmer world will necessarily be a world under greater threat from vector-borne diseases." In the case of TBEV, they report the predicted change "appears to be to our advantage."

Estrada-Peña (2003) evaluated the effects of various abiotic factors on the habitat suitability of four tick species that are major vectors of livestock pathogens in South Africa. They report "year-to-year variations in the forecasted habitat suitability over the period 1983–2000 show a clear decrease in habitat availability, which is attributed primarily to increasing temperature in the region over this period." In addition, when climate variables were projected to the year 2015, Estrada-Peña found "the simulations show a trend toward the destruction of the habitats of the four tick species," just the opposite of what is often predicted about this disease.

Randolph (2010) examined the roles played by various factors that may influence the spread of tick-borne diseases. After describing some of the outbreaks of tick-borne disease in Europe over the past couple of decades, Randolph states, "the inescapable conclusion is that the observed climate change alone cannot explain the full heterogeneity in the epidemiological change, either within the Baltic States or amongst Central and Eastern European countries," citing Sumilo *et al.* (2007). Instead, she writes, "a nexus of interrelated causal factors – abiotic, biotic and human – has been identified," and "each factor appears to operate synergistically, but with differential force in space and time, which would inevitably generate the observed epidemiological heterogeneity."

Many of these factors, she continues, "were the unintended consequences of the fall of Soviet rule and the subsequent socio-economic transition (Sumilo *et al.*, 2008b)," among which she cites "agricultural reforms resulting in changed land cover and land use, and an increased reliance on subsistence farming; reduction in the use of pesticides, and also in the emission of atmospheric pollution as industries collapsed; increased unemployment and poverty, but also wealth and leisure time in other sectors of the population as market forces took hold."

Randolph concludes "there is increasing evidence from detailed analyses that rapid changes in the incidence of tick-borne diseases are driven as much, if not more, by human behavior that determines exposure to infected ticks than by tick population biology that determines the abundance of infected ticks," as per Sumilo *et al.* (2008a) and Randolph *et al.* (2008). She ends her analysis by stating, "while nobody would deny the sensitivity of ticks and tick-borne disease systems to climatic factors that largely determine their geographical distributions, the evidence is that climate change has not been the most significant factor driving the recent temporal patterns in the epidemiology of tick-borne diseases."

Lyme disease is the most common tick-borne human disease, with an estimated annual incidence of 300,000 in the United States (Shapiro, 2014) and at least 85,000 in Europe (Lindgren and Jaensen, 2006). It is caused by the spirochete bacteria, *Borrelia burgdorferi* and sometimes by *Borrelia mayonii* (Pritt *et al.*, 2016). It is transmitted in the eastern United States and parts of Canada by the tick, *Ixodes scapularis*, and on the Pacific Coast by *I. pacificus* (Clark, 2004).

Modeling by Brownstein *et al.* (2005) "generated the current pattern of *I. scapularis* across North America with an accuracy of 89% (*P* < 0.0001). Extrapolation of the model revealed a significant expansion of *I. scapularis* north into Canada with an increase in suitable habitat and a retraction of the vector from Florida and Texas, so that the exposed population actually diminishes, by 28% in the 2020s, by 12.7% in the 2050s, and by 1.9% in the 2080s. The connection between *I.scapularis* and deciduous forest is so strong that the authors state: "recent emergence of Lyme disease throughout the northeastern and mid-Atlantic states has been linked to reforestation." The automobile may thus have contributed to the emergence of Lyme disease by converting numerous redundant horse-paddocks into woodlands and by fertilizing those woodlands with carbon dioxide. The reported incidence increased in the United States during the warming hiatus from 1998 to 2009 and then stabilized or even fell slightly despite warming in 2015–16, as shown in Figure 4.6.3.1.

In Europe and Asia, the vectors of Lyme borreliosis (LB) are *I. ricinus* (Europe) and *I. persulcatus* (Lindgren and Jaensen, 2006). Like its North American cousin, *I. ricinus* prefers forest to open land and deciduous to conifer (Zeman and Januska, 1999). Late twentieth century warming has been linked to ticks spreading into higher latitudes and altitudes (observed in the Czech Republic (Daniel *et al.*, 2004) and to higher incidences of LB (Lindgren and Gustafson, 2001), though distorted by better reporting over time in most regions. Moreover, the incidence of LB actually declined after 1995 in the Czech Republic and Lithuania (Lindgren and Jaensen, 2006).

Figure 4.6.3.1
Reported cases of Lyme disease in the United States, 1996–2016

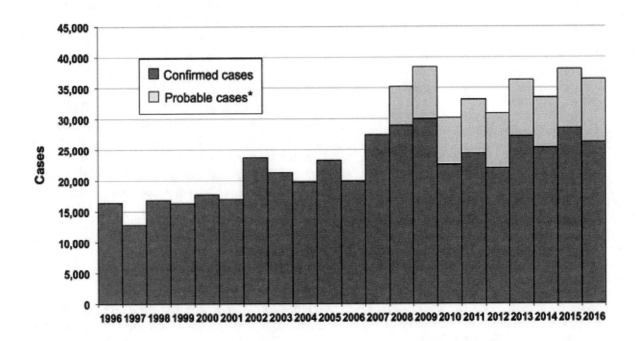

Source: Centers for Disease Control and Prevention, n.d.

References

Brownstein, J.S., Holford, T.R., and Fish, D. 2005. Effect of climate change on lyme fisease risk in North America. *Ecohealth* **2** (1): 38–46.

Centers for Disease Control and Prevention. n.d. Lyme Disease: What You Need to Know. Accessed June 12, 2018.

Clark, K. 2004. Borrelia species in host-seeking ticks and small mammals in Northern Florida. *Journal of Clinical Microbiology* **42** (11): 5076–86.

Daniel, M., *et al.* 2004. An attempt to elucidate the increased incidence of tick-borne encephalitis and its spread to higher altitudes in the Czech Republic. *International Journal of Medical Microbiology* **293** (37): 55–62.

Estrada-Peña, A. 2003. Climate change decreases habitat suitability for some tick species (*Acari: Ixodidae*) in South

Africa. *Onderstepoort Journal of Veterinary Research* **70**: 79–93.

Lindgren, E. and Gustafson, R. 2001. Tick-borne encephalitis in Sweden and climate change. *Lancet* **358**: 16–8.

Lindgren, E. and Jaensen, T. 2006. Lyme Borreliosis in Europe: Influences of Climate and Climate Change, Epidemiology, Ecology and Adaptation Measures. Copenhagen, Denmark: World Health Organization. Accessed June 12, 2018.

Pritt, B.S., *et al.* 2016. Identification of a novel pathogenic Borrelia species causing Lyme borreliosis with unusually high spirochaetaemia: a descriptive study. *Lancet Infectious Diseases* **16**: 556–64.

Randolph, S.E. 2010. To what extent has climate change contributed to the recent epidemiology of tick-borne diseases? *Veterinary Parasitology* **167**: 92–4.

Randolph, S.E., *et al.* 2008. Variable spikes in TBE incidence in 2006 independent of variable tick abundance but related to weather. *Parasites and Vectors* **1**: e44.

Randolph, S.E. and Rogers, D.J. 2000. Fragile transmission cycles of tick-borne encephalitis virus may be disrupted by predicted climate change. *Proceedings of the Royal Society of London Series B* **267**: 1741–4.

Shapiro, E.D. 2014. Clinical practice. Lyme disease. *New England Journal of Medicine* **370** (18): 1724–31.

Sumilo, D., Asokliene, L., Avsic-Zupanc, T., Bormane, A., Vasilenko, V., Lucenko, I., Golovljova, I., and Randolph, S.E. 2008a. Behavioral responses to perceived risk of tick-

borne encephalitis: vaccination and avoidance in the Baltics and Slovenia. *Vaccine* **26**: 2580–8.

Sumilo, D., Asokliene, L., Bormane, A., Vasilenko, V., Golovljova, I., and Randolph, S.E. 2007. Climate change cannot explain the upsurge of tick-borne encephalitis in the Baltics. *PLos ONE* **2**: e500.

Sumilo, D., *et al.* 2008b. Socio-economic factors in the differential upsurge of tick-borne encephalitis in Central and Eastern Europe. *Reviews in Medical Virology* **18**: 81–95.

Zeman, P. and Januska, J. 1999. Epizootic background of dissimilar distribution of human cases of Lyme borreliosis and tick-borne encephalitis in a joint endemic area. *Comparative Immunology, Microbiology and Infectious Disease* **22**: 247–60.

4.7 Conclusion

Fossil fuels directly benefit human health and longevity by powering labor-saving and life-protecting technologies and perhaps indirectly by contributing to rising surface temperatures.

Fossil fuels have benefited human health by making possible the dramatic increase in human prosperity since the first Industrial Revolution, which in turn made possible the technologies that are essential to protecting human health and prolonging human life. If, as the United Nations' Intergovernmental Panel on Climate Change (IPCC) claims, the combustion of fossil fuels leads to some global warming, then the positive as well as negative health effects of that warming should be included in any cost-benefit analysis of fossil fuels. Medical science explains why warmer temperatures are associated with health benefits. Empirical research confirms that warmer temperatures lead to a net decrease in temperature-related mortality in virtually all parts of the world, even those with tropical climates.

Climate change is likely to reduce the incidence of fatal coronary events related to low temperatures and wintry weather by a much greater degree than it increases the incidence associated with high temperatures and summer heat waves. Non-fatal myocardial infarction is also less frequent during unseasonably warm periods than during unseaonsably cold periods. Warm weather is correlated with a lower incidence of death due to respiratory disease in many parts of the world, including Canada, Shanghai,

Spain, and even on the subtropical island of Taiwan. Low minimum temperatures have been found to be a stronger risk factor than high temperatures for outpatient visits for respiratory diseases.

An extensive scientific literature contradicts the claim that malaria will expand across the globe or intensify in some regions as a result of rising global surface temperatures. Concerns over large increases in mosquito-transmitted and tick-borne diseases such as yellow fever, malaria, viral encephalitis, and dengue fever as a result of rising temperatures are unfounded and unsupported by the scientific literature. While climatic factors do influence the geographical distribution of ticks, temperature and climate change are not among the significant factors determining the incidence of tick-borne diseases.

In the face of extensive evidence of the positive effects of fossil fuels on human health, the IPCC's claims of a rising "risk of severe ill-health" and "mortality and morbidity during periods of extreme heat" ring hollow. The computer models relied on by the IPCC can be programmed to produce whatever results their sponsors want, and the IPCC has sponsored many models to predict a world filled with disease and misery. But that is not what actual medical science and empirical data allow us to predict. Thanks to fossil fuels, humanity can look forward to living longer and healthier lives than ever before.

5

Environmental Benefits

Chapter Lead Authors: Craig D. Idso, Ph.D., Patrick Moore, Ph.D.

Contributors: Tim Ball, Ph.D., Richard J. Trzupek, Steve Welcenbach

Reviewers: H. Sterling Burnett, Ph.D., Albrecht Glatzle, Dr.Sc.Agr., Kesten Green, Ph.D., Tom Harris, Tom Hennigan, Jim Petch, Willie Soon, Ph.D., James Wanliss, Ph.D.

Citation: Idso, C.D., Legates, D. and Singer, S.F. 2019. Environmental Benefits. In: *Climate Change Reconsidered II: Fossil Fuels.* Nongovernmental International Panel on Climate Change. Arlington Heights, IL: The Heartland Institute.

Key Findings

Key findings in this chapter include the following:

Fossil Fuels in the Environment

- Fossil fuels are composed mainly of carbon and hydrogen atoms (and oxygen, in the case of low-grade coal). Carbon and hydrogen appear abundantly throughout the universe and on Earth.

- In addition to mining and drilling, hydrocarbons also enter the environment through natural seepage, industrial and municipal effluent and run-off, leakage from underground storage or wells, and spills and other accidental releases.

- The chemical characteristics of fossil fuels make them uniquely potent sources of fuel. They are more abundant, compact, and reliable, and cheaper and safer to use than other energy sources.

Direct Benefits

- The greater efficiency made possible by technologies powered by fossil fuels makes it possible to meet human needs while using fewer natural resources, thereby benefiting the environment.

- Fossil fuels make it possible for humanity to flourish while still preserving much of the land needed by wildlife to survive.

- The prosperity made possible by fossil fuels has made environmental protection both highly valued and financially possible, producing a world that is cleaner and safer than it would have been in their absence.

Impact on Plants

- Elevated CO_2 improves the productivity of ecosystems both in plant tissues aboveground and in the soils beneath them.

- The effects of elevated CO_2 on plant characteristics are overwhelmingly positive, including increasing rates of photosynthesis and biomass production.

- Atmospheric CO_2 enrichment ameliorates the negative effects of a number of environmental plant stresses including high temperatures, air and soil pollutants, herbivory, nitrogen deprivation, and high levels of soil salinity.

- Exposure to elevated levels of atmospheric CO_2 prompts plants to increase the efficiency of their use of water, enabling them to grow and reproduce where it has previously been too dry for them to exist.

- The productivity of the terrestrial biosphere is increasing in large measure due to the aerial fertilization effect of rising atmospheric CO_2 concentrations.

- The benefits of CO_2 enrichment will continue even if atmospheric CO_2 rises to levels far beyond those forecast by the IPCC.

Impact on Terrestrial Animals

- The IPCC's forecasts of possible extinctions of terrestrial animals are based on computer models that have been falsified by data on temperature changes, other climatic conditions, and real-world changes in wildlife populations.

- Animal species are capable of migrating, evolving, and otherwise adapting to changes in climate that are much greater and more sudden than what is likely to result from the human impact on the global climate.

- Although there likely will be some changes in terrestrial animal population dynamics, few if any will be driven even close to extinction.

Impact on Aquatic Life

- The IPCC's forecasts of dire consequences for life in the world's oceans rely on falsified computer models and are contradicted by real-world observations.

- Aquatic life demonstrates tolerance, adaptation, and even growth and developmental improvements in response to higher temperatures and reduced water pH levels ("acidification").

- The pessimistic projections of the IPCC give way to considerable optimism with respect to the future of the planet's marine life.

Conclusion

- Combustion of fossil fuels has helped and will continue to help plants and animals thrive leading to shrinking deserts, expanded habitat for wildlife, and greater biodiversity.

Introduction

The previous two chapters considered ways the use of fossil fuels[1] benefits humanity. This chapter considers how human use of fossil fuels benefits plants and wildlife. As with the previous chapters, the focus here is on documenting the benefits rather than conducting a cost-benefit analysis. Cost-benefit analyses of climate change, fossil fuels, and regulations aimed at reducing greenhouse gas emissions are conducted in Part 3.

Why consider benefits that do not directly affect humans? Because even economists recognize the limits of a strictly utilitarian ethic. Amartya Sen, a Nobel Prize-winning economist, warned recently against taking "a strictly anthropocentric perspective on the question of the environment" (Sen, 2014). He continues,

> [W]e human beings do not only have needs. We also have values and priorities, about which we can reason. To say that worrying about other species is none of our business is not ethical reasoning, but a refusal to engage in ethical reasoning. ... It is hard to see how environmental thinking, which has many different aspects, can be reduced to a concern only with human living standards, given the

other concerns we may very reasonably have (*Ibid.*).

Fossil fuels clearly have environmental impacts beyond those directly affecting human health and well-being. Chapter 3 documented the human benefit of increased food production thanks to aerial carbon dioxide (CO_2) fertilization, but not its larger beneficial effects on the biosphere, including effects on forests, terrestrial life, and aquatic life. Chapter 4 explained how fossil fuels powered the technologies that led to great advances in human health, but did not describe how those same technologies make it possible to feed a growing global population without completely displacing wildlife habitat or how other plants and animals respond positively to elevated CO_2 in the atmosphere. This chapter fills those gaps.

In their ambition to condemn fossil fuels, the United Nations' Intergovernmental Panel on Climate Change (IPCC) and many environmental advocacy groups focus entirely on their *negative* environmental effects and studiously ignore their *beneficial* effects. For example, in a news report based on an interview with Andreas Fischlin, "an ecological modeler at the Swiss Federal Institute of Technology in Zurich," Tollefson (2015) claims, "a growing body of research suggests that ecological and economic impacts are already occurring with the 0.8°C of warming that has already occurred. These impacts will increase in severity as temperatures rise. Damage to coral reefs and Arctic ecosystems, as well as more extreme weather, can all be expected well before the 2°C threshold is reached (pp. 14–15)."

Similarly, the American Academy for the Advancement of Science (AAAS) Climate Science Panel claims,

> The overwhelming evidence of human-caused climate change documents both current impacts with significant costs and extraordinary future risks to society and natural systems. The scientific community has convened conferences, published reports, spoken out at forums and proclaimed, through statements by virtually every national scientific academic and relevant major scientific organization – including the American Academy for the Advancement of Science (AAAS) – that climate change puts the well-being of people of all nations at risk (AAAS, n.d., p. 3).

---

[1] This report follows conventional usage by using "fossil fuels" to refer to hydrocarbons, principally coal, oil, and natural gas, used by humanity to generate power. We recognize that not all hydrocarbons are derived from animal or plant sources.

In light of the alarming claims made by some scientists working on climate issues and by members of the media covering those issues, readers can be forgiven for assuming climate change produces *no* environmental benefits. This chapter demonstrates that assumption is wrong.

Section 5.1 provides background on carbon chemistry, acid precipitation, hydrogen gas, and carbon in the oceans. Section 5.2 presents the direct benefits of fossil fuels on plants and wildlife. The three main benefits are powering technologies that make it possible to use fewer resources to meet human needs; minimizing the amount of surface space needed to generate the raw minerals, fuel, and food needed to meet human needs; and bringing about the prosperity that leads to environmental protection becoming a positive social value and objective.

Fossil fuels also indirectly benefit the environment by contributing to the rise in atmospheric CO_2 levels experienced during the twentieth century and possibly the warming forecast by climate models for the twenty-first century and beyond. How much warming will occur and how much can be attributed to the combustion of fossil fuels are unsolved scientific puzzles, as explained in Chapter 2. Nevertheless, Section 5.3 considers the impacts of rising atmospheric CO_2 concentrations and *possible* warming and on plants, finding those impacts to be net positive. This extends to rates of photosynthesis and biomass production and the efficiency with which plants utilize water. Section 5.4 considers the impacts of rising CO_2 levels and temperatures on terrestrial animals and once again finds those impacts will be positive: Real-world data indicate warmer temperatures and higher atmospheric CO_2 concentrations would be beneficial, favoring a maintenance or increase in biodiversity.

Section 5.5 reviews laboratory and field studies of the impact of rising CO_2 concentrations and temperatures on aquatic life (corals and fish) and finds tolerance, adaptation, and even growth and developmental improvements. Section 5.6 provides a brief conclusion.

A previous volume in the *Climate Change Reconsidered* series subtitled "Biological Impacts" (NIPCC, 2014) contains summaries of nearly 2,000 peer-reviewed articles addressing in depth issues that are addressed only briefly in this chapter. Figure 5.1, taken from the *Summary for Policymakers* of that volume, summarizes its principal findings. Hundreds of summaries of new scientific research released since 2013 have been added to this chapter.

Figure 5.1
Summary of findings on biological impacts

- **Atmospheric carbon dioxide (CO_2) is not a pollutant.** It is a colorless, odorless, non-toxic, non-irritating, and natural component of the atmosphere. Long-term CO_2 enrichment studies confirm the findings of shorter-term experiments, demonstrating numerous growth-enhancing, water-conserving, and stress-alleviating effects of elevated atmospheric CO_2 on plants growing in both terrestrial and aquatic ecosystems.

- **The ongoing rise in the atmosphere's CO_2 content is causing a great greening of the Earth.** At locations all across the planet, the historical increase in the atmosphere's CO_2 concentration has stimulated vegetative productivity. This has occurred in spite of many real and imagined assaults on Earth's vegetation, including fires, disease, pest outbreaks, deforestation, and climatic change.

- **There is little or no risk of increasing food insecurity due to rising surface temperatures or rising atmospheric CO_2 levels.** Farmers and others who depend on rural livelihoods for income are benefitting from rising agricultural productivity throughout the world, including in parts of Asia and Africa where the need for increased food supplies is most critical. Rising temperatures and atmospheric CO_2 levels play a key role in the realization of such benefits.

- **Terrestrial ecosystems have thrived throughout the world as a result of warming temperatures and rising levels of atmospheric CO_2.** Empirical data pertaining to numerous animal species, including amphibians, birds, butterflies, other insects, reptiles, and mammals, indicate global warming and its myriad ecological effects tend to foster the expansion and preservation of animal habitats, ranges, and populations, or otherwise have no observable impacts. Multiple lines of evidence indicate animal species are adapting and in some cases evolving, to cope with climate change of the modern era.

- **Rising temperatures and atmospheric CO_2 levels do not pose a significant threat to aquatic life.** Many aquatic species have shown considerable tolerance to temperatures and CO_2 values predicted for the next few centuries and many have demonstrated a likelihood of positive responses in empirical studies. Any projected adverse impacts of rising temperatures or declining seawater and freshwater pH levels ("acidification") will be mitigated through behavioural changes during the many decades to centuries it is expected to take for pH levels to fall.

Source: Summary for Policymakers. *Climate Change Reconsidered II: Biological Impacts.* Nongovernmental International Panel on Climate Change (NIPCC). Chicago, IL: The Heartland Institute, 2014.

References

AAAS. n.d. What we know. American Academy for the Advancement of Science (website). Accessed October 12, 2018.

NIPCC. 2014. Idso, C.D, Idso, S.B., Carter, R.M., and Singer, S.F. (Eds.) *Climate Change Reconsidered II: Biological Impacts.* Nongovernmental International Panel on Climate Change. Chicago, IL: The Heartland Institute.

Sen, A. 2014. Global warming is just one of many environmental threats that demand our attention. *New Republic.* August 22.

Tollefson, J. 2015. Global-warming limit of 2°C hangs in the balance. *Nature* **520** (April 2): 14–5.

5.1 Fossil Fuels in the Environment

Fossil fuels are composed mainly of carbon and hydrogen atoms (and oxygen, in the case of low-grade coal). Carbon and hydrogen appear abundantly throughout the universe and on Earth.

As was observed at the beginning of Chapter 2, many people mistakenly believe they can address the climate change issue without understanding basic climate science. The "science tutorial" offered at the beginning of that chapter provided the science so often missing in popular and even academic writing on the subject. Section 2.1.2.2 discussed carbon chemistry in the context of the carbon cycle. Only a small part of that discussion will be repeated here, as the focus now is on fossil fuels.

5.1.1 Carbon Chemistry

Carbon and hydrogen appear abundantly throughout the universe and on the Earth. Carbon's unique function as the base element for Earth's biosphere derives from it being the lightest element capable of forming four covalent bonds with atoms of most elements in many variations. ("Covalent bonds" involve the sharing of electron pairs and are stronger than bonds involving single electrons.) The resulting molecules can contain from one to millions of carbon atoms. Carbon is so abundant and apt to bond with other atoms that the discipline of chemistry is divided into *organic chemistry,* which studies only carbon-based compounds, and *inorganic chemistry,* which studies all other compounds. Carbon-based compounds comprise the overwhelming majority of the tens of millions of compounds identified by scientists.

Compounds containing carbon atoms typically are combustible; have high melting points, low boiling points, and low solubility in water; and do not conduct electricity. All of these qualities make them good candidates for fuels that can be used to store, transport, and then release energy through combustion. Compounds containing carbon will typically produce carbon dioxide, among other byproducts, when burned.

Hydrogen is the lightest element and the most abundant substance in the universe, composing much of the mass of stars and gas giant planets. On Earth, it is rarely found in its monoatomic state due to its propensity to form covalent bonds with other elements. It is mostly present in hydrocarbons and water. At standard temperature and pressure, hydrogen is a highly combustible gas with a very high gravimetric energy density (energy per unit of weight, e.g. joules per kilogram) but a relatively low volumetric energy density (energy per unit of volume, e.g., joules per liter).

When carbon and hydrogen come together, the carbon provides the "backbone" to which hydrogen bonds, forming long and lightweight molecular chains, circles, and other complex patterns. In general, small linear hydrocarbons will be gases while medium-sized linear hydrocarbons will be liquids. Branched hydrocarbons of intermediate size tend to be waxes with low melting points. Long hydrocarbons tend to be semi-solid or solid. Figure 5.1.1.1 identifies the most common hydrocarbons and their uses.

5.1.2 Fossil Fuels

The main forms of fossil fuels are coal, oil, and natural gas (methane). Each form has in common a basis in hydrocarbons, which are molecules composed of carbon and hydrogen atoms. Types of hydrocarbons include methane, ethylene, and benzene. Coal, oil, and natural gas are made up largely of hydrocarbons, nitrogen, sulfur, and oxygen. The energy produced by burning a fossil fuel comes from breaking the carbon-hydrogen and carbon-carbon bonds and recombining them into carbon-oxygen (CO_2) and hydrogen-oxygen (H_2O) bonds. Because the hydrocarbons in coal have fewer hydrogen-carbon bonds than oil or natural gas, its gravimetric energy density (joules per kg) is less and it produces more CO_2 per unit of weight when burned. There are four types of coal according to their carbon content: anthracite has the most carbon, then bituminous, then subbituminous, then lignite. (See Figure 5.1.2.1.)

Considerable attention has been devoted to studying the possibility that some part of the world's supply of "fossil fuels" is produced by deep biospheres within the geosphere. Gold (1992, 1999) proposed that microbial life is common there and plays an important role in geochemical cycles, particularly in the carbon cycle. Kolesnikov *et al.* (2009) established experimentally that ethane and heavier hydrocarbons can be synthesized under conditions of the upper mantle, but it is as yet unknown how this may affect estimates of supplies of hydrocarbon-based fuels. According to Colman *et al.* (2017), "Despite 25 years of intense study, key questions remain on life in the deep subsurface, including whether it is endemic and the extent of its

involvement in the anaerobic formation and degradation of hydrocarbons. Emergent data from cultivation and next-generation sequencing approaches continue to provide promising new hints to answer these questions."

Hydrocarbons affect the natural environment when they are burned by releasing CO_2 and H_2O into the air. When burned, the sulfur and nitrogen in fossil fuels combine with oxygen to produce sulfur dioxide (SO_2) and nitrogen oxides (NOx). Sulfur dioxide and produce sulfuric and nitric acid, respectively, which can reduce the pH of rainwater (Cassidy and Frey, n.d.). Coal generally has more of these substances and natural gas has less. Coal also has some mineral content, typically quartz, pyrite, clay minerals, and calcite.

Hydrocarbons also enter the environment through natural seepage (Kvenvolden and Cooper, 2003), industrial and municipal effluent and run-off, leakage from underground storage or wells, and spills and other accidental releases. In some cases these releases harm plants and wildlife and endanger human health. According to Aminzadeh *et al.* (2013), "Hydrocarbon seepage can have profound local effects that may be widespread, causing vast blighted areas. The seeps that form the Buzau mounds in the Carpathian foreland of Romania are built by repeated acidic mudflows that form large blighted and barren areas," citing Baciu (2007) (p. 4). See Varjani (2017) and Chandra *et al.* (2013) for discussions of human health threats and many citations. This topic is addressed further in Chapter 8.

5.1.3 Acid Precipitation

The reduced pH of rainwater due to sulfur dioxide and nitrogen oxide emissions from the burning of fossil fuels, popularly referred to as "acid rain," was once thought to be dangerously acidifying soils and surface waters in the United States and around the world. The U.S. National Acid Precipitation Assessment Project (NAPAP, 1991), a project involving hundreds of scientists working in small groups over a period of 10 years at a cost of $550 million, found those concerns were unjustified. NAPAP found "there is no evidence of an overall or pervasive decline of forests in the United States and Canada due to acid deposition or any other stress" and "there is no case of forest decline in which acidic deposition is known to be a predominant cause" (Compendium of Summaries, p. 135).

Figure 5.1.1.1
Common hydrocarbons and their uses

Name	Number of Carbon Atoms	Uses
Methane	1	Fuel in electrical generation. Produces least amount of carbon dioxide.
Ethane	2	Used in the production of ethylene, which is utilized in various chemical applications.
Propane	3	Generally used for heating and cooking.
Butane	4	Generally used in lighters and in aerosol cans.
Pentane	5	Can be used as solvents in the laboratory and in the production of polystyrene.
Hexane	6	Used to produce glue for shoes, leather products, and in roofing.
Heptane	7	The major component of gasoline.
Octane	8	An additive to gasoline that, particularly in its branched forms, reduces knock.
Nonane	9	A component of fuel, particularly diesel.
Decane	10	A component of gasoline, but generally more important in jet fuel and diesel.

Hydrocarbons longer than 10 carbon atoms in length are generally broken down through the process known as "cracking" to yield molecules with lengths of 10 atoms or less. *Source:* Petroleum.co.uk, 2018.

Figure 5.1.2.1
Variation of selected coal properties with coal rank

	<---------- Low Rank -------><---- High Rank ------------->			
Rank:	Lignite	Subbituminous	Bituminous	Anthracite
Age:	---------------------------------- increases --------------------------------->			
% Carbon:	65-72	72-76	76-90	90-95
% Hydrogen:	~5 ---------------------- decreases ---------------------- ~2			
% Nitrogen:	<--------------------------------- ~1-2 -------------------------------->			
% Oxygen:	~30 ---------------------- decreases ---------------------- ~1			
% Sulfur:	~0 ----------- increases ------------ ~4 --- decreases --- ~0			
%Water:	70-30	30-10	10-5	~5
Heating value (BTU/lb):	~7000	~10,000	12,000–15,000	~15,000

Source: Radovic, 1997, Figure 7-3, p. 117.

NAPAP also found acidic deposition to be a threat to sensitive species of fish in only a few bodies of water and a small contributor relative to other factors, including logging and development. Remediation with lime is an inexpensive solution in such cases. A follow-up report issued in 1998 similarly found "Most forest ecosystems in the East, South, and West are not currently known to be adversely impacted by sulphur and nitrogen deposition" (NAPAP, 1998).

European researchers arrived at similar conclusions. For example, Elfving *et al.* (1996) found "in the Swedish National Forest Inventory (NFI), a steady increase in the estimated productivity of forest land has been noticed since inventory was begun in 1923. Young stands generally indicate higher site indices than old stands at equal site conditions. For spruce, this rise of site index has been estimated at 0.05–0.11 m.year^{-1}, with the highest value in the south." The authors also noted "the increasing atmospheric deposition of nitrogen is suspected to have the biggest influence" on rising forest productivity, meaning the positive effects of "acid rain" were outweighing the possible negative effects.

While "acid rain" was probably never a significant environmental threat, the dramatic reductions in SO_2 and NO_2 emissions in the United States and globally since the 1980s mean it has even less impact on the environment today. For additional commentary on the topic, see Goklany (1999), Aldrich (2003), Lomborg (2004), Menz and Seip (2004), Burns (2011), and Ridley (2012).

5.1.4 Hydrogen Gas

Pure hydrogen without carbon or the contaminants found in fossil fuels can be burned to generate energy, but it has serious disadvantages as a fuel. Hydrogen gas (H_2) is highly flammable and will explode at concentrations ranging from 4% to 75% by volume in the presence of a flame or a spark. Pure hydrogen-oxygen flames are invisible to the naked eye, making detection of a burning hydrogen leak difficult. Because hydrogen is so light, it is usually stored under pressure, introducing more cost, weight, and risk, and this is difficult to do because hydrogen embrittles many metals. While a typical automobile gas tank holds 15 gallons of gasoline weighing 90 pounds, the corresponding hydrogen tank would need to hold 60 gallons and would need to be insulated, but the fuel would weigh only 34 pounds (McCarthy, 2005).

Pure hydrogen can be obtained from methane through a process called reforming, or from water through electrolysis. However, the energy required to do either exceeds the amount of energy released when the hydrogen is burned. Current industrial electrolysis processes have effective electrical efficiency of approximately 70% to 80%, meaning they require 50 to 55 kWh of electricity to produce enough hydrogen to carry about 40 kWh of power (Christopher and Dimitrios, 2012). When the energy required to store and transport the fuel is considered, the process is even less efficient. Unless a non-fossil fuel is used to generate the electricity needed for reforming or electrolysis, using hydrogen as a fuel would not reduce carbon dioxide or other emissions generated by burning fossil fuels.

5.1.5 Carbon in the Oceans

The human contribution of oil to oceans during oil production or shipping gets extensive media attention but is small relative to natural seepage. The U.S. National Research Council found "spillage from vessels in U.S. waters during the 1990s declined significantly as compared to the prior decade and now represents less than 2% of the petroleum discharges into U.S. waters" and "only 1% of the oil discharges in North American waters is related to the extraction of petroleum" (NRC, 2003). Roberts and Feng note, "Hydrocarbons have been synonymous with the Gulf of Mexico (GOM) since early Spanish explorers wrote about the occurrence of sea surface slicks and tar balls on beaches" (Roberts and Feng, 2013, p. 43).

Because fossil fuels are carbon-based and therefore part of the carbon cycle, accidental releases or spills simply return the fuels' component parts to carbon reservoirs in different chemical forms. This often has the effect of minimizing the harm they could cause by coming into contact with plants or animals, including humans. Petroleum is typically reformed by biodegradation, dispersion, dissolution, emulsification, evaporation, photo-oxidation, resurfacing, sinking, and tar-ball formation.

Of these processes, biodegradation plays the biggest role. Hydrocarbons are energy-rich, making them inviting targets for bacteria and fungus. Atlas (1995) writes, "Hydrocarbon-utilizing micro-organisms are ubiquitously distributed in the marine environment following oil spills. These micro-organisms naturally biodegrade numerous contaminating petroleum hydrocarbons, thereby cleansing the oceans of oil pollutants" (Atlas, 1995). Aminzadeh *et al.* note, "Many marine seeps likewise have changed environments dominated by biota that can tolerate and exploit the seep. Some of these communities may be locally inhabited by very adept methanotrophs and paradoxically thrive, producing mounds similar to reefs. Fossil communities such as the Burgess Shale fauna (Friedman, 2010) have been thought to be associated with seeps (Johnston *et al.,* 2010)" (Aminzadeh, 2013.).

Varjani reported, "Petroleum hydrocarbon pollutants degradation by bacterial species has been well documented and metabolic pathways have been elucidated (Leahy and Colwell, 1990; Hendrickx *et al.,* 2006; Abbasian *et al.,* 2015; Meckenstock *et al.,* 2016; Wilkes *et al.,* 2016)" (Varjani, 2017, p. 282). Varjani's review of the literature found 38 microorganisms have been shown to biodegrade one or more of the four fractures of crude oil (saturates, aromatics, resins, and asphaltenes).

Because the efficiency and effectiveness of biodegradation is sometimes limited by the availability of indigenous colonies of bacteria and fungi or minerals needed for their replication, human intervention in the form of seeding bacterial populations and adding fertilizer can speed up and complete the biodegradation process. This process of bioremediation has been demonstrated to be successful in many different environments (Farhadian *et al.,* 2008; Chandra *et al.,* 2013; Ron and Rosenberg, 2014; Hu *et al.,* 2017).

5.1.6 Conclusion

Carbon chemistry explains why fossil fuels are preferred over other chemical compounds as sources of energy. Kiefer (2013) writes,

> Carbon transforms hydrogen from a diffuse and explosive gas that will only become liquid at ‑423° F [-253° C] into an easily handled, room-temperature liquid with 63% more hydrogen atoms per gallon than pure liquid hydrogen, 3.5 times the volumetric energy density (joules per gallon), and the ideal characteristics of a combustion fuel. ... A perfect combustion fuel possesses the desirable characteristics of easy storage and transport, inertness and low toxicity for safe handling, measured and adjustable volatility for easy mixing with air, stability across a broad range of environmental temperatures

and pressures, and high energy density. Because of sweeping advantages across all these parameters, liquid hydrocarbons have risen to dominate the global economy (p. 117).

In summary, the chemical characteristics of carbon and hydrogen, the main components of fossil fuels, make fossil fuels uniquely potent sources of fuel. They are more abundant, compact, reliable, and cheaper and safer to use than other energy sources. While it is possible to use hydrogen to transmit energy without the "backbone" provided by carbon, it is inefficient, expensive, and dangerous compared to carbon-based fuels. Acid rain, once thought to be a serious environmental threat, is no longer considered one. Human contributions of oil to the oceans via leakage and spills are trivial in relation to natural sources and quickly disperse and biodegrade. The damage caused by oil spills is a net cost of using oil, but not a major environmental problem.

References

Abbasian, F., Lockington, R., Mallavarapu, M., and Naidu, R. 2015. A comprehensive review of aliphatic hydrocarbon biodegradation by bacteria. *Applied Biochemistry and Biotechnology* **176** (3): 670–99.

Aldrich, S. 2003. *Smoke or Steam: A Guide to Environmental, Regulatory and Food Safety Concerns.* Second edition. Flo Min Publications/Keystone.

Aminzadeh, F., Berge, T.B., and Connolly, D.L. (Eds.) 2013. *Hydrocarbon Seepage: From Source to Surface.* Tulsa, OK: Society of Exploration Geophysicists and American Association of Petroleum Geologists.

Atlas, R.M. 1995. Petroleum biodegradation and oil spill bioremediation. *Marine Pollution Bulletin* **31** (4–12) (April–December): 178–82.

Baciu, C., Caracausi, A., Italiano, F., and Etiope, G. 2007. Mud volcanoes and methane seeps in Romania: main features and gas flux. *Annals of Geophysics* **50** (4).

Burns, D.A., et al. 2011. *National Acid Precipitation Assessment Program Report to Congress 2011: An Integrated Assessment.* Washington, DC: National Science and Technology Council.

Cassidy, R. and Frey, R. n.d. Acid rain (website). St. Louis, MO: Washington University, Department of Chemistry. Accessed October 17, 2018.

Chandra, S., Sharma, R., Singh, K., and Sharma, A. 2013. Application of bioremediation technology in the environment contaminated with petroleum hydrocarbon. *Annals of Microbiology* **63** (2): 417–31.

Christopher, K. and Dimitrios, R. 2012. A review on exergy comparison of hydrogen production methods from renewable energy sources. *Energy and Environmental Science* **5**: 6640.

Colman, D.R., *et al.* 2017. The deep, hot biosphere: twenty-five years of retrospection. *Proceedings of the National Academy of Sciences USA* **114** (27): 6895–903.

Elfving, B., Tegnhammar, L., and Tveite, B. 1996. Studies on growth trends of forests in Sweden and Norway. In: Spiecker, H. (Ed.) *Growth Trends in European Forests Studies from 12 Countries.* European Forest Research Institute, *Research Report* No. 5. New York, NY: Springer-Verlag Berlin Heidelberg, pp. 61–70.

Farhadian, M., *et al.* 2008. In situ bioremediation of monoaromatic pollutants in groundwater: a review. *Bioresource Technology* **99** (13): 5296–308.

Friedman, B., 2010. Burgess Shale to add another chapter. AAPG Explorer (website). Accessed November 27, 2012.

Gold, T. 1992. The deep, hot biosphere. *Proceedings of the National Academy of Sciences USA* **89**: 6045–9.

Goklany, I.M. 1999. *Clearing the Air: The Real Story of the War on Air Pollution.* Washington, DC: Cato Institute.

Gold, T. 1999. *The Deep Hot Biosphere: The Myth of Fossil Fuels.* New York, NY: Springer.

Hendrickx, B., *et al.* 2006. Alternative primer sets for PCR detection of genotypes involved in bacterial aerobic BTEX degradation: distribution of the genes in BTEX degrading isolates and in subsurface soils of a BTEX contaminated industrial site. *Journal of Microbiological Methods* **64** (2): 250–65.

Hu, P., *et al.* 2017. Simulation of Deepwater Horizon oil plume reveals substrate specialization within a complex community of hydrocarbon degraders. *Proceedings of the National Academy of Sciences USA* **114** (28): 7432–7.

Johnston, P., Johnston, K., and Keith, S. 2010. Burgess Shale tales: mud volcanism and chemosynthetic communities on the Middle Cambrian seafloor of southeastern British Columbia. *AAPG Search and Discovery Article* #90108. Tulsa, OK: American Association of Petroleum Geologists.

Kiefer, T.A. 2013. Energy insecurity: the false promise of liquid biofuels. *Strategic Studies Quarterly* (Spring): 114–51.

Kolesnikov, A., Kutcherov, V.G., and Goncharov, A.F. 2009. Methane-derived hydrocarbons produced under upper-mantle conditions. *Nature Geoscience* **2**: 566–70.

Kvenvolden, K. and Cooper, C. 2003. Natural seepage of crude oil into the marine environment. *Geo-Marine Letters* **23** (4): 140–6.

Leahy, J.G. and Colwell, R.R. 1990. Microbial degradation of hydrocarbons in the environment. *Microbiology Review* **54** (3): 305–15.

Lomborg, B. 2004. The skeptical environmentalist. In: Anderson, T. (Ed.) *You Have to Admit It's Getting Better: From economic prosperity to environmental quality.* Stanford, CA: Hoover Press, pp. 1–51.

McCarthy, J. 2005. Progress and Its Sustainability, Hydrogen (website). Accessed May 23, 2018.

Meckenstock, R.U., *et al.* 2016. Anaerobic degradation of benzene and polycyclic aromatic hydrocarbons. *Journal of Molecular Microbiology and Biotechnology* **26** (1–3): 92–118.

Menz, F. and Seip, H. 2004. Acid rain in Europe and the United States: an update. *Environmental Science & Policy*, **7**(4), 253-265.

NAPAP. 1991. National Acid Precipitation Assessment Program. *National Acid Precipitation Assessment Program 1990 Integrated Assessment Report.* Washington, DC: U.S. Government Printing Office.

NAPAP. 1998. National Acid Precipitation Assessment Program. *Biennial Report to Congress: An Integrated Assessment.* Silver Spring, MD.

NRC. 2003. U.S. National Research Council Committee on Oil in the Sea. *Oil in the Sea III: Inputs, Fates, and Effects.* Washington, DC: National Academies Press.

Petroleum.co.uk. 2018. Hydrocarbons (website). Accessed May 24, 2018.

Radovic, L. 1997. *Energy & Fuels in Society.* Course syllabus, Pennsylvania State University, Chapter 7.

Ridley, M. 2012. Apocalypse not: here's why you shouldn't worry about end times. *Wired.* August 17.

Roberts, H.H. and Feng, D. 2013. Carbonate precipitation at Gulf of Mexico hydrocarbon seeps: an overview. In: Aminzadeh, F., Berge, T.B., and Connolly, D.L. (Eds.) *Hydrocarbon Seepage: From Source to Surface.* Tulsa, OK: Society of Exploration Geophysicists and American Association of Petroleum Geologists, pp. 43–61.

Ron, E.Z. and Rosenberg, E. 2014. Enhanced bioremediation of oil spills in the sea. *Current Opinion in Biotechnology* **27**: 191–4.

Varjani, S. 2017. Microbial degradation of petroleum hydrocarbons. *Bioresource Technology* **223**: 277–86.

Wilkes, H., *et al.* 2016. Metabolism of hydrocarbons in n-Alkane utilizing anaerobic bacteria. *Journal of Molecular Microbiology and Biotechnology* **26** (1–3): 138–51.

5.2 Direct Benefits

Fossil fuels benefit the environment directly in three ways. First, they power the technologies that dramatically improve the efficiency with which natural resources are used, thereby reducing the impact of human activities on nature. One attempt to measure this benefit found the impact of global human consumption on the environment was reduced 32% from 1900 to 2006 due to technological advances (Goklany, 2009).

Second, fossil fuels save land for wildlife. They do this in three ways. The first is via the application of technology already mentioned. The use of fossil fuels to create ammonia fertilizer, to power tractors and other farm machinery, and to speed the transport of perishable food products to processing plants and markets allowed humanity's nutrition needs to be met with fewer acres under cultivation. According to Goklany (2009), technology reduced the impact of population and affluence on the amount of cropland used in the United States by 95%. In other words, fossil fuels erased all but 5% of the increased use of land that human population growth and prosperity otherwise would have required.

Fossil fuels also save land for wildlife by being more power-dense than alternative sources of energy, thus requiring less surface area than wind or solar power to produce equal amounts of energy to meet human needs. According to one estimate, using windmills to produce the same amount of energy as is currently produced globally by fossil fuels would require 14.4 million onshore turbines requiring some 570 million acres, an area equal to 25% of the entire land area of the United States (30% of the lower 48 states) (Driessen, 2017).

Fossil fuels also save land for wildlife by increasing the level of carbon dioxide (CO_2) in the atmosphere, which acts as fertilizer for crops, increasing yields and making it possible to meet the nutritional needs of a growing human population without converting yet more forests and grasslands into cropland. As documented below, assuming the 120 ppm increase in atmospheric CO_2 concentration since the beginning of the Industrial Revolution was caused by the burning of fossil fuels, fossil fuels increased agricultural production per unit of land area by 70% for C_3 cereals, 28% for C_4 cereals, 33% for fruits and melons, 62% for legumes, 67% for root and tuber crops, and 51% for vegetables (Idso *et al.*, 2003, p. 18). As the atmosphere's CO_2 content continues to rise, agricultural land use efficiency will rise with it.

The third direct benefit of fossil fuels is the impact prosperity has on the willingness of people to pay to protect the environment. Once a society attains a level of prosperity sufficient to meet its basic physical needs, the willingness of citizens to spend and sacrifice for a better environment rises more than twice as fast as per-capita income (Coursey, 1992), leading to greater investments over time in safe and clean drinking water, sanitary handling of human and animal wastes, and other measures of environmental protection. As Bryce (2014) writes,

> It's only by creating wealth that we will be able to support the scientists, tinkerers and entrepreneurs who will come up with the new technologies we need. It's only by getting richer that we will be able to afford the adaptive measures we may need to take in the decades ahead as we adjust to the Earth's ever-changing climate. It is only by using more energy, not less, that we will be able to provide more clean water and better sanitation to the poorest of the poor" (p. 54).

References

Bryce, R. 2014. *Smaller Faster Lighter Denser Cheaper: How Innovation Keeps Proving the Catastrophists Wrong.* New York, NY: PublicAffairs.

Coursey, D. 1992. *The Demand for Environmental Quality.* St. Louis, MO: John M. Olin School of Business, Washington University.

Driessen, P. 2017. Revisiting wind turbine numbers. Townhall (website). September 2. Accessed May 18, 2018.

Goklany, I.M. 2009. Have increases in population, affluence and technology worsened human and environmental well-being? *The Electronic Journal of Sustainable Development* **1** (3): 3–28.

Idso, C.D., Idso, S.B., Idso, K.E. 2003. *Enhanced or Impaired? Human Health in a CO₂-Enriched Warmer World.* Tempe, AZ: Center for the Study of Carbon Dioxide and Global Change.

5.2.1 Efficiency

The greater efficiency made possible by technologies powered by fossil fuels makes it possible to meet human needs while using fewer natural resources, thereby benefiting the environment.

Fossil fuels power the technologies used to protect the environment. Chapters 3 and 4 documented how those technologies contribute to human prosperity and human health. This chapter shows how those technologies make it possible to protect and clean the air and water of both manmade and natural pollutants, leading to benefits not only for humanity but for nearly all other forms of life.

"Without cheap supplies of electricity produced from coal, the ongoing revolution in information technology, as well as the age of biotech and nanotech, simply wouldn't be possible. Electricity accelerates the trend toward objects and systems that

are Smaller Faster Lighter Denser Cheaper," writes Bryce (2014, p. 191). "If oil didn't exist, we would have to invent it. No other substance comes close to oil when it comes to energy density, ease of handling, and flexibility. Those properties explain why oil provides more energy to the global economy than any other fuel" (*Ibid.*, p. 173). Figure 5.2.1.1 illustrates the dominance of oil, coal, and natural gas in meeting the world's energy needs.

The market system spurs innovation and efficient use of natural resources required to produce consumer goods and services, thereby indirectly leading to protection of the environment (see Chapter 1, Section 1.2.6 and Goklany, 1999; Huber, 1999, Chapter 4; Bradley, 2000; Baumol, 2002). Producers benefit when they use fewer resources because their costs decline. They also benefit, as do their customers, by developing new technologies that increase the value of the output from the resources they use.

The history of the three Industrial Revolutions briefly told in Chapter 3 reveals how central fossil fuels and especially coal were to economic progress in the past, continue to be today, and will be for the foreseeable future. McNeill (2000) writes,

Figure 5.2.1.1
Shares of global primary energy consumption by fuel

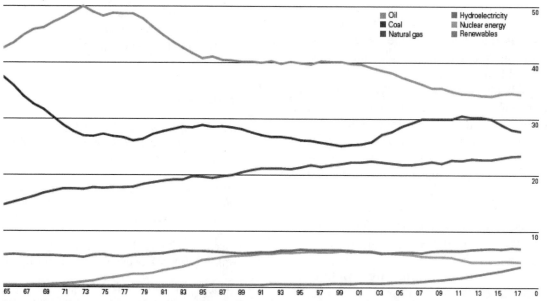

Oil remains the world's dominant fuel, making up just over a third of all energy consumed. In 2017 oil's market share declined slightly, following two years of growth. Coal's market share fell to 27.6%, the lowest level since 2004. Natural gas accounted for a record 23.4% of global primary energy consumption, while renewable power hit a new high of 3.6%.

Source: BP, 2018, p. 11.

No other century – no millennium – in human history can compare with the twentieth for its growth in energy use. We have probably deployed more energy since 1900 than in all of human history before 1900. My very rough calculation suggests that the world in the twentieth century used 10 times as much energy as in the thousand years before 1900 A.D. In the 100 centuries between the dawn of agriculture and 1900, people used only about two-thirds as much energy as in the twentieth century.

Many authors have documented the remarkable pace of growth in human population, energy use, and well-being in the twentieth century and its impact on the environment (Cronon, 1992; Schlereth, 1992; Avery, 2000; Norton Green, 2008; McNeill and Engelke, 2016; Gordon, 2016). While these authors document the negative as well as positive impacts of fossil fuels on the environment, the positive effects are dominant. Gordon (2016) observed, "When the electric elevator allowed buildings to extend vertically instead of horizontally, the very nature of land use was changed and urban density was created" (p. 4). Cities are "greener," in some ways, than less-dense population patterns due to their smaller footprint and lower per-capita use of many resources (Owen, 2004; Brand, 2010). Gordon also noted, "And so it was with motor vehicles replacing horses as a primary form of intra-urban transportation; no longer did society have to allocate a quarter of its agricultural land to support the feeding of the horses or maintain a sizable labor force for removing their waste" (*Ibid.*).

By reducing the demand for wood for use as a fuel and by increasing the productivity of land used for agriculture, fossil fuels allowed more land to remain as forests or even return to forests. Mather and Needle (1998) described the transition in the United States as follows:

Perhaps the most striking example of the process, however, is from the United States. Here, as elsewhere, the process has operated at a number of scales and is closely linked to reforestation. Within the south, for example, cropland has been increasingly concentrated on areas of high quality land. A 'process of natural selection' has led to the concentration of cropland on the better land and the vacating by agriculture of the poorer land. The areas of greatest abandonment of land

coincided with major environmental limitations, such as steep slopes and infertile soils, which limited the range of operations in which the farmers could engage. More generally, large areas of relatively poor land in New England were abandoned as better land in the Mid-West and other parts of the country was opened up. Much of the abandoned land in New England (and in the South) subsequently reverted to forest. The result was that, by 1980, the percentage of the land area of Maine under forest was 90, compared with 74 in the mid-1800s. In New Hampshire, the corresponding figures for these dates were 86 and 50%: in Vermont 76 and 35% (p. 122)

This process continues today. According to the Food and Agriculture Organization of the United Nations, in 2015 net forest area increased or was unchanged from the previous year in 12 of the agency's 15 regions and unchanged globally (FAO, 2018, Figure 26). The three regions that saw declines were Southeast Asia, North Africa, and landlocked developing countries, all areas experiencing poverty and/or civil strife. In contrast to these poor countries, Kauppi *et al.* (2018) report "a universal turnaround has been detected in many countries of the World from shrinking to expanding forests" during the 25-year period 1990 to 2015, which they depict in the figure reproduced as Figure 5.2.1.2.

According to Kauppi *et al,* the most rapid expansion of forests is occurring in nations with the highest life expectancy, education, and per-capita income indicators, as recorded in national scores on the United Nations Human Development Index. The authors say "This indicates that forest resources of nations have improved along with progress in human well-being. Highly developed countries apply modern agricultural methods on good farmlands and abandon marginal lands, which become available for forest expansion. Developed countries invest in sustainable programs of forest management and nature protection." Significantly, they add, "Our findings are significant for predicting the future of the terrestrial carbon sink. They suggest that the large sink of carbon recently observed in forests of the World will persist, if the well-being of people continues to improve" (*Ibid.*)

Jesse Ausubel, head of the Program for the Human Environment at Rockefeller University, has written extensively on how modern technology made

Figure 5.2.1.2
Change in Forest Growing Stock, 1990 – 2015

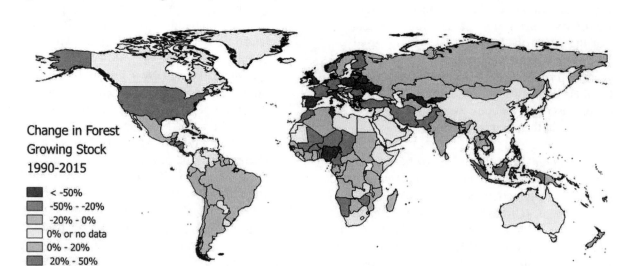

Source: Kauppi *et al.*, 2018.

possible by electricity and the fossil fuels that produce it has led to a "dematerialization" of modern civilization, the steady reduction in natural resources required to produce each unit of income or wealth (see, e.g., Ausubel, 1996; Wernick *et al.*, 1996; Wernick and Ausubel, 2014). In 2008, Ausubel and Paul E. Waggoner of the Connecticut Agricultural Experiment Station in New Haven observed, "During past years, dematerialization and declining intensity of impact have ameliorated a range of humanity's environmental impacts, from the carbon emission attending energy use to the cropland and fertilizer attending food production, and the use of wood" (Ausubel and Waggoner, 2008).

Ausubel and Waggoner asked whether the trend was ending or would continue. They found that from 1980 to 2006, the carbon intensity of the Chinese economy declined to 40% of its 1980 level. "Without the dematerialization from 1980–2006 by Chinese consumers, actual national energy use in 2006 would have been 180% greater," they write. "Reversing China's 26-year dematerialization would increase the entire global energy consumption by fully 28%." The authors found dematerialization taking place in both an early period (1980–1995) and a more recent period (1995–2006) globally and for the United States, China, and India. They write,

Although the average global consumer enjoyed 45% more affluence in 2006 than in 1980, each only consumed 22% more crops and 13% more energy. The richer consumer actually used 20% less wood, a saving of 0.67 minus 0.53 m³ per person or 39 board feet. The evidence … also shows persistently declining intensity of the impact of crop production on land and fertilizer use and persistence of declining French carbon emissions per energy production (*Ibid.*).

"The USA dematerialized steadily near 2%/year throughout the 25 years. ... Its intensity of impact did not decrease," the two authors report. In conclusion, they write,

The dematerialization of crop, fertilizer and wood use plus the decarbonization of carbon emission per GDP continue. And although a declining intensity of impact is hard to find for energy, it continues for other phenomena. The declining intensities continue assisting the journey across sustainability's dual dimensions of present prosperity without compromising the future environment (*Ibid.*).

Vaclav Smil, professor emeritus in the faculty of environment at the University of Manitoba in Winnipeg, Manitoba, Canada, also has written extensively on dematerialization. In 2013 he estimated that a dollar's worth of value produced today in the United States requires about 2.5 ounces of raw material, whereas a dollar's worth of value (adjusted for inflation) would have required 10 ounces of raw material in 1920. He estimated that since 1900, the energy required to produce a ton of steel and nitrogen fertilizer has fallen by 80% and a ton of aluminum and cement by 70% (Smil, 2013).

An example of dematerialization at work is the extraordinary energy savings made possible by the widespread use of cellphones. Tupy (2012) has documented how one smart phone saves 444 watts of power consumption by doing the work of at least nine devices previously used. A graphic illustrating his findings appears as Figure 5.2.1.3.

Figure 5.2.1.3
Dematerialization at work: One smart phone saves 444 watts of power consumption

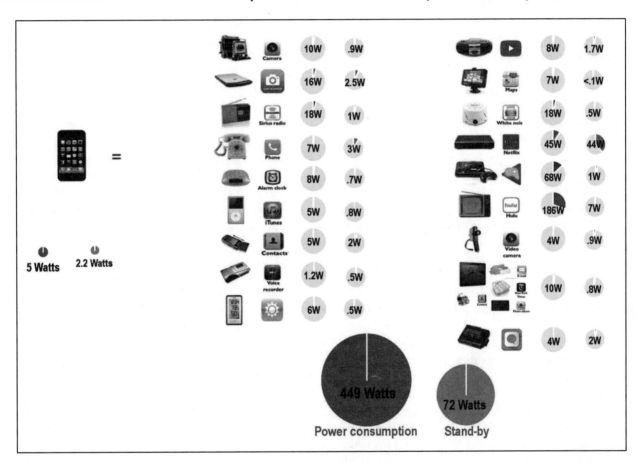

A single smart phone, pictured on the left, consumes 5 Watts of power and requires 2.2 Watts of stand-by power to produce the work of 18 devices consuming 449 Watts and requiring 72 Watts of stand-by power. Power and energy use data based on Lawrence Berkeley Laboratory standby statistics and other industry sources. Graphic courtesy of Nuno Bento, IIASA, 2017. *Source:* Adapted from Tupy, 2012.

IPAT Equation and T-Factor

A formula commonly used to estimate the environmental impact of human activities (Ehrlich and Holdren, 1971) is:

$$I = P \times A \times T$$

where I is environmental impact, P is human population, A is per-capita affluence or wealth (commonly denoted as per-capita Gross Domestic Product (GDP)) and T is technological innovation. Following Goklany (1999, 2009), we can see that since A = GDP/P, the equation can be simplified as:

$$I = GDP \times T$$

The technological change (ΔT) from an initial time (t_i) to final time (t_f) is therefore:

$$T = I/GDP$$

The impact of technological innovation is therefore:

$$\Delta T = \Delta(I/GDP)$$

If population, affluence, their product (GDP), and the technology-factor are all normalized to unity at t_i, then:

$$\Delta T = (I_f / GDP_f) - 1$$

where subscript f denotes the value at the end of the period.

Indur Goklany, a writer on technology and science who served as a contributor to and reviewer of IPCC reports as well as chief of the technical assessment division of the National Commission on Air Quality and a consultant in the Office of Policy, Planning, and Evaluation at EPA, calls this final equation the "T-factor." The smaller the T-factor, the more efficiently natural resources are being used. Goklany used this measure to show how new technology is making possible giant steps forward in environmental protection.

The T-factor for sulfur dioxide (SO_2) emissions – a pollutant produced largely from fossil fuel combustion at power plants and other industrial facilities – in the United States between 1900 and 1997 was 0.084, "which means that $1 of economic activity produced 0.084 times as much SO_2 in 1997 as it did in 1900," a dramatic reduction (Goklany,

1999, p. 72). Similarly, the T-factor for volatile organic compounds (VOCs) was 0.094 and for nitrogen oxide (NO_x), 0.374. Between 1940 and 1997, the T-factor for particulate matter (PM_{10}) was 0.034 and for carbon monoxide, 0.121.The T-factor for lead emissions between 1970 and 1997 was 0.008.Emissions levels for all of these pollutants have continued to fall since 1997.

More recently, Goklany (2009) estimated the T-factors for habitat converted to cropland, water withdrawal, air pollution, death from extreme weather events, and carbon dioxide emissions in the United States, other countries, and globally. Goklany summarized the impact of technology on carbon dioxide (CO_2) emissions:

> [F]or the U.S., despite a 27-fold increase in consumption (i.e., GDP) since 1900, CO_2 emissions increased 8-fold. This translates into a 67% reduction in impact per unit of consumption (i.e., the T-factor, which is also the carbon intensity of the economy) during this period, or a 1.1% reduction per year in the carbon intensity between 1900 and 2004. Since 1950, however, U.S. carbon intensity has declined at an annual rate of 1.7%. Arguably, CO_2 emissions might have been lower, but for the hurdles faced by nuclear power.

> Globally, consumption increased 21-fold since 1900, while CO_2 increased 13-fold because technology reduced the impact cumulatively by 32% or 0.4% per year. Both U.S. and global carbon intensity increased until the early decades of the 20th century. Since 1950, global carbon intensity has declined at the rate of 0.9% per year (Goklany, 2009, p. 18).

Some of Goklany's other findings include:

- Technological change reduced the amount of land that would have been converted from habitat to cropland globally by 84.3% from 1950 to 2005, and by 95% in the United States from 1910 to 2006.

- Technology reduced air pollution in the United States by between 70.5% and 99.8%, depending on the pollutant and time period.

■ Globally, technology reduced the number of deaths due to climate-related disasters by 95.3% from 1900/09 to 1997/2006 despite a 300% rise in world population in this period (*Ibid.*, Table 2, pp. 22–23).

The T-factor is so powerful it dominates the IPAT equation. The greater productivity, prosperity, and economic opportunities created by technological advances encourage smaller family sizes, resulting in slower population growth or even a negative population growth rate. Goklany plots total fertility rate (TFR) versus per-capita income, demonstrating the close negative correlation. (See Figure 5.2.1.4.) He concludes, "Thus, in the IPAT equation, P is not independent of A and T: sooner or later, as a nation grows richer, its population growth rate falls (e.g., World Bank 1984), which might lead to a cleaner environment (Goklany 1995, 1998, 2007b)" (Goklany, 2009, p. 7).

Figure 5.2.1.4
Total fertility rate (tfr) vs. per-capita income, 1977–2003

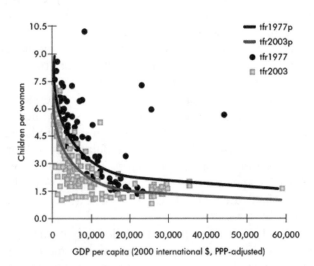

Source: Goklany, 2007a.

In summary, the human impact on the environment is smaller than it would otherwise be thanks to the technologies fueled by fossil fuels. "Dematerialization" made possible by electricity and advanced technologies means fewer raw materials must be mined and processed to meet a growing population's demand for goods and services.

References

Ausubel, J.H. 1996. Liberation of the environment. *Daedalus* **125** (3): 1–17.

Ausubel, J.H. and Waggoner, P. 2008. Dematerialization: variety, caution, and persistence. *Proceedings of the National Academy of Sciences USA* **105** (35): 12,774–9.

Avery, D. 2000. *Saving the Planet with Pesticides and Plastic.* Second Edition. Indianapolis, IN: Hudson Institute.

Baumol, W.J. 2002. *The Free-Market Innovation Machine: Analyzing the Growth Miracle of Capitalism.* Princeton, NJ: Princeton University Press.

BP. 2018. BP Statistical Review of World Energy 2018.

Bradley Jr., R.L. 2000. *Julian Simon and the Triumph of Energy Sustainability.* Washington, DC: American Legislative Exchange Council.

Brand, S. 2010. How slums can save the planet. *Prospect Magazine.* February.

Cronon, J. 1992. *Nature's Metropolis: Chicago and the Great West.* New York, NY: W.W. Norton & Company.

Ehrlich, P.R. and Holdren, J.P. 1971. Impact of population growth. *Science* **171** (3977): 1212–17.

FAO. 2018. Food and Agricultural Organization of the United Nations. *The State of the World's Forests 2018.*

Goklany, I.M. 1995. Strategies to enhance adaptability: technological change, economic growth and free trade. *Climatic Change* **30**: 427–49.

Goklany, I.M. 1998. Saving habitat and conserving biodiversity on a crowded planet. *BioScience* **48**: 941–53.

Goklany, I.M. 1999. *Cleaning the Air: The Real Story of the War on Air Pollution.* Washington, DC: Cato Institute.

Goklany, I.M. 2007a. *The Improving State of the World: Why We're Living Longer, Healthier, More Comfortable Lives on a Cleaner Planet.* Washington, DC: Cato Institute.

Goklany, I.M. 2007b. Integrated strategies to reduce vulnerability and advance adaptation, mitigation, and sustainable development. *Mitigation and Adaptation Strategies for Global Change* **12** (5): 755–86.

Goklany, I.M. 2009. Have increases in population, affluence and technology worsened human and environmental well-being? *The Electronic Journal of Sustainable Development* **1** (3): 3–28.

Gordon, R.J. 2016. *The Rise and Fall of American Growth: The U.S. Standard of Living Since the Civil War.* Princeton, NJ: Princeton University Press.

Huber, P. 1999. *Hard Green: Saving the Environment from the Environmentalists, a Conservative Manifesto.* New York, NY: Basic Books.

Kauppi, P.E., Sandström, V., and Lipponen, A. 2018. Forest resources of nations in relation to human well-being. *PLoS ONE* **13**(5): e0196248.

Mather, A. and Needle, C. 1998. The forest transition: a theoretical basis. *Area* **30** (2).

McNeill, J.R. 2000. *Something New Under the Sun: An Environmental History of the Twentieth-Century World.* New York, NY: W.W. Norton & Company.

McNeill, J.R. and Engelke, P. 2016. *The Great Acceleration: An Environmental History of the Anthropocene Since 1945.* Cambridge, MA: Harvard University Press.

Norton Greene, A. 2008. *Horses at Work: Harnessing Power in Industrial America.* Cambridge, MA: Harvard University Press.

Owen, D. 2004. Green Manhattan. *The New Yorker.* October.

Schlereth, T. 1992. *Victorian America: Transformations in Everyday Life, 1876–1915.* New York, NY: Harper Perennial.

Smil, V. 2013. *Making the Modern World: Materials and Dematerialization.* New York, NY: Wiley.

Tupy, M. 2012. Dematerialization (update). Cato at Liberty (blog). Accessed June 28, 2018.

Wernick, I. and Ausubel, J.H. 2014. *Making Nature Useless? Global Resource Trends, Innovation, and Implications for Conservation.* Washington, DC: Resources for the Future. November.

Wernick I., Herman, R., Govind, S., and Ausubel, J.H. 1996. Materialization and dematerialization: measures and trends. *Daedalus* **125** (3): 171–98.

World Bank. 1984. *World Development Report 1984.* New York, NY: Oxford University Press.

5.2.2 Saving Land for Wildlife

Fossil fuels make it possible for humanity to flourish while still preserving much of the land needed by wildlife to survive.

Fossil fuels benefit the environment by minimizing the amount of surface space needed to generate the raw materials, fuel, and food needed to meet human needs. If it were not for fossil fuels, the human need for surface space would crowd out habitat for many species of plants and animals. Fossil fuels save land for wildlife in three ways: by being more energy-dense than alternative fuels, thereby reducing the amount of surface space needed to meet the demand for energy; by making possible the "Green Revolution" dramatically reducing the acreage needed to feed the planet's growing population; and via aerial fertilization, the "greening of the Earth" that occurs when plants benefit from the CO_2 produced when fossil fuels are burned.

Fossil fuels save land for wildlife because of their exceptional power density, a concept explained in Chapter 3, Section 3.4.1. A natural gas well is nearly 50 times more power-dense than a wind turbine, more than 100 times as dense as a biomass-fueled power plant and 1,000 times as dense as corn ethanol (Bryce, 2010, p. 93). Coal has an energy density 50% to 75% that of oil, still far superior to solar, wind, and biofuels (Layton, 2008; Smil, 2010).

Power density benefits the environment because "energy sources with high power densities have the least deleterious effect on open space" (Bryce 2010, p. 92). Bryce estimates replacing U.S. coal-fired generation capacity in 2011 (300 gigawatts) with wind turbines at 1 watt per square meter would have required 300 billion square meters, or roughly 116,000 square miles (Bryce, 2014, pp. 217–218). Driessen (2017), using a number of conservative assumptions, estimated using windmills to produce the same amount of energy as is currently produced globally by fossil fuels would require 14.4 million onshore turbines requiring some 570 million acres (890,625 square miles), an area equal to 25% of the entire land area of the United States (30% of the lower 48 states).

Smil (2016) conducted a detailed tally of the land used by different energy systems around the world. He estimated that in 2010, new renewable energy sources (solar PV, wind, and liquid biofuels) required 270,000 km^2 of land to produce just 130 GW of power. Fossil fuels, thermal, and hydroelectricity generation claimed roughly 230,000 km^2 of land to deliver 14.34 TW of power, *110 times as much power on approximately 15% less land.* Fossil fuels, thermal, and hydropower required less than 0.2% of the Earth's ice-free land and nearly half that was surface area covered by water for reservoirs (pp. 211–212).

A study of the use of biofuels to replace fossil fuels conducted by the UK's Energy Research Centre and published in 2011 found that replacing half of current global primary energy supply with biofuels would require an area ranging from twice to ten times the size of France. Replacing the entire current global energy supply would require …

> an area of high yielding agricultural land the size of China. … In addition these estimates assume that an area of grassland and marginal land larger than India (>0.5Gha) is converted to energy crops. The area of land allocated to energy crops could occupy over 10% of the world's land mass, equivalent to the existing global area used to grow arable crops. For most of the estimates in this band a high meat diet could only be accommodated with extensive deforestation (Slade *et al.*, 2011, p. vii).

Kiefer (2013) calculated that replacing the energy used by the United States each year just for transportation "would require more than 700 million acres of corn. This is 37% of the total area of the continental United States, more than all 565 million acres of forest and more than triple the current amount of annually harvested cropland. Soy biodiesel would require 3.2 billion acres – one billion more than all U.S. territory including Alaska" (*Ibid.*). The figure Kiefer used to illustrate the difference power density makes in the amount of land required to produce 2,000 MW appears in Figure 5.2.2.1.

If any energy source other than fossil fuels (or nuclear) had been used to fuel the enormous growth in human population and prosperity in the twentieth century, the ecological consequences would have been disastrous. Wildlife would have been crowded out to make way for millions of windmills or millions of square miles of corn or soy planted to fuel cars, trucks, ships, and airplanes.

The second way fossil fuels save land for wildlife is by making possible the Green Revolution described in Chapter 3, Section 3.3.1. The discovery in 1909 of a process by which natural gas and atmospheric nitrogen could be converted into ammonia, now widely used as fertilizer, was only one of many technological innovations that improved farm productivity. Recall that Goklany (2009), in the

Figure 5.2.2.1
Area required by different fuels to produce 2,000 MW of power

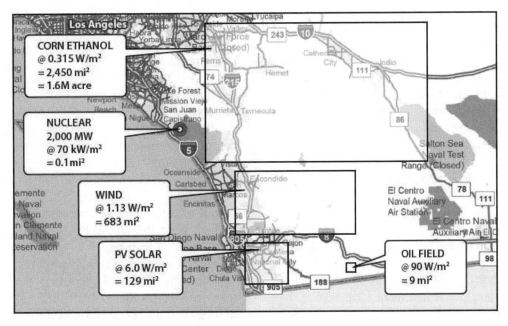

Source: Kiefer, 2013, p. 131.

T-factor analysis described in the previous section, applied his formula to cropland in the United States. He found a T-factor of 0.05 in 2006 relative to 1910, meaning technology reduced the impact of increases in population and affluence on the amount of cropland used by 95% since 1910. In other words, advances in technology alone erased all but 5% of the effect of population growth and increased affluence. Farmers in the United States were able to feed a growing and increasingly affluent population without significantly increasing the amount of land they needed

Savage (2011) estimated in 2011 that using organic farming methods to produce the 2008 U.S. yield of all crops would have required an additional 121.7 million acres of cropland, 39% more than was actually in production that year. That cropland "would be the equivalent of all the current cropland acres in Iowa, Illinois, North Dakota, Florida, Kansas, and Minnesota combined" (*Ibid.*). While not all of the superior yield of non-organic crops is due to ammonia fertilizers, much of it is and most of the pesticides and herbicides that explain the remainder of the high yield are produced from petroleum and natural gas.

Ausubel, Wernick, and Waggoner calculated the land spared in India thanks to the Green Revolution just for growing one crop, wheat, was 65 MHa (million hectares), "an area the size of France or four Iowas" (Ausubel *et al.*, 2013). Their graph showing how "the land sparing continued into the twentieth century" appears below as Figure 5.2.2.2. Similarly, they report the amount of land devoted to growing corn in China doubled from 1960 to 2010 while each harvested hectare became four-and-a-half times more productive, sparing some 120 MHa.

Ausubel and his coauthors propose a formula similar to the IPAT formula described in Section 5.2.1 to predict how many acres of land must be taken for crop production:

$$im = \textbf{Impact} = \textbf{P x A x C}_1 \textbf{ x C}_2 \textbf{ x T}$$

where

> im = cropland (in hectares) taken
> P = population (persons)
> A = affluence (in GDP per capita)
> C_1 = dietary response to affluence (in kilocalories/GDP)
> C_2 = FAO's Production Index Number/kcal)
> T = Technology (hectares divided by crop PINs)

In the ImPACT formula, rising population and affluence can increase the amount of land moved from habitat or other uses and devoted to cropland, while technology reduces that shift by increasing efficiency. Changes in consumer behavior (C_1 and C_2) can either increase or decrease the need for more land under cultivation. Declining C_1, as shown in Figure 5.2.2.3, reveals how "in country after country after calories exceed minimum levels, caloric intake rises, slows, and may eventually level off as affluence grows" (p. 226).

Ausubel and his coauthors find dematerialization of food is occurring globally and is likely to continue. In the authors' ImPACT formula, im = -0.02 for the period 2010 to 2060. The trend is driven partly by the tendency of people to reduce their consumption of meat relative to their income once a threshold of prosperity is reached and partly by the increasing productivity of the world's farmers, who are likely to increase crop outputs/hectare by about 2% per year. "[T]he number of hectares of cropland has barely changed since 1990," they report. Using conservative estimates of trends, they predict "by 2060, some 146 MHa of land could be restored to Nature, an area equal to one and a half times the size of Egypt, two and a half times France, or ten times Iowa" (Ausubel *et al.*, 2013).

The third way fossil fuels save land for wildlife is via the aerial fertilization effect described in Chapter 3, Section 3.4 and in greater detail in Section 5.3 below. As noted by Huang *et al.* (2002), human populations "have encroached on almost all of the world's frontiers, leaving little new land that is cultivatable." And in consequence of humanity's ongoing usurpation of this most basic of natural resources, Raven (2002) noted "species-area relationships, taken worldwide in relation to habitat destruction, lead to projections of the loss of fully two-thirds of all species on Earth by the end of this century." Fortunately, humanity has a powerful ally in the ongoing rise in the atmosphere's CO_2 content resulting, research shows, from the human combustion of fossil fuels. Since CO_2 is the basic "food" of essentially all terrestrial plants, the more of it there is in the atmosphere, the bigger and better they grow. Section 5.3 summarizes extensive research in support of this finding.

Since the start of the Industrial Revolution, it can be calculated on the basis of the work of Mayeux *et al.* (1997) and Idso and Idso (2000) that the 120 ppm increase in atmospheric CO_2 concentration increased agricultural production per unit land area by 70%

Figure 5.2.2.2
Actual and potential land used for wheat production in India, 1961–2010

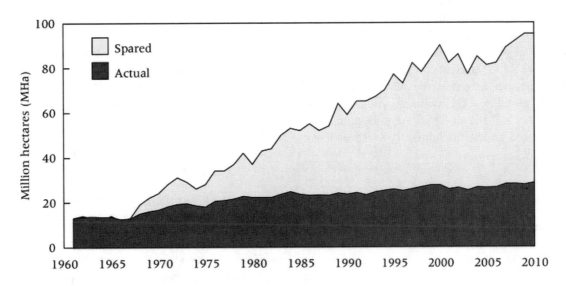

Upper segment shows the hectares farmers would have tilled to produce the actual harvest had yields stayed at the 1960 level. *Source:* Ausubel *et al.*, 2013, citing FAO, 2012.

Figure 5.2.2.3
Dematerialization of food, 1961–2007

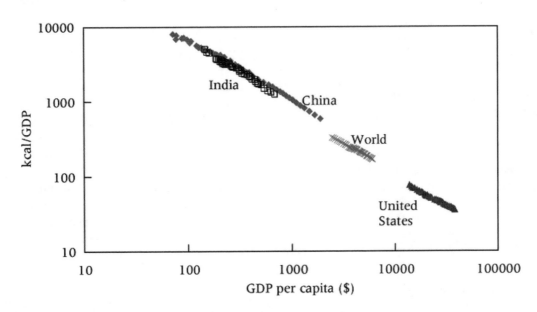

Graph shows kcal/GDP – as a function of calories consumed divided by GDP for China, India, the United States, and the world – consistently declines with rise in per-capita GDP from 1961 to 2007 over a range of incomes and cultures. *Source:* Ausubel *et al.*, 2013, Figure 6, p. 227, citing FAO, 2012 and World Bank, 2012.

for C_3 cereals, 28% for C_4 cereals, 33% for fruits and melons, 62% for legumes, 67% for root and tuber crops, and 51% for vegetables. A nominal doubling of the atmosphere's CO_2 concentration will raise the productivity of Earth's herbaceous plants by 30% to 50% (Kimball, 1983; Idso and Idso, 1994), while the productivity of its woody plants will rise by 50% to 80% (Saxe *et al.* 1998; Idso and Kimball, 2001). As the atmosphere's CO_2 content continues to rise, so too will crop yields per acre rise, meaning we will need less land to raise the food we need, giving wildlife the space it needs to live. This is a substantial and underappreciated benefit of humanity's use of fossil fuels.

References

Ausubel, J., Wernick, I., and Waggoner, P. 2013. Peak farmland and the prospect for land sparing. *Population and Development Review* **38**: 221–42.

Bryce, R. 2010. *Power Hungry: The Myths of "Green" Energy and the Real Fuels of the Future.* New York, NY: PublicAffairs.

Bryce, R. 2014. *Smaller Faster Lighter Denser Cheaper: How Innovation Keeps Proving the Catastrophists Wrong.* New York, NY: PublicAffairs.

Driessen, P. 2017. Revisiting wind turbine numbers. Townhall (website). September 2. Accessed May 18, 2018.

Goklany, I.M. 2009. Have increases in population, affluence and technology worsened human and environmental well-being? *The Electronic Journal of Sustainable Development* **1** (3): 3–28.

Huang, J., Pray, C., and Rozelle, S. 2002. Enhancing the crops to feed the poor. *Nature* **418**: 678–84.

Idso, K.E. and Idso, S.B. 1994. Plant responses to atmospheric CO_2 enrichment in the face of environmental constraints: a review of the past 10 years' research. *Agricultural and Forest Meteorology* **69**: 153–203.

Idso, C.D. and Idso, K.E. 2000. Forecasting world food supplies: the impact of the rising atmospheric CO_2 concentration. *Technology* **7S**: 33–55.

Idso, S.B. and Kimball, B.A. 2001. CO_2 enrichment of sour orange trees: 13 years and counting. *Environmental and Experimental Botany* **46**: 147–53.

Kiefer, T.A. 2013. Energy insecurity: the false promise of liquid biofuels. *Strategic Studies Quarterly* (Spring): 114–51.

Kimball, B.A. 1983. Carbon dioxide and agricultural yield: an assemblage and analysis of 430 prior observations. *Agronomy Journal* **75**: 779–88.

Layton, B.E. 2008. A comparison of energy densities of prevalent energy sources in units of joules per cubic meter. *International Journal of Green Energy* **5**: 438–55.

Mayeux, H.S., Johnson, H.B., Polley, H.W., and Malone, S.R. 1997. Yield of wheat across a subambient carbon dioxide gradient. *Global Change Biology* **3**: 269–78.

Raven, P.H. 2002. Science, sustainability, and the human prospect. *Science* **297**: 954–9.

Savage, S. 2011. A Detailed Analysis of U.S. Organic Crops (website). Accessed November 7, 2018.

Saxe, H., Ellsworth, D.S., and Heath, J. 1998. Tree and forest functioning in an enriched CO_2 atmosphere. *New Phytologist* **139**: 395–436.

Slade, R., Saunders, R., Gross, R., and Bauen, A. 2011. *Energy from Biomass: The Size of the Global Resource.* London, UK: Imperial College Centre for Energy Policy and Technology and UK Energy Research Centre.

Smil, V. 2010. *Power Density Primer: Understanding the Spatial Dimension of the Unfolding Transition to Renewable Electricity Generation (Part 1 – Definitions).* Master Resource (website). May 8.

Smil, V. 2016. *Power Density: A Key to Understanding Energy Sources and Uses.* Cambridge, MA: The MIT Press.

5.2.3 Prosperity

The prosperity made possible by fossil fuels has made environmental protection both highly valued and financially possible, producing a world that is cleaner and safer than it would have been in their absence.

The contribution of fossil fuels to human prosperity was documented in detail in Chapter 3. While there are many claims that human prosperity fueled environmental destruction (e.g., Heinberg, 2007; NRDC, 2008), data show the opposite has been true. As Bailey (2015) writes,

It is in rich democratic capitalist countries that the air and water are becoming cleaner, forests are expanding, food is abundant, education is universal, and women's rights respected. Whatever slows down economic growth also slows down environmental improvement. By vastly increasing knowledge and pursuing technological progress, past generations met their needs and vastly increased the ability of our generation to meet our needs. We should do no less for future generations (p. 72).

Similarly, Bryce (2014) writes,

The pessimistic worldview ignores an undeniable truth: more people are living longer, healthier, freer, more peaceful, lives than at any time in human history. ... The plain reality is that things are getting better, a lot better, for tens of millions of people all around the world. Dozens of factors can be cited for the improving conditions of humankind. But the simplest explanation is that innovation is allowing us to do more with less. We are continually making things and processes Smaller Faster Lighter Denser Cheaper (pp. xxi–xxii).

As fossil fuels create global prosperity, more care is taken to protect the environment and more humans are protected from air and water pollution, food contaminated with bacteria or toxic substances, contagious diseases, and accidental death from floods and other natural risks (Ausubel, 1996; Avery, 2000; Goklany, 2007; Epstein, 2014; Moore and Hartnett White, 2016). Fossil fuels may contribute to rising levels of air and water pollution in the early stages of a society's economic growth, but even during the worst period those risks pale compared to the risks of life without fossil fuels described by Goklany and other historians in Chapter 3. Over time, those same fuels make it possible to *clean* air and water of both manmade and natural pollutants, leading to a cleaner and safer environment, demonstrated by the Environmental Kuznets Curves (EKCs) described in Chapter 1, Section 1.3.3 and reproduced here as Figure 5.2.3.1 and by government data on air quality in the United States reported in Chapter 6.

Developed countries and even many developing countries are on the downward slope of the right side of EKCs as measured by emissions of pollutants that pose potential threats to human health. The trend toward a cleaner and safer world has been documented by many of the scholars previously cited in this chapter, but see specifically Julian Simon (1980, 1982, 1995, 1996); Julian Simon and Herman

Figure 5.2.3.1
A typical Environmental Kuznets Curve

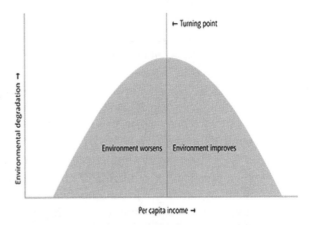

Source: Ho and Wang, 2015, p. 42.

Kahn (1984); Vaclav Smil (2005, 2006); and Indur Goklany (2007, 2012). More recent academic studies documenting EKCs for carbon dioxide emissions include Shahbaz *et al.* (2012), Tiwari *et al.* (2013), Osabuohien *et al.* (2014), Apergis and Ozturk (2015), and Sarkodie (2018).

Between 1970 and 2017, for example, U.S. emissions of six air pollutants (particulates, ozone, lead, carbon monoxide, nitrous oxide, and sulfur dioxide) declined by 73%. Those reductions occurred even as U.S. gross domestic product (GDP) grew 262%, energy consumption rose 44%, miles traveled rose 189%, and the nation's population increased 59% (EPA, 2018). The graphic used by the U.S. Environmental Protection Agency (EPA) to illustrate these trends is reproduced as Figure 5.2.3.2 below.

Steven Hayward, currently a fellow in law and economics at the American Enterprise Institute, began producing with various coauthors an annual "Index of Environmental Indicators" in 1994 reporting the latest data on environmental quality in the United States and worldwide. In 2011, observing that much of the data his team was reporting was now available online, he replaced the annual index with an "Almanac of Environmental Trends," a website that could be frequently updated. In the first (and only) print edition of Hayward's almanac (Hayward, 2011), he summarized progress on air quality in the United States as follows:

- "The improvement in air quality is the greatest public policy success story of the last generation."

- "Virtually the entire nation has achieved clean air standards for four of the six main pollutants regulated under the Clean Air Act. The exceptions are ozone and particulates."

- "In the cases of ozone and particulates, the areas of the nation with the highest pollution levels have shown the greatest magnitude of improvement."

- "The chief factor in the reduction of air pollution has been technology. Regulations played a prominent role in some innovations, but many were the result of market forces and economic growth."

- "The long-term trend of improving air quality is sure to continue."

Concerning water quality, Hayward presented the following summary:

Although water quality has improved substantially over the past 40 years, the federal government lacks good nationwide monitoring programs for assessing many basic water quality issues. Partly, this is due to the complexity and diversity of water pollution problems, which make a uniform national program methodologically difficult. Partial datasets and snapshots of particular areas provide a sense of where the main challenges remain.

Total water use in the United States has been flat for the last 30 years, even as population, food production, and the economy have continued to grow. In general the U.S. has improved water use efficiency by about 30% over the last 35 years.

On "toxic chemicals and other environmental health risks," Hayward reported:

- "The total amount of toxic chemicals used in American industry is steadily declining – a measure of resource efficiency."

- "Hazardous waste is declining. After a slow start, human exposure to toxic chemicals at Superfund sites has declined by more than 50% over the last decade."

- "Levels of most heavy metals and synthetic chemicals in human blood, tissue, and urine samples are either very low or declining."

- "Dioxin compounds in the environment have declined more than 90% over the last two decades."

- "After rising steadily for several decades, cancer rates peaked in the early 1990s and have been declining."

Hayward attributes this remarkable improvement in environmental quality mainly to technology and markets, not to government regulations. He writes,

The chief drivers of environmental improvement are economic growth, constantly increasing resource efficiency, technological innovation in pollution control, and the deepening of environmental values among the American public that have translated to changed behavior and consumer preferences. Government regulation has played a vital role, to be sure, but in the grand scheme of things regulation can be understood as a lagging indicator, often achieving results at needlessly high cost, and sometimes failing completely (p. 2).

While the environmental record of the United States stands out even among developed countries for its successes, the story is similar if not the same in all but communist and formerly communist countries. There is little doubt but that the prosperity made possible by fossil fuels has made environmental protection both highly valued and financially possible, producing a world that is cleaner and safer than it would have been in their absence.

Figure 5.2.3.2
Trends in prosperity vs. emissions of CO$_2$ and pollutants in the United States, 1970–2017.

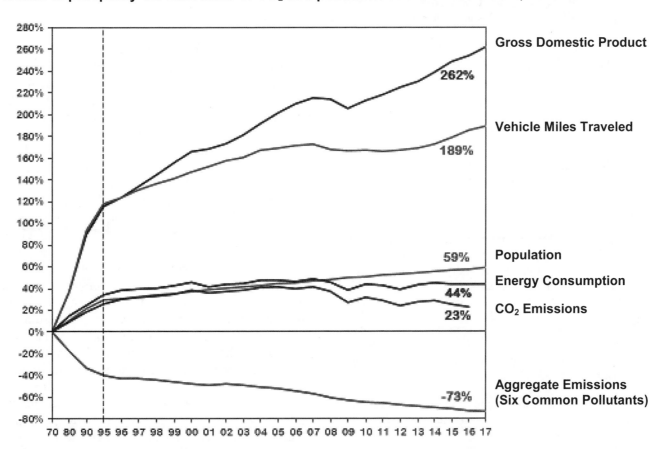

Source: EPA, 2018.

References

Apergis, N. and Ozturk, I. 2015. Testing environmental Kuznets curve hypothesis in Asian countries. *Ecological Indicators* **52** (May): 16–22.

Ausubel, J.H. 1996. Liberation of the environment. *Daedalus* **125** (3): 1–17.

Avery, D. 2000. *Saving the Planet with Pesticides and Plastic*. Second Edition. Indianapolis, IN: Hudson Institute.

Bailey, R. 2015. *The End of Doom: Environmental Renewal in the Twenty-first Century*. New York, NY: Thomas Dunne Books/St. Martin's Press.

Bryce, R. 2014. *Smaller Faster Lighter Denser Cheaper: How Innovation Keeps Proving the Catastrophists Wrong*. New York, NY: PublicAffairs.

EPA. 2018. Environmental Protection Agency. Air quality - national summary (website). Accessed October 21, 2018.

Epstein, A. 2014. *The Moral Case for Fossil Fuels*. New York, NY: Portfolio/Penguin.

Goklany, I.M. 2007. *The Improving State of the World: Why We're Living Longer, Healthier, More Comfortable Lives on a Cleaner Planet*. Washington, DC: Cato Institute.

Goklany, I.M. 2012. Humanity unbound: How fossil fuels saved humanity from nature and nature from humanity. *Cato Policy Analysis* #715. Washington, DC: Cato Institute.

Hayward, S. 2011. *Almanac of Environmental Trends*. San Francisco, CA: Pacific Institute for Public Policy Research.

Heinberg, R. 2007. *Peak Everything: Waking Up to a Century of Declines.* Gabriola Island, BC: New Society Publishers.

Ho, M. and Wang, Z. 2015. Green growth for China? *Resources.* Washington, DC: Resources for the Future.

Moore, S. and Hartnett White, K. *Fueling Freedom: Exposing the Mad War on Energy.* Washington, DC: Regnery.

NRDC. 2008. Natural Resources Defense Council. *The Cost of Climate Change: What We'll Pay if Global Warming Continues Unchecked.*

Osabuohien, E.S., Efobi, U.R., and Gitau, C.M.W. 2014. Beyond the environmental Kuznets curve in Africa: evidence from panel cointegration. *Journal of Environmental Policy and Planning* **16** (4): 517–38.

Sarkodie, A.S. 2018. The invisible hand and EKC hypothesis: what are the drivers of environmental degradation and pollution in Africa? *Environmental Science and Pollution Research International* **25** (22): 21,993–22,022.

Shahbaz, M., Lean, H.H., and Shabbir, M.S. 2012. Environmental Kuznets curve hypothesis in Pakistan: cointegration and granger causality. *Renewable and Sustainable Energy Reviews* **16** (5): 2947–53.

Simon, J. 1980. Resources, population, environment: an oversupply of false bad news. *Science* **208**: 1431–7.

Simon, J. 1982. Are we losing our farmland? *The Public Interest* **67** (Spring): 49–62.

Simon, J.L. 1995. *The State of Humanity.* Cambridge, MA: Blackwell Publishers, Inc.

Simon, J. 1996. *The Ultimate Resource.* Second edition. Princeton, NJ: Princeton University Press.

Simon, J. and Kahn, H. 1984. *The Resourceful Earth: A Response to Global 2000.* New York, NY: Basil Blackwell Publishers Inc.

Smil, V. 2005. *Creating the Twentieth Century: Technical Innovations of 1867–1914 and Their Lasting Impact.* New York, NY: Oxford University Press.

Smil, V. 2006. *Transforming the Twentieth Century: Technical Innovations and Their Consequences.* New York, NY: Oxford University Press.

Tiwari, A.K., Shahbaz, M., and Hye, Q.M.A. 2013. The environmental Kuznets curve and the role of coal consumption in India: cointegration and causality analysis in an open economy. *Renewable and Sustainable Energy Reviews* **18**: 519–27.

Yandle, B., Vijayaraghavan, M., and Bhattarai, M. 2002. The Environmental Kuznets Curve: a primer. *PERC Research Study 02-1.* Bozeman, MT: PERC.

5.3 Impact on Plants

A major environmental benefit produced by the combustion of fossil fuels is an entirely unintended consequence: the beneficial effects on plant life of elevated levels of carbon dioxide (CO_2) in the atmosphere. As reported in Chapter 2, long-term studies confirm the findings of shorter-term experiments, demonstrating numerous growth-enhancing, water-conserving, and stress-alleviating effects of elevated atmospheric CO_2 on plants growing in both terrestrial and aquatic ecosystems.

At locations across the planet, the increase in the atmosphere's CO_2 concentration has stimulated vegetative productivity in spite of many real and imagined assaults on Earth's vegetation, including fires, disease, pest outbreaks, deforestation, and climatic change. Farmers and others who depend on rural livelihoods for income are benefitting from the consequent rising agricultural productivity throughout the world, including in Africa and Asia where the need for increased food supplies is most critical.

This section presents a literature review of the effects of rising CO_2 levels on ecosystems, then plants under stress and plant water use efficiency. The final section looks at the future impacts of higher CO_2 levels on plants, including effects on food production, biodiversity, and extinction rates. The studies summarized here are nearly all based on observational data – real-world experiments and field research – and not computer models, which often are programmed to predict negative effects.

5.3.1 Introduction

As early as 1804, de Saussure showed that peas exposed to high CO_2 concentrations grew better than control plants in ambient air. Work conducted in the early 1900s significantly increased the number of species in which a growth-enhancing effect of atmospheric CO_2 enrichment was observed to occur (Demoussy, 1902–1904; Cummings and Jones, 1918). By the time a group of scientists convened at Duke University in 1977 for a workshop on Anticipated Plant Responses to Global Carbon Dioxide Enrichment, an annotated bibliography of

590 scientific studies dealing with CO_2 effects on vegetation had been prepared (Strain, 1978). This body of research demonstrated increased levels of atmospheric CO_2 generally produce increases in plant photosynthesis, decreases in plant water loss by transpiration, increases in leaf area, and increases in plant branch and fruit numbers, to name but a few of the most commonly reported benefits.

Five years later, at the International Conference on Rising Atmospheric Carbon Dioxide and Plant Productivity, it was concluded a doubling of the atmosphere's CO_2 concentration likely would lead to a 50% increase in photosynthesis in C_3 plants, a doubling of water use efficiency in both C_3 and C_4 plants, significant increases in nitrogen fixation in almost all biological systems and an increase in the ability of plants to adapt to a variety of environmental stresses (Lemon, 1983). In the years since, many other studies have been conducted on hundreds of plant species, repeatedly confirming the growth-enhancing, water-saving, and stress-alleviating advantages of elevated atmospheric CO_2 concentrations on Earth's plants and soils (Idso and Idso, 2011).

The sections below update the literature review conducted for Chapter 1 of the previous volume in the *Climate Change Reconsidered* series titled *Biological Impacts* (NIPCC, 2014). The key findings of that chapter are presented in Figure 5.3.1.1. That report also included two appendices with tables summarizing more than 5,500 individual plant photosynthetic and biomass responses to CO_2-enriched air reported in the scientific literature, finding nearly all plants experience increases in these two parameters at higher levels of CO_2.

References

Cummings, M.B. and Jones, C.H. 1918. The aerial fertilization of plants with carbon dioxide. *Vermont Agricultural Station Bulletin* No. 211.

Demoussy, E. 1902–1904. Sur la vegetation dans des atmospheres riches en acide carbonique. *Comptes Rendus Academy of Science Paris* **136**: 325–8; **138**: 291–3; **139**: 883–5.

Idso, C.D. and Idso, S.B. 2011. *The Many Benefits of Atmospheric CO_2 Enrichment*. Pueblo West, CO: Vales Lake Publishing, LLC.

Lemon, E.R. (Ed.) 1983. *CO_2 and Plants: The Response of Plants to Rising Levels of Atmospheric Carbon Dioxide*. Boulder, CO: Westview Press.

NIPCC. 2014. Idso, C.D, Idso, S.B., Carter, R.M., and Singer, S.F. (Eds.) *Climate Change Reconsidered II: Biological Impacts*. Nongovernmental International Panel on Climate Change. Chicago, IL: The Heartland Institute.

Strain, B.R. 1978. *Report of the Workshop on Anticipated Plant Responses to Global Carbon Dioxide Enrichment*. Durham, NC: Duke University, Department of Botany.

Figure 5.3.1.1
Key findings: CO_2, plants, and soils

- Results obtained under 3,586 separate sets of experimental conditions conducted on 549 plant species reveal nearly all plants experience increases in dry weight or biomass in response to atmospheric CO_2 enrichment. Additional results obtained under 2,094 separate experimental conditions conducted on 472 plant species reveal nearly all plants experience increases in their rates of photosynthesis in response to atmospheric CO_2 enrichment.

- Long-term CO_2 enrichment studies confirm the findings of shorter-term experiments, demonstrating that the growth-enhancing, water-conserving, and stress-alleviating effects of elevated atmospheric CO_2 likely persist throughout plant lifetimes.

- Forest productivity and growth rates throughout the world have increased gradually since the Industrial Revolution in concert with and in response to the historical increase in the atmosphere's CO_2 concentration. Therefore, as the atmosphere's CO_2 concentration continues to rise, forests will likely respond by exhibiting significant increases in biomass production and they likely will grow more robustly and significantly expand their ranges.

- Modest increases in air temperature tend to increase carbon storage in forests and their soils. Thus, old-growth forests can be significant carbon sinks and their capacity to sequester carbon in the future will be enhanced as the atmosphere's CO_2 content continues to rise.

- As the atmosphere's CO_2 concentration increases, the productivity of grassland species will increase even under unfavorable growing conditions characterized by less-than-adequate soil moisture, inadequate soil nutrition, elevated air temperature, and physical stress imposed by herbivory.

- The thawing of permafrost caused by increases in air temperature will likely not transform peatlands from carbon sinks to carbon sources. Instead, rapid terrestrialization likely will act to intensify carbon-sink conditions.

- Rising atmospheric CO_2 concentrations likely will enhance the productivity and carbon sequestering ability of Earth's wetlands. In addition, elevated CO_2 may help some coastal wetlands counterbalance the negative impacts of rising seas.

- Rising atmospheric CO_2 concentrations likely will allow greater numbers of beneficial bacteria (that help sequester carbon and nitrogen) to exist within soils and anaerobic water environments, thereby benefitting both terrestrial and aquatic ecosystems.

- The aerial fertilization effect of atmospheric CO_2 enrichment likely will result in greater soil carbon stores due to increased carbon input to soils, even in nutrient-poor soils and in spite of predicted increases in temperature. The carbon-sequestering capability of Earth's vegetation likely will act as a brake on the rate-of-rise of the atmosphere's CO_2 content and thereby help to mute the effects of human CO_2 emissions on global temperatures.

- The historical increase in the atmosphere's CO_2 content has significantly reduced the erosion of valuable topsoil over the past several decades; the continuing increase in atmospheric CO_2 can maintain this trend and perhaps even accelerate it for the foreseeable future.

Source: Chapter 1. "CO_2, Plants, and Soils." *Climate Change Reconsidered II: Biological Impacts.* Nongovernmental International Panel on Climate Change (NIPCC). Chicago, IL: The Heartland Institute, 2014.

5.3.2 Ecosystem Effects

Elevated CO_2 improves the productivity of ecosystems both in plant tissues aboveground and in the soils beneath them.

Zhu *et al.* (2016), in an article in *Nature Climate Change* titled "Greening of the Earth and its drivers," discussed global changes in leaf area index (LAI) associated with increasing atmospheric CO_2 concentrations. They reported,

We show a persistent and widespread increase of growing season integrated LAI (greening) [from 1982 to 2009] over 25% to 50% of the global vegetated area, whereas less than 4% of the globe shows decreasing LAI (browning). Factorial simulations with multiple global ecosystem models suggest that CO_2 fertilization effects explain 70% of

the observed greening trend, followed by nitrogen deposition (9%), climate change (8%) and land cover change (LCC) (4%).

Zhu *et al.* illustrated their findings with the figure reproduced as Figure 5.3.2.1 below. Similarly, Li *et al.* (2017) studied 2,196 globally distributed databases containing observations of net primary production (NPP) – the net carbon that is fixed (sequestered) by a given plant community or ecosystem – as well as five environmental variables thought to most impact NPP trends (precipitation, air temperature, leaf area index, fraction of photosynthetically active radiation, and atmospheric CO_2 concentration). They analyzed the spatiotemporal patterns of global NPP over the past half century (1961–2010) and found global NPP increased significantly, from 54.95 Pg C yr^{-1} in 1961 to 66.75 Pg C yr^{-1} in 2010, representing a linear increase of 21.5% over the period. They report,

Figure 5.3.2.1
Greening of the Earth, 1982 to 2009, trend in average observed leaf area index (LAI)

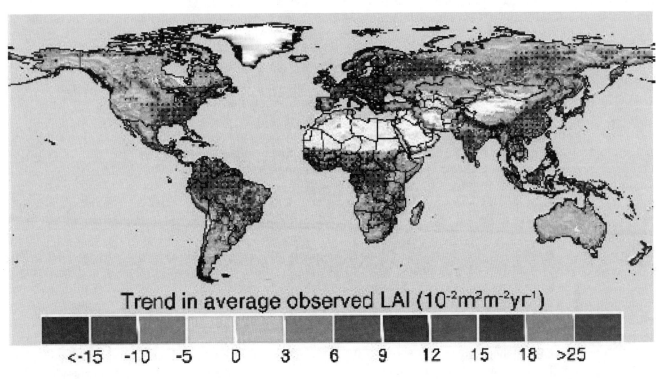

Source: Zhu *et al.*, 2016.

"atmospheric CO_2 concentration was found to be the dominant factor that controlled the interannual variability and to be the major contribution (45.3%) of global NPP." Leaf area index was the second most important factor, contributing an additional 21.8%, followed by climate change (precipitation and air temperature together) and the fraction of photosynthetically active radiation, which accounted for the remaining 18.3% and 14.6% increase in NPP, respectively.

Cheng *et al.* (2017) present similar findings with respect to global terrestrial carbon uptake (i.e., gross primary production, or GPP). Using a combination of ground-based and remotely sensed land and atmospheric observations, they estimated changes in global GPP, water use efficiency (WUE), and evapotranspiration (E) over the period 1982–2011. They estimate global GPP has increased by 0.83 ± 0.26 Pg C per year, or a total of 24.9 Pg C over the past three decades. They also report 82% of the global vegetated land area shows positive trends in GPP despite "the large-scale occurrence of droughts and disturbances over the study period." Similarly,

ecosystem WUE was found to increase in 90% of the world's vegetative areas and there was a high correlation between the spatial trends in these two parameters.

According to Cheng *et al.*, global WUE "increased at a mean rate of 13.7 ± 4.3 mg C mm^{-1} H_2O per year from 1982 to 2011 ($p < 0.001$), which is about 0.7 ± 0.2% per year of mean annual WUE." Global E experienced a non-significant very small increase of 0.06 ± 0.13% per year. Thus, both WUE and E were found to "positively contribute to the estimated increase in GPP," though the contribution from WUE accounted for 90% of the total GPP trend. Cheng *et al.* conclude the "estimated increase in global GPP under climate change and rising atmospheric CO_2 conditions over the past 30 years is taking place at no cost of using proportionally more water, but it is largely driven by the increase in carbon uptake per unit of water use, i.e. WUE."

Cheng *et al.* conclude their results show "terrestrial GPP has increased significantly and is primarily associated with [an] increase in WUE, which in turn is largely driven by rising atmospheric

CO_2 concentrations and [an] increase in leaf area index." They add, "the most important driver for the increases in GPP and WUE from 1982 to 2011 is rising atmospheric CO_2," noting a 10% increase in atmospheric CO_2 induces an approximate 8% increase in global GPP and a 14% increase in global WUE.

Numerous studies have focused on the impact of elevated CO_2 or surface temperature increases on specific ecosystems. We can begin this literature review in the alpine regions of Switzerland, where Rammig et al. (2010) monitored climatic conditions and plant growth for nearly a decade at 17 snow meteorological stations in the region. They used their empirical research to project what plant growth would be if the onset of springtime growth were to begin 17 days earlier, as predicted by the gridded output of a set of regional climate models.

Rammig et al. found "plant height and biomass production were expected to increase by 77% and 45%, respectively," evidence of a dramatic benefit from warming. In some cases "projections of biomass production over a season resulted in changes of up to two-fold." Thus, future warming, if it occurs, would likely benefit plants located in areas like the Alps, where low temperatures, snow cover, and permafrost now limit their ability to grow to their full potential.

Similarly, Kullman (2010a) monitored subalpine/alpine vegetation in the Swedish Scandes (Scandinavian Mountains), deriving "tentative projections of landscape transformations in a potentially warmer future" based on "actual observations and paleoecological data (Kullman and Kjallgren, 2006; Kullman 2006)." He notes post-Little Ice Age warming has halted "a multi-millennial trend of plant cover retrogression" and "floristic and faunal impoverishment, all imposed by progressive and deterministic neoglacial climate cooling." He reports the "upper range margin rise of trees and low-altitude (boreal) plant species, expansion of alpine grasslands and dwarf-shrub heaths are the modal biotic adjustments during the past few decades, after a century of substantial climate warming." He writes, "alpine plant life is proliferating, biodiversity is on the rise and the mountain world appears more productive and inviting than ever." In contrast to model predictions, he notes, "no single alpine plant species has become extinct, neither in Scandinavia nor in any other part of the world in response to climate warming over the past century," citing, in addition to his own studies, the work of Pauli et al. (2001, 2007), Theurillat and Guisan (2001), and Birks (2008).

Kullman concludes, "continued modest warming over the present century will likely be beneficial to alpine biodiversity, geological stability, resilience, sustainable reindeer husbandry and aesthetic landscape qualities."

In a second 2010 publication on the status of alpine communities in the Swedish Scandes, Kullman (2010b) notes in a modestly warming world "plant species diversity will further increase, both in remaining treeless alpine areas and emerging forest outliers on the former alpine tundra," and that this "new alpine landscape may come to support a previously unseen mosaic of richly flowering and luxuriant plant communities of early Holocene character," citing the works of Smith (1920), Iversen (1973), and Birks (2008).

Kullman explains "many alpine species are extremely tolerant of high temperatures per se," citing Dahl (1998) and Birks (2008), as indicated "by their prospering and spread along roadsides far below the treeline, where emerging trees and shrubs are regularly mechanically exterminated (Kullman, 2006; Westerstrom, 2008)." And he notes "another argument against the much-discussed option of pending mass-extinction of alpine species in a warmer future is that some alpine and arctic plant species contain a variety of ecotypes, pre-adapted to quite variable microclimatic and edaphic conditions, which could buffer against extinction in a possibly warmer future (Crawford, 2008)." In addition, he writes this view is supported "by the fact that in the early Holocene, alpine plants survived, reproduced and spread in accordance with higher and more rapidly rising temperatures than those projected for the future by climate models (Oldfield, 2005; Birks, 2008)."

"Over all," Kullman concludes, "continued warming throughout the present century would be potentially and predominantly advantageous for alpine flora and vegetation."

Capers and Stone (2011) "studied a community in western Maine, comparing the frequency and abundance of alpine plants in 2009 with frequency and abundance recorded in 1976" by Stone (1980). The 2009 survey, they write, "recorded an increase in total species richness of the community with the addition of four lower montane species that had not been recorded previously." They also "found no evidence that species with high-arctic distributions had declined more than other species." The changes they recorded are, they write, "consistent with those reported in tundra communities around the world."

Two teams of researchers looked at the possible

impact on arid landscapes of more periods of drought or heavy rainfall, which some computer models forecast will accompany warming surface temperatures. D'Odorico and Bhattachan (2012) note "dryland ecosystems are particularly affected by relatively intense hydroclimatic fluctuations," citing Reynolds *et al.* (2007), and they say "there is some concern that the interannual variability of precipitation in dryland regions might increase in the future thereby enhancing the occurrence of severe water stress conditions in ecosystems and societies." To explore this concern further, D'Odorico and Bhattachan studied "some of the current patterns of hydrologic variability in drylands around the world," reviewing "the major effects of hydrologic fluctuations on ecosystem resilience, maintenance of biodiversity and food security." They report the preponderance of the investigations they reviewed indicated random hydrologic fluctuations may in fact *enhance* the resilience of dryland ecosystems by eliminating threshold-like responses to external drivers. They conclude dryland ecosystem resilience is "enhanced by environmental variability through the maintenance of relatively high levels of biodiversity, which may allow dryland ecosystems to recover after severe disturbances, including those induced by extreme hydroclimatic events."

Also considering what to expect in a CO_2-enriched and warmer world in which precipitation could be more variable, Salguero-Gomez *et al.* (2012) write, "a far-too-often overlooked fact is that desert flora have evolved a set of unique structures and mechanisms to withstand extensive periods of drought," among which are "succulence (Smith *et al.*, 1997), deep roots (Canadell *et al.*, 1996), modified metabolic pathways (Dodd *et al.*, 2002), high modularity (Schenk *et al.*, 2008) and bet hedging mechanisms such as seed dormancy (Angert *et al.*, 2010) or extreme longevity (Bowers *et al.*, 1995)."

Salguero-Gomez *et al.* examined the effects of precipitation on populations of two desert plant species, coupling "robust climatic projections, including variable precipitation, with stochastic, stage-structured models constructed from long-term demographic data sets of the short-lived *Cryptantha flava* in the Colorado Plateau Desert (USA) and the annual *Carrichtera annua* in the Negev Desert (Israel)." They found "a surprising pattern of increased population growth for both study species when [they] compared population dynamics in the future to current conditions, consistent with increasing precipitation in Utah, USA and despite decreasing precipitation in Israel."

Salguero-Gomez *et al.* say their study "contributes two notable exceptions to the accepted view that short-lived species, regardless of habitat, are particularly vulnerable to climate change," emphasizing their findings "challenge the commonly held perception based on correlative approaches (e.g. bioclimatic envelope approaches) suggesting that desert organisms may be particularly vulnerable to climate change."

Polley *et al.* (2012) "grew communities of perennial forb and C_4 grass species for five years along a field CO_2 gradient (250–500 ppm) in central Texas (USA) on each of three soil types, including upland and lowland clay soils and a sandy soil." They measured a number of plant physiological properties, processes and ecosystem aboveground net primary productivity (ANPP). They found CO_2 enrichment from 280 to 480 ppm "increased community ANPP by 0–117% among years and soils and increased the contribution of the tallgrass species *Sorghastrum nutans* (Indian grass) to community ANPP on each of the three soil types," noting the "CO_2-induced changes in ANPP and *Sorghastrum* abundance were linked." They report, "by favoring a mesic C_4 tall grass, CO_2 enrichment approximately doubled the initial enhancement of community ANPP on two clay soils." As a result, they conclude "CO_2-stimulation of grassland productivity may be significantly underestimated if feedbacks from plant community change are not considered."

Kullman (2014) analyzed plant species richness on several alpine summits in the southern Swedish Scandes between 2004/2006 and 2012, which "experienced consistent summer and winter cooling and finalized with a cold and snow rich summer 2012." He reports "plant species richness on high alpine summits decreased by 25–46% between 2004/2006 and 2012" and "most of the lost species have their main distribution in subalpine forest and the low-alpine region." He also noted they "advanced upslope and colonized the summit areas in response to warmer climate between the 1950s and early 2000s." He noted "despite the reduction in species numbers, the summit floras are still richer than in the 1950s," and "substantial and consistent climate cooling (summer and winter) during a decade preceded the recent floristic demise." He concludes, "taken together," the findings "highlight a large capability of certain alpine plant species to track their ecological niches as climate fluctuates on annual to decadal scales."

In 1985, Bert Drake, a scientist at the Smithsonian Environmental Research Center in

Edgewater, Maryland, chose a Chesapeake Bay wetland sustaining both pure and mixed stands of the C_4 grass *Spartina patens* and the C_3 sedge *Scirpus olneyi* for an open-top chamber study of the effects of full-day (24-hour) atmospheric CO_2 enrichment to 340 ppm above the then-ambient concentration of the same value. In a paper published 28 years later in *Global Change Biology*, Drake (2014) summarized some of the important findings of this undertaking. The Chesapeake Bay study offers "strong evidence that shoot and root biomass and net ecosystem production increased significantly." He infers – from the fact that methane emission (Dacey *et al.*, 1994) and nitrogen fixation (Dakora and Drake, 2000) were also stimulated by elevated CO_2 and that inputs of soil carbon also increased – that "ecosystems will accumulate additional carbon as atmospheric CO_2 continues to rise, as suggested by Luo *et al.* (2006)."

Drake also writes that the long duration of the Chesapeake Bay wetland study allows for a test of "the idea that some process, such as progressive nitrogen limitation, may constrain ecosystem responses to elevated CO_2 in native ecosystems." His findings, as well as those of Norby *et al.* (2005) and Norby and Zak (2011), imply, as he notes, that quite to the contrary, Earth's ecosystems will continue to accumulate carbon as the atmosphere's CO_2 content continues to rise.

Ruzicka *et al.* (2015) studied talus slopes (the pile of rocks that accumulates at the base of a cliff, chute, or slope) "inhabited by isolated populations of boreal and alpine plants and invertebrates," citing previous research on the subject by Ruzicka (2011), Ruzicka *et al.* (2012), and Nekola (1999). They measured, over a period of five years, air temperatures of low-elevation talus slopes at three locations in the North Bohemia region of the Czech Republic. They found "the talus microclimate can be sufficiently resistant to an increase of mean annual atmospheric temperature by 3°C, retaining a sufficient number of freezing days during the winter season." They conclude, "based on our data, we can justifiably suppose that even such an extent of warming in the future (an increase of mean annual atmospheric temperature by 3°C) will not endanger the cold talus ecosystems."

In 1997, two field sites were established in alpine meadows at the Haibei Research Station in Haibei, Quinghai, China. Control and experimental plots were established to examine the impact of simulated warming (1–2°C above control plots) on plant species diversity. After four years of warming, it was determined that warm plots lost an average of 11 to 19 species (~40%) relative to control plots (Klein, 2003; Klein *et al.*, 2004). In more recent work, Zhang *et al.* (2017) contend a four-year period is "too short to detect the role colonization and re-establishment may play in community re-assembly," as such processes are known to take place over decades and not years.

Zhang *et al.* resampled the plots after 18 years of simulated warming, in order to see if the shorter-term findings were indeed premature. They found the initial warming-induced decline in species diversity "had rebounded to initial levels, on a par with control plots," concluding, "the long-term impacts of continued global warming are [likely] to result in highly dynamic processes of community reassembly and turnover that do not necessarily lead to a net decline in local diversity," adding that "short-term experiments may be insufficient to capture the temporal variability in community diversity and composition in response to climate change."

O'Leary *et al.* (2017) surveyed publications of 97 expert researchers who had studied six major types of coastal biogenic ecosystems in order to identify "bright spots of resilience" in the face of climate change. They report 80% of the researchers found resilience in the ecosystems they studied, with resilience "observed in all ecosystem types and at multiple locations worldwide." They conclude these findings suggest "coastal ecosystems may still hold great potential to persist in the face of climate change and that local- to regional-scale management can help buffer global climatic impacts."

References

Angert, A.L., Horst, J.L., Huxman, T.E., and Venable, D.L. 2010. Phenotypic plasticity and precipitation response in Sonoran desert winter annuals. *American Journal of Botany* 97: 405–11.

Birks, H.H. 2008. The late-quaternary history of arctic and alpine plants. *Plant Ecology and Diversity* 1: 135–46.

Bowers, J.E., Webb, R.H., and Rondeau, R.J. 1995. Longevity, recruitment and mortality of desert plants in Grand Canyon, Arizona, USA. *Journal of Vegetation Science* 6: 551–64.

Canadell, J., Jackson, R.B., Ehleringer, J.R., Mooney, H.A., Sala, O.E., and Schulze, E.D. 1996. Maximum rooting depth of vegetation types at the global scale. *Oecologia* 108: 583–95.

Capers, R.S. and Stone, A.D. 2011. After 33 years, trees more frequent and shrubs more abundant in northeast U.S. alpine community. *Arctic, Antarctic, and Alpine Research* **43**: 495–502.

Cheng, L., Zhang, L., Wang, Y.-P., Canadell, J.G., Chiew, F.H.S., Beringer, J., Li, L., Miralles, D.G., Piao, S., and Zhang, Y. 2017. Recent increases in terrestrial carbon uptake at little cost to the water cycle. *Nature Communications* **8**: 110.

Crawford, R.M.M. 2008. Cold climate plants in a warmer world. *Plant Ecology and Diversity* **1**: 285–97.

Dacey, J.W.H., Drake, B.G., and Klug, M.J. 1994. Simulation of methane emission by carbon dioxide enrichment of marsh vegetation. *Nature* **370**: 47–9.

Dahl, E. 1998. *The Phytogeography of Northern Europe.* Cambridge, UK: Cambridge University Press.

Dakora, F. and Drake, B.G. 2000. Elevated CO_2 stimulates associative N_2 fixation in a C_3 plant of the Chesapeake Bay wetland. *Plant, Cell and Environment* **23**: 943–53.

Dodd, A.N., Borland, A.M., Haslam, R.P., Griffiths, H., and Maxwell, K. 2002. Crassulacean acid metabolism: plastic, fantastic. *Journal of Experimental Botany* **53**: 569–80.

D'Odorico, P. and Bhattachan, A. 2012. Hydrologic variability in dryland regions: impacts on ecosystem dynamics and food security. *Philosophical Transactions of the Royal Society B* **367**: 3145–57.

Drake, B.G. 2014. Rising sea level, temperature, and precipitation impact plant and ecosystem responses to elevated CO_2 on a Chesapeake Bay wetland: review of a 28-year study. *Global Change Biology* **20**: 3329–43.

Iversen, J. 1973. The development of Denmark's nature since the last glacial. *Danmarks Geologiske Undersogelse Series V. Raeeke* **7-C**: 1–126.

Klein, J.A. 2003. *Climate Warming and Pastoral Land Use Change: Implications for Carbon Cycling, Biodiversity and Rangeland Quality on the Northeastern Tibetan Plateau.* PhD Thesis. Berkeley, CA: University of California.

Klein, J.A., Harte, J., and Zhao, X.-Q. 2004. Experimental warming causes large and rapid species loss, dampened by simulated grazing, on the Tibetan Plateau. *Ecology Letters* **7**: 1170–9.

Kullman, L. 2006. Transformation of alpine and subalpine vegetation in a potentially warmer future, the Anthropocene era: tentative projections based on long-term observations and paleovegetation records. *Current Trends in Ecology* **1**: 1–16.

Kullman, L. 2010a. A richer, greener and smaller alpine world: review and projection of warming-induced plant cover change in the Swedish Scandes. *Ambio* **39**: 159–69.

Kullman, L. 2010b. One century of treeline change and stability – experiences from the Swedish Scandes. *Landscape Online* **17**: 1–31.

Kullman, L. 2014. Recent cooling and dynamic responses of alpine summit floras in the southern Swedish Scandes. *Nordic Journal of Botany* **32**: 369–76.

Kullman, L. and Kjallgren, L. 2006. Holocene tree-line evolution in the Swedish Scandes: recent tree-line rise and climate change in a long-term perspective. *Boreas* **35**: 159–68.

Li, P., Peng, C., Wang, M., Li, W., Zhao, P., Wang, K., Yang, Y., and Zhu, Q. 2017. Quantification of the response of global terrestrial net primary production to multifactor global change. *Ecological Indicators* **76**: 245–55.

Luo, Y., Hui, D., and Zhang, D. 2006. Elevated CO_2 stimulates net accumulations of carbon and nitrogen in land ecosystems: a meta-analysis. *Ecology* **87**: 53–63.

Nekola, J.C. 1999. Paleorefugia and neorefugia: the influence of colonization history on community pattern and process. *Ecology* **80**: 2459–73.

Norby, R.J. and Zak, D.R. 2011. Ecological lessons from free-air CO_2 enrichment (FACE) experiments. *Annual Review of Ecology, Evolution, and Systematics* **42**: 181–203.

Norby, R.J., *et al.* 2005. Forest response to elevated CO_2 is conserved across a broad range of productivity. *Proceedings of the National Academy of Sciences USA* **102**: 18,052–6.

Oldfield, F. 2005. *Environmental Change. Key Issues and Alternative Perspectives.* Cambridge, UK: Cambridge University Press.

O'Leary, J.K., *et al.* 2017. The resilience of marine ecosystems to climatic disturbances. *BioScience* **67**: 208–20.

Pauli, H., Gottfried, M., and Grabherr, G. 2001. High summits of the Alps in a changing climate. In: Walther, G.-R., Burga, C.A., and Edwards, P.J. (Eds.). *Fingerprints of Climate Change.* New York, NY: Kluwer, pp. 139–49.

Pauli, H., Gottfried, M., Reiter, K., Klettner, C., and Grabherr, G. 2007. Signals of range expansions and contractions of vascular plants in the high Alps:

observations (1994–2004) at the GLORIA master site Schrankogel, Tyrol, Austria. *Global Change Biology* **13**: 147–56.

Polley, H.W., Jin, V.L., and Fay, P.A. 2012. Feedback from plant species change amplifies CO_2 enhancement of grassland productivity. *Global Change Biology* **18**: 2813–23.

Rammig, A., Jonas, T., Zimmermann, N.E., and Rixen, C. 2010. Changes in alpine plant growth under future climate conditions. *Biogeosciences* **7**: 2013–24.

Reynolds, J.F., *et al.* 2007. Global desertification: building a science for dryland development. *Science* **316**: 847–51.

Ruzicka, V. 2011. Central European habitats inhabited by spiders with disjunctive distributions. *Polish Journal of Ecology* **59**: 367–80.

Ruzicka, V., Zacharda, M., Nemcova, L., Smilauer, P., and Nekola, J.C. 2012. Periglacial microclimate in low-altitude scree slopes supports relict biodiversity. *Journal of Natural History* **46**: 2145–57.

Ruzicka, V., Zacharda, M., Smilauer, P., and Kucera, T. 2015. Can paleorefugia of cold-adapted species in talus slopes resist global warming? *Boreal Environment Research* **20**: 403–12.

Salguero-Gomez, R., Siewert, W., Casper, B.B., and Tielborger, K. 2012. A demographic approach to study effects of climate change in desert plants. *Philosophical Transactions of the Royal Society B* **367**: 3100–14.

Schenk, H.J., Espino, S., Goedhart, C.M., Nordenstahl, M., Cabrera, H.I., and Jones, C.S. 2008. Hydraulic integration and shrub growth form linked across continental aridity gradients. *Proceedings of the National Academy of Sciences USA* **105**: 11,248–53.

Smith, H. 1920. *Vegetationen och dess Utvecklingshistoria i det Centralsvenska Hogfjallsomradet*. Uppsala, Sweden: Almqvist och Wiksell.

Smith, W.K., Monson, R.K., and Anderson, J.E. 1997. *Physiological Ecology of North American Desert Plants*. New York, NY: Springer.

Stone, A. 1980. Avery Peak on Bigelow Mountain, Maine: The Flora and Vegetation Ecology of a Subalpine Heathland. M.S. Thesis. Burlington, VT: University of Vermont.

Theurillat, J.-P. and Guisan, A. 2001. Potential impacts of climate change on vegetation in the European Alps: a review. *Climatic Change* **50**: 77–109.

Westerstrom, G. 2008. Floran i tre socknar i nordvastra Angermanland. *Svensk Botanisk Tidskrift* **102**: 225–61.

Zhang, C., Willis, C.G., Klein, J.A., Ma, Z., Li, J., Zhou, H., and Zhao, X. 2017. Recovery of plant species diversity during long-term experimental warming of a species-rich alpine meadow community on the Qinghai-Tibet plateau. *Biological Conservation* **213**: 218–24.

Zhu, Z., *et al.* 2016. Greening of the Earth and its drivers. *Nature Climate Change* **6**: 791–5.

5.3.3 Plants under Stress

Atmospheric CO_2 enrichment ameliorates the negative effects of a number of environmental plant stresses including high temperatures, air and soil pollutants, herbivory, nitrogen deprivation, and high levels of soil salinity.

According to the IPCC, a warmer future will introduce new sources of stress on the biological world, including increases in forest fires, droughts, and extreme heat events. The IPCC fails to ask whether the higher levels of atmospheric CO_2 its models also predict will aid or hinder the ability of plants to cope with these challenges. Had it looked, the IPCC would have discovered an extensive body of research showing how atmospheric CO_2 enrichment ameliorates the negative effects of a number of environmental plant stresses. For example, increased ambient CO_2 improves water use efficiency (discussed in detail in Section 5.3.4) of plants by allowing more CO_2 to enter the photosynthetic tissue per unit of time, thereby enhancing the rate of photosynthesis (carboxylation) while water loss is kept at a constant level or even reduced as plants' stomata are open for less time. This well-documented biological process is absent from many computer models that assume climate change has negative effects on agriculture.

This section updates the literature review that appeared in Chapter 3 of *Climate Change Reconsidered II: Biological Impacts* (NIPCC, 2014). The key findings of the previous report appear in Figure 5.3.3.1.

Koutavas (2013) studied tree growth rings to investigate potential growth-climate relationships, developing growth indices from cores extracted from 23 living Greek fir (*Aibes cephalonica*) trees for the period AD 1820–2007. He reports the growth of the trees historically has been "limited by growing-

Figure 5.3.3.1
Key Findings: Impacts on plants under stress

■ Atmospheric CO_2 enrichment (henceforth referred to as "rising CO_2") exerts a greater positive influence on diseased as opposed to healthy plants because it significantly ameliorates the negative effects of stresses imposed on plants by pathogenic invaders.

■ Rising CO_2 helps many plants use water more efficiently, helping them overcome stressful conditions imposed by drought or other less-than-optimum soil moisture conditions.

■ Enhanced rates of plant photosynthesis and biomass production from rising CO_2 will not be diminished by any surface temperature increase that might accompany it in the future. In fact, if ambient air temperatures rise concurrently, the growth-promoting effects of atmospheric CO_2 enrichment will likely rise even more.

■ Although rising CO_2 increases the growth of many weeds, the fraction helped is not as large as that experienced by non-weeds. Thus, CO_2 enrichment of the air may provide non-weeds with greater protection against weed-induced decreases in productivity.

■ Rising CO_2 improves plants' abilities to withstand the deleterious effects of heavy metals where they are present in soils at otherwise-toxic levels.

■ Rising CO_2 reduces the frequency and severity of herbivory against crops and trees by increasing production of natural substances that repel insects, leading to the production of more symmetrical leaves that are less susceptible to attacks by herbivores and making trees more capable of surviving severe defoliation.

■ Rising CO_2 increases net photosynthesis and biomass production by many agricultural crops, grasses, and grassland species even when soil nitrogen concentrations tend to limit their growth. Additional CO_2-induced carbon input to the soil stimulates microbial decomposition and thus leads to more available soil nitrogen, thereby challenging the progressive nitrogen limitation hypothesis.

■ Rising CO_2 typically reduces and can completely override the negative effects of ozone pollution on the photosynthesis, growth, and yield of nearly all agricultural crops and trees that have been experimentally evaluated.

■ Rising CO_2 can help plants overcome stresses imposed by the buildup of soil salinity from repeated irrigation.

■ The ongoing rise in the atmosphere's CO_2 content is a powerful antidote for the deleterious biological impacts that might be caused by an increase in the flux of UV-B radiation at the surface of Earth due to depletion of the planet's stratospheric ozone layer.

Source: Chapter 3. "Plants Under Stress," *Climate Change Reconsidered II: Biological Impacts.* Nongovernmental International Panel on Climate Change. Chicago, IL: The Heartland Institute, 2014.

season moisture in late spring/early summer, most critically during June," but "by the late 20th–early 21st century, there remains no statistically significant relationship between moisture and growth."

According to Koutavas, despite the "pronounced shift to greater aridity in recent decades," tree growth in the region experienced "a net increase over the last half-century, culminating with a sharp spike in AD

1988–1990," which implies the trees have acquired a "markedly enhanced resistance to drought." Koutavas says that result is "most consistent with a significant CO_2 fertilization effect operating through restricted stomatal conductance [the rate of passage of carbon dioxide (CO_2) entering, or water vapor exiting, through the stomata of a leaf] and improved water-use efficiency."

Naudts *et al.* (2014) "assembled grassland communities in sunlit, climate-controlled greenhouses and subjected these to three stressors (drought, zinc toxicity, nitrogen limitation) and their combinations," where "half of the communities were exposed to ambient climate conditions (current climate) and the other half were continuously kept at 3°C above ambient temperatures and at 620 ppm CO_2 (future climate)." They found "across all stressors and their combinations, future climate-grown plants coped better with stress, i.e. above-ground biomass production was reduced less in future than in current climate." They identify three mechanisms driving improved stress protection and conclude, "there could be worldwide implications connected to the alleviation of the stress impact on grassland productivity under future climate conditions," noting as an example that "enhanced protection against drought could mitigate anticipated productivity losses in regions where more frequent and more intense droughts are predicted."

Zong and Shangguan (2014) hydroponically cultivated maize (*Zea mays* L. cv. Zhengdan 958) seedlings in sand within two climate-controlled chambers and exposed them to CO_2 concentrations of either 380 or 750 ppm CO_2 until the end of the study. They also irrigated the seedlings with Hoagland solutions and "different N solutions (5 mM N as the nitrogen deficiency treatment and 15 mM N as the control)." The two scientists report "maize seedlings suffering combined N limitation and drought had a better recovery of new leaf photosynthetic potential than those suffering only drought with ambient CO_2." But with elevated CO_2, "the plants were able to maintain favorable water content as well as enhance their biomass accumulation, photochemistry activity, leaf water use efficiency and new leaf growth recoveries." Zong and Shangguan conclude, "elevated CO_2 could help drought-stressed seedlings to maintain higher carbon assimilation rates under low water content," noting that was the case "even under N-limited conditions, which allow the plants to have a better performance under drought following re-watering."

Song and Huang (2014) studied Kentucky Bluegrass plants obtained from field plots in New Brunswick, New Jersey (USA) in controlled environment chambers maintained at ambient and double-ambient atmospheric CO_2 concentrations (400 and 800 ppm, respectively). They divided the plants into sub-treatments of optimum temperature and water availability, drought-stressed (D) and heat-stressed (H) conditions, and a combined D and H environment. They report "the ratio of root to shoot biomass increased by 65% to 115% under doubling ambient CO_2 across all treatments with the greatest increase under D" (see Figure 5.3.3.2, panel C). They noted "the positive carbon gain under doubling ambient CO_2 was the result of both increases in net photosynthesis rate and suppression of respiration rate." Leaf net photosynthesis "increased by 32% to 440% with doubling ambient CO_2" and there was a significant decline (by 18% to 37%) in leaf respiration rate under the different treatments "with the greatest suppression under D + H." The two scientists concluded, "the increase in carbon assimilation and the decline in respiration carbon loss could contribute to improved growth under elevated CO_2 conditions," as they note has been found to be the case with several other plants, citing the studies of Drake *et al.* (1997), Ainsworth *et al.* (2002), Long *et al.* (2004), and Reddy *et al.* (2010).

Lee *et al.* (2015) grew *Perilla frutescens* var. *japonica* 'Arum' – an herb of the mint family – from seeds for a period of 60 days in two controlled-environment chambers, where "the pots were flushed once a day and fertilized twice a week with a nutrient solution developed for leafy vegetables," and where after the first week the plants were exposed to either near-ambient or elevated atmospheric CO_2 concentrations (350 vs. 680 ppm, respectively) for the remainder of the experiment. Relative to the plants growing in near-ambient CO_2 air, as shown in Figure 5.3.3.3, they found the plants growing in the CO_2 enriched air experienced a higher photosynthetic rate, increased stomatal resistance, declining transpiration rates, and improved water-use efficiency. The elevated CO_2 concentration also reduced drought-induced oxidative damage to the plants.

Dias de Oliveira *et al.* (2015) conducted a field experiment to determine the interactive effects of CO_2, temperature, and drought on two pairs of sister lines of wheat (*Triticum aestivum* L.) over the course of a growing season. The experiment was conducted outdoors in poly-tunnels (steel frames covered in polythene) under all possible combinations of CO_2

concentration (400 or 700 ppm), temperature (ambient or +3°C above ambient daytime temperature), and water status (well-watered or terminal drought post anthesis). They found, among other things, that elevated CO_2 "increased grain yield and aboveground biomass." Terminal drought "reduced grain yield and aboveground biomass," but elevated CO_2 "was the key driver in the amelioration of [its negative] effects." They note "temperature did not have a major effect on ameliorating the effects of terminal drought."

Chen *et al.* (2015) explain "drought stress is one of the most detrimental abiotic stresses for plant growth," in that it "leads to stomatal closure and reduces photosynthesis resulting from restricted CO_2 diffusion through leaf stomata and inhibition of carboxylation activity," as described by Flexas *et al.* (2004). They note "minimizing cellular dehydration and maintaining active photosynthesis are key strategies for plant survival or persistence through dry-down periods," as is described in more detail by Nilsen and Orcutt (1996).

Figure 5.3.3.2
Shoot dry weight, root dry weight, and root/shoot dry weight ratio of Kentucky Bluegrass grown under drought stress, heat stress, and drought and heat stress, under ambient and elevated CO_2 concentrations

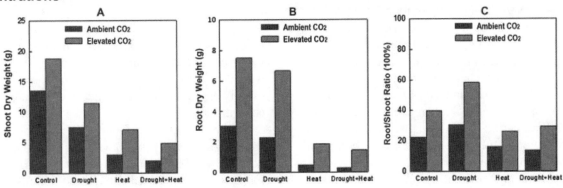

Source: Song and Huang, 2014.

Figure 5.3.3.3
Effect of elevated CO_2 on photosynthetic rate, stomatal resistance, and transpiration in *P. frutescens* under well-watered and drought-stressed conditions

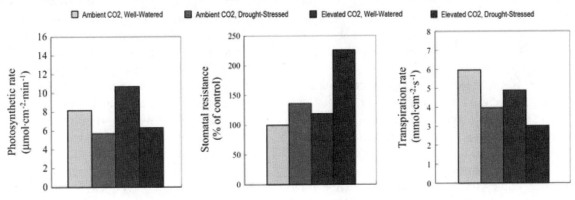

Source: Lee *et al.*, 2015.

Hypothesizing that drought stress might be alleviated by the positive effects of atmospheric CO_2 enrichment, Chen *et al.* grew a cool-season grass – tall fescue (*Festuca arundinacea* Schreb. cv. Rembrandt) – in controlled-environment chambers maintained at either 400 or 800 ppm CO_2 under both well-watered (control) conditions or subjected to drought stress followed by re-watering. This work revealed, among other things, that "elevated CO_2 reduced stomatal conductance and transpiration rate of leaves during both drought stress and re-watering" and the "elevated CO_2 enhanced net photosynthetic rate with lower stomatal conductance but higher Rubisco and Rubisco activase activities during both drought and re-watering." They conclude, "the mitigating effects of elevated CO_2 on drought inhibition of photosynthesis and the enhanced recovery in photosynthesis on re-watering were mainly the result of the elimination of metabolic limitation from drought damages associated with increased enzyme activities for carboxylation."

Using *Chrysolaena obovata* plants cultivated within four open-top chambers inside a greenhouse, Oliveira *et al.* (2016) maintained half of the plants in air of 380 ppm CO_2 and half of them in air of 760 ppm CO_2 for a period of 45 days, after which for each CO_2 concentration they separated the plants into four water replacement treatments (in which the water used by the plant, lost to the soil and evaporation, is replaced so the plant never dries out): control (100% water replacement), low drought (75% water replacement), medium drought (50% water replacement), and severe drought (25% water replacement) of the total transpired water of the previous 48 hours, as determined by the before-and-after measured weights of each plant-pot combination. They report, "under elevated CO_2, the negative effects of water restriction on physiological processes were minimized, including the maintenance of rhizophore water potential, increase in water use efficiency, maintenance of photosynthesis and fructan reserves for a longer period."

Van der Kooi *et al.* (2016) searched library archives of the scientific literature between 1979 and 2014 for CO_2 enrichment studies of agricultural plants exposed to drought. For biomass effects, they identified "62 different data entries (for both dry and well-watered conditions) from 41 different experimental studies on 30 crop species," while for yield, they identified "19 data entries (for both dry and well-watered conditions) from 17 experimental studies on 8 crop species." They found C_3 and C_4

crops responded very similarly to atmospheric CO_2 enrichment when experiencing drought conditions. They conclude, "crops grown in areas with limited water availability will benefit from future elevated CO_2, regardless of their metabolism," noting "drought leads to stomatal limitation of photosynthesis in both C_3 and C_4 crops, which is alleviated [in both cases] when the plants are grown under elevated CO_2."

Schmid *et al.* (2016) investigated the effects of elevated CO_2 and drought on two barley cultivars, Golden Promise and Bambina, growing the cultivars in controlled environment chambers at two CO_2 levels (380 and 550 ppm) and two water levels (normal and 33% less water) in order to simulate drought conditions, based on average rainfall over the course of the growing season. They found grain dry weight was enhanced by 31% and 62% in Golden Promise (GP) and Bambina (BA) cultivars, respectively, under normal water conditions. Under reduced water conditions, elevated CO_2 proved even more beneficial, enhancing BA and GP grain dry weight by 50% and 150%, respectively, coming close to fully ameliorating the impact of drought on the two cultivars.

Schmid *et al.* also found total plant biomass was enhanced by CO_2 enrichment in both plants under normal (8% biomass enhancement for GP and 34% for BA) and simulated drought conditions (52% for GP and 21% for BA). The water use efficiency of both plants also was enhanced by elevated CO_2, including a 200% increase under reduced water conditions for GP when calculated based on grain yield. The edible portion of the plant, including grain number per plant and harvest index, were significantly enhanced by elevated CO_2 under both normal and water-stressed conditions.

Wijewardana *et al.* (2016) investigated the growth response of six maize hybrids to drought, UV-B radiation, and carbon dioxide (CO_2). The maize was grown in sunlit chambers under a variety of treatment conditions, including two levels of CO_2 (400 and 800 ppm), two levels of water stress (100% and 50% irrigation treatment based on evapo-transpiration measurements), and two levels of biologically effective UV-B radiation intensities (0 and 10 kJ m^{-2} d^{-1}). Compared with ambient conditions, alone or in combination with water and/or UV-B stress, elevated CO_2 significantly increased maize hybrid plant height, leaf number, and leaf area, with the magnitudes of the responses varying by hybrid. Elevated CO_2 increased total plant dry matter as well. Averaged across the six hybrids, elevated

CO_2 induced an average 13% increase in dry matter under adequate water and normal UV-B conditions, a 19% increase under UV-B stressed conditions, a 20% increase when water stressed, and a 10% increase under combined water and UV-B stressed conditions. Thus, although drought and UV-B stress had negative effects on maize hybrid growth, elevated CO_2 "caused a positive increase in both vegetative and physiological traits."

Roy *et al.* (2016) used the Montpellier (France) CNRS Ecotron facility to simulate elevated CO_2 (eCO_2) and extreme climate events (ECEs), co-occurring heat and drought events as projected for the 2050s, the effects of which they analyzed on the ecosystem-level for both carbon and water fluxes in the Ecotron's C_3 grassland. They report "eCO_2 not only slows down the decline of ecosystem carbon uptake during the ECE but also enhances its recovery after the ECE, as mediated by increases of root growth and plant nitrogen uptake induced by the ECE." They say their findings indicate "in the predicted near-future climate, eCO_2 could mitigate the effects of extreme droughts and heat waves on ecosystem net carbon uptake."

Da Silva *et al.* (2017) examined the interactive effects of elevated CO_2 and reduced water availability on a number of key physiological and growth traits for six-week-old Concord (*Vitis labrusca*) grape plants grown in controlled environment chambers at either ambient (400 ppm) or elevated (800 ppm) levels of atmospheric CO_2 for a period of 24 days under three water management treatments: full irrigation, in which the rootzone was irrigated to saturation three days per week; partial root drying, where alternating halves of the rootzone were irrigated to saturation three days per week; and no irrigation, in which irrigation was suspended for the duration of the 24-day experiment.

Da Silva *et al.* report elevated CO_2 increased net photosynthesis by 24% and specific leaf weight by 16%, whereas it reduced stomatal density by 25% over the 24 days of enrichment. With respect to its influence on drought, "elevated CO_2 dramatically increased drought tolerance of grapevines" by enhancing plant water use efficiency, primarily because of observed CO_2-induced increases in net photosynthesis and corresponding CO_2-induced declines in stomatal conductance and transpiration rates. In the no irrigation treatment, water depletion in the root zone developed at a slower rate in the elevated CO_2 chambers, delaying the effects of drought by about four days. They concluded, "overall, elevated CO_2 improves the leaf carbon

balance and this mitigates the deleterious effects of drought on grapevines."

Kumar *et al.* (2017) grew rice (*Oryza sativa*, cv. Naveen) in open-top chambers under two moisture regimes and three CO_2 concentrations over two consecutive growing seasons. Plant moisture regimes included well-watered, where the water depth of the soil was maintained at 3 ± 2 cm, or water-deficit, where surface irrigation was applied only when the soil water potential at 15 cm reached -60 kPa. Atmospheric CO_2 concentrations were maintained at ambient (400 ppm), mid-elevation (550 ppm), or high-elevation (700 ppm) during daylight hours only.

Kumar *et al.* found atmospheric CO_2 enrichment (to both 550 and 700 ppm) "exhibited a positive response on plant growth, grain yield and [water use efficiency] of rice as compared to ambient CO_2." Elevated levels of CO_2 increased grain yield by 15% to 18% under well-watered conditions and by 39% to 43% under water-deficit conditions. Under elevated CO_2 conditions there was a decline in water use of 11% to 14% under well-watered and 5% under water-deficit conditions. The water use efficiency (ratio of grain yield to total water input) in the two CO_2-enriched chambers increased by 30% to 35% under well-watered conditions and by approximately 48% under water deficit conditions, relative to that observed in the ambient CO_2 chambers.

Kumar *et al.* also report higher levels of CO_2 significantly altered leaf tissue parameters (e.g., relative water content, leaf water potential, and electrolyte leakage) under moisture stress, so as to help mitigate the negative impacts of water deficit. They also found the concentrations of certain antioxidant metabolites were reduced in plants growing under elevated CO_2 in the moisture-stress treatment. This observation further supports the notion that elevated CO_2 helps mitigate water stress in rice: The CO_2-induced mitigation of water stress reduces the production of harmful reactive oxygen species, which subsequently reduces the need for plants to produce antioxidant enzymes to counter them.

Wang *et al.* (2017) examined the interactive effects of elevated CO_2 and drought on soybean (*Glycine max*, cv. Yu 19) by growing plants from seed for 40 days in controlled-environment greenhouses under ambient and twice ambient CO_2 concentrations and three water regimes: well-watered (80% water holding capacity of the soil), moderate drought (60% water holding capacity), and severe drought (40% water holding capacity). Drought negatively impacted the net photosynthesis of the soybean plants, but the positive influence of elevated CO_2 was so great that even under severe drought conditions, the net photosynthetic rate was 73% greater than that observed under well-watered conditions at ambient CO_2. Water use efficiency also was enhanced by elevated CO_2, where it was "almost 2.5 times larger than that under ambient CO_2."

Wang *et al.* also report that elevated CO_2 increased soil enzyme activities by stimulating plant root exudation (a below-ground response to pests) and "resulted in a longer retention time of dissolved organic carbon (DOC) in [the] soil, probably by improving the soil water effectiveness for organic decomposition and mineralization." Consequently, they conclude "drought stress had significant negative impacts on plant physiology, soil carbon, and soil enzyme activities, whereas elevated CO_2 and plant physiological feedbacks indirectly ameliorated these impacts."

Sekhar *et al.* (2017) investigated the interactive effects of elevated CO_2 and water stress on a drought-tolerant cultivar (selection-13) of mulberry (*Morus* spp.) trees. The experiment was conducted in open-top chambers where six-month-old saplings were planted into chambers of either ambient (400 ppm) or enriched (550 ppm) atmospheric CO_2. The scientists cut the plants every four months at stump height (30 cm above the soil surface) to create a coppice culture system (where new shoots grow from the stump). After one year, a subset of plants was subjected to drought stress, where all water was withheld for 30 days.

Sekhar *et al.* observed evidence of stress in the drought treatments under both elevated and ambient CO_2 conditions, but the stress was less under elevated CO_2. They report "plants grown under elevated CO_2 had more green leaves (58%) and individual leaf densities (60%) with less leaf senescence (78%) compared to their ambient controls." In addition, they note there was "a significant decrease in total chlorophyll content (45%) in ambient CO_2 grown plants" compared to plants grown in high CO_2 conditions. Plants grown in the CO_2-enriched environment had greater net photosynthesis, water use efficiency, and aboveground fresh (92%) and dry (83%) biomass, as well as reduced stomatal conductance and transpiration. Trees growing in the CO_2-enriched environment produced less reactive oxygen species and triggered more and better antioxidant systems to combat the drought-induced oxidative stress.

Sekhar *et al.* say their results "clearly demonstrate that future increases in atmospheric CO_2

enhance the photosynthetic potential and also mitigate the drought-induced oxidative stress" in mulberry.

One of the effects predicted by computer models to result from increases in the atmosphere's CO_2 content is that there will be an increase in the number of heavy precipitation events, which could lead to flooding and waterlogging of soils. Pérez-Jiménez *et al.* (2017) point out "the combined effect of waterlogging and elevated CO_2 has been scarcely studied," while additionally noting the topic "has never been studied in fruit trees." They performed the first such analysis by examining the interactive effects of elevated CO_2 and waterlogging on sweet cherry (*Prunus avium*).

Pérez-Jiménez *et al.* subjected one-year-old seedlings of a Burlat sweet cherry cultivar grafted onto one of two rootstocks (Gisela 5 and Gisela 6) to three weeks of growth in a controlled-environment chamber of either 400 or 800 ppm CO_2. After the first seven days, plants in each chamber were subjected to two additional treatments: a control treatment in which normal daily irrigation was allowed to counter daily water loss, or a waterlogging treatment where the water level was maintained at least 1 cm above the surface of the soil. Seven days later, the waterlogged plants were drained and returned to the control conditions.

Pérez-Jiménez *et al.* report the net photosynthesis of the cherry tree leaves was "markedly influenced by the CO_2 concentration," such that "it was significantly higher after every phase of the experiment, in plants grafted on both rootstocks, at elevated CO_2." Percentage increases in net photosynthesis due to elevated CO_2 in both rootstock plants ranged from 113% to 180% higher under control conditions (normal watering) and from 106% to 663% higher under waterlogged conditions. Pérez-Jiménez *et al.* also found there was a reduction in stomatal conductance in waterlogged plants under the ambient CO_2 treatment. They report "an atmosphere enriched with CO_2 improved the physiological status of waterlogged plants, reducing the need for a stomatal conductance reduction to mitigate water logging." They conclude that "elevated CO_2 was able to increase photosynthesis and thereby help plants to overcome waterlogging."

References

Ainsworth, E.A., Davey, P.A., Bernacchi, C.J., Dermody, O.C., Heaton, E.A., Moore, D.J., Morgan, P.B., Naaidu,

S.L., Ra, H.-S.Y., Zhu, X.G., Curtis, P., and Long, S.P. 2002. A meta-analysis of elevated CO_2 effects on soybean (*Glycine max*) physiology, growth and yield. *Global Change Biology* **8**: 695–709.

Chen, Y., Yu, J., and Huang, B. 2015. Effects of elevated CO_2 concentration on water relations and photosynthetic responses to drought stress and recovery during re-watering in tall fescue. *Journal of the American Society of Horticultural Science* **140**: 19–26.

da Silva, J.R., Patterson, A.E., Rodrigues, W.P., Campostrini, E., and Griffin, K.L. 2017. Photosynthetic acclimation to elevated CO_2 combined with partial rootzone drying results in improved water use efficiency, drought tolerance and leaf carbon balance of grapevines (*Vitis labrusca*). *Environmental and Experimental Botany* **134**: 82–95.

Dias de Oliveira, E.A., Siddique, K.H.M., Bramley, H., Stefanova, K., and Palta, J.A. 2015. Response of wheat restricted-tillering and vigorous growth traits to variables of climate change. *Global Change Biology* **21**: 857–73.

Drake, B.G., Gonzalez-Meler, M.A., and Long, S.P. 1997. More efficient plants: a consequence of rising atmospheric CO_2? *Annual Review of Plant Physiology and Plant Molecular Biology* **48**: 609–39.

Flexas, J., Bota, J., Loreto, F., Cornic, G., and Sharkey, T.D. 2004. Diffusive and metabolic limitations to photosynthesis under drought and salinity in C_3 plants. *Plant Biology* **6**: 269–79.

Koutavas, A. 2013. CO_2 fertilization and enhanced drought resistance in Greek firs from Cephalonia Island, Greece. *Global Change Biology* **19**: 529–39.

Kumar, A., Nayak, A.K., Sah, R.P., Sanghamitra, P., and Das, B.S. 2017. Effects of elevated CO_2 concentration on water productivity and antioxidant enzyme activities of rice (*Oryza sativa* L.) under water deficit stress. *Field Crops Research* **212**: 61–72.

Lee, S.H., Woo, S.Y., and Je, S.M. 2015. Effects of elevated CO_2 and water stress on physiological responses of *Perilla frutescens* var. *japonica* HARA. *Journal of Plant Growth Regulation* **75**: 427–34.

Long, S.P., Ainsworth, E.A., Rogers, A., and Ort, D.R. 2004. Rising atmospheric carbon dioxide: plants FACE the future. *Annual Review of Plant Biology* **55**: 591–628.

Naudts, K., van den Berge, J., Farfan, E., Rose, P., AbdElgawad, H., Ceulemans, R., Janssens, I.A., Asard, H., and Nijs, I. 2014. Future climate alleviates stress impact on grassland productivity through altered antioxidant capacity. *Environmental and Experimental Botany* **99**: 150–8.

Nilsen, E.T. and Orcutt, D.M. 1996. *The Physiology of Plants Under Stress*. New York, NY: Wiley.

NIPCC. 2014. Nongovernmental International Panel on Climate Change. Idso, C.D, Idso, S.B., Carter, R.M., and Singer, S.F. (Eds.) *Climate Change Reconsidered II: Biological Impacts*. Chicago, IL: The Heartland Institute.

Oliveira, V.F., Silva, E.A., and Carvalho, M.A.M. 2016. Elevated CO_2 atmosphere minimizes the effect of drought on the Cerrado species *Chrysolaena obovata*. *Frontiers in Plant Science* **7**: 810.

Pérez-Jiménez, M., Hernández-Munuera, M., Piñero, M.C., López-Ortega, G., and del Amor, F.M. 2017. CO_2 effects on the waterlogging response of 'Gisela 5' and 'Gisela 6' (*Prunus cerasus* x *Prunus canescens*) sweet cherry (*Prunus avium*) rootstocks. *Journal of Plant Physiology* **213**: 178–87.

Reddy, A.R., Rasineni, G.K., and Raghavendra, A.S. 2010. The impact of global elevated CO_2 concentration on photosynthesis and plant productivity. *Current Science* **99**: 1305–19.

Roy, J., *et al.* 2016. Elevated CO_2 maintains grassland net carbon uptake under a future heat and drought extreme. *Proceedings of the National Academy of Sciences USA* **113**: 6224–9.

Schmid, I., Franzaring, J., Müller, M., Brohon, N., Calvo, O.C., Högy, P., and Fangmeier, A. 2016. Effects of CO_2 enrichment and drought on photosynthesis, growth and yield of an old and a modern barley cultivar. *Journal of Agronomy and Crop Science* **202**: 81–95.

Sekhar, K.M., Reddy, K.S., and Reddy, A.R. 2017. Amelioration of drought-induced negative responses by elevated CO_2 in field grown short rotation coppice mulberry (*Morus* spp.), a potential bio-energy tree crop. *Photosynthesis Research* **132**: 151–64.

Song, Y. and Huang, B. 2014. Differential effectiveness of doubling ambient atmospheric CO_2 concentration mitigating adverse effects of drought, heat, and combined stress in Kentucky Bluegrass. *Journal of the American Society of Horticultural Science* **139**: 364–73.

van der Kooi, C.J., Reich, M., Low, M., De Kok, L.J., and Tausz, M. 2016. Growth and yield stimulation under elevated CO_2 and drought: a meta-analysis on crops. *Environmental and Experimental Botany* **122**: 150–7.

Wang, Y., Yan, D., Wang, J., Sing, Y., and Song, X. 2017. Effects of elevated CO_2 and drought on plant physiology, soil carbon and soil enzyme activities. *Pedosphere* **27**: 846–55.

Wijewardana, C., Henry, W.B., Gao, W., and Reddy, K.R. 2016. Interactive effects of CO_2, drought, and ultraviolet-B radiation on maize growth and development. *Journal of Photochemistry & Photobiology, B: Biology* **160**: 198–209.

Zong, Y. and Shangguan, Z. 2014. CO_2 enrichment improves recovery of growth and photosynthesis from drought and nitrogen stress in maize. *Pakistan Journal of Botany* **46**: 407–15.

5.3.4 Water Use Efficiency

Exposure to elevated levels of atmospheric CO_2 prompts plants to increase the efficiency of their use of water, enabling them to grow and reproduce where it previously has been too dry for them to exist.

Another major environmental benefit typically ignored in climate change cost-benefit analyses is a CO_2-induced enhancement of plant water use efficiency, the ratio of water used in plant metabolism to water lost by the plant through transpiration. Numerous studies have confirmed that plants exposed to elevated levels of atmospheric CO_2 generally do not open their leaf stomatal pores – through which they take in air (including CO_2) and emit oxygen and water vapor (transpiration) – as wide as they do at lower CO_2 concentrations (Morison, 1985; NIPCC, 2014). In addition, they sometimes produce fewer of these pores per unit area of leaf surface at higher CO_2 levels (Woodward, 1987). Both of these changes have the effect of reducing rates of plant water loss by transpiration (Overdieck and Forstreuter, 1994); and the amount of carbon they gain per unit of water lost – or water-use efficiency – therefore typically rises (Rogers *et al.*, 1983; NIPCC, 2014), greatly increasing their ability to withstand drought (Tuba *et al.*, 1998).

As the atmosphere's CO_2 content continues to rise, plants will be able to grow and reproduce where it has previously been too dry for them to exist (Johnson *et al.*, 1997). Consequently, terrestrial vegetation should become more robust and begin to win back lands previously lost to desertification (Idso and Quinn, 1983). The greater vegetative cover of the land produced by this phenomenon should reduce the adverse effects of soil erosion caused by wind and rain.

Section 5.3.4.1 summarizes recent research on the impact of elevated CO_2 and higher surface temperatures on water use efficiency by crops and Section 5.3.4.2 reviews research on trees. Section 5.3.5 addresses the *future* impact of elevated CO_2 and higher temperatures on plants.

References

Idso, S.B. and Quinn, J.A. 1983. Vegetational redistribution in Arizona and New Mexico in response to a doubling of the atmospheric CO_2 concentration. Tempe, AZ: Laboratory of Climatology, Arizona State University.

Johnson, J.D., Michelozzi, M., and Tognetti, R. 1997. Carbon physiology of *Quercus pubescens* Wild. growing at the Bossoleto CO_2 springs in central Italy. In: Raschi, A., Miglietta, F., Tognetti, R., and van Gardingen, P.R. (Eds.) *Plant Responses to Elevated* CO_2: *Evidence from Natural Springs*. New York, NY: Cambridge University Press, pp. 148–64.

Morison, J.I.L. 1985. Sensitivity of stomata and water use efficiency to high CO_2. *Plant, Cell and Environment* **8**: 467–74.

NIPCC. 2014. Nongovernmental International Panel on Climate Change. Idso, C.D, Idso, S.B., Carter, R.M., and Singer, S.F. (Eds.) *Climate Change Reconsidered II: Biological Impacts*. Chicago, IL: The Heartland Institute.

Overdieck, D. and Forstreuter, M. 1994. Evapotranspiration of beech stands and transpiration of beech leaves subject to atmospheric CO_2 enrichment. *Tree Physiology* **14**: 997–1003.

Rogers, H.H., Bingham, G.E., Cure, J.D., Smith, J.M., and Surano, K.A. 1983. Responses of selected plant species to elevated carbon dioxide in the field. *Journal of Environmental Quality* **12**: 569–74.

Tuba, Z., Csintalan, Z., Szente, K., Nagy, Z., and Grace, J. 1998. Carbon gains by desiccation-tolerant plants at elevated CO_2. *Functional Ecology* **12**: 39–44.

Woodward, F.I. 1987. Stomatal numbers are sensitive to increase in CO_2 from pre-industrial levels. *Nature* **327**: 617–8.

5.3.4.1 Agriculture

At the turn of the century, Wallace (2000) wrote that the projected increase in global population in the coming half-century (a median best-guess of 3.7 billion) is more certain to occur than is any other environmental change currently underway or reasonably anticipated. This population increase would require an increase in the food supply, which would require an increase in water for agriculture. "Over the entire globe," Wallace warned, "a staggering 67% of the future population of the world may experience some water stress," which would likely lead to food insufficiency, malnutrition, and starvation. The best response is to produce much more food per unit of available water – an improvement in plant water use efficiency that is already underway as atmospheric CO_2 levels rise.

Noting "looming water scarcity and climate change pose big challenges for China's food security," Zhao *et al*. (2014) state "previous studies have focused on the impacts of climate change either on agriculture or on water resources," while "few studies have linked water and agriculture together in the context of climate change and demonstrated how climate change will affect the amount of water used to produce per unit of crop, or virtual water content (VWC)."

Unlike the experiment-based studies reported in previous sections and below, Zhao *et al*. used a GIS-based Environmental Policy Integrated Climate (GEPIC) model to analyze the current spatial distribution of the VWC of various crops in China and the impacts of climate change on VWC in different future scenarios. They report "three general change trends exist for future VWCs of crops: continuous decline (for soybean and rice without considering CO_2 concentration changes) and continuous increase (for rice with considering CO_2 concentration changes) and first-decline-then-increase (other crop-scenario combinations)." They say the "integrated effects of precipitation, temperature and CO_2 concentration changes will benefit agricultural productivity and crop water productivity through all the future periods till the end of the century." Zhao *et al*. conclude, "climate change is likely to benefit food security and help alleviate water scarcity in China."

Pazzagli *et al*. (2016) grew two tomato (*Solanum lycopersicum*) cultivars – one potentially drought tolerant (ST 22) and one thought to be heat tolerant (ST 52) – in a controlled greenhouse environment from March to June 2014. The plants were subjected to three irrigation regimes (full irrigation, deficit irrigation, and partial root-zone drying) and two atmospheric CO_2 concentrations (380 and 590 ppm). Statistical analyses indicated there was a significant CO_2 effect on both cultivars for net photosynthetic rate, intrinsic water use efficiency (WUEi, photosynthetic rate/stomatal conductance), plant water use efficiency (WUEp, aboveground biomass/plant water use), root water potential, stem

dry weight, leaf dry weight, total dry weight, and flower number. They write, "despite large differences between the cultivars, both of them showed significant improvements in plant water use efficiency under both reduced irrigation and CO_2 enrichment, as well as under the combination of the two treatments."

Cruz *et al*. (2016) studied the effects of CO_2 enrichment on cassava, an important food staple whose tuberous roots are the third-largest source of carbohydrates in tropical regions, after rice and maize. The plant is drought-tolerant and "even under adverse soil and climatic conditions, cassava can produce a satisfactory root yield, while other annual crops barely survive (El-Sharkawy and Cock, 1987)." To "help understand the interaction between elevated CO_2 levels and water deficit on growth, physiology and dry mass accumulation in cassava," they grew two- to three-month-old cassava plantlets in a climate-controlled greenhouse for 100 days at two CO_2 concentrations (390 or 750 ppm) for 12 hours per day and two water treatments (well-watered and water-stressed).

Cruz *et al*. found "water deficits led to reductions in the Leaf Elongation Rate of plants grown at ambient as well as CO_2-enriched concentrations," but "plants grown at 750 ppm of CO_2 maintained leaf growth two days longer than plants grown at 390 ppm." They further noted "three days after withholding water, photosynthesis and stomatal conductance were reduced in plants grown under ambient CO_2, while in plants under an elevated CO_2 concentration, these physiological functions remained similar to that of control plants grown under good water availability."

Continuing, Cruz *et al*. report "five days after withholding water plants grown with 750 ppm continued to have enhanced gas exchange compared with plants grown under 390 ppm." Moreover, "under drought stress, the instantaneous transpiration efficiency was always greatest for plants grown under elevated CO_2." They also found "the positive response of elevated CO_2 levels on total dry mass was 61% in the water-stressed plants and only 20% for the plants grown under good water availability."

Deryng *et al*. (2016) combined the results obtained for networks of field experiments and global crop models in order to derive a global perspective on crop water productivity (CWP) –the ratio of crop yield to evapotranspiration – for maize, rice, soybeans, and wheat under elevated CO_2. They report, "the projected increase in the air's CO_2 concentration would likely increase global CWP by

10–27% by the 2080s, with particularly large increases in arid regions (by up to 48%, for example, in the case of rain-fed wheat)." They add, "if realized in the fields," the effects of elevated CO_2 could considerably mitigate global yield losses while reducing agricultural consumptive water use by 4% to 17%. They conclude their findings "quantify the importance of CO_2 effects on potential water savings and, in so doing, highlight key limitations of global hydrological models that do not consider effects of CO_2 on evapotranspiration."

Kumar *et al.* (2017) grew rice in naturally sunlit and irrigated (flooded) controlled-environment chambers for two dry and two wet seasons. CO_2 treatments in the chambers included 195, 390, 780, and 1,560 ppm and were maintained at these levels for 22 hours a day. The chambers were flushed of air for one hour before dawn and one hour after dusk to flush out trace gases. Kumar *et al.* report "the current level of 390 ppmv [CO_2] was distinctly sub-optimal for rice biomass production." The mean aboveground dry weight across all seasons was 1,744 g m^{-2} at the current or ambient CO_2 level (390 ppm), which value "decreased by 43% at 195 ppmv (0.5 x ambient), increased by 29% at 780 ppmv (2 x ambient) and increased by 42% at 1560 ppmv (4 x ambient)."

With respect to water use, Kumar *et al.* report whole-season crop water use under sub-ambient and current CO_2 conditions was 564 and 719 mm, rising to 928 and 803 mm at 780 and 1,560 ppmv CO_2, respectively. Although more water was used at the higher CO_2 concentrations, the amount of biomass produced per mm of water also increased (1.76, 2.43, 2.43, and 3.08 g m^{-2} mm^{-1} at 195, 390, 780 and 1,560 ppm CO_2, respectively).

Singh *et al.* (2017) examined the combined effects of elevated CO_2 and elevated ozone on the growth, biomass, and water use efficiency of chickpea (*Cicer arietinum*). Their experiment was conducted outdoors at a Free Air Ozone and Carbon Dioxide Enrichment (FAOCE) facility in New Delhi, India. Forty days after sowing, chickpea plants (cv. Pusa 5023) were subjected to one of four atmospheric treatments daily until the end of the growing season: normal air (AMB = 400 ppm CO_2, 30 ppb O_3), elevated ozone (EO = 400 ppm CO_2, 70 ppb O_3), elevated CO_2 (EC = 550 ppm CO_2, 30 ppb O_3), and elevated CO_2 and ozone (ECO = 550 ppm CO_2, 70 ppb O_3). Several growth and development parameters were measured or calculated from measurements, including crop phenology, plant height, aboveground biomass, crop growth rate, relative growth rate, and water use efficiency.

It was anticipated that elevated concentrations of ozone – a plant stressor – would negatively impact chickpea growth, whereas elevated concentrations of carbon dioxide – a plant nutrient – would enhance it. Singh *et al.* report both elevated CO_2 and elevated O_3 advanced plant phenological development, shortening the growth period by about 10 days and 14 days, respectively, compared to ambient conditions, albeit due to different mechanisms: Elevated CO_2 sped up the development and likely induced earlier senescence, whereas elevated O_3 damaged leaf chlorophyll content and nutrient status to enhance senescence.

Singh *et al.* also report chickpea plant height, growth rate, aboveground biomass, seed yield, and water use efficiency benefited from the approximate 37% increase in atmospheric CO_2. In contrast, these parameters were negatively impacted by elevated O_3 concentrations. When in combination, the positive effects of elevated CO_2 were strong enough to completely ameliorate the negative impacts of elevated O_3. Compared to ambient conditions, for example, seed yield was enhanced by 32% in the EC treatment, reduced by 22% in the EO treatment and increased by 10% in the ECO treatment. Similarly, water use efficiency increased by 44% in the EC treatment, declined by 22% in the EO treatment and experienced a 5% increase in the ECO treatment.

Wang *et al.* (2017) examined the interactive effects of elevated CO_2 and drought on soybean (*Glycine max*, cv. Yu 19), growing plants from seed for 40 days in controlled-environment greenhouses under ambient and twice ambient CO_2 concentrations and three water regimes: well-watered (80% water holding capacity of the soil), moderate drought (60% water holding capacity), and severe drought (40% water holding capacity). They found drought negatively impacted the net photosynthesis of the soybean plants, which declined by 52% and 23% in comparing the well-watered to the severe drought treatment under ambient and elevated CO_2 conditions, respectively. The positive influence of elevated CO_2 was so great that even under severe drought conditions, the net photosynthetic rate was 73% greater than that observed under well-watered conditions at ambient CO_2 (Figure 5.3.4.1.1, left panel). Water use efficiency also was enhanced by elevated CO_2 (right panel), where it was "almost 2.5 times larger than that under ambient CO_2."

Wang *et al.* also report elevated CO_2 increased soil enzyme activities and "resulted in a longer retention time of dissolved organic carbon (DOC) in [the] soil, probably by improving the soil water

Figure 5.3.4.1.1
Net photosynthesis (Np) and water use efficiency (WUE) of soybean plants grown under various treatments of drought

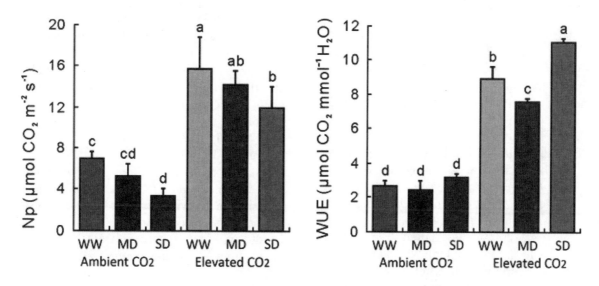

WW = well-watered; MD = moderate drought; SD = severe drought) and atmospheric CO_2 (ambient and elevated, where elevated = twice ambient). *Source*: Wang *et al.*, 2017

effectiveness for organic decomposition and mineralization." They conclude "drought stress had significant negative impacts on plant physiology, soil carbon, and soil enzyme activities, whereas elevated CO_2 and plant physiological feedbacks indirectly ameliorated these impacts."

References

Cruz, J.L., Alves, A.A.C., LeCain, D.R., Ellis, D.D., and Morgan, J.A. 2016. Elevated CO_2 concentrations alleviate the inhibitory effect of drought on physiology and growth of cassava plants. *Scientia Horticulturae* **210**: 122–9.

Deryng, D., *et al.* 2016. Regional disparities in the beneficial effects of rising CO_2 concentrations on crop water productivity. *Nature Climate Change* **6**: 786–90.

El-Sharkawy, M.A. and Cock, J.H. 1987. Response of cassava to water stress. *Plant and Soil* **100**: 345–60.

Kumar, U., Quick, W.P., Barrios, M., Sta Cruz, P.C., and Dingkuhn, M. 2017. Atmospheric CO_2 concentration effects on rice water use and biomass production. *PLoS ONE* **12**: e0169706.

Pazzagli, P.T., Weiner, J., and Liu, F. 2016. Effects of CO_2 elevation and irrigation regimes on leaf gas exchange, plant water relations, and water use efficiency of two tomato cultivars. *Agricultural Water Management* **169**: 26–33.

Singh, R.N., Mukherjee, J., Sehgal, V.K., Bhatia, A., Krishnan, P., Das, D.K., Kumar, V., and Harit, R. 2017. Effect of elevated ozone, carbon dioxide and their inter action on growth, biomass and water use efficiency of chickpea (*Cicer arietinum* L.). *Journal of Agrometeorology* **19**: 301–5.

Wallace, J.S. 2000. Increasing agricultural water use efficiency to meet future food production. *Agriculture, Ecosystems & Environment* **82**: 105–19.

Wang, Y., Yan, D., Wang, J., Sing, Y., and Song, X. 2017. Effects of elevated CO_2 and drought on plant physiology, soil carbon and soil enzyme activities. *Pedosphere* **27**: 846–55.

Zhao, Q., Liu, J., Khabarov, N., Obersteiner, M., and Westphal, M. 2014. Impacts of climate change on virtual water content of crops in China. *Ecological Informatics* **19**: 26–34.

5.3.4.2 Trees

Extensive research documents how elevated CO_2 levels improve water use efficiency by trees, enabling them to better withstand droughts and other changes in precipitation that may accompany climate change. This bodes well both for forestry and preservation of wildlife habitat.

Wang *et al.* (2012) note "empirical evidence from lab studies with a controlled CO_2 concentration and from free-air CO_2 enrichment (FACE) experiments have revealed significantly increased iWUE [intrinsic water-use efficiency] in response to rising CO_2," as demonstrated by the studies of Luo *et al.* (1996), Ainsworth and Rogers (2007), and Niu *et al.* (2011). They also note "tree-ring stable carbon isotope ratios ($\delta^{13}C$) have proven to be an effective tool for evaluating variations in iWUE around the world," citing Farquhar *et al.* (1989), Saurer *et al.* (2004), Liu *et al.* (2007), and Andreu *et al.* (2011). Working at a site in the Xinglong Mountains in the eastern part of northwestern China, Wang *et al.* extracted two cores from the trunks of each of 17 dominant living Qinghai spruce (*Picea crassifolia*) trees, from which they obtained ring-width measurements they used to calculate yearly mean basal area growth increments. Thereafter they used subsamples of the cores to conduct the analyses needed to obtain the $\delta^{13}C$ data required to calculate iWUE over the period 1800–2009. By calibrating the $\delta^{13}C$ data against climatic data obtained at the nearest weather station over the period 1954–2009, they were able to extend the histories of major meteorological parameters back to 1800. By comparing these weather data with the tree growth and water use efficiency data, they were able to interpret the impacts of climate change and atmospheric CO_2 enrichment on spruce tree growth and water use efficiency.

Wang *et al.* determined iWUE increased by approximately 40% between 1800 and 2009, rising very slowly for the first 150 years, but then more rapidly to about 1975 and then faster still until 1998, whereupon it leveled off for the remaining 11 years of the record. They say the main cause of the increasing trend in iWUE from 1800 to 1998 "is likely to be the increase in atmospheric CO_2," because "regression analysis suggested that increasing atmospheric CO_2 explained 83.0% of the variation in iWUE from 1800 to 1998 ($p<0.001$)."

Battipaglia *et al.* (2013) combined tree-ring analyses with carbon and oxygen isotope measurements made at three FACE sites to assess changes in water-use efficiency and stomatal conductance. They found elevated CO_2 increased water-use efficiency on average by 73% for sweetgum (*Liquidambar styraciflua*, +200 ppm CO_2), 77% for loblolly pine (*Pinus taeda*, +200 ppm CO_2), and 75% for poplar (*Populus sp.*, +153 ppm CO_2). They say their findings provide "a robust means of predicting water-use efficiency responses from a variety of tree species exposed to variable environmental conditions over time and species-specific relationships that can help modeling elevated CO_2 and climate impacts on forest productivity, carbon and water balances."

Keenan *et al.* (2013) documented and analyzed recent trends in the water-use efficiencies (Wei) of forest canopies, using direct and continuous long-term measurements of CO_2 and water vapor fluxes, focusing on seven sites in the midwestern and northeastern United States. They compared their results with those derived from data obtained by others from 14 additional temperate and boreal forest sites.

Keenan *et al.* found "a substantial increase in water-use efficiency in temperate and boreal forests of the Northern Hemisphere over the past two decades." They determined "the observed increase is most consistent with a strong CO_2 fertilization effect," because, as they note, "of all the potential drivers of the observed changes in Wei, the only driver that is changing sufficiently and consistently through time at all sites is atmospheric CO_2."

Keenan *et al.* additionally note "the direct tradeoff between water loss and carbon uptake through the stomata means that, as water-use efficiency increases, either evapotranspiration decreases or gross photosynthetic carbon uptake increases, or both occur simultaneously." They write "increases in Wei may account for reports of global increases in photosynthesis (Nemani *et al.*, 2003), forest growth rates (Lewis *et al.*, 2009; Salzer *et al.*, 2009; McMahon *et al.*, 2010), and carbon uptake (Ballantyne *et al.*, 2012)," leading them to suggest "rising atmospheric CO_2 is having a direct and unexpectedly strong influence on ecosystem processes and biosphere-atmosphere interactions in temperate and boreal forests."

Soulé and Knapp (2015) collected tree-ring data from ponderosa pine (*Pinus ponderosa* var. *ponderosa* - PIPO) and Douglas fir (*Pseudotsuga menziesii* var. *glauca* - PSME) at 14 locations, from which they determined yearly changes (from AD 1850 to the present) in basal area index (BAI) and intrinsic water use efficiency (iWUE). They determined both PIPO and PSME trees experienced "exponentially increasing iWUE rates during AD 1850–present, suggesting either increased net photosynthesis or decreased stomatal conductance, or both" (Figure 5.3.4.2.1, upper panel). They add "both species experienced above-average BAI in the latter half of the 20th century despite no favorable changes in climate" (lower panel), further noting "this response occurred at all sites, suggesting a pan-regional effect."

Working with four native tree species of China (*Schima superba, Ormosia pinnata, Castanopsis hystrix* and *Acmena acuminatissima*) from January 2006 to January 2010, Li *et al.* (2015) studied the effects of an approximate 300 ppm increase in the air's CO_2 concentration on the trees' WUE, which they did within open-top chambers exposed to full light and rain out-of-doors, either with (CN) or without (CC) added nitrogen fertilization. They found, compared to the control, the average increased extents of intrinsic WUE were 98% and 167% in CC and CN treatments for *S. superba*; 88% and 74% for *O. pinnata*; 234% and 194% for *C. hystrix*; and 153% and 81% for *A. acuminatissima*.

Ghini *et al.* (2015) conducted an experiment to observationally determine the response of two coffee cultivars to elevated levels of atmospheric CO_2 in the first FACE facility in Latin America. Small specimens of two coffee cultivars, Catuaí and Obatã, were sown in the field under ambient (~390 ppm) and enriched (~550 ppm) CO_2 conditions in August 2011 and allowed to grow under normal cultural growing conditions without supplemental irrigation for two years. No significant effect of CO_2 was observed on

Figure 5.3.4.2.1
Mean tree-ring iWUE values and basal-area index values for Douglas fir and Ponderosa Pine trees, 1850–2005

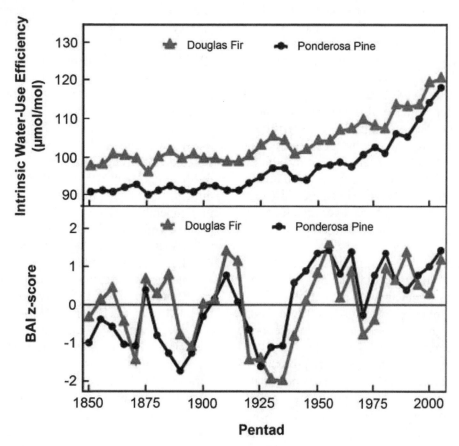

Source: Adapted from Soulé and Knapp, 2015.

the growth parameters during the first year. However, during the growing season of Year 2, net photosynthesis increased by 40% and plant water use efficiency by approximately 60%, regardless of cultivar. During the winter, when growth was limited, daily mean net photosynthesis "averaged 56% higher in the plants treated with CO_2 than in their untreated counterparts."

WUE in winter also was significantly higher (62% for Catuaí and 85% for Obatã). Such beneficial impacts resulted in significant CO_2-induced increases in plant height, stem diameter, and harvestable yield over the course of Year 2. Ghini et al. report the increased crop yield "was associated with an increased number of fruits per branch, with no differences in fruit weight."

Working in southern Chile, Urrutia-Jalabert et al. (2015) performed a series of analyses on tree-ring cores they obtained from long-lived cypress (Fitzroya cupressoides) stands, which they say "may be the slowest-growing and longest-lived high biomass forest stands in the world." Focusing on two of the more pertinent findings of their study, both the BAI and iWUE of Fitzroya experienced dramatic increases over the past century. The authors write, "the sustained positive trend in tree growth is striking in this old stand, suggesting that the giant trees in this forest have been accumulating biomass at a faster rate since the beginning of the [20th] century." Coupling that finding with the 32% increase in water use efficiency over the same time period, Urrutia-Jalabert et al. state "we believe that this increasing growth trend … has likely been driven by some combination of CO_2 and/or surface radiation increases," adding that "pronounced changes in CO_2 have occurred in parallel with changes in climate, making it difficult to distinguish between both effects."

Carles et al. (2015) subjected white spruce (Picea glauca) seedlings to a combination of two temperature regimes (ambient and ambient plus 5°C) and two levels of atmospheric CO_2 (380 and 760 ppm) over two growing seasons. They report "warmer temperatures and CO_2 elevation had a positive effect on the height and diameter growth of 2- and 3-year-old seedlings …" They also report that water use efficiency was "affected positively by the CO_2 treatment, showing a 51% increase that was consistent across families."

Wils et al. (2016) studied cores or discs extracted from five African juniper (Juniperus procera) trees of Gondar, Ethiopia, and one from the Hugumburda forest on the north-western escarpment of the

Ethiopian Rift Valley, along with discs obtained from a Mimusops caffra tree growing in South Africa's KwaZulu-Natal and an Acacia erioloba growing in the Koichab Valley of Namibia. They report, "tree-ring intrinsic water-use efficiency (iWUE) records for Africa show a 24.6% increase over the 20th century." Because a high iWUE can partly counterbalance decreases in precipitation, Wils et al. conclude this finding "has important implications for those involved in water resource management and highlights the need for climate models to take physiological forcing into account." They note "the 24.6% increase in mean iWUE confirms that African trees are already adapting to increasing atmospheric CO_2 concentrations."

Huang et al. (2017) examined the relationship between BAI and iWUE indices derived from cores of Smith fir trees (also known as Yunnan fir) (Abies georgei var. smithii) growing at a high-elevation timberline site in the southeastern Tibetan Plateau, rising atmospheric CO_2 concentration, and climate. They hypothesized "if intrinsic water use efficiency ... has increased due to rising net photosynthetic rates under rising atmospheric CO_2 concentration over the past century, tree growth should have benefitted." They found iWUE rose by 27.83% over the period 1900 to 2006. They also report "the increasing iWUE is mainly caused by the rising atmospheric CO_2 concentration," and "iWUE would continue to increase in the near future."

Huang et al. note there also has been a strong increasing trend in BAI over the past century and conclude that trend is also largely driven by the aerial fertilization effect of atmospheric CO_2, being highly influenced in the short term by interannual variations in temperature. They report finding "a significant positive correlation ($r = 0.79$, $p < 0.01$) between BAI and iWUE," which they say indicates "changes in iWUE and tree growth were likely to have had a common cause, i.e., the CO_2 fertilization effect."

Choury et al. (2017) analyzed long-term trends in the BAI and WUEi of native Aleppo pines (Pinus halepensis Mill.) growing near the northern border of the Sahara Desert. They cored multiple trees from three locations so as to evaluate trends over the period 1925–2013, during which period mean annual temperatures rose by 1.5°C and atmospheric CO_2 concentrations rose by approximately 30%. They report "the BAI patterns of natural Aleppo pine stands did not show a decreasing trend over the last century, indicating that warming-induced drought stress has not significantly affected secondary growth of pines in the area; instead, BAI trends were stable or even showed a significant increase in the case of the North slope site." Similar results were noted for the trees' WUEi, which "increased by ca. 39% across sites between 1925 and 2013." Choury *et al.* conclude their study "highlights the substantial plasticity of Aleppo pine to warming-induced drought stress," adding, "the extent of such plastic responses for Aleppo pines growing at the southernmost limit of the species distribution area is, from a physiological point of view, remarkable."

Giammarchi *et al.* (2017) assessed the changes in productivity of two similarly aged Norway pine (*Picea abies*) forests and then examined "the role of several environmental drivers, such as atmospheric CO_2 levels, temperature, and precipitation regimes on the intrinsic water-use efficiency (iWUE) temporal patterns of the above-mentioned forests." They found an increase in forest productivity at both sites since the 1860s, paralleled by a significant increase of iWUE, which they say was "mainly triggered by a CO_2-driven increase in photosynthetic capacity, rather than by a reduction of stomatal conductance."

Weiwei *et al.* (2018) cored *Platycladus orientalis* trees, an evergreen coniferous species endemic to China, to investigate trends in tree-ring carbon discrimination and iWUE over the past century. They found both iWUE and BAI have increased with time. Both variables were positively correlated with atmospheric CO_2 concentration, which findings, the authors say, "are consistent with other studies conducted on the effects of elevated CO_2 on leaf physiological activity, which demonstrate that increased CO_2 promotes water use efficiency."

References

Ainsworth, E.A. and Rogers, A. 2007. The response of photosynthesis and stomatal conductance to rising [CO_2]: mechanisms and environmental interactions. *Plant, Cell and Environment* **30**: 258–70.

Andreu, L., Planells, O., Gutierrez, E., Muntan, E., Helle, G., Anchukaitis, K.J., and Schleser, G.H. 2011. Long tree-ring chronologies reveal 20th century increases in water-use efficiency but no enhancement of tree growth at five Iberian pine forests. *Global Change Biology* **17**: 2095–112.

Ballantyne, A.P., Alden, C.B., Miller, J.B., Tans, P.P., and White, J.W.C. 2012. Increase in observed net carbon dioxide uptake by land and oceans during the past 50 years. *Nature* **488**: 70–2.

Battipaglia, G., Saurer, M., Cherubini, P., Calfapietra, C., McCarthy, H.R., Norby, R.J., and Cotrufo, M.F. 2013.

Elevated CO_2 increases tree-level intrinsic water use efficiency: insights from carbon and oxygen isotope analyses in tree rings across three forest FACE sites. *New Phytologist* **197**: 544–54.

Carles, S., Groulx, D.B., Lamhamedi, M.S., Rainville, A., Beaulieu, J., Bernier, P., Bousquet, J., Deblois, J., and Margolis, H.A. 2015. Family variation in the morphology and physiology of white spruce (*Picea glauca*) seedlings in response to elevated CO_2 and temperature. *Journal of Sustainable Forestry* **34**: 169–98.

Choury, Z., Shestakova, T.A., Himrane, H., Touchan, R., Kherchouche, D., Camarero, J.J., and Voltas, J. 2017. Quarantining the Sahara desert: growth and water-use efficiency of Aleppo pine in the Algerian Green Barrier. *European Journal of Forest Research* **136**: 139–52.

Farquhar, G.D., Ehleringer, J.R., and Hubick, K.T. 1989. Carbon isotope discrimination and photosynthesis. *Annual Reviews of Plant Physiology and Plant Molecular Biology* **40**: 503–37.

Ghini, R., Torre-Neto, A., Dentzien, A.F.M., Guerreiro-Filho, O., Iost, R., Patrício, F.R.A., Prado, J.S.M., Thomaziello, R.A., Bettiol, W., and DaMatta, F.M. 2015. Coffee growth, pest and yield responses to free-air CO_2 enrichment. *Climatic Change* **132**: 307–20.

Giammarchi, F., Cherubini, P., Pretzsch, H., and Tonon, G. 2017. The increase of atmospheric CO_2 affects growth potential and intrinsic water-use efficiency of Norway spruce forests: insights from a multi-stable isotope analysis in tree rings of two Alpine chronosequences. *Trees* **31**: 503–15.

Huang, R., Zhu, H., Liu, X., Liang, E., Griebinger, J., Wu, G., Li, X., and Bräuning, A. 2017. Does increasing intrinsic water use efficiency (iWUE) stimulate tree growth at natural alpine timberline on the southeastern Tibetan Plateau? *Global and Planetary Change* **148**: 217–26.

Keenan, T.F., Hollinger, D.Y., Bohrer, G., Dragoni, D., Munger, J.W., Schmid, H.P., and Richardson, A.D. 2013. Increase in forest water-use efficiency as atmospheric carbon dioxide concentrations rise. *Nature* **499**: 324–7.

Lewis, S.L., *et al.* 2009. Increasing carbon storage in intact African tropical forests. *Nature* **457**: 1003–6.

Li, Y., Liu, J., Chen, G., Zhou, G., Huang, W., Yin, G., Zhang, D., and Li, Y. 2015. Water-use efficiency of four native trees under CO_2 enrichment and N addition in subtropical model forest ecosystems. *Journal of Plant Ecology* **8**: 411–9.

Liu, X.H., Shao, X.M., Liang, E.Y., Zhao, L.J., Chen, T., Qin, D.H., and Ren, J.W. 2007. Species-dependent responses of juniper and spruce to increasing CO_2 concentration and to climate in semi-arid and arid areas of northwestern China. *Plant Ecology* **193**: 195–209.

Luo, Y., Sims, D.A., Thomas, R.B., Tissue, D.T., and Ball, J.T. 1996. Sensitivity of leaf photosynthesis to CO_2 concentration is an invariant function for C_3 plants: a test with experimental data and global applications. *Global Biogeochemical Cycles* **10**: 209–22.

McMahon, S.M., Parker, G.G., and Miller, D.R. 2010. Evidence for a recent increase in forest growth. *Proceedings of the National Academy of Sciences USA* **107**: 3611–15.

Nemani, R.R., Keeling, C.D., Hashimoto, H., Jolly, W.M., Piper, S.C., Tucker, C.J., Myneni, R.B., and Running, S.W. 2003. Climate-driven increases in global terrestrial net primary production from 1982 to 1999. *Science* **300**: 1560–3.

Niu, S., Xing, X., Zhang, Z., Xia, J., Zhou, X., Song, B., Li, L., and Wan, S. 2011. Water-use efficiency in response to climate change: from leaf to ecosystem in a temperate steppe. *Global Change Biology* **17**: 1073–82.

Salzer, M., Hughes, M., Bunn, A., and Kipfmueller, K. 2009. Recent unprecedented tree-ring growth in bristlecone pine at the highest elevations and possible causes. *Proceedings of the National Academy of Sciences USA* **106**: 20,346–53.

Saurer, M., Siegwolf, R.T.W., and Schweingruber, F.H. 2004. Carbon isotope discrimination indicates improving water-use efficiency of trees in northern Eurasia over the last 100 years. *Global Change Biology* **10**: 2109–20.

Soulé, P.T. and Knapp, P.A. 2015. Analyses of intrinsic water-use efficiency indicate performance differences of ponderosa pine and Douglas-fir in response to CO_2 enrichment. *Journal of Biogeography* **42**: 144–55.

Urrutia-Jalabert, R., Malhi, Y., Barichivich, J., Lara, A., Delgado-Huertas, A., Rodríguez, C.G., and Cuq, E. 2015. Increased water use efficiency but contrasting tree growth patterns in *Fitzroya cupressoides* forests of southern Chile during recent decades. *Journal of Geophysical Research, Biogeosciences* **120**: 2505–24.

Wang, W., Liu, X., An, W., Xu, G., and Zeng, X. 2012. Increased intrinsic water-use efficiency during a period with persistent decreased tree radial growth in northwestern China: causes and implications. *Forest Ecology and Management* **275**: 14–22.

Weiwei, L.U., Xinxiao, Y.U., Guodong, J.I.A., Hanzhi, L.I., and Ziqiang, L.I.U. 2018. Responses of intrinsic

water-use efficiency and tree growth to climate change in semi-arid areas of north China. *Scientific Reports* **8**: 308.

Wils, T.H.G., Robertson, I., Woodborne, S., Hall, G., Koprowski, M., and Eshetu, Z. 2016. Anthropogenic forcing increases the water-use efficiency of African trees. *Journal of Quaternary Science* **31**: 386–90.

5.3.5 Future Impacts on Plants

The productivity of the biosphere is increasing in large measure due to the aerial fertilization effect of rising atmospheric CO_2. The benefits of CO_2 enrichment will continue even if atmospheric CO_2 rises to levels far beyond those forecast by the IPCC.

Atmospheric CO_2 enrichment is boosting biospheric productivity around the world, but will it continue to do so in coming decades and centuries? Extensive research has been conducted on the possible effects on plants of elevated CO_2 levels and higher temperatures in the future. In Chapter 4 of *Climate Change Reconsidered II: Biological Impacts* (NIPCC, 2014), NIPCC presented a thorough literature review of the subject, finding ample support for its conclusion that plants will flourish if temperatures and CO_2 levels rise in the future. The key findings of that chapter are summarized in Figure 5.3.6.1.

This section updates the literature review that appeared in *Biological Impacts*. Section 5.3.5.1 summarizes new research (most of it published since 2014) on the impacts of rising atmospheric CO_2 concentrations and temperatures on plants important for food production; Section 5.3.5.2 addresses future biospheric productivity; Section 5.3.5.3 addresses future biodiversity; Section 5.3.5.4 addresses future extinction; and Section 5.3.5.5 addresses future evolution.

Reference

NIPCC. 2014. Nongovernmental International Panel on Climate Change. Idso, C.D, Idso, S.B., Carter, R.M., and Singer, S.F. (Eds.) *Climate Change Reconsidered II: Biological Impacts*. Chicago, IL: The Heartland Institute.

Figure 5.3.5.1
Key Findings: Impacts on Earth's vegetative future

- The vigor of Earth's terrestrial biosphere has been increasing with time, revealing a great post-industrial revolution greening of the Earth that extends across the entire globe. Over the past 50 years global carbon uptake has doubled from 2.4 ± 0.8 billion tons in 1960 to 5.0 ± 0.9 billion tons in 2010.

- The atmosphere's rising CO_2 content, which the IPCC considers to be the chief culprit behind all of its "reasons for concern" about the future of the biosphere, is most likely the primary cause of the observed greening trend.

- The observed greening of the Earth has occurred in spite of all the many real and imagined assaults on Earth's vegetation, including fires, disease, pest outbreaks, air pollution, deforestation, and climatic change. Rising levels of atmospheric CO_2 are making the biosphere more resilient to stress even as it becomes more lush and productive.

- Agricultural productivity in the United States and across the globe dramatically increased over the last three decades of the twentieth century, a phenomenon partly due to new cultivation techniques but also due partly to warmer temperatures and higher CO_2 levels.

- A future warming of the climate coupled with rising atmospheric CO_2 levels will further boost global agricultural production and help to meet the food needs of the planet's growing population.

- The positive direct effects of CO_2 on future crop yields are likely to dominate any hypothetical negative effects associated with changing weather conditions, just as they have during the twentieth and early twenty-first centuries.

- Plants have a demonstrated ability to adjust their physiology to accommodate a warming of both the magnitude and rate-of-rise typically predicted by climate models, should such a warming actually occur.

- Evidence continues to accumulate for substantial heritable variation of ecologically important plant traits, including root allocation, drought tolerance, and nutrient plasticity, which suggests rapid evolution is likely to occur based on epigenetic variation alone. The ongoing rise in the atmosphere's CO_2 content will exert significant selection pressure on plants, which can be expected to improve their performance in the face of various environmental stressors via the process of micro-evolution.

- As good as things currently are for world agriculture, natural selection and bioengineering could bring about additional beneficial effects. For example, highly CO_2-responsive genotypes of a wide variety of plants could be selected to take advantage of their genetic ability to optimize their growth in response to projected future increases in the atmosphere's CO_2 content.

Source: Chapter 4. "Earth's Vegetative Future," *Climate Change Reconsidered II: Biological Impacts.* Nongovernmental International Panel on Climate Change. Chicago, IL: The Heartland Institute, 2014.

5.3.5.1 Agriculture

The beneficial effects for agriculture of rising levels of CO_2 in the modern era were documented in detail in Chapter 3, Section 3.4 and earlier in this chapter in Section 5.2.2.3 and so do not need to be reported again here. But will those benefits continue? Agricultural species grown in elevated CO_2 environments often, but not always, at some point exhibit some degree of photosynthetic acclimation or down regulation, which is typically characterized by reduced rates of photosynthesis resulting from decreased activity and/or amount of rubisco, the primary plant carboxylating enzyme (Sims *et al.*, 1999; Gavito *et al.*, 2000; Ulman *et al.*, 2000).

Ziska (1998), for example, reported that soybeans grown at an atmospheric CO_2 concentration of 720 ppm initially exhibited photosynthetic rates 50% greater than those observed in control plants grown at 360 ppm. However, after the onset of photosynthetic acclimation, CO_2-enriched plants displayed subsequent photosynthetic rates only 30% greater than their ambiently grown counterparts. Nevertheless, in nearly every reported case of CO_2-induced photosynthetic acclimation, the reduced rates of photosynthesis displayed by CO_2-enriched plants are *greater* than those exhibited by plants growing at ambient CO_2 concentrations

Several studies have tried to estimate the effects on agriculture of temperatures and CO_2 concentrations forecast by the IPCC. Mariani (2017) utilized a physiological-process-based crop simulation model to estimate the change in food production under five temperature and CO_2 scenarios for four crops (wheat, maize, rice, and soybean) that account for two-thirds of total global human caloric consumption. The scenarios were identified as Today, Pre-Industrial, Glacial, Future_560, and Future_800, which correspond to respective atmospheric CO_2 concentrations of 400, 280, 180, 560, and 800 ppm, and temperatures that were -1 (Pre-Industrial), -6 (Glacial), +2 (Future_560), and +4 °C (Future_800) different from the Today scenario. The results are shown in Figure 5.3.5.1.1.

Mariani found a return to glacial period conditions would reduce global production of the four keystone crops by 51% while a return to pre-industrial conditions – the IPCC's declared objective -- would reduce food production by 18%. Looking ahead, Mariani estimates a world with double the pre-industrial level of CO_2 and temperatures 2°C higher than today's levels (Future_560) would witness food production 15% higher. A world where CO_2 levels were even higher (800 pmm) and temperatures were 4°C higher than today's levels (Future_800) would witness food production 24% above today's values.

Mariani writes, "the return of temperature and CO_2 to glacial or pre-industrial values would give rise to serious disadvantages for food security and should be as far as possible avoided, as also highlighted by the results of Sage and Coleman (2001) and Araus *et al.* (2003)."

Ruiz-Vera *et al.* (2017) write, "with the continuous increase of atmospheric CO_2, it is critical to understand the role of sink limitation in the down-regulation of photosynthetic capacity under agricultural field conditions and the capacity of N [nitrogen] availability to mitigate it if agriculture is to meet future demand (Long *et al.*, 2004; Tilman and Clark, 2015)." They wonder if down regulation can be avoided by genetically increasing plant sink size and providing sufficient N so as to capitalize on "the full potential photosynthetic benefit of rising CO_2 [in] crops."

To investigate this possibility, Ruiz-Vera *et al.* designed an experiment to assess the potential of nitrogen fertilization to mitigate photosynthetic down regulation in tobacco (*Nicotiana tabacum L.*). The experiment was performed at a Free-Air CO_2 Enrichment (FACE) facility in Champaign, Illinois (USA) in 2015. Two tobacco cultivars of different sink strength were selected for study: Petit Havana (low sink capacity, producing small leaves) and Mammoth (high sink capacity, producing large leaves). After four weeks of initial growth in a greenhouse, plants of each cultivar were transplanted outdoors at the FACE facility where they were subjected in a full factorial design to two CO_2 levels (400 or 600 ppm) and two nitrogen applications (normal, 150 Kg N/ha, or high, 300 Kg N/ha). Over the next 48 days the scientists measured gas exchange, plant height, specific leaf area, leaf carbon and nitrogen content, leaf carbohydrates, and plant dry weight.

The authors report, "high sink strength resulting from rapid growth throughout the experiment appears to have prevented down-regulation in tobacco cv. Mammoth whereas the small stature of cv. Petite Havana appears to have resulted in progressive down-regulation." Nevertheless, despite down-regulation, photosynthetic uptake averaged over the growing season in Petit Havana was significantly higher (+11%) under elevated CO_2 regardless of nitrogen treatment. Ruiz-Vera *et al.* also report that increased nitrogen "partially mitigated the down-regulation of photosynthesis in cv. Petit Havana."

Figure 5.3.5.1.1
Percent change in the combined production of wheat, maize, rice, and soybean under five temperature and CO$_2$ scenarios

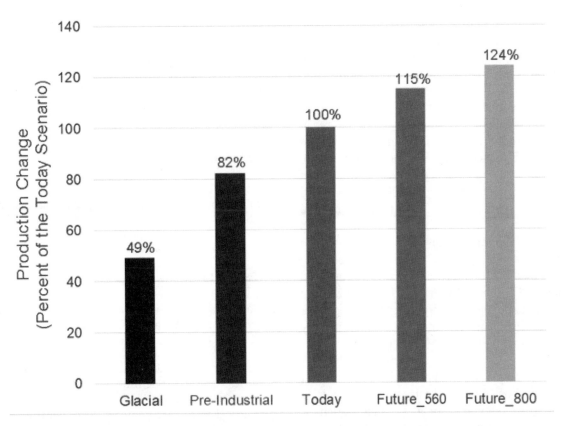

Columns from left to right are five climate scenarios: Glacial, 180 ppm CO$_2$ and -6°C relative to Today; Pre-Industrial, 280 ppm CO$_2$ and -1°C relative to Today; Today, 400 ppm CO$_2$; Future_560, 560 ppm CO$_2$ and +2°C relative to Today; and Future_800, 800 ppm CO$_2$ and +4°C relative to Today. *Source:* Mariani, 2017.

These findings and others, according to the authors, "confirm that under open-air conditions of CO$_2$ elevation in an agricultural field, down-regulation can be strongly offset in germplasm with a high sink capacity." Therefore, as they conclude, "down-regulation of photosynthetic capacity is not inevitable under field conditions where there is no limitation of rooting volume or interference with micro-climate if there is sufficient sink potential and nitrogen supply." This suggests society can capitalize on the full potential photosynthetic benefit of rising atmospheric CO$_2$ in crops by selecting cultivars with high sink capacity and/or adding supplemental nitrogen during the growing season.

Gesch *et al.* (2002) grew rice (*Oryza sativa* L.) in controlled environment chambers receiving atmospheric CO$_2$ concentrations of 350 ppm for about one month. Thereafter, plants were either maintained at 350 ppm CO$_2$ or switched to atmospheric CO$_2$ concentrations of 175 or 700 ppm for an additional 10 days to determine the effects of switching atmospheric CO$_2$ concentrations on photosynthesis, growth, and enzyme function in this important agricultural species.

Within 24 hours after the CO$_2$ concentration switch, plants placed in air of elevated CO$_2$ displayed significant increases in the activity of sucrose-phosphate synthase, a key enzyme involved in the production of sucrose. Plants moved to air of sub-ambient CO$_2$ exhibited significant reductions in the activity of this enzyme. Similarly, elevated CO$_2$ significantly increased the activity of ADP-glucose pyrophosphorylase, a key regulatory enzyme in starch synthesis, while sub-ambient CO$_2$ significantly reduced its activity.

Sucrose concentrations in mature source leaves of plants decreased following their transfer to air of high CO_2 concentration, while sucrose concentrations in sink organs (stems and sheaths) increased. At one day post-transfer, sucrose comprised only 43% of the total nonstructural carbohydrates present in these sinks. However, at 10 days post-transfer (the end of the experiment), sucrose made up 73% of the total nonstructural carbohydrates present in stems and sheaths.

Plants switched to air of elevated CO_2 concentration immediately displayed increases in their photosynthetic rates, while plants switched to sub-ambient CO_2 concentrations displayed immediate reductions in their photosynthetic rates. At the end of the experiment, plants growing at 700 ppm CO_2 still displayed photosynthetic rates 31% greater than those exhibited by unswitched controls, while plants subjected to 175 ppm CO_2 displayed photosynthetic rates 36% less than those exhibited by the same control plants. Ultimately, plants switched to atmospheric CO_2 concentrations of 700 and 175 ppm displayed total aboveground dry weights 54% greater and 18% less, respectively, than those exhibited by control plants maintained at 350 ppm CO_2.

The study by Gesch *et al.* shows that as the CO_2 content of the air rises, rice plants will likely exhibit increased rates of photosynthesis and carbohydrate production that should ultimately increase their biomass. The data acquired from this study suggest that rice plants may avoid the onset of photosynthetic acclimation by synthesizing and exporting sucrose from source leaves into sink tissues to avoid any photosynthetic end-product accumulation in source leaves. Through this mechanism, rice plants can take full advantage of the increasing atmospheric CO_2 concentration and stimulate their productivity and growth without exhibiting lower growth efficiencies resulting from photosynthetic acclimation.

In summary, many peer-reviewed studies suggest food production will continue to increase with increasing atmospheric CO_2 concentrations. Agricultural species may not necessarily exhibit photosynthetic acclimation, even under conditions of low soil nitrogen, for if a plant can maintain a balance between its sources and sinks for carbohydrates at the whole-plant level, acclimation should not be necessary. Because Earth's atmospheric CO_2 content is rising by an average of only 1.5 ppm per year, most plants should be able to adjust their relative growth rates by the small amount that would be needed to prevent low nitrogen-induced acclimation from occurring or expand their root systems by the small amount that would be needed to supply the extra nitrogen required to take full advantage of the CO_2-induced increase in leaf carbohydrate production. In the event a plant cannot initially balance its sources and sinks for carbohydrates at the whole-plant level, CO_2-induced acclimation represents a beneficial secondary mechanism for achieving that balance, redistributing resources away from the plant's photosynthetic machinery to strengthen sink development or enhance other nutrient-limiting processes.

References

Araus, J.L., Slafer, G.A., Buxó, R., and Romagosa, R. 2003. Productivity in prehistoric agriculture: physiological models for the quantification of cereal yields as an alternative to traditional approaches. *Journal of Archaeological Science* **30**: 681–93.

Gavito, M.E., Curtis, P.S., Mikkelsen, T.N., and Jakobsen, I. 2000. Atmospheric CO_2 and mycorrhiza effects on biomass allocation and nutrient uptake of nodulated pea (*Pisum sativum* L.) plants. *Journal of Experimental Botany* **52**: 1931–8.

Gesch, R.W., Vu, J.C., Boote, K.J., Allen Jr., L.H., and Bowes, G. 2002. Sucrose-phosphate synthase activity in mature rice leaves following changes in growth CO_2 is unrelated to sucrose pool size. *New Phytologist* **154**: 77–84.

Long, S.P., Ainsworth, E.A., Rogers, A., and Ort, D.R. 2004. Rising atmospheric carbon dioxide: plants FACE the future. *Annual Review of Plant Biology* **55:** 591–628.

Mariani, L. 2017. Carbon plants nutrition and global food security. *The European Physical Journal Plus* **132**: 69.

Sage, R.F. and Coleman, J.R. 2001. Effects of low atmospheric CO_2 on plants: more than a thing of the past. *Trends in Plant Science* **6**: 18–24.

Sims, D.A., Cheng, W., Luo, Y., and Seeman, J.R. 1999. Photosynthetic acclimation to elevated CO2 in a sunflower canopy. *Journal of Experimental Botany* **50**: 645–53.

Tilman, D. and Clark, M. 2015. Food, agriculture & the environment: can we feed the world and save the earth? *Daedalus* **144**: 8–23.

Ulman, P., Catsky, J. and Pospisilova, J. 2000. Photosynthetic traits in wheat grown under decreased and increased CO_2 concentration, and after transfer to natural CO_2 concentration. *Biologia Plantarum* **43**: 227–37.

Ziska, L.H. 1998. The influence of root zone temperature on photosynthetic acclimation to elevated carbon dioxide concentrations. *Annals of Botany* **81**: 717–21.

5.3.5.2 *Biospheric Productivity*

The vigor of Earth's terrestrial biosphere has been increasing with time, revealing a great post-Industrial Revolution greening of the Earth that extends across the entire globe, a phenomenon documented in Section 5.3.2 (see Zhu *et al.,* 2016; Campbell *et al.* 2017; and Cheng *et al.,* 2017). Nevertheless, it has been hypothesized that future greenhouse gas-induced climate changes could turn the terrestrial biosphere from a net carbon sink into a net carbon source (Cox *et al.,* 2000; Matthews *et al.,* 2005). Will biospheric productivity continue to improve during the twenty-first century and beyond?

Future biospheric productivity is difficult and probably impossible to predict due to our inability to forecast future local surface temperatures and other climatic conditions, poor understanding of feedbacks such as precipitation and cloud formation, and uncertainty over how much carbon is held in each of the four reservoirs (air, water, stone, and the biosphere) and the exchange rates among reservoirs. Different assumptions placed in the models used to forecast each of these variables can lead to dramatically different forecasts. In light of such uncertainty, the only scientific forecast is a continuation of past trends pointing to a continued greening of the Earth.

The physiological mechanisms whereby warmer temperatures and higher levels of CO_2 in the atmosphere lead to enhanced plant growth operate on a planetary scale. Research cited in the previous section demonstrates they are unlikely to be limited by photosynthetic acclimation or down-regulation. Computer models bear this out. Qian *et al.* (2010) analyzed the outputs of 10 models that were part of the Coupled Carbon Cycle Climate Model Intercomparison Project (C4MIP) of the International Geosphere-Biosphere Program and World Climate Research Program. All of the models, Qian *et al.* note, "used the same anthropogenic fossil fuel emissions from Marland *et al.* (2005) from the beginning of the industrial period until 2000 and the IPCC SRES A2 scenario for the 2000–2100 period."

The 10 models predicted a mean warming of 5.6°C from 1901 to 2100 in the northern high latitudes (NHL) and, Qian *et al.* found, "the NHL will be a carbon sink of 0.3 ± 0.3 PgCyr⁻¹ by 2100"

[PgC is a petagram, one billion metric tonnes.]. They also state "the cumulative land organic carbon storage is modeled to increase by 38 ± 20 PgC over 1901 levels, of which 17 ± 8 PgC comes from vegetation [a 43% increase] and 21 ± 16 PgC from the soil [an 8% increase]," noting "both CO_2 fertilization and warming enhance vegetation growth in the NHL."

Thus over the course of the current century, even the severe warming predicted by some climate models would likely not be a detriment to plant growth and productivity in the NHL. In fact, it would likely be a benefit, enhancing plant growth and soil organic carbon storage.

Friend (2010) used the Hybrid6.5 model of terrestrial primary production and "the climate change anomalies predicted by the GISS-AOM GCM under the A1B emissions scenario for the 2090s [relative] to observed modern climate, and with atmospheric CO_2 increased from 375.7 ppm to 720 ppm" – a 92% increase – to calculate the changes in terrestrial plant production that would occur throughout the world in response to the projected climate changes alone and the projected concurrent changes in climate and atmospheric CO_2 concentration.

In response to projected climate changes between 2001–2010 and 2091–2100, Friend found net primary production (NPP) of the planet as a whole was reduced by 2.5%. When both climate and atmospheric CO_2 concentration were changed concurrently, however, Friend found a mean *increase* in global NPP of 37.3%. Thus, even for the magnitude of warming predicted to occur by the models relied on by the IPCC over the remainder of the twenty-first century, biospheric productivity can be expected to increase dramatically.

Lin *et al.* (2010) conducted a meta-analysis of pertinent data they obtained from 127 studies published prior to June 2009, in order to determine if the overall impact of a substantial increase in the air's CO_2 concentration on terrestrial biomass production would likely be positive or negative. They found for the totality of terrestrial plants included in their analysis, "warming significantly increased biomass by 12.3%," while noting there was a "significantly greater stimulation of woody (+26.7%) than herbaceous species (+5.2%)." They conclude, "results in this and previous meta-analyses (Arft *et al.,* 1999; Rustad *et al.,* 2001; Dormann and Woodin, 2002; Walker *et al.,* 2006) have revealed that warming generally increases terrestrial plant biomass, indicating enhanced terrestrial carbon uptake via plant growth and net primary productivity."

New research continues to point to a positive future for Earth's terrestrial biosphere. Just one recent example is the discovery that seagrass meadows in Greenland could be emerging as a major carbon sink. Marbà *et al.* (2018) observed, "Seagrass meadows have been shown to rank amongst the most intense carbon-sink ecosystems of the biosphere with conservation and restoration programs aimed at protecting and restoring the carbon stocks and sink capacity lost with global seagrass decline." While the loss of seagrass in tropical areas has gained global attention, the expansion of seagrass meadows in Greenland has been overlooked. Seagrass meadows in Greenland "appear to be expanding and increasing their productivity," the authors write. "This is supported by the rapid growth in the contribution of seagrass-derived carbon to the sediment Corg pool, from less than 7.5% at the beginning of 1900 to 53% at present, observed in the studied meadows. Expansion and enhanced productivity of eelgrass meadows in the subarctic Greenland fjords examined here is also consistent with the on average 6.4-fold acceleration of Corg burial in sediments between 1940 and present."

According to Marbà *et al.*, "The expansion of seagrass in Greenland fjords represents a novel carbon sink, with limited significance at present due to the small size of the meadows. However, the potential for further expansion is huge, as the convoluted Greenland coastline represents about 12% of the global coastline." They conclude, "whereas the carbon sink associated with sediments under Greenland eelgrass meadows is likely to be very modest at present, it may reach significant levels along the 21st century."

In summary, the rising vitality of Earth's terrestrial biosphere observed during the twentieth and early twenty-first centuries by Zhu *et al.* (2016), Campbell *et al.* (2017), and Cheng *et al.* (2017) is very likely to continue through the twenty-first century and beyond. This is good news for humanity and for the natural world.

References

Arft, A.M., *et al.* 1999. Responses of tundra plants to experimental warming: meta-analysis of the international tundra experiment. *Ecological Monographs* **69**: 491–511.

Campbell, J.E., Berry, J.A., Seibt, U., Smith, S.J., Montzka, S.A., Launois, T., Belviso, S., Bopp, L., and Laine, M. 2017. Large historical growth in global terrestrial gross primary production. *Nature* **544**: 84–7.

Cheng, L., Zhang, L., Wang, Y.-P., Canadell, J.G., Chiew, F.H.S., Beringer, J., Li, L., Miralles, D.G., Piao, S., and Zhang, Y. 2017. Recent increases in terrestrial carbon uptake at little cost to the water cycle. *Nature Communications* **8**: 110.

Cox, P.M., Betts, R.A., Jones, C.D., Spall, S.A., and Totterdell, I.J. 2000. Acceleration of global warming due to carbon-cycle feedbacks in a coupled climate model. *Nature* **408**: 184–7.

Dormann, C.F. and Woodin, S.J. 2002. Climate change in the arctic: using plant functional types in a meta-analysis of field experiments. *Functional Ecology* **16**: 4–17.

Friend, A.D. 2010. Terrestrial plant production and climate change. *Journal of Experimental Botany* **61**: 1293–1309.

Lin, D., Xia, J., and Wan, S. 2010. Climate warming and biomass accumulation of terrestrial plants: a meta-analysis. *New Phytologist* **188**: 187–198.

Marbà, N., Krause-Jensen, D., Masqué, P., and Duarte, C.M. 2018. Expanding Greenland seagrass meadows contribute new sediment carbon sinks. *Scientific Reports* **8**: 14024.

Marland, G., Boden, T.A., and Andres, R.J. 2005. Global, regional, and national CO$_2$ emissions. In: *Trends: A Compendium of Data on Global Change*. Oak Ridge, TN: Carbon Dioxide Information Analysis Center, Oak Ridge National Laboratory, U.S. Department of Energy.

Matthews, H.D., Eby, M., Weaver, A.J., and Hawkins, B.J. 2005. Primary productivity control of simulated carbon cycle-climate feedbacks. *Geophysical Research Letters* **32**: 10.1029/2005GL022941.

Rustad, L.E., Campbell, J.L., Marion, G.M., Norby, R.J., Mitchell, M.J., Hartley, A.E., Cornelissen, J.H.C., Gurevitch, J., and GCTE-NEWS. 2001. A meta-analysis of the response of soil respiration, net nitrogen mineralization, and aboveground plant growth to experimental ecosystem warming. *Oecologia* **126**: 543–62.

Qian, H., Joseph, R., and Zeng, N. 2010. Enhanced terrestrial carbon uptake in the Northern High Latitudes in the 21st century from the Coupled Carbon Cycle Climate Model Intercomparison Project model projections. *Global Change Biology* **16**: 641–56.

Walker, M.D., *et al.* 2006. Plant community responses to experimental warming across the tundra biome.

Proceedings of the National Academy of Sciences USA **103**: 1342–6.

Zhu, Z., *et al.* 2016. Greening of the Earth and its drivers. *Nature Climate Change* **6**: 791–5.

5.3.5.3 Biodiversity

How will the ongoing rise in the air's CO_2 content affect the biodiversity of Earth's many ecosystems? Hundreds of studies have considered that question, often in the course of addressing other things, including:

- *genetic variability within species* (Hedhly, 2011; Rampino *et al.*, 2012; Hahn *et al.*, 2012; Oney *et al.*, 2013; Thilakarathne *et al.*, 2013; Marinciu *et al.*, 2013);

- C_3 *plants vs.* C_4 *plants* (Derner *et al.*, 2003; Zeng *et al.*, 2011; Hyovenen, 2011);

- *grasslands* (Ramseier *et al.*, 2005; Strengbom *et al.*, 2008; Steinbeiss *et al.*, 2008; Reich, 2009; Crain *et al.* 2012);

- *nitrogen fixers vs. non-nitrogen-fixers* (Roumet et al., 2000; Lilley et al., 2001); and

- *weeds vs. non-weeds* (Taylor and Potvin, 1997; Dukes, 2002).

The research gives good reason to believe Earth's increasing atmospheric CO_2 concentration will be *beneficial* to biodiversity by increasing niche security and expanding the ranges of nearly all the planet's many life forms. The historical record shows few cases of a negative effect of warming on diversity, even in cases where temperature increases were larger and more sudden than those forecast in coming centuries.

Jaramillo *et al.* (2010) looked back in time to the Paleocene-Eocene Thermal Maximum (PETM) of some 56 million years ago, which they noted "was one of the most abrupt global warming events of the past 65 million years." It was driven, as they described it, by "a massive release of 13C-depleted carbon" that led to "an approximate 5°C increase in mean global temperature in about 10,000 to 20,000 years." It was thought by many that Earth's tropical ecosystems "suffered extensively because mean temperatures are surmised to have exceeded the ecosystems' heat tolerance." But did the ancient warming of the world truly constitute a major problem for the planet's rainforests?

In an attempt to answer that question, Jaramillo *et al.* analyzed pollen and spore contents and the stable carbon isotopic composition of organic materials obtained from three tropical terrestrial PETM sites in eastern Colombia and western Venezuela. Their findings revealed that the onset of the PETM was "concomitant with an increase in diversity produced by the addition of many taxa (with some representing new families) to the stock of preexisting Paleocene taxa." They determined this increase in biodiversity "was permanent and not transient."

Hof *et al.* (2011) note recent and projected climate change is assumed to be exceptional because of its supposedly unprecedented velocity; they say this view has fueled the prediction that CO_2-induced increases in surface temperatures "will have unprecedented effects on earth's biodiversity," primarily by driving many species to extinction. It is widely assumed that Earth's plants and animals are unable to migrate poleward in latitude or upward in altitude fast enough to avoid extinction and also that current climate change simply outpaces evolutionary adaptation.

Hof *et al.* present evidence demonstrating "recent geophysical studies challenge the view that the speed of current and projected climate change is unprecedented." For example, they report Steffensen *et al.* (2008) showed temperatures in Greenland warmed by up to 4°C/year near the end of the last glacial period. They state this change and other rapid climate changes during the Quaternary (the last 2.5 million years) did not cause a noticeable level of broad-scale, continent-wide extinctions of species. Instead, the rapid changes appeared to "primarily affect a few specific groups, mainly large mammals (Koch and Barnosky, 2006) and European trees (Svenning, 2003)," with the result that "few taxa became extinct during the Quaternary (Botkin *et al.*, 2007)."

Hof *et al.* speculate that "species may have used strategies other than shifting their geographical distributions or changing their genetic make-up." They note, for example, that "intraspecific variation in physiological, phenological, behavioral or morphological traits may have allowed species to cope with rapid climatic changes within their ranges (Davis and Shaw, 2001; Nussey *et al.*, 2005; Skelly *et al.*, 2007)," based on "preexisting genetic variation within and among different populations, which is an important prerequisite for adaptive responses," noting "both intraspecific phenotypic variability and individual phenotypic plasticity may allow for rapid adaptation without actual microevolutionary changes."

Hof *et al.* noted, "habitat destruction and fragmentation, not climate change *per se*, are usually identified as the most severe threat to biodiversity (Pimm and Raven, 2000; Stuart *et al.*, 2004; Schipper *et al.*, 2008)." And since "species are probably more resilient to climatic changes than anticipated in most model assessments of the effect of contemporary climate change on biodiversity," addressing habitat destruction and fragmentation, rather than climate change, should take priority, since those more direct and obvious effects of mankind are more destructive, more imminent, and more easily addressed than are the less direct, less obvious, less destructive, less imminent, and less easily addressed effects of the burning of fossil fuels.

Polley *et al.* (2012) looked for the impact of CO_2 enrichment on the composition and diversity of vegetation in tallgrass prairie communities. They hypothesized that "feedbacks from species change would amplify the initial CO_2 stimulation of aboveground net primary productivity (ANPP) of tallgrass prairie communities." To test that hypothesis, they "grew communities of perennial forb and C_4 grass species for 5 years along a field CO_2 gradient (250–500 ppm) in central Texas (USA) on each of three soil types, including upland and lowland clay soils and a sandy soil," measuring a number of plant physiological properties and processes, and ecosystem ANPP.

Polley *et al.* found CO_2 enrichment from 280 to 480 ppm "increased community ANPP by 0–117% among years and soils and increased the contribution of the tallgrass species *Sorghastrum nutans* (Indian grass) to community ANPP on each of the three soil types," noting the "CO_2-induced changes in ANPP and *Sorghastrum* abundance were linked." They write, "by favoring a mesic C_4 tall grass, CO_2 enrichment approximately doubled the initial enhancement of community ANPP on two clay soils," and conclude, "CO_2-stimulation of grassland productivity may be significantly underestimated if feedbacks from plant community change are not considered."

Royer and Cheroff (2013) analyzed "how atmospheric CO_2 and temperature relate to an angiosperm-dominated record of plant diversity," based on the specific types and proportions of pollen found in central Colombia and western Venezuela

that dated back to the Palaeogene and early Neogene (65–20 million years ago), where the knowledge of pollen morphospecies richness came from Jaramillo *et al.* (2006); atmospheric CO_2 data came from the compilation of Beerling and Royer (2011), together with subsequent updates provided by Pagani *et al.* (2011) and Grein *et al.* (2011); and benthic $\delta^{18}O$ data came from the compilation of Zachos *et al.* (2008).

Royer and Cheroff report "pollen morphospecies richness from the neotropics of Colombia and Venezuela is more strongly correlated with atmospheric CO_2 than it is with temperature." In fact, "atmospheric CO_2 is the only dataset that mirrors (1) the low richness values at the beginning (Palaeocene) and end (Miocene) of the time series, (2) sustained high values during the mid-Eocene, and (3) a short-term spike in the late Palaeocene." In other words, higher atmospheric levels of CO_2 promoted plant diversity regardless of changes in global temperatures.

The 53-member research team of Steinbauer *et al.* (2018), publishing in the journal *Nature*, analyzed a massive continent-wide dataset of repeated plant surveys from 302 mountain summits across Europe dating back to 1871 in an effort to "assess the temporal trajectory of mountain biodiversity changes." Vegetation surveys were conducted predominantly on the uppermost 10 meters of elevation on each summit during the summer, with each summit being resurveyed one to six times between 1871 and 2016, for a total of 698 surveys over the period of study. Such surveys, in the words of the authors, were "optimal ... for detecting changes in plant species richness over time."

Steinbauer *et al.* report there has been "a continent-wide acceleration in the rate of increase in plant species richness, with five times as much species enrichment between 2007 and 2016 as fifty years ago, between 1957 and 1966." (See Figure 5.3.5.3.1.) They note this trend of increasing biodiversity was "consistent across all [continental regions], with no single region showing the opposite pattern."

Despite such good news rooted in real-world observations, the team of researchers just could not bring themselves to reject the CO_2-induced global warming-extinction hypothesis. They opine that "accelerating plant species richness increases are expected to be a transient phenomenon that hides the accumulation of a so-called extinction debt," where "a rapid loss of alpine-nival species may occur under accelerated global warming." But extensive research

Figure 5.3.5.3.1
Rate of change in species richness

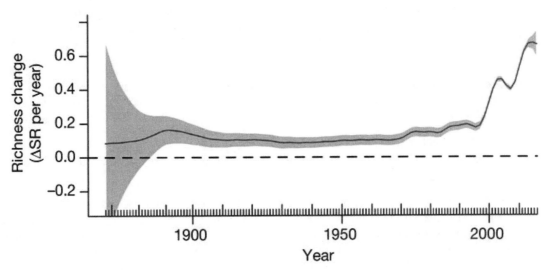

Red line is mean, shaded grey area represents ± standard error of the mean (SEM). *Source:* Steinbauer *et al.*, 2018.

on alpine ecosystems, reported earlier in Section 5.3.2, finds no basis for such concern.

Additional research summarized in the next two sections, addressing extinction and evolution, exposes the faulty assumptions underlying forecasts that climate change will reduce plant diversity. In summary, fear that climate change might reduce biodiversity is contradicted by the historical record and by what we know about the mechanisms whereby warmer temperatures and higher levels of CO_2 in the atmosphere benefit plant life.

References

Beerling, D.J. and Royer, D.L. 2011. Convergent Cenozoic CO_2 history. *Nature Geoscience* **4**: 418–20.

Botkin, D.B., *et al.* 2007. Forecasting the effects of global warming on biodiversity. *BioScience* **57**: 227–36.

Craine, J.M., Ocheltree, T.W., Nippert, J.B., Towne, E.G., Skibbe, A.M., Kembel, S.W., and Fargione, J.E. 2012. Global diversity of drought tolerance and grassland climate-change resilience. *Nature Climate Change* **3**: 63–7.

Davis, M.B. and Shaw, R.G. 2001. Range shifts and adaptive responses to Quaternary climate change. *Science* **292**: 673–9.

Derner, J.D., *et al.* 2003. Above- and below-ground responses of C_3-C_4 species mixtures to elevated CO_2 and soil water availability. *Global Change Biology* **9**: 452–60.

Dukes, J.S. 2002. Comparison of the effect of elevated CO_2 on an invasive species (*Centaurea solstitialis*) in monoculture and community settings. *Plant Ecology* **160**: 225–34.

Grein, M., Konrad, W., Wilde, V., Utescher, T., and Roth-Nebelsick, A. 2011. Reconstruction of atmospheric CO_2 during the early Middle Eocene by application of a gas exchange model to fossil plants from the Messel Formation, Germany. *Paleogeography, Paleoclimatology, Paleoecology* **309**: 383–91.

Hahn, T., Kettle, C.J., Ghazoul, J., Frei, E.R., Matter, P., and Pluess, A.R. 2012. Patterns of genetic variation across altitude in three plant species of semi-dry grasslands. *PLoS ONE* **7**: e41608.

Hedhly, A. 2011. Sensitivity of flowering plant gametophytes to temperature fluctuations. *Environmental and Experimental Botany* **74**: 9–16.

Hof, C., Levinsky, I., Araujo, M.B., and Rahbek, C. 2011. Rethinking species' ability to cope with rapid climate change. *Global Change Biology* **17**: 2987–90.

Hyvonen, T. 2011. Impact of temperature and germination time on the success of a C_4 weed in a C_3 crop: *Amaranthus retroflexus* and spring barley. *Agricultural and Food Science* **20**: 183–90.

Jaramillo, C., *et al.* 2010. Effects of rapid global warming at the Paleocene-Eocene boundary on neotropical vegetation. *Science* **330**: 957–61.

Jaramillo, C., Rueda, M.J., and Mora, G. 2006. Cenozoic plant diversity in the Neotropics. *Science* **311**: 1893–6.

Koch, P.L. and Barnosky, A.D. 2006. Late Quaternary extinctions: state of the debate. *Annual Review of Ecology, Evolution and Systematics* **37**: 215–50.

Lilley, J.M., Bolger, T.P., and Gifford, R.M. 2001. Productivity of *Trifolium subterraneum* and *Phalaris aquatica* under warmer, higher CO_2 conditions. *New Phytologist* **150**: 371–83.

Marinciu, C., Mustatea, P., Serban, G., Ittu, G., and Sauleseu, N.N. 2013. Effects of climate change and genetic progress on performance of wheat cultivars, during the last twenty years in south Romania. *Romanian Agricultural Research* No. 30, online ISSN 2067-5720.

Nussey, D.H., Postma, E., Gienapp, P., and Visser, M.E. 2005. Selection on heritable phenotypic plasticity in a wild bird population. *Science* **310**: 304–6.

Oney, B., Reineking, B., O'Neill, G., and Kreyling, J. 2013. Intraspecific variation buffers projected climate change impacts on *Pinus contorta*. *Ecology and Evolution* **3**: 437–49.

Pagani, M., Huber, M., Liu, Z., Bohaty, S.M., Hendriks, J., Sijp, W., Krishnan, S., and DeConto, R.M. 2011. The role of carbon dioxide during the onset of Antarctic glaciation. *Science* **334**: 1261–4.

Pimm, S.L. and Raven, P. 2000. Biodiversity – extinction by numbers. *Nature* **403**: 843–5.

Polley, H.W., Jin, V.L., and Fay, P.A. 2012. Feedback from plant species change amplifies CO_2 enhancement of grassland productivity. *Global Change Biology* **18**: 2813–23.

Rampino, P., Mita, G., Fasano, P., Borrelli, G.M., Aprile, A., Dalessandro, G., De Bellis, L., and Perrotta, C. 2012. Novel durum wheat genes up-regulated in response to a combination of heat and drought stress. *Plant Physiology and Biochemistry* **56**: 72–8.

Ramseier, D., Connolly, J., and Bazzaz, F.A. 2005. Carbon dioxide regime, species identity and influence of species initial abundance as determinants of change in stand biomass composition in five-species communities: an investigation using a simplex design and RGRD analysis. *Journal of Ecology* **93**: 502–11.

Reich, P.B. 2009. Elevated CO_2 reduces losses of plant diversity caused by nitrogen deposition. *Science* **326**: 1399–1402.

Roumet, C., Garnier, E., Suzor, H., Salager, J.-L., and Roy, J. 2000. Short and long-term responses of whole-plant gas exchange to elevated CO_2 in four herbaceous species. *Environmental and Experimental Botany* **43**: 155–69.

Royer, D.L. and Chernoff, B. 2013. Diversity in neotropical wet forests during the Cenozoic is linked more to atmospheric CO_2 than temperature. *Proceedings of the Royal Society B* **280**: 10.1098/rspb.2013.1024.

Schipper, J., *et al.* 2008. The status of the world's land and marine mammals: diversity, threat, and knowledge. *Science* **322**: 225–30.

Skelly, D.K., Joseph, L.N., Possingham, H.P., Freidenburg, L.K., Farrugia, T.J., Kinnison, M.T., and Hendry, A.P. 2007. Evolutionary responses to climate change. *Conservation Biology* **21**: 1353–5.

Steffensen, J.P., *et al.* 2008. High-resolution Greenland Ice Core data show abrupt climate change happens in few years. *Science* **321**: 680–4.

Steinbauer, M.J., *et al.* 2018. Accelerated increase in plant species richness on mountain summits is linked to warming. *Nature* **556**: 231–4.

Steinbeiss, S., Bessler, H., Engels, C., Temperton, V.M., Buchmann, N., Roscher, C., Kreutziger, Y., Baade, J., Habekost, M., and Gleixner, G. 2008. Plant diversity positively affects short-term soil carbon storage in experimental grasslands. *Global Change Biology* **14**: 2937–49.

Strengbom, J., Reich, P.B., and Ritchie, M.E. 2008. High plant species diversity indirectly mitigates CO_2- and N-induced effects on grasshopper growth. *Acta Oecologica* **34**: 194–201.

Stuart, S.N., Chanson, J.S., Cox, N.A., Young, B.E., Rodrigues, A.S.L., Fischman, D.L., and Waller, R.W. 2004. Status and trends of amphibian declines and extinctions worldwide. *Science* **306**: 1783–6.

Svenning, J.C. 2003. Deterministic Plio-Pleistocene extinctions in the European cool-temperate tree flora. *Ecology Letters* **6**: 646–53.

Taylor, K. and Potvin, C. 1997. Understanding the long-term effect of CO_2 enrichment on a pasture: the importance of disturbance. *Canadian Journal of Botany* **75**: 1621–7.

Thilakarathne, C.L., Tausz-Posch, S., Cane, K., Norton, R.M., Tausz, M., and Seneweera, S. 2013. Intraspecific variation in growth and yield response to elevated CO_2 in wheat depends on the differences of leaf mas per unit area. *Functional Plant Biology* **40**: 185–94.

Zachos, J.C., Dickens, G.R., and Zeebe, R.E. 2008. An early Cenozoic perspective on greenhouse warming and carbon-cycle dynamics. *Nature* **451**: 279–83.

Zeng, Q., Liu, B., Gilna, B., Zhang, Y., Zhu, C., Ma, H., Pang, J., Chen, G., and Zhu, J. 2011. Elevated CO_2 effects on nutrient competition between a C_3 crop (*Oryza sativa* L.) and a C_4 weed (*Echinochloa crusgalli* L.). *Nutrient Cycling in Agroecosystems* **89**: 93–104.

5.3.5.4 Extinction

According to the Working Group II contribution to the IPCC's Fifth Assessment Report, a "large fraction of both terrestrial and freshwater species faces increased extinction risk under projected climate change during and beyond the 21st century, especially as climate change interacts with other stressors, such as habitat modification, over-exploitation, pollution, and invasive species (*high confidence*)" (IPCC 2014, pp. 14-15). Like so many of the IPCC's other predictions, this one ignores natural variability and extensive data on plant and animal life that contradict it.

The IPCC takes advantage of the public's false perception that extinctions are often or only the result of human disturbances to the natural order. In reality, the vast majority of extinctions have nothing to do with human behavior. According to Raup (1986), "five extinctions stand out consistently as the largest and they are conventionally labeled mass extinctions. The five are terminal Ordovician (-440 million years ago), late Devonian (-365 million years ago), terminal or late Permian (-250 million years ago), terminal Triassic (-215 million years ago), and terminal Cretaceous (65 million years ago). The ranking of the five depends somewhat on database and metric, but the Permian event usually emerges as the largest, with published estimates of species kill ranging as high as 96 percent" (p. 1529).

Raup warns that our knowledge of these mass extinctions and many smaller ones in the distant past relies on a database of only about 250,000 known fossil species, "an extremely small sample of past life because of the negligible probability of preservation and discovery of any given species" (p. 1528). Nevertheless, he says "up to 4 billion species of plants and animals are estimated to have lived at some time in the geologic past, most of these in the last 600 million years (Phaneromic time). Yet there

are only a few million species living today. Thus, extinction of species has been almost as common as origination."

More recently, Jablonski (2004) observed, "Extinction is a fundamental part of nature – more than 99% of all species that ever lived are now extinct." He also warns that our understanding of past and future extinctions is limited not only by the absence of fossil records observed by Raup but by our ignorance of how many species *currently* exist. "The most daunting obstacle to assessing and responding to these problems is the lack of anything close to a full accounting of present-day biodiversity: the 1.75 million known species probably represent less than 10% of the true inventory, and the figure is surely less than 1% for genetically distinct populations."

What follows the extinction of a species or a larger group of related species (called a "clade")? Jablonski writes, "post-extinction periods are famously important in opening opportunities for once-marginal groups, for example the expansion of mammals after the dinosaurs' demise." Similarly, Erwin (2001) writes, "However much one may mourn the passing of trilobites, conodonts, ammonoids, richtofenid brachiopods, and even dinosaurs, there is no denying the profound evolutionary impetus mass extinctions have provided to the history of life. Mass extinctions create new evolutionary opportunities and redirect the course of evolution."

Extinctions, then, are not unnatural or objectively bad things: They occur constantly even in the absence of human action, they are as "natural" as the origination of species and they create opportunities (ecological niches) for new species or improve the success of other, better-adapted species. Viewed in this context, the IPCC's worries about humans causing more extinctions seem naïve. We do not even know how many species currently exist; we do not know how the number of species driven to extinction by human action in the past (e.g., the dodo and passenger pigeon) compares to the number made extinct by other causes during that same period (perhaps thousands and possibly millions), or how other species may *benefit* from the extinction of a few. The IPCC itself admits "only a few recent species extinctions have been attributed as yet to climate change (*high confidence*)" (IPCC, 2014, p. 4) and even those claims have been controversial.

The IPCC bases its forecasts of extinctions due to climate change on computer models, often referred to as *bioclimatic envelope models,* that rely on unrealistic and invalidated assumptions about the ability of plants and animals to adapt to changes in their environment, forecasts of temperature and precipitation changes, and the assumption that natural ranges are fully occupied *but for* climatic reasons, failing to account for many different reasons such as disease, competition, and more.

Focusing now on plants rather than animals (the possible extinction of animals is addressed in Section 5.4), Willis and MacDonald (2011) note that extinction scenarios are typically derived from "a suite of predictive species distribution models (e.g., Guisan and Thuiller, 2005)" that "predict current and future range shifts and estimate the distances and rates of movement required for species to track the changes in climate and move into suitable new climate space." They write that one of the most-cited studies of this type, that of Thomas *et al.* (2004), "predicts that, on the basis of mid-range climatic warming scenarios for 2050, up to 37% of plant species globally will be committed to extinction owing to lack of suitable climate space."

In contrast, Willis and MacDonald point out that "biotic adaptation to climate change has been considered much less frequently." This phenomenon – sometimes referred to as *evolutionary resilience* – is "the ability of populations to persist in their current location and to undergo evolutionary adaptation in response to changing environmental conditions (Sgro *et al.*, 2010)." They note this approach to the subject "recognizes that *ongoing change is the norm in nature* and one of the dynamic processes that generates and maintains biodiversity patterns and processes [italics added]," citing MacDonald *et al.* (2008) and Willis *et al.* (2009).

Willis and MacDonald examined the effects of significant and rapid warming on Earth's plants during several previous intervals of the planet's climatic history that were as warm as, or even warmer than, what the IPCC typically predicts for the next century. These intervals included the Paleocene-Eocene Thermal Maximum, the Eocene climatic optimum, the mid-Pliocene warm interval, and the Eemian interglacial. Their approach relies on empirical, data-based, reconstructions of the past – unlike the IPCC's approach, which is built on theoretical model-based projections of the future.

Willis and MacDonald found, "persistence and range shifts (migrations) seem to have been the predominant terrestrial biotic response (mainly of plants) to warmer intervals in Earth's history," while "the same responses also appear to have occurred during intervals of rapid climate change." In addition,

they note "evidence for global extinctions or extinctions resulting from reduction of population sizes on the scale predicted for the next century owing to loss of suitable climate space (Thomas *et al.*, 2004) is not apparent."

Also questioning the accuracy of standard bioclimatic envelope models, Feurdean *et al.* (2012) note "models run at finer scales (Trivedi *et al.*, 2008; Randin *et al.*, 2009) or including representations of plant demography (Hickler *et al.*, 2009) and more accurate dispersal capability (Engler and Guisan, 2009) appear to predict a much smaller habitat and species loss in response to climate model predictions than do more coarse-scale models (Thomas *et al.*, 2004; Thuiller *et al.*, 2005; Araujo *et al.*, 2008)."

Feurdean *et al.* analyzed seven fossil pollen sequences from Romania situated at different elevations "to examine the effects of climate change on community composition and biodiversity between 15,000 and 10,500 cal. yr BP," a period "characterized by large-amplitude global climate fluctuations occurring on decadal to millennial time scales (Johnsen *et al.*, 1992; Jouzel *et al.*, 2007)." They sought to understand "how repeated temperature changes have affected patterns of community composition and diversity" and identify "recovery processes following major disruptions of community structure."

Feurdean *et al.* report "community composition at a given time was not only the product of existing environmental conditions, but also the consequence of previous cumulative episodes of extirpation and recolonization." They note "many circumpolar woody plants were able to survive when environmental conditions became unfavorable" and "these populations acted as sources when the climate became more favorable again." That behavior, they write, "is in agreement with modeling results at the local scale, predicting the persistence of suitable habitats and species survival within large-grid cells in which they were predicted to disappear by coarse-scale models."

Bocsi *et al.* (2016) write, "projections of habitat loss due to climate change assume that many species will be unable to tolerate climate conditions outside of those found within their current distributional ranges." To explore whether that assumption is justified, they "compared the climatic conditions between occurrences in U.S. native vs. U.S. non-native ranges using 144 non-invasive plant species," quantifying "differences in January minimum temperature, July maximum temperature and annual precipitation as indicators of climatic tolerance."

They also compared "modelled potential distributions throughout the U.S. based on native and total ranges to test how expanded climatic tolerance translates into predicted geographical range."

Bocsi *et al.* report that "plants' native ranges strongly underestimate climatic tolerance, leading species distribution models to under-predict potential range," further noting "the climatic tolerance of species with narrow native ranges appears most prone to underestimation." They conclude, "many plants will be able to persist *in situ* with climate change for far longer than projected by species distribution models."

References

Araujo, M.B., Nogues-Bravo, D., Reginster, I., Rounsevell, M., and Whittaker, R.J. 2008. Exposure of European biodiversity to changes in human-induced pressures. *Environmental Science and Policy* **11**: 38–45.

Bocsi, T., Allen, J.M., Bellemare, J., Kartesz, J., Nishino, M., and Bradley, B.A. 2016. Plants' native distributions do not reflect climatic tolerance. *Diversity and Distributions* **22**: 615–24.

Erwin, D.H. 2001. Lessons from the past: biotic recoveries from mass extinctions. *Proceedings of the National Academy of Sciences USA* **98**: 5399–403.

Engler, R. and Guisan, A. 2009. MIGCLIM: predicting plant distribution and dispersal in a changing climate. *Diversity and Distributions* **15**: 590–601.

Feurdean, A., Tamas, T., Tantau, I., and Farcas, S. 2012. Elevational variation in regional vegetation responses to late-glacial climate changes in the Carpathians. *Journal of Biogeography* **39** (2): 258–71.

Guisan, A. and Thuiller, W. 2005. Predicting species distribution: offering more than simple habitat models. *Ecology Letters* **8**: 993–1009.

Hickler, T., Fronzek, S., Araujo, M.B., Schweiger, O., Thuiller, W., and Sykes, M.T. 2009. An ecosystem-model-based estimate of changes in water availability differs from water proxies that are commonly used in species distribution models. *Global Ecology and Biogeography* **18**: 304–13.

IPCC. 2014. Intergovernmental Panel on Climate Change. *Climate Change 2014: Impacts, Adaptation, and Vulnerability*. Contribution of Working Group II to the Fifth Assessment Report of the Intergovernmental Panel on Climate Change. Cambridge, UK and New York, NY: Cambridge University Press.

Jablonski, D. 2004. Extinction: past and present. *Nature* **427**: 589.

Johnsen, S.J., Clausen, H.B., Dansgaard, W., Fuhrer, K., Gundestrup, N., Hammer, C.U., Iversen, P., Jouzel, J., Stauffer, B., and Steffensen, J.P. 1992. Irregular glacial interstadials recorded in a new Greenland ice core. *Nature* **359**: 311–3.

Jouzel, J., Stievenard, M., Johnsen, S.J., Landais, A., Masson-Delmotte, V., Sveinbjornsdottir, A., Vimeux, F., von Grafenstein, U., and White, J.W.C. 2007. The GRIP deuterium-excess record. *Quaternary Science Reviews* **26**: 1–17.

MacDonald, G.M., Bennett, K.D., Jackson, S.T., Parducci, L., Smith, F.A., Smol, J.P., and Willis, K.J. 2008. Impacts of climate change on species, populations and communities: palaeobiogeographical insights and frontiers. *Progress in Physical Geography* **32**: 139–72.

Randin, C., Engler, R., Normans, S., Zappa, M., Zimmermann, N.E., Perman, P.B., Vittoz, P., Thuiller, W., and Guisan, A. 2009. Climate change and plant distribution: local models predict high-elevation persistence. *Global Change Biology* **15**: 1557–69.

Raup, D.M. 1986. Biological extinction in Earth history. *Science* **231** (4745): 1528–33.

Sgro, C.M., Lowe, A.J., and Hoffmann, A.A. 2010. Building evolutionary resilience for conserving biodiversity under climate change. *Evolutionary Applications* **4**: 326–37.

Thomas, C.D., *et al.* 2004. Extinction risk from climate change. *Nature* **427**: 145–8.

Thuiller, W., Lavorel, S., Araujo, M.B., Sykes, M.T., and Prentice, I.C. 2005. Climate change threats to plant diversity in Europe. *Proceedings of the National Academy of Sciences USA* **102**: 8245–50.

Trivedi, M.R., Berry, P.M., Morecroft, M.D., and Dawson, T.P. 2008. Spatial scale affects bioclimate model projections of climate change impacts on mountain plants. *Global Change Biology* **14**: 1089–103.

Willis, K.J., Bennett, K.D., and Birks, H.J.B. 2009. Variability in thermal and UV-B energy fluxes through time and their influence on plant diversity and speciation. *Journal of Biogeography* **36**: 1630–44.

Willis, K.J. and MacDonald, G.M. 2011. Long-term ecological records and their relevance to climate change predictions for a warmer world. *Annual Review of Ecology, Evolution, and Systematics* **42**: 267–87.

5.3.5.5 Evolution

Various researchers (e.g., Gonzalo-Turpin and Hazard, 2009; Steinbauer, *et al.,* 2018) have asserted that alpine and other ecosystems are "threatened by global warming" and the many species that comprise them "are at risk of extinction." Hansen (2006) has claimed life in alpine regions is in danger of being "pushed off the planet" by rising temperatures.

Researchers have been unable to identify any species of plants that have been "pushed off the planet" in alpine regions (Walther *et al.*, 2005; Kullman, 2007; Holzinger *et al.*, 2008; Randin *et al.*, 2009; Erschbamer *et al.*, 2009). Research continues to confirm the ability of plants (and terrestrial animals and marine life, discussed in Sections 5.4 and 5.5 below) to adjust to changes in their environment.

Stocklin *et al.* (2009) studied the highly structured alpine landscape in the Swiss Alps for evidence of evolutionary processes in four plants (*Epilobium fleischeri*, *Geum reptans*, *Campanula thyrsoides*, and *Poa alpina*), testing for whether genetic diversity within their populations was related to altitude and land use, while seeking to determine whether genetic differentiation among populations was related more to different land use or to geographic distances. They determined the within-population genetic diversity of the four species was high and mostly not related to altitude and population size, while genetic differentiation among populations was pronounced and strongly increasing with distance, implying "considerable genetic drift among populations of alpine plants."

Based on their findings and the observations of others, Stocklin *et al.* note "phenotypic plasticity is particularly pronounced in alpine plants," and "because of the high heterogeneity of the alpine landscape, the pronounced capacity of a single genotype to exhibit variable phenotypes is a clear advantage for the persistence and survival of alpine plants." Hence, they conclude "the evolutionary potential to respond to global change is mostly intact in alpine plants, even at high altitude."

Ensslin and Fischer (2015) studied how plants respond to transplantation to different elevations on Mt. Kilimanjaro, Tanzania, in order to determine whether there is sufficient quantitative genetic (among-seed family) variation in and selection on life-history traits and their phenotypic plasticity. They transplanted seed families of 15 common tropical herbaceous species of the montane and savanna vegetation zones of Mt. Kilimanjaro to watered experimental gardens in those zones and then measured species performance, reproduction, and some phenological traits.

They found "seed families within species responded differently to warming," suggesting "some genotypes may persist" and "species may subsequently adapt to warming." They also "found genetic variation in all trait means and in some trait plasticities to transplantation," which is "the prerequisite for adaptive evolution of traits and of plasticities to changes in environmental conditions." They also reported, "because selection on the measured traits did not change between gardens, it appears that the adaptive potential of these species will not be compromised by high temperatures." Ensslin and Fischer conclude "evolutionary adaptation seems a probable scenario for most of the studied common species and might alleviate the negative responses to warming."

References

Ensslin, A. and Fischer, M. 2015. Variation in life-history traits and their plasticities to elevational transplantation among seed families suggests potential for adaptive evolution of 15 tropical plant species to climate change. *American Journal of Botany* **102**: 1371–9.

Erschbamer, B., Kiebacher, T., Mallaun, M., and Unterluggauer, P. 2009. Short-term signals of climate change along an altitudinal gradient in the South Alps. *Plant Ecology* **202**: 79–89.

Gonzalo-Turpin, H. and Hazard, L. 2009. Local adaptation occurs along altitudinal gradient despite the existence of gene flow in the alpine plant species *Festuca eskia*. *Journal of Ecology* **97**: 742–51.

Hansen, J. 2006. The threat to the planet. *The New York Review of Books*. July 13.

Holzinger, B., Hulber, K., Camenisch, M., and Grabherr, G. 2008. Changes in plant species richness over the last century in the eastern Swiss Alps: elevational gradient, bedrock effects and migration rates. *Plant Ecology* **195**: 179–96.

Kullman, L. 2007. Long-term geobotanical observations of climate change impacts in the Scandes of West-Central Sweden. *Nordic Journal of Botany* **24**: 445–67.

Randin, C.F., Engler, R., Normand, S., Zappa, M., Zimmermann, N.E., Pearman, P.B., Vittoz, P., Thuiller, W., and Guisan, A. 2009. Climate change and plant distribution: local models predict high-elevation persistence. *Global Change Biology* **15**: 1557–69.

Steinbauer, M.J., *et al.* 2018. Accelerated increase in plant species richness on mountain summits is linked to warming. *Nature* **556**: 231–4.

Stocklin, J., Kuss, P., and Pluess, A.R. 2009. Genetic diversity, phenotypic variation and local adaptation in the alpine landscape: case studies with alpine plant species. *Botanica Helvetica* **119**: 125–33.

Walther, G.-R., Beissner, S., and Burga, C.A. 2005. Trends in the upward shift of alpine plants. *Journal of Vegetation Science* **16**: 541–8.

5.4 Impact on Terrestrial Animals

The IPCC's forecasts of possible extinctions of terrestrial animals are based on computer models that have been falsified by data on temperature changes, other climatic conditions, and real-world changes in wildlife populations.

The Working Group II contribution to the IPCC's Fifth Assessment Report quoted in Section 5.3.6.4, claiming "a large fraction of both terrestrial and freshwater species faces increased extinction risk under projected climate change during and beyond the 21st century," appears to be a retreat from the Fourth Assessment Report, wherein the IPCC claimed "new evidence suggests that climate-driven extinctions and range retractions are already widespread" and the "projected impacts on biodiversity are significant and of key relevance, since global losses in biodiversity are irreversible (*very high confidence*)" (IPCC, 2007). Unfortunately, the IPCC has not retreated far enough to catch up to the truth.

Before undertaking a survey of the literature on increasing surface temperatures and terrestrial animals, it is necessary to acknowledge that virtually all studies alleging to find a negative effect are based on the IPCC's climate models, which are known to exaggerate the likely warming, frequency of extreme

weather, and other possibly harmful climatic conditions in the twenty-first century and beyond. The IPCC's forecasts were rigorously critiqued in Chapter 2 citing many sources (e.g., Diffenbaugh *et al.*, 2008; Armstrong *et al.*, 2008; Sardeshmukh *et al.*, 2015; Landsea, 2015; Burn and Palmer, 2015; Camargo and Wing, 2016; Stapleton *et al.*, 2016; Crockford, 2016, 2017; Christy, 2017; and Sutton *et al.*, 2018). A previous volume in the *Climate Change Reconsidered* was devoted to debunking the IPCC's physical science findings (NIPCC, 2013). The significance of this cannot be stressed enough: Virtually all predictions of future extinctions due to climate change are invalid because researchers assume too much warming, droughts, extreme weather, and other kinds of climate change.

Section 5.4.1 presents some of the extensive research showing terrestrial animals have the ability to adapt to changes in climate as great as or even greater than those forecast by the IPCC. Section 5.4.2 looks at the most probable future impact of climate change on terrestrial animals. Although there likely will be some changes in terrestrial animal population dynamics, few if any will be driven even close to extinction. Real-world data indicate warmer temperatures and higher atmospheric CO_2 concentrations will be highly beneficial, favoring a proliferation of species.

A chapter of a previous volume in the *Climate Change Reconsidered* series, Chapter 5 of *Climate Change Reconsidered II: Biological Impacts* (NIPCC, 2014), reviewed and analyzed IPCC-based species extinction claims, highlighting many of the problems inherent in the models on which such claims are based. The model projections were then evaluated against real-world observations of various animal species and their response to what the IPCC has called the unprecedented rise in atmospheric CO_2 and temperature of the twentieth and twenty-first centuries. NIPCC's key findings regarding terrestrial animals appear in Figure 5.4.1.

**Figure 5.4.1
Key Findings: Impacts on terrestrial animals**

- The IPCC's forecast of future species extinction relies on a narrow review of the literature that is highly selective and based almost entirely on model projections as opposed to real-world observations; the latter often contradict the former.

- Numerous shortcomings are inherent in the models utilized in predicting the impact of climate on the health and distributions of animal species. Assumptions and limitations make them unreliable.

- Research suggests amphibian populations will suffer little, if any, harm from rising surface temperatures and CO_2 levels in the atmosphere and they may even benefit.

- Although some changes in bird populations and their habitat areas have been documented in the literature, linking such changes to climate change remains elusive. Also, when there have been changes, they often are positive, as many species have adapted and are thriving in response to rising temperatures of the modern era.

- Polar bears have survived historic changes in climate that have exceeded those of the twentieth century or are forecast by computer models to occur in the future. In addition, some populations of polar bears appear to be stable despite rising temperatures and summer sea ice declines. The biggest threat they face is not from climate change but hunting by humans, which historically has taken a huge toll on polar bear populations.

- The net effect of climate change on the spread of parasitic and vector-borne diseases is complex and at this time appears difficult to predict. Rising temperatures increase the mortality rates as well as the development rates of many parasites of veterinary importance and temperature is only one of many variables that influence the range of viruses and other sources of diseases.

- Existing published research indicates rising temperatures likely will not increase and may decrease plant damage from leaf-eating herbivores, as rising atmospheric CO_2 boosts the production of certain defensive compounds in plants that are detrimental to animal pests.

- Empirical data on many other animal species, including butterflies, other insects, reptiles, and mammals, indicate warmer temperatures and higher CO_2 levels in the atmosphere tend to foster the expansion and proliferation of animal habitats, ranges, and populations, or otherwise have no observable impacts one way or the other.

- Multiple lines of evidence indicate animal species are adapting and in some cases evolving to cope with climate change of the modern era.

Source: Chapter 5. "Terrestrial Animals," *Climate Change Reconsidered II: Biological Impacts.* Nongovernmental International Panel on Climate Change. Chicago, IL: The Heartland Institute, 2014.

References

Armstrong, J.S., Green, K.C., and Soon, W. 2008. Polar bear population forecasts: a public-policy forecasting audit. *Interfaces* **38**: 382–404.

Burn, M.J. and Palmer, S.E. 2015. Atlantic hurricane activity during the last millennium. *Scientific Reports* **5**: 10.1038/srep12838.

Camargo, S.J. and Wing, A.A. 2016. Tropical cyclones in climate models. *WIREs Climate Change* **7**: 211–37.

Christy, J. 2017. Testimony before the U.S. House Committee on Science, Space & Technology. March 29.

Crockford, S.J. 2016. *Polar Bears: Outstanding Survivors of Climate Change.* CreateSpace Independent Publishing Platform.

Crockford, S.J. 2017 v3. Testing the hypothesis that routine sea ice coverage of 3-5 mkm2 results in a greater than 30% decline in population size of polar bears (*Ursus maritimus*). *PeerJ Preprints.* March 2.

Diffenbaugh, N.S., Trapp, R.J., and Brooks, H. 2008. Does global warming influence tornado activity? *EOS, Transactions of the American Geophysical Union* **89**: 553–4.

IPCC. 2007. Intergovernmental Panel on Climate Change. *Climate Change 2007: Impacts, Adaptation and Vulnerability*. Contribution of Working Group II to the Fourth Assessment Report of the Intergovernmental Panel on Climate Change. Cambridge, UK and New York, NY: Cambridge University Press.

IPCC. 2014. Intergovernmental Panel on Climate Change. *Climate Change 2014: Impacts, Adaptation, and Vulnerability*. Contribution of Working Group II to the Fifth Assessment Report of the Intergovernmental Panel on Climate Change. Cambridge, UK and New York, NY: Cambridge University Press.

Landsea, C.W. 2015. Comments on "Monitoring and understanding trends in extreme storms: state of knowledge." *Bulletin of the American Meteorological Society* **96**: 1175–6.

McKitrick, R.R., McIntyre, S., and Herman, C. 2010. Panel and multivariate methods for tests of trend equivalence in climate data sets. *Atmospheric Science Letters* **11** (4): 270–7.

NIPCC. 2013. Nongovernmental International Panel on Climate Change. Idso, C.D., Carter, R.M., and Singer, S.F. (Eds.) *Climate Change Reconsidered: Physical Science.* Chicago, IL: The Heartland Institute.

NIPCC. 2014. Nongovernmental International Panel on Climate Change. Idso, C.D, Idso, S.B., Carter, R.M., and Singer, S.F. (Eds.) *Climate Change Reconsidered II: Biological Impacts*. Chicago, IL: The Heartland Institute.

Sardeshmukh, P.D., Compo, G.P., and Penland, C. 2015. Need for caution in interpreting extreme weather events. *Journal of Climate* **28**: 9166–87.

Stapleton, S., Peacock, E., and Garshelis, D. 2016. Aerial surveys suggest long-term stability in the seasonally ice-free Foxe Basin (Nunavut) polar bear population. *Marine Mammal Science* **32**: 181–201.

Sutton, R, Hoskins, B., Palmer, T., Shepherd, T., and Slingo, J. 2018. Attributing extreme weather to climate change is not a done deal. *Nature* **561**: 177.

5.4.1 Evidence of Ability to Adapt

Animal species are capable of migrating, evolving, and otherwise adapting to changes in climate that are much greater and more sudden than what is likely to result from the human impact on the global climate.

Even assuming its climate models were unbiased and reasonably accurate, the IPCC's forecast of future animal extinctions still would not be reliable because it depends on species distribution models based on assumptions about the immobility of species that are contradicted by real-world observations. The failure of those models with respect to plants was documented in Section 5.3.6.4 above. Here we turn to the adaptability of Earth's terrestrial animals, while Section 5.5.1 will consider the adaptability of aquatic life.

The IPCC improperly characterizes the adaptive responses (e.g., range shifts, phenotypic or genetic adaptations) of many species as supporting their model-based extinction claims, when in reality such adaptive responses provide evidence of species resilience. The "climate envelope" approach used to predict shifts in the ranges of Earth's many animal species – and sometimes their extinction – fails to accurately describe the way real animals respond to climate change in the real world.

Behavioral plasticity (the ability of a species to alter its behavior), developmental plasticity (changes in the timing of events in a species' development), migration, and evolutionary adaptation are mechanisms by which living organisms will successfully confront the challenges that may be presented to them by rising surface temperatures, as the recent research summarized in Sections 5.4.1.1 through 5.4.1.4 shows.

5.4.1.1 Amphibians

Li *et al.* (2013) synthesized the research literature on the influence of global climate change on amphibians. They report, "evidence is lacking on poleward shifts in amphibian distributions and on changes in body sizes and morphologies of amphibians in response to climate change." They also note "we have limited information on amphibian thermal tolerances, thermal preferences, dehydration breaths, opportunity costs of water conserving behaviors and actual temperature and moisture ranges amphibians experience." And even when the information *is* available, they say, "there remains little evidence that climate change is acutely lethal to amphibians." They conclude, "we must remember that climate change will likely have both positive and negative effects on amphibians and that geographic regions will vary in terms of both the severity of and species sensitivities to climate change."

Lindstrom *et al.* (2013) describe how species move into regions where they historically have not been present – called an "invasion front" in biology. They note the biology of species populations at an invasion front "differs from that of populations within the range core, because novel evolutionary and ecological processes come into play ..." Seeking to determine how individual members of a given species disperse at an invasion front, they analyzed an extensive dataset they derived by radio-tracking invasive cane toads (*Rhinella marina*) over the first eight years following their arrival at a site in tropical Australia.

Lindstrom *et al.* found "pioneer toads spent longer periods in dispersive mode and displayed longer, more directed movements while they were in dispersive mode." They discovered "overall displacement per year was more than twice as far for toads at the invasion front compared with those tracked a few years later at the same site."

Lindstrom *et al.* concluded "studies on established populations (or even those a few years post-establishment) thus may massively underestimate dispersal rates at the leading edge of an expanding population." They note that "this, in turn, will cause us to under-predict the rates at which native taxa can expand into newly available habitat under climate change."

Orizaola and Laurila (2016) note some organisms can respond to changing environmental conditions "through migration, plasticity and/or genetic adaptation," while others, "due to habitat fragmentation and low dispersal capacities, ... must respond to environmental change *in situ*," citing Chevin *et al.* (2010). They "examined variation in developmental plasticity to changing temperature in the pool frog (*Pelophylax lessonae*) across its distribution by studying populations from central areas (Poland), edge populations (Latvia) and northern marginal populations (Sweden)."

Orizaola and Laurila report, "plasticity in larval life-history traits was highest at the northern range margin," where when reared at induced high temperatures, "larvae from marginal populations shortened larval period and increased growth rate more than larvae from central and edge populations." They write that "the detection of high levels of developmental plasticity in isolated marginal populations suggests that they may be better able to respond to the temperature regimes expected under climate change than often predicted, reflecting the need to incorporate geographic variation in life-history traits into models forecasting responses to environmental change."

References

Chevin, L.-M., Lande, R., and Mace, G.M. 2010. Adaptation, plasticity, and extinction in a changing environment: towards a predictive theory. *PLoS Biology* **8**: e1000357.

Li, Y., Cohen, J.M., and Rohr, J.R. 2013. Review and synthesis of the effects of climate change on amphibians. *Integrative Zoology* **8**: 145–61.

Lindstrom, T., Brown, G.P., Sisson, S.A., Phillips, B.L., and Shine, R. 2013. Rapid shifts in dispersal behavior on an expanding range edge. *Proceedings of the National Academy of Sciences USA* **110**: 13,452–13,456.

Orizaola, G. and Laurila, A. 2016. Developmental plasticity increases at the northern range margin in a warm-dependent amphibian. *Evolutionary Applications* **9**: 471–8.

5.4.1.2 Birds

Smit *et al.* (2013) investigated the effects of air temperature on body temperature and the behavior of the White-browed Sparrow-Weaver (*Plocepasser mahali*) at two sites 100 kilometers apart in the southern Kalahari Desert of South Africa, over two consecutive summer seasons. Among other things, they found a relatively large variation in body temperature both within and between conspecific populations, which suggested to them that "an arid-zone passerine responds differently to prevailing weather conditions in two locations over its range and that it also responds to seasonal changes in weather conditions" – which further suggests "a species' current range may not be an accurate representation of its climatic tolerance."

"Taken together with the data of Glanville *et al.* (2012)," Smit *et al.* write, this result "suggests that the thermal physiology of endotherms [warm-blooded animals] is far more flexible than previously thought and could potentially contribute to the adaptation of populations under changing climatic conditions," citing Boyles *et al.* (2011)," so that "when predicting species' responses to climate change, their sensitivity (sensu Williams *et al.*, 2008) should be resolved at the population, rather than species, level."

Atkinson *et al.* (2013) note "Hawaiian honeycreepers are particularly susceptible to avian malaria and have survived into this century largely because of persistence of high elevation refugia on Kaua'I, Maui and Hawai'I Islands, where transmission is limited by cool temperatures." Because the long-term stability of these refugia could be threatened by future warming and "since cost effective and practical methods of vector control in many of these remote, rugged areas are lacking, adaptation through processes of natural selection may be the best long-term hope for recovery of many of these species." In a study devised to explore this possibility, Atkinson *et al.* discovered and documented what they describe as the "emergence of tolerance rather than resistance to avian malaria in a recent rapidly-expanding low-elevation population of Hawai'I 'Amakihi (*Hemignathus virens*) on the island of Hawai'i."

Atkinson *et al.* determined "experimentally infected low-elevation birds had lower mortality, lower reticulocyte counts during recovery from acute infection, lower weight loss, and no declines in food consumption relative to experimentally infected high elevation Hawai'I 'Amakihi in spite of similar intensities of infection." They state that the "emergence of this population provides an exceptional opportunity for determining physiological mechanisms and genetic markers associated with malaria tolerance that can be used to evaluate whether other, more threatened species have the capacity to adapt to this disease." Their finding "opens the possibility that other native honeycreepers may also be able to adapt to this disease through processes of natural selection."

Thompson *et al.* (2015) investigated the effects of a 4°C increase in ambient temperature – similar to that typically predicted for southern Africa by the year 2080 – on certain physiological variables of 10- to 12-gram Cape white-eye *Zosterops virens*, a passerine bird species endemic to South Africa. The scientists report "there was no significant difference in resting metabolism, body mass and intraperitoneal body temperature between birds housed indoors at 4°C above outside ambient temperature and those housed indoors at outside ambient temperature." They conclude, "the physiological flexibility of Cape white-eyes will aid them in coping with the 4°C increase [in air temperature] predicted for their range by 2080."

Nilsson *et al.* (2016) write that in a warming world, many "organisms in hot environments will not be able to passively dissipate metabolically generated heat," noting they will have to revert to evaporative cooling, which is "energetically expensive and promotes excessive water loss." They explored "the use of hyperthermia in wild birds captured during the hot and dry season in central Nigeria," revealing the presence of "pronounced hyperthermia in several species with the highest body temperatures close to predicted lethal levels." They also found "birds let their body temperature increase in direct relation to ambient temperatures, increasing body temperature by 0.22°C for each degree of increased ambient temperature." They also note that "to offset the costs of thermoregulation in ambient temperatures above the upper critical temperature, birds are willing to let their body temperatures increase by up to 5°C above normal temperatures." Nilsson *et al.* state "this flexibility in body temperatures may make birds well adapted to meet future global increases in ambient temperature," citing the similar prior conclusions of Khaliq *et al.* (2014) and Thompson *et al.* (2015).

Gladalski *et al.* (2016) investigated the response of Great Tits (*Parus major*) and Blue Tits (*Cyanistes caeruleus*) to an extreme variation in spring temperature that occurred in central Poland between 2013 and 2014; the spring of 2013 was the coldest in 40 years, whereas the spring of 2014 was the warmest in 40 years. They gathered data from two habitats (an urban parkland and a deciduous forest) in Lódź (central Poland), part of an ongoing long-term study into the "breeding biology of hole-nesting birds occupying nestboxes." By comparing their observations from the two spring temperature extremes, the authors observed the effects of extreme thermal conditions on the plasticity of breeding phenology and double broodedness of both bird species. They report, "extremely low spring temperatures in 2013 (coldest spring in 40 years) resulted in birds laying [eggs] unusually late," and this phenomenon "was followed in 2014 by the earliest breeding season on record (warmest spring in 40 years)."

Gladalski *et al.* also found "the breeding date of Great Tits and Blue Tits turned out to be a flexible trait" and that "populations of both tit species may tune their egg-laying dates to diverse weather conditions by about 3 weeks," while in some cases they have both early and late clutches. They conclude "such a buffer of plasticity may be sufficient for Blue Tits and Great Tits to adjust the timing of breeding to the upcoming climate changes."

Vengerov (2017) evaluated "changes in the phenology of breeding and reproductive output of the Song Thrush (*Turdus philomelos*) under conditions of increase in spring air temperature," examining reproductive data collected at the Voronezh Nature Reserve every four or five days over the period 1987–1990 and 2008–2012. A total of 459 nests were observed over the nine years of study, during which time there was a statistically significant increase in spring temperatures.

Vengerov determined higher temperatures lead to an "earlier arrival of the birds from wintering grounds," "earlier and more synchronous breeding of the majority of nesting pairs," "an increase in clutch size," a higher proportion of pairs producing two broods per season, and a reduction in "predation pressure on bird nests … which markedly improves reproductive success." Vengerov writes, "climate warming is conducive to increasing breeding productivity of the Song Thrush population as a whole."

References

Atkinson, C.T., Saili, K.S., Utzurrum, R.B., and Jarvi, S.I. 2013. Experimental evidence for evolved tolerance to avian malaria in a wild population of low elevation Hawai'i 'Amakihi (*Hemignathus virens*). *EcoHealth* **10**: 366–75.

Boyles, J.G., Seebacher, F., Smit, B., and McKechnie, A.E. 2011. Adaptive thermoregulation in endotherms may alter responses to climate change. *Integrative and Comparative Biology* **51**: 676–90.

Gladalski, M., Banbura, M., Kalinski, A., Markowski, M., Skwarska, J., Wawrzyniak, J., Zielinski, P., and Banbura, J. 2016. Effects of extreme thermal conditions on plasticity in breeding phenology and double-broodedness of Great Tits and Blue Tits in central Poland in 2013 and 2014. *International Journal of Biometeorology* **60**: 1795–1800.

Glanville, E.J., Murray, S.A., and Seebacher, F. 2012. Thermal adaptation in endotherms: climate and phylogeny interact to determine population-level responses in a wild rat. *Functional Ecology* **26**: 390–8.

Khaliq, I., Hof, C., Prinzinger, R., Bophning-Gaese, K., and Pfenninger, M. 2014. Global variation in thermal tolerances and vulnerability of endotherms to climate change. *Proceedings of the Royal Society B* **281** (1789):1097.

Nilsson, J.-A., Molokwu, M.N., and Olsson, O. 2016. Body temperature regulation in hot environments. *PLoS ONE* **11** (8): eO161481.

Smit, B., Harding, C.T., Hockey, P.A.R., and McKechnie, A.E. 2013. Adaptive thermoregulation during summer in two populations of an arid-zone passerine. *Ecology* **94**: 1142–54.

Thompson, L.J., Brown, M., and Downs, C.T. 2015. The potential effects of climate-change-associated temperature increases on the metabolic rate of a small Afrotropical bird. *The Journal of Experimental Biology* **218**: 1504–12.

Vengerov, P.D. 2017. Effect of rise in spring air temperature on the arrival dates and reproductive success of the Song Thrush, *Turdus philomelos* (C.L. Brehm, 1831) in the forest-steppe of the Russian plain. *Russian Journal of Ecology* **48**: 134–40.

Williams, S.E., Shoo, L.P., Isaac, J.L., Hoffman, A.A., and Langham, G. 2008. Towards an integrated framework for assessing the vulnerability of species to climate change. *PLoS Biology* **6**: 2621–6.

5.4.1.3 Mammals

Coulson *et al.* (2011) write, "environmental change has been observed to generate simultaneous responses in population dynamics, life history, gene frequencies, and morphology in a number of species." They studied these adaptive responses in Yellowstone Park wolves, using "survival and reproductive success data, body weights, and genotype at the K locus (*CBD103*, a β-defensin gene that has two alleles and determines coat color), which were collected from 280 radio-collared wolves living in the park between 1998 and 2009." They noted "body weight and genotype at the K locus vary across U.S. wolf populations" and that both traits influence fitness, citing the studies of Schmitz and Kolenosky (1985), Anderson *et al.* (2009), and MacNulty *et al.* (2009).

Coulson *et al.* say their results "reveal that, for Yellowstone wolves, (i) environmental change will inevitably generate eco-evolutionary responses; (ii) change in the mean environment will have more profound population consequences than changes in the environmental variance; and (iii) environmental change affecting different functions can generate contrasting eco-evolutionary dynamics," which suggests that "accurate prediction of the consequences of environmental change will probably prove elusive."

Maldonado-Chaparro *et al.* (2015) "aimed to characterize patterns of phenotypic change in morphological (body mass), life-history (reproductive success and litter size), and social (embeddedness) traits of female yellow-bellied marmots (*Marmota flaviventris*) in response to climatic and social variation." They used data collected over a period of 36 years on a population in Colorado, using "mixed effect models to explore phenotypically plastic responses" while testing for individual variations in mean trait values and plasticity.

Maldonado-Chaparro *et al.* report "all examined traits were plastic and the population's average plastic response often differed between spatially distinct colonies that varied systematically in timing of snowmelt, among age classes and between females with different previous reproductive experiences." In addition, they detected "individual differences in June mass and pup mass plasticity," all of which led them to conclude that in the case of yellow-bellied marmots, "plasticity plays a key role buffering the effects of continuous changes in environmental conditions."

Smith and Nagy (2015) note American pikas (*Ochotona princeps*) "have been characterized as an indicator species for the effects of global warming on animal populations," citing the works of Smith *et al.* (2004), Beever and Wilkening (2011), and Ray *et al.* (2012). They investigated the resilience of a pika metapopulation residing near Bodie, California that was exposed to several decades of natural warming, testing for a relationship between pika extinctions/recolonizations and chronic/acute temperature warming.

With respect to chronic temperature warming, Smith and Nagy report that despite a relatively high rate of patch (islands of pika-suitable habitat) turnover across the study location, there was "a near balance" of pika patch extinctions and recolonizations during the past four decades. Statistical analyses performed on the patch turnover and historic temperature data revealed there was "no evidence that warming temperatures have directly and negatively affected pika persistence at Bodie." The only significant correlation they found among the two parameters occurred between mean maximum August temperature and the number of pika recolonizations the following year, which correlation was *positive*, indicating higher August temperatures led to a greater rate of pika recolonization the next year, "in the opposite direction of the expectation that climate stress inhibits recolonizations."

With respect to acute temperature warming, defined as the number of hot summer days exceeding a temperature threshold of 25°C or 28°C (77°F or 82.4°F), Smith and Nagy write, "neither warm chronic nor acute temperatures increased the frequency of extinctions of populations on patches and relatively cooler chronic or acute temperatures did not lead to an increase in the frequency of recolonization events."

Varner *et al.* (2016) also studied American pikas (*Ochotona princeps*), these populations living in two habitat ranges in Oregon. One range comprised an elevation, landscape, and climate typical of the American pika's range, while the other was situated within an atypical low-elevation landscape and climate that "appears to be unsuitable [as a pika habitat], based on the species' previously described thermal niche." The researchers sought to quantify behavioral differences among the two populations, including differences pertaining to foraging and territorial behaviors. They collected 417 observer-hours of behavioral data in July 2011, 2012, and 2013, during which they made 5,250 pika detections.

Varner *et al.* report there were "substantial differences" in behavior between pika populations at the two habitats. They noted "low-elevation pikas do not invest as much time or energy in caching food for winter" and were more likely to spend time in forested areas off the open talus landscape around midday than pikas living at higher elevations. Pikas in the lower elevation and warmer habitat had smaller home range sizes compared to those at the higher elevation site.

Varner *et al.* write their findings "indicate that behavioral plasticity likely allows pikas to accommodate atypical conditions in this low-elevation habitat and that they may rely on critical habitat factors such as suitable microclimate refugia to behaviorally thermoregulate." They conclude, "these results suggest that behavioral adjustments are one important mechanism by which pikas can persist outside of their previously appreciated dietary and thermal niches."

Loe *et al.* (2016) analyzed "responses in space use to rain-on-snow and icing events and their fitness correlates, in wild reindeer in high-Arctic Svalbard." This work revealed that "range displacement among GPS-collared females occurred mainly in icy winters to areas with less ice, lower over-winter body mass loss, lower mortality rate, and higher subsequent fecundity, than the departure area." The researchers say their study provides "rare empirical evidence that mammals may buffer negative effects of climate change and extreme weather events by adjusting behavior in highly stochastic environments." They conclude, "under global warming, behavioral buffering may be important for the long-term population persistence in mobile species with long generation time and therefore limited ability for rapid evolutionary adaptation."

References

Anderson, T.M., *et al.* 2009. Molecular and evolutionary history of melanism in North American gray wolves. *Science* **323**: 1339–43.

Beever, E.A. and Wilkening, J.L. 2011. Playing by new rules: altered climates are affecting some pikas dramatically –and rapidly. *The Wildlife Professional* **5**: 38-41.

Coulson T., MacNulty, D.R., Stahler, D.R., vonHoldt, B., Wayne, R.K., and Smith, D.W. 2011. Modeling effects of environmental change on wolf population dynamics, trait evolution, and life history. *Science* **334**: 1275–8.

Loe, L.E., *et al.* 2016. Behavioral buffering of extreme weather events in a high-Arctic herbivore. *Ecosphere* **7**: e01374.

MacNulty, D., *et al.* 2009. Body size and predatory performance in wolves: is bigger better? *Journal of Animal Ecology* **78** (3): 532–9.

Maldonado-Chaparro, A.A., Martin, J.G.A., Armitage, K.B., Oli, M.K., and Blumstein, D.T. 2015. Environmentally induced phenotypic variation in wild yellow-bellied marmots. *Journal of Mammalogy* **96**: 269–78

Ray, C., Beever, E., and Loarie, S. 2012. Retreat of the American pika: up the mountain or into the void? In Brodie, J.F., Post, E., and Doak, D.F. (Eds.) *Wildlife Conservation in a Changing Climate*. Chicago, IL: University of Chicago Press, pp. 245–70.

Schmitz, O.J. and Kolenosky, G.B. 1985. Wolves and coyotes in Ontario: morphological relationships and origins. *Canadian Journal of Zoology* **63**: 1130–7.

Smith, A.T., Li, W., and Hik, D. 2004. Pikas as harbingers of global warming. *Species* **41**: 4–5.

Smith, A.T. and Nagy, J.D. 2015. Population resilience in an American pika (*Ochotona princeps*) metapopulation. *Journal of Mammalogy* **96**: 394–404.

Varner, J., Horns, J.J., Lambert, M.S., Westberg, E., Ruff, J.S., Wolfenberger, K., Beever, E.A., and Dearing, M.D. 2016. Plastic pikas: Behavioural flexibility in low-elevation pikas (*Ochotona princeps*). *Behavioural Processes* **125**: 63–71.

5.4.1.4 Reptiles

Logan *et al.* (2014) say "tropical ectotherms [cold-blooded animals] are thought to be especially vulnerable to climate change because they are adapted to relatively stable temperature regimes, such that even small increases in environmental temperature may lead to large decreases in physiological performance." Nevertheless, they hypothesize that tropical organisms may mitigate the detrimental effects of warming through evolutionary change in thermal physiology.

To determine whether and how thermal physiology is subject to natural selection, Logan *et al.* "measured survival as a function of the thermal sensitivity of sprint speed in two populations of *Anolis sagrei* lizards from the Bahamas," quantifying the relationship between thermal performance and survival of 85 males from a non-manipulated population in order "to test whether a simulated change in thermal environment would increase or otherwise alter selection on thermal performance." They repeated the test on a population of 80 males they transplanted from an interior forested habitat to a warmer, more thermally variable site.

Logan *et al.* report, "when we simulated a rapid change in the thermal environment by transplanting a population of lizards to a warmer and more thermally variable habitat, we observed strong natural selection on thermal physiology," which implies "rapid climate change may result in directional selection on thermal physiology, even in species whose thermoregulatory behaviors are thought to shelter them from natural selection." They warn "evolutionary change will not occur unless thermal performance traits are heritable," but ultimately conclude, "even if the amount of warming expected through the end of the century occurred during a single breeding season, this species could hypothetically compensate for as much as 30% of that environmental change through evolutionary adaptation alone."

"[B]iologists have increasingly recognized that evolutionary change can occur rapidly," Stuart *et al.* (2014) confirm, and therefore "real-time studies of evolution can be used to test classic evolutionary hypotheses," one of which is that "negative interactions between closely related species can drive phenotypic divergence." They say "an opportunity to study such real-time divergence between negatively interacting species has been provided by the recent invasion of the Cuban brown anole lizard, *Anolis sagrei*, into the southeastern United States, where *Anolis carolinensis* is the sole native anole." There, they studied "the eco-evolutionary consequences of this interaction."

Stuart *et al.* report, "on small islands in Florida, we found that the lizard *Anolis carolinensis* moved to higher perches following invasion by *Anolis sagrei* and, in response, adaptively evolved larger toepads after only 20 generations," illustrating that "interspecific interactions between closely related species can drive evolutionary change on observable time scales."

Barrows and Fisher (2014) studied congeneric lizards in southern California, noting "species and species assemblages extant today survived multiple past climate shifts throughout the Pleistocene." One potential mechanism for their survival could have been behavioral adaptation, whereby the lizards shuttle between sun and shade to maintain a preferred body temperature (Tb) that is independent of ambient temperature, as described by Dawson (1967). More recently, Lopez-Alcaide *et al.* (2014) discovered "*Sceloporus adleri* can alter its thermoregulatory behavior to maintain its preferred Tb for key physiological processes when environmental temperatures were increased by 6°C."

Barrows and Fisher (2014) constructed a set of habitat suitability models (HSMs) for an assemblage of four sympatric species of lizards within the genus *Sceloporus* – *S. magister*, *S. occidentalis*, *S. vandenburgianus*, and *S. orcutti* – in order to predict their distributions under three climate conditions: the last glacial maximum of ca 20 kya, the present, and the end of the current century as foreseen by the IPCC (2013). They say their results suggest the elevational heterogeneity of the landscape they studied "provided suitable habitat for these species throughout a past cold climate extreme and will likely continue to do so under predicted future warming."

Llewelyn *et al.* (2016) tested for intraspecific variation in climate-relevant traits in the rainforest sunskink (*Lampropholis coggeri*). They tested for four traits that are potentially important in determining a lizard species' climate sensitivity: critical thermal minimum, critical thermal maximum, thermal optimum for sprinting, and desiccation rate. Working in the Wet Tropics Bioregion of Australia, the researchers studied 12 populations of *L. coggeri*. They found "substantial variation both through time and across space in the measured traits," which the authors say suggests the lizards possess both "strong plasticity and substantial geographic variation." They conclude that if physiological variability similar to that observed in rainforest sunskinks occurs in

tropical rainforest species more generally, "these several taxa may not be as climatically specialized" and therefore "not as vulnerable to climate change, as previously thought."

References

Barrows, C.W. and Fisher, M. 2014. Past, present and future distributions of a local assemblage of congeneric lizards in southern California. *Biological Conservation* **180**: 97–107.

Dawson, W.R. 1967. Interspecific variation in physiological responses of lizards to temperature. In: Milstead, W.W. (Ed.) *Lizard Ecology Symposium*. Columbia, MO: University of Missouri Press, pp. 230–57.

IPCC. 2013. Intergovernmental Panel on Climate Change. *Climate Change 2013: The Physical Science Basis*. Contribution of Working Group I to the Fifth Assessment Report of the Intergovernmental Panel on Climate Change. New York, NY: Cambridge University Press.

Llewelyn, J., Macdonald, S.L., Hatcher, A., Moritz, C., and Phillips, B.L. 2016. Intraspecific variation in climate-relevant traits in a tropical rainforest lizard. *Diversity and Distribution*s **22**: 1000–12.

Logan, M.L., Cox, R.M., and Calsbeek, R. 2014. Natural selection on thermal performance in a novel thermal environment. *Proceedings of the National Academy of Sciences USA* **111**: 14,165–14,169.

Lopez-Alcaide, S., Nakamura, M., Macip-Rios, R., and Martinez-Meyer, E. 2014. Does behavioural thermoregulation help pregnant *Sceloporus adleri* lizards in dealing with fast environmental temperature rise? *Herpetological Journal* **24**: 41–7.

Stuart, Y.E., Campbell, T.S., Hohenlohe, P.A., Reynolds, R.G., Revell, L.J., and Losos, J.B. 2014. Rapid evolution of a native species following invasion by a congener. *Science* **346**: 463–6.

5.4.2 Future Impacts on Terrestrial Animals

Although there likely will be some changes in terrestrial animal population dynamics, few if any will be driven even close to extinction.

As noted in the introduction to Section 5.4, the IPCC's climate model simulations bear no resemblance to real-world observations of global warmth and the simulations are diverging further from reality over time (Green and Armstrong, 2014; Christy, 2017). Given that the IPCC's species-modeling research relies almost exclusively on those failed climate models, it comes as little surprise that its species extinction predictions are also failing.

Hundreds of studies, including those summarized here, have concluded that although there likely will be some changes in terrestrial animal population dynamics, few if any will be driven even close to extinction. Real-world data indicate warmer temperatures and higher atmospheric CO_2 concentrations will be beneficial, favoring a maintenance or even proliferation of species.

Anchukaitis and Evans (2010) write, "widespread amphibian extinctions in the mountains of the American tropics have been blamed on the interaction of anthropogenic climate change and a lethal pathogen," but they note that "limited meteorological records make it difficult to conclude whether current climate conditions at these sites are actually exceptional in the context of natural variability," challenging the contention that modern warming was the primary culprit in the demise of the Monteverde golden toad (*Bufo periglenes*).

Anchukaitis and Evans developed annual proxy records of hydroclimatic variability over the past century within the Monteverde Cloud Forest of Costa Rica, based on measurements of the stable oxygen isotope ratio ($\delta^{18}O$) made on trees lacking annual rings, as described in the papers of Evans and Schrag (2004) and Anchukaitis *et al.* (2008). Their work led them to conclude "the extinction of the Monteverde golden toad appears to have coincided with an exceptionally dry interval caused by the 1986–1987 El Niño event." They say their analysis suggests "the cause of the specific and well-documented extinction of the Monteverde golden toad was the combination of the abnormally strong ENSO-forced dryness and the lethality of the introduced chytrid fungus, but was not directly mediated by anthropogenic temperature trends, a finding from paleoclimatology that is in agreement with statistical reanalysis (Rohr *et al.*, 2008; Lips *et al.*, 2008) of the 'climate-linked epidemic hypothesis'."

Willis *et al.* (2010) considered the IPCC's (IPCC, 2007a) contentions that "global temperatures will increase by 2–4°C and possibly beyond, sea levels will rise (~1 m ± 0.5 m), and atmospheric CO_2 will increase by up to 1000 ppm." They note it is "widely suggested that the magnitude and rate of these changes will result in many plants and animals going extinct," citing studies that suggest "within the next century, over 35% of some biota will have gone

extinct (Thomas *et al.*, 2004; IPCC, 2007b) and there will be extensive die-back of the tropical rainforest due to climate change (e.g. Huntingford *et al.*, 2008)."

Willis *et al.* go on to note some biologists and climatologists have pointed out "many of the predicted increases in climate have happened before, in terms of both magnitude and rate of change (e.g. Royer, 2008; Zachos *et al.*, 2008), and yet biotic communities have remained remarkably resilient (Mayle and Power, 2008) and in some cases thrived (Svenning and Condit, 2008)." They report that those who mention such things are often "placed in the 'climate-change denier' category," although the purpose for pointing out these facts is simply to present "a sound scientific basis for understanding biotic responses to the magnitudes and rates of climate change predicted for the future through using the vast data resource that we can exploit in fossil records."

Willis *et al.* focus on "intervals in time in the fossil record when atmospheric CO_2 concentrations increased up to 1200 ppm, temperatures in mid- to high-latitudes increased by greater than 4°C within 60 years, and sea levels rose by up to 3 m higher than present," describing studies of past biotic responses that indicate "the scale and impact of the magnitude and rate of such climate changes on biodiversity." What emerges from those studies, they write, "is evidence for rapid community turnover, migrations, development of novel ecosystems and thresholds from one stable ecosystem state to another." And, most importantly, they report "there is very little evidence for broad-scale extinctions due to a warming world." They conclude, "based on such evidence we urge some caution in assuming broad-scale extinctions of species will occur due solely to climate changes of the magnitude and rate predicted for the next century," reiterating that "the fossil record indicates remarkable biotic resilience to wide amplitude fluctuations in climate."

Mergeay and Santamaria (2012) introduce nine papers in a special issue of *Evolutionary Applications*, all of which were based on contributions to a meeting on Evolution and Biodiversity held in Mallorca, Spain (April 12–15, 2010) and a preparatory e-conference. Shine (2012) opens the special issue by "showing how evolution can rapidly modify ecologically relevant traits in invading as well as native species." Bijlsma and Loeschcke (2012) then "tackle the interaction of drift, inbreeding and environmental stress," while Angeloni *et al.* (2012) "provide a conceptual tool-box for genomic research in conservation biology and highlight some of its possibilities for the mechanistic study of functional variation, adaptation and inbreeding."

Van Dyck (2012) shows "an organism's perception of its environment is subject to selection, a mechanism that could reduce the initial impact of environmental degradation or alleviate it over the longer run." Urban et al. (2012) contend "certain consequences of global change can only be accounted for by interactions between ecological and evolutionary processes," and Lemaire et al. (2012) highlight "the important role of evolution in predator-prey interactions."

Focusing on eco-evolutionary interactions, Palkovacs et al. (2012) "review studies on phenotypic change in response to human activities" and "show that phenotypic change can sometimes cascade across populations, communities and even entire ecosystems," while Bonduriansky et al. (2012) examine "non-genetic inheritance and its role in adaptation," dissecting "the diversity of epigenetic and other transgenerational effects." Finally, Santamaria and Mendez (2012) "build on the information reviewed in all previous papers to identify recent advances in evolutionary knowledge of particular importance to improve or complement current biodiversity policy."

"Overall," Mergeay and Santamaria conclude, "these nine papers offer compelling evidence for the role of evolutionary processes in the maintenance of biodiversity and the adaptation to global change."

In summary, terrestrial animals are able to adapt to climate change occurring on scales that surpass those forecast even by the IPCC. Claims to the contrary invariably rely on the IPCC's flawed forecasts of future climate conditions and species survival models that overlook or ignore the documented real-world responses to change by many species. Climate change is not a threat to terrestrial animals and its impact on wildlife is not a cost. History suggests it may even be a benefit.

References

Anchukaitis, K.J. and Evans, M.N. 2010. Tropical cloud forest climate variability and the demise of the Monteverde golden toad. *Proceedings of the National Academy of Sciences USA* **107**: 5036–40.

Anchukaitis, K.J., Evans, M.N., Wheelwright, N.T., and Schrag, D.P. 2008. Stable isotope chronology and climate signal in neotropical montane cloud forest trees. *Journal of Geophysical Research* **113**: G03030.

Angeloni, F., Wagemaker, C.A.M., Vergeer, P., and Ouborg, N.J. 2012. Genomic toolboxes for conservation biologists. *Evolutionary Applications* **5**: 130–43.

Bijlsma, R. and Loeschcke, V. 2012. Genetic erosion impedes adaptive responses to stressful environments. *Evolutionary Applications* **5**: 117–29.

Bonduriansky, R., Crean, A.J., and Day, D.T. 2012. The implications of nongenetic inheritance for evolution in changing environments. *Evolutionary Applications* **5**: 192–201.

Christy, J. 2017. Testimony before the U.S. House Committee on Science, Space & Technology. March 29.

Evans, M.N. and Schrag, D.P. 2004. A stable isotope-based approach to tropical dendroclimatology. *Geochimica et Cosmochimica Acta* **68**: 3295–305.

Green, K.C. and Armstrong, J.S. 2014. Forecasting global climate change. In: Moran, A. (Ed.) *Climate Change: The Facts 2014*. Melbourne, Australia: IPA, pp. 170–86.

Huntingford, C., et al. 2008. Towards quantifying uncertainty in predictions of Amazon 'dieback'. *Philosophical Transactions of the Royal Society B: Biological Sciences* **363**: 1857–64.

IPCC. 2007a. Intergovernmental Panel on Climate Change. *Climate Change 2007: Impacts, Adaptation and Vulnerability*. Contribution of Working Group II to the Fourth Assessment Report of the Intergovernmental Panel on Climate Change. Cambridge, UK and New York, NY: Cambridge University Press.

IPCC. 2007b. Intergovernmental Panel on Climate Change. *Climate Change 2007: The Physical Science Basis*. Contribution of Working Group I to the Fourth Assessment Report of the Intergovernmental Panel on Climate Change. Cambridge, UK and New York, NY: Cambridge University Press.

Lemaire, V., Bruscotti, S., Van Gremberghe, I., Vyverman, W., Vanoverbeke, J., and De Meester, L. 2012. Genotype x genotype interactions between the toxic cyanobacterium *Microcystis* and its grazer, the water flea *Daphnia*. *Evolutionary Applications* **5**: 168–82.

Lips, K.R., Diffendorfer, J., Mendelson III, J.R., and Sears, M.W. 2008. Riding the wave: reconciling the roles of disease and climate change in amphibian declines. *PLoS Biology* **6** (3): e72.

Mayle, F.E. and Power, M.J. 2008. Impact of a drier Early-Mid-Holocene climate upon Amazonian forests.

Philosophical Transactions of the Royal Society B: Biological Sciences **363**: 1829–38.

Mergeay, J. and Santamaria, L. 2012. Evolution and biodiversity: the evolutionary basis of biodiversity and its potential for adaptation to global change. *Evolutionary Applications* **5**: 103–6.

Palkovacs, E., Kinnison, M.T., Correa, C., Dalton, C.M., and Hendry, A. 2012. Ecological consequences of human-induced trait change: fates beyond traits. *Evolutionary Applications* **5**: 183–91.

Rohr, J.R., Raffel, T.R., Romansic, J.M., McCallum, H., and Hudson, P.J. 2008. Evaluating the links between climate, disease spread, and amphibian declines. *Proceedings of the National Academy of Sciences USA* **105**: 17,436–17,441.

Royer, D.L. 2008. Linkages between CO_2, climate, and evolution in deep time. *Proceedings of the National Academy of Sciences USA* **105**: 407–8.

Santamaria, L. and Mendez, P.F. 2012. Evolution in biodiversity policy: current gaps and future needs. *Evolutionary Applications* **5**: 202–18.

Shine, R. 2012. Invasive species as drivers of evolutionary change: cane toads in tropical Australia. *Evolutionary Applications* **5**: 107–16.

Svenning, J.C. and Condit, R. 2008. Biodiversity in a warmer world. *Science* **322**: 206–7.

Thomas, C.D., *et al.* 2004. Extinction risk from climate change. *Nature* **427**: 145–8.

Urban, M.C., De Meester, L., Vellend, M., Stoks, R., and Vanoverbeke, J. 2012. A crucial step towards realism: responses to climate change from an evolving metacommunity perspective. *Evolutionary Applications* **5**: 154–67.

Van Dyck, H. 2012. Changing organisms in rapidly changing anthropogenic landscapes: the significance of the "Umwelt"-concept and functional habitat for animal conservation. *Evolutionary Applications* **5**: 144–53.

Willis, K.J., Bennett, K.D., Bhagwat, S.A., and Birks, H.J.B. 2010. 4°C and beyond: what did this mean for biodiversity in the past? *Systematics and Biodiversity* **8**: 3–9.

Zachos, J.C., Dickens, G.R., and Zeebe, R.E. 2008. An early Cenozoic perspective on greenhouse warming and carbon-cycle dynamics. *Nature* **451**: 279–83.

5.5. Impact on Aquatic Life

The IPCC's forecasts of dire consequences for life in the world's oceans rely on falsified computer models and are contradicted by real-world observations.

The Working Group I contribution to the IPCC's Fifth Assessment Report (IPCC, 2013) warns that rising atmospheric CO_2 concentrations will harm aquatic life via changes in ocean temperature/heat content, salinity, and pH balance. That warning is based on the climate-model-driven claim that human emissions of CO_2 will cause Earth to warm unnaturally.

As noted in the introduction to this section, the IPCC's climate model simulations bear no resemblance to real-world observations of global warmth and the simulations are diverging further from reality over time (Green and Armstrong, 2014; Christy, 2017). Therefore, the assumptions about temperature, precipitation, weather, and other climate factors typically fed into the models used to forecast the impact of climate change on marine life are invalid, invalidating the models outputs.

Several researchers have specifically noted the sensitivity of ocean warming projections to "temperature biases associated with differing instrumentation" (Gouretski and Koltermann, 2007) and data-processing methods (Carson and Harrison, 2008). Lyman *et al.* (2006) point out that ocean temperature is highly variable and "this variability is not adequately simulated in the current generation of coupled climate models used to study the impact of anthropogenic influences on climate," which "may complicate detection and attribution of human-induced climate influences."

Natural variability is also the rule and not the exception regarding ocean pH levels. Liu *et al.* (2009) studied 18 samples of fossil and modern *Porites* corals recovered from the South China Sea, employing ^{14}C dating and positive thermal ionization mass spectrometry to generate high precision $\delta^{11}B$ (boron) data. From that data they reconstructed the paleo-pH record of the past 7,000 years that is depicted in Figure 5.5.1.

Figure 5.5.1 shows there is nothing unusual, unnatural, or unprecedented about the two most recent pH values (shown on the far right edge of the figure). Hence, there is no compelling reason to believe they were significantly influenced by the nearly 40% increase in the air's CO_2 concentration that occurred during Industrial Revolution. As for the

Figure 5.5.1
Reconstructed pH history of the South China Sea

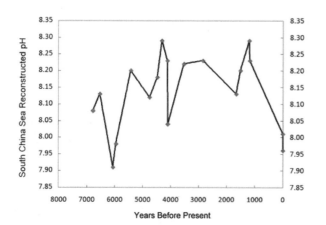

Source: Created from Table 1 of Liu *et al.*, 2009.

Figure 5.5.2
Reconstructed pH history of Arlington Reef off the northeast coast of Australia

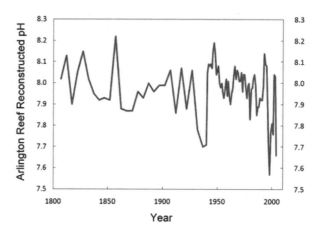

Source: Adapted from Wei *et al.*, 2009.

prior portion of the record, Liu *et al.* note there is also "no correlation between the atmospheric CO_2 concentration record from Antarctica ice cores and

Wei *et al.* (2009) derived the pH history of Arlington Reef (off the northeast coast of Australia) that is depicted in Figure 5.5.2. As can be seen there, there was a ten-year pH minimum centered at about 1935 (which obviously was not CO_2-induced) and a shorter more variable minimum at the end of the record (which also was not CO_2-induced); and apart from these two non-CO_2-related exceptions, the majority of the data once again fall within a band that exhibits no long-term trend, such as would be expected to have occurred if the gradual increase in atmospheric CO_2 concentration since the inception of the Industrial Revolution were truly making the global ocean less basic.

Coral bleaching models also are flawed, many assuming a fixed thermal tolerance not exhibited by corals in the real world, as documented below. For fish, models are often too coarse, assuming a broad species-specific response to change when in fact responses can vary by genetic lineages and even among populations within those lineages.

In contrast to the alarming projections of the IPCC's flawed computer models, real-world observations suggest aquatic species, like terrestrial plants and animals, are built to survive changes in their environment, including those that might develop in a world of increasing atmospheric CO_2. A previous

volume in the *Climate Change Reconsidered* series (Idso *et al.*, 2014) found hundreds of peer-reviewed studies suggesting a much better future is in store for Earth's aquatic life. NIPCC's 2014 key findings regarding aquatic life, which challenge the alarming and negative projections of the IPCC, are presented in Figure 5.5.3.

More recent research is summarized below in Sections 5.5.1 on the ability of corals and fish to adapt to climate change and Section 5.5.2, on the likely future impacts of climate change on aquatic life. Many laboratory and field studies demonstrate toleration, adaptation, and even growth and developmental improvements in aquatic life in response to higher temperatures and reduced water pH levels. When these observations are considered, the pessimistic projections of the IPCC give way to considerable optimism with respect to the future of the planet's marine life.

References

Gouretski, V. and Koltermann, K.P. 2007. How much is the ocean really warming? *Geophysical Research Letters* **34**: 10.1029/2006GL027834.

Green, K.C. and Armstrong, J.S. 2014. Forecasting global climate change. In: Moran, A. (Ed.) *Climate Change: The Facts 2014*. Melbourne, Australia: IPA, pp. 170–186.

IPCC. 2013. Intergovernmental Panel on Climate Change. Observations: Ocean. In: *Climate Change 2013: The Physical Science Basis*. Contribution of Working Group I to the Fifth Assessment Report of the Intergovernmental Panel on Climate Change. Cambridge, UK and New York, NY: Cambridge University Press.

Liu, Y., Liu, W., Peng, Z., Xiao, Y., Wei, G., Sun, W., He, J., Liu, G., and Chou, C.-L. 2009. Instability of seawater pH in the South China Sea during the mid-late Holocene: evidence from boron isotopic composition of corals. *Geochimica et Cosmochimica Acta* **73**: 1264–72.

Lyman, J.M., Willis, J.K., and Johnson, G.C. 2006. Recent cooling of the upper ocean. *Geophysical Research Letters* **33**: 10.1029/2006GL027033.

NIPCC. 2014. Idso, C.D, Idso, S.B., Carter, R.M., and Singer, S.F. (Eds.) *Climate Change Reconsidered II: Biological Impacts*. Nongovernmental International Panel on Climate Change. Chicago, IL: The Heartland Institute.

Tans, P. 2009. An accounting of the observed increase in oceanic and atmospheric CO_2 and an outlook for the future. *Oceanography* **22**: 26–35.

Wei, G., McCulloch, M.T., Mortimer, G., Deng, W., and Xie, L. 2009. Evidence for ocean acidification in the Great Barrier Reef of Australia. *Geochimica et Cosmochimica Acta* **73**: 2332–46.

Figure 5.5.3
Key Findings: Impacts on aquatic life

- Multiple studies from multiple ocean regions confirm ocean productivity tends to increase with temperature. Subjects of this research include phytoplankton and macroalgae, corals, crustaceans, and fish.

- Rising seawater temperature is conducive to enhanced coral calcification, leading some experts to forecast coral calcification will increase by about 35% beyond pre-industrial levels by 2100 and no extinction of coral reefs will occur in the future.

- Many aquatic species demonstrate the capability to adjust their individual critical thermal maximum (the upper temperature at which the onset of behavioral incapacitation occurs) upwards in response to temperature increases of the amount forecast by the IPCC.

- Aquatic life has survived decadal, centennial, and millennial-scale climate oscillations that have persisted for millions of years. Evidence indicates aquatic species are well-equipped to adapt to forecasted increases in temperature, if necessary.

- Caution should be applied when interpreting results from laboratory-based studies of lower seawater pH levels. Such studies often are incapable, or fall far short, of mimicking conditions in the real world and thus they frequently yield results quite different than what is observed in nature.

- Rising atmospheric CO_2 levels do not pose a significant threat to aquatic life. Many aquatic species have shown considerable tolerance to declining pH values predicted for the next few centuries and many have demonstrated a likelihood of positive responses in empirical studies.

- The projected decline in ocean pH levels in the year 2100 (as compared to preindustrial times) may be significantly overstated, amounting to only half of the 0.4 value the IPCC predicts.

- The natural variability of oceanic pH is often much greater than the change in pH levels forecast by the IPCC.

- Natural fluctuations in pH may have a large impact on the development of resilience in marine populations, as heterogeneity in the environment with regard to pH and pCO_2 exposure may result in populations that are acclimatized to variable pH or extremes in pH.

- For those aquatic species showing negative responses to pH declines in experimental studies, there are adequate reasons to conclude such responses will be largely mitigated through phenotypic adaptation or evolution during the many decades to centuries the pH concentration is projected to fall.

Source: Chapter 6. "Aquatic Life," *Climate Change Reconsidered II: Biological Impacts.* Nongovernmental International Panel on Climate Change. Chicago, IL: The Heartland Institute, 2014.

5.5.1   Evidence of Ability to Adapt

Aquatic life demonstrates tolerance, adaptation, and even growth and developmental improvements in response to higher temperatures and reduced water pH levels.

The effects of climate change on coral reefs and fish have been extensively studied. As the recent research summarized here indicates, corals and fish are capable of tolerance, acclimation, and adaptation, allowing them to successfully cope with future changes in their aquatic environment.

5.5.1.1 Corals

By inducing changes in ocean water chemistry that can lead to reductions in the calcium carbonate saturation state of seawater (Ω), which lowers the water's pH level, it has been predicted that elevated levels of atmospheric CO_2 may reduce rates of coral calcification, possibly leading to slower-growing – and, therefore, weaker – coral skeletons, and in some cases even death (Barker and Ridgwell, 2012). Such projections, however, often fail to account for the fact that coral calcification is a biologically mediated process and in the real world, living organisms tend to find a way to meet and overcome the many challenges they face. Coral calcification in response to so-called "ocean acidification" is no exception.

Pelejero *et al.* (2005) developed a reconstruction of seawater pH spanning the period 1708–1988, based on the boron isotopic composition ($\delta^{11}B$) of a long-lived massive *Porites* coral from Flinders Reef in the western Coral Sea of the southwestern Pacific. They found "no notable trend toward lower $\delta^{11}B$ values" over the 300-year period, which began "well before the start of the Industrial Revolution." Instead, they say "the dominant feature of the coral $\delta^{11}B$ record is a clear interdecadal oscillation of pH, with $\delta^{11}B$ values ranging between 23 and 25 per mil (7.9 and 8.2 pH units)," which "is synchronous with the Interdecadal Pacific Oscillation." Furthermore, they calculated aragonite saturation state values from the Flinders pH record that varied between about 3 and 4.5, which values encompass, in their words, "the lower and upper limits of aragonite saturation state within which corals can survive." Nevertheless, they report the "skeletal extension and calcification rates for the Flinders Reef coral fall within the normal range for *Porites* and *are not correlated with aragonite saturation state* [italics added]."

Working with specimens of *Montipora capitata*, Bahr *et al.* (2017) investigated the "direct and interactive effects of temperature, irradiance, and pCO_2" on the growth of this important Hawaiian reef-building coral. Their work was accomplished at the Hawaii Institute of Marine Biology, University of Hawaii, in a mesocosm system that represented present and projected conditions of climate change, including 12 experimental regimes consisting of two temperature levels (ambient and +2°C), three irradiance conditions (ambient, 50% reduction, and 90% reduction) and two pCO2 values (ambient and twice ambient). Over a period of approximately two years, several replicates of the various treatment conditions were conducted. The authors measured net coral calcification and through statistical analysis were able to untangle its relationship among these three factors.

The analysis revealed that temperature and irradiance were the primary factors driving net calcification of *M. capitata* and "the effect of pCO_2 acting alone and/or with other predictors did not contribute to the multiple regression model." Bahr *et al.* conclude, "ocean warming in shallow water environments with high irradiance poses a more immediate threat to coral growth than acidification for this dominant coral species." Indeed, ocean acidification appears to pose no threat to *M. capitata* since its influence carried no predictive power in their regression model of factors influencing net calcification.

McCulloch *et al.* (2017) developed geochemical proxies ($\delta11B$ and B/Ca) from *Porites* corals located on Davis Reef, a mid-shelf reef located east-northeast of Townsville, Queensland, Australia in the central Great Barrier Reef, and Coral Bay, which is part of the Ningaloo Reef coastal fringing system of Western Australia. They obtained seasonal records of dissolved inorganic carbon (DIC) and pH of the corals' calcifying fluid (cf) at these locations for the period 2007–2012. The records revealed that coral colonies from both reef locations "exhibit strong seasonal changes in pHcf, from ~8.3 during summer to ~8.5 during winter," which "represents an elevation in pHcf relative to ambient seawater of ~0.4 pH units together with a relatively large seasonal range in pHcf of ~0.2 units."

These observations, McCulloch *et al.* note, "are in stark contrast to the far more muted changes based on laboratory-controlled experiments," which laboratory-based values are "an order of magnitude smaller than those actually observed in reef environments." With respect to DICcf , they report that the "highest DICcf (~ x 3.2 seawater) is found during summer, consistent with thermal/light enhancement of metabolically (zooxanthellae) derived carbon, while the highest pHcf (~8.5) occurs in winter during periods of low DICcf (~ x 2 seawater)."

The proxy records also revealed that coral DICcf was inversely related (r2 ~ 0.9) to pHcf. Commenting on this relationship, McCulloch *et al.* say it "indicate[s] that the coral is actively maintaining both high (~x 4 to x 6 seawater) and relatively stable (within ± 10% of mean) levels of elevated Ωcf year-round." Or, as they explain it another way, "we have now identified the key functional characteristics of chemically controlled calcification in reef-building coral. The seasonally varying supply of summer-enhanced metabolic DICcf is accompanied by dynamic out-of-phase upregulation of coral pHcf. These parameters acting together maintain elevated but near-constant levels of carbonate saturation state (Ωcf) of the coral's calcifying fluid, the key driver of calcification."

The implications of the McCulloch *et al.* findings are enormous, for they reveal that "pHcf upregulation occurs largely independent of changes in seawater carbonate chemistry, and hence ocean acidification," demonstrating "the ability of the coral to 'control'

what is arguably one of its most fundamental physiological processes, the growth of its skeleton within which it lives." Furthermore, McCulloch *et al.* say their work presents "major ramifications for the interpretation of the large number of experiments that have reported a strong sensitivity of coral calcification to increasing ocean acidification," explaining that "an inherent limitation of many of these experiments is that they were generally conducted under conditions of fixed seawater pHsw and/or temperature, light, nutrients, and little water motion, hence conditions that are not conducive to reproducing the natural interactive effects between pHcf and DICcf that we have documented here." They conclude that "since the interactive dynamics of pHcf and DICcf upregulation do not appear to be properly simulated under the short-term conditions generally imposed by such artificial experiments, the relevance of their commonly reported finding of reduced coral calcification with reduced seawater pH must now be questioned."

Moving from ocean pH to temperature, the ability of coral reefs to survive climate change "will depend partly on the relative rates of increase of thermal tolerance and of environmental temperatures" (Bay and Palumbi, 2015). The degree of thermal tolerance depends, in large measure, on an organism's ability to adapt (evolutionary change) and/or acclimate (physiological change) to temperature stress. In long-lived organisms, acclimation generally produces the more rapid response.

A growing body of work demonstrates the strong potential for recovery from coral bleaching at various places around the world, with perhaps an evolving potential for corals to successfully recover from increasingly more intense warming episodes in the face of rising global temperatures.

Yamano *et al.* (2011) report, "although most studies of climate change effects on corals have focused on temperature-induced coral bleaching in tropical areas, poleward range shifts and/or expansions may also occur in temperate areas, as suggested by geological records and present-day eyewitnesses in several localities," citing the work of Greenstein and Pandolfi (2008) and Precht and Aronson (2004).

Yamano *et al.* collected records of coral species occurrence from eight temperate regions of Japan, where they obtained "the first large-scale evidence of the poleward range expansion of modern corals, based on 80 years of national records ... where century-long measurements of *in situ* sea-surface temperatures have shown statistically significant rises." They determined "four major coral species categories, including two key species for reef formation in tropical areas, showed poleward range expansions since the 1930s, whereas no species demonstrated southward range shrinkage or local extinction," adding "the speed of these expansions reached up to 14 km per year," which they say "is far greater than that for other species." They conclude that "temperate areas may serve as refugia for tropical corals in an era of global warming."

Carilli *et al.* (2012) write "there is evidence that corals may adapt to better withstand heat stress via a number of mechanisms," noting "corals might acquire more thermally-resistant symbionts (Buddemeier and Fautin, 1993; Rowan, 2004), or might increase their own physiological mechanisms to reduce bleaching susceptibility by producing oxidative enzymes (Coles and Brown, 2003) or photoprotective compounds (Salih *et al.*, 2000)." They point out that evidence suggests the susceptibility of a given coral or reef to bleaching depends on the thermal history of that coral or reef (Thompson and Van Woesik, 2009; Donner, 2011; Brown *et al.*, 2002).

Carilli *et al.* "collected cores from massive *Porites sp.* corals in the Gilbert Islands of Kiribati to investigate how corals along a natural gradient in temperature variability responded to recent heat stress events," examining "changes in coral skeletal growth rates and partial mortality scars (Carilli *et al.*, 2010) to investigate the impact of the bleaching event in 2004 (Donner, 2011) on corals from different temperature variability regimes."

They found the spatial patterns in skeletal growth rates and partial mortality scars found in corals from the central and northern islands suggest "corals subject to larger year-to-year fluctuations in maximum ocean temperature were more resistant to a 2004 warm-water event," and "a subsequent 2009 warm event had a disproportionately larger impact on those corals from the island with lower historical heat stress." They say their study indicates "coral reefs in locations with more frequent warm events may be more resilient to future warming."

Bellantuono *et al.* (2012) "tested the response of *Acropora millepora* to thermal preconditioning by exposing coral nubbins to 28°C (3°C below bleaching threshold) for 10 days, prior to challenging them with water temperatures of 31°C for 8 days." Additionally, "in another treatment (non-preconditioned), corals were exposed to 31°C without prior exposure to the 28°C treatment." They

discovered that short-term preconditioning to higher-than-ambient temperatures (but still 3°C below the experimentally determined bleaching threshold) for a period of ten days provided thermal tolerance for the coral and its symbionts.

Bellantuono *et al.* say their findings suggest "the physiological plasticity of the host and/or symbiotic components appears to play an important role in responding to ocean warming." They describe some real-world examples of where this phenomenon may have played a crucial role in preserving corals exposed to extreme warm temperatures in the past, citing Fang *et al.*, 1997; Middlebrook *et al.*, 2008; and Maynard *et al.*, 2008.

"To uncover the long-term impacts of elevated temperature exposure to corals from reefs that experience episodic upwelling," Mayfield *et al.* (2013) conducted a mesocosm-based experiment whereby *P. damicornis* specimens collected from an upwelling coral reef on Houbihu (a small embayment within Nanwan Bay, southern Taiwan) were exposed for nine months to nearly 30°C, a temperature the corals normally encounter *in situ* for just a few hours per year (Mayfield *et al.*, 2012).

They found, "upon nine months of exposure to nearly 30°C, all colony (mortality and surface area), polyp (*Symbiodinium* density and chlorophyll *a* content), tissue (total thickness), and molecular (gene expression and molecular composition)-level parameters were documented at similar levels between experimental corals and controls incubated at 26.5°C, suggesting that this species can readily acclimate to elevated temperatures that cause significant degrees of stress, or even bleaching and mortality, in conspecifics of other regions of the Indo-Pacific."

Mayfield *et al.* say "there is now a growing body of evidence to support the notion that corals inhabiting more thermally unstable habitats outperform conspecifics from reefs characterized by more stable temperatures when exposed to elevated temperatures," citing Coles (1975), Castillo and Helmuth (2005), and Oliver and Palumbi (2011).

Graham *et al.* (2015) document coral reef responses to the major warming-induced bleaching event of 1998 that caused unprecedented region-wide mortality of Indo-Pacific corals. They report, "following loss of more than 90% live coral cover, 12 of 21 reefs recovered towards pre-disturbance live coral states, while nine reefs underwent regime shifts to fleshy macroalgae." They determined recovery was favored when reefs were structurally complex and in deeper water; when the density of juvenile corals and herbivorous fishes was relatively high; and when nutrient loads were low. In a commentary on these findings, Pandolfi (2015) writes, "the fact that more than half of the reefs fully recovered after the bleaching event is a promising outcome for the future of coral reefs." Pandolfi writes, "put simply, many reef corals just might be capable of adapting fast enough to survive current rates of global environmental change," citing the work of Pandolfi *et al.* (2011) and Munday *et al.* (2013).

Bay and Palumbi (2015) conducted a laboratory-based experiment to investigate the temperature acclimation of a reef-building coral, *Acropora nana*. Colonies were subjected to three baseline temperature regimes, ambient (29°C), elevated (31°C), and variable (29–33°C, mimicking the diel tidal fluctuation range). After zero, two, seven, and 11 days of treatment, samples were taken and evaluated for their response to acute temperature stress (five hours of 34°C temperature). After seven days of exposure to each coral colony's respective baseline temperature regime, *A. nana* specimens subjected to acute temperature stress displayed a "striking increase in heat tolerance," which tolerance was higher in corals acclimated to elevated and variable temperature regimes as opposed to the ambient treatment.

Bay and Palumbi say their findings suggest corals "can track environmental temperatures better than previously believed," and that the observed temperature acclimation may provide "some protection for this species of coral against slow onset of warming ocean temperatures." They also note "such rapid change in heat sensitivity runs contrary to coral bleaching models based on fixed thermal tolerance that are currently used to predict coral bleaching and climate change response," citing the works of Liu *et al.* (2013) and Logan *et al.* (2013), which suggest predictions of future coral reef demise due to rising ocean temperatures are overstated.

In addition to acclimation, corals likely have another important tool in *adaptation*. Physiological acclimation, as that found by Bay and Palumbi (2015) is generally assumed to be the more common (and most rapid) mode of stress response among long-lived organisms like corals. But the work of Dixon *et al.* (2015) reveals there is also "the potential for rapid adaptation at the genetic level based on standing genetic variation." They studied *Acropora millepora* corals inhabiting two thermally divergent locations separated by 5° of latitude on the Great Barrier Reef: Princess Charlotte Bay and Orpheus Island.

Dixon *et al.* established 10 crosses according to a diallel scheme by "cross-fertilizing gametes from four adult colonies from the two locations." The heat tolerances of the larval crosses were analyzed based on their odds of survival after approximately 30 hours exposure to 35.5°C temperatures. Among many findings, Dixon *et al.* report "parents from the warmer location (Princess Charlotte Bay) conferred significantly higher thermo-tolerance to their offspring relative to parents from the cooler location (Orpheus Island); a dam from warmer Princess Charlotte Bay conferred a fivefold increase in survival odds and a sire from Princess Charlotte Bay conferred an additional twofold increase."

Dixon *et al.* conclude their study "demonstrates heritability of coral stress-related phenotypic and molecular traits and thus highlights the adaptive potential stemming from standing genetic variation in coral metapopulations." They write, "the genetic rescue scenario, therefore, emerges as a plausible mechanism of rapid coral adaptation to climate change, especially if the natural connectivity of corals across latitudes is enhanced by assisted colonization efforts."

Madeira *et al.* (2015) collected a large number of octocorals in the spring and summer of 2013 from a pristine sandy intertidal shore in Troia, Setubal, Portugal, during midday at low tide, when temperatures were normally at their daily extreme warmth, while simultaneously recording air temperature, intertidal water temperature, salinity, and pH. In the laboratory, they analyzed the activities of several antioxidant defense enzymes and other biomarkers, along with total octocoral protein content. They found "this species is able to withstand low tide conditions in warmer temperatures without evidence of thermal or oxidative stress." And they also state that, as observed by McClanahan *et al.* (2007), corals that experience the greatest temperature variability – at higher latitudes, as in this study – are also "the corals most capable of surviving in challenging conditions." Consequently, they predict "this species is likely to be quite resilient" or even to "thrive under future climate warming conditions."

References

Bahr, K.D., Jokiel, P.L., and Rodgers, K.S. 2017. Seasonal and annual calcification rates of the Hawaiian reef coral, *Montipora capitate*, under present and future climate change scenarios. *ICES Journal of Marine Science* **74**: 1083–91.

Barker, S. and Ridgwell, A. 2012. Ocean acidification. *Nature Education Knowledge* **3** (10): 21.

Bay, R.A. and Palumbi, S.R. 2015. Rapid acclimation ability mediated by transcriptome changes in reef-building corals. *Genome Biology and Evolution* **7**: 1602–12.

Bellantuono, A.J., Hoegh-Guldberg, O., and Rodriguez-Lanetty, M. 2012. Resistance to thermal stress in corals without changes in symbiont composition. *Proceedings of the Royal Society B* **279**: 1100–7.

Brown, N., Dunne, R., Goodson, M., and Douglas, A. 2002. Experience shapes the susceptibility of a reef coral to bleaching. *Coral Reefs* **21**: 119–26.

Buddemeier, R.W. and Fautin, D.G. 1993. Coral bleaching as an adaptive mechanism. *BioScience* **43**: 320–6.

Carilli, J., Donner, S.D., and Hartmann, A.C. 2012. Historical temperature variability affects coral response to heat stress. *PLoS ONE* **7**: e34418.

Carilli, J., Norris, R.D., Black, B., Walsh, S.W., and McField, M. 2010. Century-scale records of coral growth rates indicate that local stressors reduce coral thermal tolerance threshold. *Global Change Biology* **16**: 1247–57.

Castillo, K.D. and Helmuth, B.S.T. 2005. Influence of thermal history on the response of *Montastraea annularis* to short-term temperature exposure. *Marine Biology* **148**: 261–70.

Coles, S. 1975. A comparison of effects of elevated temperature versus temperature fluctuations on reef corals at Kahe Point, Oahu. *Pacific Science* **29**: 15–8.

Coles, S.L. and Brown, B.E. 2003. Coral bleaching: capacity for acclimatization and adaptation. *Advances in Marine Biology* **46**: 183–223.

Dixon, G.B., Davies, S.W., Aglyamova, G.V., Meyer, E., Bay, L.K., and Matz, M.V. 2015. Genomic determinants of coral heat tolerance across latitudes. *Science* **348**: 1460–2.

Donner, S.D. 2011. An evaluation of the effect of recent temperature variability on the prediction of coral bleaching events. *Ecological Applications* **21**: 1718–30.

Fang, L.S., Huang, S.P., and Lin, K.L. 1997. High temperature induces the synthesis of heat-shock proteins and the elevation of intracellular calcium in the coral *Acropora grandis*. *Coral Reefs* **16**: 127–31.

Graham, N.A.J., Jennings, S., MacNeil, M.A., Mouillot, D., and Wilson, S.K. 2015. Predicting climate-driven regime shifts versus rebound potential in coral reefs. *Nature* **518**: 94–7.

Greenstein, B.J. and Pandolfi, J.M. 2008. Escaping the heat: range shifts of reef coral taxa in coastal Western Australia. *Global Change Biology* **14**: 513–28.

Liu, G., Rauenzahn, J.L., Heron, S.F., Eakin, C.M., Skirving, W.J., Christensen, T.R.L., and Strong, A.E. 2013. *NOAA Coral Reef Watch 50 km Satellite Sea Surface Temperature-based Decision Support System for Coral Bleaching Management*. Silver Spring, MD: NOAA/NESDIS.

Logan, C.A., Dunne, J.P., Eakin, C.M., and Donner, S.D. 2013. Incorporating adaptive responses into future projections of coral bleaching. *Global Change Biology* **20**: 125–39.

Madeira, C., Madeira, D., Vinagre, C., and Diniz, M. 2015. Octocorals in a changing environment: seasonal response of stress biomarkers in natural populations of *Veretillum cynomorium*. *Journal of Sea Research* **103**: 120–8.

Mayfield, A.B., Chan, P.H., Putnam, H.P., Chen, C.S., and Fan, T.Y. 2012. The effects of a variable temperature regime on the physiology of the reef-building coral *Seriatopora hystrix*: results from a laboratory-based reciprocal transplant. *Journal of Experimental Biology* **215**: 4183–95.

Mayfield, A.B., Chen, M., Meng, P.J., Lin, H.J., Chen, C.S., and Liu, P.J. 2013. The physiological response of the reef coral *Pocillopora damicornis* to elevated temperature: results from coral reef mesocosm experiments in southern Taiwan. *Marine Environmental Research* **86**: 1–11.

Maynard, J.A., Anthony, K.R.N., Marshall, P.A., and Masiri, I. 2008. Major bleaching events can lead to increased thermal tolerance in corals. *Marine Biology* **155**: 173–82.

McClanahan, T.R., Atweberhan, M., Ruiz-Sebastian, C., Graham, N.A.J., Wilson, S., Bruggemann, J.H., and Guillaume, M.M.M. 2007. Predictability of coral bleaching from synoptic satellite and in situ temperature observations. *Coral Reefs* **26**: 695–701.

McCulloch, M.T., D'Olivo, J.P., Falter, J., Holcomb, M., and Trotter, J.A. 2017. Coral calcification in a changing world and the interactive dynamics of pH and DIC upregulation. *Nature Communications* **8**: 15686.

Middlebrook, R., Hoegh-Guldberg, O., and Leggat, W. 2008. The effect of thermal history on the susceptibility of reef-building corals to thermal stress. *Journal of Experimental Biology* **211**: 1050–6.

Munday, P.L., Warner, R.R., Monro, K., Pandolfi, J.M., and Marshall, D.J. 2013. Predicting evolutionary responses to climate change in the sea. *Ecology Letters* **16**: 1488–1500.

Oliver, T.A. and Palumbi, S.R. 2011. Do fluctuating temperature environments elevate coral thermal tolerance? *Coral Reefs* **30**: 429–40.

Pandolfi, J.M. 2015. Deep and complex ways to survive bleaching. *Nature* **518**: 43–4.

Pandolfi, J.M., Connolly, S.R., Marshall, D.J., and Cohen, A.L. 2011. Projecting coral reef futures under global warming and ocean acidification *Science* **333**: 418–22.

Pelejero, C., Calvo, E., McCulloch, M.T., Marshall, J.F., Gagan, M.K., Lough, J.M., and Opdyke, B.N. 2005. Preindustrial to modern interdecadal variability in coral reef pH. *Science* **309**: 2204–7.

Precht, W.F. and Aronson, R.B. 2004. Climate flickers and range shifts of reef corals. *Frontiers in Ecology and the Environment* **2**: 307–14.

Rowan, R. 2004. Coral bleaching: thermal adaptation in reef coral symbionts. *Nature* **430**: 742.

Salih, A., Larkum, A., Cox, G., Kuhl, M., and Hoegh-Guldberg, O. 2000. Fluorescent pigments in corals are photoprotective. *Nature* **408**: 850–3.

Thompson, D.M. and Van Woesik, R. 2009. Corals escape bleaching in regions that recently and historically experienced frequent thermal stress. *Proceedings of the Royal Society B* **276**: 2893.

Yamano, H., Sugihara, K., and Nomura, K. 2011. Rapid poleward range expansion of tropical reef corals in response to rising sea surface temperatures. *Geophysical Research Letters* **38**: L04601.

5.5.1.2 Fish

Thermal tolerance, acclimation, and adaptation are evident in freshwater and ocean fish species as well. They may alter their ranges or behavior; over time, they can even evolve traits, such as body size, resistance to parasitic infection, and swimming ability, that make it easier for them to cope with a changing environment. A growing body of evidence, including the recent research summarized here, shows fish are not the fragile creatures the IPCC makes them out to be.

Seebacher *et al.* (2012) analyzed six populations of mosquitofish (*Gambusia holbrooki*) from coastal and mountain environments and compared their capacity for thermal acclimation, demonstrating that mosquitofish populations "are divided into distinct genetic lineages and that populations within lineages have distinct genetic identities." They report "there were significant differences in the capacity for acclimation between traits (swimming performance, citrate synthase and lactate dehydrogenase activities), between lineages, and between populations within lineages," thereby demonstrating "there can be substantial variation in thermal plasticity between populations within species."

Noting "many predictions of the impact of climate change on biodiversity assume a species-specific response to changing environments," Seebacher *et al.* say "this resolution can be too coarse and that analysis of the impacts of climate change and other environmental variability should be resolved to a population level," since their findings suggest some populations of a species may be able to cope with a change others may not be able to tolerate.

Stitt *et al.* (2014) studied the upper thermal tolerance and capacity for acclimation in three captive populations of brook trout (*Salvelinus fontinalis*), which they obtained from three ancestral environments that differed in their upper thermal tolerance and capacity for acclimation. Building on a number of pioneering studies of thermal performance in cold-water fish (e.g., Fry *et al.*, 1946; Brett, 1952; Brett *et al.*, 1958; McCauley, 1958), they say their research revealed "populations can possess substantial thermal acclimation capacity, as well as heritable variation in thermal tolerance among populations," further citing the work of Danzmann *et al.* (1998) and Timusk *et al.* (2011).

Stitt *et al.* report the three populations they studied "differed in their upper thermal tolerance and capacity for acclimation, consistent with their ancestry," in that "the northernmost strain had the lowest thermal tolerance, while the strain with the most southern ancestry had the highest thermal tolerance." They conclude, "with changing climatic conditions, populations of brook trout may have some degree of plasticity to cope with acute and chronic thermal stressors."

Shama *et al.* (2014) write, "empirical evidence is accumulating that marine species might be able to adapt to rapid environmental change if they have sufficient standing variation (the raw material for evolutionary change) and/or phenotypic plasticity to mount fast responses," citing the studies of Munday *et al.* (2013) and Sunday *et al.* (2014). They used a combined experimental approach – transgenerational plasticity (TGP) along with quantitative genetics – to partition the relative contributions of maternal and paternal (additive genetic) effects to offspring body size, a key fitness component of marine sticklebacks.

Shama *et al.* found "TGP can buffer short-term detrimental effects of climate warming and may buy time for genetic adaptation to catch up, therefore markedly contributing to the evolutionary potential and persistence of populations under climate change."

Narum and Campbell (2015) note "thermal adaptation is a widespread phenomenon in organisms that are exposed to variable and extreme environments," adding "some organisms may alter their distribution or behavior to avoid stressors and others may acclimate through physiological plasticity, [and] many species evolve adaptive responses to local conditions over generations through natural selection," citing Dahloff and Rank (2000), Hoffman *et al.* (2003), and Kavanagh *et al.* (2010). They continue, "evolutionary adaptation to local environments has been demonstrated across a wide variety of taxa" – citing Keller and Seehausen (2012) – "and is expected to play a critical role for species with limited dispersal capabilities."

Narum and Campbell "tested for differential transcriptional response of ecologically divergent populations of redband trout (*Oncorhynthus mykiss gairdneri*) that had evolved in desert and montane climates." They reared each pure strain and their F1 cross "in a common garden environment ... exposed over four weeks to diel water temperatures that were similar to those experienced in desert climates within the species' range," after which "gill tissues were collected from the three strains of fish (desert, montane, F1 crosses) at the peak of heat stress and tested for mRNA expression differences across the transcriptome with RNA-seq."

Narum and Campbell found "redband trout from a desert climate have a much larger number of strongly differentially expressed genes than montane and F1 strains in response to heat stress, suggesting that a combination of genes has evolved for redband trout to adapt in their desert environment."

Cure *et al.* (2015) assessed the size structure and habitat associations of juvenile *Choerodon rubescens*, a popular reef fish, during the summer and autumn of 2013 (January–May) by means of an underwater visual census conducted across available shallow water habitats towards the southern range edge of their historic distribution. They report "high abundances of juveniles (up to 14 fish/40 m^2) were found in areas where they were previously absent or in low abundance." Based on the size structure of the

populations they encountered, they say "recruitment was estimated to occur during summer 2011–12 and 2012–13," which "coincides with water temperatures 1 to 2°C higher than long-term averages in the region, making conditions more favorable for recruits to survive in greater numbers." They say their finding "mirrors the well-established patterns observed on the east coast of Australia," citing the studies of Booth *et al.* (2007), Figueira *et al.* (2009), Figueira and Booth (2010), and Last *et al.* (2011).

In 1980, heated water from a nuclear power plant in Forsmark, Sweden began to be discharged into Biotest Lake, an adjacent artificial semi-enclosed lake in the Baltic Sea created in 1977. The heated water has raised the temperature of the lake by 6–10°C compared to the surrounding Baltic Sea, but other physical conditions between the lake and the sea are very similar.

A few years after the power plant began operation, scientists conducted a study to determine the effect of the lake's increased temperatures on the host-parasite dynamics between a fish parasite, the eyefluke (*Diplostomum baeri*), and its intermediate host, European perch (*Perca fluviatilis*). That analysis, performed in 1986 and 1987, revealed that perch in Biotest Lake experienced a higher degree of parasite infection compared to perch living in the cooler confines of the surrounding Baltic Sea (Höglund and Thulin, 1990), which finding is consistent with the IPCC's concerns that rising temperatures may lead to an increase in infectious diseases.

Mateos-Gonzales *et al.* (2015) returned to Biotest Lake and reexamined the host-parasite dynamic. They note Biotest Lake "provides an excellent opportunity to study the effect of a drastically changed environmental factor, water temperature, on the evolution of host-parasite interactions, in a single population recently split into two." They compared the prevalence and intensity of parasitic infection in perch populations growing in warmer Biotest Lake versus the natural population from the surrounding cooler Baltic Sea in 2013 and 2014. They also conducted a controlled laboratory experiment in which they exposed perch from both locations to *D. baeri*, comparing their infection rates.

The field results indicated the "intensity of infection in Baltic fish was on average 7.2 times higher than in the corresponding Biotest fish." In addition, Baltic fish were found to acquire "slightly more parasites as they age," whereas Biotest fish did not. With respect to the laboratory tests, Mateos-Gonzales *et al.* report exposure to parasites "did not have an effect in fish from the Biotest Lake, but it did in fish from the Baltic Sea," increasing their intensity of infection by nearly 40%.

Mateos-Gonzales *et al.* write the findings present "a dramatic contrast" to those reported nearly three decades earlier when Biotest fish were infected at a rate of "almost twice" that of Baltic fish. Compared to 1986/87, the intensity of parasitic infection in Biotest fish has fallen almost 80%, whereas it has decreased only slightly in Baltic fish. The authors conclude their results illustrate "how an increased temperature has potentially aided a dramatic change in host-parasite dynamics." They further note this adaptation has "direct implications for consequences of global climate change, as they show that fast environmental changes can lead to equally rapid evolutionary responses."

Veilleux *et al.* (2015) sequenced and assembled "*de novo* transcriptomes of adult tropical reef fish exposed developmentally or trans-generationally to projected future ocean temperatures and correlated the resulting expression profiles with acclimated metabolic traits from the same fish." They "identified 69 contigs [overlapping DNA sequences] representing 53 key genes involved in thermal acclimation of aerobic capacity," noting "metabolic genes were among the most upregulated trans-generationally, suggesting shifts in energy production for maintaining performance at elevated temperatures." They also found "immune- and stress-responsive genes were upregulated trans-generationally, indicating a new complement of genes allowing the second generation of fish to better cope with elevated temperatures."

Veilleux *et al.* conclude, "the plasticity of these genes and their strong correlation to known acclimating phenotypic traits suggests that they may be critical in aiding reef fishes and possibly other marine organisms to survive in a warmer future environment."

Madeira *et al.* (2016) examined the cellular stress response of a tropical clownfish species (*Amphiprion ocellaris*) exposed to elevated temperatures over a period of one month. Their experiment was conducted in a controlled laboratory setting in which they subjected juvenile *A. ocellaris* to either ambient (26°C) or elevated (30°C) temperatures, examining several biomarkers (e.g., stress proteins and antioxidants) in several tissue types (brain, gills, liver, intestine, and muscle) at zero, seven, 14, 21, and 28 days of temperature treatment. They write, "results showed that exposure time significantly interacted with temperature responses and tissue-

type, so in fact time influenced the organisms' reaction to elevated temperature." At Day 7 they observed significantly higher levels of biomarkers in fish in the high temperature environment, indicative of a typical thermal stress response. Thereafter, biomarker levels stabilized, showing either "a significant decrease in comparison with controls or no significant differences from the control" through the end of the experiment, which observations they suggest are indicative of temperature acclimation.

Madeira *et al.* write, "*A. ocellaris* probably lives far from its upper thermal limit and is capable of adjusting the protein quality control system and enzymes' activities to protect cell functions under elevated temperature," adding "these results suggest that this coral reef fish species presents a significant acclimation potential under ocean warming scenarios of +4°C."

Munday *et al.* (2017) reared offspring of wild-caught breeding pairs of the coral reef damselfish, *Acanthochromis polyacanthus*, for two generations at current-day and two elevated temperature treatments (+1.5 and +3.0°C), consistent with current climate change predictions, while "length, weight, body condition and metabolic traits (resting and maximum metabolic rate and net aerobic scope) were measured at four stages of juvenile development." They found "significant genotype x environment interactions indicated potential for adaptation of maximum metabolic rate and net aerobic scope at higher temperatures," noting "net aerobic scope was negatively correlated with weight," and indicating "any adaptation of metabolic traits at higher temperatures could be accompanied by a reduction in body size."

Munday *et al.* write their results suggest there is "a high potential for adaptation of aerobic scope to high temperatures, which could enable reef fish populations to maintain their performance as ocean temperatures rise." They also report "recent studies indicate that plasticity may be especially important in enabling populations of marine species to adjust to climate change," citing among others Munday (2014), Shama *et al.* (2014), and Thor and Dupont (2015), while also noting "this type of adaptive plasticity may buffer populations against the immediate effects of environmental change and give genetic adaptation time to catch up," citing the study of Chevin and Lande (2010).

Madeira *et al.* (2017) examined the acclimation potential of the common clownfish (*Amphiprion ocellaris*) to rising temperature, exposing juvenile fish to seawater temperatures of either 26° or 30°C for a period of four weeks, during which time they measured two biochemical markers – one involved in preventing protein damage (heat shock protein 70, Hsp70) and another involved in dealing with it (ubiquitin, Ub) – to determine the presence of thermal damage to cellular proteins.

Madiera *et al.* say there were no differences in survival rates among the control and elevated temperature treatments. However, they report that thermal stress was observed in the fish after one week of exposure (both biomarkers increased significantly), after which Ub levels decreased, which the authors say suggests "the animals were able to acclimate." Thereafter, "as the juveniles acclimated to the new temperature conditions, Hsp70 kept showing increased levels in order to maintain cellular homeostasis, while the degree of irreversible damage (protein denaturation) started to decrease, as shown by lower Ub levels." Thus, Madiera *et al.* conclude "*A. ocellaris* is capable of displaying a plastic response to elevated temperature by adjusting the protein quality control system to protect cell functions, without decreasing survival."

Madiera *et al.* say the observed physiological acclimation in *A. ocellaris* "may come as counterintuitive, considering that tropical species have evolved in a relatively stable thermal environment and are therefore expected to exhibit narrower thermal reaction norms," yet acclimate they did, indicating clownfish "do not seem to be in immediate danger due to direct effects of warming oceans."

References

Booth, D.J., Figueira, W.F., Gregson, M.A., Brown, L., and Beretta, G. 2007. Occurrence of tropical fishes in temperate southeastern Australia: role of the East Australian Current. *Estuarine and Coastal Shelf Science* **72**: 102–14.

Brett, J.R. 1952. Temperature tolerance in young Pacific salmon, genus *Oncorhynchus*. *Journal of the Fisheries Research Board of Canada* **9**: 265–323.

Brett, J.R., Hollands, M., and Alderdice, D.F. 1958. The effect of temperature on the cruising speed of young sockeye and coho salmon. *Journal of the Fisheries Research Board of Canada* **15**: 587–605.

Chevin, L.M. and Lande, R. 2010. When do adaptive plasticity and genetic evolution prevent extinction of a density-regulated population? *Evolution* **64**: 1143–50.

Cure, K., Hobbs, J-P.A., and Harvey, E.S. 2015. High recruitment associated with increased sea temperatures towards the southern range edge of a Western Australian endemic reef fish *Choerodon rubescens* (family Labridae). *Environmental Biology of Fishes* **98**: 1059–67.

Dahlhoff, E.P. and Rank, N.E. 2000. Functional and physiological consequences of genetic variation at phosphoglucose isomerase: Heat Shock protein expression is related to enzyme genotype in a montane beetle. *Proceedings of the National Academy of Sciences USA* **97**: 10,056–10,061.

Danzmann, R.G., Morgan, I.R.P., Jones, M.W., Bernatchez, L., and Ihssen, P.E. 1998. A major sextet of mitochondrial DNA phylogenetic assemblages extant in eastern North American brook trout (*Salvelinus fontinalis*): distribution and postglacial dispersal patterns. *Canadian Journal of Zoology* **76**: 1300–18.

Figueira, W.F. and Booth, D.J. 2010. Increasing ocean temperatures allow tropical fishes to survive overwinter in temperate waters. *Global Change Biology* **16**: 506–16.

Figueira, W., Biro, P., Booth, D., and Valenzuela, V. 2009. Performance of tropical fish recruiting to temperate habitat: role of ambient temperature and implications of climate change. *Marine Ecology Progress Series* **384**: 231–9.

Fry, F.E.J., Hart, J.S., and Walker, K.F. 1946. Lethal temperature relations for a sample of young speckled trout, *Salvelinus fontinalis*. *Publications of the Ontario Fisheries Research Laboratory* **66**: 1–35.

Hoffman, A.A., Sorensen, J.G., and Loeschcke, V. 2003. Adaptation of *Drosophila* to temperature extremes: bringing together quantitative and molecular approaches. *Journal of Thermal Biology* **28**: 175–216.

Höglund, J. and Thulin, J. 1990. The epidemiology of the metacercariae of *Diplostomum baeri* and *D. spathaceum* in perch (*Perca fluviatilis*) from the warm water effluent of a nuclear power station. *Journal of Helminthology* **64**: 139–50.

Kavanagh, K.D., Haugen, T.O., Gregersen, F., Jernvall, J., and Vollestad, L.A. 2010. Contemporary temperature-driven divergence in a Nordic freshwater fish under conditions commonly thought to hinder adaptation. *BMC Evolutionary Biology* **10**: 350.

Keller, I. and Seehausen, O. 2012. Thermal adaptation and ecological speciation. *Molecular Ecology* **21**: 782–99.

Last, P.R., White, W.T., Gledhill, D.C., Hobday, A.J., Brown, R., Edgar, G.J., and Pecl, G. 2011. Long-term shifts in abundance and distribution of a temperate fish fauna: a response to climate change and fishing practices. *Global Ecology and Biogeography* **20**: 58–72.

Madeira, C., Madeira, D., Diniz, M.S., Cabral, H.N., and Vinagre, C. 2016. Thermal acclimation in clownfish: an integrated biomarker response and multi-tissue experimental approach. *Ecological Indicators* **71**: 280–92.

Madeira, C., Madeira, D., Diniz, M.S., Cabral, H.N., and Vinagre, C. 2017. Comparing biomarker responses during thermal acclimation: a lethal vs. non-lethal approach in a tropical reef clownfish. *Comparative Biochemistry and Physiology, Part A* **204**: 104–12.

Mateos-Gonzalez, F., Sundström, L.F., Schmid, M., and Björklund, M. 2015. Rapid evolution of parasite resistance in a warmer environment: insights from a large scale field experiment. *PLoS ONE* **10**: e0128860.

McCauley, R.W. 1958. Thermal relations of geographic races of *Salvelinus*. *Canadian Journal of Zoology* **36**: 655–62.

Munday, P.L 2014. Transgenerational acclimation of fishes to climate change and ocean acidification. *F1000 Prime Reports* **6**: 99.

Munday, P.L., Donelson, J.M., and Domingos, J.A. 2017. Potential for adaptation to climate change in a coral reef fish. *Global Change Biology* **23**: 307–17.

Munday, P.L., Warner, R.R., Monro, K., Pandolfi, J.M., and Marshall, D.J. 2013. Predicting evolutionary responses to climate change in the sea. *Ecology Letters* **16**: 1488–1500.

Narum, S.R. and Campbell, N.R. 2015. Transcriptomic response to heat stress among ecologically divergent populations of redband trout. *BMC Genomics* **16**: 10.1186/s12864-015-1246-5.

Seebacher, F., Holmes, S., Roosen, N.J., Nouvian, M., Wilson, R.S., and Ward, A.J.W. 2012. Capacity for thermal acclimation differs between populations and phylogenetic lineages within a species. *Functional Ecology* **26**: 1418–28.

Shama, L.N.S., Strobel, A., Mark, F.C., and Wegner, K.M. 2014. Transgenerational plasticity in marine sticklebacks: maternal effects mediate impacts of a warming ocean. *Functional Ecology* **28**: 1482–93.

Stitt, B.C., Burness, G., Burgomaster, K.A., Currie, S., McDermid, J.L., and Wilson, C.C. 2014. Intraspecific variation in thermal tolerance and acclimation capacity in brook trout (*Salvelinus fontinalis*): physiological implications for climate change. *Physiological and Biochemical Zoology* **87**: 15–29.

Sunday, J.M., Calosi, P., Dupont, S., Munday, P.L., Stillman, J.H., and Reusch, T.B.H. 2014. Evolution in an acidifying ocean. *Trends in Ecology and Evolution* **29**: 117–25.

Thor, P. and Dupont, S. 2015. Transgenerational effects alleviate severe fecundity loss during ocean acidification in a ubiquitous planktonic copepod. *Global Change Biology* **21**: 2261–71.

Timusk, E.R., Ferguson, M.M., Moghadam, H.K., Norman, J.D., Wilson, C.C., and Danzmann, R.G. 2011. Genome evolution in the fish family *Salmonidae*: generation of a brook charr genetic map and comparisons among charrs (Arctic charr and brook charr) with rainbow trout. *BMC Genetics* **12**: 2–15.

Veilleux, H.D., Ryu, T., Donelson, J.M., van Herwerden, L., Seridi, L., Ghosheh, Y., Berumen, M.L., Leggat, W., Ravasi, T., and Munday, P.L. 2015. Molecular processes of transgenerational acclimation to a warming ocean. *Nature Climate Change* **5**: 1074–8.

5.5.2 Future Impacts on Aquatic Life

The pessimistic projections of the IPCC give way to considerable optimism with respect to the future of the planet's marine life.

The experimental and observational research cited in Section 5.5.1 suggests an optimistic outlook on the future of Earth's marine life and this is in fact what researchers predict, contradicting the IPCC's pessimistic outlook.

Starting again with acidification, Loaiciga (2006) used a mass-balance approach to "estimate the change in average seawater salinity caused by the melting of terrestrial ice and permanent snow in a warming earth." He applied "a chemical equilibrium model for the concentration of carbonate species in seawater open to the atmosphere" in order to "estimate the effect of changes in atmospheric CO_2 on the acidity of seawater." Assuming that the rise in the planet's mean surface air temperature continues unabated and that it eventually causes the melting of *all* terrestrial ice and permanent snow – an extreme assumption – Loaiciga calculated that "the average seawater salinity would be lowered not more than 0.61‰ from its current 35‰."

Loaiciga also reports that across the range of seawater temperature considered (0 to 30°C) "a doubling of CO_2 from 380 ppm to 760 ppm increases the seawater acidity [lowers its pH] approximately 0.19 pH units." He thus concludes that "on a global scale and over the time scales considered (hundreds of years), there would not be accentuated changes in either seawater salinity or acidity from the rising concentration of atmospheric CO_2."

Similarly, an analysis of Tans (2009), the results of which are included in Figure 5.5.2.1 below, estimated the decline in oceanic pH by the year 2100 is likely to be only about half of that projected by the IPCC and that this drop will begin to be ameliorated shortly after 2100, gradually returning oceanic pH to present-day values beyond AD 2500.

Turning to temperature, Brown *et al.* (2010) write, "climate change is altering the rate and distribution of primary production in the world's oceans," which in turn "plays a fundamental role in structuring marine food webs (Hunt and McKinnell, 2006; Shurin *et al.*, 2006)," which are "critical to maintaining biodiversity and supporting fishery catches." They note the "effects of climate-driven production change on marine ecosystems and fisheries can be explored using food web models that incorporate ecological interactions such as predation and competition," citing the work of Cury *et al.* (2008).

Brown *et al.* used the output of an ocean general circulation model driven by a "plausible" greenhouse gas emissions scenario (IPCC, 2007, scenario A2) to calculate changes in climate over a 50-year time horizon, the results of which were fed into a suite of models for calculating primary production of lower trophic levels (phytoplankton, macroalgae, seagrass, and benthic microalgae). Those results were used as input to "twelve existing Ecopath with Ecosim (EwE) dynamic marine food web models to describe different Australian marine ecosystems," which protocol ultimately predicted "changes in fishery catch, fishery value, biomass of animals of conservation interest, and indicators of community composition."

Brown *et al.* state that under the IPCC's "plausible climate change scenario, primary production will *increase* around Australia" with "overall positive linear responses of functional groups to primary production change," and that "generally this benefits fisheries catch and value and leads to increased biomass of threatened marine animals such as turtles and sharks." They conclude the primary production increases suggested by their work to result from future IPCC-envisioned greenhouse gas emissions and their calculated impacts on climate "will provide opportunities to recover overfished fisheries, increase profitability of fisheries and conserve threatened biodiversity."

Figure 5.5.2.1
The change in surface seawater pH vs. time and as calculated by Tans and the IPCC

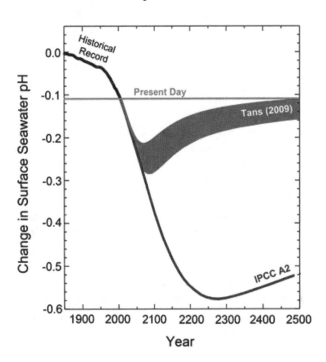

Sources: Red band is from Figure 5 of Tans, 2009 representing two emission scenarios. Blue line is the IPCC's forecast based on emission scenario A2 from IPCC, 2007.

In a comprehensive literature review published in *Science*, Pandolfi *et al.* (2011) summarize what they describe as "the most recent evidence for past, present and predicted future responses of coral reefs to environmental change, with emphasis on rapid increases in temperature and ocean acidification and their effects on reef-building corals."

Focusing here only on Pandolfi *et al.*'s findings with respect to the future of coral reefs, they write, "because bleaching-susceptible species often have faster rates of recovery from disturbances, their relative abundances will not necessarily decline." In fact, they say "such species could potentially increase in abundance, depending on how demographic characteristics and competitive ability are correlated with thermal tolerance and on the response of other benthic taxa, such as algae," while they further note "the shorter generation times typical of more-susceptible species (Baird *et al.*, 2009) may also confer faster rates of evolution of bleaching thresholds, which would further facilitate maintenance of, or increases to, the relative abundance of thermally sensitive but faster-evolving species (Baskett *et al.*, 2009)."

In summing up their analysis, Pandolfi *et al.* state emerging evidence for variability in the coral calcification response to acidification, geographical variation in bleaching susceptibility and recovery, responses to past climate change, and potential rates of adaptation to rapid warming "supports an alternative scenario in which reef degradation occurs with greater temporal and spatial heterogeneity than current projections suggest." Further noting that "non-climate-related threats already confronting coral reefs are likely to reduce the capacity of coral reefs to cope with climate change," they conclude "the best and most achievable thing we can do for coral reefs currently to deal with climate change is to seek to manage them well," by reducing more direct anthropogenic impacts such as fishing, pollution, and habitat destruction.

Gurney *et al.* (2013) also note the expected future importance of reef management, writing, "given that climate change impacts on coral reefs cannot be mitigated directly, the question arises whether reduction of stressors that originate and can be managed at a local scale (i.e. local-scale stressors) provides a tractable opportunity to increase the potential of coral reefs to cope with inevitable changes in the climate," citing Pandolfi *et al.* (2011) and Hughes *et al.* (2007). They use a simulation model validated for four sites in Bolinao, Philippines to "simulate future reef state for each site 40 years into the future under scenarios involving the cumulative impact of fishing, poor water quality and thermal bleaching-induced mortality related to climate change."

Gurney *et al.* simulated 18 scenarios, "all possible combinations of different levels of fishing pressure (three levels), coral mortality due to bleaching (three levels) and water quality (two levels)." Water quality was represented in the model through the combined effects of nutrification and sedimentation, for which they examined two scenarios: unregulated (the current situation, with high nutrification and sedimentation) and highly regulated (no impact of nutrification and sedimentation).They examined three fisheries management approaches: no reduction in fishing pressure (zero fisheries management), a 50% reduction in fishing pressure, and "no-take marine reserves" (zero fishing pressure).

Gurney *et al.* write, "our analysis of the cumulative impact of bleaching, poor water quality and fishing indicate that management of the two local stressors will significantly influence future reef state under climate change." They conclude, "our research supports the paradigm that managing local-scale stressors is critical to the persistence of coral reefs in the context of global climate change, a concept that is widely advocated," citing Pandolfi *et al.* (2011) and Donner (2009) "but still subject to debate," citing Keller *et al.* (2009) and Baker *et al.* (2008).

Baker (2014) lists some of the processes described by Bay and Palumbi (2014) that indicate "reef-building corals may have a broad repertoire of responses to deal with warming temperatures." In addition to their capacity for "maintaining diverse allelic variation," as described by Bay and Palumbi, Baker also mentions "front-loading genes involved in heat stress," citing Barshis *et al.* (2013) and Kenkel *et al.* (2013); "employing rapid acclimatization pathways," citing Palumbi *et al.* (2014), DeSalvo *et al.* (2010), and Kenkel *et al.* (2013); "changing the composition of their algal symbiont communities," citing Baker *et al.* (2004) and Berkelmans and van Oppen (2006); and "maintaining a healthy pool of microbial associates" in order "to prevent infection and disease during recovery from heat stress," citing Bourne *et al.* (2009).

Baker closes, optimistically, that "these diverse responses provide hope that the world's remaining corals may still contain the adaptive ingredients needed to survive."

References

Baird, A.H., Guest, J.R., and Willis, B.L. 2009. Systematic and biogeographical patterns in the reproductive biology of scleractinian corals. *Annual Review of Ecology, Evolution and Systematics* **40**: 551–71.

Baker, A.C. 2014. Climate change: many ways to beat the heat for reef corals. *Current Biology* **24**: 10.1016/j.cub.2014.11.014.

Baker, A.C., Glynn, P.W., and Riegl, B. 2008. Climate change and coral reef bleaching: an ecological assessment of long-term impacts, recovery trends and future outlook. *Estuarine, Coastal and Shelf Science* **80**: 435–71.

Baker, A.C., Starger, C.J., McClanahan, T.R., and Glynn, P.W. 2004. Corals' adaptive response to climate change. *Nature* **430**: 741.

Barshis, D.J., Ladner, J.T., Oliver, T.A., Seneca, F.O., Traylor-Knowles, N., and Palumbi, S.R. 2013. Genomic basis for coral resilience to climate change. *Proceedings of the National Academy of Sciences USA* **110**: 1387–92.

Baskett, M.L, Gaines, S.D., and Nisbet, R.M. 2009. Symbiont diversity may help coral reefs survive moderate climate change. *Ecological Applications* **19**: 3–17.

Bay, R.A. and Palumbi, S.R. 2014. Multi-locus adaptation associated with heat resistance in reef-building corals. *Current Biology* **24**: 2952–6.

Berkelmans, R. and van Oppen, M.J.H. 2006. The role of zooxanthellae in the thermal tolerance of corals: a 'nugget of hope' for coral reefs in an era of climate change. *Proceedings of the Royal Society B Biological Sciences* **273**: 2305–12.

Bourne, D.G., Garren, M., Work, T.M., Rosenberg, E., Smith, G.W., and Harvell, C.D. 2009. Microbial disease and the coral holobiont. *Trends in Microbiology* **17**: 554–2.

Brown, C.J., *et al.* 2010. Effects of climate-driven primary production change on marine food webs: implications for fisheries and conservation. *Global Change Biology* **16**: 1194–212.

Cury, P.M., Shin, Y.J., Planque, B., Durant, J.M., Fromentin, J.-M., Kramer-Schadt, S., Stenseth, N.C., Travers, M., and Grimm, V. 2008. Ecosystem oceanography for global change in fisheries. *Trends in Ecology and Evolution* **23**: 338–46.

DeSalvo, M.K., Sunagawa, S., Voolstra, C.R., and Medina, M. 2010. Transcriptomic responses to heat stress and bleaching in elkhorn coral *Acropora palmata*. *Marine Ecology Progress Series* **402**: 97–113.

Donner, S. 2009. Coping with commitment: projected thermal stress on coral reefs under different future scenarios. *PLoS ONE* **4**: e5712.

Gurney, G.G., Melbourne-Thomas, J., Geronimo, R.C., Aliño, P.M., and Johnson, C.R. 2013. Modelling coral reef futures to inform management: can reducing local-scale stressors conserve reefs under climate change? *PLoS ONE* **8** (11): e80137.

Hughes, T.P., Rodrigues, M.J., Bellwood, D.R., Ceccarelli, D., Hoegh-Guldberg, O., McCook, L., Moltschaniwskyj, N., Pratchett, M.S., Steneck, R.S., and Willis, B. 2007. Phase shifts, herbivory, and the resilience of coral reefs to climate change. *Current Biology* **17**: 360–5.

Hunt, G.L. and McKinnell, S. 2006. Interplay between top-down, bottom-up, and wasp-waist control in marine ecosystems. *Progress in Oceanography* **68**: 115–24.

IPCC. 2007. Intergovernmental Panel on Climate Change. *Climate Change 2007: Impacts, Adaptation and Vulnerability*. Contribution of Working Group II to the Fourth Assessment Report of the Intergovernmental Panel on Climate Change. Cambridge, UK and New York, NY: Cambridge University Press.

Keller, B.D., *et al.* 2009. Climate change, coral reef ecosystems, and management options for marine protected areas. *Environmental Management* **44**: 1069–88.

Kenkel, C.D., Meyer, E., and Matz, M.V. 2013. Gene expression under chronic heat stress in populations of the mustard hill coral (*Porites astreoides*) from different thermal environments. *Molecular Ecology* **22**: 4322–34.

Loaiciga, H.A. 2006. Modern-age buildup of CO_2 and its effects on seawater acidity and salinity. *Geophysical Research Letters* **33**: 10.1029/2006GL026305.

Palumbi, S.R., Barshis, D.J., Traylor-Knowles, N., and Bay, R.A. 2014. Mechanisms of reef coral resistance to future climate change. *Science* **344**: 895–8.

Pandolfi, J.M., Connolly, S.R., Marshall, D.J., and Cohen, A.L. 2011. Projecting coral reef futures under global warming and ocean acidification. *Science* **333**: 418–22.

Shurin, J.B., Gruner, D.S., and Hillebrand, H. 2006. All wet or dried up? Real differences between aquatic and terrestrial food webs. *Proceedings of the Royal Society B Biological Sciences* **273**: 1–9.

Tans, P. 2009. An accounting of the observed increase in oceanic and atmospheric CO_2 and an outlook for the future. *Oceanography* **22**: 26–35.

5.6 Conclusion

Combustion of fossil fuels has helped and will continue to help plants and animals thrive leading to shrinking deserts, expanded habitat for wildlife, and greater biodiversity.

Many of the scholars and advocates who write about climate change are either unfamiliar with or overlook the environmental benefits created by human use of fossil fuels. Chemists and biologists should know better: Fossil fuels are composed mainly of hydrogen and carbon atoms, two of the most abundant elements found in nature. They are not "pollutants" but share a common chemical basis with all of life on Earth.

Geologists should know better, too. The global carbon cycle acts to buffer the impact of man-made carbon dioxide (CO_2) by including it in exchange processes among carbon reservoirs that are huge compared to the human contribution. The size of the human contribution to atmospheric CO_2 concentrations is so small it may be less than the margin of error in measurements of known exchange rates among carbon reservoirs. Geologists ought to realize that current atmospheric CO_2 levels are not unprecedented and indeed are low when considered over geologic time scales. Because CO_2 is essential to plant and animal life, it is possible human use of fossil fuels may avert an ecological disaster.

Fossil fuels directly benefit the environment by making possible huge (orders of magnitude) advances in efficiency, making it possible to meet human needs while using fewer natural resources. Fossil fuels make it possible for humanity to flourish while still preserving much of the land needed by wildlife to survive. And the prosperity made possible by fossil fuels has made environmental protection both highly valued and financially possible, producing a world that is cleaner and safer than it would have been in their absence.

This chapter also finds the CO_2 released when fossil fuels are burned improves the productivity of ecosystems and has a positive effect on plant characteristics, including rates of photosynthesis and biomass production and the efficiency with which plants utilize water. Atmospheric CO_2 enrichment ameliorates the negative effects of a number of environmental plant stresses including high temperatures, air and soil pollutants, herbivory, nitrogen deprivation, and high levels of soil salinity. With the help of the ongoing rise in the air's CO_2 content, humankind should be able to meet the food needs of a growing population without occupying much of the land needed by wildlife to survive.

Although there likely will be some changes in terrestrial animal population dynamics, few (if any) will be driven even close to extinction. In a number of instances, real-world data indicate warmer temperatures and higher atmospheric CO_2 concentrations will be beneficial, favoring a proliferation of terrestrial species. Similarly, many laboratory and field studies of aquatic life demonstrate tolerance, adaptation, and even growth and developmental improvements in response to higher temperatures and reduced water pH levels. When these observations are considered, the pessimistic projections of the IPCC give way to considerable optimism with respect to the future of the planet's terrestrial and marine life.

PART III

COSTS OF
FOSSIL FUELS

Introduction to Part III

Part II documented the benefits reaped by mankind from the use of fossil fuels, including:

- Fossil fuels have vastly improved human well-being and safety by powering labor-saving and life-protecting technologies such as air conditioning, modern medicine, and cars and trucks.

- A warmer world would see a net decrease in temperature-related mortality and diseases in virtually all parts of the world, even those with tropical climates.

- The greater efficiency made possible by technologies powered by fossil fuels makes it possible to meet human needs while using fewer natural resources and land, thereby benefiting the natural environment.

Against these benefits must be balanced the costs imposed on humanity and the environment from the use of fossil fuels. The Working Group II contribution to Fifth Assessment Report of the United Nations' Intergovernmental Panel on Climate Change (IPCC) claims climate change causes a "risk of death, injury, and disrupted livelihoods" due to sea-level rise, coastal flooding, and storm surges; food insecurity, inland flooding, and negative effects on fresh water supplies, fisheries, and livestock; and "risk of mortality, morbidity, and other harms during periods of extreme heat, particularly for vulnerable urban populations" (IPCC, 2014, p. 7).

Environmental advocacy groups similarly claim the "hidden costs" of using oil and coal amount to billions and even trillions of dollars a year for the United States alone. For example, the Natural Resources Defense Council (NRDC) says "the total cost of global warming [which the NRDC attributes to fossil fuels] will be as high as 3.6% of gross domestic product (GDP). Four global warming impacts alone – hurricane damage, real estate losses, energy costs, and water costs – will come with a price of 1.8% of U.S. GDP, or almost $1.9 trillion annually (in today's dollars) by 2100. ... [T]he true cost of all aspects of global warming – including economic losses, noneconomic damages, and increased risks of catastrophe – will reach 3.6% of U.S. GDP by 2100 if business-as-usual emissions are allowed to continue" (NRDC, 2008, pp. iv, vi). See also Lovins (2011, pp. 5–6) for a similar discussion.

These claims seem disconnected from reality. The predictions of "droughts, floods, famines, [and] disease spread" were shown in Parts I and II to be without any scientific basis, so we should be skeptical when seeing them included in cost-benefit analyses. As for the economic impact of "oil dependence," just one recent innovation in energy technology – combining horizontal drilling and hydraulic fracturing (fracking) to tap oil and natural gas trapped in shale deposits – has created 1.7 million new direct and indirect jobs in the United States, with the total likely to rise to 3 million in the next eight years (IHS Global Insight, 2012). It has added $62 billion to federal and state treasuries, with that total expected to rise to $111 billion by 2020. By 2035, U.S. fracking operations could inject more than $5 trillion in cumulative capital expenditures into the economy, while generating more than $2.5 trillion in cumulative additional government revenues (*Ibid*). And this is only one of many value-creating innovations occurring in the energy sector.

The NRDC and other advocacy groups like it have several things in common. First, they accept

Citation: Idso, C.D., Legates, D. and Singer, S.F. 2019. Air Quality. In: *Climate Change Reconsidered II: Fossil Fuels.* Nongovernmental International Panel on Climate Change. Arlington Heights, IL: The Heartland Institute.

uncritically the claims of the IPCC, invariably citing the *Summaries for Policymakers* of its Fourth or Fifth Assessment Reports while overlooking the caution and uncertainties expressed in the full reports. (This is especially ironic in the case of the NRDC since the organization infiltrated the IPCC, placing its own staffers on many of the IPCC's editing and peer-review committees, and so effectively *wrote* the reports they now cite as proof of their views. See Laframboise (2011).) The IPCC's computer models fail to replicate past temperature trends, meaning they cannot produce accurate forecasts of future climate conditions (Fyfe *et al.*, 2013). According to McKitrick and Christy (2018), for the period from 1958 to 2017 the models hindcast a warming of~ 0.33° C/decade while observations show only ~ 0.17°C/decade. (With a break term for the 1979 Pacific climate shift included the models hindcast ~ 0.39°C/decade and observations show ~ 0.14° C/decade.) This fact undermines all alleged cost-benefit analyses of climate change that rely on IPCC reports for forecasts of future climate conditions.

Second, the Rocky Mountain Institute, NRDC, and groups like them invariably exclude from their accounting any of the *benefits* of fossil fuels. As documented in Chapters 3, 4, and 5, these benefits are huge relative to any damages due to fossil fuels or to climate change. Ignoring those benefits is obviously wrong. Epstein (2016) notes,

> [I]t is a mistake to look at costs in isolation from benefits, or benefits apart from costs. Yet that appears to be the approach taken in these reports. … [A] truly neutral account of the problem must be prepared to come to the conclusion that increased levels of CO_2 emissions could be, as the Carbon Dioxide Coalition has argued, a net benefit to society when a more comprehensive investigation is made. The entire process of expanding EPA regulations and other Obama administration actions feeds off this incorrect base assumption.

Environmental groups also rely heavily on economic models and simulations, called integrated assessment models (IAMs) to reach their conclusions. Like the climate models relied on by the IPCC, these models hide assumptions and uncertainties, are often invalidated by real-world data, and fail the test of genuine scientific forecasts. They are merely scenarios based on their authors' best guesses, "tuned" by their biases and political agendas, and far

from reliable. See Chapter 2 for the candid discussion of by a group of leading modelers of "the art and science of climate model tuning" (Hourdin *et al.,* 2017) and Green and Armstrong (2007) for an audit of the use of IAMs for forecasting. Real data are available to fact-check the models, but they are curiously absent from the claims of advocates and the academic literature they cite.

Chapters 6 and 7 of Part III set out an accurate accounting of the biggest alleged costs of fossil fuels, those attributable to chemical compounds released during the combustion of fossil fuels and what the IPCC calls "threats to human security" which includes famine, conflict, damage from floods and extreme weather, and forced migration. The authors find that in both cases, costs are exaggerated in the popular as well as the academic literature. Non-specialists feed these inflated cost estimates into their computer models apparently without understanding they are unsupported by real observational data and credible economic, scientific, and public health research. When these major sources of concern are addressed, any remaining costs are quite small or speculative.

Chapter 8 conducts cost-benefit analyses of climate change attributed by the IPCC to the combustion of fossil fuels, the use of fossil fuels, and regulations enacted or advocated in the name of slowing or stopping global warming. At the risk of overly simplifying what is a very complicated analysis, the conclusions of that chapter can be said to affirm the small and highly uncertain cost of man-made climate change, the net benefits of fossil fuels, and the very high cost of regulations aimed at reducing greenhouse gas emissions.

References

Epstein, R. 2016. Obsolete climate science on CO_2. Defining Ideas (blog). Hoover Institution. December 20.

Fyfe, J.C., Gillett, N.P., and Zwiers, F.W. 2013. Overestimated global warming over the past 20 years. *Nature Climate Change* **3**: 767–9.

Green, K.C. and Armstrong, J.S. 2007. Global warming: forecasts by scientists versus scientific forecasts. *Energy and Environment* **18**: 997–1021.

Hourdin, F. *et al.* 2017. The art and science of climate model tuning. *Bulletin of the American Meteorological Society* **98** (3), 589-602.

IHS Global Insight. 2012. *America's New Energy Future: The Unconventional Revolution and the Economy*. October 23.

IPCC. 2014. Intergovernmental Panel on Climate Change. Chapter 11: Human health: Impacts, adaptations, and co-benefits. In *Climate Change 2014: Impacts, Adaptation, and Vulnerabilities*. Contribution of Working Group II to the Fifth Assessment Report of the Intergovernmental Panel on Climate Change. New York, NY: Cambridge University Press.

Laframboise, D. 2011. *The Delinquent Teenager Who Was Mistaken for the World's Top Climate Expert*. Toronto, ON: Ivy Avenue Press.

Lovins, A.B. 2011. *Reinventing Fire: Bold Business Solutions for the New Energy Era*. White River Junction, VT: Chelsea Green Publishing.

McKitrick, R. and Christy, J. 2018. A Test of the tropical 200- to 300-hPa warming rate in climate models. *Earth and Space Science* **5** (9).

NRDC. 2008. Natural Resources Defense Council. *The Cost of Climate Change: What We'll Pay if Global Warming Continues Unchecked*. New York, NY: Natural Resources Defense Council.

6

Air Quality

Chapter Lead Author: John D. Dunn, M.D., J.D.

Contributors: Jerome C. Arnett, Jr., M.D., Charles Battig, M.D., M.S.E.E., James E. Enstrom, Ph.D., Steve Milloy, J.D., S. Stanley Young, Ph.D.

Reviewers: Kesten Green, Ph.D., Richard J. Trzupek, Art Viterito, Ph.D.

Key Findings

The key findings of this chapter include the following:

An Air Quality Tutorial

- The combustion of fossil fuels without air pollution abatement technology releases chemicals known to be harmful to humans, other animal life, and plants.

Citation: Idso, C.D., Legates, D. and Singer, S.F. 2019. Air Quality. In: *Climate Change Reconsidered II: Fossil Fuels.* Nongovernmental International Panel on Climate Change. Arlington Heights, IL: The Heartland Institute.

- At low levels of exposure, the chemical compounds produced by burning fossil fuels are not known to be toxic.

- Exposure to potentially harmful emissions from the burning of fossil fuels in the United States declined rapidly in recent decades and is now at nearly undetectable levels.

- Exposure to chemical compounds produced during the combustion of fossil fuels is unlikely to cause any fatalities in the United States.

Failure of the EPA

- Due to its faulty mission, flawed paradigm, and political pressures on it to chase the impossible goal of zero risk, the U.S. Environmental Protection Agency (EPA) is an unreliable source of research on air quality and its impact on human health.

- The EPA makes many assumptions about relationships between air quality and human health, often in violation of the Bradford Hill Criteria and other basic requirements of the scientific method.

- The EPA has relied on research that cannot be replicated and violates basic protocols for conflict of interest, peer review, and transparency.

Observational Studies

- Observational studies are easily manipulated, cannot prove causation, and often do not support a hypothesis of toxicity with the small associations found in uncontrolled observational studies.

- Observational studies cited by the EPA fail to show relative risks (RR) that would suggest a causal relationship between chemical compounds released during the combustion of fossil fuels and adverse human health effects.

- Real-world data and common sense contradict claims that ambient levels of particulate matter kill hundreds of thousands of Americans and millions of people around the world annually.

- By conducting human experiments involving exposure to levels of particulate matter and other pollutants it claims to be deadly, the EPA reveals it does not believe its own epidemiology-based claims of a deadly threat to public health.

Circumstantial Evidence

- Circumstantial evidence cited by the EPA, World Health Organization (WHO), and other air quality regulators is easily refuted by pointing to contradictory evidence.

- EPA cannot point to any cases of death due to inhaling particulate matter, even in environments where its National Ambient Air Quality Standard (NAAQS) is exceeded by orders of magnitude.

- Life expectancy continues to rise in the United States and globally despite what should be a huge death toll, said to be equal to the entire death toll caused by cancer, attributed by the EPA and WHO to just a single pollutant, particulate matter.

Conclusion

- It is unlikely that chemical compounds released during the combustion of fossil fuels kill or harm anyone in the United States, though it may be a legitimate health concern in third-world countries that rely on burning biofuels and fossil fuels without modern emission control technologies.

Introduction

Data cited by Simon (1995, 1996), Lomborg (2001), Anderson (2004), Hayward (2011), Goklany (2007, 2012), Epstein (2014), Pinker (2018), and many others reported in Part II, much of it compiled by the U.S. Environmental Protection Agency (EPA) and other government sources, document a dramatic improvement in public health since the beginning of the industrial revolution. *Do chemical compounds released by burning fossil fuels nevertheless pose a public health risk?*

In 2010, the EPA claimed just one kind of air pollutant, particulate matter (fine dust particles), caused approximately 360,000 and as many as 500,000 premature deaths in the United States in 2005, citing Laden *et al.* (2006) (EPA, 2010, p. G7). The high estimate would be more than one-fifth of all deaths in the United States that year and nearly as high as all deaths from cancer (Kung *et al.,* 2008). In 2011, then-EPA Administrator Lisa Jackson endorsed the highest estimate in testimony to Congress, saying, "If we could reduce particulate matter to levels that are healthy we would have an identical impact to finding a cure for cancer" (quoted in Harris and Broun, 2011, p. 2).

The World Health Organization (WHO) similarly claims air pollution is a major health problem globally, saying it caused 600,000 premature deaths in 2010 in Europe alone (WHO, 2015). A 2016 WHO report claimed "3.9 million premature deaths each year [are] attributable to outdoor air pollution" and exposure to household air pollution (HAP) "causes 4.3 million premature deaths each year" (WHO, 2016, p. ix).

These claims are reported and repeated without hesitation or scrutiny by environmental groups, the media, and even serious scholars in the climate change debate. But the EPA and WHO claims are based on weak epidemiological relationships and trends carelessly described without definition as "associations" or "trends." Much like assumptions, computer models, and circumstantial evidence are paraded by the United Nations' Intergovernmental Panel on Climate Change (IPCC) as evidence in the climate science debate, so too are these unscientific lines of reasoning presented as evidence by the EPA and WHO in the debate over air quality.

This chapter begins with a brief tutorial on air quality[1] and then explains why chemical compounds released during the combustion of fossil fuels do not present a significant human health threat in the

United States or other developed countries. In developing countries, where exposure to pollutants is greater, a health risk may be present, though fossil fuels may prove to be a solution rather than the problem in many regions. Morrison (2018), for example, describes an effort to replace old biomass cookstoves in developing countries with "stoves that use propane, a fossil fuel, the same blue-flamed byproduct of gas drilling contained in cylinders under countless American backyard grills." The solution to air quality issues in developing countries lies in the prosperity, values, and technologies used by developed countries to solve their air quality problems.

References

EPA. 2010. U.S. Environmental Protection Agency. *Quantitative Health Risk Assessment for Particulate Matter.* EPA-452/R-10-005. June. Washington, DC.

EPA. 2011. U.S. Environmental Protection Agency. Air trends in particulate matter (website).

Anderson, T. (Ed.) 2004. *You Have to Admit It's Getting Better.* Stanford, CA: Hoover Press.

Epstein, A. 2014. *The Moral Case for Fossil Fuels.* New York, NY: Portfolio/Penguin.

Goklany, I.M. 2007. *The Improving State of the World: Why We're Living Longer, Healthier, More Comfortable Lives on a Cleaner Planet.* Washington, DC: Cato Institute.

Goklany, I.M. 2012. Humanity unbound: How fossil fuels saved humanity from nature and nature from humanity. *Policy Analysis* #715. Washington, DC: Cato Institute.

Harris, A., and Broun, P. 2011. Letter to Cass R. Sunstein, Administrator, Office of Information and Regulatory Affairs. Washington, DC: U.S. House of Representatives: Committee on Science, Space, and Technology. November 15.

Hayward, S. 2011. *Almanac of Environmental Trends.* San Francisco, CA: Pacific Institute for Public Policy Research.

Krewski, D., *et al.* 2009. Extended follow-up and spatial analysis of the American Cancer Society study linking particulate air pollution and mortality. *HEI Research Report* 140. Boston, MA: Health Effects Institute.

Laden, F., Schwartz, J., Speizer, F.E., and Dockery, D.W. 2006. Reduction in fine particulate air pollution and

---

[1] We use the term "air quality" rather than "air pollution" when possible because the public policy goal is to improve air quality, not necessarily to reduce or end "air pollution." Referring to chemical compounds created during the combustion of fossil fuels as "pollution" prejudges them as harmful. Emissions are not harmful unless they are present in concentrations sufficient to endanger human health.

mortality. *American Journal of Respiratory and Critical Care Medicine* **173**: 667–72.

Lomborg, B. 2001. *The Skeptical Environmentalist: Measuring the Real State of the World.* Cambridge, MA: Cambridge University Press.

Morrison, S. 2018. Undercooked: an expensive push to save lives and protect the planet falls short. ProPublica (website). July 12.

Pinker, S. 2018. *Enlightenment Now: The Case for Reason, Science, Humanism, and Progress.* New York, NY: Viking.

Simon, J.L. 1995. *The State of Humanity.* Cambridge, MA: Blackwell Publishers, Inc.

Simon, J.L. 1996. *The Ultimate Resource.* Second edition. Princeton, NJ: Princeton University Press.

WHO. 2015. World Health Organization. *Economic Cost of the Health Impact of Air Pollution in Europe: Clean Air, Health and Wealth.* Regional Office for Europe. Copenhagen: OECD.

WHO. 2015. World Health Organization. *Burning Opportunity: Clean Household Energy for Health, Sustainable Development, and Wellbeing of Women and Children.* Copenhagen: OECD.

6.1 An Air Quality Tutorial

Critics of fossil fuels often attribute social costs to the public health consequences of emissions created by the combustion of fossil fuels without understanding basic facts about chemistry, alternative (often natural) sources of the same chemicals, evidence of human exposure and trends of the same, and how all these data are interpreted. This section offers a brief tutorial on these topics.

6.1.1 Chemistry

The combustion of fossil fuels without air pollution abatement technology releases chemicals known to be harmful to humans, other animal life, and plants.

When burned, fossil fuels release carbon dioxide (CO_2), water (H_2O), carbon monoxide (CO), sulfur dioxide (SO_2), nitrogen oxides (NO_x), and particulate matter (PM). Another pollutant, ozone (O_3), is created through photochemical reaction with the other pollutants. Carbon dioxide and water, as Moore

has observed, are "the two most essential foods for life" (Moore, 2015) and are not public health concerns, leaving five emissions of concern.

Carbon monoxide (CO) is a colorless, odorless gas formed when carbon in wood or fossil fuels is not burned completely. Approximately 80% or more of human outdoor CO emissions in the United States comes from motor vehicle exhaust while the remaining 20% comes from industrial processes and residential wood burning. CO is produced indoors by woodstoves, gas stoves, unvented gas and kerosene space heaters, and smoking.

Sulfur dioxide (SO_2) is formed when fossil fuels containing sulfur, such as coal and oil, are burned, when gasoline is extracted from crude oil, and when metals are extracted from ore. Sulfur dioxide dissolves in water, creating droplets that are less basic or alkaline than would otherwise occur, creating what is popularly and inaccurately called "acid rain."

Nitrogen oxides (NO_x) are a group of gases containing nitrogen and oxygen, most of which are colorless and odorless. Nitrogen oxides form when fuel is burned at high temperatures, as in a combustion process. Half of NO_x emissions in the United States come from motor vehicle exhaust and most of the rest from stationery generators.

Particulate matter (PM) is the general term used to describe a mixture of solid particles and liquid droplets found in the air. Some PM particles are large enough to be seen as dust or dirt. Others are so small they can be detected only with an electron microscope. $PM_{2.5}$ refers to particles less than or equal to 2.5 μm (micrometer) in diameter. PM_{10} refers to particles less than or equal to 10 μm in diameter (about one-seventh the diameter of a human hair). "Primary" PM is emitted directly into the atmosphere. Examples of primary particles are dust from roads or black carbon (soot) from burning wood or fossil fuels. "Secondary" particles, which are formed in the atmosphere from gaseous emissions, include sulfates (formed from SO_2), nitrates (formed from NO_x), and carbon (formed from CO_2).

Fossil fuels create PM in the form of soot when the supply of oxygen during combustion is insufficient to completely convert carbon to carbon oxides. This typically occurs during the combustion of coal and oil, not natural gas. PM also is produced by agriculture (plowing, planting, and harvesting activities), resuspension by wind or traffic of dust particles from roads, and many natural processes including forest fires, wind erosion, desert dust, volcanoes, sea salt aerosols (sodium chloride

(NaCl)), and biological aerosols (e.g., spores and pollen). The EPA estimates approximately 16% of U.S. PM$_{10}$ emissions and 40% of PM$_{2.5}$ emissions are anthropogenic while the rest is "fugitive dust" (dust from open fields, roadways, storage piles, and other non-point sources) and "miscellaneous and natural sources" (EPA, 2018a). See Figure 6.1.1.1.

Ozone (O₃) is a triatomic oxygen molecule gas that occurs in Earth's upper atmosphere and at ground level. Ozone is not directly emitted into the atmosphere when fossil fuels are combusted, but it can be counted as a pollutant resulting from their use because fossil fuel use produces precursors to the photochemical reaction that creates ozone at ground level. Those precursors are carbon monoxide, nitrogen oxides, and particulate matter. Trees and other plants also produce ozone precursors, in particular hydrocarbons, but primarily in rural areas where their ratio to nitrogen oxides is too large to create the conditions in which ozone is formed.

Volatile organic compounds (VOCs) are often included in lists of pollutants attributable to the use of fossil fuels. All molecules containing carbon with high vapor pressure at ordinary room temperature are classified as VOCs, meaning they readily evaporate in air. This category necessarily duplicates or overlaps with others in this list of emissions. Nature, primarily plants, produces about ten times as much VOCs, by weight, as all human activities (1,150 versus 142 teragrams per year). The combustion of fossil fuels contributes only a small fraction of man-made VOCs, with carbon monoxide, gasoline fumes, and benzene being three examples.

Lead (Pb) is often included as an emission from the combustion of fossil fuels, but it was a lead-containing compound called tetraethyllead added to petroleum to improve engine performance that was responsible for lead emissions from motor vehicles.

Lead is not found in appreciable amounts in coal or refined oil products. Due to the phase-out of leaded gasoline in the United States and other nations, lead in the air is no longer a public health hazard in the United States or other developed countries (von Storch *et al.*, 2003). The main sources of human lead emissions today are waste incinerators and lead-acid battery manufacturers.

Some trace minerals in fossil fuels also are present in ash when fossil fuels are burned. The ash can become airborne or dissolved into and transported by water. One such compound is *mercury (Hg)*, which in its organic form (methylmercury) can be poisonous to humans and other living creatures. Mercury is a naturally occurring substance, with

Figure 6.1.1.1
Sources of particulate matter (PM) in the United States

A. Relative amounts of U.S. PM$_{10}$ emissions from anthropogenic and other sources, 2011

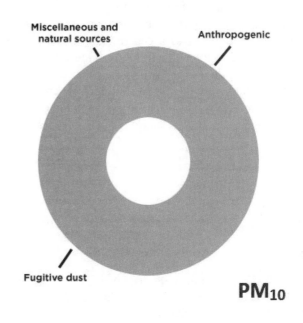

B. Relative amounts of U.S. PM$_{2.5}$ emissions from anthropogenic and other sources, 2011

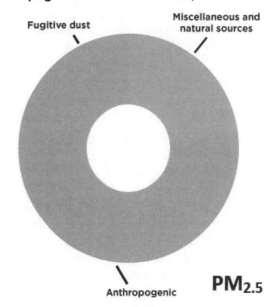

Source: EPA, 2018a, data from the 2011 National Emissions Inventory, Version 1.

some 200 million tons present in seawater. Mercury emissions from the combustion of fossil fuels in the United States are very small relative to other sources: approximately 7 tons annually (EPA, 2018b) versus 5,000 to 8,000 tons from all sources globally, including volcanoes, subsea vents, geysers, forest fires, and other natural sources.

Gasoline evaporates quickly when exposed to air, a property that leads to rapid dispersal of spills above ground, but when spilled underground (say, from leaking gas station tanks) it can remain in place for years and pose a threat to drinking water. Finally, carbon monoxide and particulate matter from incomplete fuel combustion by automobile engines and NO_x can react with sunshine to create ozone (already discussed above) and a visible haze called "smog."

References

EPA. 2018a. U.S. Environmental Protection Agency. 2018. EPA's Report on the Environment, Particulate Emissions (website). Accessed June 9, 2018.

EPA. 2018b. U.S. Environmental Protection Agency. Electric Utilities Mercury Releases in the 2016 TRI National Analysis. Washington, DC.

Moore, P. 2015. Should we celebrate CO_2? *2015 Annual GWPF Lecture*. London, UK: Global Warming Policy Foundation. October 14.

von Storch, H., *et al.* 2003. Four decades of gasoline lead emissions and control policies in Europe: a retrospective assessment. *Science of the Total Environment* **311**(1), 151-176.

6.1.2 Exposure

At low levels of exposure, the chemical compounds produced by burning fossil fuels are not known to be toxic.

The most important lesson regarding air quality is what matters most is not the toxicity of a chemical but the *level of exposure*. As Paracelsus, a Swiss physician, observed some five centuries ago, "Everything is poison. There is nothing without poison. *Only the dose makes a thing not a poison*." Without exposure there can be no harm.

Emissions from the combustion of fossil fuels increased early in the modern era due to rising population and per-capita energy consumption, but

have been falling since the 1940s. Today, most of the potentially harmful chemical compounds created during the combustion of fossil fuels for electricity generation are removed by pollution control technologies and never enter the air. According to the National Energy Technology Laboratory (NETL, 2015), an agency within the U.S. Department of Energy, pollution controls used by a "subcritical pulverized coal plant with a nominal net output of 550 MWe" reduce NO_x emissions by 83%, SO_2 emissions by 98%, mercury by 96.8%, and PM by 99.9% when compared with a similar plant with no pollution controls (p. 77). Catalytic converters on cars and trucks convert CO and unburned hydrocarbons in the combustion process into CO_2 and converts NO_x into harmless N_2.

This section begins with an explanation of the potential threat to human health posed by exposure to seven chemical compounds produced by the combustion of fossil fuels: carbon monoxide (CO), sulfur dioxide (SO_2), nitrogen oxides (NO_x), particulate matter (PM), ozone (O_3) (formed by the interactions of the previous four), volatile organic compounds (VOCs) (the vaporous state of some previous compounds), and elementary mercury (Hg). It then summarizes recent research on current levels of exposure to these chemicals. Particulate matter is reported briefly here, but is addressed in much greater detail in Section 6.3.

Potential threat to human health

Carbon monoxide (CO) can be poisonous at high levels of exposure not commonly found in ambient air. In the human body, hemoglobin (an iron compound) in the blood carries oxygen from the lungs to various tissues and transports carbon dioxide (CO_2) back to the lungs. Hemoglobin has 240 times more affinity toward CO than it does for oxygen. When the hemoglobin reacts with CO, it reduces the hemoglobin available for the transport of oxygen. This in turn reduces oxygen supply to the body's organs and tissues. Consequently, people who suffer from cardiovascular disease are most at risk from elevated levels of CO. There is also the potential for harm in pregnancy, because relative oxygen levels have a greater impact on the fetus, which depends on maternal blood oxygen. More commonly, exposure to elevated levels of CO may result in visual impairment, reduced manual dexterity, and difficulty in performing complex tasks. Figure 6.1.2.1 shows

6.1.2.1
Health effects associated with human exposure to carbon monoxide

CO concentration (parts per million)	Duration of exposure (hours)	Effect
100	10	Headache
300	10	Nausea, unconsciousness
600	10	Death
1000	1	Unconsciousness
1000	4	Death

Source: Radovic, 1992.

one estimate of the health effects associated with different levels of exposure to carbon monoxide.

Most of the sulfur in a fossil fuel combines with oxygen and forms *sulfur dioxide (SO₂)* in the combustion chamber. Unless captured by emission control technology, this SO_2 is emitted into the atmosphere where it oxidizes to sulfur trioxide (SO_3). SO_3 is soluble in water in the clouds and forms H_2SO_4 (sulfuric acid). Exposure to sulfuric acid irritates the mucous membranes of the respiratory tract, which causes airways to restrict and damages the cells of the mucous membranes, causing the release of inflammatory mediators that cause airway swelling and spasm, restricting airway size, causing an increase in work of breathing and decrease in available inspired air.

Exposure to a concentration of 1 part per million of SO_3 can cause coughing and choking; higher levels can result in temporary breathing impairment such as wheezing, chest tightness, or shortness of breath. Long-term exposure can aggravate existing cardiovascular disease and respiratory illnesses. SO_3 in the atmosphere also acts as a precursor to fine particulate matter.

Nitric oxide (NO) released during combustion of fossil fuels is oxidized in the atmosphere to *nitrogen dioxide* (NO_2). (NO_2 is also created from the nitrogen in the air during some high-temperature processes that do not involve fossil fuels.) Nitrogen dioxide is a noxious gas that can cause inflammation of the respiratory tract, similar to sulfuric acid, and, at high concentrations, even death.

NO_2 is soluble in water and forms HNO_3 (nitric acid). Like sulfuric acid, nitric acid constricts airways in humans and animals and can cause adverse health effects. Short-term exposure may lead to changes in airway responsiveness and lung function in individuals with preexisting respiratory illnesses. Long-term exposure may lead to increased susceptibility to respiratory infection and may cause irreversible alterations in lung structure.

The sulfuric and nitric acids created by SO_2 and NO_x in the atmosphere return to the surface in the form of dry deposition of particles or rain that is slightly more acidic than would otherwise occur, popularly referred to as "*acid rain.*" Pure water is neither acidic nor basic, but natural rainfall even in the absence of human use of fossil fuels is slightly acidic because it dissolves carbon dioxide from the air. Nitrogen, like carbon dioxide, is a plant fertilizer, and therefore higher levels are generally beneficial to most types of plant life and, by expanding habitats, to animal life as well. However, the addition of nitrogen to lakes and rivers can cause excessive algae growth, which contributes to eutrophication (depletion of dissolved oxygen), which can harm fish and other aquatic life. This concern is addressed in Chapter 5.

Particulate matter (PM), whether produced by the combustion of fossil fuels or by other processes described in the preceding section, can enter lungs and get trapped in the very thin air passages, reducing the air capacity of the lungs. Reduced air capacity can lead to such breathing and respiratory problems as emphysema and bronchitis, as well as increased general susceptibility to respiratory diseases. People with heart or lung disease and the elderly are especially at risk. Depending on the composition of the particles, chemical or mechanical or even allergenic, the effect is directly on the tissues, like the chemical effects described above for nitrous and sulfuric compounds, but not as toxic. The effect of particles is determined by their composition since they are not large enough to obstruct airways, even the terminal bronchioles that allow air into the air sacs that exchange oxygen and carbon dioxide. However, deposits of small particles can occur because the cleaning mechanisms in the alveoli and airways are not 100% efficient. There is no medical research establishing a mechanism for how small particles might cause death.

Exposure to ground-level *ozone (O₃)* can cause inflammation of the lining of the lungs, reduced lung function, and respiratory symptoms such as cough, wheezing, chest pain, burning in the chest, and shortness of breath. Longer-term exposure has been associated with the aggravation of respiratory illnesses such as asthma, emphysema, and bronchitis, leading to increased use of medication, absences from school, doctor and emergency department visits, and hospital admissions.

Volatile organic compounds (VOCs) related to fossil fuels include the compounds mentioned above, since the classification is determined by their ability to evaporate at relatively low temperatures. Most VOCs considered public health threats come from the use of cleaners, paints, and building materials in indoor spaces and not the combustion of fossil fuels. Outdoor levels of VOCs are monitored and regulated due to their role in the creation of ozone and smog.

Exposure to *mercury (Hg)* fumes can cause harmful effects on the nervous, digestive and immune systems, lungs and kidneys, and may be fatal. The inorganic salts of mercury are corrosive to the skin, eyes and gastrointestinal tract, and may induce kidney toxicity if ingested. Neurological and behavioral disorders may be observed after inhalation, ingestion or dermal exposure of different mercury compounds. Symptoms include tremors, insomnia, memory loss, neuromuscular effects, headaches and cognitive and motor dysfunction WHO, 2017).

Once in the environment, mercury can be transformed by bacteria into methylmercury, which bioaccumulates in fish and shellfish. Human consumption of seafood with high levels of methylmercury can cause some of the health effects described above. Methylmercury can pass through the placenta, exposing the fetus and causing birth defects, possibly manifested as lower IQ.

Current levels of exposure

The U.S. EPA was required by the Clean Air Act to establish National Ambient Air Quality Standards

(NAAQS) setting the maximum level of exposure, measured in concentration of the pollutant in the air and time of exposure for substances believed to endanger public health or the natural environment. The EPA has set NAAQS for six pollutants, which it calls "criteria air pollutants," being the five identified in the previous section as attributable to fossil fuels plus lead (EPA, 2018a). The current NAAQS appear in Figure 6.1.2.2.

In its description of the table in Figure 6.1.2.2, the EPA says "The Clean Air Act identifies two types of national ambient air quality standards. ***Primary standards*** provide public health protection, including protecting the health of 'sensitive' populations such as asthmatics, children, and the elderly. ***Secondary standards*** provide public welfare protection, including protection against decreased visibility and damage to animals, crops, vegetation, and buildings."

EPA says of its NAAQS, "The primary standards are set at a level intended to protect public health, including the health of at-risk populations, with an adequate margin of safety. In selecting a margin of safety, the EPA considers such factors as the strengths and limitations of the evidence and related uncertainties, the nature and severity of the health effects, the size of the at-risk populations, and whether discernible thresholds have been identified below which health effects do not occur. In general, for the criteria air pollutants, there is no evidence of discernible thresholds" (EPA, 2018b, p. 1). EPA's use of "safety factors" and a "linear no-threshold dose-response relation" are controversial and are explored in Section 6.2.2.

The EPA has estimated the "percentage of children living in [U.S.] counties with pollutant concentrations above the levels of the current air quality standards" for the six EPA criteria pollutants in the most recent year, 2013. Its findings are summarized in Figure 6.1.2.3.

As shown in Figure 6.1.2.3, according to the EPA carbon monoxide in ambient outdoor air is a nonexistent threat, with 0% of children living in counties in which they might be exposed to harmful levels of that pollutant. Fewer than 1% of children live in counties where lead exposure might be a threat, 2% where nitrogen dioxide is a problem, and 3% for sulfur dioxide. Particulate matter and ozone seem to pose larger problems, with between 3% and 21% of children living in counties where they might be exposed to unhealthy levels of PM and 58% threatened by ozone.

EPA also has created an "Air Quality Index" combining and weighing its measures of exposure to

Figure 6.1.2.2
National Ambient Air Quality Standards

Pollutant [links to historical tables of NAAQS reviews]		Primary/ Secondary	Averaging Time	Level	Form
Carbon Monoxide (CO)		primary	8 hours	9 ppm	Not to be exceeded more than once per year
			1 hour	35 ppm	
Lead (Pb)		primary and secondary	Rolling 3 month average	0.15 µg/m^3 [1]	Not to be exceeded
Nitrogen Dioxide (NO$_2$)		primary	1 hour	100 ppb	98th percentile of 1-hour daily maximum concentrations, averaged over 3 years
		primary and secondary	1 year	53 ppb [2]	Annual Mean
Ozone (O$_3$)		primary and secondary	8 hours	0.070 ppm [3]	Annual fourth-highest daily maximum 8-hour concentration, averaged over 3 years
Particle Pollution (PM)	PM$_{2.5}$	primary	1 year	12.0 µg/m^3	annual mean, averaged over 3 years
		secondary	1 year	15.0 µg/m^3	annual mean, averaged over 3 years
		primary and secondary	24 hours	35 µg/m^3	98th percentile, averaged over 3 years
	PM$_{10}$	primary and secondary	24 hours	150 µg/m^3	Not to be exceeded more than once per year on average over 3 years
Sulfur Dioxide (SO$_2$)		primary	1 hour	75 ppb [4]	99th percentile of 1-hour daily maximum concentrations, averaged over 3 years
		secondary	3 hours	0.5 ppm	Not to be exceeded more than once per year

(1) In areas designated nonattainment for the Pb standards prior to the promulgation of the current (2008) standards, and for which implementation plans to attain or maintain the current (2008) standards have not been submitted and approved, the previous standards (1.5 µg/m3 as a calendar quarter average) also remain in effect.

(2) The level of the annual NO2 standard is 0.053 ppm. It is shown here in terms of ppb for the purposes of clearer comparison to the 1-hour standard level.

(3) Final rule signed October 1, 2015, and effective December 28, 2015. The previous (2008) O3 standards additionally remain in effect in some areas. Revocation of the previous (2008) O3 standards and transitioning to the current (2015) standards will be addressed in the implementation rule for the current standards.

(4) The previous SO2 standards (0.14 ppm 24-hour and 0.03 ppm annual) will additionally remain in effect in certain areas: (1) any area for which it is not yet 1 year since the effective date of designation under the current (2010) standards, and (2)any area for which an implementation plan providing for attainment of the current (2010) standard has not been submitted and approved and which is designated nonattainment under the previous SO2 standards or is not meeting the requirements of a SIP call under the previous SO2 standards (40 CFR 50.4(3)). A SIP call is an EPA action requiring a state to resubmit all or part of its State Implementation Plan to demonstrate attainment of the required NAAQS.

Source: EPA, 2018a.

Figure 6.1.2.3

Percentage of children living in counties with exposures above the EPA NAAQS in 2015

Percentage of children exposed	Pollutant	Measurement of Exposure
0	Carbon monoxide	Concentrations above the level of the current standard for carbon monoxide
0.1	Lead	Ambient lead concentrations above the level of the current three-month standard for lead
2	Nitrogen dioxide	Concentrations above the level of the current one-hour standard for nitrogen dioxide at least one day per year
3	Sulfur dioxide	Sulfur dioxide concentrations above the level of the current one-hour standard for sulfur dioxide at least one day per year
3	Particulate Matter (2.5 μm)	Average concentration above the level of the current annual $PM_{2.5}$ standard
7	Particulate Matter (10 μm)	PM_{10} concentrations above the level of the current 24-hour standard for PM_{10} at least one day per year
21	Particulate Matter (2.5 μm)	$PM_{2.5}$ concentrations above the level of the current 24-hour $PM_{2.5}$ standard at least once per year
58	Ozone	Ozone concentrations above the level of the current 8-hour ozone standard at least one day during the year

Source: EPA, 2018b, from text on p. 12.

the six criteria pollutants. The percentage of children living in counties where they might be exposed to what the EPA deems "unhealthy air" was only 3% in 2015, down from 9% 16 years earlier (EPA, 2018). A graph showing the decline appears as Figure 6.1.3.3 in the next section.

EPA versus Real-World Exposure

EPA's estimates of exposure to chemical compounds released during the combustion of fossil fuels are "stylized facts," simplifications of the very complex and uncertain data collected and interpreted to meet the needs of government regulators (and perhaps newspaper headline writers). Still, they can be shown to greatly overstate the real-world exposure to pollutants experienced by people living in the United States, including children.

Start with the EPA's assumption that every child living in a county is breathing the *worst* air quality reported by *any* air-quality monitoring station in that county over the course of a year. This is why the text above summarizing EPA's findings uses the clumsy phrase "percentage of children living in counties

where they might be exposed to pollutant concentrations above the levels of the current air quality standards" instead of the percentage or number of children *actually* exposed. As Schwartz and Hayward reported in 2007,

> EPA and ALA [American Lung Association] get their inflated numbers by counting everyone in a county as breathing air that exceeds federal standards, even if most of the county has clean air. For example, only one rural area of San Diego County, with about 1% of the population, violates the EPA's 8-hour ozone standard. But the EPA and the ALA count all three million people in the county as breathing "unhealthy" air. This is akin to giving every student in a school a failing grade if just one gets an "F" on an exam (p. 7).

It gets worse. The "one day per year" appearing in Figure 6.1.2.3 is EPA shorthand for a complex way of measuring "exceedances" and "violations" (explained by Schwartz and Hayward, 2007, pp. 8-9).

If exceedances occurred one day a year, then some children living in counties where children could be exposed to a pollutant as little as 0.27% of the time (1 ÷ 365). So for the $PM_{2.5}$ standard, a one-day violation a year in counties where 21% of the children in the United States reside means the *average child* in the United States is exposed only 0.06% of the time (0.21 x 0.27), or for about five hours a year, to ambient levels of $PM_{2.5}$ above EPA's NAAQS.

When the EPA's faulty way of counting affected children is corrected, Schwartz and Hayward (2007, p. 10) found "about 11% of Americans live in areas that violate the 8-hour ozone standard, while about the same fraction live in areas that violate for $PM_{2.5}$." The authors were using data from 2006. Since then concentrations of $PM_{2.5}$ have fallen by about 24% (see Figure 6.1.3.1 below). So maybe only 8% of Americans (0.11 x (1 - 0.24)) live in areas that violate the $PM_{2.5}$ standard 0.27% of the time, so average exposure is 0.02% a year, or less than two hours a year.

EPA estimates anthropogenic emissions account for about 40% of $PM_{2.5}$ released into the air each year in the United States (EPA, 2018, see Figure 6.1.1.1 above). Fossil fuel-related activities account for approximately half of those emissions, so fossil fuels account for about 20% of human exposure to $PM_{2.5}$ in the United States. So maybe fossil fuels are responsible for exposing Americans to levels of PM_{10} that exceed EPA's NAAQS for about 24 minutes a year (0.02 x 0.2 x 60).

The same exercise could be performed for ozone and other pollutants and would arrive at similar conclusions: exposure to possibly harmful air pollutants due to the use of fossil fuels in the United States is probably too low to accurately measure or distinguish from background levels. This is according to the EPA's own monitoring stations and assuming *arguendo* that EPA's NAAQS actually are meaningful indicators of a possible threat to public health. That assumption is taken up (and refuted) in Section 6.1.4 and in later sections.

References

EPA. 2018a. U.S. Environmental Protection Agency. National ambient air quality standards (website). Accessed June 7, 2018.

EPA. 2018b. U.S. Environmental Protection Agency. 2018b. America's Children and the Environment. Third Edition. January. Washington, DC.

NETL. 2015. National Energy Technology Laboratory. Cost and performance baseline for fossil energy plants. Volume 1a: Bituminous Coal (PC) and Natural Gas to Electricity, Revision 3. Washington, DC: U.S. Department of Energy.

Radovic, L.R. 1992. *Energy and Fuels in Society*. New York, NY: McGraw-Hill.

Schwartz, J.M. and Hayward, S.F. 2007. *Air Quality in America: A Dose of Reality on Air Pollution Levels, Trends, and Health Risks*. Washington, DC: AEI Press.

WHO. 2017. World Health Organization. Mercury and health (website). Accessed November 15, 2018.

6.1.3 Trends

Exposure to potentially harmful emissions from the burning of fossil fuels in the United States declined rapidly in recent decades and is now at nearly undetectable levels.

Chemical compounds released during the combustion of fossil fuels in the United States and in developed countries around the world have fallen dramatically since the 1940s and 1950s as a result of technological change, public pressure for a cleaner environment, and government regulations. Air quality data for the United States are readily available from government agencies and are used to document these trends for the rest of this chapter. Data for Europe, readily available on the website of the European Environment Agency, show similar trends for that part of the world.

Figure 6.1.3.1 shows the trends for emissions and aerial concentrations in the United States during each of four periods: 1980 to 2016, 1990 to 2016, 2000 to 2016, and 2010 to 2016. Sulfur dioxide emissions fell by 90% since 1980, carbon monoxide emissions by 73%, and emissions of nitrogen oxides by 62%. The declines in just the most recent period, the six years from 2010 to 2016, were substantial for every pollutant except particulate matter. Aerial carbon monoxide concentrations have fallen 85% since 1980, lead 99%, and nitrogen dioxide between 61% and 62%. The trend analysis reveals much of the improvement took place in only the past 16 years, since 2000, and that major improvements occurred in the past six years.

As noted in Section 6.1.2, the EPA tracks the percentage of children in the United States living in counties where they might be exposed to pollutant

Figure 6.1.3.1
Change in criteria pollutants in the United States, 1980–2016

A. Percent change in emissions of five criterion pollutants plus VOCs in the United States, 1980-2016

	1980 vs 2016	1990 vs 2016	2000 vs 2016	2010 vs 2016
Carbon Monoxide	-73	-66	-52	-21
Lead	-99	-80	-50	-23
Nitrogen Oxides (NO$_x$)	-62	-59	-54	-30
Volatile Organic Compounds (VOC)	-55	-42	-21	-10
Direct PM$_{10}$	-57	-18	-15	-4
Direct PM$_{2.5}$	---	-25	-33	-6
Sulfur Dioxide	-90	-89	-84	-66

B. Percent change in aerial concentration of six criteria pollutants in United States, 1980–2016

	1980 vs 2016	1990 vs 2016	2000 vs 2016	2010 vs 2016
Carbon Monoxide	-85	-77	-61	-14
Lead	-99	-99	-93	-77
Nitrogen Dioxide (annual)	-62	-56	-47	-20
Nitrogen Dioxide (1-hour)	-61	-50	-33	-15
Ozone (8-hour)	-31	-22	-17	-5
PM10 (24-hour)	---	-39	-40	-9
PM2.5 (annual)	---	---	-42	-22
PM2.5 (24-hour)	---	---	-44	-23
Sulfur Dioxide (1-hour)	-87	-85	-72	-56

Source: EPA, 2018b.

concentrations higher than the levels of the current air quality standards. Its graph showing estimates for 1999–2016 appears as graph A in Figure 6.1.3.2. It shows exposure to what the EPA believes to be unsafe levels of exposure is in steep decline. For example, the percentage of children living in counties where they might be exposed to harmful levels of PM$_{2.5}$ decreased from 55% to 21%, to SO$_2$ from 31%

to 3%, and to NO$_2$, from 23% to 2%. These are dramatic declines.

The EPA's "Air Quality Index," which combines and weights its measures of exposure to the six criteria pollutants, also shows a dramatic reduction in exposure to possibly harmful pollutants from 1999 to 2015. The EPA's graph showing changes in the percentage of days with "good," "moderate," or

Figure 6.1.3.2
Trends in U.S. Air Quality

A. Percentage of children ages 0 to 17 years living in U.S. counties with pollutant concentrations above the levels of the current air quality standards, 1999–2016

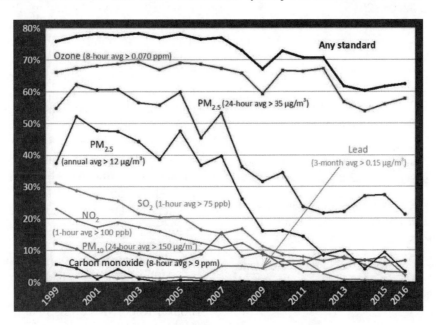

B. Percentage of days with good, moderate, or unhealthy air quality for children ages 0 to 17 years in the United States, 1999–2015

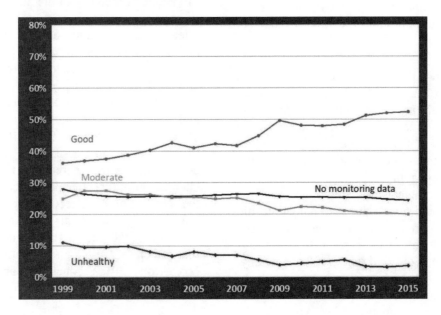

Source: EPA, 2018c.

"unhealthy" air quality for children from 1999 to 2015 appears in Figure 6.1.3.2 as graph B. The percentage of days during which children lived in counties where they might be exposed to what the EPA deems "unhealthy air" has declined from 9% in 1999 to 3% in 2015, while the percentage of children's days with "good" air quality increased from 36% in 1999 to 52% in 2015.

References

EPA. 2018a. U.S. Environmental Protection Agency. National ambient air quality standards (website). Accessed June 7, 2018.

EPA. 2018b. U.S. Environmental Protection Agency. EPA's Report on the Environment, Particulate Emissions (website). Accessed November 15, 2018.

EPA. 2018c. U.S. Environmental Protection Agency. *America's Children and the Environment.* Third edition. January. Washington, DC.

European Environment Agency. 2012. Particulate matter from natural sources and related reporting under the EU Air Quality Directive in 2008 and 2009. *EEA Technical Report* No. 10/2012.

6.1.4 Interpreting Exposure Data

Exposure to chemical compounds produced during the combustion of fossil fuels is unlikely to cause any fatalities in the United States.

The chemistry, exposure data, and trends presented in Sections 6.1.1, 6.1.2, and 6.1.3 are a necessary though not sufficient basis for rendering findings on the possible health effects of emissions caused by the combustion of fossil fuels. We can begin by ruling out negative health effects due to "acid rain" and aerial emissions of mercury.

There is no biological mechanism whereby less basic rainwater would pose a threat to human health. Benarde (1987) reported that "an exhaustive search of the pertinent literature indicates that deleterious human health effects [of "acid rain"], if there are any, remain to be established. As a consequence of pollution abatement efforts the next 15 to 20 years should witness a reduction in acid levels. Accordingly, a worsening of current levels of chemical pollutants is not anticipated. Hence, a significant threat to public health via acid rain currently or in the foreseeable future, should not be expected." More recent efforts to link acid rain with human health effects focus on the contribution of SO_2 and NO_x emissions to the formation of fine particulate matter and not to acidification *per se* (Chestnut and Mills, 2005; Menz and Seip, 2004). Even EPA says "Walking in acid rain, or even swimming in a lake affected by acid rain, is no more dangerous to humans than walking in normal rain or swimming in non-acidic lakes" (EPA, n.d.).

Mercury is a genuine threat to human health. However, *exposure* to mercury in the United States and other developed countries is well within public safety levels. The National Research Council of the National Academies of Sciences determined in 2000 that 85 micrograms of mercury per liter ($\mu g/L$) or higher in cord blood was associated with early neurodevelopmental effects. According to the Centers for Disease Control and Prevention's *Fourth National Report on Human Exposure to Environmental Chemicals,* blood samples from 8,373 people taken in 2010 (the most current test data available) found 95% had mercury levels below 4.90 $\mu g/L$ and "all blood mercury levels for persons in the Fourth Report were less than 33 $\mu g/L$" (CDC, 2018, Vol. 1, p. 319 and CDC, 2017).

The accumulation of methylmercury (MeHg) in fish tissue has been raised as a public health issue, but its accumulation depends on many environmental factors and is largely independent of concentrations of elemental mercury in the air (Mason *et al.,* 2005). Electricity generation using coal in the United States released an estimated 26.5 tons of mercury in 2011 and only 6.94 tons in 2016 (EPA, 2018). This is dwarfed by other emission sources: U.S. forest fires emit at least 44 tons per year; cremation of human remains, 26 tons; Chinese power plants, 400 tons; and volcanoes, subsea vents, geysers and other sources, approximately 9,000-10,000 tons per year (Soon and Driessen, 2011). Atmospheric concentrations of mercury do not coincide with changes in anthropogenic emissions, a reflection of the fact that humans account for less than 0.5% of all the mercury in the air and, as is the case with mercury in the oceans, the numerous natural cycles that affect its presence in the atmosphere. Soon and Monckton (2012) concluded an analysis of U.S. mercury control regulations as follows:

The scientific literature to date strongly and overwhelmingly suggests that meaningful management of mercury is likely impossible, because even a total elimination of all industrial emissions, especially those from U.S. coal-fired power plants, will almost certainly not be able to affect trace, or even high, levels of MeHg that have been found in fish tissue over century-long time periods.

Globally, emissions of mercury have plummeted since governments around the world launched campaigns to reduce industrial emissions. Since 1990, nine European countries reduced their emissions by 85% or more and five (Sweden, Denmark, Norway, Ireland, and Croatia) now report zero emissions (European Environment Agency, n.d.) U.S. emissions from electricity generation fell 89% during the same period (Oakridge National Laboratory, 2017, figure ES1, p. viii). Exposure to mercury in the air (as opposed to ingesting paint chips that might contain lead and other avenues of exposure) is not a health threat in the United States or other developed countries today.

Regarding the remaining pollutants, EPA has established National Ambient Air Quality Standards (NAAQS) that it uses to determine which states, counties, and cities are "nonattainment" areas making them subject to EPA enforcement actions, and to define and report to Congress progress toward "good" air quality. To the public, failure to attain NAAQS may appear to be evidence of "unhealthy" air, and EPA encourages this perception. However, NAAQSs are set *orders of magnitude lower* than what the best available science suggests is a level where human health and public welfare are actually threatened (Belzer, 2012). This means failing to attain EPA's NAAQS does not mean an actual threat to human health exists (Belzer, 2012).

EPA standards are based on either the dose at which no adverse effect was observed (NOAEL) or the lowest dose at which an adverse effect was observed (LOAEL). When a LOAEL is used, the default safe threshold is reduced by a factor of ten to account for the unknown distance between the observed LOAEL and the unobserved NOAEL. If the LOAEL or NOAEL comes from an animal study, the default safe threshold is reduced by another factor of 10 to account for the possibility that humans are more sensitive than the most sensitive laboratory animal tested. Together, these two "safety adjustments" can reduce the safety threshold by a factor of 100.

A third default safety factor of ten is used to make sure the most susceptible members of the population are protected. A fourth factor of ten is applied when data is obtained from studies with less-than-lifetime exposure. A fifth factor of ten is applied when the database is incomplete. When all five safety factors are used, the composite safety factor is 10,000. This means the EPA standard would be 10,000 times more strict than what the actual public health research suggests is a dose that is dangerous to human health. Mercifully – and because such extreme precaution would subject it to ridicule in the public health community – EPA has adopted a policy whereby the total safety factor applied to any particular chemical is no more than 3,000, a still remarkably high risk multiplier (EPA, 2002, pp. 4-41).

Incredibly, this is not the only way the EPA errs on the side of setting its safety standards too low. Belzer (2012) identifies the following practices:

- Extrapolating human cancer risk at very low environmental levels from very high laboratory exposures to animals;

- Using default assumptions such as daily adult inhalation, drinking water consumption, and time spent outdoors that overstate the average;

- Reliance on simulation models instead of exposure data obtained from the risk scenario of interest;

- Estimating risks and benefits using exposures to a small fraction of the population, such as the 95th percentile, rather than the mean; and

- Extrapolation of risk from each step of a risk assessment means even small over-estimations produce very large reductions in the safety standard.

The result of these default options and assumptions is "cascading bias," which Belzer (2012, p. 13, fn. 28) defines as "when each of several terms in a point estimate of risk is upwardly biased, the point estimate is biased by the product of the biases." A bureaucracy's definition of acceptable risk is not a statement of relative risk based on toxicology or observation, or even derived from epidemiological associations, but the result of a political process that balances science with institutional goals, with the latter often influenced by subjective judgements about acceptable risk. As Belzer (2017) later observed, "EPA will strive for the highest estimate of risk that does not bring upon the Agency unbearable ridicule. You simply cannot rely on the EPA risk assessment to give you an unvarnished perspective. When given an EPA risk assessment, all you know is risk can't be any worse" (p. 3). In short, the EPA's NAAQS should not be accepted as a definition of "safe" or "unsafe" air, even if they were arrived at by close attention to the science and with utmost integrity.

As the discussion of particulate matter later in this chapter demonstrates, the EPA's NAAQS were not determined by "close attention to the science and with utmost integrity." The process by which they were established demonstrates an almost shocking degree of manipulation, dishonesty, and refusal to acknowledge research findings that run counter to the agency's policy agenda. That they are still defended today by the EPA bureaucracy and the coterie of well-paid academics it has assembled to provide the appearance of scientific fact, if not by the agency's administrator, reveals a flawed culture inside a failed government agency.

Later in this chapter these issues – along with whether small-associations epidemiology is a legitimate basis for air quality standards at all, particularly when the EPA's philosophy is that there is no safe level of any primary air pollutant (the "linear no-threshold" (LNT) dose-response relationship) – are addressed in some depth. But even before those concerns are addressed, the evidence is clear that *very few people in the United States are exposed to pollutants at levels likely to pose a threat to human health.* The same is almost certainly sure for much of Europe and developed countries around the world. Further confirmation can be seen in the inability of the EPA to show any declines in mortality in the past two decades that could be attributed to the decline in particulate matter or other pollutants, a decline that should be apparent if the criteria pollutants were once a human health threat at levels higher than today's.

The very low and falling number of children who may be exposed to dangerous chemicals and the almost ridiculously low levels of exposure chosen by EPA for its NAAQS have never been reported by the press, but the EPA's highly speculative numbers of people "killed" every year by particulate matter and ozone appear countless times in headlines and the fundraising letters of environmental advocacy groups such as the American Lung Association (ALA, n.d.). They also appear in estimates of the "social cost" of fossil fuels and of future climate change and are used to justify anti-fossil fuel regulations. But as the analysis in this section shows, the real public health risks of exposure to EPA's six criteria pollutants are negligible.

References

American Lung Association. n.d. Fighting for Healthy Air (website). Accessed November 12, 2018.

Belzer, R. 2012. Risk assessment, safety assessment, and the estimation of regulatory benefits. Mercatus Research. Washington, DC: Mercatus Center.

Belzer, R. 2017. Testimony before the U.S. House of Representatives Committee on Science, Space, and Technology. February 7.

Benarde, M. 1987. Health effects of acid rain: Are there any? *Journal of the Royal Society of Health,* **107** (4), 139-145.

CDC. 2017. Centers for Disease Control and Prevention. Mercury factsheet. National Biomonitoring Program (website). Accessed June 10, 2018.

CDC. 2018. Centers for Disease Control and Prevention. Fourth National Report on Human Exposure to Environmental Chemicals. Updated Tables, March, Volume One. Accessed June 10, 2018.

Chestnut, L., and Mills, D. 2005. A fresh look at the benefits and costs of the US acid rain program. *Journal of Environmental Management,* **77**(3), 252-266.

EPA. n.d. U.S. Environmental Protection Agency. Effects of acid rain (website). Accessed November 14, 2018.

EPA. 2002. U.S. Environmental Protection Agency. A review of the reference dose and reference concentration processes. Washington, DC. Washington, DC.

EPA. 2018. U.S. Environmental Protection Agency. Electric Utilities Mercury Releases in the 2016 TRI National Analysis. Washington, DC.

European Environment Agency. n.d. Chart - changes in mercury emissions. Accessed November 14, 2018.

Mason, R.P., *et al.* 2005. Monitoring the response to mercury deposition. *Environmental Science & Technology*, vol. 39, A14-A22.

Menz, F. and Seip, H. 2004. Acid rain in Europe and the United States: an update. *Environmental Science & Policy*, **7** (4), 253-265.

Oakridge National Laboratory. 2017. Environmental Quality and the U.S. Power Sector: Air Quality, Water

Quality, Land Use and Environmental Justice. Washington, DC: U.S. Department of Energy.

Soon, W. and Driessen, P. 2011. The myth of killer mercury. *The Wall Street Journal.* May 25.

Soon, W. and Monckton, C. 2012. Reconsidering U.S. EPA's proposed NESHAP's mercury emission rule. MasterResource. Blog. June 11.

6.2 Failure of the EPA

The data presented in Section 6.1 show dramatic progress has been made in reducing emissions of possibly harmful chemical compounds produced during the use of fossil fuels, and more importantly reducing human exposure to those chemicals. While giving credit for this achievement to the government agencies most responsible for enforcing environmental protection laws might seem appropriate, this is not the case. As was shown in Chapter 1, Section 1.3.1, improvements in air quality in the United States began in the 1940s and 1950s, long before the national government and U.S. Environmental Protection Agency (EPA) got involved. Hayward writes,

> The chief drivers of environmental improvement are economic growth, constantly increasing resource efficiency, technological innovation in pollution control, and the deepening of environmental values among the American public that have translated to changed behavior and consumer preferences. Government regulation has played a vital role, to be sure, but in the grand scheme of things regulation can be understood as a lagging indicator, often achieving results at needlessly high cost, and sometimes failing completely (Hayward, 2011, p. 2).

Schwartz and Hayward (2007) note, "Improvements in air quality are not unique. Other environmental problems, such as water quality, were also improving before the federal government took over regulatory control. Likewise, other risks were dropping without federal regulation. Per mile of driving, the risk of dying in a car accident dropped 75% between 1925 and 1966 – the year Congress adopted the National Traffic and Motor Vehicle Safety Act and created the National Highway Traffic

Safety Administration. Between 1930 and 1971 – the year that the Occupational Safety and Health Administration was created – the risk of dying in a workplace accident dropped nearly 55%." In all these cases – air quality, automobile safety, and workplace safety – the rate of improvement was about the same before and after the federal government nationalized policy. Without doubt, improvements would have continued in all these areas even if the federal government had not taken the regulatory reins away from the states.

As this section will show, the EPA has often been more of a hindrance than a help in advancing the cause of environmental protection in the United States. The discussion in Chapter 1, Section 1.4.3, of how government bureaucracies work (and do not work) provides good background for this discussion. This section begins by explaining how the EPA's mission has evolved over time in response to congressional and public pressure as well as the natural tendencies of bureaucracies, creating a culture that cannot concede the possibility that human emissions of toxic substances are *not* a major public health crisis in need of the EPA's expert attention. This bias contaminates all of its scientific research, making it unreliable. Next, the EPA's repeated and flagrant violation of the basic rules of the scientific method is documented. Finally, the loss of integrity and outright corruption that have affected the agency are documented. Along the way, parallels to the mission, methodology, and corruption of the United Nations' Intergovernmental Panel on Climate Change (IPCC), documented in Chapter 2, are identified.

References

Hayward, S. 2011. *Almanac of Environmental Trends.* San Francisco, CA: Pacific Institute for Public Policy Research.

Schwartz, J.M. and Hayward, S.F. 2007. *Air Quality in America: A Dose of Reality on Air Pollution Levels, Trends, and Health Risks.* Washington, DC: AEI Press.

6.2.1 A Faulty Mission

> *Due to its faulty mission, flawed paradigm, and political pressures, the U.S. Environmental Protection Agency (EPA) is an unreliable source of research on air quality and its impact on human health.*

Chapter 2 explained how the mission of the United Nations' Intergovernmental Panel on Climate Change (IPCC) – to find and document the human impact on the global climate – blinded it to the possibility that natural variability could explain most or even all of the warming experienced in the late twentieth century and thus obviate the need for an organization tasked with solving the nonexistent problem. The EPA's mission similarly blinds it to the possibility that natural causes of cancer and other diseases may outweigh any effects of man-made chemical compounds.

EPA's website says "Born in the wake of elevated concern about environmental pollution, the EPA was established on December 2, 1970 to consolidate in one agency a variety of federal research, monitoring, standard-setting and enforcement activities to ensure environmental protection. Since its inception, the EPA has been working for a cleaner, healthier environment for the American people" (EPA, 2018). Elsewhere on its website, the EPA says its mission is simply "to protect human health and the environment."

The simple mission statement obscures profound conflicts of interest that prevent the EPA from making good on its promise. Like many government agencies, the EPA was given not one but three mandates: to identify, evaluate, and solve a social problem. But combining all three responsibilities in the same entity means the agency has no incentive to decide the social problem does not merit a significant investment of public monies to solve, or that the problem, should it exist, even could be solved. The agency is also charged with measuring its own success and then reporting it to those who control its funding and future existence. As explained in Chapter 1, Section 1.4.3, the heads of such agencies, no matter how honest or well-intended, cannot objectively evaluate their own performances (Savas, 2000, 2005). Schwartz and Hayward (2007) explained it the following way:

> The Clean Air Act charges the EPA with setting air pollution health standards. But this means that federal regulators decide when their own jobs are finished. Not surprisingly, no matter how clean the air, the EPA continues to find unacceptable risks. The EPA and state regulators' powers and budgets, as well as those of environmentalists, depend on a continued public perception that there is a serious problem to solve. Yet regulators are also

major funders of the health research intended to demonstrate the need for more regulation. They also provide millions of dollars a year to environmental groups, which use the money to augment public fear of pollution and seek increases in regulators' powers. These conflicts of interest largely explain the ubiquitous exaggeration of air pollution levels and risks, even as air quality has steadily improved (2007, pp. 11–12).

The EPA quickly grew in size and influence. Its resources and power naturally attracted the attention of interest groups. Jay Lehr, Ph.D., a scientist who was involved in the founding of the EPA, wrote in 2014, "Beginning around 1981, liberal activist groups recognized the EPA could be used to advance their political agenda by regulating virtually all human activities regardless of their impact on the environment. Politicians recognized they could win votes by posing as protectors of the public health and wildlife. Industries saw a way to use regulations to handicap competitors or help themselves to public subsidies" (Lehr, 2014).

As reported by Chase (1995), in 1993 President Bill Clinton signed the International Convention on Biological Diversity and just months later created the President's Council on Sustainable Development, making "'ecosystem protection" the EPA's highest mandate. "Under the new rules [EPA's] primary goal would no longer be to protect public health. Rather, it would seek to save nature instead" (p. 91). Evidence that the change in mission affected the EPA's research since 1993 can be found in the fact that in 1987, 1990, and 1991 the agency produced a series of reports recognizing the impacts of pollution (not only air pollution but also impacts on water and food and exposure to toxic waste) were small relative to other human health risks (EPA, 1987, 1990, 1991), but since then it has embraced a "zero risk" paradigm whereby *any* human impact on the environment, no matter how small, is regarded as justification for government regulation (e.g., EPA, 2004, 2009). Protecting public health has become a pretense for stopping any human activity that has any impact at all on the environment. Such a broad definition of "environmental protection" gives the agency license to regulate virtually every human activity.

The War on Cancer

Much as the IPCC assumes only man can cause climate change, the EPA's mission leads it to assume that natural causes of cancer and other diseases either do not exist or do not matter to the regulatory process. In both cases the assumptions are false, and they contaminate and often invalidate much of what both the IPCC and the EPA do.

The EPA ignores and even hides from the public evidence that man-made chemicals are trivial contributors to the nation's disease and mortality rates. For example, Bruce N. Ames and Lois Swirsky Gold, two distinguished medical researchers at the University of California-Berkeley, pointed out that "99.99% of all pesticides in the human diet are natural pesticides from plants" (Ames *et al.,* 1990). "All plants produce toxins to protect themselves against fungi, insects, and animal predators such as humans. Tens of thousands of these natural pesticides have been discovered, and every species of plant contains its own set of different toxins, usually a few dozen. When plants are stressed or damaged (when attacked by pests), they greatly increase their output of natural pesticides, occasionally to levels that are acutely toxic to humans" (Ames and Gold, 1993, p. 157. See also Ames, 1983, and Ames *et al.,* 1990).

The EPA's focus on man-made chemical compounds as the cause of negative health effects was reinforced by political constraints placed on the agency. According to Kent and Allen (1994), "The strong political pressures in the Congress to legislate risk levels at or near zero can have a serious impact on the costs of environmental programs. To the extent that zero risk statutes are not feasible, they also threaten the overall credibility of the nation's environmental efforts. Statutory language pursuing 'zero discharge' and extremely low cleanup standards for superfund sites could force huge social investments that would divert scarce resources from even higher-risk problems" (Kent and Allen, 1994, p. 65).

The EPA's campaign to regulate away all risks is doomed to fail since risk is inherently subjective. Lash (1994) explained why this is so:

Some people willingly die to protect their children; others abandon them. Some choose to die for religious faith, or honor, or country; others use those concepts as rhetorical symbols to achieve selfish ends. *It is the interaction of what we value with what we believe to be reality that determines how*

we act. Given identical information and alternatives, different people make different choices. The debate over what the comparative risk process is, what it should be, and whether it is essential or pernicious as a tool for public policy is a debate about decisions, who should make them, and how (p. 70).

He added, "Whether the issue is smoking or global climate change, normative questions are inextricably woven into the assessment of risk" (Lash, 1994, p. 76). Furedi (2010) noted, "frequently, worst-case thinking displaces any genuine risk-assessment process. Risk assessment is based on an attempt to calculate the probability of different outcomes. Worst-case thinking – these days known as precautionary thinking – is based on an act of imagination. It imagines the worst-case scenario and demands that we take action on that basis. ... In the absence of freedom to influence the future, how can there be human responsibility? That is why one of the principal accomplishment[s] of precautionary culture is the normalisation of irresponsibility. That is a perspective that we need to reject for a mighty dose of humanist courage."

Ames and Swirsky Gold warned, "Excessive concern for pollution will not improve public health – and, in the confusion, may cause us to neglect important hazards, such as smoking, alcohol, unbalanced diets (with too much saturated fat and cholesterol, and too few fruits and vegetables), AIDS, radon in homes, and occupational exposures to chemicals at high levels. The progress of technology and scientific research is likely to lead to a decrease in cancer death rates and incidence of birth defects, and an increase in the average human life span (Ames and Gold, 1993, p. 179).

The War on Coal

President Barack Obama understood clearly how the EPA could be used to advance his political agenda, which included penalizing manufacturers and the fossil fuel industry and rewarding high-tech companies and the alternative energy industry. When campaigning for president in January 2008, Obama told the editorial board of *The San Francisco Chronicle,* "If somebody wants to build a coal-fired power plant, they can. It's just that it will bankrupt them," and later, "Under my plan … electricity rates would necessarily skyrocket" (Martinson, 2012).

Once elected, Obama proceeded to "weaponize" the EPA against the fossil fuel industry. His administration promulgated new rules and tightened older ones in an effort to strangle the coal industry. According to Orr and Palmer (2018) those efforts included:

- Clean Power Plan

- Cross-State Air Pollution Rule

- More stringent National Ambient Air Quality Standards (NAAQS) for mercury, particulate matter, and ozone

- Cooling Water Intake Rule

- Coal Combustion Residuals Rule

- Carbon Pollution Standards for New Plants

- Effluent Limitations Guidelines

- Stream Protection Rule

- Department of the Interior bans on new mines on public lands and mountaintop mining

Many of these regulations could not be justified by cost-benefit analysis, a point that will be documented in Chapter 8. They were adopted solely as part of a "war on coal" modeled after the war on cancer to force a transition from fossil fuels to alternative energy sources (wind and solar) or mandatory energy conservation. Wrote Orr and Palmer,

> The war on coal was very real. It was led from the White House and backed by hundreds of millions of dollars in funding from left-wing foundations including the Rockefeller Brothers, the Hewlett Foundation, the MacArthur Foundation, Bloomberg Philanthropies, and even Chesapeake Energy, a natural gas drilling company seeking to grow demand for its product. These millions were funneled to environmental activist groups including Greenpeace, the Sierra Club, and Natural Resources Defense Council. Just one donor, billionaire Michael Bloomberg, has given more than $168 million to the Sierra Club to

support the effort (citing Suchecki, 2015, and Brown, 2017).

Members of the Obama administration sometimes acknowledged the real political objective of the campaign. EPA Administrator Gina McCarthy testified before the U.S. Senate Environment and Public Works Committee on July 23, 2014: "The great thing about this [Clean Power Plan] proposal is that it really is an investment opportunity. *This is not about pollution control*" (McCarthy, 2014, italics added). Secretary of State John Kerry described U.S. policy regarding coal-fueled power plants: "We're going to take a bunch of them out of commission" (Davenport, 2014). In a December 9, 2015 address at the United Nations conference where the Paris Accord was negotiated, Kerry was remarkably frank about *how the treaty was not, after all, about protecting the environment*. He said:

> The fact is that even if every American citizen biked to work, carpooled to school, used only solar panels to power their homes, if we each planted a dozen trees, if we somehow eliminated all of our domestic greenhouse gas emissions, guess what – that still wouldn't be enough to offset the carbon pollution coming from the rest of the world.

> If all the industrial nations went down to zero emissions – remember what I just said, all the industrial emissions went down to zero emissions – it wouldn't be enough, not when more than 65% of the world's carbon pollution comes from the developing world (Quoted in Watts, 2015).

The EPA was a willing accomplice in this political campaign to end the world's reliance on fossil fuels. An international climate treaty would have provided legal as well as political cover for exercising even more power over sectors of the economy that constitutionally and by tradition were the reserve of state governments or left unregulated. The Paris Accord would have been the capstone of an eight-year march to power under a president devoted to transforming the nation's energy, manufacturing, and agricultural sectors into a new system in which the agency would be empowered to regulate virtually every aspect of life in America. Today, the EPA has a budget of $8 billion and 12,000 full-time staff. Its regulations already account for more than half of the total cost of complying with federal regulations (Crews, 2018). But like all bureaucracies, it wanted to grow.

References

Ames, B.N. 1983. Dietary carcinogens and anticarcinogens. *Science* **221** (September): 1256–64.

Ames, B.N. and Gold, L.S. 1993. Environmental pollution and cancer: some misconceptions. Chapter 7 in: Foster, K.R., Bernstein, D.E., and Huber, P.W. (Eds.) *Phantom Risk: Scientific Inference and the Law*. Cambridge, MA: The MIT Press, pp. 153–81.

Ames, B.N., Profet, M., and Gold, L.S. 1990. Dietary pesticides (99.99% all natural). *Proceedings of the National Academy of Sciences USA* **87** (October): 7777–81.

Brown, D. 2017. Bloomberg puts up another $64M for 'war on coal.' Greenwire (website). October 11.

Chase, A. 1995. *In a Dark Wood: The Fight Over Forests and the Rising Tyranny of Ecology*. Boston, MA: Houghton Mifflin Company.

Crews, W. 2018. *Ten Thousand Commandments: A Policymaker's Snapshot of the Federal Regulatory State*. Washington, DC: Competitive Enterprise Institute.

Davenport, C. 2014. Strange climate event: warmth toward U.S. *The New York Times*. December 11.

EPA. 1987. U.S. Environmental Protection Agency. *Unfinished Business: A Comparative Assessment of Environmental Problems*. Washington, DC.

EPA. 1990. U.S. Environmental Protection Agency. *Reducing Risk: Setting Priorities and Strategies for Environmental Protection*. Washington, DC.

EPA. 1991. U.S. Environmental Protection Agency. *Environmental Investments: The Cost of a Clean Environment*. Washington, DC.

EPA. 2004. U.S. Environmental Protection Agency. *Air Quality Criteria for Particulate Matter*. Washington, DC.

EPA. 2009. U.S. Environmental Protection Agency. *Integrated Science Assessment for Particulate Matter*. National Center for Environmental Assessment – RTP Division, Office of Research and Development. Research Triangle Park, NC.

EPA. 2018. U.S. Environmental Protection Agency. EPA history (website). Accessed June 9, 2018.

Furedi, F. 2010. Fear is key to irresponsibility. *The Australian*. October 9.

Houser, T., Bordoff, J., and Marsters, P. 2017. *Can Coal Make a Comeback?* New York, NY: Center on Global Energy Policy. April.

Kent, C.W. and Allen, F.W. 1994. An overview of risk-based priority setting at the EPA. Chapter 4 in: Finkel, A.M. and Golding, D. (Eds.) *Worst Things First? The Debate over Risk-Based National Environmental Priorities*. Washington, DC: Resources for the Future, pp. 47–68.

Lash, J. 1994. Integrating science, values, and democracy through comparative risk assessment. Chapter 5 in: Finkel, A.M. and Golding, D. (Eds.) *Worst Things First? The Debate over Risk-Based National Environmental Priorities*. Washington, DC: Resources for the Future, pp. 69–86.

Lehr, J. 2014. Replacing the Environmental Protection Agency. *Policy Brief*. Chicago, IL: The Heartland Institute.

Martinson, E. 2012. Uttered in 2008, still haunting Obama. Politico (website). April 5.

McCarthy, G. 2014. Quoted in Pollution vs. Energy: Lacking Proper Authority, the EPA Can't Get Carbon Message Straight. *News Release*. U.S. House Energy Commerce Committee. July 23.

Orr, I. and Palmer, F. 2018. How the premature retirement of coal-fired power plants affects energy reliability, affordability. *Policy Study*. Arlington Heights, IL: The Heartland Institute. February.

Savas, E.S. 2000. *Privatization and Public Private Partnerships*. New York, NY: Chatham House Publishers.

Savas, E.S. 2005. *Privatization in the City: Successes, Failures, Lessons*. Washington, DC: CQ Press.

Schwartz, J.M. and Hayward, S.F. 2007. *Air Quality in America: A Dose of Reality on Air Pollution Levels, Trends, and Health Risks*. Washington, DC: AEI Press.

Suchecki, P. 2015. A billionaire co-founder of Yahoo and his wife quietly help fund the 'war on coal.' *Inside Philanthropy*. July 13.

Watts, A. 2015. Again, why are we there? John Kerry admits at #COP21 that US emissions cuts accomplish nothing for climate. Watts Up With That (website).

6.2.2 Violating the Bradford Hill Criteria

EPA makes many assumptions about relationships between air quality and human health, often in violation of the Bradford Hill Criteria and other basic requirements of the Scientific Method.

Belzer (1994) wrote: "Science involves a set of rigorous procedures for sorting out evidence from assertions, fact from fiction, and causation from association. Scientists develop theories of physical, biological, and human systems and craft testable hypotheses, all the while subjecting their efforts to critical review by their peers and the marketplace of ideas" (p. 176). As discussed in Chapter 2, Section 2.2.1, the scientific method requires researchers to formulate and disprove an alternative *null* hypothesis. In the case of man-made climate change, the hypothesis is that dangerous climate change is resulting, or will result, from human-related greenhouse gas emissions. A reasonable null hypothesis is that changes in global climate indices and the physical environment are the result of natural variability. Another null hypothesis could be that any hypothetical mechanism that produces some global warming will not produce a climate catastrophe. Climate scientists have failed to disprove either null hypothesis, meaning the original hypothesis has not been proven to be correct.

The scientific method imposes the same requirements on the debate over the human health effects of the chemical compounds produced during the combustion of fossil fuels. The EPA has compiled mountains of assumptions, observational studies, and circumstantial evidence in support of its implicit hypothesis that man-made chemical compounds cause measurable and harmful effects on human health, while failing to invalidate the null hypothesis that observed death rates and illnesses are the result of other causes including aging, genetics, naturally occurring carcinogens, unhealthy behaviors such as smoking and poor nutritional choices, and other forms of risky behavior. Instead of testing the elements of its hypothesis for validity, the EPA adopted the fallacies of anchoring (defending a previous decision or piece of information against new evidence), confirmation bias (interpreting all new evidence as confirmation of an existing belief), and cherry-picking arguments and information to support its hypothesis.

Bradford Hill Criteria

Much of the public concern over man-made chemicals is due to the assumption by policymakers, regulators, and advocates that evidence of an *association* between a chemical in the air or water and a human health effect is evidence that the chemical *causes* that effect. Because distinguishing between coincidence and correlation, on the one hand, and causal relationships on the other can be very difficult in matters of public health, an English epidemiologist named Sir Austin Bradford Hill (1897–1991) established in 1965 what has become known as the Bradford Hill Criteria (BHC), nine minimal conditions necessary to provide evidence of a causal relationship between an event (in this case exposure to an air pollutant) and a health effect (illness or mortality). The criteria are presented in Figure 6.2.2.1.

Similar standards have been proposed by other researchers (e.g., Henle-Koch-Evans postulates (Evans, 1976, 1977) and Susser, 1973, 1991). Commenting on the Bradford Hill Criteria, Foster *et al.* (1993) wrote,

> Most scientists would agree that they are not standards of scientific proof, or at least not the high standards that the HKE postulates are generally assumed to be. Nevertheless, Hill's criteria have been widely influential in epidemiology. The fact that epidemiologists feel it necessary to debate them at all underscores the frequent difficulty of interpreting epidemiological evidence. At the least, it points to the need for a holistic assessment of the data, and the recognition that the evidence will never be completely consistent (p. 10).

The Bradford Hill Criteria are endorsed by the Federal Judicial Center (FJC), an education and research agency of the United States federal courts established by an Act of Congress (28 U.S.C. §§ 620–629) in 1967, at the recommendation of the Judicial Conference of the United States. FJC's reference manual for judges, titled the *Reference Manual on Scientific Evidence*, provides expert advice for determining the admissibility of scientific evidence in U.S. federal courts and advises federal judges and lawyers practicing in federal courts to adhere to that advice in complying with the rules of evidence. The latest (third) edition is co-published by

Figure 6.2.2.1
Bradford Hill Criteria for establishing a causal relationship

1. *Strength of the association.* Relative risk (the incidence rate in the exposed population divided by the rate in the unexposed population) measures the strength of the association. The higher the relative risk, the greater the likelihood that the relationship is causal.

2. *Consistency of the observed association.* Has it been repeatedly observed by different persons, in different places, circumstances, and times?

3. *Specificity of the association.* Causation is most likely when the association is limited to specific occupations, particular sites, and types of diseases.

4. Temporal relationship of the association. The effect must occur after the cause.

5. *A dose-response curve.* The higher the dose, the higher the incidence of disease or mortality. A higher dose should not lead to less, rather than greater, harmful effects.

6. *Biological plausibility.* A plausible mechanism between cause and effect is helpful, but since it depends on the biological knowledge of the day, "this is a feature I am convinced we cannot demand."

7. *Coherence with current knowledge.* The cause-and-effect interpretation of the data should not seriously conflict with the generally known facts of the natural history and biology of the disease.

8. *Experimental evidence.* Before-and-after comparisons can reveal the strongest support for the causation hypothesis.

9. *Analogizing to similar known causes.* Knowing the effects of a drug such as thalidomide or a disease such as rubella on pregnant women makes it more plausible that other drugs and diseases might have similar effects.

Source: Hill, 1965.

the National Research Council of the National Academies (FJC, 2011).

The manual's chapter on epidemiology was coauthored by a distinguished legal scholar, Michael D. Green, J.D., the Bess & Walter Williams Chair in Law, Wake Forest University School of Law, and two distinguished epidemiologists: D. Michal Freedman, J.D., Ph.D., M.P.H., epidemiologist in the Division of Cancer Epidemiology and Genetics at the National Cancer Institute in Bethesda, Maryland, and Leon Gordis, M.D., M.P.H., Dr.P.H., professor emeritus of epidemiology at Johns Hopkins Bloomberg School of Public Health and professor emeritus of pediatrics at Johns Hopkins School of Medicine, Baltimore, Maryland.

The authors (on p. 566) define relative risk (the focus of BHC #1) as the ratio of the incidence rate (often referred to as incidence) of disease or mortality in exposed individuals to the incidence rate in unexposed individuals:

RR = (Incidence rate in the exposed)
(Incidence rate in the unexposed)

The FJC authors stressed, "The relative risk is one of the cornerstones for causal inferences. Relative risk measures the strength of the association. The higher the relative risk, the greater the likelihood that the relationship is causal" (p. 602). On the important question of how high a relative risk finding must be to pass the legally required threshold (in civil cases) of "more likely than not," or at least 51% probable, the FJC authors wrote:

Some courts have reasoned that when epidemiological studies find that exposure to the agent causes an incidence in the exposed group that is more than twice the incidence in the unexposed group (i.e., a relative risk greater than 2.0), the probability that exposure to the agent caused a similarly situated individual's disease is greater than 50%. These courts, accordingly, hold that when there is group-based evidence finding that exposure to an agent causes an incidence of disease in the exposed group that is more than twice the incidence in the unexposed group, the evidence is sufficient to satisfy the plaintiff's burden of production and permit submission of specific causation to a jury. In such a case, the factfinder may find that it is more likely than not that the substance caused the particular plaintiff's disease. Courts, thus, have permitted expert witnesses to testify to specific causation based on the logic of the effect of a doubling of the risk (FJC, 2011, p. 612).

Since this is an important and contentious point in the air quality debate, it is worth quoting the FJC at greater length on this question:

Having additional evidence that bears on individual causation has led a few courts to conclude that a plaintiff may satisfy his or her burden of production even if a relative risk less than 2.0 emerges from the epidemiological evidence. For example, genetics might be known to be responsible for 50% of the incidence of a disease independent of exposure to the agent. If genetics can be ruled out in an individual's case then a Relative Risk greater than 1.5 might be sufficient to support an inference that the substance was more likely than not responsible for the plaintiff's disease. ...

Eliminating other known and competing causes increases the probability that the individual's disease was caused by the exposure to the agent. ...

Similarly, an expert attempting to determine whether an individual's emphysema was caused by occupational chemical exposure would inquire whether the individual was a smoker. By ruling out (or ruling in) the possibility of other causes, the probability

that a given agent was the cause of an individual's disease can be refined. Differential etiologies are most critical when the agent at issue is relatively weak and is not responsible for a large proportion of the disease in question.

Although differential etiologies are a sound methodology in principle, this approach is only valid if general causation exists and a substantial proportion of competing causes are known. Thus, for diseases for which the causes are largely unknown, such as most birth defects, a differential etiology is of little benefit. And, like any scientific methodology, it can be performed in an unreliable manner (pp. 616–7).

The FJC's insistence on RRs of 2 (or at least 1.5) is lower than what other researchers in the field expect. Arnett (2006) wrote, "[O]bservational epidemiological studies, unless they show overwhelmingly strong associations – on the order of an increased relative risk of 3.0 or 4.0 – do not indicate causation because of the inherent systematic errors that can overwhelm the weak associations found. These errors include confounding factors, methodological weaknesses, statistical model inconsistencies, and at least 56 different biases" (p. 1).

The EPA and the voluminous research it claims in support of its regulations violate this first and most important of the Bradford Hill Criteria by relying on observational studies with RRs less than 4.0, 3.0, 2.0, and even the lowest standard of 1.5. Indeed, as shown in the next section, the studies on which the EPA relies often find zero or even negative RRs that are hidden in meta-analyses or simply left out of their reviews of the literature. The EPA simply assumes associations, even very weak ones, are proof of causation.

Another violation of the Bradford Hill Criteria is the EPA's reliance on animal experiments in which mice and rats are exposed to near-toxic doses of toxins. The EPA assumes, falsely, that such experiments produce reliable evidence of the risk to humans exposed to far lower levels of those toxins in daily life (Whelan, 1993). That assumption is contradicted by current toxicological knowledge (BHC #7). Ames and Gold (1993) wrote:

Animal cancer tests are conducted at near-toxic doses of the test chemical that cannot predict the cancer risk to humans at the much lower levels to which they are typically exposed. The prediction of cancer risk requires knowledge of the mechanisms of carcinogenesis, which is progressing rapidly. Recent understanding of these mechanisms undermines many of the assumptions of current regulatory policy regarding rodent carcinogens and requires a reevaluation of the purpose of routine animal cancer tests (p. 154).

Commenting on the use of animal testing in the search for cures to cancer rather than possible causes, Mak, Evaniew, and Lost (2014) write, "there is a growing awareness of the limitations of animal research and its inability to make reliable predictions for human clinical trials. Indeed, animal studies seem to overestimate by about 30% the likelihood that a treatment will be effective because negative results are often unpublished. Similarly, little more than a third of highly cited animal research is tested later in human trials. Of the one-third that enter into clinical trials, as little as 8% of drugs pass Phase I successfully."

A third violation of BHC and the scientific method is EPA's default assumption of a linear no-threshold (LNT) dose-response relationship. For example, EPA assumes there is no safe threshold of exposure to fine particles ($PM_{2.5}$) so that even brief exposure to extremely low levels of $PM_{2.5}$ (like those calculated in Section 6.1.3) can cause illnesses and death within hours of inhalation (i.e., "short-term" or literally "sudden death") and that long-term (i.e., years or decades) exposure to low levels of $PM_{2.5}$ also can cause premature death (EPA, 2009; Samet, 2011, p. 199). EPA reasons that if exposure to large concentrations has negative health effects, then exposure to even tiny amounts also must have negative effects, albeit smaller ones. EPA's LNT assumption for $PM_{2.5}$ and other pollutants has been vigorously disputed (e.g., Calabrese and Baldwin, 2003; Calabrese, 2005, 2015).

Calabrese and Baldwin (2003) explained, "The dose-response revolution is the changing perception that the fundamental nature of the dose response is neither linear nor threshold, but U-shaped," meaning extremely low exposures of some toxins may have *positive* health effects (called hormesis). This contradicts EPA's assumption that responses are linear all the way down to zero exposure, and if true it invalidates much of its health effects claims relying

on this assumption. Figure 6.2.2.2 shows some of the alternative dose-response curves that EPA simply assumes away.

Calabrese and Baldwin continue,

[A]cceptance that hormetic-like U-shaped dose responses are widespread and real has been difficult to achieve. The reasons for this are many, but in general include the following. First, the field of toxicology has become progressively and insidiously dependent on the role of government to set the national (and international) toxicological agenda. This agenda translates into designing and interpreting studies to fit into current risk assessment paradigms. That is, in the case of noncarcinogens, regulatory agencies design hazard assessment methodology to provide a NOAEL [no-observed-adverse-effect-level], whereas in the case of carcinogens, the study needs data that can be employed to estimate low-dose cancer risk. Such NOAEL and/or low-dose evaluations are dominating concerns. These controlling governmental regulatory perspectives have provided a seductive focus on toxicological thinking, providing the flow of financial resources and forcing private-sector and academic institutions to respond to such initiatives (*Ibid.*).

Calabrese and Baldwin's account is consistent with what we know about how government bureaucracies operate (see Chapter 1, Section 1.4) and how funding can bias research findings (see Chapter 2, Section 2.2.4). The EPA's linear no-threshold assumption means when large populations are involved, such as the population of the United States (approximately 326 million), simple math allows it to claim that even tiny amounts of an air pollutant with very small effects are responsible for thousands of deaths each year. Such claims generate favorable headlines, please political overseers, and justify a bigger research budget next year. But in fact, it is just as likely that those low levels of exposure have *positive* health effects or no effect at all. Government bureaucrats, politicians, the media, and environmental activists have no reason to let the public know that EPA's claims are implausible and even counterfactual (see Altman, 1980; Whelan, 1993; Avery, 2010; Milloy, 2001, 2016).

Figure 6.2.2.2
Alternative ways to extrapolate from high to low doses

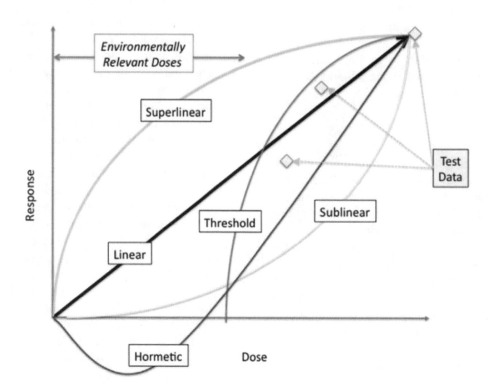

Source: Belzer, 2012, Figure A, p. 8.

Finally, the EPA's paradigm assumes that reducing potential health threats by reducing emissions is superior to making investments in health promotion, even though the latter may be far more cost effective. Focusing only on small and hypothetical health benefits, often achieved at enormous costs by further reducing already de minimus emissions of toxins, the agency misses significant opportunities for protecting public health by reforming existing policies that perversely reward harmful behavior or by making targeted public investments in improving nutrition, safety, or education. For example, Gough (1990) determined that if the EPA's estimates of cancer risks from environmental exposures were correct and if its regulatory programs were 100% successful in controlling those exposures, the agency could eliminate only between 0.25 and 1.3% of all cancers. Hattis and Goble (1994) also expressed concern that the EPA is taking resources away from solving more urgent problems (p. 125).

Just as the IPCC and its allies in the climate change debate closed ranks against distinguished climate scientists who questioned their disregard of the basic requirements of the scientific method, the EPA and its allies attacked Ames, Feinstein, Calabrese, and other highly qualified critics. Writing in 1991, Feinstein observed,

In previous eras of medical history, when major changes were proposed in customary scientific paradigms, the perceived threats to the status quo led to profound intellectual discomforts. Rational discussion of the proposed changes was sometimes replaced by passionate accusations about ethical behavior. A similar situation may arise in epidemiology today, as fundamental problems are noted in paradigmatic scientific methods, and as the available epidemiological evidence is used not only in public policy controversies, but particularly in adversarial legal conflicts. When the basic scientific quality of epidemiological evidence and methods is questioned, defenders of the status quo may respond by castigating the dissenters as fools or heretics, or by insinuating that they have been bribed with consultation honoraria (Feinstein, 1991, abstract).

* * *

In conclusion, the EPA assumes its task is to accumulate evidence in support of a self-serving hypothesis rather than disprove the null hypothesis that observed rates of death and illnesses are the result of causes other than the chemicals produced by modern industrial society. It assumes that association equals causation, administering massive doses of chemicals to laboratory animals predicts the human health impacts of much lower levels of exposure, and that even brief exposure to low levels of some pollutants can cause disease or death. All of these assumptions violate the Bradford Hill Criteria and other requirements of the scientific method, rendering EPA's science an unreliable guide for researchers and policymakers.

References

Altman, D.G. 1980. Statistics and ethics in medical research. Misuse of statistics is unethical. *British Medical Journal* 281: 1182–84.

Ames, B.N. and Gold, L.S. 1993. Environmental pollution and cancer: some misconceptions. Chapter 7 in: Foster, K.R., Bernstein, D.E., and Huber, P.W. (Eds.) *Phantom Risk: Scientific Inference and the Law*. Cambridge, MA: The MIT Press, pp. 153–81.

Arnett Jr., J.C. 2006. The EPA's fine particulate matter (PM 2.5) standards, lung disease, and mortality: a failure in epidemiology. *Issue Analysis #4*. Washington, DC: Competitive Enterprise Institute.

Avery, G. 2010. Scientific misconduct: The manipulation of evidence for political advocacy in health care and climate policy. *Cato Briefing Papers* No. 117 (February 8). Washington, DC: Cato Institute.

Belzer, R. 2012. Risk assessment, safety assessment, and the estimation of regulatory benefits. Mercatus Resarch. Washington, DC: Mercatus Center.

Belzer, R.B. 1994. Is reducing risk the real objective of risk management? Chapter 10 in: Finkel, A.M. and Golding, D. (Eds.) *Worst Things First? The Debate over Risk-Based National Environmental Priorities*. Washington, DC: Resources for the Future, pp. 167–80.

Calabrese, E.J. 2005. Paradigm lost, paradigm found: the re-emergence of hormesis as a fundamental dose response model in the toxicological sciences. *Environmental Pollution* 138 (3): 378–411.

Calabrese, E.J. 2015. On the origins of the linear no-threshold (LNT) dogma by means of untruths, artful dodges and blind faith. *Environmental Research* 142 (October): 432–42.

Calabrese, E.J. and Baldwin, L.A. 2003. Toxicology rethinks its central belief. *Nature* 421 (February): 691–2.

EPA. 2009. U.S. Environmental Protection Agency. *Integrated Science Assessment for Particulate Matter*. National Center for Environmental Assessment – RTP Division, Office of Research and Development. Research Triangle Park, NC.

Evans, A.S. 1976. Causation and disease: The Henle-Koch postulates revisited. *Yale Journal of Biology and Medicine* 49: 175–95.

Evans, A.S. 1977. Limitation of Koch's postulates. *Lancet* 2 (8051): 1277–8.

Feinstein, A.R. 1991 Scientific paradigms and ethical problems in epidemiological research. *Journal of Clinical Epidemiology* 44 (Supplement 1): 119–23.

FJC. 2011. Federal Judicial Center. *Reference Manual on Scientific Evidence*. Third Edition. Federal Judicial Center and National Research Council of the National Academies. Washington, DC: National Academies Press.

Foster, K.R., Bernstein, D.E., and Huber, P.W. (Eds.) 1993. *Phantom Risk: Scientific Inference and the Law*. Cambridge, MA: The MIT Press.

Gough, M. 1990. How much cancer can the EPA regulate away? *Risk Analysis* 10: 1–6.

Hattis, D. and Goble, R.L. 1994. Current priority-setting methodology: too little rationality or too much? Chapter 7 in: Finkel, A.M. and Golding, D., eds., *Worst Things First? The Debate over Risk-Based National Environmental Priorities*. Washington, DC: Resources for the Future, pp. 107–32.

Hill, A.B. 1965. The environment and disease: Association or causation? *Proceedings of the Royal Society of Medicine* 58 (5): 295–300.

Mak, I.W., Evaniew, N., and Ghert, M. 2014. Lost in translation: animal models and clinical trials in cancer treatment. *American Journal of Translational Research* 6(2):114-8.

Milloy, S. 2001. *Junk Science Judo: Self-defense against Health Scares and Scams*. Washington, DC: Cato Institute.

Milloy, S. 2016. *Scare Pollution: Why and How to Fix the EPA*. Lexington, KY: Bench Press, Inc.

Samet, J. 2011. The Clean Air Act and health – a clearer view from 2011. *New England Journal of Medicine* **365**: 198–201.

Susser, M. 1973.*Causal Thinking in the Health Sciences: Concepts and Strategies in Epidemiology*. London, UK: Oxford University Press.

Susser, M. 1991. What is a cause and how do we know one? A grammar for pragmatic epidemiology. *American Journal of Epidemiology* **133**: 635–48.

Whelan, E. 1993. *Toxic Terror, Second Edition*. New York, NY: Prometheus Books.

6.2.3 Lack of Integrity and Transparency

The EPA has relied on research that cannot be replicated and violates basic protocols for conflict of interest, peer review, and transparency.

In 2018, EPA Administrator Scott Pruitt acknowledged his agency had been "weaponized" by the previous administration to wage a "war on fossil fuels." "The key to me," Pruitt told a reporter for *The Daily Signal*, "is that weaponization of the agency that took place in the Obama administration, where the agency was used to pick winners and losers. Those days are over" (Bluey, 2018).

Pruitt went on to say, "Can you imagine, in the first instance, an agency of the federal government, a department of the U.S. government, declaring war on a sector of your economy? Where is that in the statute? Where does that authority exist? It doesn't. And so to restore process and restore commitment to doing things the right way, I think we've seen tremendous success this past year" (Bluey, 2018) About the Paris Accord, which President Donald Trump had said the United States would exit, Pruitt said, "What was decided in Paris under the past administration was not about carbon reduction. It was about penalties to our own economy because China and India, under that accord, didn't have to take any steps to reduce CO_2 until the year 2030. So, if it's really about CO_2 reduction, why do you let that happen?"

As described in Section 6.2.1, the "war on coal" was real. Burnett (2018) writes, "Nearly a year into his presidency, Obama's Environmental Protection Agency (EPA) issued an endangerment finding ruling carbon dioxide, the gas plants need for life and every human and animal exhales, a danger to human health or the environment. Never before had EPA found a naturally occurring chemical dangerous at levels that have no toxic effect. During his tenure, Obama also successfully pressured Congress to increase the subsidies to wind and solar power plants and directed agencies such as EPA to expand their regulatory authority to tighten regulations on coal-fired power plants. Combined with competition from natural gas, these regulations and subsidies caused the premature closure of more than 250 coal-fired power plants nationwide."

Major regulatory decisions, such as the agency's finding that carbon dioxide endangered public health and therefore could be regulated by the EPA under the Clean Air Act, were rushed through without the documentation required for a major rule and even without approval by the EPA's Science Advisory Board. This follows a long history of the EPA refusing to respond to outside criticism, reliance on a small cabal of favored researchers, refusal to consider research that contradicts its findings, and general lack of transparency (see Expert Panel, 1992; GAO, 2008, 2011; NRC, 2011; Smith, 2014; Carna, 2015b).

Many authors have reported the lack of integrity and often outright corruption that have characterized the EPA. Lehr (2014) wrote, "The vague language of the federal environmental statutes and the corresponding massive delegation of authority to the EPA to make law, enforce law, and adjudicate violations concentrate tremendous power in the hands of the agency, breeding insensitivity, zealotry, and abuse. Experience has shown that regulatory agencies will tend to expand until checked, and the potential for regulatory expansion at the EPA, unbounded as it is by congressional language, is vast."

A sample of books documenting corruption inside the EPA appears in Figure 6.2.3.1. Following the table are brief reports of some especially egregious examples of corruption inside the agency.

The John Beale Case

One of the highest-salaried EPA officials responsible for setting NAAQS for particulate matter and ozone in the 1990s and for the "Endangerment Finding" for carbon dioxide in 2009 "is a convicted felon who

Figure 6.2.3.1
Exposés of lack of integrity and corruption inside the EPA

Ron Arnold, *Freezing in the Dark: Money, Power, Politics and the Vast Left Wing Conspiracy,* 2007.

Wilfred Beckerman, *Through Green-Colored Glasses: Environmentalism Reconsidered,* 1996.

Larry Bell, *Climate of Corruption: Politics and Power Behind the Global Warming Hoax,* 2011.

James T. Bennett and Thomas J. DiLorenzo, *Cancer Scam: Diversion of Federal Cancer Funds to Politics,* 1998.

Alex B. Berezow and Hank Campbell, *Science Left Behind: Feel-Good Fallacies and the Rise of the Anti-Scientific Left,* 2012.

Rupert Darwall, *The Age of Global Warming: A History,* 2013.

James V. DeLong, *Out of Bounds, Out of Control: Regulatory Enforcement at the EPA,* 2002.

Jeff Gillman and Eric Heberlig, *How the Government Got In Your Backyard,* 2011.

Indur M. Goklany, *The Precautionary Principle: A Critical Appraisal of Environmental Risk Assessment,* 2001.

Geoffrey C. Kabat, *Hyping Health Risks: Environmental Hazards in Daily Life and the Science of Epidemiology,* 2008.

Wallace Kaufman, *No Turning Back: Dismantling the Fantasies of Environmental Thinking,* 1994.

Aynsley Kellow, *Science and Public Policy: The Virtuous Corruption of Virtual Environmental Science,* 2007.

Jay H. Lehr, ed., *Rational Readings on Environmental Concerns,* 1992.

S. Robert Lichter and Stanley Rothman, *Environmental Cancer – A Political Disease?* 1999.

Christopher Manes, *Green Rage: Radical Environmentalism and the Unmaking of Civilization,* 1990.

A.W. Montford, *The Hockey Stick Illusion: Climategate and the Corruption of Science,* 2010.

Daniel T. Oliver, *Animal Rights: The Inhumane Crusade,* 1999.

James M. Sheehan, *Global Greens: Inside the International Environmental Establishment,* 1998.

Julian Simon, *Hoodwinking the Nation,* 1999.

Rich Trzupek, *Regulators Gone Wild: How the EPA Is Ruining American Industry,* 2011.

Source: Lehr, 2014.

---

went to great lengths to deceive and defraud the U.S. government over the span of more than a decade," according to Alisha Johnson, press secretary to Gina McCarthy, the EPA administrator at the time (Isikoff, 2013).

John C. Beale, a high-ranking career bureaucrat in the EPA's Office of Air and Radiation and said to be the person most responsible for the EPA's rulings on ozone, particulate matter, and carbon dioxide, was convicted of felony theft of government property in 2014 and sentenced to 32 months in prison for fraud and stealing nearly $900,000 from American taxpayers (*Wall Street Journal,* 2013). Mark Kaminsky, an investigator for the Office of the Inspector General, testified that Beal is a pathological liar who "lied across all aspects of his life" (Gaynor, 2014). During his deposition, Beale said he lied to his friends and colleagues because he felt "an excitement about manipulating people or convincing them of something that's not true" (Hayward, 2014).

Patrick Sullivan, assistant inspector general for investigations at the EPA, told NBC News "he doubted Beale's fraud could occur at any federal agency other than the EPA. 'There's a certain culture here at the EPA where the mission is the most important thing,' he said. "They don't think like criminal investigators. They tend to be very trusting and accepting'" (Isikoff, 2013). According to NBC News, the scandal was "what some officials describe as one of the most audacious, and creative, federal frauds they have ever encountered."

Much of Beale's work at the EPA was in furtherance of agendas promoted by liberal environmental organizations, the use of collusive lawsuits with a result of sue and settle for new environmental regulations, and promotion of more burdensome air regulations with the objective of imposing maximum harm on industry in general and the coal industry in particular. A minority report issued by the U.S. Senate Committee on Environment and Public Works painted a vivid picture of manipulation and corruption:

> Before his best friend Robert Brenner [Deputy Director of the Office of Policy, Analysis, and Review (OPAR) within the office of Air and Radiation (OAR)] hired him to work at the EPA, Beale had no legislative or environmental policy experience and wandered between jobs at a small-town law firm, a political campaign, and an apple farm. Yet at the time he was recruited to the EPA, Brenner arranged to place him in the highest pay scale for general service employees, a post that typically is earned by those with significant experience.
>
> What most Americans do not know is that Beale and Brenner were not obscure no-name bureaucrats housed in the bowels of the Agency. Through his position as head of the Office of Policy, Analysis, and Review, Brenner built a "fiefdom" that allowed him to insert himself into a number of important policy issues and to influence the direction of the Agency. Beale was one of Brenner's acolytes – who owed his career and hefty salary to his best friend.
>
> During the Clinton Administration, Beale and Brenner were very powerful members of the EPA's senior leadership team within the Office of Air and Radiation, the office responsible for issuing the most expensive and onerous federal regulations. Beale

himself was the lead EPA official for one of the most controversial and far reaching regulations ever issued by the Agency, the 1997 National Ambient Air Quality Standards (NAAQS) for Ozone and Particulate Matter (PM). These standards marked a turning point for the EPA air regulations and set the stage for the exponential growth of the Agency's power over the American economy. Delegating the NAAQS to Beale was the result of Brenner's facilitating the confidence of the EPA elites, making Beale the gatekeeper for critical information throughout the process. Beale accomplished this coup based on his charisma and steadfast application of the belief that the ends justify the means (U.S. Senate Committee on Environment and Public Works, 2014, p. i).

According to Reynolds (2018), "Beale had his hand in another fishy tactic utilized by the EPA. The creation of the particulate matter regulations came about as a result of the first instance of "sue and settle," in which friendly bureaucrats negotiate settlements with activist groups. In the case of the particulate matter regulations, the American Lung Association had sued the EPA to expedite the creation of the regulations, and a court order imposed a deadline on the agency. The Obama administration stuffed the EPA with former employees of radical environmental organizations, and then put them in charge of negotiating settlements when those organizations sued. This allowed the EPA to bypass the normal rulemaking process with congressional oversight because they were under court order. Incidentally, Pruitt put an end to this practice in October 2017, another reason he's been targeted for destruction by the Left."

Anyone who claims the EPA's ozone and PM NAAQS are based on the scientific method and "best available science" should read this account carefully and reconsider.

Richard Windsor

While conducting research for a book, Christopher Horner, an attorney and author affiliated with the Competitive Enterprise Institute, found an EPA memo from 2008 describing "alias" email accounts created by former EPA Administrator Carol Browner

(1993–2001). Those accounts created a "dual account structure" used by high-level officials inside the EPA to correspond with one another and with outside environmental groups without fear that the messages would be "leaked" to the public. Many of the accounts were apparently set to "auto-delete" (Horner, 2012a).

More recently, EPA Administrator Lisa Jackson (2009–2013) invented the name "Richard Windsor" for emails sent and received to evade federal transparency laws. The scandal resulted in her abrupt resignation in December, 2013 just days after the Justice Department announced it would begin releasing the secret emails. She was never formally charged with a crime.

Federal law requires all government employees to use only official email accounts. If they use a private account to do official business, they are required to make those accounts available to their employing department or agency. Why would two EPA administrators and their senior staff seek to hide their professional (not personal) emails from the public? The Competitive Enterprise Institute, which filed Freedom of Information Act (FOIA) requests and eventually launched a lawsuit leading to a judge's decision to order the release of Jackson's emails, said in a news release on the date of her resignation announcement, "the emails relate to the war on coal Jackson was orchestrating on behalf of President Obama outside the appropriate democratic process" (Hall, 2012). The news release continued,

But this scandal cannot end with Jackson's resignation. She appears to have illegally evaded deliberative procedures and transparency requirements set in law – as did the federal appointees and career employees with whom she communicated through her alias email account. She must be held to account, as must those others – both to assure the peoples' business is done in public and to send a signal to other high-level government officials this conduct cannot and will not be tolerated.

Meanwhile, CEI will continue to try to get to the bottom of Jackson's efforts to evade public scrutiny of her actions. We have and will continue to pursue what we have determined to be widespread similar behavior including private email accounts, private computers and privately owned computer servers used to hide discussions that, by law,

must be open to scrutiny and be part of the public record. The administration has admitted the agency has destroyed documents in apparent violation of the federal criminal code, and we intend to continue to investigate and expose these attempts to hide the agency's actions.

Regarding Carol Browner, the Clinton-era EPA administrator, Horner wrote: "You remember Ms. Browner? She's the lady who suddenly ordered her computer hard drive be reformatted and backup tapes be erased, just hours after a federal court issued a 'preserve' order that her lawyers at the Clinton Justice Department insisted they hadn't yet told her about? She's the one who said it didn't matter because she didn't use her computer for email anyway?" (Horner, 2012b).

Regrettably, the corruption didn't end with Jackson's resignation. In 2015, EPA Administrator Gina McCarthy (2013–2017) repeatedly refused to turn over to congressional investigators records of the agency's interactions with environmental advocacy groups, leading the chairman of the House Science, Space and Technology Committee to issue a subpoena for the records in March (Carna, 2015a). In October, the EPA again refused to turn over records to congressional investigators, this time concerning its collaboration with environmental groups to alter global temperature records, leading to another subpoena (Warrick, 2015).

McCarthy also was subpoenaed for hiding and deleting text messages just days after being told by a House committee that she may have been violating federal document retention laws (Miller, 2015). Twenty-one members of Congress introduced legislation to impeach her, saying "Administrator McCarthy committed perjury and made several false statements at multiple congressional hearings, and as a result, is guilty of high crimes and misdemeanors – an impeachable offense" (Gosar, 2015).

Human Experiments

By conducting human experiments involving exposure to levels of particulate matter and other pollutants it claims to be deadly, the EPA reveals it doesn't believe its own epidemiology-based claims of a deadly threat to public health.

Another EPA scandal pertains to life-endangering experiments performed on human subjects in violation of international standards and medical ethics (Bell, 2013; Dunn, 2012, 2015; Milloy, 2013, 2016; Milloy and Dunn, 2012, 2016). The EPA has tested a variety of air pollutants – including very high exposures to $PM_{2.5}$ – on more than 6,000 human volunteers. Many of these volunteers were elderly or already health-compromised – the very groups the EPA claims are most susceptible to death from $PM_{2.5}$ exposure. $PM_{2.5}$ exposures in these experiments have been as high as 21 times greater than allowed by the EPA's own air quality rules (Milloy, 2012).

It is illegal, unethical, and immoral to expose experimental subjects to harmful or lethal toxins. *The Reference Manual on Scientific Evidence* (FJC, 2011), published by the United States Federal Judicial Center and cited previously in Section 6.2.2, on page 555 declares that exposing human subjects to toxic substances is "proscribed" by law and cites case law. The Nuremberg Code and the Helsinki Accords on Human Experimentation by the World Medical Association prohibit human experiments that might cause harm to the subjects. The EPA's internal policy guidance on experimental protocols prohibits, under United States law (the "Common Rule"), experiments that expose human subjects to any harm, including exposure to lethal or toxic substances.

The EPA human experiments were conducted from January 2010 to June 2011, according to information obtained by JunkScience.com from a Freedom of Information Act request, and ended three months before then-EPA Administrator Lisa Jackson testified to Congress claiming $PM_{2.5}$ was possibly the most deadly substance known to mankind, killing as many people as die from cancer in the United States every year. If the EPA believed its own rhetoric about the health threats of $PM_{2.5}$, then it also should have believed these experiments could have resulted in serious injury or death, and so were illegal and unethical.

What could have possessed these EPA researchers to conduct these illegal experiments? Robert Devlin, a senior EPA research official who supervised human experiments at the University of North Carolina School of Medicine, said in an affidavit, "Controlled human exposure studies conducted by the EPA scientists and the EPA funded scientists at multiple universities in the United States fill an information gap that cannot be filled by large population studies. ... These studies are done under conditions that are controlled to ensure safety, with measurable, reversible physiological responses. They

are not meant to cause clinically significant adverse health effects, but rather reversible physiological responses can be indicators of the potential for more serious outcomes (Devlin, 2012).

Devlin either did not believe EPA Administrator Jackson's claims that exposure to even low levels of $PM_{2.5}$ could cause instant death, or he knowingly violated the provisions of the Nuremberg Code, the Helsinki Accords on Human Experimentation, and the U.S. Common Rule. Either Jackson is wrong, or Devlin and scores of other doctors and researchers who participated in these illegal experiments should be in prison.

The EPA refused to respond to FOIA requests filed by medical researchers Steve Milloy and John Dale Dunn, M.D. (note both are contributors to this chapter). When sued, it claimed the EPA-funded researchers were immunized from any requirement to produce their data because the data were the private property of the researchers. Then the EPA's inspector general took up the case in October 2012. Eighteen months later, the inspector general concluded the agency had indeed failed to warn study subjects that it believed the experiments could kill them – but the inspector general inexplicably ignored the issue of whether the experiments were fundamentally illegal and unethical (EPA, 2014).

Embarrassed by negative publicity from the case, the EPA quietly paid the National Research Council of the National Academies of Sciences to produce a report that it expected would exonerate the agency. A committee of mostly academics, many of them recipients of government grants to find evidence favoring the government's hypothesis that man-made chemicals threaten human health, was formed and began meeting on June 1, 2015. There was no public notice of the formation of the committee or its meeting, so the legally required "public" meeting was attended only by the committee members and EPA and NRC staff.

In June 2016, Milloy and Dunn learned of the NRC investigation for the first time from a congressional aide who just happened to see information about it. They learned five meetings had been held, the last one in April 2016, none open to the public. Milloy and Dunn hurriedly provided comments to the committee docket (record) and requested an opportunity to present oral and written information to the committee. They were allowed to participate remotely in one meeting (Milloy *et al.*, 2016).

The NRC released its report in March 2017 (NRC, 2017a). As Milloy and Dunn had feared, it was a whitewash. From NRC's announcement of the report's release:

> The committee concluded that the societal benefits of CHIE [controlled human inhalation exposure] studies are greater than the risks posed to the participants in the eight studies considered, which are unlikely to be large enough to be of concern. EPA applies a broad set of health-evaluation criteria when selecting participants to determine that there is no reason to believe that their participation in the study will lead to an adverse health response (NRC, 2017b).

The first sentence in the NRC's statement ought to be shocking to all readers. Since at least the end of World War II, the ethics of human experimentation was never about balancing "societal benefits" against individual risks. The consensus of ethicists around the world is that *no* societal benefit can justify human experimentation where serious physical harm is a possibility. Even informed consent is not a permission slip to conduct such experiments. This sentence demonstrates how the NRC failed to properly frame its investigation from the very start.

The second sentence from the NRC's summary directly contradicts the EPA's claims about the health effects of exposure to low concentrations of $PM_{2.5}$. Whereas the EPA repeatedly claims there is "no safe level of exposure" to $PM_{2.5}$, that even tiny exposures raise the risk of adverse health effects up to and including sudden death, the NRC says experiments exposing volunteers to such levels do not "lead to an adverse health response." To avoid having scores of medical doctors and researchers working under its management go to jail for violating medical ethics, the EPA apparently admitted to the NRC that PM is not the deadly pollutant it has been saying it is to the public, Congress, and the public health research community.

The EPA's response to the concerns expressed by Milloy and Dunn illustrates the same aversion to transparency, defiance of the law, and opposition to transparency that were demonstrated in the previous examples in this section. EPA's motive for conducting the experiments, from Devlin's testimony and the circumstances, seems clear. The EPA knew its claims about the health effects of $PM_{2.5}$ and other pollutants are vulnerable to challenge because the underlying studies – all dubious epidemiological statistical correlation studies – do not actually show that particulate matter kills *anyone.* Neither do animal toxicology studies, no matter how much PM the laboratory animals inhale. So the EPA decided to break the rules – of the international community as well as of the agency itself – and bolster its claims about particulate matter by conducting human experiments.

The Current Administration

> *While the new administration has pledged to improve matters, some current regulations and ambient air standards are based on flawed data.*

On February 17, 2017, Scott Pruitt became EPA Administrator, although he resigned effective July 9, 2018. (At the time of this writing, a permanent replacement has not been named.) During his years as attorney general for the State of Arkansas, Pruitt grew familiar with the EPA's misuse of science, lack of transparency, and outright corruption of the regulatory process. With other state attorneys general, he sued the EPA 14 times for exceeding its constitutional authority by attempting to federalize state environment and energy regulation. As administrator, Pruitt proposed a 2018 budget for EPA that was $2.6 billion below the agency's 2017 funding level. The opening pages of the proposed budget state:

> This resource level and the agency FTE [full-time equivalent] level of 11,611 supports the agency's return to a focus on core statutory work and recognizes the appropriate federal role in environmental protection. The budget addresses our highest environmental priorities and refocuses efforts toward streamlining and reducing burden. Responsibility for funding local environmental efforts and programs is returned to state and local entities, while federal funding supports priority national work (EPA, 2017a, pp. 1–2).

Under Pruitt's leadership, the EPA began to unravel the "war on coal" waged by his predecessors. Specific regulatory changes are discussed in some detail in Chapter 8, as part of the cost-benefit analysis of regulations, and so won't be raised here. However, in light of the abuses of transparency and process documented above, three Pruitt initiatives should be mentioned here. First, on October 16, 2017, Pruitt issued an agency-wide directive designed to end the "sue and settle" practice that was used to set the PM and ozone NAAQSs. In the announcement of the directive, Pruitt is quoted as saying,

> The days of regulation through litigation are over. We will no longer go behind closed doors and use consent decrees and settlement agreements to resolve lawsuits filed against the Agency by special interest groups where doing so would circumvent the regulatory process set forth by Congress. Additionally, gone are the days of routinely paying tens of thousands of dollars in attorney's fees to these groups with which we swiftly settle (EPA, 2017b).

Also in October 2017, Pruitt announced the EPA would no longer appoint to its advisory boards individuals who receive funding from the agency. According to the directive, "members [of advisory committees] shall be independent from EPA, which shall include a requirement that no member of any of EPA's federal advisory committees be currently in receipt of EPA grants, either as principal investigator or co-investigator, or in a position that otherwise would reap substantial direct benefit from an EPA grant" (EPA, 2017c). "It is very, very important to ensure independence, to ensure that we're getting advice and counsel independent of the EPA," Pruitt told *The New York Times*. He pointed out that members of just three boards – Scientific Advisory Board, Clean Air Science Advisory Committee, and Board of Scientific Counselors – had collectively accepted $77 million in EPA grants over the previous three years. "He noted that researchers will have the option of ending their grant or continuing to advise EPA, 'but they can't do both'" (Dennis and Eilperin, 2017).

On April 30, 2018, the EPA issued a notice of a proposed rule for "strengthening transparency in regulatory science." That notice said, "Today, EPA is proposing to establish a clear policy for the transparency of the scientific information used for significant regulations: Specifically, the dose response data and models that underlie what we are calling 'pivotal regulatory science'" (EPA, 2018). The proposed rule calls for ending the use of "secret science" – research utilizing databases that are not made available to independent scholars to replicate findings – and challenges the EPA's most controversial assumption, the linear no-threshold dose-response. The rule also calls for more complete disclosure of confounding factors and model uncertainty.

These three initiatives are bold departures from "business as usual" at the EPA, and if successful they would address the most important reasons the agency has lost nearly all its credibility in the air quality debate (Johnston, 2018). It will take years for these reforms to change the agency's culture and lead to corrections of its faulty scientific and public health claims. Until that time, no one should rely on any public health research conducted by the EPA in justification of its regulations.

References

Bell, L. 2013. EPA charged with lethal experiments on hundreds of unsuspecting subjects. Forbes (website). November 13.

Bluey, R. 2018. The weaponization of the EPA is over: an exclusive interview with Scott Pruitt. The Daily Signal (website). February 25. Accessed July 31, 2018.

Burnett, S. 2018. Coal facts. *Issues in Science and Technology* **34** (3).

Carna, T. 2015a. GOP chairman subpoenas EPA on texts. *The Hill*. March 25.

Carna, T. 2015b. Senators vote to block EPA's use of 'secret science.' *The Hill*. April 28.

Dennis, B. and Eilperin, J. 2017. Scott Pruitt blocks scientists with EPA funding from serving as agency advisers. *The New York Times*. October 31.

Devlin, R. 2012. Affidavit in *American Traditions Institute* v. *US EPA*.

Dunn, J.D. 2012. EPA's unethical air pollution experiments. AmericanThinker (website). June 6.

Dunn, J.D. 2015. The EPA uses children (and adults) as guinea pigs. AmericanThinker (website). January 26.

EPA. 2014. U.S. Environmental Protection Agency. Improvements to EPA policies and guidance could enhance protection of human study subjects. Report No. 14-P-0154. Office of Inspector General. March 31. Washington DC.

EPA. 2017a. U.S. Environmental Protection Agency. FY 2018 EPA budget in brief. EPA-190-K-17-001. May. Washington, DC.

EPA. 2017b. U.S. Environmental Protection Agency. Administrator Pruitt issues directive to end EPA "sue & settle." *News Release*. October 16. Washington, DC.

EPA. 2017c. U.S. Environmental Protection Agency. Administrator Pruitt issues directive to ensure independence, geographic diversity & integrity in EPA science committees. *News Release*. October 31. Washington, DC.

EPA. 2018. U.S. Environmental Protection Agency. Strengthening transparency in regulatory science, a proposed rule by the Environmental Protection Agency. April 30. Washington, DC.

Expert Panel. 1992. *Safeguarding the Future: Credible Science, Credible Decisions*. Report of the Expert Panel on the Role of Science at the EPA to William K. Reilly, Administrator. Environmental Protection Agency. March.

FJC. 2011. Federal Judicial Center. *Reference Manual on Scientific Evidence*. Third Edition. Federal Judicial Center and National Research Council of the National Academies. Washington, DC: National Academies Press.

GAO. 2008. U.S. Government Accountability Office. *Chemical Assessments: Low Productivity and New Interagency Review Process Limit the Usefulness and Credibility of EPA's Integrated Risk Information System*. GAO-08-440. Washington, DC. March.

GAO. 2011. U.S. Government Accountability Office. *Chemical Assessments: Challenges Remain with EPA's Integrated Risk Information System Program*. GAO-12-42. Washington, DC: December.

Gaynor, M. 2014. The suit who spooked the EPA. *Washingtonian*. March 4.

Gosar, S. 2015. Rep. Gosar files articles of impeachment against EPA administrator Gina McCarthy. *News Release*. September 11.

Hall, C. 2012. CEI: EPA administrator's resignation a good first step. Competitive Enterprise Institute (website). December 27.

Hayward, S. 2014. Media incuriosity about Obama administration scandals is something to behold. Forbes (website). January 28.

Horner, C. 2012a. *The Liberal War on Transparency: Confessions of a Freedom of Information "Criminal."* New York, NY: Threshold Editions.

Horner, C. 2012b. Who is 'Richard Windsor'? *National Review*. November 12.

Isikoff, M. 2013. Climate change expert's fraud was "crime of massive proportion," say feds. NBC News (website). December 16.

Johnston, J.S. 2018. Restoring science and economics to EPA's benefit calculation. *The Regulatory Review* (website). Accessed November 15, 2018.

Lehr, J. 2014. Replacing the Environmental Protection Agency. *Policy Brief*. Chicago, IL: The Heartland Institute.

Miller, S.A. 2015. Subpoena proper after EPA chief Gina McCarthy deletes nearly 6,000 text messages, experts testify. *Washington Times*. March 26.

Milloy, S. 2012. EPA Inspector General asked to investigate illegal human experimentation. JunkScience (website). May 14.

Milloy, S. 2013. Federal judge overturns EPA human experiments case. *Washington Times*. February 13.

Milloy, S. 2016. *Scare Pollution: Why and How to Fix the EPA*. Lexington, KY: Bench Press Inc.

Milloy, S., Dunn, J., Young, S., Enstrom, J., and Donnay, A. 2016. Report: NAS meeting on EPA's illegal human experiments. JunkScience (website). August 27.

Milloy, S. and Dunn, J. 2012. Environmental Protection Agency's air pollution research: unethical and illegal? *Journal of American Physicians and Surgeons* **17** (4): 109–10.

Milloy, S. and Dunn, J. 2016. EPA whitewashes illegal human experiments. AmericanThinker (website). August 10.

NRC. 2011. National Research Council. *Review of the Environmental Protection Agency's Draft IRIS Assessment of Formaldehyde*. Washington, DC: The National Academies Press.

NRC. 2017a. National Research Council. *Controlled Human Inhalation-Exposure Studies at the EPA*. Washington, DC: The National Academies Press.

NRC. 2017b. National Research Council. New report finds EPA's controlled human exposure studies of air pollution are warranted. *News Release*. Washington, DC: The National Academies Press.

Reynolds, J. 2018. Leftist double standards: bashing Scott Pruitt while ignoring Obama's EPA scandals. PJ Media (website). April 9. Accessed November 15, 2018.

Smith, L. 2014. What is the EPA hiding from the public? *The Wall Street Journal*. June 23.

U.S. Senate Committee on Environment and Public Works. 2014. *EPA's Playbook Unveiled: A Story of Fraud, Deceit, and Secret Science*. March 19.

Wall Street Journal. 2013. An amazing fraud by an architect of government climate policies. (Editorial). December 20.

Warrick, J. 2015. Congressional skeptic on global warming demands records from U.S. climate scientists. *Washington Post*. October 23.

6.3. Observational Studies

When EPA-sponsored toxicological research fails to prove a particular exposure is harmful, the agency turns to observational studies in which the researcher is not able to control how subjects are assigned to the "treated" group or the "control" groups or the treatments each group receives. Such studies are frequently used in the field of epidemiology, a branch of medicine that studies the incidence and distribution of diseases. Section 6.3.1 describes EPA's history of relying on such studies and their shortcomings, and Section 6.3.2 presents a case studying featuring EPA's reliance on flawed studies to justify regulation of fine particulate matter.

6.3.1 Reliance on Observational Studies

Observational studies are easily manipulated, cannot prove causation, and often do not support a hypothesis of toxicity with the small associations found in uncontrolled observational studies.

Just as the academic literature on climate science is clogged with multi-author reports based on unreliable computer models generally aimed at supporting the federal government's "war on fossil fuels,' so too is the literature on air quality is clogged by government-funded observational studies, sometimes called epidemiological studies or simply epidemiology. Such studies generally compare the observed health outcomes of subjects thought to have been exposed to a relatively high level of a chemical compound in an uncontrolled setting, typically determined by air quality monitors located in or near the area where the subjects live or work, to a control group that is either larger (e.g., all residents of the country) or whose members live or work in an area with lower levels of exposure. Observational studies

differ from experiments, in which subjects are randomly assigned to a treated group or a control group. Wolff and Heuss (2012) reported EPA's increased reliance on such studies beginning in 1996:

> In considering the establishment of NAAQS, EPA relies on three types of health effect studies: controlled human exposures ("clinical"), animal toxicology ("toxicology") and epidemiology studies. In all NAAQS reviews prior to the 1996 PM review, EPA relied most heavily on controlled human exposures, which establish health effect endpoints as a function of exposure and demonstrate causality, and the toxicology studies which provide insights as to the mode of the damage caused by an exposure. Epidemiology studies were used if they supported the findings in the other two types of studies because epidemiology studies can only identify statistical associations between air pollutant concentrations and health endpoint incidence and cannot be used to demonstrate causality (cause-effect relationships).

For the PM NAAQS review that ended in 1996, Wolff and Heuss (2012) wrote, the EPA for the first time subordinated human exposure and toxicological studies to epidemiological studies "because they [the toxicological studies] showed no evidence of effects at concentrations near the level of the existing NAAQS." To make a case for a lower NAAQS for PM_{10} and a new NAAQS for $PM_{2.5}$, the EPA had to turn to epidemiology studies that found "very weak statistical associations" between exposure and mortality. "EPA promulgated new annual and 24-hour $PM_{2.5}$ NAAQS based on the epidemiology findings" (*Ibid.*).

EPA's reliance on epidemiology represented a major step away from sound science. The Federal Judicial Center, whose authors were introduced and quoted earlier in this chapter, stress, "epidemiology cannot prove causation; rather, causation is a judgment for epidemiologists and others interpreting the epidemiological data. Moreover, scientific determinations of causation are inherently tentative. The scientific enterprise must always remain open to reassessing the validity of past judgments as new evidence develops" (FJC, 2011, p. 598).

Foster *et al.* (1993), commenting on more than a dozen cases of what they call "phantom risks" ("cause-and-effect relationships whose very existence is unproven and perhaps unprovable" (p. 1)), concluded, "The epidemiological studies are frequently inconsistent; the animal studies often show clear toxic effects, but at levels that vastly exceed any reasonable human exposure. Although the issues vary, similar themes constantly reappear" (p. 13) and "despite hysterical claims that were widely publicized during the 1970s, typical environmental exposures to most chemicals are too low to be a major (or even detectable) source of illness" (p. 14).

A major shortcoming of observational studies is their failure to replicate results, a violation of BHC #2 requiring consistency of the observed association. Young and Karr (2011) wrote,

It may not be appreciated how often observational claims fail to replicate. In a small sample in 2005 [citing Ioannidis, 2005], of 49 claims coming from highly cited studies, 14 either failed to replicate entirely or the magnitude of the claimed effect was greatly reduced (a regression to the mean). Six of these 49 studies were observational studies, and in these six, in effect, randomly chosen observational studies, five failed to replicate. This last is an 83% failure rate. In an ideal world in which well-studied questions are addressed and statistical issues are accounted for properly, few statistically significant claims are false positives. Reality for observational studies is quite different (p. 117).

Young and Karr continued,

We ourselves carried out an informal but comprehensive accounting of 12 randomised clinical trials that tested observational claims. … The 12 clinical trials tested 52 observational claims. They all confirmed no claims in the direction of the observational claims. We repeat that figure: 0 out of 52. To put it another way, 100% of the observational claims failed to replicate. In fact, five claims (9.6%) are statistically significant in the clinical trials *in the opposite direction* to the observational claim. To us, a false discovery rate of over 80% is potent evidence that the observational study process is not in control. The problem, which has been recognised at least since 1988, is systemic (*Ibid.*).

Alvan R. Feinstein, a Yale epidemiologist, produced a series of devastating critiques of research relied on by the EPA and other regulatory agencies (Feinstein, 1988, 1991; Feinstein and Massa, 1997). In a 1988 article published in *Science* he observed:

Many substances used in daily life, such as coffee, alcohol, and pharmaceutical treatment for hypertension, have been accused of "menace" in causing cancer or other major diseases. Although some of the accusations have subsequently been refuted or withdrawn, they have usually been based on statistical associations in epidemiological studies that could not be done with the customary experimental methods of science. With these epidemiological methods, however, the fundamental scientific standards used to specify hypotheses and groups, get high-quality data, analyze attributable actions, and avoid detection bias may also be omitted. Despite peer-review approval, the current methods need substantial improvement to produce trustworthy scientific evidence (Feinstein, 1988, abstract).

James Enstrom, an epidemiologist long associated with the Jonsson Comprehensive Cancer Center at the University of California, Los Angeles and now head of the Scientific Integrity Institute, observed the following flaws in epidemiological studies relied on by the EPA and air quality regulatory agencies in California to estimate the health effects of particulate matter (PM):

- mobile populations

- unreliable, non-continuous, and fixed monitor information

- no monitor information on some pollutants all the time (2.5 micron particulate matter, for example) or part of the time (10 micron and others)

- an attempt to assess long-term chronic health effects of air quality by death studies, an acute phenomenon

- death certificates and raw death data used without autopsies

- inside air quality ignored for populations living indoors, particularly during old age, advanced medical illness, and terminal illness

- no biological plausibility because the deaths are in the setting of non-toxic levels of air pollution (Enstrom, 2005)

Each of these flaws can lead to violations of BHC standards and make such studies unreliable guides for public policy. Observational studies are easily manipulated, cannot prove causation, and often do not support a hypothesis of toxicity with the small associations in uncontrolled observational studies. And yet, an important part of the case against fossil fuels – that they produce emissions that threaten human health – relies entirely on such research. The flawed results are often fed, without criticism or skepticism, into the computer models used to predict future health effects and the "social cost of carbon" (see, e.g., Bosello *et al.*, 2006). This is a critical mistake that careful researchers should avoid.

References

Bosello, F., Roson, R., and Tol, R. 2006. Economy-wide estimates of the implications of climate change: Human health. *Ecological Economics* **58** (3): 579–91.

Enstrom, J.E. 2005. Fine particulate air pollution and total mortality among elderly Californians, 1973–2002. *Inhalation Toxicology* **17**: 803–16.

Feinstein, A.R. 1988. Scientific standards in epidemiological studies of the menace of daily life. *Science* **242**: 1257–63.

Feinstein, A.R. 1991 Scientific paradigms and ethical problems in epidemiological research. *Journal of Clinical Epidemiology* **44** (Supplement 1): 119–23.

Feinstein, A.R. and Massa, R.D. 1997. Problems in the "evidence" of "evidence-based medicine." *American Journal of Medicine* **103**: 529–35.

FJC. 2011. Federal Judicial Center. *Reference Manual on Scientific Evidence*. Third edition. Federal Judicial Center and the National Research Council of the National Academies. Washington, DC: The National Academies Press.

Foster, K.R., Bernstein, D.E., and Huber, P.W. (Eds.) 1993. *Phantom Risk: Scientific Inference and the Law*. Cambridge, MA: The MIT Press.

Ioannidis, J.P.A. 2005. Contradicted and initially stronger effects in highly cited clinical research. *Journal of the American Medical Association* **294**: 218–28.

Wolff, G.T. and Heuss, J.M. 2012. *Review and Critique of U.S. EPA's Assessment of the Health Effects of Particulate Matter (PM)*. Air Improvement Resource, Inc., prepared for the American Coalition for Clean Coal Electricity. August 28.

Young, S.S. and Karr, A. 2011. Deming, data and observational studies: a process out of control and needing fixing. *Significance* **8** (3): 116–20.

6.3.2 The Particulate Matter Scare

Real-world data and common sense contradict claims that ambient levels of particulate matter kill hundreds of thousands of Americans and millions of people around the world annually.

The studies relied on by the EPA to support its "war on coal" frequently fail to show relative risks (RR) that would suggest a causal relationship between the chemical compounds released during the combustion of fossil fuels and adverse human health effects. Particularly egregious is the agency's claim, against real-world data and common sense, that small particles in the air kill hundreds of thousands of Americans annually.

EPA's Research

The EPA first asserted authority to regulate fine particulate matter ($PM_{2.5}$) as a pollutant in 1997. The U.S. Senate Committee on the Environment and Public Works reported how John Beale, now a convicted felon, played a major role in the decision:

> In the case of the 1997 NAAQS, the Playbook started with a sue-and-settle agreement with the American Lung Association, which established a compressed timeline to draft and issue PM standards. This timeline was further compressed when EPA made the unprecedented decision to simultaneously issue new standards for both PM and Ozone. Issuing these standards in tandem and under the pressure of the sue-and-settle deadline, Beale had the mechanism he needed to ignore opposition to the

standards – EPA simply did not have the time to consider dissenting opinions.

The techniques of the Playbook were on full display in the "Beale Memo," a confidential document that was leaked to Congress during the controversy, which revealed how he pressured the Office of Information and Regulatory Affairs to back off its criticism of the NAAQS and forced them to alter their response to Congress in 1997. EPA also brushed aside objections raised by Congress, the Office of Management and Budget, the Department of Energy, the White House Council of Economic Advisors, the White House Office of Science and Technology Policy, the National Academy of Sciences, and EPA's own scientific advisers – the Clean Air Science Advisory Committee.

These circumstances were compounded by EPA's "policy call" to regulate $PM_{2.5}$ for the first time in 1997. $PM_{2.5}$ are ubiquitous tiny particles, the reduction of which the EPA used to support both the PM and Ozone NAAQS. In doing so, the Playbook also addressed Beale's approach to EPA's economic analysis: overstate the benefits and underrepresent the costs of federal regulations. This technique has been applied over the years and burdens the American people today, as up to 80% of the benefits associated with all federal regulations are attributed to supposed $PM_{2.5}$ reductions (U.S. Senate Committee on the Environment and Public Works, 2014, p. ii).

Fourteen years later, in 2011, EPA Administrator Lisa Jackson claimed in testimony before Congress, "If we could reduce particulate matter to levels that are healthy we would have an identical impact to finding a cure for cancer" (quoted in Harris and Broun, 2011, p. 2; see also Congressional Record, 2011). Cancer kills approximately 570,000 people in the United States annually, making this an astounding and incredible claim.

In 2014, then EPA Administrator Gina McCarthy told reporters, "John Beale walked on water at EPA." The U.S. Senate Committee on Environment and Public Works commented on that remark: "This unusual culture of idolatry has led EPA officials to blind themselves to Beale's wrongdoing and caused them to neglect their duty to act as public servants. As such, to this day EPA continues to protect Beale's work product and the secret science behind the Agency's NAAQS and PM claims" (U.S. Senate Committee on Environment and Public Works, 2014, p. iii).

As reported in the introduction to this chapter, in 2010, the EPA claimed PM caused approximately 360,000 and as many as 500,000 premature deaths in the United States in 2005, citing Laden *et al.* (2006) (EPA, 2010, p. G7). Figure 6.3.2.1 reproduces Table G-1 from the EPA report supporting the agency's claim. In 2012, approximately the same team of authors who produced the estimates relied on by the EPA for the estimates in Figure 6.3.2.1 updated their analysis to account for changes in the cohort population and air quality up to and including 2009. They reported, "Each 10-µg/m3 increase in $PM_{2.5}$ was associated with a 14% increased risk of all-cause death [95% confidence interval (CI): 7%, 22%], a 26% increase in cardiovascular death (95% CI: 14%, 40%), and a 37% increase in lung-cancer death (95% CI: 7%, 75%)" (Lepeule *et al.,* 2012). They went on to report, "Given that there were 2,423,712 deaths in the United States in 2007 (Xu *et al.* 2010) and that the average $PM_{2.5}$ level was 11.9 µg/m^3 (U.S. EPA 2011), our estimated association between $PM_{2.5}$ and all-cause mortality implies that a decrease of 1 µg/m^3 in population-average $PM_{2.5}$ would result in approximately 34,000 fewer deaths per year" (*Ibid.*).

The EPA's claim that $PM_{2.5}$ causes long-term death is grounded in two long-term epidemiological studies: the Harvard Six Cities study (Dockery *et al.*, 1993; Pope *et al.*, 2002) and the American Cancer Society (ACS) study (Pope *et al.*, 1995, 2002, 2009). The original Harvard Six Cities study tracked the health of 8,111 subjects in six cities between 1974 and 1991 and found an RR of 1.26 for those living in cities with the highest reported levels of air pollution compared to those living in the city with the lowest reported level of air pollution. The authors concluded, "fine particulate air pollution … contributes to excess mortality in certain U.S. cities."

Besides the obvious problem of a small sample size and failing to consider many possible confounding factors, the study found subjects with more than a high school education showed no association of PM exposure with mortality and even found for that group a slight *decrease* in mortality rates due to respiratory disease (Arnett, 2006, p. 5). This finding violates BHC #2 requiring consistency of the observed association and #3 requiring specificity of the association.

Figure 6.3.2.1
Estimated PM₂.₅-related premature mortality associated with incremental air quality differences between 2005 ambient mean PM₂.₅ levels and lowest measured level from the epidemiology studies or policy relevant background

Air Quality Level	Estimates Based on Krewski et al. (2009)		Estimates Based on Laden et al. (2006) *(90th percentile confidence interval)*
	'79-'83 estimate (90th percentile confidence interval)	*'99-'00 estimate (90th percentile confidence interval)*	
10 µg/m³ (LML for Laden et al., 2006)	26,000 (16,000—36,000)	33,000 (22,000—44,000)	**88,000** **(49,000—130,000)**
5.8 µg/m³ (LML for Krewski et al., 2009)	**63,000** **(39,000—87,000)**	**80,000** **(54,000—110,000)**	210,000 (120,000—300,000)
Policy-Relevant Background	110,000 (68,000—150,000)	140,000 (94,000—180,000)	360,000 (200,000—500,000)

Bold indicates that the minimum air quality level used to calculate this estimate corresponds to the lowest measured level identified in the epidemiological study

Source: EPA, 2010, Table G-1.

The original ACS study compared air quality levels with mortality in more than 500,000 people from 151 U.S. metropolitan areas between 1982 and 1989. It found RRs of 1.17 for PM and 1.15 for sulfate, once again comparing the most polluted city with the least polluted city. Even this weak association – far below the RR of 2 or even 1.5 required by the Federal Judicial Center – is suspect. According to Arnett (2006), "health information was obtained only once, at entry into the study in 1982 and it considered only a few of the 300 known risk factors that have been associated with cardiovascular disease. None of the data obtained was verified by review of medical records or by other means" (p. 6).

Not surprisingly, given the small associations they found and lack of supporting science, the EPA's own scientific advisory committee refused to approve a PM standard. In 1995, in response to a request from the agency, researchers for the National Institute of Statistical Science investigated the possible relationship between airborne particulate matter and mortality in Cook County, Illinois, and Salt Lake County, Utah. "We found no evidence that particulate matter < or = 10 microns (PM₁₀) contributes to excess mortality in Salt Lake County, Utah. In Cook County, Illinois, we found evidence of a positive PM₁₀ effect in spring and autumn, but not in winter and summer," they reported. "We conclude that the reported effects of particulates on mortality are unconfirmed" (Styer *et al.,* 1995).

In its 2013 estimate of the "social cost of carbon," which has since been rescinded, the EPA claimed public health is endangered by chemical compounds released during the combustion of fossil fuel, principally particulate matter, ozone, nitrogen dioxide (NO₂), sulfur dioxide (SO₂), mercury, and hydrogen chloride (HCl) (EPA, 2013). Other harms it cited included visibility impairment (haze), corrosion of building materials, negative effects on vegetation due to ozone, acid rain, nitrogen deposition, and negative effects on ecosystems from methylmercury

In 2014 and 2015, the EPA relied on the same sources (Laden *et al.,* 2006 and Lepeule *et al.,* 2012) for its regulatory impact statement regarding the proposed Clean Power Plan (EPA, 2014, 2015), which has since been rescinded (EPA, 2018a). The EPA claimed benefits of the new regulations would be worth an estimated $55 billion to $93 billion in

2030. Virtually all the benefits would come from reducing particulate matter emissions and exposure to ozone, which the EPA said would avoid 2,700 to 6,600 premature deaths and 140,000 to 150,000 asthma attacks in children annually (EPA, 2015). From 2009 to 2011, EPA claimed reducing PM emissions amounted to 99% or more of the benefits of eight of twelve new rules (Smith, 2011).

Despite much research, there is no generally accepted medical or biological explanation for how $PM_{2.5}$ at concentrations close to U.S. ambient levels could cause disease or death. No laboratory animal has ever died from $PM_{2.5}$ in an experimental setting, even though animals have been exposed to levels of $PM_{2.5}$ as much as 100+ times greater than human exposures to $PM_{2.5}$ in outdoor air (Arnett, 2006). The EPA assumes without providing clinical evidence that exposure to ambient levels of $PM_{2.5}$ causes disease and mortality. This violates BHC #6, requiring biological plausibility, and #8, requiring experimental evidence.

As reported earlier in Section 6.2, the EPA *has* tested a variety of air pollutants – including very high exposures to $PM_{2.5}$ – on more than 6,000 human volunteers. Many of these volunteers were elderly or already health-compromised – the very groups the agency claims are most susceptible to death from $PM_{2.5}$ exposure. The agency has admitted there have been no deaths or any dangerous adverse events clearly caused by these $PM_{2.5}$ exposures, which were as high as 21 times greater than the exposures allowed by the agency's own air quality rules (Milloy, 2012).

Recently EPA reduced the size of claims made in the past with regards to the small particle and ozone co-benefits of the Clean Power Plan (Saiyid, 2018), but a more far-reaching review of the EPA's methodology and integrity is needed. Such a review would likely result in dramatic changes in NAAQS and other EPA policies. According to the EPA, average exposure in the United States to both PM_{10} and $PM_{2.5}$ has fallen steeply since the 1990s and is now below the agency's NAAQS (EPA, 2018b). Figure 6.3.2.2 reproduces the EPA's graphs for $PM_{2.5}$ and PM_{10} concentrations for the period 2000–2016.

Independent Research

The Health Effects Institute (HEI), a nonprofit research organization jointly funded by the EPA and the automobile industry, has conducted several studies on the health effects of air quality,

reanalyzing data from the Harvard Six Cities and ACS studies as well as a newer database called the National Morbidity, Mortality and Air Pollution Study (NMMAPS) (Krewski *et al.*, 2000; Krewski *et al.*, 2005; HEI, 2008). While generally confirming the findings of the earlier reports, they also reported considerable heterogeneity in the data, indicating exposure to identical levels of particulate matter was correlated with different health outcomes in different parts of the country, a violation of BHC #2, which requires consistency of the observed association across different places. In 2008, HEI reported, "We have re-done our analyses with more stringent convergence criteria for the GAM [generalized additive models] estimation procedure and found that estimates for individual cities changed by small amounts and that the estimate of the average particulate pollution effect across the 90 largest U.S. cities changed from a 0.41% increase to a 0.27% increase in daily mortality per 10 micrograms per cubic meter of PM10," a significant reduction.

Enstrom (2005) surveyed observational studies on the health effects of PM in the United States up to that year. His table summarizing the findings appears as Figure 6.3.2.3.

None of the studies in Enstrom's table found an RR for $PM_{2.5}$ greater than 1.15 (at the 95% confidence level), far below the Federal Judicial Center requirement of an RR of 2 or more to pass the legal requirement for evidence showing exposure to a chemical compound is "more likely than not" to cause an adverse health effect (FJC, 2011). Recall that an RR = 1 means no association at all, and a negative RR means a possible *positive* effect on health outcomes. In the same article, Enstrom presented the results of his original study of the health effects of $PM_{2.5}$ in California. He described his methodology as follows:

> [T]he long-term relation between fine particulate air pollution and total mortality was examined in a cohort of 49,975 elderly Californians, with a mean age of 65 [years] as of 1973. These subjects, who resided in 25 California counties, were enrolled in 1959, recontacted in 1972, and followed from 1973 through 2002; 39,846 deaths were identified. Proportional hazards regression models were used to determine their relative risk of death (RR) and 95% confidence interval (CI) during 1973–2002 by county of residence. The models adjusted for age, sex, cigarette smoking, race, education, marital status,

Figure 6.3.2.2
Declining aerial concentrations of PM$_{2.5}$ and PM$_{10}$ in the United States, 2000–2016

A. PM2.5 seasonally weighted average annual concentration in the United States, 2000–2016

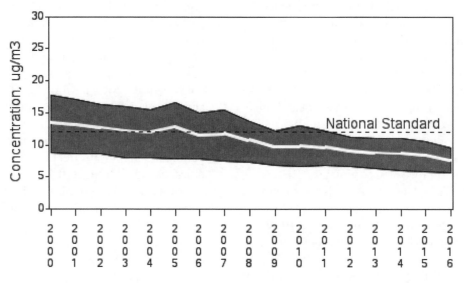

2000 to 2016 :   42% decrease in National Average

B. PM$_{10}$ seasonally weighted average annual concentration in the United States, 1990–2016.

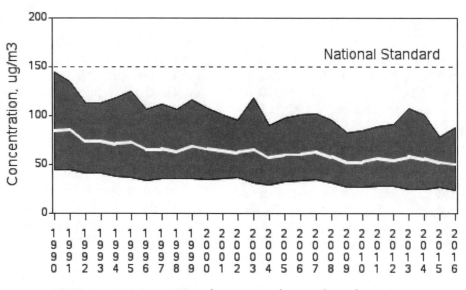

1990 to 2016 :   39% decrease in National Average

Source: EPA, 2018b.

Figure 6.3.2.3
Relative risk and 95% confidence interval (CI) for long-term all-cause mortality per 10-ug/m³ increase in PM₂.₅ for U.S. cohort studies based on PM₂.₅

TABLE 10
Relative risk (RR) and 95% confidence interval (CI) for long-term all-cause mortality per 10-μg/m³ increase in PM₂.₅ for U.S. cohort studies based on PM₂.₅ data, circa 1980

Study (author, year)	PM₂.₅ Data period	PM₂.₅ Mean (range) (μg/m³)	Cohort geographic definition	Follow-up period	Mean entry age for period	Number entered in cohort	Deaths in follow-up period	RR (95% CI)
Males								
Dockery et al., 1993	1979–1985	19 (11–30)	6 U.S. cities	1975–1989	~50	3671[a]	830[a]	1.15 (1.02–1.30)[b]
Pope et al., 1995	1979–1981	18 (9–34)	50 U.S. SMSAs	1982–1989	57	130,310[a]	~12,400[a]	1.07 (1.03–1.11)[b]
McDonnell et al., 2000	1973–1977	32 (17–45)	9 CA airsheds	1976–1992	58	≤1347	≤375	1.09 (0.98–1.21)[b]
Lipfert et al., 2000	1979–1981	24 (6–42)	42 U.S. counties	1975–1981	51	26,067	~4600[c]	0.95 (0.89–1.01)[c]
	1982–1984	22 (8–41)		1982–1988	57	~21,467	~6100[c]	0.94 (0.90–0.98)[c]
	1982–1984	22 (8–41)		1989–1996	63	~15,367	~5765[c]	0.89 (0.85–0.95)[c]
Pope et al., 2002	1979–1983	21 (10–30)	61 U.S. SMSAs	1982–1998	57	~159,000[a]	~36,000[a]	1.05 (1.01–1.10)
Enstrom, 2005	1979–1983	24 (11–42)	11 CA counties	1973–1982	66	15,573	4701	1.03 (0.99–1.07)
	1979–1983	24 (11–42)		1983–2002	74	10,872	8831	0.97 (0.95–1.00)
Females								
Dockery et al., 1993	1979–1985	19 (11–30)	6 U.S. cities	1975–1989	~50	4440[a]	599[a]	1.12 (0.96–1.30)[b]
Pope et al., 1995	1979–1981	18 (9–34)	50 U.S. SMSAs	1982–1989	57	164,913[a]	~8365[a]	1.06 (1.01–1.12)[b]
McDonnell et al., 2000	1973–1977	32 (17–45)	9 CA airsheds	1976–1992	58	≤2422	≤568	~1.00 (assumed)
Pope et al., 2002	1979–1983	21 (10–30)	61 U.S. SMSAs	1982–1998	57	~200,000[a]	~24,000[a]	1.02 (0.98–1.06)
Enstrom, 2005	1979–1983	24 (11–42)	11 CA counties	1973–1982	65	20,210	4094	1.05 (1.01–1.10)
	1979–1983	24 (11–42)		1983–2002	73	16,116	10,815	1.02 (0.99–1.04)
Both Sexes								
Dockery et al., 1993	1979–1985	19 (11–30)	6 U.S. cities	1975–1989	~50	8111	1430	1.13 (1.04–1.23)[b]
Pope et al., 1995	1979–1981	18 (9–34)	50 U.S. SMSAs	1982–1989	57	295,223	20,765	1.07 (1.04–1.10)[b]
Pope et al., 2002	1979–1983	21 (10–30)	61 U.S. SMSAs	1982–1998	57	~359,000	~60,000	1.04 (1.01–1.08)
Enstrom, 2005	1979–1983	24 (11–42)	11 CA counties	1973–1982	65	35,783	8795	1.04 (1.01–1.07)
	1979–1983	24 (11–42)		1983–2002	73	26,988	19,646	1.00 (0.98–1.02)

[a] Obtained from supplementary data (Krewski et al., 2000).
[b] Recalculated from published data (US EPA, 2004).
[c] Obtained from the author.

813

Source: Enstrom, 2005.

body mass index, occupational exposure, exercise, and a dietary factor. For the 35,789 subjects residing in 11 of these counties, county-wide exposure to fine particles was estimated from outdoor ambient concentrations measured during 1979–1983 and RRs were calculated as a function of these $PM_{2.5}$ levels (mean of 23.4 μg/m^3) (abstract).

Enstrom (2005) described his findings as follows:

For the initial period, 1973–1982, a small positive risk was found: RR was 1.04 (1.01–1.07) for a 10-μg/m^3 increase in $PM_{2.5}$. For the subsequent period, 1983–2002, this risk was no longer present: RR was 1.00 (0.98–1.02). For the entire follow-up period, RR was 1.01 (0.99–1.03). The RRs varied somewhat among major subgroups defined by sex, age, education level, smoking status, and health status. None of the subgroups that had significantly elevated RRs during 1973–1982 had significantly elevated RRs during 1983–2002. The RRs showed no substantial variation by county of residence during any of the three follow-up periods. Subjects in the two counties with the highest $PM_{2.5}$ levels (mean of 36.1 μg/m^3) had no greater risk of death than those in the two counties with the lowest $PM_{2.5}$ levels (mean of 13.1 μg/m^3). These epidemiological results do not support a current relationship between fine particulate pollution and total mortality in elderly Californians, but they do not rule out a small effect, particularly before 1983.

In later writing on this study, Enstrom (2006) said, "The methodology used in my study is completely consistent with the methodology used in the 2002 Pope study. For instance, my study controlled for smoking at entry and presented results for never smokers. Furthermore, fully adjusted relative risks hardly differed from age-adjusted relative risks. My study used the same 1979–1983 $PM_{2.5}$ data that was used in the Pope studies." Enstrom also noted his findings were consistent with those of Krewski *et al.* (2005) who found "no excess mortality risk in California due to $PM_{2.5}$ among the ACS CPS II cohort during 1982–1989."

Moolgavkar (2005) wrote a lengthy review and criticism of the EPA's reliance on epidemiology in *Regulatory Toxicology and Pharmacology*. He wrote,

"the results of observational epidemiology studies can be seriously biased, particularly when estimated risks are small, as is the case with studies of air pollution. The Agency [EPA] has largely ignored these issues." He continued, "I conclude that a particle mass standard is not defensible on the basis of a causal association between ambient particle mass and adverse effects on human health."

Smith *et al.* (2009) conducted a reanalysis of data from the NMMAPS to test intercity variability and sensitivity of the ozone-mortality associations to modeling assumptions and choice of daily ozone metric, reasoning that such analysis could reveal confounders and "effect modifiers." They report finding "substantial sensitivity. We examined ozone-mortality associations in different concentration ranges, finding a larger incremental effect in higher ranges, but also larger uncertainty. Alternative ozone exposure metrics defined by maximum 8-hour averages. Smith *et al.* concluded, "Our view is that ozone-mortality associations, based on time-series epidemiological analyses of daily data from multiple cities, reveal still-unexplained inconsistencies and show sensitivity to modeling choices and data selection that contribute to serious uncertainties when epidemiological results are used to discern the nature and magnitude of possible ozone-mortality relationships or are applied to risk assessment" (*Ibid.*).

Enstrom returned to the issue with a paper presented in 2012 at a meeting of the American Statistical Association (Enstrom, 2012). Part of that presentation included a new table summarizing more recent California-specific studies of $PM_{2.5}$ and total mortality in California. That table appears below in Figure 6.3.2.4.

While one study in Enstrom's table shows an RR of 1.84 it is clearly an outlier: None of the other studies shows an RR greater than 1.11 and several show RRs *less* than 1.0, suggesting a *positive* health effect from PM. Recent research plainly shows no support for claims by the EPA and other air quality regulators that PM poses a threat to human health. Commenting on his findings, Enstrom wrote, "There is now overwhelming epidemiological evidence that particulate matter (PM), both fine particulate matter ($PM_{2.5}$) and coarse particulate matter (PM_{10}), is not related to total mortality in California" (p. 2324).

Krstic (2013) conducted a reanalysis on the dataset used by Pope *et al.* (2009) of 51 metropolitan regions. He found "a visual analysis of Figure 4 presented on page 382 of their article indicates that data-point number 46 (Topeka, Kansas) is a

Figure 6.3.2.4
Epidemiological cohort studies of PM$_{2.5}$ and total mortality in California

Relative risk of death from all causes (RR and 95% CI) associated with increase of
10 μg/m³ in PM2.5

Krewski 2000 & 2010   CA CPS II Cohort   RR = 0.872 (0.805-0.944)   1982-1989
(N=40,408 [18,000 M + 22,408 F]; 4 MSAs;
 1979-1983 PM2.5; 44 covariates)

McDonnell 2000   CA AHSMOG Cohort   RR ~ 1.00   (0.95 – 1.05)   1977-1992
(N~3,800 [1,347 M + 2,422 F]; SC&SD&SF AB;
 M RR=1.09(0.98-1.21) & F RR~0.98(0.92-1.03))

Jerrett 2005   CPS II Cohort in Los Angeles Basin
(N=22,905; 267 zip code areas;   RR = 1.11   (0.99 - 1.25)   1982-2000
 1999-2000 PM2.5; 44 cov + max confounders)

Enstrom 2005   CA CPS I Cohort   RR = 1.039 (1.010-1.069)   1973-1982
(N=35,783 [15,573 M + 20,210 F]; 11 counties;   RR = 0.997 (0.978-1.016)   1983-2002
 1979-1983 PM2.5; 25 county internal comparison)

Enstrom 2006   CA CPS I Cohort   RR = 1.061 (1.017-1.106)   1973-1982
(N=35,783 [15,573 M + 20,210 F]; 11 counties;   RR = 0.995 (0.968-1.024)   1983-2002
 1979-1983 & 1999-2001 PM2.5)

Zeger 2008   MCAPS Cohort "West"   RR = 0.989 (0.970-1.008)   2000-2005
(3.1 M [1.5 M M + 1.6 M F]; Medicare enrollees
 in CA+OR+WA (CA=73%); 2000-2005 PM2.5)

Jerrett 2010   CA CPS II Cohort   RR ~ 0.994 (0.965-1.025)   1982-2000
(N=77,767 [34,367 M + 43,400 F]; 54 counties;
 2000 PM2.5; KRG ZIP; 20 ind cov+7 eco var; Slide 12)

Krewski 2010   CA CPS II Cohort
(N=40,408; 4 MSAs; 1979-1983 PM2.5; 44 cov) RR = 0.960 (0.920-1.002)   1982-2000
(N=50,930; 7 MSAs; 1999-2000 PM2.5; 44 cov) RR = 0.968 (0.916-1.022)   1982-2000

Jerrett 2011   CA CPS II Cohort   RR = 0.994 (0.965-1.024)   1982-2000
(N=73,609 [32,509 M + 41,100 F]; 54 counties;
 2000 PM2.5; KRG ZIP Model; 20 ind cov+7 eco var; Table 28)

Jerrett 2011   CA CPS II Cohort   RR = 1.002 (0.992-1.012)   1982-2000
(N=73,609 [32,509 M + 41,100 F]; 54 counties;
 2000 PM2.5; Nine Model Ave; 20 ic+7 ev; Fig 22 & Tab 27-32)

Lipsett 2011   CA Teachers Cohort   RR = 1.01   (0.95 – 1.09)   2000-2005
(N=73,489 [73,489 F]; 2000-2005 PM2.5)

Ostro 2011   CA Teachers Cohort   RR = 1.06   (0.96 – 1.16)   2002-2007
(N=43,220 [43,220 F]; 2002-2007 PM2.5)
 replaced Ostro 2010   Incorrect 2010 Result: RR = 1.84   (1.66 – 2.05)   2002-2007

Source: Enstrom, 2012.

potentially influential statistical outlier when the 51 metropolitan areas only are considered" and "the statistical significance of the correlation between the reduction in $PM_{2.5}$ and population-weighted life expectancy in the 51 largest U.S. metropolitan areas should not be affected by the removal of a single data point. Unfortunately, it appears that the statistical significance of the correlation is lost after removing Topeka, Kansas, from the regression analysis" (p. 133).

Specifically, Krstic found "removing data point number 46 (Topeka, Kansas), as an observed potentially influential statistical outlier, yields weak and not statistically significant correlation (i.e., ~0.35 years per 10 mg/m^3; r2 = 0.022; p = 0.31) between the studied variables" (*Ibid.*). He further reported, "Similar and statistically not significant results are obtained on the basis of the complete data kindly provided by the authors for the 211 counties from the 51 metropolitan areas." Krstic's scatter diagrams, shown in Figure 6.3.2.5, clearly show the outlier (Topeka, in the bottom left of the first scatter diagram) and the lack of correlation when it is removed. Krstic concluded:

The results of the presented reanalysis on the basis of the data from Pope *et al.* (2009) show that the statistical significance of the association between the reduction in $PM_{2.5}$ and the change in life expectancy in the United States is lost after removing one of the metropolitan areas from the regression analysis. Hence, the observed weak and statistically not significant correlation between the studied variables does not appear to provide the basis for a meaningful and reliable inference regarding potential public health benefits from air pollution emission reductions, which may raise concern for policymakers in decisions regarding further reductions in permitted levels of air pollution emissions (p. 135).

Young and Xia (2013) observed, "At one point or another, the Environmental Protection Agency (EPA) and the California Air Resources Board (CARB) speak of thousands or more than 160,000 deaths attributable to $PM_{2.5}$. ... The EPA and CARB base their case on statistical analysis of observational data. But if that analysis is not correct, and small-particle air pollution is not causing excess statistical deaths, then the faulty science is punishing society through increased costs and unnecessary regulation" (p. 375).

They reported the results of their reanalysis of data used in Pope *et al.* (2009) as follows: "We compute multiple analyses sweeping across the county from west to east and show that one can 'cut' along the longitude passing just west of Chicago and find no effect of $PM_{2.5}$ to the west and a small effect of $PM_{2.5}$ on statistical deaths to the east. Both Styer *et al.* and Smith *et al.* make the point if the effect of the pollutant is not consistent, then it is unlikely that you have a causative agent. We agree" (p. 376).

Beyond their finding of heterogeneity, Young and Xia reported, "The association between $PM_{2.5}$ with mortality, when compared to the associations between other variables and mortality, shows that the importance of $PM_{2.5}$ is relatively small. There is no measurable association in the western United States, although it accounts for about 11% of the variance in the eastern United States. The Pratt regression analysis across the entire United States has $PM_{2.5}$ explaining about 4% of the standard deviation" (p. 383). The authors conclude, "All analysis indicates that changes in income and several other variables are more influential than $PM_{2.5}$, so policymakers might better focus on improving the economy, reducing cigarette smoking, and encouraging people to pursue education" (p. 384).

Milloy (2013) reported the results of his analysis of daily air quality and daily death data in California for 2007–2010. According to the author's executive summary, "Based on a comparison of air quality data from the California Air Resources Board and death certificate data for 854,109 deaths from the California Department of Public Health for the years 2007–2010, no correlation was identified between changes in ambient $PM_{2.5}$ and daily deaths, including when the analysis was limited to the deaths among the elderly, heart and/or lung deaths only, and heart and/or lung deaths among the elderly."

Milloy concluded, "Although this is only an epidemiological or statistical study that cannot absolutely exclude the possibility that $PM_{2.5}$ actually affects mortality in some small and as yet unknown way, these results also illustrate that it would be virtually impossible to demonstrate through epidemiological study that such an effect actually exists" (*Ibid.*).

* * *

Observational studies funded by and relied upon by the EPA and other air quality regulators fail to show relative risks (RR) that would suggest a causal

Figure 6.3.2.5
Change in life expectancy vs. reduction in PM$_{2.5}$ concentration with and without Topeka, Kansas as an influential outlier

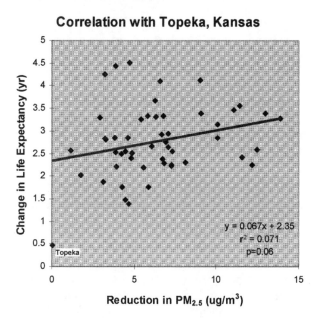
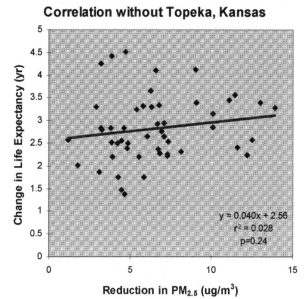

Source: Krstic, 2013. Data from Pope *et al.*, 2009.

relationship between the chemical compounds created during the combustion of fossil fuels and adverse human health effects. Independent researchers have examined the data and found no such relationship exists, meaning tens and even hundreds of billions of dollars have been wasted trying to solve a problem that did not exist. Objective research reveals aerial particulate matter poses little or no threat to public health. Similar analyses of EPA's other NAAQS and regulatory initiatives reach similar conclusions.

References

Arnett Jr., J.C. 2006. The EPA's fine particulate matter (PM$_{2.5}$) standards, lung disease, and mortality: A failure in epidemiology. *Issue Analysis #4.* Washington, DC: Competitive Enterprise Institute.

Congressional Record. 2011. Transparency in Regulatory Analysis of Impacts on the Nation Act of 2011. U.S. House of Representatives. September 11.

Dockery, D.W., *et al.* 1993. An association between air pollution and mortality in six U.S. cities. *New England Journal of Medicine* **329**: 1753–9.

Enstrom, J.E. 2005. Fine particulate air pollution and total mortality among elderly Californians, 1973–2002. *Inhalation Toxicology* **17**: 803–16.

Enstrom, J.E. 2006. Response to "A critique of fine particulate air pollution and total mortality among elderly Californians, 1973–2002" by Bert Brunekreef, PhD, and Gerard Hoek, PhD. *Inhalation Toxicology* **18**: 509514.

Enstrom, J.E. 2012. Are fine particulates killing Californians? Presentation to the American Statistical Association (August 1).

EPA. 2010. U.S. Environmental Protection Agency. *Quantitative Health Risk Assessment for Particulate Matter.* EPA-452/R-10-005. June. Washington, DC.

EPA. 2013. U.S. Environmental Protection Agency. *Technical Support Document: Technical Update of the Social Cost of Carbon for Regulatory Impact Analysis under Executive Order 12866.* Washington, DC.

EPA. 2014. U.S. Environmental Protection Agency. *Regulatory Impact Analysis for the Proposed Carbon Pollution Guidelines for Existing Power Plants and Emission Standards for Modified and Reconstructed Power Plants.* EPA-452/R-14-002. June. Washington, DC.

EPA. 2015. U.S. Environmental Protection Agency. Fact sheet: Clean Power Plan overview, cutting carbon pollution from power plants (website). Accessed November 15, 2018.

EPA. 2018a. U.S. Environmental Protection Agency. Proposal: Affordable Clean Energy (ACE) Rule. (website). Accessed November 15, 2018.

EPA. 2018b. U.S. Environmental Protection Agency. Particulate Matter ($PM_{2.5}$) Trends (website). Accessed November 13, 2018.

FJC. 2011. Federal Judicial Center. *Reference Manual on Scientific Evidence.* Third edition. Federal Judicial Center and the National Research Council of the National Academies. Washington, DC: The National Academies Press.

Harris, A., and Broun, P. 2011. Letter to Cass R. Sunstein, Administrator, Office of Information and Regulatory Affairs. Washington, DC: U.S. House of Representatives: Committee on Science, Space, and Technology. November 15.

HEI. 2008. Health Effects Institute. Frequently asked questions. (website). Accessed November 13, 2018.

Krewski, D., *et al.* 2000. Reanalysis of the Harvard six cities study and the American Cancer Society study of particulate air pollution and mortality. *Special Report.* Cambridge, MA: Health Effects Institute.

Krewski, D., *et al.* 2005. Reanalysis of the Harvard six cities study, part I: validation and replication. *Inhalation Toxicology* **17:** 7–8.

Krstic, G. 2013. A reanalysis of fine particulate matter air pollution versus life expectancy in the United States. *Journal of the Air and Waste Management Association* **62** (9): 989–91.

Kung, H.-C., *et al.* 2008. Deaths: Final Data for 2005. *National Vital Statistics Reports.* Hyattsville, MD: National Center for Health Statistics.

Laden, F., Schwartz, J., Speizer, F.E., and Dockery, D.W. 2006. Reduction in fine particulate air pollution and mortality. *American Journal of Respiratory and Critical Care Medicine* **173**: 667–72.

Lepeule, J., Laden, F., Dockery, D., and Schwartz, J. 2012. Chronic exposure to fine particles and mortality: an extended follow-up of the Harvard six cities study from 1974 to 2009. *Environmental Health Perspectives* **120** (7): 965–70.

Milloy, S. 2012. EPA Inspector General asked to investigate illegal human experimentation. JunkScience (website). May 14.

Milloy, S. 2013. Airborne fine particulate matter and short-term mortality: exploring the California experience, 2007–2010. JunkScience (website).

Moolgavkar, S. 2005. A review and critique of the EPA's rationale for a fine particle standard. *Regulatory Toxicology and Pharmacology* **42**: 123–44.

Pope III, C.A., Burnett, R.T., Thun, M.J., Calle, E.E., Krewski, D., Ito, K., and Thurston, G.D. 2002. Lung cancer, cardiopulmonary mortality, and long-term exposure to fine particulate air pollution. *Journal of the American Medical Association* **287**: 1132–41.

Pope III, C.A., Ezzati, E., and Dockery, D.W. 2009. Fine particulate air pollution and life expectancy in the United States. *New England Journal of Medicine* **360**: 376–86.

Pope III, C.A., Thun, M.J., Namboodiri, M.M., Dockery, D.W., Evans, J.S., Speizer, F.E., and Heath, J.C.W. 1995. Particulate air pollution as a predictor of mortality in a prospective study of U.S. adults. *American Journal of Respiratory and Critical Care Medicine* **151**: 669–74.

Saiyid, A.H. 2018. EPA Lays Groundwork for Avoiding Future Power Plant Mercury Rules. Bloomberg. September 7.

Smith, A.E. 2011. *An Evaluation of the $PM_{2.5}$ Health Benefits Estimates in Regulatory Impact Analyses for Recent Air Regulations*, NERA Economic Consulting. December.

Smith, R.L., Xu, B., and Switzer, P. 2009. Reassessing the relationship between ozone and short-term mortality in US urban communities. *Inhalation Toxicology* **29** (S2): 37–61.

Styer, P., McMillan, N., Gao, F., Davis, J., and Sacks, J. 1995. Effect of outdoor airborne particulate matter on daily death counts. *Environmental Health Perspectives* **103**: 490–7.

U.S. Senate Committee on Environment and Public Works. 2014. *EPA's Playbook Unveiled: A Story of Fraud, Deceit, and Secret Science*. March 19.

Xu, J., *et al.* 2010. Deaths: Final Data for 2007. *National Vital Statistics Reports.* Hyattsville, MD: National Center for Health Statistics.

Young, S.S. and Xia, J. 2013. Assessing geographic heterogeneity and variable importance in an air pollution data set. *Statistical Analysis and Data Mining* **6** (4): 375–86.

6.4 Circumstantial Evidence

Circumstantial evidence cited by the EPA and other air quality regulators is easily refuted by pointing to contradictory evidence.

The EPA and other air quality regulators cite observational studies with small sample sizes (such as the Harvard Six Cities report), historical incidents where cases of extremely poor air quality appeared to have caused a spike in illness or mortality, and laboratory experiments showing physiological responses to high levels of exposure that might be indicative of human health effects in the real world. These are all examples of circumstantial evidence being cast as proof of causation and are easily refuted by contradictory evidence.

6.4.1 Sudden Death

Real-world evidence that fine particulate matter ($PM_{2.5}$) does not cause sudden death is readily available. Everyone is constantly and unavoidably exposed to $PM_{2.5}$ from both natural and manmade sources. Natural sources include dust, pollen, mold, pet dander, forest fires, sea spray, and volcanoes. Manmade sources primarily are smoking, fossil fuel combustion, industrial processes, wood stoves, fireplaces, and indoor cooking. Indoor exposures to $PM_{2.5}$ can easily exceed outdoor exposures by as much as a factor of 100. Although the EPA claims almost 25% of annual U.S. deaths are caused by ambient levels of $PM_{2.5}$, no death has ever been medically attributed to such exposure.

Much higher exposures to $PM_{2.5}$ than exist even in the "worst" outdoor air are not associated with sudden death. The level of $PM_{2.5}$ in average U.S. outdoor air – air the EPA claims can cause sudden death – is about 10 millionths of a gram (microgram) per cubic meter. In one day, a person breathing such air would inhale about 240 micrograms of $PM_{2.5}$. In contrast, a cigarette smoker inhales approximately 10,000 to 40,000 micrograms of $PM_{2.5}$ *per cigarette*. A pack-a-day smoker inhales 200,000 to 800,000 micrograms every day.

A marijuana smoker inhales 3.5 to 4.5 times more $PM_{2.5}$ than a cigarette smoker – i.e., 35,000 to 180,000 micrograms of $PM_{2.5}$ per joint (Gettman, 2015). Typical water pipe or "hookah" smokers inhale the equivalent $PM_{2.5}$ of 100 cigarettes per session. Yet there is no example in published

medical literature of these various types of short-term smoking causing sudden death despite the very high exposures to $PM_{2.5}$ (Goldenberg, 2003). Sudden deaths due to high $PM_{2.5}$ exposures were not reported when Beijing experienced $PM_{2.5}$ levels of 886 micrograms per cubic meter – some 89 times greater than the U.S. daily average (Milloy, 2013).

References

Gettman, J. 2015. The effects of marijuana smoke. Drug Science (website). Accessed November 13, 2018.

Goldenberg, I., *et al.* 2003. Current smoking, smoking cessation, and the risk of sudden cardiac death in patients with coronary artery disease. *Archives of Internal Medicine* **63** (19): 2301–5.

Milloy, S. 2013. China's bad air puts the lie to EPA's scare tactics. *Washington Times.* January 22.

6.4.2 Life Expectancy

The sources cited in the introduction to this chapter leave no doubt that the chemical compounds created during the combustion of fossil fuels are not causing an epidemic of illnesses. Further evidence is easy to find. Life expectancy in the world's wealthiest countries – all of them with the highest levels of energy consumption and fossil-fuel use in the world – rose rapidly since the beginning of the Industrial Revolution, as shown in Figure 6.4.2.1. According to the U.S. Census Bureau:

- "The world average age of death has increased by 35 years since 1970, with declines in death rates in all age groups, including those aged 60 and older (Institute for Health Metrics and Evaluation, 2013; Mathers *et al.,* 2015).

- "From 1970 to 2010, the average age of death increased by 30 years in East Asia and 32 years in tropical Latin America, and in contrast, by less than 10 years in western, southern, and central Sub-Saharan Africa (Institute for Health Metrics and Evaluation, 2013; Figure 4-1).

- "In the mean age at death between 1970 and 2010 across different WHO regions, all regions have had increases in mean age at death, particularly East Asia and tropical Latin America.

Figure 6.4.2.1
Expected life expectancy for five rich countries, 1742–2002

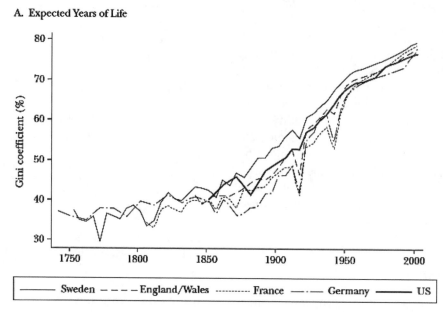

A. Expected Years of Life

Source: Peltzman, 2009, p. 180, Figure 2.

- "Global life expectancy at birth reached 68.6 years in 2015 (Table 4-2). A female born today is expected to live 70.7 years on average and a male 66.6 years. The global life expectancy at birth is projected to increase almost 8 years, reaching 76.2 years in 2050.

- "Northern America currently has the highest life expectancy at 79.9 years and is projected to continue to lead the world with an average regional life expectancy of 84.1 years in 2050." (U.S. Census Bureau, 2016, pp. 31–33).

Life expectancy in the United States rose from 47 years in 1900 to 77 years in 1998 (Moore and Simon, 2000, p. 26). Life expectancy rose for every age group in the United States during that time, as shown in Figure 6.4.2.2.

According to a landmark study on the causes of cancer commissioned by the U.S. Office of Technology Assessment and published by the National Cancer Institute in 1981, comparisons of cancer rates in countries with different levels of air quality as well as of urban and rural residents found "little or no effect of air pollution. To distinguish

between 'little' and 'no' from such direct comparisons is not of course possible, as any real effects will probably be undetectably small, while even if there are no real effects it is impossible to prove a negative" (Doll and Peto, 1981). The authors estimated "combustion products of fossil fuels in ambient air ... would ... account for about 10% of lung cancer in big cities or 1% of all cancer. These crude estimates probably provide the best basis for the formation of policy." In the three decades since the Doll and Peto report, air quality in the United States and in many other developed countries has improved dramatically, with aerial concentrations of potentially harmful man-made chemical compounds often falling to background (natural) levels. It is logical therefore to assume that the health risks of exposure to such chemicals, barely detectible when Doll and Peto were writing, are indistinguishable from zero today. This in fact is what more recent researchers have found. According to Ames and Gold (1993), "cancer death rates in the United States (after adjusting the rates for age and smoking) are steady or decreasing" and "In the United States and other industrial countries life expectancy has been steadily increasing, while infant mortality is decreasing. Although the data are less adequate, there is no evidence

Figure 6.4.2.2
Life expectancy in the United States has increased at every age

Age	1901	1954	1968	1977	1990	2014
0	40	70	70	73	76	78.9
15	62	72	72	75	77	79.5
45	70	74	75	77	79	81.1
65	77	79	80	81	82	84.4
75	82	84	84	85	86	87.3

Source: Data through 1990 from Moore and Simon, 2000, p. 27. 2014 data from Arias, 2017, Table A, p. 3.

that birth defects are increasing. Americans, on average, are healthier now than ever" (pp. 153, 154). Ames and Gold specifically reject popular claims that man-made toxins are responsible for significant human health risks:

> Epidemiology and toxicology provide no persuasive evidence that pollution is a significant cause of birth defects or cancer. Epidemiological studies of the Love Canal toxic waste dump in Niagara Falls, New York, or of dioxin in Agent Orange, or of air pollutants from refineries in Contra Costa County, California, or of contaminated well water in Silicon Valley, California or Woburn, Massachusetts, or of the pesticide DDT, provide no persuasive evidence that such forms of pollution cause human cancer. In most of these cases, the people involved appear to have been exposed to levels of chemicals that were much too low (relative to the background of rodent carcinogens occurring naturally or produced from cooking food) to be credible sources of increased cancer (p. 175).

Michael Gough (1997) wrote, "When EPA was established in 1970, there was a clear expectation that removing chemicals from the air, water, and soil would reduce cancer rates. Now, almost three decades later, scientists are almost uniform in their opinion that chemicals in the environment are associated with only a tiny proportion of cancer. Moreover, there is no evidence that EPA's efforts have had any effect on cancer rates." He concludes,

> There is no cancer epidemic. Cancer mortality from all cancers other than lung cancers has been dropping since the early 1970s, and lung cancer mortality began dropping in 1990. The contribution of environmental exposures to cancer is small – two percent or less – and regulation of those exposures can reduce cancer mortality by no more than one percent. Some of that reduction, even if realized, might be offset by increased food prices that would decrease consumption of fresh fruits and vegetables that are known to protect against cancer.

References

Ames, B.N. and Gold, L.S. 1993. Environmental pollution and cancer: Some misconceptions. Chapter 7 in: Foster, K.R., Bernstein, D.E., and Huber, P.W. (Eds.) *Phantom Risk: Scientific Inference and the Law*. Cambridge, MA: The MIT Press, pp. 153–81.

Arias, E. *et al.*, 2017. United States Life Tables, 2014. *National Vital Statistics Reports* **66** (4). U.S. Department of Health and Human Services, Centers for Disease Control and Prevention. Washington, DC. August 14.

Doll, R. and Peto, R. 1981. The causes of cancer: quantitative estimates of avoidable risks of cancer in the United States today. *Journal of the National Cancer Institute* **66** (6): 1191–308.

Gough, M. 1997. "Environmental cancer" isn't what we thought or were told. Congressional testimony presented on March 6, 1997.

Institute for Health Metrics and Evaluation. 2013. *The Global Burden of Disease: Generating Evidence, Guiding Policy*. Seattle, WA: IHME.

Mathers, C.A., Stevens, G.A., Boerma, T., White, R.A., and Tobias, M.I. 2015. Causes of international increases in older age life expectancy. *The Lancet* **385** (9967): 540–8.

Moore, S. and Simon, J. 2000. *It's Getting Better All the Time: 100 Greatest Trends of the Last 100 Years*. Washington, DC: Cato Institute.

Peltzman, S. 2009. Mortality inequality. *Journal of Economic Perspectives*. **32** (4): 175–90.

U.S. Census Bureau. 2016. *An Aging World: 2015. International Population Reports P95/16-1*. Washington, DC: U.S. Department of Commerce.

6.5 Conclusion

It is unlikely that the chemical compounds created during the combustion of fossil fuels kill or harm anyone in the United States, though it may be a legitimate health concern in third-world countries that rely on burning biofuels and fossil fuels without modern emission control technologies.

Ray Hyman's Categorical Directive, mentioned in Chapter 1, says "before we try to explain something, we should be sure it actually happened." Scientists, economists, and others attempting to incorporate the damages caused by chemical compounds created during the combustion of fossil fuels should pause and consider whether such damages exist at all. Economists and computer modelers dutifully enter stylized facts provided by the U.S. Environmental Protection Agency and World Health Organization into their integrated assessment models to calculate a "social cost of carbon," but it is unlikely such emissions kill or harm *anyone* in the United States or in other developed countries, though it may have been a concern at one time. Air quality may still be a legitimate health concern in developing countries that

rely on burning biofuels and fossil fuels without emission control technologies.

The best available evidence suggests levels of exposure to the chemicals created by the combustion of fossil fuels is too low in the United States, and higher but still too low in other developed countries, to produce the public health effects alleged by environmentalists and many government agencies. Even those low levels of exposure are falling fast in the United States thanks to prosperity, technological change, and government regulation. As Paracelsus said many centuries ago, the dose makes the poison. Without exposure, there can be no harm.

The most influential source of claims that air pollution is a public health hazard in the United States, the U.S. Environmental Protection Agency (EPA), is simply not credible on this issue. Given its constrained mission, flawed paradigm, political pressures, and evidence of actual corruption, there is no reason to believe any science produced by the EPA in justification of its regulations. The EPA makes many assumptions about relationships between air quality and human health often in violation of the Bradford Hill Criteria and other basic requirements of the scientific method. These assumptions, such as a linear no-threshold dose-response relationship and that injecting mice and rats with massive doses of chemicals can accurately forecast human health effects at ambient levels of air pollution, have been debunked again and again.

Observational studies cited by the EPA and other air quality regulators are not designed to test hypotheses and cannot establish causation. Most observational studies cannot be replicated, and in nearly one case in ten, efforts to replicate results find *benefits* where the previous study found harms or vice versa. In any case, observational studies fail to show relative risk (RR) ratios that would suggest a causal relationship between air quality and adverse human health effects.

Finally, the circumstantial evidence cited by the EPA and other air quality regulators is easily refuted by pointing to contradictory evidence. From everyday experience where we do not see people dropping dead in the street from exposure to cigarette fumes that contain orders of magnitude more particulates than ambient air, to the extensive evidence that human health and longevity are increasing over time, the circumstantial evidence is overwhelming that emissions produced from the use of fossil fuels are *not* a threat to human health.

7

Human Security

Chapter Lead Authors: Dennis Avery, Craig D. Idso, Ph.D., Aaron Stover

Contributors: Vice Admiral Edward S. Briggs, Captain Donald K. Forbes, Admiral Thomas B. Hayward

Reviewers: Donn Dears, Kesten Green, Ph.D., Marlo Lewis, Jr., Ph.D., Ronald Rychlak, J.D.

Key Findings

Fossil Fuels and Human Security

- As the world has grown more prosperous, threats to human security have become less common. The prosperity that fossil fuels make possible, including helping produce sufficient food for a growing global population, is a major reason the world is safer than ever before.

- Prosperity is closely correlated with democracy, and democracies have lower rates of violence and go to war less frequently than any other form of government. Because fossil fuels make the spread of democracy possible, they contribute to human security.

- The cost of wars fought in the Middle East is not properly counted as one of the "social costs of carbon" as those conflicts have origins and justifications unrelated to oil.

Citation: Idso, C.D., Legates, D. and Singer, S.F. 2019. Human Security. In: *Climate Change Reconsidered II: Fossil Fuels.* Nongovernmental International Panel on Climate Change. Arlington Heights, IL: The Heartland Institute.

- Limiting access to affordable energy threatens to prolong and exacerbate poverty in developing countries, increasing the likelihood of domestic violence, state failure, and regional conflict.

Climate Change

- The IPCC claims global warming threatens "the vital core of human lives" in multiple ways, many of them unquantifiable, unproven, and uncertain. The narrative in Chapter 12 of the Fifth Assessment Report illustrates the IPCC's misuse of language to hide uncertainty and exaggerate risks.

- Real-world data offer little support for predictions that CO_2-induced global warming will increase either the frequency or intensity of extreme weather events.

- Little real-world evidence supports the claim that global sea level is currently affected by atmospheric CO_2 concentrations, and there is little reason to believe future impacts would be distinguishable from local changes in sea level due to non-climate related factors.

- Alleged threats to agriculture and food security are contradicted by biological science and empirical data regarding crop yields and human hunger.

- Alleged threats to human capital – human health, education, and longevity – are almost entirely speculative and undocumented. There is no evidence climate change has eroded or will erode livelihoods or human progress.

Violent Conflict

- Empirical research shows no direct association between climate change and violent conflicts.

- The climate-conflict hypothesis is a series of arguments linked together in a chain, so if any one of the links is disproven, the hypothesis is invalidated. The academic literature on the relationship between climate and social conflict

reveals at least six methodological problems that affect efforts to connect the two.

- There is little evidence that climate change intensifies alleged sources of violent conflict including abrupt climate changes, access to water, famine, resource scarcity, and refugee flows.

- Climate change does not pose a military threat to the United States. President Donald Trump was right to remove it from the Pentagon's list of threats to national security.

- Predictions that climate change will lead directly or indirectly to violent conflict presume mediating institutions and human capital will not resolve conflicts before they escalate to violence.

Human History

- Extensive historical research in China reveals a close and positive relationship between a warmer climate and peace and prosperity, and between a cooler climate and war and poverty.

- The IPCC relies on second- or third-hand information with little empirical backing when commenting on the implications of climate change for conflict.

Introduction

The United Nations' Intergovernmental Panel on Climate Change (IPCC) refers to damages caused by climate change as "threats to human security," hence the title of this chapter. Among the topics addressed in this chapter are the role played by fossil fuels in prosperity, democracy, and wars in the Middle East, and the possible harms caused by climate change including more frequent or severe extreme weather events, sea-level rise, and damage to agriculture. The possible link between climate change and violent conflict is given particularly close attention. The final section of this chapter reviews academic literature on the role of climate in human history.

Most of the IPCC's discussion of this topic appears in Chapter 12 of the Working Group II contribution to the Fifth Assessment Report (AR5) (IPCC, 2014a, p. 759), where human security "in the context of climate change" is defined as "a condition that exists when the vital core of human lives is protected, and when people have the freedom and the capacity to live with dignity. In this assessment, the vital core of human lives includes the universal and culturally specific, material and non-material elements necessary for people to act on behalf of their interests." "The concept [of human security] was developed in parallel by UN institutions, and by scholars and advocates in every region of the world," the IPCC reports, citing many conference and committee reports and edited books.

One supposes the definition of "human security" was carefully chosen by a task force of "scholars and advocates," but all of the words in it seem derived from philosophy, ethics, and perhaps anthropology, sociology, and law, but not science or economics. While not meaningless, the standard nevertheless is incapable of quantification. As Gleditsch and Nordås (2014) comment, "the definition in the Human Security chapter is too wide to allow serious attempts to assess the secular trend. ... There is a real danger that any kind of social change disliked by some group becomes a threat to someone's human security" (pp. 85–86).

The IPCC alleges, "Climate change threatens human security because it undermines livelihoods, compromises culture and individual identity, increases migration that people would rather have avoided, and because it can undermine the ability of states to provide the conditions necessary for human security. Changes in climate may influence some or all of the factors at the same time. Situations of acute insecurity, such as famine, conflict, and sociopolitical instability, almost always emerge from the interaction of multiple factors. For many populations that are already socially marginalized, resource dependent, and have limited capital assets, human security will be progressively undermined as the climate changes (IPCC, 2014a, FAQ 12.1, p. 762).

In its *Summary for Policymakers* (SPM) for the Working Group II contribution to AR5, the IPCC claims,

> Climate change indirectly increases risks from violent conflict in the form of civil war, inter-group violence, and violent protests by exacerbating well-established drivers of these conflicts such as poverty and economic shocks (*medium confidence*). Statistical studies show that climate variability is

significantly related to these forms of conflict. ... Climate change over the 21st century will lead to new challenges to states and will increasingly shape national security policies (*medium evidence, medium agreement*) (IPCC, 2014b, p. 12).

As emphatic as these declarations seem to be, the IPCC is nevertheless deeply conflicted over whether global warming contributes to violence and other kinds of social conflicts. In Chapter 18 of the same report, on "Detection and Attribution of Observed Impacts," the IPCC found,

> ... both the detection of a climate change effect [on social conflict] and an assessment of the importance of its role can be made only with *low confidence* owing to limitations on both historical understanding and data. Some studies have suggested that levels of warfare in Europe and Asia were relatively high during the Little Ice Age (Parker, 2008; Brook, 2010; Tol and Wagner, 2010; White, 2011; Zhang *et al.*, 2011), but for the same reasons the detection of the effect of climate change and an assessment of its importance can be made only with *low confidence*. There is no evidence of a climate change effect on interstate conflict in the post-World War II period (IPCC, 2014a, p. 1001).

That this dramatic admission of uncertainty did not make it into the SPM of the Fifth Assessment Report is one of many examples of how the IPCC's editorial process, described in Chapter 2, Section 2.3.3, ensures its widely cited SPMs exaggerate the possible dangers posed by climate change, whether natural or man-made, while uncertainties and even contradictory evidence are hidden deep in its almost impenetrable tomes (Stavins, 2014; Tol, 2014).

Citing the IPCC's AR5 and its preceding Fourth Assessment Report as his scientific basis, U.S. President Barack Obama deemed climate change to be an immediate threat to the security of the United States and the entire world. Two *National Security Strategies* (White House, 2010, 2015) made that case, and two *Quadrennial Defense Reviews* (Department of Defense, 2010, 2014) discussed how the U.S. military would need to change to address the new alleged threats. When releasing the 2015 *National Security Strategy*, Obama said, "Today, there is no greater threat to our planet than climate change" (Obama, 2015).

The United States national government quickly and dramatically changed course following the election of President Donald Trump. Climate change no longer appears in the list of national security threats facing the United States (White House, 2017). In March 2017, Trump signed an executive order scrapping the Obama administration's "social cost of carbon" calculations (Trump, 2017a) and in June 2017 he announced his intention to withdraw the United States from the Paris Accord (Trump, 2017b).

Who is right, IPCC (Chapter 12) and Barack Obama, or IPCC (Chapter 18) and Donald Trump? As this chapter will show, it is not a close call. IPCC (Chapter 18) correctly describes the lack of scientific evidence supporting claims that global warming causes violence and other threats to human security and President Donald Trump was right to remove climate change from the list of threats to national security.

Similar to previous chapters, this chapter first examines the direct impact of the use of fossil fuels, in this case on human security, and then the hypothetical indirect impact of fossil fuels if they are contributing to climate change. Parts of this chapter originally appeared in reports published by the George C. Marshall Institute titled *Climate and National Security: Exploring the Connection* (Kueter, 2012) and by The Heartland Institute titled *Climate Change, Energy Policy, and National Power* (Hayward *et al.*, 2014) and *Critique of "Climate Change Adaptation: DOD Can Improve Infrastructure Planning and Processes to Better Account for Potential Impacts"* (Smith, 2015). Those reports have been extensively revised with the authors' and publishers' approval.

References

Brook, T. 2010. *The Troubled Empire: China in the Yuan and Ming Dynasties.* Cambridge, MA: Belknap Press.

Department of Defense. 2010. *Quadrennial Defense Review* (February).

Department of Defense. 2014. *Quadrennial Defense Review* (March).

Gleditsch, N.P. and Nordås, R. 2014. Conflicting messages? The IPCC on conflict and human security. *Political Geography* **43**: 82–90.

Hayward, T.B., Briggs, E.S., and Forbes, D.K. 2014. *Climate Change, Energy Policy, and National Power.* Chicago, IL: The Heartland Institute.

IPCC. 2014a. Intergovernmental Panel on Climate Change. *Climate Change 2014: Impacts, Adaptation, and Vulnerability. Part A: Global and Sectoral Aspects.* Contribution of Working Group II to the Fifth Assessment Report of the Intergovernmental Panel on Climate Change. Cambridge, UK and New York, NY: Cambridge University Press.

IPCC. 2014b. Intergovernmental Panel on Climate Change. *Summary for Policymakers.* In: *Climate Change 2014: Mitigation of Climate Change.* Contribution of Working Group III to the Fifth Assessment Report of the Intergovernmental Panel on Climate Change. New York, NY and Cambridge, UK: Cambridge University Press.

Kueter, J. 2012. *Climate and National Security: Exploring the Connection.* Arlington, VA: George C. Marshall Institute.

Obama, B. 2015. Weekly address: Climate change can no longer be ignored. Washington, DC: The White House. April 18.

Parker, G. 2008. Crisis and catastrophe: the global crisis of the seventeenth century reconsidered. *American Historical Review* **113** (4): 1053–79.

Smith, T. 2015. Critique of "Climate Change Adaptation: DOD Can Improve Infrastructure Planning and Processes to Better Account for Potential Impacts." *Policy Brief.* Chicago, IL: The Heartland Institute.

Stavins, R. 2014. Is the IPCC government approval process broken? An Economic View of the Environment (blog). April 25.

Tol, R.S.J. 2014. Bogus prophecies of doom will not fix the climate. *Financial Times.* March 31.

Tol, R.S.J. and Wagner, S. 2010. Climate change and violent conflict in Europe over the last millennium. *Climatic Change* **99**: 65–79.

Trump, D. 2017a. Presidential Executive Order on Promoting Energy Independence and Economic Growth. Washington, DC: White House. March 28.

Trump, D. 2017b. Statement by President Trump on the Paris Climate Accord. Washington, DC: White House. June 1.

White, S. 2011. *The Climate of Rebellion in the Early Modern Ottoman Empire.* New York, NY: Cambridge University Press.

White House. 2010. *National Security Strategy.* May.

White House. 2015. *National Security Strategy*. February.

White House. 2017. *National Security Strategy*. December.

Zhang, D.D., Lee, H.F., Wang, C., Li, B., Pei, Q., Zhang, J., and An, Y. 2011: The causality analysis of climate change and large-scale human crisis. *Proceedings of the National Academy of Sciences USA* **108** (42): 17296–301.

7.1. Fossil Fuels

Sections 7.1.1 and 7.1.2 show how the use of fossil fuels has contributed to human security, as defined by the IPCC, in two principal ways: by making possible the immense rise in human prosperity that resulted from the Industrial Revolution and by supporting the spread of democracy to many parts of the world. Since it is often argued that the developed nations' reliance on oil from the Middle East threatens human security by fomenting war and a huge investment in troops and arms sent to the region (e.g., Lovins, 2011, p. 5), Section 7.1.3 shows how those wars and battles for at least the past four decades, and perhaps in the more distant past, were not fought over oil but had their origins and justifications in matters unrelated to fossil fuels.

7.1.1 Prosperity

As the world has grown more prosperous, threats to human security have become less common. The prosperity that fossil fuels make possible, including helping produce sufficient food for a growing global population, is a major reason the world is safer than ever before.

Fossil fuels, as documented in Chapters 3 and 4, have unquestionably made humanity more prosperous and healthier. They have even benefited nature, as documented in Chapter 5. History reveals that cold temperatures are more dangerous to our societies than warm temperatures. The Holocene Optimum from 9000 to 6000 years ago was significantly warmer than today and humans flourished. Siberia was 3°C to 9°C warmer then than it is today, and the seas around the Great Barrier Reef were warmer by about 1°C. The Minoan, Roman, and Medieval Warmings were also warmer than today and human societies flourished during those periods as well. In contrast, during the Last Glacial Maximum, temperatures frequently dipped below minus 40°C. The latest Cambridge studies say the desperate Ice Age cold left only about 100,000 human survivors scattered in tiny refuges worldwide when the warming before the Younger Dryas Event began perhaps 14,000 years ago (Davies and Gollop, 2003).

The Dark Ages and Little Ice Age saw huge proportions of their human populations die, mostly in famines because the weather was too cold and chaotic for farmers to feed their cities. Growing seasons were shorter, colder, and cloudier with chaotic events such as killing frosts in mid-summer. The "little ice ages" also suffered centuries-long droughts, massive floods, hunger-driven combat, and hunger-related disease epidemics. Vast storms lashed the seas and lands. Northern Europe became too wet for grains, southern Europe too dry, and the vast Eurasian steppes were abandoned to drought. Their nomadic herders attacked neighboring sedentary peoples in all directions, seeking more grass for their herds. The Eastern Mediterranean nations were essentially depopulated, over and over, by extended droughts. China was ravaged by droughts, floods, wars, rebellions, and dynastic collapses during each of its cold, chaotic weather periods (Fagan, 2000).

In North America, the vegetation underwent nine major transformations in 14,000 years (Viau *et al.*, 2002). Trees, grasses, berries, and roots shifted their ranges in the cold and chaotic weather, forcing the hunter-gatherers to shift their patterns, and often their habitats too. Archaeology from North America's Little Ice Age tells us warfare was the constant and inevitable result (Rice, 2009, pp. 136–60). This pattern of cold-climate human failure continued until the seventeenth century. Then, in the continuing cold of the Little Ice Age, human technology made possible by the use of fossil fuels became effective enough to feed larger populations despite that awful weather.

Global temperatures have risen since the Little Ice Age, a warming that began before the human use of fossil fuels could have been responsible and may be continuing in the modern era. Even in today's relative warmth and with our advanced technologies and wealth, though, far more humans die during cold events than during heat events (see Gasparrini *et al.*, 2015, and the many references in Chapter 4, Section 4.2). Since a low and falling mortality rate is of fundamental importance to human security, however that term is defined, it can hardly be doubted that a warmer world would be a net improvement for the human condition. Nevertheless, some scholars worry about the possible negative "side-effects" of prosperity. Friedman (2006) writes,

> We are also increasingly aware that economic development – industrialization in particular, and more recently globalization – often brings undesirable side effects, like damage to the environment or the homogenization of what used to be distinctive cultures, and we have come to regard these matters, too, in moral terms. On both counts, we therefore think of economic growth in terms of material considerations versus moral ones: Do we have the right to burden future generations, or even other species, for our own material advantage? (p. 15)

But Friedman goes on to say, "I believe this thinking is seriously, in some circumstances dangerously, incomplete." He writes,

> The value of a rising standard of living lies not just in the concrete improvements it brings to how individuals live but in how it shapes the social, political and, ultimately, the moral character of a people. Economic growth – meaning a rising standard of living for the clear majority of citizens – *more often than not fosters greater opportunity, tolerance of diversity, social mobility, commitment to fairness, and dedication to democracy.* Ever since the Enlightenment, Western thinking has regarded each of these tendencies positively, and in explicitly moral terms (*Ibid.*, italics added).

In *The Moral Consequences of Economic Growth* (2005), Friedman showed from international studies that periods of higher economic growth tend to be accompanied historically by more tolerance, optimism, and egalitarian perspectives, while periods of declining economic growth are characterized by pessimism, nostalgia, xenophobia, and violence.

Similarly, LeBlanc and Register (2003) asked, "Has 'progress' – that escalating desire to be bigger, better, faster, stronger – totally extinguished our ancestral instincts to grow everything we consume and hunt only what we need to sustain us? Many view the march of civilization not as a blessing but as a curse, bringing with it escalating warfare and spiraling environmental destruction unlike anything in our human past" (p. xii). But also like Friedman,

LeBlanc and Register say this popular point of view is wrong: "Contrary to exceedingly popular opinion, and as bad as our problems may be today, none of this is true. The common notion of humankind's blissful past, populated with noble savages living in a pristine and peaceful world, is held by those who do not understand our past and who have failed to see the course of human history for what it is."

As the world has grown more prosperous, deaths from wars have plummeted. See Figure 7.1.1.

According to Gleditsch and Nordås (2004), "Globally, in the first decade after World War II, an average of some 300,000 people per year died in battle-related violence. In the first decade in the new Millennium the figure had shrunk to around 44,000" (p. 82). If prosperity fueled rather than discouraged war, these figures would be difficult to explain.

Focusing specifically on the threat to human security posed by civil wars, Hegre and Sambanis (2006) report, "there is now consensus that the risk of war decreases as average income increases and the size of a country's population decreases" (pp. 508–9). Revealing the distance between the explanatory power of these two variables and all others, the authors add, "Beyond these two results, however, there is little agreement."

Hegre and Sambanis conducted an empirical analysis of the role played by prosperity and other factors in the incidence of civil war, isolating causation "by using the same definition of civil war and analyzing the same time period while systematically exploring the sensitivity of 88 variables used to explain civil war in the literature." They used both the PRIO [Peace Research Institute Oslo] definition of "internal armed conflict" and their own definition of civil wars as "an armed conflict between an internationally recognized state and (mainly) domestic challengers able to mount an organized military opposition to the state. The war must have caused more than 1,000 deaths in total and in at least a three-year period" (p. 523). They included per-capita income as a variable because other researchers reasoned that higher incomes raise the opportunity cost of civil wars, citing Fearon and Laitin (2003).

For both definitions of civil war, Hegre and Sambanis found "robust" relationships between the onset of civil wars and low income levels as well as low rates of economic growth (p. 508). They found "decreasing income by one standard deviation increases the risk of civil war by 65%," and "income is substantially more important than population" (p. 524).

Figure 7.1.1
Battle-related deaths in state-based conflicts since 1946, by world region

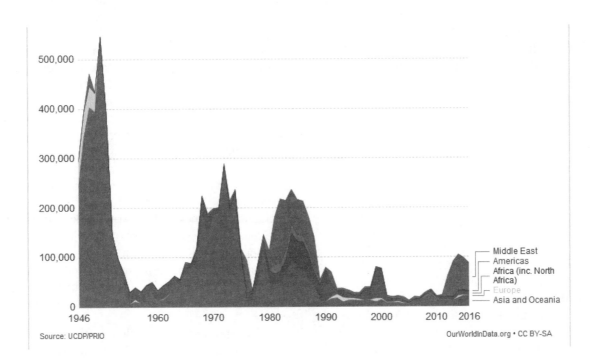

Source: UCDP/PRIO   OurWorldInData.org • CC BY-SA

Source: Our World in Data, n.d.

Other researchers have arrived at similar conclusions (see Collier and Hoeffler, 2004 for citations).

Driving much of this movement toward world peace is the rising abundance of food and other necessities made possible by the use of fossil fuels. Fagan (2000) described a world *without* fossil fuels in a book titled *The Little Ice Age*. He wrote,

Wine harvests were generally late between 1687 and 1703, when cold, wet springs and summers were commonplace. These were barren years, with cold summer temperatures that would not be equaled for the next century. The depressing weather continued as the Nine Years War engulfed the Spanish Netherlands and the Palatinate and Louis XIV's armies battled the League of Augsburg. The campaigning armies of both sides consumed grain stocks that might have fed the poor. As always, taxes were increased to pay for the war, so the peasants had little money to buy seed when they could not produce enough of their own in poor harvest years (p. 132).

As Fagan's description shows, bad weather was enough to cause starvation and wars over limited supplies of food. Armies were raised to commandeer the meager output of low-productivity peasants, which further increased social unrest. Fagan notes "there was little excess food in Europe" during the Irish famine in 1740–1741 because of poor harvests and the War of Austrian Succession. Instead, he observes, quoting Austin Bourke, help came from Britain's peaceful and prosperous North American colonies: "large supplies of provisions arrived from America" (*Ibid.*, p. 183).

Conflicts within nations can likewise arise over scarcity, especially food shortages. Noting "it is implausible to suppose that famines and massive dislocations of poor populations will be unaccompanied by civil unrest and disobedience," Fagan documents such an occurrence in sixteenth century England: "The 1520s produced five exceptional English harvests in a row, when people adapted readily to greater plenty. A spike of sudden cold weather in 1527 brought immediate threats of social unrest. In that year, the mayor's register at Norwich in eastern England noted "there was so great scarceness of corne that aboute Christmas the commons of the cyttye were ready to rise upon the ryche men" (*Ibid.*, p. 84).

As these and countless other examples attest, in centuries past, natural changes in weather as well as climate continuously pushed people into conflict with one another in the pursuit of scarce resources. Reducing this dependency on fair weather is one of the keystones of the development of civilization and the reduction of conflict among peoples. Goklany (2012) noted,

> Until the last quarter of a millennium, mankind depended on living nature for all its food and clothing, most of its energy, and much of its material and medicines. She dictated mankind's numbers, well-being, and living standards. But she has never been constant. She would smile on some, but not on others. Her smiles, always temporary, would inevitably be replaced by frowns. Her Malthusian checks – hunger, famine, disease, or conflict – ensured that there was little or no progress in the human condition. Many people did not even survive into their 20s, populations grew very slowly, and living standards were generally constrained to subsistence levels.
>
> Gradually, with the accumulation of human capital, exchange of ideas, and hard work, mankind started to commandeer more land to meet its needs and develop technologies that, in some cases, amplified Nature's bounty but, in other cases, bypassed her altogether. These led to higher food production, better health, longer lifespans, and larger populations with better living standards, which then reinforced human capital and the exchange of ideas, which begat yet more and better technologies. Thus was the cycle of progress born and set in motion (p. 26).

Fossil fuels, Goklany notes, made possible this cycle of prosperity and progress. Fossil fuels are responsible for at least 60% of mankind's food, and they provide 81% of our energy supply (with nature contributing only 10%). Worldwide, 60% of the fiber used for clothing and other textiles is synthetic, produced mainly from fossil fuels such as petroleum. Even the production of so-called natural fibers, which constitute 30% of the clothing and textile supply,

relies heavily on the use of fossil fuel-based fertilizers and pesticides.

Fossil fuels and the technologies they make possible, Goklany notes, lower our reliance on "living nature," thus reducing the effect of "the whims of nature" on human well-being and reducing the amount of land converted to human use. The reduction of "mankind's footprint on the world" makes land – and hence political sovereignty over increasing amounts of it – less important even as populations increase. A critical element of that progress was the huge increase in crop yields achieved in the twentieth century, a story told in some detail in Chapters 3, 4, and 5.

In addition to nitrogen fertilizer – mass-produced through the use of fossil fuels and delivered efficiently by fossil-fuel-powered vehicles – irrigation and pesticides have further increased crop yields, with fossil fuels playing critical roles in the production and transportation of these goods. In addition, fossil-fuel-powered transportation plays a central role in increasing the availability of food and other necessities of life. Again Goklany reports:

> Beyond increasing yields on the farm, fossil fuels have increased food availability in other ways. The food and agricultural system depends on trade within and between countries to move agricultural inputs to farms and farm outputs to markets. In particular, trade allows food surpluses to be moved to areas experiencing food deficits. But transporting these inputs and outputs in the quantities needed and with the speed necessary for such trade to be an integral part of the global food system depends on relatively cheap fossil fuels (p. 10).

Fagan also noted the importance of transportation in reducing the vulnerability of Europe to crop failures in the fourteenth century: "Vulnerability was a reality of daily life: however adaptable farmers were, Europe still lacked an effective infrastructure for moving large quantities of grain and other commodities at short notice" (Fagan, 2000, p. 80). The Industrial Revolution and rapid increase in the use of fossil fuels have eliminated that technological constraint and spread wealth across the face of the Earth.

Also critical in reducing conflict within and between nations is international trade. Greaves (1995) reported that when Britain repealed its tariffs on imported grain in the nineteenth century (known as the Corn Laws), "Free trade lowered the price of bread and improved the diet of the poor. Living standards improved. With more to eat, people lived longer and healthier lives" (Greaves, 1995, p. 13). They were also more productive, producing more goods and services for themselves and everyone around them. Improved transportation and communication shrank the world and allowed the division of labor to develop internationally, further increasing productivity, as did the global movement of capital: "Production was shifted to areas where the marginal productivity per worker was greater. New trade channels were developed." The increasing international trade, in turn, "brought peoples in different parts of the world closer together. It fostered mutual respect and friendship. People came to realize that voluntary transactions brought gains to both parties and benefits to nation and state. The way to wealth was through trade, not conquest or war." As a result, "peace and good will reigned in most of the world throughout the nineteenth century" (*Ibid.*).

While free trade encouraged peace, high tariffs and blockades encouraged war. Greaves noted the importance of resource scarcity in the rise of Adolf Hitler and Nazi Germany:

In Germany after World War I, rampant inflation had wiped out all savings, completely destroying the middle class. The people were hungry. Adolf Hitler, a rabble rouser with dramatic flair, had attracted a few misfits and malcontents to his movement. The depression added to the distress. ... Hitler made the Jews scapegoats and reached out for "Lebensraum" (living space) to obtain the food and other resources needed to make Germany self-sufficient. Hence the occupation of Austria (March 1938), the Czech Sudetenland (October 1938), and the invasion of Poland (September 1, 1939), also of Belgium, Denmark, Norway, Netherlands, Luxembourg, and Russia.

Although Hitler had grandiose reasons for at least some of these invasions, Greaves is correct to observe that economic scarcity fostered his rise to political power and the German people's acceptance of his program of occupation. Greaves quotes Ludwig von Mises as having written during World War II, "Germany does not aim at autarky because it is eager to wage war. It aims at war because it wants autarky – because it wants to live in economic self-sufficiency" (*Ibid.*, p. 15).

Regarding Germany's fellow Axis power Japan, Greaves notes: "Japan too needed 'lebensraum.' Its population was increasing." Japan's inability to produce enough food and other needed resources drove a fervor for conquest. "Japan was becoming a modern industrial state and depended on imports more than most countries. Yet Japan's attempts to buy food and resources abroad were blocked." Japan's expansion into Korea and Manchuria and its war with China spurred the United States, Britain, and Netherlands to impose trade restrictions on the island nation in the late 1930s, further increasing Japan's need for self-sufficiency. As a result, "Japan attacked Pearl Harbor to protect its flank as she struck the Dutch East Indies and British Malaya to obtain needed food, oil, rubber, and other resources (*Ibid.*)."

The dire consequences of the forced isolation of Germany and Japan led nations away from free trade in the early years of the twentieth century, and conflicts increased. Later in the century, after the brutality of two world wars and a worldwide depression, governments once again turned to freer trade, with a big boost from a technological advance: fossil fuels. Productivity worldwide began to rise rapidly once again.

The increase in trade among nations, made possible by the efficiency of fossil fuels, both alleviates hunger crises in nations hit by natural disasters or poor crop years and allows surpluses in successful nations to be sent to those suffering long-term productivity problems. Trade also increases the stock of human knowledge and inspires the spread of ideas. Consider, for example, the rapid rise of electronics production in Japan in the 1970s and 1980s, computer software in India in the 2000s, electronics in Korea in the 1990s and 2000s, and computer production in China in the 2000s. This fossil-fuel-accelerated process further increases the pace of trade. Goklany notes:

Without relatively cheap fossil fuels, the volume and speed with which goods are traded would be much lower. But trade is one of the fastest methods of disseminating technologies. Introducing new technologies to new places also helps generates new ideas. Or, as Matt Ridley has noted, ideas have "sex," which then propagates new ideas. Absent trade, such devices as personal computers, notebooks, and cell phones may not have been available outside of a handful of industrialized countries, and their prices would have been higher everywhere. This would translate into lower human capital per capita. These products also contain substantial amounts of polycarbonate and other petroleum-based plastics (Goklany, 2012, p. 25).

The argument has been made that income inequality accompanying rising prosperity results in violent conflicts and even war (Piketty, 2014; Scheidel, 2017). An analysis by Goklany (2002), however, finds rich nations are not advancing at the expense of the poor: "Gaps in these critical measures of well-being between the rich countries and the middle- or low-income countries have generally shrunk dramatically since the mid-1900s irrespective of trends in income inequality" (p. 14). Where there have been losses in well-being in the poorest nations, "the problem is not too much globalization but too little," Goklany writes. Specifically, the cycle of prosperity has been inhibited by government policies. Pinkovskiy and Sala-i-Martin (2009) estimated the income distribution for 191 countries between 1970 and 2006 and confirmed Goklany's analysis. They found,

Using the official $1/day line [the United Nations' definition of poverty], we estimate that world poverty rates have fallen by 80% from 0.268 in 1970 to 0.054 in 2006. The corresponding total number of poor has fallen from 403 million in 1970 to 152 million in 2006. Our estimates of the global poverty count in 2006 are much smaller than found by other researchers. We also find similar reductions in poverty if we use other poverty lines. *We find that various measures of global inequality have declined substantially* and measures of global welfare increased by somewhere between 128% and 145% (italics added).

In the 1990s, the gap in life expectancy between sub-Saharan Africa and the rest of the world grew due largely to government policies prohibiting the use of DDT and the subsequent return of malaria to that region of the world. Even with the AIDS epidemic, sub-Saharan mortality rates might have held their own if not for the resurgence of malaria. Thus, "the fact that life expectancy in the Sub-Saharan countries still exceeds the 20–30 years that was typical prior to globalization indicates that, despite the AIDS epidemic and the resurgence of malaria, the net effect of globalization has been positive as far as life expectancy is concerned," Goklany concludes. (*Ibid.*)

Lichbach (2000) observes that the "global political order" has not eliminated conflicts among nations and in fact encourages countries to band together to wage war against others: "The so-called global order makes overt war in Kosovo, continues unnoticed bombing in Iraq, and does nothing about genocide in East Timor," he writes. Markets, by contrast, create social order not only on the local, regional, and national level but also on a global scale. He goes on to say,

[T]he globalization problem is a perfect example of how markets can create rather than destroy social order. Global social order will come, if at all, from international markets (that is, international trade), which will lead to social contracts about international markets that, in turn, will require more general global political order. ... Given the global pluralism of values, only rationally arrived at social contracts can produce predictability, cooperation, and the absence of violence (p. 148).

This cycle of progress is entirely dependent on fossil fuels, Goklany argues:

Although fossil fuels did not initiate the cycle of progress and are imperfect, they are critical for maintaining the current level of progress. It may be possible to replace fossil fuels in the future. Nuclear energy is waiting in the wings but, as the high subsidies and mandates for renewables attest, renewables are unable to sustain themselves today. Perhaps, with help from fossil fuels, new ideas will foster technologies that will enable a natural transition away from such fuels (Goklany, 2012, p. 27).

More recent research on the economics of renewable energies – mainly wind power and solar photovoltaic cells – reported in Chapter 3, Section 3.5, shows renewables indeed have been unable to replace fossil fuels in most applications and particularly with regards to generating "dispatchable" (always available) electricity. Other research suggests there is a strong positive linkage between cheap energy, the economic growth it enables, and international stability. A report commissioned by the

U.S. Agency for International Development surveyed 93 countries to test a model attempting to show the relationships between energy consumption, gross domestic product, life expectancy, and probability of stability (Vasudeva *et al.*, 2005). Access to cheap, affordable energy and economic growth were found to increase the odds of peace by a factor of 2.5. By raising energy consumption, "the occurrence of peace is now 1.5 times more likely than the occurrence of instability in any given country," the study found. (*Ibid.*, p. 32)

The cycle of progress increases prosperity, alleviates resource scarcity crises, and fosters international trade and cooperation, all made possible by the widespread and increasing use of fossil fuels.

References

Collier, P. and Hoeffler, A. 2004. Greed and grievance in civil war. *Oxford Economic Papers* **56**: 563–95.

Davies, W. and Gollop, P. 2003. Chapter 8. The human presence in Europe during the last glacial period II: climate tolerance and climate preferences of mid-and late glacial hominids. In: van Andel, T.H. and Davies, T. (Eds.) *Neanderthals and Modern Humans in the European Landscape During the Last Glaciation: Archaeological Results of the Stage 3 Project.* Cambridge, UK: McDonald Institute for Archaeological Research, pp.131–46.

Fagan, B. 2000. *The Little Ice Age.* New York, NY: Basic Books.

Fearon, J.D. and Laitin, D.D. 2003. Ethnicity, insurgency, and civil war. *American Political Science Review* **97**: 75–90.

Friedman, B. 2005. *The Moral Consequences of Economic Growth.* New York, NY: Vintage Books.

Friedman, B. 2006. The moral consequences of economic growth. *Society* **43** (January/February): 15–22.

Gasparrini, A., *et al.* 2015. Mortality risk attributable to high and low ambient temperature: a multi-country observational study. *Lancet* **386**: 369–75.

Gleditsch, N.P. and Nordås, R. 2014. Conflicting messages? The IPCC on conflict and human security. *Political Geography* **43**: 82–90.

Goklany, I.M. 2002. The globalization of human well-being. *Cato Policy Analysis* No. 447. Washington, DC: Cato Institute. August 22.

Goklany, I.M. 2012. Humanity unbound: how fossil fuels saved humanity from nature and nature from humanity. *Cato Policy Analysis* No. 715. Washington, DC: Cato Institute. December 20.

Greaves, B. 1995. Why war. In: Opitz, E. (Ed.) *Leviathan at War.* Irvington-on-Hudson, NY: The Foundation for Economic Education.

Hegre, H. and Sambanis, N. 2006. Sensitivity analysis of empirical results on civil war onset. *Journal of Conflict Resolution* **50** (4): 508–35.

LeBlanc, S. and Register, K.E. 2003. *Constant Battles: The Myth of the Noble Savage and a Peaceful Past.* New York, NY: St. Martin's/Griffin.

Lichbach, M. 2000. A dialogic conclusion. In: Lichbach, M. and Seligman, A. *Market and Community: The Basis of Social Order, Revolution, and Relegitimation.* University Park, PA: The Pennsylvania State University Press.

Lovins, A.B. 2011. *Reinventing Fire: Bold Business Solutions for the New Energy Era.* White River Junction, VT: Chelsea Green Publishing.

Our World in Data. n.d. War and Peace (website). Accessed July 6, 2018.

Piketty, T. 2014. *Capital in the Twenty-First Century.* Cambridge, MA: Harvard University Press.

Pinkovskiy, M. and Sala-i-Martin, X. 2009. Parametric estimations of the world distribution of income. *Working Paper* No. 15433. Cambridge, MA: National Bureau of Economic Research.

Rice, J. 2009. *Nature & History in the Potomac Country.* Baltimore, MD: Johns Hopkins University Press.

Scheidel, W. 2017. *The Great Leveler: Violence and the History of Inequality from the Stone Age to the Twenty-First Century.* Princeton, NJ: Princeton University Press.

Vasudeva, G., Siegel, S., and Mandrugina, O. 2005. *Energy and Country Instability Project Report.* Reston, VA: Energy and Security Group.

Viau, A.E., *et al.* 2002. Widespread evidence of 1500 yr climate variability in North America during the past 14 000 yr. *Geology* **30** (5): 455–8.

7.1.2 Democracy

Prosperity is closely correlated with democracy, and democracies have lower rates of violence and go to war less frequently than any other form of government. Because fossil fuels make the spread of democracy possible, they contribute to human security.

Democracy can be defined as a system for selecting political leadership characterized by popular participation, broad access by candidates to the ballot, and institutional checks on the power of officials once elected (Gurr *et al.*, 1990). The rise of democracy has been called the "preeminent development" of the twentieth century (Sen, 1999). Samuel Huntington identified three "waves of democratization" in his important book titled *The Third Wave: Democratization in the Late Twentieth Century* (Huntington, 1991). Some of his findings are summarized in Figure 7.1.2.1.

The association between democracies and human security has been extensively studied. Halperin *et al.* (2004) surveyed the literature and found:

Counter to the expectations of the prevailing school, a great deal of research in the 1990s on the political dimension of conflict has revealed a powerful pattern of a "democratic peace." Democracies rarely, if ever, go to war with each other. This pattern has held from the establishment of the first modern democracies in the nineteenth century to the present. As an ever-greater share of the world's states become democratic, the implications for global peace are profound. Indeed, as the number of democracies has been increasing, major conflicts around the world (including civil wars) have declined sharply. Since 1992, they have fallen by two-thirds, numbering just 13 as of 2003 (p. 12).

According to Siegle *et al.*, 80% of all interstate conflicts are instigated by autocracies and 95% of the worst economic performances over the past 40 years were overseen by nondemocratic governments, as well as "virtually all contemporary refugee crises." They write, "Over the past 40 years, autocracies have been twice as likely to experience economic collapse as democracies." Citing Nobel laureate Amartya Sen, they report there has never been a democracy with a free press that has experienced a famine (*Ibid.*, pp. 17–18).

Writing in *Foreign Affairs* in 2004, Halperin, Siegle, and Weinstein documented how low-income democracies do a superior job advancing human security than their autocratic counterparts, observing that "development can also be measured by social indicators such as life expectancy, access to clean drinking water, literacy rates, agricultural yields, and the quality of public-health services. On nearly all of these quality-of-life measures, low-income democracies dramatically outdo their autocratic counterparts" (Siegle *et al.*, 2004). They also report:

People in low-income democracies live, on average, nine years longer than their counterparts in low-income autocracies, have a 40 percent greater chance of attending secondary school, and benefit from agricultural yields that are 25 percent higher.

… Poor democracies also suffer 20 percent fewer infant deaths than poor autocracies (*Ibid.*).

Lipset and Lakin (2004) observe "there is an extremely high correlation between civil and political liberties" (p. 32), though civil liberties may not be part of a "minimalist" definition of democracy. On the association of democracy and violence, they write:

Democracy promotes the institutionalization of nonviolent forms of social conflict and the substitution of nonviolent for violent struggle. While its inception may be the result of rational choice rather than any deep moral commitment, the institutionalization of nonviolent conflict through repeated practice eventually cultivates abiding moral support. Likewise, out-groups that have to fight for entrance into the political game often develop democratic ideologies that suit their purposes, but upon seizing power, they find that the democratic ideal has rooted itself in society, that many adherents genuinely believe in it. Thus what began as instrumental support becomes culturally entrenched (*Ibid.*, p. 35).

diZerega (2000) noted, "Unlike other forms of government, liberal democracies have never fought wars *with others of their own kind*" (p. 1). He suggests

Figure 7.1.2.1
Comparing waves of democratization

Wave	Percentage-point increase in the number of democratic states	Approximate duration (Years)
First	45	100
Second	13	20
Third	35	25

Source: Huntington, 1991, p. 26.

the reason is "democracies are spontaneous orders in [Friedrich] Hayek's sense of the term. Consequently democracies are not states in the usual sense, and often do not act like them." According to diZerega (and Hayek, 1973, 1977, 1979), a spontaneous order does not have a single purpose or an individual who can impose such a purpose on the system. Consequently,

In a democracy all specific policy goals are subordinated to democratic procedures, with the partial exception of wartime. It is only during wartime that democracies can come to resemble instrumental organizations, that is, typical states. Even here, any suspension of democratic procedures such as Britain's suspending elections during WWII, is justified as necessary in order to win the war *and return to democratic procedures*. No general agreement as to the polity's specific goals (beyond survival) need exist.

Fossil fuels and the Industrial Revolution they brought about empowered the common man relative to governments and elites by enabling even poor workers and members of their households to replace their labor with machine labor, dramatically improving their productivity and so their personal consumption or ability to trade with others. The effect was broadly egalitarian, allowing ordinary people to attain what just a generation earlier could be had only by the very rich or very privileged. Lomborg (2001) likened the productivity-boosting effect of technology to giving everyone multiple "virtual servants," each able to do the work of a

person without the assistance of machines. "[E]ach person in Western Europe today has access to 150 virtual servants, in the U.S. about 300, and even in India each person has about 15 servants to help along," he reports (p. 119).

The prosperity made possible by fossil fuels can take some but not all of the credit for the spread of democracy around the world. The relationship between democracy and prosperity has been closely studied, starting with the pioneering empirical research conducted by Lipset (Lipset, 1959). More recently, Lipset and Lakin observed "democracy is supported by a variety of non-political factors including, and *preeminent among them, economic well-being*" (Lipset and Lakin, 2004, p. 12, italics added).

Friedman (2006), cited in the previous section of this chapter, said "the evidence suggests that economic growth usually fosters democracy and all that it entails." He goes on to say, "The main story of the last two decades throughout the developing world, including many countries that were formerly either member states of the Soviet Union or close Soviet dependencies, has been *the parallel advance of economic growth and political democracy*" (p. 18, italics added).

Friedman argues the close correlation between economic growth and democracy is not a coincidence, but that the values and institutions that create economic growth are similar to those that make democracies possible. "While economic growth makes a society more open, tolerant, and democratic, such societies are, in turn, better able to encourage enterprise and creativity and hence to achieve ever greater economic prosperity" (p. 21). "[T]aken as a whole," he concludes, "the experience of the developing world during the last two decades, indeed since World War II, is clearly more consistent with a positive connection between economic growth and democratization" (p. 18; see also Friedman, 2005).

Siegel *et al.* (2004) make the important distinction that economic growth by itself does not lead to democracy. Their objective is to dispel what they call the "development first, democracy later" argument in economic development circles, which justifies massive transfers of income from developed countries to less-developed autocracies in hopes that improved economic well-being will lead to the emergence of democratic institutions. In reality, the authors say, such policies serve only to reinforce the political power of autocrats and undermine market-based economic growth. Economic aid to autocracies, they write, "has led to atrocious policies – indeed, policies that have undermined international efforts to improve the lives of hundreds of millions of people in the developing world."

Affluence may not be necessary for democracies to arise, but affluence does ensure their survival. Pzeworski (2004), widely regarded as one of the world's leading experts on democracy, notes:

[N]o democracy ever, including the period before World War II, fell in a country with a per capita income higher than that of Argentina in 1975, $6,055. This is a startling fact, given that since 1946 alone 47 democracies collapsed in poorer countries. In contrast, 35 democracies spent 1,046 years in wealthier countries and not one died. Affluent democracies survived wars, riots, scandals, economic and governmental crises, hell or high water.

Pzeworski's statistical analysis found:

[T]he probability that democracy survives increases monotonically with per capita income. In countries with per capita income under $1,000, the probability that a democracy would die during a particular year was 0.0845, which implies that their expected life was about twelve years. Between $1,001 and $3,000, this probability was 0.0362, for an expected duration of twenty-seven years. Between $3,001 and $6,055, the probability was 0.0166, which translates into about sixty years of expected life. And what happens above $6,055 we

already know: democracy lasts forever" (*Ibid.*).

Pzeworski explains the association between democracy and prosperity this way:

The reason everyone opts for democracy in affluent societies is that too much is at stake in turning against it. In poor societies there is little to distribute, so that a group that moves against democracy and is defeated has little income to lose: in poor countries, incomes of people suffering from a dictatorship are not much lower than of those living under democracy, whether they won or lost an election. But in affluent societies, the gap between incomes of electoral losers and of people oppressed by a dictatorship is large (*Ibid.*).

Finally and in summarizing his findings, Pzeworski observes, "We know that democracies are frequent among the economically developed countries and rare among the very poor ones. The reason we observe this pattern is not that democracies are more likely to emerge as a consequence of economic development but that they are much more likely to survive if they happen to emerge in more developed countries" (*Ibid.*).

The research cited above makes it clear that fossil fuels, by making possible the dramatic rise in global prosperity since the great expansion of their use starting in the eighteenth century, have created the conditions necessary for democracies to survive. Democracies, in turn, promote world peace and create other conditions needed to ensure human security. Rather than being a net cost to society in terms of human security, fossil fuels clearly have been human security's surest guarantor.

References

diZerega, G. 2000. Democracy, spontaneous order and peace: implications for the classical liberal critique of democratic politics. *Working Paper* #20. Independent Institute. January.

Friedman, B. 2005. The moral case for growth. *The International Economy* **40** (Fall).

Friedman, B. 2006. The moral consequences of economic growth. *Society* **43** (January/February): 15–22.

Gurr, T.R., Jaggers, K., and Moore, W. 1990. The transformation of the western state: the growth of democracy, autocracy, and state power since 1800. *Studies in Comparative International Development* **25**: 73–108.

Halperin, M.H., Siegle, J.T., and Weinstein, M.M. 2004. *The Democracy Advantage: How Democracies Promote Prosperity and Peace.* New York, NY: Routledge.

Hayek, F. 1973. *Law, Legislation, and Liberty, Vol. 1. Rules and Order.* Chicago, IL: University of Chicago Press.

Hayek, F. 1977. *Law, Legislation, and Liberty, Vol. 2. The Mirage of Social Justice.* Chicago, IL: University of Chicago Press.

Hayek, F. 1979. *Law, Legislation, and Liberty, Vol. 3. The Political Order of a Free People.* Chicago, IL: University of Chicago Press.

Huntington, S. 1991. *The Third Wave: Democratization in the Late Twentieth Century.* Norman, OK: University of Oklahoma Press.

Lipset, S.M. 1959. Some social requisites of democracy: economic development and political legitimacy. *American Political Science Review* **53** (1): 69–105.

Lipset, S.M. and Lakin, J.M. 2004. *The Democratic Century.* Norman, OK: University of Oklahoma Press.

Lomborg, B. 2001. *The Skeptical Environmentalist: Measuring the Real State of the World.* Cambridge, UK: Cambridge University Press.

Pzeworski, A. 2004. Democracy and economic development. In: Mansfield, E.D. and Sisson, R. (Eds.) *The Evolution of Political Knowledge: Democracy, Autonomy, and Conflict in Comparative and International Politics.* Columbus, OH: Ohio State University Press, pp. 300–24.

Sen, A. 1999. Democracy as a universal value. *Journal of Democracy* **10** (3): 3–17.

Siegle, J.T., Weinstein, M.M., and Halperin, M.H. 2004. Why democracies excel. *Foreign Affairs* **83** (5): 57–71.

7.1.3 Wars for Oil

The cost of wars fought in the Middle East is not properly counted as one of the "social costs of carbon" as those conflicts have origins and justifications unrelated to oil.

According to the Rocky Mountain Institute, quoting Alan Greenspan, "the Iraq War 'is largely about oil.'

That war has already cost more than 4,400 U.S. lives, plus one to several trillion borrowed dollars" (Lovins, 2011, p. 5, no citation to Greenspan). The author continues: "in 2010, a Princeton study pegged the cost of U.S. forces just in the Persian Gulf in just one year (2007) at half a trillion dollars, or about three-fourths of the nation's total military expenditures" (*Ibid.*, citing Stern, 2010).

In the wake of President George W. Bush's invasion of Iraq in 2003, countless other commentators claimed the effort was undertaken to ensure the availability of oil for U.S. consumers. This "blood for oil" argument has a long history. More than three decades ago, Husbands (1983) noted, "One frequently hears that our presence in the Middle East is necessary to protect 'our' oil. The implication is that in our absence, the oil would necessarily fall into unfriendly hands and those parties would then embargo exports to the United States" (Husbands, 1983). It is an implausible claim, Husbands argued, given that U.S. oil companies at the time were making efforts to minimize their purchases of oil from Saudi Arabia in favor of cheaper oil from Russia and Mexico.

Although resource scarcity historically has been a common factor in war, the "blood for oil" thesis relies on several premises, all of which are dubious: that the United States suffers from a scarcity of oil; that the U.S. government could reasonably expect that invading Iraq would reduce scarcity by an amount great enough to provide a larger return than the amount of resources and human lives it would cost; that there were no less-expensive (in money and lives) ways to achieve a similar increase in the supply of oil; and that there were no other, more compelling reasons for the intervention in Afghanistan, Iraq, and Pakistan.

The notion that the United States has a scarcity of oil is a value judgment, not a factual statement. The amount of oil people use depends on its price and its value to the consumer: People will use oil as long as the money spent on it brings them greater benefits than the same amount of money spent on something else. Hence, the issue is not whether there is "enough" oil but whether people can afford it. The latter is visible in consumption numbers: U.S. crude oil consumption reached a peak of 20,800 billion barrels a day in 2005, which stayed stable until 2007, at just under 20,700 billion barrels, according to the U.S. Energy Information Administration. When the United States invaded Iraq in March 2003, consumption was at 20,000 barrels a day, up from just under 17,000 in 1990 and 19,700 in 2000. As a result of the 2008 recession, daily consumption fell to 18,700 in 2009, and then 18,400 in 2012.

As those figures indicate, U.S. crude oil consumption tracks with the strength of the nation's economy. We use more when the economy is strong, and we use less when it is weak. The notion that oil is so scarce that the United States had to go to war to ensure supplies is not supported by the facts. Glaser (2017) noted,

Indeed, the United States today is far less reliant on foreign oil supplies than it once was. In 2015, only about 24 percent of the petroleum consumed by the United States was imported from foreign countries (the lowest level since 1970), and only about 16 percent of that was imported from the Middle East. This is largely because U.S. domestic production has significantly increased thanks to technological advances in exploiting shale reserve areas. Since 2008, annual U.S. crude production has grown by about 75 percent and net import volumes are projected to decline by 55 percent by 2020. Canadian oil output is also expected to double by 2040, meaning North America is on track to be a net oil exporter by 2020 and to remain so through 2040.

Glaser also points out:

[O]il is a fungible commodity traded on global markets and subject to the laws of supply and demand. Supply disruptions from one source impact the overall price, but can quickly be offset by an increase in output from another source. In every oil shock since 1973, global energy markets adapted quickly, by increasing production from other sources, rerouting existing supplies and putting both private and government-held stockpiles around the world into use. These market adjustments mitigated the ramifications of the shocks and stabilized prices and supply. U.S. military presence in the Persian Gulf did not prevent the disruptions, nor did it ease the resulting economic pain (*Ibid.*).

As to the costs and presumed oil-supply benefits of the War in Iraq, that war alone had cost $1.7 trillion by 2013, and the nation owed another almost half-billion dollars in benefits to veterans of wars in Afghanistan, Iraq, and Pakistan, according to the

Costs of War Project by the Watson Institute for International Studies at Brown University, as reported by Reuters (2013). The combined cost of the wars in Afghanistan, Iraq, and Pakistan was estimated at nearly $4 trillion. In addition, the interest costs for paying off the U.S. government debt incurred in the wars were expected to tally another $4 trillion over the next 40 years. "The report concluded the United States gained little from the war while Iraq was traumatized by it," Reuters noted. (*Ibid.*)

At the approximate 2017 price of crude oil of about $50 per barrel, the United States could have purchased *160 billion* barrels of oil for the $8 trillion the wars in Afghanistan, Iraq, and Pakistan cost. In 2014, the United States consumed just under seven billion barrels of petroleum products in total. That means the United States could have purchased almost 23 years of its *total* oil consumption with what the wars in the Middle East cost. If ensuring access to cheap oil were the rationale for the U.S. presence in the Middle East, then it has been a spectacularly bad investment. Taylor and Peter Van Doren (2008) remarked, "The U.S. 'oil mission' is thus best thought of as a taxpayer-financed gift to oil regimes and, perhaps, the Israeli government that has little, if any, effect on the security of oil production facilities. One may support or oppose such a gift, but our military expenditures in the Middle East are not necessary to remedy a market failure."

Instead of ensuring a greater flow of oil from the Middle East to the United States, the years since the War in Iraq have brought a decreasing dependency on oil from the Middle East. As Figure 7.1.3.1 indicates, U.S. imports of OPEC oil have been falling since 2008, and non-OPEC sources have supplied more U.S. oil imports since the early 1990s, with the gap widening. Canada now supplies the lion's share of U.S. oil imports.

In February 2018, the Persian Gulf region, led by Saudi Arabia at 8%, provided the United States with just 18% of its imported oil. Iraq itself has not been a major supplier of oil to the United States for quite some time. The International Energy Agency projects Iraq will raise production to 6.1 million barrels by 2020, but most of that oil will be exported to China and other Asian markets. Although it is possible to argue the war backfired in ensuring an adequate supply of oil from Iraq, the important thing to note is that Iraq was not, and never had been, a significant supplier of U.S. oil. And even if other Middle Eastern oil-producing nations wanted the United States to

Figure 7.1.3.1
U.S. crude oil and petroleum products imports by year and by nation of origin, 1950–2017 (million barrels per day)

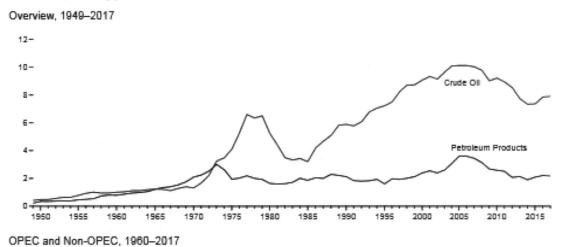

Overview, 1949–2017

OPEC and Non-OPEC, 1960–2017

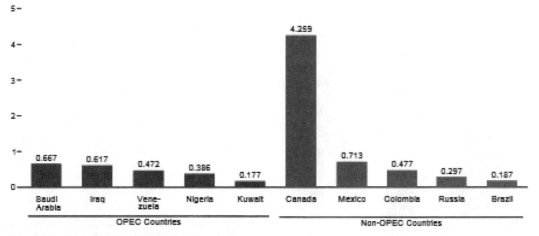

From Selected Countries, February 2018

Note: OPEC–Organization of the Petroleum Exporting Countries.
Web Page: http://www.eia.gov/totalenergy/data/monthly/#petroleum.
Sources: Tables 3.3b–3.3d.

Source: EIA, 2018b, Figure 3.3.b, p. 56.

Figure 7.1.3.2
U.S. crude oil and natural gas liquids imports, exports, and production, 1950-2017
(million barrels per day)

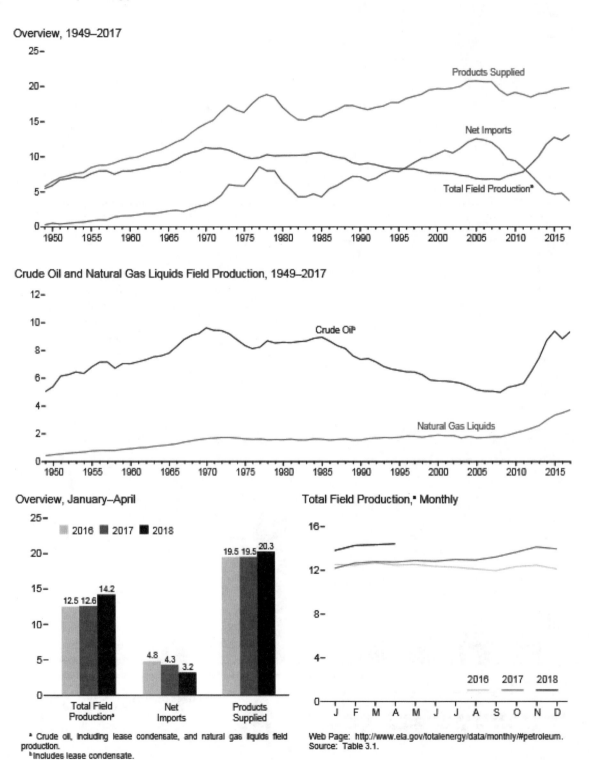

Source: EIA 2018b, Figure 3.1, p. 50.

invade Iraq, their production was already of decreasing importance at the time, as noted above.

While imports of Middle Eastern oil have fallen, domestic U.S. oil production has risen since the War in Iraq. When the U.S. forces invaded in March 2003, U.S. domestic crude oil and natural gas production was just under 18 million barrels per day (b/d) (see Figure 7.1.3.2 on the previous page). Production fell to about 14.6 million b/d in November 2005, but it has been rising steadily since then. The U.S. Energy Information Administration (EIA, 2018a) estimates U.S. crude oil production will average 10.9 million b/d in 2018, up from 9.4 million b/d in 2017, and will average 12.1 million b/d in 2019. Note U.S. oil production declined when OPEC imposed its embargo in 1973. The United States did not see fit to increase oil production at that time, much less go to war in the Middle East to ensure a resumption of supplies.

The rising U.S. consumption of non-Middle Easter oil production shows the wars in Afghanistan, Iraq, and Pakistan did not increase U.S. consumption of Middle Eastern oil but instead accompanied a rise in the use of oil from other suppliers. Instead of spending $8 trillion on wars, the United States, were it intent on increasing oil supplies, could simply have continued to develop these other sources, especially domestic production. That was indeed much less expensive, in money and lives, as a way of ensuring an ample supply of oil.

The U.S. government had, and repeatedly stated, other significant reasons for its interventions in Afghanistan, Iraq, and Pakistan. U.S. intervention in the Middle East has long been based on geopolitical and humanitarian concerns that have nothing to do with oil supplies. Defending Israel, the lone stable democracy in the region, has been a high U.S. priority since May 14, 1948, when Israel declared its existence and President Harry S. Truman recognized the new nation on the same day. Since the collapse of the former Soviet Union, it has been learned that Israel figured more prominently in the communist regime's cold war stratagems than was publicly known at the time, providing another justification for the U.S. presence in the region (Ginor and Remez, 2007).

Testifying before the U.S. Senate Armed Services Committee on U.S. policy in the Middle East in September 2015, former Obama administration CIA Director David Petraeus urged the government to intervene in Syria by threatening to destroy President Bashar Assad's air force if the Syrian forces continued to bomb the Syrian people.

Petraeus also recommended "the establishment of enclaves in Syria protected by coalition air power, where a moderate Sunni force could be supported and where additional forces could be trained, internally displaced persons could find refuge, and the Syrian opposition could organize" (Wong, 2015). Syria is not an oil-exporting country, so there are obviously other reasons for the U.S. government to be so concerned about its affairs.

In 1991, the United States established safe havens and enforced no-fly zones under Operation Provide Comfort in an effort to stop Iraqi leader Saddam Hussein from massacring Kurds in northern Iraq after his suppression and killings of Shiites in the southern part of the nation. That effort resulted in self-rule for Iraqi Kurdistan. These interventions and many others have no bearing on U.S. oil supplies and appear to be the result of humanitarian and geopolitical concerns, not U.S. economic interests, and certainly not the flow of imported oil from the Middle East.

If claims of a humanitarian mission in the Middle East are not persuasive, then perhaps the explanation lies in American hubris. Bacevich (2017) summarized U.S. involvement in the Middle East as follows:

> From day one, the larger purpose of America's War for the Greater Middle East has been to affirm that we are a people to whom limits do not apply. The advertised purpose has been to liberate, defend, or deter. Yet the actual purpose has been far more ambitious in my view. The real mission has been to sustain the claims of American exceptionalism that have long since become central to our self-identity – to bring into compliance with American purposes the revolutionaries, warlords, terrorists, despots, or bad actors of various stripes given to defiance. To employ the kind of jargon that's popular in this city, back in 1980, the United States set out in willy-nilly fashion to "shape" the greater Middle East. Given the conditions existing there, employing military means to bring the region into conformity with American purposes has resulted in an undertaking of breathtaking scope.

Bacevich's views on U.S. military involvement in the Middle East are echoed by many other military experts (Glaser and Kelanic, 2016; Cohen, 2011; Glaser, 2017; Codevilla, 2018). It is safe to say none of them believes U.S. military forces are in the Middle East to protect American access to oil.

Finally, it is unclear whether a forced transition from fossil fuels would reduce violence in the Middle East. Indeed, the opposite is more likely to be the case. Pipes (2018) observed, "yes, the demise of oil and gas will bring some good news: More water desalination plants, less Islamism (petrodollars basically fund it), and Israel's enemies weakened. But the negative implications of a gas and oil price collapse will be much greater" (p. 21). He explained:

> Foreign direct investment will shrivel. The majority of Middle Eastern economies will convulse. Regimes such as the Islamic Republic of Iran or the People's Democratic Republic of Algeria will not survive, leading to more anarchy (already rampant in Afghanistan, Egypt, Iraq, Lebanon, Libya, Somalia, Syria, the West Bank, and Yemen). Disagreements over access to scarce resources will spur new conflicts. Guest workers will return home in droves, upsetting those economies. Economic and other migrants will pour out of the region, headed mostly to the West, further upsetting the politics of Europe. Key airline and shipping routes will be disrupted. U.S. disengagement will enable nuclear weapons programs. In brief, the world's chief trouble spot will retain its role, only more so. Attention to the Middle East, still the world's premier irritant, will continue long after the decline of oil and gas (p. 21).

* * *

In summary, fossil fuels have made the world a safer place than ever before. Prosperity has led to more tolerance, optimism, and egalitarian perspectives and less xenophobia, pessimism, and violence. Driving this movement toward world peace is the rising abundance of food and other necessities as well as international trade that brings people together in their common pursuit of happiness. Prosperous countries are more likely to be (and remain) democracies, and democracies have lower rates of violence and go to war less frequently than any other form of government. Limiting access to affordable energy threatens to prolong and exacerbate poverty in developing countries, increasing the likelihood of domestic violence, state failure, and regional conflict. Wars will continue to be fought in

the Middle East even if, and perhaps especially if, the world reduces its reliance on fossil fuels.

References

Bacevich, A. 2016. Endless war in the Middle East. *Cato's Letter* **14** (3).

Bacevich, A. 2017. *America's War for the Greater Middle East: A Military History.* New York, NY: Random House.

Codevilla, A.M. 2018. On the natural law of war and peace. *Claremont Review of Books* **18** (2): 25–30.

Cohen, D. 2011. Myth-telling: the cult of energy insecurity and China-US relations. *Global Asia* (June).

EIA. 2018a. U.S. Energy Information Administration. U.S. Liquid Fuels. Short-term Energy Outlook. November 6.

EIA. 2018b. U.S. Energy Information Administration. *Monthly Energy Review.* May.

Ginor, I. and Remez, G. 2007. *Foxbats over Dimona: The Soviets' Nuclear Gamble in the Six-Day War.* New Haven, CT: Yale University Press.

Glaser, C.L. and Kelanic, R.A. (Eds.) 2016. *Crude Strategy: Rethinking the U.S. Military Commitment to Defend Persian Gulf Oil.* Washington, DC: Georgetown University Press.

Glaser, J. 2017. Does the U.S. military actually protect Middle East oil? The National Interest (website). January 9.

Husbands, S. 1983. Free trade and foreign wars. *The Freeman* **33** (4): 195–204.

Lovins, A.B. 2011. *Reinventing Fire: Bold Business Solutions for the New Energy Era.* White River Junction, VT: Chelsea Green Publishing.

Pipes, D. 2018. The end of carbon fuels? A symposium of views. *The International Economy* (Spring): 10–21.

Reuters. 2013. Iraq war costs U.S. more than $2 trillion: study. March 14.

Stern, R.J. 2010. United States cost of military force projection in the Persian Gulf, 1976–2007. *Energy Policy* **38** (6): 2816–25.

Taylor, J. and Van Doren, P. 2008. The energy security obsession. *The Georgetown Journal of Law & Public Policy* **6** (2).

Wong, K. 2015. Petraeus: ISIS war progress "inadequate." *The Hill*. September 22.

7.2 Climate Change

Section 7.1 makes clear that far from threatening human security, fossil fuels are actually its best guarantor. *But what if fossil fuels cause or contribute to climate change?* Would higher global surface temperatures trigger floods, droughts, more violent weather, and other climate effects described in vivid detail in IPCC reports? Would those climate changes reduce human security? The survey of climate science presented in Chapter 2 concludes such an outcome is highly unlikely, but the IPCC and its followers plainly disagree. The rest of the current chapter assumes *arguendo* that the IPCC is right and man-made climate change is a genuine possibility.

Section 7.2.1 describes how the IPCC frames the discussion of "human security and points out the major problems with it. Section 7.2.2 addresses sea-level rise, Section 7.2.3 addresses impacts on agriculture, and Section 7.2.4 addresses other impacts on human security.

7.2.1 The IPCC's Perspective

The IPCC claims global warming threatens "the vital core of human lives" in multiple ways, many of them unquantifiable, unproven, and uncertain. The narrative in Chapter 12 of the Fifth Assessment Report illustrates the IPCC's misuse of language to hide uncertainty and exaggerate risks.

The introduction to this chapter discussed the elastic definition of "threats to human security" used by Working Group II in Chapter 12 of IPCC's Fifth Assessment Report (AR5) to characterize the alleged damages caused by man-made climate change. The IPCC sorts these damages into "deprivation of human needs" and "erosion of livelihood and human capabilities," provides a table, reproduced below as Figure 7.2.1.1, presenting additional "dimensions of impact" and examples of observed and potential impacts of climate change.

Gleditsch and Nordås (2014), quoted in the introduction, observed how "the definition in the Human Security chapter is too wide to allow serious attempts to assess the secular trend" pp. 85–86). A

second general problem is the confusion of impacts due to natural causes and those that could be attributable to and an impact on climate due to the human presence. In the *Summary for Policymakers* of the Working Group II contribution to the Fifth Assessment Report, IPCC says:

> Climate change refers to a change in the state of the climate that can be identified (e.g., by using statistical tests) by changes in the mean and/or the variability of its properties, and that persists for an extended period, typically decades or longer. Climate change may be due to natural internal processes or external forcings such as modulations of the solar cycles, volcanic eruptions, and persistent anthropogenic changes in the composition of the atmosphere or in land use. (IPCC, 2014a, p. 5, Background Box SPM.2).

A third problem is IPCC's frequent assertion of unproven causal links between climate change and factors contributing to or detracting from human security. The IPCC admits to some uncertainty, saying "Given the many and complex links between climate change and human security, uncertainties in the research on the biophysical dimensions of climate change, and the nature of the social science, highly confident statements about the influence of climate change on human security are not possible," citing Scheffran *et al.* (2012). But then in characteristic IPCC fashion, the very next sentence tries to deny the uncertainty it just confessed: "Yet there is good evidence about many of the discrete links in the chains of causality between climate change and human insecurity" (IPCC, 2014b, p. 760).

At issue is not whether "highly confident statements" can be made, but whether declarative statements with *any* degree of confidence can be made. The assertion of unproven causal links pervades Chapter 12, more so than any other chapter of AR5, partly because of the nature of the issues being addressed. The fact that many alleged consequences cannot be measured and are largely subjective already has been mentioned. But it also is due to the methodology the IPCC chose. Much of the "evidence" is based on futurology – attempts to predict the future – woven from historic anecdotes and expert opinions. There is an extensive literature on scientific forecasting demonstrating that such an approach is no more likely to produce accurate forecasts than uneducated guesses (see the literature surveys in Armstrong, 2001, 2006). But there are no references in AR5 either to the general literature on scientific forecasting or to its application to climate change.

A fourth problem with AR5 Chapter 12 is what is called, in statistics, propagation of error. The errors or uncertainty in one variable, due perhaps to measurement limitations or confounding factors, are compounded (propagated) when that variable becomes part of a function involving other variables that are similarly uncertain. For example, there is a range of uncertainty regarding surface temperatures due to the placement of temperature stations and changes in technology over time. The human impact on global average temperature is uncertain due to incomplete understanding of the climate (e.g., exchange rates between CO_2 reservoirs and the

behavior of clouds). There are also ranges of uncertainty regarding human emissions of CO_2 in the past, present, and future. There are also ranges of uncertainty as to how to measure an alleged effect (e.g., loss of livelihood, loss of personal property, forced migration) and how much of the effect to attribute to a specific weather-related event (e.g., flood, hurricane, drought) or to some other variable (e.g., poverty, civil war, mismanagement of infrastructure). The more variables in a function, the wider the "uncertainty bars" surrounding the outcome must be. Even a formula with few variables is subject to propagation error if it attempts to forecast events far into the future, e.g., a century or more in the case of climate models (Frank, 2015).

Figure 7.2.1.1
IPCC's list of threats to human security due to climate change

Dimensions of impact		Illustrative examples of observed impacts due to aggravating climate stresses	Illustrative examples of potential changes in livelihoods and capabilities as a consequence of climate variability and climate change
Deprivation of basic needs	Livelihood assets	Household assets such as livestock sold or lost during drought: documented examples are the 1999–2000 drought, Ethiopia, and 1999–2004 drought, Afghanistan (Carter et al., 2007; de Weijer 2007). Riverbank erosion, floods, and groundwater depletion and salinization are associated with changed hydrological regimes and cause loss of agricultural land (Paul and Routray, 2010; Taylor et al., 2013).	Simulated future climate volatility leads to reduced future production of staple grains and increases in poverty (Ahmed et al., 2009). Changes in the viability of livestock feed crops have an impact on smallholder farmers: maize yields are projected to decline in many regions (Jones and Thornton, 2003; Section 7.4). Projections of land loss, riverbank erosion, and groundwater depletion, in combination with environmental change and human interventions, suggest future stress on livelihood assets (Le et al., 2007; Taylor et al., 2013).
	Water stress and scarcity	Glacier retreat leads to lower river flows and hence affects water stress and livelihoods, representing a cultural loss (Orlove et al., 2008). For example, glacier recession in the Cordillera Blanca in Peru has altered the hydrological regime with implications for local livelihoods and water availability downstream (Mark et al., 2010).	Projected stresses to water availability show increased populations without sustainable access to safe drinking water (Hadipuro, 2007). Projected reduction in glacier extent and the associated loss of a hydrological buffer is expected to increase (Vuille, 2008; Section 3.4.4).
	Loss of property and residence	Floods destroy shelter and properties and curtail ability to meet basic needs. For example, the Fiji flood in 2009 resulted in economic losses of F$24 million affecting at least 15% of farm households (Lal, 2010). Sea level rise and increased frequency of extreme events increases the risk of loss of lives, homes, and properties and damages infrastructure and transport systems (Adrianto and Matsuda, 2002; Suarez et al., 2005; Philips and Jones, 2006; Ashton et al., 2008; Von Storch et al., 2008).	Changes in flood risk may increase and cause economic damages: in the Netherlands, the total amount of urban area that can potentially be flooded has increased sixfold during the 20th century and may double again during the 21st century (de Moel et al., 2011). In England and Wales, projected changes in flood risk mean economic damages may increase up to 20 times by the 2080s (Hall et al., 2003).
Erosion of livelihood and human capabilities	Agriculture and food security	Interaction of climate change with poverty and other political, social, institutional, and environmental factors may adversely affect agriculture production and exacerbate the problem of food insecurity (Downing, 2002; Saldana-Zorrilla, 2008; Trotman et al., 2009). Examples include in Kenya (Oluoko-Odingo, 2011); in Southern Africa (Drimie and Gillespie, 2010); in Zimbabwe and Zambia (Mubaya et al., 2012).	Studies of African agriculture using diverse climate scenarios indicate increasing temperature and rainfall variation have negative impacts on crops and livestock production and lead to increased poverty, vulnerability, and loss of livelihoods. Examples include Ethiopia (Deressa and Hassan, 2009); Kenya (Kabubo-Mariara, 2009); Burkina Faso, Egypt, Kenya, and South Africa (Molua et al., 2010); and sub-Saharan Africa (Jones and Thornton, 2009). Potential livelihood insecurity among small-scale rain-fed maize farmers in Mexico is projected owing to potential loss of traditional seed sources in periods of climate stress (Bellona et al., 2011).
	Human capital (health, education, loss of lives)	Food shortage, absence of safe and reliable access to clean water and good sanitary conditions, and destruction of shelters and displacements all have a negative bearing on human health (Costello et al., 2009; Sections 11.4 and 11.8). Droughts and floods can intensify the pressure to transfer children to the labor market (Ethiopia and Malawi; UNDP, 2007). Indian women born during a drought or flood in the 1970s were 19% less likely to ever attend primary school, when compared with women of the same age who were not affected by natural disasters (UNDP, 2007).	Analysis of the economic and climatic impacts of three emission scenarios and three tax scenarios estimates the impacts on food productivity and malaria infection to be very severe in some Asian countries (Kainuma et al., 2004). Studies of the impacts of future floods using a combination of socioeconomic and climate change scenarios for developed countries show an increase in mortality. For example, in the Netherlands, sea level rise, combined with other factors, potentially increases the number of fatalities four times by 2040 (Maaskant et al., 2009).

Source: IPCC, 2014a, Table 12-1, p. 761.

Propagation of error means it is likely to be impossible to attribute to climate change *any* impacts on human security. Deaths and loss of income due to storms, flooding, and other weather-related phenomena are and always have been part of the human condition. We can at best document trends in the frequency of storms, the number of deaths, and the value of property losses, but these statistics are meaningless for a discussion of public policy if they cannot be reliably correlated with long-term climate change, and climate change with human greenhouse gas emissions.

A final problem with AR5 Chapter 12 is the language the IPCC uses to hide uncertainty and exaggerate risks. Statements seeming to express certainty are often followed immediately by sentences expressing uncertainty, or vice versa. To some extent this is the result of editing by committees seeking consensus. Advocates of making strong statements are allowed to use their language on the condition that doubters and skeptics can follow with sentences that begin with "However..." or "Nevertheless... ." This dynamic produces reports that journalists and advocates can use to justify dramatic headlines, but it misleads serious researchers, policymakers, and the public.

Gleditsch and Nordås (2014) describe the many "expressions of uncertainty" that appear in IPCC reports, including such words as may, might, can, and could, and such phrases as "has a potential to," "is a potential cause of," and "is sensitive to." The terms that are most vague appear more frequently in Working Group II reports on "impacts, adaptation, and vulnerability" than in Working Group I reports on "the physical science basis." Gleditsch and Nordås write,

> The frequent use of "may" terms might have been justified as a way of indicating that "under certain circumstances, a relationship is likely." But this does not work well if those circumstances are not specified. On the whole, it would probably be best to avoid the use of terms like "may" in academic writing except to state conjectures. Misrepresentation of the scientific basis is a real hazard when using such terminology (p. 88).

In conclusion, the IPCC claims climate change threatens "the vital core of human lives" in multiple ways, many of them unquantifiable, unproven, and uncertain. The narrative in AR5 Chapter 12 illustrates the IPCC's misuse of language to hide uncertainty and exaggerate risks.

References

Armstrong, J.S. 2001. *Principles of Forecasting – A Handbook for Researchers and Practitioners*. Norwell, MA: Kluwer Academic Publishers.

Armstrong, J.S. 2006. Findings from evidence-based forecasting: methods for reducing forecast error. *International Journal of Forecasting* **22**: 583–98.

Frank, P. 2015. Negligence, non-science, and consensus climatology. *Energy & Environment* **26** (3): 391–415.

Gleditsch, N.P. and Nordås, R. 2014. Conflicting messages? The IPCC on conflict and human security. *Political Geography* **43**: 82–90.

IPCC. 2014a. Intergovernmental Panel on Climate Change. *Summary for Policymakers*. In: *Climate Change 2014: Impacts, Adaptation, and Vulnerability*. Contribution of Working Group II to the Fifth Assessment Report of the Intergovernmental Panel on Climate Change. New York, NY: Cambridge University Press.

IPCC. 2014b. Intergovernmental Panel on Climate Change. *Climate Change 2014: Impacts, Adaptation, and Vulnerability. Part A: Global and Sectoral Aspects*. Contribution of Working Group II to the Fifth Assessment Report of the Intergovernmental Panel on Climate Change. New York, NY: Cambridge University Press.

Scheffran, J., Brzoska, M., Kominek, J., Link, P.M., and Schilling, J. 2012. Disentangling the climate-conflict nexus: empirical and theoretical assessment of vulnerabilities and pathways. *Review of European Studies* **4** (5): 1–13.

7.2.2 Extreme Weather

Real-world data offer little support for predictions that CO_2-induced global warming will increase either the frequency or intensity of extreme weather events.

According to the IPCC (quoting from Figure 7.2.1.1), "sea level rise and increased frequency of extreme events increases the risk of loss of lives, homes, and properties, and damages infrastructure and transport systems." Extrapolating from isolated incidents, damages associated with these calamities are imagined to cost billions or hundreds of billions of dollars a year. But the link is very tenuous. Though often cited as the source of alarming projections of violent weather, the IPCC has been quite cautious on the topic. In a special report on the issue published in 2012, it found only mixed and weak evidence of a trend toward more extreme weather:

> There is evidence from observations gathered since 1950 of change in some extremes. Confidence in observed changes in extremes depends on the quality and quantity of data and the availability of studies analyzing these data, which vary across regions and for different extremes. Assigning "low confidence" in observed changes in a specific extreme on regional or global scales neither implies nor excludes the possibility of changes in this extreme (IPCC, 2012).

In Working Group II's contribution to IPCC's Fifth Assessment Report, in Chapter 10, the IPCC admits "The impact of natural disasters on economic growth in the long-term is disputed, with studies reporting positive effects (Skidmore and Toya, 2002), negative effects (Raddatz, 2009), and no discernible effects (Cavallo *et al.*, 2013)" (*Ibid.*, p. 692). The IPCC authors conclude, "The literature on the impact of climate and climate change on economic growth and development has yet to reach firm conclusions. There is agreement that climate change would slow economic growth, by a little according to some studies and by a lot according to other studies" (IPCC, 2014, p. 693).

However, the *Summary for Policymakers* of the Working Group I contribution to the IPCC's AR5, as usual more alarmist than the underlying report itself, claims, "Extreme precipitation events over most of the mid-latitude land masses and over wet tropical regions will *very likely* become more intense and more frequent by the end of this century, as global mean surface temperature increases" (IPCC, 2013, p. 23).

Literature reviews conducted in Chapter 2 of this volume and for previous volumes in the *Climate Change Reconsidered* series failed to find a convincing relationship between global warming over the past 100 years and increases in any of these extreme weather events. Other authors have reached the same conclusion (Maue, 2011; Alexander *et al.*, 2006; Khandekar, 2013; Pielke Jr., 2013, 2014). Instead, the number and intensity of extreme events wax and wane often in parallel with natural decadal

or multidecadal climate oscillations. Basic meteorological science suggests a warmer world would experience fewer storms and weather extremes, as indeed has been the case in recent years.

Globally, there has been no detectable long-term trend in the amount or intensity of tropical storm activity. The trend in the number of storms making landfall in the United States has been relatively flat since the 1850s. Before the active 2017 hurricane season in the United States, there was a lull in the number of major hurricane landfalls that lasted nearly 12 years, the longest such drought in the United States since the 1860s (Landsea, 2018).

Hurricane activity varies year to year and over multidecadal periods. Activity is affected by numerous factors, including ocean cycles and the El Niño and La Niña oscillations. Data show multidecadal cycles in the Atlantic and Pacific Oceans favor some basins over others. An evaluation of the Accumulated Cyclone Energy (ACE) Index – which takes into account the number, duration, and strength of all tropical storms in a season – shows over the past 45 years there has been variability but no trend in tropical storms, both in the Northern Hemisphere and globally. See Figure 7.2.2.1, which is the same as Figure 2.7.5.1 in Chapter 2

Khandekar and Idso summarized their extensive survey of the literature in 2013 as follows:

> Air temperature variability decreases as mean air temperature rises, on all time scales. Therefore the claim that global warming will lead to more extremes of climate and weather, including of temperature itself, seems theoretically unsound; the claim is also unsupported by empirical evidence. Although specific regions have experienced significant changes in the intensity or number of extreme events over the twentieth century, for the globe as a whole no relationship exists between such events and global warming over the past 100 years.

> Observations from across the planet demonstrate droughts have not become more extreme or erratic in response to global warming. In most cases, the worst droughts in recorded meteorological history were much milder than droughts that occurred periodically during much colder times. There is little or no evidence that precipitation will become more variable and intense in a

Figure 7.2.2.1
Cyclonic energy, globally and northern hemisphere, from 1970 through October 2018

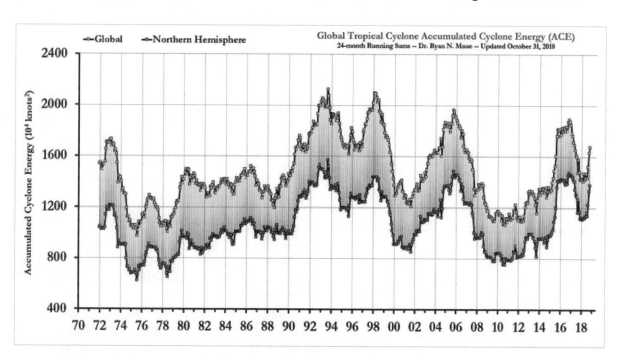

Last 4 decades of Global and Northern Hemisphere Accumulated Cyclone Energy: 24 month running sums. Note that the year indicated represents the value of ACE through the previous 24-months for the Northern Hemisphere (bottom line/gray boxes) and the entire global (top line/blue boxes). The area in between represents the Southern Hemisphere total ACE. *Source:* Maue, 2018.

warming world; indeed, some observations show just the opposite. There has been no significant increase in either the frequency or intensity of stormy weather in the modern era.

Despite the supposedly "unprecedented" warming of the twentieth century, there has been no increase in the intensity or frequency of tropical cyclones globally or in any of the specific ocean basins (Khandekar and Idso, 2013, pp. 809–810).

Khandekar and Idso conclude, "It is clear in almost every instance of each extreme weather event examined, there is little support for predictions that CO_2-induced global warming will increase either the frequency or intensity of those events. The real-world data overwhelmingly support an opposite conclusion: Weather will more likely be less extreme in a warmer world (*Ibid.,* p. 810).

References

Alexander, L.V., *et al.* 2006. Global observed changes in daily climate extremes of temperature and precipitation. *Journal of Geophysical Research* **111** (D5): 10.1029/2005JD006290.

Cavallo, E., Galiani, E., Noy, I., and Pantano, J. 2013. Catastrophic natural disasters and economic growth. *Review of Economics and Statistics* **95** (5): 1549–61.

IPCC. 2012. Intergovernmental Panel on Climate Change. *Managing the Risks of Extreme Events and Disasters to Advance Climate Change Adaptation. A Special Report of Working Groups I and II of the Intergovernmental Panel on Climate Change.* Cambridge University Press, Cambridge, UK, and New York, NY.

IPCC. 2013. Intergovernmental Panel on Climate Change. *Summary for Policymakers.* In: *Climate Change 2013: the Physical Science Basis.* Contribution of Working Group I to the Fifth Assessment Report of the Intergovernmental Panel on Climate Change. Cambridge, UK: Cambridge University Press.

IPCC. 2014. Intergovernmental Panel on Climate Change. *Climate Change 2014: Impacts, Adaptation, and Vulnerability. Part A: Global and Sectoral Aspects*. Contribution of Working Group II to the Fifth Assessment Report of the Intergovernmental Panel on Climate Change. New York, NY: Cambridge University Press.

Khandekar, M.L. 2013. Are extreme weather events on the rise? *Energy & Environment* **24**: 537–49.

Khandekar, M. and Idso, C. 2013. Chapter 7. Observations: extreme weather. In: Idso, C.D., Carter, R.M., and Singer, S.F. (Eds.) *Climate Change Reconsidered II: Physical Science*. Nongovernmental International Panel on Climate Change. Chicago, IL: The Heartland Institute.

Landsea, C. 2018. What is the complete list of continental U.S. landfalling hurricanes? (website). Hurricane Research Division, Atlantic Oceanographic and Meteorological Laboratory, National Oceanic and Atmospheric Administration. Accessed June 29, 2018.

Maue, R.N. 2018. Global Tropical Cyclone Activity (website). Accessed November 16, 2018.

Maue, R.N. 2011. Recent historically low global tropical cyclone activity. *Geophysical Research Letters* **38**: L14803.

Pielke Jr., R.A. 2013. Testimony before the Committee on Environment and Public Works of the U.S. Senate. July 18.

Pielke Jr., R.A. 2014. *The Rightful Place of Science: Disasters and Climate Change*. Tempe, AZ: Arizona State University Consortium for Science, Policy & Outcomes.

Raddatz, C. 2009. The wrath of god – macroeconomic costs of natural disasters. *Policy Research Working Paper* 5039. The Word Bank Development Research Group, Macroeconomics and Growth Team. Washington, DC: The World Bank.

Skidmore, M. and Toya, H. 2002. Do natural disasters promote long-term growth? *Economic Inquiry* **40** (4): 664–87.

7.2.3 Sea-level Rise

Little real-world evidence supports the claim that global sea level is currently affected by atmospheric CO_2 concentrations, and there is little reason to believe future impacts would be distinguishable from local changes in sea level due to non-climate related factors.

The IPCC claims, in the *Summary for Policymakers* for the Working Group I contribution to AR5, that "Under all RCP scenarios, the rate of sea level rise will *very likely* exceed that observed during 1971 to 2010 due to increased ocean warming and increased loss of mass from glaciers and ice sheets" (IPCC, 2013, p. 25). Most IAMs duly incorporate estimates of damages due to flooding and "climate refugees" forced to retreat from shorelines.

IPCC and IAM modelers can find support for their forecasts in studies relying on computer models and manipulation of recent satellite data purporting to be able to measure small changes in global sea level and to be sufficiently comparable to observational data from tidal gauges to justify being grafted onto past trends despite the different methodologies (e.g., Nerem *et al.,* 2018). But many experts observe the new model-derived estimates do not agree with tidal gauges located in geologically stable areas of the world and conclude any recent warming trend they claim to reveal is an artifact of the change in methodologies.

Tidal gauges continue to show local and regional sea levels exhibit typical natural variability – in some places rising and in others falling. Parker and Ollier reported in 2017,

> Sea levels are oscillating, with well-known inter-annual, decadal and multi-decadal oscillations well evidenced in the measurements collected by tidal gauges. There are oscillations of synchronous and non-synchronous phases moving from one location to another. Furthermore, it is well known that local sea-level changes occur also because of local factors such as subsidence due to groundwater or oil extraction, or tectonic movements that may be either up or down. *Relative sea-level changes due to subsidence or uplift are sometimes far larger than the global average sea-level changes* (Parker and Ollier, 2017, italics added).

Parker and Ollier report "the loud divergence between sea-level reality and climate change theory – the climate models predict an accelerated sea-level rise driven by the anthropogenic CO_2 emission – has been also evidenced in other works such as Boretti (2012a, b), Boretti and Watson (2012), Douglas (1992), Douglas and Peltier (2002), Fasullo *et al.* (2016), Jevrejeva *et al.* (2006), Holgate (2007), Houston and Dean (2011), Mörner (2010a, b, 2016), Mörner and Parker (2013), Scafetta (2014), Wenzel and Schröter (2010) and Wunsch *et al.* (2007) reporting on the recent lack of any detectable acceleration in the rate of sea-level rise" (*Ibid.*) To which we would add Wöppelmann *et al.* (2009) and Frederikse *et al.* (2018).

If unusual sea-level rise were occurring, it has not forced significant numbers of people to migrate. Andrew Baldwin *et al.*, writing in 2014, observed:

> The origins of climate change-induced migration discourse go back to the 1980s, when concerned scientists and environmental activists argued that unchecked environmental and climate change could lead to mass displacement (Mathews 1989; Myers 1989). However, at that time, hardly any actual climate or environmental refugees could be detected. *Even today, almost three decades later, the term as such remains merely a theoretical possibility but not an actually existing, clearly defined group of people* (Baldwin *et al.*, 2014, p. 121, italics added).

Similarly, the British government's *Foresight Report on Migration and Global Environmental Change,* widely regarded as the most authoritative study of the issue of environmental migration, found "the range and complexity of the interactions between these drivers [of migration] means that it will rarely be possible to distinguish individuals for whom environmental factors are the sole driver" (Foresight, 2011, p. 9) and "Environmental change is equally likely to make migration less possible as more probable. This is because migration is expensive and requires forms of capital, yet populations who experience the impacts of environmental change may see a reduction in the very capital required to enable a move" (*Ibid.*). In other words, there may be *no net increase* in the number of environmental refugees.

The best available data show dynamic variations in Pacific sea level in accord with El Niño-La Niña cycles, superimposed on a natural long-term eustatic rise (Australian Bureau of Meteorology, 2011). Island coastal flooding results not from sea-level rise, but from spring tides or storm surges in combination with development pressures such as borrow pit digging or groundwater withdrawal. Persons emigrating from the islands are doing so for social and economic reasons rather than in response to environmental threat.

Another claim concerning the effect of climate change on oceans is that increases in freshwater runoff into the oceans will disrupt the global thermohaline circulation system. But the range of natural fluctuation in the global ocean circulation system has yet to be fully delineated (Srokosz *et al.*, 2012). Research to date shows no evidence of changes that lie outside previous natural variability, nor of any malign influence from increases in human-related CO_2 emissions. Singer summarized the state of current knowledge on sea-level rise in 2018,

> Currently, sea-level rise does not seem to depend on ocean temperature, and certainly not on CO_2. We can expect the sea to continue rising at about the present rate for the foreseeable future. By 2100 the seas will rise another 6 inches or so – a far cry from Al Gore's alarming numbers. There is nothing we can do about rising sea levels in the meantime. We'd better build dikes and sea walls a little bit higher (Singer, 2018).

See the data on sea-level rise graphed in Chapter 2, Section 2.1 for further evidence that sea-level rise is unrelated to CO_2 levels, and Figure 2.7.3.1 in that same chapter for more findings about climate change and oceans from Chapter 6 of *Climate Change Reconsidered II: Physical Science*. The myth of "climate refugees" is addressed again later in the chapter as part of the discussion of whether climate change causes violent conflicts.

In conclusion, there is too little scientific evidence to support the contention that changes in global sea level are being affected by CO_2 concentrations in the atmosphere. Further, there is little scientific effort to support the contention that any future impact would be distinguishable from local changes in sea level due to groundwater or oil extraction, tectonic movements, sedimentation, and other non-climate related factors.

References

Australian Bureau of Meteorology. 2011. The South Pacific sea-level and climate monitoring program. Sea-level summary data report. July 2010–June 2011.

Baldwin, A., Methmann, C., and Rothe, D. 2014. Securitizing 'climate refugees': the futurology of climate-induced migration. *Critical Studies on Security* 2 (2): 121–30.

Boretti, A. 2012a. Short term comparison of climate model predictions and satellite altimeter measurements of sea levels. *Coastal Engineering* **60**: 319–22.

Boretti, A. 2012b. Is there any support in the long term tide gauge data to the claims that parts of Sydney will be swamped by rising sea levels? *Coastal Engineering* **64**: 161–7.

Boretti, A. and Watson, T. 2012. The inconvenient truth: ocean level not rising in Australia. *Energy & Environment* **23** (5): 801–17.

Douglas, B.C. 1992. Global sea level acceleration. *Journal of Geophysical Research Oceans* **97** (C8): 12699–706.

Douglas, B.C. and Peltier, W.R. 2002. The puzzle of global sea-level rise. *Physics Today* **55** (3): 35–41.

Fasullo, J.T., Nerem, R.S., and Hamlington, B. 2016. Is the detection of accelerated sea level rise imminent? *Scientific Reports* **6**: 31245

Foresight. 2011. *Migration and Global Environmental Change.* London, UK: The Government Office for Science.

Frederikse, T., Jevrejeva, S., Riva, R.E.M., and Dangendorf, S. 2018. A consistent sea-level reconstruction and its budget on basin and global scales over 1958–2014. *Journal of Climate* **31** (3): 1267–80.

Holgate, S.J. 2007. On the decadal rates of sea level change during the twentieth century. *Geophysical Research Letters* **34** (1).

Houston, J.R. and Dean, R.G. 2011. Sea-level acceleration based on U.S. tide gauges and extensions of previous global-gauge analyses. *Journal of Coastal Research* **27**: 409–17.

IPCC. 2013. Intergovernmental Panel on Climate Change. *Climate Change 2013: The Physical Science Basis.* Contribution of Working Group I to the Fifth Assessment Report of the Intergovernmental Panel on Climate Change. Cambridge, United Kingdom: Cambridge University Press.

Jevrejeva, S., Grinsted, A., Moore, J.C., and Holgate, S. 2006. Nonlinear trends and multiyear cycles in sea level records. *Journal of Geophysical Research Oceans* **111** (C9).

Mathews, J.T. 1989. Redefining security. *Foreign Affairs* **68**: 162–77.

Mörner, N-A. 2010a. Sea level changes in Bangladesh: new observational facts. *Energy & Environment* **21** (3): 235–49.

Mörner, N-A. 2010b. Some problems in the reconstruction of mean sea level and its changes with time. *Quaternary International* **221** (1–2): 3–8.

Mörner, N-A. 2016. Rates of sea level changes—a clarifying note. *International Journal of Geoscience* **7** (11): 1318.

Mörner, N-A. and Parker, A. 2013. Present-to-future sea level changes: the Australian case. *Environmental Science: An Indian Journal* **8** (2): 43–51.

Myers, N. 1989. Environment and security. *Foreign Policy* **74**: 23–41.

Nerem, R.S., Beckley, B.D, Fasullo, J.T., Hamblington, B.D., Masters, D., and Mitchum, G.T. 2018. Climate-change–driven accelerated sea-level rise detected in the altimeter era. *Proceedings of the National Academy of Sciences,* pre-print publication online, February 12.

Parker, A. and Ollier, C.D. 2017. Short-term tide gauge records from one location are inadequate to infer global sea-level acceleration. *Earth Systems and Environment* (December): 1–17.

Scafetta, N. 2014. Multi-scale dynamical analysis (MSDA) of sea level records versus PDO, AMO, and NAO indexes. *Climate Dynamics* **43** (1–2): 175–92.

Singer. S.F. 2018. The sea is rising, but not because of climate change. *The Wall Street Journal.* May 15.

Srokosz, M., Baringer, M., Bryden, H., Cunningham, S., Delworth, T., Lozier, S., Marotzke, J., and Sutton, R. 2012. Past, present, and future changes in the Atlantic Meridional Overturning Circulation. *Bulletin of the American Meteorological Society* **93**: 1663–76.

Wenzel, M. and Schröter, J. 2010. Reconstruction of regional mean sea level anomalies from tide gauges using neural networks. *Journal of Geophysical Research: Oceans* **115**: C08013.

Wöppelmann, G., Letetrel, C., Santamaria, A., Bouin, M.-N., Collilieux, X., Altamimi, Z., Williams, S.D.P., and Miguez, B.M. 2009. Rates of sea-level change over the past century in a geocentric reference frame. *Geophysical Research Letters* **36**.

Wunsch, C., Ponte, R.M., and Heimbach, P. 2007 Decadal trends in sea level patterns: 1993–2004. *Journal of Climate* **20** (24): 5889–911.

7.2.4 Agriculture

Alleged threats to agriculture and food security are contradicted by biological science and empirical data regarding crop yields and human hunger.

Another alleged threat to human security is harms to agriculture and food security caused by extreme heat and drought. According to the IPCC, "illustrative examples of observed impacts due to aggravating climate stresses" on agriculture and food security can be found in Kenya, Southern Africa, Zambia, and Zimbabwe. Illustrative examples the IPCC says are a consequence of climate change also come from Africa and one reference to "small-scale rain-fed maize farmers in Mexico" (IPCC, 2014a, p. 761). In the *Summary for Policymakers* for the WGII contribution to AR5, the IPCC says:

> For the major crops (wheat, rice, and maize) in tropical and temperate regions, climate change without adaptation is projected to negatively impact production for local temperature increases of 2°C or more above late-20th-century levels, although individual locations may benefit (*medium confidence*). Projected impacts vary across crops and regions and adaptation scenarios, with about 10% of projections for the period 2030–2049 showing yield gains of more than 10%, and about 10% of projections showing yield losses of more than 25%, compared to the late 20th century. (IPCC, 2014b, pp. 17–18, italics in original).

There is much to question here. The examples cited do not support a broad projection onto world food production. The forecast focuses oddly on two ends of a probability distribution, which implies that 80% of studies find little or no impact of climate change on agriculture. Why are not they more likely to be true? The assumption that there would be no adaptation is plainly wrong. The human capability to produce food in the face of climate change has been on display since at least the last Ice Age. At that time, a nomadic people we call the Grevettians used mammoth-skin tents instead of living in caves. That permitted them to pursue the mammoths and other game animals that had to migrate because their grass had turned to less-nourishing tundra.

The Grevettians also used atlatls (spear-throwers) to kill mammoths from a safe distance. Perhaps most importantly, they tamed wolves and bred them into dogs, to help find game on the trackless steppes. The dogs also protected their communities where campfires were inadequate. Language evolved into writing, writing evolved into printing and libraries and then into today's research laboratories and digital communications. All this has allowed humans to learn collectively and thus evolve better survival strategies than our forebears could have imagined. There is no reason to expect the collective learning that has given us books, libraries, computers, and space travel would somehow fail to meet humanity's most basic need – adequate food production techniques – in the years ahead.

The application of technology to agriculture makes adaptation far easier and faster than it has ever been before (Waggoner, 1995; Goklany, 2009). During the twentieth and early twenty-first centuries, when the IPCC claims the world's temperatures rose at an "unprecedented" pace, increases in agricultural output rose even faster. Despite global population growth, "the number of hungry people in the world has dropped to 795 million – 216 million fewer than in 1990–92 – or around one person out of every nine" (FAO, 2015). In developing countries, under-nourishment (having insufficient food to live an active and healthy life) fell from 23.3% 25 years earlier to 12.9%. A majority of the 129 countries monitored by FAO reduced under-nourishment by half or more since 1996 (*Ibid.*). This is not evidence of a negative effect of climate change on food security in the world today, but evidence of just the opposite.

Extensive evidence reviewed in Chapters 3, 4, and 5 showed rising ambient CO_2 concentrations and higher temperatures *benefit* and do not harm food crops and nearly all other plant life on Earth, and why shouldn't they? Most plants on Earth today evolved during times when research shows the planet was much warmer and CO_2 levels were much higher than they are today.

The IPCC admits "food security is determined by a range of interacting factors including poverty, water availability, food policy agreements and regulations, and the demand for productive land for alternative uses (Barrett, 2010, 2013)." Blurring the issue of causation, the IPCC says "many of these factors are themselves *sensitive* to climate variability and climate change" (IPCC, 2014a, p. 763, italics added). The IPCC identifies incidents where "food price spikes have been associated with food riots," but then cites literature attributing those riots to other factors. It says "there are complex pathways between climate, food production, and human security and hence this area requires further concentrated research as an area of concern" (*Ibid.*). Why, then, does IPCC say in Figure 7.2.1.1 that climate change "may adversely affect agriculture production and exacerbate the problem of food insecurity"?

References

Barrett, C.B. 2010. Measuring food security. *Science* **327**: 825–8.

Barrett, C.B. 2013. Food or consequences: food security and its implications for global sociopolitical stability. In: Barrett, C.B. (Ed.) *Food Security and Sociopolitical Stability*. Oxford, UK: Oxford University Press, pp. 1–34.

FAO. 2015. Food and Agriculture Organization of the United Nations. *The State of Food Insecurity in the World 2015*. Rome, Italy.

Goklany, I.M. 2009. Discounting the future. *Regulation* **32** (1): 37–40.

IPCC. 2014a. Intergovernmental Panel on Climate Change. *Climate Change 2014: Impacts, Adaptation, and Vulnerability. Part A: Global and Sectoral Aspects*. Contribution of Working Group II to the Fifth Assessment Report of the Intergovernmental Panel on Climate Change. Cambridge, UK and New York, NY: Cambridge University Press.

IPCC. 2014b. Intergovernmental Panel on Climate Change. *Summary for Policymakers*. In: *Climate Change 2014: Impacts, Adaptation, and Vulnerability*. Contribution of Working Group II to the Fifth Assessment Report of the Intergovernmental Panel on Climate Change. New York, NY: Cambridge University Press.

Waggoner, P.E. 1995. How much land can ten billion people spare for nature? Does technology make a difference? *Technology in Society* **17**: 17–34.

7.2.5 Human Capital

Alleged threats to human capital – human health, education, and longevity – are almost entirely speculative and undocumented. There is no evidence climate change has eroded or will erode livelihoods or human progress.

The final "dimension of impact" described by the IPCC in its Table 12-1, reprinted as Figure 7.2.1.1

above, is "human capital (health, education, loss of lives)." As "illustrative examples of observed impacts due to aggravating climate stresses" it includes examples that duplicate those offered in its description of "deprivation of basic needs," such as "food shortage, absence of safe and reliable access to clean water and good sanitary conditions, and destruction of shelters and displacements" (IPCC, 2014, p. 761). Examples specifically attributed to climate change are computer projections of falling food productivity and increased malaria infection and fatalities due to floods.

The IPCC's labeling of these possible effects as threats to "human capital" is curious at best and likely misleading. Human capital is more typically and usefully defined as "intangible collective resources possessed by individuals and groups within a given population. These resources include all the knowledge, talents, skills, abilities, experience, intelligence, training, judgment, and wisdom possessed individually and collectively, the cumulative total of which represents a form of wealth available to nations and organizations to accomplish their goals" (Encyclopedia Britannica, n.d.). In economics, the term has come to refer more narrowly to the knowledge, skills, health, and values people possess that enable them to be productive, produce earnings, and live a comfortable life. Becker (n.d.) wrote,

> Schooling, a computer training course, expenditures on medical care, and lectures on the virtues of punctuality and honesty are also capital. That is because they raise earnings, improve health, or add to a person's good habits over much of his lifetime. Therefore, economists regard expenditures on education, training, medical care, and so on as investments in human capital. They are called human capital because people cannot be separated from their knowledge, skills, health, or values in the way they can be separated from their financial and physical assets.

Does climate change threaten "human capital" as *Encyclopedia Britannica* or Becker defines it? The case, as has been shown to be true with every other "dimension of impact" in the IPCC's list, seems tenuous. Climate change might cause extreme weather events or flooding, although this assertion is not supported by the climate science and data presented in Chapter 2. Such events might interrupt people's educations or training or their ability to pass knowledge and skills on to others, but only if one assumes no adaptation, no response by civil and political institutions, and no long-term recovery. But this only rarely happens. More often the effects of even natural catastrophes are short-term, and over time they severely affect shrinking numbers of people thanks to the mobility, technologies, and resiliency made possible by fossil fuels.

Available evidence on crop yields and hunger in the world shows rising productivity and a trend that is likely to continue, boosted rather than hurt by rising temperatures and carbon dioxide levels in the atmosphere (Waggoner, 1995; Epstein, 2014). Fear that warmer temperatures will lead to the spread of malaria and other diseases is entirely speculative and contradicted by extensive real-world research, much of it summarized in Chapter 4. To date, global warming's main effects appear to be *increasing* food supplies and food security and a greening of Earth that is much more beneficial than harmful (Zhu *et al.,* 2016). Violent weather has become less common, not more common, as the world has warmed. Each of these points was made and documented in previous chapters.

As also was demonstrated in previous chapters, the fossil fuels the IPCC holds responsible for some part of global warming in the late twentieth and early twenty-first centuries were clearly a boon to human capital. They provided the prosperity that made possible huge investments in schooling, health care, and technologies that in turn boosted human productivity. They helped protect human capital from nature by providing technologies that made it possible to survive hot or cold weather and periods of heavy rain or drought, and even to escape the paths of floods or hurricanes (Goklany, 2002, 2012). This positive trend since the beginning of the fossil fuel era has overwhelmed any negative effects that might be attributed to a slight and gradual rise in average global surface temperatures.

Human capital is the solution to whatever problems climate change might present to humanity (Simon, 1996). The IPCC's claim that climate change threatens human capital is almost entirely speculative and undocumented. There is no evidence global warming has eroded or will erode livelihoods or human progress.

References

Becker, G. n.d. Human capital. Econlib (website). Accessed June 24, 2018.

Encyclopedia Britannica. n.d. Human capital (website). Accessed June 24, 2018.

Epstein, A. 2014. *The Moral Case for Fossil Fuels.* New York, NY: Portfolio/Penguin.

Goklany, I.M. 2002. The globalization of human well-being. *Cato Policy Analysis* No. 447. Washington, DC: Cato Institute. August 22.

Goklany, I.M. 2012. Humanity unbound: how fossil fuels saved humanity from nature and nature from humanity. *Cato Policy Analysis* No. 715. Washington, DC: Cato Institute. December 20.

IPCC. 2014. Intergovernmental Panel on Climate Change. *Climate Change 2014: Impacts, Adaptation, and Vulnerability. Part A: Global and Sectoral Aspects.* Contribution of Working Group II to the Fifth Assessment Report of the Intergovernmental Panel on Climate Change. New York, NY: Cambridge University Press.

Simon, J. 1996. *The Ultimate Resource 2.* Princeton, NJ: Princeton University Press.

Waggoner, P.E. 1995. How much land can ten billion people spare for nature? Does technology make a difference? *Technology in Society* 17: 17–34.

Zhu, Z., *et al.* 2016. Greening of the Earth and its drivers. *Nature Climate Change* 6: 791–5.

7.3 Violent Conflict

According to the IPCC, "Climate change has the potential to increase rivalry between countries over shared resources. For example, there is concern about rivalry over changing access to the resources in the Arctic and in transboundary river basins. Climate changes represent a challenge to the effectiveness of the diverse institutions that already exist to manage relations over these resources. However, *there is high scientific agreement that this increased rivalry is unlikely to lead directly to warfare between states*" (IPCC, 2014a, p. 772, italics added).

The IPCC reviews the literature on "the relationship between short-term warming and armed conflict" and concludes: "Some of these find a weak relationship, some find no relationship, and *collectively the research does not conclude that there is a strong positive relationship between warming and armed conflict*" (*Ibid.*, italics added).

As is typical of the IPCC *Summaries for Policymakers*, the uncertainty made so clear in the full report is dropped from the much more widely read summary: "Climate change can indirectly increase risks of violent conflicts in the form of civil war and inter-group violence by amplifying well-documented drivers of these conflicts such as poverty and economic shocks (*medium confidence*). Multiple lines of evidence relate climate variability to these forms of conflict" (IPCC, 2014b, p. 20). This is certainly the message politicians and the media took from the Fifth Assessment Report.

In 2015, U.S. President Barack Obama issued an executive statement echoing those claims, but with much more than "medium confidence." According to Obama, "A changing climate will act as an accelerant of instability around the world, exacerbating tensions related to water scarcity and food shortages, natural resource competition, underdevelopment, and overpopulation" (Executive Office of the President, 2015, p. 8). These effects, he said, "are threat multipliers that will aggravate stressors abroad such as poverty, environmental degradation, political instability, and social tensions – conditions that enable terrorist activity and other forms of violence. The risk of conflict may increase" (*Ibid.*).

Reliance by the U.S. government on the IPCC for the "scientific consensus" on climate change reached its apex during the Obama administration, but it predated Obama's election. Dr. Thomas Fingar, deputy director of National Intelligence for Analysis and chairman of the National Intelligence Council, testified to Congress in 2008 that "our primary source for climate science was the United Nations Intergovernmental Panel on Climate Change (IPCC) Fourth Assessment Report" and "we relied predominantly upon a mid-range projection from among a range of authoritative scenario trajectories provided by the IPCC. ... In the study, we assume that the climate will change as forecast by the IPCC" (Fingar, 2008, pp. 2–3). Apparently no one at the IPCC told Fingar the IPCC does not issue "forecasts," only scenarios.

Environmental groups endorse and promote the climate-conflict hypothesis without reviewing the data in part because their leaders believe it is an argument that appeals to conservatives and Republicans in the United States (Ungar, 2007; Baldwin *et al.*, 2014). The motivation of members of the defense and intelligence communities and some retired senior military officials is different. They see

in climate change a justification for investments in new military equipment and force planning. Like economists who say they support "market-based solutions to climate change" yet know little about climate science, these military experts accept the findings of the IPCC without critical review and then limit their own contributions to the debate to planning efficient responses to scenarios derived from the IPCC's computer models, misunderstood to be forecasts or predictions. By doing so, they create the appearance of validating or endorsing the IPCC's exaggerated and implausible claims.

A robust set of studies has emerged in recent years examining the climate-conflict hypothesis. These studies cast much doubt on the central links of the argument and, in turn, undermine support for the notion that a warming planet will give rise to future conflict. Section 7.3.1 summarizes some of the scholarly research on the association between climate and armed conflict. (A much larger literature review appears later, in Section 7.4, where the historical relationship between climate and conflict is reported.) Section 7.3.2 addresses methodological problems with the climate-conflict theory, helping to explain why the hypothesis fails in the real world. Section 7.3.3 reviews evidence on five specific alleged sources of conflict: abrupt climate change, water, famine, resource scarcity, and refugee flows.

References

Baldwin, A., Methmann, C., and Rothe, D. 2014. Securitizing 'climate refugees': the futurology of climate-induced migration. *Critical Studies on Security* **2** (2): 121–30.

Executive Office of the President. 2015. *The National Security Implications of a Changing Climate*. Washington, DC: White House.

Fingar, D.T. 2008. National intelligence assessment on the national security implications of global climate change to 2030. Testimony before the House Permanent Select Committee on Intelligence and the House Select Committee on Energy Independence and Global Warming. Washington, DC.

IPCC. 2014a. Intergovernmental Panel on Climate Change. *Climate Change 2014: Impacts, Adaptation, and Vulnerability. Part A: Global and Sectoral Aspects*. Contribution of Working Group II to the Fifth Assessment Report of the Intergovernmental Panel on Climate Change. New York, NY: Cambridge University Press.

IPCC.2014b. Intergovernmental Panel on Climate Change. *Summary for Policymakers.* In: *Climate Change 2014: Impacts, Adaptation, and Vulnerability. Part A: Global and Sectoral Aspects.* Contribution of Working Group II to the Fifth Assessment Report of the Intergovernmental Panel on Climate Change. New York, NY: Cambridge University Press.

Ungar, S. 2007. Public scares: changing the issue culture. Chapter 4 in: Moser, S.C. and Dilling, L. (Eds.) *Creating a Climate for Change: Communicating Climate Change and Facilitating Social Change.* Cambridge, MA: Cambridge University Press, pp. 81–8.

7.3.1 Empirical Research

Empirical research shows no direct association between climate change and violent conflicts.

There is no empirical evidence that natural disasters have tended to lead directly to violent conflict in the years since the end of the Little Ice Age. But then the weather has been wonderfully supportive of humans, no matter how we decry our comparatively feeble storms, floods, and droughts. In addition, food productivity has soared through technology. The outstanding example was Dr. Norman Borlaug's Agricultural Green Revolution, which tripled most of the world's crop yields with disease-resistant seed varieties, modern pesticides, and chemical fertilizers. Borlaug's own father had dealt with Norman's departure for college by buying an early model of a gasoline tractor with his brother. The Borlaugs' tractor quadrupled the farm's productivity, in no small part because no land was needed any longer for horse feed.

Hunger-driven conflicts had been characteristic of "little ice ages" from the dawn of time until the Colombian Exchange of the fifteenth and sixteenth centuries, but no longer. The modern world relies on a vastly successful pattern of research and engineering to support history's most effective food production system. The modern world also typically offers food aid (and the vital transportation to carry it) to nations stricken by droughts, floods, and other natural impacts.

Gleditsch and Nordås observed, "none of the studies on climate and conflict, with the possible exception of literature on heat and individual aggression, assume that climate has a direct influence on violence. The assumption, usually if not always made explicit, is that climate change (be it increasing heat or changes in precipitation) influences other factors, which in turn lead to conflict" (Gleditsch and Nordås, 2014, p. 85). The attribution of violent conflict to global warming does not rest on empirical data, but is a hypothesis (see Hsiang and Burke, 2014), and as the following sections will show, a very complicated and unlikely one at that.

The research summarized in this section consists of only a few recent studies in the literature often referred to as "peace studies." A much larger literature exists, primarily found in academic history journals, concerning the historical association between climate and conflict reaching back centuries and including findings from nearly every country in the world. That literature appears to be largely unknown to the climate science community, and in particular the IPCC. Since that literature is so voluminous, it is reviewed in its own section, Section 7.4, below.

Raleigh and Kniveton (2012) observed "the climate-conflict literature suffers from a lack of theoretical connections between its main driver (climate) and its possible consequence (conflict)." Concluding an extensive review of the literature, Theisen *et al.* (2013) similarly found, "Taken together, extant studies provide mostly inconclusive insights, with contradictory or weak demonstrated effects of climate variability and change on armed conflict" (Theisen *et al.*, 2013).

Like Homer-Dixon (1999) and Nel and Righarts (2008) before him, Slettebak (2012) focused primarily on how natural disasters might cause the breakdown of social structures or scarcity of important resources. His analysis addressed the environmental impacts frequently alleged to be associated with rising temperatures, including storms, droughts, floods, landslides, wildfires, and extreme temperatures. He tested six models incorporating a host of socioeconomic and environmental variables, concluding:

> I set out to test whether natural disasters can add explanatory power to an established model of civil conflict. The results indicate that they can, but that *their effect on conflict is the opposite of popular perception*. To the extent that climate-related natural disasters affect the risk of conflict, they contribute to *reducing* it. This holds for measures of climate-related natural disasters in general as well as drought in particular (p. 174, italics added).

629

Another approach hypothesizes that climate change-driven natural disasters will slow economic growth in the affected area, increasing the likelihood of social unrest. Bergholt and Lujala (2012) tested that possibility for the period 1980–2007, developing a dataset covering 171 countries with a total of more than 4,000 country-year observations. Finding natural disasters do in fact slow economic growth, they nevertheless conclude "climate-related natural disasters do not have any direct effect on conflict onset," nor did "economic shocks caused by climate-related disasters have an effect on conflict onset" (Bergholt and Lujala, 2012, p. 148).

Similarly, Koubi *et al.* (2012) tested how deviations in precipitation and temperature trends from their long-run averages relate to economic growth and civil conflict. For the period 1980–2004, they conclude, "climate variability … does not affect violent intrastate conflict through economic growth" (Koubi *et al.*, 2012).

The IPCC, as noted earlier, has been cautious in declaring a direct causal relationship between climate change and armed conflict. In a special report released in 2012 titled "Managing the Risks of Extreme Events and Disasters to Advance Climate Change Adaptation," the IPCC admitted great uncertainty over forecasts of more extreme weather events as a result of climate change. It notes,

> Confidence in projecting changes in the direction and magnitude of climate extremes depends on many factors, including the type of extreme, the region and season, the amount and quality of observational data, the level of understanding of the underlying processes, and the reliability of their simulation in models. Projected changes in climate extremes under different emissions scenarios generally do not strongly diverge in the coming two to three decades, but these signals are relatively small compared to natural climate variability over this time frame. Even the sign of projected changes in some climate extremes over this time frame is uncertain. For projected changes by the end of the 21st century, either model uncertainty or uncertainties associated with emissions scenarios used becomes dominant, depending on the extreme (IPCC, 2012, p. 11).

The statement is significant for its admission that natural forces will exert dominant influence over "climate extremes" over the period of 10 to 20 years and that, in some instances, the models are unable to state whether the purported human impact is positive or negative. The IPCC also expresses caution about the climate-conflict link in AR5 Chapter 18, on "Detection and attribution of observed impacts," saying "the detection of the effect of climate change [on warfare] and an assessment of its importance can be made only with *low confidence*. There is no evidence of a climate change effect on interstate conflict in the post-World War II period. … [N]either the detection of an effect of climate change on civil conflict nor an assessment of the magnitude of such an effect can currently be made with a degree of confidence" (IPCC, 2014, p. 1001).

Also in 2014, in the introduction to a 2014 special issue of *Political Geography* devoted to climate and conflict, Idean Salehyan, a professor in the department of political science at the University of North Texas, wrote,

> The relationship between climate, climate change, and conflict has been empirically tested in a wide variety of studies, but the literature has yet to converge on a commonly accepted set of results. This is mainly due to poor conceptualization of research designs and empirical measurements. Data are often collected at different temporal, geographic, and social scales. In addition, "climate" and "conflict" are rather elusive concepts and scholars have utilized different measures of each. The choice of measures and empirical tests is not a trivial one, but reflects different theoretical frameworks for understanding environmental influences on conflict. Therefore, results from different analyses are often not commensurable with one another and readers should be wary of broad, sweeping characterizations of the literature (Salehyan, 2014, abstract).

Gleditsch and Nordås (2014) wrote, "there is no consensus in the scholarly community about such dire projections of future climate wars; in fact most observers conclude that there is no robust and consistent evidence for an important relationship between climate change and conflict (Bernauer, Bohmelt, & Koubi, 2012; Scheffran, Brzoska, Kominek, Link, & Schilling, 2012; Theisen, Gleditsch, & Buhaug, 2013)" (pp. 1–2).

In conclusion, empirical research shows no direct association between climate change and violent conflicts, and we should be surprised if it did. A warmer world is a safer and more prosperous world in which there is less cause for conflict.

References

Bergholt, D. and Lujala, P. 2012. Climate-related natural disasters, economic growth, and armed civil conflict. *Journal of Peace Research* **49** (1): 147–62.

Bernauer, T., Bohmelt, T., and Koubi, V. 2012. Environmental changes and violent conflict. *Environmental Research Letters* **7** (1): 1–8.

Gleditsch, N.P. and Nordås, R. 2014. Conflicting messages? The IPCC on conflict and human security. *Political Geography* **43**: 82–90.

Homer-Dixon, T. 1999. *Environment, Scarcity and Violence*. Princeton, NJ: Princeton University Press.

Hsiang, S. and Burke, M. 2014. Climate, conflict, and social stability: what does the evidence say? *Climatic Change* **123** (1): 39–55.

Hsiang, S., Burke, M., and Miguel, E. 2013. Quantifying the influence of climate on human conflict. *Science* **341** (6151): 123657-1–12.

IPCC. 2012. Intergovernmental Panel on Climate Change. *Managing the Risks of Extreme Events and Disasters to Advance Climate Change Adaptation. A Special Report of Working Groups I and II of the Intergovernmental Panel on Climate Change*. New York, NY: Cambridge University Press.

IPCC. 2014. Intergovernmental Panel on Climate Change. *Climate Change 2014: Impacts, Adaptation, and Vulnerability. Part A: Global and Sectoral Aspects*. Contribution of Working Group II to the Fifth Assessment Report of the Intergovernmental Panel on Climate Change. New York, NY: Cambridge University Press.

Koubi, V., Bernauer, T., Kalbhenn, A., and Spilker, G. 2012. Climate variability, economic growth, and civil conflict. *Journal of Peace Research* **49** (1): 113–27.

Nel, P. and Righarts, M. 2008. Natural disasters and the risk of violent armed conflict. *International Studies Quarterly* **52** (1): 159–85.

Raleigh, C. and Kniveton, D. 2012. Come rain or shine: an analysis of conflict and climate variability in East Africa. *Journal of Peace Research* **49** (1): 51–64.

Salehyan, I. 2014. Climate change and conflict: making sense of disparate findings. *Political Geography* **43**: 1–5.

Scheffran, J., Brzoska, M., Kominek, J., Link, P.M., and Schilling, J. 2012. Climate change and violent conflict. *Science* **336** (6083): 869–71.

Slettebak, R. 2012. Don't blame the weather! Climate-related natural disasters and civil conflict. *Journal of Peace Research* **49** (1): 163–76.

Theisen, O.M., Gleditsch, N.P., and Buhaug, H. 2013. Is climate change a driver of armed conflict? *Climatic Change* **117**: 613–25.

7.3.2. Methodological Problems

The climate-conflict hypothesis is a series of arguments linked together in a chain, so if any one of the links is disproven, the hypothesis is invalidated. The academic literature on the relationship between climate and social conflict reveals at least six methodological problems that affect efforts to connect the two.

Why do nearly all empirical studies invalidate the climate-conflict hypothesis? The climate-conflict hypothesis is driven by a number of unproven assumptions, many of which have been challenged in previous chapters of this volume and previous volumes of the *Climate Change Reconsidered* series. The hypothesis assumes not only that climate models are accurate on a global scale but also that these models can accurately move from global to regional scales. The hypothesis also assumes the accuracy of computer-model-generated scenarios projecting economic growth, demand for energy, and consumer behavior (among other factors) even though the flaws of such projections are well known.

Using President Barack Obama's language quoted at the beginning of Section 7.3 (Executive Office of the President, 2015), the hypothesis can be expressed like this:

Any changes in climate ("a changing climate") will result in changes to the weather, *all of them negative* (droughts, floods, hurricanes or storms, etc. etc.), which in turn will exacerbate *and never alleviate* "tensions" that already exist due to other causes (water scarcity and food shortages, underdevelopment, etc.), which in turn will always create *and never relieve* "social tensions" (poverty, environmental degradation, etc.), which in turn will "enable" *and never handicap* terrorists and other armed combatants, thereby increasing *and never reducing* the "risk of conflict."

How plausible is this hypothesis? On its face, not very. Consider only the text in italics and see how brittle the hypothesis is:

- Climate is always changing, it did so before and without the human presence, so there is no way to test the hypothesis by "stopping climate change" for, say, a few decades, and seeing what impact that might have on the frequency of violent conflicts.

- Some of the impacts of a warmer planet would clearly be good: expanded ranges for wildlife, forestry, and agriculture, longer growing seasons, lower winter heating bills, and fewer deaths due to cold weather. Climatology and the historical record also suggest there are *fewer* extreme weather events, not more, in a warmer world.

- More precipitation and a greening Earth, two well-documented trends occurring during the twentieth and early twenty-first centuries, result in more food production and more food security, not less, which likely alleviate social tensions arising from poverty and hunger.

- Civil wars are statistically most closely associated with low per-capita income and slow economic growth and are not related at all to average global surface temperature, so the effect of global warming on terrorists and other armed combatants must be ambiguous at best.

- The actual *rate* of conflict around the world has been falling, as reflected in the rapid decline in the number of deaths arising from armed conflicts around the world reported in Section 7.1.1. So the effect of climate change on the "risk of conflict" is either negative – meaning the world grows safer as temperatures rise – or too small to detect.

The climate-conflict hypothesis, like all the alleged threats to human security in Chapter 12 of the Fifth Assessment Report, is an argument linked together in a chain, so if any one of the links is disproven, the hypothesis is invalidated. For example, if it can be demonstrated that the human impact on climate is probably too small to measure, then the entire chain of reasoning ends with falsification of the first assumption. If a human impact on climate is found and thought to be statistically significant, then its negative impacts on food, water, housing, or other basic needs must be found to be so large as to not only cancel out its positive impacts but also to cause natural disasters that can "exacerbate social tensions." If the benefits of modest warming to human prosperity, health, and even to the environment previously documented in Part II outweigh the costs, then the chain of reasoning ends with that link.

If the small human impact on climate is nevertheless causing natural disasters, what evidence is there that these disasters lead to civil war or other forms of violence? As reported above and again below, there is no consistent association between natural disasters and war or civil conflicts, so the chain of reasoning ends again. How often might such conflicts, should they occur, rise to the level where they affect the security of other countries? If they are rare, this will probably not rank high on a list of priorities for more than a few undeveloped countries. It certainly would not justify placing climate change at the top of a list of priorities for the U.S. military, as called for by U.S. President Obama. Finally, to what extent do these new security threats require investments in new military equipment or changes to force planning? Would such changes even require a net increase in spending, rather than only small shifts in resources?

The academic literature on the relationship between climate and social conflict reveals at least six methodological problems that affect efforts to connect the two.

A. Untestable Models

The case studies used to construct the proofs typically rely on multiple independent variables acting through intervening variables, such as changing rainfall patterns creating droughts that reduce food supplies, leading to group manipulation of food supplies and social unrest. Many of the dependent variables used are imprecise as well, such as social unrest or health problems, meaning they defy measurement in a meaningful fashion. Without greater specificity in the dependent variable, tests for causal connections are imprecise.

B. Lack of a Control Group

The case study approach by its nature is anecdotal, and scholars must take care to construct their research designs in ways that enable variation of the factors under examination. A defense of biased case selections for environmental scenarios has been offered by Homer-Dixon (1999) and others, claiming environmental scenarios offer greater complexity than other sources of conflict. Not only is that untrue, but accepting that view requires the concession that environmental scenarios cannot be tested in a qualitative format with variable variation. Empirical work done subsequently reveals such tests are possible.

C. Reverse Causality

In many of the regions examined by the literature, ongoing conflicts have destroyed and damaged local environments resulting in lost food supplies and dislocated populations. In turn, that damage decreases a community's resiliency in the face of natural disasters, resulting in more damage caused by climate change. In the context of the climate-conflict debate, these ongoing conflicts cut against the explanatory power of climate change as the source of local environmental degradation and potential causation of local or regional tension or conflict.

D. Using the Future as Evidence

Much of the literature presents environmental variables as a cause of future, rather than past, conflicts. The environment may be a causal element in conflict, but reliance on the future is an appeal to argument, rather than evidence, as proof of the causal relationship. All the environmental variables cited in the climate-conflict literature are documentable and therefore testable against known instances of conflict. A review of that evidence should show a positive link between past floods, droughts, or other environmental degradation with intra- or interstate conflict when other explanatory variables are accounted for. If it does not, then the hypothesis is not proven and the conclusion that environmental conditions breed conflict is not supported.

E. Drawing Lessons from Foreign and Domestic Conflict

The resource wars literature draws lessons from interstate war, but most warfare in the post-World War II period is internal to states. Internal conflicts have very different characteristics and causes. Generalizing lessons from interstate to intrastate conflict is problematic, and the climate-conflict literature generally fails to reflect those lessons. As was documented in Section 7.1.1, empirical data show civil war is most strongly correlated with low income and slow economic growth, not with climate (Hegre and Sambanis, 2006).

F. Changing Levels of Analysis

The climate-conflict literature freely jumps between systems, nations, and individual levels of analysis when developing theories and examining empirical evidence. Hypotheses appropriate for one level of analysis may not follow to another or even be logically consistent with the other levels. In their study of the effects of changing rainfall patterns on rates of rebel and communal violence in Africa, Raleigh and Kniveton (2012) offer an illustration of how these concerns can manifest themselves and confound the resulting interpretations. As noted, in order for social disorder or conflict to emerge from an environmental cause, a number of intervening actions and reactions have to occur in sequence. Raleigh and Kniveton observed that alternative, and sometimes competing, hypotheses can emerge during careful consideration of those sequences. In their case, the key intervening variable between climate and conflict is rainfall pattern change. Raleigh and Kniveton offer four competing hypotheses to illustrate this point:

- Increased conflict is likely to follow periods of above-average decreases in rainfall as groups compete over a scarce resource;

- Decreases in conflict are likely to be correlated with decreased rainfall because there is little to fight for because the gains to be had from conflict do not justify the costs of conflict;

- Increases in political violence will follow periods of higher than average rainfall as agricultural abundance spurs greed; and

■ Political violence is less following increases in rainfall because agricultural abundance breeds contentment and self-sufficiency (p. 54).

In this example, climatic variables are hypothesized to have positive and negative influences on the likelihood of conflict, further highlighting the methodological critiques. Prevailing public argumentation on the issue has all tended in the same direction, but the variances in the intervening variables can generate alternative outcomes. Careful examination shows these critiques have persisted in study after study, decades after Gleditsch (1998) published the first substantive review of the literature. Combined, they cast doubt on the explanatory power of the central claim and undermine the generalizability of the argument.

References

Executive Office of the President. 2015. *The National Security Implications of a Changing Climate.* Washington, DC: White House.

Gleditsch, N.P. 1998. Armed conflict and the environment: a critique of the literature. *Journal of Peace Research* **35** (3): 381–400.

Hegre, H. and Sambanis, N. 2006. Sensitivity analysis of empirical results on civil war onset. *Journal of Conflict Resolution* **50** (4): 508–35.

Homer-Dixon, T. 1999. *Environment, Scarcity and Violence.* Princeton, NJ: Princeton University Press.

Raleigh, C. and Kniveton, D. 2012. Come rain or shine: an analysis of conflict and climate variability in East Africa. *Journal of Peace Research* **49** (1): 51–64.

7.3.3 Alleged Sources of Conflict

There is little evidence that climate change intensifies alleged sources of violent conflict including abrupt climate changes, access to water, famine, resource scarcity, and refugee flows.

The literature on the climate-conflict hypothesis, including Chapter 12 of the IPCC's Fifth Assessment Report, cites five sources of violent conflict allegedly intensified by climate change: abrupt climate changes, access to water, famine, resource scarcity, and refugee flows. Yet the literature on each of these alleged sources of conflict does not support claims of a causal relationship for any one of them.

7.3.3.1 Abrupt Climate Change

The possibility that climate change could occur suddenly rather than gradually is clear from the geological record. By happening too suddenly for plants, humans, and other animals to adapt, abrupt climate change could result in sudden losses of livelihood and residences, famines, mass migrations, and other conditions that could, in turn, lead to violent conflict. That is the theory, but how credible is it?

In 2002, the National Research Council of the U.S. National Academies of Sciences published a report titled *Abrupt Climate Change: Inevitable Surprises* (NRC, 2002). The report quickly became the most frequently cited source said to support the claim that abrupt climate change could lead to violent conflicts. In fact, conflict is hardly mentioned in the report, and only once regarding conflicts over water. It actually makes the opposite case, that adaptation is likely:

> It is important not to be fatalistic about the threats posed by abrupt climate change. Societies have faced both gradual and abrupt climate changes for millennia and have learned to adapt through various mechanisms, such as moving indoors, developing irrigation for crops, and migrating away from inhospitable regions. Nevertheless, because climate change will likely continue in the coming decades, denying the likelihood or downplaying the relevance of past abrupt events could be costly. Societies can take steps to face the potential for abrupt climate change.

> The committee believes that increased knowledge is the best way to improve the effectiveness of response, and thus that research into the causes, patterns, and likelihood of abrupt climate change can help reduce vulnerabilities and increase our adaptive capabilities. The committee's research recommendations fall into two broad categories: (1) implementation of targeted research to expand instrumental and paleoclimatic observations and (2)

implementation of modeling and associated analysis of abrupt climate change and its potential ecological, economic, and social impacts (NRC, 2002, p. 2).

This nuanced approach was quickly forgotten when, in response to the NRC report, the U.S. Pentagon commissioned a report by two consultants, Peter Schwartz and Doug Randall, on the national security implications of abrupt climate change (Schwartz and Randall, 2003). The resulting report released in 2003, titled "An Abrupt Climate Change Scenario and Its Implications for United States National Security," is still one of the most frequently cited sources on the subject. The saliency of the topic and the paper were not hurt by the debut in 2004 of a movie, *The Day After Tomorrow,* whose premise was a nearly instantaneous return to a global ice age.

Schwartz and Randall illustrated what they believed to be the association between abrupt climate change and national security in the graphic reproduced in Figure 7.3.3.1.1 below. As close inspection of the figure might suggest, Schwartz and Randall is not a scholarly report, and it is surprising it was ever treated as though it were. It is a 22-page essay with only two footnotes. The authors, both affiliated at the time with a consulting firm called Global Business Network, are "futurists" without any background or publications in climate science or warfare. They made no effort to document any part of their narrative by referring to any authoritative article or book. As befits consultants to Hollywood moviemakers, they say, "Rather than predicting how climate change will happen, our intent is to dramatize the impact climate change could have on society if

we are unprepared for it." "Dramatize," "could," and "if" are the key words in this sentence. The authors do not over-sell their work. The following disclaimer of sorts appears on the first page in a large font:

The purpose of this report is to imagine the unthinkable – to push the boundaries of current research on climate change so we may better understand the potential implications on United States national security.

We have interviewed leading climate change scientists, conducted additional research, and reviewed several iterations of the scenario with these experts. The scientists support this project, but caution that the scenario depicted is extreme in two fundamental ways. First, they suggest the occurrences we outline would most likely happen in a few regions, rather than on globally [sic]. Second, they say the magnitude of the event may be considerably smaller.

We have created a climate change scenario that although not the most likely, is plausible, and would challenge United States national security in ways that should be considered immediately (Schwartz and Randall, 2003, p. 1).

The methodology used by the authors, interviewing "leading climate change scientists," is not promising. As mentioned in Section 7.2.1, basing

Figure 7.3.3.1.1
Association of abrupt climate change and national security

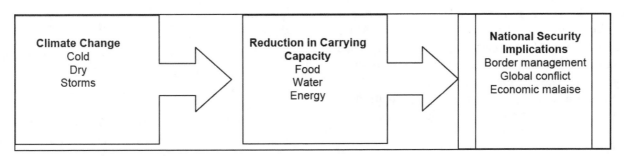

Source: Schwartz and Randall, 2003, p. 3.

forecasts on the opinions of experts is no more likely to be accurate than making uneducated guesses (Armstrong, 2001, 2006). When done scientifically, the most accurate forecasts concerning climate science virtually rule out the possibility of an abrupt climate change resembling Schwartz and Randall's scenario in the coming century (Green *et al.*, 2009).

The scenario presented by the authors is hardly plausible. They imagine "the thermohaline collapse begins in 2010, disrupting the temperate climate of Europe," whereas the IPCC has "*low confidence* in projections of when an anthropogenic influence on the AMOC [Atlantic meridianal overturning circulation] might be detected" (IPCC, 2013, p. 995). Schwartz and Randall assume that over the course of a decade rapid temperature declines of 5°F *per year* occur over Asia and North America and 6°F in northern Europe, and annual temperature increases up to 4°F in "key areas throughout Australia, South America, and southern Africa." Drought would strike "critical agricultural regions and in the water resource regions for major population centers in Europe and eastern North America." Winter storms and winds would intensifySchwartz and Randall then *assume to be true* every link in the chain of association that must be proven to make the rest of their scenario credible: food shortages due to decreases in net global agricultural production, decreased availability and quality of fresh water in key regions due to shifted precipitation patterns, and disrupted access to energy supplies due to extensive sea ice and storminess. The literature on the associations between climate change and all of these variables, and then these variables and violent conflict, is reviewed in the sections above and below, but it needs to be said here that real experts on these subjects are nearly unanimous that violent conflicts only rarely arise from these conditions and when they do, they are invariably the result of the failure of civil and political institutions to address public needs.

The Schwartz and Randall report cannot be taken seriously. It more closely resembles a movie script or hurriedly composed college term paper than a serious research paper. Nevertheless, the paper influenced the public debate and set the stage for a more alarmist report by the U.S. Climate Change Science Program (2008) and a new report from NRC issued in 2013 titled "Abrupt Impacts of Climate Change: Anticipating Surprises" (NRC, 2013).

There is no plausible scenario under which small increases in carbon dioxide in the atmosphere lead to abrupt climate changes like those observed in the geologic record. To plan for possible violent conflicts

that might arise from such a scenario is a waste of public resources and human capital.

References

Armstrong, J.S. 2001. *Principles of Forecasting – A Handbook for Researchers and Practitioners*. Norwell, MA: Kluwer Academic Publishers.

Armstrong, J.S. 2006. Findings from evidence-based forecasting: Methods for reducing forecast error. *International Journal of Forecasting* **22**: 583–98.

Green, K.C., Armstrong, J.S., and Soon, W. 2009. Validity of climate change forecasting for public policy decision making. *International Journal of Forecasting* **25**: 826–32.

IPCC. 2013. Intergovernmental Panel on Climate Change. *Climate Change 2013: The Physical Science Basis*. Contribution of Working Group I to the Fifth Assessment Report of the Intergovernmental Panel on Climate Change. New York, NY: Cambridge University Press.

NRC. 2002. National Research Council. *Abrupt Climate Change: Inevitable Surprises. Committee on Abrupt Climate Change*, Ocean Studies Board, Polar Research Board, Board on Atmospheric Sciences and Climate, Division on Earth and Life Studies. Washington, DC: The National Academies Press.

NRC. 2013. National Research Council. *Abrupt Impacts of Climate Change: Anticipating Surprises*. Washington, DC: The National Academies Press.

Schwartz, P. and Randall, D. 2003. *An Abrupt Climate Change Scenario and Its Implications for United States National Security*. Pasadena, CA: Jet Propulsion Laboratory.

U.S. Climate Change Science Program. 2008. *Abrupt Climate Change. A report by the U.S. Climate Change Science Program and the Subcommittee on Global Change Research*. U.S. Geological Survey. Reston, VA.

7.3.3.2 Water as a Source of Conflict

According to the *Summary for Policymakers* for the Working Group II contribution to AR5, "Freshwater-related risks of climate change increase significantly with increasing greenhouse gas concentrations (*robust evidence, high agreement*). The fraction of global population experiencing water scarcity and the fraction affected by major river floods increase with the level of warming in the 21st century. Climate change over the 21st century is projected to reduce renewable surface water and groundwater resources significantly in most dry subtropical regions (*robust evidence, high agreement*), intensifying competition for water among sectors (*limited evidence, medium agreement*) (IPCC, 2014, p. 14).

Water, whether too much or too little, is a main variable in the climate-conflict argument. An Intelligence Community Assessment published in February 2012 by the Office of the Director of National Intelligence asserts as its "bottom line" that "during the next 10 years, many countries important to the United States will experience water problems – shortages, poor water quality, or floods – that will risk instability and state failure, increase regional tensions, and distract them from working with the United States in important U.S. policy objectives" (Intelligence Community Assessment, 2012, p. iii).

"Tensions" over water were cited as a source of conflict by the Center for Naval Analyses in 2007. John Podesta (who served in the Clinton and Obama administrations) and Peter Ogden of the liberal Center for American Progress predicted in 2008 that "increasing water scarcity due to climate change will contribute to instability throughout the world ... water scarcity also shapes the geopolitical order when states engage in direct competition with neighbors over shrinking water supplies" (Podesta and Ogden, 2008, pp. 104–5). The Obama administration repeatedly claimed water scarcity and floods would exacerbate tensions and flooding could harm U.S. military bases and installations at home and abroad (e.g., Executive Office of the President, 2015).

The empirical evidence strongly refutes these claims. A thorough analysis of 412 crises during the period 1918–1994 reveals only seven where water was even a partial cause (Wolf, 1999). "As we see, the actual history of armed water conflict is somewhat less dramatic than the water wars literature would lead one to believe. ... As near as we can find, there has never been a single war fought over water," Wolf concluded. Writing in the pages of *International Security*, a preeminent security studies journal, three Norwegian scholars examined the linkages between water scarcity, drought, and incidence of civil wars. Factors other than the environment were much more significant in explaining the onset of conflict. They conclude:

The results presented in this article demonstrate that there is no direct, short-term relationship between drought and civil war onset, even within contexts presumed most conducive to violence. ... Ethnopolitical

exclusion is strongly and robustly related to the local risk of civil war. These findings contrast with efforts to blame violent conflict and atrocities on exogenous non-anthropogenic events, such as droughts or desertification. The primary causes of intrastate armed conflict and civil war are political, not environmental (Theisen *et al.*, 2011, p. 105).

Salehyan and Hendrix (2014) examined civil conflict, defined as confrontation between organized, armed groups as well as terrorism, and confirmed the absence of a positive relationship between water scarcity and conflict. They summarized their findings:

Most importantly, we have shown that analysts and policy planners should not look for significant increases in armed violence during periods of acute water scarcity. Climate change may cause certain regions of the world to be more drought-prone, but such droughts are not likely to cause fighting to erupt – at least in the short term. It would be more appropriate to focus on humanitarian concerns, capacity building, and development needs in order to assure that drought-stricken communities are able to adapt to a more uncertain climate (p. 249).

A war over water is difficult to imagine. A downstream state may have high motivation to secure greater supplies, but unless it could exert control over the entire watershed, it would be continually subject to manipulation by upstream sources. The costs of ensuring complete control would be quite high with little guarantee of short- or long-term success. This explains why the opposite result – peaceful cooperation to manage a shared resource – is the more likely consequence of water scarcity. International cooperation over transboundary water sources is much more common than conflict over the same resources (Yoffe *et al.*, 2003). Tir and Stinnett (2012) tested whether the pressures exerted by climate change will weaken transboundary river treaties and encourage non-compliance. By testing historical data on water availability between 1950 and 2000, they found the slightly increased risk of military conflict was offset by institutionalized agreements. The length of time over which the effects of climate change will be felt offers sufficient time to strengthen and institutionalize international treaties governing use of water.

Of course, treaties and agreements that have limited conflict in the past may not do so in the future. Climate-conflict proponents imply that states would ignore those agreements and move to protect their interests by any means necessary. Proponents of the "water wars" view appeal to the future and contend past trends will be overwhelmed by the enormity of the problems to come; they point to specific hot-spots where water-induced conflicts seem most probable. Podesta and Ogden (2008), for example, viewed the Middle East as the primary location where a water conflict could emerge, as have a number of others (see Trondalen, 2009, and Brown and Crawford, 2009). CNA (2007) pointed to water as a source of interstate and intrastate tension in the region and a contributor to terrorism.

Feitelson *et al.* (2012) tested these claims using four scenarios of climate change, along with varying assumptions about refugee return, in the Israeli-Palestinian context projected to 2030. They conclude:

> … based on analysis of extreme scenarios, we find that the likely direct effects of climate change per se are limited. While climate change may affect the livelihood of Palestinian farmers and semi-nomads, particularly in remote areas, it is unlikely to affect the welfare of the urban population substantially if some water re-allocation occurs, even under extreme scenarios (Feitelson *et al.*, 2012, p. 253).

The authors conclude "climate change does not seem to pose a major direct security risk in the Israeli-Palestinian context" (*Ibid.*, p. 254). They do note a danger in characterizing water as a security problem. "However, the framing of water issues and of climate change as security issues, and the subservience of water and environmental issues to the 'high politics' of conflict may hinder the ability to undertake adaptive measures that may mitigate the effects of climate change" (*Ibid.*). Adding a security dimension to environmental or shared resource concerns, when other factors have created conditions of mistrust and tension among the parties, is expected to greatly reduce the probability of an amicable resolution. As Feitelson *et al.* show, water shortage is not a sufficiently robust condition to generate conflict on its own. Ironically, the climate-conflict literature may do more than climate change itself to militarize environmental crises by characterizing them as security challenges, thereby prompting decision-makers to turn away from cooperative or diplomatic solutions and towards military options.

In Central Asia, the Syr Darya river basin is cited as another area where a transboundary dispute over water could spark conflict (see Swarup, 2009 and Hodgson, 2010). The region is comprised of poor, undemocratic states with weak international water management agreements. It is a perfect test case for the claim that the introduction of new supply pressures borne out of climate change will incite conflict and tension. Bernauer and Siegfried (2012) tested this proposition using IPCC climate models projected to 2050. They conclude that even though climate change is expected to make water supplies scarcer in the region (not a surprising conclusion given the previous discussion of the IPCC modeling approach), "such shifts are likely to occur only in the medium to long term" (Bernauer and Siegfried, 2012, p. 237). Rather than conflict, which they judge as "unlikely," Bernauer and Siegfried believe the countries in the region will respond by strengthening the international agreements governing water; a response consistent with past experiences, globally and regionally (Deudney, 1990).

Examining the relationship between precipitation, temperature, and drought on the incidence of civil war in Asia, Wischnath and Buhaug (2014) found climatic events play only a "trivial role" in explaining the risk of conflict.

Africa is frequently cited as a case where rainfall and changing water patterns could elicit greater risk of conflict. Darfur was called the first climate conflict by Jan Egeland, former United Nations Undersecretary General for Humanitarian Affairs, and U.N. Secretary General Ban Ki-Moon (see Salehyan, 2008, and Mazo, 2010). A strong relationship between rising temperature and civil war has been suggested to exist in Africa (Burke *et al.*, 2009). A subsequent analysis, however, shows Burke *et al.*'s findings are not supported when tested using different methods, notably a different set of armed conflict data (Buhaug, 2010).

Raleigh and Kniveton (2012) look at the Africa case from the perspective of small-scale conflict, rather than interstate conflict. Since a major hypothesis of the climate-conflict literature is that changing water dynamics create conditions within states that weaken social structures and government institutions, their examination of rainfall variability on rebel and communal violence is highly informative. Most studies that have examined the causes of civil wars have shown little statistical significance for environmental variables when other standard political and economic variables are controlled for (see Nordås and Gleditsch, 2007, and Raleigh and Urdal, 2007).

Detailed examination of rebel and communal conflicts in East Africa shows rainfall patterns emerge as an explanation for conflict only when other socioeconomic conditions exist. Then, the outcome that emerges is one where communal violence has a tendency to increase during wet periods, when the abundance of resources provides the motives and opportunities for inter-group violence. In contrast, during dry periods, communal violence is suppressed and the conditions for rebel conflicts emerge (Raleigh and Kniveton, 2012).

Other examinations of the impact of climate variability on social unrest and conflict in Africa show less connection between the two. Looking at the Sahel, which under climate change scenarios will become drier as rainfall is reduced through the effects of rising temperatures, a team of researchers from the Peace Research Institute in Oslo studied land use conflicts using both statistical and case study approaches. Both methods "provide little evidence supporting the notion that water scarcity and rapid environmental change are important drivers of intercommunal conflict in the Sahel" (Benjaminsen *et al.*, 2012). They judge political and economic forces as more significant than climate variability. Similarly, an examination of the Kenyan range found drought conditions suppress conflict and encourage groups to share resources (see Butler and Gates, 2012, and Eaton, 2008), further reinforcing the finding of cooperation rather than conflict arising out of environmental pressures.

Examining Kenyan armed conflict below the common civil conflict level, Theisen (2012) determined that years with below-average rainfall were generally more peaceful, concluding, "Tests of the hypotheses on resource scarcity lend most support to those that argue that resource scarcity does not fuel violence and seems even to favor those that see droughts as temporarily cooling tensions" (Theisen, 2012, p. 93).

In conclusion, the notion that global warming's effect on access to water might lead to more armed conflict around the world has been repeatedly tested and invalidated by a wide range of researchers using data from many parts of the world.

References

Benjaminsen, T., Alinon, K., Buhaug, H., and Buseth, J.T. 2012. Does climate change drive land-use conflicts in the Sahel? *Journal of Peace Research* **49** (1): 97–111.

Bernauer, T. and Siegfried, T. 2012. Climate change and international water conflict in central Asia. *Journal of Peace Research* **49** (1): 227–39.

Brown, O. and Crawford, A. 2009. *Rising Temperatures, Rising Tensions: Climate Change and the Risk of Violent Conflict in the Middle East.* Winnipeg, MB: International Institute for Sustainable Development.

Buhaug, H. 2010. Climate not to blame for African civil wars. *Proceedings of the National Academy of Sciences USA* **197**: 16,477–82.

Burke, M., Miguel, E., Satyanah, S., Dyekema, J., and Lobell, D. 2009. Warming increases the risk of civil war in Africa. *PNAS* **106** (49): 20670–4.

Butler, C. and Gates, S. 2012. African range wars: climate, conflict, and property rights. *Journal of Peace Research* **49** (1): 23–34.

CNA. 2007. Center for Naval Analyses. *National Security and the Threat of Climate Change.* Washington, DC: Center for Naval Analyses.

Deudney, D. 1990. The case against linking environmental degradation and national security. *Millennium* **19** (3): 461–76.

Eaton, D. 2008. The business of peace: raiding and peace work along the Kenya-Uganda border (Part I). *African Affairs* **107** (426): 89–110.

Executive Office of the President. 2015. *The National Security Implications of a Changing Climate.* Washington, DC: White House.

Feitelson, E., Tamimi, A., and Rosenthal, G. 2012. Climate change and security in the Israeli-Palestinian context. *Journal of Peace Research* **49** (1): 241–57.

Hodgson, S. 2010. *Strategic Water Resources in Central Asia.* Brussels, Belgium: Centre for European Policy Studies.

Intelligence Community Assessment. 2012. *Global Water Security.* Washington, DC: Office of the Director of National Intelligence.

IPCC. 2014. Intergovernmental Panel on Climate Change. *Summary for Policymakers.* In: *Climate Change 2014: Impacts, Adaptation, and Vulnerability.* Contribution of Working Group II to the Fifth Assessment Report of the

Intergovernmental Panel on Climate Change. New York, NY: Cambridge University Press.

Mazo, J. 2010. *Climate Conflict*. New York, NY: Routledge.

Nordås, R. and Gleditsch, N.P. 2007. Climate change and conflict: a critical overview. *Political Geography* **26** (6): 627–38.

Podesta, J. and Ogden, P. 2008. Security implications of climate scenario 1. In: Campbell, K. (Ed.) *Climatic Cataclysm*. Washington, DC: Brookings Institution, pp. 97–132.

Raleigh, C. and Kniveton, D. 2012. Come rain or shine: an analysis of conflict and climate variability in East Africa. *Journal of Peace Research* **49** (1): 51–64.

Raleigh, C. and Urdal, H. 2007. Climate change, environmental degradation, and armed conflict. *Political Geography* **26** (6): 674–94.

Salehyan, I. 2008. From climate change to conflict? No consensus yet. *Journal of Peace Research* **45** (3): 315–26.

Salehyan, I. and Hendrix, C. 2014. Climate shocks and political violence. *Global Environmental Change* **28**: 239–50.

Swarup, A. 2009. *Reaching Tipping Point? Climate Change and Poverty in Tajikistan*. Dushanbe, Tajikistan: Oxfam International.

Theisen, O.M. 2012. Climate clashes? Weather variability, land pressure, and organized violence in Kenya, 1989–2004. *Journal of Peace Research* **49** (1): 81–96.

Theisen, O.M., Holtermann, H., and Buhaug, H. 2011. Climate wars? Assessing the claim that drought breeds conflict. *International Security* **36** (3): 79–106.

Tir, J. and Stinnett, D. 2012. Weathering climate change: can international institutions mitigate international water conflict? *Journal of Peace Research* **49** (1): 211–25.

Trondalen, J.M. 2009. *Climate Change, Water Security and Possible Remedies for the Middle East*. Paris, France: UNESCO.

Wischnath, G. and Buhaug, H. 2014. On climate variability and civil war in Asia. *Climatic Change* **122**: 709–21.

Wolf, A. 1999. 'Water wars' and water reality. In: Lonergan, S. (Ed.) *Environmental Change, Adaptation, and Human Security*. Dordrecht, Netherlands: Kluwer Academic, pp. 251–65.

Yoffe, S., Wolf, A., and Giordano, M. 2003. Conflict and cooperation over international freshwater resources: indicators of basins at risk. *Journal of the American Water Resources Association* **39** (5): 1109–26.

7.3.3.3 *Famine as a Source of Conflict*

Famine does not appear in Chapter 12 of the Working Group II contribution to the Fifth Assessment Report as one of the factors that increase the risk of violent conflicts and are "sensitive" to climate change, but it was featured in the previously discussed report by Schwartz and Randall commissioned by the U.S. Pentagon (Schwartz and Randall, 2003) and made regular appearances in declarations by President Barack Obama and federal agencies during his two terms in office (see Executive Office of the President, 2015). It frequently appears in the popular media, as illustrated by a *Newsweek* story in 2017 titled "Famine Isn't Just a Result of Conflict – It's a Cause" (Hopma, 2017).

Yet according to Nobel laureate Amartya Sen, there has never been a democracy with a free press that has experienced a famine (Sen, 1999, p. 178). While Sen's statement has been criticized as being overly broad and dependent on the definition of "famine," it has withstood the test of time (see Halperin *et al.*, 2004, p. 18). Sen's observation is significant because it illustrates a huge confounding factor in the climate-famine-conflict theory. If climate drives famines, why are democracies somehow immune? Given the close association between prosperity and democracy documented in Section 7.1.2, the solution to famines would seem to be to promote prosperity and democracy by making energy more abundant and affordable, rather than attempt to control the weather by increasing the cost of energy and impoverishing people.

While famines still occur in the world today, they invariably are the result of government mismanagement of food supplies or use of starvation by autocracies to oppress their people. Worldwide, food production outpaced population growth during the past century, with production per capita rising along with significant increases in world production of maize (203%), wheat (122%), rice (131%), vegetables (251%), cassava (146%), and soybeans (431%) between 1969 and 2009 (Hofstrand, 2011). Food production has "never been higher than it is today, largely due to fertilizers, pesticides, irrigation and farm machinery" (Goklany, 2011, p. 168).

According to the Food and Agriculture Organization of the United Nations (FAO), "the number of hungry people in the world has dropped to 795 million – 216 million fewer than in 1990–92 – or around one person out of every nine" (FAO, 2015). In developing countries, the share of population that is undernourished (having insufficient food to live an active and healthy life) fell from 23.3% 25 years earlier to 12.9%. A majority of the 129 countries monitored by FAO reduced undernourishment by half or more since 1996 (*Ibid.*).

Claims that climate change will reduce global food output are frequently made (e.g., Challinor *et al.*, 2014), but these forecasts invariably are based on computer models not validated by real-world data. Biological science, some of it summarized in Chapter 5, Section 3, conclusively shows plants thrive in a warmer world with higher-than-current levels of carbon dioxide (CO_2). Since aerial fertilization by CO_2 helps plants thrive even in hot and dry conditions, there is no scientific reason to believe those benefits will not continue even into the distant future.

In Chapter 3, Section 3.3.4, the graph below was presented and explained. It shows improvement in yields of one representative crop, sugar cane, due to improvements in technology ("techno-intel") and CO_2 fertilization continuing to 2050 and beyond (Idso, 2013).

The climate scenarios used by the IPCC improperly discount the adaptive capacity of modern agriculture and the large beneficial impacts of atmospheric CO_2 on crop productivity and food production. Idso and Idso (2000) identified the 45 crops that at the turn of the century supplied 95% of the world's food needs and projected historical trends in the productivities of these crops 50 years into the future, after which they evaluated the growth-enhancing effects of atmospheric CO_2 enrichment on these plants and made similar yield projections based on the increase in atmospheric CO_2 concentration likely to have occurred by that future date. While world population would likely be 51% greater in the year

Figure 7.3.3.3.1

Historical and projected increases in total yield and the portion of the total yield due to the techno-intel and CO_2 effects, 2012–2050

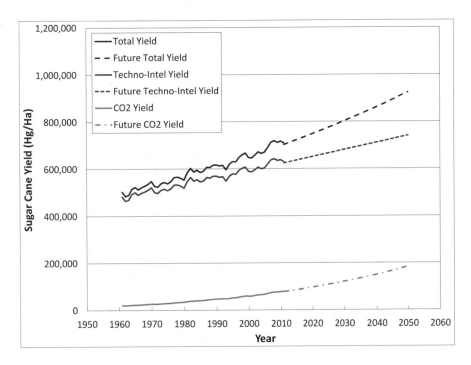

Source: Idso, 2013.

2050 than it was in 1998, Idsos' exercise revealed that as a consequence of anticipated improvements in agricultural technology and expertise and the aerial fertilization effect of anthropogenic CO_2 emissions, farm production would keep pace with population growth.

Norman Borlaug, father of the Green Revolution and recipient of the 1970 Nobel Peace Prize, wrote about the need to vastly increase the world's agricultural productivity. In an article published at the turn of the twenty-first century he wrote, "agricultural scientists and leaders have a moral obligation to warn political, educational, and religious leaders about the magnitude and seriousness of the arable land, food, and population problems that lie ahead, *even with breakthroughs in biotechnology* [italics added]." In fact, "if we fail to do so," as he described it, "we will be negligent in our duty and inadvertently may be contributing to the pending chaos of incalculable millions of deaths by starvation" (Borlaug, 2000).

References

Borlaug, N.E. 2000. Ending world hunger: the promise of biotechnology and the threat of antiscience zealotry. *Plant Physiology* **124**: 487–90.

Challinor, A.J., Watson, J., Lobell, D.B., Howden, S.M., Smith, D.R., and Chhetri, N. 2014. A meta-analysis of crop yield under climate change and adaptation. *Nature Climate Change* **4**: 287–91.

Executive Office of the President. 2015. *The National Security Implications of a Changing Climate.* Washington, DC: White House.

FAO. 2015. Food and Agriculture Organization of the United Nations. *The State of Food Insecurity in the World 2015.* Rome, Italy.

Goklany, I.M. 2011. Economic development in developing countries. In: Michaels, P. (Ed.) *Climate Coup.* Washington, DC: Cato Institute, pp. 157–84.

Halperin, M.H., Siegle, J.T., and Weinstein, M.M. 2004. *The Democracy Advantage: How Democracies Promote Prosperity and Peace.* New York: Routledge.

Hofstrand, D. 2011. Can the World Feed Nine Billion People by 2050? Agricultural Marketing Resource Center (website). Accessed June 24, 2018.

Hopma, J. 2017. Famine isn't just a result of conflict – it's a cause. *Newsweek.* January 25.

Idso, C.D. 2013. *The Positive Externalities of Carbon Dioxide.* Center for the Study of Carbon Dioxide and Global Change.

Idso, C.D. and Idso, K.E. 2000. Forecasting world food supplies: the impact of the rising atmospheric CO_2 concentration. *Technology* **7S**: 33–56.

Schwartz, P. and Randall, D. 2003. *An Abrupt Climate Change Scenario and Its Implications for United States National Security.* Pasadena, CA: Jet Propulsion Laboratory.

Sen, A. 1999. *Development as Freedom.* New York, NY: Random House.

7.3.3.4 Resource Scarcity as a Source of Conflict

The authors of Chapter 12 of the Working Group II contribution to AR5 say "Climate change has the potential to increase rivalry between countries over shared resources," but as reported earlier, they stop short of claiming any empirical evidence to support such a link (IPCC, 2014, p. 772). The authors of Chapter 22 of AR5, on Africa, are more assertive, claiming "the degradation of natural resources as a result of both overexploitation and climate change will contribute to increased conflicts over the distribution of those resources" (*Ibid.,* p. 1204).

The possibility of armed conflicts over scarce resources caused by abrupt climate change was raised by Schwartz and Randall in their 2003 report for the U.S. Pentagon. More credibly, Rune Slettebak, a Norwegian researcher affiliated with the Norwegian University of Science and Technology and the Peace Research Institute Oslo, writes, "Within the current debate on how environmental factors may affect the risk of conflict, scarcity of important resources holds a prominent place" (Slettebak, 2012). Similarly, Barnett and Adger write, "Acute scarcities, caused by reduced supply, increased demand or skewed distribution, are suggested as a significant current and future source of violent conflict" (Barnett and Adger, 2007).

That resource scarcity might lead to instability, state collapse, civil strife, or international conflict is a familiar argument in international security affairs. Under the "resource war" framework, nations are said to fight over territory, raw materials, energy, water, and food (Gleditsch, 1998). Deteriorating environmental conditions create resource scarcity and competition, thus creating conditions conducive to violence, the argument goes. Therefore, to the extent that climate change contributes to deteriorating environmental conditions, it is viewed in this framework as one of many possible causal factors.

These perspectives became popular in the 1970s and gained prominence with the end of the Cold War. The first Gulf War appeared to offer an excellent case supporting the view that the United States would go to war to secure a vital resource – petroleum (see Klare, 2001). More recently Kahl argued resource scarcity can result in the collapse of a state's ability to operate effectively, thereby undermining social structures and the cohesion of the state. He also identified another possible outcome: cooption of the state by groups that exploit the power of government to disperse resources selectively (Kahl, 2006).

Drawing on archaeological data, LeBlanc and Register (2003) argue warfare was "quite common in the past" and "not a fluke but the norm" throughout human history. Humans often fight when population growth outstrips the "carrying capacity" of their natural environment, they say, while peace occurs when carrying capacity increases faster than population growth thanks to the invention of agriculture, the discovery of new energy sources and technologies, and the expansion of trade with other regions. According to LeBlanc and Register, modernity has broken the pattern of "constant battles," though a war-free future is not guaranteed, human nature being what it is. They write, "In spite of the pronounced impact industrialized states make on the environment, their technology and slow [population] growth rates enable them to live well below the carrying capacity. The decline in warfare among those countries is incredibly strong" (*Ibid.*, p. 228) and "For the first time in history, technology and science enable us to understand Earth's ecology and our impact on it, to control population growth, and to increase the carrying capacity in ways never before imagined. The opportunity for humans to live in long term balance with nature is within our grasp if we do it right" (*Ibid.*, p. 229).

Much of the argument and evidence presented in the debate over resource scarcity-conflict are the same as that presented in the climate-conflict debate. A recent review of the literature by Shields and Solar (2011) provides a nuanced view of the scarcity-conflict hypothesis. Conflicts over minerals do occur, they say, but they are dependent upon the existence of other social factors (weak rule of law, inequitable distribution of revenue) and not the depletion of the supply. In fact, "in modern times, no interstate conflicts have been driven by depletion," the review concludes (p. 261).

Four critiques of the resource scarcity-conflict hypothesis have been advanced:

- Human inventiveness and technological innovation enhance agricultural output and improve resource extraction abilities.

- International trade enables the reallocation of resources that are plentiful in one location to those areas where they are scarcer.

- Many raw materials have substitutes that are cheaper or more plentiful.

- Under conditions of scarcity, prices will rise which in turn encourages innovation, trade, and incentives to substitute (Simon, 1996).

Since the resource scarcity argument grew into prominence during the 1970s, actual experience shows the concerns to be overstated. The *Limits to Growth* report (Meadows *et al.,* 1972), for instance, predicted aluminum, copper, gold, lead, tin, zinc, and many other materials would be exhausted by the 1990s–2000s. All remain in widespread production today. Further illustration of the absence of predictive foresight were the expectations that natural gas supplies would be exhausted by 1994 and petroleum by 1992. The application of new technologies has greatly expanded known and recoverable supplies of both natural gas and petroleum in recent years.

Scarcity may give rise to cooperation, rather than conflict. Deudney argued, "analysts of environmental conflict do not systematically consider ways in which environmental scarcity or change can stimulate cooperation" (Deudney, 1999). As discussed in Section 7.3.3.2, water scarcity more often gives rise to cooperation than to conflict (Dinar, 2011).

The logic behind cooperation, trade, or innovation as the preferred strategy for addressing resource scarcity is simple and compelling. The costs of military action are always high, the probability of success (in either the short or long run) is not guaranteed, and the costs of holding the gains from military action undermine the benefits of securing supplies of the desired resource. The German and Japanese experiences during World War II are instructive for these purposes. Both nations were strongly incentivized to secure supplies of resources before the onset of conflict and during the course of the war. Neither succeeded – obviously at great cost.

Institutions, international markets, and diplomatic solutions offer options short of conflict for resolving natural resource disputes. Trading on the international market expands supply options, as does investment in efficiency or substitutions. For these reasons, few wars in the modern era were fought over natural resources, and that is likely to continue to be the case.

References

Barnett, J. and Adger, W.N. 2007. Climate, security, and human conflict. *Political Geography* **26** (6): 639–55.

Deudney, D. 1999. Environmental security: a critique. In: Deudney, D. and Matthew, R. (Eds.) *Contested Grounds: Security and Conflict in the New Environmental Politics*. Albany, NY: State University of New York Press, pp. 187–219.

Dinar, S. 2011. Conflict and cooperation along international rivers. In: Dinar, S. *Beyond Resource Wars*. Cambridge, MA: MIT Press, pp. 166–99.

Gleditsch, N.P. 1998. Armed conflict and the environment: a critique of the literature. *Journal of Peace Research* **35** (3): 381–400.

IPCC. 2014. Intergovernmental Panel on Climate Change. *Summary for Policymakers*. In: *Climate Change 2014: Impacts, Adaptation, and Vulnerability*. Contribution of Working Group II to the Fifth Assessment Report of the Intergovernmental Panel on Climate Change. New York, NY: Cambridge University Press.

Kahl, C. 2006. *States, Scarcity, and Civil Strife in the Developing World*. Princeton, NJ: Princeton University Press.

Klare, M. 2001. *Resource Wars*. New York, NY: Metropolitan Books.

LeBlanc, S. and Register, K.E. 2003. *Constant Battles: The Myth of the Noble Savage and a Peaceful Past*. New York, NY: St. Martin's/Griffin.

Meadows, D., Meadows, D., Randers, J., and Behrens III, W. 1972. *The Limits to Growth*. New York, NY: Signet.

Schwartz, P. and Randall, D. 2003. *An Abrupt Climate Change Scenario and Its Implications for United States National Security*. Pasadena, CA: Jet Propulsion Laboratory.

Shields, D. and Solar, S. 2011. Responses to alternative forms of mineral scarcity. In: Dinar, S. *Beyond Resource Wars*. Cambridge, MA: MIT Press, pp. 239–85.

Simon, J. 1996. *The Ultimate Resource 2*. Princeton, NJ: Princeton University Press.

Slettebak, R. 2012. Don't blame the weather! Climate-related natural disasters and civil conflict. *Journal of Peace Research* **49** (1): 163–76.

7.3.3.5. Refugee Flows as a Source of Conflict

Flows of environmental refugees are another source of concern raised by the climate-conflict argument. According to the *Summary for Policymakers* of the Working Group II contribution to AR5,

> Climate change over the 21st century is projected to increase displacement of people (*medium evidence, high agreement*). Displacement risk increases when populations that lack the resources for planned migration experience higher exposure to extreme weather events, in both rural and urban areas, particularly in developing countries with low income. Expanding opportunities for mobility can reduce vulnerability for such populations. Changes in migration patterns can be responses to both extreme weather events and longer-term climate variability and change, and migration can also be an effective adaptation strategy. There is low confidence in quantitative projections of changes in mobility, due to its complex, multi-causal nature (IPCC, 2014, p. 20).

These migrations of displaced peoples, driven from their homes out of necessity because of drought, flood, or famine, or driven out intentionally by more powerful groups looking to secure greater shares of scarcer resources for themselves, are regularly cited. CNA (2007), for example, warns of unwelcomed migrations in Africa, Asia, Europe, and North America. Fingar (2008) cites migration concerns as well.

A widely cited figure for the number of possible "climate refugees" is 200 million, often attributed to a 1993 book by British environmentalist Norman Myers (Myers, 1993, and see Environmental Justice Foundation, 2009). The figure was cited by the IPCC in its Third Assessment Report, but not in AR5. Of Myers, Gleditsch and Nordås write, "it is generally recognized that this figure represents guesswork rather than a scientifically-based estimate" (Gleditsch and Nordås, 2014). The number in fact is pure speculation and detached from any current real-world estimates of the actual number of people forced to move by climate change. The United Nations endorsed the prediction of 50 million environmental refugees by 2010, a claim subsequently discredited by reality (Atkins, 2011).

Like conflicts over water, the environmental refugee problem is a future one, conditioned on the assumption that things will be worse than ever observed. All forecasts are based on anecdotal accounts of natural disasters causing migration, and then computer models predicting increased incidences of such disasters and no human adaptation. The models have not been validated and the best global data show declining, not increasing, frequency of extreme weather events. (See Chapter 2 for citations.)

Any cases of "environmental refugees" in the world today are either model predictions with no real-world data to confirm them, or the result of naturally occurring disasters (hurricanes, tornadoes, floods) with no evidence of a connection to long-term climate change, whether caused by the human presence or by natural cycles. Baldwin *et al.*, writing in 2014, observed:

> The origins of climate change-induced migration discourse go back to the 1980s, when concerned scientists and environmental activists argued that unchecked environmental and climate change could lead to mass displacement (Mathews 1989; Myers 1989). However, at that time, hardly any actual climate or environmental refugees could be detected. *Even today, almost three decades later, the term as such remains merely a theoretical possibility but not an actually existing, clearly defined group of people* (Baldwin *et al.,* 2014, p. 121, italics added).

In 2011, the British Government Office for Science published the *Foresight Report on Migration and Global Environmental Change,* the work of some "350 experts and stakeholders from 30 countries across the world" and referred to by Baldwin *et al.* as "by far the most authoritative scientific account of the relationship between climate change and human migration." According to the report, "the range and complexity of the interactions between these drivers [of migration] means that it will rarely be possible to distinguish individuals for whom environmental factors are the sole driver" (Foresight, 2011, p. 9).

After pointing out that "17 million people were displaced by natural hazards in 2009 and 42 million in 2010," the authors say, "Environmental change is equally likely to make migration less possible as more probable. This is because migration is expensive and requires forms of capital, yet populations who experience the impacts of environmental change may see a reduction in the very capital required to enable a move" (*Ibid.*). In other words, there may be *no net increase* in the number of environmental refugees.

While it is certainly possible to speculate about scenarios wherein displaced peoples create conflict, directly or indirectly, the empirical evidence suggests that is highly unlikely (Salehyan, 2005). The research shows "there are few, if any, cases of environmental refugees leading to violent conflict in receiving areas and while there are certainly examples of sporadic violence, such violence is generally small-scale, interpersonal and disorganized" (Buckland, 2007, p. 9).

According to a 2017 Reuters news story, "Statements by such public voices as Britain's Prince Charles and former U.S. Vice President Al Gore have linked the violence in Syria with global warming, saying the 2006 drought played a key role in urban migration that helped spark the civil war." But according to University of Sussex Professor Jan Selby, the coauthor of a study of the matter published in the journal *Political Geography,* "There is no sound evidence that global climate change was a factor in sparking the Syrian civil war. ... It is extraordinary that this claim has been so widely accepted when the scientific evidence is so thin" (Reuters, 2017). In their journal article, Selby *et al.* (2017) report,

> This article provides a systematic interrogation of these claims, and finds little merit to them. Amongst other things it shows that there is no clear and reliable evidence that anthropogenic climate change was a factor in Syria's pre-civil war drought; that this drought did not cause anywhere near the scale of migration that is often alleged; and that there exists no solid evidence that drought migration pressures in Syria contributed to civil war onset. The Syria case, the article finds, does not support 'threat multiplier' views of the impacts of climate change; to the contrary, we conclude, policymakers, commentators and scholars alike should exercise far greater caution

when drawing such linkages or when securitizing climate change.

After examining many environmental refugee claims, Tertrais (2011) concluded, "Such are the reasons why experts of environmental migrations generally agree that climate change in itself is rarely a root cause of migration. Major population displacements due to environmental and/or climatic factors will remain exceptional except in the case of a sudden natural disaster. And most importantly for the sake of this analysis, they are rarely a cause of violent conflict" (p. 24).

References

Atkins, G. 2011. The origins of the 50 million climate refugees prediction. *Asian Correspondent*. April 23.

Baldwin, A., Methmann, C., and Rothe, D. 2014. Securitizing 'climate refugees': the futurology of climate-induced migration. *Critical Studies on Security* **2** (2): 121–30.

Buckland, B. 2007. *A Climate of War? Stopping the Securitization of Global Climate Change*. Berlin, Germany: International Peace Bureau, pp. 1–15.

CNA. 2007. Center for Naval Analyses. *National Security and the Threat of Climate Change*. Washington, DC: Center for Naval Analyses.

Environmental Justice Foundation. 2009. *No Place Like Home: Where Next for Climate Refugees*. London, UK: Environmental Justice Foundation.

Fingar, D.T. 2008. National intelligence assessment on the national security implications of global climate change to 2030. Testimony before the House Permanent Select Committee on Intelligence and the House Select Committee on Energy Independence and Global Warming. Washington, DC.

Foresight. 2011. *Migration and Global Environmental Change*. London, UK: The Government Office for Science.

Gleditsch, N.P. and Nordås, R. 2014. Conflicting messages? The IPCC on conflict and human security. *Political Geography* **43**: 82–90.

IPCC. 2014. Intergovernmental Panel on Climate Change. *Summary for Policymakers*. In: *Climate Change 2014: Impacts, Adaptation, and Vulnerability*. Contribution of Working Group II to the Fifth Assessment Report of the Intergovernmental Panel on Climate Change. New York, NY: Cambridge University Press.

Mathews, J.T. 1989. Redefining security. *Foreign Affairs* **68**: 162–77.

Myers, N. 1989. Environment and security. *Foreign Policy* **74**: 23–41.

Myers, N. 1993. *Ultimate Security: The Environmental Basis of Political Stability*. New York, NY: W.W. Norton and Co.

Reuters. 2017. Claims that climate change fueled Syria's civil war questioned in new study. September 7.

Salehyan, I. 2005. Refugees, climate change and instability. In: *Human Security and Climate Change*. Oslo, Norway: International Peace Research Institute, pp. 1–20.

Selby, J., *et al.* 2017. Climate change and the Syrian Civil War revisited. *Political Geography* **60**: 251–2.

Tertrais, B. 2011. The climate wars myth. *Washington Quarterly* **34** (3): 17–29.

7.3.4 U.S. Military Policy

Climate change does not pose a military threat to the United States. President Donald Trump was right to remove it from the Pentagon's list of threats to national security.

Throughout his two terms in office, President Barack Obama tried to frame climate change as a matter of United States national security. In May 2015, the White House issued a report saying,

Climate change is an urgent and growing threat to U.S. national security, contributing to increased weather extremes which worsen refugee flows and conflicts over basic resources like food and water. The national security implications of climate change reach far beyond U.S. coastlines, further threatening already fragile regions of the world. Increased sea levels and storm surges threaten coastal regions, infrastructure, and property. A changing climate will act as an accelerant of instability around the world, exacerbating tensions related to water scarcity and food shortages, natural resource competition, underdevelopment, and overpopulation (Executive Office of the President, 2015).

Obama did not invent the idea that climate change would threaten U.S. national security; he inherited it from the previous administration. Recall from Section 7.3 that Dr. Thomas Fingar, deputy director of National Intelligence for Analysis and chairman of the National Intelligence Council, testified to Congress in 2008 that "our primary source for climate science was the United Nations Intergovernmental Panel on Climate Change (IPCC) Fourth Assessment Report" and "we relied predominantly upon a mid-range projection from among a range of authoritative scenario trajectories provided by the IPCC" (Fingar, 2008, pp. 2–3). This was *before* Obama took office.

Obama and activists in the environmental movement apparently thought by casting climate change as a security issue, they could win over conservatives and Republicans who prioritized national defense and spending on the military. It may have worked: Congress, even when controlled by Republicans, approved virtually all of Obama's spending requests involving military programs advancing his climate change agenda. Secretary of Defense Chuck Hagel, a Republican appointed to the position by Obama, said,

> Among the future trends that will impact our national security is climate change. Rising global temperatures, changing precipitation patterns, climbing sea levels, and more extreme weather events will intensify the challenges of global instability, hunger, poverty, and conflict. By taking a proactive, flexible approach to assessment, analysis, and adaptation, the Defense Department will keep pace with a changing climate, minimize its impacts on our missions, and continue to protect our national security (DoD, 2014b).

The Obama administration used the Department of Defense (DoD) to help wage its "war on coal," part of its announced strategy of weaning the nation away from fossil fuels. DoD, like other executive agencies, made public statements that seemed to validate the claims and predictions of climate change alarmists. The department's "2014 Climate Change Adaptation Roadmap" illustrates the acceptance of this view. Its preface reads like a news release from Greenpeace:

> Among the future trends that will impact our national security is climate change. Rising global temperatures, changing precipitation patterns, climbing sea levels, and more extreme weather events will intensify the challenges of global instability, hunger, poverty, and conflict. They will likely lead to food and water shortages, pandemic disease, disputes over refugees and resources, and destruction by natural disasters in regions across the globe. In our defense strategy, we refer to climate change as a "threat multiplier" because it has the potential to exacerbate many of the challenges we are dealing with today – from infectious disease to terrorism. We are already beginning to see some of these impacts (DoD, 2014a).

The U.S. military, with its abundant technological, scientific, and financial resources, has a massive platform from which to steward energy innovation. Research and development is a legitimate function of DoD and other government agencies. However, investing in unreliable renewable energy resources for purposes other than those supporting the department's mission is wasteful, unnecessary, and potentially dangerous when it diverts funding from higher priorities. Unfortunately, such diversion seems to be the goal of the various environmental advocacy groups and consultants paid to produce reports on how DoD can "accommodate" or "respond" to climate change (e.g, Busby, 2007; Center for a New American Security, 2008; McGrady *et al.*, 2010; CNA and Oxfam America, 2011; CNA, 2014). Most of these reports are little more than illustrated versions of the superficial Schwartz and Randall report commissioned by the Pentagon in 2003 (Schwartz and Randall, 2003).

During the Obama administrations, DoD was directed to spend scarce funds on expensive alternative energy projects to help pave the way to commercialization. In 2011, the U.S. Army Corps of Engineers issued a power purchase agreement (PPA) authorizing $7 billion in spending on alternative energy sources (biomass, geothermal, solar, and wind). In 2014, the program had 79 contracts to purchase power from third parties (Casey, 2014). Fossil fuel resources are far more affordable and reliable than alternatives available to DoD. Research reported in Chapter 3 found electricity generated by wind turbines and solar PV cells cost approximately three times as much as fossil fuels (Stacy and Taylor, 2016).

In 2009, the U.S. Navy purchased 40,000 gallons of jet fuel derived from camelina (wild flax) at $67.50 per gallon and 20,055 gallons of algae-derived diesel-like fuel at a hefty $424 per gallon (Biello, 2009). Conventional jet fuel cost less than $2 a gallon in 1999. *Scientific American* also reported, "The Defense Advanced Research Projects Agency has spent $35 million to sponsor research into oil from algae and the Air Force is also looking for cleaner ways to fly and fight" (*Ibid.*).

Attempting to transition the U.S. military away from fossil fuels to biofuels, solar, and wind cannot be done without compromising military power and preparedness. T.A. "Ike" Kiefer, a captain in the U.S. Navy in addition to having degrees in physics and strategy, explained the trade-off as follows:

No materials other than very exotic and toxic substances like lithium borohydride (LiBH4) or expensive rare metals like beryllium surpass the energy density of diesel and jet fuel. Biodiesel and ethanol both fall short. Hydrogen fuel cells, electrical storage batteries, and capacitors miss by a much greater margin. Other alternatives, such as wind, solar, geo-thermal, or waste-to-energy devices, can power some laptops and light some fixed facilities but simply cannot harvest enough energy to propel the tanks, jets, helos, and trucks that are by far the major battlefield fuel consumers. These can offer only an incidental decrease in overall fuel requirements for mechanized forces and then only in low-hostility circumstances where they can be set up and safeguarded (Kiefer, 2013, pp. 117–8).

According to Kiefer, "the US Navy directly rejected a RAND study conducted at the direction of Congress and delivered to the secretary of defense in January of 2011 that unambiguously found biofuels of 'no benefit to the military' (Bartis and Van Bibber, 2011; Maron, 2011). A second RAND study and a report by the U.S. National Academy of Sciences, both severely questioning the wisdom and efficacy of current U.S. biofuels policies, also resulted in no adjustments to U.S. biofuels programs (Bartis, 2012; NRC, 2011)" (*Ibid.*, p. 116).

Another unnecessary expense is "hardening" military installations for unrealistic forecasts of sea level rise or the increased probability of intense storms. According to Obama, "Installations near the coastlines are threatened by coastal erosion and sea level rise, damaging infrastructure and reducing the land available for operations" (Executive Office of the President, 2015, p. 9). But as reported in Chapter 2, Section 2.1.2, globally averaged sea-level change has been stable and less than seven inches per century for the past 1,000 years, a rate that is functionally negligible because it is frequently exceeded by coastal processes like erosion, sedimentation, and subsidence unrelated to climate.

What matters to military bases and military strategy is not global average sea level – itself an abstract concept and not an empirical finding – but actual *local* changes in sea level. Local sea-level trends vary considerably depending on tectonic movements of adjacent land and other factors. In many places vertical land motion, either up or down, exceeds the very slow global sea-level trend. Efforts to document an accelerated sea-level rise, to the extent they are made rather than simply assumed by relying on secondary sources and television documentaries, typically use very short measurement records or short, low-quality, satellite altimetry measurements rather than long, high-quality, coastal measurements. Church and White (2006), for example, spliced together measurements from different locations at different times and claimed to find (from the study's title) "A 20th Century Acceleration in Global Sea-Level Rise." Later researchers found all of the (very slight) acceleration Church and White measured occurred prior to 1930 – when atmospheric carbon dioxide levels were under 310 ppm (Burton, 2012).

More frequent or more intense storms could become a concern for military bases, but empirical data do not show a long-term trend in either measure (Alexander *et al.*, 2006; Khandekar, 2013; Pielke Jr., 2013, 2014; Landsea, 2018). The IPCC's computer models cannot produce reliable regional results, much less forecast the weather near existing military installations, so a global average is meaningless for military purposes. The best practice is to measure real-world weather conditions on-the-ground and determine if trends justify taking action.

Another unnecessary expense is making preparations for the U.S. military to respond to humanitarian crises. Natural disasters occur around the world on a nearly daily basis. In most cases, local governments, civic institutions, and private enterprise rise to the challenge by providing medical aid to the injured and rebuilding damaged homes and infrastructure. International aid organizations such as

Red Cross also arrive to help. Under Obama, DoD was told to anticipate conditions where the U.S. military would be called upon to provide disaster relief and humanitarian assistance on an ever-increasing basis; to consider how to alter force plans, training, and acquisition strategies; and to contemplate alterations and adaptations in DoD's bases and physical infrastructure to accommodate expected environmental challenges.

The United States is a generous nation. Natural disasters generally elicit an outpouring of money and assistance from U.S. citizens, philanthropic organizations, and the government, but not for every disaster and not in every circumstance. Using public concern and interest in climate change as a way to divert public resources intended for national defense to foreign aid missions, without congressional appropriations or express public approval, seems an improper use of presidential power. Choices must be made about when and how extensively to respond. In a world where such demands on U.S. resources might increase, policymakers and defense officials need to make choices based on solid science and real-world situations, not United Nations computer models (Hayward *et al.,* 2014).

Development of a credible national energy policy would help support national strategy that defines our role in international affairs. Where timing is of the essence, it would direct distribution of needed resources when circumstances warrant. Rather than burden the U.S. military with unnecessary and costly preparations for international assistance based upon unrealistic predictions of global warming, military planning ought to reflect national interests and strategic policies, and certainly our humanitarian values, and engender diplomatic and geopolitical advantage. DoD is never the sole repository of disaster relief capabilities. As noted above, various institutions also assist. Nor should key military resources be diverted for ill-conceived and premature infrastructure adaptations or altering basic force requirements, as was proposed by the Obama administration. A national energy policy brings unity to disparate public and private agencies involved with international assistance.

Among the choices to be made is whether to continue U.S. military engagements in the Middle East. Section 7.1.3 of this chapter addressed "wars for oil" in some depth, and concluded the United States is not in the Middle East to ensure access to cheap oil, since many of our interventions had other (among them humanitarianism and national pride) justifications, oil is hardly a scarce resource, and the United States is no longer dependent on the Middle East for a significant part of its oil supplies. With the United States about to become a net oil exporter thanks to the shale revolution (EIA, 2018), public support for maintaining so many troops in the region (approximately 35,000, with 13,000 in Kuwait and 5,000 in Bahrain, where energy security is the stated purpose (see Glaser, 2017)), may be expected to fall.

The election of Donald Trump as president of the United States marked a decisive turning point in climate change policy in the United States. Immediately after taking office, Trump approved the Keystone XL and Dakota Access natural gas pipeline projects that had been blocked by the Obama administration for years (Cama, 2017). In March 2017, Trump issued an "Executive Order on Promoting Energy Independence and Economic Growth" revoking and beginning the process of rescinding many Obama-era policies, including Obama's Climate Action Plan and Clean Power Plan, and disbanding the Interagency Working Group on Social Cost of Greenhouse Gases (Trump, 2017a).

In June 2017, Trump announced he would withdraw the United States from the Paris Climate Accord (Trump, 2017b). In December, he announced the administration would remove "climate change" from its list of threats to national security (Trump, 2017c). Indeed, the phrase appears nowhere in the *National Security Strategy* released that month; it says only, "The United States will continue to advance an approach that balances energy security, economic development, and environmental protection" (Executive Office of the President, 2017).

Under Trump, the U.S. Department of Energy, Department of the Interior, and the Environmental Protection Agency have taken steps to remove punitive regulations imposed on coal, oil, and natural gas producers during the Obama era, and recently announced plans to protect the nation's coal generation plants in the name of ensuring a reliable energy supply in the event of cyberattacks that could disable gas pipelines (Colman, 2018). These seem to be reasonable steps toward restoring balance to U.S. energy policy as well as military policy.

References

Alexander, L.V., *et al.* 2006. Global observed changes in daily climate extremes of temperature and precipitation. *Journal of Geophysical Research* **111** (D5): 10.1029/2005JD006290.

Bartis, J.T. 2012. *Promoting International Energy Security*. Santa Monica, CA: RAND Corp.

Bartis, J.T. and Van Bibber, L. 2011. *Alternative Fuels for Military Applications*. Santa Monica, CA: RAND Corp.

Biello, D. 2009. Navy green: military investigates biofuels to power its ships and planes. Scientific American (website). September 14.

Burton, D. 2012. Comments on 'Assessing future risk: quantifying the effects of sea level rise on storm surge risk for the southern shores of Long Island, New York.' *Natural Hazards* **63**: 1219.

Busby, J. 2007. *Climate Change and National Security: An Agenda for Action*. New York, NY: Council on Foreign Relations.

Cama, T. 2017. Trump approves Keystone pipeline. *The Hill*. March 14.

Casey, T. 2014. Department of Defense goes big on wind, solar, and biomass. Clean Technica (website). February 21.

Center for a New American Security. 2008. Uncharted waters: the U.S. Navy and navigating climate change. *Working Paper*. December.

Church, J.A. and White, N.J. 2006. A 20th century acceleration in global sea-level rise. *Geophysical Research Letters* **33**: L01602.

CNA. 2014. CNA Corporation. *National Security and the Accelerating Risks of Climate Change*. CNA Military Advisory Board. Alexandria, VA: CNA Corporation.

CNA. 2011. Center for Naval Analyses and Oxfam America. A*n Ounce of Prevention: Preparing for the Impact of a Changing Climate on U.S. Humanitarian and Disaster Response*. Washington, DC: Center for Naval Analyses.

Colman, Z. 2018. Foes of Trump coal plan worry it will work. *E&E News*. June 11.

DoD. 2014a. U.S. Department of Defense. *2014 Climate Change Adaptation Roadmap*.

DoD. 2014b. U.S. Department of Defense. DoD releases 2014 climate change adaptation roadmap. *News Release*. October 13.

EIA. 2018. U.S. Energy Information Administration. The United States is projected to become a net energy exporter in most AEO2018 cases. Today in Energy (website). February 12.

Executive Office of the President. 2015. *The National Security Implications of a Changing Climate*. Washington, DC: White House. May 20.

Executive Office of the President. 2017. *National Security Strategy of the United States of America*. Washington, DC: White House. December.

Fingar, D.T. 2008. National intelligence assessment on the national security implications of global climate change to 2030. Testimony before the House Permanent Select Committee on Intelligence and the House Select Committee on Energy Independence and Global Warming. Washington, DC.

Glaser, J. 2017. Does the U.S. military actually protect Middle East oil? The National Interest (website). January 9.

Hayward, T.B., Briggs, E.S., and Forbes, D.K. 2014. *Climate Change, Energy Policy, and National Power*. Chicago, IL: The Heartland Institute.

Khandekar, M.L. 2013. Are extreme weather events on the rise? *Energy & Environment* **24**: 537–49.

Kiefer, T.A. 2013. Energy insecurity: the false promise of liquid biofuels. *Strategic Studies Quarterly* (Spring): 114–51.

Landsea, C. 2018. What is the complete list of continental U.S. landfalling hurricanes? (website). Hurricane Research Division, Atlantic Oceanographic and Meteorological Laboratory, National Oceanic and Atmospheric Administration. Accessed June 29, 2018.

Maron, D.F. 2011. Biofuels of no benefit to military—RAND. *The New York Times*. January 25.

McGrady, E., Kingsley, M., and Stewart, J. 2010. *Climate Change: Potential Effects on Demands for U.S. Military Humanitarian Assistance and Disaster Response*. Washington, DC: Center for Naval Analyses.

NRC. 2011. National Research Council. Renewable fuel standard: potential economic and environmental effects of U.S. biofuel policy. Washington, DC: National Academies Press.

Pielke Jr., R.A. 2013. Testimony before the Committee on Environment and Public Works of the U.S. Senate. July 18.

Pielke Jr., R.A. 2014. *The Rightful Place of Science: Disasters and Climate Change*. Tempe, AZ: Arizona State University Consortium for Science, Policy & Outcomes.

Schwartz, P. and Randall, D. 2003. *An Abrupt Climate Change Scenario and Its Implications for United States*

National Security. Pasadena, CA: Jet Propulsion Laboratory.

Stacy, T.F. and Taylor, G.S. 2016. *The Levelized Cost of Electricity from Existing Generation Resources*. Washington, DC: Institute for Energy Research.

Trump, D. 2017a. Presidential Executive Order on Promoting Energy Independence and Economic Growth. Washington, DC: White House. March 28.

Trump, D. 2017b. Statement by President Trump on the Paris Climate Accord. Washington, DC: White House. June 1.

Trump, D. 2017c. Remarks by President Trump on the administration's national security strategy. Washington, DC: White House. December 18.

7.3.5 Conclusion

Predictions that climate change will lead directly or indirectly to violent conflict presume mediating institutions and human capital will not resolve conflicts before they escalate to violence.

Empirical research does not support the IPCC's contention that climate change will lead to violent conflicts, a failure easily explained by the methodological flaws in the argument. Each of the five alleged sources of conflict examined in this section – abrupt climate change, water shortages, famine, resource scarcity, and refugee flows – are revealed to be lacking in proof and plausibility.

One way in which proponents of the climate-conflict argument have responded to the lack of empirical support for their position is to suggest that climate-induced change will cause future conflicts because the problems will be *so much worse* than anything that has been experienced previously. This logic allows proponents to dismiss the lack of empirical evidence in support of the causal linkages, because the argument is purely concerned with the prospects for future conflict. Environmental factors then become an additive fuel to a combustible mixture. Statements like that offered by President Barack Obama's 2010 *National Security Strategy*, "The change wrought by a warming planet will lead to new conflicts over refugees and resources," are deterministic and predictive, but ultimately not testable.

The deterministic interpretation artificially assumes limits on the adaptability of the actors involved or other institutions that can play stabilizing roles. The countries and groups affected by an environmental phenomenon may not react in a manner consistent with the expectations of computer modelers or "futurists." The mediating effects of other nations, nongovernmental organizations, new technology, and the output of human capital can all defuse a crisis. These dynamics are impossible to model or incorporate into a testable hypothesis, and yet experience shows they exist and are important. As Tir and Stinnett observed, "Forecasts that do not account for the important conflict management potential of international institutions will produce overly pessimistic scenarios regarding the impact of climate change on international security" (Tir and Stinnett, 2012). Those agreements and institutions provide a means to seek reconciliation and adjudication of interests before conflict escalates to violence and offer a venue for the appropriate expression of tension. The conflict scenarios all presume these elements fail or are not present, and so they are wrong.

References

Executive Office of the President. 2010. *National Security Strategy*. Washington, DC: The White House.

Tir, J. and Stinnett, D. 2012. Weathering climate change: can international institutions mitigate international water conflict? *Journal of Peace Research* **49** (1): 211–25.

7.4 Human History

A large literature exists on the historical relationship between climate and human security. Much of it shows humanity enjoyed periods of peace during warmer periods or periods of rising temperatures, while cooler periods or periods of falling temperatures have been accompanied by human suffering and often armed conflict. This research contradicts the narrative of the IPCC and its supporters, and for that reason it is seldom referenced in the IPCC assessment reports or by those who advocate for immediate action to address climate change.

Section 7.4.1 summarizes recent research on the relationship between climate and human security in China, the world's most populous nation and the one with the longest and most detailed historical records.

Section 7.4.2 presents research from other parts of the world.

7.4.1 China

Extensive historical research in China reveals a close and positive relationship between a warmer climate and peace and prosperity, and between a cooler climate and war and poverty.

China is a good test case for the relationship between global warming and violent conflict because it has been a well-populated, primarily agricultural country for millennia, and it has a relatively well-recorded history over this period. Accordingly, several researchers have conducted analyses of factors influencing social stability in China.

Zhang *et al.* (2005) noted historians typically identify political, economic, cultural, and ethnic unrest as the chief causes of war and civil strife in China. However, the five Chinese scientists contend climate plays a key role as well, and to examine their thesis they compared proxy climate records with historical data on wars, social unrest, and dynastic transitions in China from the late Tang to Qing Dynasties (mid-ninth century to early twentieth century). Their research revealed war frequencies, peak war clusters, nationwide periods of social unrest, and dynastic transitions were all significantly associated with cold, not warm, phases of China's oscillating climate. Specifically, all three distinctive peak war clusters (defined as more than 50 wars in a 10-year period) occurred during cold climate phases, as did all seven periods of nationwide social unrest and nearly 90% of all dynastic changes that decimated this largely agrarian society. They conclude climate change was "one of the most important factors in determining the dynastic cycle and alternation of war and peace in ancient China," with warmer climates having been immensely more effective than cooler climates in terms of helping "keep the peace."

Zhang *et al.* (2007a) utilized high-resolution paleoclimate data to explore the effects of climate change on the outbreak of war and population decline at a global and continental scale in the pre-industrial era, as discerned by analyses of historical socioeconomic and demographic data over the period AD 1400–AD 1900. In describing their findings, they report "cooling impeded agricultural production,

which brought about a series of serious social problems, including price inflation, then successively war outbreak, famine, and population decline." And they suggest that "worldwide and synchronistic war-peace, population, and price cycles in recent centuries have been driven mainly by long-term climate change," wherein warm periods were supportive of good times and cooling led to bad times.

In response to "the gradual temperature drop and the increase in size of the cold area from the 'Medieval Warm Period' to the Little Ice Age," for example (when Zhang *et al.* found that every sudden temperature drop would induce a "demographic shock"), population growth rate "reached its lowest level in the 13–14th centuries, primarily because of epidemics, wars, and famines." In providing more detail, they say "the invasion by the Mongols in the 13–14th centuries was related to the ecological stress caused by cooling, which reduced China's total population nearly by half (~55 million decline)," while in Europe they report the Black Death held sway, "accompanied by massive social unrest and economic collapse, which wiped out a quarter to one-third of the population in AD 1347–1353, the coldest period in the last several hundred years." Then, in the seventeenth century, which was the longest cold period of the Little Ice Age, they report "more wars of great magnitude and the associated population declines in Europe and Asia followed." More specifically, they state "the European population was devastated by possibly the worst war in its history in terms of the share of the population killed in AD 1618–1648, starvation, and epidemics." Likewise, they report "in China, the population plummeted 43 percent (~70 million) because of wars, starvation and epidemics in AD 1620–1650."

Liu *et al.* (2009) derived a 2,485-year mean annual temperature history of the mid-eastern Tibetan Plateau based on Qilian juniper (*Sabina przewalskii*) tree-ring width chronologies spanning the time period 484 BC–AD 2000, which they demonstrated to be well correlated with several temperature histories of the Northern Hemisphere. The eight researchers report there were four periods of average temperatures in their record similar to "or even higher than" the mean of AD 1970–2000. Liu *et al.* also report the high-temperature intervals during the first millennium were what could be described as relatively good times. The downfalls of most major dynasties in China coincided with intervals of low temperature, or at least the beginnings of their downfalls did, citing the demise of the Qin, Three

Kingdoms, Tang, Song (North and South), Yuan, Ming, and Qing Dynasties.

Lee and Zhang (2010) examined data on Chinese history, including temperature, wars and rebellions, epidemics, famines, and population for the past millennium. Over their study interval of 911 years, they found nomad migrations, rebellions, wars, epidemics, floods, and droughts were all higher during cold periods. All of these factors tended to disrupt population growth or increase mortality. Overall, five of six population contractions, constituting losses of 11.4% to 49.4% of peak population, were associated with a cooling climate. The sixth cool period evinced a great reduction in population growth rate during a cool phase, but not a collapse. None of the population contractions was associated with a warming climate.

Zhang *et al.* (2010) note "climatic fluctuation may be a significant factor interacting with social structures in affecting the rise and fall of cultures and dynasties," citing Cowie (1998) and Hsu (1998). When the climate worsens beyond what the available technology and economic system can accommodate – that is, beyond the society's adaptive capacity – they state, "people are forced to move or starve." Zhang *et al.* also note "climate cooling has had a huge impact on the production of crops and herds in pre-industrial Europe and China (Hinsch, 1998; Atwell, 2002; Zhang *et al.*, 2007a), even triggering mass southward migration of northern nomadic societies (Fang and Liu, 1992; Wang, 1996; Hsu, 1998)," and "this ecological and agricultural stress is likely to result in wars and social unrest, often followed by dynastic transitions (Zhang *et al.*, 2005)." In fact, they write, "recent studies have demonstrated that wars and social unrests in the past often were associated with cold climate phases (Zhang *et al.*, 2005, 2007a,b)," and "climate cooling may have increased locust plagues through temperature-driven droughts or floods in ancient China (Stige *et al.*, 2007; Zhang *et al.*, 2009)."

In a study designed to explore the subject further, Zhang *et al.* employed "historical data on war frequency, drought frequency and flood frequency" compiled by Chen (1939), and "a multi-proxy temperature reconstruction for the whole of China reported by Yang *et al.* (2002), air temperature data for the Northern Hemisphere (Mann and Jones, 2003), proxy temperature data for Beijing (Tan *et al.*, 2003), and a historical locust dataset reported by Stige *et al.* (2007)," plus "historical data of rice price variations reported by Peng (2007)." In analyzing the linkages among these factors, the researchers report

"food production during the last two millennia has been more unstable during cooler periods, resulting in more social conflicts." They specifically note "cooling shows direct positive association with the frequency of external aggression war to the Chinese dynasties mostly from the northern pastoral nomadic societies, and indirect positive association with the frequency of internal war within the Chinese dynasties through drought and locust plagues," which typically have been more pronounced during cooler as opposed to warmer times.

Zhang *et al.* conclude "it is very probable that cool temperature may be the driving force in causing high frequencies of meteorological, agricultural disasters and then man-made disasters (wars) in ancient China," noting "cool temperature could not only reduce agricultural and livestock production directly, but also reduce agricultural production by producing more droughts, floods and locust plagues." They also observe the subsequent "collapses of agricultural and livestock production would cause wars within or among different societies." Consequently, although "it is generally believed that global warming is a threat to human societies in many ways (IPCC, 2007)," Zhang *et al.* arrive at a different conclusion, stating some countries or regions might actually "benefit from increasing temperatures," citing the work of Nemani *et al.* (2003), Stige *et al.* (2007), and Zhang *et al.* (2009), while restating the fact that "during the last two millennia, food production in ancient China was more stable during warm periods owing to fewer agricultural disasters, resulting in fewer social conflicts."

In their study of widespread crises in China, Lee and Zhang (2013) write "the fall of the Ming dynasty in the first half of the 17th century and the Taiping Rebellion from 1851–1865 were two of the most chaotic periods in Chinese history," each of which "was accompanied by large-scale population collapses." Utilizing "high-resolution empirical data, qualitative survey, statistical comparison and time-series analysis" to investigate how climate change and population growth "worked synergistically to drive population cycles in 1600–1899," they found that "recurrences of population crises were largely determined by the combination of population growth and climate change." More specifically, "in China in the past millennium, the clustering of natural calamities and human catastrophes in times of cold climate was found not only in one or two cold phases, but in all of the cold phases (Lee and Zhang, 2010)."

China is not different from the rest of the world

in this regard. During what is known as the General Crisis of the Seventeenth Century, for example, Lee and Zhang note "the crown of the Holy Roman Empire was unsettled by the Thirty Years' War," "civil war devastated France," "in London, Charles I was condemned to death by his own subjects," and Spain's Philip IV "lost almost all his possessions in Asia." In addition, Lee and Zhang mention the Puritan Revolution in England, the revolts of Scotland and Ireland, the insurrections in the Spanish monarchy – including Catalonia and Portugal in 1640 and Naples and Palermo in 1647 – the Fronde in France between 1648 and 1653, the bloodless revolt of 1650 that displaced the stadholderate in the Netherlands, the revolt of the Ukraine from 1648 to 1654, as well as "a string of peasant risings across the [European] continent (Parker and Smith, 1978)."

After analyzing these situations and others, Lee and Zhang conclude "both natural calamities and human catastrophes are clustered in periods of cold climate," primarily because cooling "generates a devastating impact on agricultural production everywhere," citing the work of Atwell (2001, 2002), while also noting "declines in temperatures often have had catastrophic consequences for the world's food supply."

Wei *et al.* (2014) point out "climate change has long been suggested as a factor of great importance in facilitating the rise or fall of culture," citing Issar and Zohar (2007), but they note "this type of study still faces the lack of high-resolution data of long-term socio-economic processes." In research designed to overcome this deficiency, they found more than 1,100 such sets of information in 24 Chinese fiscal and economic history books, plus other well-preserved historical documents, from which they constructed a 2,130-year (220 BC to AD 1910) fiscal-state sequence with decadal resolution that is representative of the phase transition history of China's fiscal soundness.

Wei *et al.* found "the fiscal balance of dynasties from 220 BC to AD 1910 experienced seven large stages." More specifically, "the relatively sufficient periods dominated from 220 to 31 BC, AD 581–1020, AD 1381–1520 and from AD 1681–1910," whereas the relatively deficient periods were the three intervening time intervals. The three Chinese researchers discovered that "fiscal crisis was more likely to occur in cold-dry climatic scenarios," noting that "both temperature and precipitation displayed more significant effects on the fiscal fluctuation within the long term, particularly for temperature."

Jia (2014) notes China is "a good testing ground for the link between weather shocks and civil conflict, as there is detailed information on abnormal weather conditions and the occurrence of peasant revolts at the prefecture level going back to the 15th century," which data indicate a peasant revolt occurred in 0.22% of all prefecture-years. However, when focusing only on prefecture-years when there was an exceptional drought, Jia says "there was a peasant revolt in 0.58 percent of prefecture-years," such that "a peasant revolt at the prefecture level was almost three times more likely in a drought year." In addition, Jia found the price effect of droughts was nearly three times that of floods, and droughts thus had more severe negative effects on local food production, consistent "with historians' argument that droughts were the most important natural disasters driving historical peasant revolts," citing Xia (2010).

With respect to how the introduction of drought-resistant sweet potatoes helped mitigate civil conflict, Jia collected data on their adoption and diffusion across different provinces or collections of prefectures, finding that before the introduction of sweet potatoes "there was a peasant revolt in 0.78 percent of prefecture-years with an exceptional drought," but that "after the introduction of sweet potatoes, there was a peasant revolt in only 0.26 percent of prefecture-years with an exceptional drought."

Wei *et al.* (2015) investigated the long-term relationship between the climate and economy of China, returning to the 2,130-year record of the Chinese economy they developed in previous research. This proxy was statistically analyzed in conjunction with historical proxies of Chinese temperature and precipitation previously compiled by Ge *et al.* (2013) and Zheng *et al.* (2006), respectively. Wei *et al.* found that warm and wet climate periods coincided with more prosperous and robust economic phases (above-average mean economic level, higher ratio of economic prosperity, and less intense variations), whereas opposite economic conditions ensued during cold and dry periods, where the possibility of economic crisis was "greatly increased" (see Figure 7.4.1.1). They also report temperature was "more influential than precipitation in explaining the long-term economic fluctuations, whereas precipitation displayed more significant effects on the short-term macro-economic cycle."

In their study of climate change impacts on dynastic well-being in China over the period 210 BC to AD 1910, Yin *et al.* (2016) focused on relationships among dynastic transition and prosper-

Figure 7.4.1.1
Series comparison between economic fluctuations and climate changes in China from BC 220 to AD 1910

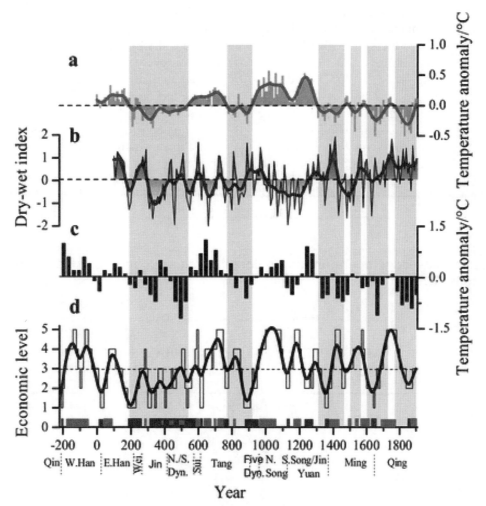

Panel a: Decadal temperature anomaly for all of China during the period AD 1–1910 (Ge *et al.*, 2013); the red curve is the low-pass filtered series. Panel b: Decadal precipitation over eastern China during the period 101–1910 (Zheng *et al.*, 2006); the blue curve is the low-pass filtered series. Panel c: Winter half-year temperature anomaly series for eastern China during the period BC 210–AD 1920 with a 30-year resolution (Ge, 2011). Panel d: Decadal macro-economic series during the period BC 220–AD 1910 in China; the black curve is the low-pass filtered series. The red and blue bars indicate typical episodes of prosperity and crisis periods (respectively). The gray and white areas delineate cold and warm phases, respectively.

Source: Wei *et al.,* 2015.

ity and how they were affected by historical climate change and its impacts on grain harvests. The three Chinese researchers report that from 210 BC to AD 1910, unfavorable dynastic transitions mostly coincided with changes from warm-to-cold and wet-to-dry periods, when there were relatively poor harvests, noting "dynastic prosperity mostly coincided with warm ages or the periods that changed from cold to warm and wet or dry-to-wet periods," when they report there were bumper grain harvests.

Yin *et al*. note "dynastic prosperity tended to appear in warm periods or cold-to-warm periods, wet or dry-to-wet periods, and crop abundance periods," further noting "transitions from chaos to unity tended to occur at the ends of centuries-long cold periods and at the beginning of warm periods." They say "collapse of the Tang Dynasty was haunted by colder weather and declining grain harvests."

Lee *et al.* (2017) analyzed the association between climate change and health-related epidemics recorded in China over the period 1370–1909 AD. For climate data, they utilized the temperature reconstruction of Yang *et al.* (2002) and the precipitation reconstruction of Zhang *et al.* (2015). Epidemic data were aggregated from three independently derived datasets, *Collection of Meteorological Records in China over the Past Three Thousands Years* (Zhang, 2004), *Historical Records of Infectious Diseases in China* (Li, 2004), and *Epidemic Records in Historical China* (Zhang, 2007).

All data and the relationships among them were analyzed on three spatial scales (national, regional, and provincial).There were a total of 5,961 epidemic incidents across China during the study period. Statistical analyses revealed that precipitation was not significantly correlated with epidemic count.

Temperature, on the other hand, was found to be "negatively correlated with epidemic incidents" (see Figure 7.4.1.2). Additionally, Lee *et al.* calculated that for every one standard deviation decrease in temperature at the country, regional, or provincial level, increases of 162, 34, and 3.4 epidemic outbreaks were observed, respectively. Consequently, Lee *et al.* conclude their analysis "supports the notion that climate change, be it the ultimate cause or direct trigger, acts as a driver of historical epidemics," but that global *cooling, not warming,* is to be feared.

Figure 7.4.1.2
The relationship between temperature and epidemic incident count for all of China over the period 1370-1909 AD

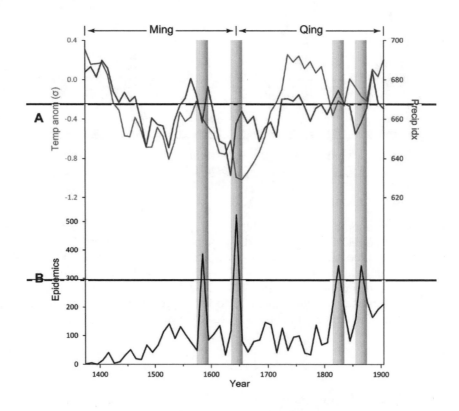

Panel (A) temperature anomaly (°C) (blue line) and precipitation index (red line). Panel (B) count of epidemics incidents.

Source: Lee *et al.,* 2017.

Wang *et al.* (2018) developed a 4,000-year proxy temperature reconstruction based on chironomid (midge) assemblages in a sediment core from Gonghai Lake (38.9°N, 112.23°E), an alpine freshwater lake located on the northeastern margin of the Chinese Loess Plateau in the Shanxi Province. The reconstruction was then compared with published war and population records for the Shanxi Province to explore the relationship between climate change and human societal changes for this region. Their findings are presented in Figure 7.4.1.3.

As shown in the figure, there have been multiple centennial-scale fluctuations but an overall decline in temperature over the 4,000 year record. That finding is not surprising since the record begins at the warmest interval of the current interglacial period. Notable warm events in the record include the Sui-Tang Warm Period (1270–1040 cal yr BP), the Medieval Warm Period (~970–570 cal yr BP), and the modern warm period. Notable cold events include the Chinese Period of Disunity (~1700–1270 cal yr BP), the Era of the Five Dynasties and Ten Kingdoms (~1040–970 cal yr BP), and the Little Ice Age (~570–270 cal yr BP).

In examining the relationship between climate (their chironomid temperature proxy and an independent pollen-based reconstruction from the same lake by Chen *et al.*, 2015) and societal change, Wang *et al.* report wars "occurred more frequently when temperature and precipitation decreased abruptly," noting that war events were more strongly correlated with temperature than precipitation. The most severe era of war events occurred during the coldest period of the record, i.e., the Little Ice Age.

With respect to population, Wang *et al.* report "an increase [in population] often occurred during warm periods," which provided relief from the harsh economic pressures brought about by poor crop harvests during colder periods, when yields were reduced by as much as 50%. Not surprisingly, reduced crop yields during cold eras would trigger higher food prices and famine, creating "large numbers of homeless refugees and outbreaks of plague," eventually resulting in "wars and social unrest which acted to reduce the population size."

References

Atwell, W.S. 2001. Volcanism and short-term climatic change in East Asian and world history, c. 1200–1699. *Journal of World History* 12: 29–98.

Atwell, W.S. 2002. Time, money, and the weather: Ming China and the 'great depression' of the mid-fifteenth century. *Journal of Asian Studies* 61: 83–113.

Chen, F.H., *et al.* 2015 East Asian summer monsoon precipitation variability since the last deglaciation. *Scientific Reports* 5: 11186.

Chen, G.Y. 1939. *China Successive Natural and Manmade Disasters Table.* Guangzhou, China: Jinan University Book Series.

Cowie, J. 1998. *Climate and Human Change: Disaster or Opportunity?* New York, NY: Parthenon Publishing Group.

Fang, J. and Liu, G. 1992. Relationship between climatic change and the nomadic southward migrations in eastern Asia during historical times. *Climatic Change* 22: 151–68.

Ge, Q., Hao, Z., Zheng, J., and Shao, X. 2013. Temperature changes over the past 2000 yr in China and comparison with the Northern Hemisphere. *Climate of the Past* 9: 1153–60.

Ge, Q.S. 2011. *Climate Change in Chinese Dynasties.* Beijing, China: Science Press.

Hinsch, B. 1998. Climate change and history in China. *Journal of Asian History* 22: 131–59.

Hsu, K.J. 1998. Sun, climate, hunger and mass migration. *Science in China Series D – Earth Sciences* 41: 449–72.

IPCC. 2007. Intergovernmental Panel on Climate Change. *Climate Change 2007: Synthesis Report.* Contribution of Working Groups I, II and III to the Fourth Assessment Report of the Intergovernmental Panel on Climate Change. Geneva, Switzerland.

Issar, A.S. and Zohar, M. 2007. *Climate Change: Environment and History of the Near East.* Berlin/Heidelberg, Germany: Springer.

Jia, R. 2014. Weather shocks, sweet potatoes and peasant revolts in historical China. *The Economic Journal* 124: 92–118.

Lee, H.F. and Zhang, D.D. 2010. Changes in climate and secular population cycles in China, 1000 CE to 1911. *Climate Research* 42: 235–46.

Lee, H.F. and Zhang, D.D. 2013. A tale of two population crises in recent Chinese history. *Climatic Change* 116: 285–308.

Lee, H.F., Fei, J., Chan, C.Y.S., Pei, Q., Jia, X., and Yue, R.P.H. 2017. Climate change and epidemics in Chinese history: a multi-scalar analysis. *Social Science & Medicine* 174: 53–63.

Figure 7.4.1.3
Temperature proxy, number of wars, and population of the Shanxi Province of China, from 4,000 years BP to current

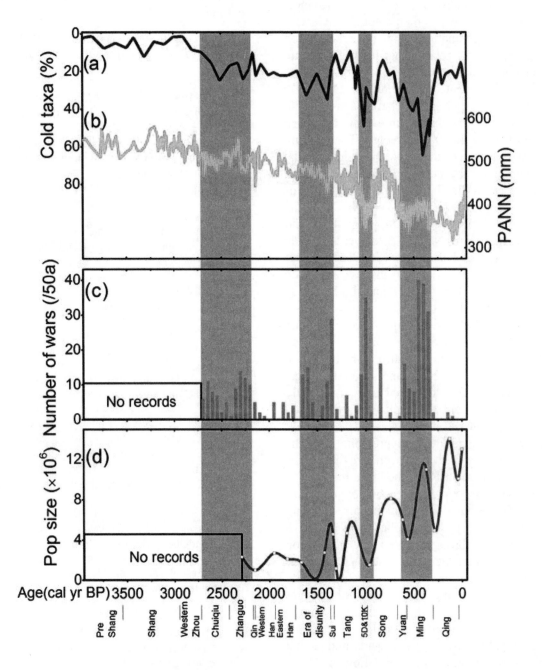

Comparison of (a) cold-preferring taxa percentages and (b) reconstructed precipitation at Gonghai Lake (Chen *et al.*, 2015) with (c) frequencies of wars in Shanxi Province, China, and (d) population size (in units of 1 million, square dots) of Shanxi Province during the past 2300 years; the data are spline connected. Grey shaded areas indicate cold events. *Source:* Wang *et al.,* 2018.

Li, W. 2004. *Zhongguo Chuanranbing Shiliao* [Historical Records of Infectious Diseases in China]. Beijing, China: Chemical Industry Press.

Liu, Y., An, Z.S., Linderholm, H.W., Chen, D.L., Song, M.H., Cai, Q.F., Sun, J.S., and Tian, H. 2009. Annual temperatures during the last 2485 years in the mid-eastern Tibetan Plateau inferred from tree rings. *Science in China Series D Earth Science* **52**: 348–59.

Mann, M.E. and Jones, P.D. 2003. Global surface temperatures over the past two millennia. *Geophysical Research Letters* **30**: 10.1029/2003GL017814.

Nemani, R.R., Keeling, C.D., Hashimoto, H., Jolly, W.M., Piper, S.C., Tucker, C.J., Myneni, R.B., and Running, S.W. 2003. Climate-driven increases in global terrestrial net primary production from 1982 to 1999. *Science* **300**: 1560–3.

Parker, G. and Smith, L.M. 1978. Introduction. In: Parker, G. and Smith, L.M. (Eds.) *The General Crisis of the Seventeenth Century*. London, UK: Routledge and Kegan Paul, pp. 1–25.

Peng, X.W. 2007. *History of Chinese Currency*. Shanghai, China: Shanghai People's Press.

Stige, L.C., Chan, K.S., Zhang, Z.B., Frank, D., and Stenseth, N.C. 2007. Thousand-year-long Chinese time series reveals climatic forcing of decadal locust dynamics. *Proceedings of the National Academy of Sciences USA* **104**: 16,188–93.

Tan, M., Liu, T., Hou, J., Qin, X., Zhang, H., and Li, T. 2003. Cyclic rapid warming on centennial scale revealed by a 2650-year stalagmite record of warm season temperature. *Geophysical Research Letters* **30**: 10.1029/2003GL017352.

Wang, H., Chen, J., Zhang, S., Zhang, D.D., Wang, Z., Xu, Q., Chen, S., Wang, S., Kang, S., and Chen, F. 2018. A chironomid-based record of temperature variability during the past 4000 years in northern China and its possible societal implications. *Climate of the Past* **14**: 383–96.

Wang, H.C. 1996. The relationship between the southern migrations of north nomadic tribes and climate change in China. *Scientia Geographica Sinica* **16**: 274–9.

Wei, Z., Fang, X., and Su, Y. 2014. Climate change and fiscal balance in China over the past two millennia. *The Holocene* **24**: 1771–84.

Wei, Z., Fang, X., and Su, Y. 2015. A preliminary analysis of economic fluctuations and climate changes in China from BC 220 to AD 1910. *Regional Environmental Change* **15**: 1773–85.

Xia, M. 2010. Zhongguo lishishangde hanzai jiqi chengyin (Droughts in historical China and their causes. *Guangming Daily*. April 21, 2004.

Yang, B., Braeuning, A., Johnson, K.R., and Shi, Y.F. 2002. General characteristics of temperature variation in China during the last two millennia. *Geophysical Research Letters* **29**: 38–42.

Yin, J., Fang, X., and Su, Y. 2016. Correlation between climate and grain harvest fluctuations and the dynastic transitions and prosperity in China over the past two millennia. *The Holocene* **26**: 1914–23.

Zhang, D. 2004. *Zhongguo Sanqiannian Qixiang Jilu Zongji* [Collection of Meteorological Records in China over the Past Three Thousands Years]. Nanjing, China: Fenghuang Chubanshe.

Zhang, D.D., Brecke, P., Lee, H.F., He, Y.Q., and Zhang, J. 2007a. Global climate change, war, and population decline in recent human history. *Proceedings of the National Academy of Sciences USA* **104**: 19,214–9.

Zhang, D.D., Jim, C.Y., Lin, C.S., He, Y.Q., and Lee, F. 2005. Climate change, social unrest and dynastic transition in ancient China. *Chinese Science Bulletin* **50**: 137–44.

Zhang, D.D., Pei, Q., Lee, H.F., Zhang, J., Chang, C., Li, B., Li, J., and Zhang, X. 2015. The pulse of imperial China: a quantitative analysis of long-term geopolitical and climatic cycles. *Global Ecology and Biogeography* **24**: 87–96.

Zhang, D.D., Zhang, J., Lee, H.F., and He, Y. 2007b. Climate change and war frequency in eastern China over the last millennium. *Human Ecology* **35**: 403–14.

Zhang, Z. 2007. *Zhongguo Gudai Yibing Liuxing Nianbiao* [Epidemic Records in Historical China]. Fujian, China: Fujian Science & Technology Publishing.

Zhang, Z., Tian, H., Cazelles, B., Kausrud, K.L., Brauning, A., Guo, F., and Stenseth, N.C. 2010. Periodic climate cooling enhanced natural disasters and wars in China during AD 10–1900. *Proceedings of the Royal Society B* **277**: 3745–53.

Zhang, Z.B., Cazelles, B., Tian, H.D., Stige, L.C., Brauning, A., and Stenseth, N.C. 2009. Periodic temperature-associated drought/flood drives locust plagues in China. *Proceedings of the Royal Society B* **276**: 823–31.

Zheng, J.Y., Wang, W.C., Ge, Q.S., Man, Z.M., and Zhang, P.Y. 2006. Precipitation variability and extreme events in eastern China during the past 1500 years. *Terrestrial Atmospheric and Oceanic Sciences* **17**: 579–92.

7.4.2 Rest of the World

The IPCC relies on second- or third-hand information with little empirical backing when commenting on the implications of climate change for conflict.

Focusing on Europe, Tol and Wagner (2010) write that in "gloomier scenarios of climate change, violent conflict plays a key part," noting that in such visions of the future "war would break out over declining water resources, and millions of refugees would cause mayhem." They note "the Nobel Peace Prize of 2007 was partly awarded to the IPCC and Al Gore for their contribution to slowing climate change and thus preventing war." However, they warn "scenarios of climate-change-induced violence can be painted with abandon," citing the example of Schwartz and Randall (2003), because, as they continue, "there is little research to either support or refute such claims."

Tol and Wagner proceeded to do for Europe what Zhang et al. (2005, 2006) had done for China. Their results indicate "periods with lower temperatures in the pre-industrial era are accompanied by violent conflicts." They further determined "this effect is much weaker in the modern world than it was in pre-industrial times," which implies, in their words, "that future global warming is not likely to lead to (civil) war between (within) European countries." Therefore, they conclude, "should anyone ever seriously have believed that, this paper does put that idea to rest."

Buntgen et al. (2011) developed a set of tree ring-based reconstructions of central European summer precipitation and temperature variability over the past 2,500 years. In the abstract of their paper, the 12 researchers state, "wet and warm summers occurred during periods of Roman and medieval prosperity," and in the body of their paper they write, "average precipitation and temperature showed fewer fluctuations during the period of peak medieval and economic growth, ~1000 to 1200 C.E. (Kaplan et al., 2009; McCormick, 2001)," which suggests a warmer climate is better than a colder one for humanity.

Support for this point of view is provided by Buntgen et al.'s description of what happened as temperatures declined and the Medieval Warm Period gave way to the Little Ice Age, with its onset "likely contributing," in their words, "to widespread famine across central Europe," when they say "unfavorable climate may have even played a role in debilitating the underlying health conditions that contributed to the devastating economic crisis that arose from the second plague pandemic, the Black Death, which reduced the central European population after 1347 C.E. by 40 to 60 percent (Buntgen et al., 2010; Kaplan et al., 2009; Kausrud et al., 2010)." In addition, they note this period "is also associated with a temperature decline in the North Atlantic and the abrupt desertion of former Greenland settlements (Patterson et al., 2010)," and "temperature minima in the early 17th and 19th centuries accompanied sustained settlement abandonment during the Thirty Years' War and the modern migrations from Europe to America."

Chen et al. (2011) developed a high temporal resolution (four-year) sea surface temperature (SST) history based on a dinoflagelate cyst record obtained from a well-dated sediment core retrieved from a site in the Gulf of Taranto located at the distal end of the Po River discharge plume (39°50.07'N, 17°48.05'E) in the southern Italian region of the Mediterranean Sea. According to the authors, SST reconstructions based on the composition of dinoflagellate cysts recovered from the sediment core "suggest high stable temperatures between 60 BC and 90 AD followed by a decreasing trend between 90 AD and 200 AD." They also observed their "reconstruction of relatively warm stable climatic conditions corresponds to the time of the 'Pax Romana'," i.e., the long period of relative peace and minimal expansion by military force experienced by the Roman Empire in the first and second centuries AD.

Zhang et al. (2011a) preface their work by noting early paleo-temperature reconstructions suggested "massive social disturbance, societal collapse, and population collapse often coincided with great climate change in America, the Middle East, China, and many other countries in preindustrial times (Bryson and Murray, 1977; Atwell, 2001; deMenocal, 2001; Weiss and Bradley, 2001; Atwell, 2002)." They also say it has been shown more recently that "climate change was responsible for the outbreak of war, dynastic transition, and population decline in China, Europe, and around the world because of climate-induced shrinkage of agricultural production (Zhang et al., 2005, 2006, 2007a,b; Lee et al., 2008; Lee et al., 2009; Lee and Zhang, 2010; Tol and Wagner, 2010; Zhang et al., 2010; Zhang et al., 2011b)."

In a study designed to provide still greater support for this general relationship, Zhang et al. (2011a) "examined the climate-crisis causal mechanism in a period [AD 1500–1800] that contained both periods of harmony and times of crisis," the most prominent of the latter of which was the General Crisis of the Seventeenth Century (GCSC) in Europe, which was marked by widespread economic distress, social unrest, and population decline. The researchers examined linkages between temperature data and climate-driven economic variables that defined the "golden" and "dark" ages in Europe and North America.

Zhang et al. demonstrated that "climate change was the ultimate cause, and climate-driven economic downturn was the direct cause, of large-scale human crises in pre-industrial Europe and the Northern Hemisphere." In addition, they say it was *cooling* that triggered the chain of negative responses in variables pertaining to physical and human systems. Initially, for example, they found agricultural production "decreased or stagnated in a cold climate and increased rapidly in a mild climate at the multi-decadal timescale," while the time course of crisis development was such that "bio-productivity, agricultural production and food supply per capita (FSPC) sectors responded to temperature change immediately, whereas the social disturbance, war, migration, nutritional status, epidemics, famine and population sectors responded to the drop in FSPC with a 5- to 30-year time lag." Thus, the dark ages they delineated by these means were AD 1212–1381 (the Crisis of Late Middle Ages) and AD 1568–1665 (the GCSC), whereas the golden ages were the tenth to twelfth centuries (the High Middle Ages), the late fourteenth to early sixteenth centuries (the Renaissance), and the late seventeenth to eighteenth centuries (the Enlightenment). It thus can be concluded from several centuries of European and Northern Hemispheric data that warming and warmth beget human wellness, while cooling and cold produce human misery.

Cleaveland et al. (2003) developed a history of winter–spring (November–March) precipitation for the period 1386–1993 for the area around Durango, Mexico, based on earlywood width chronologies of Douglas-fir tree rings collected at two sites in the Sierra Madre Occidental. This reconstruction, in their words, "shows droughts of greater magnitude and longer duration than the worst historical drought," and none of them occurred during a period of unusual warmth, as some researchers claim they should; instead, they occurred during the Little Ice Age. They also note, "Florescano et al. (1995) make a connection between drought, food scarcity, social upheaval and political instability, especially in the revolutions of 1810 and 1910," and they note the great megadrought that lasted from 1540 to 1579

"may be related to the Chicimeca war (Stahle et al., 2000), the most protracted and bitterly fought of the many conflicts of natives with the Spanish settlers." If these concurrent events were indeed related, they too suggest warmer is better than cooler for maintaining social stability.

Working in East Africa, Nicholson and Yin (2001) analyzed climatic and hydrologic conditions from the late 1700s to close to the present, based on histories of the levels of 10 major African lakes and a water balance model they used to infer changes in rainfall associated with the different conditions, concentrating on Lake Victoria. The results they obtained were indicative of "two starkly contrasting climatic episodes." The first, which began sometime prior to 1800 during the Little Ice Age, was one of "drought and desiccation throughout Africa." This arid episode, which was most intense during the 1820s and 1830s, was accompanied by extremely low lake levels. As the two researchers describe it, "Lake Naivash was reduced to a puddle. ... Lake Chad was desiccated. ... Lake Malawi was so low that local inhabitants traversed dry land where a deep lake now resides. ... Lake Rukwa [was] completely desiccated. ... Lake Chilwa, at its southern end, was very low and nearby Lake Chiuta almost dried up."

Nicholson and Yin state that throughout this period "intense droughts were ubiquitous." Some, in fact, were "long and severe enough to force the migration of peoples and create warfare among various tribes." As the Little Ice Age's grip on the world began to loosen in the middle to latter part of the 1800s, however, things began to change for the better. The two researchers report, "semi-arid regions of Mauritania and Mali experienced agricultural prosperity and abundant harvests; floods of the Niger and Senegal Rivers were continually high; and wheat was grown in and exported from the Niger Bend region." Then, as the nineteenth century came to an end and the twentieth began, there was a slight lowering of lake levels, but nothing like what had occurred a century earlier; and in the latter half of the twentieth century, things once again improved, with the levels of some lakes rivaling high water characteristic of the years of transition to the Modern Warm Period.

According to Benjaminsen et al. (2012), "during the last few years, violent land-use conflict in the Sahel has become the most popular example of the alleged link between global climate change and conflict," noting "many politicians and international civil servants seem particularly attracted to this idea," as described in the study of Benjaminsen (2009).

They indicate this idea "was also at the core of the decision to award the 2007 Nobel Peace Prize to former US vice-president Al Gore and the Intergovernmental Panel on Climate Change (IPCC)."

Focusing on an area in the heart of the Sahel (the inland delta of the Niger River in the Mopti region of Mali), Benjaminsen *et al.* collected from the regional Court of Appeal in Mopti data on land-use conflicts that occurred within that region between 1992 and 2009, after which they compared the court data with contemporaneous climatic data. They also conducted a qualitative analysis of one of the many land-use conflicts in the region: a farmer-herder conflict, where young men from the village of Karbaye fired on a group of herders from the neighboring village of Guirowel, who were bringing livestock to a pond close to their homes, killing as many as five of them and injuring some 15 to 30 others.

With respect to the findings of the initial thrust of their study, the four Norwegian researchers found "a comparison of the conflict data with statistics on contemporaneous climatic conditions gives little substance to claims that climate variability is an important driver of these conflicts." And they go on to say they "interpret this finding as indicative evidence that land-use conflicts in the delta region are shaped by political and economic texts (e.g., confidence in the judicial system, economic opportunities, and learning) rather than climate variability." As for the second part of their study, they also conclude "factors other than those directly related to environmental conditions and resource scarcity dominate as plausible explanations of the violent conflict," arguing "three structural factors are the main drivers behind these conflicts: agricultural encroachment that obstructed the mobility of herders and livestock, opportunistic behavior of rural actors as a consequence of an increasing political vacuum, and corruption and rent seeking among government officials."

The findings of Benjaminsen *et al.*, and those of many others whom they cite (Grandin, 1987; Bassett, 1988; Ellis and Swift, 1988; Bonfiglioli and Watson, 1992; Behnke *et al.*, 1993; Turner, 1998, 2004; Hagberg, 2005; Hesse and MacGregor, 2006; Moritz, 2006; Nordås and Gleditsch, 2007; Benjaminsen, 2008; Benjaminsen *et al.*, 2009; Benjaminsen and Ba, 2009), give further credence to the conclusion of Nordås and Gleditsch (2007) that even the IPCC, which "prides itself on being a synthesis of the best peer-reviewed science, has fallen prey to relying on second- or third-hand information with little

empirical backing when commenting on the implications of climate change for conflict." Real-world evidence for their climate-change-causes-conflict claim is just not there – at least in the case where the climate change involves warming.

In another study from Africa, O'Loughlin *et al.* (2014) write, "continued public and academic interest in the topic of global climate change consequences for political instability and the risk of conflict has generated a growing but inconclusive literature, especially about the effects in sub-Saharan Africa." They note many of the studies supporting that hypothesis "do not elaborate on nor test the causal mechanisms." So "using a new disaggregated dataset of violence and climate anomaly measures (temperature and precipitation variations from normal) for sub-Saharan Africa 1980–2012, we consider political, economic and geographic factors, not only climate metrics, in assessing the chances of increased violence."

O'Loughlin *et al.* found "the location and timing of violence are influenced less by climate anomalies than by key political, economic and geographic factors," such that "overall, the temperature effect is statistically significant, but important inconsistencies in the relationship between temperature extremes and conflict are evident in more nuanced relationships than have been previously identified." They cite several independent studies that reached a similar conclusion, including those of Buhaug (2010), Bergholt and Lujala (2012), Koubi *et al.* (2012), Raleigh and Kniveton (2012), and Wischnath and Buhaug (2014).

Field and Lape (2010) note it has been repeatedly suggested that in many parts of the world climate change has "encouraged conflict and territorialism," as this response "serves as an immediate means of gaining resources and alleviating shortfalls," such as those that occur when the climate change is detrimental to agriculture and the production of food. To investigate this hypothesis, they compared "periods of cooling and warming related to hemispheric-level transitions (namely the Medieval Warm Period and the Little Ice Age) in sub-regions of the Pacific with the occurrence of fortifications at the century-level." Their study revealed "the comparison of fortification chronologies with paleoclimatic data indicate that fortification construction was significantly correlated with periods of cooling, which in the tropical Pacific is also associated with drying." In addition, "the correlation was most significant in the Indo-Pacific Warm Pool, the Southwestern Pacific and New Zealand," where "people constructed more fortifications during periods that match the chronology for the Little Ice Age (AD 1450–1850)," as opposed to the Medieval Warm Period (AD 800–1300) when the Indo-Pacific Warm Pool was both warm and saline "with temperatures approximating current conditions (Newton *et al.*, 2006)." Field and Lape's study provides additional evidence that periods of greater warmth have generally led to more peaceful times throughout the world, whereas periods of lesser warmth have typically led to greater warfare.

Zhang *et al.* (2011b) note it has long been assumed that "deteriorating climate" – defined as either cooling or warming – "could shrink the carrying capacity of agrarian lands, depriving the human population of sufficient food," with "population collapses (i.e., negative population growth)" the unavoidable consequence. They further note "this human-ecological relationship has rarely been verified scientifically," pointing out that at the high end of the temperature spectrum, "evidence of warming-caused disaster has never been found."

Zhang *et al.* performed time-series analyses to examine the association between temperature change and country-wide/region-wide population collapses in different climatic zones of the Northern Hemisphere (NH), focusing on all known population collapses over the period AD 800–1900. In addition, they computed regressions to estimate the relative sensitivity of population growth in the NH to climate change, where the independent variables employed were time and temperature anomalies. Of the 88 NH population collapses they identified, fully 80% were caused by cooling, while 12% occurred during what the six scientists called "mild conditions," and only 8% of them were caused by warming. They found "temperature was positive and highly significant in the regressions in which a 10 percent increase in temperature produced on average a 3.1 percent increase in population growth rate."

Historically, and for the Northern Hemisphere as a whole, warming and warmer times have most often been prosperous times for humanity, as exemplified by the greater numbers of people the Earth supports under such conditions, while cooling and colder times are typically just the opposite, with many significant population collapses caused by what Zhang *et al.* describe as "Malthusian checks (i.e., famines, wars and epidemics)."

Koubi *et al.* (2012) state "despite many claims by high-ranking policymakers and some scientists that climate change breeds violent conflict, the existing empirical literature has so far not been able to identify a systematic, causal relationship of this kind" – see, for example, Bruckner and Ciccone (2007, 2010), Buhaug (2010), Ciccone (2011), Theisen *et al.* (2011), and Bergholt and Lujala (2012) – which failure "may either reflect de facto absence of such a relationship, or it may be the consequence of theoretical and methodological limitations of existing work." In a study designed to explore these two possibilities, Koubi *et al.* "examine the causal pathway linking climatic conditions to economic growth and to armed conflict," as well as the degree to which this pathway is contingent upon the political systems of the potential conflict participants, using data "from all countries of the world in the period 1980–2004."

Koubi *et al.* say their results suggest "climate variability, measured as deviations in temperature and precipitation from their past, long-run levels (a 30-year moving average), does not affect violent intrastate conflict through economic growth." This finding, in their words, "is important because the causal pathway leading from climate variability via (deteriorating) economic growth to conflict is a key part of most theoretical models of the climate-conflict nexus." They further note there is "some, albeit weak, support for the hypothesis that non-democratic [i.e., 'autocratic'] countries are more likely to experience civil conflict when economic conditions deteriorate," but they add that even this weak connection "is fragile with regard to model specification."

Focusing on nearly the same time period, Bergholt and Lujala (2012) examined "how climate-related natural disasters, including flash floods, surges, cyclones, blizzards, and severe storms, affect economic growth and peace," after which they focused on the question of "whether climate-related disasters have an indirect effect on conflict onset via slowdown in economic growth." They utilized climate-related disaster data for the period 1980–2007 found in the Emergency Events Database developed by the Centre for Research on the Epidemiology of Disasters, economic growth data found in the Penn World Table Version 6.3 (Heston *et al.*, 2009), and armed civil conflict data tabulated in the annually updated UCDP/PRIO Armed Conflict Dataset (Gleditsch *et al.*, 2002; Harbom and Wallensteen, 2010).

In the first stage of their analysis, Bergholt and Lujala found "climate-related disasters have a negative impact on growth," but they say their analysis of disaster data and conflict onset shows "climate-related natural disasters do not have any direct effect on conflict onset." They also report they "did not find any evidence that economic shocks caused by climate-related disasters have an effect on conflict onset," noting their findings "are similar to those in the recent cross-country study by Ciccone (2011)." They conclude "storms and floods adversely affect people and production inputs such as land, infrastructure, and factories, which in turn have a negative impact on the aggregate economy," but "these negative income shocks do not increase the risk of armed civil conflict as predicted by prominent studies in the field (Collier and Hoeffler, 2004; Fearon and Laitin, 2003; Miguel et al., 2004)."

In another large-scale study, Slettebak (2012) writes, "academic, policy, and popular discussions that surround the issue of climate change predict changing weather patterns to increase natural disasters," and he states that many of the discussants "expect these disasters to increase the risk of violent conflict." In a test of this hypothesis, Slettebak examined "whether natural disasters can add explanatory power to an established model of civil conflict." Results "indicate that they can, but that their effect on conflict is the opposite of popular perception." He explains, "to the extent that climate-related natural disasters affect the risk of conflict, they contribute to reducing it." This result holds "for a measure of climate-related natural disasters in general, as well as drought in particular," adding these findings are "consistent with a large amount of research ... on the relation between disasters and the risk of anti-social behavior," going back to the work of Fritz (1961), which was not made public until some 35 years later (Fritz, 1996).

In commenting on his findings, Slettebak says his primary result "underscores the importance of being cautious about assuming that adversity will automatically translate into increased levels of conflict – a perception that appears frequent among a number of vocal actors in the debate around the political consequences of climate change." Thus he emphasizes "one worrying facet of the claims that environmental factors cause conflict is that they may contribute to directing attention away from more important conflict-promoting factors, such as poor governance and poverty," noting "there is a serious risk of misguided policy to prevent civil conflict if the assumption that disasters have a significant effect on war is allowed to overshadow more important causes."

According to Gartzke (2012), "while anecdote and some focused statistical research suggests that civil conflict may have worsened in response to recent climate change in developing regions, these claims have been severely criticized by other studies," citing Nordås and Gleditsch (2007), Buhaug (2010), and Buhaug et al. (2010). In addition, he states "the few long-term macro statistical studies actually find that conflict increases in periods of climatic chill (Zhang et al., 2006, 2007[a]; Tol and Wagner, 2010)." He reports "research on the modern era reveals that interstate conflict has declined in the second half of the 20th century, the very period during which global warming has begun to make itself felt (Goldstein, 2011; Hensel, 2002; Levy et al., 2001; Luard, 1986, 1988; Mueller, 2009; Pinker, 2011; Sarkees et al., 2003)."

Gartzke explored "the relationship between climate change, liberal processes fueled by industrialization (development, democracy, international institutions) and interstate conflict," based on information gleaned from the Correlates of War (COR) Militarized Interstate Dispute (MID) dataset (Gochman and Maoz, 1984; Ghosn et al., 2004) and annual average temperature data provided by NASA's Goddard Institute for Space Studies and the United Kingdom's Meteorological Office Hadley Centre and the Climatic Research Unit of the University of East Anglia, while measures of regime type come from the Polity IV project described by Gurr et al. (1989) and Marshall and Jaggers (2002).

"Surprisingly," Gartzke writes, "analysis at the system level suggests that global warming is associated with a reduction in interstate conflict," and "incorporating measures of development, democracy, cross-border trade, and international institutions reveals that systemic trends toward peace are actually best accounted for by the increase in average international income," which in turn is driven by "the processes that are widely seen by experts as responsible for global warming." Furthermore, in the concluding sentence of his paper's abstract Gartzke writes, "ironically, stagnating economic development in middle-income states caused by efforts to combat climate change could actually realize fears of climate-induced warfare." And thus he states in the concluding section of his paper that "we must add to the advantages of economic development that it appears to make countries more peaceful," and we must therefore ask if environmental objectives should be "modified by the prospect that combating climate change could prolong the process of transition from warlike to peaceful polities."

Buhaug et al. (2015) note earlier research has suggested there is "a correlational pattern between climate anomalies and violent conflict" due to "drought-induced agricultural shocks and adverse economic spillover effects as a key causal mechanism linking the two phenomena." They compared half a century of statistics on climate variability, food production, and political violence across Sub-Saharan Africa, which effort "offers the most precise and theoretically consistent empirical assessment to date of the purported indirect relationship." Their analysis "reveals a robust link between weather patterns and food production where more rainfall generally is associated with higher yields." However, they also report "the second step in the causal model is not supported," noting "agricultural output and violent conflict are only weakly and inconsistently connected, even in the specific contexts where production shocks are believed to have particularly devastating social consequences," which leads them to suggest "the wider socioeconomic and political context is much more important than drought and crop failures in explaining violent conflict in contemporary Africa."

Buhaug *et al.* continue, "social protest and rebellion during times of food price spikes may be better understood as reactions to poor and unjust government policies, corruption, repression and market failure," citing the studies of Bush (2010), Buhaug and Urdal (2013), Sneyd *et al.* (2013), and Chenoweth and Ulfelder (2015). They note that even the IPCC's Fifth Assessment Report concludes "it is *likely* that socioeconomic and technological trends, including changes in institutions and policies, will remain a relatively stronger driver of food security over the next few decades than climate change," citing Porter *et al.* (2014).

References

Atwell, W.S. 2001. Volcanism and short-term climatic change in East Asian and world history, c. 1200–1699. *Journal of World History* 12: 29–98.

Atwell, W.S. 2002. Time, money, and the weather: Ming China and the 'great depression' of the mid-fifteenth century. *Journal of Asian Studies* 61: 83–113.

Bassett, T.J. 1988. The political ecology of peasant-herder conflicts in the northern Ivory Coast. *Annals of the Association of American Geographers* 78: 453–72.

Behnke, R.H., Scoones, I., and Kerven, C. (Eds.). 1993. *Range Ecology at Disequilibrium: New Models of Natural Variability and Pastoral Adaptation in African Savannas.* London, UK: Overseas Development Institute and International Institute for Environment and Development.

Benjaminsen, T.A. 2008. Does supply-induced scarcity drive violent conflicts in the African Sahel? The case of the Tuareg rebellion in northern Mali. *Journal of Peace Research* 45: 831–48.

Benjaminsen, T.A. 2009. Klima og konflikter I Sahel - eller politikk og vitenskap ved klimaets nullpunkt [Climate and conflicts in the Sahel - or politics and science at the Ground Zero of climate change]. *Internasjonal politikk* 67: 151–72.

Benjaminsen, T.A., Alinon, K., Buhaug, H., and Buseth, J.T. 2012. Does climate change drive land-use conflicts in the Sahel? *Journal of Peace Research* 49: 97–111.

Benjaminsen, T.A. and Ba, B. 2009. Farmer-herder conflicts, pastoral marginalization and corruption: a case study from the inland Niger delta of Mali. *Geographical Journal* 174: 71–81.

Benjaminsen, TA., Maganga, F., and Abdallah, J.M. 2009. The Kilosa killings: political ecology of a farmer-herder conflict in Tanzania. *Development and Change* 40: 423–45.

Bergholt, D. and Lujala, P. 2012. Climate-related natural disasters, economic growth, and armed civil conflict. *Journal of Peace Research* 49: 147–62.

Bonfiglioli, A.M. and Watson, C. 1992. *Pastoralists at a Crossroads: Survival and Development Issues in African Pastoralism.* Nairobi, Kenya: United Nations Children's Fund, United Nations Sudano-Sahelian Office.

Bruckner, M. and Ciccone, A. 2007. Growth, democracy, and civil war. *CEPR Discussion Paper* No. DP6568.

Bruckner, M. and Ciccone, A. 2010. International commodity prices, growth, and civil war in sub-Saharan Africa. *Economic Journal* 120: 519–34.

Bryson, R.A. and Murray, T.J. 1977. *Climates of Hunger: Mankind and the World's Changing Weather.* Madison, WI: University of Wisconsin Press.

Buhaug, H. 2010. Climate not to blame for African civil wars. *Proceedings of the National Academy of Sciences USA* 197: 16,477–82.

Buhaug, H. and Urdal, H. 2013. An urbanization bomb? Population growth and social disorder in cities. *Global Environmental Change* 23: 1–10.

Buhaug, H., Benjaminsen, T.A., Sjaastad, E., and Theisen, O.M. 2015. Climate variability, food production shocks, and violent conflict in Sub-Saharan Africa. *Environmental Research Letters* 10: 10.1088/1748-9326/10/12/125015.

Buhaug, H., Gleditsch, N.P., and Theisen, O.M. 2010. Implications of climate change for armed conflict. In:

Mearns, R. and Norton, A. (Eds.) *The Social Dimensions of Climate Change: Equity and Vulnerability in a Warming World.* Washington, DC: World Bank, pp. 75–101.

Buntgen, U., *et al.* 2011. 2500 years of European climate variability and human susceptibility. *Science* **331**: 578–82.

Buntgen, U., Trouet, V., Frank, D., Leuschner, H.H., Friedrichs, D., Luterbacher, J., and Esper, J. 2010. Tree-ring indicators of German summer drought over the last millennium. *Quaternary Science Reviews* **29**: 1005–16.

Bush, R. 2010. Food riots: poverty, power and protest. *Journal of Agrarian Change* **10**: 119–29.

Chen, L., Zonneveld, K.A.F., and Versteegh, G.J.M. 2011. Short term climate variability during the "Roman Classical Period" in the eastern Mediterranean. *Quaternary Science Reviews* **30**: 3880–91.

Chenoweth, E. and Ulfelder, J. 2015. Can structural conditions explain the onset of nonviolent uprisings? *Journal of Conflict Resolution* **61** (2): 298–324.

Ciccone, A. 2011. Economic shocks and civil conflict: a comment. *American Economic Journal: Applied Economics* **3**: 215–27.

Cleaveland, M.K., Stahle, D.W., Therrell, M.D., Villanueva-Diaz, J., and Burns, B.T. 2003. Tree-ring reconstructed winter precipitation and tropical teleconnections in Durango, Mexico. *Climatic Change* **59**: 369–88.

Collier, P. and Hoeffler, A. 2004. Greed and grievance in civil war. *Oxford Economic Papers* **56**: 563–95.

deMenocal, P.B. 2001. Cultural responses to climate change during the late Holocene. *Science* **292**: 667–73.

Ellis, J.E. and Swift, D.M. 1988. Stability of African pastoral ecosystems: alternate paradigms and implications for development. *Journal of Range Management* **41**: 450–9.

Fearon, J.D. and Laitin, D.D. 2003. Ethnicity, insurgency, and civil war. *American Political Science Review* **97**: 75–90.

Field, J.S. and Lape, P.V. 2010. Paleoclimates and the emergence of fortifications in the tropical Pacific islands. *Journal of Anthropological Archaeology* **29**: 113–24.

Florescano, E., Swan, S., Menegus, M., and Galindo, I. 1995. *Breve Historia de la Sequia en Mexico.* Veracruz, Mexico: Universidad Veracruzana.

Fritz, C.E. 1961. Disaster. In: Merton, R.K. and Nisbet, R.A. (Eds.) *Contemporary Social Problems: An Introduction to the Sociology of Deviant Behavior and Social Disorganization.* New York, NY: Harcourt, Brace and World, pp. 651–94.

Fritz, C.E. 1996. Disasters and mental health: therapeutic principles drawn from disaster studies. *Historical and Comparative Disaster Series 10.* Newark, DE: University of Delaware Disaster Research Center.

Gartzke, E. 2012. Could climate change precipitate peace? *Journal of Peace Research* **49**: 177–92.

Ghosn, F., Palmer, G., and Bremer, S. 2004. The MID 3 data set, 1993–2001: procedures, coding rules, and description. *Conflict Management and Peace Science* **21**: 133–54.

Gleditsch, N.P., Wallensteen, P., Eriksson, M., Sollenberg, M., and Strand, H. 2002. Armed conflict 1946–2001: a new dataset. *Journal of Peace Research* **39**: 615–37.

Gochman, C.S. and Maoz, Z. 1984. Militarized interstate disputes, 1816–1976: procedure, patterns, and insights. *Journal of Conflict Resolution* **28**: 585–615.

Goldstein, J.S. 2011. *Winning the War on War: The Decline of Armed Conflict Worldwide.* New York, NY: Dutton.

Grandin, B.E. 1987. Pastoral Culture and Range Management: Recent Lessons from Maasailand. *ILCA Bulletin* 28. Addis Ababa, Ethiopia: International Livestock Center for Africa.

Gurr, T.R., Jaggers, K., and Moore, W.H. 1989. *Polity II: Political Structures and Regime Change, 1800–1986.* Boulder, CO: University of Colorado.

Hagberg, S. 2005. Dealing with dilemmas: violent farmer-pastoralist conflicts in Burkina Faso. In: Richards, P. (Ed.) *No Peace, No War: An Anthropology of Contemporary Armed Conflicts.* Oxford, UK: James Currey Publishing.

Harbom, L. and Wallensteen, P. 2010. Armed conflicts, 1946–2009. *Journal of Peace Research* **47**: 501–9.

Hensel, P. 2002. The more things change ...: recognizing and responding to trends in armed conflict. *Conflict Management and Peace Science* **19**: 27–53.

Hesse, C. and MacGregor, J. 2006. Pastoralism: Drylands' Invisible Asset? Developing a Framework for Assessing the Value of Pastoralism in East Africa. *IIED Issue Paper* 142. London, UK: International Institute for Environment and Development.

Heston, A., Summers, R., and Aten, B. 2009. *Penn World Table, Version 6.3.* Center for International Comparisons, University of Pennsylvania (CICUP).

Kaplan, J.O., Krumhardt, K.M., and Zimmermann, N. 2009. The prehistoric and preindustrial deforestation of Europe. *Quaternary Science Reviews* **28**: 3016–34.

Kausrud, K.L., *et al.* 2010. Modeling the epidemiological history of plague in Central Asia: palaeoclimatic forcing on a disease system over the past millennium. *BMC Biology* **8**: 10.1186/1741-7007-8-112.

Koubi, V., Bernauer, T., Kalbhenn, A., and Spilker, G. 2012. Climate variability, economic growth, and civil conflict. *Journal of Peace Research* **49**: 113–27.

Lee, H.F., Fok, L., and Zhang, D.D. 2008. Climatic change and Chinese population growth dynamics over the last millennium. *Climatic Change* **88**: 131–56.

Lee, H.F. and Zhang, D.D. 2010. Changes in climate and secular population cycles in China, 1000 CE to 1911. *Climate Research* **42**: 235–46.

Lee, H.F., Zhang, D.D., and Fok, L. 2009. Temperature, aridity thresholds, and population growth dynamics in China over the last millennium. *Climate Research* **39**: 131–47.

Levy, J.S., Walker, T.C., and Edwards, M.S. 2011. Continuity and change in the evolution of warfare. In: Maoz, Z. and Gat, A. (Eds.) *War in a Changing World*. Ann Arbor, MI: University of Michigan Press, pp. 15–48.

Luard, E. 1986. *War in International Society: A Study in International Sociology*. London, UK: Tauris.

Luard, E. 1988. *Conflict and Peace in the Modern International System: A Study of the Principles of International Order*. Albany, NY: State University of New York Press.

Marshall, M. and Jaggers, K. 2002. Polity IV: Political regime characteristics and transitions, 1800–2002 (website). Accessed June 25, 2018.

McCormick, M. 2001. *Origins of the European Economy: Communications and Commerce, A.D. 300–900*. Cambridge, UK: Cambridge University Press.

Miguel, E., Satyanath, S., and Sergenti, E. 2004. Economic shocks and civil conflict: an instrumental variables approach. *Journal of Political Economy* **112**: 725–54.

Moritz, M. 2006. The politics of permanent conflict: farmer-herder conflicts in northern Cameroon. *Canadian Journal of African Studies* **40**: 101–26.

Mueller, J. 2009. War has almost ceased to exist: an assessment. *Political Science Quarterly* **124**: 297–321.

Newton, A., Thunnell, R., and Stott, L. 2006. Climate and hydrographic variability in the Indo-Pacific warm pool during the last millennium. *Geophysical Research Letters* **33**: 10.1029/2006GL027234.

Nicholson, S.E. and Yin, X. 2001. Rainfall conditions in equatorial East Africa during the nineteenth century as inferred from the record of Lake Victoria. *Climatic Change* **48**: 387–98.

Nordås, R. and Gleditsch, N.P. 2007. Climate change and conflict: a critical overview. *Political Geography* **26** (6): 627–38.

O'Loughlin, J., Linke, A.M., and Witmer, F.D.W. 2014. Effects of temperature and precipitation variability on the risk of violence in sub-Saharan Africa, 1980–2012. *Proceedings of the National Academy of Sciences USA* **111**: 16,712–7.

Patterson, W.P., Dietrich, K.A., Holmden, C., and Andrews, J.T. 2010. Two millennia of North Atlantic seasonality and implications for Norse colonies. *Proceedings of the National Academy of Sciences USA* **107**: 5306–10.

Pinker, S. 2011. *The Better Angels of Our Nature*. New York, NY: Viking.

Porter, J.R., *et al.* 2014. Food security and food production systems. In: *Climate Change 2014: Impacts, Adaptation, and Vulnerability. Part A: Global and Sectoral Aspects.* Contribution of Working Group II to the Fifth Assessment Report of the Intergovernmental Panel on Climate Change. Cambridge. MA: Cambridge University Press, pp. 485–533.

Raleigh, C. and Kniveton, D. 2012. Come rain or shine: an analysis of conflict and climate variability in East Africa. *Journal of Peace Research* **49**: 51–64.

Sarkees, M.R., Wayman, F.W., and Singer, J.D. 2003. Inter-state, intra-state, and extra-state wars: a comprehensive look at their distribution over time, 1816-1997. *International Studies Quarterly* **47**: 49–70.

Schwartz, P. and Randall, D. 2003. *An Abrupt Climate Change Scenario and Its Implications for United States National Security*. Pasadena, CA: Jet Propulsion Laboratory.

Slettebak, R.T. 2012. Don't blame the weather! Climate-related natural disasters and civil conflict. *Journal of Peace Research* **49**: 163–76.

Sneyd, I.Q., Legwegoh, A., and Fraser, E.D.G. 2013. Food riots: media perspectives on the causes of food protest in Africa. *Food Security* **5**: 485–97.

Stahle, D.W., Cook, E.R., Cleaveland, M.K., Therrell, M.D., Meko, D.M., Grissino-Mayer, H.D., Watson, E., and Luckman, B.H. 2000. Tree-ring data document 16th century megadrought over North America. *EOS, American Geophysical Union, Transactions* **81**: 121, 125.

Theisen, O.M., Holtermann, H., and Buhaug, H. 2011. Climate wars? Assessing the claim that drought breeds conflict. *International Security* **36** (3): 79–106.

Tol, R.S.J. and Wagner, S. 2010. Climate change and violent conflict in Europe over the last millennium. *Climatic Change* **99**: 65–79.

Turner, M. 1998. The interaction of grazing history with rainfall and its influence on annual rangeland dynamics in the Sahel. In: Zimmerer, K. and Young, K.R. (Eds.) *Nature's Geography: New Lessons for Conservation in Developing Countries.* Madison, WI: University of Wisconsin Press, pp. 237–51.

Turner, M. 2004. Political ecology and the moral dimensions of 'resource conflicts': the case of farmer-herder conflicts in the Sahel. *Political Geography* **23**: 863–89.

Weiss, H. and Bradley, R.S. 2001. Archaeology. What drives societal collapse? *Science* **291**: 609–10.

Wischnath, G. and Buhaug, H. 2014. On climate variability and civil war in Asia. *Climatic Change* **122**: 709–21.

Zhang, D.D., Brecke, P., Lee, H.F., He, Y.Q., and Zhang, J. 2007a. Global climate change, war, and population decline in recent human history. *Proceedings of the National Academy of Sciences USA* **104**: 19,214–9.

Zhang, D.D., Jim, C.Y., Lin, C.S., He, Y.Q., and Lee, F. 2005. Climate change, social unrest and dynastic transition in ancient China. *Chinese Science Bulletin* **50**: 137–44.

Zhang, D.D., Jim, C.Y., Lin, G.C.-S., He, Y.-Q., Wang, J.J., and Lee, H.F. 2006. Climatic change, wars and dynastic cycles in China over the last millennium. *Climatic Change* **76**: 459–77.

Zhang, D.D., Lee, H.F., Wang, C., Li, B., Pei, Q., Zhang, J., and An, Y. 2011a. The causality analysis of climate change and large-scale human crisis. *Proceedings of the National Academy of Sciences USA* **108** (42): 17,296–301.

Zhang, D.D., Lee, H.F., Wang, C., Li, B., Zhang, J., Pei, Q., and Chen, J. 2011b. Climate change and large scale human population collapses in the pre-industrial era. *Global Ecology and Biogeography* **20**: 520–31.

Zhang, D.D., Zhang, J., Lee, H.F., and He, Y. 2007b. Climate change and war frequency in eastern China over the last millennium. *Human Ecology* **35**: 403–14.

Zhang, Z., Tian, H., Cazelles, B., Kausrud, K.L., Brauning, A., Guo, F., and Stenseth, N.C. 2010. Periodic climate cooling enhanced natural disasters and wars in China during AD 10–1900. *Proceedings of the Royal Society B* **277**: 3745–53.

7.5 Conclusion

The IPCC relies on second- or third-hand information with little empirical backing when commenting on the implications of climate change for conflict.

This chapter makes a strong case that citizens and many policymakers around the world have been misled into believing the use of fossil fuels poses a threat to their security. The truth is just the opposite: The prosperity fossil fuels make possible, including helping produce sufficient food for a growing global population, is a major reason the world is safer today than ever before. And since prosperity is closely correlated with democracy, and democracies have lower rates of violence and go to war less frequently than any other form of government, it follows that fossil fuels contribute to human security by making the spread of democracy possible.

Some commentators set against this record of achievement the cost of wars "fought for oil" in the Middle East. While it is true that the presence in that region of troops from the United States and other nations has sometimes been justified by the desire to keep oil flowing from the region, those conflicts have origins and justifications unrelated to oil. The extraordinarily high cost of fighting those wars – in lost lives as well as the trillions of dollars spent on arms, equipment, and logistics – far exceed whatever benefits might have been obtained by keeping the global price of oil low, and likely did not even succeed in achieving that.

The IPCC claims climate change threatens "the vital core of human lives" in multiple ways, many of them unquantifiable, unproven, and uncertain. The narrative in Chapter 12 of the Working Group II contribution to the Fifth Assessment Report illustrates how the IPCC misuses language to hide uncertainty and exaggerate risks. The alleged threats to human security due to "deprivation of basic needs" are speculative, not supported by real-world evidence, and contradicted by the IPCC's own survey of the economic literature. Alleged threats to agriculture and food security are contradicted by biological science and empirical data regarding crop yields and human hunger. Alleged threats to human capital – human health, education, and longevity –

are almost entirely speculative and undocumented. There is no evidence global warming has eroded or will erode livelihoods or human progress.

Even though the IPCC is often cited as the scientific basis for the claim that climate change increases the risk of violent conflicts around the world, its reports express deep uncertainty over the matter. Recall the admission in Chapter 18 of the Working Group II contribution to AR5, on "Detection and Attribution of Observed Impacts," that "both the detection of a climate change effect and an assessment of the importance of its role can be made only with *low confidence* owing to limitations on both historical understanding and data" (IPCC, 2014, p. 1001). But the IPCC's spokespersons rarely mention these doubts and they may have been inconvenient truths for the politicians, interest groups, and journalists who have done so much to confuse the public.

While some politicians and the news media profess absolute certainty that global warming increases the risk of warfare, the academic community has produce extensive research pointing in the opposite direction. Empirical research shows no direct association between climate change and armed conflicts. The climate-conflict hypothesis is an argument linked together in a chain, and if any one of these links is disproven, the hypothesis is invalidated. The academic literature on the relationship between climate and social conflict reveals at least six methodological problems affecting efforts to connect the two.

The IPCC relies on second- or third-hand information with little empirical backing when commenting on the implications of climate change for conflict. Real-world evidence demonstrates warmer weather is closely associated with peace and prosperity, and cooler weather with war and poverty. A warmer world, should it occur, is therefore more likely to bring about peace and prosperity than war and poverty.

When Harvard archaeologist and history of war expert Steven LeBlanc looked to the future, he concluded "the decline in warfare among those countries is incredibly strong" and "for the first time in history, technology and science enable us to understand Earth's ecology and our impact on it, to control population growth, and to increase the carrying capacity in ways never before imagined. The opportunity for humans to live in long term balance with nature is within our grasp if we do it right" (LeBlanc and Register, 2003, p. 229).

References

IPCC. 2014. Intergovernmental Panel on Climate Change. *Climate Change 2014: Impacts, Adaptation, and Vulnerability*. Contribution of Working Group II to the Fifth Assessment Report of the Intergovernmental Panel on Climate Change. New York, NY: Cambridge University Press.

LeBlanc, S. and Register, K.E. 2003. *Constant Battles: The Myth of the Nobel Savage and a Peaceful Past*. New York, NY: St. Martin's/Griffin.

8

Cost-Benefit Analysis

Chapter Lead Authors: Roger Bezdek, Ph.D., Christopher Monckton of Brenchley

Contributors: Joseph L. Bast, Barry Brill, OBE, JP, Kevin Dayaratna, Ph.D., Bryan Leyland, M.Sc.

Reviewers: David Archibald, Patrick Frank, Ph.D., Donald Hertzmark, Ph.D., Paul McFadyen, Ph.D., Patrick Michaels, Ph.D., Robert Murphy, Ph.D., Charles N. Steele, Ph.D., David Stevenson

Citation: Idso, C.D., Legates, D. and Singer, S.F. 2019. Cost-Benefit Analysis. In: *Climate Change Reconsidered II: Fossil Fuels.* Nongovernmental International Panel on Climate Change. Arlington Heights, IL: The Heartland Institute.

Key Findings

Key findings of this chapter include the following:

CBA Basics

- Cost-benefit analysis (CBA) is an economic tool that can help determine if the social benefits over the lifetime of a government project exceed its social costs.

- In the climate change debate, CBA is used to answer questions about the costs and benefits of climate change, the use of fossil fuels, and specific measures to mitigate, rather than adapt to, climate change.

- Integrated assessment models (IAMs) are a key element of cost-benefit analysis in the climate change debate. They are enormously complex and can be programmed to arrive at widely varying conclusions.

- A typical IAM has four steps: emission scenarios, future CO_2 concentrations, climate projections and impacts, and economic impacts.

- IAMs suffer from propagation of error, sometimes called cascading uncertainties, whereby uncertainty in each stage of the analysis compounds, resulting in wide uncertainty bars surrounding any eventual results.

- The widely cited "social cost of carbon" calculations produced during the Obama administration by the Interagency Working Group on the Social Cost of Carbon have been withdrawn and are not reliable guides for policymakers.

- The widely cited "Stern Review" was an important early attempt to apply cost-benefit analysis to climate change. Its authors focused on worst-case scenarios and failed to report profound uncertainties.

Assumptions and Controversies

- Most IAMs rely on emission scenarios that are little more than guesses and speculative "storylines." Even current greenhouse gas emissions cannot be measured accurately, and technology is likely to change future emissions in ways that cannot be predicted.

- IAMs falsely assume the carbon cycle is sufficiently understood and measured with sufficient accuracy as to make possible precise predictions of future levels of carbon dioxide (CO_2) in the atmosphere.

- Many IAMs rely on estimates of climate sensitivity – the amount of warming likely to occur from a doubling of the concentration of atmospheric carbon dioxide – that are too high, resulting in inflated estimates of future temperature change.

- Many IAMs ignore the extensive scholarly research showing climate change will not lead to more extreme weather, flooding, droughts, or heat waves.

- The "social cost of carbon" (SCC) derived from IAMs is an accounting fiction created to justify regulation of fossil fuels. It should not be used in serious conversations about how to address the possible threat of man-made climate change.

- The IPCC acknowledges great uncertainty over estimates of the "social cost of carbon" and admits the impact of climate change on human welfare is small relative to many other factors.

- Many IAMs apply discount rates to future costs and benefits that are much lower than the rates conventionally used in cost-benefit analysis.

Climate Change

- By the IPCC's own estimates, the cost of reducing emissions in 2050 by enough to avoid a warming of ~2°C would be 6.8 times as much as the benefits would be worth.

- Changing only three assumptions in two leading IAMs – the DICE and FUND models – reduces the SCC by an order of magnitude for the first and changes the sign from positive to negative for the second.

- Under very reasonable assumptions, IAMs can suggest the SCC is more likely than not to be negative, even though they have many assumptions and biases that tend to exaggerate the negative effects of GHG emissions.

Fossil Fuels

- Sixteen of 25 possible impacts of fossil fuels on human well-being are net benefits, only one is a net cost, and the rest are either unknown or likely to have no net impact.

- Wind and solar cannot generate enough dispatchable energy (available 24/7) to replace fossil fuels, so energy consumption must fall in order for emissions to fall.

- Transitioning from a world energy system dependent on fossil fuels to one relying on alternative energies would cost trillions of dollars and take decades to implement.

- The evidence seems compelling that the costs of restricting use of fossil fuels greatly exceed the benefits, even accepting many of the IPCC's very questionable assumptions.

- Reducing greenhouse gas emissions to levels suggested by the IPCC or the goal set by the European Union would be prohibitively expensive.

Regulations

- Cost-benefit analysis applied to greenhouse gas mitigation programs can produce like-to-like comparisons of their cost-effectiveness.

- The cap-and-trade bill considered by the U.S. Congress in 2009 would have cost 7.4 times more than its benefits, even assuming all of the IPCC's assumptions and claims about climate science were correct.

- Other bills and programs already in effect have costs exceeding benefits by factors up to 7,000. In short, even accepting the IPCC's flawed science and scenarios, there is no justification for adopting expensive emission mitigation programs.

- The benefits of fossil fuels far outweigh their costs. Various scenarios of reducing greenhouse gas emissions have costs that exceed benefits by ratios ranging from 6.8:1 to 162:1.

Introduction

The debate over climate change would be advanced if it were possible to weigh, in an even-handed and precise manner, the costs imposed by the use of fossil fuels on humanity and the environment, on the one hand, and the benefits produced by their use on the other. If the costs exceed the benefits, then efforts to force a transition away from fossil fuels are justified and ought to continue. If, on the other hand, the benefits are found to exceed the costs, then the right path forward would be the *energy freedom* path described in Chapter 1 rather than more restrictions on the use of fossil fuels.

Cost-benefit analysis (CBA) can be used to conduct such an investigation. CBA is an economic tool that is widely used in the private and public sectors to determine if the benefits of an investment or spending on a government program exceed its costs (Singer, 1979; Hahn and Tetlock, 2008; Wolka, 2000; Ellig, McLaughlin and Morrall, 2013; OMB, 2013). The history of CBA in shaping public policy was briefly surveyed in Chapter 1, Section 1.2.9.

We apologize in advance to the many researchers and reviewers, especially in the UK, who prefer "benefit-cost analysis" or BCA to "cost-benefit analysis" or CBA. Some researchers distinguish between the two, using CBA to refer only to analyses that rely on the potential compensation test (PCT) and BCA for analyses that rely on willingness to pay (WTP) or willingness to accept (WTA) (see Zerbe, 2008, 2017) but others do not. A Google search for both terms suggests CBA is preferred over BCA by a margin of about 17:1. In keeping with this choice, the two approaches are not distinguished here and results are reported as the ratio of costs to benefits rather than benefits to costs. Except for the final section, where the editors defer to the wishes of a chapter lead author.

Cost-benefit analysis is a complex endeavor typically involving subjective choices about what data to include and what to leave out, how to weigh evidence, and how to interpret results. The discipline is complicated enough to merit its own society, the Society for Benefit-Cost Analysis, and its own journal, *Journal of Benefit-Cost Analysis*. Section 8.1 begins with a brief tutorial on the application of CBA to the climate change debate. It is followed by an introduction to integrated assessment models (IAMs), an explanation of their biggest shortcoming (the "propagation of error" or cascading uncertainty), and reviews of CBAs of global warming produced by the Interagency Working Group on the Social Cost of Carbon (since disbanded) and the British Stern Review.

Section 8.2 examines the assumptions and biases that underlie IAMs. Tracking the order of "blocks" or "modules" in IAMs and drawing on research presented in previous chapters, it shows how errors or uncertainties in choosing emission scenarios, estimating the amount of carbon dioxide that stays in the atmosphere, the likelihood of increases in flooding and extreme weather, and other inputs render IAMs too unreliable to be of any use to policymakers.

Section 8.3 shows how two leading IAMs – the DICE and FUND models – rely on inaccurate equilibrium climate sensitivity rates, low discount rates, and a too-long time horizon (300 years). Correcting only these errors reveals the SCC is most likely negative, even accepting all of the IPCC's other errors and faulty assumptions. In other words, the social benefits of anthropogenic GHG emissions exceed their social cost.

Sections 8.4 summarizes the extensive literature reviews on the impacts of fossil fuels on human well-being conducted for earlier chapters in a single table. It reveals 16 of 25 possible impacts are positive (net benefits), only one is negative (net cost), and the rest are unknown or produce benefits and costs that are likely to offset each other. It presents cost-benefit analyses showing the cost of ending humanity's reliance on fossil fuels would be between 32 and 162 times as much as the hypothetical benefits of a slightly cooler world in 2050 and beyond.

Section 8.5 presents a formula for calculating the cost-effectiveness of GHG mitigation programs using the IPCC's own data and assumptions to produce like-to-like comparisons. The formula reveals a sample of proposed and existing programs has cost-benefit ratios ranging from 7.4:1 to 7,000:1, suggesting that current regulations, subsidies, and tax schemes aimed at reducing GHG emissions are not justified by their social benefits.

Section 8.6 offers a brief conclusion. According to the authors, CBA reveals the global war on energy freedom, which commenced in earnest in the 1980s and reached a fever pitch in the second decade of the twenty-first century, was never founded on sound science or economics. They urge the world's policymakers to acknowledge this truth and end that war.

References

Ellig, J., McLaughlin, P.A., and Morrall, J. 2013. Continuity, change, and priorities: The quality and use of regulatory analysis across US administrations. *Regulation and Governance* **7** (2).

Hahn, R.W. and Tetlock, P.C. 2008. Has economic analysis improved regulatory decisions? *Journal of Economic Perspectives* **22**: 1. 67–84.

OMB. 2013. U.S. Office of Management and Budget. Draft report to Congress on the benefits and costs of federal regulations and agency compliance with the Unfunded Mandates Reform Act.

Singer, S.F. 1979. *Cost Benefit Analysis as an Aid to Environmental Decision Making*. McLean, VA: The MITRE Corporation.

Wolka, K. 2000. Chapter 8. Analysis and modeling, Section 8.9 economics, in Lehr, J.H. (Ed.) *Standard Handbook of Environmental, Science, Health, and Technology*. New York, NY: McGraw-Hill, 8.122–8.133.

Zerbe, R.O. 2008. Ethical benefit cost analysis as art and science: ten rules for benefit-cost analysis. *University of Pennsylvania Journal of Law and Social Change* **12** (1).

Zerbe, R.O. 2017. A distinction between benefit-cost analysis and cost benefit analysis, moral reasoning and a justification for benefit-cost analysis. Seattle, WA: Evans School of Public Policy and Governance, University of Washington.

Society for Cost Benefit Analysis, n.d. (website). Accessed November 17, 2018.

8.1 CBA Basics

Cost-benefit analysis (CBA) is an economic tool that can help determine if the social benefits over the lifetime of a government project exceed its social costs.

Section 8.1.1 describes how cost-benefit analysis can be used to answer four key questions in the climate change debate. Section 8.1.2 provides background and an overview of the structure of integrated assessment models (IAMs) and describes how the "propagation of error" or cascading uncertainty renders their outputs unreliable. Sections 8.1.3 and 8.1.4 critique two of the best known attempts to apply CBA to climate change, the U.S. Interagency Working Group on the Social Cost of Carbon (since disbanded) and the British Stern Review. Section 8.1.5 presents a brief conclusion.

8.1.1 Use in the Climate Change Debate

In the climate change debate, CBA is used to answer four distinct questions:

1. Do the benefits from the use of fossil fuels, such as the increase in per-capita income made possible by affordable energy and higher agricultural output due to higher carbon dioxide (CO_2) levels in the atmosphere, exceed the costs it may have imposed, such as reduced air quality and, if they contribute to climate change, damage and harm from floods, droughts, or other severe weather events? (Bezdek, 2014)

2. Do the *social* benefits of either fossil fuels or climate change exceed the *social* cost – that is, do the positive externalities produced by the private use of fossil fuels exceed the negative externalities imposed on others? This is often called the "social cost of carbon" (SCC), calculated as the welfare loss associated with each additional metric ton of CO_2 emitted. (Tol, 2011)

3. Will the benefits of *a particular program* to reduce greenhouse gas emissions or sequester CO_2 by planting trees or injecting the gas into wells for underground storage exceed the costs incurred in implementing that program? (Monckton, 2016)

4. Is the cost-benefit ratio of a particular program to mitigate climate change higher or lower than the cost-benefit ratio of adapting to climate change by investing in stronger levees and dams, finding alternative sources of water, or "hardening" critical infrastructure? This is the *"mitigate versus adapt"* question that is frequently referenced in the Working Group II contribution to the IPCC's Fifth Assessment Report (AR5) (e.g., IPCC, 2014a, Chapter 10, pp. 665–666, 669, 679).

Regarding the first question, about the total private and social costs and benefits of the use of fossil fuels, Chapters 3 and 4 showed how fossil fuels made possible three Industrial Revolutions which in turn made possible large increases in human population, per-capita income, and lifespan (Bradley, 2000; Smil, 2005, 2006; Goklany, 2007; Bryce, 2010, 2014; Gordon, 2016). The benefits continue to accumulate today as cleaner-burning fossil fuels bring electricity to third-world countries and replace wood and dung as sources of heat in homes (Yadama, 2013; Bezdek, 2014). How much of the benefits of that economic transformation should be counted as "private" versus "social" benefits is not immediately apparent, but those benefits cannot be ignored entirely.

Rising atmospheric carbon dioxide concentrations and higher temperatures produce other benefits such as higher agricultural productivity, expanded ranges for most terrestrial animals having economic value such as livestock, and lower levels of human mortality and morbidity traditionally caused by exposure to cold temperatures (see Chapters 4 and 5 and Idso, 2013 for a detailed review of this literature). These well-known and observable benefits must be compared and weighed against cost estimates appearing in CBAs that are much less certain or well documented, many of which could even be judged conjectural.

Forward-looking CBAs must be based on reasonably accurate forecasts of future climate conditions. This requires climate models that take explicit, quantitative account of the principal relevant results in climatology, notably the radiative-forcing functions of CO_2 and other greenhouse gases and the various values of the climate sensitivity parameter. Current climate models have not shown much promise in this regard, as deomonstrated in Chapter 2, Section 2.2.2 (and see Fyfe *et al.*, 2013; McKitrick and Christy, 2018). CBAs also require economic models that can predict future changes in per-capita income, energy supply and demand, rate of technological innovation, economic growth rates in the developed and developing worlds, demographic trends, changes in land use and lifestyles, greenhouse gases other than carbon dioxide, and even political trends such as whether civil and economic freedoms are likely to expand or contract in various parts of the world (van Kooten, 2013).

The IPCC claims it can resolve all these uncertainties. In the Working Group III contribution to AR5, the IPCC says "a *likely* chance to keep average global temperature change below 2°C relative to pre-industrial levels" would require "lower global GHG emissions in 2050 than in 2010, 40% to 70% lower globally, and emissions levels near zero GtCO$_2$eq or below in 2100" (IPCC, 2014b, pp. 10, 12). Since fossil fuels are responsible for approximately 80% of anthropogenic greenhouse gas emissions, this would require gradually phasing out the use of fossil fuels and banning their use entirely by 2100.

Any effort to calculate the costs and benefits of future climate change confronts fundamental problems inherent in making forecasts in the absence of complete understanding of underlying causes and effects. In such cases, the most reliable method of forecasting is to project a simple linear continuation of past trends (Armstrong, 2001), but this plainly is not what is done by the IPCC or the authors of the models on which it relies. An audit of the IPCC's Fourth Assessment Report conducted by experts in scientific forecasting found "the forecasting procedures that were described [in sufficient detail to be evaluated] violated 72 principles" of scientific forecasting (Green and Armstrong, 2007). The authors found no evidence the scientists involved in making the IPCC's forecasts were even aware of the literature on scientific forecasting.

Cost-benefit analysis of future climate change also must address the effects of dematerialization. As the research by Wernick and Ausubel (2014), Smil (2013), and others cited in Chapter 5 demonstrates, technological change is lowering the energy- and carbon-intensity of manufacturing and goods and services generally in the United States and globally, meaning future emission levels may be lower or less certain than is presently assumed. The cost of reducing emissions is likely to be lower in the future as well, as new technologies emerge to capture and sequester carbon dioxide or generate energy or consumer goods without emissions. As Mendelsohn (2004) writes, "there is no question but that we will learn a great deal about controlling greenhouse gases

and about climate change over even the next few decades. The optimal policy is to commit to only what one will do in the near term. Every decade, this policy should be reexamined in light of new evidence. Once the international community has a viable program in place, it is easy to imagine the community being able to adjust their policies based on what new information is forthcoming" (p. 47).

Comparing the costs and benefits of specific mitigation efforts, the third question, requires CBA methodologies that are case-specific, which means they can be applied to specific mitigation projects such as a carbon tax, a carbon trading program, investment in solar photo-voltaic systems, or subsidizing electric cars (Monckton, 2014, 2016). Conducting CBAs of mitigation strategies is complicated by the fact that the possible benefits from mitigation will not be apparent until many years into the future – the models used often claim to be accurate and policy-relevant 100 years and even longer – even though the costs will be incurred immediately and will be ongoing. This makes choosing an appropriate discount rate – the subject of Section 8.2.5.2 – critical to producing an accurate evaluation.

There is considerable uncertainty regarding whether man-made emissions will ever cause a contribution to atmospheric warming of more than 1° or 2°C. Some experts believe costs may begin to exceed benefits if the contribution of man-made emissions is a temperature rise exceeding 2.5°C above pre-industrial levels (Mendelsohn and Williams, 2004; Tol, 2009; Doiron, 2014). If temperatures stop rising before or around that point, due to natural feedbacks or simply because man no longer is producing large quantities of greenhouse gases or because the climate sensitivity to greenhouse gases is lower than the IPCC projects, then enormous expenditures spanning generations will have been entirely wasted.

Because forcing a transition away from fossil fuels to alternative fuels requires raising the price of energy, and the price of energy is closely related to the rate of economic growth, actions taken today to reduce emissions will reduce the wealth of future generations. Thus, investing today to avoid or delay a future hazard that may or may not even materialize may undermine the ability of future generations to cope with climate change (whether natural or man-made) or make further progress in protecting the natural environment from other, real, threats.

The fourth question, which asks if mitigation is preferable to adaptation, is often overlooked by scientists and policymakers alike. Environmentalists who are predisposed to oppose initiatives that shape or alter the natural world view adaptation strategies as insufficient and likely to result in doubling down on past bad behavior that could make the situation worse rather than better (Orr, 2012). However, if future climate change is gradual, unlikely to reach the levels feared by some proponents of the hypothesis, or unlikely to be accompanied by many of the negative impacts thought to occur, then adaptation would indeed be the preferred strategy.

While the cost of adaptation to unmitigated warming is not always case-specific, since it may consist of countless choices made by similarly countless individuals over long periods of time, the cost of mitigation projects can be assessed case by case. This could make direct cost-benefit ratio comparisons of mitigation strategies with adaptation difficult, unless the cost of adaptation to unmitigated global warming can be shown to be lower than even the best mitigation strategies.

The complexity of climate science and economics makes conducting any of these CBAs a difficult and perhaps even impossible challenge (Ceronsky *et al.,* 2011; Pindyck, 2013). In a candid statement alluding to the many difficulties associated with determining the "social cost of carbon," Weitzman remarked, "the economics of climate change is a problem from hell," adding that "trying to do a benefit-cost analysis (BCA) of climate change policies bends and stretches the capability of our standard economist's toolkit up to, and perhaps beyond, the breaking point" (Weitzman, 2015).

References

Armstrong, J.S. 2001. *Principles of Forecasting – A Handbook for Researchers and Practitioners.* Norwell, MA: Kluwer Academic Publishers.

Bezdek, R.H. 2014. *The Social Costs of Carbon? No, the Social Benefits of Carbon.* Oakton, VA: Management Information Services, Inc.

Bradley, R.L. 2000. *Julian Simon and the Triumph of Energy Sustainability.* Washington, DC: American Legislative Exchange Council.

Bryce, R. 2010. *Power Hungry: The Myths of 'Green' Energy and the Real Fuels of the Future.* New York, NY: PublicAffairs.

Bryce, R. 2014. *Smaller Faster Lighter Denser Cheaper.* New York, NY: PublicAffairs.

Ceronsky, M., Anthoff, D., Hepburn, C., and Tol, R.S.J. 2011. Checking the price tag on catastrophe: the social cost of carbon under non-linear climate response. *ESRI Working Paper* No. 392. Dublin, Ireland: Economic and Social Research Institute.

Doiron, H.H. 2014. Bounding GHG climate sensitivity for use in regulatory decisions. The Right Climate Stuff (website). February.

Fyfe, J.C., Gillett, N.P., and Zwiers, F.W. 2013. Overestimated global warming over the past 20 years. *Nature Climate Change* **3**: 767–9.

Goklany, I.M. 2007. *The Improving State of the World: Why We're Living Longer, Healthier, More Comfortable Lives on a Cleaner Planet.* Washington, DC: Cato Institute.

Gordon, R.J. 2016. *The Rise and Fall of American Growth.* Princeton, NJ: Princeton University Press.

Green, K.C. and Armstrong, J.S. 2007. Global warming: forecasts by scientists versus scientific forecasts. *Energy and Environment* **18:** 997–1021.

Idso, C.D. 2013. *The Positive Externalities of Carbon Dioxide.* Tempe, AZ: Center for the Study of Carbon Dioxide and Global Change.

IPCC. 2014a. Intergovernmental Panel on Climate Change. *Climate Change 2014: Impacts, Adaptation, and Vulnerability.* Contribution of Working Group II to the Fifth Assessment Report of the Intergovernmental Panel on Climate Change. New York, NY: Cambridge University Press.

IPCC. 2014b. Intergovernmental Panel on Climate Change. *Climate Change 2014: Mitigation of Climate Change.* Contribution of Working Group III to the Fifth Assessment Report of the Intergovernmental Panel on Climate Change. New York, NY: Cambridge University Press.

McKitrick, R. and Christy, J. 2018. A test of the tropical 200- to 300-hPa warming rate in climate models. *Earth and Space Science* **5** (9).

Mendelsohn, R. 2004. The challenge of global warming. Opponent paper on climate change. In: Lomborg, B. (Ed.) *Global Crises, Global Solutions.* Cambridge, MA: Cambridge University Press.

Mendelsohn, R. and Williams, L. 2004. Comparing forecasts of the global impacts of climate change. *Mitigation and Adaptation Strategies for Global Change* **9**: 315–33.

Monckton, C. 2014. *To Mitigate or to Adapt? Inputs from Mainstream Climate Physics to Mitigation Benefit-Cost Appraisal.* Haymarket, VA: Science and Public Policy Institute.

Monckton, C. 2016. Chapter 10. Is CO_2 mitigation cost effective? In: Easterbrook, D. (Ed.) *Evidence-Based Climate Science.* Second Edition. Cambridge, MA: Elsevier, Inc.

Orr, D.W. 2012. *Down to the Wire: Confronting Climate Collapse.* New York, NY: Oxford University Press.

Pindyck, R.S. 2013. Pricing carbon when we don't know the right price. *Regulation* **36** (2): 43–6.

Smil, V. 2005. *Creating the Twentieth Century: Technical Innovations of 1867–1914 and Their Lasting Impact.* New York, NY: Oxford University Press, Inc.

Smil, V. 2006. *Transforming the Twentieth Century: Technical Innovations and Their Consequences.* New York, NY: Oxford University Press, Inc.

Smil, V. 2013. *Making the Modern World: Materials and Dematerialization.* New York, NY: Wiley.

Tol, R.S.J. 2009. The economic effects of climate change. *Journal of Economic Perspectives* **23** (2): 29–51.

Tol, R.S.J. 2011. The social cost of carbon. *Annual Review of Resource Economics* **3**: 419–43.

van Kooten, G.C. 2013. *Climate Change, Climate Science and Economics: Prospects for an Alternative Energy Future.* New York, NY: Springer.

Weitzman, M.L. 2015. A review of William Nordhaus' "The climate casino: risk, uncertainty, and economics for a warming world." *Review of Environmental Economics and Policy* **9**: 145–56.

Wernick, I. and Ausubel, J.H. 2014. *Making Nature Useless? Global Resource Trends, Innovation, and Implications for Conservation.* Washington, DC: Resources for the Future. November.

Yadama, G.N. 2013. *Fires, Fuels, and the Fate of 3 Billion: The State of the Energy Impoverished.* New York, NY: Oxford University Press.

8.1.2 Integrated Assessment Models

Integrated assessment models (IAMs) are a key element of cost-benefit analysis in the climate change debate. They are enormously complex and can be programmed to arrive at widely varying conclusions.

Integrated assessment models (IAMs), as they are used in the climate change debate, are mathematical constructs that provide a framework for combining knowledge from a wide range of disciplines, in particular climate science and economics, to measure economic damages associated with carbon dioxide (CO_2)-induced climate change. In public discourse as well as academic research, this measure is often referred to as the "social cost of carbon" (SCC) which Nordhaus (2011) defines as "the economic cost caused by an additional ton of carbon dioxide emissions (or more succinctly carbon) or its equivalent. In a more precise definition, it is the change in the discounted value of the utility of consumption denominated in terms of current consumption per unit of additional emissions. In the language of mathematical programming, the SCC is the shadow price of carbon emissions along a reference path of output, emissions, and climate change" (p. 2).

The SCC label is regretfully inaccurate since "carbon" exists in several states in the natural environment (including in the human body and in the breath we exhale), it is a basic building block of life on Earth, and the "cost" being estimated is typically only the cost of the effects of climate change attributed to CO_2 and other greenhouse gases emitted by humanity, not the net social and environmental costs *and benefits* of the activities that produce greenhouse gases. Since that is quite a mouthful, the brief but inaccurate moniker "social cost of carbon" has been adopted generally by researchers and is used here.

The building and tweaking of IAMs has become so complex its practitioners, like those who specialize in cost-benefit analysis, have formed their own society, The Integrated Assessment Society, and publish their own academic journal, titled *Integrated Assessment Journal,* dedicated to "issues in how to calibrate and validate complex integrated assessment models" (IAJ, 2018). As noted by Wilkerson *et al.* (2015), there are "dozens of IAMs to choose from when evaluating policy options and each has different strengths and weaknesses, solves using different techniques, and has different levels of technological and regional aggregation. So it is critical that consumers of model results (*e.g.,* scientists, policymakers, leaders in emerging technologies) know how a particular model behaves (and why) before making decisions based on the results."

The three models used by the U.S. government for policymaking prior to 2017 were the Dynamic Integrated Climate-Economy (DICE) model, developed by Yale University economist William Nordhaus (Newbold, 2010; Nordhaus, 2017); the Climate Framework for Uncertainty, Negotiation and Distribution, referred to as the FUND model, originally developed by Richard Tol, an economist at the University of Sussex and now co-developed by Tol and David Anthoff, an assistant professor in the Energy and Resources Group at the University of California at Berkeley (Anthoff and Tol, 2014; Waldhoff *et al.*, 2014); and the Policy Analysis of the Greenhouse Effect (PAGE) model created by Chris Hope and other researchers affiliated with the Judge Business School at the University of Cambridge (Hope, 2006, 2013).

Because of their prominent role in producing SCC estimates, the bulk of the present chapter focuses on IAMs. Section 8.1.2.1 discusses their background and structure and 8.1.2.2 discusses perhaps their biggest problem, the propagation of error (sometimes referred to as the "cascade of uncertainty" due to the chained logic of the computer programming on which they rely). Descriptions of the IAMs used by the U.S. Interagency Working Group on the Social Cost of Carbon and by a UK report – the Stern Review – are presented in Sections 8.1.3 and 8.1.4, and a brief conclusion appears in Section 8.1.5.

References

Anthoff, D. and Tol, R.S.J. 2014. The income elasticity of the impact of climate change. In: Tiezzi, S. and Martini, C. (Eds.) *Is the Environment a Luxury? An Inquiry into the Relationship between Environment and Income.* Abingdon, UK: Routledge, pp. 34–47.

Hope, C.W. 2006. The marginal impact of CO_2 from PAGE 2002. *Integrated Assessment Journal* **6** (1): 9–56.

Hope, C.W. 2013. Critical issues for the calculation of the social cost of CO_2: why the estimates from PAGE09 are higher than those from PAGE2002. *Climatic Change* **117**: 531–43.

IAJ. 2018. IAJ integrated assessment: bridging science and policy (website). Accessed June 28, 2018.

Newbold, S.C. 2010. Summary of the DICE Model. Prepared for the EPA/DOE workshop, Improving the Assessment and Valuation of Climate Change Impacts for Policy and Regulatory Analysis, Washington DC, November 18–19.

Nordhaus, W.D. 2011. Estimates of the social cost of Carbon: background and results from the RICE-2011 Model. *NBER Working Paper* No. 17540. Cambridge, MA: National Bureau of Economic Research. October.

Nordhaus, W. 2017. Evolution of assessments of the economics of global warming: Changes in the DICE model, 1992–2017. *Discussion Paper* No. w23319. Cambridge, MA: National Bureau of Economic Research and Cowles Foundation.

Waldhoff, S., Anthoff, D., Rose, S., and Tol, R.S.J. 2014. The marginal damage costs of different greenhouse gases: an application of FUND. *Economics: The Open-Access, Open Assessment E-Journal* **8**: 2014–31.

Wilkerson, J.T., Leibowicz, B.D., Turner, D.D., and Weyant, J.P. 2015. Comparison of integrated assessment models: carbon price impacts on U.S. energy. *Energy Policy* **76**: 18–31.

8.1.2.1 Background and Structure

A typical IAM has four steps: emission scenarios, future CO_2 concentrations, climate projections and impacts, and economic impacts.

Prior to the widespread use of modern-day mathematical computer models, questions involving cross-disciplinary issues were generally addressed by scientific panels or commissions convened to bring together a group of experts from different disciplines who would provide their collective wisdom and judgment on the issue at hand. The first formal application of an IAM in global environmental issues was the Climate Impacts Assessment Program (CIAP) of the U.S. Department of Transportation, which examined the potential environmental impacts of supersonic flight in the early 1970s. Other efforts to address global challenges using IAMs followed, but it was not until the 1990s that IAMs proliferated and became commonplace in studies of global climate change.

In an early description and review of these models, appearing as a chapter in the IPCC's Second Assessment Report, Weyant *et al.* explained how "Integrated Assessment Models (IAMs) use a computer program to link an array of component models based on mathematical representations of information from the various contributing disciplines.

This approach makes it easier to ensure consistency among the assumptions input to the various components of the models, but may tend to constrain the type of information that can be used to what is explicitly represented in the model" (Weyant *et al.*, 1996, p. 371). Today there are hundreds of IAMs investigating multiple aspects of the global climate change debate, including the calculation of SCC estimates (Stanton *et al.*, 2009; Wilkerson *et al.*, 2015).

As shown in Figure 8.1.2.1.1, there are four basic steps to calculating the SCC in an IAM: (1) projecting future CO_2 emissions based on various socioeconomic conditions, (2) calculating future atmospheric CO_2 concentrations based on the predicted emission streams, (3) determining how future CO_2 concentrations will change global temperature and weather, and what impact such changes would have on society, and (4) calculating the economic impact ("monetizing the damages") of weather-related events.

The *Emission Scenarios* block, called *Economic Dynamics* in some models, encompasses the impact = population x affluence x technology (IPAT) equation discussed in Chapter 5, Section 5.2.1. This block usually contains a fairly robust energy module as well as a component representing agriculture, forestry, and livestock.

The *Future CO_2 Concentration* block, also called the *Carbon Cycle* module, contains a model of the

carbon cycle that estimates the net increase of carbon in the atmosphere based on what we know of carbon reservoirs, exchange rates, and the residence time of CO_2 in the atmosphere. See Chapter 5, Section 5.1.2 for a tutorial on the carbon cycle.

Changes in carbon concentrations are used as inputs into a *Climate Projections and Impacts* block, sometimes called a *Climate Dynamics* module, which attempts to predict changes in global average surface temperature based on an estimate of "climate sensitivity." See Chapter 2, Section 2.5.3 for a discussion of climate sensitivity, and Chapters 1, 2 and 3 of Idso *et al.* (2013) for hundreds of source citations on this issue. In that same block, changes in temperature are determined or assumed to cause specific effects such as extreme weather events and sea-level rise, which in turn are determined or assumed to have adverse effects on agriculture, human health, and human security. See Chapter 2, Sections 2.1 and 2.7, and Chapter 7, Section 7.2, for discussions of these associations and chains of impacts, and more generally NIPCC (2013, 2014) for thousands of source citations on the subject.

Finally, the postulated changes to weather and then damage to property and livelihood are fed into an *Economic Impacts* block, often called the *Damage Function* module, which monetizes the effects, usually expressing them as a change in per-capita income or gross national product (GNP) or economic growth rates, discounts them to account for the length

Figure 8.1.2.1.1
Simplified linear causal chain of an IAM illustrating the basic steps required to obtain SCC estimates

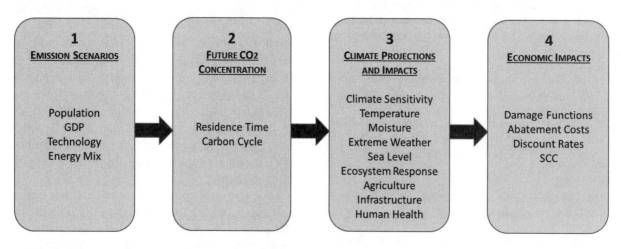

Source: Modified from Parson *et al.*, 2007, Figure ES-1, p. 1.

of time that passes before the effects are experienced, calculates the total (global) net social cost (or benefit), divides it by the number of tons of carbon dioxide emitted according to the *Emission Scenarios* block, and produces a "social cost of carbon" typically expressed in USD per metric ton of CO_2-equivalent greenhouse gases.

Models can be more complex than the one shown in Figure 8.1.2.1.1. For example, the model illustrated in Figure 8.1.2.1.2 incorporates an Ocean Carbon Cycle model as well as a terrestrial and atmospheric model. Conceptually, there is no limit to the degree of sophistication that can be built into the IAMs. Computational limits, however, are another matter, and these weigh heavily in optimization models based on computable general equilibrium (CGE) economic modules, such as the DICE model, which compute optimal growth paths by computing thousands of iterations over hundreds of periods.

Model complexity does not necessarily equate to model accuracy or reliability. Illustrating this point,

Risbey *et al.* (1996) compared IAMs to a home built from bricks, where the bricks represent the substantive knowledge found in the different disciplines represented in the various IAM modules, and the mortar or "glue" is the modelers' subjective judgements linking the disparate blocks of knowledge together. They wrote,

Unfortunately, while the bricks may be quite sound and well described, the subjective judgments (glue) are often never made explicit. As a result, it is difficult to judge the stability of the structure that has been constructed. Thus, in the case of integrated assessment, not only do we need criteria for assessing the quality of the individual components of the analysis, we also need criteria that are applicable to the glue or the subjective judgments of the analyst, as also for the analysis as a whole. While criteria for adequacy for the individual components may be obtained from the individual disciplines, a

Figure 8.1.2.1.2
Wiring diagram for integrated assessment models of climate change

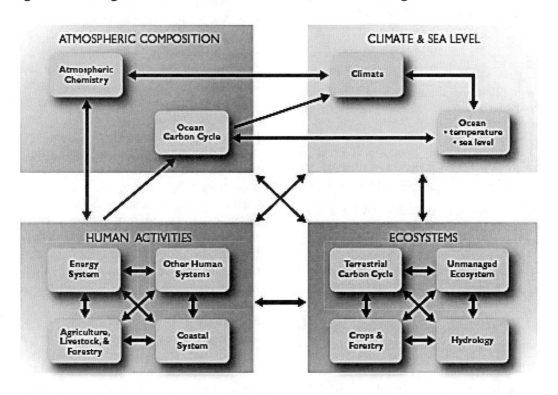

Source: Parson *et al.*, 2007, Figure 2.1, p. 23, citing Wyant *et al.*, 1996.

similar situation does not exist for the 'glue' in the analysis (Risbey *et al.*, 1996, p. 383).

Not only is the "glue" suspect in IAMs, but the blocks themselves are also questionable. Major module limitations include the simplicity of their approach, using only one or two equations associating aggregate damage to one climate variable – in most cases temperature change – which does not recognize interactions among different impacts. More problems include the ability to capture only a limited number of impacts and often omitting those impacts that may be large but are difficult to quantify or show high levels of uncertainty, and presenting damage in terms of loss of income without recognizing capital implications. A particularly difficult problem to solve is the application of "willingness to pay" or "stated preference" quantifications that frequently overstate values relative to observed behavior, since people responding to surveys face no real consequences in terms of required payment for the good or service. This positive "hypothetical bias" is widely noted and discussed in economic literature (e.g., Murphy *et al.*, 2004; Vossler and Evans, 2009; Penn and Hu, 2018). These and other weaknesses described below erode confidence in the ability of IAMs to accurately estimate the "social cost of carbon" in CBAs.

References

Murphy, J., *et al.* 2004. A meta-analysis of hypothetical bias in stated preference valuation. *Environmental and Resource Economics* **30** (3): 313–25.

NIPCC. 2013. Nongovernmental International Panel on Climate Change. *Climate Change Reconsidered II: Physical Science*. Chicago, IL: The Heartland Institute.

NIPCC 2014. Nongovernmental International Panel on Climate Change. *Climate Change Reconsidered II: Biological Impacts*. Chicago, IL: The Heartland Institute.

Parson, E., *et al.* 2007. Global-change scenarios: their development and use. *U.S. Department of Energy Publications* 7. Washington, DC.

Penn, J. and Hu, W. 2018. Understanding hypothetical bias: an enhanced meta-analysis. *American Journal of Agricultural Economics* **100** (4): 1186–1206.

Risbey, J., Kandlikar, M., and Patwardhan, A. 1996. Assessing integrated assessments. *Climatic Change* **34**: 369–95.

Stanton, E.A., Ackerman, F., and Kartha, S. 2009. Inside the integrated assessment models: four issues in climate economics. *Climate and Development* **1**: 166–84.

Vossler, C. and Evans, M. 2009. Bridging the gap between the field and the lab: environmental goods, policy maker input, and consequentiality. *Journal of Environmental Economics and Management* **58** (3): 338–45.

Weyant, J.P., *et al.* 1996. Chapter 10. Integrated assessment of climate change: an overview and comparison of approaches and results. In: Bruce, J.P., *et al.* (Eds.) *Climate Change 1995: Economic and Social Dimensions of Climate Change*. Intergovernmental Panel on Climate Change. Cambridge, UK: Cambridge University Press.

Wilkerson, J.T., *et al.* 2015. Comparison of integrated assessment models: carbon price impacts on U.S. energy. *Energy Policy* **76**: 18–31.

8.1.2.2 Propagation of Error

IAMs suffer from propagation of error, sometimes called cascading uncertainties, whereby uncertainty in each stage of the analysis compounds, resulting in wide uncertainty bars surrounding any eventual results.

"Propagation of error" is a term introduced in Chapter 2, Section 2.1.1.3, used in statistics to refer to how errors or uncertainty in one variable, due perhaps to measurement limitations or confounding factors, are compounded (or propagated) when that variable becomes part of a function involving other variables that are similarly uncertain. Error propagation through sequential calculations is widely used in the physical sciences to reveal the reliability of an experimental result or of a calculation from theory. As the number of variables or steps in a function increases, uncertainties multiply until there can be no confidence in the calculational outcomes. In academic literature this is sometimes referred to as "cascading uncertainties" or "uncertainty explosions."

Uncertainties in climate science, described in Chapter 2, create major difficulties for IAMs. Although considerable progress has been made in climate science and in the understanding of how human activity interacts with and impacts the biosphere and economy, significant uncertainties persist in each block or module of an IAM. As the model progresses through each of these phases, uncertainties surrounding each variable in the chain of computations are compounded one upon another, creating a cascade of uncertainties that peaks upon completion of the final calculation.

An interesting example of the uncertainty and arbitrariness of damage functions can be shown in a comparison conducted by Aldy *et al.* (2009) of the results of IAM impact studies. They found there was a significant amount of consistency among several disparate studies of the economic impact of a 2.5°C warming by 2100 of average global temperatures compared to pre-industrial levels. Five models predicted economic damages between 1% and 2% of global GDP. However, although the gross damage estimates were similar, *there were huge differences in the estimates of the sources of the damages within each study.* The similar results for the gross damage estimates could have occurred by remarkable coincidence. More likely, the modelers "tuned" their models to arrive at total damage values they knew to be in the range of what other researchers have reported. This is an example of the "herding" behavior documented in Chapter 2, Section 2.2.2.1, and the "tuning" of climate models reported by Voosen (2016) and Hourdin *et al.* (2017).

When confronted with the fact that their models include only a limited number of sectors of the economy, the modelers typically argue any unrepresented sectors would result in even greater damage assessment if included. For example, the IPCC says "Different studies include different aspects of the impacts of climate change, but no estimate is complete; most experts speculate that excluded impacts are on balance negative" (IPCC, 2014, p. 690). However, little evidence is presented to support these claims. In contrast, as shown in previous chapters of this volume, the opposite is more likely to be true. Tunnel vision prevents bureaucracies from searching for evidence that might seem to lower the risk of the problem they are responsible for solving, so the "excluded impacts" are likely to be exculpatory rather than reinforce the government's theory. Publication bias (the tendency of academic journals to publish research that finds associations and not to publish those that do not) means more research is likely to reveal that relationships between climate change and alleged impacts is weaker than currently thought.

IAMs increasingly address the issue of uncertainty by including probability distributions – a range of values around a norm – of the parameters to explicitly address the issue of uncertainty. While this serves to acknowledge that we have no real scientific evidence to support one value over another, their use introduces another bias into IAM results. Since the structure of the damage function is made up of quadratic equations, the results of using probability distributions of equation parameters results in so-called "fat tail" impacts that are larger for higher temperature increases than for lower increases. Multiplying a series of upper-bound estimates results in a phenomenon called "cascading conservatism" (Council of Economic Advisors, 2004, p. 179) or what Belzer (2012, p. 13) calls "cascading bias," leading to risk assessments that are orders of magnitude higher than what observational data suggest.

Many experts have concluded the uncertainty problem affecting IAMs makes them too unreliable to form the basis of public policy decisions. Payne (2014) noted "the activist policy [of reducing CO$_2$ emissions] depends on a teetering chain of improbabilities" and represents "an extensive chain of assumptions, every one of which has to be true in order for carbon-dioxide-limiting policies to be justified." Pindyck wrote in the *Journal of Economic Literature* in 2013,

> [IAMs] have crucial flaws that make them close to useless as tools for policy analysis: certain inputs (e.g., the discount rate) are arbitrary, but have huge effects on the SCC estimates the models produce; the models' descriptions of the impact of climate change are completely ad hoc, with no theoretical or empirical foundation; and the models can tell us nothing about the most important driver of the SCC, the possibility of a catastrophic climate outcome. IAM-based analyses of climate policy create a perception of knowledge and precision, but that perception is illusory and misleading (Pindyck, 2013a, abstract).

Writing that same year in the *Review of Environmental Economics and Policy,* Pindyck (2013b, p. 6) also observed:

> IAM damage functions are completely made up, with no theoretical or empirical foundation. They simply reflect common beliefs (which might be wrong) regarding the impact of 2°C or 3°C of warming, and can tell us nothing about what might happen if the temperature increases by 5°C or more.

And yet those damage functions are taken seriously when IAMs are used to analyze climate policy.

Harvard University's Martin Weitzman (2015, pp. 145–6) has commented,

[D]isconcertingly large uncertainties are everywhere, including the most challenging kinds of deep structural uncertainties. The climate change problem unfolds over centuries and millennia, a long intergenerational human time frame that most people are entirely unaccustomed to thinking about. With such long time frames, discounting becomes ultra-decisive for BCA, and there is much debate and confusion about which long-run discount rate should be chosen.

According to Tapia Granados and Carpintero (2013, p. 40), "The lack of robustness of results of different IAMs indicates the limitations of the neoclassical approach, which constitutes the theoretical base of most IAMs; the variety of so-called ad hoc assumptions (often qualified as 'heroic' by their own authors), and the controversial nature of the methods to estimate the monetary value of non-market costs and benefits (mortality, morbidity, damage to ecosystems, etc.). These features explain why many contributions of this type of macroeconomics-oriented IAMs have been criticized for their dubious political usefulness and limited scientific soundness."

Tapia Granados and Carpintero then presented several important shortcomings of IAMs, most of which have been discussed previously: (1) a lack of transparency to explain and justify the assumptions behind the estimates, (2) questionable treatment of uncertainty and discounting of the future, (3) assumption of perfect substitutability between manufactured capital and "natural" capital in the production of goods and services, and (4) problems in the way IAMs estimate monetary costs of non-market effects, which can lead to skepticism about policies based on the results of the models. In another blunt assessment, Ackerman *et al.* (2009, pp. 131–2) wrote:

[P]olicy makers and scientists should be skeptical of efforts by economists to specify optimal policy paths using the current generation of IAMs. These models do not embody the state of the art in the economic theory of uncertainty, and the foundations of the IAMs are much shakier than the general circulation models that represent our best current understanding of physical climate processes. Not only do the IAMs entail an implicit philosophical stance that is highly contestable, they suffer from technical deficiencies that are widely recognized within economics.

Even the latest contributors to the IPCC's assessment reports agree. According to the Working Group II contribution to Chapter 10 of the Fifth Assessment Report (AR5), "Uncertainty in SCC estimates is high due to the uncertainty in underlying total damage estimates (see Section 10.9.2), uncertainty about future emissions, future climate change, future vulnerability and future valuation. The spread in estimates is also high due to disagreement regarding the appropriate framework for aggregating impacts over time (discounting), regions (equity weighing), and states of the world (risk aversion)." As the result of such uncertainties, they say,

Quantitative analyses have shown that SCC estimates can vary by at least approximately two times depending on assumptions about future demographic conditions (Interagency Working Group on the Social Cost of Carbon, 2010), at least approximately three times owing to the incorporation of uncertainty (Kopp *et al.,* 2012), and at least approximately four times owing to differences in discounting (Tol, 2011) or alternative damage functions (Ackerman and Stanton, 2012) (IPCC, 2014, p. 691).

According to the IPCC, "In sum, estimates of the aggregate economic impact of climate change are relatively small but with a large downside risk. Estimates of the incremental damage per tonne of CO_2 emitted vary *by two orders of magnitude,* with the assumed discount rate the main driver of the differences between estimates. The literature on the impact of climate and climate change on economic growth and development has yet to reach firm conclusions. There is agreement that climate change would slow economic growth, by a little according to some studies and by a lot according to other studies. Different economies will be affected differently. Some studies suggest that climate change may trap more people in poverty" (*Ibid.,* p. 692–693, italics added).

For the foreseeable future, IAM analyses will be saddled with the fact that the degree of uncertainty within the various computational stages is immense – especially when the most significant input is subjective (*i.e.*, the discount rate). For all practical purposes the errors inherent to IAMs render their use as policy tools highly questionable, if not irresponsible. They are simply not capable of providing realistic estimates of the SCC, nor can they justify GHG emission reduction policies.

References

Ackerman, F., *et al.* 2009. Limitations of integrated assessment models of climate change. *Climatic Change* **95**: 297–315.

Ackerman, F. and Stanton, E.A. 2012. Climate risks and climate prices: revisiting the social cost of carbon. *Economics: The Open-Access, Open-Assessment E-Journal* **6** (10).

Aldy, J.E., *et al.* 2009. Designing climate mitigation policy. *RFF DP* 08-16. Washington, DC: Resources for the Future. May.

Belzer, R. 2012. *Risk Assessment, Safety Assessment, and the Estimation of Regulatory Benefits*. Mercatus Research. Washington, DC: Mercatus Center.

Council of Economic Advisors. 2004. *Economic Report of the President*. Washington, DC: Government Printing Office.

Hourdin, F., *et al.* 2017. The art and science of climate model tuning. *Bulletin of the American Meteorological Society* **98** (3): 589–602.

IPCC. 2001. Intergovernmental Panel on Climate Change. *Climate Change 2001: Impacts, Adaptation and Vulnerability*. Contribution of Working Group II to the Third Assessment Report of the Intergovernmental Panel on Climate Change. New York, NY: Cambridge University Press.

IPCC. 2014. Intergovernmental Panel on Climate Change. *Climate Change 2014: Impacts, Adaptation, and Vulnerability. Part A: Global and Sectoral Aspects*. Contribution of Working Group II to the Fifth Assessment Report of the Intergovernmental Panel on Climate Change. New York, NY: Cambridge University Press.

IWG. 2010. Interagency Working Group on the Social Cost of Carbon. *Social Cost of Carbon for Regulatory Impact Analysis under Executive Order 12866*. Washington, DC. February.

Kopp, R.E., Golub, A., Keohane, N.O., and Onda, C. 2012: The influence of the specification of climate change damages on the social cost of carbon. *Economics: The Open-Access, Open-Assessment E-Journal* **6**: 2012–3.

Payne, J.L. 2014. The real case against activist global warming policy. *The Independent Review* **19**: 265–70.

Pindyck, R.S. 2013a. Climate change policy: what do the models tell us? *Journal of Economic Literature* **51**: 860–72.

Pindyck, R.S. 2013b. The climate policy dilemma. *Review of Environmental Economics and Policy* **7**: 219–37.

Tapia Granados, J.A. and Carpintero, O. 2013. Dynamics and economic aspects of climate change. In: Kang, M.S. and Banga, S.S. (Eds.) *Combating Climate Change: An Agricultural Perspective*. Boca Raton, FL: CRC Press.

Tol, R.S.J. 2011: The social cost of carbon. *Annual Review of Resource Economics* **3**: 419–43.

Voosen, P. 2016. Climate scientists open up their black boxes to scrutiny. *Science* **354**: 401–2.

Weitzman, M.L. 2015. A review of William Nordhaus' "The climate casino: risk, uncertainty, and economics for a warming world. *Review of Environmental Economics and Policy* **9**: 145–56.

8.1.3 IWG Reports

The widely cited "social cost of carbon" calculations produced during the Obama administration by the Interagency Working Group on the Social Cost of Carbon have been withdrawn and are not reliable guides for policymakers.

On March 28, 2017, President Donald Trump issued an executive order ending the U.S. government's endorsement of estimates of the "social cost of carbon" (SCC) (Trump, 2017). The executive order, which also rescinded other legacies of the Obama administration's environmental agenda, read in part:

Section 5. Review of Estimates of the Social Cost of Carbon, Nitrous Oxide, and Methane for Regulatory Impact Analysis.

(a) In order to ensure sound regulatory decision making, it is essential that agencies use estimates of costs and benefits in their regulatory analyses that are based on the best available science and economics.

(b) The Interagency Working Group on the Social Cost of Greenhouse Gases (IWG), which was convened by the Council of Economic Advisers and the OMB Director, *shall be disbanded*, and the following documents issued by the IWG shall be withdrawn as no longer representative of governmental policy:

(i) Technical Support Document: Social Cost of Carbon for Regulatory Impact Analysis Under Executive Order 12866 (February 2010);

(ii) Technical Update of the Social Cost of Carbon for Regulatory Impact Analysis (May 2013);

(iii) Technical Update of the Social Cost of Carbon for Regulatory Impact Analysis (November 2013);

(iv) Technical Update of the Social Cost of Carbon for Regulatory Impact Analysis (July 2015);

(v) Addendum to the Technical Support Document for Social Cost of Carbon: Application of the Methodology to Estimate the Social Cost of Methane and the Social Cost of Nitrous Oxide (August 2016); and

(vi) Technical Update of the Social Cost of Carbon for Regulatory Impact Analysis (August 2016).

(c) Effective immediately, when monetizing the value of changes in greenhouse gas emissions resulting from regulations, including with respect to the consideration of domestic versus international impacts and the consideration of appropriate discount rates, agencies shall ensure, to the extent permitted by law, that any such estimates are consistent with the guidance contained in OMB Circular A-4 of September 17, 2003 (Regulatory Analysis), which was issued after peer review and public comment and has been widely accepted for more than a decade as embodying the best practices for conducting regulatory cost-benefit analysis (Trump, 2017).

It is not unusual for a president to rescind his predecessor's executive orders, and Trump's predecessor relied heavily on executive orders to implement his anti-fossil-fuel agenda. Disbanding the Interagency Working Group (IWG) sent a clear signal that the president did not want to see the "social cost of carbon" concept kept alive by agency bureaucrats.

The IWG was comprised of representatives from 12 federal agencies brought together specifically to come up with a number – the alleged damages due to climate change caused by each ton of CO_2 emitted by the use of fossil fuels – that could be used to support President Barack Obama's war on fossil fuels (IER, 2014, p. 2). It was an example of the "seeing like a state" phenomenon reported by Scott (1998) and discussed in Chapter 1, Section 1.3.4, when government agencies succumb to pressure to find what they believe their overseers want them to find. IWG utilized experts from numerous agencies who explored technical literature in relevant fields, discussed key model inputs and assumptions, considered public comments, and then duly produced some stylized facts to meet the government's needs.

The first IWG report, issued in 2010, put the social cost of carbon in 2010 at between $4.70 and $35.10 per metric ton of CO_2, depending on the discount rate used (5% for the lower estimate and 2.5% for the higher estimate) (IWG, 2010). The numbers were based on the average SCC calculated by three IAMs (DICE, PAGE, and FUND) and three discount rates (2.5%, 3%, and 5%). A fourth value was calculated as the 95th percentile SCC estimate across all three models at a 3% discount rate and was included to characterize higher-than-expected impacts from temperature change in the tails of the SCC distribution. See Figure 8.1.3.1.

New versions of the three IAMs prompted IWG to recalculate and publish revised SCC estimates in 2013, shown in Figure 8.1.3.2 below (IWG, 2013). In this follow-up exercise, IWG did not revisit other methodological decisions so no changes were made to the discount rate, reference case socioeconomic and emission scenarios, or equilibrium climate sensitivity. Changes in the way damages are modeled were confined to those that had been incorporated into the latest versions of the models by the developers themselves and reported in the peer-reviewed literature.

The IWG's new estimates for the SCC in 2010 ranged from $11 to $52 per metric ton of CO_2, once again depending on the discount rate used, considerably higher than its previous estimate. The

Figure 8.1.3.1
Estimates of the social cost of carbon in 2007 dollars per metric ton of CO_2 from the IWG's 2010 report

Discount Rate Year	5.0% Avg	3.0% Avg	2.5% Avg	3.0% 95th
2010	4.7	21.4	35.1	64.9
2015	5.7	23.8	38.4	72.8
2020	6.8	26.3	41.7	80.7
2025	8.2	29.6	45.9	90.4
2030	9.7	32.8	50.0	100.0
2035	11.2	36.0	54.2	109.7
2040	12.7	39.2	58.4	119.3
2045	14.2	42.1	61.7	127.8
2050	15.7	44.9	65.0	136.2

Source: IWG, 2010.

Figure 8.1.3.2
Estimates of the social cost of carbon in 2007 dollars per metric ton of CO_2 from the IWG's 2013 report

Discount Rate Year	5.0% Avg	3.0% Avg	2.5% Avg	3.0% 95th
2010	11	33	52	90
2015	12	38	58	109
2020	12	43	65	129
2025	14	48	70	144
2030	16	52	76	159
2035	19	57	81	176
2040	21	62	87	192
2045	24	66	92	206
2050	27	71	98	221

Source: IWG, 2013.

new, higher SCC estimates were used by the U.S. government for the first time in a June 2013 rule on efficiency standards for microwave ovens (U.S. Department of Energy, 2013). IWG's SCC estimates were fiercely criticized by experts in the climate change debate. Much of the criticism focused on the IAMs it used as the basis of its estimates – an average of the DICE, FUND, and PAGE models – and since those models are critiqued later in this chapter (see Section 8.3), there is no need to repeat that analysis here.

The Institute for Energy Research (IER), in comments submitted to the Office of Management and Budget in 2014, offered a stinging critique of SCCs in general, making many of the points made in the previous section, and then focused specifically on IWG's process for arriving at an SCC estimate. The IER authors wrote:

> The most obvious example of the dubious implementation of the SCC in federal cost/benefit analyses is the ignoring of clear [Office of Management and Budget (OMB)] guidelines on how such analyses are to be quantified. Specifically, OMB requires that the costs and benefits of proposed policies be quantified at discount rates of 3% and 7% (with additional rates being optional), and OMB also requires that the costs and benefits be quantified at the domestic (not global) level. In practice, the Working Group and agencies that have relied on its estimates of the SCC have simply ignored these two clear OMB guidelines (IER, 2014, p. 12).

Similar points were made in comments submitted by Michaels and Knappenberger (2014). When Heritage Foundation researchers re-ran two of the three IAMs using the 7% discount rate, the SCC dropped by more than 80 percent in one of the models and actually went negative in the other (Dayaratna and Kreutzer, 2013, 2014). The authors of the IER comment went on to say, "No one is arguing that the Working Group or federal agencies should be prohibited from reporting results using a low discount rate. Rather, the public deserves to know what the results would be, were the cost/benefit calculations performed at a 7% discount rate, as OMB guidelines clearly require," and "This omission of a 7% figure masks just how dependent the SCC is on discount rates" (IER, 2014, p. 12) The importance of choosing proper discount rates is discussed in detail later in this chapter (see Section 8.2.5.2).

The IWG's decision to include in its cost-benefit analysis estimates of the global costs (and presumably benefits) of climate change reflected the fact that the three IAMs it chose to rely on attempt to find a global cost rather than a cost specific to the United States. But not only does this violate the purpose of CBA as set forth in national policy guidelines, it also produces false results by disregarding the "leakage" problem reported in Chapter 1, Section 1.2.10, which found reducing emissions in the United States by 10 metric tons could cause emissions by other countries to increase between 1.2 and 13 tons (Brown, 1999; Babiker, 2005).

A net reduction of 10 tons assuming the lower of the two estimates would require an emissions reduction by the United States of 11.4 tons, so the IWG estimate of the SCC is too low. The second estimate means no reductions by the United States, no matter how high, will lead to a net reduction in global emissions since emissions in other countries rise faster than reductions in the United States. In choosing to use a global estimate of damages in its SCC, the IWG disregarded an extensive body of literature on leakage rates by industry, by type of program, and by country (Fischer *et al.*, 2010).

Finally, the IER researchers also observe that "According to Cass Sunstein, the man who convened the SCC Working Group, 'Neither the 2010 TSD [Technical Support Document] nor the 2013 update was subject to peer review in advance, though an interim version was subject to public comment in 2009' [Sunstein, 2013]. This is a direct violation of the administration's stance on 'Transparency and Open Government' [Obama, 2009]" (IER, 2014, p. 19).

For all these reasons, the Trump administration was right to withdraw the social cost of carbon calculations produced during the Obama administration by the Interagency Working Group on the Social Cost of Carbon.

References

Babiker, M.H. 2005. Climate change policy, market structure, and carbon leakage. *Journal of International Economics* **65**: 421.

Brown, S.P.A. 1999. Global Warming Policy: Some Economic Implications. *Policy Report #224*. Dallas, TX: National Center for Policy Analysis.

Dayaratna, K. and Kreutzer, D. 2013. Loaded DICE: an EPA model not ready for the big game. *Backgrounder* No. 2860. Washington, DC: The Heritage Foundation. November 21.

Dayaratna, K. and Kreutzer, D. 2014. Unfounded FUND: yet another EPA model not ready for the big game. *Backgrounder* No. 2897. Washington, DC: The Heritage foundation. April 29.

Fischer, C., Moore, E., Morgenstern, R., and Arimura, T. 2010. *Carbon Policies, Competitiveness, and Emissions Leakage: An International Perspective*. Washington, DC: Resources for the Future.

IER. 2014. Institute for Energy Research. Comment on technical support document: technical update of the social cost of carbon for regulatory impact analysis under executive order no. 12866. February 24.

IWG. 2010. Interagency Working Group on the Social Cost of Carbon. *Social Cost of Carbon for Regulatory Impact Analysis under Executive Order 12866*. Washington, DC. February.

IWG. 2013. Interagency Working Group on the Social Cost of Carbon. *Technical Update of the Social Cost of Carbon for Regulatory Impact Analysis Under Executive Order 12866*. Washington, DC. November.

Michaels, P. and Knappenberger, P. 2014. Comment for Cato Institute on "Office of Management and Budget's request for comments on the technical support document entitled technical update of the social cost of carbon for regulatory impact analysis under executive order 12866." January 27.

Obama, B. 2009. Memorandum for the heads of executive departments and agencies on transparency and open government. Washington, DC: White House. January 21.

Scott, J.C. 1998. *Seeing Like a State: How Certain Schemes to Improve the Human Condition Have Failed*. New Haven, CT: Yale University Press.

Sunstein, C.R. 2013. On not revisiting official discount rates: institutional inertia and the social cost of carbon. *Regulatory Policy Program Working Paper* RPP-2013-21. Cambridge, MA: Mossavar-Rahmani Center for Business and Government, Harvard Kennedy School, Harvard University.

Trump, D. 2017. Presidential Executive Order on Promoting Energy Independence and Economic Growth. March 28.

U.S. Department of Energy. 2013. Energy conservation program: energy conservation standards for standby mode and off mode for microwave ovens. *Federal Register* **78** (116): 10 CFR Parts 429 and 430.

8.1.4 The Stern Review

The widely cited "Stern Review" was an important early attempt to apply cost-benefit analysis to climate change. Its its authors focused on worst-case scenarios and failed to report profound uncertainties.

The Economics of Climate Change: The Stern Review was prepared for the British government by Nicholas Stern, professor of economics and government at the London School of Economics, released in October 2006, and published by Cambridge University Press in 2007 (Stern *et al.*, 2007). Commonly known as the Stern Review, it claimed "using the results from formal economic models, the Review estimates that if we don't act, the overall costs and risks of climate change will be equivalent to losing at least 5% of global GDP each year, now and forever. If a wider range of risks and impacts is taken into account, the estimates of damage could rise to 20% of GDP or more."

The Stern Review's findings were markedly different from prior works on the subject and thus led to questions as to how and why its authors came to such a radically different conclusion. It did not take long for researchers to determine the disparity and the report was quickly refuted (Byatt *et al.*, 2006; Mendelsohn, 2006; Nordhaus, 2006; Tol and Yohe, 2006). Nevertheless, the report was heralded throughout the policy world and continues to be frequently cited as justification for enacting CO_2 emission reduction policies. Following are some key shortcomings of the report.

Uncertainty. Uncertainties all along the chain of calculations are poorly expressed or not acknowledged at all, creating the appearance of a specific and certain finding (such as the numbers cited above) even though its conclusions could be off by an order of magnitude or even reverse sign. This sort of rhetoric is tolerated in the "gray literature" – policy research and commentary that is not peer-reviewed – but it should not then be presented as an authoritative scientific report.

Emission Scenarios. The Stern Review uncritically endorses the IPCC's future emissions scenarios, which have been widely criticized as problematic and based on flawed economic analyses to which no probabilities have been assigned

(Henderson, 2005). As an example, the Stern Review considers only one baseline of demographic change over the next two centuries, which assumes rapid population growth in lower latitudes. Further, the scenario assumes an anemic growth in per-capita income of only 1.3% per year instead of recent growth rates of approximately 3%. This blend of assumptions creates a future full of billions of poor people living in regions deemed most sensitive to warming. Had the Stern Review assumed economic growth to continue at just 2%, and if population growth rates continued to slow, there would actually be a reduction in the poorest and most vulnerable rural populations in these lower latitudes.

Climate Impacts. The Stern Review consistently exaggerates the potential impacts of climate change, giving much more weight and credence to worst-case future climate scenarios. The report assumes powerful positive feedbacks will cause temperatures to increase more rapidly than previously thought, especially throughout the twenty-second century. The central assumption is temperatures might rise 2° to 5°C by 2100, and then by another 2°C by 2200. But the report also raises the possibility warming might be as high as 10° to 11°C by 2100, in which event the global cost is estimated to be as high as 5% to 20% of GDP.

Much of the economic damages are expected to result from increasing extreme weather events, which gain in magnitude and frequency and time in the Stern Review. Observational evidence, in contrast, shows no conclusive relationship between extreme weather events and global warming, with much of the literature suggestive that such events will decline as temperatures warm (see Chapter 2 and references in NIPCC, 2013, Chapter 7). Estimated annual climate-related damages in the Stern Review amount to only 0.2% of GDP at present but rise to 5% of GDP in 2200. This translates to around $70 billion in damages in 2000 to a staggering $23 trillion per year by 2200. There is no evidence to suggest climate impacts could possibly reach this height.

Discount Rate. The Stern Review utilized an extremely low value for the discount rate, just 1.4%. As discussed in some detail in Section 8.2.5.2, the application of such a low value will inherently produce a very high SCC (one dollar of damage in 2200 is worth six cents in 2000 if discounted at 1.4%, but worth only 0.03 cents if discounted at 4%). The authors of the Stern Review argue for using the 1.4% value, which is only 0.1% above the rate of growth of consumption in their analysis, saying it is "ethically proper" – they consider using a higher discount rate to be unfair to future generations. In fact, using a low discount rate to justify draconian reductions on CO_2 emissions today would reduce the welfare of all generations by slowing economic growth today and for decades to come. Further, by using a low discount rate, the Stern Review placed too much near-term significance on events that may occur only far into the future.

Another problem concerning the discount rate was pointed out by Mendelsohn (2006, p. 43), who wrote

> Despite arguing for the low discount rate in the impact analysis, the report does not use it when evaluating the cost of mitigation. To be consistent, the opportunity cost of investing in mitigation must also be valued using the same discount rate as was used to determine the cost of climate change. Because investing in mitigation substitutes for investing in other activities that can earn the market rate of interest, society loses the income that it could have gained from other valuable projects. Assuming that we use the historic rate of return of 4% (that the mitigation program does not drive up interest rates), the value of $1 of abatement is $2.9 when evaluated at a discount rate of 1.4%. The mitigation costs reported in the study need to be multiplied by a factor of three to be consistent with how the damages are calculated.

No Adaptation. Despite discussing the importance of human adaptation to climate change at various points in the report, the influence of adaptation on welfare damages is not taken into account. As Mendelsohn (2006, p. 44) once again critiqued: "[T]he report's estimates of flood damage costs from earlier spring thaws do not consider the probability that people will build dams to control the flooding. Farmers are envisioned as continuing to grow crops that are ill suited for new climates. People do not adjust to the warmer temperatures they experience year after year, and they thus die from heat stroke. Protective structures are not built along the coasts to stop rising sea levels from flooding cities. No public health measures are taken to stop infectious diseases from spreading." The result is that "compared to studies that include adaptation, the [Stern] report overestimates damages by more than an order of magnitude" (*Ibid.*).

Emission Abatement. The Stern Review concedes a present-day high cost of abatement. To reach the stabilization goal of 550 ppm, which corresponds to a two-thirds reduction in emissions by 2050, a carbon tax on the order of $168 per ton would need to be implemented, amounting to a rough estimate of $8.9 trillion per year, which is 6.5% of GDP, or a displaced investment worth about 20% of GDP. The Stern Review reassures its readers these costs will be reduced over time by technological advancements that will drive the costs to only 3% of GDP in 2020 and 1% in 2050. But the costs of technologies do not always fall over time.

Mendelsohn (2006, p. 46) wrote, "Many technologies have been abandoned precisely because their costs have not fallen. Moreover, one must be careful projecting how far costs will fall because one will eventually exhaust all the possible improvements that can be made. One of the critical linchpins of the Stern Report is that technical change will drive down the cost of abatement six-fold by 2050."

Carbon recapture remains a costly, unproven technology and there are multiple problems with renewable technology (see Chapter 3). To meet the Stern Review's goals, Mendelsohn notes, an area covering some 5 million to 10 million hectares of land would be needed for solar panels (in sunny locations), 33 million hectares would be needed to install two million additional wind turbines, and a whopping 500 million additional hectares of land would be needed to increase energy production from biofuels (Mendelsohn, 2006, p. 45). And, despite the increased pressure these actions would place on land, the Stern Review assumes they would have no impact on the price of land, nor on the industries from which the land presumably would be taken (agriculture, timber, and tourism).

Economic Impact. Because of its errors in emission scenarios, estimates of climate impact, use of improper discount rates, and failure to consider adaptation, the Stern Review's claim that unabated global warming would produce large negative economic impacts "now and forever" is not credible. Unlike most other studies, the Stern Review attempts to account for non-market (*i.e.*, environmental) impacts as well as the risk of catastrophe (see, *e.g.*, Freeman and Guzman, 2009, p. 127). Tol observes, "[The Stern Review's] impact estimates are pessimistic even when compared to other studies in the gray literature and other estimates that use low discount rates" (Tol, 2008, p. 9).

Goklany (2009) used the Stern Review's four emission scenarios (taken from the IPCC's Third Assessment Report (TAR)) and its inflated estimates of the damages caused by global warming (expressed as a loss of GDP) for developing and industrialized countries to produce estimates of per-capita income in 2100 and 2200 *net of the cost of global warming –* in other words, subtracting the Stern Review's estimate of income loss attributed to unabated global warming – and compared them to actual 1990 per-capita income (both Stern's and the IPCC's baseline year). His findings appear in Figure 8.1.4.1.

In Figure 8.1.4.1, the net GDP per capita for 1990 is the same as the actual (unadjusted) GDP per capita (in 1990 US dollars, using market exchange rates, per the IPCC's practice). This is consistent with using 1990 as the base year for estimating changes in globally averaged temperatures. The average global temperature increases from 1990 to 2085 for the scenarios are as follows: 4°C for AIFI, 3.3°C for A2, 2.4°C for B2, and 2.1°C for B1. For context, in 2006, GDP per capita for industrialized countries was $19,300; the United States, $30,100; and developing countries, $1,500.

For 2100, the unadjusted GDP per capita accounts for any population and economic growth assumed in the IPCC scenarios from 1990 (the base year) to 2100. For 2200, Goklany assumed the unadjusted GDP per capita is double that in 2100, which is equivalent to a compounded annual growth rate of 0.7%, less than the Stern Review's assumed annual growth rate of 1.3%. Thus, Goklany's calculation substantially understates the unadjusted GDP per capita and, therefore, also the net per-capita GDP in 2200. The costs of global warming are taken from the Stern Review's 95th percentile estimates under the "high climate change" scenario, which is equivalent to the IPCC's warmest scenario (A1F1). Per the Stern Review, these costs amount to 7.5% of global GDP in 2100 and 35.2% in 2200. These losses are adjusted downwards for the cooler scenarios per Goklany (2007).

Figure 8.1.4.1
Net GDP per capita, 1990–2200, after accounting for losses due to global warming as estimated by the Stern Review, for four IPCC emission and climate scenarios

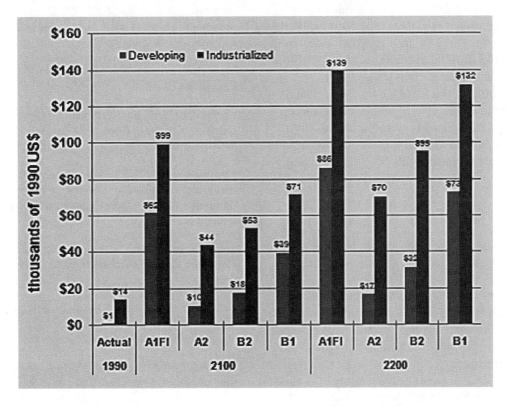

A1FI, A2, B2, and B1 are four emission scenarios for the years 2100 and 2200 as postulated by IPCC TAR arranged from the warmest (A1FI) on the left to the coolest (B1) on the right. Per-capita income growth rate is explained in the text. *Source:* Goklany, 2009.

Figure 8.1.4.1 shows that even accepting the Stern Review's unrealistic assumptions and highest damage estimates, and despite assuming an economic growth rate in the absence of global warming that is *less* than the already low rate assumed by the Stern Report, for populations living in countries currently classified as "developing," net GDP per capita (after accounting for global warming) will be 11 to 65 times higher in 2100 than it was in the base year. It will be even higher (18 to 95 times) in 2200.

Goklany's calculation also found that industrialized countries will have net GDP per capita three to seven times higher in 2100 than in 1990. In 2200 it will be five to 10 times higher. Net GDP per capita in today's developing countries will be higher in 2200 than it was in industrialized countries in the base year (1990) under all scenarios, despite any global warming. *That is, regardless of any global warming, populations living in today's developing countries will be far better off in the future than people currently inhabiting today's industrialized countries.* This is also true for 2100 for all but the "poorest" (A2) scenario.

Under the warmest scenario (A1FI), the scenario that prompts much of the apocalyptic warnings about global warming, net GDP per capita of inhabitants of developing countries in 2100 ($61,500) will be double that of the United States in 2006 ($30,100), and almost triple in 2200 ($86,200 versus $30,100). (All dollar estimates are in 1990 US dollars.)

In other words, if the Stern Review's pessimistic scenario were to come about, people everywhere – even in developing countries – would be wealthy by today's standards, and their ability to cope with and adapt to climate change will be correspondingly higher.

* * *

In conclusion, at the time of its publication in 2007 the Stern Review was a serious attempt to estimate the social costs and benefits of climate change. Written by a distinguished author (and team of assistants), it was and still is accepted as being at least close to the mark by many policymakers and activists around the world. But the report was not even close to the mark.

Admitting and reporting uncertainty is what separates scholarship from propaganda. For the authors of the Stern Review not to admit or reveal the cascade of uncertainties that render their predictions wholly implausible is not excusable. To manipulate economic growth and discount rates to arrive at headline-grabbing numbers that invoke fear is not the best way to advance an informed public debate. Painting a picture of widespread poverty and despair a century from now, when its own data show people living in developing countries would be *11 to 65 times better off in 2100 than they were in 1990*, suggests at best sleight of hand rather than transparency. In short, the Stern Review set back, rather than advanced, global understanding of the consequences of climate change.

References

Byatt, I., *et al.* 2006. The Stern Review: a dual critique. Part II: economic aspects. *World Economics* 7: 199–229.

Freeman, J. and Guzman, A. 2009. *Climate Change & U.S. Interests*. Version 3.1. September 13.

Goklany, I.M. 2007. Is a richer-but-warmer world better than poorer-but-cooler worlds? *Energy & Environment* 18 (7 and 8): 1023–48.

Goklany, I.M. 2009. Discounting the future. *Regulation* 32 (Spring): 36–40.

Henderson, D. 2005. SRES, IPCC and the treatment of economic issues: what has emerged? *Energy and Environment* 16: 549–78.

Mendelsohn, R.O. 2006. A critique of the Stern Report. *Regulation* 29: 42–6.

NIPCC. 2013. Nongovernmental International Panel on Climate Change. *Climate Change Reconsidered II: Physical Science*. Chicago, IL: The Heartland Institute.

Nordhaus, W. 2006. *The "Stern Review" on the Economics of Climate Change. Working Paper* 12741. Cambridge, MA: National Bureau of Economic Research. December.

Stern, N., *et al.* 2007. *The Economics of Climate Change: The Stern Review*. Cambridge, UK: Cambridge University Press.

Tol, Richard S.J. 2008. The social cost of carbon: trends, outliers and catastrophes. *Economics: The Open-Access, Open-Assessment E-Journal* 2: 2008–25.

Tol, R.S.J. and Yohe, G. 2006. A review of the Stern Review. *World Economics* 7 (4): 233–50.

8.1.5 Conclusion

Cost-benefit analysis (CBA) is a well-established practice in finance and economics, and its application to fossil fuels, climate change, and environmental regulations should be welcomed. Decisions need to be made, but they are being made without a full appreciation of the costs and benefits involved, who will bear the costs and when they might arrive, and other key factors that need to be considered. Integrated assessment models (IAMs) attempt to fill this gap by combining what is known about climate change – the carbon cycle, climate sensitivity, and the residence time of CO_2 in the atmosphere, for example – with economic models that monetize the possible consequences of climate change. These models can be simple or complex; complexity is no guarantee of a superior ability to forecast the future, and may have the opposite effect.

The problem with IAMs is that they are not reliable. This is the result of cascading uncertainties in each block or module of the models, a problem that cannot be solved by more computer power, more data, or averaging the outputs of multiple models. Even small amounts of uncertainty in, say, the residence time of CO_2 in the atmosphere or the cost of adaptation in future years, produces false signals that get amplified year after year as the model is run, making predictions 50 years or 100 years distant purely speculative.

A second problem with IAMs is the judgments that act as the "glue" between the modules. Those judgments are subjective and can be "tuned" to produce practically any result their modelers like: a low "social cost of carbon" estimate if the intent is to avoid having to pay for polluting the commons with greenhouse gases that may injure future generations, or a higher estimate if the intent is to justify punitive regulations on fossil fuel producers and users. Certainly this latter was the case with the SCC estimates produced by the now-disbanded Interagency Working Group and the Stern Review.

Later in this chapter, Section 8.3 reports two attempts made to correct some of the biggest mistakes that appear in IAMs. The exercise is useful if only to reveal how unrealistic current models are and to give policymakers a basis for rejecting calls that they act on the current models' flawed and exaggerated forecasts.

8.2 Assumptions and Controversies

Each of the four modules of a typical integrated assessment model (IAM) relies on assumptions about and controversial estimates of key data, processes, and trends. Efforts to use the models to apply cost-benefit analysis (CBA) to climate change are consequently deeply compromised. In this section the major assumptions and controversies in each model are identified and errors documented.

8.2.1 Emission Scenarios

Most IAMs rely on emission scenarios that are little more than guesses and speculative "storylines." Even current greenhouse gas emissions cannot be measured accurately, and technology is likely to change future emissions in ways that cannot be predicted.

As illustrated in Figure 8.1.2.1.1, the first step in an IAM is to project future changes in human greenhouse gas emissions. The *Emission Scenarios* block, called *Economic Dynamics* in some models, encompasses the impact = population x affluence x technology (IPAT) equation discussed in Chapter 5, Section 5.2.1.

Scenarios (or "storylines," as the IPCC has called them in the past) of future CO_2 emissions are generated by forecasting economic growth rates and their related emissions. Prior to 2013, most IAMs and "gray literature" such as the Stern Review relied on emission scenarios called "SRESs," named after the IPCC's 2000 *Special Report on Emissions Scenarios* that proposed them (IPCC, 2000). In 2013, those scenarios were superseded by "representative concentration pathways" (RCPs) used in the Fifth Assessment Report (IPCC, 2013, Chapters 1 and 12). As the IPCC explains, "Representative Concentration Pathways are referred to as pathways in order to emphasize that they are not definitive scenarios, but rather internally consistent sets of time-dependent forcing projections that could potentially be realized with more than one underlying socioeconomic scenario. The primary products of the RCPs are concentrations but they also provide [estimates of] gas emissions" (IPCC, 2013, p. 1045).

Each RCP starts with projections of emissions of four greenhouse gases (carbon dioxide, methane, nitrous oxide, and chlorofluorocarbons (CFCs)) obtained from the Coupled Model Intercomparison Project (CMIP), an international effort to achieve

Figure 8.2.1.1

Greenhouse gas emissions in IPCC's four representative concentration pathways (RCPs) from 1765 to 2300

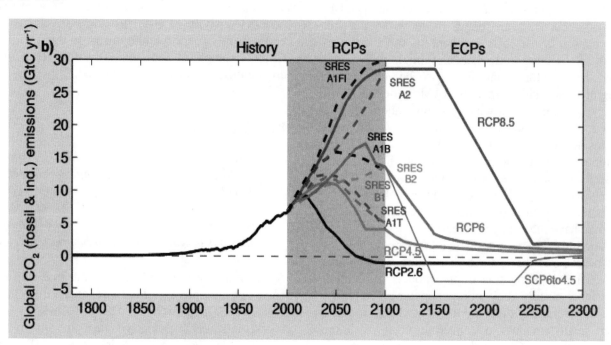

Source: IPCC, 2013, Box 1.1, Figure 3b, p. 149.

consensus on inputs to IAMs. According to CMIP's website, "CMIP provides a community-based infrastructure in support of climate model diagnosis, validation, intercomparison, documentation and data access. This framework enables a diverse community of scientists to analyze [global climate models] in a systematic fashion, a process which serves to facilitate model improvement. Virtually the entire international climate modeling community has participated in this project since its inception in 1995" (CMIP, 2018). CMIP is funded by the U.S. Department of Energy.

Total global annual anthropogenic emissions of greenhouse gases, measured as CO_2 equivalents, are estimated for the past two centuries (starting "around 1765") and forecast from the present for approximately 300 years (to the year 2300). The IPCC's forecast appears as Figure 8.2.1.1, reprinted from Working Group I's contribution to AR5 (IPCC, 2013, p. 149).

The IPCC's four RCPs are titled RCP2.6, RCP4.5, RCP6, and RCP8.5, named after radiative forcing (RF) values of cumulative anthropogenic CO_2 and CO_2 equivalents by the year 2100 relative to pre-

industrial values (2.6, 4.5, 6.0, and 8.5 W/m², respectively). Whether these emissions would actually have these radiative forcing values is discussed in Section 8.2.3 below. IPCC's switch from SRESs to RCPs seems designed to direct attention away from the complexity and uncertainty of its estimates of past, current, and future emissions. Accepting an RCP amounts to accepting a black box that produces easy and "consistent" answers to questions about concentrations and radiative forcing, questions IAM modelers have great difficulty answering without admitting to great uncertainty.

The IPCC's RCPs allow it to say none of its scenarios depends on "one underlying socioeconomic scenario" (IPCC, 2013, p. 1045). This makes challenging their credibility more difficult, but doesn't make any scenario more credible. Apparently, no change in population or economic (consumption) growth, war, natural disaster, or appearance of a new technology for emissions control or efficiency can discredit any one of the RCPs because they no longer rely explicitly on real-world events that might affect these variables. Instead, they rest on an amalgam of IAMs reported in 23 published reports (see sources cited for Table 12.1 on pp. 1048–49). This resembles the IPCC's decision to rely on the DICE, FUND, and PAGE models for its cost-benefit analysis. But what is convenient for modelers is not necessarily good science. Once again, averaging the outputs of models that share common flaws does not produce results more accurate than any one model can produce.

Forecasting future emissions is no easy task given the crude and often simplified parameterizations utilized by IAMs to mimic the global economy (Lemoine and Rudik, 2017). The key is to accurately portray the quantity of current and annual future emissions by properly estimating future population and economic growth and changes in technology and productivity, and predicting seemingly unpredictable events such as changes in government policies, wars, scientific and medical breakthroughs, and more. Few self-described "futurists" can accurately predict such things a year or two in advance. None can predict them across decades or even centuries, as the IPCC attempts to do.

Even measuring *current* emissions is an extremely complex and difficult exercise. Because natural sources (oceans and vegetation) produce massive amounts of CO_2 relative to human emissions, their background presence makes the measurement of anthropogenic emissions on the ground impossible. As reported in Chapter 5, Section 5.1.2, human CO_2 emissions account for only about 3.5% (7.8 Gt divided by 220 Gt) of the carbon entering the atmosphere each year and so, with about 0.5% (1.1 Gt divided by 220 Gt) from net land use change, natural sources account for the remaining 96.0%. The residual of the human contribution the IPCC believes remains in the atmosphere after natural processes move the rest to other reservoirs is as little as 1.17 Gt per year (15% of 7.8 Gt), just 0.53% of the carbon entering the atmosphere each year (IPCC, 2013, p. 471, Figure 6.1). This is less than two-tenths of 1% (0.195%) of the total amount of carbon thought to be in the atmosphere.

Of course, the IPCC does not actually measure CO_2 emissions, since there are millions of sources (billions if humans are counted). Virtually all emission estimates coming from CMIP and therefore the IPCC are not observational data, but stylized facts standing in for unknown quantities that can only be estimated by models and formulas homogenizing disparate and often poorly maintained databases. As the IPCC says, "the final RCP data sets comprise land use data, harmonized GHG emissions and concentrations, gridded reactive gas and aerosol emissions, as well as ozone and aerosol abundance fields" (p. 1046).

Not all countries are able to keep accurate records of economic activity, much less emissions of a dozen gases. "Informal economies" constitute a large part of the economies of many developing and even industrial countries, and little is known about their use of natural resources or emissions. It is thought that 50% of the world's workforce works in informal markets and are likely to escape government regulations and reporting requirements (Jutting and de Laiglesia, 2009). If their use of energy and emissions are comparable to use in the formal economy, then an economy as large as that of China and Japan *combined* is largely invisible to government data collectors. This source of uncertainty is never reported by IAM modelers.

Some countries, including major emitters such as China and Russia, routinely manipulate data regarding economic growth and investment to hide economic woes from their citizens or exaggerate their success to other world leaders. Martinez (2018) observes that totalitarian regimes have "a stronger incentive to exaggerate economic performance (years of low growth, before elections, after becoming ineligible for foreign aid)" which might be observed in their reporting of "GDP sub-components that rely on government information and have low third-party verification." To measure this deception, he compared satellite images of changes over time in electric lights in free and authoritarian countries to their reported economic growth rates during the same period. He found that "yearly GDP growth rates are inflated by a factor of between 1.15 and 1.3 in the most authoritarian regimes" (Martinez, 2018; see also Ingraham, 2018). According to Freedom House, an organization that monitors democracy and authoritarianism around the world, countries designated as "free" in 2013 represented only 40% of the global population (Freedom House, 2014).

With respect to economic growth, IAMs typically assume compound annual global economic growth rates for the period 1995 to 2100 that range between 1.48% and 2.45%, with an average baseline rate of growth of 2.17%. Spreadsheets with the various parameters for the created scenarios can be found on a website maintained by the Energy Modeling Forum at Stanford University (Energy Modeling Forum, 2018). These rates of growth are not particularly high, especially when compared to global growth rates over the past 50 years. The International Energy Agency (IEA) assumes that world GDP, in purchasing power parity (PPP), will grow by an average of 3.4% annually over the period 2010–2030 (IEA, 2017). Similarly, the U.S. Energy Information Agency (EIA) forecasts that from 2015 to 2040, real world GDP growth averages 3.0% in its Reference case (EIA, 2017). If a warmer world is also a more prosperous world, as countless historians have documented was the case in the past (see the review in Chapter 7), then even these projected rates will be too low and along with them, forecasts of future greenhouse gas emissions.

More rapid-than-expected technological change would have the opposite effect of faster-than-expected economic growth. Thanks to the spread of electrification, technology-induced energy efficiency, and the emergence of natural gas as a vigorous competitor to coal for electricity production in the United States (and likely in other countries in the future), the correlation between economic growth (consumption) and greenhouse gas emissions has weakened since the end of the twentieth century (Handrich *et al.,* 2015). Emissions in industrialized countries generally, and especially in the United States, have slowed or even fallen despite population and consumption growth, evidence of the "dematerialization" reported by Ausubel and Waggoner (2008), Goklany (2009), Smil (2013), and others cited in Chapter 5.

In conclusion, most IAMs rely on emission scenarios that are little more than guesses and speculative "storylines." Even current greenhouse gas emissions cannot be measured accurately, and technology is likely to change future emissions in ways that cannot be predicted.

References

Ausubel, J.H. and Waggoner, P. 2008. Dematerialization: variety, caution, and persistence. *Proceedings of the National Academy of Sciences USA* **105** (35): 12,774–12,779.

CMIP. 2018. About CMIP (website). Accessed June 27, 2018.

EIA. 2017. U.S. Energy Information Administration. *2017 International Energy Outlook.*

Energy Modeling Forum. 2018. About (website). Stanford, CA: Stanford University. Accessed June 29, 2018.

Freedom House. 2014. Freedom in the world: 2014. Fact Sheet (website). Accessed July 4, 2018.

Goklany, I.M. 2009. Have increases in population, affluence and technology worsened human and environmental well-being? *The Electronic Journal of Sustainable Development* **1** (3): 3–28.

Handrich, L., Kemfert, C., Mattes, A., Pavel, F., and Traber, T. 2015. *Turning Point: Decoupling Greenhouse Gas Emissions from Economic Growth.* A study by DIW Econ. Berlin, Germany: Heinrich Böll Stiftung.

IEA. 2017. International Energy Agency. *World Energy Outlook.*

Ingraham, C. 2018. Satellite data strongly suggests that China, Russia and other authoritarian countries are fudging their GDP reports. *Washington Post.* May 15.

IPCC. 2000. Intergovernmental Panel on Climate Change. *IPCC Special Report: Emissions Scenarios. A Special Report of IPCC Working Group III.* Cambridge, UK: Cambridge University Press.

IPCC. 2013. Intergovernmental Panel on Climate Change. *Climate Change 2013: The Physical Science Basis.* Contribution of Working Group I to the Fifth Assessment Report of the Intergovernmental Panel on Climate Change. Cambridge, UK: Cambridge University Press.

Jutting, J. and de Laiglesia, J.R. (Eds.) 2009. *Is Informal Normal? Towards More and Better Jobs in Developing Countries.* Paris, France: Organization for Economic Cooperation and Development.

Lemoine, D.M. and Rudik, I. 2017. Steering the climate system: using inertia to lower the cost of policy. *American Economic Review* **107** (10): 2947–57.

Martinez, L.R. 2018. How much should we trust the dictator's GDP estimates? University of Chicago Irving B. Harris Graduate School of Public Policy Studies, May 1. Available at SSRN: https://ssrn.com/abstract=3093296.

Smil, V. 2013. *Making the Modern World: Materials and Dematerialization.* New York, NY: Wiley.

8.2.2 Carbon Cycle

IAMs falsely assume the carbon cycle is sufficiently understood and measured with sufficient accuracy as to make possible precise predictions of future levels of carbon dioxide (CO_2) in the atmosphere.

As illustrated in Figure 8.1.2.1.1, the second step in an IAM is to compute the trajectory of global atmospheric CO$_2$ concentrations based on the emission scenarios calculated in Step 1. The *Future CO$_2$ Concentration* block, also called the *Carbon Cycle* module, contains a model of the carbon cycle that estimates the net increase of carbon in the atmosphere based on what is known about carbon reservoirs, exchange rates, and the residence time of CO$_2$ in the atmosphere.

The IPCC describes the carbon cycle in some detail in Chapter 6 of the Working Group I contribution to AR5 (IPCC, 2013, pp. 465–570), but for its cost-benefit analysis it relies on a single carbon cycle model provided by CMIP. The IPCC uses it to estimate how much anthropogenic carbon dioxide remains in the atmosphere and how it affects future atmospheric concentrations. Figure 8.2.2.1, reprinted from Working Group I's contribution to AR5, illustrates historical and projected estimated atmospheric CO$_2$ concentrations for the four RCPs from 1800 to 2300 (IPCC, 2013, p. 149).

How accurate or certain is the carbon cycle model provided by CMIP for this part of the IPCC's cost-benefit analysis? The IPCC itself says "a single carbon cycle model with a representation of carbon-climate feedbacks was used in order to provide consistent values of CO$_2$ concentration for the CO$_2$

emission provided by a different IAM for each of the scenarios. This methodology was used to produce consistent data sets across scenarios *but does not provide uncertainty estimates for them*" (*Ibid.*, p. 1046, italics added). Estimates without uncertainty estimates should be a red flag for all serious researchers.

As described in more detail in Chapter 5, Section 5.1.2, carbon is stored in four reservoirs: soil, rocks, and sediments, oceans and lakes, plants and animals, and the air. The amount of carbon in each reservoir and the rates of exchange among reservoirs are not known with certainty. Estimates vary in the literature (e.g., Ruddiman, 2008; Falkowski *et al.*, 2000; IPCC, 2013, p. 471). Falkowski *et al.* admitted, "Our knowledge is insufficient to describe the interactions between the components of the Earth system and the relationship between the carbon cycle and other biogeochemical and climatological processes" (Falkowski *et al.*, 2000).

Carbon moves from soil, rocks, and sediment into the air via natural oxidation, bacterial processing, degassing from midocean ridges and hotspot volcanoes, seepage of crude oil and natural gas from land and the ocean floor, the weathering of rocks, and

Figure 8.2.2.1
Historical and projected estimated atmospheric CO$_2$ concentrations, 1765–2300

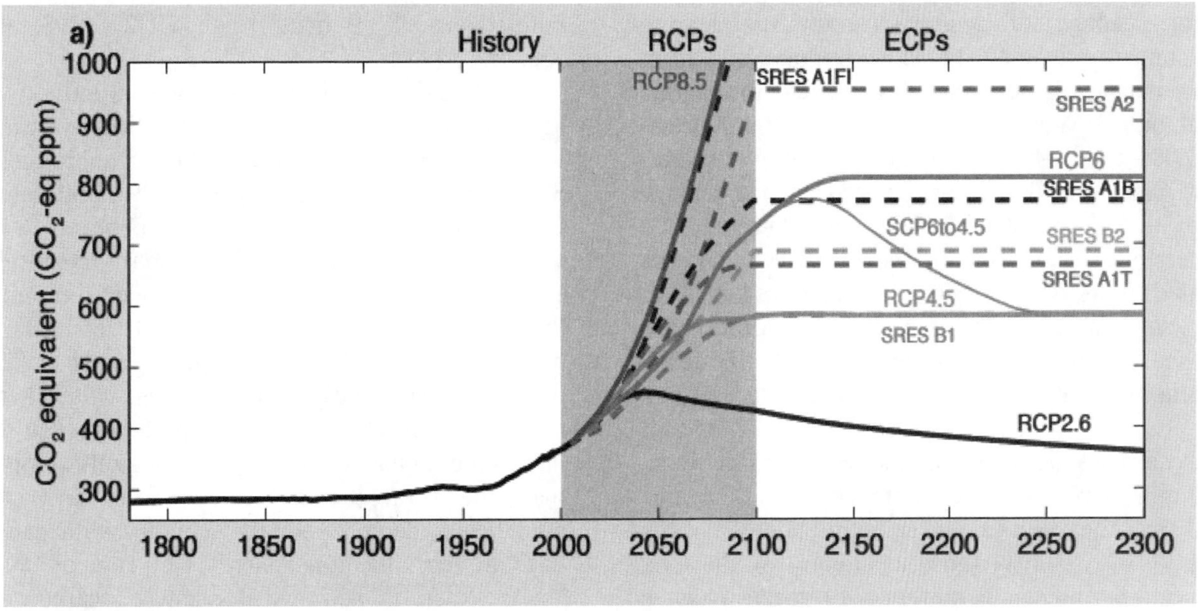

Source: IPCC, 2013, Box 1.1, Figure 3, p. 149.

burning of fossil fuels. Just how much carbon is released naturally from the lithosphere in any given year is uncertain. According to Burton *et al.* (2013, p. 323), "the role of CO_2 degassing from the Earth is clearly fundamental to the stability of the climate, and therefore to life on Earth. Notwithstanding this importance, the flux of CO_2 from the Earth is poorly constrained. The uncertainty in our knowledge of this critical input into the geological carbon cycle led Berner and Lagasa (1989) to state that it is the most vexing problem facing us in understanding that cycle."

According to Wylie (2013), estimates of volcanic degassing rose from around 100 million metric tons of CO_2 per year in 1992 to 600 million metric tons in 2013, a six-fold increase in two decades. According to Aminzadeh *et al.* (2013, p. 4), "What is the volume of hydrocarbon seepage worldwide? The Coal Oil Point seeps are a large source of air pollution in Santa Barbara County, California. Those seeps are similar in many ways to the seeps discussed in this volume. When multiplied by any reasonable assumption of seep numbers worldwide, it is easy to imagine that natural seepage of oil in the range of thousands of barrels per day and gas leakage of hundreds of millions of cubic feet per day is not unreasonable."

In 2003, a U.S. National Research Council report titled *Oil in the Sea III* acknowledged "the inputs from land-based sources are poorly understood, and therefore estimates of these inputs have a high degree of uncertainty," and "estimating the amount of natural seepage of crude oil into the marine environment involves broad extrapolations from minimal data." It nevertheless estimated the annual global oil seepage rate to be between 200,000 and 2,000,000 tons (60 and 600 million gallons) (NRC, 2003). Since the NRC report was produced, extensive use by the oil industry of 3D seismic data, manned submersibles, and remotely operated vehicles has revealed more seeps than previously assumed to exist, suggesting natural seepage of hydrocarbons from the ocean floor may be understated by the IPCC and other research bodies (Roberts and Feng, 2013, p. 56).

Oceans are the second largest reservoir of carbon, containing about 65 times as much as the air. The IPCC and other political and scientific bodies assume roughly 50% to 70% (note the range) of the CO_2 produced by human combustion of fossil fuels is absorbed and sequestered by the oceans, most of the remainder is taken up by plants and animals (terrestrial as well as aquatic), and what's left remains in the air, contributing to the slow increase in atmospheric concentrations of CO_2 during the modern era.

Earth's atmosphere (air) is the fourth and smallest reservoir, estimated to hold approximately 870 gigatons of carbon (GtC). (Note this estimate is generated by mathematical formulas and is not observational data.) As mentioned in the previous section, the total human contribution, including net land use change (primarily agriculture and forestry), is only about 4.3% of total annual releases of carbon into the atmosphere (IPCC, 2013, p. 471, Figure 6.1). The residual of the human contribution that the IPCC believes remains in the atmosphere after natural processes move the rest to other reservoirs is just 0.53% of the carbon entering the air each year. It is less than two-tenths of 1% (0.195%) of the total amount of carbon thought to be in the atmosphere, per Ruddiman (2008). Given uncertainties in the sizes of the reservoirs and the exchange rates among them, it is proper to ask if this residual is measurable, and if not, if it exists at all.

The IPCC apparently assumes atmospheric CO_2 concentrations would be stable, decade after decade and century after century, *but for* anthropogenic emissions. Yet research suggests 500 million years ago the atmosphere's CO_2 concentration was approximately 20 times higher than it is today, at around 7,500 ppm. Two hundred million years later it declined to close to the air's current CO_2 concentration of just over 400 ppm, after which it rose to four times that amount at 220 million years before present (Berner, 1990, 1992, 1993, 1997; Kasting, 1993).

During the middle Eocene, some 43 million years ago, the atmospheric CO_2 concentration is estimated to have dropped to a mean value of approximately 385 ppm (Pearson and Palmer, 1999), and between 25 to nine million years ago, it is believed to have varied between 180ppm and 290 ppm (Pagani *et al.*, 1999). This latter concentration range is essentially the same in which the air's CO_2 concentration oscillated during the 100,000-year glacial cycles of the past 420,000 years (Fischer *et al.*, 1999; Petit *et al.*, 1999). While the natural processes that have driven these changes in CO_2 are not likely to operate over the shorter time scales of an IAM, they nonetheless demonstrate the natural world can and does influence the atmosphere's CO_2 content.

But there is also evidence nature's carbon cycle can impact atmospheric CO_2 at the shorter time periods that matter to IAMs. Joos and Bruno (1998) used ice core data and direct observations of atmospheric CO_2 and ^{13}C to reconstruct the histories of terrestrial and oceanic uptake of anthropogenic carbon over the past two centuries. They discovered that, whereas the land and ocean biosphere typically acted as a source of CO_2 to the atmosphere during the nineteenth century and the first decades of the twentieth century, it subsequently "turned into a sink." In another study, Tans (2009) employed measurements of atmospheric and oceanic carbon contents, along with reasonably constrained estimates of global anthropogenic CO_2 emissions, to calculate the residual fluxes of carbon (in the form of CO_2) from the terrestrial biosphere to the atmosphere (+) or from the atmosphere to the terrestrial biosphere (-), obtaining the results depicted in Figure 8.2.2.2.

As Figure 8.2.2.2 illustrates, Earth's land surfaces were a net *source* of CO_2-carbon to the atmosphere until about 1940, primarily because of the felling of forests and the plowing of grasslands to make way for expanded agricultural activities. From 1940 onward, however, the terrestrial biosphere has become, in the mean, an increasingly greater *sink* for CO_2-carbon, and it has done so despite all the many real and imagined assaults on Earth's vegetation that have occurred over the past several decades, including wildfires, disease, pest outbreaks, deforestation, and climatic changes in temperature

and precipitation, more than compensating for any of the negative effects these phenomena may have had on the global biosphere.

Such findings, which do "not depend on models" but "only on the observed atmospheric increase and estimates of fossil fuel emissions," led Tans (2009) to conclude, "suggestions that the carbon cycle is becoming less effective in removing CO_2 from the atmosphere (*e.g.*, LeQuere *et al.*, 2007; Canadell *et al.*, 2007) can perhaps be true locally, but they do not apply globally, not over the 50-year atmospheric record, and not in recent years." Tans continues, "to the contrary," and "despite global fossil fuel emissions increasing from 6.57 GtC in 1999 to 8.23 in 2006, the five-year smoothed global atmospheric growth rate has not increased during that time, which requires more effective uptake [of CO_2] either by the ocean or by the terrestrial biosphere, or both, to satisfy atmospheric observations."

Confirming evidence has come from Ballantyne *et al.* (2012), who used "global-scale atmospheric CO_2 measurements, CO_2 emission inventories and their full range of uncertainties to calculate changes in global CO_2 sources and sinks during the past fifty years." The five U.S. scientists say their mass balance

Figure 8.2.2.2
Five-year smoothed rates of carbon transfer from land to air (+) or from air to land (-) vs. time

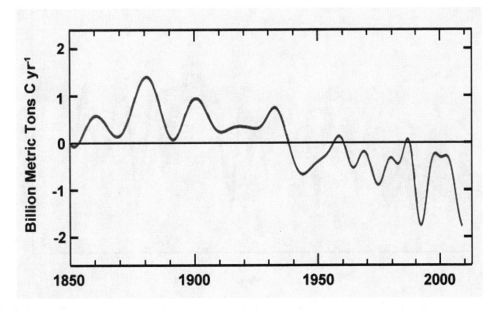

Source: Adapted from Tans (2009).

analysis shows "net global carbon uptake has increased significantly by about 0.05 billion tonnes of carbon per year and that global carbon uptake doubled, from 2.4 ± 0.8 to 5.0 ± 0.9 billion tonnes per year, between 1960 and 2010." See Figure 8.2.2.3 for the authors' plot of their findings.

Commenting on the significance of their findings, Ballantyne *et al.* (2012) wrote in the concluding paragraph of their *Nature* article, "although present predictions indicate diminished C uptake by the land and oceans in the coming century, with potentially serious consequences for the global climate, as of 2010 there is no empirical evidence that C uptake has started to diminish on the global scale." In fact, as their results clearly indicate, just the *opposite* appears to be the case, with global carbon uptake actually doubling over the past half-century. When estimating future concentrations of atmospheric CO_2, IAMs must reconcile model projections of diminished future C uptake by the land and oceans with past observations that indicate land and ocean uptake is being enhanced.

As for the *cause* of this increased removal of CO_2 from the atmosphere, it is primarily the product of Earth's rising atmospheric CO_2 content itself. Thousands of studies demonstrate the photosynthetic response of terrestrial and aquatic plants is enhanced at higher CO_2 concentrations via a phenomenon known as the *aerial fertilization effect* of CO_2 (Idso *et al.*, 2014; Idso, 2018). As Earth's atmospheric CO_2 content has risen since the beginning of the Industrial Revolution, so too has the magnitude of its aerial fertilization effect. This enhancement of terrestrial and oceanic productivity, in turn, has led to an increase in the average amount of CO_2 annually being sequestered from the atmosphere into the land and ocean biosphere, as illustrated in Figures 8.2.2.2 and 8.2.2.3. And that upsurge in sequestration impacts the atmosphere's CO_2 concentration, reducing it from what it would have been without the fertilization effect.

The carbon cycle modules utilized within IAMs must correctly capture all the detailed workings of the global carbon cycle – and how those workings are influenced by both natural and anthropogenic factors – or their estimates of future atmospheric CO_2 concentrations will be wrong. And if those estimates

Figure 8.2.2.3

Annual global net carbon (C) uptake by Earth's lands and oceans (solid blue line) for 1959–2010

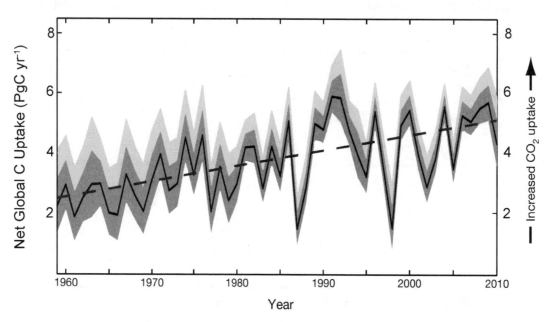

The linear trend (dashed red line) and 1σ (dark shaded bands) and 2σ (light shaded bands) uncertainties are also shown. *Source:* Adapted from Ballantyne *et al.* (2012).

are inaccurate, so will be the projected impacts of future climate that depend on them. No current IAM incorporates this moderating influence of the aerial fertilization effect on future CO_2 concentrations.

In conclusion, IAMs falsely assume the carbon cycle is sufficiently understood and measured with sufficient accuracy as to make possible precise predictions of future levels of CO_2 in the atmosphere.

References

Aminzadeh, F., Berge, T.B., and Connolly, D.L. (Eds.) 2013. *Hydrocarbon Seepage: From Source to Surface.* Geophysical Developments Series No. 16. Tulsa, OK: Society of Exploration Geophysicists and American Association of Petroleum Geologists.

Ballantyne, A.P., Alden, C.B., Miller, J.B., Tans, P.P., and White, J.W. 2012. Increase in observed net carbon dioxide uptake by land and oceans during the past 50 years. *Nature* **488**: 70–2.

Berner, R.A. 1990. Atmospheric carbon dioxide levels over Phanerozoic time. *Science* **249**: 1382–6.

Berner, R.A. 1992. Paleo-CO_2 and climate. *Nature* **358**: 114.

Berner, R.A. 1993. Paleozoic atmospheric CO_2: importance of solar radiation and plant evolution. *Science* **261**: 68–70.

Berner, R.A. 1997. The rise of plants and their effect on weathering and atmospheric CO_2. *Science* **276**: 544–6.

Berner R.A. and Lasaga, A.C. 1989. Modeling the geochemical carbon cycle. *Scientific American* **260**:74–81.

Burton, M.R., Sawyer, G.M., and Granieri, D. 2013. Deep carbon emissions from volcanoes. *Reviews in Mineralogy and Geochemistry* **75** (1): 323–54.

Canadell, J.G., *et al.* 2007. Contributions to accelerating atmospheric CO_2 growth from economic activity, carbon intensity, and efficiency of natural sinks. *Proceedings of the National Academy of Sciences USA* **104**: 18,866–70.

Falkowski, P., *et al.* 2000. The global carbon cycle: a test of our knowledge of Earth as a system. *Science* **290** (5490): 291–96.

Fischer, H., Wahlen, M., Smith, J., Mastroianni, D., and Deck, B. 1999. Ice core records of atmospheric CO_2 around the last three glacial terminations. *Science* **283**: 1712–4.

Idso, C.D. 2018. CO_2 Science Plant Growth Study Database (website). Accessed June 29.

Idso, C.D., Idso, S.B., Carter, R.M., and S.F. Singer. 2014. (Eds.) Nongovernmental International Panel on Climate Change. *Climate Change Reconsidered II: Biological Impacts.* Chicago, IL: The Heartland Institute.

IPCC. 2013. Intergovernmental Panel on Climate Change. *Climate Change 2013: The Physical Science Basis.* Contribution of Working Group I to the Fifth Assessment Report of the Intergovernmental Panel on Climate Change. Cambridge, UK: Cambridge University Press.

Joos, F. and Bruno, M. 1998. Long-term variability of the terrestrial and oceanic carbon sinks and the budgets of the carbon isotopes ^{13}C and ^{14}C. *Global Biogeochemical Cycles* **12**: 277–95.

Kasting, J.F. 1993. Earth's early atmosphere. *Science* **259**: 920–6.

Le Quere, C., *et al.* 2007. Saturation of the Southern Ocean CO_2 sink due to recent climate change. *Science* **316**: 1735–8.

NRC. 2003. U.S. National Research Council Committee on Oil in the Sea. *Oil in the Sea III: Inputs, Fates, and Effects.* Washington, DC: National Academies Press.

Pagani, M., Arthur, M.A., and Freeman, K.H. 1999. Miocene evolution of atmospheric carbon dioxide. *Paleoceanography* **14**: 273–92.

Pearson, P.N. and Palmer, M.R. 1999. Middle Eocene seawater pH and atmospheric carbon dioxide concentrations. *Science* **284**: 1824–6.

Petit, J.R., *et al.* 1999. Climate and atmospheric history of the past 420,000 years from the Vostok ice core, Antarctica. *Nature* **399**: 429–36.

Roberts, H.H. and Feng, D. 2013. Carbonate precipitation at Gulf of Mexico hydrocarbon seeps: an overview. In: Aminzadeh, F., Berge, T.B., and Connolly, D.L. (Eds.). *Hydrocarbon Seepage: From Source to Surface.* Geophysical Developments Series No. 16. Tulsa, OK: Society of Exploration Geophysicists and American Association of Petroleum Geologists, pp. 43–61.

Ruddiman, W.F. 2008. *Earth's Climate: Past and Future.* Second edition. New York, NY: W.H. Freeman and Company.

Tans, P. 2009. An accounting of the observed increase in oceanic and atmospheric CO_2 and an outlook for the future. *Oceanography* **22**: 26–35.

Wylie, R. 2013. Long Invisible, Research Shows Volcanic CO_2 Levels Are Staggering. Live Science (website). October 15.

8.2.3 Climate Sensitivity

Many IAMs rely on estimates of climate sensitivity – the amount of warming likely to occur from a doubling of the concentration of atmospheric carbon dioxide – that are too high, resulting in inflated estimates of future temperature change.

As illustrated in Figure 8.1.2.1.1, the third step in an IAM is to project future changes in global surface temperatures and weather for a given atmospheric CO_2 concentration. Changes in carbon concentrations are used as inputs into a *Climate Projections and Impacts* block, sometimes called a *Climate Dynamics* module, which attempts to predict changes in global average surface temperature.

Equilibrium climate sensitivity (ECS) was discussed in some detail in Chapter 2. It is broadly defined as the equilibrium global mean surface temperature change following a doubling of atmospheric CO_2 concentration. In its Fifth Assessment Report (AR5), the IPCC decided on "a range of 2°C to 4.5°C, with the CMIP5 model mean at 3.2°C" (IPCC, 2013, p. 83). Having estimated the increase in atmospheric CO_2 concentrations caused by emissions in its four Representative Concentration Pathways (RCP), the IPCC used its climate sensitivity estimate to calculate recent past and future radiative forcing, and then the resulting changes to average global surface temperatures, for each RCP. Figure 8.2.3.1 shows the IPCC's estimates for 1950 to 2100. Figure 8.2.3.2 shows the IPCC's RCP estimates for 1765 to 2500. (That is not a typo: The IPCC believes it can hindcast to before the American Revolutionary War and forecast the impact of human greenhouse gas emissions 600 years in the future.)

The IPCC predicts the increase in global average surface temperature by the end of the twenty-first century, relative to the average from year 1850 to 1900, due to human greenhouse gas emissions is "*likely* to exceed 1.5°C for RCP4.5, RCP6.0, and RCP8.5 (*high confidence*). Warming is likely to exceed 2°C for RCP6.0 and RCP8.5 (*high confidence*), more likely than not to exceed 2°C for RCP4.5 (*high confidence*), but *unlikely* to exceed 2°C for RCP2.6 (*medium confidence*). Warming is

unlikely to exceed 4°C for RCP2.6, RCP4.5, and RCP6.0 (*high confidence*) and is *about as likely as not* to exceed 4°C for RCP8.5 (*medium confidence*)" (IPCC, 2013, p. 20). The IPCC illustrates its forecast with the graph reprinted as Figure 8.2.3.3 below.

A better rendering of predicted future global average surface temperatures, in this case forecast by the DICE Model, one of the IAMs relied on by the IPCC, is shown in Figure 8.2.3.4.

How credible are these estimates of climate sensitivity and the temperature changes attributed to them? The Nongovernmental International Panel on Climate Change (NIPCC, 2013) says its best guess of ECS is 0.3°C to 1.1°C, about two-thirds lower than the IPCC's. Figure 2.1.4.1 in Chapter 2 presented a visual representation of estimates of climate sensitivity appearing in scientific research papers published between 2011 and 2016. According to Michaels (2017), the climate sensitivities reported in that figure average ~2.0°C (median) with a range of ~1.1°C (5th percentile) and ~3.5°C (95th percentile). The median is high than NIPCC's 2013 estimate but still more than one-third lower than the estimate used by the IPCC.

Also reported in Chapter 2, Christy and McNider (2017), relying on the latest satellite temperature data, put the transient climate response (ΔT_{LT} at the time CO_2 doubles) at +1.10 ± 0.26 K, which they say "is about half of the average of the IPCC AR5 climate models of 2.31 ± 0.20 K. Assuming that the net remaining unknown internal and external natural forcing over this period is near zero, the mismatch since 1979 between observations and CMIP-5 model values suggests that excessive sensitivity to enhanced radiative forcing in the models can be appreciable."

The fact that the climate models relied on by the IPCC tend to "run hot" is demonstrated in Figure 8.2.3.5, showing the results of 108 climate model runs during the 20-year and 30-year periods ending in 2014 (Michaels and Knappenberger, 2014). The blue bars show the number of runs that predicted a specific maximum trend in °C/decade, while the red and yellow lines point to the actual observed trend during those periods. Remarkably, every model predicted maximum temperature increases higher than the observed 20-year trend and nearly all of them ran "hotter" than the observed 30-year trend. All of these models were specifically tuned to reproduce the twentieth century air temperature trend, an exercise at which they clearly failed.

Figure 8.2.3.1
Historical and projected total anthropogenic radiative forcing (RF) (W/m²) relative to preindustrial (around 1765) between 1950 and 2100

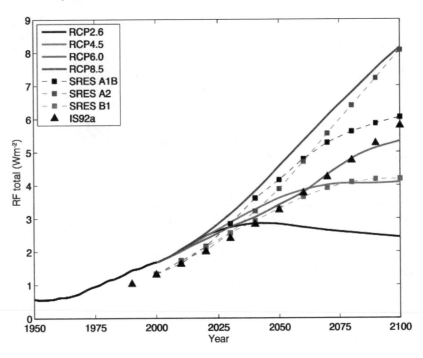

Source: IPCC, 2013, p. 146.

Figure 8.2.3.2
Estimated total radiative forcing (RF) (W/m²) (anthropogenic plus natural) for four RCPs and extended concentration pathways (ECPs) from around 1765 to 2500

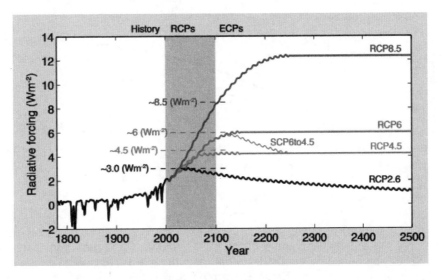

Source: IPCC, 2013, p. 147, citing Meinshausen *et al.,* 2011.

Figure 8.2.3.3

IPCC estimated historical and global average surface temperature changes for four RCPs, 1950–2100

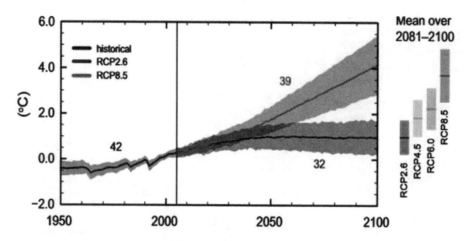

CMIP5 multi-model simulated time series from 1950 to 2100 for change in global annual mean surface temperature relative to 1986-2005. The mean and associated uncertainties for 2085-2100 are given for all RCP scenarios as colored vertical bars. *Source:* IPCC, 2013, p. 21.

Figure 8.2.3.4

Future temperature changes for the years 2000–2300, projected by the DICE model for each of the five emissions scenarios used by the 2013 IWG social cost of carbon estimate

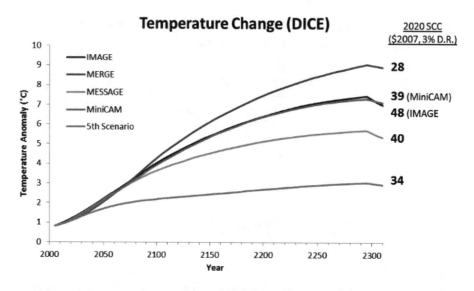

Temperature changes are the arithmetic average of the 10,000 Monte Carlo runs from each scenario. The 2020 value of the SCC (in $2007) produced by the DICE model (assuming a 3% discount rate) is included in the upper right of the figure. DICE data provided by Kevin Dayaratna and David Kreutzer of The Heritage Foundation. *Source:* Michaels and Knappenberger, 2014, p. 4.

Figure 8.2.3.5
20- and 30-year trend distributions from 108 climate model runs versus observed change in temperature

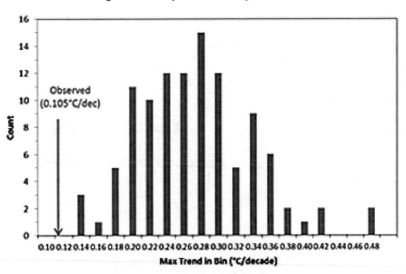

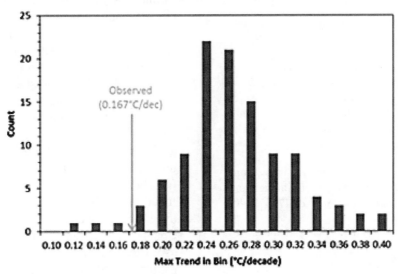

Source: Michaels and Knappenberger, 2014

Critics of the models used by the IPCC have produced their own estimates (e.g., Spencer and Braswell, 2008; Lindzen and Choi, 2011; Monckton *et al.* 2015). Monkton *et al.* (2015) cited 27 peer-reviewed articles "that report climate sensitivity to be below [IPCC's] current central estimates." Their

list of sources appears in Chapter 2 of the present volume as Figure 2.5.3.1.

No one actually knows what the "true" climate sensitivity value is because it is, like so many numbers in the climate change debate, a stylized fact: a single number chosen for the sake of convenience for those who make their living modeling climate change. The number is inherently uncertain for much the same reason it is impossible to know how much CO_2 is emitted into the air every year or how much of it stays there, which is the enormous size of natural processes relative to the "human signal" caused by our CO_2 emissions (Frank, 2015). See Chapter 2, Section 2.2.3 for a discussion of climate sensitivity, and more generally NIPCC (2013) for hundreds of source citations on this complex matter.

The IPCC's estimate of climate sensitivity is very likely to be too high, which invalidates its temperature forecasts and consequently any IAMs that rely on its forecasts. But the IPCC is not the only participant in the climate debate that is wrong. Deep uncertainty about the dynamics of climate means it is probably impossible to reliably estimate climate sensitivity.

References

Christy, J.R. and McNider, R.T. 2017. Satellite bulk tropospheric temperatures as a metric for climate sensitivity. *Asia-Pacific Journal of Atmospheric Sciences* **53** (4): 511–518.

Frank, P. 2015. Negligence, non-science, and consensus climatology. *Energy & Environment* **26** (3): 391–415.

IPCC. 2013. Intergovernmental Panel on Climate Change. *Climate Change 2013: The Physical Science Basis.* Contribution of Working Group I to the Fifth Assessment Report of the Intergovernmental Panel on Climate Change. Cambridge, United UK: Cambridge University Press.

Lindzen, R.S. and Choi, Y-S. 2011. On the observational determination of climate sensitivity and its implications. *Asia-Pacific Journal of Atmospheric Science* **47**: 377–90.

Meinshausen, M., *et al.* 2011. The RCP greenhouse gas concentrations and their extensions from 1765 to 2300. *Climate Change* **109**: 213–41.

Michaels, P. 2017. Testimony. Hearing on At What Cost? Examining the Social Cost of Carbon, before the U.S. House of Representatives Committee on Science, Space, and Technology, Subcommittee on Environment, Subcommittee on Oversight. February 28.

Michaels, P. and Knappenberger, 2014. Comment submitted on Standards of Performance for Greenhouse Gas Emissions from Existing Sources: Electric Utility Generating Units. Docket ID: EPA-HQ-OAR-2013-0602-0001-OAR-2013-0602-0001. Federal Register Number: 2014-13726. Environmental Protection Agency. December 1.

Monckton, C., Soon, W.W.-H, Legates, D.R., and Briggs, W.M. 2015. Keeping it simple: the value of an irreducibly simple climate model. *Science Bulletin* **60** (15): 1378–90.

NIPCC. 2013. Nongovernmental International Panel on Climate Change. *Climate Change Reconsidered II: Physical Science.* Chicago, IL: The Heartland Institute.

Spencer, R.W. and Braswell, W.D. 2008. Potential biases in cloud feedback diagnosis: a simple model demonstration. *Journal of Climate* **23**: 5624–8.

8.2.4 Climate Impacts

Many IAMs ignore the extensive scholarly research showing climate change will not lead to more extreme weather, flooding, droughts, or heat waves.

The *Climate Projections and Impacts* module of Integrated Assessment Models (IAMs) that contains the estimate of climate sensitivity also contains formulas linking changes in temperature to specific climate impacts such as extreme weather events and sea-level rise, violent conflict over water or other scarce resources, negative health effects caused by exposure to heat or diseases spread by mosquitoes, ticks, and other parasites, loss of livelihoods (economic displacement), and more.

Efforts to link global warming to these alleged harms are crippled by cascading uncertainty, described in Section 8.1.2.2, whereby the errors or uncertainty in one variable are compounded (propagated) when that variable becomes part of a function involving other variables that are similarly uncertain, as well as cascading bias, explained in Chapter 6, Section 6.3.1, whereby upper-limit risk estimates are multiplied by each other resulting in estimates orders of magnitude greater than what empirical data suggest. Coming late in the sequence of calculations in an IAM, the *Climate Projections and Impacts* module already has to manage the uncertainties infecting earlier modules, so modelers cannot say with confidence that one additional metric ton of CO_2 released into the air will result in any warming at all. When they try to document an association, they add even more links and more uncertainties to a logical chain that already defies reason.

Since all the climate impacts alleged by the IPCC were addressed in previous chapters, we avoid repetition by only briefly discussing them here. Monetizing the impacts of climate change, the last nodule of an IAM, is addressed in Section 8.2.5.

Extreme Weather

According to the IPCC, "sea level rise and increased frequency of extreme events increases the risk of loss of lives, homes, and properties, and damages infrastructure and transport systems" (IPCC, 2014, Table 12-1, p. 761). But as reported in Chapter 2, Section 2.3.1, researchers have failed to find a convincing relationship between higher surface temperatures over the past 100 years and increases in the frequency or severity of extreme weather events (Maue, 2011; Alexander *et al.*, 2006; Khandekar, 2013; Pielke Jr., 2013, 2014). Instead, the number and intensity of extreme events wax and wane often in parallel with natural decadal or multidecadal climate oscillations.

Legates (2014) writes, "Current state-of-the-art General Circulation Models (GCMs) do not simulate precipitation well because they do not include the full range of precipitation-forming mechanisms that occur in the real world. It is demonstrated here that the impact of these errors are not trivial – an error of only 1 mm in simulating liquid rainfall is equivalent to the energy required to heat the entire troposphere by $0.3°C$. Given that models exhibit differences between the observed and modeled precipitation that often exceed 1 mm day, this lost energy is not trivial. Thus, models and their prognostications are largely unreliable" (abstract).

Basic meteorological science suggests a warmer world would experience fewer storms and weather extremes, as indeed has been the case in recent years. Khandekar and Idso concluded, "It is clear in almost every instance of each extreme weather event examined, there is little support for predictions that CO_2-induced global warming will increase either the frequency or intensity of those events. The real-world data overwhelmingly support an opposite conclusion: Weather will more likely be less extreme in a warmer world (Khandekar and Idso, p. 810).

Sea-level Rise

The IPCC says "for countries made up entirely of low-lying atolls, sea level rise, ocean acidification, and increase in episodes of extreme sea surface temperatures compromise human security for present or future higher populations. With projected high levels of sea level rise beyond the end of this century, the physical integrity of low-lying islands is under threat" (IPCC, 2014, p. 775). But as was documented in Chapter 2, Section 2.3.3, sea level rise for the past thousand years it is generally believed to have averaged less than seven inches per century, a rate that is functionally negligible because it is frequently exceeded by coastal processes like erosion and sedimentation (Parker and Ollier, 2017; Burton, 2012). Local sea-level trends vary considerably because they depend on tectonic movements of adjacent land and other local factors. In many places vertical land motion, either up or down, exceeds the very slow global sea-level trend. Consequently, at some locations sea level is rising much faster than the global rate, and at other locations sea level is falling.

Curry (2018) writes, "Tide gauges show that sea levels began to rise during the 19th century, after several centuries associated with cooling and sea level decline. Tide gauges also show that rates of global mean sea level rise between 1920 and 1950 were comparable to recent rates." Her review of recent research found "there is no consistent or compelling evidence that recent rates of sea level rise are abnormal in the context of the historical records back to the 19th century that are available across Europe" and "There is not yet convincing evidence of a fingerprint on sea level rise associated with human-caused global warming."

Agriculture

In its *Summary for Policymakers* for the Working Group II contribution to AR5, the IPCC says "For the major crops (wheat, rice, and maize) in tropical and temperate regions, climate change without adaptation is projected to negatively impact production for local temperature increases of $2°C$ or more above late-20th-century levels, although individual locations may benefit (*medium confidence*) (IPCC, 2014b, pp. 17–18). But as explained in Chapter 2, Section 2.3.5 as well as in great depth in Chapter 5, this forecast is at odds with the fact that CO_2 is plant food and most plants benefit from warmer surface temperatures. Food production has been growing faster than population growth thanks to the technologies of the Green

Revolution and the Gene Revolution and the aerial fertilization effect caused by the combustion of fossil fuels (FAO, 2015; Idso, 2013).

The IPCC acknowledges that "food security is determined by a range of interacting factors including poverty, water availability, food policy agreements and regulations, and the demand for productive land for alternative uses (Barrett, 2010, 2013)." Blurring the issue of causation by using one of the "expressions of uncertainty" identified in Section 7.2, the IPCC says "many of these factors are themselves *sensitive to* climate variability and climate change" (IPCC, 2014a, p. 763, italics added). The IPCC identifies incidents where "food price spikes have been associated with food riots," but then cites literature attributing those riots to other factors. It ends with a remarkable example of combining words that seem to convey certainty ("critical elements," "robust evidence," and "associated with") with an admission of complete uncertainty: "Food prices, food access, and food availability are critical elements of human security. There is robust evidence that food security affects basic-needs elements of human security and, in some circumstances, is associated with political stability and climate stresses. But there are complex pathways between climate, food production, and human security and hence this area requires further concentrated research as an area of concern" (IPCC, 2014a).

In other words, the relationship between climate and food supply and security is so nuanced there likely is no causal relationship between them. Why, then, does it appear in the IPCC's table purporting to show climate impacts on human security?

Public Health

The IPCC claims "Until mid-century, projected climate change will impact human health mainly by exacerbating health problems that already exist (*very high confidence*). Throughout the twenty-first century, climate change is expected to lead to increase in ill-health in many regions and especially in developing countries with low income, as compared to a baseline without climate change (*high confidence*)" (IPCC, 2014b, p. 19). Chapter 4 explains how medical science and empirical data both contradict that forecast. Warmer temperatures are associated with net health benefits, as is confirmed by empirical research in virtually all parts of the world, even those with tropical climates

(Gasparrini *et al.,* 2015; Seltenrich (2015).

An extensive medical literature contradicts the claim that malaria will expand across the globe or intensify in some regions as a result of rising global surface temperatures (Reiter, 2008; Zhao *et al.,* 2016). Concerns over large increases in mosquito-transmitted and tick-borne diseases such as yellow fever, malaria, viral encephalitis, and dengue fever as a result of rising temperatures are similarly unfounded. While climatic factors do influence the geographical distribution of ticks, temperature and climate change are not among the significant factors determining the incidence of tick-borne diseases (Gething, 2010).

Fossil fuels have been an essential part of the campaign to reduce diseases and extend human life since the start of the Industrial Revolution. While somehow avoiding or slowing rising global temperatures would almost assuredly not improve public health, it is certain that restricting access to fossil fuels would *harm* public health.

Violent Conflict

In the *Summary for Policymakers* for the Working Group II contribution AR5, the IPCC claims "Climate change indirectly increases risks from violent conflict in the form of civil war, inter-group violence, and violent protests by exacerbating well-established drivers of these conflicts such as poverty and economic shocks (*medium confidence*). Statistical studies show that climate variability is significantly related to these forms of conflict. … Climate change over the 21st century will lead to new challenges to states and will increasingly shape national security policies (*medium evidence, medium agreement*) (IPCC, 2014, p. 12).

This strong language, common in the IPCC's summaries for policymakers, is not repeated in Chapter 12 of the Working Group II contribution to AR5. There, one reads:

> [B]oth the detection of a climate change effect [on the incidence of violent conflicts] and an assessment of the importance of its role can be made only with *low confidence* owing to limitations on both historical understanding and data. Some studies have suggested that levels of warfare in Europe and Asia were relatively high during the Little Ice Age (Parker, 2008; Brook, 2010; Tol and Wagner, 2010; White, 2011; Zhang

et al., 2011), but for the same reasons the detection of the effect of climate change and an assessment of its importance can be made only with *low confidence*. There is no evidence of a climate change effect on interstate conflict in the post-World War II period (IPCC, 2014, p. 1001).

The extensive literature review presented earlier in the current volume, in Chapter 7, Sections 7.3 and 7.4, demonstrates a consensus among historians that warmer temperatures in the past clearly *reduced* the incidence of violent conflict by resulting in more food production, food security, and faster income growth (increasing the opportunity cost of wars), and facilitating more trade. Gleditsch and Nordås (2014) write, "there is no consensus in the scholarly community about such dire projections of future climate wars; in fact most observers conclude that there is no robust and consistent evidence for an important relationship between climate change and conflict."

Conflicts over scarce resources most frequently arise when they are treated as common property without the sort of management described by Ostrom (1990, 2005, 2010) and her international network of researchers. The way to reduce such conflicts is not to try to control the weather, but to empower people with technologies and wealth so they can turn such "tragedies of the commons" into "opportunities of the commons" (Boettke, 2009).

References

Alexander, L.V., *et al.* 2006. Global observed changes in daily climate extremes of temperature and precipitation. *Journal of Geophysical Research* **111** (D5): 10.1029/2005JD006290.

Barrett, C.B. 2010. Measuring food security. *Science* **327**: 825–8.

Barrett, C.B. 2013. Food or consequences: food security and its implications for global sociopolitical stability. In: Barrett, C.B. (Ed.). *Food Security and Sociopolitical Stability*. Oxford, UK: Oxford University Press, pp. 1–34.

Boettke, P. 2009. Liberty should rejoice: Elinor Ostrom's Nobel Prize. *The Freeman*. November 18.

Brook, T. 2010. *The Troubled Empire: China in the Yuan and Ming Dynasties*. Cambridge, MA: Belknap Press.

Burton, D. 2012. Comments on 'Assessing future risk: quantifying the effects of sea level rise on storm surge risk for the southern shores of Long Island, New York. *Natural Hazards* **63**: 1219.

Curry, J. 2018. Special report on sea level rise. Climate Etc. (website). Accessed November 27, 2018.

FAO. 2015. Food and Agriculture Organization of the United Nations. *The State of Food Insecurity in the World 2015*. Rome, Italy.

Gasparrini, A., *et al.* 2015. Mortality risk attributable to high and low ambient temperature: a multi-country observational study. *Lancet* **386**: 369–75.

Gething, P.W., Smith, D.L., Patil, A.P., Tatem, A.J., Snow, R.W., and Hay, S.I. 2010. Climate change and the global malaria recession. *Nature* **465**: 342–5.

Legates, D. 2014. Climate models and their simulation of precipitation. *Energy & Environment* **25** (6–7): 1163–75.

Idso, C.D. 2013. *The Positive Externalities of Carbon Dioxide*. Tempe, AZ: Center for the Study of Carbon Dioxide and Global Change.

IPCC. 2014a. Intergovernmental Panel on Climate Change. *Climate Change 2014: Impacts, Adaptation, and Vulnerability. Part A: Global and Sectoral Aspects*. Contribution of Working Group II to the Fifth Assessment Report of the Intergovernmental Panel on Climate Change. New York, NY: Cambridge University Press.

IPCC. 2014b. Intergovernmental Panel on Climate Change. *Summary for policymakers*. In: *Climate Change 2014: Impacts, Adaptation, and Vulnerability. Part A: Global and Sectoral Aspects*. Contribution of Working Group II to the Fifth Assessment Report of the Intergovernmental Panel on Climate Change. New York, NY: Cambridge University Press.

Khandekar, M.L. 2013. Are extreme weather events on the rise? *Energy & Environment* **24**: 537–49.

Khandekar, M. and Idso, C. 2013. Chapter 7. Observations: extreme weather. In: Idso, C.D., Carter, R.M., and Singer, S.F. (Eds.) *Climate Change Reconsidered II: Physical Science*. Nongovernmental International Panel on Climate Change. Chicago, IL: The Heartland Institute.

Maue, R.N. 2011. Recent historically low global tropical cyclone activity. *Geophysical Research Letters* **38**: L14803.

Ostrom, E. 1990. *Governing the Commons: The Evolution of Institutions for Collective Action*. Cambridge, UK: Cambridge University Press.

Ostrom, E. 2005. *Understanding Institutional Diversity.* Princeton, NJ: Princeton University Press.

Ostrom, E. 2010. Polycentric systems for coping with collective action and global environmental change. *Global Environmental Change* 20: 550–7.

Parker, G. 2008. Crisis and catastrophe: the global crisis of the seventeenth century reconsidered. *American Historical Review* **113** (4): 1053–79.

Parker, A. and Ollier, C.D. 2017. Short-term tide gauge records from one location are inadequate to infer global sea-level acceleration. *Earth Systems and Environment* (December): 1–17.

Pielke Jr., R.A. 2013. Testimony before the Committee on Environment and Public Works of the U.S. Senate. July 18.

Pielke Jr., R.A. 2014. *The Rightful Place of Science: Disasters and Climate Change.* Tempe, AZ: Arizona State University Consortium for Science, Policy & Outcomes.

Reiter, P. 2008. Global warming and malaria: knowing the horse before hitching the cart. *Malaria Journal* **7** (Supplement 1): S1–S3.

Seltenrich, N. 2015. Between extremes: health effects of heat and cold. *Environmental Health Perspectives* **123**: A276–A280.

Tol, R.S.J. and Wagner, S. 2010. Climate change and violent conflict in Europe over the last millennium. *Climatic Change* **99**: 65–79.

White, S. 2011. *The Climate of Rebellion in the Early Modern Ottoman Empire.* New York, NY: Cambridge University Press.

Zhang, D.D., Lee, H.F., Wang, C., Li, B., Pei, Q., Zhang, J., and An, Y. 2011: The causality analysis of climate change and large-scale human crisis. *Proceedings of the National Academy of Sciences USA* **108** (42): 17296–301.

Zhao, X., Smith, D.L., and Tatem, A.J. 2016. Exploring the spatiotemporal drivers of malaria elimination in Europe. *Malaria Journal* **15**: 122.

8.2.5 Economic Impacts

The "social cost of carbon" (SCC) derived from IAMs is an accounting fiction created to justify regulation of fossil fuels. It should not be used in serious conversations about how to address the possible threat of man-made climate change.

As illustrated in Figure 8.1.2.1.1, the final step in an integrated assessment model (IAM) is to project the economic impacts of climate change due to anthropogenic greenhouse gas emissions. The *Economic Impacts* block, also called the *Damage Function* module, monetizes the damages fed to it by the *Climate Projections and Impacts* block or module. The effects, usually expressed as a change in per-capita income, gross national product (GNP), or economic growth rates, are discounted to account for the length of time that passes before the effects are experienced. Formulas aggregate, weigh, and calculate the total (global) net social cost (or benefit), divide it by the number of tons of carbon dioxide emitted according to the *Emission Scenarios* block, and produce a "social cost of carbon" (SCC) typically expressed in USD per metric ton of CO_2-equivalent greenhouse gas.

Coming at the very end of the sequence of calculations in an IAM, the *Economic Impacts* module is most affected by uncertainties infecting earlier modules. By now, the propagation of error first described in Section 8.1.2.2 is so great that modelers cannot say with confidence whether supposed impacts having to do with weather, sea-level rise, agriculture, and human security will be positive, negative, or nonexistent. Nevertheless, dollar figures are assigned and the models are run in classic GIGO ("garbage in, garbage out") style.

Whereas *climate* impacts have been addressed in great depth in previous chapters of this volume and in previous volumes in the *Climate Change Reconsidered* series, *economic* impacts have not. Therefore, there are fewer references in this section to previous chapters or books. Section 8.2.5.1 addresses the IPCC's findings concerning economic impacts, and Section 8.2.5.2 addresses the issue of choosing a discount rate.

8.2.5.1 The IPCC's Findings

The IPCC's effort to monetize the impacts of climate change appears mainly in Chapter 10 of Working Group II's contribution to AR5 titled "Key Economic Sectors and Services." That chapter, as Gleditsch and Nordås note, "is quite modest when it comes to the global economic effects expected to result from global warming" (Gleditsch and Nordås, 2014, p. 85). From the chapter's executive summary (with two paragraph breaks added to facilitate reading):

> Global economic impacts from climate change are difficult to estimate. Economic impact estimates completed over the past 20 years vary in their coverage of subsets of economic sectors and depend on a large number of assumptions, many of which are disputable, and many estimates do not account for catastrophic changes, tipping points, and many other factors.
>
> With these recognized limitations, the incomplete estimates of global annual economic losses for additional temperature increases of ~2°C are between 0.2 and 2.0% of income (±1 standard deviation around the mean) (*medium evidence, medium agreement*). Losses are more likely than not to be greater, rather than smaller, than this range (*limited evidence, high agreement*). Additionally, there are large differences between and within countries.
>
> Losses accelerate with greater warming (*limited evidence, high agreement*), but few quantitative estimates have been completed for additional warming around 3°C or above. Estimates of the incremental economic impact of emitting carbon dioxide lie between a few dollars and several hundreds of dollars per tonne of carbon (*robust evidence, medium agreement*). Estimates vary strongly with the assumed damage function and discount rate (IPCC, 2014, p. 663).

The IPCC adds "for most economic sectors, the impact of climate change will be small relative to the impacts of other drivers (*medium evidence, high agreement*). Changes in population, age, income, technology, relative prices, lifestyle, regulation, governance, and many other aspects of socioeconomic development will have an impact on the supply and demand of economic goods and services that is large relative to the impact of climate change" (IPCC, 2014, p. 662).

Saying "the impact of climate change will be small relative to the impacts of other drivers," even "changes in … relative prices, lifestyle," is a major concession to what real data show. It is at odds with

the tone and narrative of every IPCC *Summary for Policymakers* since publication of the first IPCC assessment report in 1990. It is certainly at odds with the spin put on the release of AR5 by the IPCC and the breathless headlines it generated (e.g., "UN Panel Issues Its Starkest Warning Yet on Global Warming" (Gillis, 2014), "Threat from Global Warming Heightened in Latest U.N. Report" (Reuters, 2014), and "Fossil Fuels Should be 'Phased Out by 2100' says IPCC" (BBC, 2014).

There is another, even bigger, admission in AR5 that undermines its narrative of an impending climate crisis. The authors of the Working Group II contribution admit climate change "may be due to natural internal processes or external forcings such as modulations of the solar cycles, volcanic eruptions, and persistent anthropogenic changes in the composition of the atmosphere or in land use" (IPCC, 2014, Background Box SPM.2). While this may be obvious to all climate scientists, the IPCC Working Group I has *defined* "climate change" as referring only to changes attributable to human activities, either the release of greenhouse gas emissions (primarily by the use of fossil fuels) or by changes in land use (primarily agriculture and forestry). But Working Group II says "attribution of observed impacts in the WGII AR5 generally links responses of natural and human systems to observed climate change, *regardless of its cause*" (IPCC, 2014, p. 4, italics added). In a footnote, they add, "the term attribution is used differently in WGI and WGII. Attribution in WGII considers the links between impacts on natural and human systems and observed climate change, regardless of its cause. By comparison, attribution in WGI quantifies the links between observed climate change and human activity, as well as other external climate drivers" (*Ibid.*).

This is an important clarification with considerable consequences for IAM modelers. It means the "climate impacts" IPCC describes, often at great length and most likely to be reported by media outlets and featured by environmental advocacy groups in their fundraising appeals, may be due to natural causes ("solar cycles, volcanic eruptions") and not be attributable to human activities. Why, then, would IAM modelers incorporate any of them in models intended to forecast "the social cost" of human carbon emissions? Nearly all IAMs make a major error by relying on IPCC data for their inputs.

Rather than produce its own IAM to estimate economic impacts, the IPCC surveyed the IAM literature and, in a fashion similar to what the Interagency Working Group did in the United States, reported an average of the findings. Its estimates of the social cost of carbon (SCC), reported in Table 10-9 of the Working Group II contribution to AR5, are reproduced as Figure 8.2.5.1.1 below. Note that *all these IAMs use IPCC anecdotes and scenarios as inputs into their damage function modules,* so these IAMs are not independent research or confirmation of IPCC's findings.

The IPCC chose to report a range of discount rates, from 0% to 3%, starting lower than and not extending as high as what was used by the IWG (2.5% to 5%) and lower than many experts in the field recommend. (This is the topic of Section 8.2.5.2 below.) Its estimate of the SCC at the 3% discount rate is $40 per metric ton for all studies and $33 for studies published since the Fourth Assessment Report (AR4) published in 2007, figures within IWG's range of $11 to $52 reported in 2013, and the second figure is even a perfect match with IWG's estimate assuming a 3% discount rate. The proximity is neither coincidence nor evidence of accuracy, however, given the herding tendency of model builders and their shared assumptions (Park *et al.,* 2014).

The economic impact of global warming also can be expressed as a measure of lost income or consumption over time, typically expressed as per-capita gross domestic product (GDP). The IPCC says "the incomplete estimates of global annual economic losses for additional temperature increases of ~2°C are between 0.2 and 2.0% of income," with only "*medium evidence, medium agreement*" (IPCC, 2014, p. 663). Oddly, this estimate appears in the executive summary of the chapter but nowhere in the body of the chapter. Presumably this is the lost income growth over a 50-year period (the time required for temperatures to increase ~2°C). (This interpretation of the IPCC's very terse statement of its finding is from Gleditsch and Nordås (2014, p. 85), who cite Tol (2014a) for support.)

The order of magnitude separating the IPCC's low and high estimates is proof that this is little more than a guess. The IPCC admits this, saying "The literature on the impact of climate and climate change on economic growth and development has yet to reach firm conclusions. There is agreement that climate change would slow economic growth, by a little according to some studies and by a lot according to other studies. Different economies will be affected differently" (IPCC, 2014, p. 693).

The economic impacts forecast by the three main IAMs the IPCC uses were plotted by the Interagency Working Group in 2010 in a figure that is reproduced below as Figure 8.2.5.1.2. For a 4°C increase in temperatures by the end of the century – a midpoint in the IPCC's range "from 3.7°C to 4.8°C compared to pre-industrial levels" (IPCC, 2014b., p. 8) – the three IAMs find an annual consumption loss of about 1%, 3%, and 4.5%. For a 2°C warming – IPCC's estimate for the year 2050 – the PAGE and DICE models forecast consumption losses of about 0.5% and 1% while the FUND model forecasts a consumption *benefit* of about 1% of GDP. See Figure 8.2.5.1.2. The models average about a 0.5% consumption loss, the number we can use for a cost-benefit ratio.

The IPCC's attempt to conduct a cost-benefit analysis of global warming illustrates the profound difficulty confronting such endeavors. The IPCC's admissions of uncertainty are explicit and could hardly be more emphatic; from the executive summary previously cited, "Global economic impacts from climate change are difficult to estimate. ... [They] depend on a large number of assumptions, many of which are disputable, and many estimates do not account for catastrophic changes, tipping points, and many other factors" (IPCC, 2014). The IPCC's decision not to build its own IAM speaks volumes as well. The IPCC reports many efforts to monetize the impact of climate change on specific sectors of the economies of many nations, including energy (supply, demand, transport and transmission, and macroeconomic impacts), water services, transportation, recreation and tourism, insurance and financial services, and "other primary and secondary economic activities" including agriculture, forestry, fisheries, and mining. The estimates come from hundreds of sources, many of them "gray literature" meaning they were not peer reviewed. Most estimates are country-specific and would need to be extrapolated to produce global estimates, an exercise fraught with uncertainties. All estimates cover different time periods (long, short, decades ago, or more recent) and use different methodologies (often formulas applied to limited sets of observational data). Most have not been replicated.

To perform a cost-benefit analysis, the IPCC would need to aggregate these extensive but disparate and often unreliable data on these individual economic sectors, an impossible task. The

Figure 8.2.5.1.1
Social cost of carbon estimates reported in AR5

PRTP	Post-AR4			Pre-AR4			All studies		
	Avg	SD	*N*	Avg	SD	*N*	Avg	SD	*N*
0%	270	233	97	745	774	89	585	655	142
1%	181	260	88	231	300	49	209	284	137
3%	33	29	35	45	39	42	40	36	186
All	241	233	462 (35)	565	822	323 (49)	428	665	785 (84)

"PRTP" is pure rate of time preference (discount rate). Columns titled "N" report the number of findings using each of three discount rates (0%, 1%, and 3%). The number of studies surveyed before and after publication of AR4 and the total number of unique studies is reported in parenthesis at the bottom of the "N" columns. "Avg" is the average social cost of carbon in dollars per metric ton of carbon dioxide equivalent greenhouse gas emissions as reported by the studies, "SD" is standard deviation (a measure of variability around the mean). *Source:* IPCC, 2014, Table 10-9, p. 691, citing Section SM10.2 of the on-line supplementary material.

Figure 8.2.5.1.2
Annual consumption loss as a fraction of global GDP in 2100 due to an increase in annual global temperature in the DICE, FUND, and PAGE models

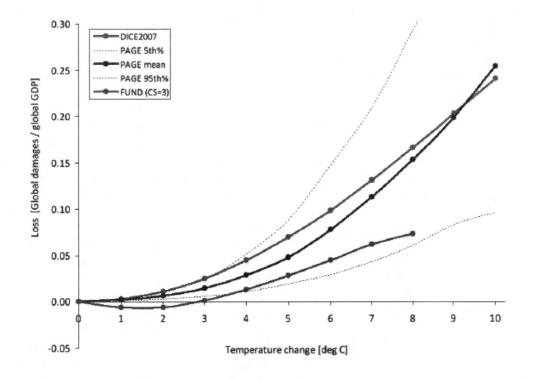

Source: IWG, 2010, Figure 1A, p. 9.

insurmountable problems it would have faced did not disappear when it decided to rely on IAMs created by others. *The DICE, PAGE, and FUND modelers faced the same challenges* but went ahead and produced unreliable estimates anyway.

Averaging the results of multiple IAMs does not raise the probability of finding an SCC estimate or impact on economic growth that is accurate. As Frank observed, "systematic error does not average away with repeated measurements. Repetition can even increase error. When systematic error cannot be eliminated and is known to be present, uncertainty statements must be reported along with the data" (Frank, 2016, p. 338).

In a review of IAMs, Warren *et al.* (2006) concluded, "The assumption of a quadratic dependence of damage on temperature rise is even less grounded in any empirical evidence. Our review of the literature uncovered no rationale, whether empirical or theoretical, for adopting a quadratic form for the damage function – although the practice is endemic in IAMs." Similarly, Pindyck has lamented,

> IAM damage functions are completely made up, with no theoretical or empirical foundation. They simply reflect common beliefs (which might be wrong) regarding the impact of 2°C or 3°C of warming, and can tell us nothing about what might happen if the temperature increases by 5°C or more. And yet those damage functions are taken seriously when IAMs are used to analyze climate policy (Pindyck, 2013a, p. 16).

Also troubling is that these functions are usually based on only one country or region because the literature on the topic of environmentally induced costs is very limited, except in agriculture. For example, as described by Mastrandrea (2009):

> Market and non-market damages in DICE are based on studies of impacts on the United States that are then scaled up or down for application to other regions. Many of the estimates to which market damages in PAGE are calibrated are also based on an extrapolation of studies of the United States. Only FUND uses regional and sector-specific estimates. However, in some sectors these estimates also originate in one country, or may be dominated by estimates from one region. For example, in the energy sector,

the sector which accounts for most of the economic damages in FUND, estimates for the UK are scaled across the world.

Summing up the cumulative effects of the many shortcomings that prevent IAMs from being able to accurately determine the economic impacts of climate change, Pindyck writes:

> … the greatest area of uncertainty concerns the economic impact (including health and social impacts) of climate change. The economic loss functions that are part of most IAMs are essentially ad hoc. This is not surprising given how little we know – in terms of both theory and data – about the ways and extent to which changes in temperature and other climate variables are likely to affect the economy. In fact, the economic impact of climate change may well be in the realm of the "unknowable." This in turn means that IAM-based analyses of climate change may not take us very far, and the models may be of very limited use as a policy tool (Pindyck, 2013b, p. 17).

Assuming *arguendo* that the IPCC's estimate of the economic impacts of global warming in 50 or 100 years is accurate, how should it be interpreted? The IPCC's estimate of the impact of a surface temperature increase of ~2°C (from pre-industrial levels), a loss of 1% of GDP around the year 2050, is *less than the expected global economic growth rate in about four months*. A single recession, even a very short and mild recession, would have a larger impact, and several are likely to occur before 2050.

Other than their choice of a low discount rate, the authors of Chapter 10 of Working Group II's contribution to AR5 may be out-of-step with the rhetoric and tone of other chapters of the WGII contribution to AR5, *but that is a good thing*. While the authors of other chapters seemed to think it their duty to compile anecdotes of human suffering due to extreme weather and natural disasters and to speculate that such events will become more frequent in the future due to human interference in the climate, the authors of Chapter 10 took more seriously their duty to prove the links in the logical chain behind such claims (even while accepting the IPCC's distorted views in the emission scenarios and carbon cycle modules), and then to monetize the harms.

One of the two lead authors of Chapter 10, Richard S.J. Tol, is the creator of the FUND model, one of the three IAMs most prominent in the climate change literature. Significantly, Tol resigned from the IPCC shortly before AR5 was released. He explained why in a blogpost on April 25, 2014:

In the earlier drafts of the SPM, there was a key message that was new, snappy and relevant: **Many of the more worrying impacts of climate change really are symptoms of mismanagement and under-development.** This message does not support the political agenda for greenhouse gas emission reduction. Later drafts put more and more emphasis on the reasons for concern about climate change, a concept I had helped to develop for AR3. Raising the alarm about climate change has been tried before, many times in fact, but it has not had an appreciable effect on greenhouse gas emissions. I reckoned that putting my name on such a document would not be credible – my opinions are well-known – and I withdrew (Tol, 2014b, boldface in original).

Economics, as explained in Chapter 1, uses data about prices and investment returns to make objective what are otherwise only subjective impressions, preferences, and anecdotes. When applied to the impacts of climate change, economics can reveal the true net costs of climate change, should it occur and provided the data that enter earlier modules in the IAMs are accurate. Even with the IPCC's thumb on the scale in this respect, it is remarkable to see how small the economic consequences of climate change would be.

References

BBC. 2014. Fossil fuels should be "phased out by 2100" says IPCC. November 2. Accessed November 17, 2018.

Frank, P. 2016. Systematic error in climate measurements: the surface air temperature record. International Seminars on Nuclear War and Planetary Emergencies 48th Session. Presentation in Erice, Italy on August 19–25, pp. 337–51.

Gillis, J. 2014. UN panel issues its starkest warning yet on global warming. *The New York Times,* November 2. Accessed November 17, 2018.

Gleditsch, N.P. and Nordås, R. 2014. Conflicting messages? The IPCC on conflict and human security. *Political Geography* **43**: 82–90.

IPCC. 2014. Intergovernmental Panel on Climate Change. *Climate Change 2014: Impacts, Adaptation, and Vulnerability. Part A: Global and Sectoral Aspects.* Contribution of Working Group II to the Fifth Assessment Report of the Intergovernmental Panel on Climate Change. New York, NY: Cambridge University Press.

IWG. 2010. Interagency Working Group on the Social Cost of Carbon. *Technical Support Document: Social Cost of Carbon for Regulatory Impact Analysis Under Executive Order 12866.* Washington, DC. February.

Mastrandrea, M.D. 2009. *Calculating the Benefits of Climate Policy: Examining the Assumptions of Integrated Assessment Models.* Arlington, VA: The Pew Center on Global Climate Change.

Park, I.-U., Peacey, M.W., and Munafo, M.R. 2014. Modelling the effects of subjective and objective decision making in scientific peer review. *Nature* **506** (7486): 93–6.

Pindyck, R.S. 2013a. Climate change policy: what do the models tell us? *Journal of Economic Literature* **51**: 860–72.

Pindyck, R.S. 2013b. The climate policy dilemma. *Review of Environmental Economics and Policy* **7**: 219–37.

Reuters. 2014. Threat from global warming heightened in latest U.N. report. March 31. Accessed on November 17, 2018.

Tol, R.S.J. 2014a. Bogus prophecies of doom will not fix the climate. *Financial Times.* March 31.

Tol, R.S.J. 2014b. IPCC again. Richard Tol Occasional thoughts on all sorts (blog). Accessed June 26, 2018.

Warren, R., Hope, C., Mastrandrea, M., Tol, R.S.J., Adger, N., and Lorenzoni, I. 2006. Spotlighting impacts functions in integrated assessment. Research Report Prepared for the Stern Review on the Economics of Climate Change. Tyndall Centre for Climate Change Research. *Working Paper* **91**. September.

8.2.5.2 Discount Rates

As discussed in Chapter 1, Section 1.2.8, the selection of a discount rate (referred to in the U.K. as the "social time preference rate" or STPR) is one of the most controversial issues in the climate change debate (Heal and Millner, 2014; Weitzman, 2015).

According to the U.K. Treasury's *Green Book*, a STPR has two components:

- "time preference" – the rate at which consumption and public spending are discounted over time, assuming no change in per capita consumption. This captures the preference for value now rather than later.

- "wealth effect" – this reflects expected growth in per capita consumption over time, where future consumption will be higher relative to current consumption and is expected to have a lower utility (H.M. Treasury, 2018, p. 101).

The STPR is expressed as an equation, $r = \rho + \mu g$, where r is the STPR, ρ (rho) is time preference comprising pure time preference (δ, delta) and catastrophic risk (L), and μg is the wealth effect, derived as the marginal utility of consumption (μ, mu), multiplied by expected growth rate of future real per capita consumption g. In 2018, the *Green Book* put the three variables at $\rho = 1.5\%$; $\mu = 1.0$; and $g = 2\%$, so $0.015 + 1 \times 0.02 = 3.5\%$. However, the *Green Book* recommends a lower rate of 1.5% for "risk to health and life values" because "the 'wealth effect', or real per capita consumption growth element of the discount rate, is excluded." The STPR also should "decline over the long term," says the *Green Book,* "due to uncertainty about future values of its components." The result is a range of STPRs which it summarizes in the table reproduced as Figure 8.2.5.2.1 below.

Many IAMs and reports in the "gray literature" use rates similar to the *Green Book's* long-term

health rates – 0.71% to 1.07% – which are much lower than those used in any other area of public policy. While different rates are appropriate for different kinds of analysis, it seems the practice of using extremely low rates (and even zero) was adopted early on in the climate change debate to draw attention to what was thought to be an under-appreciated long-term problem. Over time, much of the urgency about the issue has been removed as temperatures have risen less than expected and the predicted climate impacts have failed to materialize. The high cost of mitigation has become better understood, strengthening the case that investments in emissions mitigation should compete on equal footing with spending on other long-term public needs such as education, health care, and infrastructure.

The IPCC originally endorsed discount rates much higher than those recommended by the *Green Book.* The IPCC's Third Assessment Report (IPCC, 2001) said the following about discount rates:

> For climate change the assessment of mitigation programmes and the analysis of impacts caused by climate change need to be distinguished. The choice of discount rates applied in cost assessment should depend on whether the perspective taken is the social or private case.
>
> *For mitigation effects,* the country must base its decisions at least partly on discount rates that reflect the opportunity cost of capital. In developed countries rates around 4%–6% are probably justified. Rates of this level are

Figure 8.2.5.2.1
Declining long term social time preference rate (STPR)

Year	0 – 30	31 – 75	76 – 125
STPR (standard)	3.50%	3.00%	2.50%
STPR (reduced rate where pure STP = 0)	3.00%	2.57%	2.14%
Health	1.50%	1.29%	1.07%
Health (reduced rate where pure STP = 0)	1.00%	0.86%	0.71%

Source: H.M. Treasury, 2018, Table 8, p. 104.

in fact used for the appraisal of public sector projects in the European Union (EU) (Watts, 1999). In developing countries the rate could be as high as 10%–12%. The international banks use these rates, for example, in appraising investment projects in developing countries. It is more of a challenge, therefore, to argue that climate change mitigation projects should face different rates, unless the mitigation project is of very long duration. These rates do not reflect private rates of return, which typically need to be considerably higher to justify the project, potentially between 10% and 25%.

For climate change impacts, the long-term nature of the problem is the key issue. The benefits of reduced [greenhouse gas (GHG)] emissions vary with the time of emissions reduction, with the atmospheric GHG concentration at the reduction time, and with the total GHG concentrations more than 100 years after the emissions reduction. Any "realistic" discount rate used to discount the impacts of increased climate change impacts would render the damages, which occur over long periods of time, very small. With a horizon of around 200 years, a discount rate of 4% implies that damages of USD1 at the end of the period are valued at 0.04 cents today. At 8% the same damages are worth 0.00002 cents today. Hence, at discount rates in this range the damages associated with climate change become very small and even disappear (Cline, 1993)" (IPCC, 2001, p. 466).

There are two main points to be taken from this passage. First, investments in mitigation should be held to the same standard as other investments, public or private, to ensure capital flows to its highest and best use. For developing countries, the IPCC suggests using discount rates as high as 10% to 12%. Second, "the range of dangers associated with climate change become very small and even disappear" as the chosen discount rate increases. It should therefore come as no surprise that governments and other proponents of immediate action to slow or stop climate change favor the use of lower discount rates. At higher (and likely more appropriate) discount rates, there is no economic rationale for immediate action.

In 2001, the IPCC cited a survey by Weitzman (1998) of 1,700 professional economists suggesting they believe "lower rates should be applied to problems with long time horizons, such as that being discussed here," and Weitzman "suggests the appropriate discount rate for long-lived projects is less than 2%" (IPCC, 2001, p. 467). In the eyes of some, discounting at all is unethical (Broome, 2004, 2012; Heal, 2009; Stern, 2014). They claim it violates intergenerational neutrality, causing future generations to be held as less valuable than the current one. But this logic seems flawed since the cost of reducing greenhouse gas emissions to benefit future generations must be compared to other investments *that would also benefit future generations.* Nearly any investment in capital and services that raises productivity and produces wealth will benefit future generations.

Weitzman (2007) and a team of other economists (Arrow *et al.,* 2013) have sided with a declining discount rate based on a formula called the Ramsey discounting formula, in which benefits realized in the immediate future (one to five years) might be discounted at 4%, those in the medium future (26–75 years) at 2%, and those in the distant future (76–300 years) at 1%. But once again, this seems counterintuitive. Making investments in emission reductions that yield less than the return on alternative investments impoverishes future generations (Birdsall and Steer, 1993; Klaus, 2012). As Robert Mendelsohn wrote in 2004, "if climate change can only earn a 1.5% return each year, there are many more deserving social activities that we must fund before we get to climate. Although climate impacts are long term, that does not justify using a different price for time" (Mendelsohn, 2004).

Other economists argue for discount rates higher than the Ramsey formula. Carter *et al.* wrote, "because our knowledge of future events becomes more uncertain as the time horizon is extended, discount rates should if anything increase rather than diminish with time" (Carter *et al.,* 2006). The passage of time diminishes the odds that any specific event, whether harmful (cost) or desirable (benefit), will come to pass. It is therefore logical to discount the possibility of ever seeing a benefit whose delivery is decades or even a century distant. In the climate debate, delivery of the benefit can be foiled by even small changes in population, consumption, technology, politics, and international affairs that can (following the IPCC's chain of logic) change emission scenarios, hence atmospheric concentrations of CO_2, hence climate impacts, and hence economic impacts.

Another reason to believe discount rates should be high rather than low for benefits realized in the far future is because future generations will be much wealthier than people are today and therefore better able to cope with the risks that might accompany climate change. "There is a general consensus among economists that future generations will be able to deal with the average impacts of climate change relatively uneventfully," writes Litterman (2013, p. 38). At an annual per-capita income growth rate of 2.8% (the average over the past 50 years), average personal income will be four times as high as today in 50 years and 16 times as high in 100 years. In the latter case, even the world's poor will be wealthier than middle-income wage earners today, giving them access to mobility, air conditioning, and other forms of adaptation to climate hazards that currently may be beyond their reach (Goklany, 2009).

Nigel Lawson reports the rate the British Treasury set for public-sector projects was 6% during his time as U.K. Chancellor, and he is skeptical of the justification for a subsequent reduction to 3.5%, pointing out the private-sector rate is considerably higher (Lawson, 2008, p. 84). The issue, he observes, is not what would be an appropriate rate for developed countries, but what rate should be applied to a global project, and as the IPCC admits in the excerpt above, normal rates in developing countries are considerably higher.

The U.S. Office of Management and Budget (OMB) guidelines for base-line analysis state, "Constant-dollar benefit-cost analyses of proposed investments and regulations should report net present value and other outcomes determined using a real discount rate of 7%. This rate approximates the marginal pretax rate of return on an average investment in the private sector in recent years" (OMB, 1992, p. 9). Another commonly referenced benchmark is the return on U.S. Treasury notes, which at the time of this writing was 3.14% (Bankrates.com, 2018).

Economists generally reject the notion that climate change should be singled out for unique treatment, arguing the assessment of present values of future benefits/costs rests on principles that are rational and immutable (*e.g.*, Mendelsohn, 2004). Although expenditures can be viewed very differently in terms of diverse politics, moral philosophy, or ethics, they contend discount rates used for inter-temporal calculations should be around the real rate of return on capital, because only that rate represents the true opportunity cost of investments in climate mitigation (Nordhaus, 1998; Murphy, 2008). According to Kreutzer (2016),

> What, then, is the best reasonable return on investment? While one cannot predict what future rates will be, past rates of return on broad indexes are an excellent guide. The return on the Standard & Poor's 500 from 1928 to 2014 was 9.60 percent. Over this time inflation was a compounded 3.1 percent. The real rate of return would be the difference, 6.5 percent per year. Another source estimates the return for all stocks in the U.S. from 1802 to 2002 and gets the same 6.5 percent real return on capital. Yet another source calculates the real return on stocks between 1802 and 2002 to be 6.8 percent per year. These estimates reflect the returns after corporate income taxes are paid. Adjusting for corporate profits taxes increases these rates to between 7.5 percent and 9.9 percent.

Kreutzer concludes, "In any event, the 7 percent discount rate that is part of the Office of Management and Budget's guidance does not seem too high" (*Ibid.*).

The exception that seems to draw many researchers away from this consensus is Sir Nicholas Stern, whose 2007 Stern Review based its analysis on a discount rate of roughly 1.4% or even as low as 0.1% (Stern, 2007; Stern Review team, 2006). Stern justifies his rate as follows:

> The most straightforward and defensible interpretation (as argued in the Review) of [the utility discount factor] δ is the probability of existence of the world. In the Review, we took as our base case δ = 0.1%/year, which gives roughly a one-in-ten chance of the planet not seeing out this century. [Annual per-capita consumption growth] is on average ~1.3% in a world without climate change, giving an average consumption or social discount rate across the entire period of 1.4% (being lower where the impacts of climate change depress consumption growth) (Dietz *et al.*, 2007).

Stern assumes a one-in-ten probability that anthropogenic global warming will bring the world to an end by 2100, the social discount rate would indeed be vanishingly different from zero. But that doomsday scenario defies logic as well as climate science and economics. Carbon dioxide's effect on climate and then climate change's effect on human well-being are likely to be small relative to other human needs and priorities, even well past the end of the twenty-first century. Investing in efforts to mitigate their effects ought not be raised above other needs without sound scientific and economic justification. Stern's focus on an utterly implausible scenario makes his advice on a discount rate unreliable.

The detailed analyses of the risk of anthropogenic climate change presented earlier in this chapter and in previous chapters make a strong case that there is nothing special or unique about climate change that would justify an exceptional discount rate. Estimates of future costs and benefits and investments in emission reductions should be discounted at the same rate as other costs, benefits, and investment opportunities that face similar uncertainties. Special pleading or exception-making opens the door for bad public policy choices, thereby undermining the goals of CBA in the first place.

Finding the right discount rate has major consequences for estimating the human welfare impacts of climate change. The debate over choosing an appropriate discount rate is certainly worth having, but opponents of using a constant discount rate of approximately 7%, as recommended by OMB, Kreutzer, and others, have a tough position to defend.

References

Arrow, K.J., *et al.* 2013. How should benefits and costs be discounted in an intergenerational context? *Economics Department Working Paper Series* No. 56-2013. Colchester, UK: University of Essex, Business, Management, and Economics.

Bankrates.com. 2018. Ten-year Treasury constant maturity (website). Accessed November 17, 2018.

Birdsall, N. and Steer, A. 1993. Act now on global warming – but don't cook the books. *Finance and Development* **30** (1): 6–8.

Broome, J. 2004. *Weighing Lives.* Oxford, UK: Oxford University Press.

Broome, J. 2012. *Climate Matters: Ethics in a Warming World.* New York, NY: W.W. Norton & Company.

Carter, R.M., *et al.* 2006. The Stern Review: A dual critique. *World Economics* **7** (4).

Cline, W.M. 1993. Give greenhouse abatement a chance. *Finance and Development.* March.

Dietz, S., Hope, C., Stern, N., and Zenghelis, D. 2007. Reflections on the Stern Review (1): a robust case for strong action to reduce the risks of climate change. *World Economics* **8** (1): 121–68.

Goklany, I.M. 2009. Discounting the future. *Regulation* **32** (1): 37–40.

Heal, G. 2009. Climate economics: a meta-review and some suggestions. *Review of Environmental Economics and Policy* **3**: 4–21.

Heal, G.M. and Millner, A. 2014. Agreeing to disagree on climate policy. *Proceedings of the National Academy of Sciences USA* **111**: 3695–8.

H.M. Treasury. 2018. *Green Book: Central Governance Guidance on Appraisal and Evaluation.* London, UK.

IPCC. 2001. Intergovernmental Panel on Climate Change. *Climate Change 2001: Mitigation.* Report of Working Group III to the Third Assessment Report of the Intergovernmental Panel on Climate Change. New York, NY: Cambridge University Press.

Klaus, V. 2012. Magistral Lecture on Climate Economics, 45th Annual Seminars on Planetary Emergencies. World Federation of Scientists, Erice, Sicily. Washington, DC: Science and Public Policy Institute.

Kreutzer, D. 2016. Discounting climate costs. The Heritage Foundation (website). Accessed July 19, 2018.

Lawson, N. 2008. *An Appeal to Reason: A Cool Look at Global Warming.* New York, NY: Overlook.

Litterman, B. 2013. What is the right price for carbon emissions? *Regulation* **36** (2): 38–42.

Mendelsohn, R. 2004. The challenge of global warming. Opponent paper on climate change. In: Lomborg, B. (Ed.) *Global Crises, Global Solutions.* Cambridge, MA: Cambridge University Press.

Murphy, K.M. 2008. *Some Simple Economics of Climate Changes.* Paper presented to the Mont Pelerin Society General Meeting, Tokyo. September 8.

Nordhaus, W.D. 1998. *Economics and Policy Issues in Climate Change.* Washington, DC: Resources for the Future.

OMB. 1992. U.S. Office of Management and Budget. Circular A-94 Guidelines and discount rates for benefit-cost analysis of federal programs. Washington, DC.

Stern, N. 2007. *The Economics of Climate Change: The Stern Review*. Cambridge, UK: Cambridge University Press.

Stern, N. 2014. Ethics, equity and the economics of climate change paper 1: science and philosophy. *Economics and Philosophy* **30**: 397–444.

Stern Review Team. 2006. Personal communication to Christopher Monckton of Brenchley.

Tol, R.S.J. 2010. Carbon dioxide mitigation. In: Lomborg, B. (Ed.) *Smart Solutions to Climate Change: Comparing Costs and Benefits*. New York, NY: Cambridge University Press, pp. 74–105.

Watts, W. 1999. *Discounting and Sustainability*. Brussels, Belgium: The European Commission, Directorate General of Democracy.

Weitzman, M.L. 1998. Why the far-distant future should be discounted at its lowest possible rate. *Journal of Environmental Economics and Management* **36**: 201–8.

Weitzman, M.L. 2007. A review of the Stern Review on the Economics of Climate Change. *Journal of Economic Literature* **XLV** (September).

Weitzman, M.L. 2015. A review of William Nordhaus' The Climate Casino: Risk, Uncertainty, and Economics for a Warming World. *Review of Environmental Economics and Policy* **9**: 145–56.

8.3 Climate Change

Previous sections of this chapter have shown how cascading uncertainty cripples integrated assessment models (IAMs). All five steps in an IAM – emission scenarios, carbon cycle, climate sensitivity, climate impacts, and economic impacts – rely on assumptions and controversial assertions that undermine the credibility of these academic exercises. They are, as Pindyck (2013) wrote, "close to useless as tools for policy analysis."

Assuming *arguendo* that IAMs get some aspects of the climate change problem right, this section begins with a summary of what the IPCC in its Fifth Assessment Report says the models show. It is seldom noted that the IPCC's estimates of the cost of reducing greenhouse gas emissions is reported in the Working Group III report while the benefits appear in the Working Group II report. What happens when those two estimates are compared? Section 8.3.1 answers that question.

Sections 8.3.2 and 8.3.3 report what happened when Dayaratna *et al.* (2017) re-ran two of the three IAMs relied upon by the IPCC to estimate the "social cost of carbon" using different assumptions regarding climate sensitivity, discount rates, and number of years being forecast. (The researchers also were interested in examining the robustness of the IPCC's third model, the PAGE model (Hope, 2013, 2018), but the author of that model, Chris Hope, insisted on co-authorship of any publications that would be written in exchange for providing his codes, so that model was not studied.)

References

Dayaratna, K., McKitrick, R., and Kreutzer, D. 2017. Empirically-constrained climate sensitivity and the social cost of carbon. *Climate Change Economics* **8**: 2.

Doiron, H.H. 2016. *Recommendations to the Trump Transition Team Investigating Actions to Take at the Environmental Protection Agency (EPA): A Report of The Right Climate Stuff Research Team*. November 30, p. 20.

Hope, C.W. 2013. Critical issues for the calculation of the social cost of CO_2: why the estimates from PAGE09 are higher than those from PAGE2002. *Climatic Change* **117**: 531–43.

Hope, C.W. 2018. PAGE. Climate Colab (website). Accessed July 19, 2018.

Pindyck, R.S. 2013. Climate change policy: what do the models tell us? *Journal of Economic Literature* **51**: 860–72.

8.3.1 The IPCC's Findings

By the IPCC's own estimates, the cost of reducing emissions in 2050 by enough to avoid a warming of ~2°C would be 6.8 times as much as the benefits would be worth.

The IPCC's estimate of the economic impact of unmitigated climate change was discussed in some detail in Section 8.2.5.1. Working Group II's contribution to AR5 put the cost of unmitigated climate change at between 0.2% and 2.0% of annual global GDP for a warming of approximately 2°C by 2050 (IPCC, 2014a, p. 663). Presumably this is the lost income growth over a 50-year period (the time required for temperatures to increase ~2°C) (Gleditsch and Nordås, 2014, p. 85). A mean cost estimate might be 1% (2.2 / 2), but this is higher than what the IPCC's IAMs forecast (see Figure 8.2.5.1.2). For a 2°C warming the PAGE and DICE models forecast consumption losses of about 0.5% and 1% while the FUND model forecasts a consumption *benefit* of about 1% of GDP. The models average about a 0.5% consumption loss. Avoiding this cost would be the *benefit* of reducing emissions sufficiently to keep the warming from occurring.

The Working Group III contribution to AR5 puts the *cost* of reducing greenhouse gas emissions enough to avoid more than 2°C warming by 2100 at 1.7% of global GDP in 2030, 3.4% in 2050, and 4.8% in 2100 (IPCC, 2014b, Table SPM.2, p. 15). These are "global mitigation costs" discounted at 5% per year and do not include the possible benefits or costs of climate impacts.

Working Group III says without mitigation, "global mean surface temperature increases in 2100 from 3.7°C to 4.8°C compared to pre-industrial levels" (IPCC, 2014b, p. 8). But Working Group II doesn't offer an estimate of the cost of unmitigated climate change much higher than ~2°C, saying "losses accelerate with greater warming (*limited evidence, high agreement*), but few quantitative estimates have been completed for additional warming around 3°C or above. ... Estimates vary strongly with the assumed damage function and discount rate" (IPCC, 2014a, p. 663). On this point we can agree with the IPCC: Accurately forecasting economic costs and benefits more than 40 or 50 years distant is impossible.

The ratio of the IPCC's estimates of the costs and benefits of reducing emissions sufficiently to prevent more than 2°C warming by 2050 is 6.8:1 (3.4/0.5). This seems as close to a cost-benefit ratio as one can derive from the IPCC's voluminous research and commentary on impacts and mitigation. Reducing emissions would cost approximately seven times as much as any possible benefits that might come from a slightly cooler world in 2050 and beyond. This means the IPCC itself makes a strong case *against* reducing emissions before 2050. But given all the errors in the IPCC's analysis documented in this and earlier chapters, a better cost-benefit ratio is in order.

References

Gleditsch, N.P. and Nordås, R. 2014. Conflicting messages? The IPCC on conflict and human security. *Political Geography* **43**: 82–90.

IPCC. 2014a. Intergovernmental Panel on Climate Change. *Climate Change 2014: Impacts, Adaptation, and Vulnerability. Part A: Global and Sectoral Aspects.* Contribution of Working Group II to the Fifth Assessment Report of the Intergovernmental Panel on Climate Change. New York, NY: Cambridge University Press.

IPCC. 2014b. Intergovernmental Panel on Climate Change. *Climate Change 2014: Mitigation of Climate Change.* Contribution of Working Group III to the Fifth Assessment Report of the Intergovernmental Panel on Climate Change. New York, NY: Cambridge University Press.

8.3.2 DICE and FUND Models

Changing only three assumptions in two leading IAMs – the DICE and FUND models – reduces the SCC by an order of magnitude for the first and changes the sign from positive to negative for the second.

The two publicly available models used by the U.S. Interagency Working Group (IWG) for policymaking prior to 2017 were the Dynamic Integrated Climate-Economy (DICE) model (Newbold, 2010; Nordhaus, 2017), and the Climate Framework for Uncertainty, Negotiation and Distribution (FUND) model (Anthoff and Tol, 2014; Waldhoff *et al.*, 2014; Tol and Anthoff, 2018). Examination of the DICE and FUND models by Dayaratna *et al.* (2017) revealed they are especially sensitive to three parameters chosen by IWG: discount rates, equilibrium climate sensitivity, and the number of years being forecast. IWG simply chose not to run the models with the 7% discount rate required by the U.S. Office of Management and Budget (OMB, 1992) and recommended by many economists as recounted in Section 8.2.5.2. So Dayaratna *et al.* ran the models themselves. As previously mentioned, a third model, PAGE, was not used due to the author's insistence of co-authorship, precluding independent analysis.

Equilibrium climate sensitivity (ECS) was discussed in detail in Chapter 2 and in Section 8.2.3. The ECS distribution used by IWG was published in the journal *Science* 11 years ago (Roe and Baker, 2007). Rather than being based on empirical data, this distribution was calibrated to assumptions made by IWG. Since it was published, studies regarding ECS distributions have found a significantly lower probability of extreme global warming (see Figure 8.2.3.5 and Otto *et al.*, 2013; Lewis, 2013; and Lewis and Curry, 2015). Dayaratna *et al.* (2017) re-ran the

DICE and FUND models with these new ECS estimates.

The IWG also chose to run the DICE and FUND models with time horizons of 300 years, which defies credibility. The "cascade of uncertainty" identified earlier in this chapter grows greater with every year, making predictions beyond even one or a few decades speculative. Three centuries is far beyond the horizon of any credible scientific or economic model. As seen in the outputs reported below, reducing the horizon by half, to a still-unbelievable 150 years, dramatically changes the SCC.

When Dayaratna *et al.* (2017) ran the DICE model using a 7% discount rate but retaining the Roe and Baker ECS estimate and 300-year horizon, the social cost of carbon (SCC) estimates ranged from $4.02 per marginal ton of CO_2eq generated in 2010 to $12.25 in 2050, dramatically less than the estimates produced when lower discount rates are assumed. For example, between a 2.5% and a 7% discount rate, the SCC falls by more than 80% in 2050. The reductions in SCC for other years are also quite substantial. The results appear in (A) in Figure 8.3.2.1.

Figure 8.3.2.1
Re-running the DICE model with truncated time horizon

Year	Discount Rate			
	2.50%	3%	5%	7%
(A) DICE model SCC estimates using outdated Roe-Baker (2007) ECS distribution and 300 year time horizon				
2010	$46.57	$30.04	$8.81	$4.02
2020	$56.92	$37.79	$12.10	$5.87
2030	$66.52	$45.14	$15.33	$7.70
2040	$76.95	$53.25	$19.02	$9.85
2050	$87.69	$61.72	$23.06	$12.25
(B) DICE model SCC estimates using outdated Roe-Baker (2007) ECS distribution with time horizon truncated at 150 years				
2010	$36.78	$26.01	$8.66	$4.01
2020	$44.41	$32.38	$11.85	$5.85
2030	$50.82	$38.00	$14.92	$7.67
2040	$57.17	$43.79	$18.36	$9.79
2050	$62.81	$49.20	$22.00	$12.13
(C) Percentage change in DICE model's SCC estimates using outdated Roe-Baker (2007) ECS distribution after truncating time horizon to 150 years				
2010	-21.04%	-13.43%	-1.77%	-0.20%
2020	-21.98%	-14.32%	-2.10%	-0.27%
2030	-23.60%	-15.82%	-2.66%	-0.39%
2040	-25.71%	-17.78%	-3.45%	-0.60%
2050	-28.37%	-20.28%	-4.58%	-0.94%

Running the DICE model using the 7% discount rate and truncating the time horizon to 150 years instead of 300 years significantly reduced SCC estimates for model runs using low discount rates while leaving the SCC estimates for the 7% discount rate relatively unchanged. The absolute values appear in (B) and the percentage change from (A) to (B) appears in (C) in Figure 8.3.2.1.

Dayaratna *et al.* (2017) also found the DICE model is sensitive to the choice of its equilibrium climate sensitivity distribution. Running the model with the Otto *et al.* (2013) ECS instead of the out-of-date Roe-Baker (2007) ECS revealed an SCC with a 7% discount rate of between $2.80 (2010) and $8.29 (2050), a decline by some 30%. (D) in Figure 8.3.2.2 presents the absolute values and (E) shows the percentage change from (A) in Figure 8.3.2.1.

Dayaratna *et al.* (2017) also ran the DICE model using the Lewis and Curry (2015) ECS distribution instead of the outdated Roe-Backer (2007) ECS distribution and found similar lower SCC results and large percentage changes at all discount rates as shown in (F) and (G) in Figure 8.3.2.3.

These reductions in SCC estimates are due to a very simple aspect of the ECS distribution used. The outdated Roe-Baker distribution has a significantly higher probability of high-end global warming than these more up-to-date distributions. For example, the probability of a temperature increase greater than 4° Celsius is slightly above 0.25 under the outdated Roe-Baker distribution; under the Otto *et al.* (2013) and Lewis and Curry (2015) distributions, this probability is less than 0.05. As a result, model simulations draw more from such extreme cases of global warming using the Roe-Baker distribution, and those extreme cases manifest themselves in higher estimates of the SCC.

Similarly, Dayaratna *et al.* (2017) re-ran the FUND model using the 7% discount rate and replacing the outdated Roe-Baker (2007) ECS distribution with the more recent Otto *et al.* (2013) and Lewis and Curry (2015) ECS distributions. The FUND model's estimates of SCC start out slightly lower than the DICE model because it includes some social benefits attributable to enhanced agricultural productivity due to increased CO_2 fertilization. With a 7% discount rate and updated ECS distributions, the FUND model reports a slightly negative SCC for all years from 2010 to 2050 ranging from $-0.14 per metric ton to -$1.12. See (H), (I), and (J) in Figure 8.3.2.4 for the SCC estimates for all four discount rates and three ECS distributions.

Figure 8.3.2.2
Re-running the DICE model with Otto et al. (2013) ECS distribution

Year	Discount Rate			
	2.5%	3%	5%	7%
(D) DICE model SCC estimates using Otto *et al.* (2013) ECS distribution				
2010	$26.64	$17.72	$5.73	$2.80
2020	$32.65	$22.32	$7.82	$4.04
2030	$38.33	$26.74	$9.88	$5.26
2040	$44.54	$31.63	$12.24	$6.69
2050	$51.19	$36.91	$14.84	$8.29
(E) Percentage change in DICE model's SCC estimates after switching from the outdated Roe-Baker (2007) to Otto *et al.* (2013) ECS distribution				
2010	-42.79%	-41.00%	-35.02%	-30.39%
2020	-42.63%	-40.93%	-35.37%	-31.20%
2030	-42.38%	-40.77%	-35.52%	-31.71%
2040	-42.12%	-40.61%	-35.65%	-32.13%
2050	-41.62%	-40.20%	-35.62%	-32.33%

Figure 8.3.2.3
Re-running the DICE model with the Lewis and Curry (2015) ECS distribution

Year	Discount Rate			
	2.5%	3%	5%	7%
(F) DICE model SCC estimates using Lewis and Curry (2015) ECS distribution				
2010	$23.62	$15.62	$5.03	$2.48
2020	$28.92	$19.66	$6.86	$3.57
2030	$33.95	$23.56	$8.67	$4.65
2040	$39.47	$27.88	$10.74	$5.91
2050	$45.34	$32.51	$13.03	$7.32
(G) Percentage change in DICE model's SCC estimates after switching from the outdated Roe-Baker (2007) to Lewis and Curry (2015) ECS distribution				
2010	-49.28%	-48.00%	-42.91%	-38.31%
2020	-49.19%	-47.98%	-43.31%	-39.18%
2030	-48.96%	-47.81%	-43.44%	-39.61%
2040	-48.71%	-47.64%	-43.53%	-40.00%
2050	-48.30%	-47.33%	-43.50%	-40.24%

Re-running the DICE and FUND models with these reasonable changes to discount rates and equilibrium climate sensitivity reveals several things:

(a) The models relied on by the IPCC, EPA, and other government agencies depend on factors whose values violate conventional cost-benefit analysis (low discount rates), rely on outdated and invalidated data (the Roe-Baker (2007) ECS estimate), or lie outside the range of plausibility (the 300-year horizon);

(b) Altering only these three variables is sufficient to reduce the SCC to less than $10 in the DICE model (e.g from $87.69 to $7.32 in 2050) and to change its sign from positive to negative in the FUND model (e.g. from $42.98 to -$0.53 in 2050);

(c) Using the FUND model – the only model that takes into account potential benefits from CO2 emissions – the estimates of the SCC are close to zero or even negative under very reasonable assumptions, suggesting that climate change may offer more benefits than costs to society.

References

Anthoff, D. and Tol, R.S.J. 2014. The income elasticity of the impact of climate change. In: Tiezzi, S. and Martini, C. (Eds.) *Is the Environment a Luxury? An inquiry into the relationship between environment and income.* Abingdon, UK: Routledge, pp. 34–47.

Dayaratna, K., McKitrick, R., and Kreutzer, D. 2017. Empirically-constrained climate sensitivity and the social cost of carbon. *Climate Change Economics* **8**: 2.

Lewis, N. 2013. An objective Bayesian improved approach for applying optimal fingerprint techniques to estimate climate sensitivity. *Journal of Climate* **26** (19): 7414–29.

Lewis, N. and Curry, J. 2015. The implications for climate sensitivity of AR5 forcing and heat uptake estimates. *Climate Dynamics* **45** (3-4): 1009–23.

Newbold, S.C. 2010. Summary of the DICE Model. Prepared for the EPA/DOE workshop, Improving the Assessment and Valuation of Climate Change Impacts for Policy and Regulatory Analysis, Washington DC, November 18–19.

Nordhaus, W. 2017. Evolution of assessments of the economics of global warming: Changes in the DICE model, 1992–2017. *Discussion Paper* No. w23319. Cambridge, MA: National Bureau of Economic Research and Cowles Foundation.

Figure 8.3.2.4
Re-running the FUND model using Roe-Baker (2007), Otto *et al.* (2013), and Lewis and Curry (2015)

Year	Discount Rate			
	2.50%	3%	5%	7%
(H) FUND model SCC estimates using outdated Roe-Baker (2007) ECS distribution				
2010	$29.69	$16.98	$1.87	-$0.53
2020	$32.90	$19.33	$2.54	-$0.37
2030	$36.16	$21.78	$3.31	-$0.13
2040	$39.53	$24.36	$4.21	$0.19
2050	$42.98	$27.06	$5.25	$0.63
(I) FUND model SCC estimates using Otto *et al.* (2013) ECS distribution				
2010	$11.28	$6.27	$0.05	-$0.93
2020	$12.66	$7.30	$0.36	-$0.87
2030	$14.01	$8.35	$0.74	-$0.75
2040	$17.94	$11.08	$1.50	-$0.49
2050	$19.94	$12.69	$2.21	-$0.14
(J) FUND model SCC estimates using Lewis and Curry (2015) ECS distribution				
2010	$5.25	$2.78	-$0.65	-$1.12
2020	$5.86	$3.33	-$0.47	-$1.10
2030	$6.45	$3.90	-$0.19	-$1.01
2040	$7.02	$4.49	-$0.18	-$0.82
2050	$7.53	$5.09	$0.64	-$0.53

OMB. 1992. U.S. Office of Management and Budget. Circular A-94 Guidelines and discount rates for benefit-cost analysis of federal programs. Washington, DC.

Otto, A., *et al.* 2013. Energy budget constraints on climate response. *Nature Geoscience* 6: 415–416.

Roe, G. and Baker, M.B. 2007. Why is climate sensitivity so unpredictable? *Science* 318 (5850): 629–32.

Tol, R.S.J. and Anthoff, D. 2018. FUND—climate framework for uncertainty, negotiation and distribution (website). Accessed July 20, 2018.

Waldhoff, S., Anthoff, D., Rose, S., and Tol, R.S.J. 2014. The marginal damage costs of different greenhouse gases: an application of FUND. *Economics: The Open-Access, Open Assessment E-Journal* 8: 2014–31

8.3.3 A Negative SCC

Under very reasonable assumptions, IAMs can suggest the SCC is more likely than not to be negative, even though they have many assumptions and biases that tend to exaggerate the negative effects of GHG emissions.

The negative SCC estimates produced by the FUND model are interesting and warrant further discussion. Since SCC is presented as a cost, a negative estimate signifies more social benefits than social costs associated with greenhouse gas emissions, and therefore such emissions are net beneficial for the planet. As these models are estimated via Monte Carlo simulation, Dayaratna *et al.* (2017) were able to compute the probability of a negative SCC. Their findings are summarized in (A), (B), and (C) in Figure 8.3.3.1.

There are a few noteworthy points from these results. First, with a 7% discount rate and updated ECS range, the probability ranges from 54% to 73% that the SCC is negative. Even with lower discount rates the probability of a negative SCC ranges from 22.8% to 60.1%. Even using the outdated Roe-Baker distribution, with a 7% discount rate there is a greater probability of a negative SCC than a positive SCC through 2040.

These results may be one of the reasons the IWG researchers chose not to report a 7% discount rate in their analysis. Acknowledging that the combustion of fossil fuels – the main source of anthropogenic CO_2 emissions – likely causes more social benefits than social harms would hardly have aided the Obama administration in its "war on coal." That result would more plausibly support efforts to protect the nation's coal-powered electric generation capacity, something Obama's successor is pursuing (Cama, 2017; Dlouhy, 2018).

The analysis by Dayaratna *et al.* (2017) makes clear that estimates of the social cost of carbon are sensitive to changes to assumptions and a few key variables. Although these models are interesting to explore in academic research, they are not robust enough for use in setting regulatory policy. Fortunately, the Trump administration disbanded the IWG and halted use of SCC estimates in regulatory policy (Trump, 2017). Future administrations, both in the United States and elsewhere in the world, would benefit from doing the same.

References

Cama. T. 2017. Trump proposes higher payments for coal, nuclear power. *The Hill,* September 29.

Dayaratna, K., McKitrick, R., and Kreutzer, D. 2017. Empirically-constrained climate sensitivity and the social cost of carbon. *Climate Change Economics* **8**: 2

Dlouhy, J.A. 2018. Trump prepares lifeline for money-losing coal plants. Bloomberg, May 31.

Otto, A., *et al.* 2013. Energy budget constraints on climate response. *Nature Geoscience* **6**: 415–416.

Roe, G. and Baker, M.B. 2007. Why is climate sensitivity so unpredictable? *Science* **318** (5850): 629–32.

Trump, D. 2017. Presidential executive order on promoting energy independence and economic growth. White House, Washington, DC. March 28.

8.4 Fossil Fuels

Efforts to calculate the "social cost of carbon" (SCC) routinely underestimate the cost of reducing humanity's reliance on fossil fuels by excluding the private benefits of fossil fuels and then the opportunity cost of foregoing those benefits. As was mentioned at the start of this chapter, in Section 8.1.2, the SCC label is typically applied only to the cost of the net effects of climate change attributed to CO_2 and other greenhouse gases emitted by humanity.

But to ignore this opportunity cost is obviously wrong. In its 2017 report to Congress, the U.S. Office of Management and Budget (OMB) said "cost-benefit analysis as required by EO 12866 remains the primary analytical tool to inform specific regulatory decisions. Accordingly, except where prohibited by law, agencies must continue to assess and consider *both the benefits and costs* of regulatory and deregulatory actions, and issue such actions only upon a reasoned determination that benefits justify costs" (OMB, 2018, p. 51, italics added).

It should have occurred to the IWG economists that the integrated assessment models (IAMs) they chose to rely on for the SCC estimates failed to meet OMB's requirement, and not only by failing to report costs using a 7% discount rate and by comparing domestic costs with global benefits, as reported in Section 8.1.4. IAMs *by design* monetize only the costs of climate change attributable to anthropogenic greenhouse gas emissions. The DICE model

Figure 8.3.3.1
Probability of a negative Social Cost of Carbon (SCC) estimate

Year	Discount Rate			
	2.5%	3%	5%	7%
(A) Probability of negative SCC estimates for DICE and FUND models using outdated Roe-Baker (2007) ECS distribution				
2010	0.087	0.121	0.372	0.642
2020	0.084	0.115	0.344	0.601
2030	0.08	0.108	0.312	0.555
2040	0.075	0.101	0.282	0.507
2050	0.071	0.093	0.251	0.455
B. Probability of negative SCC estimates for DICE and FUND models using Otto *et al.* (2013) ECS distribution				
2010	0.278	0.321	0.529	0.701
2020	0.268	0.306	0.496	0.661
2030	0.255	0.291	0.461	0.619
2040	0.244	0.274	0.425	0.571
2050	0.228	0.256	0.386	0.517
(C) Probability of negative SCC estimates for DICE and FUND models using Lewis and Curry (2015) ECS distribution				
2010	0.416	0.450	0.601	0.730
2020	0.402	0.432	0.570	0.690
2030	0.388	0.414	0.536	0.646
2040	0.371	0.394	0.496	0.597
2050	0.354	0.372	0.456	0.542

deliberately excludes *any benefits* from climate change, while the FUND model includes only the benefits from aerial CO_2 fertilization (Dayaratna and Kreutzer, 2013, 2014). They omit entirely the extensive benefits produced by the use of fossil fuels, and hence the opportunity cost of losing those benefits. Consequently, while IAMs might be used to monetize one or a few of the many costs and benefits

arising from the use of fossil fuels, they are not a true CBA (Pindyck, 2013).

The rest of this section attempts to produce more accurate cost-benefit ratios for the use of fossil fuels. Section 8.4.1 reviews all the impacts of fossil fuels identified earlier in this chapter and in other chapters of this book and finds 16 benefits and only one net cost. Section 8.4.2 produces realistic estimates of the cost of reducing GHG emissions by the amounts recommended by the IPCC and according to a goal set by the European Union. Section 8.4.3 produces new cost-benefit ratios using the findings from the IPCC, the Interagency Working Group, and Bezdek (2014, 2015). The authors find the cost of reducing humanity's reliance on fossil exceeds the benefits by ratios as low as 6.8:1 to as high as 160:1.

References

Bezdek, R.H. 2014. *The Social Costs of Carbon? No, the Social Benefits of Carbon*. Oakton, VA: Management Information Services, Inc.

Bezdek, R.H. 2015. Economic and social implications of potential UN Paris 2015 global GHG reduction mandates. Oakton, VA: Management Information Services, Inc.

Dayaratna, K. and Kreutzer, D. 2013. Loaded DICE: an EPA model not ready for the big game. *Backgrounder* No. 2860. The Heritage Foundation. November 21.

Dayaratna, K. and Kreutzer, D. 2014. Unfounded FUND: yet another EPA model not ready for the big game. *Backgrounder* No. 2897. The Heritage Foundation. April 29.

OMB, 2018. U.S. Office of Management and Budget. Draft report to Congress on the benefits and costs of federal regulations and agency compliance with the Unfunded Mandates Reform Act. February.

Pindyck, R.S. 2013. Climate change policy: what do the models tell us? *Journal of Economic Literature* **51** (3): 860–72.

Section 8.4.1 Impacts of Fossil Fuels

Sixteen of 25 possible impacts of fossil fuels on human well-being are net benefits, only one is a net cost, and the rest are either unknown or likely to have no net impact.

The authors of the Working Group II contribution to the IPCC's Fifth Assessment Report (AR5) reported hundreds of studies allegedly documenting the impacts of climate change on humanity, but they did not attempt to aggregate those impacts, observing that differences in methodology, geographical areas, time periods, and outputs made such a meta-analysis impossible. Instead, they opted to summarize the possible impacts in a table (Assessment Box SPM.2 Table 1 in the Summary for Policymakers (IPCC, 2014a, pp. 21–25).

The authors of the current volume follow the IPCC's lead by producing the table shown in Figure 8.4.1.1 summarizing the findings of previous chapters regarding the impacts of fossil fuels on human well-being. Possible impacts appear in alphabetical order, their net impact (benefit, cost, no net impact, or unknown) appear in the second column, brief observations on the impacts appear in the third column, and chapters and sections of chapters in which the topics are addressed appear in the fourth column of the table.

Figure 8.4.1.1
Impact of fossil fuels on human well-being

Impact	Benefit or Cost	Observations	Chapter References
Acid rain	No net impact	Once feared to be a major environmental threat, the deposition of sulfuric and nitric acid due to smokestack emissions, so-called "acid rain," was later found not to be a threat to forest health and to affect only a few bodies of water, where remediation with lime is an inexpensive solution. The fertilizing effect of nitrogen deposition more than offsets its harms to vegetation. Dramatic reductions in SO$_2$ and NO$_2$ emissions since the 1980s mean "acid rain" has no net impact on human well-being today.	5.1, 6.1
Agriculture	Benefit	Fossil fuels have contributed to the enormous improvement in crop yields by making artificial fertilizers, mechanization, and modern food processing techniques possible. Higher atmospheric CO$_2$ levels are causing plants to grow better and require less water. Numerous	3.4, 4.1, 5.2 5.3, 7.2, 8.2

		studies show the aerial fertilization effect of CO_2 is improving global agricultural productivity, on average by at least 15%.	
Air quality	Benefit	Exposure to potentially harmful chemicals in the air has fallen dramatically during the modern era thanks to the prosperity, technologies, and values made possible by fossil fuels. Safe and clean fossil fuels made it possible to rapidly increase energy consumption while improving air quality.	5.2, Chapter 6
Catastrophes	Unknown	No scientific forecasts of possible catastrophes triggered by global warming have been made. CO_2 is not a "trigger" for abrupt climate change. Inexpensive fossil fuel energy greatly facilitates recovery.	8.1
Conflict	Benefit	The occurrence of violent conflicts around the world has fallen dramatically thanks to prosperity and the spread of democracy made possibly by affordable and reliable energy and a secure food supply.	7.1, 7.3, 8.2
Democracy	Benefit	Prosperity is closely correlated with the values and institutions that sustain democratic governments. Tyranny promoted by zero- sum wealth is eliminated. Without fossil fuels, there would be fewer democracies in the world.	7.1
Drought	No net impact	There has been no increase in the frequency or intensity of drought in the modern era. Rising CO_2 lets plants use water more efficiently, helping them overcome stressful conditions imposed by drought.	2.7, 5.3
Economic growth (consumption)	Benefit	Affordable and reliable energy is positively correlated with economic growth rates everywhere in the world. Fossil fuels were indispensable to the three Industrial Revolutions that produced the unprecedented global rise in human prosperity.	Chapter 3, 4.1, 5.2, 7.1, 7.2, 8.1, 8.2
Electrification	Benefit	Transmitted electricity, one of the greatest inventions in human history, protects human health in many ways. Fossil fuels directly produce some 80% of electric power in the world. Without fossil fuels, alternative energies could not be built or relied on for continuous power.	Chapter 3, 4.1
Environmental protection	Benefit	Fossil fuels power the technologies that make it possible to meet human needs while using fewer natural resources and less surface space. The aerial CO_2 fertilization effect has produced a substantial net greening of the planet, especially in arid areas, that has been measured using satellites.	1.3, Chapter 5
Extreme weather	No net impact	There has been no increase in the frequency or intensity of extreme weather in the modern era, and therefore no reason to expect any economic damages to result from CO_2 emissions.	2.7, 8.2
Forestry	Benefit	Fossil fuels made it possible to replace horses as the primary means of transportation, saving millions of acres of land for forests. Elevated CO_2 concentrations have positive effects on forest growth and health, including efficiency of water use. Rising CO_2 has reduced and overridden the negative effects of ozone pollution on the photosynthesis, growth, and yield of nearly all the trees that have been evaluated experimentally.	5.3
Human development	Benefit	Affordable energy and electrification, better derived from fossil fuels than from renewable energies, are closely correlated with the United Nations' Human Development Index and advances what the IPCC labels "human capital."	3.1, 4.1, 7.2
Human health	Benefit	Fossil fuels contribute strongly to the dramatic lengthening of average lifespans in all parts of the world by improving nutrition, health care, and human safety and welfare. (See also "Air quality.")	3.1, Chapter 4, 5.2
Human settlements /migration	Unknown	Forced migrations due to sea-level rise or hydrological changes attributable to man-made climate change have yet to be documented and are unlikely since the global average rate of sea-level rise has not accelerated. Global warming is as likely to decrease as increase the number of people forced to migrate.	7.3, 8.2

Ocean acidification	Unknown	Many laboratory and field studies demonstrate growth and developmental improvements in aquatic life in response to higher temperatures and reduced water pH levels. Other research illustrates the capability of both marine and freshwater species to tolerate and adapt to the rising temperature and pH decline of the planet's water bodies.	5.5
Oil spills	Cost	Oil spills can harm fish and other aquatic life and contaminate drinking water. The harm is minimized because petroleum is typically reformed by dispersion, evaporation, sinking, dissolution, emulsification, photo-oxidation, resurfacing, tar-ball formation, and biodegradation.	5.1
Other market sectors	No net impact	The losses incurred by some businesses due to climate change, whether man-made or natural, will be offset by profits made by other businesses taking advantage of new opportunities to meet consumer wants. Institutional adaptation, including of markets, to a small and slow warming is likely.	1.2, 7.2
Polar ice melting	Unknown	What melting is occurring in mountain glaciers, Arctic sea ice, and polar icecaps is not occurring at "unnatural" rates and does not constitute evidence of a human impact on the climate. Global sea-ice cover remains similar in area to that at the start of satellite observations in 1979, with ice shrinkage in the Arctic Ocean offset by growth around Antarctica.	2.7
Sea-level rise	No net impact	There has been no increase in the rate of increase in global average sea level in the modern era, and therefore no reason to expect any economic damages to result from it. Local sea levels change in response to factors other than climate.	2.7, 8.2
Sustainability	Benefit	Fossil fuels are a sustainable source of energy for future generations. The technology they support makes sustainable development possible. Rising prosperity and market forces also are working to ensure a practically endless supply of fossil fuels.	1.5, 5.2
Temperature-related mortality	Benefit	Extreme cold kills more people than extreme heat, and fossil fuels enable people to protect themselves from temperature extremes. A world made warmer and more prosperous by fossil fuels would see a net decrease in temperature-related mortality.	4.2
Transportation	Benefit	Fossil fuels revolutionized society by making transportation faster, less expensive, and safer for everyone. The increase in human, raw material, and product mobility was a huge boon for humanity, with implications for agriculture, education, health care, and economic development.	4.1
Vector-borne diseases	No net impact	Warming will have no impact on insect-borne diseases because temperature plays only a small role in the spread of these diseases. The technologies and prosperity made possible by fossil fuels eliminated the threat of malaria in developed countries and could do the same in developing countries regardless of climate change.	4.6
Water resources	Benefit	While access to water is limited by climate and other factors in many locations around the world, there is little evidence warming would have a net negative effect on the situation. Fossil fuels made it possible for water quality in the United States and other industrial countries to improve substantially while improving water use efficiency by about 30% over the past 35 years. Aerial CO_2 fertilization improves plant water use efficiency, reducing the demand for irrigation.	5.2, 5.3

Twenty-five climate impacts appear in Figure 8.4.1.1. Some general observations are possible:

- *Net benefits:* 14 impacts (agriculture, air quality, conflict, democracy, economic growth (consumption), electrification, environmental protection, forestry, heat-related mortality, human development, human health, sustainability, transportation, and water resources) are benefits, meaning their net social benefits exceed their social costs.

- *No net impact:* Six impacts (acid rain, drought, extreme weather, other market sectors, sea-level rise, and vector-borne diseases) are either not being intensified or made more harmful by anthropogenic climate change or are likely to have offsetting benefits resulting in no net impacts.

- *Unknown costs and benefits:* Four impacts (catastrophes, human settlements/migration, ocean acidification, and polar ice melting) are not sufficiently understood to determine if net costs exceed benefits.

- *Net cost:* Only one impact (oil spills) is likely to have costs that exceed benefits. Although accidental releases of oil into bodies of water do occur and cause damage, their harm is unlikely to be great. Natural seepage from ocean floors exceeds the human contribution by nearly ten-fold and biodegradation quickly diminishes the threat to human health or wildlife (see Atlas, 1995; NRC, 2003; Aminzadeh *et al.,* 2013). Still, we count this as a net cost.

A visualization of the findings in Figure 8.4.1.1 appears in Figure 8.4.1.2. This image is modeled after, but is quite different from, one produced by the U.S. National Oceanic and Atmospheric Administration (NOAA, 2016).

This summary differs dramatically from the opinions expressed by the IPCC, but the reason should be clear: Working Group II did not conduct a cost-benefit analysis of fossil fuels. It was tasked with producing a catalogue of every possible negative consequence of climate change, whether natural or man-made (see Section 8.2.5.1 for a brief comment on that), and did its job with superb attention to detail. But since the chains of causality linking human activity to temperature changes, and then to climate impacts, and finally to human impacts decades and even centuries in the future are long, tenuous, and little more than speculation, WGII's conclusions are necessarily ambiguous: "Global economic impacts from climate change are difficult to estimate. ... Estimates vary strongly with the assumed damage function and discount rate ... the impact of climate change will be small relative to the impacts of other drivers" (IPCC, 2014, p. 663).

The authors of the current volume asked a different question: "What does observational data show to be the real impacts of the use of fossil fuels on human well-being?" and so reached a different conclusion. Extensive literature reviews have found 14 impacts of the use of fossil fuels are beneficial, meaning their net benefits to society exceed their costs. Six impacts are likely to have neither net benefits nor net costs (benefits offset costs). The net costs or benefits of four impacts are unknown due to our lack of scientific understanding of the processes involved. Only one impact of fossil fuels, oil spills, is likely to be net negative, and it is small relative to natural sources of hydrocarbons in the oceans.

In economic terms, our calculation of net benefits combines private benefits – those enjoyed by individuals and paid for by them – and net social benefits – the benefits enjoyed by people who do not pay for them minus any negative costs imposed on them. This is not a "social cost of carbon" calculation, which by design ignores private costs and benefits. Like the IPCC, we do not attempt to aggregate widely different databases on such diverse impacts. However, private benefits are easier to estimate than social costs thanks to the prices and investment data created by market exchanges, a point explained in Chapter 1, Section 1.2.3. This means the *opportunity cost* of doing without fossil fuels, calculated as a loss of per-capita income or GDP, can be estimated. This calculation is performed in the next section.

Figures 8.4.1.1 and 8.4.1.2 make it clear that the benefits of fossil fuels exceed their cost by a wide margin.

Figure 8.4.1.2
Impact of Fossil Fuels on Human Health

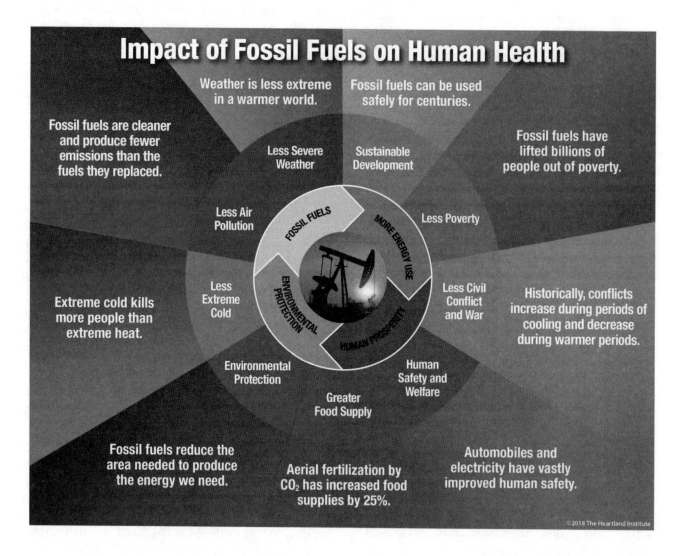

References

Aminzadeh, F., Berge, T.B., and Connolly, D.L. (Eds.) 2013. *Hydrocarbon Seepage: From Source to Surface.* Geophysical Developments Series No. 16. Tulsa, OK: Society of Exploration Geophysicists and American Association of Petroleum Geologists.

Atlas, R.M. 1995. Petroleum biodegradation and oil spill bioremediation. *Marine Pollution Bulletin* **31** (4–12): 178–82.

IPCC. 2014. Intergovernmental Panel on Climate Change. *Climate Change 2014: Impacts, Adaptation, and Vulnerability.* Contribution of Working Group II to the Fifth Assessment Report of the Intergovernmental Panel on Climate Change. New York, NY: Cambridge University Press.

NOAA. 2016. National Oceanic and Atmospheric Administration. Impact of climate change on human health. U.S. climate resilience toolkit (website). Accessed July 13, 2018.

NRC. 2003. U.S. National Research Council, Committee on Oil in the Sea. *Oil in the Sea III: Inputs, Fates, and Effects.* Washington, DC: National Academies Press

8.4.2 Cost of Mitigation

Wind and solar cannot generate enough dispatchable energy (available 24/7) to replace fossil fuels, so energy consumption must fall in order for emissions to fall.

According to the Working Group III contribution to the IPCC's Fifth Assessment Report, keeping average global surface temperature change to less than 2°C above its pre-industrial level by 2100 requires limiting atmospheric concentrations of CO_2 in 2100 to "about 450 ppm CO_2eq (*high confidence*)," which would require "substantial cuts in anthropogenic GHG emissions by mid-century through large scale changes in energy systems and potentially land use (*high confidence*). Scenarios reaching these concentrations by 2100 are characterized by lower global GHG emissions in 2050 than in 2010, 40% to 70% lower globally, and emissions levels near zero $GtCO_2$eq or below in 2100" (IPCC, 2014b, pp. 10, 12). Emissions can supposedly fall to below zero through the use of "carbon dioxide removal technologies" (*Ibid.*).

Also according to Working Group III, the cost of reducing emissions to meet these goals in the IPCC's best-case scenario – where all countries immediately begin mitigation efforts, adopt a single global carbon tax, and impose no regulations favoring some technologies over others – expressed as a percentage of baseline global gross domestic product (GDP) without climate policies, would be 1% to 4% (median: 1.7%) in 2030, 2% to 6% (median: 3.4%) in 2050, and 3% to 11% (median: 4.8%) in 2100 relative to consumption in baseline scenarios (IPCC, 2014b, pp. 15–16, text and Table SPM.2).

The following sections explain why IPCC's estimate of the cost of a forced transition away from fossil fuels to "near zero ... or below in 2100" is too low for two reasons. First, replacing a world energy system currently dependent on fossil fuels to provide more than 80% of primary energy with one relying mostly or entirely on alternative energies would cost far greater sums and take decades to implement. Second, wind and solar face physical limits that prevent them from generating enough dispatchable energy (available 24/7) to replace fossil fuels, so energy consumption must fall in order for emissions to fall. Energy demand is forecast to grow significantly in the twenty-first century, and the opportunity cost of reversing that trend – of reducing rather than increasing per-capita energy consumption

– is enormous. Section 8.4.2.1 addresses the first concern, and Section 8.4.2.2 addresses the second.

8.4.2.1 High Cost of Reducing Emissions

Transitioning from a world energy system dependent on fossil fuels to one relying on alternative energies would cost trillions of dollars and take decades to implement.

Chapter 3, Section 3.5, documented at great length the inherent limitations on alternative energy sources and the history of past transitions to new energy sources suggesting the cost of forcing a transition from fossil fuels would be very costly (Smil, 2010; Morriss *et al.*, 2011; Clack *et al.*, 2017). The sheer size of the global energy market makes replacing it massively expensive and time consuming. Smil (2010) notes the global oil industry "handles about 30 billion barrels annually or 4 billion tons" and operates about 3,000 large tankers and more than 300,000 miles of pipelines. "Even if an immediate alternative were available, writing off this colossal infrastructure that took more than a century to build would amount to discarding an investment worth well over $5 trillion – and it is quite obvious that its energy output could not be replaced by any alternative in a decade or two" (p. 140). Later, Smil (2010, p. 148) writes the cost of a transition "would be easily equal to the total value of U.S. gross domestic product (GDP), or close to a quarter of the global economic product."

Wind and solar power face cost, scale, and intermittency problems that make extremely expensive any efforts to increase their share of total energy production to more than 10% or 15% of total production. In particular, their low power density means scaling them up to replace fossil fuels would require alarming amounts of surface space, crowding out agriculture and wildlife habitat with harmful effects on food production and the natural environment. See Chapter 3, Section 3.2, and Chapter 5, Section 5.2, for discussions of these problems and many references there (e.g., Rasmussen, 2010; Hansen, 2011; Kelly, 2014; Bryce, 2014; Smil, 2016; Stacy and Taylor, 2016; Driessen, 2017).

Advocates of rapid decarbonization underestimate the negative consequences of the intermittency of solar and wind power. In a critique of Jacobson *et al.* (2015) and an earlier paper also by Jacobson and a coauthor (Jacobson and Delucchi,

2009) claiming a transition to a 100% renewables future is possible, Clack *et al.* (2017) observe,

Wind and solar are variable energy sources, and some way must be found to address the issue of how to provide energy if their immediate output cannot continuously meet instantaneous demand. The main options are to (*i*) curtail load (i.e., modify or fail to satisfy demand) at times when energy is not available, (*ii*) deploy very large amounts of energy storage, or (*iii*) provide supplemental energy sources that can be dispatched when needed. It is not yet clear how much it is possible to curtail loads, especially over long durations, without incurring large economic costs. There are no electric storage systems available today that can affordably and dependably store the vast amounts of energy needed over weeks to reliably satisfy demand using expanded wind and solar power generation alone. These facts have led many U.S. and global energy system analyses to recognize the importance of a broad portfolio of electricity generation technologies, including sources that can be dispatched when needed.

Modern economies require a constant supply of electricity 24/7, not just when the sun shines and the wind blows. The grid needs to be continuously balanced – energy fed into the grid must equal energy leaving the grid – which requires dispatchable (on-demand) energy and spinning reserves (Backhaus and Chertkov, 2013; Dears, 2015). This effectively requires that approximately 90% of the energy produced by wind turbines and solar PV cells be backed up by rotating turbines powered by fossil fuels (E.ON Netz, 2005). Today, only fossil fuels and nuclear can provide dispatchable power in sufficient quantities to keep grids balanced.

Similarly, and as explained in Chapter 3, the technology to safely and economically store large amounts of electricity does not exist (Clack *et al.*, 2017), at least not outside the few areas where large bodies of water and existing dams make pumped-storage hydroelectricity possible. The frequent announcements of "breakthroughs" in battery technology have not resulted in commercial products capable of even a small fraction of the storage needs of a transition from fossil fuels (Fildes, 2018). Scholars have even developed a "hype curve" to track how far the claims about new battery technologies overstate their potential and how long it takes for them to achieve commercial success (Sapunkov *et al.,* 2015). See Figure 8.4.2.1.1.

There is no question that fuels superior to coal, oil, and natural gas for some applications already exist or will be found and that their use will increase as new technologies are discovered and commercialized. *Energy freedom* – relying on markets to balance the interests and needs of today with those of tomorrow and to access the local knowledge needed to find efficient win-win responses to climate change – should be permitted to dictate the pace of this transition, not fears of a climate catastrophe and hope for technological breakthroughs.

References

Backhaus, S. and Chertkov, M. 2013. Getting a grip on the electrical grid. *Physics Today* **66**:42–48.

Bryce, R. 2014. *Smaller Faster Lighter Denser Cheaper.* New York, NY: PublicAffairs.

Clack, C.T.M., *et al.* 2017. Evaluation of a proposal for reliable low-cost grid power with 100% wind, water, and solar. *Proceedings of the National Academy of Sciences USA* **114** (26): 6722–7.

Dears, D. 2015. *Nothing to Fear: A Bright Future for Fossil Fuels.* The Villages, FL: Critical Thinking Press.

Driessen, P. 2017. Revisiting wind turbine numbers. Townhall (website). September 2. Accessed May 18, 2018.

E.ON Netz. 2005. *Wind Power Report 2005.* Bayreuth, Germany.

Fildes, N. 2018. Beyond lithium – the search for a better battery. *Financial Times.* January 7.

Hansen, J. 2011. Baby Lauren and the Kool-Aid. Climate Science, Awareness, and Solutions (blog). July 29.

Jacobson, M.Z. and Delucchi, M.A. 2009. A plan to power 100 percent of the planet with renewables. *Scientific American* **303** (5): 58–65.

Jacobson, M.Z., Delucchi, M.A., Cameron, M.A., and Frew, B.A. 2015. Low-cost solution to the grid reliability problem with 100% penetration of intermittent wind, water, and solar for all purposes. *Proceedings of the National Academy of Sciences USA* **112** (49):15060–5.

Figure 8.4.2.1.1
The new battery technology "hype cycle"

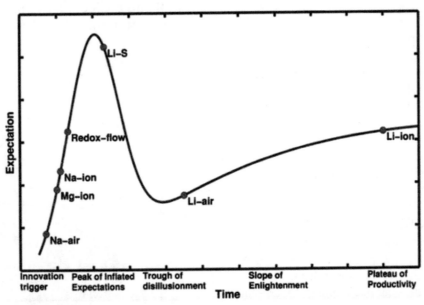

Source: Sapunkov *et al., 2015.*

Kelly, M.J. 2014. Technology introductions in the context of decarbonisation: Lessons from recent history. *GWPF Note 7.* London, UK: The Global Warming Policy Foundation.

Kiefer, T.A. 2013. Energy insecurity: the false promise of liquid biofuels. *Strategic Studies Quarterly* (Spring): 114–51.

Morriss, A.P., Bogart, W.T., Meiners, R.E., and Dorchak, A. 2011. *The False Promise of Green Energy.* Washington, DC: Cato Institute.

Rasmussen, K. 2010. *A Rational Look at Renewable Energy and the Implications of Intermittent Power. Edition 2.0.* South Jordan, UT: Deseret Power. November.

Sapunkov, O., *et al.* 2015. Quantifying the promise of 'beyond' Li–ion batteries. *Translational Materials Research* **2** (4).

Smil, V. 2010. *Energy Myths and Realities: Bringing Science to the Energy Policy Debate.* Washington, DC: American Enterprise Institute.

Smil, V. 2016. *Power Density: A Key to Understanding Energy Sources and Uses.* Cambridge, MA: The MIT Press.

Stacy, T.F. and Taylor, G.S. 2016. *The Levelized Cost of Electricity from Existing Generation Resources.* Washington, DC: Institute for Energy Research.

8.4.2.2 High Cost of Reducing Energy Consumption

Reducing greenhouse gas emissions to levels suggested by the IPCC or the goal set by the European Union would be prohibitively expensive.

If a rapid transition away from fossil fuels is physically impossible due to intermittency and the lack of surface space to accommodate wind turbines, solar panels, and biofuels, or too expensive owing to the trillions of dollars required to replace an energy system delivering energy 24/7 to a global population of 7.4 billion people and the higher levelized cost of electricity (LCOE) produced by alternatives to nuclear power and fossil fuels, what is the alternative? It is, as Clack *et al.* (2007) noted in the previous section, to "curtail load (i.e., modify or fail to satisfy demand) at times when energy is not available." As this section shows, reducing energy consumption would impose even larger social costs than substituting expensive alternatives for inexpensive fossil fuels.

According to *BP Energy Outlook 2035* (BP, 2014[1]), primary energy demand is expected to increase by 41% between 2012 and 2035, with growth averaging 1.5% per annum (p.a.). Growth slows from 2.2% p.a. for 2005–15 to 1.7% p.a. 2015–25 and to just 1.1% p.a. in the final decade. Fossil fuels lose share but they are still the dominant form of energy in 2035 with a share of 81%, compared to 86% in 2012. See Figure 8.4.2.2.1.

Driving this growth in energy demand are rising global population and per-capita consumption. BP forecasts GDP growth (expressed in purchasing power parity (PPP)) averaging 3.5% p.a. from 2012 to 2035. Due to rising energy efficiency and the "dematerialization" trend described in Chapter 5, Section 5.2, energy intensity (the amount of energy required per unit of GDP) declines by 1.9% p.a., and about 36% between 2012 and 2035. BP forecasts the rate of decline in energy intensity post 2020 will be more than double the rate achieved from 2000 to 2010, resulting in a growing decoupling of GDP and energy consumption, as depicted in Figure 8.4.2.2.2.

Despite declining energy intensity, BP projects carbon dioxide (CO_2) emissions will continue to grow at approximately 1.1% p.a., only slightly slower than energy consumption, as shown in Figure 8.4.2.2.3. Figure 8.4.2.2.4 combines the trends shown in the three earlier figures with a common index (1990 = 100) for the x-axis.

The U.S. Energy Information Administration (EIA) similarly forecasts the world's real GDP will increase 3.5% per year from 2010 to 2040 and world energy consumption will increase 56% between 2010 and 2040 (EIA, 2013, 2014). Like BP, EIA forecasts fossil fuels will continue to supply most of the energy used worldwide. See Figure 8.4.2.2.5.

---

[1] This section cites the 2014 edition of BP's annual *Energy Outlook* even though more recent editions are available partly because it was the source cited in source material for this section (Bezdek, 2015) but also because subsequent editions incorporate assumptions about taxes and subsidies that recent political developments show are unlikely to be true. BP management apparently assumes international agreements such as the Paris Accord and national policies such as the U.S. Clean Power Plan will be implemented and massive subsidies to wind and solar power generation by China and Germany will continue, even though they already are being reduced. As described later in this section, even the 2014 edition used for this analysis assumes very optimistic rates of technological progress and decarbonization.

Figure 8.4.2.2.1
Global energy consumption by type of fuel, actual and projected, in billion tons of oil equivalent (toe), 1965–2035

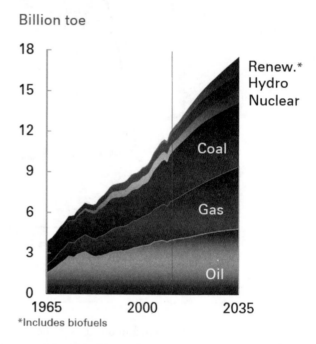

Source: BP, 2014, p. 12.

Figure 8.4.2.2.2
GDP and energy consumption, actual and projected, 1965–2035

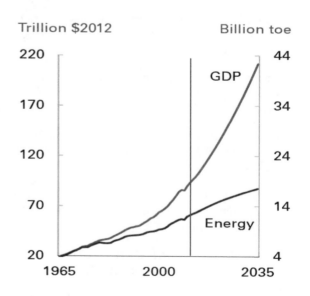

Source: BP, 2014, p. 16.

Figure 8.4.2.2.3
Energy consumption and CO_2 emissions, actual and projected, 1965–2035

Index: 1965 = 100

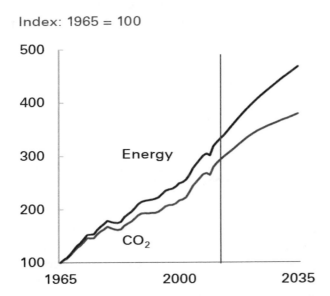

Source: BP, 2014, p. 20.

Figure 8.4.2.2.4
GDP, energy consumption, and CO_2 emissions, actual and projected, from 1990–2035

Index: 1990 = 100

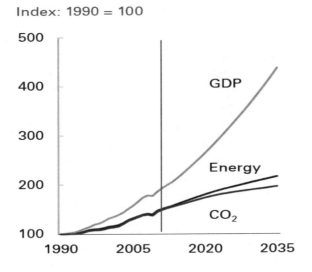

Source: BP, 2014, p. 88.

Figure 8.4.2.2.5
World energy consumption by fuel type, in quadrillion Btu, actual and projected, 1990–2040

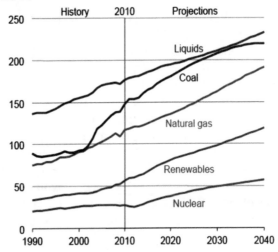

Source: EIA, 2013, Figure 2, p. 2.

What does this tell us about the cost of reducing energy consumption as a way to reduce global GHG emissions in 2050 by 40% to 70% below 2010 levels and to "near zero GtCO$_2$eq or below in 2100" (IPCC, 2014, p. 10, 12)? The relationship between world GDP and CO_2 emissions over the past century is illustrated in Figure 8.4.2.2.6. In 2010, expressed in 2007 dollars, a ton of CO_2 resulting from the use of fossil fuels "created" about $2,400 in world GDP.

Using BP and EIA's forecasts of GDP, energy use, and CO_2 emissions, Bezdek (2015) extended the relationship between world GDP and CO_2 emissions in the EIA reference case through 2050, with results shown in Figure 8.4.2.2.7. The relationship is forecast to be roughly linear, with an elasticity of 0.254 from 2020 to 2050. This is the CO_2-GDP elasticity rate, meaning reducing CO_2 emissions by 1% reduces GDP by 0.254%.

It merits emphasis that the EIA forecast already assumes world GDP will increase at a faster rate than primary energy consumption, and CO_2 emissions will increase at a lower rate than either GDP or energy consumption thanks to continued and even escalating government subsidies, favorable regulatory treatment, and tax breaks. Specifically, EIA projects:

■ world GDP increases 3.6% annually,

- world primary energy consumption increases only 1.5% annually, and

- world CO_2 emissions increase only 1.3% annually.

This implies ambitious goals for technological advancements and public policies favorable to alternative energies are already incorporated into these forecasts, meaning even more programs aimed at speeding a transition to alternative fuels would be increasingly difficult and expensive. It also assumes, contrary to the analysis presented in Chapter 3 and earlier in this chapter, that alternative energies are able to produce enough dispatchable energy to replace fossil fuels at such an ambitious pace and beyond the 10% or 20% level beyond which the addition of intermittent energy begins to destabilize grids and impose large grid-management expenses. Different assumptions in the baseline projections would increase the cost estimates this model predicts, making this a very conservative model.

Figure 8.4.2.2.8 presents the independent variables and constants and calculates the impact on GDP and per-capita GDP of the IPCC's two reduction scenarios (of 40% and 70% below 2010 levels) and applying the European Union's goal of reducing emissions to 90% below 1990 levels to global emissions, rather than only to EU nations. Sources are presented in the note under the table. The impact on GDP can be summarized as follows:

- Reducing CO_2 emissions to 40% below 2010 levels by 2050 would reduce global GDP by 16%, to $245 trillion instead of the benchmark $292 trillion, a loss of $47 trillion.

- Reducing CO_2 emissions to 70% below 2010 levels by 2050 would reduce global GDP by 21%, to $231 trillion, a loss of $61 trillion.

- Reducing CO_2 emissions to 90% below 1990 levels would reduce global GDP by 24%, to $220 trillion, a loss of $72 trillion.

GDP losses can be converted into per-capita GDP numbers using the United Nations' 2017 population forecast for world population in 2050 of 9.8 billion (UN, 2017). The reference case forecast of

world per-capita GDP in 2050 is about $29,800. As Figure 8.4.2.2.8 shows,

- Reducing CO_2 emissions to 40% below 2010 levels by 2050 would reduce average annual global per-capita GDP by 16%, to $24,959 instead of the benchmark $29,796, a loss of income of $4,837.

- Reducing CO_2 emissions to 70% below 2010 levels by 2050 would reduce global per-capita GDP by 21%, to $23,587, a loss of $6,209.

- Reducing CO_2 emissions to 90% below 1990 levels would reduce global per-capita GDP by 24%, to $22,531, a loss of $7,265.

These estimates assume alternatives to fossil fuels will be found that can supply enough energy, albeit at a higher cost, to meet the needs of a growing global population, albeit it once again at a lower level of prosperity than is currently being forecast. The analysis presented in Chapter 3 and again in the section preceding this one suggests this assumption is wrong. The need by intermittent energy sources such as wind and solar power for back-up power generation, which today can be provided in sufficient quantities only by fossil fuels, is unlikely to change enough to avert energy shortages, particularly in those countries that have chosen to abandon their coal, natural gas, and nuclear energy generation capacity.

Recall from Chapter 3, Section 3.6.2, the calculation by Tverberg (2012), who sought to measure the lost GDP in 2050 resulting from the failure of renewable energies to offset the loss of 80% of the energy produced by fossil fuels, requiring a decrease in global energy consumption of 50%. She estimated the long-term elasticity of energy consumption (not fossil fuel use, the metric used in the preceding analysis) and GDP was 0.89. Among her findings: world per-capita energy consumption in 2050 would fall to what it was in 1905 and global per-capita GDP would decline by 42% from its 2010 level. Converting Tverberg's estimates into the outputs specified by the model developed in this section shows she forecast a reduction in GDP from our baseline projection of 81%; GDP in 2050 would be $54 trillion, a loss of $238 trillion; and per-capita income would be approximately $5,518. These figures appear in the bottom row of Figure 8.4.2.2.8.

Figure 8.4.2.2.6
Historical relationship between world GDP and CO₂ emissions, 1900–2010

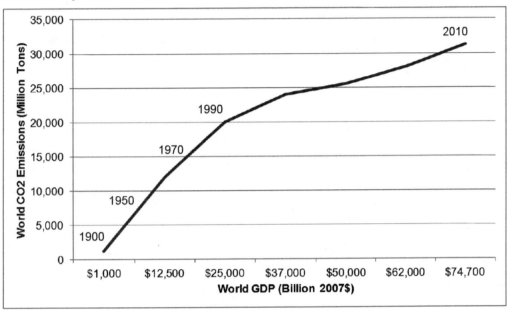

Source: Bezdek, 2014.

Figure 8.4.2.2.7
Projected relationship between world GDP and CO₂ emissions, 2010–2050

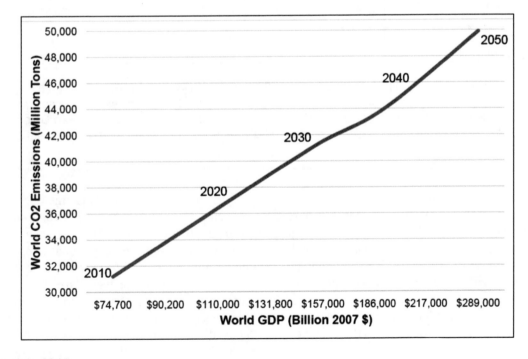

Source: Bezdek, 2015.

Figure 8.4.2.2.8
Independent variables and constants for IPCC and EU 2050 mitigation targets

	Variables	Beginning values	Δ from 2050 baseline	% less than 2050 baseline (BMT)	% reduction in 2050 GDP	2050 GDP (billion $)	Per-capita GDP
UN	2050 population estimate (billion)	9.80	--	--	--	--	--
y_2	CO2 in 2020 (billion metric tons (BMT)	36.40	15	29.62%	--	--	--
y_1	CO2 in 2050 (BMT)	51.72	0	0.00%	--	--	--
x_2	GDP in 2020 (billion $)	$110,000	$182,000	62.33%	--	--	--
x_1	GDP in 2050 (billion $)	$292,000	0	0.00%	--	--	$29,796
Rise	$(y_2 - y_1) / y_2$	-42.09%	--	--	--	--	--
Run	$(x_2 - x_1) / y_2$	-165.45%	--	--	--	--	--
	Elasticity (rise / run)	0.254	--	--	--	--	--
	$/ton in 2050	$5,645.57	--	--	--	--	--
EIA	CO_2 emissions in 2010 (BMT)	31.20	20.52	39.68%	--	--	--
	$7/ton in 2010	$2,400	--	--	--	--	--
IPCC	CO_2 emissions 40% below 2010 CO_2 (BMT)	18.72	33.00	63.81%	16.23%	$244,599	$24,959
IPCC	CO_2 emissions 70% below 2010 CO_2 (BMT)	9.36	42.36	81.90%	20.84%	$231,156	$23,587
EIA	CO_2 emissions in 1990 (BMT)	21.50	30.22	58.43%	--	--	--
EU	CO_2 emissions 90% below 1990 CO_2 (BMT)	2.15	49.57	95.84%	24.38%	$220,800	$22,531
	50% reduction in global energy consumption	--	--	--	81.48%	$54,072	$5,518

2050 population estimate is from UN (2017); 2010 and 2020 CO_2 emission from EIA (2013); GDP forecast from World Bank (n.d.); 2050 CO_2 emission forecast from Bezdek (2015); IPCC emission reduction scenarios from Working Group III SPM pp. 10, 12 (IPCC 2014); EU emission reduction scenario is European Union nations only presented in EU (2012), projected to a global scenario. 50% reduction in global energy consumption scenario is from Tverberg (2012).

As catastrophic as these numbers appear to be, Tverberg believes her estimate is conservative, writing, "it assumes that financial systems will continue to operate as today, international trade will continue as in the past, and that there will not be major problems with overthrown governments or interruptions to electrical power. It also assumes that we will continue to transition to a service economy, and that there will be continued growth in energy efficiency."

In conclusion, achieving the IPCC's goal of reducing CO_2 emissions by between 40% and 70% from 2010 levels by 2050, the amount it believes would be necessary to keep global temperatures from increasing by more than 2°C above pre-industrial levels, would cost the world's energy consumers at least $47 trillion to $61 trillion in lost goods and services (GDP) in the year 2050. Achieving the EU's goal of reducing CO_2 emissions to 90% below 1990 levels by 2050 would cost at least $72 trillion that year. If renewables cannot completely replace the energy expected to be produced by fossil fuels in coming decades, the cost would skyrocket to $238 trillion. These are enormous numbers: the entire U.S. GDP in 2017 was only $19.4 trillion, and China's GDP was only $12.2 trillion. Would such a great loss of wealth be worthwhile? That is the topic of the next section.

References

Bezdek, R.H. 2014. *The Social Costs of Carbon? No, the Social Benefits of Carbon*. Oakton, VA: Management Information Services, Inc.

Bezdek, R.H. 2015. Economic and social implications of potential UN Paris 2015 global GHG reduction mandates. Oakton, VA: Management Information Services, Inc.

BP. 2014. *BP Energy Outlook 2035*. London, UK: BP p.l.c., January.

Clack, C.T.M., *et al.* 2017. Evaluation of a proposal for reliable low-cost grid power with 100% wind, water, and solar. *Proceedings of the National Academy of Sciences USA* **114** (26): 6722–7.

Deming, D. 2013. What if Atlas Shrugged? LewRockwell.com (website). Accessed October 24, 2018.

EIA. 2013. U.S. Energy Information Administration *International Energy Outlook 2013, With Projections to 2040*.

EU. 2012. European Union. *Energy Roadmap 2050*. Luxembourg: Publications Office of the European Union.

IPCC. 2014. Intergovernmental Panel on Climate Change. *Climate Change 2014: Mitigation of Climate Change*. Contribution of Working Group III to the Fifth Assessment Report of the Intergovernmental Panel on Climate Change. New York, NY: Cambridge University Press.

Tverberg, G. 2012. An energy/GDP forecast to 2050. Our Finite World (website). July 26.

UN. 2017. United Nations. *World Population Prospects: The 2017 Revision*. New York, NY.

World Bank. n.d. GDP per-capita, all countries and economies. Website. Accessed November 26, 2018.

8.4.3 New Cost-Benefit Ratios

The evidence seems compelling that the costs of restricting use of fossil fuels greatly exceed the benefits, even accepting many of the IPCC's very questionable assumptions.

According to the Working Group III contribution to the IPCC's Fifth Assessment Report, keeping average global surface temperature change to less than 2°C above its pre-industrial level by 2100 requires "lower global GHG emissions in 2050 than in 2010, 40% to 70% lower globally, and emissions levels near zero $GtCO_2eq$ or below in 2100" (IPCC, 2014a, pp. 10, 12). Without such reductions, "global mean surface temperature increases in 2100 from 3.7°C to 4.8°C compared to pre-industrial levels" (p. 8). Working Group III put the cost of reducing greenhouse gas emissions at 3.4% of GDP in 2050 (Table SPM.2, p. 15). Working Group II estimated the benefit of avoiding unmitigated climate change would be worth approximately 0.5% of GDP in 2050 (2050 (IPCC, 2014b, p. 663, and see Section 8.3.1 for discussion). The ratio of the two values suggests a cost-benefit ratio of 6.8:1 (3.4/0.5). This means reducing emissions would cost approximately seven times as much as the possible benefits of a slightly cooler world in 2050 and beyond. The IPCC itself makes a strong case *against* reducing emissions before 2050.

The IPCC's estimate of the benefit of avoiding ~ 2° C warming by 2050 can be compared to other estimates of the cost of reducing emissions to produce additional and more reliable cost-benefit ratios. Section 8.4.2.2 found a CO_2-GDP elasticity of 0.254, meaning for every 1% reduction in CO_{2eq} emissions between 2020 and 2050, GDP falls by 0.254%. This is the case even assuming rapid progress in technologies and a "de-coupling" of economic growth and per-capita energy consumption. Reducing emissions to 40% below 2010 levels would lower global GDP in 2050 by 16%; a 70% reduction would lower GDP by 21%; and a 90% below 1990 levels (the EU goal) would lower GDP by 24%. Tverberg found that if alternative energy sources were able to produce only 50% of the energy that would have been produced with the help of fossil fuels in 2050 – a reasonable scenario given the low density, intermittency, and high cost of fossil fuels – GDP would decline by a catastrophic 81%. Comparing these cost estimates to the possible benefit of avoiding a loss of 0.5% of annual global GDP in 2050, the IPCC's estimate of the benefits of mitigation, yields cost-benefit ratios of 32:1, 42:1, 48:1, and 162:1 respectively.

A third way to construct a cost-benefit ratio is to compare Bezdek's estimate of the GDP "created" by fossil fuels to the IPCC's and IWG's estimates of the "social cost of carbon" (SCC). According to Bezdek, in 2010 every ton of CO_2eq emitted "created" about $2,400 in world GDP. The IPCC and IWG converge on an SCC of $33 per ton of CO_2eq for 2010 assuming a 3% discount rate (IWG, 2013). Bezdek's estimate constitutes the opportunity cost of reducing emissions by one ton, while the IPCC's and IWG's estimate constitutes the possible social benefit of avoiding a 2°C warming by 2050. The ratio is 73:1 (2,400/33).

Bezdek estimates every ton of CO_2eq emitted in 2050 will "create" approximately $5,645 in GDP. The IPCC doesn't offer an estimate of SCC for 2050, but the IWG does: $71 at the 3% discount rate. (See Figure 8.1.3.2.) The ratio is 79:1 (5,645/71). In other words, the benefits of fossil fuels will exceed their social costs by a factor of 79 in 2050, *using the IWG's own SCC numbers.*

Replacing the IPCC's and IWG's inflated SCC estimates with the corrected estimates derived by Dayaratna *et al.* (2017) and reported in Section 8.3 would create even more lopsided cost-benefit ratios. An SCC near zero causes the cost-benefit ratio to become meaninglessly large and the net benefit of every ton of CO_2eq is approximately equal to $2,400 in 2010 and $5,645 in 2050.

Figure 8.4.3.1 summarizes the seven cost-benefit ratio analyses presented in this section.

Summarizing, the IPCC itself says the cost of reducing emissions enough to avoid more than a 2°C warming in 2050 will exceed the benefits by a ratio of approximately 6.8:1. The linear relationship between GDP and CO_2 emissions means attempting to avoid climate change by reducing the use of fossil fuels would cost between 32 and 48 times more than the IPCC's estimate of the possible benefit (measured as a percentage of GDP) of a cooler climate. If renewable energies are unable to entirely replace

Figure 8.4.3.1
Cost-benefit ratios of reducing CO_2eq emissions by 2050 sufficiently, according to IPCC, to prevent more than 2°C warming

	IPCC cost estimate (% of GDP)[a]	IPCC benefit estimate (% of GDP)[b]	NIPCC cost estimate (% of GDP)[c]	Bezdek cost estimate (SCC)[d]	IWG benefit estimate (SCC)[e]	Cost-benefit ratio
IPCC's cost-benefit ratio	3.4%	0.5%	--	--	--	6.8:1
NIPCC/IPCC cost-benefit ratio, 40% by 2050	--	0.5%	16%	--	--	32:1
NIPCC/IPCC cost-benefit ratio, 70% by 2050	--	0.5%	21%	--	--	42:1
NIPCC/IPCC cost-benefit ratio, 90% by 2050[f]	--	0.5%	24%	--	--	48:1
50% reduction in global energy consumption[g]	--	0.5%	81%	--	--	162:1
Bezdek/IWG SCC cost-benefit ratio 2010	--	--	--	$2,400	$33	73:1
Bezdek/IWG SCC cost-benefit ratio 2050	--	--	--	$5,645	$71	79:1

Sources: (a) IPCC, 2014b, cost of emission mitigation in 2050; (b) IPCC, 2014a, consumption loss avoided through emission mitigation, 1% is a single point estimate for model range of -1% – 1%, see Section 8.3.1 for discussion; (c) Lost GDP due to reduced energy consumption per Figure 8.4.2.2.8 (d) Bezdek, 2015, lost GDP divided by tons of CO_2eq mitigated; (e) IWG, 2013, estimate of SCC in 2010 and 2050 assuming 3% discount rate; (f) EU target using 1990 as baseline, extrapolated to reductions needed to achieve a national target; (g) Tverberg (2012).

fossil fuels, the cost could be 162 times more than the benefits. If we accept the IPCC's and IWG's estimates of the "social cost of carbon," the *benefits* of fossil fuels still exceed their social cost by ratios of 73:1 in 2010 and 79:1 in 2050.

In conclusion, the evidence seems compelling that the costs of restricting use of fossil fuels greatly exceed the benefits, even accepting many of the IPCC's very questionable assumptions.

References

Bezdek, R.H. 2014. *The Social Costs of Carbon? No, the Social Benefits of Carbon*. Oakton, VA: Management Information Services, Inc.

Bezdek, R.H. 2015. Economic and social implications of potential UN Paris 2015 global GHG reduction mandates. Oakton, VA: Management Information Services, Inc.

Dayaratna, K., McKitrick, R., and Kreutzer, D. 2017. Empirically-constrained climate sensitivity and the social cost of carbon. *Climate Change Economics* **8**: 2.

IWG. 2013. Interagency Working Group on the Social Cost of Carbon. *Technical Support Document: Technical Update of the Social Cost of Carbon for Regulatory Impact Analysis Under Executive Order 12866*. Washington, DC. May.

IPCC. 2014a. Intergovernmental Panel on Climate Change. *Climate Change 2014: Mitigation of Climate Change*. Contribution of Working Group III to the Fifth Assessment Report of the Intergovernmental Panel on Climate Change. New York, NY: Cambridge University Press.

IPCC. 2014b. Intergovernmental Panel on Climate Change. *Climate Change 2014: Impacts, Adaptation, and Vulnerability*. Contribution of Working Group II to the Fifth Assessment Report of the Intergovernmental Panel on Climate Change. New York, NY: Cambridge University Press.

8.5 Regulations

Mitigation of global warming is attempted by many competing methods, including regulating or taxing profitable activities or commodities and subsidizing otherwise unprofitable activities or commodities. Cost-benefit analysis (CBA) can determine whether and by how much the benefits of a particular program exceed the costs incurred in implementing that program. Regulations can be ranked according to their cost-benefit ratios, revealing which programs

produce the most benefits per dollar invested (Singer, 1979).

Since it is usually (and incorrectly) assumed there is an immediate need to act to save the planet from catastrophic global warming, real cost-benefit analyses of existing or proposed global warming mitigation strategies are seldom carried out. When they are, they generally rely on global climate models (GCMs) and integrated assessment models (IAMs), the flaws of which are described at length in Chapter 2, Section 2.5, and earlier in the current chapter.

Data and a methodology do exist, however, to produce fair, like-for-like, and transparent cost-effectiveness comparisons among competing mitigation strategies. A practicable metric was proposed by Christopher Monckton of Brenchley at international conferences in 2010, 2011, and 2012 and published in 2013 after peer review by the World Federation of Scientists and in 2016 as a chapter in *Evidence-based Climate Science* (Monckton, 2013, 2016). The remainder of this section presents a slightly revised and updated version of the 2016 publication.

References

Monckton of Brenchley, C. 2013. Is CO_2 mitigation cost-effective? In: Zichichi, A. and Ragaini, R. (Eds.) *Proceedings of the 45th Annual International Seminar on Nuclear War and Planetary Emergencies*. World Federation of Scientists. London, UK: World Scientific, pp.167–85.

Monckton of Brenchley, C. 2016. Is CO_2 mitigation cost effective? In: Easterbrook, D. (Ed.) *Evidence-Based Climate Science*. Second Edition. Amsterdam, Netherlands: Elsevier, pp. 175–87.

Singer, S.F. 1979. *Cost Benefit Analysis as an Aid to Environmental Decision Making*. McLean, VA: The MITRE Corporation.

8.5.1 Common Variables

Nine independent variables and constants common to all mitigation-cost assessments appear in Figure 8.5.1.1. Their derivations are presented in the sections below.

The observed rate of global warming from 1979 to 2017, taken as the least-squares linear-regression trend on the mean of the HadCRUT4 terrestrial surface and UAH satellite lower-troposphere datasets

(Morice *et al.*, 2012, updated; UAH, 2018), is equivalent to 1.5 K per century, little more than one-quarter of the 5.7 K per century upper-bound warming rate imagined by Stern. (See Figure 8.5.1.2.) This implies the notion of a 10% probability that global warming will destroy the Earth by 2100 may be dismissed as mere rhetoric. Application of such low discount rates on the grounds of "intergenerational equity" unduly favors investment in mitigation as against doing nothing now and adapting later. As President Vaclav Klaus of the Czech Republic explained in a lecture at Cambridge, "By assuming a very low (near-zero) discount rate, the proponents of the global-warming doctrine neglect the issues of time and of alternative opportunities. Using a low discount rate in global warming models means harming current generations vis-à-vis future generations. Undermining current

economic development harms future generations as well" (Klaus 2011).

The U.S. Environmental Protection Agency (EPA) adopts as its starting point a somewhat more reasonable 3% intertemporal discount rate, still less than half the U.S. Office of Management and Budget (OMB) recommended discount rate of 7% for intertemporal appraisal of public-sector investments. However, the EPA also adopts a device for which there is little rational justification: It arbitrarily assumes the value of every human life on the planet increases by 2.75% per year. This allows the EPA to conduct its intertemporal appraisals on the basis of a 0.25% discount rate. At the time of writing, EPA is inviting expert comment on its investment appraisal method.

Figure 8.5.1.1
Independent variables and constants for all mitigation strategies

y	2000	2010	2020	2030	2040	2050	2060	2070	2080	2090	2100
Annual real global GDP ($ trillion) assuming 3% p.a. growth											
g_m	44.7	60.0	80.6	108	146	196	263	354	475	638	858
Cumulative real global GDP ($ trillion) assuming 3% p.a. growth											
$g_{m_{cum}}$	44.7	409	658	828	944	1023	1077	1114	1140	1157	1169
Discounted cost of inaction (% GDP) to year y											
Z_m	3.0	2.05	1.40	0.96	0.65	0.45	0.31	0.21	0.14	0.10	0.07
Transient sensitivity parameter ($K\,W^{-1}\,m^2$) including feedback											
λ_{tra_y}	0.30	0.31	0.32	0.33	0.34	0.35	0.36	0.37	0.38	0.39	0.40
Predicted business-as-usual CO_2 concentration (ppmv) in year y											
C_y	368	392	418	446	476	508	541	577	616	656	700
Global population (billions)											
w	7.0	7.5	8.0	8.6	9.2	9.9	10.6	11.4	12.2	13.1	14.0

Constant	Value	Description
ΔT_{Cha}	3.35 K	Charney sensitivity to doubled CO_2 (CMIP5)
d	7%	OMB required discount rate
q	0.734	Fraction of anthropogenic forcing arising from CO_2
k	5	Coefficient in the CO_2 forcing function
ΔQ_C	$3.464\,W\,m^{-2}$	Radiative forcing from doubled CO_2
λ_P	$0.3\,K\,W^{-1}\,m^2$	Planck (zero-feedback) sensitivity parameter

Figure 8.5.1.2
Global surface temperature anomalies and trend, 1979–2017 (mean of HadCRUT4 & UAH)

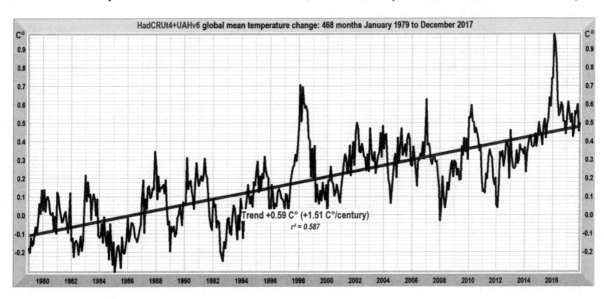

On the grounds of intergenerational equity one should not harm future generations by adopting the device of a sub-market discount rate such as that of Stern, still less that of the EPA. Instead, one should adopt, as Murphy (2008) and Nordhaus (2008) conclude, the minimum reasonable discount rate of 5% or better, the OMB's 7% rate, just as one would adopt it for any commercial investment appraisal.

8.5.1.2 Global GDP g and its growth rate

Since the cost of mitigating future anthropogenic warming is very large, it is generally expressed as a fraction of global gross domestic product (GDP), the total annual output of all of humanity's endeavors, enterprises, and industries. Global GDP g is taken as $60 trillion in 2010 (World Bank, 2011), growing at $3\%\,year^{-1}$ (3% per year). See Figure 8.5.1.1 for twenty-first-century values.

8.5.1.3 The cost Z of climate inaction

A predicted cost of climate inaction Z over a term t of years $1 \leq a \leq t$ is generally expressed as a percentage of GDP. Eq. 8.5.1 converts such a cost Z_s derived on the basis of a suspect or submarket discount rate d_s to the equivalent mitigation cost Z_m derived on the basis of a mainstream or midmarket

discount rate d_m. Here, the annual percentage GDP growth rate r will be assumed to be 3%.

$$Z_m = Z_s \frac{\sum_{a=1}^{t} \left(\frac{1 + r/100}{1 + d_m/100} \right)^a}{\sum_{a=1}^{t} \left(\frac{1 + r/100}{1 + d_s/100} \right)^a} \qquad \text{(Eq. 8.5.1)}$$

For instance, Stern's mid-range inaction cost Z_s, amounting to 3% of GDP across the entire twenty-first century derived on the basis of his 1.4% submarket discount rate and the assumption of $3\,K$ global warming by 2100, falls by nine-tenths to just 0.3% of GDP when rebased on the U.S. OMB's 7% discount rate using Eq. 8.5.1.

Furthermore, Stern made no allowance for the fact that no welfare loss arises from global warming of less than 2 K above pre-industrial temperature, equivalent to 1.1 K above the temperature in 2000. On Stern's mid-range assumption of 3 K twenty-first-century warming, and assuming a uniform twenty-first-century warming rate, no welfare loss would arise until 2038, so that at the U.S. OMB's 7% discount rate the cumulative welfare loss arising from total climate inaction would fall to less than 0.1% of GDP.

8.5.1.4 Charney sensitivity ΔT_{Cha}

The standard metric for projecting anthropogenic global warming is Charney sensitivity ΔT_{Cha}; i.e., climate sensitivity to a doubling of atmospheric CO_2 concentration. Charney (1979) held that his eponymous sensitivity was 3.0 [1.5, 4.5] K, the value adopted by IPCC (1990) and, with little change, in all subsequent Assessment Reports. In the third-generation (CMIP3) and fifth-generation (CMIP5) models of the Climate Model Intercomparison Project, Charney sensitivity was thought to be 3.35 [2.1, 4.7] K (Andrews *et al.*, 2012). Predicted global warming from all anthropogenic sources over the twenty-first century tends to be approximately equal to Charney sensitivity. Here, the CMIP3/5 mid-range estimate 3.35 K will be assumed *arguendo* to be normative.

8.5.1.5 The CO_2 fraction q

IPCC (2013, Fig. SPM.5) finds that, of the 2.29 W m^{-2} net anthropogenic forcing to 2011, 1.68 W m^{-2} is attributable to CO_2. Accordingly, the CO_2 fraction $q = 1.68/2.29 = 0.734$.

8.5.1.6 The CO_2 radiative forcing ΔQ_C

Andrews *et al.* (2012), reviewing an ensemble of two dozen CMIP5 models, provides data on the basis of which one may conclude that the radiative forcing ΔQ_C in response to doubled CO_2 concentration is 3.464 W m^{-2}. On the interval of interest, the CO_2 forcing function is approximately logarithmic. Thus, $\Delta Q_C = k \ln(C_1/C_0)$, where C_0 is the unperturbed concentration (Myhre *et al.*, 1998; IPCC, 2001, Section 6.1). Thus, $k = 3.464/\ln 2 = 5$.

8.5.1.7 The Planck sensitivity parameter λ_P

The Planck sensitivity parameter λ_P, the quantity by which a radiative forcing ΔQ_E in Watts per square meter is multiplied to yield a temperature change ΔT_S before accounting for temperature feedback, is the first derivative $\Delta T_S/\Delta Q_E = T_S/(4Q_E)$ of the fundamental equation of radiative transfer. Surface temperature $T_S = 288.4$ K (ISCCP, 2018). Given total solar irradiance $S_0 = 1364.625$ W m^{-2} (Mekaoui *et al.*, 2010) and albedo $\alpha = 0.293$ (Loeb *et al.*, 2009), radiative flux density $Q_E = S_0(1 -$

$\alpha)/4 = 241.2$ W m^{-2} at the mean emission altitude. Therefore, λ_P is today equal to 0.30 K W^{-1} m^2 (Schlesinger, 1985).

8.5.1.8 The transient-sensitivity parameter λ_{tra}

IPCC (2007, Table SPM.3) gives predicted transient anthropogenic forcings ΔQ_{tra} and warmings ΔT_{tra} from 1900 to 2100 for six scenarios. On all six scenarios, the bicentennial transient-sensitivity parameter λ_{bi}, which exceeds λ_P to the extent that some temperature feedbacks have acted, is 0.5 K W^{-1} m^2 (see IPCC 2001, p. 354, citing Ramanathan, 1985).

An appropriate twenty-first-century centennial value λ_{tra} is the mean of λ_P and λ_{bi}; i.e., 0.4 K W^{-1} m^2, in agreement with Garnaut (2008), who wrote of keeping greenhouse-gas increases to 450 ppmv CO_2 equivalent above the 280 ppmv prevalent in 1750 with the aim of holding twenty-first-century global warming to 2 K, implying $\lambda_{tra} = 0.4$ K W^{-1} m^2. Values of ΔT_{tra} implicit in this value of λ_{tra} are shown in Figure 8.5.1.1.

8.5.1.9 Global population w

Global population w is here taken as 7 billion (bn) in 2000, rising exponentially to 14 bn in 2100.

References

Andrews, T., Gregory, J.M., Webb, M.J., and Taylor, K.E. 2012. Forcing, feedbacks and climate sensitivity in CMIP5 coupled atmosphere-ocean climate models. *Geophysical Research Letters* **39**: L09712.

Charney, J.G., *et al.* 1979. *Carbon Dioxide and Climate: A Scientific Assessment*. Report of an Ad-Hoc Study Group on Carbon Dioxide and Climate, Climate Research Board, Assembly of Mathematical and Physical Sciences, and National Research Council. Washington, DC: National Academies of Sciences.

Dietz, S., Hope, C., Stern, N., and Zenghelis, D. 2007. Reflections on the Stern Review (1): a robust case for strong action to reduce the risks of climate change. *World Economics* **8** (1): 121–68.

Garnaut, R. 2008. *The Garnaut Climate Change Review: Final Report*. Port Melbourne, Australia: Cambridge University Press.

Garnaut, R. 2011. *The Garnaut Review 2011: Australia in the Global Response to Climate Change*. Cambridge, UK: Cambridge University Press.

Green Book. 2018. *The Green Book: Appraisal and Evaluation in Central Government: Treasury Guidance*. London, UK: HM Treasury.

IPCC. 1990. Intergovernmental Panel on Climate Change. *Climate Change – The IPCC Assessment (1990)*. Report prepared for the Intergovernmental Panel on Climate Change by Working Group I. Cambridge, , UK: Cambridge University Press.

IPCC. 2001. Intergovernmental Panel on Climate Change. *Climate Change 2001: The Scientific Basis*. Contribution of Working Group I to the Third Assessment Report of the Intergovernmental Panel on Climate Change. Cambridge, UK: Cambridge University Press.

IPCC. 2007. Intergovernmental Panel on Climate Change. *Climate Change 2007: the Physical Science Basis.* Contribution of Working Group I to the Fourth Assessment Report of the Intergovernmental Panel on Climate Change. Cambridge, UK: Cambridge University Press.

IPCC. 2013. Intergovernmental Panel on Climate Change. *Climate Change 2013: The Physical Science Basis.* Contribution of Working Group I to the Fifth Assessment Report of the Intergovernmental Panel on Climate Change. Cambridge, UK: Cambridge University Press.

ISCCP. 2018. Cloud analysis (part I): Climatology of global cloud and surface properties (website). Accessed May 25, 2018.

Klaus, V. 2011. *The Global Warming Doctrine Is Not a Science*. Conference on The Science and Economics of Climate Change. Cambridge, UK: Downing College. May.

Loeb, G.N., Wielicki, B.A., Dölling, D.R., Smith, G.L., Keyes, D.F., Kato, S., Manalo-Smith, N., and Wong, T., 2009. Toward optimal closure of the Earth's top-of-atmosphere radiation budget. *Journal of Climate* **22**: 748–66.

Mekaoui, S., Dewitte S., Conscience, C., and Chevalier A. 2010. Total solar irradiance absolute level from DIRAD/SOVIM on the International Space Station. *Advanced Space Research* **45**: 1393–1406.

Murphy, K.M. 2008. *Some Simple Economics of Climate Changes*. Paper presented to the Mont Pelerin Society General Meeting, Tokyo. September 8.

Lowe, J. 2008. Intergenerational Wealth Transfers and Social Discounting: Supplementary *Green Book* Guidance. London, UK: HM Treasury.

Morice, C.P., Kennedy, J.J., Rayner, N., and Jones, P.D. 2012 Quantifying uncertainties in global and regional temperature change using an ensemble of observational estimates: the HadCRUT4 dataset. *Journal of Geophysical Research* **117**:D08101.

Myhre, G., *et al.*, 1998. New estimates of radiative forcing due to well mixed greenhouse gases. *Geophysical Research Letters* **25** (14): 2715–8.

Nordhaus, W.D. 2008. *A Question of Balance: Weighing the Options on Global Warming Policies*. New Haven, CT: Yale University Press.

Ramanathan, V., *et al.* 1985. Trace gas trends and their potential role in climate change. *Journal of Geophysical Research* **90**: 5547–66.

Schlesinger, M.E. 1985. Feedback analysis of results from energy balance and radiative-convective models. In: *The Potential Climatic Effects of Increasing Carbon Dioxide.* MacCracken, M.C. and Luther, F.M. (Eds,). Washington, DC: U.S. Department of Energy, 280–319.

Stern, N., *et al.* 2007. *The Economics of Climate Change: The Stern Review.* Cambridge, UK: Cambridge University Press.

UAH. 2018. Monthly global mean land and sea lower-troposphere temperature anomalies. (website). Accessed November 19, 2018.

World Bank. 2011. Gross Domestic Product 2009, World Development Indicators.

8.5.2 Case-specific variables

Only three case-specific inputs are required and are described below.

8.5.2.1 The discounted cost X_m

The cost X_m of a given mitigation strategy is discounted to present value at the chosen market intertemporal discount rate d_m.

8.5.2.2 The business-as-usual CO_2 concentration C_y

The currently predicted business-as-usual CO_2 concentration C_y in the target final year y_2 of any existing or proposed mitigation strategy is given in Figure 8.5.1.1, allowing ready derivation of an appropriate value for any year of the twenty-first century. CO_2 concentration in the twenty-first century is extrapolated from trends in Tans (2011) and Conway & Tans (2011) according to the mid-range estimates in IPCC (2007, 2013), by which CO_2 increases exponentially from 368 ppmv in 2000 to 700 ppmv in 2100.

This value is the benchmark against which any foreseeable reduction in CO_2 concentration achieved by the mitigation strategy is measured. It will be seen from the case studies that such reductions are in practice negligible.

8.5.2.3 The fraction p of global business-as-usual CO_2 emissions the strategy will abate

The fraction p of projected global business-as-usual CO_2 emissions until year y_2 that will be abated under the (usually generous) assumption that the strategy will work as advertised is an essential quantity that is seldom derived in any integrated assessment model.

References

Conway, T. and Tans, P. 2011, Recent trends in globally-averaged CO_2 concentration, NOAA/ESRL (website). Accessed November 19, 2018.

IPCC. 2007. Intergovernmental Panel on Climate Change. *Climate Change 2007: the Physical Science Basis.* Contribution of Working Group I to the Fourth Assessment Report of the Intergovernmental Panel on Climate Change. Cambridge, UK: Cambridge University Press.

IPCC. 2013. Intergovernmental Panel on Climate Change. *Climate Change 2013: The Physical Science Basis.* Contribution of Working Group I to the Fifth Assessment Report of the Intergovernmental Panel on Climate Change. Cambridge, UK: Cambridge University Press.

Tans, P. 2011. NOAA global monthly mean CO2 concentration trends dataset (website). Accessed November 19, 2018.

8.5.3 Outputs

A robust cost-benefit model comprising a system of simple equations informed by the independent variables described in Sections 8.5.1 and 8.5.2 may readily be applied to any given mitigation strategy. The model can produce three outputs.

8.5.3.1 The unit mitigation cost M

Unit mitigation cost M is here defined as the cost of abating 1 K global warming on the assumption that all measures to abate anthropogenic warming to year y have a cost-effectiveness identical to the strategy under consideration.

8.5.3.2 Global abatement cost H per capita and J as % GDP.

On the same assumption, the strategy's global abatement cost is the total cost of abating all predicted global warming ΔT_{C21} (see Figure 8.5.1.1) over the term t from year y_1 to year y_2. This global abatement cost may be expressed in three ways: as a global cash cost X_d, as a cost H per head of global population, and as a percentage J of global GDP over the term t.

8.5.3.3 The benefit/cost or action/inaction ratio A

Finally, the benefit/cost or action/inaction ratio A of the chosen mitigation strategy is the ratio of the GDP cost J of implementation to the GDP cost Z_d of inaction now and adaptation later.

8.5.4 Cost-benefit Model

The purpose of the model is to give policymakers unfamiliar with climatology a simple but focused and robust method of answering two questions: how the cost of an existing or proposed mitigation strategy compares with those of competing strategies, and whether that cost exceeds the cost of not mitigating global warming at all.

The model comprises the following sequence of equations designed to be readily programmable. The model is so simple that it can be run on a pocket calculator. Yet, because it is rooted in mainstream climate science, it will give a more focused and reliable indication of the costs and benefits of individual mitigation strategies than any integrated assessment model.

Where p, on $[0, 1]$, is the fraction of future global emissions that a mitigation strategy is projected to abate by a target calendar year y_2, and C_{y2} is the IPCC's projected unmitigated CO_2 concentration in year y_2, model Eq. M1 gives C_{mit}, the somewhat lesser concentration in ppmv that is expected to obtain in year y_2 if the strategy is successfully followed.

$$C_{mit} = C_{y2} - p(C_{y2} - C_{y1}) \qquad \text{M1}$$

The CO_2 forcing equation is model Eq. M2, where C_0 is the unperturbed concentration.

$$\Delta Q_c = k \ln(C_1/C_0) = 5 \ln(C_1/C_0). \qquad \text{M2}$$

Accordingly, the CO_2 forcing ΔQ_{aba} abated by the chosen strategy is given by Eq. M3.

$$\Delta Q_{aba} = 5 \ln(C_{y2}/C_{mit}). \qquad \text{M3}$$

Then the global warming abated by the mitigation strategy to the final year y_2 is given by Eq. M4.

$$\Delta T_{aba} = \lambda_{tra_{y2}} k \, [\ln(C_{y2}/C_{y1}) \qquad \text{M4}$$
$$- \ln(C_{mit}/C_{y1})]$$
$$= 5 \, \lambda_{tra_{y2}} \ln(C_{y2}/C_{mit}).$$

The unit mitigation cost M is given by Eq. M5, and the predicted global warming ΔT_{y2} over the term is given by Eq. M6,

$$M = X_m/\Delta T_{aba}; \qquad \text{M5}$$
$$\Delta T_{y2} = 5 \, \lambda_{tra_{y2}} \ln(C_{y2}/C_{y1}). \qquad \text{M6}$$

The global abatement costs G in cash, H per capita and J as a percentage of global GDP of abating all predicted global warming ΔT_{y2} over the term t of the strategy are given by Eqs. M7–M9, where w is global population, G is the real cumulative discounted cost of abating ΔT_{y2}, and Z_m is the real cumulative discounted cost of inaction over the same term.

$$G = M \, \Delta T_{y2}. \qquad \text{M7}$$
$$H = G/w. \qquad \text{M8}$$
$$J = 100 \, G/Z_m. \qquad \text{M9}$$

Finally, the benefit-cost or action-inaction ratio A is given by Eq. M10, where A is the ratio of the cumulative discounted GDP cost J of the strategy over the term to the cumulative discounted cost of inaction as a percentage of GDP over the term.

$$A = J/Z_m. \qquad \text{M10}$$

8.5.5 Model Applied

8.5.5.1 2009 U.S. Cap-and-Trade Bill

In the United States in 2009, Democrats tried and failed to pass a "cap-and-trade" bill (HR 2454, SB 311) its sponsors said would cost \$180 bn per year for 40 years, or \$7.2 tn in all, which is here discounted by 7% yr^{-1} to \$2.6 tn at present value. The stated aim of the bill was to abate 83% of U.S. CO_2 emissions by 2050. Since the U.S. emitted 17% of global CO_2 at the time (derived from Olivier and Peters, 2010, Table A1), the fraction of global CO_2 emissions abated would have been 0.14. The business-as-usual CO_2 concentration in 2050 would be 508 ppmv without the bill and 492 ppmv (from Eq. M1) with it, whereupon radiative forcing abated would have been 0.16 W m^{-2} (Eq. M3) and global warming abated over the 40-year term would have been less than 0.06 K (Eq. M4).

Accordingly, the unit cost of abating 1 K global warming by measures of cost-ineffectiveness equivalent to the bill would have been equal to the ratio of the discounted cost to the warming averted; i.e., \$46 tn K^{-1} per Kelvin (Eq. M5). Therefore, the cash cost of abating all of the predicted 0.44 K global warming (Eq. M6) from 2011 to 2050 would have been more than \$20 tn; or more than \$2000 per head of global population, man, woman, and child; or 3.3% of cumulative discounted global GDP over the term. The action/inaction ratio would then be the ratio of the 3.3% GDP cost of action to the 0.45% GDP cost of inaction to 2050, i.e. 7.4 : 1. Implementing the cap-and-trade bill, or measures of equivalent unit mitigation cost, would have been almost seven and a half times costlier than the cost of doing nothing now and adapting to global warming later – always supposing that the cost of any such adaptation were to exceed the benefit of warmer weather and more CO_2 fertilization worldwide.

Additional case studies are briefly summarized in Sections 8.5.5.2–8.5.5.9. Section 8.5.6 draws lessons from the results delivered by the cost-benefit model.

8.5.5.2 The UK's Climate Change Act

In 2008, the British parliament approved the Climate Change Act of 2008. The cost stated in the government's case was \$39.4 bn yr^{-1} for 40 years, which, discounted at 7% yr^{-1} to present value, would be \$526 bn. The aim was to cut national emissions by 80% over the term.

Since UK emissions are only 1.5% of global emissions, the fraction of global emissions abated will be 0.012. CO_2 concentration by 2050 would have been 508 ppmv without the bill and will be 506 ppmv with it. Anthropogenic forcing abated over the term will be 0.013 W m^{-2}, and warming abated will be 0.005 K. Unit mitigation cost will thus be $526 bn / 0.005, or 112 tn K^{-1}. The cash cost of abating all of the predicted 0.44 K warming over the term will be $49.5 tn; the cost per head of global population will be $5000, or 8% of GDP over the term. The action-inaction ratio is 18 : 1.

8.5.5.3 The European Union's carbon trading scheme

EU carbon trading costs 92 bn yr^{-1} (World Bank 2009, p. 1), here multiplied by 2.5 (implicit in Lomborg, 2007) to allow for numerous non-trading mitigation measures. Total cost is $2.3 tn over the 10-year term to 2020, or $1.6 tn at present value. The declared aim of the EU scheme was to abate 20% of member states' emissions, which were 13% of global emissions (from Boden *et al.,* 2010a, 2010b). Thus, the fraction of global emissions abated will be 0.026. CO_2 emissions in 2020 will be 419 ppmv without the EU scheme and 418 ppmv with it. Radiative forcing abated is thus just 0.007 W m^{-2}, and warming abated is 0.002 K.

Accordingly, the unit mitigation cost of the EU's carbon trading scheme is 690 tn K^{-1}, and the cash cost of abating the < 0.1 K warming predicted to occur over the 10-year term is $64 tn; or $8000 per head of global population; or 26% of global GDP. Acting on global warming by measures of equivalent unit mitigation cost would be 18 times costlier than doing nothing now and adapting later.

8.5.5.4 California's cap-and-trade Act

Under AB32 (2006), which came fully into effect in 2012, some 182 bn yr^{-1} (Varshney and Tootelian, 2009) will be spent in the 10 years to 2021 on cap and trade and related measures. The gross cost is thus $1.8 tn, discounted to $1.3 tn. California's stated aim was to reduce its emissions, which represent 8% of U.S. emissions, by 25%. U.S. emissions at the beginning of the scheme were 17% of global emissions: thus, the fraction of global emissions

abated will be 0.0033. CO_2 concentration will fall from a business-as-usual 421 ppmv to just under 421 ppmv by 2021. Anthropogenic forcing abated will be 0.001 W m^{-2}, and warming abated will be less than one-thousandth of a Kelvin.

Accordingly, unit mitigation cost of California's cap-and-trade program will approach $4 quadrillion per °K of global warming avoided; cost per head will be $43,000, or almost 150% of global GDP over the term. It will be well more than 100 times costlier to mitigate global warming by measures such as this than to take no measures at all and adapt to such warming as may occur.

8.5.5.5 The Thanet wind array

Subsidy to one of the world's largest wind turbine installations, off the English coast, is guaranteed at $100 million annually for its 20-year lifetime; i.e. $1.06 bn at present value. Rated output of the 100 turbines is 300 MW, but such installations yield only 24% of rated capacity (Young, 2011, p. 1), so total output, at 72 MW, is only 1/600 of mean 43.2 GW UK electricity demand (Department for Energy and Climate Change, 2011). Electricity accounts for one-third of U.K. emissions, which represent 1.5% of global emissions. Therefore, the fraction of global emissions abated over the 20-year period will be 8.333×10^{-6}. Business-as-usual CO_2 concentration in 2030 would be 446.296 ppmv without the array, falling to 446.2955 ppmv with it. Forcing abated is $0.000005 \text{ W m}^{-2}$, so that warming abated is less than 0.000002 K.

Accordingly, the unit mitigation cost of the Thanet wind array is 670 tn K^{-1}. To abate the predicted 0.2 K warming over the 20-year term, the cost in cash would be $135 tn; per head $16,000; and almost one-third of global GDP over the term. It would be 34 times costlier to act on global warming than to do nothing today and adapt later.

8.5.5.6 Australia's carbon trading scheme

Australia's 2011 Clean Energy Act cost 10.1 bn yr^{-1}, plus 1.6 bn yr^{-1} for administration (Wong 2010, p. 5), plus 1.2 bn yr^{-1} for renewables and other costs, a total of 13 bn yr^{-1}, rising at $5\% \text{ yr}^{-1}$, giving a total discounted cost of $117 bn at present value. The stated aim of the legislation was to reduce Australia's CO_2 emissions by 5% over the term. Australia's emissions represent 1.2% of global emissions (derived from Boden and Marland, 2010; Boden *et al.*, 2010). Thus the fraction of global emissions abated is 0.0006. CO_2 concentration would fall from a business-as-usual 418.5 ppmv without the Australian scheme to 418.49 ppmv with it, so that warming of 0.00005 K would be abated.

The unit mitigation cost of Australia's carbon trading scheme would be $2.2 quadrillion per Kelvin. To abate the 0.1 K warming predicted for the period (the warming did not occur) would be $201 tn. The cost per head would exceed $25,000, or 81% of global GDP throughout the term. The cost of acting on climate change by measures such as Australia's scheme would approach 60 times that of inaction.

8.5.5.7 Gesture politics 1: a wind turbine on an elementary school roof in England

On March 31, 2010, Sandwell Council in England answered a freedom-of-information request (McCauley, 2011) by disclosing that it had spent £5875 ($7730) on buying and installing a small wind turbine capable of generating 209 KWh in a year – enough to power a single 100 W reading-lamp for less than three months. The fraction of UK emissions abated by the wind turbine is 33% of 209 KWh / 365 days / 24 hr / 43.2 GW, or 0.00000002%, and, since UK emissions are only 1.5% of global emissions, the fraction of global emissions abated is less than 0.000000000003. Forcing abated is < 0.000000000002 W m^{-2}. Warming abated is 0.0000000000005 K.

The unit mitigation cost of a wind turbine on an elementary school roof in England is $14.5 quadrillion per Kelvin. To abate all of the 0.2 K global warming predicted to occur over the 20-year life of the wind turbine, the cost is $3 quadrillion, or $340 tn per head, or 700% of GDP. Action costs more than 730 times inaction.

8.5.5.8 Gesture politics 2: Maryland's 90% cut in CO_2 emissions

In the United States, the state of Maryland's government decided that from 2011 to 2050 it would reduce its CO_2 emissions by 90% at a discounted cost of $7.3 tn, about three times the discounted cost of the rejected national cap-and-tax scheme over the same period. The reduction would have amounted to

1.5% of national emissions, which are 17% of global emissions. Therefore, the fraction of global emissions abated is 0.0025. The predicted business-as-usual CO_2 concentration of 507.55 ppmv would fall to 507.25 ppmv. Radiative forcing abated is less than 0.003 W m^{-2}, and warming abated is 0.001 K.

The unit mitigation cost is $7.3 quadrillion. The cost of abating the predicted 0.44 K global warming over the period is $3 quadrillion, or $320,000 per head of global population, or well more than 500% of global GDP over the period. Attempted mitigation by measures as costly as Maryland's scheme would be 1150 times costlier than inaction today and adaptation later.

8.5.5.9 Gesture politics 3: The London bicycle-hire scheme

Perhaps the costliest measure ever adopted in the name of abating global warming was the London bicycle-hire scheme, which cost $130 bn upfront, together with large annual maintenance costs that are not included here, for just 5000 bicycles – a cost of $26,000 per bicycle. Transport represents 15.2% of UK emissions (from Office for National Statistics, 2010, Table C). Cycling represents 3.1 bn of the 316.3 bn vehicle miles traveled on UK roads annually (Department for Transport, 2011). There are 23 million bicycles in use in Britain (Cyclists' Touring Club, 2011).

Global emissions will be cut by 1.5% of 15.2% of 3.1/316.3 x 5000/23,000,000. Thus the fraction of global emissions abated will be 4.886 x 10^{-9}. If the lifetime of bicycles and docking stations is 20 years, business-as-usual CO_2 concentration of 446.296 ppmv will fall to 446.2989 ppmv through the scheme. Forcing abated is 0.000000003 W m^{-2}; warming abated is 0.000000001 K; unit mitigation cost exceeds $141 quadrillion per Kelvin abated; and the cash global abatement cost of $28.5 quadrillion is $3.3 million per head, or almost 7000% of global GDP to 2030. Action costs more than 7000 times inaction.

References

AB 32. 2006. Global Warming Solutions Act. Sacramento, CA: California State Legislature.

Boden, T. and Marland, G. 2010. *Global CO$_2$ Emissions from Fossil-Fuel Burning, Cement Manufacture, and Gas*

Flaring, 1751-2007. Oak Ridge, TN: Carbon Dioxide Information and Analysis Center.

Boden, T. *et al.* 2010. Ranking of the world's countries by 2007 total CO2 emissions from fossil-fuel burning, cement production, and gas flaring. Oak Ridge, TN: Carbon Dioxide Information and Analysis Center.

Clean Energy Bill. 2011. Exposure Draft. Parliament of the Commonwealth of Australia, 2011.

Climate Change Act. 2008. London, UK: HM Stationery Office.

Cyclists' Touring Club. 2011. Cyclists' Touring Club Facts and Figures (website).

Department for Energy and Climate Change. 2011. UK Energy Statistics: Electricity (website).

Department for Transport. 2011. Table TRA0101, Road Traffic by Vehicle Type, Great Britain, 1950–2009 (website).

HR 2454. 2009. American Clean Energy & Security Bill. 111th Congress, Washington DC.

Lomborg, B. 2007. Perspective on Climate Change. Testimony before the Committee on Science and Technology of the U.S. House of Representatives. March 21.

McCauley, R. 2011. Letter of response to a request under the Freedom of Information Act received from Peter Day. Oldbury, UK: Sandwell Borough Council. March 31.

Office for National Statistics. 2010. *Statistical Bulletin: Greenhouse Gas Emissions Intensity Falls in 2008.* Newport, UK: Office for National Statistics.

Olivier, J.G.J. and Peters, J.A.H.W. 2010. Mondiale emissies koolstofdioxide door gebruik fossiele brandstoffen en cementproductie [Global emissions of carbon dioxide by using fossil fuels and cement production], 1990–2009. The Hague, Netherlands: PBL Netherlands Environmental Assessment Agency.

Varshney, S.B. and Tootelian, D.H. 2009. *Cost of AB 32 on California Small Businesses*. Sacramento, CA: California Small Business Roundtable.

Wong, P. 2010. Portfolio Budget Statements 2010–11: Budget-Related Paper No. 1.4. Climate Change and Energy Efficiency Portfolio. Canberra, Australia: Commonwealth of Australia.

World Bank. 2009. State and Trends of the Carbon Market. Washington, DC.

Young, S. 2011. *Analysis of UK Wind Power Generation, November 2008 to December 2010.* Edinburgh, Scotland: John Muir Trust. March.

8.5.6 Results Discussed

For the sake of simplicity and accessibility, the focus of the method is deliberately narrow. Potential benefits external to CO_2 mitigation, changes in global warming potentials, variability in the global GDP growth rate, or relatively higher mitigation costs in regions with lower emission intensities are ignored, for little error arises. GDP growth rates and climate-inaction costs are assumed uniform, though in practice little climate-related damage would arise unless global temperature rose by at least 1 K above today's temperatures. Given the small amount of warming abated by CO_2-reduction strategies, as well as the breadth of the intervals of published estimates of inaction and mitigation costs, modeling non-uniform GDP growth rates and climate-inaction costs may in any event prove irrelevant.

Government predictions of abatement costs (cases 1 and 2) are of the same order as those in Stern (2006) and Garnaut (2008) and the reviewed literature. However, the costs of specific measures (cases 3 through 6) prove significantly higher than official predictions. Gesture policies (cases 7 through 9) are absurdly costly and are studied here because there are so many of them. These results indicate there is no rational economic case for global warming mitigation. The arguments for mitigation are, therefore, solely political.

Mitigation is so much costlier than adaptation that real and substantial damage is being done to Western economic interests. Though the cost-benefit model concentrates exclusively on the direct costs and benefits of specific mitigation strategies, it should be understood that there are very heavy costs (but very few and very small benefits) not included in this analysis. All industries suffer by the doubling and tripling of electricity and gasoline prices allegedly in the name of abating global warming. Given that the raw material costs of coal, oil, and gas have halved in recent decades, electricity prices should have fallen commensurately. Instead, almost entirely owing to global warming mitigation policy, they have risen, and are now perhaps five times what they would be in a free market. Likewise, no account has been taken of such real and substantial indirect costs as the need for spinning-reserve backup for wind turbines and solar panels.

Further, no account has been taken of the considerable economic damage done by excessive electricity prices. Certain energy-intensive industries, such as aluminum smelting, will soon be extinct in the West. Britain's last aluminum smelter was closed some years ago owing to the government-mandated, global-warming-policy-driven cost of power, even though the facility was powered by its own hydroelectric generating station. Aluminum smelters in Australia are also under direct threat.

The results from the cost-benefit model show very clearly why mitigation of global warming is economically unjustifiable. The impact of any individual mitigation strategy on CO_2 emissions is so minuscule as to be in most cases undetectable, yet the cost of any such strategy is very large. Accordingly, CO_2 mitigation strategies inexpensive enough to be affordable will be ineffective, while strategies costly enough to be effective will be unaffordable.

The results from the model would lead us to expect that, notwithstanding the squandering of trillions in taxpayers' and energy-users' funds by national governments and, increasingly, by global warming profiteers, mitigation strategies have had so little effect that the global mix of primary energy sources for power generation is unlikely to have changed much. Sure enough, coal, the primary target of the war on fossil fuels, had a 38% global market share in 1997 and has a 38% market share today, not least because, while the West cripples its economies by closing coal-fired power stations, China alone opens at least as many new power stations per month as the West closes. (See Figure 8.5.6.1.) China's CO_2 emissions per capita are now as high as in Western countries: yet China, unlike the West, is counted as a "developing country" and is, therefore, exempted from any obligations under the Paris Climate Accord.

References

BP. 2018. BP Statistical Review of World Energy 2018. London, UK: BP p.l.c.

Garnaut, R. 2008. The Garnaut Climate Change Review: Final Report. Port Melbourne, Australia: Cambridge University Press.

Stern, N., et al. 2007. The Economics of Climate Change: The Stern Review. Cambridge, UK: Cambridge University Press.

Figure 8.5.6.1
Fuel sources as percentages of global power generation, 1997-2017

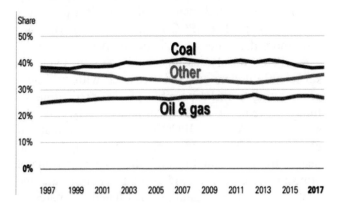

Source: BP, 2018, p. 6

8.6 Conclusion

The benefits of fossil fuels far outweigh their costs. Various scenarios of reducing greenhouse gas emissions have costs that exceed benefits by ratios ranging from 6.8:1 to 162:1.

Cost-benefit analysis (CBA) offers a methodology for weighing, in an even-handed and precise manner, the costs imposed by the use of fossil fuels on humanity and the environment, on the one hand, and the benefits produced by their use on the other. If the costs, including all the damages associated with toxic emissions and anthropogenic climate change, exceed the benefits, then efforts to force a transition from fossil fuels to alternatives such as wind turbines and solar photovoltaic (PV) cells are justified and ought to continue. If, on the other hand, the benefits of fossil fuels are found to exceed the costs, even after proper discounting of costs and risks far in the future, then the right path forward would be *energy freedom* rather than more restrictions on the use of fossil fuels.

A focus of this chapter was on the use of integrated assessment models (IAMs) in the climate change debate. While relied on by governments around the world to "put a price on carbon," they are unreliable, suffering from "cascading uncertainty" whereby uncertainties in each stage of the analysis propagate forward (Frank, 2016) and "cascading bias," whereby high-end estimates are multiplied together resulting in risk estimates that are orders of magnitude greater than what empirical data suggest (Belzer, 2012).

This chapter closely examined the IAMs used by the IPCC and found major methodological problems at every step. The IPCC's emission scenarios are little more than guesses and speculative "storylines." The IPCC falsely assumes the carbon cycle is sufficiently understood and measured with sufficient accuracy as to make possible precise predictions of future levels of carbon dioxide (CO_2) in the atmosphere. And the IPCC relies on an estimate of climate sensitivity – the amount of warming likely to occur from a doubling of the concentration of atmospheric carbon dioxide – that is out-of-date and probably too high, resulting in inflated estimates of future temperature change (Michaels, 2017).

The IPCC has done a masterful job compiling research on the impacts of climate change, but it does not distinguish between those that might be due to the human presence and those caused by such natural events as changes in solar radiation, volcanic activity, or ocean currents (IPCC, 2014a, pp. 4–5). It makes no attempt to aggregate the many studies it cites, and since they utilize different methodologies and definitions of terms, cover different periods of time, and often focus on small geographic areas, such an effort would prove impossible. The IPCC is left saying "the impact of climate change will be small relative to the impacts of other drivers" " (IPCC, 2014b, p. 662), and when measured in terms of lost gross national product (GDP), its estimate of -0.2 to -2.0% (IPCC, 2014a, p. 663) should put climate change near the bottom of the agenda for governments around the world.

The "social cost of carbon" (SCC) derived from IAMs is little more than an accounting fiction created to justify regulation of fossil fuels (Pindyck, 2013). Model-generated numbers are without any physical meaning, being so far removed from any empirical data that, using the analogy provided by Risbey et al. (1996), there are hardly any "bricks" in this edifice, only "glue," being the subjective judgements of the modelers. Changing only three assumptions in two leading IAMs – the DICE and FUND models – reduces the SCC by an order of magnitude for the first and changes the sign from positive to negative for the second (Dayaratna et al., 2017). With reasonable assumptions, IAMs show the benefits of future climate change probably exceed its cost, even though such models have many other assumptions

and biases that tend to exaggerate the negative effects of greenhouse gas emissions.

The literature review conducted in earlier chapters of this book and summarized in Figure 8.4.1.1 identified 25 impacts of fossil fuels on human well-being. Sixteen are net benefits, only one (oil spills) is a net cost, and the rest are either unknown or likely to have no net impact. This finding presents a serious challenge to any calls to restrict access to fossil fuels. While the decisions about how to classify each impact may be somewhat subjective, they are no more so than those made by the IPCC when it composed its own similar table, which appears in the Summary for Policymakers for the Working Group II contribution to the Fifth Assessment Report (IPCC, 2014a, pp. 21–25).

Cost-benefit analyses conducted for this chapter and summarized in Figure 8.4.3.1 show the IPCC's own cost and benefit estimates put the cost of restricting the use of fossil fuels at approximately 6.8 *times* greater than the benefits. Replacing the IPCC's unrealistically low cost estimate with ones originally produced by Bezdek (2014, 2015) and updated for this chapter show reducing the use of fossil fuels costs between 32 and 48 times as much as the IPCC's estimate of the benefits of a slightly cooler world. If renewable energy sources are unable to entirely replace fossil fuels, the cost could soar to 162 times the possible benefit. The ratio of Bezdek's cost estimate per ton of CO_{2eq} and the SCC produced by the Interagency Working Group in 2015 is 73:1 for fossil fuel used in 2010 and 79:1 for fossil fuels used in 2050: the cost of stopping climate change by restricting the use of fossil fuels would be 73 to 79 times greater than the benefits, and this assumes there are benefits.

Why is the case for fossil fuels so strong? Because wind and solar power cannot generate enough dispatchable energy (available 24/7) to replace fossil fuels. Energy consumption would have to fall to attain the IPCC's stated goal to lower global greenhouse emissions 40% to 70% by 2050 and to "near zero GtCO2eq or below in 2100" (IPCC, 2014c, pp. 10, 12). Less use of fossil fuels means slower economic growth and lower per-capita income for billions of people around the world, even assuming rapidly advancing technologies and a "decoupling" of economic growth from energy consumption. Reducing greenhouse gas emissions to 90% below 1990 levels by 2050, the goal of the European Union, would lower world GDP in 2050 by 24%, a loss of some $72 trillion, the equivalent of losing eight times the entire GDP of the United States.

The major losses of per-capita income that would be caused by achieving either the IPCC's or the EU's goals for reducing greenhouse gas emissions would be most harmful to the poor living in developed countries and many people living in developing countries. Recall from Chapter 3, Section 3.2.3, the comparison of life in Ethiopia and the Netherlands, two countries with similar natural endowments but dramatically different in per-capita energy consumption and per-capita incomes. People living in Ethiopia are 10.5 times more likely to have HIV/AIDS, 17 times more likely to die in infancy, die 24 years sooner, and spend 99% less money on health care (ifitweremyhome.com, n.d.). Implementing the IPCC's plan would slow or prevent the arrival of life-saving energy and technologies to millions of people living in similar conditions around the world.

Chapter 5 described how replacing fossil fuels with wind turbines and solar PV cells would cause devastating damage to the environment by vastly increasing the amount of surface area required to grow food and generate energy. Figure 5.2.2.1 showed how using oil to produce 2,000 MW of power requires about nine square miles, solar panels require 129 square miles, wind turbines require 683 square miles, and ethanol an incredible 2,450 square miles (Kiefer, 2013). Wildlife would be pushed to extinction nearly everywhere in the world were alternative energies relied on for most of our energy needs.

Finally, Chapter 7 documented the close associations between prosperity and peace, prosperity and democracy, and democracy and peace. By slowing economic growth around the world, a dramatic reduction in the use of fossil fuels could undermine some of the world's democracies and make wars over food and other scarce resources a reality again. The dramatic fall in battle-related deaths in state-based conflicts since 1946, shown in Figure 7.1.1, could be reversed as wars broke out around the world. Recall LeBlanc and Register (2003, p. 229) saying the history of humanity was "constant battles" until the prosperity, technology, and freedom made possible by the Industrial Revolution made peace possible. They thought "the opportunity for humans to live in long term balance with nature is within our grasp if we do it right." That vision would be lost as humanity is plunged back into never-ending warfare by attempts to restrict access to fossil fuels.

The cost-benefit analyses conducted in this chapter confirm that the benefits of using fossil fuels far outweigh their costs. More than that, continued reliance on fossil fuels is essential if we are to feed a growing world population and still preserve space for nature. The process that allowed humanity to discover and put to use the tremendous energy trapped inside fossil fuels – relying on markets to find efficient win-win responses to climate change and balance the interests and needs of today with those of tomorrow – should be permitted to continue to dictate the pace of a transition to alternatives to fossil fuels, not irrational fears of a climate catastrophe or hopes of technological miracles.

The global war on energy freedom, which commenced in earnest in the 1980s and reached a fever pitch in the second decade of the twenty-first century, was never founded on sound science or economics. The world's policymakers ought to acknowledge this truth and end that war.

References

Belzer, R. 2012. *Risk Assessment, Safety Assessment, and the Estimation of Regulatory Benefits*. Mercatus Research. Washington, DC: Mercatus Center.

Bezdek, R.H. 2014. *The Social Costs of Carbon? No, the Social Benefits of Carbon*. Oakton, VA: Management Information Services, Inc.

Dayaratna, K., McKitrick, R., and Kreutzer, D. 2017. Empirically-constrained climate sensitivity and the social cost of carbon. *Climate Change Economics* **8**: 2.

Frank, P. 2016. Systematic error in climate measurements: the surface air temperature record. International Seminars on Nuclear War and Planetary Emergencies 48th Session. Presentation in Erice, Italy on August 19–25, pp. 337–51.

Idso, C.D. 2013. *The Positive Externalities of Carbon Dioxide: Estimating the Monetary Benefits of Rising Atmospheric CO_2 Concentrations on Global Food Production*. Tempe, AZ: Center for the Study of Carbon Dioxide and Global Change. October 21.

ifitweremyhome.com. n.d. Compare Netherlands to Ethiopia (website). Accessed November 17, 2018.

IPCC. 2013. Intergovernmental Panel on Climate Change. *Climate Change 2013: The Physical Science Basis*. Contribution of Working Group I to the Fifth Assessment Report of the Intergovernmental Panel on Climate Change. Cambridge, UK: Cambridge University Press.

IPCC. 2014a. Intergovernmental Panel on Climate Change. Summary for Policymakers. *Climate Change 2014: Impacts, Adaptation, and Vulnerability*. Contribution of

Working Group II to the Fifth Assessment Report of the Intergovernmental Panel on Climate Change. New York, NY: Cambridge University Press.

IPCC. 2014b. Intergovernmental Panel on Climate Change. *Climate Change 2014: Impacts, Adaptation, and Vulnerability. Part A: Global and Sectoral Aspects.* Contribution of Working Group II to the Fifth Assessment Report of the Intergovernmental Panel on Climate Change. New York, NY: Cambridge University Press.

IPCC. 2014c. Intergovernmental Panel on Climate Change. *Climate Change 2014: Mitigation of Climate Change.* Contribution of Working Group III to the Fifth Assessment Report of the Intergovernmental Panel on Climate Change. New York, NY: Cambridge University Press.

Kiefer, T.A. 2013. Energy insecurity: the false promise of liquid biofuels. *Strategic Studies Quarterly* (Spring): 114–51.

LeBlanc, S. and Register, K.E. 2003. *Constant Battles: The Myth of the Noble Savage and a Peaceful Past.* New York, NY: St. Martin's/Griffin.

Michaels, P. 2017. Testimony. Hearing on At What Cost? Examining the Social Cost of Carbon, before the U.S. House of Representatives Committee on Science, Space, and Technology, Subcommittee on Environment, Subcommittee on Oversight. February 28.

Moore, S. and Simon, J. 2000. *It's Getting Better All the Time: 100 Greatest Trends of the Last 100 Years.* Washington, DC: Cato Institute.

Risbey, J., Kandlikar, M., and Patwardhan, A. 1996. Assessing integrated assessments. *Climatic Change* **34**: 369–95.

Pindyck, R.S. 2013. Climate change policy: what do the models tell us? *Journal of Economic Literature* **52** (3).

Appendix 1

Acronyms

AAAS	American Association for the Advancement of Science		**BEA**	U.S. Bureau of Economic Analysis
			BHC	Bradford Hill Criteria
AC	alternating current		**BLM**	U.S. Bureau of Land Management
ACC	American Chemistry Council *also* anthropogenic climate change		**BOD**	biological oxygen demand
			BP	before present
ACE	accumulated cyclone energy		**BR**	brassinosteriod
ACRIM	Active Cavity Radiometer Irradiance Monitor satellite		**BTU**	British thermal unit
ACS	American Cancer Society		**C**	carbon; also Celsius
AD	*anno Domini*		**C₃**	C_3 carbon fixation pathway for photosynthesis
AEMO	Australian Energy Market Operator		**C₄**	C_4 carbon fixation pathway for photosynthesis
AGW	anthropogenic global warming			
AMI	acute myocardial infarction		**C4MIP**	Coupled Carbon Cycle Climate Model Intercomparison Project
AMO	Atlantic Multidecadal Oscillation		**Ca**	CO_2 concentration in the air
AMOC	Atlantic Meridional Overturning Circulation		**CAFE**	corporate average fuel economy
AMS	American Meteorological Society		**CAM**	crassulacean acid metabolism
ANN	artificial neural network		**CARB**	California Air Resources Board
ANPP	aboveground net primary production		**CBA**	cost-benefit analysis
AO/NAO	Arctic Oscillation/North Atlantic Oscillation		**CBD**	cerebrovascular disease
			CC	combined cycle (natural gas)
APS	atmospheric pollen season		**CCS**	carbon capture and sequestration (storage)
AR5	Fifth Assessment Report			
aSAH	aneurysmal subarachnoid hemorrhage		**CDC**	U.S. Centers for Disease Control and Prevention
BA	burned area			
BAI	basal area index		**CE**	Common Era
BC	before Christ *also* black carbon		**CER**	cumulative excess [mortality] risk
BCA	benefit-cost analysis		**CERN**	European Institute for Nuclear Research
BCE	before Common Era		**CF**	capacity factor

Citation: Idso, C.D., Legates, D. and Singer, S.F. 2019. Appendix 1. In: *Climate Change Reconsidered II: Fossil Fuels.* Nongovernmental International Panel on Climate Change. Arlington Heights, IL: The Heartland Institute.

CFC	chlorofluorocarbon
CGCM	coupled general circulation model
CGE	computable general equilibrium
CH_4	methane
CHD	coronary heart disease
CLOUD	Cosmics Leaving Outdoor Droplets experiment
CME	coronal mass ejection
CMIP	Coupled Model Intercomparison Project
CNRS	National Center for Scientific Research (France)
CO	carbon monoxide
CO_2	carbon dioxide
CO_{2e}	carbon dioxide equivalent
CONUS	continental United States
COP	Conference of the Parties
COPD	chronic obstructive pulmonary disease
C_{org}	organic carbon
CPP	Clean Power Plan
CRF	cosmic ray flux
CRU	Climatic Research Unit at the University of East Anglia
CT	combustion turbine (natural gas)
CVD	cardiovascular disease
CWP	crop water productivity
DBD	dry bulk density
DC	direct current
DIC	dissolved organic carbon
DM	dry matter
DMS	dimethyl sulfide
DNA	deoxyribonucleic acid
DOC	dissolved organic carbon
DoD	U.S. Department of Defense
DICE	Dynamic Integrated Climate-Economy
DOI	U.S. Department of Interior
DTR	diurnal temperature range
ECE	extreme climate event
ECS	equilibrium climate sensitivity
EDF	Environmental Defense Fund
EIA	U.S. Energy Information Administration
EJ	exajoule

EKC	Environmental Kuznets Curve
EM	electromagnetic
ENSO	El Niño-Southern Oscillation
EPA	U.S. Environmental Protection Agency
EROI	energy return on investment
ESA	Endangered Species Act
ETS	extratropical storm
EU	European Union
EVI	Enhanced Vegetation Index
F	Fahrenheit
FACE	free-air CO_2 enrichment
FAOCE	free-air ozone and CO_2 enrichment
FAO	Food and Agriculture Organization of the United Nations
Fe	iron
FERC	U.S. Federal Energy Regulatory Commission
FJC	Federal Judicial Center
FME	free-market environmentalism
FOIA	Freedom of Information Act
FUND	Climate Framework for Uncertainty, Negotiation, and Distribution
FWS	U.S. Fish and Wildlife Service
FY	fiscal year
GAM	generalized additive model
GAO	U.S. Government Accountability Office
GCM	General Circulation Models
GCR	galactic cosmic ray
GCSC	General Crisis of the Seventeenth Century
GDP	Gross Domestic Product
GDP_{pc}	Gross Domestic Product per capita
GHG	greenhouse gas
GIGO	garbage in, garbage out
GIMMS	Global Inventory Modeling and Mapping Studies
GIS	geographic information system
GISP	Greenland Ice Sheet Project
GISS	Goddard Institute for Space Studies
GJ	gigajoule
$GMCO_2$	global mean carbon dioxide
GMST	global mean surface temperature

gNDVI	Normalized Difference Vegetation Index over the Growing Season	**IPCC 2007-II**	Intergovernmental Panel on Climate Change – Group II Contribution
GNP	Gross National Product	**IPCC 2007-III**	Intergovernmental Panel on Climate Change – Group III Contribution
GPP	gross primary production		
gr	gram(s)	**IPCC-FAR**	Intergovernmental Panel on Climate Change – First Assessment Report
GRACE	Gravity Recovery and Climate Experiment	**IPCC-SAR**	Intergovernmental Panel on Climate Change – Second Assessment Report
GREB	Globally Resolved Energy Balance model	**IPCC-TAR**	Intergovernmental Panel on Climate Change – Third Assessment Report
GrIS	Greenland Ice Sheet	**IPCC-AR4**	Intergovernmental Panel on Climate Change – Fourth Assessment Report
GSM	grand solar minimum		
GSP	gross state product	**IPCC-AR5**	Intergovernmental Panel on Climate Change – Fifth Assessment Report
GtC	gigatons of carbon		
GW	gigawatt	**IR**	infrared
H$_2$O$_2$	hydrogen peroxide	**IRR**	incidence rate ratio
H$_2$SO$_4$	sulfuric acid	**IS**	ischemic stroke
H$_2$S	hydrogen sulfide	**ISEE**	International Society for Ecological Economics
HadCRUT	Hadley Center/Climatic Research Unit dataset of monthly instrumental temperature records	**ITS**	Investment Tax Credit
		IWG	Interagency Working Group on the Social Cost of Carbon
HAMOCC	Hamburg Ocean Carbon Cycle model	**IWUE (iWUE)**	intrinsic water use efficiency
HCFC	hydrochlorofluorocarbon	**JAMA**	Journal of the American Medical Association
HEI	Health Effects Institute		
HEM	Heritage Energy Model	**K**	Kelvin
HFC	hydrofluorocarbon	**ka**	thousand years
HNO$_3$	nitric acid	**kcal**	kilocalorie
HU	hydrologic unit	**kPa**	kilopascal, a measure of pressure
HURdat	hurricane database	**kWh**	kilowatt hour
IAC	InterAcademy Council	**kWhe**	kilowatt hour of electricity
IAM	integrated assessment model	**kWht**	thermal kilowatt hour
IEA	International Energy Agency	**J**	joule
IER	Institute for Energy Research	**JI**	joint implementation
IHD	ischaemic heart disease	**LAI**	leaf area index
INM-CM4	Institute for Numerical Mathematics (Russian Academy of Sciences) climate model	**LCOE**	levelized cost of electricity
		LED	light-emitting diode
IPAT	equation where I is environmental impact, P is human population, A is per-capita affluence, and T is technological innovation	**LGM**	Last Glacial Maximum
		LIA	Little Ice Age
		LIG	last interglacial
		LIM	linear inverse model
IPCC	United Nations' Intergovernmental Panel on Climate Change	**LNT**	linear-no threshold
		LOAEL	lowest observed adverse effect level
IPCC 2007-I	Intergovernmental Panel on Climate Change – Group 1 Contribution	**LOI**	loss on ignition

LST	land surface temperature	**NMMAPS**	National Morbidity, Mortality and Air Pollution Study
LWIR	longwave infrared	**NOAA**	U.S. National Oceanic and Atmospheric Administration
m	meter		
Ma	million years	**NOAEL**	no observed adverse effect level
Ma BP	million years before present	**NO**	nitric oxide
MAAT	mean annual air temperature	**NO$_x$**	nitrogen oxides
MBtu	million British thermal units	**NPP**	net primary productivity (production)
MCA	Medieval Climate Anomaly	**NRDC**	Natural Resources Defense Council
MeHg	methylmercury	**NSPS**	New Source Performance Standards
MMT	minimum mortality temperature; also million metric tons	**NUE**	nitrogen use efficiency
		NWS	U.S. National Weather Service
MMTS	maximum-minimum temperature sensor	**O$_3$**	ozone
MODIS	MODerate resolution Imaging Spectroradiometer	**OECD**	Organisation for Economic Co-operation and Development
MS	magnetic susceptibility	**OHCA**	out-of-hospital cardiac arrest
MSL	mean sea level	**OIRA**	U.S. Office of Information and Regulatory Affairs
MTCE	million tons of carbon equivalent		
MW	megawatt	**OMB**	U.S. Office of Management and Budget
MWP	Medieval Warm Period	**OTC**	open-top chamber
N	nitrogen	**PAGE**	Policy Analysis of the Greenhouse Effect
N$_2$O	nitrous oxide	**PAR**	populations at risk
NAAQS	National Ambient Air Quality Standards	**Pb**	lead
NaCl	sodium chloride	**PCT**	potential compensation test
NAE	National Academy of Engineering	**PDI**	power dissipation index
NAO	North Atlantic Oscillation	**PDO**	Pacific Decadal Oscillation
NAPAP	National Acid Precipitation Assessment Program	**PET**	physiologically equivalent temperature
		PETM	Palaeocene-Eocene Thermal Maximum
NARR	North American Regional Reanalysis	**Pg**	petagram (one billion metric tonnes)
NAS	National Academy of Sciences (USA)	**PGD**	Plant Growth Database
NASA	National Aeronautics and Space Administration (USA)	**PGR**	post-glacial rebound
		pH	a measure of acidity or basicity
NCDC	National Climatic Data Center	**PM**	particulate matter
NDVI	Normalized Difference Vegetation Index	**PM$_{2.5}$**	particles less than or equal to 2.5 μm (micrometer)
NEMS	National Energy Modeling System		
NETL	National Energy Technology Laboratory	**PM$_{10}$**	particles less than or equal to 10 μm in diameter (about one-seventh the diameter of a human hair)
NFI	National Forest Inventory		
NH$_4$	ammonium	**PPA**	power purchase agreement
NH	Northern Hemisphere	**ppb**	parts per billion
NHL	northern high latitudes	**ppm**	parts per million
NIPCC	Nongovernmental International Panel on Climate Change	**ppmv**	parts per million by volume

PPP	purchasing power parity
PRIO	Peace Research Institute of Oslo
PSMSL	Permanent Service for Mean Sea Level
PTC	Production Tax Credit
PV	photovoltaic
PVC	polyvinyl chloride
rBC	refractory black carbon
RCP	representative concentration pathway
RD	respiratory disease
RF	radiative forcing
RIA	regulatory impact analysis
RPM	revolutions per minute
RR	relative risk
RSV	respiratory syncytial virus
RTI	respiratory tract infection
RWP	Roman Warm Period
SARS	sudden acute respiratory disease
SCC	social cost of carbon
SCD	sudden cardiac death
SEPP	Science and Environmental Policy Project
SLR	sea-level rise
SNEP	Sierra Nevada Ecosystem Project
SO$_2$	sulfur dioxide
SO$_3$	sulfur trioxide
SOC	soil organic carbon
SOM	soil organic matter
SPM	Summary for Policymakers
SRES	Special Report on Emission Scenarios
SST	sea surface temperature
STPR	social time preference rate
TAR	Third Assessment Report
TBE	tick-borne encephalitis
TBEV	tick-borne encephalitis virus
TC	tropical cyclone
Tcf	trillion cubic feet
tCO$_{2e}$	ton of CO_2 equivalent
TCPI	Tropical Cycle Potential Impact index
TCS	transient climate sensitivity
TFR	total fertility rate
Tg	teragram
TLT	temperature lower troposphere
Tmax	maximum temperature
Tmin	minimum temperature
TOA	top of atmosphere
Topt	optimum temperature
TRCS	The Right Climate Stuff
TRW	tree-ring width
TSI	total solar irradiance
TW	terawatt
TWh	terawatt hour
UAH	University of Alabama Huntsville
UCD	unusually cold day
UN	United Nations
UNCED	United Nations Conference on Environment and Development
UNEP	United Nations Environment Program
UNFCCC	United Nations Framework Convention on Climate Change
USDM	U.S. Drought Monitor
USDOE	U.S. Department of Energy
USHCN	U.S. Historical Climatology Network
UV	ultraviolet
VOC	volatile organic compound
VWC	virtual water content
W	watt
WCED	World Commission on Economic Development
WGI	Working Group I of IPC
WGII	Working Group II of IPCC
WGIII	Working Group III of IPCC
WHO	World Health Organization
WMO	World Meteorological Organization
WNP	Western North Pacific
WTA	willingness to accept
WTP	willingness to pay
WUE	water use efficiency
YLL	years of life lost
Zn	zinc

Appendix 2

Authors, Contributors, and Reviewers

Lead Authors/Editors

Roger Bezdek, Ph.D.
Management Information Services, Inc.
USA

Craig D. Idso, Ph.D.
Center for the Study of Carbon Dioxide and Global Change
USA

David Legates, Ph.D.
University of Delaware
USA

S. Fred Singer, Ph.D.
Science and Environmental Policy Project
USA

Chapter Lead Authors

Dennis Avery
The Heartland Institute
USA

Roger Bezdek, Ph.D.
Management Information Services, Inc.
USA

John D. Dunn, M.D., J.D.
Emergency Physician
USA

Craig D. Idso, Ph.D.
Center for the Study of Carbon Dioxide and Global
    Change
USA

David Legates, Ph.D.
University of Delaware
USA

Christopher Monckton
Third Viscount Monckton of Brenchley
UNITED KINGDOM

Citation: Idso, C.D., Legates, D. and Singer, S.F. 2019. Appendix 2. In: *Climate Change Reconsidered II: Fossil Fuels.* Nongovernmental International Panel on Climate Change. Arlington Heights, IL: The Heartland Institute.

Patrick Moore, Ph.D.
GreenSpirit
CANADA

S. Fred Singer, Ph.D.
Science and Environmental Policy Project
USA

Charles N. Steele, Ph.D.
Hillsdale College
USA

Aaron Stover
The Heartland Institute
USA

Richard L. Stroup, Ph.D.
North Carolina State University
USA

Chapter Contributing Authors

Jerome C. Arnett, Jr., M.D.
American Council on Science and Health
USA

John Baden, Ph.D.
Foundation for Research on Economics and the
    Environment
USA

Timothy Ball, Ph.D.
University of Winnipeg (ret.)
CANADA

Joseph L. Bast
The Heartland Institute
USA

Charles Battig, M.D., M.S.E.E.
Private Practice Medicine
USA

Vice Admiral Edward S. Briggs
Commander, Naval Surface Force
U.S. Atlantic Fleet (ret.)
USA

Barry Brill, OPE, JP
Associate Minister of Energy & Science (ret.)
NEW ZEALAND

Kevin Dayaratna, Ph.D.
The Heritage Foundation
USA

John D. Dunn, M.D., J.D.
Emergency Physician
USA

James Enstrom, Ph.D.
Scientific Integrity Institute
USA

Captain Donald K. Forbes
Commandant of Midshipmen
U.S. Naval Academy (ret.)
USA

Patrick Frank, Ph.D.
SLAC National Accelerator Center
Stanford University
USA

Kenneth Haapala
Science and Environmental Policy Project
USA

Howard Hayden, Ph.D.
University of Connecticut (emeritus)
USA

Admiral Thomas B. Hayward
Commander-in-Chief
U.S. Pacific Fleet (ret.)
USA

Jay Lehr, Ph.D.
The Heartland Institute
USA

Bryan Leyland, M.Sc.
LCL Ltd.
NEW ZEALAND

Steve Milloy, J.D.
JunkScience
USA

Patrick Moore, Ph.D.
GreenSpirit
CANADA

Willie Soon, Ph.D.
Astrophysicist
USA

Richard J. Trzupek
Chemist and Author
USA

Steve Welcenbach
Alchemical Ventures, Inc.
USA

S. Stanley Young, Ph.D.
National Institute of Statistical Sciences (ret.)
USA

Chapter Reviewers

D. Weston Allen, MBBS, FRACGP, Dip.Phys.Med.
Kingscliff Family Medical Services
AUSTRALIA

Mark Alliegro, Ph.D.
Brown University
USA

Charles Anderson, Ph.D.
Anderson Materials Evaluation, Inc.
USA

David Archibald
Backreef Oil Pty Ltd
AUSTRALIA

Dennis T. Avery
The Heartland Institute
USA

Timothy Ball, Ph.D.
University of Winnipeg (ret.)
CANADA

David Bowen, Ph.D.
Cardiff University (emeritus)
UNITED KINGDOM

Barry Brill, OPE, JP
Associate Minister of Energy & Science (ret.)
NEW ZEALAND

H. Sterling Burnett, Ph.D.
The Heartland Institute
USA

David Burton
SeaLevelInfo
USA

William N. Butos, Ph.D.
Trinity College
USA

Mark Campbell, Ph.D.
United States Naval Academy
USA

Jorge David Chapas
Red de Amigos de la Naturaleza / Universidad
Francisco Marroquín
GUATEMALA

Ian D. Clark, Ph.D.
University of Ottawa
CANADA

Donald R. Crowe
Absaroke Corporation
USA

Weihong Cui
Institute of Remote Sensing and Digital Earth
Chinese Academy of Sciences
CHINA

Donn Dears
GE Company (ret.)
USA

David Deming, Ph.D.
University of Oklahoma
USA

Terry W. Donze
Missouri University of Science & Technology
USA

Paul Driessen, J.D.
Committee for a Constructive Tomorrow
USA

John Droz, Jr.
Independent Scientist
USA

James Enstrom, Ph.D.
Scientific Integrity Institute
USA

Rex J. Fleming, Ph.D.
Global Aerospace, LLC (ret.)
USA

Vivian Richard Forbes
Carbon Sense Coalition
AUSTRALIA

Patrick Frank, Ph.D.
SLAC National Accelerator Center
Stanford University
USA

Lee C. Gerhard, Ph.D.
Kansas Geological Survey, University of Kansas (ret.)
USA

François Gervais, Ph.D.
University of Tours (emeritus)
FRANCE

Albrecht Glatzle, Dr.Sc.Agr.
Iniciativa para la Investigación y Transferencia de
    Technología Agraria Sostenible (INTTAS)
PARAGUAY

Steve Goreham
Climate Science Coalition of America
USA

Pierre Gosselin
NoTricksZone
EUROPE

Laurence Gould, Ph.D.
University of Hartford
USA

Kesten Green, Ph.D.
University of South Australia
AUSTRALIA

Kenneth Haapala
Science and Environmental Policy Project
USA

Hermann Harde, Ph.D.
Helmut-Schmidt-University Hamburg (ret.)
GERMANY

Tom Harris
International Climate Science Coalition
CANADA

Howard Hayden, Ph.D.
University of Connecticut (emeritus)
USA

Tom Hennigan
Truett McConnell University
USA

Donald Hertzmark, Ph.D.
DMP Resources
USA

Ole Humlum, Ph.D.
University of Oslo (emeritus)
NORWAY

Mary Hutzler
Institute for Energy Research
USA

Hans Konrad Johnsen, Ph.D.
Det Norske Oljeselskap ASA (ret.)
NORWAY

Brian Joondeph, M.D., M.P.S.
Colorado Retina Associates, PC
USA

Richard A. Keen, Ph.D.
University of Colorado, Boulder (emeritus)
USA

William Kininmonth, M.Sc.
National Climate Centre (ret.)
AUSTRALIA

Joseph Leimkuhler
LLOG Exploration L.L.C.
USA

Marlo Lewis, Jr., Ph.D.
Competitive Enterprise Institute
USA

Bryan Leyland
LCL Ltd.
NEW ZEALAND

Anthony R. Lupo, Ph.D.
University of Missouri
USA

Paul McFadyen, Ph.D.
Biologist (ret.)
AUSTRALIA

John Merrifield, Ph.D.
University of Texas At San Antonio
USA

Patrick J. Michaels, Ph.D.
Cato Institute
USA

Alan Moran, Ph.D.
Regulation Economics
AUSTRALIA

Robert Murphy, Ph.D.
Institute for Energy Research
USA

Daniel W. Nebert, M.D.
University of Cincinnati Medical Center (emeritus)
USA

Norman J. Page, Ph.D.
ClimateSense
USA

Fred Palmer, J.D.
The Heartland Institute
USA

Garth Paltridge, Ph.D., D.Sc., FAA
Australian National University
AUSTRALIA

Jim Petch
University of Manchester Trican
Manchester Metropolitan University (retired)
UNITED KINGDOM

Charles T. Rombough, Ph.D.
CTR Technical Services, Inc.
USA

Ronald Rychlak, J.D.
University of Mississippi, School of Law
USA

Tom V. Segalstad, Ph.D.
University of Oslo
NORWAY

Gary D. Sharp, Ph.D.
Center for Climate/Ocean Resources Study (ret.)
USA

Jan-Erik Solheim
Arctic University of Norway
NORWAY

Willie Soon, Ph.D.
Astrophysicist
USA

Charles N. Steele, Ph.D.
Hillsdale College
USA

David Stevenson
Caesar Rodney Institute
USA